About the Cover

The cover image illustrates a dodeca-spidroball formed by applying spidrons onto the surface of a dodecahedron. A spidron is a plane figure composed entirely of triangles, where, for every pair of triangles, each has a leg of the other as one of its legs, and neither has any point inside the interior of the other.

▶ **Meet the Artist**

The cover art was generated by Paul Nylander, a mechanical engineer with strong programming and mathematical skills that he uses to design complex engineering systems. Nylander says that he always enjoyed science and art as a hobby. When he was in high school, he had some aptitude for math, but programming was difficult for him. However, he became much more interested in programming when he began studying computer graphics. Most of Nylander's artwork is created in Mathematica, POV-Ray, and C++.

You can find a short bio and description of his work at:
http://virtualmathmuseum.org/mathart/ArtGalleryNylander/Nylanderindex.html.

ALGEBRA 2

GLENCOE

McGraw Hill Education

mheducation.com/prek-12

Copyright © 2018 McGraw-Hill Education

All rights reserved. No part of this publication may be reproduced or distributed in any form or by any means, or stored in a database or retrieval system, without the prior written consent of McGraw-Hill Education, including, but not limited to, network storage or transmission, or broadcast for distance learning.

Send all inquiries to:
McGraw-Hill Education
8787 Orion Place
Columbus, OH 43240

ISBN: 978-0-07-903990-3
MHID: 0-07-903990-1

Printed in the United States of America.

4 5 6 7 8 9 LWI 21 20 19 18

Understanding by Design® is a registered trademark of the Association for Supervision and Curriculum Development ("ASCD").

McGraw-Hill is committed to providing instructional materials in Science, Technology, Engineering, and Mathematics (STEM) that give all students a solid foundation, one that prepares them for college and careers in the 21st century.

Contents in Brief

Chapter 0
Preparing for Advanced Algebra

Chapter 1
Linear Equations

Chapter 2
Relations and Functions

Chapter 3
Quadratic Functions

Chapter 4
Polynomials and Polynomial Functions

Chapter 5
Inverses and Radical Functions

Chapter 6
Exponential and Logarithmic Functions

Chapter 7
Rational Functions

Chapter 8
Statistics and Probability

Chapter 9
Trigonometric Functions

Chapter 10
Trigonometric Identities and Equations

Authors

Our lead authors ensure that the Macmillan/McGraw-Hill and Glencoe/McGraw-Hill mathematics programs are truly vertically aligned by beginning with the end in mind — success in Algebra 1 and beyond. By "backmapping" the content from the high school programs, all of our mathematics programs are well articulated in their scope and sequence.

LEAD AUTHORS

John A. Carter, Ph.D.

Mathematics Teacher
WINNETKA, ILLINOIS

Areas of Expertise:
Using technology and manipulatives to visualize concepts; mathematics achievement of English-language learners

Gilbert J. Cuevas, Ph.D.

Professor of Mathematics Education, Texas State University—San Marcos
SAN MARCOS, TEXAS

Areas of Expertise:
Applying concepts and skills in mathematically rich contexts; mathematical representations; use of technology in the development of geometric thinking

Roger Day, Ph.D., NBCT

Mathematics Department Chairperson, Pontiac Township High School
PONTIAC, ILLINOIS

Areas of Expertise:
Understanding and applying probability and statistics; mathematics teacher education

In Memoriam
Carol Malloy, Ph.D.

Dr. Carol Malloy was a fervent supporter of mathematics education. She was a Professor at the University of North Carolina, Chapel Hill, NCTM Board of Directors member, President of the Benjamin Banneker Association (BBA), and 2013 BBA Lifetime Achievement Award for Mathematics winner. She joined McGraw-Hill in 1996. Her influence significantly improved our programs' focus on real-world problem solving and equity. We will miss her inspiration and passion for education.

PROGRAM AUTHORS

Berchie Holliday, Ed.D.
National Mathematics Consultant
SILVER SPRING, MARYLAND

Areas of Expertise:
Using mathematics to model and understand real-world data; the effect of graphics on mathematical understanding

Ruth Casey
Regional Teacher Partner
UNIVERSITY OF KENTUCKY

Areas of Expertise:
Graphing technology and mathematics

CONTRIBUTING AUTHORS

Dinah Zike FOLDABLES
Educational Consultant
Dinah-Might Activities, Inc.
SAN ANTONIO, TEXAS

Jay McTighe
Educational Author and Consultant
COLUMBIA, MARYLAND

connectED.mcgraw-hill.com

Consultants and Reviewers

These professionals were instrumental in providing valuable input and suggestions for improving the effectiveness of the mathematics instruction.

LEAD CONSULTANT

Viken Hovsepian
Professor of Mathematics
Rio Hondo College
WHITTIER, CALIFORNIA

CONSULTANTS

MATHEMATICAL CONTENT

Grant A. Fraser, Ph.D.
Professor of Mathematics
California State University, Los Angeles
LOS ANGELES, CALIFORNIA

Arthur K. Wayman, Ph.D.
Professor of Mathematics Emeritus
California State University, Long Beach
LONG BEACH, CALIFORNIA

GIFTED AND TALENTED

Shelbi K. Cole
Research Assistant
University of Connecticut
STORRS, CONNECTICUT

COLLEGE READINESS

Robert Lee Kimball, Jr.
Department Head, Math and Physics
Wake Technical Community College
RALEIGH, NORTH CAROLINA

DIFFERENTIATION FOR ENGLISH-LANGUAGE LEARNERS

Susana Davidenko
State University of New York
CORTLAND, NEW YORK

Alfredo Gómez
Mathematics/ESL Teacher
George W. Fowler High School
SYRACUSE, NEW YORK

GRAPHING CALCULATOR

Jerry Cummins
Former President
National Council of Supervisors of Mathematics
WESTERN SPRINGS, ILLINOIS

MATHEMATICAL FLUENCY

Robert M. Capraro
Associate Professor
Texas A&M University
COLLEGE STATION, TEXAS

PRE-AP

Dixie Ross
Lead Teacher for Advanced Placement Mathematics
Pflugerville High School
PFLUGERVILLE, TEXAS

READING AND WRITING

ReLeah Cossett Lent
Author and Educational Consultant
MORGANTON, GEORGIA

Lynn T. Havens
Director of Project CRISS
KALISPELL, MONTANA

REVIEWERS

Corey Andreasen
Mathematics Teacher
North High School
SHEBOYGAN, WISCONSIN

Mark B. Baetz
Mathematics Coordinating Teacher
Salem City Schools
SALEM, VIRGINIA

Kathryn Ballin
Mathematics Supervisor
Newark Public Schools
NEWARK, NEW JERSEY

Kevin C. Barhorst
Mathematics Department Chair
Independence High School
COLUMBUS, OHIO

Brenda S. Berg
Mathematics Teacher
Carbondale Community High School
CARBONDALE, ILLINOIS

Dawn Brown
Mathematics Department Chair
Kenmore West High School
BUFFALO, NEW YORK

Sheryl Pernell Clayton
Mathematics Teacher
Hume Fogg Magnet School
NASHVILLE, TENNESSEE

Bob Coleman
Mathematics Teacher
Cobb Middle School
TALLAHASSEE, FLORIDA

Jane E. Cotts
Mathematics Teacher
O'Fallon Township High School
O'FALLON, ILLINOIS

Michael D. Cuddy
Mathematics Instructor
Zypherhills High School
ZYPHERHILLS, FLORIDA

Melissa M. Dalton, NBCT
Mathematics Instructor
Rural Retreat High School
RURAL RETREAT, VIRGINIA

Tina S. Dohm
Mathematics Teacher
Naperville Central High School
NAPERVILLE, ILLINOIS

Laurie L.E. Ferrari
Mathematics Teacher
L'Anse Creuse High School–North
MACOMB, MICHIGAN

Steve Freshour
Mathematics Teacher
Parkersburg South High School
PARKERSBURG, WEST VIRGINIA

Shirley D. Glover
Mathematics Teacher
TC Roberson High School
ASHEVILLE, NORTH CAROLINA

Caroline W. Greenough
Mathematics Teacher
Cape Fear Academy
WILMINGTON, NORTH CAROLINA

Susan Hack, NBCT
Mathematics Teacher
Oldham County High School
BUCKNER, KENTUCKY

Michelle Hanneman
Mathematics Teacher
Moore High School
MOORE, OKLAHOMA

Theresalynn Haynes
Mathematics Teacher
Glenbard East High School
LOMBARD, ILLINOIS

Sandra Hester
Mathematics Teacher/AIG Specialist
North Henderson High School
HENDERSONVILLE, NORTH CAROLINA

Jacob K. Holloway
Mathematics Teacher
Capitol Heights Junior High School
MONTGOMERY, ALABAMA

Robert Hopp
Mathematics Teacher
Harrison High School
HARRISON, MICHIGAN

Eileen Howanitz
Mathematics Teacher/Department Chairperson
Valley View High School
ARCHBALD, PENNSYLVANIA

Charles R. Howard, NBCT
Mathematics Teacher
Tuscola High School
WAYNESVILLE, NORTH CAROLINA

Sue Hvizdos
Mathematics Department Chairperson
Wheeling Park High School
WHEELING, WEST VIRGINIA

Elaine Keller
Mathematics Teacher
Mathematics Curriculum Director K-12
Northwest Local Schools
CANAL FULTON, OHIO

Sheila A. Kotter
Mathematics Educator
River Ridge High School
NEW PORT RICHEY, FLORIDA

Frank Lear
Mathematics Department Chair
Cleveland High School
CLEVELAND, TENNESSEE

Jennifer Lewis
Mathematics Teacher
Triad High School
TROY, ILLINOIS

Catherine McCarthy
Mathematics Teacher
Glen Ridge High School
GLEN RIDGE, NEW JERSEY

Jacqueline Palmquist
Mathematics Department Chair
Waubonsie Valley High School
AURORA, ILLINOIS

Thom Schacher
Mathematics Teacher
Otsego High School
OTSEGO, MICHIGAN

Laurie Shappee
Teacher/Mathematics Coordinator
Larson Middle School
TROY, MICHIGAN

Jennifer J. Southers
Mathematics Teacher
Hillcrest High School
SIMPSONVILLE, SOUTH CAROLINA

Sue Steinbeck
Mathematics Department Chair
Parkersburg High School
PARKERSBURG, WEST VIRGINIA

Kathleen D. Van Sise
Mathematics Teacher
Mandarin High School
JACKSONVILLE, FLORIDA

Karen Wiedman
Mathematics Teacher
Taylorville High School
TAYLORVILLE, ILLINOIS

Digital Tools to Enhance Your Learning

Students today have an unprecedented access to and appetite for technology and new media. You see technology as your friend and rely on it to study, work, play, relax, and communicate.

You are accustomed to the role that computers play in today's world. You are the first generation whose primary educational tool is a computer or a cell phone. The eStudentEdition gives you access to your math curriculum anytime, anywhere.

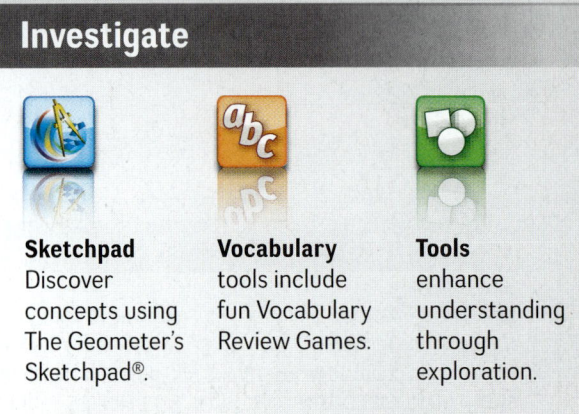

Investigate

Sketchpad Discover concepts using The Geometer's Sketchpad®.

Vocabulary tools include fun Vocabulary Review Games.

Tools enhance understanding through exploration.

The Geometer's Sketchpad

The Geometer's Sketchpad gives you a tangible, visual way to see the math in action through dynamic model manipulation of lines, shapes, and functions.

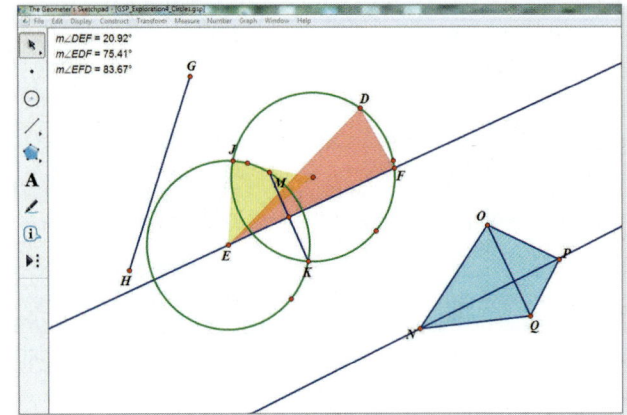

eToolkit

eToolkit helps you to use virtual manipulative's to extend your learning beyond the classroom by modifying concrete models in a real-time, interactive format focused on problem-based learning.

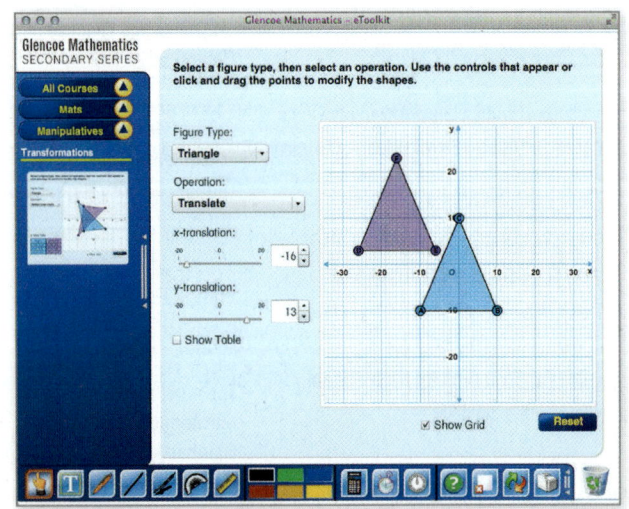

Learn

LearnSmart Topic based online assessment

Animations illustrate key concepts through step-by-step tutorials and videos.

Tutors See and hear a teacher explain how to solve problems.

Calculator Resources provides other calculator keystrokes for each Graphing Technology Lab.

Practice

Self-Check Practice allows students to check their understanding and send results to their teacher.

eBook Interactive learning experience with links directly to assets

eStudentEdition

Use your eStudentEdition to access your print text 24/7. This interactive eBook gives you access to all the resources that help you learn the concepts. You can take notes, highlight, digitally write on the pages, and bookmark where you are.

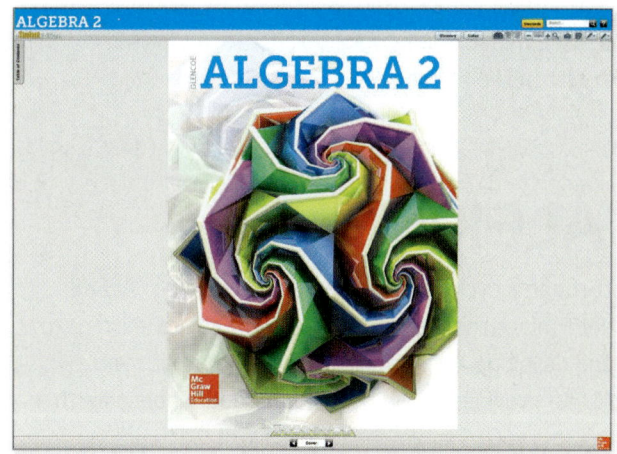

Interactive Student Guide

Interactive Student Guide is a dynamic resource to help you meet the challenges of content standards.

This guide works together with the student edition to ensure that you can reflect on comprehension and application, apply math concepts to the real world, and internalize concepts to develop "second nature" recall.

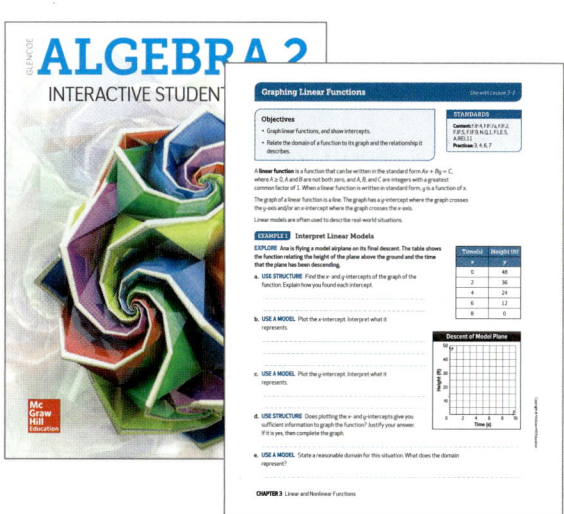

connectED.mcgraw-hill.com

Adaptive Learning

LEARNSMART®

LearnSmart is your online test-prep solution with adaptive capability and resources for end-of-course assessment. Through a series of adaptive questions, LearnSmart identifies areas where you need to focus your learning. It provides you with learning resources such as slides, videos, kaleidoscopes, and label games to encourage you to review material again. This adaptive learning will help increase your likelihood that you will retain your new knowledge.

LearnSmart can be used as a self-study tool or it can be assigned to you by your teacher through your ConnectED platform.

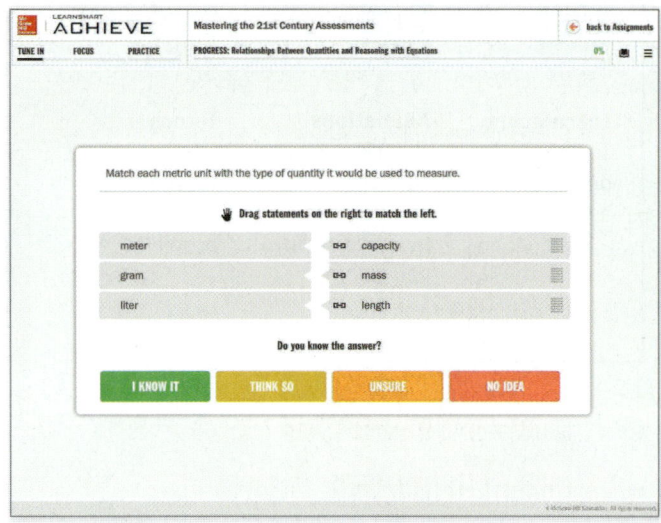

ALEKS®

Through a cycle of assessment and learning, *ALEKS determines the topics that you are ready to learn next and develops a personalized learning path for you. ALEKS provides you with real-time, actionable data that informs you what you need to learn. It will help you review and master the skills needed to be successful in your math class. Use your ALEKS Pie to see a snapshot of your current progress of the course. Click on the Pie slice to see the number of topics mastered and ready to learn next.

Use the timeline to watch your progress toward your learning goals, and if needed you can toggle between English and Spanish translations of the content and interface.

*Ask your teacher if you have access to ALEKS.

CHAPTER 0
Preparing for Advanced Algebra

	Get Started on Chapter 0	P2
	Pretest	P3
0-1	Representing Functions	P4
0-2	FOIL	P6
0-3	Factoring Polynomials	P7
0-4	Counting Techniques	P9
0-5	Adding Probabilities	P13
0-6	Multiplying Probabilities	P16
0-7	Congruent and Similar Figures	P20
0-8	The Pythagorean Theorem	P22
0-9	Measures of Center, Spread, and Position	P24
	Posttest	P29

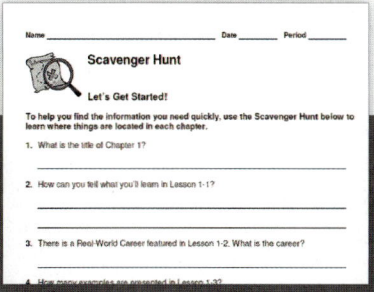

Worksheets help to explain key concepts, let you practice your skills, and offer opportunities for extending the lessons. Find them in the Resources in ConnectED.

Go Online!
connectED.mcgraw-hill.com

CHAPTER 1
Linear Equations

	Get Ready for Chapter 1	2
1-1	Solving Linear Equations	5
1-2	Solving Linear Inequalities	13
	Extend 1-2 Algebra Lab: Interval and Set-Builder Notation	20
1-3	Rate of Change and Slope	21
1-4	Writing Linear Equations	28
1-5	Graphing Linear Inequalities	35
	Mid-Chapter Quiz	40
	Explore 1-6 Graphing Technology Lab: Intersections of Graphs	41
1-6	Solving Systems of Equations	42
1-7	Solving Systems of Inequalities by Graphing	52
	Extend 1-7 Graphing Technology Lab: Systems of Linear Inequalities	59
1-8	Optimization with Linear Programming	60
1-9	Solving Systems of Equations in Three Variables	67

ASSESSMENT

Study Guide and Review	74
Practice Test	79
Performance Task	80
Preparing for Assessment, Test-Taking Strategy	81
Preparing for Assessment, Cumulative Review	82

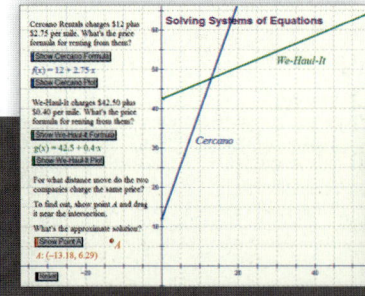

Geometer's Sketchpad® allows you to interact with functions in a visual way. Investigate systems of equations and inequalities with sketches in ConnectED.

CHAPTER 2
Relations and Functions

	Get Ready for Chapter 2	84
2-1	Functions and Continuity	87
2-2	Linearity and Symmetry	95
2-3	Extrema and End Behavior	103
2-4	Sketching Graphs of Functions	111
	Mid-Chapter Quiz	117
2-5	Graphing Special Functions	118
2-6	Transformations of Functions	125
2-7	Solving Equations by Graphing	133

ASSESSMENT

Study Guide and Review	138
Practice Test	143
Performance Task	144
Preparing for Assessment, Test-Taking Strategy	145
Preparing for Assessment, Cumulative Review	146

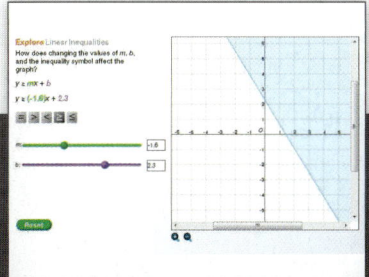

 With the **Graphing Tools** in ConnectED, you can explore how changing parameters affects the graph of a function or an inequality.

connectED.mcgraw-hill.com   xiii

CHAPTER 3
Quadratic Functions

	Get Ready for Chapter 3	148
3-1	Graphing Quadratic Functions	151
	Extend 3-1 Graphing Technology Lab: Modeling Real-World Data	160
3-2	Solving Quadratic Equations by Graphing	163
	Extend 3-2 Graphing Technology Lab: Solving Quadratic Equations by Graphing	171
3-3	Complex Numbers	172
3-4	Solving Quadratic Equations by Factoring	179
	Extend 3-4 Algebra Lab: The Complex Plane	188
	Mid-Chapter Quiz	190
3-5	Solving Quadratic Equations by Completing the Square	191
	Extend 3-5 Graphing Technology Lab: Solving Quadratic Equations	198
3-6	The Quadratic Formula and the Discriminant	199
	Extend 3-6 Algebra Lab: Quadratics and Rate of Change	208
3-7	Quadratic Inequalities	209

ASSESSMENT

Study Guide and Review	216
Practice Test	221
Performance Task	222
Preparing for Assessment, Test-Taking Strategy	223
Preparing for Assessment, Cumulative Review	224

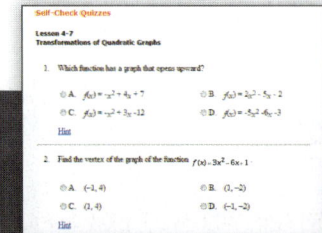

 Review concepts with quick **Self-Check Quizzes** in ConnectED. Use them to check your own progress as you complete each lesson.

connectED.mcgraw-hill.com

CHAPTER 4
Polynomials and Polynomial Functions

	Get Ready for Chapter 4	226
4-1	**Operations with Polynomials**	229
	Extend 4-1 Algebra Lab: Dimensional Analysis	236
4-2	**Powers of Binomials**	237
4-3	**Dividing Polynomials**	242
	Extend 4-3 Graphing Technology Lab: Dividing Polynomials	250
	Explore 4-4 Graphing Technology: Power Functions	251
4-4	**Graphing Polynomial Functions**	253
4-5	**Analyzing Graphs of Polynomial Functions**	262
	Extend 4-5 Graphing Technology Lab: Modeling Data Using Polynomial Functions	270
	Mid-Chapter Quiz	272
	Explore 4-6 Graphing Technology Lab: Solving Polynomial Equations by Graphing	273
4-6	**Solving Polynomial Equations**	274
4-7	**Proving Polynomial Identities**	282
4-8	**The Remainder and Factor Theorem**	287
4-9	**Roots and Zeros**	293
	Extend 4-9 Graphing Technology Lab: Analyzing Polynomial Functions	301

ASSESSMENT

Study Guide and Review	302
Practice Test	307
Performance Task	308
Preparing for Assessment, Test-Taking Strategy	309
Preparing for Assessment, Cumulative Review	310

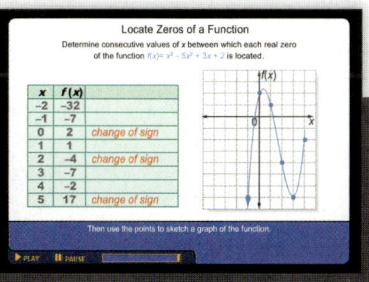

Animations demonstrate Key Concepts and topics from the chapter. Click to watch animations in ConnectED.

Go Online!
connectED.mcgraw-hill.com

CHAPTER 5
Inverses and Radical Functions

	Get Ready for Chapter 5	312
5-1	Operations with Functions	315
5-2	Composition of Functions	322
5-3	Inverse Functions and Relations	329
	Extend 5-3 Graphing Technology Lab: Inverse Functions and Relations	336
	Mid Chapter Quiz	337
5-4	Graphing Square Root Functions	338
5-5	Graphing Cube Root Functions	345
	Extend 5-5 Graphing Technology Lab: Graphing nth Root Functions	351
5-6	Solving Radical Equations	352
	Extend 5-6 Graphing Technology Lab: Solving Radical Equations	359

ASSESSMENT

Study Guide and Review	361
Practice Test	365
Performance Task	366
Preparing for Assessment, Test-Taking Strategy	367
Preparing for Assessment, Cumulative Review	368

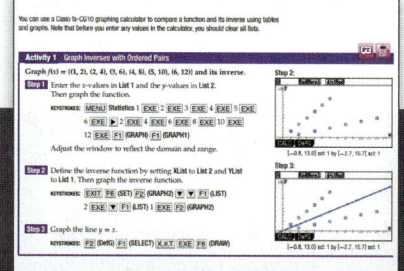

 A graphing calculator is a powerful tool for studying algebra. **Graphing Calculator Keystrokes** help you harness that power. Find the keystrokes for your calculator in the Resources in ConnectED.

CHAPTER 6
Exponential and Logarithmic Functions

	Get Ready for Chapter 6	370
6-1	Graphing Exponential Functions	373
	Explore 6-2 Graphing Technology Lab: Solving Exponential Equations and Inequalities	381
6-2	Solving Exponential Equations and Inequalities	383
6-3	Geometric Sequences and Series	390
6-4	Logarithms and Logarithmic Functions	397
6-5	Modeling Data	405
	Mid Chapter Quiz	415
6-6	Properties of Logarithms	416
6-7	Common Logarithms	423
6-8	Natural Logarithms	430
6-9	Solving Logarithmic Equations and Inequalities	437
	Extend 6-9 Graphing Technology Lab: Solving Logarithmic Equations and Inequalities	443
6-10	Using Logarithms to Solve Exponential Problems	445
	Extend 6-10 Graphing Technology Lab: Cooling	452

ASSESSMENT

Study Guide and Review	453
Practice Test	459
Performance Test	460
Preparing for Assessment, Test-Taking Strategy	461
Preparing for Assessment, Cumulative Review	462

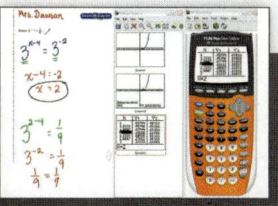

Personal Tutors that use graphing calculator technology show you every step to solving problems with this powerful tool. Find them in the Resources in ConnectED.

Go Online!
connectED.mcgraw-hill.com

CHAPTER 7
Rational Functions

	Get Ready for Chapter 7	464
7-1	Multiplying and Dividing Rational Expressions	467
7-2	Adding and Subtracting Rational Expressions	476
7-3	Graphing Reciprocal Functions	483
	Mid Chapter Quiz	490
7-4	Graphing Rational Functions	491
	Extend 7-4 Graphing Technology Lab: Graphing Rational Functions	499
7-5	Variation Functions	500
7-6	Solving Rational Equations and Inequalities	508
	Extend 7-6 Graphing Technology Lab: Solving Rational Equations and Inequalities	517

ASSESSMENT

Study Guide and Review	519
Practice Test	523
Performance Task	524
Preparing for Assessment, Test-Taking Strategy	525
Preparing for Assessment, Cumulative Review	526

Worksheets help to explain key concepts, let you practice your skills, and offer opportunities for extending the lessons. Find them in the Resources in ConnectED.

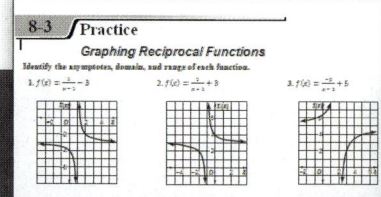

CHAPTER 8
Statistics and Probability

	Get Ready for Chapter 8	528
8-1	Random Sampling	531
8-2	Using Statistical Experiments	538
8-3	Population Parameters	545
8-4	Distributions of Data	551
	Mid Chapter Quiz	560
8-5	Evaluating Published Data	561
8-6	Normal Distributions	566
8-7	Using Probability to Make Decisions	573

ASSESSMENT

Study Guide and Review	579
Practice Test	585
Performance Task	586
Preparing for Assessment, Test-Taking Strategy	587
Preparing for Assessment, Cumulative Review	588

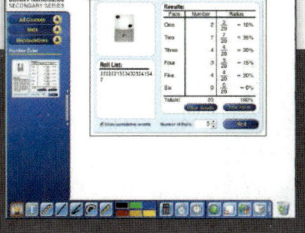

The number cube and coin toss tools can be helpful as you study this chapter. Find them in the **eToolkit** in ConnectED.

Go Online!
connectED.mcgraw-hill.com

CHAPTER 9
Trigonometric Functions

	Get Ready for Chapter 9	590
	Explore 9-1 Spreadsheet Lab: Investigating Special Right Triangles	593
9-1	**Trigonometric Functions in Right Triangles**	594
	Extend 9-1 Geometry Lab: Regular Polygons	603
9-2	**Angles and Angle Measure**	604
	Extend 9-2 Geometry Lab: Areas of Parallelograms	611
9-3	**Trigonometric Functions of General Angles**	612
	Mid Chapter Quiz	619
9-4	**Circular and Periodic Functions**	620
9-5	**Graphing Trigonometric Functions**	627
	Extend 9-5 Graphing Technology Lab: Trigonometric Graphs	634
9-6	**Translations of Trigonometric Graphs**	635

ASSESSMENT

Study Guide and Review	643
Practice Test	647
Performance Task	648
Preparing for Assessment, Test-Taking Strategy	649
Preparing for Assessment, Cumulative Review	650

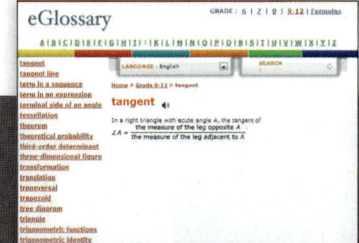

connectED.mcgraw-hill.com

Vocabulary is important to learning the Key Concepts in this chapter. Find all the terms with animations, English pronunciations, and translations to 13 languages in the eGlossary in ConnectED.

CHAPTER 10
Trigonometric Identities and Equations

	Get Ready for Chapter 10	652
10-1	Trigonometric Identities	655
10-2	Verifying Trigonometric Identities	662
10-3	Sum and Difference Identities	668
	Mid Chapter Quiz	674
10-4	Double-Angle and Half-Angle Identities	675
	Explore 10-5 Graphing Technology Lab: Solving Trigonometric Equations	682
10-5	Solving Trigonometric Equations	683

ASSESSMENT

Study Guide and Review	690
Practice Test	693
Performance Task	694
Preparing for Assessment, Test-Taking Strategy	695
Preparing for Assessment, Cumulative Review	696

Use **Self-Check Quizzes** in ConnectED to check your understanding of trigonometric identities and equations.

Go Online!
connectED.mcgraw-hill.com

connectED.mcgraw-hill.com   xxi

Student Handbook

Built-In Workbook

Extra Practice R1

Reference

Selected Answers and Solutions R11

Glossary/Glosario R100

Index R118

Formulas and Measures .... Inside Back Cover

Symbols and Properties .... Inside Back Cover

Standards for Mathematical Practice

Glencoe Algebra 2 exhibits these practices throughout the entire program. All of the Standards for Mathematical Practice will be covered in each chapter. The MP icon notes specific areas of coverage.

Mathematical Practices	What does it mean?
1. Make sense of problems and persevere in solving them.	Solving a mathematical problem takes time. Use a logical process to make sense of problems, understand that there may be more than one way to solve a problem, and alter the process if needed.
2. Reason abstractly and quantitatively.	You can start with a concrete or real-world context and then represent it with abstract numbers or symbols (decontextualize), find a solution, then refer back to the context to check that the solution makes sense (contextualize).
3. Construct viable arguments and critique the reasoning of others.	Sound mathematical arguments require a logical progression of statements and reasons. Mathematically proficient students can clearly communicate their thoughts and defend them.
4. Model with mathematics.	Modeling links classroom mathematics and statistics to everyday life, work, and decision-making. High school students at this level are expected to apply key takeaways from earlier grades to high-school level problems.
5. Use appropriate tools strategically.	Certain tools, including estimation and virtual tools are more appropriate than others. You should understand the benefits and limitations of each tool.
6. Attend to precision.	Precision in mathematics is more than accurate calculations. It is also the ability to communicate with the language of mathematics. In high school mathematics, precise language makes for effective communication and serves as a tool for understanding and solving problems.
7. Look for and make use of structure.	Mathematics is based on a well-defined structure. Mathematically proficient students look for that structure to find easier ways to solve problems.
8. Look for and express regularity in repeated reasoning.	Mathematics has been described as the study of patterns. Recognizing a pattern can lead to results more quickly and efficiently.

Folding Instructions

The following pages offer step-by-step instructions to make the Foldables® study guides.

Layered-Look Book

1. Collect three sheets of paper and layer them about 1 cm apart vertically. Keep the edges level.
2. Fold up the bottom edges of the paper to form 6 equal tabs.
3. Fold the papers and crease well to hold the tabs in place. Staple along the fold. Label each tab.

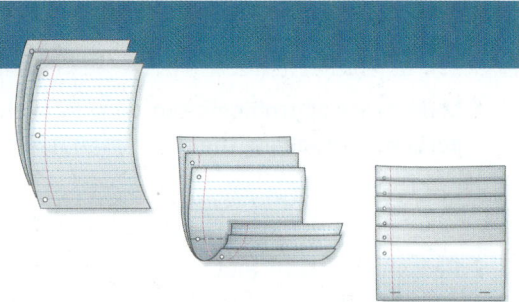

Shutter-Fold and Four-Door Books

1. Find the middle of a horizontal sheet of paper. Fold both edges to the middle and crease the folds. Stop here if making a shutter-fold book. For a four-door book, complete the steps below.
2. Fold the folded paper in half, from top to bottom.
3. Unfold and cut along the fold lines to make four tabs. Label each tab.

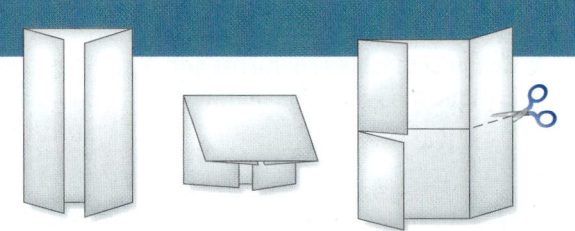

Concept-Map Book

1. Fold a horizontal sheet of paper from top to bottom. Make the top edge about 2 cm shorter than the bottom edge.
2. Fold width-wise into thirds.
3. Unfold and cut only the top layer along both folds to make three tabs. Label the top and each tab.

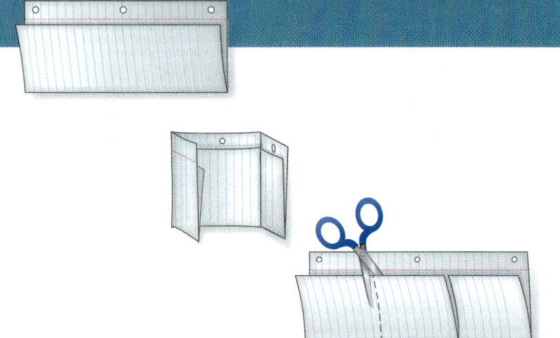

Vocabulary Book

1. Fold a vertical sheet of notebook paper in half.
2. Cut along every third line of only the top layer to form tabs. Label each tab.

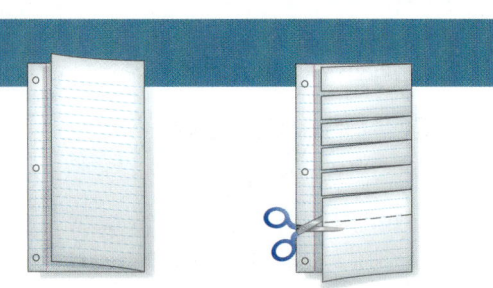

Pocket Book

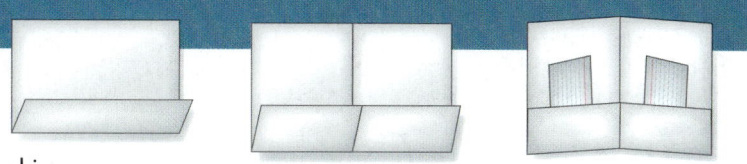

1. Fold the bottom of a horizontal sheet of paper up about 3 cm.

2. If making a two-pocket book, fold in half. If making a three-pocket book, fold in thirds.

3. Unfold once and dot with glue or staple to make pockets. Label each pocket.

Bound Book

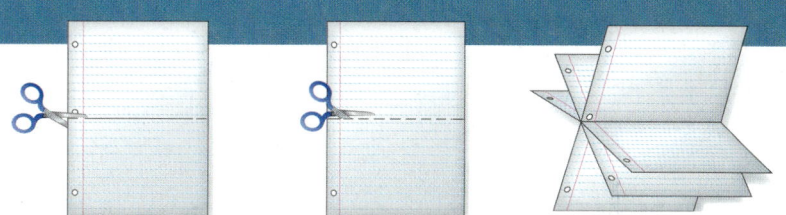

1. Fold several sheets of paper in half to find the middle. Hold all but one sheet together and make a 3-cm cut at the fold line on each side of the paper.

2. On the final page, cut along the fold line to within 3-cm of each edge.

3. Slip the first few sheets through the cut in the final sheet to make a multi-page book.

Top-Tab Book

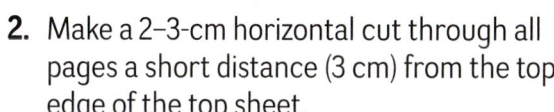

1. Layer multiple sheets of paper so that about 2–3 cm of each can be seen.

2. Make a 2–3-cm horizontal cut through all pages a short distance (3 cm) from the top edge of the top sheet.

3. Make a vertical cut up from the bottom to meet the horizontal cut.

4. Place the sheets on top of an uncut sheet and align the tops and sides of all sheets. Label each tab.

Accordion Book

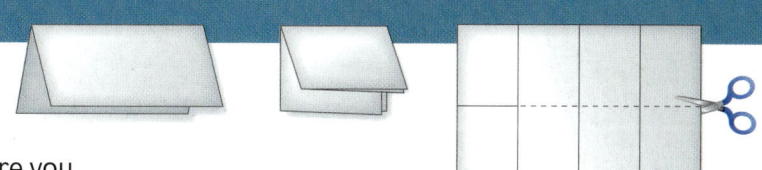

1. Fold a sheet of paper in half. Fold in half and in half again to form eight sections.

2. Cut along the long fold line, stopping before you reach the last two sections.

3. Refold the paper into an accordion book. You may want to glue the double pages together.

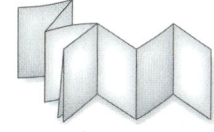

CHAPTER 0

Preparing for Advanced Algebra

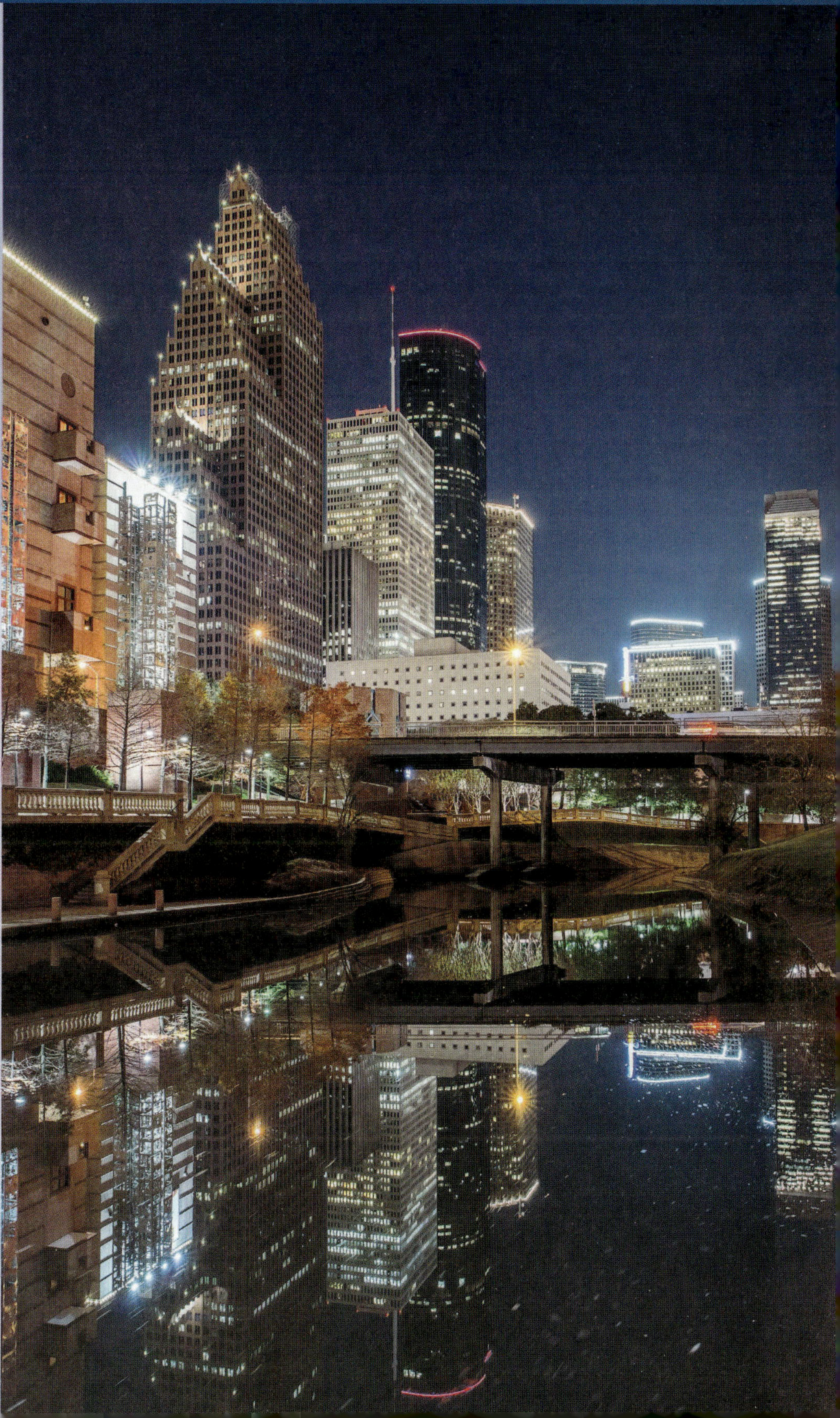

NOW

Chapter 0 contains lessons on topics from previous courses. You can use this chapter in various ways.

- Begin the school year by taking the Pretest. If you need additional review, complete the lessons in this chapter. To verify that you have successfully reviewed the topics, take the Posttest.
- As you work through the text, you may find that there are topics you need to review. When this happens, complete the individual lessons that you need.
- Use this chapter for reference. When you have questions about any of these topics, flip back to this chapter to review definitions or key concepts.

MP WHY

CITY LIFE Regardless of the size of the city, math is used every day in a myriad of ways by those who live and work in a city, including city planners.

Use the Mathematical Practices to complete the activity.

1. **Use Tools** Use the Internet or another source to find out about the demographics of a nearby city.
2. **Model with Mathematics** Use the Line Graph tool to show the change in population over time for the city you selected.

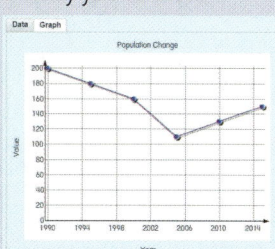

3. **Discuss** How could a city planner use this information in decisions about roads, public transportation, or schools?

jld3 Photography/Getty Images

 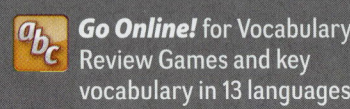

Go Online to Guide Your Learning

Go Online! for Vocabulary Review Games and key vocabulary in 13 languages.

Organize

Throughout this text, you will be invited to use **FOLDABLES** to organize your notes.

Why should you use them?
- They help you organize, display, and arrange information.
- They make great study guides, specifically designed for you.
- They give you a chance to improve your math vocabulary.

How should you use them?
- Write general information—titles, vocabulary terms, concepts, questions, and main ideas—on the front tabs of your Foldable.
- Write specific information—ideas, your thoughts, answers to questions, steps, notes, and definitions—under the tabs.

When should you use them?
- Set up your Foldable as you begin a chapter, or when you start learning a new concept.
- Write in your Foldable every day.
- Use your Foldable to review for homework, quizzes, and tests.

New Vocabulary

English		Español
domain	p. P4	dominio
range	p. P4	rango
quadrants	p. P4	cuadrantes
mapping	p. P4	transformaciones
function	p. P4	función
outcome	p. P9	resultados
sample space	p. P9	espacio muestral
permutation	p. P9	permutación
factorial	p. P10	factorial
combination	p. P11	combinación
theoretical probability	p. P13	probabilidad teórica
experimental probability	p. P13	probabilidad experimental
simple event	p. P14	evento simple
compound event	p. P14	suceso compuesto
mutually exclusive	p. P14	mutuamente exclusiva
independent events	p. P16	eventos independientes
dependent events	p. P16	eventos dependientes
conditional probability	p. P16	probabilidad condicional
two-way frequency table	p. P18	tabla de doble entrada o de contingencia
population	p. P24	población
sample	p. P24	muestra
mean	p. P24	media
median	p. P24	mediana
mode	p. P24	moda
range	p. P25	rango
variance	p. P25	varianza
standard deviation	p. P25	desviación estándar
five-number summary	p. P26	resumen de cinco números
outlier	p. P27	valor atípico

Explore & Explain

 Mapping Tool

Use the **Mapping** tool to explore relations and functions in Lesson 0-1.

CHAPTER 0
Pretest

State the domain and range of each relation. Then determine whether each relation is a function. Write *yes* or *no*.

1. $\{(14, 1), (-3, 6), (8, 4)\}$
2.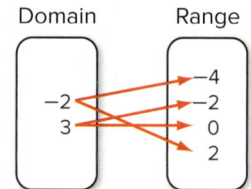

Name the quadrant in which each point is located.

3. $(-6, -2)$ 4. $(4, -3)$ 5. $(-5, 7)$

Find each product.

6. $(x + 1)(x + 4)$
7. $(a - 3)(a + 6)$
8. $(m - 2)(m - 5)$
9. $(c + 8)(c - 8)$

10. **NUMBER THEORY** There are two integers. One is 5 more than a number, and the other is 1 less than the same number.

   a. Write expressions for the two numbers.
   b. Write a polynomial expression for the product of the numbers.

Factor each polynomial.

11. $10ab^2 + 5b$
12. $15d - 12cd^2$
13. $y^2 + 6y - 7$
14. $a^2 - 13a + 36$

15. **ICE CREAM** How many different 1-scoop ice cream cones can you order if you have a choice of two cone types and 15 flavors of ice cream?

Determine whether each situation involves *permutations* or *combinations*. Then solve.

16. How many ways are there to place an algebra book, a geometry book, a chemistry book, an English book, and a health book on a shelf?

17. How many ways are there to select 3 of 15 flavors of juice at the grocery store?

A card is drawn at random from a standard deck. Determine whether the events are *mutually exclusive* or *not mutually exclusive*. Then find each probability.

18. $P(2 \text{ or black card})$
19. $P(\text{jack or } 4)$

Determine whether the events are *independent* or *dependent*. Then find the probability.

20. Two cards are selected at random one after the other from a standard deck without replacement. What is the probability that both cards are aces?

21. A coin is tossed and a dot cube is rolled. What is the probability that the coin lands on heads and the number rolled is greater than 4?

22. **SHOPPING** During one month, 18% of INET Clothing customers purchased item #345A. If 5% of customers order this item and item #345B, find the probability that someone who ordered item #345A also ordered item #345B.

23. Determine whether the triangles are *similar*, *congruent*, or *neither*.

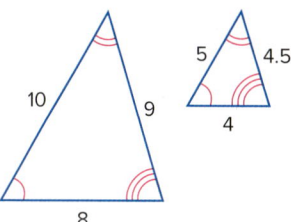

24. **GARDENS** A garden is 3 feet wide by 5 feet long. Susan enlarges the garden so that it is 12 feet long. If she wants to increase the width by the same proportion, how wide should she make the garden?

Find each missing measure. Round to the nearest tenth, if necessary.

25.

26. $a = 6$ yd, $b = 9$ yd, $c = ?$

The lengths of three sides of a triangle are given. Determine whether each triangle is a right triangle.

27. 12 yd, 14 yd, 16 yd
28. 15 km, 20 km, 25 km

Find the mean, median, mode, range, and standard deviation of each data set. Then identify any outliers.

29. scores for 16 students on a math test: 85, 100, 92, 36, 74, 88, 92, 86, 88, 82, 98, 70, 78, 92, 84, 100

30. weight to the nearest pound of 17 cats: 10, 15, 9, 8, 11, 10, 9, 8, 13, 9, 12, 10, 13, 11, 9, 12, 10

LESSON 1
Representing Functions

Objective
- Identify the domain and range of functions.

New Vocabulary
domain
range
quadrants
mapping
function

Recall that a *relation* is a set of ordered pairs. The **domain** of a relation is the set of all first coordinates (*x*-coordinates) from the ordered pairs, and the **range** is the set of all second coordinates (*y*-coordinates) from the ordered pairs.

Example 1 Domain and Range

State the domain and the range of the relation.
$\{(-3, 3), (0, -7), (1, -5), (2, 4)\}$

The domain is the set of *x*-coordinates.
$D = \{-3, 0, 1, 2\}$

The range is the set of *y*-coordinates.
$R = \{-7, -5, 3, 4\}$

A relation can be graphed on a coordinate plane. A coordinate plane is formed by the intersection of the horizontal axis, or *x*-axis, and the vertical axis, or *y*-axis. The axes meet at the origin (0, 0) and divide the plane into four **quadrants**. Any ordered pair in the coordinate plane can be written in the form (x, y).

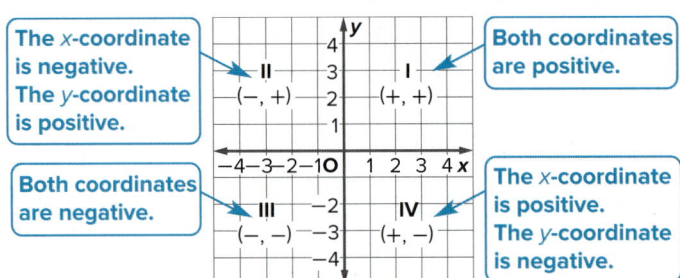

Example 2 Locate Coordinates

Name the quadrant in which $T(-8, 5)$ is located.

Point T has a negative *x*-coordinate and a positive *y*-coordinate. The point is located in Quadrant II.

A relation can also be represented by a table or a mapping. A **mapping** illustrates how each element of the domain is paired with an element in the range.

Ordered Pairs	Table		Graph	Mapping
(1, 2)	**x**	**y**		Domain → Range
(−2, 3)	1	2		1 → 2
(0, −3)	−2	3		−2 → 3
	0	−3		0 → −3

A **function** is a relation in which each element of the domain is paired with *exactly one* element of the range.

Example 3 Identify Domain and Range

State the domain and range of each relation. Then determine whether each relation is a function.

Watch Out!

Functions Remember that in a function, an element of the range can be paired with more than one element of the domain. But an element of the domain cannot be paired with more than one element of the range.

a. {(10, 3), (6, −2), (7, 4), (−8, −9)}

D = {−8, 6, 7, 10}
R = {−9, −2, 3, 4}
For each element of the domain, there is only one corresponding element in the range. So, this relation is a function.

b.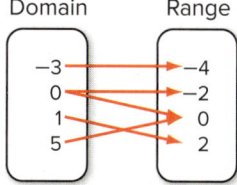

D = {1, 2, 3}
R = {3, 4, 7}

Because 1 is paired with 3 and 4, this is not a function.

c. Domain   Range

[mapping diagram: −3, 0, 1, 5 → −4, −2, 0, 2]

D = {−3, 0, 1, 5}
R = {−4, −2, 0, 2}
Because 0 is paired with −2 and 0, this is not a function.

d.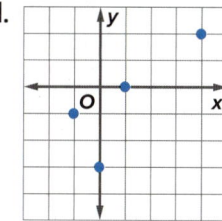

D = {−1, 0, 1, 4}
R = {−3, −1, 0, 2}
This is a function.

Exercises

State the domain and range of each relation. Then determine whether each relation is a function. Write *yes* or *no*.

1. {(2, 7), (3, 10), (1, 6)}

2. {(−6, 0), (5, 5), (9, −2), (−2, −9)}

3.
x	y
1	5
2	7
1	9

4.
x	y
−12	0
−10	1
−8	2
−6	4

5. Domain   Range

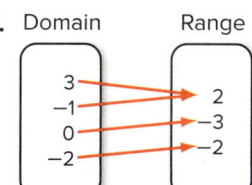

6. Domain   Range

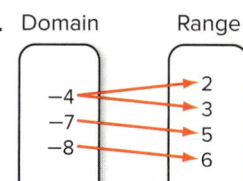

7.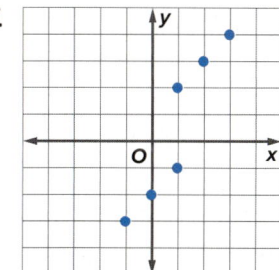

8.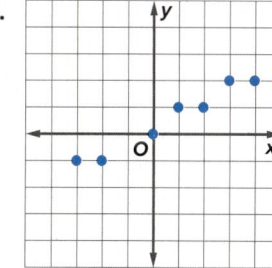

Name the quadrant in which each point is located.

9. (5, 3)   10. (8, −6)   11. (2, 0)   12. (−7, −1)

LESSON 2
FOIL

Objective
- Use the FOIL method to multiply binomials.

The product of two binomials is the sum of the products of the *first* terms, the *outer* terms, the *inner* terms, and the *last* terms.

F   O   I   L

Example 1   Use the FOIL Method

Find each product.

a. $(x + 3)(x - 5)$

$$\underset{\text{F}\ \ \ \ \text{L}}{(x + 3)(x - 5)} = \underset{\text{First}}{x \cdot x} + \underset{\text{Outer}}{x \cdot (-5)} + \underset{\text{Inner}}{3 \cdot x} + \underset{\text{Last}}{3 \cdot (-5)}$$

$$= x^2 - 5x + 3x - 15$$
$$= x^2 - 2x - 15$$

b. $(3y + 2)(5y + 4)$

$$(3y + 2)(5y + 4) = 3y \cdot 5y + 3y \cdot 4 + 2 \cdot 5y + 2 \cdot 4$$
$$= 15y^2 + 12y + 10y + 8$$
$$= 15y^2 + 22y + 8$$

Exercises

Find each product.

1. $(a + 2)(a + 4)$
2. $(v - 7)(v - 1)$
3. $(h + 4)(h - 4)$
4. $(d - 1)(d + 1)$
5. $(b + 4)(b - 3)$
6. $(t - 9)(t + 11)$
7. $(r + 3)(r - 8)$
8. $(k - 2)(k + 5)$
9. $(p + 8)(p + 8)$
10. $(x - 15)(x - 15)$
11. $(2c + 1)(c - 5)$
12. $(7n - 2)(n + 3)$
13. $(3m + 4)(2m - 5)$
14. $(5g + 1)(6g + 9)$
15. $(2q - 17)(q + 2)$
16. $(4t - 7)(3t - 12)$

17. **NUMBERS** I am thinking of two integers. One is 7 less than a number, and the other is 2 greater than the same number.
    a. Write expressions for the two numbers.
    b. Write a polynomial expression for the product of the numbers.

18. **OFFICE SPACE** Monica's current office is square. Her office in the company's new building will be 3 feet wider and 5 feet longer.
    a. Write expressions for the dimensions of Monica's new office.
    b. Write a polynomial expression for the area of Monica's new office.
    c. Suppose Monica's current office is 7 feet by 7 feet. How much larger will her new office be?

LESSON 3
Factoring Polynomials

Objective
- Use various techniques to factor polynomials.

Some polynomials can be factored using the Distributive Property.

Example 1 — Use the Distributive Property

Factor $4a^2 + 8a$.

Find the GCF of $4a^2$ and $8a$.
$4a^2 = 2 \cdot 2 \cdot a \cdot a \qquad 8a = 2 \cdot 2 \cdot 2 \cdot a \quad \rightarrow \quad$ GCF: $2 \cdot 2 \cdot a$ or $4a$

$4a^2 + 8a = 4a(a) + 4a(2)$ Rewrite each term using the GCF.
$ = 4a(a + 2)$ Distributive Property

Thus, the completely factored form of $4a^2 + 8a$ is $4a(a + 2)$.

To factor trinomials of the form $x^2 + bx + c$, find two integers m and p with a product equal to c and a sum equal to b. Then write $x^2 + bx + c$ using the pattern $(x + m)(x + p)$.

Example 2 — Use Factors and Sums

Factor each polynomial.

a. $x^2 + 5x + 6$ ← Both *b* and *c* are positive.

b is 5 and c is 6. Find two numbers with a product of 6 and a sum of 5.

Factors of 6	Sum of factors
1, 6	7
2, 3	5

The correct factors are 2 and 3.

$x^2 + 5x + 6 = (x + m)(x + p)$ Write the pattern.
$ = (x + 2)(x + 3)$ $m = 2$ and $p = 3$

b. $x^2 - 8x + 12$ ← *b* is negative, and *c* is positive.

$b = -8$ and $c = 12$. This means that $m + p$ is negative and mp is positive. So m and p must both be negative.

Factors of 12	Sum of factors
−1, −12	−13
−2, −6	−8

The correct factors are −2 and −6.

$x^2 - 8x + 12 = (x + m)(x + p)$ Write the pattern.
$ = [x + (-2)][x + (-6)]$ $m = -2$ and $p = -6$
$ = (x - 2)(x - 6)$ Simplify.

c. $x^2 + 14x - 15$ ← *b* is positive, and *c* is negative.

$b = 14$ and $c = -15$. This means that $m + p$ is positive and mp is negative. So either m or p must be negative, but not both.

Factors of −15	Sum of factors
1, −15	−14
−1, 15	14

The correct factors are −1 and 15.

$x^2 + 14x - 15 = (x + m)(x + p)$ Write the pattern.
$ = [x + (-1)](x + 15)$ $m = -1$ and $p = 15$
$ = (x - 1)(x + 15)$ Simplify.

connectED.mcgraw-hill.com

To factor quadratic trinomials of the form $ax^2 + bx + c$, find two integers m and p with a product equal to ac and a sum equal to b. Write $ax^2 + bx + c$ using the pattern $ax^2 + mx + px + c$. Then factor by grouping.

Example 3 — Use Factors and Sums

Factor $6x^2 + 7x - 3$.

$a = 6$, $b = 7$, and $c = -3$. This means that $m + p$ is positive and mp is negative. So either m or p must be negative, but not both.

Factors of −18	Sum of factors
1, −18	−17
−1, 18	17
2, −9	−7
−2, 9	7

The correct factors are −2 and 9.

$$\begin{aligned} 6x^2 + 7x - 3 &= 6x^2 + mx + px - 3 && \text{Write the pattern.}\\ &= 6x^2 + (-2)x + 9x - 3 && m = -2 \text{ and } p = 9\\ &= (6x^2 - 2x) + (9x - 3) && \text{Group terms with common factors.}\\ &= 2x(3x - 1) + 3(3x - 1) && \text{Factor the GCF from each group.}\\ &= (2x + 3)(3x - 1) && \text{Distributive Property} \end{aligned}$$

Study Tip

Checking Solutions You can check to see that you have factored correctly by multiplying the factors and comparing the product to the original polynomial.

Here are some special products.

Perfect Square Trinomials

$(a + b)^2 = (a + b)(a + b)$
$\qquad\quad = a^2 + 2ab + b^2$

$(a - b)^2 = (a - b)(a - b)$
$\qquad\quad = a^2 - 2ab + b^2$

Difference of Squares

$a^2 - b^2 = (a + b)(a - b)$

Example 4 — Use Special Products

Factor each polynomial.

a. $4x^2 + 20x + 25$

The first and last terms are perfect squares. The middle term is equal to $2(2x)(5)$. This is a perfect square trinomial of the form $(a + b)^2$.

$$\begin{aligned} 4x^2 + 20x + 25 &= (2x)^2 + 2(2x)(5) + 5^2 && \text{Write as } a^2 + 2ab + b^2.\\ &= (2x + 5)^2 && \text{Factor using the pattern.} \end{aligned}$$

b. $x^2 - 4$

This is a difference of squares.

$$\begin{aligned} x^2 - 4 &= x^2 - (2)^2 && \text{Write in the form } a^2 - b^2.\\ &= (x + 2)(x - 2) && \text{Factor the difference of squares.} \end{aligned}$$

Exercises

Factor each polynomial.

1. $12x^2 + 4x$
2. $6x^2y + 2x$
3. $8ab^2 - 12ab$
4. $x^2 + 5x + 4$
5. $y^2 + 12y + 27$
6. $x^2 + 6x + 8$
7. $3y^2 + 13y + 4$
8. $7x^2 + 51x + 14$
9. $3x^2 + 28x + 32$
10. $x^2 - 5x + 6$
11. $y^2 - 5y + 4$
12. $6x^2 - 13x + 5$
13. $6a^2 - 50ab + 16b^2$
14. $11x^2 - 78x + 7$
15. $18x^2 - 31xy + 6y^2$
16. $x^2 + 4xy + 4y^2$
17. $9x^2 - 24x + 16$
18. $4a^2 + 12ab + 9b^2$
19. $x^2 - 144$
20. $4c^2 - 9$
21. $16y^2 - 1$
22. $25x^2 - 4y^2$
23. $36y^2 - 16$
24. $9a^2 - 49b^2$

LESSON 4
Counting Techniques

:·Objective

- Use various techniques to factor polynomials.

New Vocabulary

outcome
probability experiment
sample space
tree diagram
permutation
factorial
combination

Recall that an **outcome** is the result of a single trial of a process involving chance, called a **probability experiment**. The set of all possible outcomes of such an experiment is called a **sample space**. A **tree diagram**, like the one shown, can be used to systematically list the outcomes in a sample space. Notice that there are 8 total outcomes in this sample space.

When you only need to know the total number of outcomes, you can use the **Fundamental Counting Principle**.

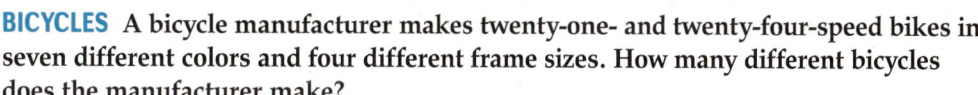

	Key Concept	Fundamental Counting Principle
Words		The number of possible outcomes in a sample space can be found by multiplying the number of possible outcomes for each event.
Symbols		If event M can occur in m ways and is followed by event N that can occur in n ways, then the event M followed by N can occur in m · n ways.

This rule can be extended to three or more events.

If there are 4 possible blood types and 2 possible Rh-factors, there are 4 · 2 or 8 possible ways to label blood donations.

Example 1 Fundamental Counting Principle

BICYCLES A bicycle manufacturer makes twenty-one- and twenty-four-speed bikes in seven different colors and four different frame sizes. How many different bicycles does the manufacturer make?

There are 2 gear choices, 7 color choices, and 4 frame choices. Apply the Fundamental Counting Principle.

gear		color		frame		number of bicycles
2	·	7	·	4	=	56

56 different bicycles can be made.

Counting problems can also involve determining the number of different arrangements of objects. An arrangement of a group of distinct objects in a certain order is called a **permutation**. For example, there are 6 permutations of the letters A, B, and C.

ABC   ACB   BAC   BCA   CAB   CBA

To determine this number mathematically, apply the Fundamental Counting Principle. There are 3 choices for the first letter. Once this letter is chosen, 2 choices remain for the second letter. After the first two letters are chosen, only one choice remains.

first letter		second letter		third letter		permutations
3	·	2	·	1	=	6

The product 3 · 2 · 1 can also be written as 3!, which is read *3 factorial*.

connectED.mcgraw-hill.com   P9

> **Key Concept** Factorial
>
> **Words** *n* factorial is the product of all counting numbers beginning with *n* and counting backward to 1. Zero factorial is defined to equal 1.
>
> **Symbols** $n! = n \cdot (n-1) \cdot (n-2) \cdot (n-3) \cdots 3 \cdot 2 \cdot 1$ and $0! = 1$

The number of permutations of *n* distinct objects is *n*!.

> **Example 2** Permutations of *n* Objects
>
> **BAND** There are 8 finalists in a band competition. In how many different ways can the bands be ranked if they cannot receive the same ranking?
>
> $8! = 8 \cdot 7 \cdot 6 \cdot 5 \cdot 4 \cdot 3 \cdot 2 \cdot 1$ or 40,320   Permutations of *n* distinct objects
>
> The bands can be ranked in 40,320 different ways.

Suppose in Example 2 you wanted to know how many different ways the first-, second-, and third-place rankings in the competition could be awarded. This also involves permutations. It is the number of permutations of 8 finalists taken 3 at a time.

first place		second place		third place		permutations
8	•	7	•	6	=	336

The number of permutations can also be found using factorials by finding the number of permutations of all 8 bands and dividing out the number of arrangements of bands that finished with rankings other than first, second, or third.

$$\frac{8!}{(8-3)!} = \frac{8!}{5!} = \frac{8 \cdot 7 \cdot 6 \cdot 5 \cdot 4 \cdot 3 \cdot 2 \cdot 1}{5 \cdot 4 \cdot 3 \cdot 2 \cdot 1} = 8 \cdot 7 \cdot 6$$

This result is generalized below.

> **Key Concept** Permutations of *n* Objects Taken *r* at a Time
>
> The number of permutations of *r* objects taken from a group of *n* distinct objects is given by
>
> $$_nP_r = \frac{n!}{(n-r)!}.$$

Reading Math

Permutations The number of permutations of *n* objects taken *r* at a time can also be denoted as $P(n, r)$.

> **Example 3** Permutation of *n* Objects Taken *r* at a Time
>
> **TUTORING** How many different ways can two students be assigned to five tutors if only one student is assigned to each tutor?
>
> Because the order in which the students are arranged determines the tutor to which a student is assigned, this situation involves permutations.
>
> $_5P_2 = \dfrac{5!}{(5-2)!}$ Permutations of 5 things taken 2 at a time
>
> $= \dfrac{5!}{3!}$ Subtract.
>
> $= \dfrac{5 \cdot 4 \cdot 3 \cdot 2 \cdot 1}{3 \cdot 2 \cdot 1}$ or 20   Divide out common factors and simplify.
>
> The students can be assigned in 20 different ways.

Study Tip

With Repetition The number of permutations of *n* objects in which one object is repeated *p* times and another object is repeated *q* times is $\dfrac{n!}{p!q!}$. This rule can be extended to permutations involving three or more repeated objects.

A selection of distinct objects in which the order of the objects selected is *not* important is called a **combination**. Suppose you select 3 out of 4 books, A, B, C, and D, for a book report. Selecting books A, B, and C is not different from selecting books C, B, A, or any other arrangements of these three books.

To find the number of combinations of *n* objects taken *r* at a time, divide the number of permutations by the number of ways to arrange each selection of *r* objects.

> **Key Concept** Combinations of *n* Objects Taken *r* at a Time
>
> The number of combinations of *r* objects taken from a group of *n* distinct objects is given by
> $$_nC_r = \frac{_nP_r}{r!} = \frac{n!}{(n-r)!\,r!}.$$

Example 4   Combinations of *n* Objects Taken *r* at a Time

CARDS How many ways are there to choose 5 cards from a standard deck of 52 playing cards?

Because the order in which the cards are chosen is not important, this situation involves combinations.

$$_{52}C_5 = \frac{52!}{(52-5)!\,5!} \quad \text{Combinations of 52 things taken 5 at a time}$$

$$= \frac{52!}{47!\,5!} \quad \text{Subtract.}$$

$$= 2{,}598{,}960 \quad \text{Use a calculator.}$$

There are 2,598,960 ways to choose 5 cards from a standard deck of playing cards.

Study Tip

Evaluating Permutations and Combinations You can also use the *nPr* and *nCr* features on your graphing calculator to evaluate permutations and combinations, respectively.

A critical first step in solving a counting problem is deciding whether order is important. If the order of objects selected is important, then the problem involves permutations. If objects are selected without regard to order, the problem involves combinations.

Example 5   Permutations or Combinations?

Twenty-five students write their names on slips of paper. Then three different names are chosen at random to receive prizes. Determine whether each situation involves *permutations* or *combinations*.

a. choosing 3 people to each receive a "no homework" coupon

Because the prizes are identical, the order in which the students' names are drawn *is not* important. This situation involves combinations.

b. choosing 3 people to each receive one of the following prizes: 1st prize, a new graphing calculator; 2nd prize, a "no homework" coupon; 3rd prize, a new pencil

Because the prizes are different, the order in which the students' names are drawn *is* important. This situation involves permutations.

Exercises

Use the Fundamental Counting Principle to determine the number of outcomes.

1. **FOOD** How many different combinations of sandwich, side, and beverage are possible?

Sandwiches	Sides	Beverages
hot dog	chips	bottled water
hamburger	apple	soda
veggie burger	pasta salad	juice
bratwurst		milk
grilled chicken		

2. **QUIZZES** Each question on a five-question multiple-choice quiz has answer choices labeled A, B, C, and D. How many different ways can a student answer the five questions?

3. **DANCES** Dane is renting a tuxedo for prom. Once he has chosen his jacket, he must choose from three types of pants and six colors of vests. How many different ways can he select his attire for prom?

4. **MANUFACTURING** A baseball glove manufacturer makes a glove with the different options shown in the table. How many different gloves are possible?

Option	Number of Choices
sizes	4
types by position	3
materials	2
levels of quality	2

Evaluate each permutation or combination.

5. $_6P_3$
6. $_7P_5$
7. $_4C_2$
8. $_{12}C_7$
9. $_6C_1$
10. $_9P_5$

Determine whether each situation involves *permutations* or *combinations*. Then solve the problem.

11. **SCHOOL** Charlita wants to take 6 different classes next year. Assuming that each class is offered each period, how many different schedules could she have?

12. **BALLOONS** How many 4-colored groups can be selected from 13 different colored balloons?

13. **CONTEST** How many ways are there to choose the winner and first, second, and third runners-up in a contest with 10 finalists?

14. **BANDS** A band is choosing 3 new backup singers from a group of 18 who try out. How many ways can they choose the new singers?

15. **PIZZA** How many different two-topping pizzas can be made if there are 6 options for toppings?

16. **SOFTBALL** How many ways can the manager of a softball team choose players for the top 4 spots in the lineup if she has 7 possible players in mind?

17. **EMPLOYEES** A department store has 9 employees available to cover 4 different sections of the store. How many ways can the employees be assigned?

18. **READING** Jack has a reading list of 12 books. How many ways can he select 9 books from the list to check out of the library?

19. **CHALLENGE** Abby is registering at a Web site and must select a six-character password. The password can contain either letters or digits.

    a. How many passwords are possible if characters can be repeated? if no characters can be repeated?

    b. How many passwords are possible if all characters are letters that can be repeated? if the password must contain exactly one digit? Which type of password is more secure? Explain.

LESSON 5
Adding Probabilities

Objectives

- Compute theoretical and experimental probabilities.
- Compute probabilities of compound events.

New Vocabulary
probability
probability model
uniform or simple probability model
theoretical probability
experimental probability
simple event
compound event
mutually exclusive events
odds

Probability is a measure of the chance that a given event E will occur. The probability $P(E)$ that an event will occur is always given as a ratio between 0 and 1, inclusive.

E will not occur. E is equally likely to occur or not occur. E is certain to occur.

$P(E) = 0$ $P(E) = \frac{1}{2}$ $P(E) = 1$

A **probability model** is a mathematical model used to represent the outcomes of an experiment. A **uniform** or **simple probability model** is used to describe an experiment for which all outcomes are equally likely or have the same probability of occurring.

If we assume that all outcomes of an experiment are equally likely, then we can calculate the **theoretical probability** that an event will occur using the sample space of possible outcomes. We can also calculate the empirical or **experimental probability** using outcomes obtained by actually performing trials of the experiment.

Key Concept Theoretical and Experimental Probability

If each outcome is assumed to be equally likely, the theoretical probability P of an event E is given by

$$P(E) = \frac{\text{number of favorable outcomes}}{\text{number of possible outcomes}}.$$

Given the frequency of outcomes from a certain number of trials of an experiment, the experimental probability P of an event E is given by

$$P(E) = \frac{\text{number of favorable trials}}{\text{number of trials}}.$$

Example 1 Theoretical and Experimental Probability

The graph shows the results of several trials of an experiment in which a single number cube is rolled.

a. What is the experimental probability of rolling a 6?

From the graph we can see that there were 10 trials that resulted in 6. The total number of trials was $9 + 6 + 12 + 6 + 7 + 10$ or 50.

$P(6) = \frac{\text{number of favorable trials}}{\text{number of trials}}$

$= \frac{10}{50}$ or $\frac{1}{5}$, which is 20%

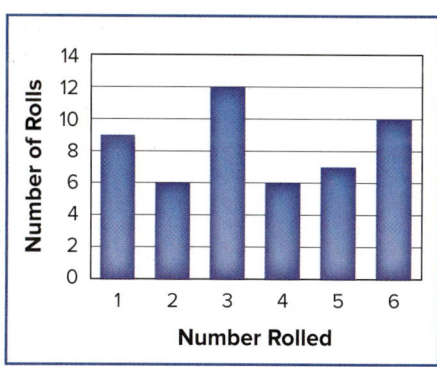

b. What is the theoretical probability of rolling a 6?

List the sample space. The possible outcomes for this experiment are 1, 2, 3, 4, 5, or 6. Of these 6 possible outcomes, only one is a 6.

$P(6) = \frac{\text{number of favorable outcomes}}{\text{number of possible outcomes}} = \frac{1}{6}$ or about 17%

connectED.mcgraw-hill.com **P13**

An event that has a single outcome, such as rolling a 6 on a number cube, is called a **simple event**. Many problems involve finding the probability of a composite or **compound event**, which consists of two or more simple events. To find the probability of a compound event, you must consider whether the events can occur at the same time.

Events that cannot occur at the same time are said to be **mutually exclusive**. Mutually exclusive events have no outcomes in common. For example, because it is not possible to draw a card from a standard deck that is both a king and a queen, these two events are mutually exclusive. It is possible to draw a card that is both a king and a spade, so these events are *not* mutually exclusive.

> **Study Tip**
> **Mutually and Not Mutually Exclusive** Events that are mutually exclusive are sometimes said to be disjoint. Events that are not mutually exclusive are sometimes said to be inclusive events.

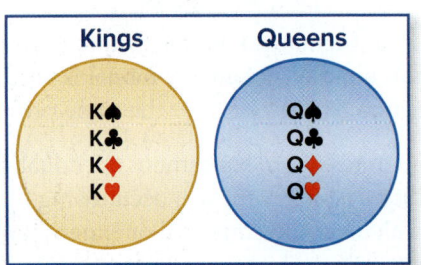

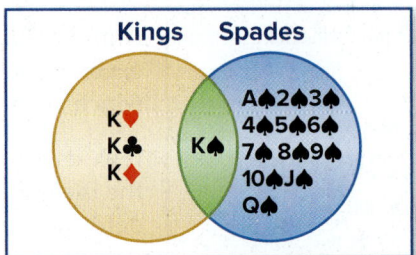

> **Key Concept** Addition Rules for Probability
>
> If two events *A* and *B* are **mutually exclusive**, the probability that *A* or *B* will occur is
>
> $$P(A \text{ or } B) = P(A) + P(B).$$
>
> If two events *A* and *B* are **not mutually exclusive**, the probability that *A* or *B* will occur is
>
> $$P(A \text{ or } B) = P(A) + P(B) - P(A \text{ and } B).$$

> **Study Tip**
> **Why Subtract?** Notice that when two events are not mutually exclusive, you add the probabilities and then subtract the probabilities of the outcomes that are common to both. If these common probabilities are not subtracted, you have actually counted them twice.

Example 2 Add Probabilities

Determine whether the events are *mutually exclusive* or *not mutually exclusive*. Then find the probability.

a. **Tasheika's teacher is randomly assigning students a day to present their projects. If 8 students present on Tuesday, 5 on Wednesday, and 6 on Thursday, what is the probability that Tasheika will present on Tuesday or Thursday?**

These events are mutually exclusive, because Tasheika will not present on both Tuesday *and* Thursday. There are a total of 8 + 5 + 6 or 19 different spots for presentations.

$P(\text{Tuesday or Thursday}) = P(\text{Tuesday}) + P(\text{Thursday})$ Mutually exclusive events

$= \frac{8}{19} + \frac{6}{19}$ or $\frac{14}{19}$ Substitute and add.

b. **Suppose that of 1400 students, 550 take Spanish, 700 take biology, and 400 take both Spanish and biology. What is the probability that a student selected at random takes Spanish or biology?**

Because some students take both Spanish *S* and biology *B*, the events are not mutually exclusive.

$P(S) = \frac{550}{1400}$ or $\frac{11}{28}$ $P(B) = \frac{700}{1400}$ or $\frac{14}{28}$ $P(S \text{ and } B) = \frac{400}{1400}$ or $\frac{8}{28}$

$P(S \text{ or } B) = P(S) + P(B) - P(S \text{ and } B)$ Not mutually exclusive events

$= \frac{11}{28} + \frac{14}{28} - \frac{8}{28}$ or $\frac{17}{28}$ Substitute and simplify.

Exercises

1. **CARNIVAL GAMES** A spinner has sections of equal size. The table shows the results of several spins.

Color	Frequency
red	6
blue	7
yellow	9
orange	12
purple	5
green	11

   a. Copy the table and add a column to show the experimental probability of the spinner landing on each of the colors with the next spin.
   b. Create a bar graph that shows these experimental probabilities.
   c. Add a column to your table that shows the theoretical probability of the spinner landing on each of the colors with the next spin.
   d. Create a bar graph that shows these theoretical probabilities.
   e. Interpret and compare the graphs you created in parts b and d.

Determine whether the events are *mutually exclusive* or *not mutually exclusive*. Then find the probability.

2. Two number cubes are rolled.
   a. P(sum of 10 or doubles)   b. P(sum of 6 or 7)   c. P(sum < 3 or sum > 10)

3. A card is drawn at random from a standard deck of cards.
   a. P(club or diamond)   b. P(ace or spade)   c. P(jack or red card)

4. In a French class, there are 10 freshmen, 8 sophomores, and 2 juniors. Of these students, 9 freshmen, 2 sophomores, and 1 junior are female. A student is selected at random.
   a. P(freshman or female)   b. P(sophomore or male)   c. P(freshman or sophomore)

5. There are 40 vehicles on a rental car lot. All are either sedans or SUVs. There are 18 red vehicles, and 3 of them are sedans. There are 15 blue vehicles, and 9 of them are SUVs. Of the remaining vehicles, all are black and 2 are SUVs. A vehicle is selected at random.
   a. P(blue or black)   b. P(red or SUV)   c. P(black or sedan)

6. **DRIVING** A survey of Longview High School students found that the probability of a student driving while texting was 0.16, the probability of a student getting into an accident while driving was 0.07, and the probability of a student getting into an accident while driving and texting was 0.05. What is the probability of a student driving while texting or getting into an accident while driving?

7. **REASONING** Explain why the rule $P(A \text{ or } B) = P(A) + P(B) - P(A \text{ and } B)$ can be used for both mutually exclusive and not mutually exclusive events.

ODDS Another measure of the chance that an event will occur is called odds. The odds of an event occurring is a ratio that compares the number of ways an event can occur s (successes) to the number of ways it cannot occur f (failure), or s to f. The sum of the number of success and failures equals the number of possible outcomes.

8. A card is drawn from a standard deck of 52 cards. Find the odds in favor of drawing a heart. Then find the odds against drawing an ace.

9. Two fair coins are tossed. Find the odds in favor of both landing on heads. Then find the odds in favor of exactly one landing on tails.

10. The results of rolling a dot cube 120 times are shown.

Roll	1	2	3	4	5	6
Frequency	16	24	17	25	30	8

    Find the experimental odds against rolling a 1 or a 6. Then find the experimental odds in favor of rolling a number less than 3.

LESSON 6
Multiplying Probabilities

Objectives
- Find probabilities of independent and dependent events.
- Use two-way frequency tables to find conditional probabilities.

 **New Vocabulary**
independent events
dependent events
conditional probability
two-way frequency table

If the occurrence of one event does not affect the probability of a second event occurring, then the two events are **independent events**.

Key Concept Probability of Independent Events

If two events A and B are independent, then the probability that A and B will occur is
$$P(A \text{ and } B) = P(A) \cdot P(B).$$

This rule can be extended to three or more independent events.

Example 1   Probability of Independent Events

A coin is tossed and a number cube is rolled. What is the probability of the coin landing on tails and rolling a 3?

Because the outcome of tossing the coin does not affect the outcome of rolling the cube, these events are independent.

$P(\text{tails and } 3) = P(\text{tails}) \cdot P(3)$ Probability of independent events

$= \dfrac{1}{2} \cdot \dfrac{1}{6}$ or $\dfrac{1}{12}$ $P(\text{tails}) = \dfrac{1}{2}$ and $P(3) = \dfrac{1}{6}$

CHECK Use a listing of the possible outcomes from this compound event.

H1, H2, H3, H4, H5, H6, T1, T2, **T3**, T4, T5, T6

There is only 1 outcome out of the 12 possible outcomes that indicates landing on tails and rolling a 3, so $P(\text{tails and } 3)$ is $\dfrac{1}{12}$. ✓

If the occurrence of the first event *does* affect the probability of the second event occurring, then the events are **dependent events**. An example of dependent events is drawing a card from a standard deck of cards, not putting it back, and then drawing a second card. To find the probability of drawing a jack and then a queen in this situation, we modify the multiplication rule.

$P(\text{jack and queen}) = \underset{\text{Out of 52 cards, 4 are jacks.}}{\underset{\text{jack on the first draw}}{\underset{\text{Probability of drawing a}}{\dfrac{4}{52}}}} \cdot \underset{\text{Out of the 51 cards left, 4 are queens.}}{\underset{\text{draw given that a jack was drawn first}}{\underset{\text{Probability of drawing a queen on second}}{\dfrac{4}{51}}}}$

The probability of an event A occurring given that event B has already occurred is called a **conditional probability** and is represented by $P(B|A)$, read *the probability of B given A*. This notation is used in the rule for the probability of two dependent events.

Key Concept Probability of Dependent Events

If two events A and B are dependent, then the probability that A and B will occur is
$$P(A \text{ and } B) = P(A) \cdot P(B|A).$$

This rule can be extended to three or more dependent events.

Example 2 Probability of Dependent Events

A bag contains 12 red, 9 blue, 11 yellow, and 8 green marbles. If two marbles are drawn at random and not replaced, what is the probability that a red and then a blue marble are drawn?

The event of drawing the first marble affects the probability of drawing the second marble, because there is one fewer marble from which to choose. So, the events are dependent.

$P(\text{red}) = \dfrac{12}{40}$ ← number of red marbles / total number of marbles

$P(\text{blue} \mid \text{red}) = \dfrac{9}{39}$ ← number of blue marbles / number of marbles remaining

$= \dfrac{3}{10}$

$= \dfrac{3}{13}$

$P(\text{red and blue}) = P(\text{red}) \cdot P(\text{blue} \mid \text{red})$ Probability of dependent events

$= \dfrac{3}{10} \cdot \dfrac{3}{13}$ Substitute.

$= \dfrac{9}{130}$ Multiply.

The probability that a red and then a blue marble are drawn is $\dfrac{9}{130}$.

Solving the equation for the probability of two dependent events for $P(A \mid B)$ gives a rule for computing the conditional probability of an event.

$P(A \text{ and } B) = P(A) \cdot P(B \mid A)$ Probability of dependent events

$\dfrac{P(A \text{ and } B)}{P(A)} = P(B \mid A)$ Divide each side by $P(A)$.

Study Tip

Independent Events If A and B are independent events, then $P(B \mid A) = P(B)$.

Key Concept Conditional Probability

If A and B are dependent events, then the conditional probability of event B occurring, given that event A has already occurred, is

$$P(B \mid A) = \dfrac{P(A \text{ and } B)}{P(A)}, \text{ where } P(A) \neq 0.$$

Example 3 Conditional Probability

FOOD At a restaurant, 25% of customers order chili. If 4% of customers order chili and a baked potato, find the probability that someone who orders chili also orders a baked potato.

$P(\text{baked potato} \mid \text{chili}) = \dfrac{P(\text{chili and baked potato})}{P(\text{chili})}$ Conditional probability

$= \dfrac{0.04}{0.25}$ $P(\text{chili and baked potato}) = 0.04$ and $P(\text{chili}) = 0.25$

$= 0.16$ Simplify.

The probability that someone who orders chili also orders a baked potato is 16%.

Review Vocabulary

relative frequency the ratio of the number of observations in a statistical category to the total number of observations

A *contingency* or two-way frequency table is often used to show the observed or relative frequencies of data from an experiment classified according to two variables, with the rows indicating one variable and the columns indicating the other. These tables can be used to find conditional probabilities.

Real-World Example 4   Two-Way Frequency Table

MEDICINE A drug company conducted an experiment to determine the effectiveness of a certain new drug. Test subjects were randomly assigned to one of two groups: a treatment group, which received the drug, or a control group, which received a placebo instead of the drug. The contingency table below shows the results.

Group	Condition Improves (Y)	Condition Does Not Improve (N)
Treatment (T)	1600	400
Control (C)	1200	800

a. Find the probability that a test subject's condition improved given that he or she was in the treatment group.

Step 1 Add a row and a column to the table, and calculate the totals.

Group	Condition Improves (Y)	Condition Does Not Improve (N)	Totals
Treatment (T)	1600	400	2000
Control (C)	1200	800	2000
Totals	2800	1200	4000

Step 2 Divide each observed frequency by the total number of people in the study, 4000, to obtain a table of relative frequencies.

Group	Condition Improves (Y)	Condition Does Not Improve (N)	Totals
Treatment (T)	0.4	0.1	0.5
Control (C)	0.3	0.2	0.5
Totals	0.7	0.3	1

Study Tip

Joint and Marginal Frequencies The totals row and totals column in a two-way table report the *marginal frequencies*, while the cells in the interior of the table report the *joint frequencies*. The total number of observations is reported in the bottom right corner of the table.

Step 3 Find the conditional probability $P(Y|T)$.

$P(Y|T) = \dfrac{P(T \text{ and } Y)}{P(T)}$ Conditional probability

$= \dfrac{0.4}{0.5}$ or 0.8 $P(T \text{ and } Y) = 0.4$ and $P(T) = 0.5$

The probability that a subject's condition improved given that he or she was in the treatment group is 0.8 or 80%.

b. Find the probability that a test subject was in the control group given that his or her condition did not improve.

$P(C|N) = \dfrac{P(N \text{ and } C)}{P(N)}$ Conditional probability

$= \dfrac{0.2}{0.3}$ or about 0.667 $P(N \text{ and } C) = 0.2$ and $P(N) = 0.3$

The probability that a test subject was in the control group given that his or her condition did not improve is about 0.667 or about 66.7%.

Exercises

Determine whether the events are *independent* or *dependent*. Then find the probability.

1. A red dot cube and a blue dot cube are rolled. What is the probability of getting the result shown?

2. Yana has 4 black socks, 6 blue socks, and 8 white socks in his drawer. If he selects three socks at random with no replacement, what is the probability that he will first select a blue sock, then a black sock, and then another blue sock?

A number cube is rolled twice. Find each probability.

3. P(2 and 3)   4. P(two 4s)   5. P(no 6s)   6. P(two of the same number)

A bag contains 8 blue marbles, 6 red marbles, and 5 green marbles. Three marbles are drawn one at a time. Find each probability.

7. The second marble is green, given that the first marble is blue and not replaced.

8. The second marble is red, given that the first marble is green and is replaced.

9. The third marble is red, given that the first two are red and blue and not replaced.

10. The third marble is green, given that the first two are red and are replaced.

CLOTHING There are 8 black, 3 red, and 5 white shirts in a closet. Suppose three shirts are selected at random from the closet. Find each probability.

11. P(3 black), with replacement

12. P(2 black, then red), without replacement

13. **CARDS** You draw a card from a standard deck of cards and show it to a friend. The friend tells you that the card is red. What is the probability that you correctly guess that the card is the ace of diamonds?

14. **HONOR ROLL** Suppose the probability that a student takes AP Calculus and is on the honor roll is 0.0035, and the probability that a student is on the honor roll is 0.23. Find the probability that a student takes AP Calculus given that he or she is on the honor roll.

15. **DRIVING TESTS** The table shows how students in Mr. Diaz's class fared on their first driving test. Some took a class to prepare, while others did not. Find each probability.

Status	Class	No Class
passed	64	48
failed	18	32

    a. Paige passed, given that she took the class.
    b. Madison failed, given that she did not take the class.
    c. Jamal did not take the class, given that he passed.

16. **SCHOOL CLUBS** King High School tallied the number of males and females that were members of at least one after school club. Find each probability.

Gender	Clubs	No Clubs
male	156	242
female	312	108

    a. A student is a member of a club given that he is male.
    b. A student is not a member of a club given that she is female.
    c. A student is a male given that he is not a member of a club.

17. **FOOTBALL ATTENDANCE** The number of students who have attended a football game at North Coast High School is shown. Find each probability.

Class	Freshman	Sophomore	Junior	Senior
attended	48	90	224	254
not attended	182	141	36	8

    a. Given that a student is a freshman, the student has not attended a game.
    b. Given that a student has attended a game, the student is an upperclassman (a junior or senior).

LESSON 7
Congruent and Similar Figures

:: Objective
- Identify and use congruent and similar figures.

New Vocabulary
congruent
similar

Congruent figures have the same size and the same shape. Two polygons are congruent if their corresponding sides and their corresponding angles are congruent.

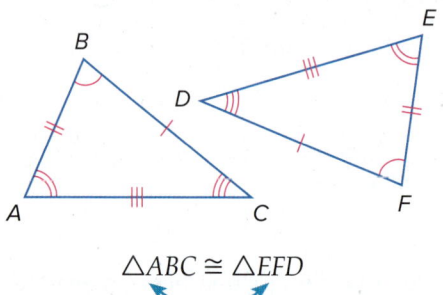

△ABC ≅ △EFD

The order of the vertices indicates the corresponding parts.

Congruent Angles	Congruent Sides
∠A ≅ ∠E	$\overline{AB} \cong \overline{EF}$
∠B ≅ ∠F	$\overline{BC} \cong \overline{FD}$
∠C ≅ ∠D	$\overline{AC} \cong \overline{ED}$

Read the symbol ≅ as *is congruent to*.

Example 1 Congruence Statements

The corresponding parts of two congruent triangles are marked on the figure. Write a congruence statement for the two triangles.

List the congruent angles and sides.

∠A ≅ ∠D   ∠B ≅ ∠E   ∠ACB ≅ ∠DCE

$\overline{AB} \cong \overline{DE}$ $\overline{AC} \cong \overline{DC}$ $\overline{BC} \cong \overline{EC}$

Match the vertices of the congruent angles. Therefore, △ABC ≅ △DEC.

Similar figures have the same shape, but not necessarily the same size.

In similar figures, corresponding angles are congruent, and the measures of corresponding sides are proportional. (They have equivalent ratios.)

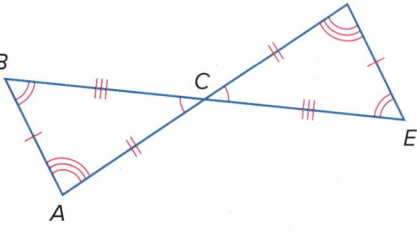

△ABC ~ △DEF

Congruent Angles
∠A ≅ ∠D, ∠B ≅ ∠E, ∠C ≅ ∠F

Proportional Sides
$\dfrac{AB}{DE} = \dfrac{BC}{EF} = \dfrac{AC}{DF}$

Read the symbol ~ as *is similar to*.

Example 2 Determine Similarity

Determine whether the polygons are similar. Justify your answer.

Because $\dfrac{4}{3} = \dfrac{8}{6} = \dfrac{4}{3} = \dfrac{8}{6}$, the measures of the sides are proportional. However, the corresponding angles are not congruent. The polygons are not similar.

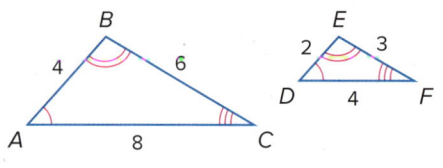

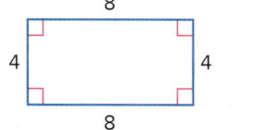
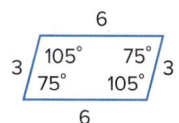

Example 3 Solve a Problem Involving Similarity

CIVIL ENGINEERING The city of Mansfield plans to build a bridge across Pine Lake. Use the information in the diagram at the right to find the distance across Pine Lake.

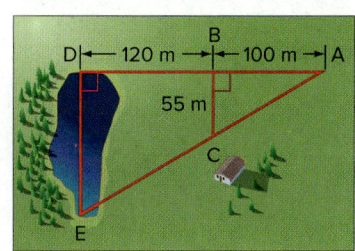

$\triangle ABC \sim \triangle ADE$

$\dfrac{AB}{AD} = \dfrac{BC}{DE}$ Definition of similar polygons

$\dfrac{100}{220} = \dfrac{55}{DE}$ $AB = 100$, $AD = 100 + 120$ or 220, $BC = 55$

$100DE = 220(55)$ Cross products
$100DE = 12{,}100$ Simplify.
$DE = 121$ Divide each side by 100.

The distance across the lake is 121 meters.

Study Tip

Reasonableness When solving a problem using a proportion, examine your solution for reasonableness. In Example 3, *DA* is more than twice *BA*, so *DE* should be more than twice *BC*. The solution is reasonable.

Exercises

Determine whether each pair of figures is *similar*, *congruent*, or *neither*.

1.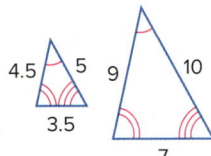
2.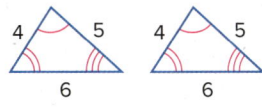
3.
4.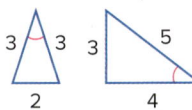
5.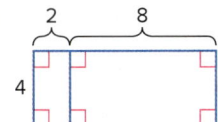
6.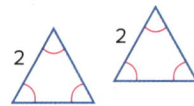

Each pair of polygons is similar. Find the values of *x* and *y*.

7.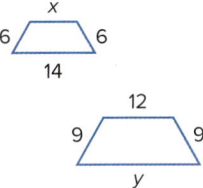
8.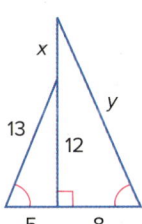
9.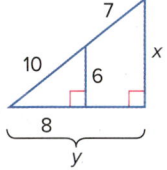

10. **SHADOWS** On a sunny day, Jason measures the length of his shadow and the length of a tree's shadow. Use the figures at the right to find the height of the tree.

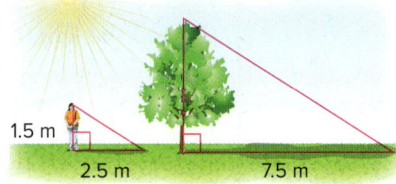

11. **DECORATING** Amy wants to put carpeting in part of her living room. Her living room is 12 feet wide by 18 feet long. If she wants the carpeted part of the living room to have the same proportions as the entire room and the carpet is 5 feet wide, find its length.

12. **SURVEYING** Surveyors use instruments to measure objects that are too large or too far away to measure by hand. They can use the shadows that objects cast to find the height of the objects without measuring them. A surveyor finds that a telephone pole that is 25 feet tall is casting a shadow 20 feet long. A nearby building is casting a shadow 52 feet long. What is the height of the building?

LESSON 8
The Pythagorean Theorem

Objective
- Use the Pythagorean Theorem and its converse.

The **Pythagorean Theorem** states that in a right triangle, the square of the length of the hypotenuse c is equal to the sum of the squares of the lengths of the legs a and b.

That is, in any right triangle, $c^2 = a^2 + b^2$.

Example 1 — Find Hypotenuse Measures

Find the length of the hypotenuse of each right triangle.

a. (triangle with legs 5 in. and 12 in., hypotenuse c in.)

$c^2 = a^2 + b^2$ — Pythagorean Theorem
$c^2 = 5^2 + 12^2$ — $a = 5$ and $b = 12$
$c^2 = 25 + 144$ — Simplify.
$c^2 = 169$ — Add.
$c = \sqrt{169}$ — Take the positive square root of each side.
$c = 13$ — The length of the hypotenuse is 13 inches.

b. (triangle with legs 10 cm and 6 cm, hypotenuse c cm)

$c^2 = a^2 + b^2$ — Pythagorean Theorem
$c^2 = 6^2 + 10^2$ — $a = 6$ and $b = 10$
$c^2 = 36 + 100$ — Simplify.
$c^2 = 136$ — Add.
$c = \sqrt{136}$ — Take the positive square root of each side.
$c \approx 11.7$ — Use a calculator.

To the nearest tenth, the length of the hypotenuse is 11.7 centimeters.

You can also find the length of a leg of a right triangle given the lengths of the hypotenuse and the other leg.

Example 2 — Find Leg Measures

Find the length of the missing leg in each right triangle.

a. (right triangle with leg 7 ft, hypotenuse 25 ft, missing leg a ft)

$c^2 = a^2 + b^2$ — Pythagorean Theorem
$25^2 = a^2 + 7^2$ — $c = 25$ and $b = 7$
$625 = a^2 + 49$ — Simplify.
$625 - 49 = a^2 + 49 - 49$ — Subtract 49 from each side.
$576 = a^2$ — Simplify.
$\sqrt{576} = a$ — Take the positive square root of each side.
$24 = a$ — The length of the leg is 24 feet.

Watch Out!

Positive Square Root When finding the length of a side using the Pythagorean Theorem, use only the positive and not the negative square root, since length cannot be negative.

b.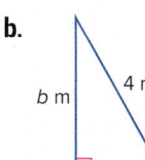

$$c^2 = a^2 + b^2 \quad \text{Pythagorean Theorem}$$
$$4^2 = 2^2 + b^2 \quad c = 4 \text{ and } a = 2$$
$$16 = 4 + b^2 \quad \text{Simplify.}$$
$$16 - 4 = 4 - 4 + b^2 \quad \text{Subtract 4 from each side.}$$
$$12 = b^2 \quad \text{Simplify.}$$
$$\sqrt{12} = b \quad \text{Take the positive square root of each side.}$$
$$3.5 \approx b \quad \text{Use a calculator.}$$

To the nearest tenth, the length of the leg is 3.5 meters.

The **converse of the Pythagorean Theorem** states that if the sides of a triangle have lengths a, b, and c, and $c^2 = a^2 + b^2$, then the triangle is a right triangle.

Example 3   Identify a Right Triangle

The lengths of the three sides of a triangle are 5, 7, and 9 inches. Determine whether this triangle is a right triangle.

Because the longest side is 9 inches, use 9 as c, the measure of the hypotenuse.

$$c^2 = a^2 + b^2 \quad \text{Pythagorean Theorem}$$
$$9^2 \stackrel{?}{=} 5^2 + 7^2 \quad c = 9, a = 5, \text{ and } b = 7$$
$$81 \stackrel{?}{=} 25 + 49 \quad \text{Evaluate } 9^2, 5^2, \text{ and } 7^2.$$
$$81 \neq 74 \quad \text{Simplify.}$$

Because $c^2 \neq a^2 + b^2$, the triangle is *not* a right triangle.

Exercises

Find each missing measure. Round to the nearest tenth, if necessary.

1.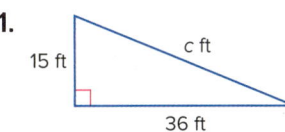
2.
3.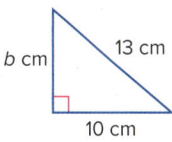

4. $a = 3, b = 4, c = ?$
5. $a = ?, b = 12, c = 13$
6. $a = 14, b = ?, c = 50$
7. $a = 2, b = 9, c = ?$
8. $a = 6, b = ?, c = 13$
9. $a = ?, b = 7, c = 11$

The lengths of three sides of a triangle are given. Determine whether each triangle is a right triangle.

10. 5 in., 7 in., 8 in.
11. 9 m, 12 m, 15 m
12. 6 cm, 7 cm, 12 cm
13. 11 ft, 12 ft, 16 ft
14. 10 yd, 24 yd, 26 yd
15. 11 km, 60 km, 61 km

16. **FLAGPOLES** If a flagpole is 30 feet tall and Mai-Lin is standing a distance of 15 feet from the flagpole, what is the distance from her feet to the top of the flagpole?

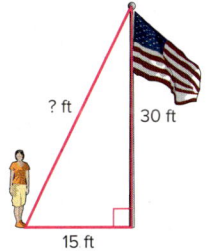

17. **CONSTRUCTION** The walls of a recreation center are being covered with paneling. A doorway is 0.9 meter wide and 2.5 meters high. What is the length of the widest rectangular panel that can be taken through this doorway?

18. **OPEN-ENDED** Create an application problem involving right triangles and the Pythagorean Theorem. Then solve your problem, drawing diagrams if necessary.

LESSON 9
Measures of Center, Spread, and Position

Objective
- Find measures of center, spread, and position.

New Vocabulary
statistics
descriptive statistics
population
variable
data
univariate data
sample
measures of center or central tendency
mean
median
mode
measures of spread or variation
range
variance
standard deviation
quartile
lower quartile
upper quartile
five-number summary
interquartile range
outlier

Statistics is the science of collecting, organizing, displaying, and analyzing data in order to draw conclusions and make predictions. The branch of statistics that focuses on collecting, summarizing, and displaying data is called **descriptive statistics**.

The entire group of interest to a statistician is called a **population**. A **variable** is a characteristic of a population that can assume different values called **data**. Data that include only one variable are called **univariate data**. When it is not possible to obtain data about every member of a population, a representative **sample** or subset of the population is selected.

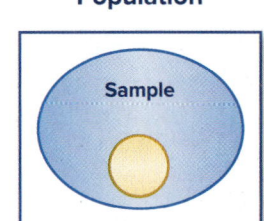
Population

Univariate data are often summarized using a single number to represent what is average or typical. Measures of average are also called **measures of center** or **central tendency**. The most common measures of center are mean, median, and mode.

Key Concept Measures of Center

- The **mean** is the sum of the values in a set of data $x_1, x_2, \ldots, x_n$ divided by the total number of values n in the set. The formula for population mean μ is $\mu = \dfrac{x_1 + x_2 + \ldots + x_n}{n}$.

- The **median** is the middle value or the mean of the two middle values in a set of data when the data are arranged in numerical order.

- The **mode** is the value or values that appear most often in a set of data. A set of data can have no mode, one mode, or more than one mode.

For sample mean, replace μ with $\bar{x}$.

Example 1 Measures of Center

SOFT DRINKS The number of milligrams of sodium in a 12-ounce can of ten different brands of regular cola are shown below. Find the mean, median, and mode.

50, 30, 25, 20, 40, 35, 35, 10, 15, 35

Mean $\bar{x} = \dfrac{50 + 30 + 25 + 20 + 40 + 35 + 35 + 10 + 15 + 35}{10}$

$= \dfrac{295}{10}$ or 29.5 milligrams

Find the sum of the data values and divide by the number of values, 10.

Median 10, 15, 20, 25, 30, 35, 35, 35, 40, 50

$\dfrac{30 + 35}{2}$ or 32.5 milligrams

Arrange the data in order. Find the mean of the middle two values.

Mode The value that occurs most often in the set is 35, so the mode of the data set is 35 milligrams.

Because two very different data sets can have the same mean, statisticians also use **measures of spread** or **variation** to describe how widely the data values vary and how much the values differ from what is typical. The most common of these measures are listed below.

> ### 🔑 Key Concept Measures of Spread
>
> - The **range** is the difference between the greatest and least values in a set of data.
> - The **variance** in a set of data $x_1, x_2, \ldots, x_n$ is the mean of the squares of the deviations or differences from the mean. The formula for population variance σ^2 is
>
> $$\sigma^2 = \frac{(x_1 - \mu)^2 + (x_2 - \mu)^2 + \ldots + (x_n - \mu)^2}{n}.$$
>
> - The **standard deviation** in a set of data $x_1, x_2, \ldots, x_n$ is the average amount by which each individual value deviates or differs from the mean. It is the square root of the variance. The formula for population standard deviation σ is
>
> $$\sigma = \sqrt{\sigma^2} \text{ or } \sqrt{\frac{(x_1 - \mu)^2 + (x_2 - \mu)^2 + \ldots + (x_n - \mu)^2}{n}}.$$

Study Tip

Unbiased Estimator Dividing by $n - 1$ instead of n when finding a sample variance and standard deviation gives a slightly larger value, and therefore an unbiased estimate, of the population standard deviation or variance.

For sample variance and standard deviation, replace σ^2 with s^2, μ with $\bar{x}$, and n with $n - 1$.

Example 2 Measures of Spread

MIDTERM EXAMS Two classes took the same midterm exam. The scores of five students from each class are shown. Both sets of scores have a mean of 84.2.

Class A
85, 76, 92, 88, 80

Class B
75, 85, 95, 98, 68

a. Find the range, variance, and standard deviation for the sample scores from Class A.

Range $92 - 76$ or 16 range = greatest value − least value

Variance   Subtract the mean of the data, 84.2, from each value. Then square each result and find the sum of these squares. Finally, since these data were obtained from a sample of all scores in Class A, divide the sum by $n - 1$.

Study Tip

Squared Differences The sum of the differences of the data values from the mean of the set will always be zero. For this reason, variance is calculated using the sum of the *squares* of the differences.

X	X − $\bar{X}$	(X − $\bar{X}$)²
85	85 − 84.2 = 0.8	0.8² = 0.64
76	76 − 84.2 = −8.2	(−8.2)² = 67.24
92	92 − 84.2 = 7.8	7.8² = 60.84
88	88 − 84.2 = 3.8	3.8² = 14.44
80	80 − 84.2 = −4.2	(−4.2)² = 17.64
		Sum = 160.8

$$s^2 = \frac{(x_1 - \bar{x})^2 + \ldots + (x_n - \bar{x})^2}{n - 1} \quad \text{Sample variance}$$

$$= \frac{160.8}{5 - 1} = 40.2 \quad (x_1 - \bar{x})^2 + \ldots + (x_5 - \bar{x})^2 = 160.8 \text{ and } n = 5$$

Standard Deviation $s = \sqrt{s^2} = \sqrt{\dfrac{160.8}{5 - 1}}$ or about 6.3   Sample standard deviation

(continued on the next page)

b. Use a calculator to find the range, variance, and standard deviation for the sample scores from Class B.

To find measures of center and spread using a graphing calculator, first clear all lists. Press STAT ENTER to enter each data value. Then press STAT and select **1-Var Stats** from the **CALC** menu. Scroll to see all one-variable statistics.

The range **maxX − minX** for Class B is 98 − 68 or 30. The sample standard deviation **Sx** is about 12.8, so the variance is about 12.8^2 or 163.8.

c. Compare the sample standard deviations of Class A and Class B.

Because the sample standard deviation of Class A is 6.3 and the sample standard deviation of Class B is 12.8, there is more variability in the sample scores from Class B than from Class A.

Study Tip

Comparing Standard Deviations Standard deviation can only be compared in this way when the units of both data sets are the same.

Statisticians use measures of position to describe where specific values fall within a data set. **Quartiles** are three position measures that divide a data set arranged in ascending order into four groups, each containing about 25% of the data. The median marks the second quartile Q_2 and separates the data into upper and lower halves. The first or **lower quartile** Q_1 is the median of the lower half, while the third or **upper quartile** Q_3 is the median of the upper half.

Study Tip

Calculating Quartiles When the number of values in a set of data is odd, the median is not included in either half of the data when calculating Q_1 or Q_3.

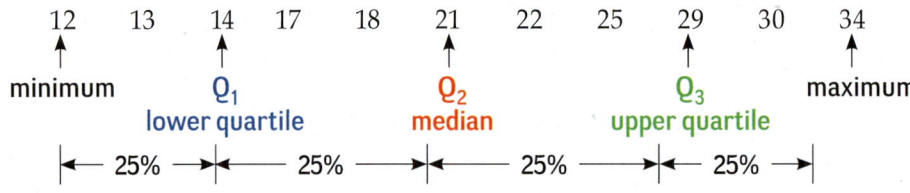

The three quartiles, along with the minimum and maximum values, are called a **five-number summary** of a data set.

Example 3 Five-Number Summary

PART-TIME JOB The number of hours Liana worked each week for the last 12 weeks were 21, 10, 18, 12, 15, 13, 20, 20, 19, 16, 18, and 14.

a. Find the minimum, lower quartile, median, upper quartile, and maximum of the data set.

Use a calculator to find the one-variable statistics for the data set. The minimum **minX** of the data set is 10, the lower quartile Q_1 is 13.5, the median **Med** is 17, the upper quartile Q_3 is 19.5, and the maximum **maxX** is 21.

b. Interpret this five-number summary.

Over the last 12 weeks, Liana worked a minimum of 10 hours and a maximum of 21 hours. She worked less than 13.5 hours 25% of the time, less than 17 hours 50% of the time, and less than 19.5 hours 75% of the time.

> **Study Tip**
>
> **Interquartile Range** A large interquartile range means that the data are spread out.

The difference between Q_3 and Q_1, is called the **interquartile range** or IQR. The interquartile range contains about 50% of the values. Before deciding on which measures best describe a set of data, check the data set for outliers. An **outlier** is an extremely high or extremely low value when compared with the rest of the values in the set. To check for outliers, look for data values that are beyond the upper or lower quartiles by more than 1.5 times the interquartile range.

Example 4   Effect of an Outlier

HOMEWORK The number of minutes each of the 22 students in a class spent working on the same algebra assignment is shown below.

15, 12, 25, 15, 27, 10, 16, 18, 30, 35, 22, 25, 65, 20, 18, 25, 15, 13, 25, 22, 15, 28

a. Identify any outliers in the data.

Calculate the interquartile range IQR by using a graphing calculator to find Q_1 and Q_3.

$$\text{IQR} = Q_3 - Q_1 = 25 - 15 \text{ or } 10$$

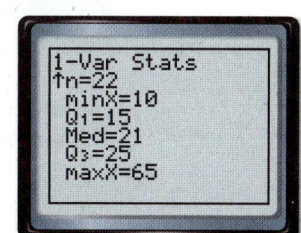

Use the interquartile range to find the values beyond which any outliers would lie.

$Q_1 - 1.5(\text{IQR})$ and $Q_3 + 1.5(\text{IQR})$		Values beyond which outliers lie
$15 - 1.5(10)$ $25 + 1.5(10)$		$Q_1 = 15, Q_3 = 25, \text{IQR} = 10$
0 40		Simplify.

There are no values less than 0, but there is one value greater than 40. The value 65 can be considered an outlier for this data set.

b. Find the mean, median, mode, range, and standard deviation of the data set with and without the outlier. Describe the effect on each measure.

You can use a graphing calculator. The data set represents the entire population of the class, so use the population standard deviation σ.

Data Set	Mean	Median	Mode	Range	Standard Deviation
with outlier	≈22.5	21	15, 25	55	≈11.2
without outlier	≈20.5	20	15, 25	25	≈6.5

Removal of the outlier causes the mean, median, range, and standard deviation to decrease. Notice that the mean is affected more by the outlier than the median.

Exercises

Find the mean, median, and mode for each set of data.

1. number of pages in each novel assigned for summer reading:
   224, 272, 374, 478, 960, 394, 404, 308, 480, 624

2. height in centimeters of bean plants at the end of an experiment:
   14.5, 12, 16, 11, 14, 11, 10.5, 14, 11.5, 15, 13.5

3. number of text messages sent each day during the last two weeks:
   63, 112, 166, 158, 84, 201, 78, 83, 125, 97, 21, 114, 125, 131

State whether the data in sets A and B represent *sample* or *population* data. Then find the range, variance, and standard deviation of each set. Use the standard deviations to compare the variability between the data sets.

4.
Wait Times (min)					
Ride A			Ride B		
45	22	40	35	50	32
48	11	51	31	35	45
36	55	60	45	49	40
32	24	37	43	37	45

5.
Number of Sponsors Obtained by Participants					
Charity Walk A			Charity Walk B		
44	14	61	8	28	15
22	27	25	100	42	19
38	50	49	25	75	82

6.
Number of Days Each Student Missed This Year													
Class A													
10	8	5	9	7	3	6	8	14	11	8	4	7	8	2
5	13	0	15	9	7	9	10	9	11	14	8	12	10	1
Class B													
5	8	13	7	9	4	10	2	12	9	6	11	3	8	5
12	6	7	8	11	12	8	9	3	10	5	13	9	1	8

Find the minimum, lower quartile, median, upper quartile, and maximum of each data set. Then interpret this five-number summary.

7.
Number of Students in Each Math Class at Central High													
25	27	26	26	19	27	24	23	19	28	25	24	20	22	22
24	26	18	28	29	29	26	24	24	23	23	25	25	29	28

8.
State Mean ACT Scores									
20.2	21.3	21.5	20.4	21.6	20.3	22.5	21.5	17.8	20.5
20.0	21.7	21.3	20.2	21.6	22.0	21.6	20.3	19.8	22.6
20.8	22.4	21.4	22.2	18.8	21.5	21.7	21.2	22.5	21.2
20.1	22.3	20.3	21.2	21.4	20.6	22.5	21.8	21.9	19.3
21.5	20.5	20.3	21.5	22.7	20.9	22.5	22.2	21.4	20.7

Identify any outliers in each data set, and explain your reasoning. Then find the mean, median, mode, range, and standard deviation of the data set with and without the outlier. Describe the effect on each measure.

9. fuel efficiency in miles per gallon of 16 randomly selected automobiles:
40, 36, 29, 45, 51, 36, 48, 34, 36, 22, 13, 42, 31, 44, 32, 34

10. number of posts to a certain blog each month during a particular year:
25, 23, 21, 27, 29, 19, 10, 21, 20, 18, 26, 23

11. **CEREAL** The weights, in ounces, of 20 randomly selected boxes of a certain brand of cereal are shown.
16.7, 16.8, 15.9, 16.1, 16.5, 16.6, 16.5, 15.9, 16.7, 16.5,
16.6, 14.9, 16.5, 16.1, 15.8, 16.7, 16.2, 16.5, 16.4, 16.6

   a. Identify any outliers in the data set, and explain your reasoning.
   b. If the outlier was removed and an additional cereal box that was 17.35 ounces was added, would this value be an outlier of the new data set? Explain.
   c. What are some possible causes of outliers in this situation?

CHAPTER 0
Posttest

State the domain and range of each relation. Then determine whether each relation is a function. Write *yes* or *no*.

1. $\{(4, 5), (5, -1), (0, 12), (0, -2), (7, 9)\}$

2.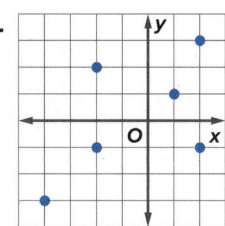

Name the quadrant in which each point is located.

3. $(-3, 7)$
4. $(10, -11)$
5. $(-15, -3)$

Find each product.

6. $(4n - 3)(2n + 2)$
7. $(5p - 1)(6p - 10)$
8. $(7x + 4)(7x + 4)$
9. $(3k - 2)(6k + 9)$

10. **GEOMETRY** The height of a rectangle is 3 millimeters less than twice the width.
    a. Write an expression for each measure.
    b. Write a polynomial expression for the area of the rectangle.

Factor each polynomial.

11. $4x^2 + 4xy + y^2$
12. $25a^2 - 20a + 4$
13. $4a^2 + 16ab + 16b^2$
14. $81t^2 - 36$

15. **STUDENT COUNCIL** A student council has 6 seniors, 5 juniors, and 1 sophomore as members. How many ways can a 3-member committee be formed that includes one member from each class?

Determine whether each situation involves *permutations* or *combinations*. Then solve.

16. How many ways are there to select one competitor and one alternate out of 8 students?

17. How many ways are there to form a team of 7 athletes from a group of 15 who try out?

RESTAURANT There are 24 male and 36 female patrons in a restaurant. Of the 11 patrons under 10 years old, 6 are male. Of the 14 patrons over 55 years old, 9 are female. A patron is selected at random. Determine whether the events are *mutually exclusive* or *not mutually exclusive*. Then find each probability.

18. P(female or under 10)
19. P(under 10 or over 55)

Determine whether the events are *independent* or *dependent*. Then find the probability.

20. Slips of paper numbered 1 through 10 are placed into a bag. What is the probability of drawing the number 10 three times in a row if a slip is drawn at random and then replaced?

21. Two students are selected at random from a class that consists of 13 males and 7 females. What is the probability that both students are female?

22. **TESTING** Of the students who took both the Mid-Chapter 4 Quiz and the Chapter 4 Test, 56% passed the quiz and 48% passed both the quiz and the test. If a student passed the quiz, find the probability that he or she also passed the test.

23. Determine whether the rectangles are *similar*, *congruent*, or *neither*.

     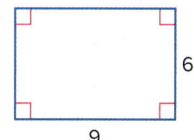

24. **TABLETS** An image of a painting on a tablet is 320 pixels wide by 240 pixels high. If the actual painting is 42 inches wide, how high is it?

Find each missing measure. Round to the nearest tenth, if necessary.

25.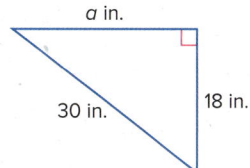
26. $a = 33$ cm, $b = ?$ cm, $c = 45$ cm

The lengths of three sides of a triangle are given. Determine whether each triangle is a right triangle.

27. 6 in., 8 in., 12 in.
28. 30 m, 34 m, 16 m

Find the mean, median, mode, range, and standard deviation of each data set. Then identify any outliers.

29. number of students present at 8 student council meetings: 23, 45, 16, 75, 32, 35, 28, 35

30. running time in minutes for 17 movies: 95, 102, 148, 140, 110, 103, 107, 104, 99, 111, 109, 124, 109, 90, 92, 110, 129

CHAPTER 1

Linear Equations

THEN

You wrote expressions with variables.

NOW

You will:

- Solve and write linear equations.
- Solve and graph linear inequalities.
- Solve systems of linear equations and linear inequalities.
- Solve problems by using linear programming.

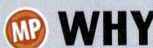

MP WHY

BUSINESS Often, being successful in business means that you have good math skills. To make the most money possible, you need to maximize your profits and minimize your costs.

Use the Mathematical Practices to complete the activity.

1. **Sense Making** In a business, you need to know the cost and quantity of all materials that go into a product. What materials might be needed in the business shown?
2. **Reasoning** What would the quantity and price of the materials be used to determine? How would this impact your profits or losses?
3. **Modeling** Use the Equation Chart Mat and the Algebra Tiles tool to model a business equation.

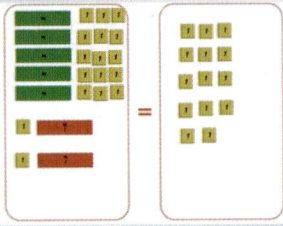

4. **Discuss** Can this equation be expanded to include other factors that might impact a company's profitability?

Westend61 Premium/Shutterstock.com

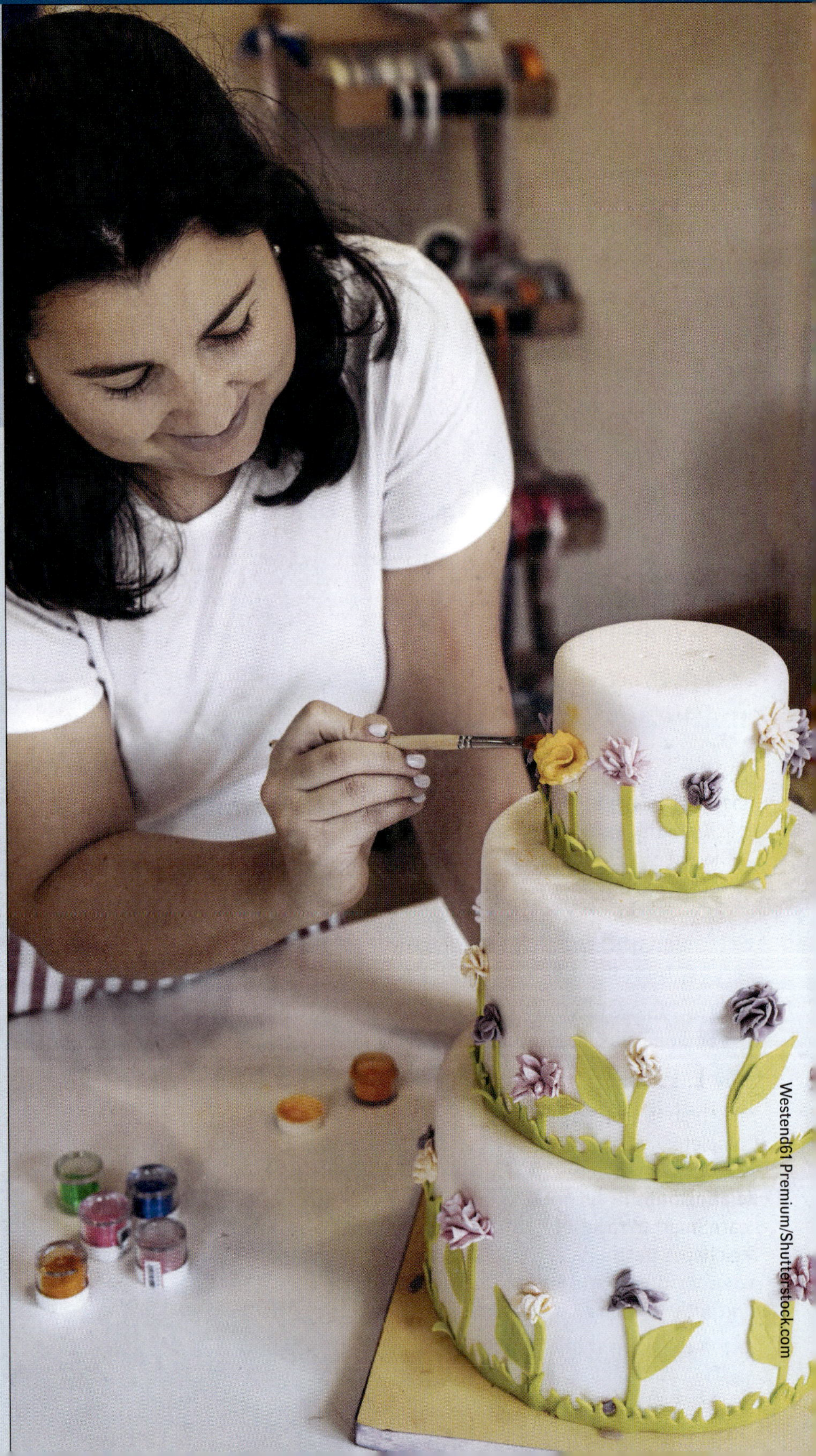

 ## Go Online to Guide Your Learning

Explore & Explain

 The Geometer's Sketchpad

The Geometer's Sketchpad allows you to interact with functions in a visual way. You can investigate systems of equations and inequalities with sketches in ConnectED as you learn about them in Lessons 1-4 through 1-8.

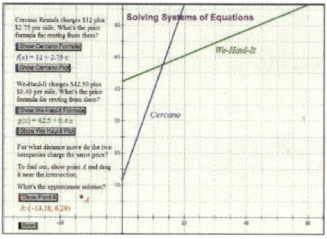

 Graphing Tools

Use the **Graphing Tools** to enhance your understanding of the topics discussed in Lessons 1-3 through 1-6.

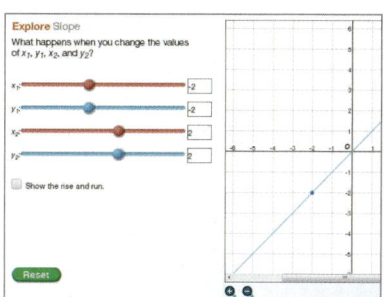

eBook
Interactive Student Guide

Before starting the chapter, answer the **Chapter Focus** preview questions. Check your answers as you complete each lesson. At the end of the chapter, try the **Performance Task**.

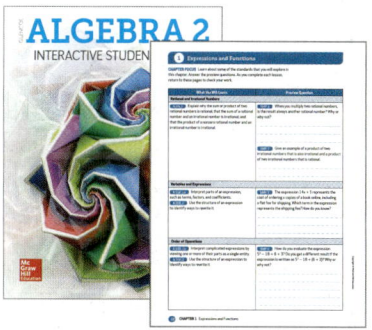

Organize

 Foldables

Get organized! Create an **Equations and Inequalities Foldable** before you start the chapter to arrange your notes about linear equations.

Collaborate

 Chapter Project

In the **Take Me Out to the Ball Game** project, you will use what you have learned about equations and inequalities to complete a project that addresses business literacy.

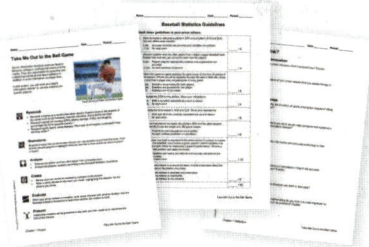

Focus

LEARNSMART

Need help studying? Complete the **Polynomial, Rational, and Radical Relationships** domain in LearnSmart to review for the chapter test.

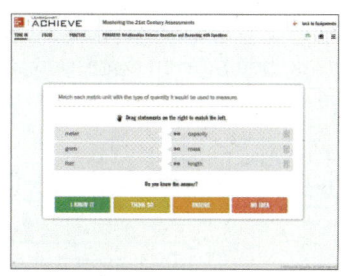

ALEKS

You can use the **Real Numbers and Linear Equations** topic in ALEKS to explore what you know about linear equations and what you are ready to learn.*

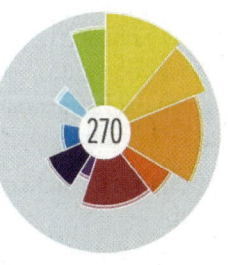

*Ask your teacher if this is part of your program.

connectED.mcgraw-hill.com

Get Ready for the Chapter

Connecting Concepts

Concept Check
Review the concepts used in this chapter by answering the questions below.

1. Given $\left(\frac{3}{6}\right)\left(\frac{4}{5}\right)$, what steps would you take to simplify?
2. Given $\frac{12}{80}$, what step would you take to simplify?
3. How would you evaluate $(-1.5)^3$?
4. Would the answer to question 3 be positive or negative?
5. How would you evaluate $\left(\frac{7}{10}\right)^2$?
6. Is $\frac{1}{7} \leq \frac{1}{9}$?
7. Is $\frac{5}{6} \leq \frac{25}{30}$?
8. The graph of $2y + 5x = -10$ is shown. How do you determine the x-intercept?

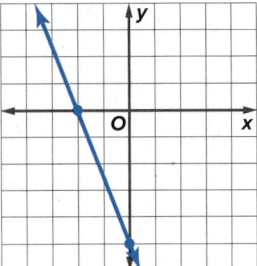

9. For the equation in question 8, how do you determine the y intercept?

Performance Task Preview
You can use the concepts and skills in the chapter to solve problems for a catering company. Knowing how to work with linear equations will help you finish the Performance Task at the end of the chapter.

MP In this Performance Task you will:
- make sense of problems and persevere in solving them
- reason abstractly and quantitatively

New Vocabulary

English		Español
open sentence	p. 5	enunciado abierto
equation	p. 5	ecuación
solution	p. 5	solución
set-builder notation	p. 15	notación de construcción de conjuntos
rate of change	p. 21	tasa de cambio
slope	p. 22	pendiente
slope-intercept form	p. 28	forma pendiente-intersección
point-slope form	p. 29	forma punto-pendiente
parallel lines	p. 30	rectas paralelas
perpendicular lines	p. 30	rectas perpendiculars
linear inequality	p. 35	desigualdad lineal
boundary	p. 35	frontera
constraint	p. 36	restriccion
break-even point	p. 42	el punto de equilbrio
system of equations	p. 42	sistema de ecuaciones
substitution method	p. 44	método de sustitución
elimination method	p. 45	método de eliminación
system of inequalities	p. 52	sistema de desigualdades
linear programming	p. 60	programación lineal
feasible region	p. 60	región viable
bounded	p. 60	acotada
optimize	p. 62	optimice
ordered triple	p. 67	triple ordenado

Review Vocabulary

evaluate *evaluar* to find the value of an expression

inequality *desigualdad* an open sentence that contains the symbol $<$, $\leq$, $>$, or $\geq$

power *potencia* an expression of the form x^n, read x to the nth power

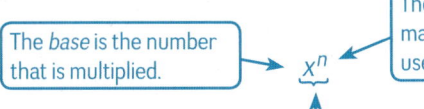

The *base* is the number that is multiplied.

The *exponent* tells how many times the base is used as a factor.

The number that can be expressed using an exponent is called a *power*.

LESSON 1
Solving Linear Equations

::Then
- You used properties of real numbers to evaluate expressions.

::Now
1. Translate verbal expressions into algebraic expressions and equations, and vice versa.
2. Solve equations using the properties of equality.

::Why?
- The United States is one of the few countries in the world that measures distances in miles. When traveling by car in different countries, it is often useful to convert miles to kilometers. To find the approximate number of kilometers k in miles m, divide the number of miles by 0.62137.

$$m \text{ miles} \times \frac{1 \text{ kilometer}}{0.62137 \text{ mile}} \approx k \text{ kilometers}$$

$$\frac{m}{0.62137} \approx k \text{ kilometers}$$

New Vocabulary
open sentence
equation
solution

Mathematical Practices
3 Construct viable arguments and critique the reasoning of others.
8 Look for and express regularity in repeated reasoning.

1 Verbal Expressions and Algebraic Expressions Verbal expressions can be translated into algebraic expressions by using the language of algebra.

Example 1   Verbal to Algebraic Expression

Write an algebraic expression to represent each verbal expression.

a. 2 more than 4 times the cube of a number $4x^3 + 2$

b. the quotient of 5 less than a number and 12 $\dfrac{n-5}{12}$

▶ **Guided Practice**

1A. the cube of a number increased by 4 times the same number

1B. three times the difference of a number and 8

A mathematical sentence containing one or more variables is called an **open sentence**. A mathematical sentence stating that two mathematical expressions are equal is called an **equation**.

Example 2   Algebraic to Verbal Sentence

Write a verbal sentence to represent each equation.

a. $6x = 72$ The product of 6 and a number is 72.

b. $n + 15 = 91$ The sum of a number and 15 is ninety-one.

▶ **Guided Practice**

2A. $g - 5 = -2$ **2B.** $2c = c^2 - 4$

Open sentences are neither true nor false until the variables have been replaced by numbers. Each replacement that results in a true sentence is called a **solution** of the open sentence.

connectED.mcgraw-hill.com   5

2 Properties of Equality

To solve equations, we can use properties of equality. Some of these properties are listed below.

Math History Link

Diophantus of Alexandria (c. 200–284) Diophantus was famous for his work in algebra. His main work was titled *Arithmetica* and introduced symbolism to Greek algebra as well as propositions in number theory and polygonal numbers.

Key Concept Properties of Equality

Property	Symbols	Examples
Reflexive	For any real number a, $a = a$.	$b + 12 = b + 12$
Symmetric	For all real numbers a and b, if $a = b$, then $b = a$.	If $18 = -2n + 4$, then $-2n + 4 = 18$.
Transitive	For all real numbers a, b, and c, if $a = b$ and $b = c$, then $a = c$.	If $5p + 3 = 48$ and $48 = 7p - 15$, then $5p + 3 = 7p - 15$.
Substitution	If $a = b$, then a may be replaced by b and b may be replaced by a.	If $(6 + 1)x = 21$, then $7x = 21$.

Example 3   Identify Properties of Equality

Name the property illustrated by each statement.

a. If $3a - 4 = b$, and $b = a + 17$, then $3a - 4 = a + 17$.

  Transitive Property of Equality

b. If $2g - h = 62$, and $h = 24$, then $2g - 24 = 62$.

  Substitution Property of Equality

Guided Practice

3. If $-11a + 2 = -3a$, then $-3a = -11a + 2$.

Solving most equations requires assuming that the original equation has a solution, and performing the same operations on each side of the equal sign. The properties of equality allow for the equation to be solved in this way.

Study Tip

Using Your Text Look for Key Concepts to learn important properties, definitions, and concepts.

Key Concept Properties of Equality

Addition and Subtraction Properties of Equality	
Symbols	For any real numbers, a, b, and c, if $a = b$, then $a + c = b + c$ and $a - c = b - c$.
Examples	If $x - 6 = 14$, then $x - 6 + 6 = 14 + 6$. If $n + 5 = -32$, then $n + 5 - 5 = -32 - 5$.
Multiplication and Division Properties of Equality	
Symbols	For any real numbers, a, b, and c, $c \neq 0$, if $a = b$, then $a \cdot c = b \cdot c$ and $\frac{a}{c} = \frac{b}{c}$.
Examples	If $\frac{m}{8} = -7$, then $8 \cdot \frac{m}{8} = 8 \cdot (-7)$. If $-2y = 12$, then $\frac{-2y}{-2} = \frac{12}{-2}$.

Example 4 Solve One-Step Equations

Solve each equation. Check your solution.

a. $n - 3.24 = 42.1$

$n - 3.24 = 42.1$ Original equation

$n - 3.24 + 3.24 = 42.1 + 3.24$ Add 3.24 to each side.

$n = 45.34$ Simplify.

The solution is 45.34.

CHECK $n - 3.24 = 42.1$ Original equation

$45.34 - 3.24 \stackrel{?}{=} 42.1$ Substitute 45.34 for n.

$42.1 = 42.1$ ✓   Simplify.

b. $-\frac{5}{8}x = 20$

$-\frac{5}{8}x = 20$ Original equation

$-\frac{8}{5}\left(-\frac{5}{8}\right)x = -\frac{8}{5}(20)$ Multiply each side by $-\frac{8}{5}$.

$x = -32$ Simplify.

The solution is -32.

CHECK $-\frac{5}{8}x = 20$ Original equation

$-\frac{5}{8}(-32) \stackrel{?}{=} 20$ Replace x with -32.

$20 = 20$ ✓   Simplify.

> **Study Tip**
>
> **MP Reasoning** In Example 4b, notice that multiplying both sides of the equation by $-\frac{8}{5}$ is the same as dividing both sides by $-\frac{5}{8}$.

Guided Practice

4A. $x - 14.29 = 25$ **4B.** $\frac{2}{3}y = -18$

To solve an equation with more than one operation, undo operations by working backward.

Example 5 Solve a Multi-Step Equation

Solve $5(x + 3) + 2(1 - x) = 14$.

$5(x + 3) + 2(1 - x) = 14$ Original equation

$5x + 15 + 2 - 2x = 14$ Apply the Distributive Property.

$3x + 17 = 14$ Simplify the left side.

$3x = -3$ Subtract 17 from each side.

$x = -1$ Divide each side by 3.

> **Study Tip**
>
> **MP Precision** When solving for a variable, you can use substitution to check your answer by replacing the variable in the original equation with your answer.

Guided Practice

Solve each equation.

5A. $-10x + 3(4x - 2) = 6$ **5B.** $2(2x - 1) - 4(3x + 1) = 2$

connectED.mcgraw-hill.com   7

You can use properties to solve an equation for a variable.

Example 6 — Solve for a Variable

GEOMETRY The formula for the area A of a trapezoid is $A = \frac{1}{2}h(b_1 + b_2)$, where h represents the height, and b_1 and b_2 represent the measures of the bases. Solve the formula for b_2.

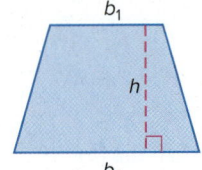

$A = \frac{1}{2}h(b_1 + b_2)$ Area formula

$2A = 2\left[\frac{1}{2}h(b_1 + b_2)\right]$ Multiply each side by 2.

$2A = h(b_1 + b_2)$ Simplify.

$\dfrac{2A}{h} = \dfrac{h(b_1 + b_2)}{h}$ Divide each side by h.

$\dfrac{2A}{h} = b_1 + b_2$ Simplify.

$\dfrac{2A}{h} - b_1 = b_1 + b_2 - b_1$ Subtract b_1 from each side.

$\dfrac{2A}{h} - b_1 = b_2$ Simplify.

Guided Practice

6. The formula for the surface area S of a cylinder is $S = 2\pi r^2 + 2\pi rh$, where r is the radius of the base and h is the height of the cylinder. Solve the formula for h.

> **Go Online!**
> **Personal Tutors** for each example let you follow along as a teacher solves a problem. Pause and rewind as you need.
>
>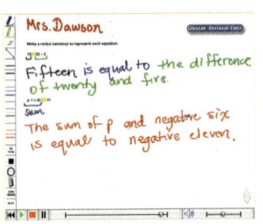

There are often many ways to solve a problem. Using the properties of equality can help you find a simpler way.

Example 7 — Use Properties of Equality

If $6x - 12 = 18$, what is the value of $6x + 5$?

A 5   **B** 11   **C** 35   **D** 41

Read the Item

You are asked to find the value of $6x + 5$. Note that you do not have to find the value of x. Instead, you can use the Addition Property of Equality to make the left side of the equation $6x + 5$.

Solve the Item

$6x - 12 = 18$ Original equation

$6x - 12 + 17 = 18 + 17$ Add 17 to each side because $-12 + 17 = 5$.

$6x + 5 = 35$ Simplify.

The answer is C.

> **Study Tip**
> **Read the Question**
> Read the question carefully before solving the equation. In Example 7, you are to find the value of $6x + 5$, not the value of x.

Guided Practice

7. If $5y + 2 = \frac{8}{3}$, what is the value of $5y - 6$?

    F $\dfrac{-20}{3}$ **G** $\dfrac{-16}{3}$ **H** $\dfrac{16}{3}$ **J** $\dfrac{32}{3}$

Check Your Understanding

= Step-by-Step Solutions begin on page R11.

Go Online! for a Self-Check Quiz

Example 1 Write an algebraic expression to represent each verbal expression.

1. the product of 12 and the sum of a number and negative 3
2. the difference between the product of 4 and a number and the square of the number

Example 2 Write a verbal sentence to represent each equation.

3. $5x + 7 = 18$
4. $x^2 - 9 = 27$
5. $5y - y^3 = 12$
6. $\frac{x}{4} + 8 = -16$

Example 3 Name the property illustrated by each statement.

7. $(8x - 3) + 12 = (8x - 3) + 12$
8. If $a = -3$ and $-3 = d$, then $a = d$.

Examples 4–5 **MP PRECISION** Solve each equation. Check your solution.

9. $z - 19 = 34$
10. $x + 13 = 7$
11. $-y = 8$
12. $-6x = 42$
13. $5x - 3 = -33$
14. $-6y - 8 = 16$
15. $3(2a + 3) - 4(3a - 6) = 15$
16. $5(c - 8) - 3(2c + 12) = -84$
17. $-3(-2x + 20) + 8(x + 12) = 92$
18. $-4(3m - 10) - 6(-7m - 6) = -74$

Example 6 Solve each equation or formula for the specified variable.

19. $8r - 5q = 3$, for q
20. $Pv = nrt$, for n

Example 7

21. **MULTIPLE CHOICE** If $\frac{y}{5} + 8 = 7$, what is the value of $\frac{y}{5} - 2$?
   A -10
   B -3
   C 1
   D 5

Practice and Problem Solving

Extra Practice is on page R1.

Example 1 Write an algebraic expression to represent each verbal expression.

22. the difference between the product of four and a number and 6
23. the product of the square of a number and 8
24. fifteen less than the cube of a number
25. five more than the quotient of a number and 4

Example 2 Write a verbal sentence to represent each equation.

26. $8x - 4 = 16$
27. $\frac{x + 3}{4} = 5$
28. $4y^2 - 3 = 13$

29. **BASEBALL** During a recent season, Miguel Cabrera and Mike Jacobs hit a combined total of 46 home runs. Cabrera hit 6 more home runs than Jacobs. How many home runs did each player hit? Define a variable, write an equation, and solve the problem.

Example 3 Name the property illustrated by each statement.

30. If $x + 9 = 2$, then $x + 9 - 9 = 2 - 9$
31. If $y = -3$, then $7y = 7(-3)$
32. If $g = 3h$ and $3h = 16$, then $g = 16$
33. If $-y = 13$, then $-(-y) = -13$

34. MONEY Aiko and Kendra arrive at the state fair with $51.50. What is the total number of rides they can go on if they each pay the entrance fee?

Examples 4–5 **PRECISION** Solve each equation. Check your solution.

35. $3y + 4 = 19$

36. $-9x - 8 = 55$

37. $7y - 2y + 4 + 3y = -20$

38. $5g + 18 - 7g + 4g = 8$

39. $5(-2x - 4) - 3(4x + 5) = 97$

40. $-2(3y - 6) + 4(5y - 8) = 92$

41. $\frac{2}{3}(6c - 18) + \frac{3}{4}(8c + 32) = -18$

42. $\frac{3}{5}(15d + 20) - \frac{1}{6}(18d - 12) = 38$

43. GEOMETRY The perimeter of a regular pentagon is 100 inches. Find the length of each side.

44. MEDICINE For Nina's illness her doctor gives her a prescription for 28 pills. The doctor says that she should take 4 pills the first day and then 2 pills each day until her prescription runs out. For how many days does she take 2 pills?

Example 6 Solve each equation or formula for the specified variable.

45. $E = mc^2$, for m

46. $c(a + b) - d = f$, for a

47. $z = \pi q^3 h$, for h

48. $\frac{x + y}{z} - a = b$, for y

49. $y = ax^2 + bx + c$, for a

50. $wx + yz = bc$, for z

51. GEOMETRY The formula for the volume of a cylinder with radius r and height h is π times the radius times the radius times the height.

  a. Write this as an algebraic expression.

  b. Solve the expression in part **a** for h.

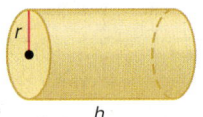

52. AWARDS BANQUET A banquet room can seat a maximum of 69 people. The coach, principal, and vice principal have invited the award winning girls' tennis team to the banquet. If the tennis team consists of 22 girls, how many guests can each student bring?

PRECISION Solve each equation. Check your solution.

53. $5x - 9 = 11x + 3$

54. $\frac{1}{x} + \frac{1}{4} = \frac{7}{12}$

55. $5.4(3k - 12) + 3.2(2k + 6) = -136$

56. $8.2p - 33.4 = 1.7 - 3.5p$

57. $\frac{4}{9}y + 5 = -\frac{7}{9}y - 8$

58. $\frac{3}{4}z - \frac{1}{3} = \frac{2}{3}z + \frac{1}{5}$

59. FINANCIAL LITERACY Benjamin spent $11,216 on his living expenses last year. Most of these expenses are listed at the right. Benjamin's only other expense last year was rent. If he paid rent 12 times last year, how much was Benjamin's rent each month?

Expense	Annual Cost
Electric	$864
Gas	$428
Water	$480
Renter's Insurance	$144

60 CHALLENGE The Queen Isabella Memorial Bridge connects South Padre Island, Texas, to Port Isabel on the mainland. Suppose one crew began building from South Padre Island, and another crew began building from Port Isabel. The two crews met 6256.5 feet from the South Padre Island side of the bridge approximately 2.5 years after construction began.

   a. Suppose the South Padre Island crew built an average of 208.5 feet per month. Together the two crews built 12,513 feet of bridge. Determine the average number of feet built per month by the Port Isabel crew.

   b. About how many miles of bridge did each crew build?

   c. Is this answer reasonable? Explain.

61 MULTIPLE REPRESENTATIONS The absolute value of a number describes the distance of the number from zero.

   a. Geometric Draw a number line. Label the integers from −5 to 5.

   b. Tabular Create a table of the integers on the number line and their distance from zero.

   c. Graphical Make a graph of each integer x and its distance from zero y using the data points in the table.

   d. Verbal Make a conjecture about the integer and its distance from zero. Explain the reason for any changes in sign.

H.O.T. Problems   Use Higher-Order Thinking Skills

62. ERROR ANALYSIS Steven and Jade are solving $A = \frac{1}{2}h(b_1 + b_2)$ for b_2. Is either of them correct? Explain your reasoning.

Steven
$$A = \frac{1}{2}h(b_1 + b_2)$$
$$\frac{2A}{h} = (b_1 + b_2)$$
$$\frac{2A - b_1}{h} = b_2$$

Jade
$$A = \frac{1}{2}h(b_1 + b_2)$$
$$\frac{2A}{h} = (b_1 + b_2)$$
$$\frac{2A}{h} - b_1 = b_2$$

63. PERSEVERANCE Solve $d = \sqrt{(x_2 - x_1)^2 + (y_2 - y_1)^2}$ for y_1.

64. CHALLENGE Use what you have learned in this lesson to explain why the following number trick works.
- Take any number.
- Multiply it by ten.
- Subtract 30 from the result.
- Divide the new result by 5.
- Add 6 to the result.
- Your new number is twice your original.

65. CHALLENGE Provide one example of an equation involving the Distributive Property that has no solution and another example that has infinitely many solutions.

66. WRITING IN MATH Compare and contrast the Substitution Property of Equality and the Transitive Property of Equality.

Preparing for Assessment

67. If $mn + p = m + 3n$, what is the value of m when $n = 4$ and $p = 6$? **MP 2**

- A -2
- B 2
- C 3
- D 3.6
- E 6

68. Given $\frac{y}{2} = \frac{z}{w} + \frac{x}{3}$, what is x in terms of w, y, and z if w, x, y, and z are positive numbers? **MP 2**

- A $\frac{y-z}{6w} = x$
- B $\frac{3y-3z}{2-w} = x$
- C $\frac{3y-3z}{2w} = x$
- D $\frac{wy-2z}{6w} = x$
- E $\frac{3wy-6z}{2w} = x$

69. What is the value of x in the figure below? **MP 2**

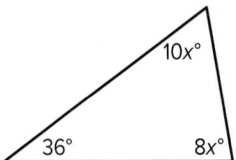

- A 2
- B 8
- C 12
- D 18
- E 63

70. The average of 3, 3, 16, 19, and y is 10. What is the value of y? **MP 2**

- A 3
- B 9
- C 10.2
- D 16

71. If $4x + 18 = 25$, what is the value of $4x - 8$? **MP 2**

- A 4
- B 2
- C -1
- D -2

72. Solve the Celsius temperature formula for Fahrenheit (F). **MP 2**

$C = \frac{5}{9}(F - 32)$

Solve for F.

- A $\frac{9}{5}C - 32$
- B $\frac{5}{9}C + 32$
- C $\frac{5}{9}C - 32$
- D $\frac{9}{5}C + 32$

73. MULTI-STEP Suppose that the sum of three numbers is 100 and the lowest number is n. **MP 2**

a. If the second highest number is two more than twice the lowest number, write an expression for the second number.

b. If the highest number is six less than five times the lowest number, write an expression for the highest number.

c. Write an equation using the expressions for each number that express the sum of the three numbers to be 100.

d. Solve the equation for n.

e. Name the three numbers.

f. Check the solution to the equation.

12 | Lesson 1-1 | Solving Linear Equations

LESSON 2
Solving Linear Inequalities

::Then
- You solved equations involving absolute values.

::Now
1. Solve one-step inequalities.
2. Solve multi-step inequalities.

::Why?
- Josh is trying to decide between two data plans offered by a wireless telephone company.

To compare these two rate plans, we can use inequalities. The monthly access fee for Plan 1 is less than the fee for Plan 2, $60 < $75. However, the additional data fee for Plan 1 is greater than that of Plan 2, $20 > $10.

	Plan 1	Plan 2
Monthly Fee	$60	$75
Data Included	2 GB	4 GB
Additional 1 GB/month	$20	$10

New Vocabulary
set-builder notation

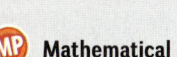
Mathematical Practices
4 Model with mathematics.

1 One-Step Inequalities For any two real numbers a and b, exactly one of the following statements is true.
$$a < b \qquad a = b \qquad a > b$$

Adding the same number to, or subtracting the same number from, each side of an inequality does not change the truth of the inequality.

Key Concept

Addition Property of Inequality

Words

For any real numbers, a, b, and c:

If $a > b$, then $a + c > b + c$.

If $a < b$, then $a + c < b + c$.

Models

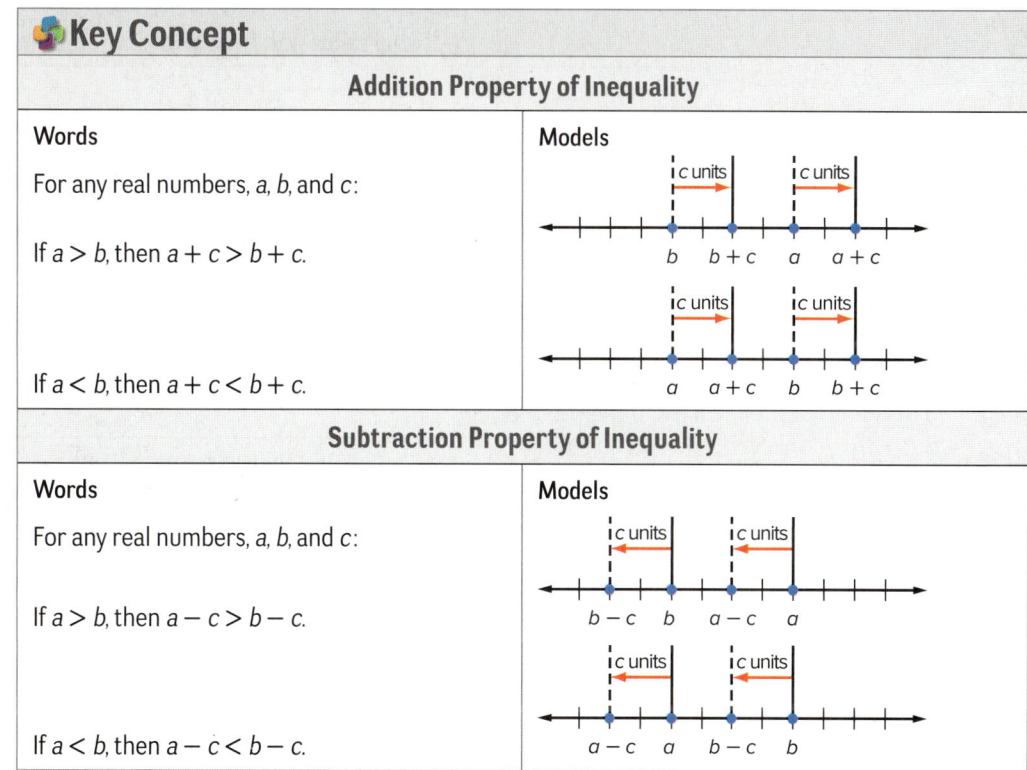

Subtraction Property of Inequality

Words

For any real numbers, a, b, and c:

If $a > b$, then $a - c > b - c$.

If $a < b$, then $a - c < b - c$.

Models

These properties are also true for $\leq$, $\geq$, and $\neq$.

These properties can be used to solve inequalities. The solution sets of inequalities in one variable can then be graphed on number lines.

connectED.mcgraw-hill.com

Review Vocabulary

Inequality Symbols

> greater than; is more than

< less than; is fewer than

≥ greater than or equal to; is at least; is no less than

≤ less than or equal to; is at most; is no more than

Study Tip

Graphing Inequalities
A circle is used for < and >.
A dot is used for ≤ and ≥.

Example 1 Solve an Inequality Using Addition or Subtraction

Solve $y - 6 < 3$. Graph the solution set on a number line.

$y - 6 < 3$ Original inequality

$y - 6 + 6 < 3 + 6$ Add 6 to each side.

$y < 9$ Simplify.

Any real number less than 9 is a solution of this inequality. The graph of the solution set is shown at the right.

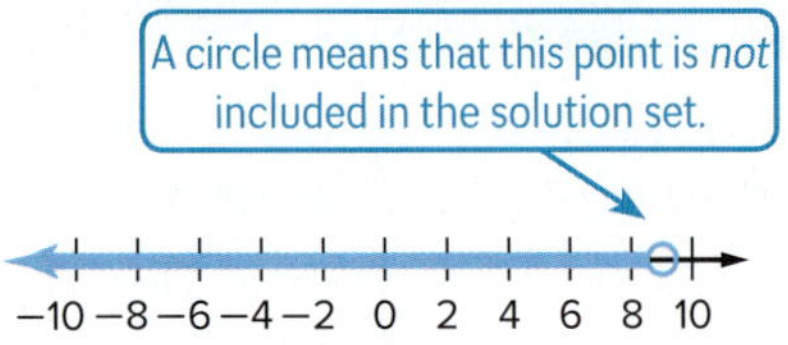

A circle means that this point is *not* included in the solution set.

CHECK Substitute 8 and then 10 for y in $y - 6 < 3$. The inequality should be true for $y = 8$ and false for $y = 10$. ✓

Guided Practice

Solve each inequality. Graph the solution set on a number line.

1A. $5w + 3 > 4w + 9$

1B. $5x - 3 > 4x + 2$

Multiplying or dividing each side of an inequality by a positive number does not change the truth of the inequality. However, multiplying or dividing each side of an inequality by a *negative* number requires that the order of the inequality be *reversed*. For example, to reverse ≤, replace it with ≥.

Key Concept

Multiplication Property of Inequality

Words	Examples
For any real numbers, *a*, *b*, and *c*,	$-5 < -3$
where *c* is positive:	$-5(6) < -3(6)$
If $a > b$, then $ac > bc$.	$-30 < -18$
If $a < b$, then $ac < bc$.	
where *c* is negative:	$12 > -7$
If $a > b$, then $ac < bc$.	$12(-4) < -7(-4)$
If $a < b$, then $ac > bc$.	$-48 < 28$

Division Property of Inequality

Words	Examples
For any real numbers, *a*, *b*, and *c*,	$-12 < -8$
where *c* is positive:	$\frac{-12}{4} < \frac{-8}{4}$
If $a > b$, then $\frac{a}{c} > \frac{b}{c}$.	$-3 < -2$
If $a < b$, then $\frac{a}{c} < \frac{b}{c}$.	
where *c* is negative:	$-21 < -14$
If $a > b$, then $\frac{a}{c} < \frac{b}{c}$.	$\frac{-21}{-7} > \frac{-14}{-7}$
If $a < b$, then $\frac{a}{c} > \frac{b}{c}$.	$3 > 2$

These properties are also true for ≤, ≥, and ≠.

Reading Math

Set-Builder Notation
$\{y \mid y < 9\}$ is read *the set of all numbers y such that y is less than 9.*

The solution set of an inequality can be expressed by using **set-builder notation**. For example, the solution set in Example 1 can be expressed as $\{y \mid y < 9\}$.

Example 2 Solve an Inequality Using Multiplication or Division

Solve $-4.2x \leq -29.4$. Graph the solution set on a number line.

$-4.2x \leq -29.4$ Original inequality

$\dfrac{-4.2x}{-4.2} \geq \dfrac{-29.4}{-4.2}$ Divide each side by -4.2, reversing the inequality symbol.

$x \geq 7$ Simplify.

The solution set is $\{x \mid x \geq 7\}$. The graph of the solution is shown below.

A dot means that this point is included in the solution set.

CHECK Substitute 6 and then 8 for x in $-4.2x \leq -29.4$. The inequality should be true for $x = 8$ and false for $x = 6$. ✓

Guided Practice

Solve each inequality. Graph the solution set on a number line.

2A. $-4x \geq -24$ **2B.** $-9.2y < 23$

2 Multi-Step Inequalities
Solving multi-step inequalities is similar to solving multi-step equations.

Go Online!

Look for the **Tools** icon for places where the tools in the eToolkit may be useful. Log into ConnectED to use the tools.

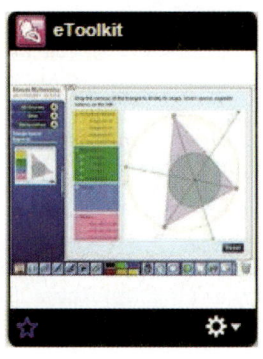

Example 3 Solve Multi-Step Inequalities

Solve $-4c \leq \dfrac{5c + 58}{6}$. Graph the solution set on a number line.

$-4c \leq \dfrac{5c + 58}{6}$ Original inequality

$-24c \leq 5c + 58$ Multiply each side by 6.

$-29c \leq 58$ Add $-5c$ to each side.

$c \geq -2$ Divide each side by -29, reversing the inequality symbol.

The solution set is $\{c \mid c \geq -2\}$ and is graphed below.

CHECK Substitute -3 and then -1 for x in $-4c \leq \dfrac{5c + 58}{6}$. The inequality should be true for $x = -1$ and false for $x = -3$. ✓

Guided Practice

Solve each inequality. Graph the solution set on a number line.

3A. $-3x \leq \dfrac{-4x + 22}{5}$ **3B.** $8y \geq \dfrac{-5y + 9}{-4}$

3C. $-6(-4v + 3) \leq 2(10v + 3)$ **3D.** $-5(3d - 7) > 3(2d + 14)$

Real-World Example 4 Write and Solve an Inequality

WEBSITES Enrique's company pays Salim to advertise on Salim's website. Salim's website earns $15 per month plus $0.05 every time a visitor clicks on the advertisement. What is the least number of clicks per month that Salim needs in order to earn at least $50 per month?

Real-World Link
Recently, the Netcraft Web Server Survey found over 108,000,000 distinct websites.
Source: Netcraft

Understand Let c = the number of clicks on the advertisement. Salim earns $15 per month and $0.05 per click, and he wants to earn a minimum of $50.

Plan Write an inequality.

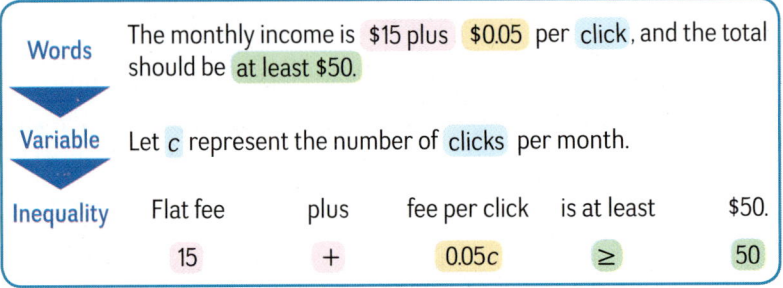

Solve
$15 + 0.05c \geq 50$ Original inequality
$0.05c \geq 35$ Subtract 15 from each side.
$c \geq 700$ Divide each side by 0.05.

Visitors to Salim's website need to click on Enrique's advertisement at least 700 times per month in order for Salim to earn $50 or more from Enrique's company.

Check
$15 + 0.05c \geq 50$ Original inequality
$5 + 0.05(700) \stackrel{?}{\geq} 50$ Replace c with 700.
$15 + 35 \stackrel{?}{\geq} 50$ Multiply.
$50 \geq 50$ ✓ Add.

Translating a real-world problem into a mathematical expression allows developed tools to be used.

Guided Practice

4. Rosa's Internet radio plan costs her $9.95 per month plus $0.88 for each song she buys. How many songs can she buy and still pay less than a total of $20?

Check Your Understanding

 = Step-by-Step Solutions begin on page R11.

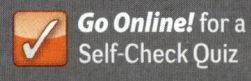 Go Online! for a Self-Check Quiz

Examples 1–3 Solve each inequality. Then graph the solution set on a number line.

1. $b + 6 < 14$
2. $12 - d > -8$
3. $18 \leq -3x$
4. $-5y \geq -35$
5. $-4w - 13 > -21$
6. $8z - 9 \geq -15$
7. $s \geq \dfrac{s+6}{5}$
8. $\dfrac{2x-9}{4} \leq x + 2$

Example 4

9. **MODELING** Tara is delivering bags of mulch. Each bag weighs 48 pounds, and the push cart weighs 65 pounds. If her flat-bed truck is capable of hauling 2000 pounds, how many bags of mulch can Tara safely take on each trip?

Practice and Problem Solving

Extra Practice is on page R1.

Examples 1–3 Solve each inequality. Then graph the solution set on a number line.

10. $m - 8 > -12$
11. $n + 6 \leq 3$
12. $6r < -36$
13. $-12t \geq -6$
14. $-\frac{w}{4} \leq -7$
15. $\frac{k}{3} - 14 < -5$
16. $4x - 15 \leq 21$
17. $-6z - 14 > -32$
18. $-16 \geq 5(2z - 11)$
19. $12 < -4(3c - 6)$
20. $\frac{3y - 4}{0.2} - 8 > 12$
21. $\frac{9z + 5}{4} + 18 < 26$

Example 4

22. **GYMNASTICS** In a gymnastics competition, an athlete's final score is calculated by taking 75% of the average technical score and adding 25% of the artistic score. All scores are out of 10, and one gymnast has a 7.6 average technical score. What artistic score does the gymnast need to have a final score of at least 8.0?

Define a variable and write an inequality for each problem. Then solve.

23. Twelve less than the product of three and a number is less than 21.
24. The quotient of three times a number and 4 is at least -16.
25. The difference of 5 times a number and 6 is greater than the number.
26. The quotient of the sum of 3 and a number and 6 is less than -2.
27. **HIKING** Danielle can hike 3 miles in an hour, but she has to take a one-hour break for lunch and a one-hour break for dinner. If Danielle wants to hike at least 18 miles, solve $3(x - 2) \geq 18$ to determine how many hours the hike should take.

Solve each inequality. Then graph the solution set on a number line.

28. $18 - 3x < 12$
29. $-8(4x + 6) < -24$
30. $\frac{1}{4}n + 12 \geq \frac{3}{4}n - 4$
31. $0.24y - 0.64 > 3.86$
32. $10x - 6 \leq 4x + 42$
33. $-6v + 8 > -14v - 28$
34. $n > \frac{-3n - 15}{8}$
35. $-2r < \frac{6 - 2r}{5}$
36. $\frac{9z - 4}{5} \leq \frac{7z + 2}{4}$

37. **MONEY** Jin is selling advertising space on the *Central City Blog* to local businesses. Jin earns 3% commission for every advertisement he sells plus a salary of $250 a week. If the average amount of money that a business spends on an advertisement is $500, how many advertisements must he sell each week to make a salary of at least $700 that week?

   a. Write an inequality to describe this situation.
   b. Solve the inequality and interpret the solution.

Define a variable and write an inequality for each problem. Then solve.

38. One third of the sum of 5 times a number and 3 is less than one fourth the sum of six times that number and 5.
39. The sum of one third a number and 4 is at most the sum of twice that number and 12.
40. **CHALLENGE** The sides of square *ABCD* are extended to form rectangle *DEFG*. If the perimeter of the rectangle is at least twice the perimeter of the square, what is the maximum length of a side of square *ABCD*?

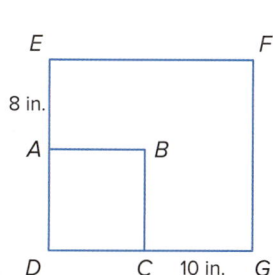

41. MARATHONS Jamie wants to be able to run at least the standard marathon distance of 26.2 miles. A good rule for training is that runners generally have enough endurance to finish a race that is up to 3 times his or her average daily distance.

  a. If the length of her current daily run is 5 miles, write an inequality to find the amount by which she needs to increase her daily run to have enough endurance to finish a marathon.

  b. Solve the inequality and interpret the solution.

42. MP MODELING The costs for renting a car from Ace Car Rental and from Basic Car Rental are shown in the table. For what mileage does Basic have the better deal? Use the inequality $49 + 0.1x > 55 + 0.05x$. Explain why this inequality works.

Rental Car Costs		
Company	Cost per Day	Cost per Mile
Ace	$49	$0.10
Basic	$55	$0.05

43. MULTIPLE REPRESENTATIONS In this exercise, you will explore graphing inequalities on a coordinate plane.

  a. **Tabular** Organize the following into a table. Substitute 5 points into the inequality $y \geq -\frac{1}{2}x + 3$. State whether the resulting statement is *true* or *false*.

  b. **Graphical** Graph $y = -\frac{1}{2}x + 3$. Also graph the 5 points from the table. Label all points that resulted in a true statement with a T. Label all points that resulted in a false statement with an F.

  c. **Verbal** Describe the pattern produced by the points you have labeled. Make a conjecture about which points on the coordinate plane would result in true and false statements.

H.O.T. Problems   Use Higher-Order Thinking Skills

44. MP MODELING If $-4 < x < 5$ and $0.25 < y < 4$, then $a < \frac{x}{y} < b$. What is $a + b$?

45. ERROR ANALYSIS Madlynn and Emilie were comparing their homework. Is either of them correct? Explain your reasoning.

Madlynn	Emilie
$\frac{4x+5}{-2} - 1 > -3$	$\frac{4x+5}{-2} - 1 > -3$
$\frac{4x+5}{-2} < -2$	$\frac{4x+5}{-2} > -2$
$4x + 5 > 4$	$4x + 5 > 4$
$4x > -1$	$4x > -1$
$x > -\frac{1}{4}$	$x > -\frac{1}{4}$

46. MP CONSTRUCT ARGUMENTS Determine whether the following statement is *sometimes*, *always*, or *never* true. Explain your reasoning.

The opposite of the absolute value of a negative number is less than the opposite of that number.

47. MP MODELING Given $\triangle ABC$ with sides $AB = 3x + 4$, $BC = 2x + 5$, and $AC = 4x$, determine the values of x such that $\triangle ABC$ exists.

48. OPEN-ENDED Write an inequality for which the solution is all real numbers in the form $ax + b > c(x + d)$. Explain how you know this.

49. WRITING IN MATH Why does the inequality symbol need to be reversed when multiplying or dividing by a negative number?

Preparing for Assessment

50. Which graph represents the complete solution set of the inequality $x + 6 < 2x - 9$? MP 5

- A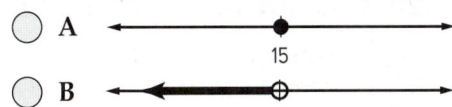
- B
- C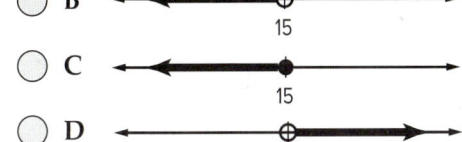
- D
- E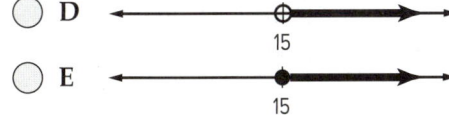

51. The solution set of which inequality is represented by this graph? MP 2

- A $6 - \frac{x}{2} < 7$
- B $\frac{x - 10}{3} < -4$
- C $\frac{2 - x}{4} \geq 1$
- D $x + \frac{5 + x}{2} < -1$
- E $x + \frac{2x - 9}{3} < -5$

52. A kennel has enough space to hold at most 80 cats and dogs. If c represents the number of cats in the kennel and d represents the number of dogs, which inequality describes the capacity of the kennel? MP 4

- A $c + d > 80$
- B $c + d \geq 80$
- C $c + d < 80$
- D $c + d \leq 80$

53. MULTI-STEP The quotient of 5 times a number and -4 is at least 20. MP 1

a. Write an inequality for the problem.

b. Solve the inequality.

c. Graph the solution set on a number line.

54. The solution to $-x > 15$ is: MP 2

- A all numbers greater than 15.
- B all numbers greater than -15.
- C all numbers less than 15.
- D all numbers less than -15.

55. Solve the inequality. MP 6

$-3x + 5 \leq 20$

- A $x \leq -5$
- B $x \leq 5$
- C $x \geq 5$
- D $x \geq -5$

56. Solve the inequality $1 < 3x - 2 \leq 7$. MP 1

- A $-1 < x \leq 3$
- B $1 < x \leq 3$
- C $x < 1$ or $x \geq 3$
- D $x < 1$ or $x \geq 9$

57. Solve the inequality $3x - 2 \leq \frac{7x - 1}{2}$. MP 1

58. Solve the inequality $x < 3x$. MP 1

- A $x > 0$
- B $x > 3$
- C $x < 0$
- D $x < 3$
- E no solutions

EXTEND 1-2
Algebra Lab
Interval and Set-Builder Notation

The solution set of an inequality can be described by using **interval notation**. The **infinity** symbols below are used to indicate that a set is unbounded in the positive or negative direction, respectively.

Mathematical Practices
MP 5 Use appropriate tools strategically.

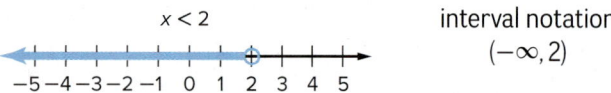

To indicate that an endpoint is *not* included in the set, a parenthesis, (or), is used. Parentheses are always used with the symbols $+\infty$ and $-\infty$, because they do not include endpoints.

$x < 2$ interval notation
$(-\infty, 2)$

A bracket is used to indicate that the endpoint, -2, *is* included in the solution set below.

$x \geq -2$ interval notation
$[-2, +\infty)$

In interval notation, the symbol for the union of the two sets is $\cup$. The compound inequality $y \leq -7$ or $y > -1$ is written as $(-\infty, -7] \cup (-1, +\infty)$.

Exercises

Work cooperatively. Write each inequality using interval notation.

1. $\{a \mid a \leq -3\}$
2. $\{n \mid n > -8\}$
3. $\{y \mid y < 2 \text{ or } y \geq 14\}$
4. $\{b \mid b \leq -9 \text{ or } b > 1\}$
5. $\{t \mid 1 < t < 3\}$
6. $\{m \mid m \geq 4 \text{ or } m \leq -7\}$
7. $\{x \mid x \neq -4\}$
8. $\{r \mid 3 < r < 4\}$

9. [number line graph]
10. [number line graph]
11. [number line graph]
12. [number line graph]
13. [number line graph]
14. [number line graph]

Graph each solution set on a number line.

15. $(-1, +\infty)$
16. $(-\infty, 4]$
17. $(-\infty, 5] \cup (7, +\infty)$

18. **WRITING IN MATH** Write in words the meaning of $(-\infty, 3) \cup [10, +\infty)$. Then write the compound inequality that this notation represents.

19. **WRITING IN MATH** How are symbols used to write solution sets for inequalities? Explain.

LESSON 3
Rate of Change and Slope

::Then::
- You graphed linear relations.

::Now::
1. Find rate of change.
2. Determine the slope of a line.

::Why?::
- The table shows the total distance a car traveled over various time intervals. The distance formula, $rt = d$ or $r = \frac{d}{t}$, relates time and distance.

Time (h)	Distance (mi)
1	68
2.5	170
3	204
4.5	306
5	340

New Vocabulary
rate of change
slope

Mathematical Practices
7 Look for and make use of structure.
8 Look for and express regularity in repeated reasoning.

1 Rate of Change
Rate of change is a ratio that compares how much one quantity changes, on average, relative to the change in another quantity. If x is the independent variable and y is the dependent variable, then rate of change $= \frac{\text{change in } y}{\text{change in } x}$. This is sometimes referred to as $\frac{\Delta y}{\Delta x}$.

Real-World Example 1 Constant Rate of Change

CHEMISTRY The table shows the temperature of a solution after it has been removed from a heat source. Find the rate of change in temperature for the solution.

Use the ordered pairs (2, 139.4) and (5, 133.1).

$$\text{rate of change} = \frac{\text{change in } y}{\text{change in } x}$$
$$= \frac{\text{change in temperature}}{\text{change in time}}$$
$$= \frac{133.1 - 139.4}{5 - 2}$$
$$= \frac{-6.3}{3} \text{ or } -2.1$$

Time (min)	Temperature (°C)
0	143.6
2	139.4
5	133.1
8	126.8
12	118.4

The rate of change is -2.1. This means that the temperature is decreasing by 2.1°C each minute.

Guided Practice

1. **RECREATION** The graph at the right shows the number of gallons of water in a swimming pool as it is being filled. At what rate is the pool being filled?

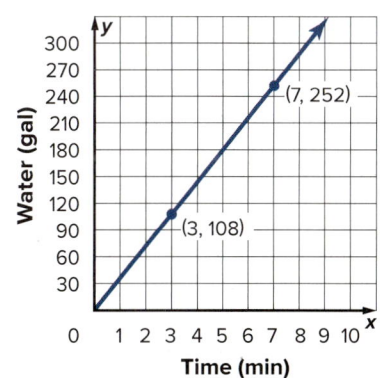

Study Tip
Independent Quantities Rates of change often include a measure of *time* as the independent variable.

Up to this point, you have used rates of change that are constant. Many real-world situations involve rates of change that are not constant. These situations are often described using an average rate of change over a specified interval.

Real-World Example 2 Average Rate of Change

MUSIC Refer to the graph at the right. Find the average rate of change of the sales for both digital music and ebooks from 2004 to 2009. Compare the rates.

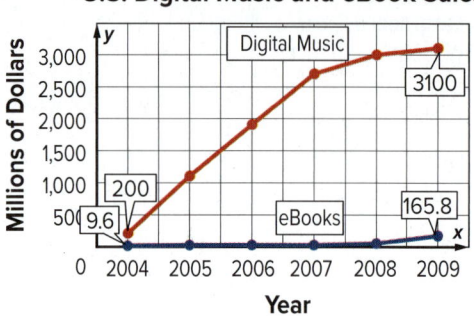

U.S. Digital Music and eBook Sales

Digital Musics:

$$\text{rate of change} = \frac{\text{change in } y}{\text{change in } x}$$

$$= \frac{\text{change in sales}}{\text{change in time}}$$

$$= \frac{3100 - 200}{2009 - 2004}$$

$$= \frac{2900}{5} \text{ or } 580$$

eBooks:

$$\text{rate of change} = \frac{\text{change in } y}{\text{change in } x}$$

$$= \frac{\text{change in sales}}{\text{change in time}}$$

$$= \frac{165.8 - 9.6}{2009 - 2004}$$

$$= \frac{156.2}{5} \text{ or } 31.24$$

The sales of digital music increased at an average rate of $580 million per year, while the sales of ebooks increased at an average rate of $31.24 million per year.

Guided Practice

2. **EDUCATION** In 2007, 23,142 students applied to State College, and 34,689 students applied to Central University. In 2015, 29,563 students applied to State College, and 36,107 applied to Central University. Determine the average rate of change in applicants for both schools from 2007 to 2015.

Real-World Link
According to a recent online survey, 36% of teens in the United States purchased at least one CD in the past year, while 51% paid to download at least one song.

Source: CNN

2 Slope
The **slope** of a line is the ratio of the change in the y-coordinates to the corresponding change in the x-coordinates. The slope of a line is the same as its rate of change.

Suppose a line passes through points at (x_1, y_1) and (x_2, y_2).

$$\text{Slope} = \frac{\text{change in } y\text{-coordinates}}{\text{change in } x\text{-coordinates}} = \frac{y_2 - y_1}{x_2 - x_1}$$

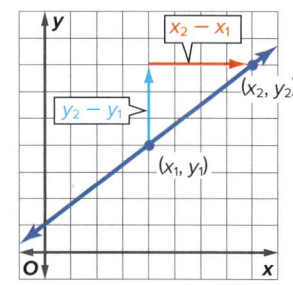

Go Online!

Investigate slope by using the slope tool in the **Graphing Tools** in ConnectED.

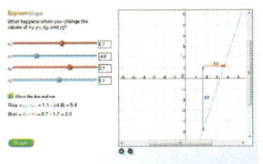

Key Concept Slope of a Line

Words The slope of a line is the ratio of the change in *y*-coordinates to the change in *x*-coordinates.

Symbols The slope *m* of a line passing through (x_1, y_1) and (x_2, y_2) is given by $m = \frac{y_2 - y_1}{x_2 - x_1}$, where $x_1 \neq x_2$.

SLOPE The formula for slope is often remembered as *rise over run*, where the rise is the difference in *y*-coordinates and the run is the difference in *x*-coordinates.

Example 3   Find Slope Using Coordinates

Find the slope of the line that passes through $(-4, 3)$ and $(2, 5)$.

$m = \dfrac{y_2 - y_1}{x_2 - x_1}$ Slope Formula

$ = \dfrac{5 - 3}{2 - (-4)}$ $(x_1, y_1) = (-4, 3), (x_2, y_2) = (2, 5)$

$ = \dfrac{2}{6}$ or $\dfrac{1}{3}$ Simplify.

▶ Guided Practice

Find the slope of the line that passes through each pair of points.

3A. $(1, -3)$ and $(3, 5)$ **3B.** $(-8, 11)$ and $(24, -9)$

You can choose any two points from the graph of a line to find the slope.

Study Tip

Slope is Constant The slope of a line is the same, no matter what two points on the line are used.

Example 4   Find Slope Using a Graph

Find the slope of the line shown at the right.

The line passes through $(-2, 0)$ and $(0, -3)$.

$m = \dfrac{y_2 - y_1}{x_2 - x_1}$ Slope Formula

$ = \dfrac{-3 - 0}{0 - (-2)}$ $(x_1, y_1) = (-2, 0), (x_2, y_2) = (0, -3)$

$ = \dfrac{-3}{2}$ or $-\dfrac{3}{2}$ Simplify.

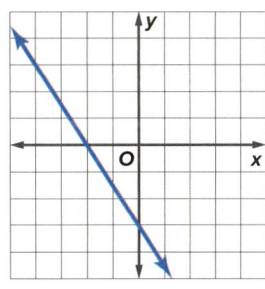

▶ Guided Practice

Find the slope of each line.

4A. 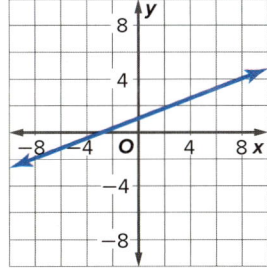   **4B.**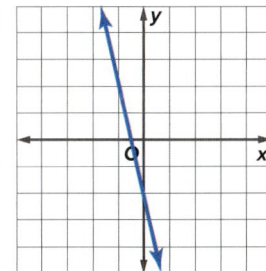

connectED.mcgraw-hill.com

Check Your Understanding

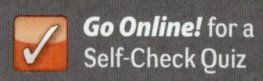

Example 1 **MP REASONING** Find the rate of change for each set of data.

1.
Time (min)	2	4	6	8	10
Distance (ft)	12	24	36	48	60

2.
Time (sec)	5	10	15	20	25
Volume (cm³)	16	32	48	64	80

Example 2

3. **CAMERAS** The graph shows the number of digital still cameras and film cameras sold by Yellow Camera Stores in recent years.

   a. Find the average rate of change of the number of digital cameras sold from 2006 to 2015.

   b. Find the average rate of change of the number of film cameras sold from 2006 to 2015.

   c. What do the signs of each rate of change represent?

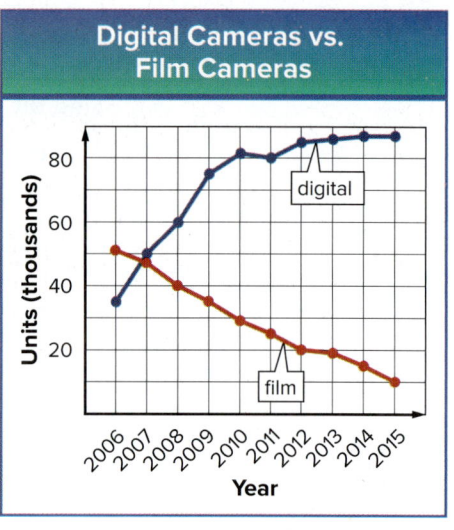

Example 3 Find the slope of the line that passes through each pair of points.

4. (3, 2), (8, 12)   5. (−1, 4), (3, −8)   6. (−2, −5), (−7, 10)

Example 4 Determine the rate of change of each graph.

7. 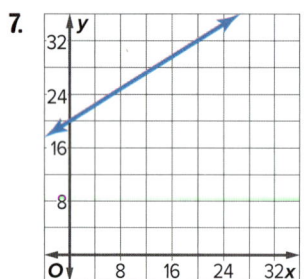   8.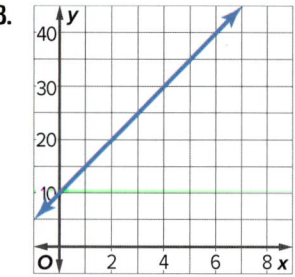

Practice and Problem Solving

Extra Practice is on page R1.

Example 1 Find the rate of change for each set of data.

9.
Time (day)	3	6	9	12	15
Height (mm)	20	40	60	80	100

10.
Weight (lb)	11	22	33	44	55
Cost ($)	8	16	24	32	40

24 | Lesson 1-3 | Rate of Change and Slope

Example 2

11. HEALTH The table below shows Lisa's temperature during an illness over a 3-day period.

Day	Monday		Tuesday		Wednesday	
Time	8:00 A.M.	8:00 P.M.	8:00 A.M.	8:00 P.M.	8:00 A.M.	8:00 P.M.
Temp (°F)	100.5	102.3	103.1	100.7	99.9	98.6

a. What was the average rate of change in Lisa's temperature from 8:00 A.M. on Monday to 8:00 P.M. on Monday?

b. What was the average rate of change in Lisa's temperature from 8:00 A.M. on Tuesday to 8:00 P.M. on Wednesday? Is your answer reasonable? What does the sign of the rate mean?

c. During which 12-hour period was the average rate of change in Lisa's temperature the greatest?

Example 3 Find the slope of the line that passes through each pair of points. Express as a fraction in simplest form.

12. $(-2, 11), (5, 6)$

13. $(-9, -11), (6, 3)$

14. $(-1.5, 3.5), (4.5, 6)$

15. $(-4.5, 9.5), (-1, 2.5)$

16. $(-8, -0.5), (-4, 5)$

17. $(-6, -2), (-1.5, 5.5)$

Example 4 Determine the rate of change of each graph.

18.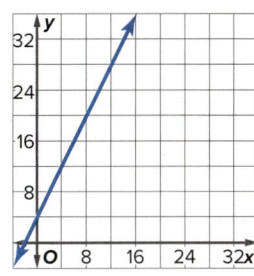

19.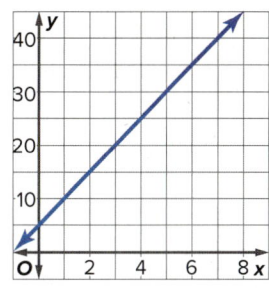

20.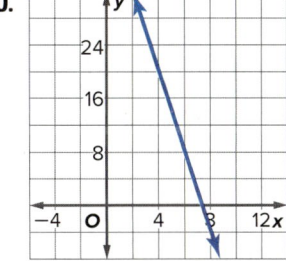

21.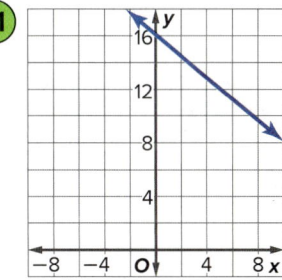

22. STRUCTURE The table shows your height on a water slide at various time intervals.

Time (s)	Height (ft)
0	120
1	90
2	60
3	30
4	0
5	0

a. Graph the height versus the time on the water slide.

b. Find the average rate of change in a rider's height between 1 and 3 seconds.

c. Find the average rate of change in a rider's height between 0 and 5 seconds.

d. What is another word for *rate of change* in this situation?

Determine the rate of change for each equation.

23. $6y = 8x - 40$

24. $-2y - 16x = 41$

25. $12x - 4y + 5 = 18$

26. $20x + 85y = 120$

27. $\frac{3}{2}x - \frac{5}{4}y = 15$

28. $\frac{1}{6}y + \frac{3}{8}x = 24$

29. **WASHINGTON MONUMENT** The Washington Monument is 555 feet $5\frac{1}{8}$ inches tall and weighs 90,854 tons. The monument is topped by an aluminum square pyramid. The sides of the pyramid's base measure 5.6 inches, and the pyramid is 8.9 inches tall. Estimate the slope that a face of the pyramid makes with its base.

30. **MARINE LIFE** The illustrations show the growth of a starfish over time.

    a. Find the average rate of change in the measure over time.

    b. Predict the size of the starfish in 2019.

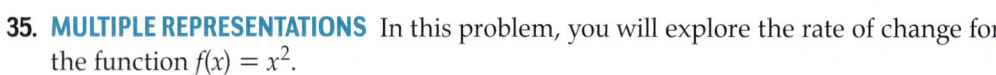

Find the value of r so that the line that passes through each pair of points has the given slope.

31. $(6, r), (3, 3), m = 2$

32. $(8, 1), (5, r), m = \frac{1}{3}$

33. $(10, r), (4, -3), m = \frac{4}{3}$

34. $(8, -2), (r, -6), m = -4$

35. **MULTIPLE REPRESENTATIONS** In this problem, you will explore the rate of change for the function $f(x) = x^2$.

    a. **Graphical** Graph $f(x) = x^2$.

    b. **Tabular** Copy and complete the table. To complete the slope row, find the slope of the line containing two consecutive points such as $(-4, 16)$ and $(-3, 9)$. The first one is completed for you.

x	−4	−3	−2	−1	0	1	2	3	4
f(x)	16	9							
slope		−7							

    c. **Verbal** Describe what happens to the rate of change for $f(x) = x^2$ as x increases.

H.O.T. Problems — Use Higher-Order Thinking Skills

36. **ERROR ANALYSIS** Patty and Tim are asked to find the slope of the line passing through the points (4, 3) and (7, 9). Is either of them correct? Explain.

    Patty
    $m = \dfrac{9-3}{7-4}$
    $= \dfrac{6}{3}$ or 2

    Tim
    $m = \dfrac{7-4}{9-3}$
    $= \dfrac{3}{6}$ or $\dfrac{1}{2}$

37. **MODELING** The graph of a line passes through the points (2, 3) and (5, 8). Explain how you would find the y-coordinate of the point (11, y) on the same line. Then find y.

38. **WRITING IN MATH** In what ways can change be represented mathematically?

39. **REASONING** Determine whether the statement *A line has a slope that is a real number* is *sometimes*, *always*, or *never* true. Explain your reasoning.

40. **REASONING** Describe the process of finding the rate of change for each.

    a. a table of values
    b. a graph
    c. an equation

Preparing for Assessment

41. Line ℓ contains the points $(8, a)$ and $(6, -4)$. Line m contains the points $(-2, 7)$ and $(4, 2)$. For what value of a are the slopes of lines ℓ and m the same? **MP 2**

- A -17
- B $-\frac{3}{7}$
- C $\frac{7}{3}$
- D 7
- E $-\frac{17}{3}$

42. What is the rate of change for the data set in the table? **MP 6**

Time (min)	2	5	8	11	14
Distance (ft)	10	25	40	55	70

- A 3 ft/min
- B 5 ft/min
- C 8 ft/min
- D 10 ft/min
- E 15 ft/min

43. What is the slope of the line $-6x - 2y = -10$? **MP 8**

- A -6
- B -5
- C -3
- D -2
- E 3

44. What is the slope of the line pictured below? **MP 5**

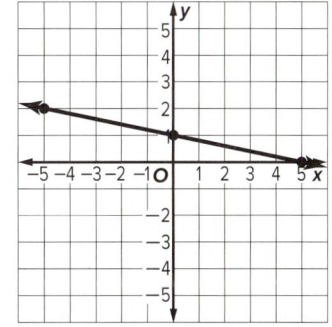

- A -5
- B $-\frac{1}{5}$
- C $\frac{1}{5}$
- D 1
- E 5

45. A family took a road trip across many U.S. states in 20 hours. What was their average rate of change in miles per hour? **MP 6**

- A 49 mi/h
- B 48 mi/h
- C 49 h/mi
- D 48 h/m

Time (hours)	Distance (miles)
10	490
15	735
20	980

46. Find the slope of the linear function with the following x and y coordinates. **MP 7**

- A -1
- B 2
- C $-1/2$
- D -2

x	y
1	-12
2	-14
3	-16

47. MULTI-STEP A baseball team's total wins were plotted on the graph below. Each point represents the cumulative season total of wins at the end of the month. For example, at the end of April, the team had 18 wins as shown by $(4,18)$. At the end of May, the total wins for both April and May was 33 wins. **MP 3**

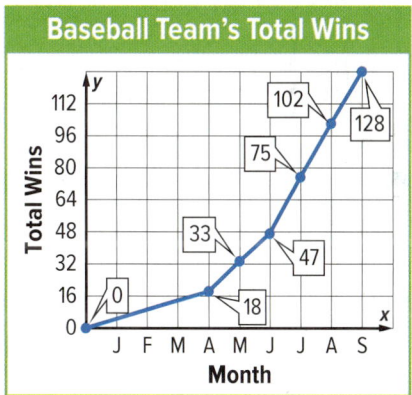

a. What is the average rate of change (wins per month) from the end of April to the end of May?

b. What is the average rate of change (wins per month) from the end of May to the end of June?

c. By the end of the season, the team had won 128 games and had won the league. What was their average rate of change from the end of June to the end of September?

d. Compare the average rate of change during the hot streak to the rate during the earlier months. Explain how the rates relate to the graph.

LESSON 4

Writing Linear Equations

::Then
- You determined slopes of lines.

::Now
1. Write an equation of a line given the slope and a point on the line.
2. Write an equation of a line parallel or perpendicular to a given line.

::Why?
- Medical insurance companies often require their customers to make a co-payment for every doctor's office visit in addition to an annual insurance premium.

  If an insurance company charges $2280 annually and requires a copayment of $35 per doctor's office visit, then the linear equation $y = 35x + 2280$ represents the total annual cost y for x doctor's office visits.

New Vocabulary
slope-intercept form
point-slope form
parallel
perpendicular

Mathematical Practices
2 Reason abstractly and quantitatively.

1 Forms of Equations Consider the line through $A(0, b)$ and $C(x, y)$. Notice that b is the y-intercept. You can use these two points to find the slope of $\overleftrightarrow{AC}$.

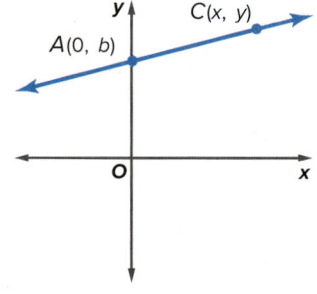

$m = \dfrac{y_2 - y_1}{x_2 - x_1}$ Slope Formula

$ = \dfrac{y - b}{x - 0}$ $(x_1, y_1) = (0, b), (x_2, y_2) = (x, y)$

$ = \dfrac{y - b}{x}$ Simplify.

Now solve the equation for y.

$mx = y - b$ Multiply each side by x.

$mx + b = y$ Add b to each side.

$y = mx + b$ Symmetric Property of Equality

Equations written in this format are in **slope-intercept form**.

Key Concept Slope-Intercept Form

Words   The slope-intercept form of the equation of a line is $y = mx + b$, where m is the slope and b is the y-intercept.

Model

Symbols $y = mx + b$

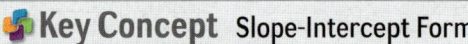
slope ———↑   ↑— y-intercept

If you are given the slope and y-intercept of a line, you can find an equation of the line by substituting the values of m and b into the slope-intercept form.

Sometimes it is necessary to calculate the slope before you can write an equation.

> **Watch Out!**
>
> **MP Reasoning** The equation of a vertical line cannot be written in slope-intercept form because its slope is undefined.

Example 1   Write an Equation in Slope-Intercept Form

Write an equation in slope-intercept form for the line.

The graph intersects the y-axis at -2. So $b = -2$.

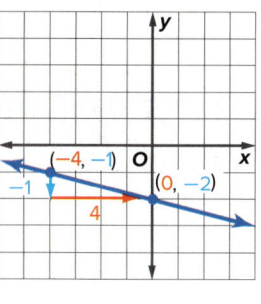

Step 1 Find the slope.

$$m = \frac{y_2 - y_1}{x_2 - x_1} \quad \text{Slope Formula}$$

$$= \frac{-2 - (-1)}{0 - (-4)} \quad (x_1, y_1) = (-4, -1), (x_2, y_2) = (0, -2)$$

$$= \frac{-1}{4} \text{ or } -\frac{1}{4} \quad \text{Simplify.}$$

Step 2 Substitute the values into the slope-intercept equation.

$$y = mx + b \quad \text{Slope-intercept form}$$

$$y = -\frac{1}{4}x - 2 \quad m = -\frac{1}{4}, b = -2$$

> **Study Tip**
>
> **MP Precision** You can check that your equation satisfies the conditions by graphing it.

▶ **Guided Practice**

Write an equation in slope-intercept form for the line described.

1A. slope $\frac{4}{3}$, passes through $(0, 4)$ **1B.** passes through $(0, -6)$ and $(-4, 10)$

If you know the slope of a line and the coordinates of a point on the line, you can use the **point-slope form** to find an equation of the line.

Key Concept   Point-Slope Form

Words	Symbols
The point-slope form of the equation of a line is $y - y_1 = m(x - x_1)$, where (x_1, y_1) are the coordinates of a point on the line and m is the slope of the line.	slope ↓ $y - y_1 = m(x - x_1)$ coordinates of a point on the line

Example 2   Write an Equation Given Slope and One Point

Write an equation of the line through $(6, -2)$ with a slope of -4.

$$y - y_1 = m(x - x_1) \quad \text{Point-slope form}$$

$$y - (-2) = -4(x - 6) \quad (x_1, y_1) = (6, -2), m = -4$$

$$y + 2 = -4x + 24 \quad \text{Simplify.}$$

$$y = -4x + 22 \quad \text{Subtract 2 from each side.}$$

▶ **Guided Practice**

Write an equation in slope-intercept form for the line described.

2A. passes through $(2, 3)$; $m = \frac{1}{2}$ **2B.** passes through $(-2, -1)$; $m = -3$

connectED.mcgraw-hill.com

You can use any two points on a line to write an equation.

Example 3   Write an Equation Given Two Points

Write an equation of the line passing through each pair of points.

a. $(-2, 7), (3, -3)$

Step 1 Determine the slope of the line.

$m = \dfrac{y_2 - y_1}{x_2 - x_1}$ Slope Formula

$= \dfrac{-3 - 7}{3 - (-2)}$ $(x_1, y_1) = (-2, 7)$
$(x_2, y_2) = (3, -3)$

$= -\dfrac{10}{5}$ Subtract.

$= -2$ Divide.

Step 2 Write an equation. Use either ordered pair for (x_1, y_1).

$y - y_1 = m(x - x_1)$ Point-slope form

$y - (-3) = -2(x - 3)$ $(x_1, y_1) = (3, -3),$
$m = -2$

$y + 3 = -2x + 6$ Simplify.

$y = -2x + 3$ Subtract 3 from each side.

An equation of the line is $y = -2x + 3$.

b. $(4, -7), (4, 6)$

Step 1 Determine the slope of the line.

$m = \dfrac{y_2 - y_1}{x_2 - x_1}$ Slope Formula

$= \dfrac{6 - (-7)}{4 - 4}$ $(x_1, y_1) = (4, -7)$
$(x_2, y_2) = (4, 6)$

$= \dfrac{13}{0}$ Subtract.

The slope is undefined.

Step 2 Write an equation.

Because the slope is undefined, the line is a vertical line.

Because any point on the line has an x-cooridinate of 4, an equation of the line is $x = 4$.

Guided Practice

3A. $(4, -9), (2, -4)$ **3B.** $(-2, 0), (1, 5)$

3C. $(-3, -1), (2, -1)$ **3D.** $(6, -2), (8, 3)$

2 Parallel and Perpendicular Lines
Slopes can help you determine whether two lines are parallel or perpendicular.

Key Concept   Parallel and Perpendicular Lines

Parallel Lines	Perpendicular Lines
Two nonvertical lines are **parallel** if and only if they have the same slope. All vertical lines are parallel.	Two nonvertical lines are **perpendicular** if and only if the product of the slopes is −1. Vertical lines and horizontal lines are perpendicular.

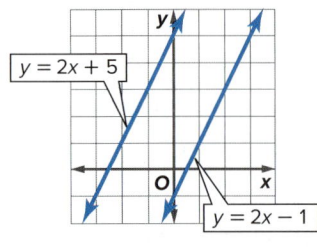

$y = 2x + 5$ and $y = 2x - 1$

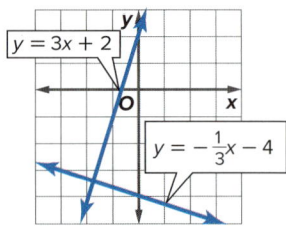

$y = 3x + 2$ and $y = -\dfrac{1}{3}x - 4$

Go Online!

Be sure to review key **vocabulary**, such as *y*-intercept, so that you understand what is being asked in a problem. Find all the terms with animations, English pronunciations, and translations into 13 languages in the eGlossary in ConnectED.

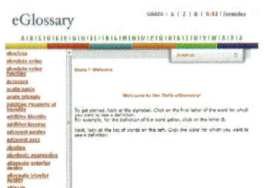

Example 4 Write an Equation of a Parallel or Perpendicular Line

Write an equation in slope-intercept form for the line that passes through $(5, -6)$ and is perpendicular to the line with equation $y = -\frac{3}{2}x + 7$.

The slope of the given line is $-\frac{3}{2}$. Because the slopes of perpendicular lines are opposite reciprocals, the slope of the line perpendicular to the given line is $\frac{2}{3}$.

Use the point-slope form and the ordered pair $(5, -6)$.

$y - y_1 = m(x - x_1)$ Point-slope form

$y - (-6) = \frac{2}{3}(x - 5)$ $(x_1, y_1) = (5, -6)$ and $m = \frac{2}{3}$

$y + 6 = \frac{2}{3}x - \frac{10}{3}$ Distributive Property

$y = \frac{2}{3}x - \frac{28}{3}$ Subtract 6 from each side and simplify.

CHECK Graph both equations to verify the solution.

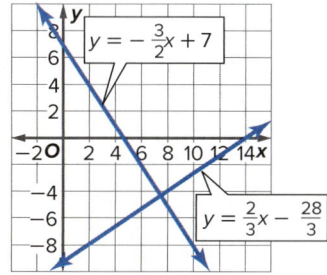

Guided Practice

4. Write an equation in slope-intercept form for the line that passes through $(3, 7)$ and is parallel to the line with equation $y = \frac{3}{4}x - 5$.

Check Your Understanding

= Step-by-Step Solutions begin on page R11.

Go Online! for a Self-Check Quiz

Example 1 Write an equation in slope-intercept form for the line described.

1. slope 1.5, passes through $(0, 5)$

2. passes through $(-2, 3)$ and $(0, 1)$

Example 2

3. passes through $(3, 5)$; $m = -2$

4. passes through $(-8, -2)$; $m = \frac{5}{2}$

Example 3 Write an equation of the line passing through each pair of points.

5. $(-9, 11), (-6, -1)$

6. $(4, 3), (-1, 6)$

7. $(8, 10), (-4, -6)$

Example 4 **MP MODELING** Write an equation in slope-intercept form for the line that satisfies each set of conditions.

8. passes through $(-9, -3)$, perpendicular to $y = -\frac{5}{3}x - 8$

9 passes through $(4, -10)$, parallel to $y = \frac{7}{8}x - 3$

connectED.mcgraw-hill.com **31**

Practice and Problem Solving

Extra Practice is on page R1.

Example 1 Write an equation in slope-intercept form for the line described.

10. slope 3, passes through $(0, -2)$
11. slope $-\frac{1}{2}$, passes through $(0, 5)$
12. slope $-\frac{6}{5}$, passes through $(0, 8)$
13. slope $\frac{9}{2}$, passes through $\left(0, -\frac{13}{2}\right)$

Example 2
14. slope -2, passes through $(-3, 14)$
15. slope 4, passes through $(6, 9)$
16. slope $\frac{3}{5}$, passes through $(-6, -8)$
17. slope $-\frac{1}{4}$, passes through $(12, -4)$

18. **PART-TIME JOB** Each week, Carmen earns a base pay of $35 plus $0.17 for every pamphlet that she delivers. Write an equation that can be used to find how much Carmen earns each week. How much will she earn the week that she delivers 300 pamphlets?

Example 3 Write an equation of the line passing through each pair of points.

19. $(-2, -6), (4, 6)$
20. $(-8, -5), (-3, 10)$
21. $(-4, 12), (-2, -4)$
22. $(4.6, 3.4), (2.2, 2.8)$
23. $(5.5, 0.6), (1.1, 2.8)$
24. $(-25, -16), (-29, 12)$

Example 4 **MP PRECISION** Write an equation in slope-intercept form for the line that satisfies each set of conditions.

25. passes through $(4, 2)$, perpendicular to $y = -2x + 3$
26. passes through $(-6, -6)$, parallel to $y = \frac{4}{3}x + 8$
27. passes through $(12, 0)$, parallel to $y = -\frac{1}{2}x - 3$
28. passes through $(10, 2)$, perpendicular to $y = 4x + 6$

29. **FINANCIAL LITERACY** Julio buys a used car for $5900. Monthly expenses for the car—which include insurance, maintenance, and gas—average $180 per month. Write an equation that represents the total cost of buying and owning the car for x months.

30. **DELI** The sales of a sandwich store increased approximately linearly from $52,000 to $116,000 during the first five years of business. Write an equation that models the sales y after x years. Determine what the sales will be at the end of 12 years if the pattern continues.

31. **WHALES** In 2015, it was estimated that there were 300 northern right whales in existence in a specific area. The population of northern right whales is expected to decline by at least 25 whales each generation. Write an equation that represents the number of northern right whales that will be in existence in x generations.

Write an equation in slope-intercept form for each graph.

32.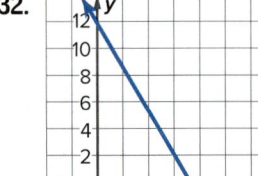
33.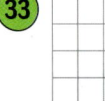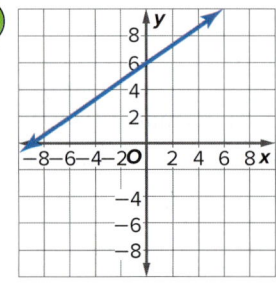
34.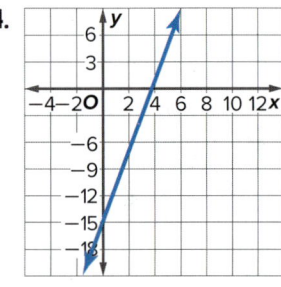

35. **ROSES** Brad wants to send his girlfriend Kelli a dozen roses. He visits two stores. For what distance do the two stores charge the same amount to deliver a dozen roses?

Full Bloom	Flowers R US
Dozen roses $30	Dozen roses $40
Delivery: $3 per mile	Delivery: $2 per mile

36. **KEY BOARDING** The equation $y = 55(23 - x)$ can be used to model the number of words y you have left to type after x minutes.
    a. Write this equation in slope-intercept form.
    b. Identify the slope and y-intercept.
    c. Find the number of words you have left to type after 20 minutes.

37. **MULTI-STEP** Sadie is organizing her craft room. She has 142 spools of thread, 61 balls of yarn, and 35 bottles of paint. She can buy thread racks that hold 60 spools of thread for $14.99, baskets that hold 20 balls of yarn for $15 each, and a paint storage rack that holds 81 bottles of paint for $49.00.
    a. She currently has $32 saved. If she wants to have all of the necessary storage to organize her craft room in 2 months (9 weeks), how much money should she save every week in order to afford the supplies she needs?
    b. Explain your solution process.

38. **MODELING** Refer to the table at the right.

Miles	Kilometers
100	161
50	80.5

    a. Write and graph the linear equation that gives the distance y in kilometers in terms of the number x in miles.
    b. What distance in kilometers corresponds to 20 miles?
    c. What number is the same in kilometers and miles? Explain your reasoning.

H.O.T. Problems   Use Higher-Order Thinking Skills

39. **REASONING** Determine whether the following statement is *always*, *sometimes*, or *never* true. Explain your reasoning.

    The quadrilateral formed by any two parallel lines and two lines perpendicular to those lines is a square.

40. **MODELING** Given $\square ABCD$ with vertices $A(a, b)$, $B(c - a, d)$, $C(c + a, d)$, and $D(c, b)$, write an equation of a line perpendicular to diagonal $\overline{BD}$ that contains A.

41. **STRUCTURE** Write $y = ax + b$ in point-slope form.

42. **OPEN ENDED** Write the equations of two parallel lines with negative slopes.

43. **OPEN-ENDED** Write an equation in point-slope form of a line with an x-intercept of c and y-intercept of d.

44. **WRITING IN MATH** Why do we represent linear equations in more than one form?

Preparing for Assessment

45. The points $Q(1, 3)$, $R(4, -1)$, and $S(10, 4)$ are three vertices of a parallelogram $QRST$. What are the equations of the lines containing the two missing sides of the parallelogram, $\overline{QT}$ and $\overline{ST}$? **MP 1**

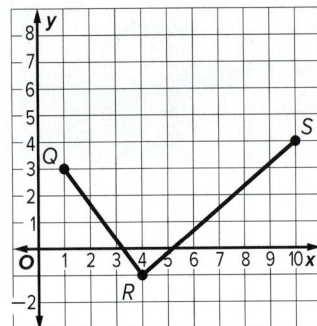

- A $y = -\frac{6}{5}x + \frac{21}{5}$ and $y = -\frac{4}{3}x + \frac{52}{3}$
- B $y = -\frac{6}{5}x + \frac{21}{5}$ and $y = \frac{3}{4}x - \frac{7}{3}$
- C $y = \frac{5}{6}x + \frac{13}{6}$ and $y = -\frac{4}{3}x + \frac{52}{3}$
- D $y = -\frac{5}{6}x + \frac{23}{6}$ and $y = \frac{4}{3}x - \frac{24}{3}$
- E $y = \frac{6}{5}x + \frac{9}{5}$ and $y = -\frac{3}{4}x + \frac{23}{2}$

46. What is the equation of the line containing the point $(3, 7)$ that is perpendicular to the line containing the points $(0, 5)$ and $(-1, -3)$? **MP 6**

- A $y = 8x - 17$
- B $y = -8x + 31$
- C $y = \frac{1}{8}x + \frac{53}{8}$
- D $y = -\frac{1}{8}x + \frac{31}{8}$
- E $y = -\frac{1}{8}x + \frac{59}{8}$

47. Which of the following lines has an undefined slope? **MP 2**

- A $x - y = 1$
- B $x + y = 1$
- C $x = 1$
- D $y = 1$
- E $y = x$

48. What is the slope and y-intercept of a line with an equation of $y = -x + 1$? **MP 7**

- A $m = -1$ and $b = 1$
- B $m = 1$ and $b = -1$
- C $m = 0$ and $b = -1$
- D $m = 1$ and $b = 0$

49. Write an equation of a line with the following points. **MP 6**

- A $y = 2x - 1$
- B $y = -12x - 1$
- C $y = -12x + 2$
- D $y = -2x - 1$

x	y
2	-5
3	-7
4	-9

50. MULTI-STEP A chemistry student converted temperatures to Celsius (C) and Fahrenheit (F) and then graphed the two temperature equations below in her chemistry lab. **MP 4**

(1) $F = \frac{9}{5}C + 32$

(2) $C = \frac{5}{9}(F - 32)$

a. What is the slope of the Fahrenheit equation (1)?

b. When the student graphed equation (1), where should the line intersect the F-axis?

c. What is the slope of the Celsius equation (2)?

d. When the student graphed equation 2, approximately where should the line intersect the C-axis?

51a. Write an equation of the line passing through the points $(3, 2)$ and $(-3, 2)$. **MP 2**

b. Write an equation of the line that is the perpendicular bisector of $(3, 2)$ and $(-3, 2)$.

LESSON 5
Graphing Linear Inequalities

Then
- You wrote linear equations.

Now
1. Graph linear inequalities.
2. Apply linear inequalities.

Why?
- Randy is planning to treat his lacrosse team to a pizza party after the championship game, but he does not want to spend more than $200.

  Randy can use the inequality $11p + 2.25d \leq 200$, where p represents the number of pizzas and d represents the number of soft drinks, to check whether certain combinations of pizzas and drinks will fall within his budget.

 New Vocabulary
linear inequality
boundary
constraint

 Mathematical Practices
4 Model with mathematics.
6 Attend to precision.

1 Graph Linear Inequalities
A **linear inequality** resembles a linear equation, but with an inequality symbol instead of an equals symbol. For example, $y > -3x - 2$ is a linear inequality and $y = -3x - 2$ is the related linear equation.

The graph of the inequality $y > -3x - 2$ is shown at the right as a shaded region. Every point in the shaded region satisfies the inequality. The graph of $y = -3x - 2$ is the **boundary** of the region. It is drawn as a dashed line to show that points on the line do not satisfy the inequality. If the symbol was $\leq$ or $\geq$, then points on the boundary would satisfy the inequality, so the boundary would be drawn as a solid line.

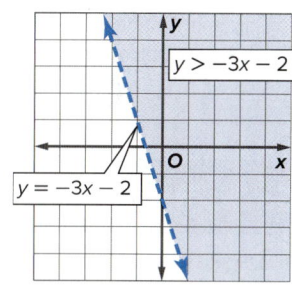

Example 1 Graph an Inequality

Graph $x + 4y > 2$.

Step 1 The boundary of the graph is the graph of $x + 4y = 2$. Because the inequality symbol is $>$, the boundary will be dashed.

Step 2 Test the point (0, 0) because it is not on the boundary.

$x + 4y > 2$ Original inequality

$0 + 4(0) \stackrel{?}{>} 2$ $(x, y) = (0, 0)$

$0 > 2$ ✗   False

The region that does *not* contain (0, 0) is shaded.

CHECK The graph indicates that (0, 3) is a solution.

$x + 4y > 2$ Original inequality

$0 + 4(3) \stackrel{?}{>} 2$ $(x, y) = (0, 3)$

$12 > 2$ ✓   True

The solution checks.

▶ **Guided Practice**

1A. Graph $3x + \frac{1}{2}y < 2$.

1B. Graph $-x + 2y > 4$.

2 Apply Linear Inequalities

A **constraint** is a condition that the solution of a problem must satisfy. You can think of a constraint as a limitation or a restriction. Sometimes a constraint requires a tradeoff: if you choose to do one thing, it will have an effect on something else you are trying to do.

Real-World Career

Recreation Workers Recreation workers plan, organize, and manage recreational activities to meet the needs of a variety of people. Educational requirements range from a high school diploma to a graduate degree.

Go Online!

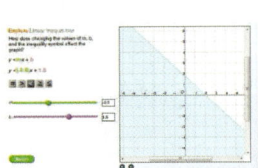

Investigate the graphs of linear inequalities by using the tool in the **Graphing Tools** in ConnectED.

Real-World Example 2 Solve a Problem with a Constraint

RECREATION A recreation center offers various 30-minute and 60-minute art classes. The recreation director has allotted up to 20 hours per week for art classes.

a. Write an inequality to represent the number of classes that can be offered per week. Graph the inequality.

Let x represent the number of 30-minute or $\frac{1}{2}$-hour art classes, and let y represent the number of 60-minute or 1-hour art classes. Because the sum can equal the maximum, the inequality symbol is ≤, and the boundary is solid. The inequality is $\frac{1}{2}x + y \leq 20$.

Step 1 Graph the boundary $\frac{1}{2}x + y = 20$.

Step 2 Test the point (0, 0).

$\frac{1}{2}x + y \leq 20$ Original inequality

$\frac{1}{2}(0) + (0) \stackrel{?}{\leq} 20$ $(x, y) = (0, 0)$

$0 \leq 20$ ✓   True

The region that contains (0, 0) is shaded.

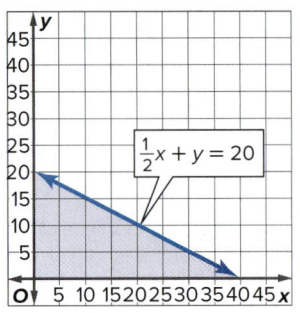

b. Can the recreation director schedule 25 of the 30-minute classes and 15 of the 60-minute classes during a given week? Explain your reasoning.

The point (25, 15) lies outside the shaded region, so it does not satisfy the inequality. Thus, the recreation director cannot schedule twenty-five 30-minute and fifteen 60-minute classes.

Guided Practice

2. Manuel has $25 to spend at the county fair. The fair costs $7.50 for admission, $1.25 for each ride ticket, and $1.00 for each game ticket. Write an inequality, and draw a graph that represents the number of r rides and g game tickets that Manuel can buy.

Check Your Understanding

 = Step-by-Step Solutions begin on page R11.

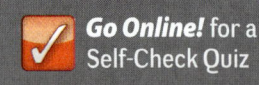
Go Online! for a Self-Check Quiz

Example 1 Graph each inequality.

1. $y \leq 4$
2. $x \geq -6$
3. $x + 4y \leq 2$
4. $3x + y > -8$

Example 2

5. **MP MODELING** Gregg needs to buy gas and oil for his car. Gas costs $3.85 a gallon, and oil costs $6.41 a quart. He has $65 to spend.

   a. Write an inequality to represent the situation, where g is the number of gallons of gas he buys and q is the number of quarts of oil.

   b. Graph the inequality.

   c. Can Gregg buy 10 gallons of gasoline and 8 quarts of oil? Explain.

36 | Lesson 1-5 | Graphing Linear Inequalities

Practice and Problem Solving

Extra Practice is on page R1.

Example 1 Graph each inequality.

6. $x + 2y > 6$
7. $y \geq -3x - 2$
8. $2y + 3 \leq 11$
9. $4x - 3y > 12$
10. $6x + 4y \leq -24$
11. $y \geq \frac{3}{4}x + 6$

Example 2

12. **COLLEGE** April's guidance counselor says that she needs a combined score of at least 1700 on her college entrance exams to be eligible for the college of her choice. The highest possible score is 2400. There is a math portion and a verbal portion.

    a. The inequality $x + y \geq 1700$ represents this situation, where x is the verbal score and y is the math score. Graph this inequality.

    b. Refer to your graph. If she scores a 680 on the math portion of the test and 910 on the verbal portion of the test, will April be eligible for the college of her choice?

13. **SCHOOL DANCE** Carlos estimates that he will need to earn at least $700 to take his girlfriend to the prom. Carlos works two jobs as shown in the table.

    a. Write an inequality to represent this situation.

    b. Graph the inequality.

    c. Will he make enough money if he works 50 hours at each job?

Job	Pay
Main St. Deli	$8 per hour
Babysitting	$6 per hour

Write an inequality for each graph.

14.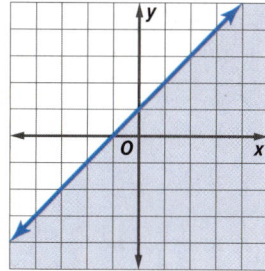

15.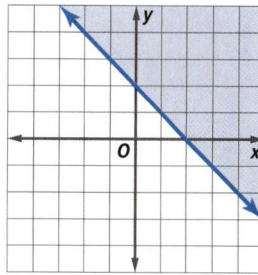

16.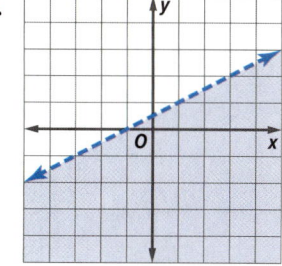

17.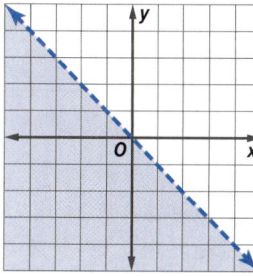

18.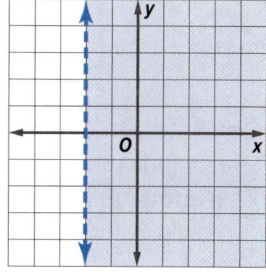

19.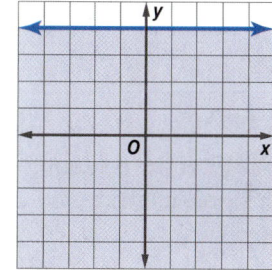

20. **MODELING** Mei is making necklaces and bracelets to sell at a craft show. She has enough beads to make 50 pieces. Let x represent the number of bracelets and y represent the number of necklaces.
    a. Write an inequality that shows the possible number of necklaces and bracelets Mei can make.
    b. Graph the inequality.
    c. Give three possible solutions for the number of necklaces and bracelets that can be made.

21. **MULTI-STEP** Susan won a $250 gift certificate from an online store. She wants to use it to buy e-books and movies. The average price of an e-book is $7.99, and the average price of a movie is $15. Because the movies must be shipped, there is an additional $2.00 charge for shipping and handling for every movie that she buys.
    a. List three possible ways in which she can buy books and movies if she spends between $200 and $250.
    b. Explain your solution process.
    c. What assumptions did you make?

Graph each inequality.

22. $y < |x|$ 23. $y \geq |x + 1|$ 24. $y \leq |x| - 2$

H.O.T. Problems — Use Higher-Order Thinking Skills

25. **OPEN-ENDED** Create a linear inequality in which none of the possible solutions fall in the second or third quadrant.

26. **CHALLENGE** Graph the following inequality.
$$g(x) > \begin{cases} x + 2 & \text{if } x < -2 \\ x & \text{if } -2 \leq x \leq 3 \\ -x & \text{if } x > 3 \end{cases}$$

27. **ERROR ANALYSIS** Paulo and Janette are graphing $x - y \geq 2$. Is either of them correct? Explain your reasoning.

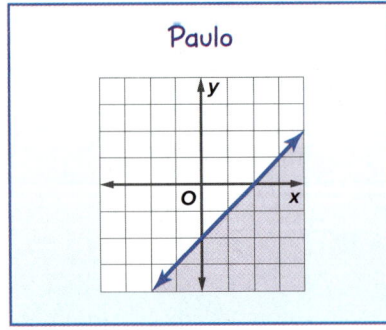
Paulo

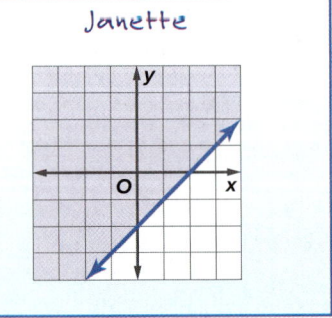
Janette

28. **REASONING** If the solution of the inequality $y < mx + b$ includes the point $(0, 0)$, what conclusions, if any, can you make about the values of m and b? Explain.

29. **WRITING IN MATH** Describe a situation in which the solution to a linear inequality includes all points along the x-axis. Explain your reasoning.

Preparing for Assessment

30. Jenna has $24 to spend on tomato and pepper plants for her garden. Each tomato plant costs $3.00 and each pepper plant costs $4.00. Which could be the graph of the inequality that models the problem? **MP** 4

○ **A** 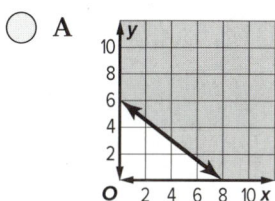 ○ **B**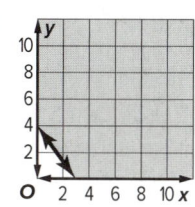

○ **C** 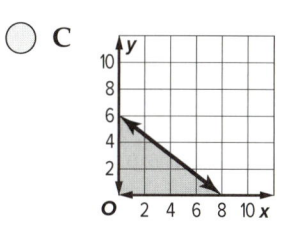 ○ **D**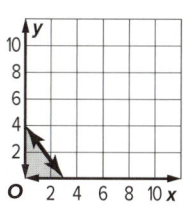

31. Which points do not lie in the solution set of the inequality? **MP** 6

$$x - y < 2$$

☐ **A** $(0, -1)$
☐ **B** $(-6, 1)$
☐ **C** $(9, 2)$
☐ **D** $(5, -4)$
☐ **E** $(-8, 7)$

32. Which statement best describes the inequality? **MP** 6

$$x + 8y > 16$$

○ **A** The boundary line of its graph is solid.
○ **B** It has an infinite number of solutions.
○ **C** The point $(0, 2)$ lies in the solution set.
○ **D** Its x-intercept is 2.

33. Which inequality has a solution whose graph includes the origin? **MP** 6

○ **A** $x + y > 0$
○ **B** $2x - 3y < 0$
○ **C** $x - 3y \geq -1$
○ **D** $2x + y \geq 1$

34. MULTI-STEP Jamal is a camp counselor. He is buying T-shirts and caps for the children at the camp. The T-shirts cost $8.00 each and the caps cost $12.00 each. He can spend at most $480.00. **MP** 4

a. Write an inequality to represent the number of T-shirts x and the number of caps y that Jamal can buy.

b. Graph the inequality.

c. Can Jamal buy 30 T-shirts and 30 caps? Explain how you can use your graph to decide.

d. What is the greatest number of caps Jamal can buy? Explain.

35. Which inequality is shown in the graph? **MP** 6

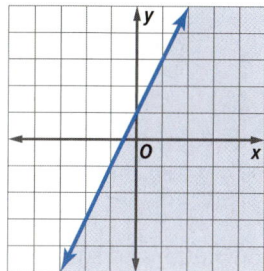

○ **A** $y < 2x + 1$
○ **B** $y \leq 2x + 1$
○ **C** $y < \frac{1}{2}x + 1$
○ **D** $y \leq \frac{1}{2}x + 1$

36. Graph the inequality $y \leq 2x - 4$. **MP** 1

37. Celia and June want to donate some money to a local food pantry. To raise funds, they are selling necklaces and earrings that they have made themselves. Necklaces cost $8 and earrings cost $5. By selling these, they want to donate at least $100. **MP** 1

a. Write an inequality to model the situation.

b. Determine a number of earrings and necklaces that will let them accomplish their donation goal.

c. Determine a number of earrings and necklaces that will let them accomplish their donation goal, if Celia and June make 15 pieces of jewelry.

CHAPTER 1
Mid-Chapter Quiz
Lessons 1-1 through 1-5

1. Solve $6x + 4y = -1$ for x. (Lesson 1-1)

2. **MULTIPLE CHOICE** Which algebraic sentence represents the verbal statement, *the product of 4 and the difference of a number and 13 equals 18*? (Lesson 1-1)

    A $4n - 13 = 18$
    B $4(n - 13) = 18$
    C $\frac{4}{n - 13} = 18$
    D $\frac{4n}{13} = 18$

3. Solve $-3(6x + 5) + 2(4x) = 20$. (Lesson 1-1)

4. What is the height of the trapezoid below? (Lesson 1-1)

    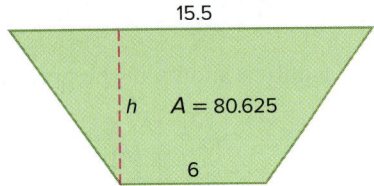

5. **GEOMETRY** The formula for the surface area of a sphere is $SA = 4\pi r^2$, and the formula for the volume of a sphere is $V = \frac{4}{3}\pi r^3$. (Lesson 1-1)

    a. Find the volume and surface area of a sphere with radius 2 inches. Write your answers in terms of π.

    b. Is it possible for a sphere to have the same numerical value for the surface area and volume? If so, find the radius of such a sphere.

6. **MONEY** Marilyn is going on a class outing to an amusement park. She has $40 to spend. The admission price is $20. She plans to buy a sandwich for $5. A cup of lemonade costs $3.00. Write an inequality to show how many cups of lemonade, c, she can buy. (Lesson 1-2)

7. **MULTIPLE CHOICE** For which equation(s) is the following graph a solution set? (Lesson 1-2)

    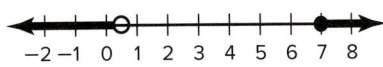

    A $2x + 3 < 4$ or $2x \geq 14$
    B $2x + 3 \leq 4$ or $2x > 14$
    C $4 \geq 2x + 3 > 17$
    D $4 > 2x + 3 \geq 17$

8. Solve $-5c + 7 \geq -16c - 15$. Graph the solution set on a number line. (Lesson 1-2)

9. Write an inequality for the following statement and then solve: *Fifteen less than the product of two and a number x is less than 31.* (Lesson 1-2)

10. **SPEED** The table shows the distance traveled by a car after each time given in minutes. Find the rate of change in distance for the car. (Lesson 1-3)

Time (min)	Distance (mi)
15	20
30	40
45	60

Find the slope of the line that passes through each pair of points. Express as a fraction in simplest form. (Lesson 1-3)

11. $(-2, 6), (1, 15)$

12. $(-2.5, 4), (1.5, -2)$

13. Find the slope of the line shown. (Lesson 1-3)

    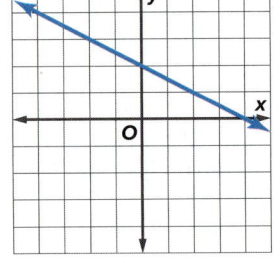

14. Write an equation in standard form for the line with slope $\frac{2}{3}$ that passes through $(3, -4)$. (Lesson 1-4)

15. Write an equation in slope-intercept form for the line that passes through the points $(4.2, 3.6)$ and $(1.8, -1.2)$. (Lesson 1-4)

16. **PART-TIME JOB** Each day Jesse delivers pizzas his employer pays him $20 plus $0.50 for every pizza he delivers. (Lesson 1-4)

    a. Write an equation to represent his daily earnings if he delivers x pizzas.

    b. How much will he earn the day he delivers 20 pizzas?

    c. **MP** What mathematical practice did you use to solve this problem?

Graph each inequality.

17. $y \leq 3x - 2$

18. $3x + 6y > 12$

19. Is the point $(-1, 5)$ in the solution set of the inequality $y > 3x + 8$? Explain.

20. What is the boundary line of the graph of $5x + 2y \leq 3x - 4$? Is the line solid or dashed?

EXPLORE 1-6

Graphing Technology Lab
Intersections of Graphs

You can use a TI-83/84 Plus graphing calculator to find where two graphs intersect. You can use the Y= menu to graph each equation on the same set of axes.

Example Intersection of Two Graphs

Work cooperatively. Graph both equations in the standard viewing window.

$3x + y = 9$
$x - y = -1$

Step 1 Write each equation in the form $y = mx + b$.

$3x + y = 9$ $\qquad$ $x - y = 1$
$\quad y = -3x + 9$ $\qquad$ $-y = -x - 1$
$\qquad\qquad\qquad\qquad\qquad y = x + 1$

Step 2 Enter $y = -3x + 9$ as **Y1** and $y = x + 1$ as **Y2**. Then graph the lines.

KEYSTROKES: Y= (−) 3 X,T,θ,n + 9 ENTER
X,T,θ,n + 1 ENTER ZOOM 6

Step 3 Find the intersection of the lines.

KEYSTROKES: 2nd [CALC] 5 ENTER ENTER ENTER

The intersection is at (2, 3).

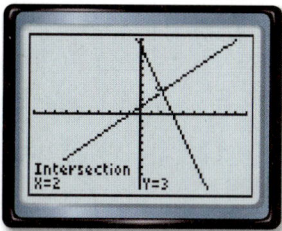

[-10, 10] scl: 1 by [-10, 10] scl: 1

Exercises

Work cooperatively. Use a graphing calculator to find where each pair of graphs intersect.

1. $2x + 4y = 36$
   $10y - 5x = 0$

2. $2y - 3x = 7$
   $5x = 4y - 12$

3. $4x - 2y = 16$
   $7x + 3y = 15$

4. $2x + 4y = 4$
   $x + 3y = 13$

5. $5x + y = 13$
   $3x = 15 - 3y$

6. $4y - 5 = 20 - 3x$
   $4x - 7y + 16 = 0$

7. $\frac{1}{4}x + y = \frac{11}{4}$
   $x - \frac{1}{2}y = 2$

8. $3x + 2y = -3$
   $x + \frac{1}{3}y = -4$

9. $3x - 6y = 6$
   $2x - 4y = 4$

10. $6x + 8y = -16$
    $3x + 4y = 12$

LESSON 6
Solving Systems of Equations

::Then
- You graphed and solved linear equations.

::Now
1. Solve systems of linear equations graphically.
2. Solve systems of linear equations algebraically.

::Why?
- Libby borrowed $450 to start a lawn-mowing business. She charges $35 per lawn and incurs $8 in operating costs per lawn. A system of equations can be used to determine the break-even point. The break-even point is the point at which income equals cost.

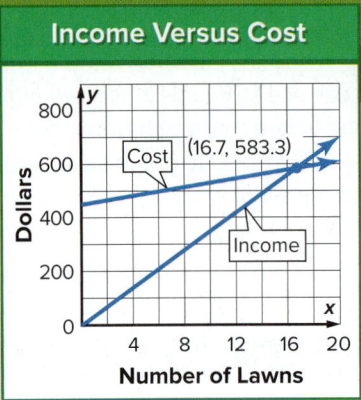

 New Vocabulary
system of equations
consistent
inconsistent
independent
dependent
substitution method
elimination method

1 Solve Systems Graphically

A **system of equations** is two or more equations with the same variables. To solve a system of equations with two variables, find the ordered pair that satisfies all of the equations.

To solve a system of equations by using a table, first write each equation in slope-intercept form. Then substitute different values for x and solve for the corresponding y-values. For ease of use, choose 0 and 1 as your first x-values.

$y_1 = -2x + 8$
$y_2 = 4x - 7$

x	y_1	y_2	Difference
0	8	−7	15
1	6	−3	9
2	4	1	3
3	2	5	−3

Because the difference between the y-values is closer to 0 for $x = 1$ than $x = 0$, a value greater than 1 should be tried next.

Because the difference between the y-values changed signs from $x = 2$ to $x = 3$, a value between these should be tried next.

The solution is between 2 and 3.

Example 1 Solve by Using a Table

Solve the system of equations.

$3x + 2y = -2$
$-4x + 5y = -28$

Write each equation in slope-intercept form.
$3x + 2y = -2 \rightarrow y = -1.5x - 1$
$-4x + 5y = -28 \rightarrow y = 0.8x - 5.6$

Use a table to find the solution that satisfies both equations.

x	y_1	y_2	(x, y_1)	(x, y_2)
0	−1	−5.6	(0, −1)	(0, −5.6)
1	−2.5	−4.8	(1, −2.5)	(1, −4.8)
2	−4	−4	(2, −4)	(2, −4)

The solution of the system is $(2, -4)$.

Guided Practice

1A. $2x - 5y = 11$
$-3x + 4y = -13$

1B. $4x + 3y = -17$
$-7x - 2y = 20$

42 | Lesson 1-6

Another method for solving a system of equations is to graph the equations on the same coordinate plane. The point of intersection represents the solution.

Example 2 Solve by Graphing

Solve the system of equations by graphing.

$2x - y = -1 \qquad 2y + 5x = -16$

Write each equation in slope-intercept form.
$2x - y = -1 \quad \rightarrow \quad y = 2x + 1$
$2y + 5x = -16 \quad \rightarrow \quad y = -2.5x - 8$

The graphs of the lines appear to intersect at $(-2, -3)$.

CHECK Substitute the coordinates into each original equation.

$$2x - y = -1 \qquad\qquad 2y + 5x = -16 \qquad \text{Original equations}$$
$$2(-2) - (-3) \stackrel{?}{=} -1 \qquad 2(-3) + 5(-2) \stackrel{?}{=} -16 \qquad x = -2 \text{ and } y = -3$$
$$-1 = -1 \checkmark \qquad\qquad -16 = -16 \checkmark \qquad \text{Simplify.}$$

The solution of the system is $(-2, -3)$.

Study Tip

MP Precision Always check to see if the values work for **both** of the original equations.

Guided Practice

2A. $4x + 3y = 12$
$-6x + 4y = -1$

2B. $-3x + 8y = 36$
$6x + y = -21$

Systems of equations can be classified by the number of solutions. A system is **consistent** if it has at least one solution and **inconsistent** if it has no solutions. If it has exactly one solution, it is **independent**, and if it has an infinite number of solutions, it is **dependent**.

Study Tip

Slope and Classifying Systems If the equations have different slopes, then the system is consistent and independent.

Example 3 Classify Systems

Graph each system of equations and describe them as *consistent and independent, consistent and dependent,* or *inconsistent.*

a. $4x + 3y = 24$
$-3x + 5y = 30$

b. $-2x + 5y = 10$
$4x - 10y = -20$

The graphs of the lines intersect at one point, so there is one solution. The system is consistent and independent.

Because the equations are equivalent, their graphs are the same line. The system is consistent and dependent.

Guided Practice

3A. $6x - 4y = 15$
$-6x + 4y = 18$

3B. $-4x + 5y = -17$
$-4x - 2y = 15$

The relationship between the graph and the solutions of a system is summarized below.

Concept Summary Characteristics of Linear Systems

Consistent and Independent	Consistent and Dependent	Inconsistent
intersecting lines; one solution	same line; infinitely many solutions	parallel lines; no solution

2 Solve Systems Algebraically Algebraic methods are used to find exact solutions of systems of equations. One algebraic method is called the **substitution method**.

Key Concept Substitution Methods

Step 1 Solve one equation for one of the variables.

Step 2 Substitute the resulting expression into the other equation to replace the variable. Then solve the equation.

Step 3 Substitute to solve for the other variable.

Systems of equations can be used to solve many real-world problems involving constraints modeled by two or more different functions.

Go Online!

Interact with the graphs of systems of linear equations by using the **Graphing Tools** in ConnectED.

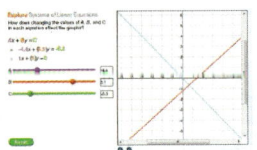

Real-World Example 4 Use the Substitution Method

BUSINESS Alejandro has a computer support business. He estimates that the cost to run his business can be represented by $y = 48x + 500$, where x is the number of customers. He also estimates that his income can be represented by $y = 65x - 145$. How many customers will Alejandro need in order to break even? What will his profit be if he has 60 customers?

$y = 65x - 145$	Income equation
$48x + 500 = 65x - 145$	Substitute $48x + 500$ for y.
$500 = 17x - 145$	Subtract $48x$ from each side.
$645 = 17x$	Add 145 to each side.
$37.9 \approx x$	Divide each side by 17.

Alejandro needs 38 customers to break even. If he has 60 customers, his income will be $65(60) - 145$ or $3755, and his costs will be $48(60) + 500$ or $3380, so his profit will be $3755 - 3380$ or $375.

CHECK You can use a graphing calculator to check this solution. The break-even point is near (37.9, 2321.2). Use the **CALC** function to find the cost and income for 60 customers.

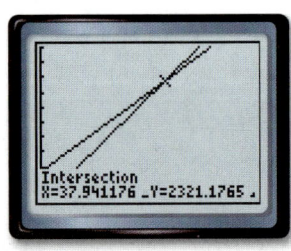

[0, 65] scl: 5 by [0, 3000] scl: 300

44 | Lesson 1-6 | Solving Systems of Equations

Guided Practice

Use substitution to solve each system of equations.

4A. $5x - 3y = 23$
$2x + y = 7$

4B. $x - 7y = 11$
$5x + 4y = -23$

4C. $-6x - y = 27$
$3x + 8y = 9$

You can use the **elimination method** to solve a system when one of the variables has the same coefficient in both equations.

> **Key Concept** Elimination Method
>
> **Step 1** Multiply one or both equations by a number to result in two equations that contain opposite terms.
>
> **Step 2** Add the equations, eliminating one variable. Then solve the equation.
>
> **Step 3** Substitute to solve for the other variable.

Variables can be eliminated by addition or subtraction.

Study Tip

MP Tools Remember when you add or subtract one equation from another to add or subtract every term, including the constant on the other side of the equal sign.

Example 5 Solve by Using Elimination

Use the elimination method to solve the system of equations.

$5x + 3y = -19$
$8x + 3y = -25$ ← Notice that solving by substitution would involve fractions.

Step 1 Multiply one equation by -1 so the equations contain $3y$ and $-3y$.

$8x + 3y = -25$ — Multiply by -1. → $-8x - 3y = 25$

Step 2 Add the equations to eliminate one variable.

$5x + 3y = -19$ — Equation 1
$(+)\,-8x - 3y = 25$ — Equation 2 × (−1)
$\overline{-3x \; = 6}$ — Add the equations.
$x = -2$ — Divide each side by -3.

Step 3 Substitute -2 for x into either original equation.

$8x + 3y = -25$ — Equation 2
$8(-2) + 3y = -25$ — $x = -2$
$-16 + 3y = -25$ — Multiply.
$3y = -9$ — Add 16 to each side.
$y = -3$ — Divide each side by 3.

The solution is $(-2, -3)$.

Guided Practice

5A. $4x - 3y = -22$
$2x + 3y = 16$

5B. $6x - 5y = -8$
$4x - 5y = -12$

5C. $2x - 9y = 34$
$-2x + 6y = -28$

Review Vocabulary

Least Common Multiple the least number that is a common multiple of two or more numbers

Sometimes, adding or subtracting equations will not eliminate either variable. You can use multiplication and least common multiples to find a common coefficient.

Example 6 No Solution and Infinite Solutions

Solve each system of equations.

a. $5x + 3y = 52$
$15x + 9y = 54$

Neither variable has a common coefficient. The coefficients of the y-variables are 3 and 9 and their least common multiple is 9, so multiply each equation by the value that will make the y-coefficient 9.

$5x + 3y = 52$ **Multiply by 3.** $15x + 9y = 156$
$15x + 9y = 54$ $(-)\ 15x + 9y = 54$
                                      $0 = 102$ Subtract the equations.

Because $0 = 102$ is not true, the system has no solution.

b. $2x + 3y = 15$
$-4x - 6y = -30$

The coefficients of the x-variables are 2 and -4 and their least common multiple is -4, so multiply each equation by the value that will make the x-coefficient -4.

$2x + 3y = 15$ **Multiply by -2.** $-4x - 6y = -30$
$-4x - 6y = -30$ $(-)\ -4x - 6y = -30$
                                         $0 = 0$ Subtract the equations.

Because $0 = 0$, the equations represent lines that coincide. So, there are infinitely many solutions.

Guided Practice

6A. $2x + 3y = 5$
$6x + 9y = 15$

6B. $2x - 12y = 3$
$-x + 6y = -9$

6C. $3x + 6y = 30$
$2x + 4y = -12$

Study Tip

MP Reasoning If you add or subtract two equations in a system and the result is an equation that is never true, then the system is inconsistent. When you add or subtract two equations in a system and the result is an equation that is always true, then the system is dependent.

Math History Link

Nina Karlovna Bari (1901–1961) Russian mathematician Nina Karlovna Bari was considered the principal leader of mathematics at Moscow State University, shown above. She is best known for her textbooks *Higher Algebra* and *The Theory of Series*.

The following summarizes the various methods for solving systems.

Concept Summary Solving Systems of Equations

Method	The Best Time to Use
Table	to estimate the solution, since a table may not provide an exact solution
Graphing	to estimate the solution, since graphing usually does not give an exact solution
Substitution	if one of the variables in either equation has a coefficient of 1 or -1
Elimination Using Addition	if one of the variables has opposite coefficients in the two equations
Elimination Using Subtraction	if one of the variables has the same coefficient in the two equations
Elimination Using Multiplication	if none of the coefficients are 1 or -1 and neither of the variables can be eliminated by simply adding or subtracting the equations

Lesson 1-6 | Solving Systems of Equations

Check Your Understanding

= Step-by-Step Solutions begin on page R11.

Example 1 Solve each system of equations by using a table.

1. $y = 3x - 4$
   $y = -2x + 11$

2. $4x - y = 1$
   $5x + 2y = 24$

Example 2 Solve each system of equations by graphing.

3. $y = -3x + 6$
   $2y = 10x - 36$

4. $y = -x - 9$
   $3y = 5x + 5$

5. $y = 0.5x + 4$
   $3y = 4x - 3$

6. $-3y = 4x + 11$
   $2x + 3y = -7$

7. $4x + 5y = -41$
   $3y - 5x = 5$

8. $8x - y = 50$
   $x + 4y = -2$

9. **MODELING** Refer to the table at the right.

   a. Write equations that represent the cost of printing digital photos at each lab.

   b. Under what conditions is the cost to print digital photos the same at both stores?

   c. When is it best to order prints online, and when is it best to use the local store?

Digital Photos
Online Store
$0.15 per photo + $2.70 shipping
Local Store
$0.25 per photo

Example 3 Graph each system of equations and describe it as *consistent and independent*, *consistent and dependent*, or *inconsistent*.

10. $y + 4x = 12$
    $3y = 8 - 12x$

11. $-2x - 3y = 9$
    $4x + 6y = -18$

12. $9x - 2y = 11$
    $5x + 4y = 13$

Example 4 Solve each system of equations by using substitution.

13. $x + 5y = 3$
    $3x - 2y = -8$

14. $y = 2x - 10$
    $y = -4x + 8$

15. $2a + 8b = -8$
    $3a - 5b = 22$

16. $a - 3b = -22$
    $4a + 2b = -4$

17. $6x - 7y = 23$
    $8x + 4y = 44$

18. $9c - 3d = -33$
    $6c + 5d = -8$

Examples 5–6 Solve each system of equations by using elimination.

19. $-6w - 8z = -44$
    $3w + 6z = 36$

20. $4x - 3y = 29$
    $4x + 3y = 35$

21. $3a + 5b = -27$
    $4a + 10b = -46$

22. $8a - 3b = -11$
    $5a + 2b = -3$

23. $5a + 15b = -24$
    $-2a - 6b = 28$

24. $6x - 4y = 30$
    $12x + 5y = -18$

25. $4x + 3y = 2$
    $4x - 2y = 12$

26. $-35x + 40y = 55$
    $7x - 8y = -11$

27. $3x + 2y = -6$
    $x + y = -2$

Practice and Problem Solving

Extra Practice is on page R1.

Example 1 Solve each system of equations by using a table.

28. $y = 5x + 3$
$y = x - 9$

29. $3x - 4y = 16$
$-6x + 5y = -29$

30. $2x - 5 = y$
$-3x + 4y = 0$

31. FUNDRAISER To raise money for new uniforms, the band boosters sell T-shirts and hats. The cost and sale price of each item is shown. The boosters spend a total of $2000 on T-shirts and hats. They sell all of the merchandise, and make $3375. How many T-shirts did they sell?

Item	Cost	Sale Price
T-Shirt	$6	$10
Hat	$4	$7

Example 2 Solve each system of equations by graphing.

32. $-3x + 2y = -6$
$-5x + 10y = 30$

33. $4x + 3y = -24$
$8x - 2y = -16$

34. $6x - 5y = 17$
$6x + 2y = 31$

35. $-3x - 8y = 12$
$12x + 32y = -48$

36. $y - 3x = -29$
$9x - 6y = 102$

37. $-10x + 4y = 7$
$2x - 5y = 7$

38. MODELING Jerilyn has a $10 coupon and a 15% discount coupon for her favorite store. The store has a policy that only one coupon may be used per purchase. When is it best for Jerilyn to use the $10 coupon, and when is it best for her to use the 15% discount coupon?

Example 3 Graph each system of equations and describe it as *consistent and independent*, *consistent and dependent*, or *inconsistent*.

39. $y = 3x - 4$
$y = 6x - 8$

40. $y = 2x - 1$
$y = 2x + 6$

41. $2x + 5y = 10$
$-4x - 10y = 20$

42. $x - 6y = 12$
$3x + 18y = 14$

43. $-5x - 6y = 13$
$12y + 10x = -26$

44. $8y - 3x = 15$
$-16y + 6x = -30$

Example 4 Solve each system of equations by using substitution.

45. $9y + 3x = 18$
$-3y - x = -6$

46. $5x - 20y = 70$
$6x + 5y = -32$

47. $-4x - 16y = -96$
$7x + 3y = 68$

48. $-4a - 5b = 14$
$9a + 3b = -48$

49. $-9c - 4d = 31$
$6c + 6d = -24$

50. $8f + 3g = 12$
$-32f - 12g = 48$

51. TENNIS At a park, there are 38 people playing tennis. Some are playing doubles, and some are playing singles. There are 13 matches in progress. A doubles match requires 4 players, and a singles match requires 2 players.

  a. Write a system of two equations that represents the number of singles and doubles matches going on.

  b. How many matches of each kind are in progress?

Examples 5–6 Solve each system of equations by using elimination.

52. $8x + y = 27$
$-3x + 4y = 3$

53. $2a - 5b = -20$
$2a + 5b = 20$

54. $6j + 4k = -46$
$2j + 4k = -26$

55. $3x - 8y = 24$
$-12x + 32y = 96$

56. $5a - 2b = -19$
$8a + 5b = -55$

57. $r - 6t = 44$
$9r + 12t = 0$

58. $6d + 5f = -32$
$5d - 9f = 26$

59. $11u = 5v + 35$
$8v = -6u + 62$

60. $-1.2c + 3.4d = 6$
$6c = -30 + 17d$

Use a graphing calculator to solve each system of equations. Round the coordinates of the intersection to the nearest hundredth.

61. $12y = 5x - 15$
$4.2y + 6.1x = 11$

62. $-3.8x + 2.9y = 19$
$6.6x - 5.4y = -23$

63. $5.8x - 6.3y = 18$
$-4.3x + 8.8y = 32$

Solve each system of equations.

64. $11p + 3q = 6$
$-0.75q - 2.75p = -1.5$

65. $8r - 5t = -60$
$6r + 3t = -18$

66. $10t + 4v = 13$
$-4t - 7v = 11$

67. $6w = 12 - 4x$
$6x = -9w + 18$

68. $\frac{3}{2}y + z = 3$
$-y - \frac{2}{3}z = -2$

69. $\frac{5}{2}a - \frac{3}{4}b = 46$
$-\frac{7}{8}a - 3b = 10$

70. ROWING Allison can row a boat 1 mile upstream (against the current) in 24 minutes. She can row the same distance downstream in 13 minutes. Assume that both the rowing speed and the speed of the current are constant.

  a. Find the speed at which Allison is rowing and the speed of the current.

  b. If Allison plans to meet her friends 3 miles upstream one hour from now, will she be on time? Explain.

71. MODELING The table shows the winning times in seconds for the 100-meter dash at the Olympics between 1968 and 2012.

Years Since 1968, x	Men's Gold Medal Time	Women's Gold Medal Time
0	9.90	11.0
4	10.14	11.07
8	10.06	11.08
12	10.25	11.06
16	9.99	10.97
20	9.92	10.54
24	9.96	10.82
28	9.84	10.94
32	9.87	10.75
36	9.85	10.93
40	9.69	10.78
44	9.63	10.75

  a. Write equations that represent the winning times for men and women since 1968. Assume that both times continue along the same trend.

  b. Graph both equations. Estimate when the women's performance will catch up to the men's performance. Do you think that your prediction is reasonable? Explain.

72. JOBS Levi has a job offer in which he will receive $800 per month plus a commission of 2% of the total price of the cars that he sells. At his current job, he receives $1200 per month plus a commission of 1.5% of his total sales. How much must he sell per month to make the new job a better deal?

73. TRAVEL A youth group went on a trip to an amusement park, traveling in two vans. The number of people in each van and the total cost of admission are shown in the table. Find the adult price and student price of admission.

Van	Adults	Students	Total Cost
A	2	5	$77
B	2	7	$95

GEOMETRY Find the point at which the diagonals of the quadrilaterals intersect.

74.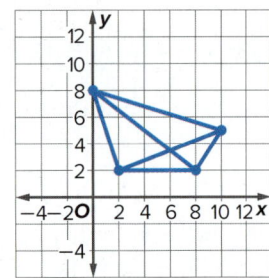

75.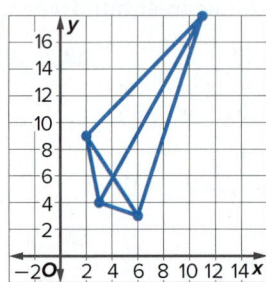

76. **ELECTIONS** In the election for student council, Candidate A received 55% of the total votes, while Candidate B received 1541 votes. If Candidate C received 40% of the votes that Candidate A received, how many total votes were cast?

77. **MULTIPLE REPRESENTATIONS** In this problem, you will explore systems of three linear equations and two variables.

$$3y + x = 16$$
$$y - 2x = -4$$
$$y + 5x = 10$$

   a. **Tabular** Make a table of x- and y-values for each equation.
   b. **Analytical** Which values from the table indicate intersections? Is there a solution that satisfies all three equations?
   c. **Graphical** Graph the three equations on a single coordinate plane.
   d. **Verbal** What conditions must be met for a system of three equations with two variables to have a solution? What conditions result in no solution?

H.O.T. Problems   Use Higher-Order Thinking Skills

78. **ERROR ANALYSIS** Gloria and Syreeta are solving the system $6x - 4y = 26$ and $-3x + 4y = -17$. Is either of them correct? Explain your reasoning.

Gloria	Syreeta
$6x - 4y = 26$ $\quad 6(3) - 4y = 26$	$6x - 4y = 26$ $\quad 6(-3) - 4y = 26$
$-3x + 4y = -17$ $\quad 18 - 4y = 26$	$-3x + 4y = -17$ $\quad -18 - 4y = 26$
$3x = 9$ $\quad\quad\quad\quad -4y = 8$	$3x = -9$ $\quad\quad\quad\quad -4y = 44$
$x = 3$ $\quad\quad\quad\quad y = -2$	$x = -3$ $\quad\quad\quad\quad y = -11$
The solution is $(3, -2)$.	The solution is $(-3, -11)$.

79. **MODELING** Find values of a and b for which the following system has a solution of $(b - 1, b - 2)$.

$$-8ax + 4ay = -12a$$
$$2bx - by = 9$$

80. **REASONING** If a is consistent and dependent with b, b is inconsistent with c, and c is consistent and independent with d, then a will *sometimes*, *always*, or *never* be consistent and independent with d. Explain your reasoning.

81. **CHALLENGE** Write a system of equations in which one equation needs to be multiplied by 3 and the other needs to be multiplied by 4 in order to solve the system with elimination. Then solve your system.

82. **WRITING IN MATH** Why is substitution sometimes more helpful than elimination, and vice versa?

Preparing for Assessment

83. At East Hills High School, there are 80 students in the school band. There are 25 fewer female students than twice the number of male students in the band. Which system of equations represents the number of male and female students participating in the school band? **MP 2, 4**

- A $m + f = 80, f - 25 = 2m$
- B $m + f = 80, f + 25 = m$
- C $m + f = 80, f + 2m = 25$
- D $m + f = 80, f = 2m - 25$

84. Lee worked a total of 30 hours at two jobs last week. Her combined pay from the jobs was $240. She earned $9 per hour working at the movie theater and $6 per hour working in her aunt's store. How many hours did she spend working in her aunt's store last week? **MP 2, 4**

85. Hiram has a total of 20 nickels and quarters. The value of the coins is $4.00. Which system of equations represents the number of coins he has? **MP 2, 4**

- A $n + q = 20, n + q = 400$
- B $5n + 25q = 20, n + q = 400$
- C $n + q = 20, 5n + 25q = 400$
- D $5n + 25q = 20, 5n + 25q = 400$

86. The Saxena family paid $180 for 2 adult tickets and 3 child tickets to an amusement park. The Hopkins family paid $105 for 1 adult and 2 child tickets to the same park. What is the price of a child ticket? **MP 2, 4**

- A $15
- B $30
- C $59
- D $75
- E $90

87. The sum of the digits of a two-digit number is 7. When the digits are reversed, the number is increased by 27. Find the number. **MP 2, 4**

88. Using g for the number of field goals and t for free throws, write two equations that could be solved to find the answer to this problem: Kim scored 21 points in a basketball game. Some baskets were field goals, worth 2 points each, and some were free throws, worth 1 point each. She made three times as many field goals as free throws. How many of each did she make? **MP 2, 4**

89. MULTI-STEP Tristan has a total of 27 dimes and nickels. The total value of the coins is $1.80. Show two ways to determine how many dimes and nickels he has. **MP 2, 4**

a. Let $d =$ the number of dimes Tristan has.

Let $n =$ the number of nickels Tristan has.

Write an equation to show the total number of coins he has.

b. Write an equation to show the total value of all the coins in cents.

c. Solve the first equation for n and substitute the value into the second equation. Then solve for d.

d. Substitute your value for d into the first equation to find n.

e. Check your work by solving the system using elimination. Multiply the equation in part **a.** by 5.

f. Subtract the equation in part **e** from the equation in part **b** and solve for d.

g. Substitute the value for d into the equation in part **b** to verify your value for n.

LESSON 7
Solving Systems of Inequalities by Graphing

∷Then
- You solved systems of linear equations graphically and algebraically.

∷Now
1. Solve systems of inequalities by graphing.
2. Determine the coordinates of the vertices of a region formed by the graph of a system of inequalities.

∷Why?
- Many weather conditions need to be met before a rocket can launch. The temperature must be greater than 35°F and less than 100°F, and the wind speed cannot exceed 30 knots. A system of inequalities can be used to show these three conditions.

New Vocabulary
system of inequalities

MP Mathematical Practices
1. Make sense of problems and persevere in solving them.

1 Systems of Inequalities
Solving a **system of inequalities** means finding the ordered pairs that satisfy all of the inequalities in the system.

> **Key Concept** Solving Systems of Inequalities
>
> **Step 1** Graph each inequality, shading the correct area.
>
> **Step 2** Identify the region that is shaded for all of the inequalities. This is the solution of the system.

Example 1 — Intersecting Regions

Solve the system of inequalities.
$y > 2x - 4$
$y \leq -0.5x + 3$

Solution of $y > 2x - 4 \rightarrow$ Regions 1 and 3

Solution of $y \leq -0.5x + 3 \rightarrow$ Regions 2 and 3

Region 3 is part of the solution of both inequalities, so it is the solution of the system.

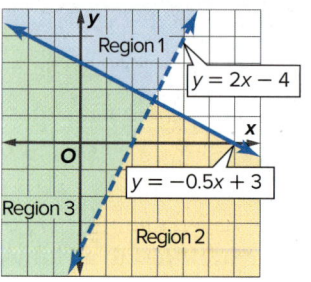

CHECK Notice that the origin is part of the solution of the system. The origin can be used as a test point. You can test the solution by substituting (0, 0) for x and y in each equation.

$y > 2x - 4$
$0 \stackrel{?}{>} 2(0) - 4$
$0 \stackrel{?}{>} 0 - 4$
$0 > -4$ ✓

$y \leq -0.5x + 3$
$0 \stackrel{?}{\leq} -0.5(0) + 3$
$0 \stackrel{?}{\leq} 0 + 3$
$0 \leq 3$ ✓

▶ Guided Practice

1A. $y \leq -2x + 5$
$y > -\frac{1}{4}x - 6$

1B. $y \geq |x|$
$y < \frac{4}{3}x + 5$

Go Online!

Watch and listen to the **animation** to learn about systems of inequalities. Summarize what you hear for a partner, and have them ask you questions to help your understanding.

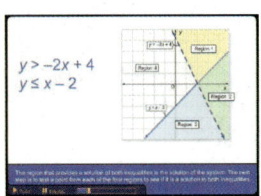

It is possible that the regions do not intersect. When this occurs, there is no solution of the system or the solution set is the *empty set*. The empty set is also called the *null set*. It can be represented by ∅ or { }.

Example 2 Separate Regions

Solve the system of inequalities by graphing.
$y \geq x + 5$
$y < x - 4$

Graph both inequalities.

Because the graphs of the inequalities do not overlap, there are no points in common and there is no solution to the system.

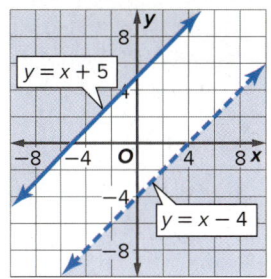

The solution set is the empty set.

▶ **Guided Practice**

2A. $y \geq -4x + 8$
$\ y < -4x + 4$

2B. $y \geq |2x|$
$\ y < 2x - 24$

Real-World Example 3 Write and Use a System of Inequalities

TIME MANAGEMENT Chelsea has final exams in calculus, physics, and history. She has up to 25 hours to study for the exams. She plans to study history for 2 hours. She needs to spend at least 7 hours studying for calculus, but over 14 is too much. She hopes to spend between 8 and 12 hours on physics. Write and graph a system of inequalities to represent the situation.

Calculus: at least 7 hours, but no more than 14
$$7 \leq c \leq 14$$

Physics: at least 8 hours, but no more than 12
$$8 \leq p \leq 12$$

Chelsea has 25 hours available, and 2 of those will be spent on history. She has up to 23 hours left for calculus and physics.
$$c + p \leq 23$$

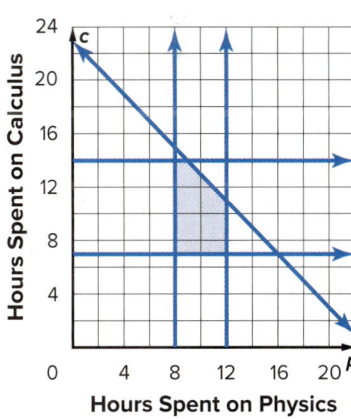

Graph all of the inequalities. Any ordered pair in the intersection is a solution of the system. One solution is 10 hours on physics and 12 hours on calculus.

Real-World Career
Typical incoming freshmen will spend more than 3 times as many hours studying in college as in high school.
Source: *National Survey of Student Engagement*

▶ **Guided Practice**

3. TRAVEL Mr. and Mrs. Rodriguez are driving across the country with their two children. They plan on driving a maximum of 10 hours each day. Mr. Rodriguez wants to drive at least 4 hours a day but no more than 8 hours a day. Mrs. Rodriguez can drive between 2 and 5 hours per day. Write and graph a system of inequalities that represents this information.

connectED.mcgraw-hill.com 53

Study Tip

Boundaries If the inequality that forms the boundary is < or >, then the boundary is not included in the solution, and the line should be dashed.

2 Find Vertices of an Enclosed Region

Sometimes the graph of a system of inequalities produces an enclosed region in the form of a polygon. To find the vertices of the region, determine the coordinates of the points at which the boundaries intersect.

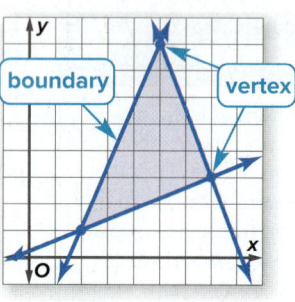

Example 4 Find Vertices

Find the coordinates of the vertices of the triangle formed by $y \geq 2x - 8$, $y \leq -\frac{1}{4}x + 6$, and $4y \geq -15x - 32$.

Step 1 Graph each inequality.

The coordinates $(-4, 7)$ and $(0, -8)$ can be determined from the graph. To find the coordinates of the third vertex, solve the system of equations $y = 2x - 8$ and $y = -\frac{1}{4}x + 6$.

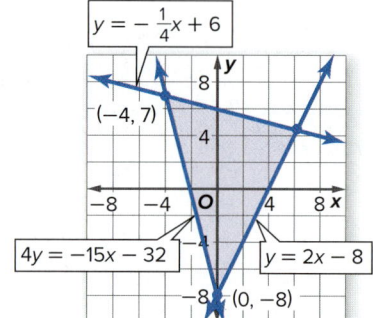

Step 2 Substitute for y in the second equation.

$2x - 8 = -\frac{1}{4}x + 6$ Replace y with $2x - 8$.

$2x = -\frac{1}{4}x + 14$ Add 8 to each side.

$\frac{9}{4}x = 14$ Add $\frac{1}{4}x$ to each side.

$x = \frac{56}{9}$ or $6\frac{2}{9}$ Divide each side by $\frac{9}{4}$.

Step 3 Find y.

$y = 2\left(6\frac{2}{9}\right) - 8$ Replace x with $6\frac{2}{9}$.

$= 12\frac{4}{9} - 8$ Distributive Property

$= 4\frac{4}{9}$ Simplify.

CHECK Compare the coordinates to the coordinates on the graph. The x-coordinate of the third vertex is between 6 and 7, so $6\frac{2}{9}$ is reasonable. The y-coordinate of the third vertex is between 4 and 5, so $4\frac{4}{9}$ is reasonable.

The vertices of the triangle are at $(-4, 7)$, $(0, -8)$, and $\left(6\frac{2}{9}, 4\frac{4}{9}\right)$.

Guided Practice

Find the coordinates of the vertices of the triangle formed by each system of inequalities.

4A. $y \geq -3x - 6$
$2y \geq x - 16$
$11y + 7x \leq 12$

4B. $5y \leq 2x + 9$
$y \leq -x + 6$
$9y \geq -2x + 5$

Check Your Understanding

= Step-by-Step Solutions begin on page R11.

Go Online! for a Self-Check Quiz

Examples 1–2 Solve each system of inequalities by graphing.

1. $y \leq 6$
   $y > -3 + x$

2. $y \leq -3x + 4$
   $y \geq 2x - 1$

3. $y > -2x + 4$
   $y \leq -3x - 3$

Example 3

4. **CHALLENGE** The most Kala can spend on hot dogs and buns for her cookout is $35. A package of 10 hot dogs costs $3.50. A package of buns costs $2.50 and contains 8 buns. She needs to buy at least 40 hot dogs and 40 buns.

   a. Graph the region that shows how many packages of each item she can purchase.

   b. Give an example of three different purchases she can make.

Example 4 Find the coordinates of the vertices of the triangle formed by each system of inequalities.

5. $y \geq 2x + 1$
   $y \leq 8$
   $4x + 3y \geq 8$

6. $3y \geq -7x - 16$
   $7y \leq x + 32$
   $y \geq 15x - 40$

Practice and Problem Solving

Extra Practice is on page R1.

Examples 1–2 Solve each system of inequalities by graphing.

7. $x < 3$
   $y \geq -4$

8. $y > 3x - 5$
   $y \leq 4$

9. $y < -3x + 4$
   $3y + x > -6$

10. $y \geq 0$
    $y < x$

11. $6x - 2y \geq 12$
    $3x + 4y > 12$

12. $-8x > -2y - 1$
    $-4y \geq 2x - 5$

13. $5y < 2x + 10$
    $y - 4x > 8$

14. $3y - 2x \leq -24$
    $y \geq \frac{2}{3}x - 1$

15. $y > -\frac{2}{5}x + 2$
    $5y \leq -2x - 15$

Example 3

16. **RECORDING** Jane's band wants to spend no more than $575 recording their first album. The studio charges at least $35 per hour to record. Graph a system of inequalities to represent this situation. Then list three possible solutions.

17. **SUMMER TRIP** Rondell has to save at least $925 to go to Rome with his Latin class in 8 weeks. He earns $9 an hour working at the Pizza Palace and $12 an hour working at a car wash. By law, he cannot work more than 25 hours per week. Graph two inequalities that Rondell can use to determine the number of hours he needs to work at each job if he wants to make the trip. Then list three possible solutions.

Example 4 Find the coordinates of the vertices of the triangle formed by each system of inequalities.

18. $x \geq 0$
    $y \geq 0$
    $x + 2y < 4$

19. $y \geq 3x - 7$
    $y \leq 8$
    $x + y > 1$

20. $x \leq 4$
    $y > -3x + 12$
    $y \leq 9$

21. $-3x + 4y \leq 15$
    $2y + 5x > -12$
    $10y + 60 \geq 27x$

22. $8y - 19x < 74$
    $38y + 26x \leq 119$
    $54y - 12x \geq -198$

23. $6y - 24x \geq -168$
    $8y + 7x > 10$
    $20y - 2x \leq 64$

24. **BAKING** Rebecca wants to bake cookies and cupcakes for a bake sale. She can bake 15 cookies at a time and 12 cupcakes at a time. She needs to make at least 120 baked goods, but no more than 360, and she wants to have at least three times as many cookies as cupcakes. What combination of batches of each could Rebecca make?

25. CELL PHONES Dale has a maximum of 800 minutes on his cell phone plan that he can use each month. Daytime minutes cost $0.15, and nighttime minutes cost $0.10. Dale plans to use at least twice as many daytime minutes as nighttime minutes. If Dale uses at least 200 nighttime minutes and does not go over his limit, what is his maximum bill? his minimum bill?

26. TREES Trees are divided into four categories according to height and trunk circumference. Data for the trees in a forest are described in the table.

Crown Class	dominant	codominant	intermediate	suppressed
Height (in feet)	over 72	56–72	40–55	under 39
Trunk Circumference (in inches)	over 60	48–60	34–48	under 33

Source: USDA Forest Service

a. Write and graph the system of inequalities that represents the range of heights h and circumferences c for a codominant tree.

b. Determine the crown class of a basswood that is 48 feet tall. Find the expected trunk circumference.

27. MODELING On a camping trip, Jessica needs at least 3 pounds of food and 0.5 gallon of water per day. Marc needs at least 5 pounds of food and 0.5 gallon of water per day. Jessica's equipment weighs 10 pounds, and Marc's equipment weighs 20 pounds. A gallon of water weighs approximately 8 pounds. Each of them carries their own supplies, and Jessica is capable of carrying 35 pounds while Marc can carry 50 pounds.

a. Graph the inequalities that represent how much they can carry.

b. How many days can they camp, assuming that they bring all their supplies in at once?

c. Who will run out of supplies first?

Solve each system of inequalities by graphing.

28. $y \geq |2x + 4| - 2$
$3y + x \leq 15$

29. $y \geq |6 - x|$
$|y| \leq 4$

30. $|y| \geq x$
$y < 2x$

31. $y > -3x + 1$
$4y \leq x - 8$
$3x - 5y < 20$

32. $6y + 2x \leq 9$
$2y + 18 \geq 5x$
$y > -4x - 9$

33. $|x| > y$
$y \leq 6$
$y \geq -2$

34. $2x + 3y \geq 6$
$y \leq |x - 6|$

35. $8x + 4y < 10$
$y > |2x - 1|$

36. $y \geq |x - 2| + 4$
$y \leq [\![x]\!] - 3$

37. MEDIA Steve is trying to decide what to put on his smartphone. Audio books are about 3 hours long and songs are about 3.5 minutes long. Steve wants no more than 4 audio books on his smartphone, but at least ten songs and one audio book. Each book costs $15.00 and each song costs $0.95. Steve has $63 to spend on books and music. Graph the inequalities to show possible combinations of books and songs that Steve can have.

38. JOBS Louie has two jobs and can work no more than 25 total hours per week. He wants to earn at least $150 per week. Graph the inequalities to show possible combinations of hours worked at each job that will help him reach his goal. Then list three possible solutions.

Job	Pay
Busboy	$6.50
Clerk	$8.00

39. **TIME MANAGEMENT** Ramir uses his spare time to write a novel and to exercise. He has budgeted 35 hours per week. He wants to exercise at least 7 hours a week but no more than 15. He also hopes to write between 20 and 25 hours per week. Write and graph a system of inequalities that represents this situation.

Find the coordinates of the vertices of the figure formed by each system of inequalities.

40. $y \geq 2x - 12$
    $y \leq -4x + 20$
    $4y - x \leq 8$
    $y \geq -3x + 2$

41. $y \geq -x - 8$
    $2y \geq 3x - 20$
    $4y + x \leq 24$
    $y \leq 4x + 22$

42. $2y - x \geq -20$
    $y \geq -3x - 6$
    $y \leq -2x + 2$
    $y \leq 2x + 14$

43. **FINANCIAL LITERACY** Mr. Hoffman is investing $10,000 in two funds. One fund will pay 6% interest, and a riskier second fund will pay 10% interest. What is the least amount he can invest in the risky fund and still earn at least $740 after one year?

44. **DODGEBALL** A high school is selecting a dodgeball team to play in a fund-raising exhibition against their rival. There can be from 10 to 15 players on the team and there must be more girls than boys on the team.

   a. Write and graph a system of inequalities to represent the situation.
   b. List all of the possible combinations of boys and girls for the team.
   c. Explain why there is not an infinite number of possibilities.

H.O.T. Problems Use Higher-Order Thinking Skills

45. **MODELING** Find the area of the region defined by the following inequalities.
$$y \geq -4x - 16$$
$$4y \leq 26 - x$$
$$3y + 6x \leq 30$$
$$4y - 2x \geq -10$$

46. **OPEN-ENDED** Write a system of two inequalities in which the solution:
   a. lies only in the third quadrant.
   b. does not exist.
   c. lies only on a line.
   d. lies on exactly one point.

47. **MODELING** Write a system of inequalities to represent the solution shown at the right. How many points with integer coordinates are solutions of the system?

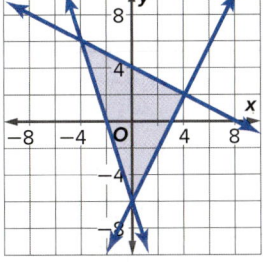

48. **CRITIQUE ARGUMENTS** Determine whether the statement is *true* or *false*. If false, give a counterexample.

   A system of two linear inequalities has either no points or infinitely many points in its solution.

49. **WRITING IN MATH** Write a how-to manual for determining where to shade when graphing a system of inequalities.

50. **WRITING IN MATH** Explain how you would test to see whether (−4, 6) is a solution of a system of inequalities.

Preparing for Assessment

51. Chloe has two jobs and can work no more than 28 total hours per week. Which system of inequalities can be solved to find the possible numbers of hours worked per week at each job for which Chloe will earn at least $160? **MP** 2, 4

Job	Cashier	Florist
Hourly Pay	$8.50	$10.00

○ **A** $x + y \geq 28$
       $8.5x + 10y \leq 160$

○ **B** $x + y \leq 28$
       $8.5x + 10y \geq 160$

○ **C** $x + y \geq 160$
       $8.5x + 10y \leq 28$

○ **D** $x + y \leq 160$
       $8.5x + 10y \geq 28$

○ **E** $x + y < 28$
       $8.5x + 10y > 160$

52. Which system of inequalities is graphed below? **MP** 2, 4

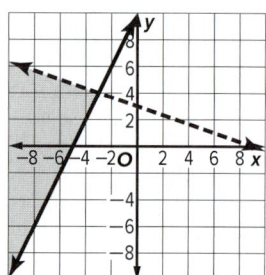

○ **A** $x + 3y < 9$ ○ **C** $x + 3y > 9$
       $2x \leq y - 10$ $2x \geq y - 10$

○ **B** $x + 3y \leq 9$ ○ **D** $x + 3y \geq 9$
       $2x < y - 10$ $2x > y - 10$

53. Which graph shows the solution of the following system of inequalities? **MP** 2, 4

$$x + 2y \geq 3$$
$$3y - 2x \leq 1$$

○ **A**

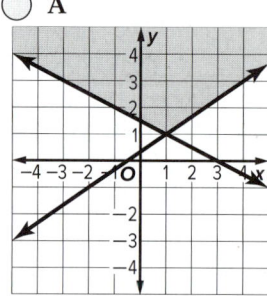

○ **B**

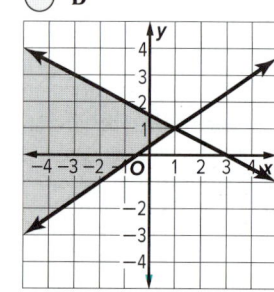

○ **C**

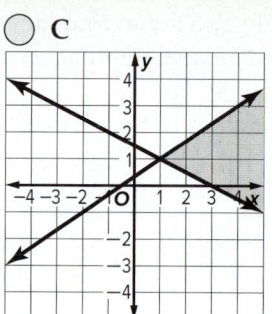

○ **D**

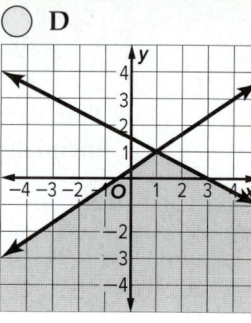

54. MULTI-STEP Daniel has $72 left on a gift card that he plans to use to buy at least 4 new shirts. Some of the shirts cost $12 and others cost $18. **MP** 2, 4

   a. Using x for the number of $12 shirts and y for the number of $18 shirts, write an inequality to show how many shirts Daniel wants to buy.

   b. Write an inequality to show how much he can spend on the shirts.

   c. Make a first quadrant graph for the system of these two inequalities. Why is it only necessary to draw the graph in the first quadrant?

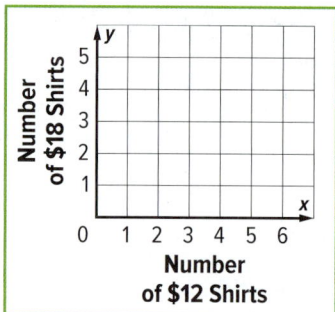

   d. Use the shaded region of the graph to fill in the table with the whole number pairs of each type of shirt that Daniel could buy. Complete the column for the total cost of each combination. Are your combinations reasonable?

$12 Shirts	$18 Shirts	Total Cost

58 | Lesson 1-7 | Solving Systems of Inequalities by Graphing

EXTEND 1-7

Graphing Technology Lab
Systems of Linear Inequalities

You can graph systems of linear inequalities with a graphing calculator by using the **Y=** menu. You can choose different graphing styles to shade above or below a line.

Example Intersection of Two Graphs

Work cooperatively. Graph the system of inequalities in the standard viewing window.

$y \geq -3x + 4$
$y \leq 2x - 1$

Step 1 Enter $-3x + 4$ as **Y1**. Because y is greater than $-3x + 4$, shade above the line.

KEYSTROKES: Y= ◄ ◄ ENTER ENTER ► ►
(−) 3 X,T,θ,n + 4 ENTER

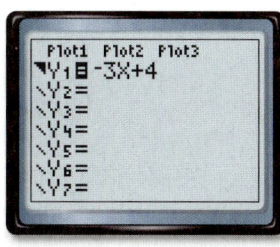

Step 2 Enter $2x - 1$ as **Y2**. Because y is less than $2x - 1$, shade below the line.

KEYSTROKES: ◄ ◄ ENTER ENTER ENTER ► ► 2
X,T,θ,n − 1 ENTER

Step 3 Display the graphs in the standard viewing window.

KEYSTROKES: ZOOM 6

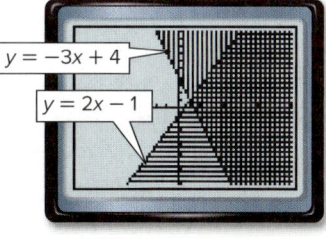

[−10, 10] scl: 1 by [−10, 10] scl: 1

Notice the shading pattern above the line $y = -3x + 4$ and the shading pattern below the line $y = 2x - 1$. The intersection of the graphs is the region where the patterns overlap. This region includes all the points that satisfy the system $y \geq -3x + 4$ and $y \leq 2x - 1$.

Exercises

Work cooperatively. Use a graphing calculator to solve each system of inequalities.

1. $y \geq 3$
   $y \leq -x + 1$

2. $y \geq -4x$
   $y \leq -5$

3. $y \geq 2 - x$
   $y \leq x + 3$

4. $y \geq 2x + 1$
   $y \leq -x - 1$

5. $2y \geq 3x - 1$
   $3y \leq -x + 7$

6. $y + 5x \geq 12$
   $y - 3 \leq 10$

7. $5y + 3x \geq 11$
   $3y - x \leq -8$

8. $10y - 7x \geq -19$
   $7y - 5x \leq 11$

9. $\frac{1}{6}y - x \geq -3$
   $\frac{1}{5}y + x \leq 7$

connectED.mcgraw-hill.com **59**

LESSON 8
Optimization with Linear Programming

::Then
- You solved systems of linear inequalities by graphing.

::Now
1. Find the maximum and minimum values of a function over a region.
2. Solve real-world optimization problems using linear programming.

::Why?
An electronics company produces two types of smart phone docking stations. A sign on the company bulletin board is shown.

If at least 2000 items must be produced per shift, how many of each type should be made to minimize costs?

The company is experiencing limitations, or constraints, on production caused by customer demand, shipping, and the productivity of their factory. A system of inequalities can be used to represent these constraints.

Keeping Costs Down: We Can Do It!

Our Goal: Production per Shift			
Unit	Minimum	Maximum	Cost per Unit
basic	600	1500	$55
multifunction	800	1700	$95

 New Vocabulary
linear programming
feasible region
bounded
unbounded
optimize

 Mathematical Practices
4 Model with mathematics.
8 Look for and express regularity in repeated reasoning.

1 Maximum and Minimum Values Situations often occur in business in which a company hopes to either maximize profits or minimize costs, and many constraints need to be considered. These issues can often be addressed by the use of systems of inequalities in linear programming.

Linear programming is a method for finding maximum or minimum values of a function over a given system of inequalities with each inequality representing a constraint. After the system is graphed and the vertices of the solution set, called the **feasible region**, are substituted into the function, you can determine the maximum or minimum value.

Key Concept Feasible Regions

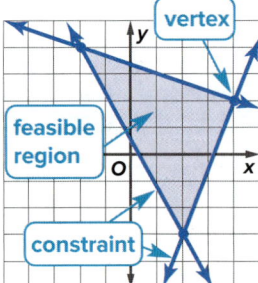

The feasible region is enclosed, or **bounded**, by the constraints. The maximum or minimum value of the related function *always* occurs at a vertex of the feasible region.

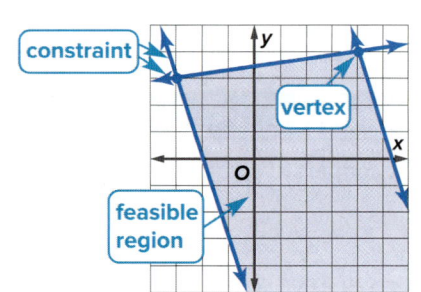

The feasible region is open and can go on forever. It is **unbounded**. Unbounded regions have either a maximum or a minimum.

60 | Lesson 1-8

Example 1 Bounded Region

Graph the system of inequalities. Name the coordinates of the vertices of the feasible region. Find the maximum and minimum values of the function for this region.

$3 \leq y \leq 6$
$y \leq 3x + 12$
$y \leq -2x + 6$
$f(x, y) = 4x - 2y$

Reading Math
Function Notation The notation $f(x, y)$ is used to represent a function with two variables, x and y. It is read f of x and y.

Step 1 Graph the inequalities and locate the vertices.

Step 2 Evaluate the function at each vertex.

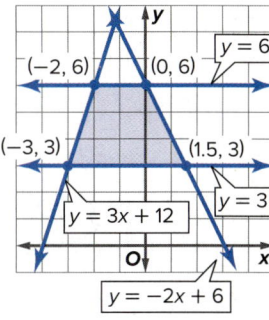

(x, y)	4x − 2y	f(x, y)	
(−3, 3)	4(−3) − 2(3)	−18	
(1.5, 3)	4(1.5) − 2(3)	0	← maximum
(0, 6)	4(0) − 2(6)	−12	
(−2, 6)	4(−2) − 2(6)	−20	← minimum

The maximum value is 0 at (1.5, 3). The minimum value is −20 at (−2, 6).

Guided Practice

1A. $-2 \leq x \leq 6$
$1 \leq y \leq 5$
$y \leq x + 3$
$f(x, y) = -5x + 2y$

1B. $-6 \leq y \leq -2$
$y \leq -x + 2$
$y \leq 2x + 2$
$f(x, y) = 6x + 4y$

When a system of inequalities does not form a closed region, it is unbounded.

Example 2 Unbounded Region

Reading Math
MP Perseverance Do not assume that there is no maximum if the feasible region is unbounded above the vertices. Test points are needed to determine if there is a minimum or maximum.

Graph the system of inequalities. Name the coordinates of the vertices of the feasible region. Find the maximum and minimum values of the function for this region.

$2y + 3x \geq -12$
$y \leq 3x + 12$
$y \geq 3x - 6$
$f(x, y) = 9x - 6y$

Evaluate the function at each vertex.

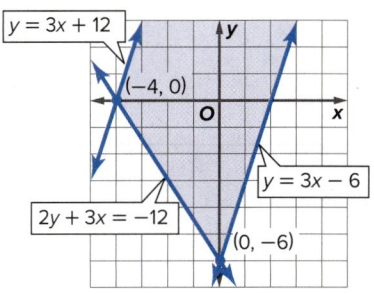

(x, y)	9x − 6y	f(x, y)
(−4, 0)	9(−4) − 6(0)	−36
(0, −6)	9(0) − 6(−6)	36

The maximum value is 36 at (0, −6). There is no minimum value. Notice that another point in the feasible region, (0, 8), yields a value of −48, which is less than −36.

Guided Practice

2A. $y \leq 8$
$y \geq -x + 4$
$y \leq -x + 10$
$f(x, y) = -6x + 8y$

2B. $y \geq x - 9$
$y \leq -4x + 16$
$y \geq -4x - 4$
$f(x, y) = 10x + 7y$

2 Optimization

To **optimize** means to seek the best price or amount to minimize costs or maximize profits. This is often obtained with the use of linear programming.

Key Concept Optimization with Linear Programming

Step 1 Define the variables.
Step 2 Write a system of inequalities.
Step 3 Graph the system of inequalities.
Step 4 Find the coordinates of the vertices of the feasible region.
Step 5 Write a linear function to be maximized or minimized.
Step 6 Substitute the coordinates of the vertices into the function.
Step 7 Select the greatest or least result. Answer the problem.

When using a system of inequalities to describe constraints in real-world problems, often only whole-number solutions will make sense.

Real-World Example 3 Optimization with Linear Programming

BUSINESS Refer to the application at the beginning of the lesson. Determine how many of each type of device should be made per shift.

Step 1 Let b = number of basic docking stations produced. Let m = number of multifunction docking stations produced.

Step 2
$600 \leq b \leq 1500$
$800 \leq m \leq 1700$
$b + m \geq 2000$

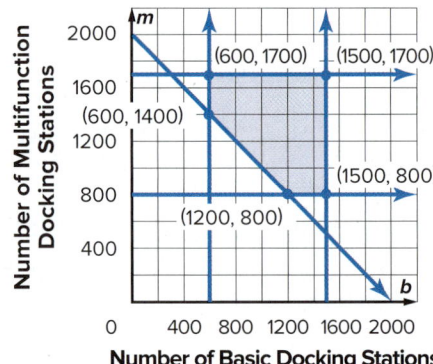

Steps 3 and 4 The system is graphed at the right. Note the vertices of the feasible region.

Step 5 The function to be minimized is $f(b, m) = 55b + 95m$.

Step 6

(b, m)	$55b + 95m$	$f(b, m)$	
(600, 1700)	55(600) + 95(1700)	194,500	
(600, 1400)	55(600) + 95(1400)	166,000	
(1500, 1700)	55(1500) + 95(1700)	244,000	← maximum
(1500, 800)	55(1500) + 95(800)	158,500	
(1200, 800)	55(1200) + 95(800)	142,000	← minimum

Step 7 Produce 1200 basic and 800 multifunction docking stations to minimize costs.

Guided Practice

3. JEWELRY Each week, Mackenzie can make 10 to 25 necklaces and 15 to 40 pairs of earrings. If she earns profits of $3 on each pair of earrings and $5 on each necklace, and she plans to sell at least 30 pieces of jewelry, how can she maximize profit?

Real-World Career

Operations Manager Operations management is an area of business that is concerned with the production of goods and services, and involves the responsibility of ensuring that business operations are efficient and effective. A master's degree in business and experience in operations are preferred.

Go Online!

Explore linear programming using **The Geometer's Sketchpad®** activity in ConnectED.

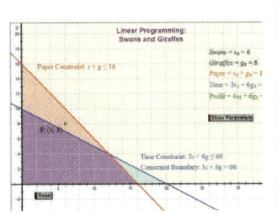

Check Your Understanding

= Step-by-Step Solutions begin on page R11.

Go Online! for a Self-Check Quiz

Examples 1–2 Graph each system of inequalities. Name the coordinates of the vertices of the feasible region. Find the maximum and minimum values of the given function for this region.

1. $y \leq 5$
   $x \leq 4$
   $y \geq -x$
   $f(x, y) = 5x - 2y$

2. $y \leq -3x + 6$
   $-y \leq x$
   $y \leq 3$
   $f(x, y) = 8x + 4y$

3. $y \geq -3x + 2$
   $9x + 3y \leq 24$
   $y \geq -4$
   $f(x, y) = 2x + 14y$

4. $-2 \leq y \leq 6$
   $3y \leq 4x + 26$
   $y \leq -2x + 2$
   $f(x, y) = -3x - 6y$

5. $-3 \leq y \leq 7$
   $4y \geq 4x - 8$
   $6y + 3x \leq 24$
   $f(x, y) = -12x + 9y$

6. $y \leq 2x + 6$
   $y \geq 2x - 8$
   $y \geq -2x - 18$
   $f(x, y) = 5x - 4y$

Example 3

7. **TRANSPORTATION** A package delivery service has a truck that can hold 4200 pounds and has a capacity of 480 cubic feet. Deliveries will only be made if there are at least 10 of each type of package.

Package Type	Weight (lb)	Volume (ft³)	Shipping Fee
Standard	Up to 25	up to 3	12.50
Oversize	Over 25 up to 50	over 3 up to 5	20.00

a. Write the constraints for this situation if all packages are at the maximum weight and volume.
b. Graph the feasible region and determine its vertices.
c. Write a function to represent the revenue per truckload.
d. Determine the number of each type of package needed per truckload to maximize revenue. What is the maximum revenue? Explain your reasoning.

Practice and Problem Solving

Extra Practice is on page R1.

Examples 1–2 Graph each system of inequalities. Name the coordinates of the vertices of the feasible region. Find the maximum and minimum values of the given function for this region.

8. $1 \leq y \leq 4$
   $4y - 6x \geq -32$
   $2y \geq -x + 4$
   $f(x, y) = -6x + 3y$

9. $2 \geq x \geq -3$
   $y \geq -2x - 6$
   $4y \leq 2x + 32$
   $f(x, y) = -4x - 9y$

10. $-2 \leq x \leq 4$
    $5 \leq y \leq 8$
    $2x + 3y \leq 26$
    $f(x, y) = 8x - 10y$

11. $-8 \leq y \leq -2$
    $y \leq x$
    $y \leq -3x + 10$
    $f(x, y) = 5x + 14y$

12. $x + 4y \geq 2$
    $2x + 4y \leq 24$
    $2 \leq x \leq 6$
    $f(x, y) = 6x + 7y$

13. $3 \leq y \leq 7$
    $2y + x \leq 8$
    $y - 2x \leq 23$
    $f(x, y) = -3x + 5y$

Examples 1–2 Graph each system of inequalities. Name the coordinates of the vertices of the feasible region. Find the maximum and minimum values of the given function for this region.

14. $-9 \leq x \leq -3$
    $-9 \leq y \leq -5$
    $3y + 12x \leq -75$
    $f(x, y) = 20x + 8y$

15. $x \geq -8$
    $3x + 6y \leq 36$
    $2y + 12 \geq 3x$
    $f(x, y) = 10x - 6y$

16. $y \geq |x - 2|$
    $y \leq 8$
    $8y + 5x \leq 49$
    $f(x, y) = -5x - 15y$

17. $x \geq -6$
    $y + x \leq -1$
    $2x + 3y \geq -9$
    $f(x, y) = -10x - 12y$

18. $-5 \geq y \geq -17$
    $y \leq 3x + 19$
    $y \leq -4x + 15$
    $f(x, y) = 8x - 3y$

19. $-8 \leq x \leq 16$
    $y \geq 2x - 10$
    $2y + x \leq 80$
    $f(x, y) = 12x + 15y$

20. $y \leq x + 4$
    $y \geq x - 4$
    $y \leq -x + 10$
    $y \geq -x - 10$
    $f(x, y) = -10x + 9y$

21. $-4 \leq x \leq 8$
    $-8 \leq y \leq 6$
    $y \geq x - 6$
    $4y + 7x \leq 31$
    $f(x, y) = 12x + 8y$

22. $y \geq |x + 1| - 2$
    $0 \leq y \leq 6$
    $-6 \leq x \leq 2$
    $x + 3y \leq 14$
    $f(x, y) = 5x + 4y$

Example 3

23. **COOKING** Jenny's Bakery makes two types of birthday cakes: yellow cake, which sells for $25, and strawberry cake, which sells for $35. Both cakes are the same size, but the decorating and assembly time required for the yellow cake is 2 hours, while the time is 3 hours for the strawberry cake. There are 450 hours of labor available for production. How many of each type of cake should be made to maximize revenue?

24. **MULTI-STEP** A landscaper is offering a 25% discount on her $150 lawn care package to new customers. She has a budget of $200 for brochures and mailers and 3 hours to distribute the materials. She estimates that 5% of the people who get a brochure and 12% of the people who get a mailer will sign up for the service. Brochures cost $0.20 each, and mailers cost $0.99 each including postage. It takes 1 hour to pass out 100 brochures, but it only takes 40 minutes to stamp 100 mailers.

    a. What is the maximum revenue that she can make from new customers less the marketing costs?
    b. Explain your solution method.

25. **MP PRECISION** Sean has a maximum of 20 days each month to paint sheds and playhouses. He can paint 2.5 sheds per day or 2 playhouses per day. Each month he has access to at most 45 structures that need to be painted.

    a. Write a system of inequalities to represent the constraints on the number of each structure Sean can paint each month.
    b. Draw a graph showing the feasible region and list the coordinates of the vertices of the feasible region.
    c. If Sean is paid $26 per shed and $30 per playhouse, how many of each should he paint to maximize his monthly earnings?
    d. What is his maximum monthly earnings?

26. **MOVIES** Employees at a local movie theater work overlapping 8-hour shifts from noon to 8 P.M. or from 4 P.M. to midnight. The table below shows the number of employees needed and their corresponding pay. Find the numbers of day-shift workers and night-shift workers that should be scheduled to minimize the cost. What is the minimal cost?

Time	noon to 4 P.M.	4 P.M. to 8 P.M.	8 P.M. to midnight
Number of Employees Needed	at least 5	at least 14	6
Rate per Hour	$5.50	$7.50	$7.50

27. **BUSINESS** Each car on a freight train can hold 4200 pounds of cargo and has a capacity of 480 cubic feet. The freight service handles two types of packages: small—which weigh 25 pounds and are 3 cubic feet each, and large—which are 50 pounds and are 5 cubic feet each. The freight service charges $5 for each small package and $8 for each large package.

   a. Find the number of each type of package that should be placed on a train car to maximize revenue.

   b. What is the maximum revenue per train car?

   c. In this situation, is maximizing the revenue necessarily the best thing for the company to do? Explain.

28. **RECYCLING** A recycling plant processes used plastic into food or drink containers. The plant processes up to 1200 tons of plastic per week. At least 300 tons must be processed for food containers, while at least 450 tons must be processed for drink containers. The profit is $17.50 per ton for processing food containers and $20 per ton for processing drink containers. What is the profit if the plant maximizes processing?

H.O.T. Problems   Use Higher-Order Thinking Skills

29. **CHALLENGE** Create a set of inequalities that forms a bounded region with an area of 20 units^2 and lies only in the fourth quadrant.

30. **MODELING** Find the area of the bounded region formed by the following constraints: $y \geq |x| - 3$, $y \leq -|x| + 3$, and $x \geq |y|$.

31. **CONSTRUCT ARGUMENTS** Identify the system of inequalities that is not the same as the other three. Explain your reasoning.

   a.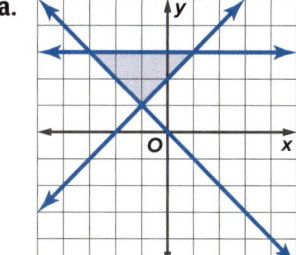

   c.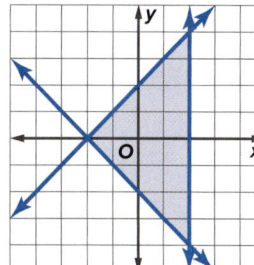

   b.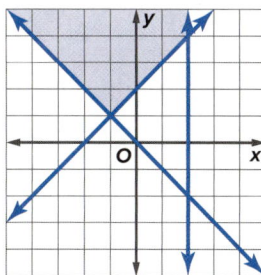

   d.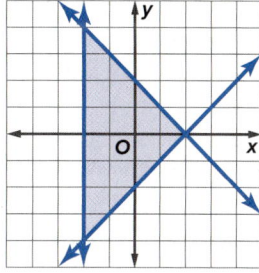

32. **REASONING** Determine whether the following statement is *sometimes*, *always*, or *never* true. Explain your reasoning.

   An unbounded region will not have both a maximum and minimum value.

33. **WRITING IN MATH** Upon determining a bounded feasible region, Ayumi noticed that vertices $A(-3, 4)$ and $B(5, 2)$ yielded the same maximum value for $f(x, y) = 16y + 4x$. Kelvin confirmed that his constraints were graphed correctly and his vertices were correct. Then he said that those two points were not the only maximum values in the feasible region. Explain how this could have happened.

Preparing for Assessment

34. What is the minimum value of $f(x, y) = 2y - 3x$ on the feasible region for this system of inequalities? **MP 1, 2**

$$x + y \geq -5$$
$$y \leq 8 - 2x$$
$$1 \leq x \leq 3$$

- A -25
- B -15
- C -5
- D 0
- E 9

35. A botanist needs at least 50 plants for an experiment. She cannot use more than 30 cacti or more than 40 ferns. Each cactus costs $3 and each fern costs $2. What is the minimum number of dollars the botanist will spend on cacti and ferns for the experiment? **MP 2, 4**

36. A factory assembles hockey helmets and lacrosse helmets. To assemble a hockey helmet, $\frac{1}{4}$ hour is required and $\frac{1}{3}$ hour is required to assemble a lacrosse helmet. There are 36 hours of labor available each day for helmet assembly. The factory makes a profit of $10 on each hockey helmet and $12 on each lacrosse helmet. What is the least number of hockey helmets the factory can produce per day in order to make a profit of at least $1400? **MP 2, 4**

- A 26
- B 30
- C 36
- D 40
- E 104
- F 144

37. MULTI-STEP Consider this system of inequalities: **MP 1, 2**

$$x + y \geq 10$$
$$x \leq 5$$
$$x + 4 \geq y$$

a. Graph the inequalities and shade the feasible region.

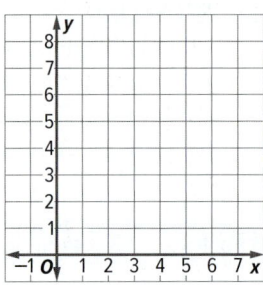

b. Identify the vertices of the feasible region for this system of inequalities.

c. Fill in the table to find the values of $f(x, y) = 5x - 2y$ for the vertices. Identify the maximum value of $f(x, y)$.

(x, y)	5x − 2y	f(x, y)

38. Find the maximum and minimum value of $z = 3x + 4y$ subject to the following constraints: **MP 2**

$$x + 2y \leq 14$$
$$3x - y \geq 0$$
$$x - y \leq 2$$

39. At a certain refinery, the refining process requires the production of at least two gallons of gasoline for each gallon of fuel oil. To meet the demands of winter, at least three million gallons of fuel oil a day will need to be produced. The demand for gasoline, on the other hand, is not more than 6.4 million gallons a day.

If gasoline is selling for $1.90 per gallon and fuel oil sells for $1.50 per gallon, how much of each should be produced in order to maximize revenue? **MP 2**

LESSON 9
Solving Systems of Equations in Three Variables

::Then
- You solved systems of linear equations in two variables.

::Now
1. Solve systems of linear equations in three variables.
2. Solve real-world problems using systems of linear equations in three variables.

::Why?
- Seats closest to the stage at an ampitheater cost $120. The seats in the next section cost $45, and lawn seats are $35. There are twice as many seats in section B as in section A. When all 20,000 seats are sold, the amphitheater makes $962,500.

- A system of equations in three variables can be used to determine the number of seats in each section.

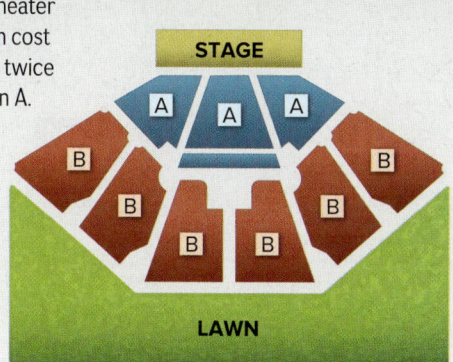

New Vocabulary
ordered triple

Mathematical Practices
3 Construct viable arguments and critique the reasoning of others.

1 Systems in Three Variables
Like systems of equations in two variables, systems in three variables can have one solution, infinite solutions, or no solution. A solution of such a system is an **ordered triple** (x, y, z).

The graph of an equation in three variables is a three-dimensional graph in the shape of a plane. The graphs of a system of equations in three variables form a system of planes.

One Solution
The three individual planes intersect at a specific point.

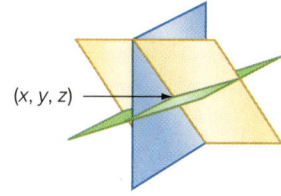
(x, y, z)

Infinitely Many Solutions

The planes intersect in a line.

Every coordinate on the line represents a solution of the system.

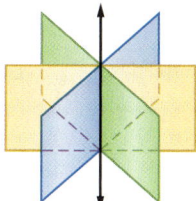

The planes intersect in the same plane.

Every equation is equivalent.
Every coordinate in the plane represents a solution of the system.

No Solution
There are no points in common with all three planes.

Similar to solving systems of equations in two variables, use the strategies of substitution and elimination to find the ordered triple that represents the solution of the system.

Example 1   A System with One Solution

Solve the system of equations.

$3x - 2y + 4z = 35$
$-4x + y - 5z = -36$ ← The coefficient of 1 in the second equation makes y a good choice for elimination.
$5x - 3y + 3z = 31$

Step 1 Eliminate one variable by using two pairs of equations.

$3x - 2y + 4z = 35$
$-4x + y - 5z = -36$ → Multiply by 2.

$3x - 2y + 4z = 35$ Equation 1
$(+) -8x + 2y - 10z = -72$ Equation 2 × 2
$\overline{-5x - 6z = -37}$

$-4x + y - 5z = -36$ → Multiply by 3.
$5x - 3y + 3z = 31$

$-12x + 3y - 15z = -108$ Equation 2 × 3
$(+) 5x - 3y + 3z = 31$ Equation 3
$\overline{-7x - 12z = -77}$

The y-terms have been eliminated. There now is a system of two equations in two variables.

Step 2 Solve the system of two equations.

$-5x - 6z = -37$ Multiply by −2.
$-7x - 12z = -77$

$10x + 12z = 74$
$(+) -7x - 12z = -77$
$\overline{3x = -3}$ Eliminate z.
$x = -1$ Divide by 3.

Use substitution to solve for z.

$-5x - 6z = -37$ Equation with two variables
$-5(-1) - 6z = -37$ Substitution
$5 - 6z = -37$ Multiply.
$-6z = -42$ Subtract 5 from each side.
$z = 7$ Divide each side by −6.

The result is $x = -1$ and $z = 7$.

Step 3 Substitute the two values into one of the original equations to find y.

$-4x + y - 5z = -36$ Equation 2
$-4(-1) + y - 5(7) = -36$ Substitution
$4 + y - 35 = -36$ Multiply.
$y = -5$ Add 31 to each side.

CHECK $3x - 2y + 4z = 35$ Equation 1
$3(-1) - 2(-5) + 4(7) \stackrel{?}{=} 35$ $x = -1, y = -5, z = 7$
$-3 + 10 + 28 \stackrel{?}{=} 35$ Simplify.
$35 = 35$ ✓

The solution is $(-1, -5, 7)$.

> **Study Tip**
> **MP Precision** Always substitute your answer into all of the original equations to confirm your answer.

Guided Practice

1A. $2x + 4y - 5z = 18$
$-3x + 5y + 2z = -27$
$-5x + 3y - z = -17$

1B. $4x - 3y + 6z = 18$
$-x + 5y + 4z = 48$
$6x - 2y + 5z = 0$

68 | Lesson 1-9 | Solving Systems of Equations in Three Variables

When solving a system of three linear equations with three variables, it is important to check your answer using all three of the original equations. This is necessary because it is possible for a solution to work for two of the equations but not the third.

Example 2 No Solution and Infinite Solutions

Solve each system of equations.

a. $5x + 4y - 5z = -10$
$-4x - 10y - 8z = -16$
$6x + 15y + 12z = 24$

Eliminate x in the second two equations.

$-4x - 10y - 8z = -16$ — Multiply by 3. → $-12x - 30y - 24z = -48$
$6x + 15y + 12z = 24$ — Multiply by 2. → $(+)\ 12x + 30y + 24z = 48$
$0 = 0$

> **Study Tip**
> **Infinite and No Solutions**
> When solving a system of more than two equations, $0 = 5$ will always yield no solution, while $0 = 0$ may not always yield infinite solutions.

The equation $0 = 0$ is always true. This indicates that the last two equations represent the same plane. Check to see if this plane intersects the first plane.

$5x + 4y - 5z = -10$ — Multiply by 4. → $20x + 16y - 20z = -40$
$-4x - 10y - 8z = -16$ — Multiply by 5. → $(+)\ -20x - 50y - 40z = -80$
$-34y - 60z = -120$

The planes intersect in a line. So, there are an infinite number of solutions.

b. $-6a + 9b - 12c = 21$
$-2a + 3b - 4c = 7$
$10a - 15b + 20c = -30$

Eliminate a in the first two equations.

$-6a + 9b - 12c = 21$ $-6a + 9b - 12c = 21$
$-2a + 3b - 4c = 7$ — Multiply by −3. → $(+)\ 6a - 9b + 12c = -21$
$0 = 0$

The equation $0 = 0$ is always true. This indicates that the first two equations represent the same plane. Check to see if this plane intersects the last plane.

$-2a + 3b - 4c = 7$ — Multiply by 5. → $-10a + 15b - 20c = 35$
$10a - 15b + 20c = -30$ $(+)\ 10a - 15b + 20c = -30$
$\phantom{10a - 15b + 20c = -30 (+)\ 10a - 15b + 20}0 = 5$

The equation $0 = 5$ is never true. So, there is no solution of this system.

Guided Practice

2A. $-4x - 2y - z = 15$
$12x + 6y + 3z = 45$
$2x + 5y + 7z = -29$

2B. $3x + 5y - 2z = 13$
$-5x - 2y - 4z = 20$
$-14x - 17y + 2z = -19$

2 Real-World Problems
When solving problems involving three variables, use the five-step plan to help organize the information. Identify the three variables and what they represent. Then use the information in the problem to form equations using the variables. Once you have three equations and all three variables are represented, you can solve the problem.

Math History Career

Music Management Music management involves acting as the talent manager to the artists. Other duties include negotiating with record labels, music promoters, and tour promoters. Managers usually earn a percentage of the artist's income. A bachelor's degree is usually required.

Go Online!

Completing the **Self-Check Quiz** with a partner can help your understanding. Take turns describing each step as you solve each problem. Ask for clarification as you need it.

Real-World Example 3   Write and Solve a System of Equations

CONCERTS Refer to the beginning of the lesson. Write and solve a system of equations to determine how many seats are in each section of the amphitheater.

Understand Define the variables. x = section A seats; y = section B seats; z = lawn seats

Plan There are 20,000 seats.
$$x + y + z = 20{,}000 \qquad \text{Equation 1}$$

The total revenue is $962,500.
$$120x + 45y + 35z = 962{,}500 \qquad \text{Equation 2}$$

There are twice as many seats in section B as in section A.
$$y = 2x \qquad \text{Equation 3}$$

Solve Solve the system.

Step 1 Substitute $y = 2x$ in the first two equations.

$x + y + z = 20{,}000$	Equation 1
$x + 2x + z = 20{,}000$	$y = 2x$
$3x + z = 20{,}000$	Add.
$120x + 45y + 35z = 962{,}500$	Equation 2
$120x + 45(2x) + 35z = 962{,}500$	$y = 2x$
$210x + 35z = 962{,}500$	Simplify.

Step 2 Solve the system of two equations in two variables.

$$3x + z = 20{,}000 \quad \text{Multiply by } -35. \quad -105x - 35z = -700{,}000$$
$$210x + 35z = 962{,}500 \qquad\qquad\qquad (+)\ 210x + 35z = 962{,}500$$
$$\overline{\qquad\qquad 105x \qquad\quad = 262{,}500}$$
$$x = 2500$$

Step 3 Substitute to find z.

$3x + z = 20{,}000$	Remaining equation in two variables
$3(2500) + z = 20{,}000$	$x = 3600$
$7500 + z = 20{,}000$	Multiply.
$z = 12{,}500$	Simplify.

Step 4 Substitute to find y.

$y = 2x$	Equation 3
$y = 2(2500)$ or 5000	$x = 3600$

The solution is (2500, 5000, 12,500). There are 2500 seats in section A, 5000 in section B, and 12,500 lawn seats.

Check Substitute the values into either of the first two equations.

In Step 1, $y = 2x$ was substituted into the other two equations because it resulted in the easiest two equation system.

Guided Practice

3. Ms. Garza invested $50,000 in three different accounts. She invested three times as much money in an account that paid 8% interest as in an account that paid 10% interest. The third account earned 12% interest. If she earned a total of $5160 in interest in a year, how much did she invest in each account?

Check Your Understanding

= Step-by-Step Solutions begin on page R11.

Example 1

1. What is the solution for the system of equations?
   $3a + 6b - 2c = -6$
   $2a + b + 4c = 19$
   $-5a - 2b + 8c = 62$

   A $(-4, 3, 6)$ **C** $(0, 0, 3)$
   B $(-1, -3, 6)$ **D** $(4, -2, 3)$

Examples 1–2 Solve each system of equations.

2. $-4r - s + 3t = -9$
   $3r + 2s - t = 3$
   $r + 3s - 5t = 29$

3. $3x + 5y - z = 12$
   $-2x - 3y + 5z = 14$
   $4x + 7y + 3z = 38$

4. $2a - 3b + 5c = 58$
   $-5a + b - 4c = -51$
   $-6a - 8b + c = 22$

Example 3

5. **DOWNLOADING** Heather downloaded some television shows. A sitcom is 0.3 gigabyte; a drama, 0.6 gigabyte; and a talk show, 0.6 gigabyte. She downloaded 7 programs totaling 3.6 gigabytes. There were twice as many episodes of the drama as the sitcom. How many episodes of each show did she download?

Practice and Problem Solving

Extra Practice is on page R1.

Examples 1–2 Solve each system of equations.

6. $-5x + y - 4z = 60$
   $2x + 4y + 3z = -12$
   $6x - 3y - 2z = -52$

7. $4a + 5b - 6c = 2$
   $-3a - 2b + 7c = -15$
   $-a + 4b + 2c = -13$

8. $-2x + 5y + 3z = -25$
   $-4x - 3y - 8z = -39$
   $6x + 8y - 5z = 14$

9. $4r + 6s - t = -18$
   $3r + 2s - 4t = -24$
   $-5r + 3s + 2t = 15$

10. $-2x + 15y + z = 44$
    $4x + 3y + 3z = 18$
    $-3x + 6y - z = 8$

11. $4x + 2y + 6z = 13$
    $-12x + 3y - 5z = 8$
    $-4x + 7y + 7z = 34$

12. $8x + 3y + 6z = 43$
    $-3x + 5y + 2z = 32$
    $5x - 2y + 5z = 24$

13. $-6x - 5y + 4z = 53$
    $5x + 3y + 2z = -11$
    $8x - 6y + 5z = 4$

14. $-9a + 3b - 2c = 61$
    $8a + 7b + 5c = -138$
    $5a - 5b + 8c = -45$

15. $2x - y + z = 1$
    $x + 2y - 4z = 3$
    $4x + 3y - 7z = -8$

16. $x + 2y = 12$
    $3y - 4z = 25$
    $x + 6y + z = 20$

17. $r - 3s + t = 4$
    $3r - 6s + 9t = 5$
    $4r - 9s + 10t = 9$

Example 3

18. **CHALLENGE** The posted results of a recent high school swim meet state that 24 individuals placed, earning a combined total of 53 points. First place earned 3 points, second place earned 2 points, and third place earned 1 point. There were as many first-place finishers as second- and third-place finishers combined.

    a. Write a system of three equations that represents how many people finished in each place.

    b. How many swimmers finished in first place, in second place, and in third place?

    c. Suppose that the athletes scored a combined total of 47 points. Explain why this statement is false and the solution is unreasonable.

19. **AMUSEMENT PARKS** Nick goes to the amusement park to ride roller coasters, bumper cars, and water slides. The wait for the roller coasters is 1 hour, the wait for the bumper cars is 20 minutes long, and the wait for the water slides is only 15 minutes long. Nick rode 10 total rides during his visit. Because he enjoys roller coasters the most, the number of times he rode the roller coasters was the sum of the times he rode the other two rides. If Nick waited in line for a total of 6 hours and 20 minutes, how many of each ride did he go on?

20. **BUSINESS** Ramón usually gets one of the routine maintenance options at Annie's Garage. Today, however, he needs a different combination of work than what is listed.

    a. Assume that the price of an option is the same price as purchasing each item separately. Find the prices for an oil change, a radiator flush, and a brake pad replacement.
    b. If Ramón wants his brake pads replaced and his radiator flushed, how much should he plan to spend?

21. **FINANCIAL LITERACY** Kate invested $100,000 in three different accounts. If she invested $30,000 more in account A than account C and is expected to earn $6300 in interest, how much did she invest in each account?

Account	Expected Interest
A	4%
B	8%
C	10%

H.O.T. Problems   Use Higher-Order Thinking Skills

22. **CHALLENGE** Write a system of equations to represent the three rows of figures below. Use the system to find the number of red triangles that will balance one green circle.

23. **MP MODELING** The general form of an equation for a parabola is $y = ax^2 + bx + c$, where (x, y) is a point on the parabola. If three points on a parabola are $(2, -10)$, $(-5, -101)$, and $(6, -90)$, determine the values of a, b, and c and write the general form of the equation.

24. **PROOF** Consider the following system and prove that if $b = c = -a$, then $ty = a$.

    $$rx + ty + vz = a$$
    $$rx - ty + vz = b$$
    $$rx + ty - vz = c$$

25. **OPEN-ENDED** Write a system of three linear equations that has a solution of $(-5, -2, 6)$. Show that the ordered triple satisfies all three equations.

26. **OPEN-ENDED** Use the diagrams of solutions of systems of equations on page 67 to consider a system of inequalities in three variables. Describe the solution of such a system.

27. **WRITING IN MATH** Use your knowledge of solving a system of three linear equations with three variables to explain how to solve a system of four equations with four variables.

Preparing for Assessment

28. What is the solution of the system of equations?

$$5x - 3y - z = -1$$
$$2x + y + 3z = 2$$
$$x + y - 2z = -7$$

- A $(-3, -4, -2)$
- B $(-3, 2, 3)$
- C $(-1, -2, 2)$
- D $(1, 2, 5)$
- E $(1, 2, 7)$

29. Three families bought food at a baseball game. The Davis family spent $16.50 on 3 hot dogs, 2 drinks, and 1 pretzel. The Ruiz family spent $26 on 4 hot dogs, 4 drinks, and 2 pretzels. The Cho family spent $13.75 on 2 hot dogs, 1 drink and 3 pretzels. What is the price of a pretzel?

- A $1.50
- B $1.75
- C $2.25
- D $3.50

30. Three trucks are being loaded with small, medium, and large crates. Each crate of the same size has the same weight, and each size has a different weight. Use x for the weight of a small crate, y for the weight of a medium crate, and z for the weight of a large crate. Write three equations for the weights. Solve to find the weight of each size crate.

- The first truck has 18 small crates, 10 medium, and 16 large, for a total weight of 1460 pounds.
- The second truck has 25 small crates, 20 medium, and 8 large, for a total weight of 1500 pounds.
- The third truck has 7 small crates, 40 medium, and 8 large, for a total weight of 1740 pounds.

31. MULTI-STEP Consider this system of equations. Follow the steps to find the solution. Show your work.

$$x + 2y - z = -1$$
$$2x + y + z = 7$$
$$3x + 3y - 2z = 2$$

a. Add the first two equations to eliminate z.

b. Eliminate z from the first and third equations. Explain your work.

c. Explain how to solve the two equations in x and y.

d. Use one of the equations with three variables to solve for z. Then write the ordered triple that is the solution.

32. Solve the following system of equations:

$$-3x + 2y - 6z - 6 = 0$$
$$5x + 7y - 5z = 6$$
$$x + 4y - 2z = 8$$

33. Solve the following system of equations:

$$x + 2y - z = 4$$
$$5x + 7y - 4z = 15$$
$$x + 4y - 2z = 7$$

34. Last weekend, the local cinema sold a total of 8000 movie tickets. Proceeds totaled $57000. Tickets can be bought in one of three ways: a matinee admission costs $5, student admission is $6 all day, and regular admissions are $8. How many of each type of ticket were sold if twice as many student tickets were sold as matinee tickets?

CHAPTER 1
Study Guide and Review

Go Online! for Vocabulary Review Games and key vocabulary in 13 languages

Study Guide

Key Concepts

Linear Equations and Inequalities (Lessons 1-1 and 1-2)
- Verbal expressions can be translated into algebraic expressions.
- Apply inverse operations to solve linear equations and inequalities.
- When you multiply or divide each side of an inequality by a negative number, the direction of the inequality symbol must be reversed.

Linear Equations and Slope (Lessons 1-3 and 1-4)
- Slope: $\dfrac{\text{change in } y}{\text{change in } x} = \dfrac{y^2 - y^1}{x^2 - x^1}$
- Slope-Intercept Form: $y = mx + b$
- Point-Slope Form: $y - y_1 = m(x - x_1)$

Graphing Linear Inequalities (Lesson 1-5)
- Determine whether the boundary is solid or dashed. Graph the boundary. Then determine which region to shade by choosing a point and testing it in the inequality.

Systems of Equations and Inequalities (Lessons 1-6 and 1-7)
- In the substitution method, one equation is solved for a variable and substituted to find the value of another variable. In the elimination method, one variable is eliminated by adding or subtracting the equations.
- The solution of a system of inequalities is found by graphing the inequalities and determining the intersection of the graphs.

Linear Programming (Lesson 1-8)
- Linear programming is a method for finding maximum or minimum values of a function over a given system of inequalities with each inequality representing a constraint.

Systems of Equations in Three Variables (Lesson 1-9)
- A system of equations in three variables can be solved algebraically by using substitution or elimination.

FOLDABLES Study Organizer

Use your Foldable to review the chapter. Working with a partner can be helpful. Ask for clarification of concepts as needed.

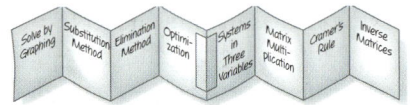

Key Vocabulary

bounded (p. 60)
break-even point (p. 42)
consistent (p. 43)
dependent (p. 43)
elimination method (p. 45)
equation (p. 5)
feasible region (p. 60)
inconsistent (p. 43)
independent (p. 43)
linear equation (p. 5)
linear inequality (p. 5)
linear programming (p. 60)
open sentence (p. 5)

optimize (p. 62)
ordered triple (p. 67)
point-slope (p. 21)
rate of change (p. 21)
set-builder notation (p. 35)
slope (p. 21)
slope-intercept form (p. 21)
solution (p. 5)
substitution method (p. 44)
system of equations (p. 42)
system of inequalities (p. 52)
unbounded (p. 60)

Vocabulary Check

Choose the term from above to complete each sentence.

1. A feasible region that is open and can go on forever is called _____.
2. To _____ means to seek the best price or profit using linear programming.
3. A mathematical sentence containing one or more variables is called a(n) _____.
4. If you are given the coordinates of two points on a line, you can use the _____ to find the equation of the line that passes through them.
5. A(n) _____ of an equation is a value that makes the equation true.
6. A system of equations is _____ if it has at least one solution.

Concept Check

7. Explain the difference between an independent system of equations and a dependent system of equations.
8. Explain how to determine which region of the coordinate plane to shade when graphing a linear inequality.

Lesson-by-Lesson Review

1-1 Solving Linear Equations

Solve each equation. Check your solution.

9. $8 + 5r = -27$
10. $4w + 10 = 6w - 13$
11. $\frac{x}{6} + \frac{x}{3} = \frac{3}{4}$
12. $6b - 5 = 3(b + 2)$
13. **MONEY** It cost Lori $19.75 to go to the movies. She bought popcorn for $5.50 and a soda for $4.75. How much was her ticket?

Solve each equation or formula for the specified variable.

14. $2k - 3m = 16$ for k
15. $\frac{r+5}{mn} = p$ for m
16. $A = \frac{1}{2}h(a + b)$ for h

17. **GEOMETRY** Yu-Jun wants to fill the water container at the right. He knows that the radius is 2 inches and the volume is 100.48 cubic inches. What is the height of the water bottle? Use the formula for the volume of a cylinder, $V = \pi r^2 h$, to find the height of the bottle.

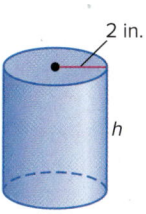

Example 1

Solve $-3(a - 3) + 2(3a - 2) = 14$.

$-3(a - 3) + 2(3a - 2) = 14$	Original equation
$-3a + 9 + 6a - 4 = 14$	Distributive Property
$-3a + 6a + 9 - 4 = 14$	Commutative Property
$3a + 5 = 14$	Substitution Property
$3a = 9$	Subtraction Property
$a = 3$	Division Property

Example 2

Solve each equation or formula for the specified variable.

a. $y = 2x + 3z$ for x

$y = 2x + 3z$	Original equation
$y - 3z = 2x$	Subtract $3z$ from each side.
$\frac{y - 3z}{2} = x$	Divide each side by 2.

b. $V = \frac{\pi r^2 h}{3}$ for h

$V = \frac{\pi r^2 h}{3}$	Original equation
$3V = \pi r^2 h$	Multiply each side by 3.
$\frac{3V}{\pi r^2} = h$	Divide each side by πr^2.

1-2 Solving Linear Inequalities

Solve each inequality. Then graph the solution set on a number line.

18. $-4a \leq 24$
19. $\frac{r}{5} - 8 > 3$
20. $4 - 7x \geq 2(x + 3)$
21. $-p - 13 < 3(5 + 4p) - 2$

22. **MONEY** Ms. Hawkins is taking her science class on a field trip to a museum. She has $884 to spend on the trip. There are 52 students that will go to the museum. The museum charges $5 per student, and Ms. Hawkins gets in for free. If the students will have slices of pizza for lunch that cost $4 each, how many slices can each student have?

Example 3

Solve $2m - 7 < -11$. Graph the solution set on a number line.

$2m - 7 < -11$	Original inequality
$2m < -4$	Add 7 to each side.
$m < -2$	Divide each side by 2.

The solution set is $\{m \mid m < -2\}$.

The graph of the solution set is shown below.

-5 -4 -3 -2 -1 0 1 2 3 4 5

CHAPTER 1
Study Guide and Review Continued

1-3 Rate of Change and Slope

23. RETAIL The table shows the number of backpacks sold each week at Bags & More. Find the average rate of change of the number of backpacks sold from week 2 to week 5.

Week	1	2	3	4	5
Backpacks Sold	76	58	94	83	112

Find the slope of the line that passes through each pair of points.

24. $(2, 5), (6, -3)$ 25. $(8, 2), (2, 8)$

Example 4

Find the slope of the line that passes through points $(-2, 9)$ and $(1, 4)$.

$m = \dfrac{y_2 - y_1}{x_2 - x_1}$ Slope Formula

$= \dfrac{4 - 9}{1 - (-2)}$ $(x_1, y_1) = (-2, 9), (x_2, y_2) = (1, 4)$

$= -\dfrac{5}{3}$ Simplify.

1-4 Writing Linear Equations

Write an equation in slope-intercept form for the line that satisfies each set of conditions.

26. slope -2, passes through $(-3, -5)$

27. slope $\dfrac{2}{3}$, passes through $(4, -1)$

28. passes through $(-2, 4)$ and $(0, 8)$

29. passes through $(3, 5)$ and $(-1, 5)$

Write an equation of the line passing through each pair of points.

30. $(6, 1), (4, 9)$ 31. $(-4, 2), (6, 8)$

Write an equation in slope-intercept form for the line that satisfies each set of conditions.

32. through $(1, 2)$, parallel to $y = 4x - 3$

33. through $(-3, 5)$, perpendicular to $y = \dfrac{2}{3}x - 8$

34. PETS Drew paid a $250 fee when he adopted a puppy. The average monthly cost of feeding and caring for the puppy is $32. Write an equation that represents the total cost of adopting and caring for the puppy for x months.

Example 5

Write an equation of the line through $(-2, 5)$ and $(0, -9)$.

Find the slope of the line.

$m = \dfrac{y_2 - y_1}{x_2 - x_1}$ Slope Formula

$= \dfrac{-9 - 5}{0 - (-2)}$ $(x_1, y_1) = (-2, 5),$ $(x_2, y_2) = (0, -9)$

$= \dfrac{-14}{2}$ or -7 Simplify.

Write an equation.

$y - y_1 = m(x - x_1)$ Point-slope form

$y - 5 = -7(x - (-2))$ Substitute.

$y - 5 = -7(x + 2)$ Simplify.

$y - 5 = -7x - 14$ Distributive Property

$y = -7x - 9$ Add 5 to each side.

The equation is $y = -7x - 9$.

1-5 Graphing Linear Inequalities

Graph each inequality.

35. $x - 3y < 6$ 36. $y \geq 2x + 1$

37. $2x + 4y \leq 12$ 38. $y < -3x - 5$

39. SHOPPING Spencer has saved $30 to spend at his favorite online store. Each song costs $2 and each game costs $3. Write and graph an inequality that shows the number of songs and games Spencer can download.

Example 6

Graph $x - 2y > 6$.

Because the inequality symbol is $>$, the graph of the boundary is dashed.

Test $x - 2y > 6$ at $(0, 0)$.

$x - 2y > 6$

$0 - 2(0) \stackrel{?}{>} 6$

$0 > 6$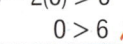

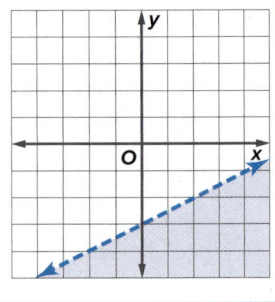

Lesson-by-Lesson Review

1-6 Solving Systems of Equations

Solve each system of equations by graphing.

40. $3x + 4y = 8$
$x - 3y = -6$

41. $x + \frac{8}{3}y = 12$
$\frac{1}{2}x + \frac{4}{3}y = 6$

42. $y - 3x = 13$
$y = \frac{1}{3}x + 5$

43. $6x - 14y = 5$
$3x - 7y = 5$

44. LAWN CARE André and Paul each mow lawns. André charges a $30 service fee and $10 per hour. Paul charges a $10 service fee and $15 per hour. After how many hours will André and Paul charge the same amount?

Solve each system of equations by using either substitution or elimination.

45. $x + y = 6$
$3x - 2y = -2$

46. $5x - 2y = 4$
$-2y + x = 12$

47. $x + y = 3.5$
$x - y = 7$

48. $3y - 5x = 0$
$2y - 4x = -2$

49. SCHOOL SUPPLIES At an office supply store, Emilio bought 3 notebooks and 5 pens for $13.75. If a notebook costs $1.25 more than a pen, how much does a notebook cost? How much does a pen cost?

Example 7

Solve the system of equations by graphing.
$x + y = 4$
$x + 2y = 5$

Graph both equations on the coordinate plane.

The solution of the system is (3, 1).

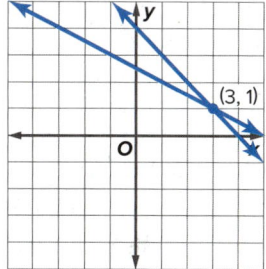

Example 8

Solve the system of equations by using either substitution or elimination.
$3x + 2y = 1$
$y = -x + 1$

Substitute $-x + 1$ for y in the first equation. Then solve for y.

$3x + 2y = 1$ $y = -x + 1$
$3x + 2(-x + 1) = 1$ $= -(-1) + 1$
$3x - 2x + 2 = 1$ $= 2$
$x + 2 = 1$
$x = -1$ The solution is (−1, 2).

1-7 Solving Systems of Inequalities by Graphing

Solve each system of inequalities by graphing.

50. $y < 2x - 3$
$y \geq 4$

51. $|y| > 2$
$x > 3$

52. $y \geq x + 3$
$2y \leq x - 5$

53. $y > x + 1$
$x < -2$

54. JEWELRY Payton makes jewelry to sell at her mother's clothing store. She spends no more than 3 hours making jewelry on Saturdays. It takes her 15 minutes to set up her supplies and 25 minutes to make each bracelet. Draw a graph that represents this. Then list three possible solutions.

Example 9

Solve the system of inequalities by graphing.
$y \geq \frac{3}{2}x - 3$
$y < 4 - 2x$

The solution of the system is the region that satisfies both inequalities. The solution of this system is the shaded region.

CHAPTER 1
Study Guide and Review Continued

1-8 Optimization with Linear Programming

55. FLOWERS A florist can make a grand arrangement in 18 minutes or a simple arrangement in 10 minutes. The florist makes at least twice as many of the simple arrangements as the grand arrangements. The florist can work only 40 hours per week. The profit on the simple arrangements is $10 and the profit on the grand arrangements is $25. Find the number and type of arrangements that the florist should produce to maximize profit.

56. MANUFACTURING A shoe manufacturer makes outdoor and indoor soccer shoes. There is a two-step process for both kinds of shoes. Each pair of outdoor shoes requires 2 hours in step one and 1 hour in step two, and produces a profit of $20. Each pair of indoor shoes requires 1 hour in step one and 3 hours in step two, and produces a profit of $15. The company has 40 hours of labor available per day for step one and 60 hours available for step two. What is the manufacturer's maximum profit? What is the combination of shoes for this profit?

Example 10

A gardener is planting two types of herbs in a 5184-square-inch garden. Herb A requires 6 square inches of space, and herb B requires 24 square inches of space. The gardener will plant no more than 300 plants. If herb A can be sold for $8 and herb B can be sold for $12, how many of each herb should be sold to maximize income?

Let a = the number of herb A and b = the number of herb B.

$a \geq 0, b \geq 0,$
$6a + 24b \leq 5184,$
and $a + b \leq 300$

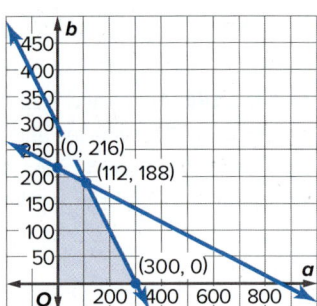

Graph the inequalities. The vertices of the feasible region are (0, 0), (300, 0), (0, 216), and (112, 188).

The profit function is $f(a, b) = 8a + 12b$.

The maximum value of $3152 occurs at (112, 188). So the gardener should plant 112 of herb A and 188 of herb B.

1-9 Solving Systems of Equations in Three Variables

Solve each system of equations.

57. $a - 4b + c = 3$
$b - 3c = 10$
$3b - 8c = 24$

58. $2x - z = 14$
$3x - y + 5z = 0$
$4x + 2y + 3z = -2$

59. AMUSEMENT PARKS Dustin, Luis, and Marci went to an amusement park. They purchased snacks from the same vendor. Their snacks and how much they paid are listed in the table. How much did each snack cost?

Name	Hot Dogs	Popcorn	Soda	Price
Dustin	1	2	3	$15.25
Luis	2	0	3	$14.00
Marci	1	2	1	$10.25

Example 11

Solve the system of equations.
$x + y + 2z = 6$
$2x + 5z = 12$
$x + 2y + 3z = 9$

$2x + 2y + 4z = 12$ Equation 1 × 2
$\underline{(-) \; x + 2y + 3z = 9}$ Equation 3
$x + z = 3$ Subtract.

Solve the system of two equations.

$2x + 5z = 12$ Equation 2
$\underline{(-) \; 2x + 2z = 6}$ 2 × (x + z = 3)
$3z = 6$ Subtract.
$z = 2$ Divide each side by 3.

Substitute 2 for z in one of the equations with two variables, and solve for y. Then, substitute 2 for z and the value you got for y into an equation from the original system to solve for x.

The solution is (1, 1, 2).

CHAPTER 1
Practice Test

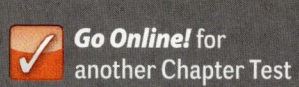

Write an algebraic expression to represent each verbal expression.

1. twice the difference of a number and 11
2. the product of the square of a number and 5

Solve each equation.

3. $8.5(3y + 4) = 3.5(y - 28)$
4. $12g - 9g + 24 - g - 9 = 13$
5. $\frac{1}{2}p - 12 = \frac{3}{4}p - 18$
6. $5(4x - 9) - 2x = 6x + 15$

7. **MULTIPLE CHOICE** If $3m + 5 = 23$, what is the value of $2m - 3$?

   A 105
   B 9
   C $\frac{47}{3}$
   D 6

8. Solve $r = \frac{1}{2}m^2 p$ for p.

Solve each inequality and graph the solution set on a number line.

9. $-2b > \frac{18 - b}{5}$
10. $-3b - 5 \geq -6b - 13$

11. **MONEY** Carson has $35 to spend at the water park. The admission price is $25 and each soda is $2.50. Write an inequality to show how many sodas he can buy.

Find the slope of the line that passes through each pair of points.

12. $(1, 6), (3, 10)$
13. $(-2, 7), (3, -1)$

14. Write an equation in slope-intercept form for the line that has slope -2 and passes through the point $(3, -4)$.

15. Write an equation of the line that passes through the points $(2, -4)$ and $(1, 6)$.

16. Write an equation in slope-intercept form for the line that passes through $(-3, 5)$ and is parallel to $y = -6x + 1$.

Graph each inequality.

17. $y \geq 4x - 1$
18. $2x + 6y < -12$

Solve each system of equations by using either substitution or elimination.

19. $y = x + 4$
    $x + y = -12$
20. $3x + 5y = -7$
    $6x - 4y = 0$

21. $5x + 2y = 4$
    $3y - 4x = -40$
22. $8x - 3y = -13$
    $-3x + 5y = 1$

23. **MULTIPLE CHOICE** Which graph shows the solution of the system of inequalities?

$$y \leq 2x + 3$$
$$y < \frac{1}{3}x + 5$$

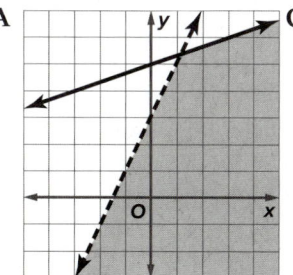
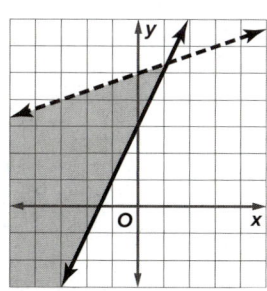
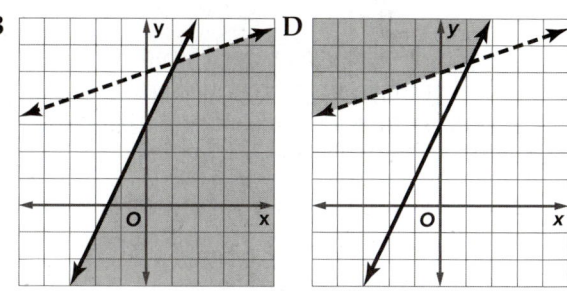

Solve each system of inequalities by graphing.

24. $x + y > 6$
    $x - y < 0$
25. $y \geq 2x - 5$
    $y \leq x + 4$
26. $3x + 4y \leq 12$
    $6x - 3y \geq 18$
27. $5y + 2x \leq 20$
    $4x + 3y > 12$

28. **SALONS** Sierra King is a nail technician. She allots 20 minutes for a manicure and 45 minutes for a pedicure in her 7-hour work day. No more than 5 pedicures can be scheduled each day. The prices are $25 for a manicure and $45 for a pedicure. If she must schedule both procedures, how many of each should Ms. King schedule to maximize her daily income? What is her maximum daily income?

29. **COLLEGE FOOTBALL** Johnny Manziel of Texas A&M is the first freshman to win the Heisman Trophy. Players are given 3 points for every first-place, 2 points for every second-place vote, and 1 point for every third-place vote. Manziel received 892 total votes for first, second, and third place, for a total of 2029 points. If he had 4 fewer than twice as many second-place votes as first-place votes, how many votes did he receive for each place?

CHAPTER 1
Preparing for Assessment

Performance Task

Provide a clear solution to each part of the task. Be sure to show all of your work, include all relevant drawings, and justify your answers.

FINANCIAL LITERACY A catering company offers various services to clients.

Part A

A corporate client requests catering service for a business event with 125 attendees. For an all-inclusive service, the catering company gives them a quote of $1512.50, which includes a flat fee of $200 and the cost per attendee.

1. Determine the cost per attendee for the client.

Part B

Another client is planning a wedding and is in the process of establishing a budget. The catering company gives her two options. The all-inclusive option costs $16.50 per person, with a flat fee of $200, while the budget-friendly option costs $12.00 per person, with a flat fee of $350. The client's catering budget is $4500.

2. **Tools** Write two inequalities that model the situation above.

3. Determine the greatest number of guests the client can have if they select the all-inclusive option.

4. Determine the greatest number of guests the client can have if they select the budget-friendly option.

Part C

For a smaller event, the company has given a client a quote of $643.75 for 35 people. The per-person rate is $11.25.

5. Write a linear equation that models the situation above.

6. **Sense-Making** Determine the flat fee the catering company charges.

Part D

Another client is planning a high-school reunion, to which alumni are welcome to bring their children. The catering company gives them a quote of $11.75 per adult and $7.50 per child. The number of children who attended was 231 less than the number of adults who attended. The total cost of catering was $4369.75.

7. Write a system of equations that models the situation above.

8. Determine the number of adults and the number of children who attended.

Test-Taking Strategy

Example

Read the problem. Identify what you need to know. Then use the information in the problem to solve.

Company A charges a monthly fee of $14.50 plus $0.05 per minute for cell phone service. Company B charges $20.00 per month plus $0.04 per minute. For what number of minutes would the total monthly charge be the same with each company?

Step 1 **What is the problem asking you to find? What do you need to do to solve the problem?** I need to find the number of minutes for which the total monthly charge will be the same with Companies A and B. I need to set up a system of equations, graph the system, and then find where the graphs intersect.

Step 2 **How can you use the values given in the problem to create a system of equations?** I am given the monthly fee and rate per minute for the two companies, so the equations that represent Company A's and Company B's total monthly charges will make up the system. Each equation will be linear, with the monthly fee representing the y-intercept and the fee per minute representing the slope. Company A: $y = 0.05x + 14.50$; Company B: $y = 0.04x + 20$

Step 3 **How can you use a graph to solve the problem? What is the correct answer?** I can find the intersection of the graphs of the two equations and determine the x-coordinate, which is the solution.

The correct answer is 550 minutes.

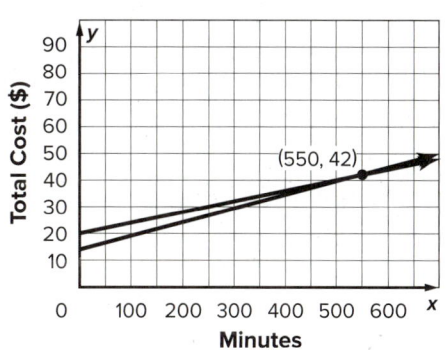

> **Test-Taking Tip**
>
> **Strategies for Solving Systems of Equations**
> For a system of linear equations in two variables, you can solve the system algebraically or by graphing. When you solve it graphically, you can substitute the solution for x in each equation to ensure your answer is correct.

Apply the Strategy

Read the problem. Identify what you need to know. Then use the information in the problem to solve.

Raul had $395 in a savings account and started adding $25 a week. At the same time, his sister Tina had $365 in her account and began adding $30 a week. After how many weeks will Raul and Tina have the same amount in their savings accounts?

Answer the questions below.
a. What is the problem asking you to find? What do you need to do to solve the problem?
b. How can you use the values given in the problem to create a system of equations?
c. How can you use a graph to solve the problem? What is the correct answer?

CHAPTER 1
Preparing for Assessment
Cumulative Review

Read each question. Then fill in the correct answer on the answer document provided by your teacher or on a sheet of paper.

1. Which system of inequalities is graphed below?

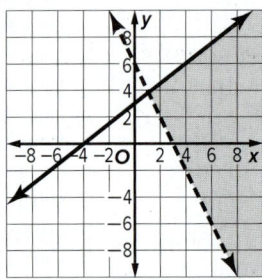

- A $2x + y \le 6$ and $4y - 12 < 3x$
- B $2x + y > 6$ and $4y - 12 \ge 3x$
- C $2x + y > 6$ and $4y - 12 \le 3x$
- D $2x + y \ge 6$ and $4y - 12 > 3x$

2. The table below shows the money spent by three families at a baseball stadium concession stand. What is the cost of a hot dog?

Concession Stand Purchases

Name	Hot Dogs	Drinks	Pretzels	Total Cost
Booth	2	2	1	$9.25
Chan	3	4	1	$14.50
McCline	1	3	3	$12.00

- A $1.50
- B $1.75
- C $2.25
- D $2.50

3. What is the maximum value of the function $f(x, y) = x^2 + 2y$ for the polygonal region bounded by this system of inequalities?

$$4y - 20 \le x$$
$$x \le 4$$
$$x + y \ge 0$$

4. Graph the system of equations. Then use the graph to find the solution of the system.
$$y = 2.5x - 4 \qquad y = -0.5x + 8$$

5. Solve the system of equations: $\begin{array}{l} 3x - 2y = 4 \\ x + y = 3 \end{array}$

- A $(2, 0)$
- B $(2, 1)$
- C $(2, -1)$
- D $(-2, -1)$

6. The school music department sold a total of 225 tickets to the holiday concert. They sold adult tickets for $5 each and student tickets for $2 each and collected a total of $4726. Which system of equations can be solved to find the number each type of ticket sold?

- A $a + s = 225$, $a + s = 4726$
- B $a + s = 225$, $5a + 2s = 4726$
- C $5a + 2s = 225$, $a + s = 4726$
- D $5a + 2s = 225$, $5a + 2s = 4726$

7. Elton has two jobs and can work at most 32 total hours per week. He wants to earn a minimum of $200 per week. Which system of inequalities can be solved to find the possible combinations of hours worked at each job that will help him reach his goal?

Job	Dog Walker	Store Clerk
Hourly Pay	$9.00	$10.50

- A $x + y \le 32$
       $9x + 10.5y \ge 200$
- B $9x + 10.5y \le 32$
       $x + y \ge 200$
- C $x + y \ge 32$
       $9x + 10.5y \le 200$
- D $9x + 10.5y \ge 32$
       $x + y \le 200$

82 | Chapter 1 | Preparing for Assessment

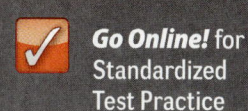
Go Online! for Standardized Test Practice

8. Which of the following equations represent lines that are either parallel or perpendicular to the line represented by the equation $y = 2x + 6$?
   - [ ] A $x = -\frac{1}{2}y - 6$
   - [ ] B $x = -2y + 6$
   - [ ] C $-\frac{1}{2}y = 4 - x$
   - [ ] D $2y + 2x = 4$
   - [ ] E $4y = 2x + 1$

9. What is the slope of a line passing through the points $(0, -5)$ and $(3, 1)$?

10. Which of the following scenarios could be represented by the equation $y = 3.5x + 10$?
    - [ ] A A car has 100 miles worth of gas left in its tank and consumes gas at a rate of 35 miles per gallon.
    - [ ] B A jogger jogs at an average rate of 3.5 miles per hour for 10 days while training for a marathon.
    - [ ] C A fabric company charges a $10 shipping fee, plus $3.50 per linear yard of fabric.
    - [ ] D A teacher makes customized T-shirts for each of her 35 students in 10 hours.
    - [ ] E A child receives a weekly allowance of $10 and gets an additional $3.50 for each time they wash the dishes after meals.

11. A factory makes two types of office chairs, wheeled and stationary. It takes 2.1 hours to assemble a wheeled chair and 1.4 hours to assemble a stationary chair. There are 84 hours of labor available each day. The factory makes a profit of $23 on each wheeled chair and $14 on each stationary chair. What is the maximum profit the factory will generate if they produce at least 12 stationary chairs?
    - ( ) A $864
    - ( ) B $888
    - ( ) C $904
    - ( ) D $920

12. Solve the inequality $-\frac{2}{3}x + 1 > 5$ and graph the solution on a number line.

13. Solve the system of equations:
    $2y + 3x = 38$
    $y - 2x = 12$

14. What is the slope of a line perpendicular to the line with equation $2y + 3x = 38$?

15. The velocity of an object fired directly upward is given by $V = 80 - 32t$, where t is in seconds.

    When will the velocity be between 32 and 64 feet per second?

Need Extra Help?

If you missed Question...	1	2	3	4	5	6	7	8	9	10	11	12	13	14	15
Go to Lesson...	1-7	1-9	1-8	1-7	1-9	1-6	1-7	1-4	1-3	1-1	1-8	1-9	1-6	1-4	1-5

connectED.mcgraw-hill.com

CHAPTER 2

Relations and Functions

THEN

You solved equations and inequalities.

NOW

You will:

- Use equations of relations and functions.
- Determine the slope of a line.
- Use scatter plots and prediction equations.
- Graph linear inequalities.

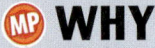

MP WHY

RECREATION Linear functions can be used to model many aspects of recreational activities, including the amount it would cost a group of people to attend the state fair.

Use the Mathematical Practices to complete the activity.

1. **Use Tools** You want to know how much it would cost your FFA club to go to the state fair. Use the KWL Chart tool to record and organize what you know, what you want to know, and what you learn.

Topic FFA State Fair Trip

What I Know	What I Want to Know	What I Learned
Total going =50	Total cost for group, including admission and parking	

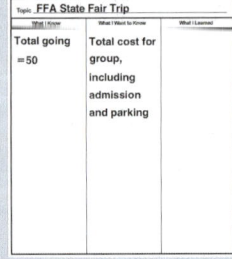

2. **Use Tools** Use the Internet to find the cost of attending the state fair. Is there a group discount? What is the cost of parking?
3. **Applying Math** Write an equation to determine the total cost of your group attending the state fair.
4. **Reasoning** What fee should you charge each person attending?

Glow Images

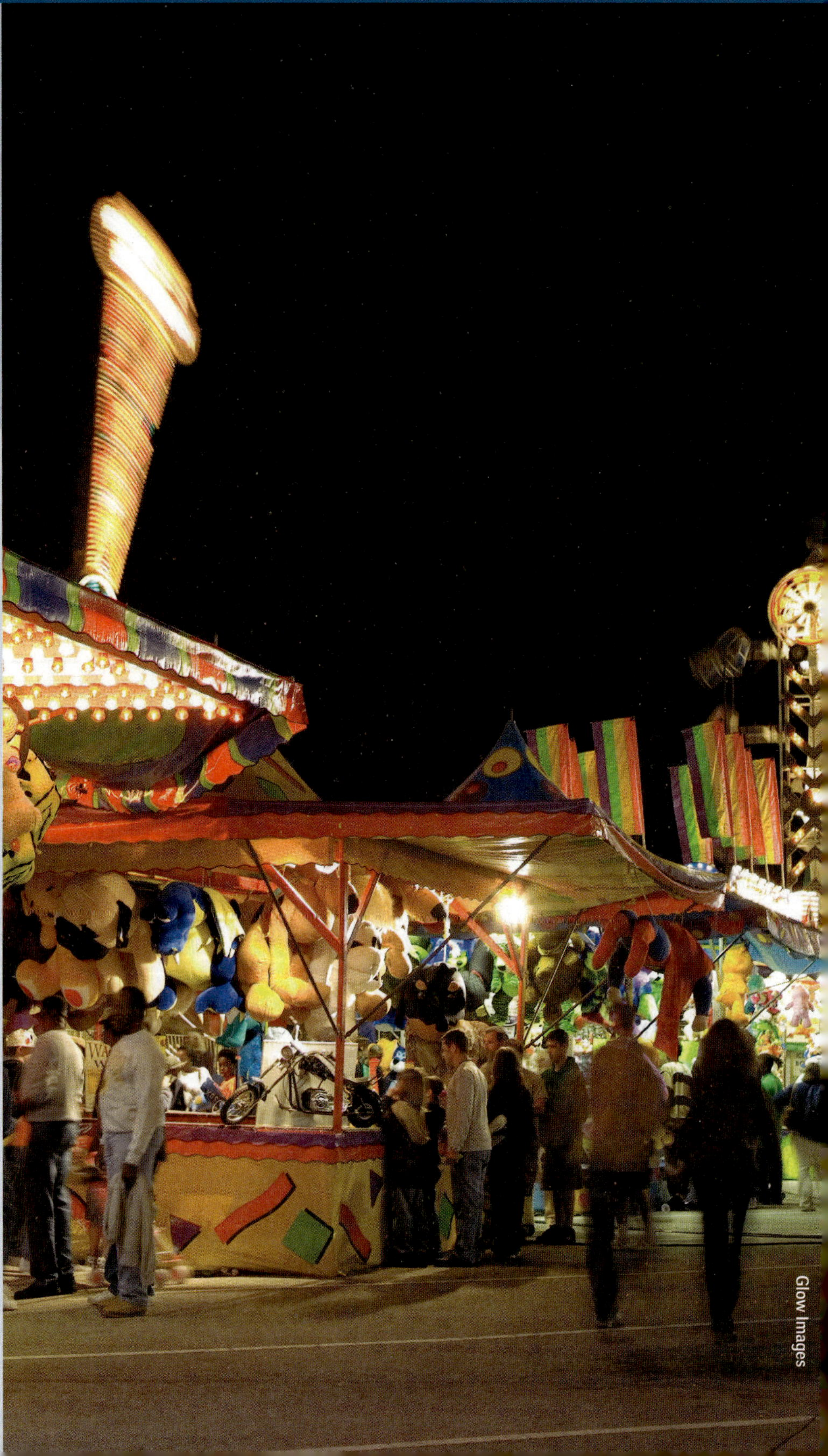

 ## *Go Online* to Guide Your Learning

Explore & Explain

 Graphing Tools

Use the **Graphing Tools** in ConnectED to explore how changing parameters affects the graph of a function or an inequality.

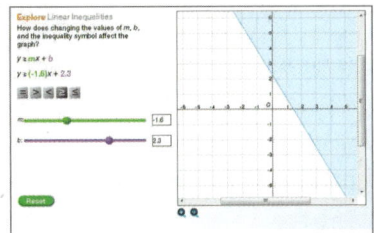

The Geometer's Sketchpad

Use **The Geometer's Sketchpad** to illustrate how to graph relations and functions. Use it to graph and analyze polynomial functions in Lesson 2-3 and to explore the piecewise, step, and absolute value functions discussed in Lesson 2-5.

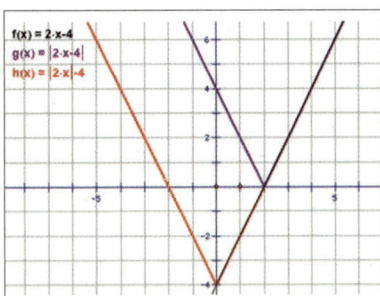

Organize

 Foldables

Get organized! Create a **Relations and Functions Foldable** before you start the chapter to organize your notes about relations and functions.

Collaborate

Chapter Project

In the **Getting Healthy** project, you will use what you have learned about rate of change and writing equations to complete a project that addresses health literacy.

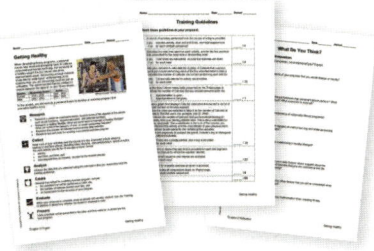

eBook

Interactive Student Guide

Before starting the chapter, answer the **Chapter Focus** preview questions. Check your answers as you complete each lesson. At the end of the chapter, try the **Performance Task**.

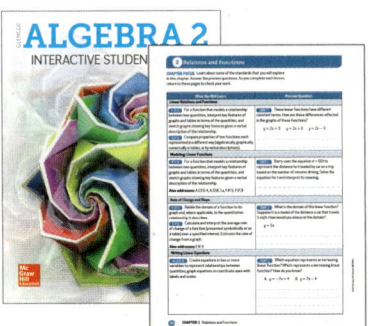

Focus

 LEARNSMART®

Need help studying? Complete the **Modeling with Functions** domain in LearnSmart to review for the chapter test.

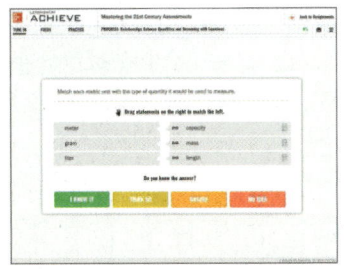

 ALEKS®

You can use the **Lines and Functions** topic in ALEKS to explore what you know about relations and functions and what you are ready to learn.*

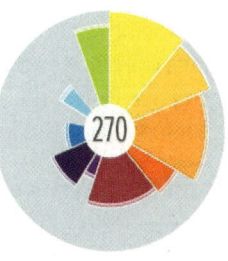

* Ask your teacher if this is part of your program.

connectED.mcgraw-hill.com

Get Ready for the Chapter

Connecting Concepts

Concept Check
Review the concepts used in this chapter by answering the questions below.

1. Given the coordinate plane shown, how would you find the *x*-coordinate of the ordered pair represented by the point *D*?

2. Given the coordinate plane shown, how would you find the *y*-coordinate of the ordered pair represented by the point *D*?

3. In the coordinate plane shown, in what quadrant is point *E*?

4. In the coordinate plane shown, what is the name for the location of point *C*?

5. In the coordinate plane shown, in what quadrant is point *B*?

6. In the coordinate plane shown, are the coordinates in the ordered pair representing point *A* positive, negative, or a combination?

7. Given $3x + 6y = 24$, what would be the first step to solve for *y*?

8. Given $\frac{2a + 4b}{c}$, where $a = -3$, $b = 4$, and $c = -2$, what would be the first step to evaluate the expression?

9. After you evaluate the expression in question 8, would you have a positive or negative number?

Performance Task Preview

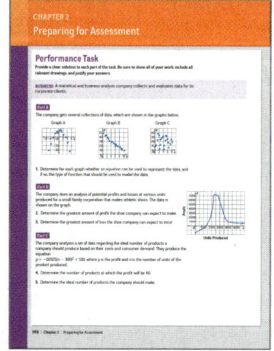

You can use the concepts and skills in the chapter to evaluate data for the corporate clients of a statistical and business analysis company. Understanding relations and functions will help you finish the Performance Task at the end of the chapter.

MP In this Performance Task you will:
- make sense of problems and persevere in solving them
- model with mathematics
- attend to precision

New Vocabulary

English		Español
one-to-one function	p. 87	función biunívoca
onto function	p. 87	sobre la función
discrete relation	p. 88	relación discreta
continuous relation	p. 88	relación continua
vertical line test	p. 88	prueba de la recta vertical
independent variable	p. 89	variable independiente
dependent variable	p. 89	variable dependiente
function notation	p. 89	notación funcional
linear equation	p. 95	ecuación lineal
linear function	p. 95	función lineal
end behavior	p. 103	comportamiento final
relative maximum	p. 105	comportamiento final
turning point	p. 105	máximo relativo
piecewise-defined function	p. 118	función por trozos-definida
piecewise-linear function	p. 119	función a intervalos lineal
step function	p. 119	función etapa
greatest integer function	p. 119	función del máximo entero
absolute value function	p. 120	función del valor absoluto
parent graph	p. 125	gráfica madre
parent function	p. 125	función del padre
line of reflection	p. 126	linea de la reflexión
dilation	p. 126	homotecia
x-intercept	p. 133	intersección *x*
y-intercept	p. 133	intersección *y*

Review Vocabulary

equation ecuación a mathematical sentence stating that two mathematical expressions are equal

function función a relation in which each *x*-coordinate is paired with exactly one *y*-coordinate

relation relación a set of ordered pairs

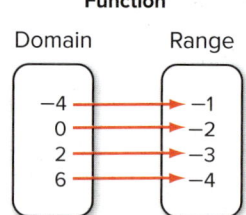

LESSON 1
Functions and Continuity

:Then
- You identified domains and ranges for given situations.

:Now
1. Determine whether functions are one-to-one and/or onto.
2. Determine whether functions are discrete or continuous.

:Why?
- The table shows the monthly average low and high temperatures for Charlotte, North Carolina. Each month's average temperatures can be represented by the ordered pair (average low, average high). For example, January's average temperatures can be expressed as (32, 51).

Monthly Average Temperature (°F) Charlotte, NC						
Month	Jan	Feb	Mar	Apr	May	June
Low	32	34	42	49	58	66
High	51	56	64	73	80	87
Month	Jul	Aug	Sep	Oct	Nov	Dec
Low	71	69	63	51	42	35
High	90	88	82	73	63	54

Source: The Weather Channel

New Vocabulary
one-to-one function
onto function
discrete relation
continuous relation
vertical line test
independent variable
dependent variable
function notation
codomain

Mathematical Practices
1 Make sense of problems and persevere in solving them.
2 Reason abstractly and quantitatively.
4 Model with mathematics.
7 Look for and make use of structure.

1 Relations and Functions
Recall that a function is a relation in which each element of the domain is paired with exactly one element in the range. All functions map elements of the domain to elements of the range, but they may differ in the way the elements of the domain and range are paired.

Key Concept Functions

one-to-one function

Each element of the domain pairs to exactly one unique element of the range.

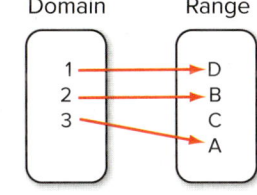

onto function

Each element of the range corresponds to an element of the domain.

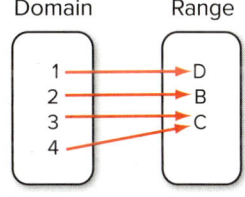

both one-to-one and onto

Each element of the domain is paired to exactly one element of the range, and each element of the range corresponds to a unique element of the domain.

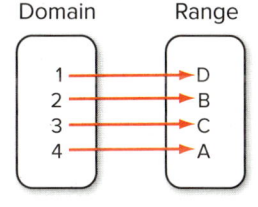

Example 1 Domain and Range

State the domain and range of each relation. Then determine whether each relation is a *function*. If it is a function, determine if it is *one-to-one*, *onto*, *both*, or *neither*.

a. $\{(-6, -1), (-5, -9), (-3, -7), (-1, 7), (6, -9)\}$

Domain: $\{-6, -5, -3, -1, 6\}$ Range: $\{-9, -7, -1, 7\}$

function: Yes, because each element of the domain is paired with one element of the range.

one-to-one: No, because each element of the domain is not paired with a unique element of the range.

onto: Yes, because each element of the range corresponds to an element of the domain.

connectED.mcgraw-hill.com   87

b.

x	2	−1	−2	−1	2
y	−2	−1	0	1	2

Domain: {−2, −1, 2}   Range: {−2, −1, 0, 1, 2}

The relation is not a function because 2 is mapped to both −2 and 2, and −1 is mapped to both −1 and 1.

▶ **Guided Practice**

State the domain and range of each relation. Then determine whether each relation is a *function*. If it is a function, determine if it is *one-to-one*, *onto*, *both*, or *neither*.

1A. 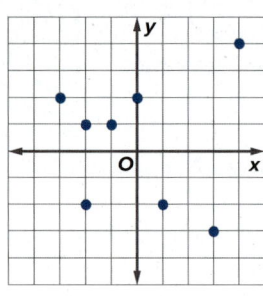   1B.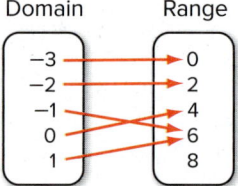

2 Discrete and Continuous Functions

A relation in which the domain consists of a set of individual elements and the graph is a set of unconnected points is said to be a **discrete relation**. When the domain of a relation has an infinite number of elements and the relation can be graphed with a line or smooth curve, the relation is a **continuous relation**.

Study Tip

MP Structure Notice that Graph A is composed of individual, or discrete, points while Graph B continues from one point to the next with no gaps.

Graph A 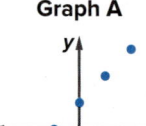   Graph B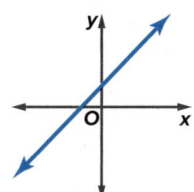

discrete relation   continuous relation

With both discrete and continuous graphs, you can use the **vertical line test** to determine whether the relation is a function.

Go Online!

Explore relations and how to identify functions using **The Geometer's Sketchpad®** activity in ConnectED.

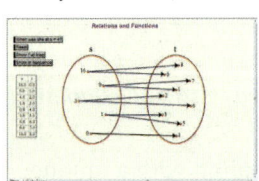

Key Concept Vertical Line Test

Words If no vertical line intersects a graph in more than one point, the graph represents a function.   If a vertical line intersects a graph in two or more points, the graph does not represent a function.

Models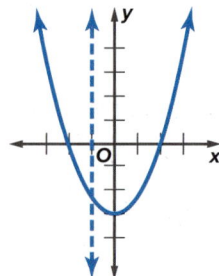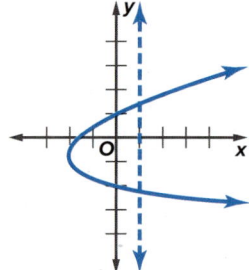

88 | Lesson 2-1 | Functions and Continuity

Relations and functions can also be represented by equations. The solutions of an equation in x and y are the set of ordered pairs (x, y) that make the equation true. To determine whether an equation represents a function, it is often simplest to look at the graph of the relation.

Example 2 Graph a Relation

Graph $y = \frac{1}{2}x - 3$, and determine the domain and range. Then determine whether the equation is a *function*, is *one-to-one*, *onto*, *both*, or *neither*. State whether it is *discrete* or *continuous*.

Make a table of values that satisfy the equation. Then graph the equation.

Every real number is the x-coordinate of some point on the line, and every real number is the y-coordinate of some point on the line. So the domain and range are both all real numbers.

x	y
−4	−5
−2	−4
0	−3
2	−2
4	−1

The graph passes the vertical line test, so the equation is a function. Every x-value is paired with exactly one unique y-value, and every y-value corresponds to an x-value. Thus, the function is both one-to-one and onto.

Because the graph is a solid line without breaks, the function is continuous.

▶ **Guided Practice**

2. Graph $y = x^2 + 1$, and determine the domain and range. Then determine whether the equation is a *function*, is *one-to-one*, *onto*, *both*, or *neither*. State whether it is *discrete* or *continuous*.

When an equation represents a function, the variable, often x, with values making up the domain is called the **independent variable**. The other variable, often y, is called the **dependent variable** because its values depend on x.

Reading Math

MP Structure The symbol $f(x)$ replaces the y and is read "f of x." The f is just the name of the function. It is not a variable that is multiplied by x.

Equations that represent functions are often written in **function notation**. The equation $y = 5x - 1$ can be written as $f(x) = 5x - 1$. Suppose you want to find the value in the range that corresponds to the element -6 in the domain of the function. The value $f(-6)$ is found by substituting -6 for each x in the equation. Therefore, $f(-6) = 5(-6) - 1$ or -31.

Example 3 Evaluate a Function

Given $f(x) = 2x^2 - 8$, find each value.

a. $f(6)$

$f(x) = 2x^2 - 8$ Original function
$f(6) = 2(6)^2 - 8$ Substitute.
$ = 2(36) - 8$ Evaluate 6^2.
$ = 72 - 8$ or 64 Simplify.

b. $f(2y)$

$f(x) = 2x^2 - 8$ Original function
$f(2y) = 2(2y)^2 - 8$ Substitute.
$ = 2(4y^2) - 8$ $(2y)^2 = 2^2 y^2$
$ = 8y^2 - 8$ Simplify.

▶ **Guided Practice**

Given $g(x) = 0.5x^2 - 5x + 3.5$, find each value.

3A. $g(2.8)$ **3B.** $g(4a)$

Real-World Link
The first Tour de France, held in 1903, was 2428 kilometers long and was completed at an average speed of 25.679 km/h. In 2013, the 100th Tour stretched 3403.5 kilometers and was finished at an average of 40.545 km/h.

Source: letour.fr

Real-World Example 4   Discrete and Continuous Functions

BICYCLING The graph shows the length of the Tour de France in kilometers each year from 2000 through 2013. Is the relation *discrete* or *continuous*? Does the graph represent a function?

Because the graph consists of distinct points, the function is discrete. Use the vertical line test. No vertical line can be drawn that contains more than one of the data points. Therefore, the relation is a function.

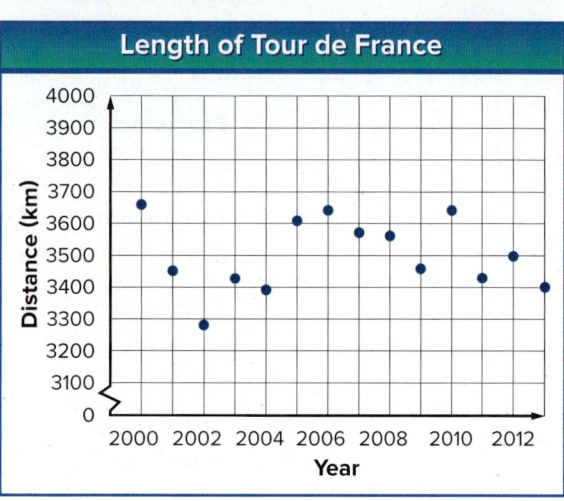

Guided Practice

4. The number of employees a company had in each year from 2010 to 2015 were 25, 28, 34, 31, 27, and 29. Graph this information and determine whether the relation is *discrete* or *continuous*. Does the graph represent a function?

A function that represents real-world data may be more accurately modeled as a discrete graph, or as a continuous graph, but usually not both.

Example 5   Choose the Correct Model

The Yogurt Shack sells prepacked cups of frozen yogurt for $2. The cost of x cups can be described by the function $y = 2x$, where y is the total cost in dollars. Determine whether the function is correctly modeled by a discrete or continuous function. Explain your reasoning.

If a person buys 2 cups of frozen yogurt, it will cost $4. Three cups will cost $6, 4 cups will cost $8, and so on. You cannot go into the Yogurt Shack and purchase 1.5 cups or 2.25 cups if the store is selling prepacked cups. Since the domain consists only of whole number, this is correctly modeled by a discrete function and the graph will consist of the set of unconnected points (0, 0), (1, 2), (2, 4), (3, 6), (4, 8), ... as shown.

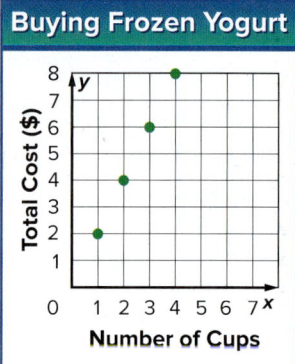

When deciding whether a real-world situation is modeled by a discrete or continuous function, consider whether an interval of all real numbers makes sense as part of the domain.

Guided Practice

5A. A graph consists of the number of hours x a car travels and the distance y the car travels. Determine whether the function is correctly modeled by a discrete or continuous function. Explain your reasoning.

5B. A graph consists of the number of rides x on a roller coaster and the number of riders who have ridden on the roller coaster y. Determine whether the function is correctly modeled by a discrete or continuous function. Explain your reasoning.

Check Your Understanding

= Step—by-Step Solutions begin on page R11.

Example 1 **STRUCTURE** State the domain and range of each relation. Then determine whether each relation is a *function*. If it is a function, determine if it is *one-to-one*, *onto*, *both*, or *neither*.

1.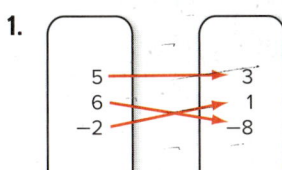

2.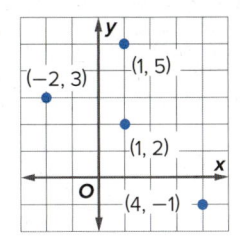

3.
x	y
−2	−4
1	−4
4	−2
8	6

4. {(−3, 4), (2, −1), (−2, −1), (6, 2), (5, 4)}

Example 2 Graph each equation, and determine the domain and range. Determine whether the equation is a *function*, is *one-to-one*, *onto*, *both*, or *neither*. Then state whether it is *discrete* or *continuous*.

5. $y = 5x + 4$ 6. $y = -4x - 2$ 7. $y = 3x^2$ 8. $x = 7$

Example 3 9. **BASKETBALL** The table shows the average points per game for a professional basketball player for four seasons.

a. Assume the ages are the domain. Identify the domain and range.

b. Write a relation of ordered pairs for the data.

c. State whether the relation is *discrete* or *continuous*.

d. Graph the relation. Is this relation a function?

Season	Player's Age	Average Points per Game
2012–2013	25	22.9
2013–2014	26	24.0
2014–2015	27	23.8
2015–2016	28	30.1

Example 4 Evaluate each function.

10. $f(-3)$ if $f(x) = -4x - 8$

11. $g(5)$ if $g(x) = -2x^2 - 4x + 1$

Example 5 Determine whether each function is correctly modeled using a discrete or continuous function. Explain your reasoning.

12. **Converting Units**

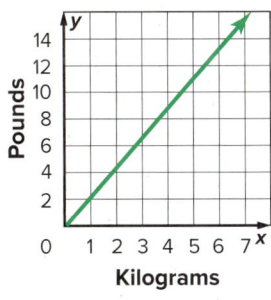

13. **T-shirts Received**

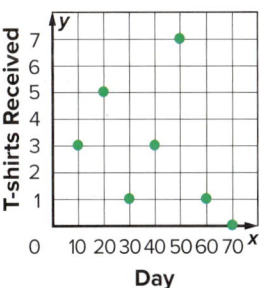

Practice and Problem Solving

Extra Practice is on page R2.

Example 1 State the domain and range of each relation. Then determine whether each relation is a *function*. If it is a function, determine if it is *one-to-one*, *onto*, *both*, or *neither*.

14.
x	y
−0.3	−6
0.4	−3
1.2	−1
1.2	4

15.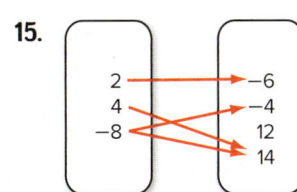

16. {(−3, −4), (−1, 0), (3, 0), (5, 3)}

Example 2 **MP STRUCTURE** Graph each equation, and determine the domain and range. Determine whether the equation is a *function*, is *one-to-one*, *onto*, *both*, or *neither*. Then state whether it is *discrete* or *continuous*.

17. $y = -3x + 2$
18. $y = 0.5x - 3$
19. $y = 2x^2$
20. $y = -5x^2$
21. $y = 4x^2 - 8$
22. $y = -3x^3 - 1$

Example 3 Evaluate each function.

23. $f(-8)$ if $f(x) = 5x^3 + 1$
24. $f(2.5)$ if $f(x) = 16x^2$

Example 4 25. **POLITICS** The table below shows the population of several states and the number of U.S. representatives from those states.

   a. Make a graph of the data with population on the horizontal axis and representatives on the vertical axis.
   b. Identify the domain and range.
   c. Is the relation *discrete* or *continuous*?
   d. Does the graph represent a function? Explain your reasoning.

State	Population (millions)	Number of Representatives
California	37.69	53
Florida	19.06	27
Illinois	12.87	18
New York	19.47	27
North Carolina	9.66	13
Texas	25.67	36

Source: U.S. Bureau of the Census

Example 5 26. **WRITING IN MATH** Give an example of a real-world function that is discrete and a real-world function that is continuous. Explain your reasoning.

27. **DIVING** The table below shows the pressure on a diver at various depths.

Depth (ft)	0	20	40	60	80	100
Pressure (atm)	1	1.6	2.2	2.8	3.4	4

   a. Write a relation to represent the data.
   b. Graph the relation.
   c. Identify the domain and range. Is the relation *discrete* or *continuous*?
   d. Is the relation a function? Explain your reasoning.

Find each value if $f(x) = 3x + 2$, $g(x) = -2x^2$, and $h(x) = -4x^2 - 2x + 5$.

28. $f(-5)$
29. $f(9)$
30. $g(-3)$
31. $g(-6)$
32. $h(3)$
33. $h(8)$
34. $f\left(\dfrac{2}{3}\right)$
35. $g\left(\dfrac{3}{2}\right)$
36. $h\left(\dfrac{1}{5}\right)$

37. A thrift store sells all clothing items for $3 each. The cost of x items can be described by the equation $y = 3x$.

   a. Should a graph of y be displayed as a continuous or discrete function? Explain.
   b. Graph the function using the domain {0, 1, 3, 6, 10}.

38. **MOVIES** Chaz has a collection of 15 movies downloaded on his media player. He decides to download 3 more movies each month. The function $M(t) = 15 + 3t$ counts the number of movies $M(t)$ he has after t months. How many movies will he have after 8 months?

39. Suppose $f(x)$ is a discrete function. Is the domain always the set of integers? Explain.

40. If the domain of a function is the set of real numbers, can the function be discrete? Explain.

41. **MULTIPLE REPRESENTATIONS** In this problem you will investigate one-to-one and onto functions.

   a. **Graphical** Graph each function on a separate graphing calculator screen.
   $$f(x) = x^2 \qquad g(x) = 2^x \qquad h(x) = x^3 - 3x^2 - 5x + 6 \qquad j(x) = x^3$$

   b. **Tabular** Use the graphs to create a table showing the number of times a horizontal line could intersect the graph of each function. List all possibilities.

   c. **Analytical** For a function to be one-to-one, a horizontal line on the graph of the function can intersect the function at most once. Which functions meet this condition? Which do not? Explain your reasoning.

   d. **Analytical** For a function to be onto, every possible horizontal line on the graph of the function must intersect the function at least once. Which functions meet this condition? Which do not? Explain your reasoning.

   e. **Graphical** Create a table showing whether each function is one-to-one and/or onto.

H.O.T. Problems   Use Higher-Order Thinking Skills

42. **CRITIQUE ARGUMENTS** Omar and Madison are finding $f(3d)$ for the function $f(x) = -4x^2 - 2x + 1$. Is either of them correct? Explain your reasoning.

Omar	Madison
$f(3d) = -4(3d)^2 - 2(3d) + 1$	$f(3d) = -4(3d)^2 - 2(3d) + 1$
$= -4(9d^2) - 6d + 1$	$= 12d^2 - 6d + 1$
$= -36d^2 - 6d + 1$	

43. **CHALLENGE** Consider the functions $f(x)$ and $g(x)$. $f(a) = 19$ and $g(a) = 33$, while $f(b) = 31$ and $g(b) = 51$. If $a = 5$ and $b = 8$, find two possible functions to represent $f(x)$ and $g(x)$.

44. **REASONING** If the graph of a relation crosses the y-axis at more than one point, is the relation *sometimes*, *always*, or *never* a function? Explain your reasoning.

45. **OPEN-ENDED** Graph a relation that can be used to represent each of the following.

   a. the height of a baseball that is hit into the outfield
   b. the speed of a car that travels to the store, stopping at two lights along the way
   c. the height of a person from age 5 to age 80
   d. the temperature on a typical day from 6 A.M. to 11 P.M.

46. **REASONING** Determine whether the following statement is *true* or *false*. Explain your reasoning.

   If a function is onto, then it must be one-to-one as well.

47. The range is part of the *codomain*, which is the set of all possible y-values, or the set of all possible values to which the function could map the domain. The codomain is assumed to be all real numbers unless otherwise stated. The diagram at the right shows one pairing of domain and codomain that is one-to-one and onto. Draw another pairing of the domain and codomain so that the pairing is neither one-to-one nor onto.

Preparing for Assessment

48. The graph below shows the function $f(x) = -x^2 + 4x + 5$. **MP** 1,2

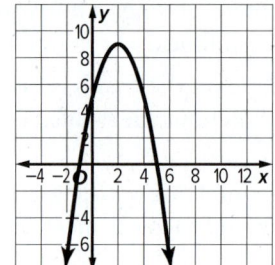

a. What are the domain and range of the function shown in the graph?

- A D = {all real numbers}, R = {all real numbers}
- B D = {all real numbers}, R = {y | y ≤ 9}
- C D = {x | x ≤ 2}, R = {y | y ≤ 9}
- D D = {x | x ≥ 2}, R = {y | y ≥ 9}

b. Tell whether the function is *one-to-one*, *onto*, or *both*.

c. Tell whether the graph of the function is continuous or discrete. Explain.

d. **MULTI-SELECT** Which of the following statements are true? Select all solutions.

- A $f(0) = 0$
- B $f(4) = 5$
- C $f(6) = 7$
- D $f(-1) = 0$
- E $f(2) = 9$
- F $f(-4) = 10$

49. Let $f(x) = -2x^2 + 2x$. What is $f(2a)$? **MP** 1

- A $-16a^2 + 4a$
- B $-8a^2 + 4a$
- C $-4a^2 + 4a$
- D $-2a^2 + 2a$
- E $16a^2 + 4a$

50. The table shows the amounts that five customers paid for bags of apples of various weights. **MP** 4

Weight (lb)	Cost ($)
2	1.60
4.5	3.60
6	4.80
1.8	1.44
3.4	2.72

Mitul noted the function $c = 0.8w$ can be used to find the cost c, in dollars, for w pounds of apples. What are the domain and range of Mitul's function?

- A D = {1.8, 2, 3.4, 4.5, 6}, R = {1.44, 1.6, 2.72, 3.6, 4.8}
- B D = {w | 1.8 ≤ w ≤ 4.8}, R = {c | 1.44 ≤ c ≤ 6}
- C D = {w | w ≥ 0}, R = {c | c ≥ 0}
- D D = {all real numbers}, R = {all real numbers}

51. Let the domain of $g(x)$ be the interval (p, q). Given $g(-3) = 2$, which of the following must be true? **MP** 1

  I. $p < -3 < q$
  II. $p ≤ 2 ≤ q$
  III. $g(2) = -3$

- A I only
- B II only
- C III only
- D I and II only
- E I, II, and III

52. Which pricing system could be modeled by a discrete function? Select all solutions. Explain your choice(s). **MP** 1, 4

- A An ice cream shop sells any amount of ice cream for $0.95 per ounce.
- B An ice cream shop sells ice cream and toppings for $5.00 per cup.
- C An ice cream shop sells ice cream for $1.20 per ounce and toppings for $0.55 per ounce.
- D none of these

53. For which domain would $f(x) = x$ be discrete? **MP** 1

- A D = {1, 2, 3, ...}
- B D = {x | x ≥ 2}
- C D = {all real numbers}
- D D = {x | 0 < x < 10}

54. Draw a graph of a real-life situation that is modeled by a discrete function. Explain why the function is discrete. **MP** 1

94 | Lesson 2-1 | Functions and Continuity

LESSON 2
Linearity and Symmetry

::Then
- You analyzed continuity of functions.

::Now
1. Identify linear and nonlinear functions by examining equations or graphs.
2. Determine whether graphs of functions have line or point symmetry.

::Why?
- Laura does yard work to earn money during the summer. She either cuts grass x or does general gardening y, and she schedules 5 jobs per day. The equation $x + y = 5$ can be used to relate how many of each task Laura can do in a day.

Yard Work

New Vocabulary

linear equation
linear function
nonlinear function
line symmetry
line of symmetry
point symmetry
point of symmetry

Mathematical Practices
3 Construct viable arguments and critique the reasoning of others.
7 Look for and make use of structure.

1 Identify Linear Functions
In a **linear function**, no variable is raised to a power other than 1. Any linear function can be written in the form $f(x) = mx + b$, where m and b are real numbers. Linear functions can be modeled by linear equations, which can be written in the form $Ax + By = C$. The graph of a linear equation is a straight line.

A function that is not linear is called a **nonlinear function**. The graph of a nonlinear function includes a set of points that cannot all lie on the same line. A nonlinear function cannot be written in the form $f(x) = mx + b$. A parabola is the graph of a quadratic function, which is a type of nonlinear function.

Key Concept Linear Functions

Words A linear function can be written in the form $f(x) = mx + b$ or $y = mx + b$, where m and b are real numbers.

Examples
$f(x) = -\frac{2}{3}x - 1$
$f(x) = \frac{1}{2}x$
$4x - 5y = 16$

Nonexamples
$f(x) = \frac{1}{x}$
$2x + 6y^2 = -25$
$x + xy = -\frac{5}{8}$

Example 1 Identify Linear Functions from Equations

State whether each function is a linear function. Write *yes* or *no*. Explain.

a. $3x + 2y = 8$

$3x + 2y = 8$ Original function
$2y = -3x + 8$ Subtract 3x from each side.
$y = \frac{-3x}{2} + \frac{8}{2}$ Divide each side by 2.
$y = -\frac{3}{2}x + 4$ Simplify.

Yes; the equation can be written as $y = -\frac{3}{2}x + 4$, where $m = -\frac{3}{2}$, and $b = 4$.

b. $f(x) = \dfrac{2}{x}$

No; the expression includes division by the variable.

c. $y = 3xy - 4$

No; the two variables are multiplied together.

Guided Practice

1A. $f(x) = \dfrac{5}{x+6}$

1B. $g(x) = -\dfrac{3}{2}x + \dfrac{1}{3}$

The graph of a linear function consists of a single straight line or a set of points that lie along a single straight line.

Real-World Example 2 Identify Linear Functions from Graphs

ECOLOGY The graph models one prediction about the population of the northern snakehead fish in a lake over time. The data in the table models a second prediction. State whether each is a linear function. Explain.

Real-World Link
The northern snakehead is an invasive species of fish that was first reported in the United States in 1997.
Source: United States Geological Survey

Model A (graph showing Population vs Time (years), curve through points)

Model B	
Time (years)	Population
0	100
1	125
2	150
3	175
4	200

For model A, draw a line through the first two points marked on the graph.

For model B, plot ordered pairs from the table. Then try to draw a line through all of the points.

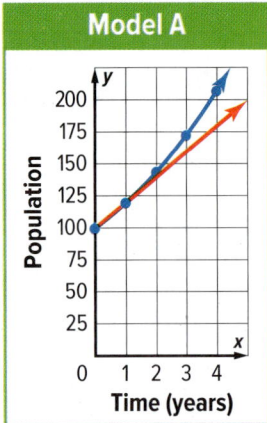

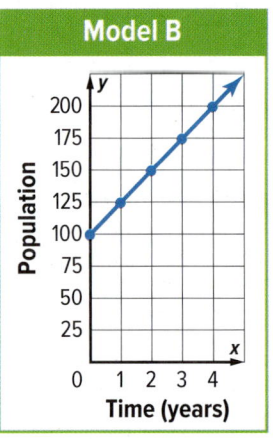

The line shows that the points on the graph of the function do not all lie single straight line. Model A is a nonlinear function.

All of the points lie on a single straight line. Model B is a linear function.

▶ **Guided Practice**

2. Shayla and her brother Malcolm both left home on their bikes at the same time. The graph shows a function modeling Shayla's distance from home over time. The table shows a function modeling Malcolm's distance from home over time.

   State whether each function is a linear function. Explain.

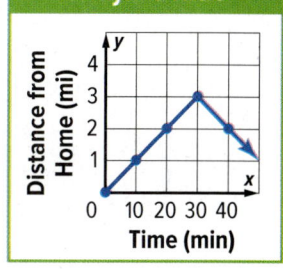

Malcolm's Ride					
Time (min)	0	10	20	30	40
Distance from Home (mi)	0	1.5	3	4.5	6

2 Identify Symmetry The graphs of some functions exhibit symmetry. The graph of a function has **line symmetry** if each half of the graph maps exactly to the other half. The line dividing the graphs into matching halves is called a **line of symmetry**.

Each point on one side is reflected in the line to a point equidistant from the line on the opposite side.

Example 3   Identify Line Symmetry

State whether the graph of each function has line symmetry. If so, identify the line of symmetry.

a.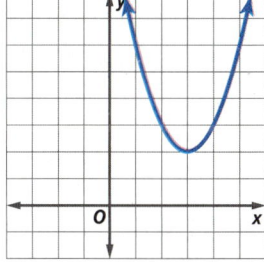

A vertical line divides the graph into two halves that are mirror images.

b.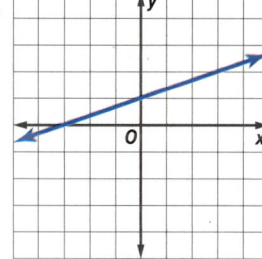

There is no vertical line that divides the graph into two halves that are mirror images.

> **Study Tip**
> **Structure** For a function with line symmetry, each point on one side of the line of symmetry has a corresponding point on the other side that is equidistant to the line of symmetry.

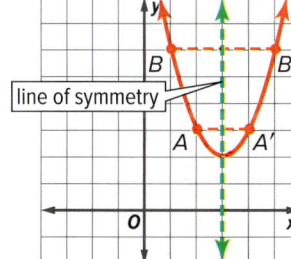

The graph has line symmetry with $x = 3$ as the line of symmetry.

The graph does not have line symmetry about a vertical line.

connectED.mcgraw-hill.com   97

▶ **Guided Practice**

State whether the graph of each function has line symmetry. If so, identify the line of symmetry.

3A.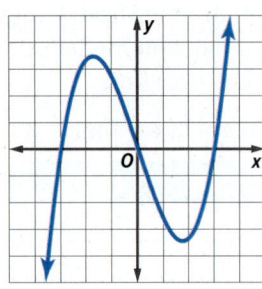

3B.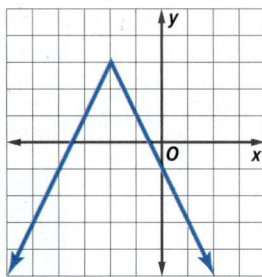

The graph of a function may also exhibit a type of symmetry called *point symmetry*.

The graph of a function has **point symmetry** if it can be rotated 180° about a point so that its rotation maps exactly onto itself. The point that acts as the center of the rotation is called the **point of symmetry**.

Example 4 **Identify Point Symmetry**

State whether the graph of each function has point symmetry. If so, identify the point or points of symmetry.

a.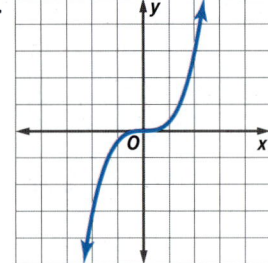

b.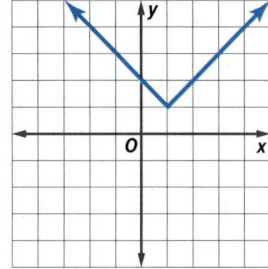

When the graph is rotated 180° about the origin, the rotation maps the original graph onto itself.

There is no point about which the graph can be rotated 180° so that it maps onto itself.

Study Tip

Structure For a function with point symmetry, each point on the graph on one side of the point of symmetry has a corresponding point on the other side so that the point of symmetry is the midpoint of the segment connecting the two.

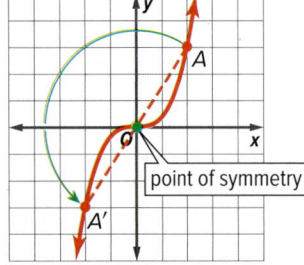

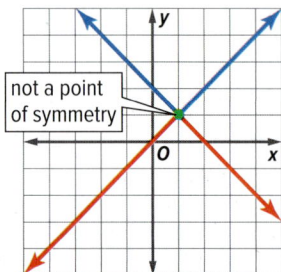

The graph has point symmetry with (0, 0) as the point of symmetry.

The graph does not have point symmetry.

98 | Lesson 2-2 | Linearity and Symmetry

Guided Practice

State whether the graph of each function has point symmetry. If so, identify the point or points of symmetry.

4A.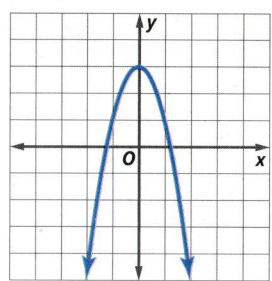

4B.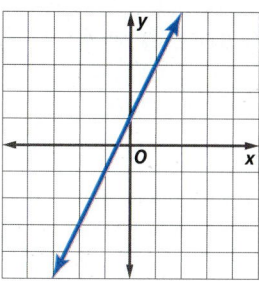

Check Your Understanding

= Step-by-Step Solutions begin on page R11.

Go Online! for a Self-Check Quiz

Example 1 State whether each function is a linear function. Write *yes* or *no*. Explain.

1. $f(x) = \frac{x + 12}{5}$
2. $g(x) = \frac{7 - x}{x}$
3. $y = 3x^2 - 4$
4. $12y = 4x + 8$

Example 2

5. **SAVINGS** Selena and Fiona are each saving money to buy a bicycle. The table shows how the total amount of Selena's savings has changed over time, and the graph shows how the total amount of Fiona's savings has changed over time. State whether each function is a linear function. Explain.

Selena's Savings					
Week	1	2	3	4	5
Total Saved ($)	10	25	40	55	70

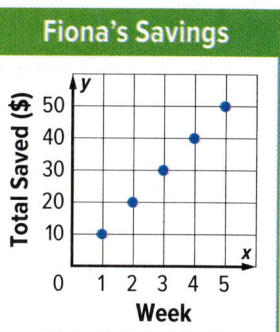

Fiona's Savings

Examples 3–4 State whether each graph has line symmetry or point symmetry. If so, identify any lines of symmetry or points of symmetry.

6.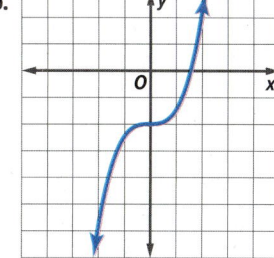

7.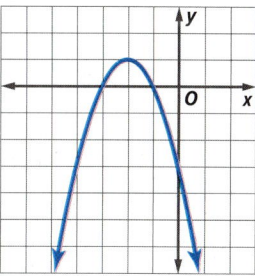

8.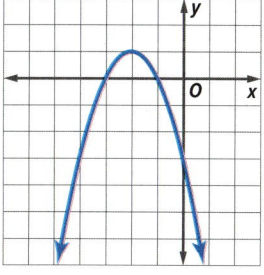

9.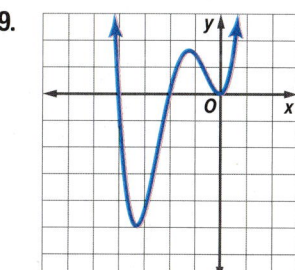

connectED.mcgraw-hill.com 99

Practice and Problem Solving

Extra Practice is on page R2.

Example 1 State whether each equation or function is a linear function. Write *yes* or *no*. Explain.

10. $3y - 4x = 20$
11. $y = x^2 - 6$
12. $h(x) = 6$
13. $j(x) = 2x^2 + 4x + 1$
14. $g(x) = 5 + \frac{6}{x}$
15. $f(x) = \sqrt{7 - x}$
16. $4x + \sqrt{y} = 12$
17. $\frac{1}{x} + \frac{1}{y} = 1$
18. $f(x) = \frac{4x}{5} + \frac{8}{3}$

Example 2

19. **CARS** Two cars are waiting at a stoplight. The graph and the table show how the speeds of the two cars change over time after the light turns green. State whether each function is a linear function. Explain.

Car B						
Time (s)	0	1	2	3	4	5
Speed (mi/h)	0	8	16	24	32	40

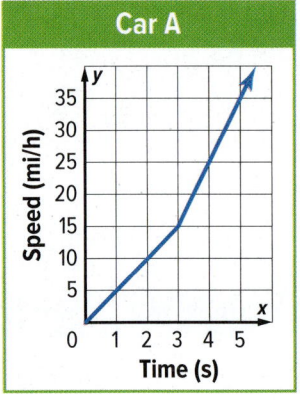

Car A

Examples 3-4 State whether each graph has line symmetry or point symmetry. If so, identify any lines of symmetry or points of symmetry.

20.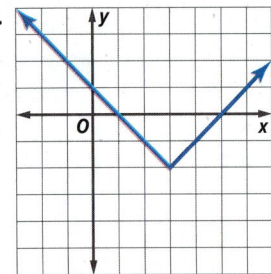

21.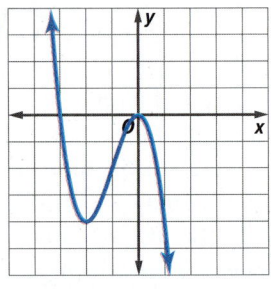

22.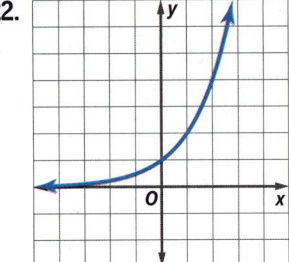

23.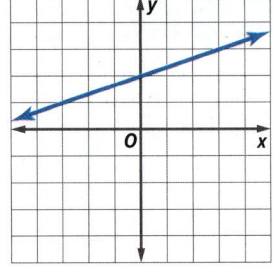

24. **MODELING** Latonya earns a commission of $25 for each 6-month gym membership that she sells and $45 for each one-year gym membership that she sells. Her goal is to earn a total of $1125 in commissions in the next two months. The equation $25s + 45\ell = 1125$ models this situation, where s is the number of 6-month memberships and ℓ is the number of one-year memberships Latonya sells.

   a. Graph the equation. Does this equation represent a linear function? Explain.
   b. If Latonya sells twelve 6-month memberships and sixteen 1-year memberships, will she meet her goal? Explain.

25. **ARCHITECTURE** The function $f(x) = -\frac{1}{3}|x - 12| + 4$ for the domain $0 \leq x \leq 24$ models the shape of a roof.

   a. Is the function linear? Justify your answer without graphing the function.
   b. Identify the line of symmetry of the function. What does the line of symmetry represent in this situation?

26. **SPORTS** The table shows how the height of a kicked football changes over time.

Height of Football									
Time (s), x	0	0.25	0.5	0.75	1	1.25	1.5	1.75	2
Height (ft), y	0	7	12	15	16	15	12	7	0

    a. Is the function shown in the table linear? Justify your answer.

    b. Describe any symmetry in the graph of the function. What does the symmetry indicate about the behavior of the ball?

Write each equation in the form $y = mx + b$. Identify m and b.

27. $\frac{x+5}{3} = -2y + 4$ 28. $\frac{4x-1}{5} = 8y - 12$ 29. $\frac{-2x-8}{3} = -12y + 18$

30. **FUND-RAISING** The Freshman Class Student Council wanted to raise money by giving car washes. The students spent $10 on supplies and charged $2 per car wash.

    a. Write an equation to model the situation.

    b. Graph the equation, and state whether it represents a linear function.

    c. How much money did they earn after 20 car washes?

    d. How many car washes are needed for them to earn $100?

Each coordinate plane represents half the graph of a function. On your own paper, sketch the other half of the graph based on the description of the symmetry of the function.

31. The origin is a point of symmetry.

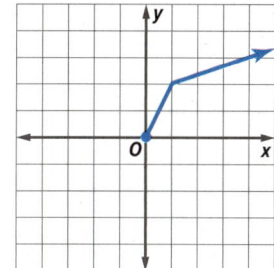

32. The y-axis is a line of symmetry.

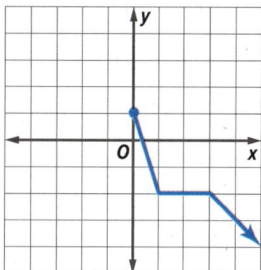

H.O.T. Problems   Use Higher-Order Thinking Skills

33. **REASONING** Determine whether an equation of the form $x = a$, where a is a constant, is *sometimes*, *always*, or *never* a function. Explain your reasoning.

34. **ERROR ANALYSIS** Of the four equations shown, identify the one that does not belong. Explain your reasoning.

$y = 2x + 3$

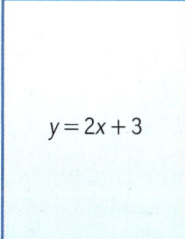

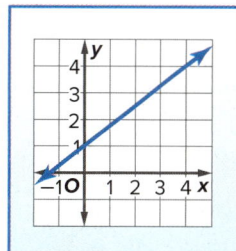

$y = 2xy$

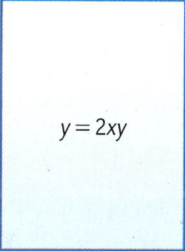

35. **WRITING IN MATH** Consider the graph of the relationship between hours worked and earnings.

    a. When would this graph represent a linear relationship? Explain your reasoning.

    b. Provide another example of a linear relationship in a real-world situation.

Preparing for Assessment

36. Which equations define a nonlinear function? Select all that apply. **MP 7**

- [ ] **A** $y = 5$
- [ ] **B** $y = \frac{7}{x} + 4$
- [ ] **C** $y = 2^x$
- [ ] **D** $y = 15 + 3x$
- [ ] **E** $y = \frac{x+3}{2}$

37. A swimming pool holds 1800 gallons of water. At the end of the summer, the pool must be drained for 3 hours. While the pool is drained, the number of gallons of water left in the pool g, after t minutes of draining, is modeled by the function $g = 1800 - 10t$. What are the domain and range of this function? **MP 2, 4**

- ○ **A** $D = \{t \mid t \leq 180\}$, $R = \{g \mid g \leq 1800\}$
- ○ **B** $D = \{t \mid t \geq 0\}$, $R = \{g \mid g \geq 0\}$
- ○ **C** $D = \{t \mid 0 \leq t \leq 180\}$, $R = \{g \mid 0 \leq g \leq 1800\}$
- ○ **D** $D = \{$all real numbers$\}$, $R = \{$all real numbers$\}$

38. Write the equation of the line of symmetry of the function shown in the graph. **MP 7**

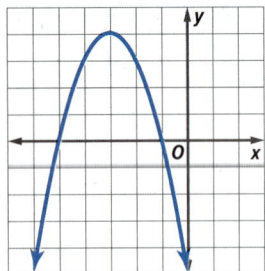

- ○ **A** $x = -5$
- ○ **B** $x = -3$
- ○ **C** $x = -1$
- ○ **D** $x = 0$
- ○ **E** $y = 0$
- ○ **F** $y = 4$

39. Which is a point of symmetry of the function shown in the graph? **MP 7**

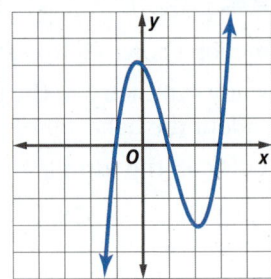

- ○ **A** $(-1, 0)$
- ○ **B** $(0, 3)$
- ○ **C** $(1, 0)$
- ○ **D** $(3, 0)$

40. MULTI-STEP Nick played in two tennis tournaments. The graph shows the number of players left after each round in the first tournament. The table shows the number of players left after each round in the second tournament. **MP 2, 3**

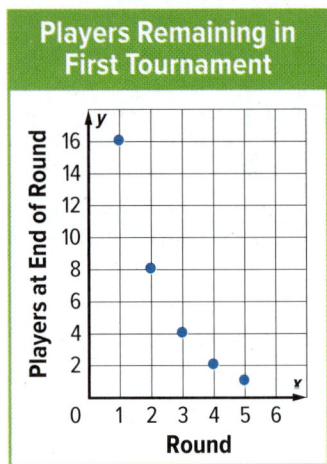

Players Remaining in Second Tournament						
Round	1	2	3	4	5	6
Players at End of Round	16	13	10	7	4	1

a. Are the functions shown in the graph and the table linear or nonlinear? Explain.

b. Compare and contrast the behavior of the two functions.

LESSON 3
Extrema and End Behavior

::Then
- You graphed continuous functions.

::Now
1. Identify the end behavior of graphs.
2. Identify extrema of functions.

::Why?
- The relationship between the speed of a car in miles per hour and the distance in feet it takes for the car to stop after braking can be modeled by the graph of a nonlinear function. Do you think the speed of the car in miles per hour and the braking distance can have unlimited values?

New Vocabulary
end behavior
relative maximum
relative minimum
turning points
extrema

Mathematical Practices
1 Make sense of problems.
2 Reason abstractly and quantitatively.
4 Model with mathematics.
5 Use appropriate tools strategically.
7 Make use of structure.

1 End Behavior of Graphs of Functions

End behavior is the behavior of a graph as x approaches positive or negative infinity. At the right end, the values of x are increasing toward infinity. This is denoted as $x \to +\infty$. At the left end, the values of x are decreasing toward negative infinity. This is denoted as $x \to -\infty$. For the graphs below, as $x \to +\infty$, $f(x)$ increases, so $f(x) \to +\infty$, and as $x \to -\infty$, $f(x)$ decreases, so $f(x) \to -\infty$. These graphs all have the same end behavior.

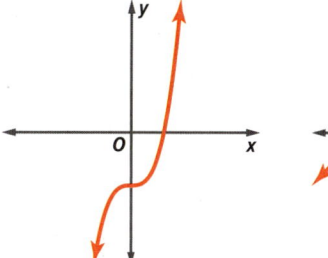 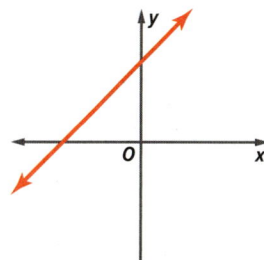 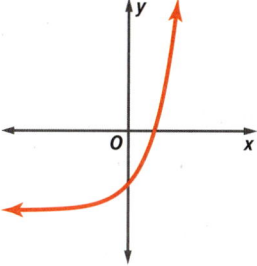

Example 1 End Behavior of Linear Functions

Describe the end behavior of each linear function graph.

a.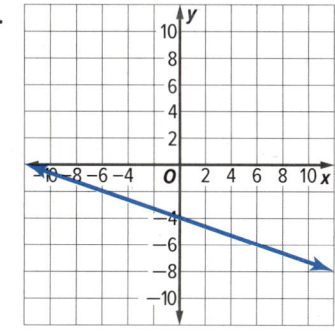

b.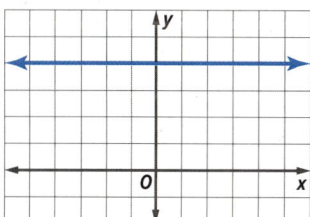

Step 1 Study the behavior on the right.
As x increases, $f(x)$ decreases.
As $x \to +\infty$, $f(x) \to -\infty$.

Step 2 Study the behavior on the left.
As x decreases, $f(x)$ increases.
As $x \to -\infty$, $f(x) \to +\infty$.

Step 3 Describe the end behavior.
As $x \to +\infty$, $f(x) \to -\infty$, and as $x \to -\infty$, $f(x) \to +\infty$.

Step 1 Study the behavior on the right.
As x increases, $g(x)$ is constant.
As $x \to +\infty$, $g(x) = 4$.

Step 2 Study the behavior on the left.
As x decreases, $g(x)$ is constant.
As $x \to -\infty$, $g(x) = 4$.

Step 3 Describe the end behavior.
As $x \to +\infty$, $g(x) = 4$, and as $x \to -\infty$, $g(x) = 4$.

connectED.mcgraw-hill.com 103

▶ **Guided Practice**

Describe the end behavior of each linear function graph.

1A.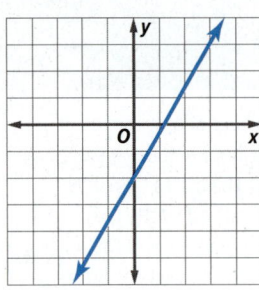

1B.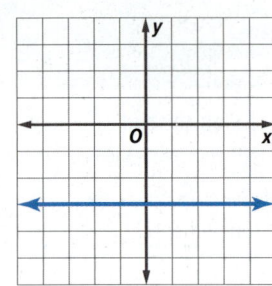

To determine end behavior, you do not need to consider the values of the graph closer to the origin. Instead, focus on the graph's behavior at the ends of the graph. Even though two functions have very different graphic representations, they can have the same end behavior, as shown in the graphs to the right.

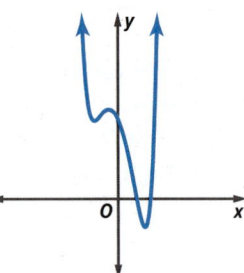 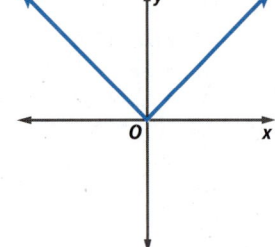

Review Vocabulary
infinity endless or boundless

Example 2 End Behavior of Nonlinear Functions

Describe the end behavior of each nonlinear function graph.

a.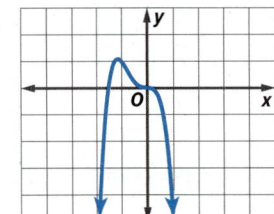

b.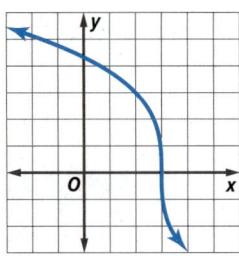

Step 1 Study the behavior on the right.
As x increases, $f(x)$ decreases.
As $x \to +\infty$, $f(x) \to -\infty$.

Step 2 Study the behavior on the left.
As x decreases, $f(x)$ decreases.
As $x \to -\infty$, $f(x) \to -\infty$.

Step 3 Describe the end behavior.
As $x \to +\infty$, $f(x) \to -\infty$ and as $x \to -\infty$, $f(x) \to -\infty$.

Step 1 Study the behavior on the right.
As x increases, $f(x)$ decreases.
As $x \to +\infty$, $f(x) \to -\infty$.

Step 2 Study the behavior on the left.
As x decreases, $f(x)$ increases.
As $x \to -\infty$, $f(x) \to +\infty$.

Step 3 Describe the end behavior.
As $x \to +\infty$, $f(x) \to -\infty$ and as $x \to -\infty$, $f(x) \to +\infty$.

Study Tip

MP Sense-Making End behavior refers to the behavior of a graph as the x-values approach infinity or negative infinity. There may be no actual maximum or minimum value, but a relative maximum or minimum value can be found within a certain interval of x-values.

▶ **Guided Practice**

Describe the end behavior of each nonlinear function graph.

2A.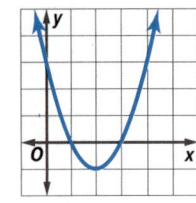

2B.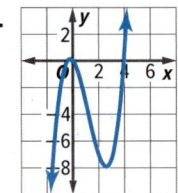

2 Extrema of Functions

Sometimes the behavior of a function closer to the origin is important to know. The graph below shows a nonlinear function with coordinates close to the origin.

The point A is a **relative maximum** of the function since no other nearby points have a greater y-coordinate. The graph is increasing as it approaches A and decreasing as it moves away from A.

The point B is a **relative minimum** of the function since no other nearby points have a lesser y-coordinate. The graph is decreasing as it approaches B and increasing as it moves away from B.

The relative maxima and relative minima are often referred to as **turning points**, or the points where a curve completely changes direction from either down to up or up to down. For some functions, you can predict how many turning points to expect by knowing the degree of the function. Knowing the number of turning points is helpful when sketching the graphs of some functions, which you will do in Chapter 4.

The relative minima, relative maxima, and turning points are known as the **extrema** of the graph.

Example 3 Zeros and Extrema of a Graph

The table and graph below are of a function with extrema. Estimate the zeros. Then estimate the coordinates at which relative maxima and minima occur.

Study Tip

MP Use Tools You can use a graphing calculator to explore the graphs of a variety of functions. Use the CALC menu to find the relative maximum and relative minimum values of a function as well as the zeros and the points where graphs intersect.

x	f(x)
−2	7
−1	2
0	3
1	−2
2	−1
3	42

- zero between 0 and 1
- indicates a relative minimum
- indicates a relative maximum
- zero between 2 and 3

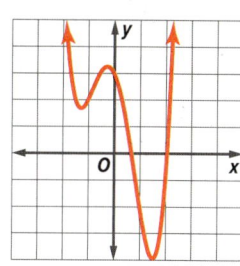

Look at the table of values and the graph.

The value of f(x) changes signs between x = 0 and x = 1, indicating a zero of the function. There must be a zero *near x = 1*.

The value of f(x) changes signs between x = 2 and x = 3, indicating another zero of the function. There must be a zero *near x = 2*.

The values of f(x) at x = −2 and x = −1 are greater than at a point between them, indicating a relative minimum between −2 and −1. There must be a relative minimum *near x = −1*.

The values of f(x) at x = −1 and x = 0 are less than at a point between them, indicating a relative maximum between −1 and 0. There must be a relative maximum *near x = 0*.

The values of f(x) at x = 1 and x = 2 are greater than at a point between them, indicating a relative minimum between 1 and 2. There must be a relative minimum *near x = 2*.

Guided Practice

3.

x	f(x)
−2	14
−1	0
0	−2
1	2
2	6
3	4
4	−10

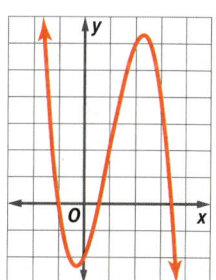

> **Study Tip**
> **Rational Values** Zeros and turning points will not always occur at integral values of *x*.

The value of $f(x)$ changes sign between $x = -1$ and $x = 0$, indicating a zero of the function. From the table, you can see that there is a zero *at* $x = -1$.

The values of $f(x)$ at $x = -1$ and $x = 0$ are greater than at a point between them, indicating a relative minimum between -1 and 0. There must be a relative minimum *near* $x = 0$.

The value of $f(x)$ changes sign between $x = 0$ and $x = 1$, indicating a zero of the function. There must be a zero *near* $x = 1$.

The values of $f(x)$ at $x = 2$ and $x = 3$ are less than at a point between them, indicating a relative maximum between 2 and 3. There must be a relative maximum *near* $x = 2$.

The value of $f(x)$ changes sign between $x = 3$ and $x = 4$, indicating a zero of the function. There must be a zero *near* $x = 3$.

Real-World Example 4   Find End Behavior and Extrema

FOOD PRICES The cost of an orange changes over the course of a year, as shown in the table for each month starting in January and in the graph of cost over time as a function. Use the table and graph to estimate the extrema for this function. Then explain the extrema in the context of the situation.

Month	Cost ($)
1	0.60
2	0.70
3	1.00
4	0.90
5	1.00
6	0.90
7	0.80
8	0.70
9	0.70
10	0.40
11	0.50
12	0.60

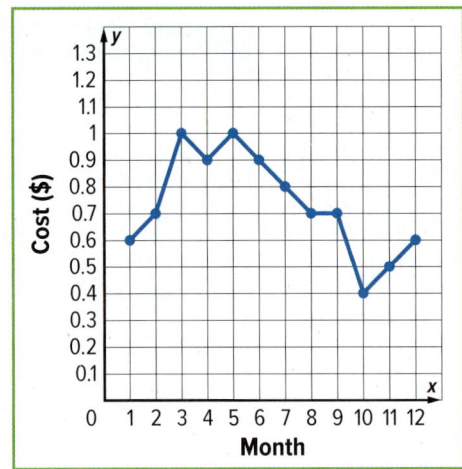

Extrema for Table
The costs in February and April are less than the cost in March, so March, which is represented by the point (3, 1.00), is a relative maximum. Similarly, the point (5, 1.00) for May is a relative maximum.
The costs in March and May are greater than the cost in April, so April, which is represented by the point (4, 0.90), is a relative minimum. Similarly, the point (10, 0.40) for October is a relative minimum.

Extrema for Graph
The function is increasing from $x = 1$ to $x = 3$ and decreasing from $x = 3$ to $x = 4$. Therefore, the value of $f(x)$ where $x = 3$ represents a relative maximum. Similarly, the value of $f(x)$ where $x = 5$ represents a relative maximum.
The function is decreasing from $x = 3$ to $x = 4$ and increasing from $x = 4$ to $x = 5$. Therefore, the value of $f(x)$ where $x = 4$ represents a relative minimum. Similarly, the value of $f(x)$ where $x = 10$ represents a relative minimum.
Because March and May are relative maxima, they are the months when oranges are the most expensive. Because October is a relative minimum, it is the month oranges are the cheapest.

▶ **Guided Practice**

4. **ENTERTAINMENT** The average cost of a theater ticket at the local cinema changes over the course of a year, as shown in the table for the months starting in January and in the graph of cost over time as a function. Use the table and graph to estimate the extrema for this function. Then explain the extrema in the context of the situation.

Month	Cost ($)
1	10
2	12
3	12
4	14
5	11
6	12
7	13
8	14
9	9
10	10
11	11
12	12

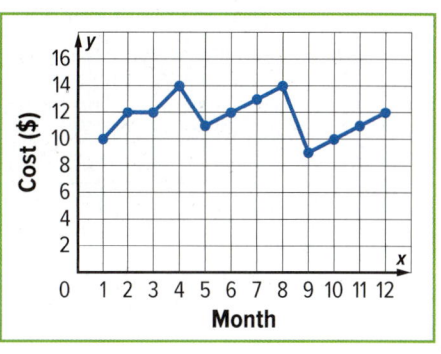

Check Your Understanding

● = Step-by-Step Solutions begin on page R11.

Go Online! for a Self-Check Quiz

Example 1 Describe the end behavior of each linear function.

1. 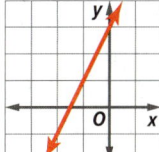   2. 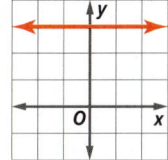   3.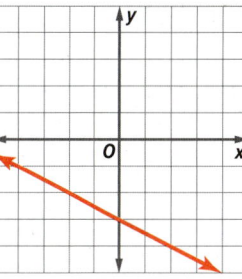

Example 2 Describe the end behavior of each nonlinear function.

4. 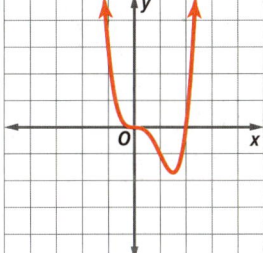   5. 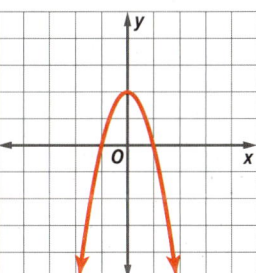   6.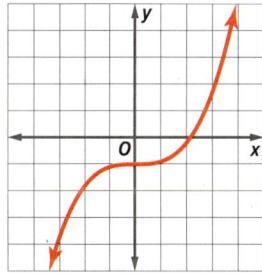

Example 3

7. The table and graph below are of a function with extrema. Estimate the zeros. Then estimate the coordinates at which relative maxima and minima occur.

Example 4

8. Harold and Marnie are selling produce for one day at a farmers' market. The table and graph model the expected number of pounds of produce sold per hour. Use the table and graph to estimate the extrema for this function. Then explain the extrema in the context of the situation.

x	f(x)
−3	−18
−2	−1
−1	4
0	3
1	2
2	7
3	24

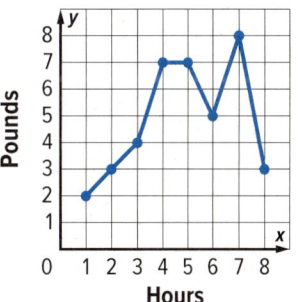

Hour	Pounds
1	2
2	3
3	4
4	7
5	7
6	5
7	8
8	3

connectED.mcgraw-hill.com 107

Practice and Problem Solving

Extra Practice is on page R2.

Example 1 Use the graphs to describe the end behavior of each linear function.

9.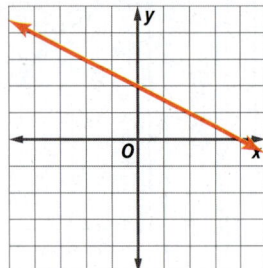
10.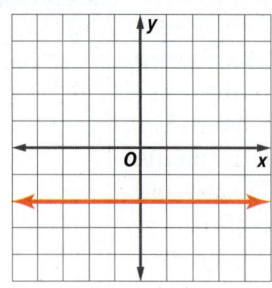
11.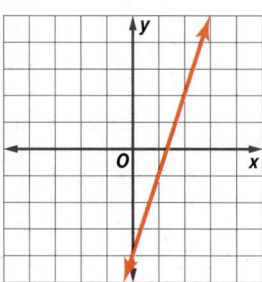

Example 2 Use the graphs to describe the end behavior of each nonlinear function.

12.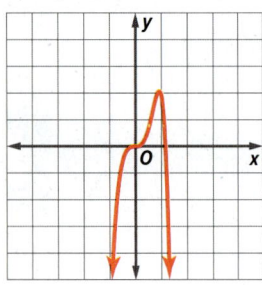
13.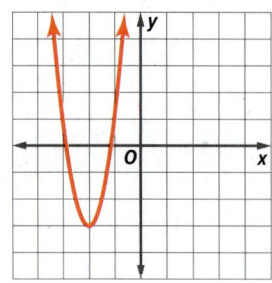
14.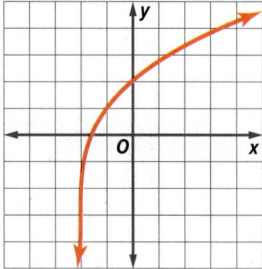

Example 3

15. The table and graph are of a function with extrema. Estimate the zeros. Then estimate the coordinates at which relative maxima and/or minima occur.

x	f(x)
−3	−18
−2	−1
−1	4
0	3
1	2
2	7
3	24

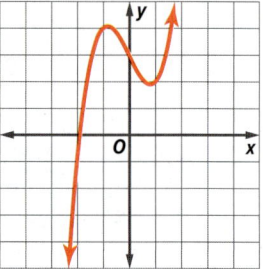

Example 4

16. **SAVINGS** Amira's parents have a college savings account for her. The graph represents the balance in the account over time.

    a. What was the initial balance of the account?
    b. Describe the end behavior of the graph as $x \to +\infty$.
    c. Is there a turning point of the graph? Explain.
    d. What trend(s) do the graph suggest?
    e. Will the trend(s) continue indefinitely? Explain.

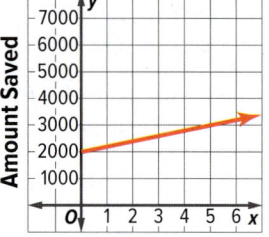

17. **ERROR ANALYSIS** Finn thinks that the graph shows a function with no extrema. Mina disagrees. Who is correct? Explain your reasoning.

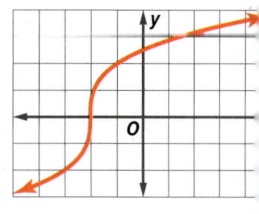

18. Consider the function graphs below.

A 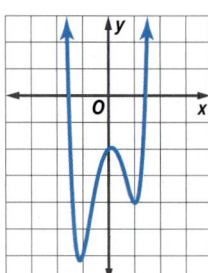   B 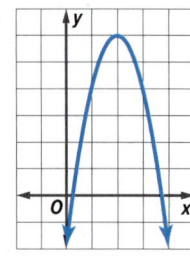   C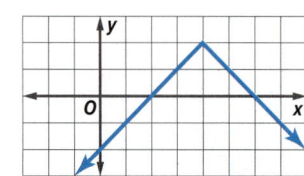

   a. Which of the graphs appear to have the same end behavior? Justify your reasoning.
   b. Do any of the graphs appear to have a relative minimum (minima)? If so, approximate the value of the relative minimum (minima).
   c. Do any of the graphs appear to have turning point(s)? If so, approximate the value(s) of the turning points. Justify your reasoning.

108 | Lesson 2-3 | Extrema and End Behavior

19. For which function is it true that, as $x \to \infty$ and as $x \to -\infty$, the values of $f(x)$ remain constant? Explain.

   A B C

20. **LIFE EXPECTANCY** The life expectancy at birth in the United States changes each year. The table and graph show the life expectancy figures, with year 0 representing 1980.

Year	Age
0	62.8
10	65.5
20	67.7
30	70.6
40	75.4

   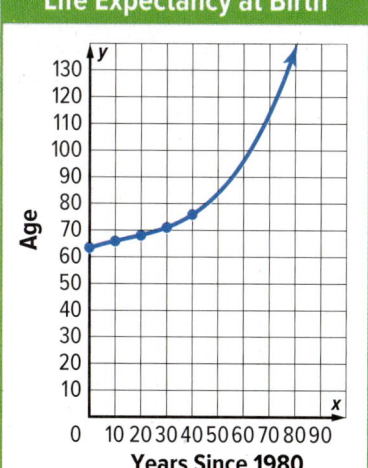

   Life Expectancy at Birth

   a. What is the y-intercept of the graph, and what does it represent?
   b. Describe the end behavior of the graph.
   c. Is there a turning point for the graph? Explain.
   d. What trends in life expectancy do the graph and table suggest? Is it reasonable that the trends will continue indefinitely?

21. **OPEN-ENDED** Sketch at least four different examples of linear and nonlinear functions that have the same end behavior as each other.

H.O.T. Problems   Use Higher-Order Thinking Skills

22. **REASONING** Carlo and Jin are selling their photographs at an art fair. The function models the expected earnings given the number of photographs sold.

    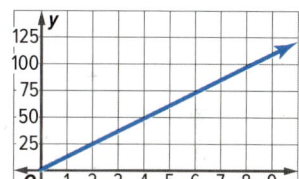

    a. Describe the end behavior of $f(x)$ as $x \to +\infty$.
    b. Describe the end behavior of $f(x)$ as $x \to -\infty$. Explain.
    c. Carlo and Jin are currently selling each photograph for $12. Would the end behavior of the function change if they decided to sell the photographs for $10 each instead? Explain your reasoning.

23. **REASONING** If a function is symmetric about a vertical line, what must be true about the end behavior of $f(x)$ as $x \to \infty$ and as $x \to -\infty$? Explain.

24. **REASONING** Do all linear functions have the opposite end behavior on the right side and on the left side of the graph? Explain.

25. **WRITING IN MATH** How would you describe point P of function $g(x)$ if g is increasing as it approaches point P from the left and decreasing as g moves away from P on the right? Explain.

26. **ERROR ANALYSIS** Damon claims that he can draw a function with two relative minima, but no relative maxima. Saul says this is impossible. Who is correct? Explain your reasoning.

27. **CHALLENGE** Use a graphing calculator to explore the graphs of $f(x) = x$, $g(x) = x^2$, $h(x) = x^3$, $k(x) = x^4$, and $j(x) = x^5$. How does the end behavior of the graphs change as the power of the x increases?

28. **OPEN-ENDED** Sketch at least three different examples of linear and nonlinear functions that do not have any extrema.

Preparing for Assessment

29. Use the graph of the function. **MP** 2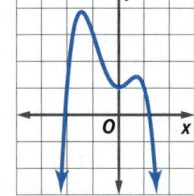

  a. Which statement describes the relative extrema of the function? Select all that apply.

    ☐ **A** relative minimum between $x = -2$ and $x = -1$

    ☐ **B** relative minimum between $x = -1$ and $x = 1$

    ☐ **C** relative maximum between $x = -2$ and $x = -1$

    ☐ **D** relative maximum between $x = 0$ and $x = 1$

  b. Which of the following statements are true for the graph in part **a**? Select all solutions.

    ☐ **A** As $x \to +\infty$, $f(x) \to +\infty$.

    ☐ **B** As $x \to -\infty$, $f(x) \to -\infty$.

    ☐ **C** As $x \to +\infty$, $f(x) \to -\infty$.

    ☐ **D** As $x \to -\infty$, $f(x) \to +\infty$.

30. To prepare for college, Randall saves $100 per week from his paycheck. The graph shows the amount he has saved over time. **MP** 2

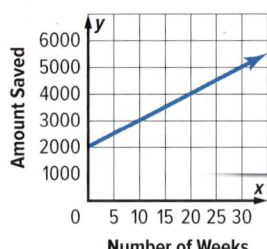

Which choice describes the end behavior of the function in the graph?

A As $x \to +\infty$, $f(x) \to +\infty$ and as $x \to -\infty$, $f(x) \to -\infty$.

B As $x \to +\infty$, $f(x) \to +\infty$ and as $x \to -\infty$, $f(x) \to +\infty$.

C As $x \to +\infty$, $f(x) \to +\infty$ and $f(x)$ is not defined for $x \leq 0$.

D As $x \to +\infty$, $f(x) \to -\infty$ and as $x \to -\infty$, $f(x) \to +\infty$.

31. Use the graph of the function. **MP** 2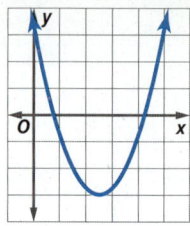

  a. Identify the extrema of the graph. Justify your reasoning.

  b. Describe the end behavior of the graph.

  c. Estimate the zeros of the graph.

32. MULTI-STEP The average costs for full-time undergraduate students at a four-year college in the United States can be modeled by the graph and table below, where x is the number of years since 2008, and y is the cost. **MP** 4

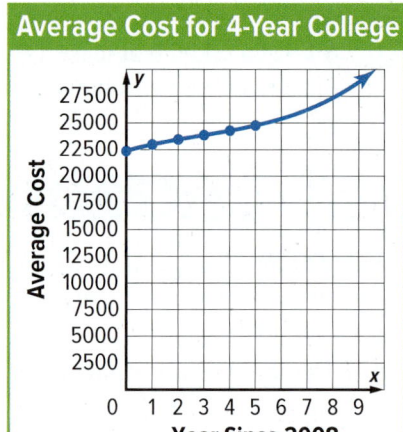

x	y ($)
0	22,340
1	22,903
2	23,460
3	23,759
4	24,245
5	24,706

  a. What is the y-intercept of the graph?

    ○ **A** 0

    ○ **B** 22,340

    ○ **C** 22,500

    ○ **D** 30,000

  b. What does the y-intercept represent?

  c. Describe the end behavior of the graph.

  d. Is there a turning point for the graph? Explain.

  e. What trends in college costs does the graph and table suggest?

  f. Is it reasonable that the trends will continue indefinitely?

LESSON 4
Sketching Graphs of Functions

::Then
- You analyzed characteristics of functions.

::Now
1. Use the key features of functions to sketch graphs of linear functions.
2. Use the key features of functions to sketch graphs of nonlinear functions.

::Why?
A group of hikers walk steadily uphill, then stop for lunch at a vista point. After lunch, they hike back downhill. You can use this information to sketch a graph of the hikers' elevation as a function of time.

MP Mathematical Practices
2. Reason abstractly and quantitatively.
7. Look for and make use of structure.

1 Sketch Graphs of Linear Functions
You can use key features of a function to sketch the graph of the function. Key features include intercepts, symmetry, end behavior, intervals where the function is increasing or decreasing, and intervals where the function is positive or negative. Not all features are needed to create a graph, and the appearance of a graph may vary depending on the scale you choose for the axes.

Key Concept Key Features for Graphing a Linear Function

x-intercept	the x-coordinate of a point where a graph crosses the x-axis
y-intercept	the y-coordinate of a point where a graph crosses the y-axis
Positive	A function is positive on a portion of its domain where its graph lies above the x-axis.
Negative	A function is negative on a portion of its domain where its graph lies below the x-axis.
Increasing	As the x-values increase, the value of the function increases.
Decreasing	As the x-values increase, the value of the function decreases.
End behavior	describes how the values of the function behave as $x \to \infty$ and as $x \to -\infty$

Example 1   Sketch a Linear Graph

Use the given key features to sketch a linear graph. The y-intercept is -70. The function is positive for $x < -30$. The function is decreasing for all values of x. As $x \to \infty$, $f(x) \to -\infty$ and as $x \to -\infty$, $f(x) \to \infty$.

Use the y-intercept to plot the point $(0, -70)$.

Because the sign of the function changes when $x = -30$, the x-intercept must be -30. Plot $(-30, 0)$.

Use these points to sketch the graph. The graph is a straight line passing through the x- and y-intercepts.

Check that the graph matches the other given key features.

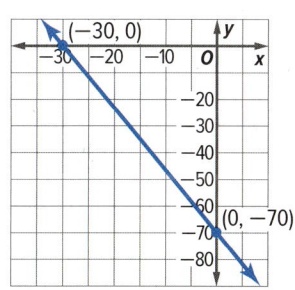

▶ **Guided Practice**

1. Use the given key features to sketch a linear graph. The y-intercept is -3. The function is positive for $x > 1.5$. The function is increasing for all values of x. The domain is $-2 \leq x \leq 3$. The range is $-7 \leq y \leq 3$.

connectED.mcgraw-hill.com 111

2 Sketch Graphs of Nonlinear Functions
The key features for sketching the graph of a nonlinear function include all of the key features for linear functions, as well as some additional information that may be provided.

Key Concept   Key Features for Graphing a Nonlinear Function

Continuity   A function is continuous if it can be graphed with connected lines or connected smooth curves.

Extrema   the maximum and minimum values of a function

Symmetry   A function has reflectional symmetry if there is a vertical line that divides the graph of the function into two halves that are mirror images of each other.

Sketching the graph of a nonlinear function is similar to sketching the graph of a linear function. First plot points based on the known key features of the graph. Then draw lines or curves to connect these points if the graph is known to be continuous.

Example 2   Sketch a Nonlinear Graph

Use the given key features to sketch a nonlinear graph.

a. The y-intercept is -5. The function is continuous. The function is positive for $x < -3$ and $x > 4$. The function has a minimum at $(2, -6)$. The function is increasing for $x > 2$. As $x \to \infty$, $f(x) \to \infty$ and as $x \to -\infty$, $f(x) \to \infty$.

Use the y-intercept to plot the point $(0, -5)$.

Because the function has a minimum at $(2, -6)$, this point is also on the graph. Plot the point $(2, -6)$.

Because the sign of the function changes when $x = -3$, there is an x-intercept at $x = -3$. Plot $(-3, 0)$.

Because the sign of the function changes when $x = 4$, there is an x-intercept at $x = 4$. Plot $(4, 0)$.

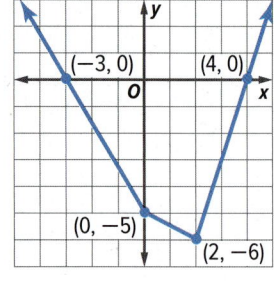

Sketch a graph passing through the points. Note that more than one graph may be possible. Check that the graph matches the other given key features.

Study Tip

MP Sense-Making Before sketching the graph of a function, consider the scales on the axes that best fit the given information.

b. The function is continuous and symmetric about the line $x = -1$. The function has a maximum at $(-1, 4)$. As $x \to \infty$, $f(x) \to -\infty$ and as $x \to -\infty$, $f(x) \to -\infty$.

Plot the maximum at $(-1, 4)$.

Draw the line of symmetry, $x = -1$.

A downward-opening parabola with vertex at $(-1, 4)$ satisfies all of the given key features.

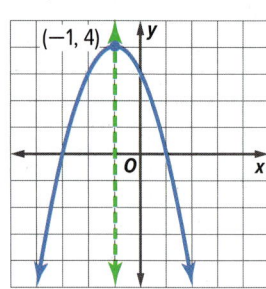

Guided Practice

2. Use the given key features to sketch a nonlinear graph.

   A. The y-intercept is 2. The function is negative for $x < -1$ and $x > 4$. The function has a maximum at $(3, 3)$.

   B. The function is symmetric about the line $x = 2$. The function has a minimum and an x-intercept at $x = 2$. As $x \to \infty$, $f(x) \to \infty$ and as $x \to -\infty$, $f(x) \to \infty$.

You can use what you know about sketching graphs based on their key features to sketch a graph based on real-world information.

Real-World Example 3 Sketch a Real-World Function

RACING Use the given key features to sketch a graph.

Jose decides to go to a test track to drive a racing car. He drives the car for a little over 5 minutes.

y-intercept: Jose starts driving at 0 miles per hour.

Linear or nonlinear: The function that models the situation is nonlinear.

Extrema: Jose's maximum speed is 160 miles per hour, which he reaches 30 seconds after he starts driving.

Increasing: Jose's speed increases steadily for the first 30 seconds.

Decreasing: At the 2-minute mark, Jose decreases his speed for 0.25 minute, then he stays at 100 miles per hour for 2.75 minutes. At the 5-minute mark, he again decreases his speed for 30 seconds until he reaches a stop.

End Behavior: The graph has no end behavior since Jose starts at 0 miles per hour and ends at 0 miles per hour.

> **Watch Out!**
>
> **Graphs** The graph showing speed as a function of time does *not* represent the path of the car.

Jose drives for a total of 5.5 minutes. His speed ranges from 0 miles per hour to 150 miles per hour. Use this information to set up the scales on the x- and y-axes of the graph.

The y-intercept is 0, so plot (0, 0).

There is an x-intercept that represents the end of the drive at (5.5, 0).

There is an increasing interval for $0 \leq x \leq 0.5$.

There are decreasing intervals for $2 \leq x \leq 2.25$ and $5 \leq x \leq 5.5$.

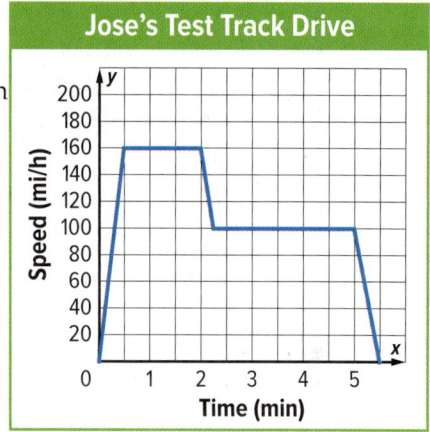

During times when the car's speed is not increasing or decreasing, the speed is constant. These periods are represented by horizontal segments in the graph.

Guided Practice

3A. WALKING TOUR Kenisha follows a walking tour for 18 minutes.

> **y-intercept:** Kenisha starts at 0 miles per hour.
>
> **Linear or nonlinear:** The function that models the situation is nonlinear.
>
> **Extrema:** Kenisha's maximum speed is 2 miles per hour, which she reaches at 1 minute. She maintains that speed for the next 15 minutes.
>
> **Decreasing:** For the last 2 minutes, Kenisha slows down at a constant rate until she comes to a stop.

3B. ELEVATOR An elevator rises from the lobby of a building to the 12th floor, which is at a height of 120 feet. It makes no stops along the way.

> **y-intercept:** The elevator starts at a height of 0 feet.
>
> **Linear or nonlinear:** The function that models the situation is linear.
>
> **Increasing:** The elevator reaches its maximum height of 120 feet after 6 seconds.

Check Your Understanding

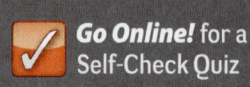

Example 1 Use the given key features to sketch a linear graph.

1. The x-intercept is 6. The y-intercept is 4. The function is decreasing for all values of x. The function is positive for $x < 6$. As $x \to \infty$, $f(x) \to -\infty$ and as $x \to -\infty$, $f(x) \to \infty$.

2. The y-intercept is -7. As $x \to \infty$, $f(x) \to \infty$ and as $x \to -\infty$, $f(x) \to -\infty$.

Example 2 Use the given key features to sketch a nonlinear graph.

3. The y-intercept is 2. The function is continuous. The function is positive for $-2 < x < 4$. The function has a maximum at $(2, 4)$. As $x \to \infty$, $f(x) \to -\infty$ and as $x \to -\infty$, $f(x) \to -\infty$.

4. The y-intercept is 3. The function is continuous and is positive for all values of x. The function has a minimum at $(0, 3)$. The y-axis is a line of symmetry. As $x \to \infty$, $f(x) \to \infty$ and as $x \to -\infty$, $f(x) \to \infty$.

Example 3

5. **DRIVING** Samantha is driving from her apartment to a nearby mall to do some errands. Use the given key features to sketch a graph.

   y-intercept: Samantha starts at a speed of 0 miles per hour.

   Linear or nonlinear: The function that models the situation is nonlinear.

   Extrema: Samantha's maximum speed is 45 miles per hour, which she reaches after 30 seconds.

   Increasing: At the start of the trip, Samantha increases her speed at a steady rate for 30 seconds. Then she maintains her speed for another 3 minutes.

   Decreasing: Samantha decreases her speed at a steady rate for the next 30 seconds until she comes to a stop at the mall.

   End behavior: Samantha starts at 0 miles per hour and ends at 0 miles per hour, so there is no end behavior.

Practice and Problem Solving

Extra Practice is found on page R2.

Use the given key features to sketch a linear graph.

Example 1

6. The y-intercept is 7. The function is increasing for all values of x. The function is positive for $x > -3$. As $x \to \infty$, $f(x) \to \infty$ and as $x \to -\infty$, $f(x) \to -\infty$.

7. The y-intercept is -1. The function is decreasing for all values of x. The domain of the function is $-4 \leq x \leq 3$. The range of the function is $-4 \leq y \leq 3$.

Use the given key features to sketch a nonlinear graph.

Example 2

8. The y-intercept is -4. The function has no x-intercepts. As $x \to \infty$, $f(x) \to -\infty$ and as $x \to -\infty$, $f(x) \to -\infty$.

9. The y-intercept is 1. The function is positive for all values of x. The function is decreasing for $x < -1$. The function is increasing for $x > 1$. As $x \to \infty$, $f(x) \to \infty$ and as $x \to -\infty$, $f(x) \to \infty$.

Example 3

10. **HIKING** Andrew is hiking in a mountainous area of a state park. Use the given key features to sketch a graph.

    y-intercept: Andrew starts at an elevation of 0 feet.

    Extrema: Andrew's maximum elevation is 800 feet.

    Increasing: At the start of the hike, Andrew's elevation increases steadily for 30 minutes. Then he stops to rest for 5 minutes.

    Decreasing: Andrew hikes back to his starting point for the final 20 minutes.

State three key features that could be given so that a sketch of the graph results in the graph shown.

11.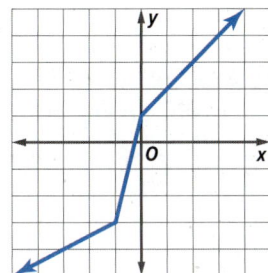

12.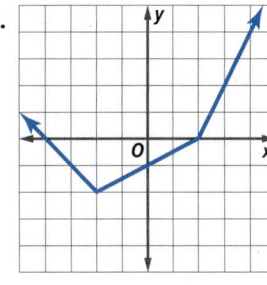

13. **CHARITY** A neighborhood charity is collecting cans of food. They start with 56 cans and collect 10 cans every 20 minutes. They continue for 4 hours. Sketch a graph to represent this situation and tell whether the function is linear or nonlinear.

14. The function $f(x)$ has a maximum at $(0, -3)$, is increasing for $x < 0$ and decreasing for $x > 0$, and is negative for all values of x. The graph of $g(x)$ is the graph of $f(x)$ after a translation 2 units right and 3 units up. Sketch a possible graph of $g(x)$.

15. Sketch a possible graph of $f(x)$, given the following key features. The function is not continuous. The function has no y-intercept. The function is increasing for all values of x. The domain of the function is $x \leq -1$ and $x \geq 2$.

16. **MP STRUCTURE** What are all the possible end behaviors for a linear function? Explain.

17. **MP REASONING** Is it possible for a function to be continuous, to have no x-intercepts, and to have the following end behavior as $x \to \infty$, $f(x) \to \infty$ and as $x \to -\infty$, $f(x) \to -\infty$? If so, sketch the graph of such a function. If not, explain why not.

H.O.T. Problems   Use Higher-Order Thinking Skills

18. **OPEN-ENDED** Sketch the graph of a function that has symmetry about the line $x = 2$, no x-intercepts, and the following end behavior: as $x \to \infty$, $f(x) \to -\infty$ and as $x \to -\infty$, $f(x) \to -\infty$.

19. **ERROR ANALYSIS** Sophia was asked to graph a function with the following key features. The y-intercept is 3. As $x \to \infty$, $f(x) \to -\infty$ and as $x \to -\infty$, $f(x) \to \infty$. Sophia's graph is shown here. Is Sophia correct or incorrect? Explain. If she is incorrect, provide a possible graph of the function.

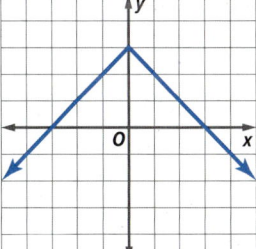

20. **MP REASONING** A continuous function is negative for $x < m$ and positive for $x > n$, where m and n are constants such that $m < n$. For what values of x is $f(x) = 0$?

21. **CHALLENGE** Sketch the graph of a nonlinear function with y-intercept -4, increasing function values for $x < 2$ and $0 < x < 2$, decreasing function values for $-2 < x < 0$ and $x > 2$, and the following end behavior: as $x \to \infty$, $f(x) \to -\infty$ and as $x \to -\infty$, $f(x) \to -\infty$.

22. **WRITING IN MATH** Explain how you can sketch the graph of a linear function if you know only the y-intercept and an interval on which the function is positive or negative.

Preparing for Assessment

23. Leshawn was asked to sketch the graph of a linear function that is increasing for all values of x. Which end behavior should his graph show? MP 7

- A As $x \to \infty$, $f(x) \to \infty$ and as $x \to -\infty$, $f(x) \to -\infty$.
- B As $x \to \infty$, $f(x) \to \infty$ and as $x \to -\infty$, $f(x) \to \infty$.
- C As $x \to \infty$, $f(x) \to -\infty$ and as $x \to -\infty$, $f(x) \to -\infty$.
- D As $x \to \infty$, $f(x) \to -\infty$ and as $x \to -\infty$, $f(x) \to \infty$.

24. Crystal correctly sketched a graph based on some given key features. Her graph was a downward-opening parabola. Which key features could have been given? MP 2, 6

- A The function is nonlinear.
- B The function is decreasing for all values of x.
- C The function is continuous.
- D The x-intercepts are 1 and 3.
- E There is no y-intercept.
- F As $x \to \infty$, $f(x) \to -\infty$ and as $x \to -\infty$, $f(x) \to -\infty$.

25. Ivan drives from his home to his office. He starts at 0 miles per hour and steadily increases his speed until he is driving at 50 miles per hour. He drives at this speed for 10 minutes, then decreases his speed until he reaches a speed of 45 miles per hour. He drives at this speed for 10 more minutes. Then he decreases his speed until he stops at his office. Which of the following must be a key feature of the graph of this situation? MP 4

- A The function is decreasing for all x.
- B The function is continuous.
- C There is only one x-intercept at (0, 0).
- D The y-intercept is 50.

26. A linear function is positive for $x < 5$ and has a y-intercept of 3. If you correctly sketch a graph of the function, at what point will the graph intersect the x-axis? MP 6, 7

27. Anita was given some key features of a graph. Then she drew the graph shown here. Assuming she made a correct graph, which of the following could not be one of the given key features? MP 7

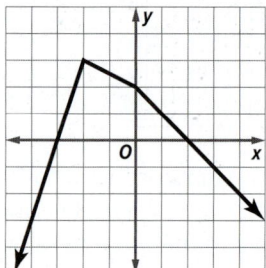

- A The function is positive for $-3 < x < 2$.
- B The function has a maximum at $(-2, 3)$.
- C The function is decreasing for $x > -2$.
- D As $x \to \infty$, $f(x) \to -\infty$ and as $x \to -\infty$, $f(x) \to \infty$.

28. The function $f(x)$ is linear. Based on the table of values, which key feature will the graph of $f(x)$ show? MP 7

x	f(x)
1	−2
3	−4
5	−6

- A The function is increasing.
- B The y-intercept is -1.
- C The function is positive for $x > -1$.
- D The graph is symmetric about the y-axis.

29. MULTI-STEP The y-intercept of a continuous nonlinear function is -1. The function is negative for $-1 < x < 3$. The function has a minimum at $(2, -2)$. As $x \to \infty$, $f(x) \to \infty$ and as $x \to -\infty$, $f(x) \to \infty$. MP 1

a. Sketch a possible graph of the function.

b. Suppose the graph of the function is translated 2 units down. Which of the given key features would remain the same for the new graph? Which of the key features would change?

c. Sketch the graph of the transformed function.

116 | Lesson 2-4 | Sketching Graphs of Functions

CHAPTER 2
Mid-Chapter Quiz
Lessons 2-1 through 2-4

State the domain and range of each relation. Then determine whether the relation is a function. If it is a function, determine if it is *one-to-one*, *onto*, *both*, or *neither*. (Lesson 2-1)

1. {(−3, 2), (4, 1), (0, 3), (5, −2), (2, 7)}

2.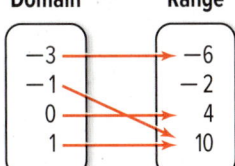

3.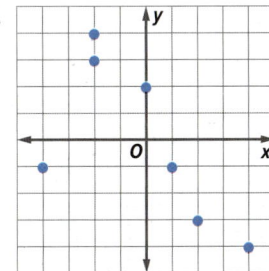

Graph each equation and determine whether the equation is a *function*, is *one-to-one*, *onto*, *both*, or *neither*. State whether it is *discrete* or *continuous*. (Lesson 2-1)

4. $y = 2x - 3$

5. $y = 4 - \frac{1}{2}x^2$

6. State whether $f(x) = 2x^2 - 9$ is a linear function. Explain. (Lesson 2-2)

7. The daily pricing for renting a mid-sized car is given by the function $f(x) = 0.35x + 49$, where $f(x)$ is the total rental price for a car driven x miles. (Lesson 2-2)

   a. State whether $f(x)$ is a linear function. Explain.

   b. **MP** What mathematical practice did you use to solve this problem?

8. State whether the graph represents a linear function. Explain. (Lesson 2-2)

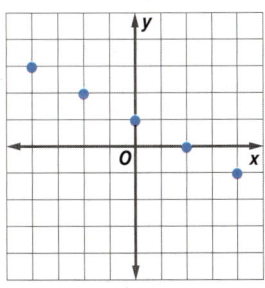

Identify the type of symmetry in each graph. (Lesson 2-2)

9.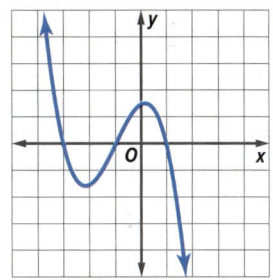

10.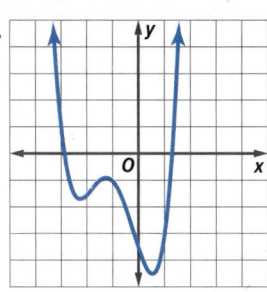

For each graph:
a. describe the end behavior.
b. state the number of real zeros. (Lesson 2-3)

11.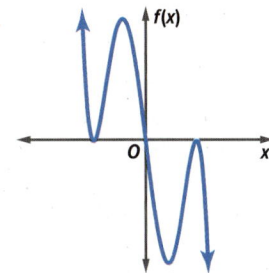

12.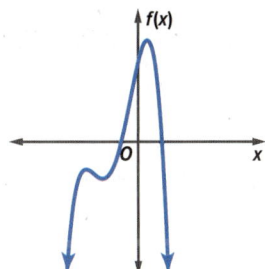

Use a graphing calculator to estimate the *x*-coordinates at which the maxima and minima of each function occur. Round to the nearest hundredth. (Lesson 2-3)

13. $f(x) = x^3 - 4x^2 + 3x - 1$

14. $f(x) = -3x^4 - 5x^3 + 1$

Use the key features of each function to sketch its graph. (Lesson 2-4)

15. *y*-intercept at (0, 4); linear function; positive for values of *x* such that $x > -3$; increasing for all values of *x*

16. *y*-intercept at (0, −3); nonlinear function; negative for values of *x* such that $-1 < x < 3$; increasing for values of *x* such that $x > 1$; relative minimum at (1, −4)

LESSON 5

Graphing Special Functions

::Then   ::Now   ::Why?

- You modeled data using lines of regression.
- 1 Graph and analyze piecewise-defined functions.
- 2 Graph and analyze step and absolute value functions.
- The table shows a recent federal income tax rate schedule. The amount of federal income tax an individual is required to pay is a function of income.

Federal Tax Rate Schedule – Filing Single		
If taxable income is over	But not over	The tax is:
$0	$8925	10% of the amount over $0
$8925	$36,250	$892.5 + 15% over $8925
$36,250	$87,850	$4991.25 + 25% over $36,250
$87,850	$183,250	$17,891.25 + 28% over $87,850
$183,250	$398,350	$44,603.25 + 33% over $183,250
$398,350	$400,000	$115,586.25 + 35% over $398,350
$400,000	No limit	$116,163.75 + 39.6% over $400,000

Source: Internal Revenue Service

New Vocabulary
piecewise-defined function
piecewise-linear function
step function
greatest integer function
absolute value function

Mathematical Practices
1 Make sense of problems and persevere in solving them.

1 Piecewise-Defined Functions
The function relating income and tax is not a linear function because each interval, or piece, of the function is defined by a different expression. A function that is written using two or more expressions is called a **piecewise-defined function.** On the graph of a piecewise-defined function, a dot indicates that the point is included in the graph. A circle indicates that the point is not included in the graph

Example 1   Piecewise-Defined Function

Graph $f(x) = \begin{cases} x - 2 \text{ if } x < -1 \\ x + 3 \text{ if } x \geq -1 \end{cases}$. Identify the domain and range.

Step 1 Graph $f(x) = x - 2$ for $x < -1$.
$f(x) = x - 2$
$= (-1) - 2$
$= -3$
Because -1 does not satisfy the inequality, begin with a circle at $(-1, -3)$.

Step 2 Graph $f(x) = x + 3$ for $x \geq -1$.
$f(x) = x + 3$
$= (-1) + 3$
$= 2$
Because -1 satisfies the inequality, begin with a dot at $(-1, 2)$.

The function is defined for all values of x, so the domain is all real numbers.

The $f(x)$-coordinates of points on the graph are all real numbers less than -3 and all real numbers greater than or equal to 2, so the range is $\{f(x) | f(x) < -3 \text{ or } f(x) \geq 2\}$.

▶ **Guided Practice**

1. Graph $f(x) = \begin{cases} x + 2 \text{ if } x < 0 \\ x \quad\quad \text{ if } x \geq 0 \end{cases}$. Identify the domain and range.

118 | Lesson 2-5

Study Tip

MP Sense-Making The graphs of different parts of a piecewise-defined function may or may not connect. A graph may also stop at a given *x*-value and then begin at a different *y*-value for the same value of *x*.

Piecewise-defined functions are often defined by several linear functions.

Example 2 Write a Piecewise-Defined Function

Write the piecewise-defined function shown in the graph.

Examine and write a function for each portion of the graph.

The left portion of the graph is the graph of $f(x) = 2x + 3$. There is a circle at (1, 5), so the linear function is defined for $\{x | x < 1\}$.

The center portion of the graph is the graph of $f(x) = -x + 2$. There are dots at (1, 1) and (2, 0), so the linear function is defined for $\{x | 1 \leq x \leq 2\}$.

The right portion of the graph is the constant function $f(x) = 3$. There is a circle at (2, 3), so the constant function is defined for $\{x | x > 2\}$.

Write the piecewise-defined function.

$$f(x) = \begin{cases} 2x + 3 & \text{if } x < 1 \\ -x + 2 & \text{if } 1 \leq x \leq 2 \\ 3 & \text{if } x > 2 \end{cases}$$

CHECK The graph shows a portion of a line with positive slope for $x < 1$. The graph has negative slope for $1 \leq x \leq 2$ and constant slope for $x > 2$. The function is reasonable for the graph.

▶ **Guided Practice**

Write the piecewise-defined function shown in each graph.

2A.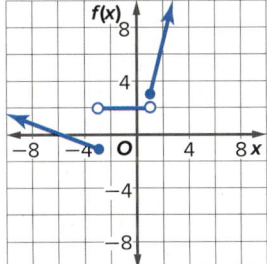

2B.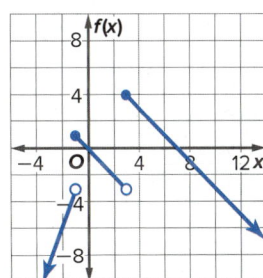

2 Step Functions and Absolute Value Functions Unlike a piecewise-defined function, a **piecewise-linear function** contains a single expression. A common piecewise-linear function is the step function. The graph of a **step function** consists of line segments.

Study Tip

Greatest Integer Function Notice that the domain of this step function is all real numbers and the range is all integers.

The **greatest integer function**, written $f(x) = [\![x]\!]$, is one kind of step function. The symbol $[\![x]\!]$ means *the greatest integer less than or equal to x*. For example, $[\![3.25]\!] = 3$ and $[\![-4.6]\!] = -5$.

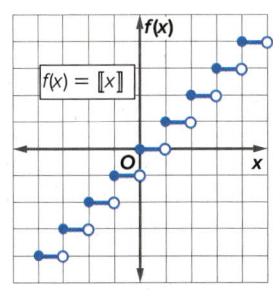

connectED.mcgraw-hill.com   **119**

Go Online!

Watch and listen to the animation to learn about step functions. Summarize what you hear for a partner, and have them ask you questions to help your understanding.

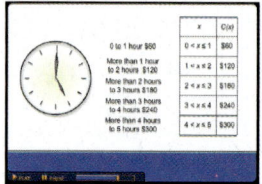

Real-World Example 3   Use a Step Function

BUSINESS An automotive repair center charges $50 for any part of the first hour of labor, and $35 for any part of each additional hour. Draw a graph that represents this situation.

Understand The total labor charge is $50 for the first hour plus $35 for each additional fraction of an hour, so the graph will be a step function.

Plan If the time spent on labor is greater than 0 hours, but less than or equal to 1 hour, then the labor charge is $50. If the time is greater than 1 hour but less than or equal to 2 hours, then the labor charge is $85, and so on.

Solve Use the pattern of times and costs to make a table, where x is the number of hours of labor and $T(x)$ is the total labor charge. Then graph.

x	$T(x)$
$0 < x \leq 1$	$50
$1 < x \leq 2$	$85
$2 < x \leq 3$	$120
$3 < x \leq 4$	$155
$4 < x \leq 5$	$190

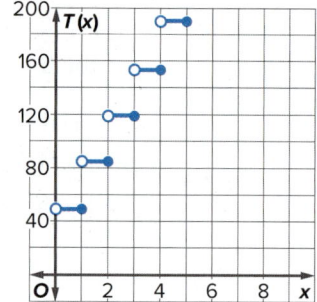

Check Because the repair center rounds any fraction of an hour up to the next whole number, each segment of the graph has a circle at the left endpoint and a dot at the right endpoint.

Though a step function can be represented by a single expression, the table allows us to break the expression into pieces based on the time intervals. The table resembles a piecewise-defined function and can be graphed in the same way. Our graph accurately represents the described situation.

▶ Guided Practice

3. RECYCLING A recycling company pays $5 for every full box of newspaper. They do not give any money for partial boxes. Draw a graph that shows the amount of money $P(x)$ for the number of boxes x brought to the recycling center.

Another piecewise-linear function is the absolute value function. An **absolute value function** is a function that contains an algebraic expression within absolute value symbols.

Math History Link

Karl Weierstrass
(1815–1897) At the wishes of his father, Weierstrass studied law, economics, and finance at the University of Bonn, but then dropped out to study his true interest, mathematics, at the University of Münster. In an 1841 essay, Weierstrass first used | | to denote absolute value.

Photo: The Granger Collection

🔑 Key Concept   Parent Function of Absolute Value Functions

Parent function: $f(x) = |x|$, defined as

$$f(x) = \begin{cases} x & \text{if } x > 0 \\ 0 & \text{if } x = 0 \\ -x & \text{if } x < 0 \end{cases}$$

Type of graph: V-shaped
Domain: all real numbers
Range: $[0, +\infty)$
Intercepts: $x = 0, f(x) = 0$
Not defined: $f(x) < 0$

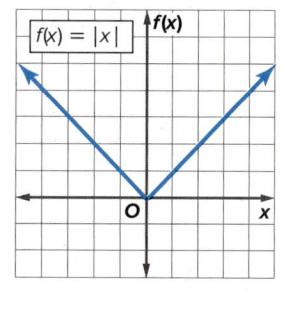

120 | Lesson 2-5 | Graphing Special Functions

Example 4 Absolute Value Functions

Graph $f(x) = |2x| - 4$. Identify the domain and range.

Create a table of values.

x	$|2x| - 4$
−3	2
−2	0
−1	−2
0	−4
1	−2
2	0
3	2

Graph the points and connect them.

D = $(-\infty, +\infty)$, {all real numbers}, or $-\infty < x < +\infty$; R = $[-4, +\infty)$, $\{f(x) \mid f(x) \geq -4\}$, or $\{-4 < x < +\infty\}$

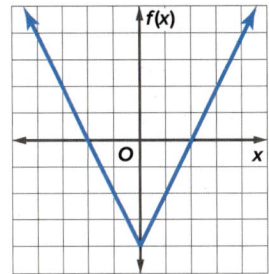

Guided Practice

Graph each function. Identify the domain and range.

4A. $f(x) = |x - 2|$

4B. $f(x) = -|x| + 1$

Check Your Understanding

= Step-by-Step Solutions begin on page R11.

Go Online! for a Self-Check Quiz

Example 1 Graph each function. Identify the domain and range.

1. $g(x) = \begin{cases} -3 & \text{if } x \leq -4 \\ x & \text{if } -4 < x < 2 \\ -x + 6 & \text{if } x \geq 2 \end{cases}$

2. $f(x) = \begin{cases} 8 & \text{if } x \leq -1 \\ 2x & \text{if } -1 < x < 4 \\ -4 - x & \text{if } x \geq 4 \end{cases}$

Example 2 Write the piecewise-defined function shown in each graph.

3.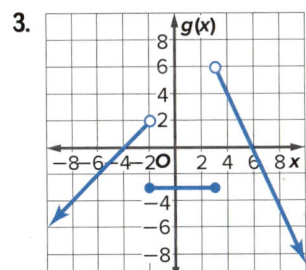

4.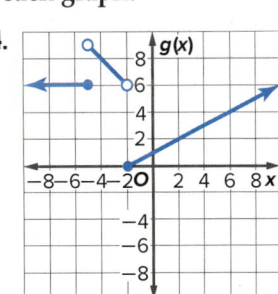

Example 3

5. **MP REASONING** Springfield High School's theater can hold 250 students. The drama club is performing a play in the theater. Draw a graph of a step function that shows the relationship between the number of tickets sold x and the minimum number of plays y that the drama club must perform.

Graph each function. Identify the domain and range.

6. $g(x) = -2[\![x]\!]$

7. $h(x) = [\![x - 5]\!]$

Example 4 Graph each function. Identify the domain and range.

8. $g(x) = |-3x|$

9. $f(x) = 2|x|$

10. $h(x) = |x + 4|$

11. $s(x) = |-2x| + 6$

Practice and Problem Solving

Extra Practice is on page R2.

Example 1 Graph each function. Identify the domain and range.

12. $f(x) = \begin{cases} -3x & \text{if } x \leq -4 \\ x & \text{if } 0 < x \leq 3 \\ 8 & \text{if } x > 3 \end{cases}$

13. $f(x) = \begin{cases} 2x & \text{if } x \leq -6 \\ 5 & \text{if } -6 < x \leq 2 \\ -2x + 1 & \text{if } x > 4 \end{cases}$

14. $g(x) = \begin{cases} 2x + 2 & \text{if } x < -6 \\ x & \text{if } -6 \leq x \leq 2 \\ -3 & \text{if } x > 2 \end{cases}$

15. $g(x) = \begin{cases} -2 & \text{if } x < -4 \\ x - 3 & \text{if } -1 \leq x \leq 5 \\ 2x - 15 & \text{if } x > 7 \end{cases}$

Example 2 Write the piecewise-defined function shown in each graph.

16.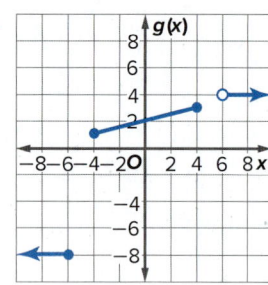

17.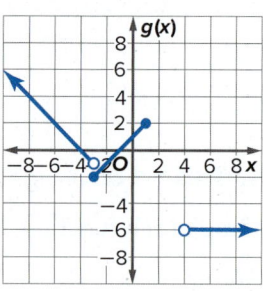

18.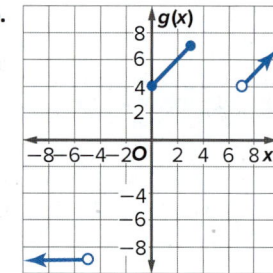

19.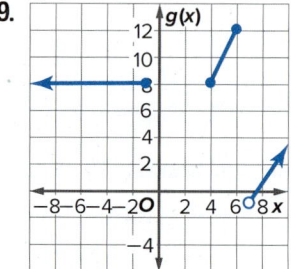

Example 3 Graph each function. Identify the domain and range.

20. $f(x) = [\![x]\!] - 6$

21. $h(x) = [\![3x]\!] - 8$

22. $f(x) = [\![3x + 2]\!]$

23. $g(x) = 2[\![0.5x + 4]\!]$

Example 4 Graph each function. Identify the domain and range.

24. $f(x) = |x - 5|$

25. $g(x) = |x + 2|$

26. $h(x) = |2x| - 8$

27. $k(x) = |-3x| + 3$

28. $f(x) = 2|x - 4| + 6$

29. $h(x) = -3|0.5x + 1| - 2$

30. **GIVING** Patrick is donating money and volunteering his time to an organization that restores homes. His employer will match his monetary donations up to $100.

   a. Identify the type of function that models the total amount of money received by the organization when Patrick donates x dollars.

   b. Write and graph a function for the situation.

31. **SENSE-MAKING** A car's speedometer reads 60 miles per hour.

   a. Write an absolute value function for the difference between the car's actual speed a and the reading on the speedometer.

   b. What is an appropriate domain for the function? Explain your reasoning.

   c. Use the domain to graph the function.

Lesson 2-5 | Graphing Special Functions

32. **RECREATION** The charge for renting a bicycle from a rental shop for different amounts of time is shown at the right.

   a. Identify the type of function that models this situation.

   b. Write and graph a function for the situation.

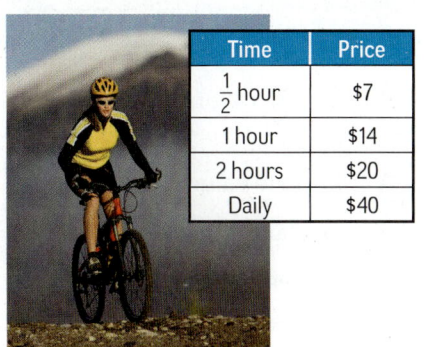

Time	Price
$\frac{1}{2}$ hour	$7
1 hour	$14
2 hours	$20
Daily	$40

Use each graph to write the absolute value function.

33.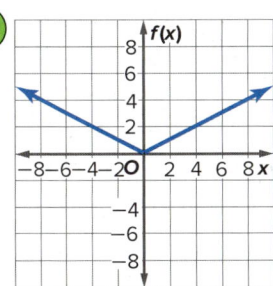

34.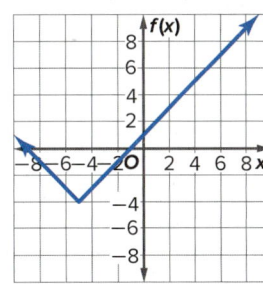

Graph each function. Identify the domain and range.

35. $f(x) = [\![\,|0.5x|\,]\!]$

36. $g(x) = |\,[\![2x]\!]\,|$

37. $g(x) = \begin{cases} [\![x]\!] & \text{if } x < -4 \\ x + 1 & \text{if } -4 \leq x \leq 5 \\ -|x| & \text{if } x > 3 \end{cases}$

38. $h(x) = \begin{cases} -|x| & \text{if } x < -6 \\ |x| & \text{if } -6 \leq x \leq 2 \\ |-x| & \text{if } x > 2 \end{cases}$

39. **MULTIPLE REPRESENTATIONS** Consider the following sets of absolute value functions.

   $f(x) = a|x|$ for $a = \{-2, -1, -0.5, 0.5, 2, 4\}$ $g(x) = |bx|$ for $b = \{-2, -1, -0.5, 0.5, 2, 2\}$

   $h(x) = |x - c|$ for $c = \{-4, -2, -0.5, 0.5, 1, 3\}$ $k(x) = |x| + d$ for $d = \{-4, -2, -0.5, 0.5, 1, 3\}$

   a. **Graphical** Graph each set of functions on a graphing calculator.

   b. **Verbal** Compare and contrast the graphs of the functions in each set.

   c. **Tabular** Choose any two functions. Find the slope between consecutive points on the graphs.

   d. **Verbal** Describe how the slopes of the two sections of an absolute value graph are related.

H.O.T. Problems   Use Higher-Order Thinking Skills

40. **OPEN-ENDED** Write an absolute value relation in which the domain is all nonnegative numbers and the range is all real numbers.

41. **CHALLENGES** Graph $|y| = 2|x + 3| - 5$.

42. **CRITIQUE ARGUMENTS** Find a counterexample to the statement and explain your reasoning.

   In order to find the greatest integer function of x when x is not an integer, round x to the nearest integer.

43. **OPEN-ENDED** Write an absolute value function in which $f(5) = -3$.

44. **WRITING IN MATH** Explain how piecewise functions can be used to accurately represent real-world problems.

Preparing for Assessment

45. Which type of function is a linear function? MP 6

A absolute value
B constant
C greatest integer
D step

46. What is the equation of the function shown in the graph? MP 7

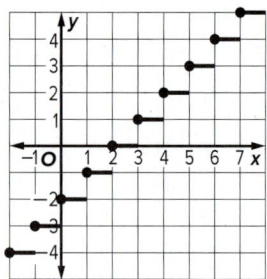

A $f(x) = [\![x - 2]\!]$
B $f(x) = [\![x]\!] - 2$
C $f(x) = |x| - 2$
D $f(x) = |x - 2|$
E $f(x) = [\![x]\!]$

47. MULTI-STEP Consider the function described below.

$$f(x) = \begin{cases} 5 & \text{if } x \leq -2 \\ 2x + 1 & \text{if } -2 < x < 0 \\ 5x + 10 & \text{if } x \geq 0 \end{cases}$$

a. Which of the following statements are true? Choose all that apply. MP 6

☐ A The range of the first part of the function and the second part of the function overlap.
☐ B There are two points on this function with the same domain.
☐ C The number 5 is included in the domain of this function.
☐ D The number 4 is not included in the range of this function.
☐ E The range ends at 100.
☐ F The graph of this piecewise function would include parts of two lines.
☐ G The graph of this piecewise function would include parts of three lines.

b. Which of the following options correctly describes the range of this function? MP 1

○ A $\{f(x) \mid f(x) = -3 < f(x) < 1$ or 5 or $f(x) \geq 10\}$
○ B $\{f(x) \mid f(x) = -3 < f(x) < 1$ or $f(x) \geq 10\}$
○ C $\{f(x) \mid f(x) = -5 < f(x) < 1$ or 5 or $f(x) > 100\}$
○ D $\{f(x) \mid f(x) = -9 < f(x) < 10$ or $f(x) > 10\}$

c. What is the y-intercept of this function? MP 6, 2

d. What is true about the x-intercept of this function? MP 2

○ A It has multiple values.
○ B It cannot be determined.
○ C It can be determined and is 2.
○ D It can be determined and is −0.5.

48. Consider the following function.

$$f(x) = \begin{cases} |x - 2| & \text{if } x > -3 \\ -x & \text{if } x < -4 \end{cases}$$

a. Which of the following statements are true? Choose all that apply. MP 6

☐ A The function bends once.
☐ B The function bends three times.
☐ C The function has no y-intercept.
☐ D The value of the function at $x = -3$ is undefined.
☐ E The value of the function at $x = 1$ is undefined.
☐ F The graph of this piecewise function would include parts of two lines.
☐ G The graph of this piecewise function would include parts of three lines.

b. What is the range of this function? MP 1

c. If it has one, what is the vertex of this function? MP 1

LESSON 6

Transformations of Functions

:Then
- You analyzed and used relations and functions.

:Now
1. Identify the effects on the graphs of functions by replacing $f(x)$ with $f(x) + k$ and $f(x - h)$ for positive and negative values.
2. Identify the effect on the graphs of functions by replacing $f(x)$ with $af(x)$, $f(ax)$, $-af(x)$ and $f(-ax)$.

:Why?
Nick makes $8 an hour working at a pizza shop. The blue line represents his wages. If he agrees to be the closing supervisor, he will be paid an extra $20 per shift. The green line represents his wages when he is the closing supervisor. These graphs are an example of a transformation.

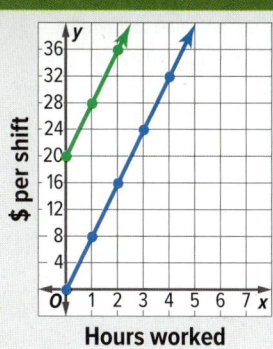

 New Vocabulary
parent graph
parent function
quadratic function
translation
reflection
line of reflection
dilation

 Mathematical Practices
6 Attend to precision.
8 Look for and express regularity in repeated reasoning.

1 Translations In a family of graphs, which have at least one common characteristic, the **parent graph** is the graph of the **parent function**, the simplest function of the family. It is transformed to create other members in the family of graphs. **Transformations** on parent graphs occur when the graph is slid, reflected in an axis, stretched, or compressed.

A translation is a transformation in which a figure is slid from one position to another without being turned.

- When a constant k is added to or subtracted from a parent function, the result $f(x) \pm k$ is a translation of the graph up or down.
- When a constant h is added to or subtracted from x before evaluating a parent function, the result, $f(x \pm h)$, is a translation left or right.

Example 1 Describe and Graph Translations

Describe the translation in each function as it relates to its parent graph. Then graph the function.

a. $y = |x| + 2$

The graph of $y = |x| + 2$ is a translation of the graph of $y = |x|$ up 2 units.

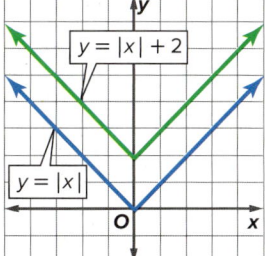

b. $y = x^2 - 7$

The graph of $y = x^2 - 7$ is a translation of the graph of $y = x^2$ down 7 units.

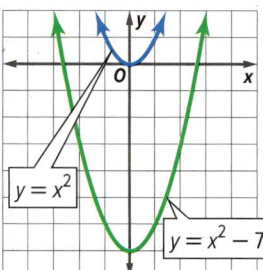

▶ **Guided Practice**

Describe the translation in each function as it relates to its parent graph.. Then graph the function.

1A. $y = |x + 3|$

1B. $y = x^2 - 4$

connectED.mcgraw-hill.com 125

Reading Math

translation Everyday use - to change from one language to another;

Math meaning - to change from one location to another; can also be called a slide, a shift, or a glide.

2 Reflections and Dilations
Two other types of transformations are *reflections* and *dilations*.

A **reflection** flips a figure in a line called the **line of reflection**.

- When a parent function is multiplied by -1, the result $-f(x)$ is a reflection of the graph in the x-axis.

- When only the variable is multiplied by -1, the result $f(-x)$ is a reflection of the graph in the y-axis.

Example 2 Describe and Graph Reflections

Describe the reflection in $y = -x^2$ as it relates to the parent graph. Then graph the function.

The graph of $y = -x^2$ is a reflection of the graph of $y = x^2$ in the x-axis.

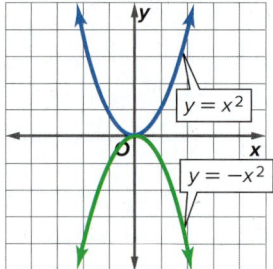

▶ **Guided Practice**

Describe the reflection in each function as it relates to its parent graph. Then graph the function.

2A. $y = -|x|$ **2B.** $y = -x$

A **dilation** is a transformation that alters the size of a figure but not its shape.

- When a nonlinear parent function is multiplied by a nonzero constant, the function is stretched or compressed vertically.

- When only the variable is multiplied by a nonzero constant, the function is stretched or compressed horizontally.

Study Tip

Transformations A transformation could include a reflection and a dilation at the same time. When $f(x)$ becomes $-af(x)$ or $f(-ax)$, it is both dilated and reflected in the x-axis.

Example 3 Describe and Graph Dilations

Describe the dilation in each equation as it relates to the graph of the parent function. Then graph the function.

a. $y = 4x^2$

The graph of $y = 4x^2$ is a dilation of the graph of $y = x^2$ compressed vertically.

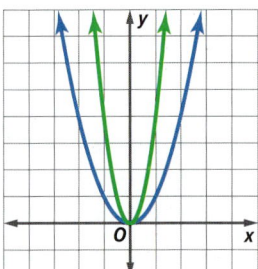

b. $y = \left|\frac{1}{3}x\right|$

The graph of $y = \left|\frac{1}{3}x\right|$ is a dilation of the graph of $y = |x|$ stretched horizontally.

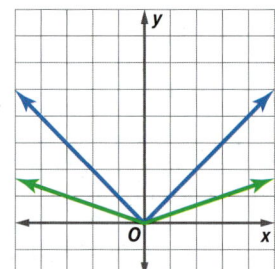

▶ **Guided Practice**

Describe the dilation in each function as it relates to its parent graph. Then graph the function.

3A. $y = 2x^2$

3B. $y = \left| \dfrac{1}{3}x \right|$

Go Online!

Parent graphs and transformations are key concepts in Algebra 2. Look for **worksheets** in the Resources if you need extra practice.

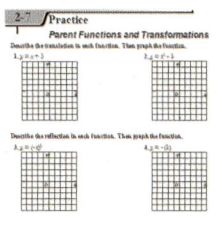

Example 4   Identify Transformations

The graph of the quadratic function $f(x)$ is shown on the grid. If the graph of $f(x)$ is translated 5 units to the right and 3 units down to create a new graph, which function best represents this new graph?

A $g(x) = -(x + 4)^2 - 1$

B $g(x) = -(x - 4)^2 - 1$

C $g(x) = (x - 1)^2 + 4$

D $g(x) = (x - 1)^2 - 4$

Read the Item

You are given the graph of a function and asked to find the equation of a translated graph.

Solve the Item

The graph of $f(x)$ is a reflection of the parent graph in the x-axis, so a is negative. The vertex of the graph is at $(-1, 2)$ so $h = -1$ and $k = 2$. The translation of the graph of $f(x)$ is 5 units to the right. So, $-1 + 5$ or 4 is the translated value of h. The graph is translated down 3 units, so $2 - 3$ or -1 is the translated value of k. Thus, an equation for the translated graph is $g(x) = -(x - 4)^2 - 1$. B is the correct choice.

▶ **Guided Practice**

4. The graph of $g(x)$ was obtained by translating the graph of $f(x)$. Based on the graph, which equation can be used to describe $g(x)$ in terms of $f(x)$?

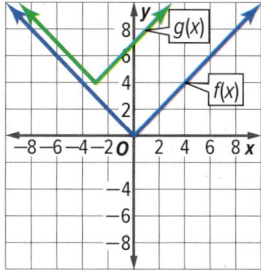

A $g(x) = f(x - 3) + 4$

B $g(x) = f(x - 3) - 4$

C $g(x) = f(x + 3) + 4$

D $g(x) = f(x + 3) - 4$

The general form of a function is $g(x) = a \cdot f(x - h) + k$, where $f(x)$ is the parent function. Each constant in the equation affects the parent graph.
- The value of $|a|$ stretches or compresses (dilates) the parent graph.
- When the value of a is negative, the graph is reflected in the x-axis.
- The value of k shifts (translates) the parent graph up or down.
- The value of h shifts (translates) the parent graph left or right.

Study Tip

MP Regularity Ask yourself these questions to help you identify transformations.
1. What type of function is it?
2. Does the graph open up or down?
3. Does the vertex lie on an axis?

Concept Summary   Transformations of Functions

Transformation	Change to Parent Graph
horizontal translation, h	If $h > 0$, the graph of $f(x)$ is translated h units right.
	If $h < 0$, the graph of $f(x)$ is translated $|h|$ units left.
vertical translation, k	If $k > 0$, the graph of $f(x)$ is translated k units up.
	If $k < 0$, the graph of $f(x)$ is translated $|k|$ units down.
reflection, a	If $a > 0$, the graph of $f(x)$ is reflected in the x-axis.
	If $a < 0$, the graph of $f(x)$ is reflected in the y-axis.
dilation, a	If $|a| > 1$, the graph of $f(x)$ is stretched vertically.
	If $0 < |a| < 1$, the graph of $f(x)$ is compressed vertically.

Check Your Understanding

= Step-by-Step Solutions begin on page R11.

Go Online! for a Self-Check Quiz

Example 1   **MP SENSE-MAKING** Describe the translation in each function as it relates to its parent graph. Then graph the function.

1. $y = x^2 - 4$
2. $y = |x + 1|$

Example 2   Describe the reflection in each function as it relates to its parent graph. Then graph the function.

3. $y = -|x|$
4. $y = (-x)^2$

Example 3   Describe the dilation in each function as it relates to its parent graph. Then graph the function.

5. $y = \frac{3}{5}x$
6. $y = 3x^2$

Example 4

7. The graph of $g(x)$ was obtained by translating the graph of $f(x)$. Based on the graph, which equation can be used to describe $g(x)$ in terms of $f(x)$?

   A $g(x) = -f(x + 2) + 1$
   B $g(x) = -f(x - 2) + 1$
   C $g(x) = f(x + 2) + 1$
   D $g(x) = f(x - 2) + 1$

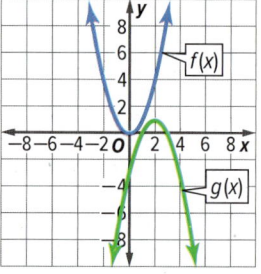

Practice and Problem Solving

Extra Practice is on page R2.

Example 1 Describe the translation in each function as it relates to its parent graph. Then graph the function.

8. $y = x^2 + 4$
9. $y = |x| - 3$
10. $y = x - 1$
11. $y = x + 2$
12. $y = (x - 5)^2$
13. $y = |x + 6|$

Example 2 Describe the reflection in each function as it relates to its parent graph. Then graph the function.

14. $y = -x$
15. $y = -x^2$
16. $y = (-x)^2$
17. $y = |-x|$
18. $y = -|x|$
19. $y = (-x)$

Example 3 Describe the dilation in each function as it relates to its parent graph. Then graph the function.

20. $y = (3x)^2$
21. $y = 6x$
22. $y = 4|x|$
23. $y = |2x|$
24. $y = \frac{2}{3}x$
25. $y = \frac{1}{2}x^2$

Example 4

26. **MP SENSE-MAKING** A non-impact workout can burn up to 7.5 Calories per minute. The equation to represent how many Calories a person burns after m minutes of the workout is $C(m) = 7.5m$.

    a. Identify the transformation in the function.

    b. Graph the function.

Describe the transformation in the graph of the function in green as it relates to its parent graph in blue.

27.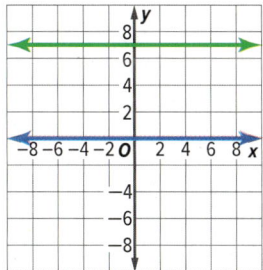

28.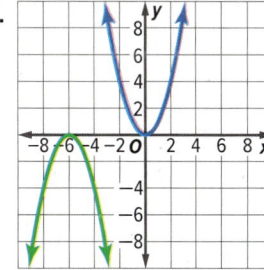

29.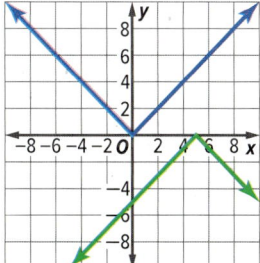

30.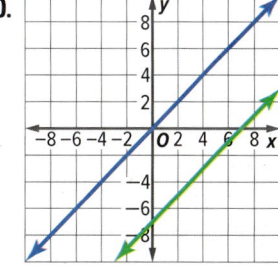

Describe the translation in the graph of the function in green as it relates to the graph of the function in blue.

31.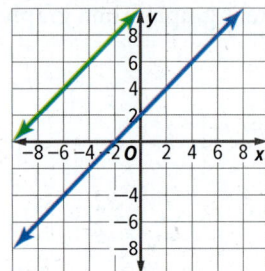

32.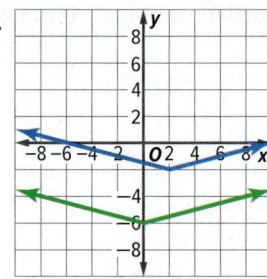

Use the transformation of a parent graph to write an equation for the function represented by each graph.

33.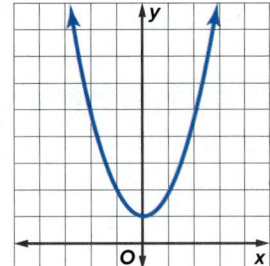

34.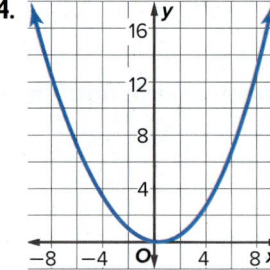

35.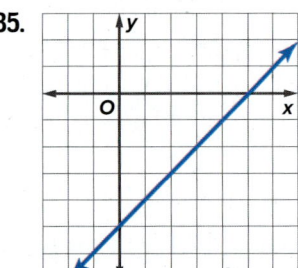

36.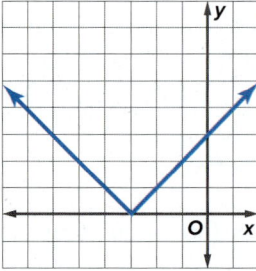

37.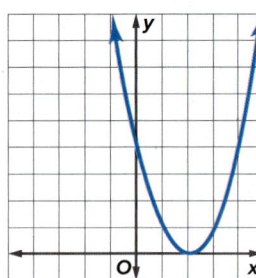

38.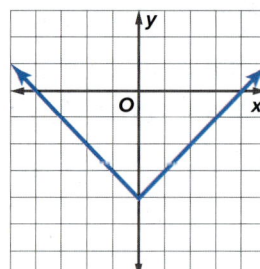

39. **BUSINESS** The previous cost for a company to produce x widgets is represented by the blue line on the graph. The green line represents the new cost after the company hired a consultant to make their production more efficient.

Describe the transformation in the graph of the green line as it relates to the graph of the blue line.

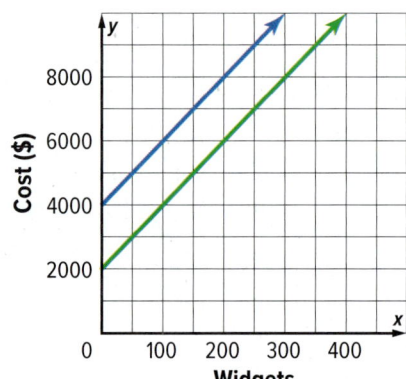

130 | Lesson 2-6 | Transformations of Functions

40. **ROCKETRY** Kenji launched a toy rocket from ground level. The height $h(t)$ of Kenji's rocket after t seconds is shown in blue. Emily believed that her rocket could fly higher and longer than Kenji's. The flight of Emily's rocket is shown in green.

   a. How much longer than Kenji's rocket did Emily's rocket stay in the air?

   b. How much higher than Kenji's rocket did Emily's rocket go?

   c. Describe the type of transformation between the two graphs.

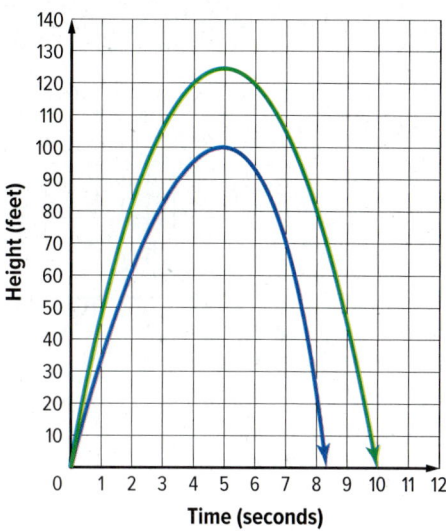

Write an equation for each function.

 41.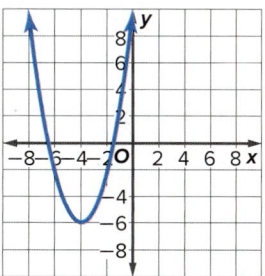

42.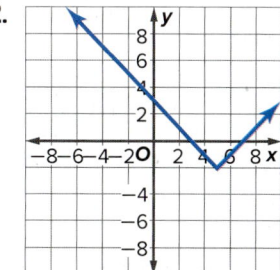

H.O.T. Problems Use Higher-Order Thinking Skills

43. **CHALLENGE** Explain why performing a horizontal translation followed by a vertical translation ends up being the same transformation as performing a vertical translation followed by a horizontal translation.

44. **ERROR ANALYSIS** Kimi thinks that the graph and table below are representations of the same linear relation. Carla disagrees. Who is correct? Explain your reasoning.

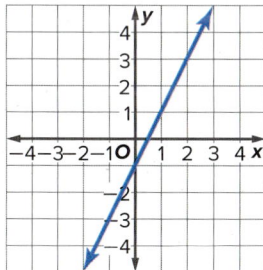

x	y
0	−1
1	1
2	3
3	5

45. **MP SENSE-MAKING** Draw a figure in Quadrant II. Use any of the transformations you learned in this lesson to move your figure to Quadrant IV. Describe your transformation.

46. **MP REASONING** Study the parent graphs at the beginning of this lesson. Select a parent graph with positive y-values at its leftmost points and positive y-values at its rightmost points.

47. **WRITING IN MATH** Explain why the reflection of the graph of $f(x) = x^2$ in the y-axis is the same as the graph of $f(x) = x^2$. Is this true for all reflections of quadratic equations? If not, describe a case when it is false.

Preparing for Assessment

48. What is the equation of the function graphed below? **MP 1**

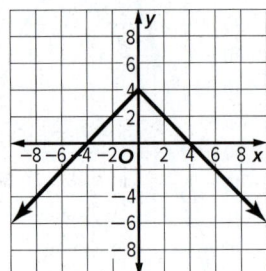

- A $g(x) = -4|x|$
- B $g(x) = -|4x|$
- C $g(x) = -|x + 4|$
- D $g(x) = -|x - 4|$
- E $g(x) = -|x| + 4$

49. Which function is a reflection of $f(x)$ in the x-axis? **MP 7**

- A $g(x) = f(-x)$
- B $g(x) = f(x) - 1$
- C $g(x) = -f(x)$
- D $g(x) = f(x - 1)$

50. Which statement correctly describes the transformation of the function $g(x) = x^2$ to the function $h(x) = -x^2 - 8$? **MP 2**

- A a reflection of $g(x)$ in the y-axis and a translation 8 units up
- B a reflection of $g(x)$ in the y-axis and a translation 8 units right
- C a reflection of $g(x)$ in the y-axis and a translation 8 units down
- D a reflection of $g(x)$ in the x-axis and a translation 8 units right
- E a reflection of $g(x)$ in the x-axis and a translation 8 units down

51. MULTI-STEP Consider the function $g(x)$ shown on the graph. **MP 7**

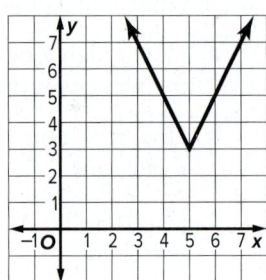

a. Which function is the parent function of $g(x)$?
- A $f(x) = |x|$
- B $f(x) = |0.5x|$
- C $f(x) = 2|x|$
- D $f(x) = -|x|$

b. What is the vertex of the graph of $g(x)$?
- A (4, 3)
- B (3, 4)
- C (5, 3)
- D (5, 4)

c. Which of the statements about $g(x)$ are true?
- A It's graph is a dilation of $f(x)$ compressed vertically.
- B It's graph is a dilation of $f(x)$ stretched vertically.
- C It's graph is a reflection of the graph of $f(x)$ in the y-axis.
- D The slopes of the graph on the left and right sides of $x = 5$ are the same.
- E It's graph is a translation of the graph of $f(x)$ 5 units right and 3 units up.
- F The function $g(x)$ is a step function.

d. Which is the equation for $g(x)$?
- A $g(x) = |x| - 5 + 3$
- B $g(x) = 2|x + 5| - 3$
- C $g(x) = 2|x - 5| + 3$
- D $g(x) = |x - 5| - 3$

LESSON 7
Solving Equations by Graphing

::Then
- You solved linear equations and you graphed linear functions.

::Now
1. Find x- and y-intercepts of functions.
2. Solve equations by examining graphs of the related functions.

::Why?
- The height of a punted football can be modeled by a function like $y = -16x^2 + 10x + 2$. The graph of the function can tell you how high the ball is during its rise and fall and when it hits the ground.

 New Vocabulary
root of an equation
x-intercept
y-intercept
zero of a function

 Mathematical Practices
3 Construct viable arguments and critique the reasoning of others.

1 Find Intercepts Important features of a graph of most functions include the intercepts of the graph. The **intercepts** of a graph are points where the graph touches or crosses an axis. A **y-intercept** of a graph is the y-coordinate of any point where the graph intersects the y-axis. Similarly, an **x-intercept** is the x-coordinate of any point where the graph intersects the x-axis.

Because the y-axis is where $x = 0$ and the x-axis is where $y = 0$, the coordinates of any y-intercept are $(0, y)$ and the coordinates of any x-intercept are $(x, 0)$.

Example 1   Find x- and y-intercepts

Find the x- and y-intercepts of the graph of $y = \frac{1}{2}x^2 + \frac{1}{2}x - 6$, which is shown at the right.

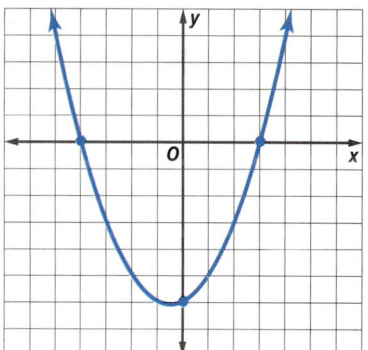

Step 1 Find the x-intercepts. Look for the point(s) where the graph intersects the x-axis.

The graph intersects the x-axis at two points, $(-4, 0)$ and $(3, 0)$. The x-intercepts are -4 and 3.

Step 2 Find the y-intercept. Look for the point where the graph intersects the y-axis.

The graph intersects the y-axis at $(0, -6)$. The y-intercept is -6.

▶ **Guided Practice**

1. Find the x- and y-intercepts of the graph of $y = -\frac{1}{3}x^2 + \frac{4}{3}x + 4$.

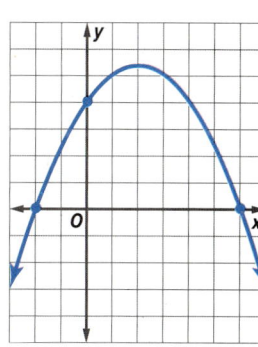

connectED.mcgraw-hill.com   133

2 Solve Equations by Using Graphs of Related Functions

The solution of an equation is called the **root of the equation**. The root of an equation is related to the zero of a function. The **zero of a function** f is the value of x for which $f(x) = 0$. Because $y = f(x)$ gives the graph of the function f, $f(x) = 0$ translates to $y = 0$ and the zero of the function is the same as the x-intercept of the graph.

Example 2 Find the Zero of a Function

Find the root of the linear equation, $-2x + 10 = 0$, by graphing the related function.

Step 1 Write the related function.

Equation: $\quad 0 = -2x + 10$

Related Function: $f(x) = -2x + 10$

Step 2 Graph the function.

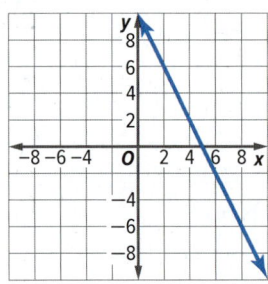

Step 3 Find the zero of the function.
The zero of the function is the x-intercept, 5.

5 is the root, or solution, of the original equation, $0 = -2x + 10$.

> **Reading Math**
> **Precision** The related function intersects the x-axis at (5, 0). The x-coordinate of the ordered pair is the x-intercept, which is the zero of the function.

Guided Practice

2A. $0 = 4x - 6$

2B. $0 = \frac{2}{3}x + 4$

Real-World Example 3 Solving an Equation by Graphing

CLASS PARTY Sara stopped at a bakery to pick up $1\frac{1}{2}$ dozen cupcakes for a class party. She was charged $7.88, which included $.38 tax. How much does one dozen cupcakes cost?

Step 1 Write an equation. Let x represent the cost of a dozen cupcakes.

$$7.88 = \frac{3}{2}x + 0.38$$

Step 2 Rewrite the equation.

$$7.88 = \frac{3}{2}x + 0.38$$

$$7.88 - 7.88 = \frac{3}{2}x + 0.38 - 7.88$$

$$0 = \frac{3}{2}x - 7.50$$

Step 3 Write and graph the related function. Describe what the zero of the function means.

$$f(x) = \frac{3}{2}x - 7.50$$

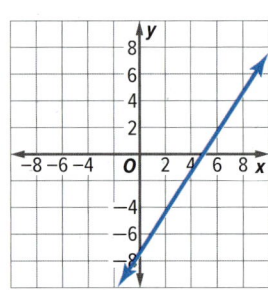

The zero of the function is 5. So, the cost of one dozen cupcakes is $5.

> **Study Tip**
> **Related Linear Function** To write a related function, the original equation must be written in the form of
> 0 = variable + constant or
> 0 = constant + variable.

Guided Practice

3. Two small pizzas each with one topping cost a total of $6.83, which includes $0.33 tax. What is the cost of each pizza? Solve by graphing the related function.

Check Your Understanding

○ = Step-by-Step Solutions begin on page R11.

Go Online! for a Self-Check Quiz

Example 1 Find the *x*- and *y*-intercepts of the graphs of each function.

1. $y = 2x - 3$
2. $f(x) = -\frac{1}{2}x^2 - x + 4$

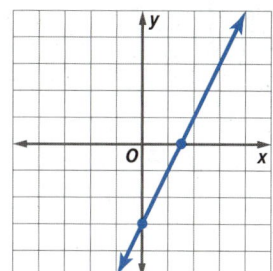
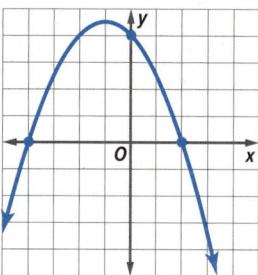

3. $y = 3x - 3$
4. $f(x) = -3x + 7$
5. $y = 5x + 8$
6. $y = -4x - 8$
7. $y = -\frac{2}{3}x + 4$
8. $f(x) = 6x + 3$

Examples 2–3 Solve each equation by graphing its related function.

9. $0 = 6x - 4$
10. $0 = -2x + 5$
11. $0 = -\frac{3}{4}x - 3$

12. **BUDGETING** Sanjit earned $84 last month mowing lawns. He saved $40 and budgeted the rest for *x* numbers of meals at $4 each.

   a. Write an equation for this situation.
   b. Write the related function.
   c. Graph the function and find its zero. State the meaning of the zero in the context of the situation.

Practice and Problem Solving

Extra Practice is on page R2.

Example 1 Find the *x*- and *y*-intercepts of the graphs of each function.

13. $f(x) = \frac{3}{4}x - 6$
14. $f(x) = -2x + 7$
15. $f(x) = -4x - 8$
16. $f(x) = -\frac{1}{3}x + 1$
17. $f(x) = 4x - 2.8$
18. $f(x) = \frac{2}{3}x + 4$
19. $f(x) = -x + 4.8$
20. $f(x) = \frac{1}{2}x - \frac{3}{4}$
21. $f(x) = -8x + 2$

Examples 2–3 Solve each equation by using the graph of its related function.

22. $0 = 2x - 11$
23. $0 = 4x + 12$
24. $0 = -\frac{5}{4}x + 5$
25. $0 = 4x + 2$
26. $0 = -6x + 18$
27. $0 = \frac{5}{8}x - 10$

28. **ELEVATOR CAPACITY** The maximum capacity of a service elevator in the new town hall is 1200 pounds. A delivery person, who weighs 180 pounds, is bringing boxes of paper to the fourth floor. Each box of paper weighs 60 pounds.

   a. Write an equation for this situation.
   b. Write the related function.
   c. Graph the function and find its zero. State the meaning of the zero in the context of the situation.

29. **EARNINGS** Francesca bought $25 worth of face painting supplies. At a local fair, she charges $5 to paint a child's face. How many faces must she paint to earn $80? Solve the equation by finding the zero of the related function.

connectED.mcgraw-hill.com 135

Find the x- and y-intercepts of the graphs.

30. $y = -\frac{3}{8}x^2 + \frac{3}{4}x + 9$

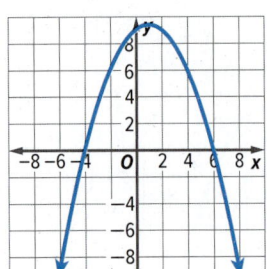

31. $f(x) = x^3 - 2x^2 - 5x + 6$

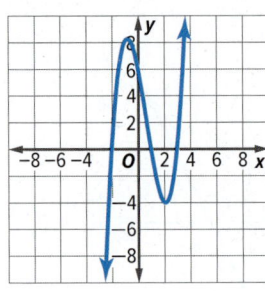

Find the zeros of each function.

32. $f(x) = 5x + 12$

33. $f(x) = 3x - 8$

34. $f(x) = -\frac{5}{8}x - 2$

35. $f(x) = \frac{2}{3}x + 4$

36. $f(x) = -6x + 3$

37. $f(x) = -\frac{5}{4}x + \frac{15}{4}$

38. **MODELING** The graph shows the function $y = -16x^2 + 10x + 2$, which models the path of a football kicked during a game. The y-axis shows the distance the ball is above the ground in feet. The x-axis shows the number of seconds from the kick. Estimate the intercepts and interpret them in the context of the problem.

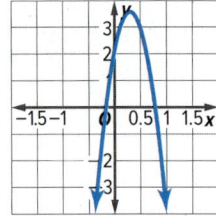

39. MULTI-STEP Hernando is saving for a school trip. He has $40 saved. He knows he can earn $15 a week mowing a neighbor's yard. He can also earn $30 helping his father at his father's shop. The trip costs $130.

a. How much money will Hernando have, if he helps his father at his shop?

b. How much additional money must he earn by mowing yards?

c. Write a function to find how many weeks he must mow yards in order to have enough money for the trip. Find the zero of the function.

d. What does the zero mean in the context of this situation?

H.O.T. Problems Use Higher-Order Thinking Skills

40. ERROR ANALYSIS Alexander and Sandra solved the equation $7 = \frac{5}{8}x + 4$ by graphing the related function. Which student's solution is correct? What error may the other student have made?

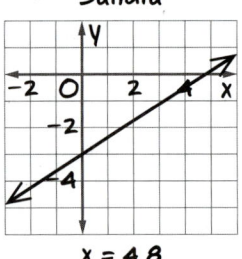

41. WRITING IN MATH Use $0 = 4x + 10$ and $f(x) = 4x + 10$ to distinguish among roots, solutions, and zeros.

42. REASONING How can you confirm that a zero of a related function is the root of the original equation?

43. REASONING Does $f(x) = 6$ have a zero? Explain.

44. OPEN-ENDED The root of a linear equation is $-\frac{43}{8}$. Write an equation and its related function.

Preparing for Assessment

45. What are the *x*- and *y*-intercepts of the graph?
MP 2

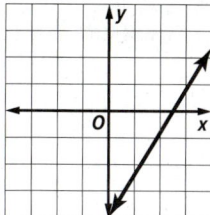

- A *x*-intercept is 4; *y*-intercept is −2.5.
- B *x*-intercept is −2.5; *y*-intercept is 4.
- C *x*-intercept is 2.5; *y*-intercept is −4.
- D *x*-intercept is −4; *y*-intercept is 2.5.

46. What are the *x*- and *y*-intercepts of the graph?
MP 2

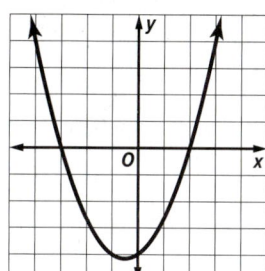

- A *x*-intercepts are −3, 2; *y*-intercept is −4.
- B *x*-intercepts are 3, −2; *y*-intercept is 4.
- C *x*-intercept is −4; *y*-intercept are −3, 2.
- D *x*-intercept is 4; *y*-intercept is 3, −2.

47. A baker buys supplies to make 5 loaves of bread. The supplies cost $8. He would like to know what price to put on each loaf to make a profit of $12. Which is the equation of the related function that can be used to find the solution? MP 4

- A $f(x) = 5x - 4$
- B $f(x) = 5x - 8$
- C $f(x) = 5x - 12$
- D $f(x) = 5x - 20$

48. MULTI-STEP Below is the graph of a function that models the height of a baseball each second it is in the air. The height in feet is on the vertical axis and the time in seconds is on the horizontal axis.
MP 1, 2, 8

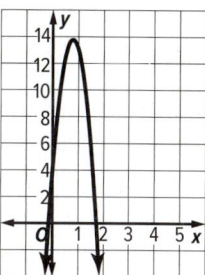

a. Which of the following statements is true?
- A The graph is linear.
- B The graph is nonlinear.
- C There is more than one *x*-intercept.
- D There is more than one *y*-intercept.
- E In terms of the context of the problem, only the first quadrant is needed.

b. What is the *x*-intercept of the graph?

c. What is the *y*-intercept of the graph?

d. What do the intercepts represent?

49. The graph of a function *f* passes through the points (5.8, 0) and (0, −7.5). What is the root of $f(x) = 0$?
MP 7

- A 0
- B 5.8
- C −7.5
- D (5.8, −7.5)

50. What is the related function for $19 = 7x + 14$? MP 7

CHAPTER 2
Study Guide and Review

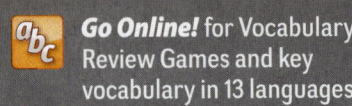 **Go Online!** for Vocabulary Review Games and key vocabulary in 13 languages

Study Guide

Key Concepts

Functions and Continuity (Lesson 2-1)
- A function is a relation where each member of the domain is paired with exactly one member of the range.
- A function is *one-to-one* if each element of the domain is paired with exactly one unique element of the range.
- A function is *onto* if each element of the range corresponds to an element of the domain.

Linearity and Symmetry (Lesson 2-2)
- A linear function can be written in the form $f(x) = mx + b$.
- A graph has line symmetry if each half of the graph on either side of a line matches the other side exactly.
- A graph has point symmetry when a figure is rotated 180° about a point and maps exactly onto the other part.

Extrema and End Behavior (Lesson 2-3)
- The degree and leading coefficient of a polynomial determine its end behavior.
- Polynomials with an even degree have end behavior in the same direction, while polynomials with an odd degree have end behavior in opposite directions.
- Extrema are the relative maximum and minimum values of a function.

Sketching Graphs of Functions (Lesson 2-4)
- Key features provide information for sketching a graph.

Special Functions and Parent Functions (Lessons 2-5 and 2-6)
- A piecewise defined function is made up of two or more expressions.
- Translations, reflections, and dilations of a parent graph form a family of graphs.

Solving Equations by Graphing (Lesson 2-7)
- To solve equations by graphing: **Step 1** Graph each side of the equation separately. **Step 2** Find the points of intersection. **Step 3** Examine the graphs. The *x*-coordinate of each point of intersection is a solution to the equation.

FOLDABLES Study Organizer

Use your Foldable to review the chapter. Working with a partner can be helpful. Ask for clarification of concepts as needed.

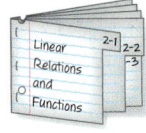

Key Vocabulary

absolute value function (p. 120) linear function (p. 95)
continuous relation (p. 88) one-to-one function (p. 87)
dependent variable (p. 89) onto function (p. 87)
dilation (p. 126) piecewise-defined function (p. 118)
discrete relation (p. 88)
end behavior (p. 103) point symmetry (p. 98)
extrema (p. 105) reflection (p. 126)
function notation (p. 89) relative maximum (p. 105)
greatest integer function (p. 119) relative minimum (p. 105)
 root of an equation (p. 134)
independent variable (p. 89) step function (p. 119)
intercepts (p. 133) translation (p. 125)
line symmetry (p. 97) vertical line test (p. 88)
linear equation (p. 95) zero of a function (p. 134)

Vocabulary Check

Choose the correct term to complete each sentence.

1. A function is (discrete, one-to-one) if each element of the domain is paired to exactly one unique element of the range.
2. The (domain, range) of a relation is the set of all first coordinates from the ordered pairs which determine the relation.
3. The (constant, identity) function is a linear function described by $f(x) = x$.
4. The (leading coefficient, constant term) and degree of a polynomial function determines the end behavior of its graph.
5. A function that is written using two or more expressions is called a (linear, piecewise-defined) function.

Concept Check

6. Explain the difference between a discrete and continuous relation.
7. Explain the difference in end behavior of polynomials with an even degree and an odd degree.

Lesson-by-Lesson Review

2-1 Functions and Continuity

State the domain and range of each relation. Then determine whether the relation is a function. If it is a function, determine if it is *one-to-one*, *onto*, *both*, or *neither*.

8. {(−3, 0), (0, 2), (2, 4), (4, 5), (5, 2)}
9. {(−4, 1), (3, 3), (1, 1), (−2, 5), (3, −4)}
10. {(7, −4), (5, −2), (3, 0), (1, 2), (−1, 4)}

11. **BOWLING** A bowling alley charges $3.50 for shoe rental and $5.25 per game bowled. The amount a bowler is charged can be expressed as $y = 3.50 + 5.25x$, when $1 \leq x \leq 5$, and x is an integer. Find the domain and range. Then determine whether the equation is a function. Is the equation discrete or continuous?

Example 1

State the domain and range of the relation {(−4, 3), (−1, 0), (−2, 4), (3, −1), (2, 6)}. Then determine whether the relation is a function. If it is a function, determine if it is *one-to-one*, *onto*, *both*, or *neither*. State whether it is *discrete* or *continuous*.

Domain: {−4, −2, −1, 2, 3}
Range: {−1, 0, 3, 4, 6}

Each element of the domain is paired with one element of the range, so the relation is a function. The function is both because each element of the domain is paired with a unique element of the range and each element of the range is paired with a unique element of the domain. The function is discrete as its domain consists of individual points.

2-2 Linearity and Symmetry

State whether each function is a linear function. Write *yes* or *no*. Explain.

12. $3x + 4y = 12$
13. $x^2 + y^2 = 4$
14. $y = x^3 − 6$
15. $y = 6x − 19$
16. $f(x) = −2x + 9$
17. $\frac{1}{x} + 3y = −5$

Identify the type of symmetry in the graph of each function.

18. $y = -\frac{1}{2}x + 1$
19. $y = x^2 − 2x − 3$

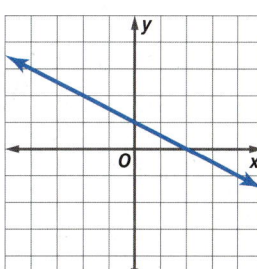
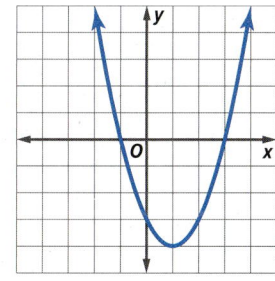

Example 2

State whether $f(x) = 3x^2$ is a linear function. Write *yes* or *no*. Explain.

No, because the expression includes a variable raised to the second power.

Example 3

Identify the type of symmetry in the graph.

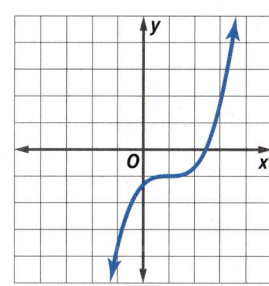

There is no vertical line in which the graph can be reflected onto itself. The graph does not display line symmetry.

The graph displays point symmetry at (1, −1) and a 180° rotation about that point gives the original graph.

CHAPTER 2
Study Guide and Review Continued

2-3 Extrema and End Behavior

For each graph:
a. describe the end behavior.
b. state the number of real zeros.

20. 21.
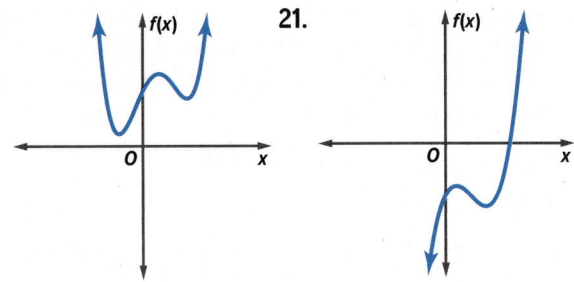

Use a graphing calculator to estimate the x-coordinates at which the maxima and minima of each function occur. Round to the nearest hundredth.

22. $f(x) = 2x^3 - 3x^2 - 4x + 5$
23. $f(x) = x^4 + 3x^3 - x^2 + 8x + 1$

Example 4

Use the graph to answer the questions below.

a. Describe the end behavior.
$f(x) \to +\infty$ as $x \to -\infty$,
$f(x) \to -\infty$ as $x \to +\infty$

b. State the number of real zeros.

The graph intersects the x-axis at three points, so there are three real zeros.

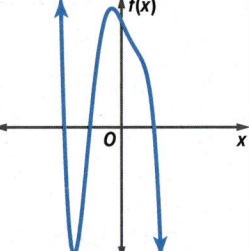

2-4 Sketching Graphs of Functions

Use the key features of the function to sketch its graph.

24. y-intercept: (0, −3)
    Linearity: linear
    Positive: for all values of x such that $x > 2$
    Increasing: for all values of x
    End Behavior: as $x \to \infty$, $f(x) \to \infty$;
    as $x \to -\infty$, $f(x) \to -\infty$

25. y-intercept: (0, 0)
    Linearity: nonlinear
    Continuity: continuous
    Positive: for all x such that $-4 < x < 0$
    Decreasing: for all x such that $x > -2$
    Extrema: relative maximum at (−2, 4)
    End Behavior: as $x \to \infty$, $f(x) \to -\infty$;
    as $x \to -\infty$, $f(x) \to -\infty$

Example 5

Use the key features of the function to sketch its graph.

y-intercept: (0, 10)
Linearity: linear
Negative: for all values of x such that $x > 15$
Decreasing: for all values of x

To include the y-intercept and all key features, set axes from −25 to 25.

Plot the y-intercept at (0, 10).

The graph is a decreasing straight line that is negative for all values of x such that $x > 15$.

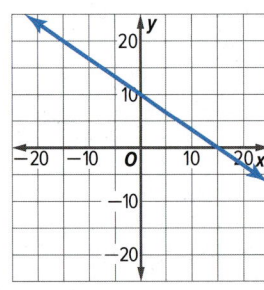

2-5 Graphing Special Functions

Graph each function. Identify the domain and range.

26. $f(x) = \begin{cases} -2x & \text{if } x \leq -1 \\ x + 1 & \text{if } -1 < x < 3 \\ x & \text{if } x \geq 3 \end{cases}$

27. $f(x) = \begin{cases} -3 & \text{if } x < -1 \\ 4x - 3 & \text{if } -1 \leq x \leq 3 \\ x & \text{if } x > 3 \end{cases}$

28. Write the piecewise-defined function shown in the graph.

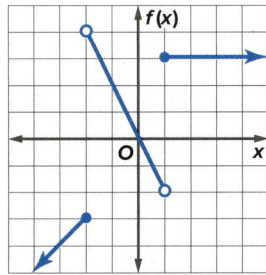

Example 6

Write the piecewise-defined function shown in the graph.

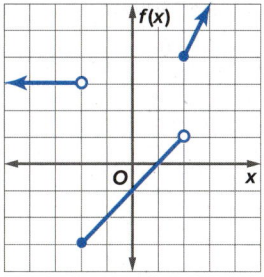

The left portion of the graph is the graph of $f(x) = 3$. There is a circle at $(-2, 3)$, so the linear function is defined for $x < -2$.

The center portion of the graph is the graph of $f(x) = x - 1$. There is a dot at $(-2, -3)$ and a circle at $(2, 1)$, so the linear function is defined for $-2 \leq x < 2$.

The right portion of the graph is the graph of $f(x) = 2x$. There is a dot at $(2, 4)$, so the linear function is defined for $x \geq 2$.

$f(x) = \begin{cases} 3 & \text{if } x < -2 \\ x - 1 & \text{if } -2 \leq x < 2 \\ 2x & \text{if } x \geq 2 \end{cases}$

Graph each function. Identify the domain and range.

29. $f(x) = [\![x]\!] + 2$
30. $f(x) = [\![x + 3]\!]$

Graph each function. Identify the domain and range.

31. $f(x) = |-x| + 1$
32. $f(x) = 2|x - 1|$

2-6 Transformations of Functions

Describe the translation in each function.

33. $y = x^2 - 3$
34. $y = x - 4$

Describe the reflection in each function.

35. $y = |-x|$
36. $y = -x^2$

Describe the dilation in each function.

37. $y = 6|x|$
38. $y = (2x)^2$

39. The graph of $g(x)$ was obtained by transforming the graph of $f(x)$. Which equation can be used to describe $g(x)$ in terms of $f(x)$?

   A $g(x) = -f(x + 2)$
   B $g(x) = -f(x - 1) - 1$
   C $g(x) = -f(x + 1) - 1$
   D $g(x) = -f(x - 1) + 1$

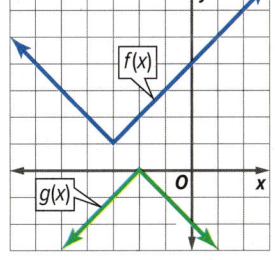

Example 7

Describe the translation in $y = |x + 6|$.

The graph of $y = |x + 6|$ is a translation of the graph of $y = |x|$ left 6 units.

Example 8

Describe the reflection in $y = (-x)^2$.

The graph of $y = (-x)^2$ is a reflection of the graph of $y = x^2$ in the y-axis.

Example 9

Describe the dilation in $y = \frac{1}{2}|x|$.

The graph of $y = \frac{1}{2}|x|$ is a dilation of the graph of $y = |x|$ compressed vertically.

CHAPTER 2
Study Guide and Review *Continued*

2-7 Solving Equations by Graphing

Find the *x*- and *y*-intercepts of the graph of each function.

40. $y = 3x - 4$

41. $f(x) = -\frac{1}{3}x + 2$

42. $f(x) = \frac{2}{5}x + \frac{4}{5}$

Solve each equation by using the graph of the related function.

43. $0 = -\frac{1}{2}x + 3$

44. $\frac{7}{3}x - \frac{14}{3} = 0$

45. $0 = -4x - 1$

Example 10

Find the *x*- and *y*-intercepts of the graph of $y = x^2 - 2$, which is shown below.

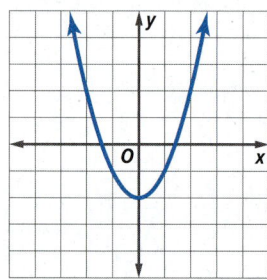

The graph intersects the *x*-axis at about the points (−1.4, 0) and (1.4, 0). The *x*-intercepts are about −1.4 and 1.4.

The graph intersects the *y*-axis at (0, −2). The *y*-intercept is −2.

CHAPTER 2
Practice Test

1. State the domain and range of the relation shown in the table. Then determine whether it is a function. If it is a function, determine if it is *one-to-one*, *onto*, *both*, or *neither*.

x	y
−2	3
4	−1
3	2
6	3

2. Graph each equation and determine whether the equation is a *function*, is *one-to-one*, *onto*, *both*, or *neither*. State whether it is *discrete* or *continuous*.

State whether each function is a linear function. Write *yes* or *no*. Explain.

3. $-4x + y = -1$

4. $y = x^2 - 2x + 1$

5. Identify the type of symmetry in the graph.

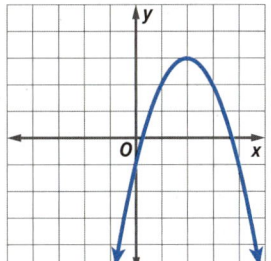

6. For the graph below:
   a. describe the end behavior.
   b. state the number of real zeros.

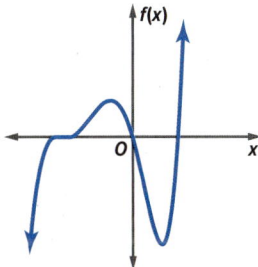

7. Analyze the function $f(x) = 2x^5 - 3x^4 + x^3 - 2x^2 - x - 1$ and describe its end behavior.

8. Use a graphing calculator to estimate the *x*-coordinates at which the maxima and minima of the function in exercise 7 occur. Round to the nearest hundredth.

9. Use the key features of the function to sketch its graph.

   y-intercept: (0, −2)
   Linearity: nonlinear
   Continuity: continuous
   Positive: for all values of *x* such that $x < -2$
   Decreasing: for all values of *x*
   Extrema: no relative maximum or minimum
   End Behavior: As $x \to \infty, f(x) \to -\infty$; as $x \to -\infty, f(x) \to \infty$

10. Graph $f(x) = \begin{cases} -x & \text{if } x < -2 \\ x + 2 & \text{if } -2 \leq x \leq 2 \\ 5 & \text{if } x > 2 \end{cases}$

11. Write the piecewise-defined function shown.

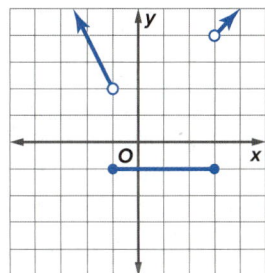

12. Graph $f(x) = 2|x| - 4$. Identify the domain and range.

13. Identify the domain and range of $y = [\![x]\!] + 2$.

14. Describe the translation to $y = x^2 + 5$.

15. Describe the reflection in $y = -|x|$.

16. Describe the dilation in $y = (5x)^2$.

Find the *x*- and *y*-intercepts of the graph of each function.

17. $f(x) = |x + 3| - 2$

18. $g(x) = x^2 + x - 6$

CHAPTER 2
Preparing for Assessment

Performance Task

Provide a clear solution to each part of the task. Be sure to show all of your work, include all relevant drawings, and justify your answers.

BUSINESS A statistical and business analysis company collects and evaluates data for its corporate clients.

Part A

The company gets several collections of data, which are shown in the graphs below.

Graph A
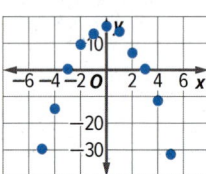

Graph B
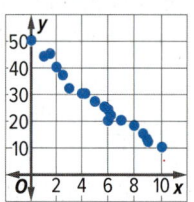

Graph C
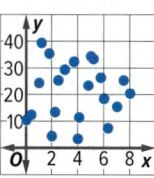

1. Determine for each graph whether an equation can be used to represent the data, and if so, the type of function that should be used to model the data.

Part B

The company does an analysis for a small family corporation that makes athletic shoes of potential profits and losses at various units produced. The data is shown on the graph.

2. Determine the greatest amount of profit the shoe company can expect to make.

3. Determine the greatest amount of loss the shoe company can expect to incur.

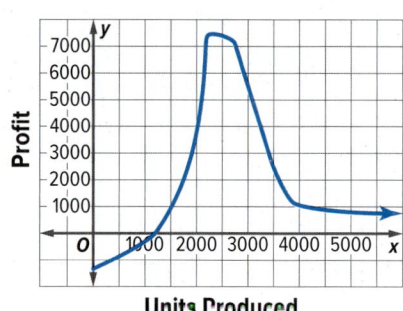

Part C

The company analyzes a set of data regarding the ideal number of products a company should produce based on their costs and consumer demand. They produce the equation
$p = -0.0125(n - 300)^2 + 500$, where p is the profit and n is the number of units of the product produced.

4. Determine the number of products at which the profit will be $0.

5. Determine the ideal number of products the company should make.

Test-Taking Strategy

Example

Read the problem. Identify what you need to know. Then use the information in the problem to solve.

Which of the following best describes the graph of $y - 2x \leq 6$?

A a solid line with shading above the line

B a solid line with shading below the line

C a dashed line with shading above the line

D a dashed line with shading below the line

Step 1 What do you know? What do you need to know? Is there enough information? Is there extra information?
The question gives the equation of the line. I need to put the equation in slope-intercept form and determine where it should be shaded.

Step 2 How are the facts given related? Are there any keywords that give you a clue on how to solve?
The answer choices all mention solid versus dashed lines and shading, so I know I need to determine what the graph of an inequality looks like.

Step 3 Can you eliminate any answer choices?
I can eliminate choices C and D because they are dashed lines, and an inequality with an "or equal to" sign is a solid line.

Step 4 What is the correct answer?
Choice B is correct.

> **Test-Taking Tip**
> **Strategies for Solving Reading Math Problems**
> The first step to solving any math problem is to read the problem. When reading a math problem to get the information you need to solve, it is helpful to use special reading strategies.

Apply the Strategy

Read the question. Then use the information in the problem to solve.

Which equation best represents the graph shown below?

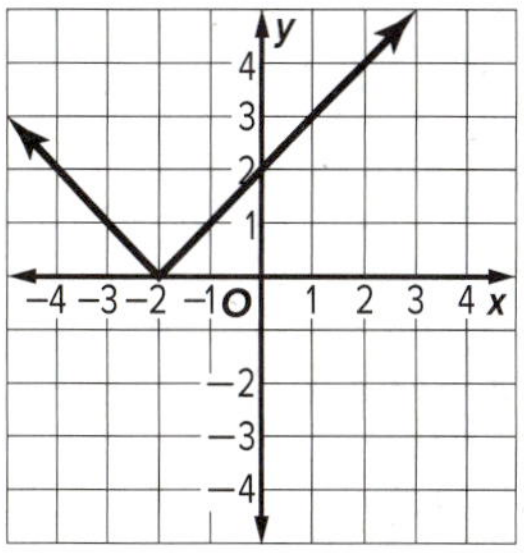

A $y = |x| + 2$ C $y = |-2x + 1|$

B $y = |2x + 1|$ D $y = |x + 2|$

Answer the questions below.

a. What do you know? What do you need to know? Is there enough information? Is there extra information?

b. How are the facts given related? Are there any keywords that give you a clue on how to solve?

c. Can you eliminate any answer choices?

d. What is the correct answer?

CHAPTER 2
Preparing for Assessment
Cumulative Review

Read each question. Then fill in the correct answer on the answer document provided by your teacher or on a sheet of paper.

1. Let $f(x) = -3x^2 - 2x$. What is $f\left(-\frac{1}{2}\right)$?
   - A 4
   - B $-\frac{7}{4}$
   - C $\frac{1}{4}$
   - D 2
   - E $\frac{13}{4}$

2. What is the y-intercept of the graph of $y = |x - 12| + 6$?

3. Which of the following scenarios should be represented by continuous functions? Select all that apply.
   - A A train with c cars can hold p people.
   - B A car uses g gallons of gas per m miles.
   - C A tree grows c centimeters for every n inches of rain.
   - D A baseball team gives away h hats for every t tickets bought.
   - E A person's femur bone is n inches for every f feet the person is tall.

4.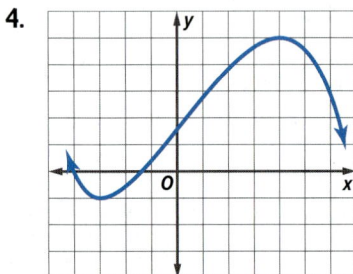

   What are the extrema of the graph shown above?

5. Graph the absolute value function that has x-intercepts of -6 and -2 and a minimum of $(-4, -4)$.

6. Which of the following functions have the end behavior $f(x) \to -\infty$, as $x \to -\infty$? Select all that apply.
   - A $y = -x^2$
   - B $y = -x^5 + 2$
   - C $y = -3x^4 - 9$
   - D $y = 2x^3 + 6$
   - E $y = 0.5x^7 - 2.5$

7. What is the value of the function when $x = 0$?
$$f(x) = \begin{cases} -\frac{1}{2}x - 4 & \text{if } x < 0 \\ 5x + 8 & \text{if } 0 \leq x < 9 \\ -2x + 10 & \text{if } x \geq 9 \end{cases}$$
   - A -4
   - B 8
   - C 9
   - D 10
   - E undefined

8. Determine whether each function is *one-to-one*, *onto*, *both*, or *neither*.

   a.
Domain	-1	-2	-3	-4
Range	8	4	-4	-8

   b.
Domain	5	10	15	20
Range	0	1	2	3

   c.
Domain	-2	0	0	2
Range	-4	-2	2	4

   d.
Domain	0.5	1.0	1.5	2.0
Range	0	0	-1	-2

Go Online! for Standardized Test Practice

9. Determine whether each graph has a *line of symmetry*, a *point of symmetry*, *both*, or *neither*.

a.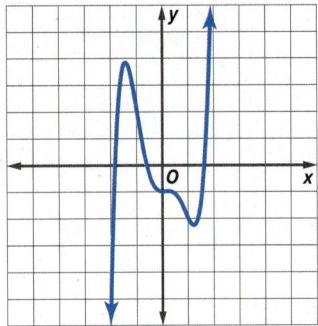

b.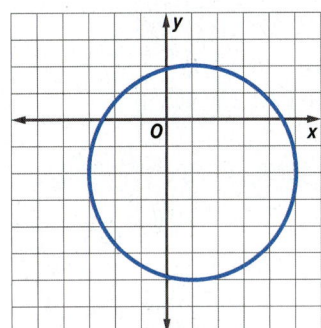

c.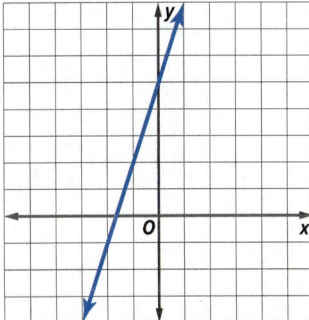

d.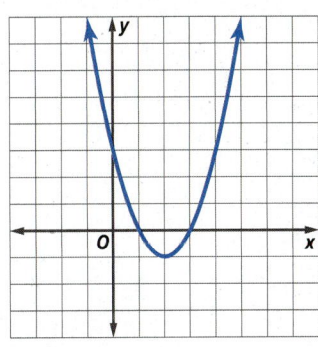

10. Which inequality has no solutions in the first quadrant?

○ A $y > -|x|$

○ B $y < |20 - x|$

○ C $y < -|x|$

○ D $y < |x| - 20$

11. What is an equation of the function shown in the graph?

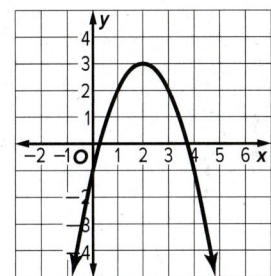

○ A $y = (x - 2)^2 + 3$

○ B $y = -(x + 2)^2 + 3$

○ C $y = -(x - 3)^2 + 2$

○ D $y = -(x - 2)^2 + 3$

12. Let $f(x) = |x + 42| + 35$. What is the minimum value of the function?

Test-Taking Tip

Question 12 The minimum value of a quadratic function like $f(x)$ occurs at its vertex.

Need Extra Help?

If you missed Question…	1	2	3	4	5	6	7	8	9	10	11	12
Go to Lesson…	2-1	2-6	2-1	2-3	2-4	2-3	2-5	2-1	2-2	2-5	2-6	2-3

CHAPTER 3
Quadratic Functions

THEN
You graphed linear equations and inequalities.

NOW
You will:
- Graph quadratic functions.
- Solve quadratic equations.
- Perform operations with complex numbers.
- Graph and solve quadratic inequalities.

 WHY

MOTION In soccer, it is important to have a plan for where the ball will go and how fast it will get there. Linear functions can be used to model an object, like a soccer ball, in motion.

Use the Mathematical Practices to complete the activity.

1. **Using Tools** Use the Internet to learn about linear motion and how it could apply to ball movement.
2. **Sense Making** What equation can be applied to determine the distance traveled by a ball? What equation can be used to determine average velocity of the ball?
3. **Reasoning** Why do you need to know the distance traveled to calculate the average velocity?
4. **Applying Math** If you know the distance a ball traveled and the time it took, use the Equation Chart Mat to illustrate the equations needed to find the average velocity.

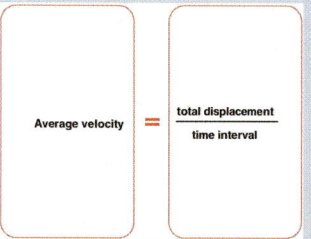

Average velocity = total displacement / time interval

 ## Go Online to Guide Your Learning

Explore & Explain

 Product Mat and Algebra Tiles
Use the **Product Mat** and the **Algebra Tiles** in ConnectED to enhance your understanding of factoring trinomials, as explored in Lesson 3-4.

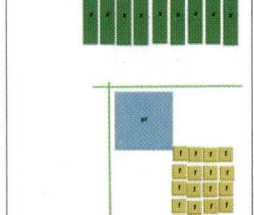

The Geometer's Sketchpad
Use **The Geometer's Sketchpad** to illustrate how to graph quadratic functions, how to solve quadratic equations graphically, and to relate the graphs of quadratic equations to their solutions. The Geometer's Sketchpad can also be used to illustrate how to recognize the number of roots for quadratic equations and to illustrate how to graph the solution sets of quadratic inequalities, as discussed in Lesson 3-7.

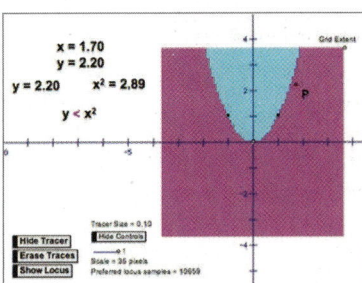

eBook
Interactive Student Guide
Before starting the chapter, answer the **Chapter Focus** preview questions. Check your answers as you complete each lesson. At the end of the chapter, try the **Performance Task**.

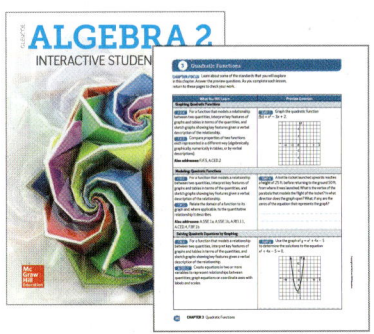

Organize

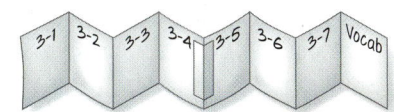 **Foldables**
Get organized! Create a **Quadratic Functions Foldable** before you start the chatper to organize your notes about quadratic functions and relations.

Collaborate

 Chapter Project
In the **Just Drop It** project, you will use what you have learned about quadratic functions to complete a project that addresses global awareness.

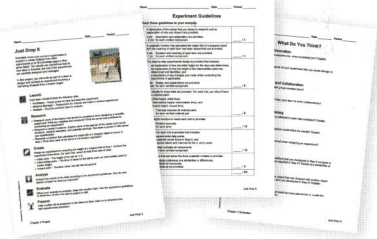

Focus

 LEARNSMART
Need help studying? Complete the **Polynomial, Rational, and Radical Relationships** and **Modeling with Functions** domains in LearnSmart to review for the chapter test.

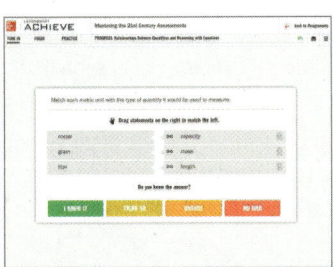

ALEKS
You can use the **Quadratic and Polynomial Functions** and **Exponents and Polynomial Expressions** topics in ALEKS to explore what you know about quadratic functions and what you are ready to learn.*

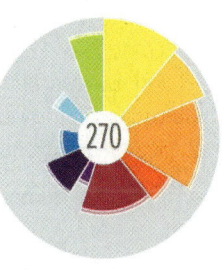

*Ask your teacher if this is part of your program.

Get Ready for the Chapter

Connecting Concepts

Concept Check
Review the concepts used in this chapter by answering the questions below.

1. Given $f(x) = -2x^2 + 3x - 1$ and $f(2)$, what would be your first step to find the value of the equation?

2. Marla has budgeted $65 per day on food during a business trip. Write a function that is a model for this situation.

3. If Marla, from question 2, is gone on a business trip for 2 weeks, how would you determine how much she has budgeted to spend on food? What is her budget?

4. Is the polynomial $2x^2 - x - 3$ factorable or prime?

5. Is the polynomial $x^2 - 11x + 15$ factorable or prime?

6. The rectangular room shown has an area of $x^2 + 14x + 48$ square feet. If the width of the room is $(x + 6)$ feet, how would you determine the length of the room?

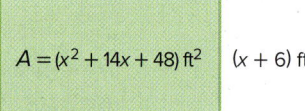
$A = (x^2 + 14x + 48)$ ft² $(x + 6)$ ft

7. What is the length of the room described in question 6?

8. Given $= 2x^2 + 2x - 3x - 3$, what property can you apply to begin to simplify the equation?

9. Given $= 2x(x + 1) + -3(x + 1)$, what property can you apply to simplify the equation?

New Vocabulary

English		Español
quadratic term	p. 151	término cuadrático
linear term	p. 151	término lineal
constant term	p. 151	término constante
vertex	p. 152	vértice
maximum value	p. 154	valor máximo
minimum value	p. 154	valor mínimo
quadratic equation	p. 163	ecuación cuadrática
standard form	p. 163	forma estándar
root	p. 163	raíz
zero	p. 163	cero
imaginary unit	p. 172	unidad imaginaria
pure imaginary number	p. 172	número imaginario puro
complex number	p. 173	número complejo
complex conjugates	p. 175	conjugados complejos
factored form	p. 179	forma reducida
FOIL method	p. 179	método FOIL
completing the square	p. 192	completar el cuadrado
Quadratic Formula	p. 199	fórmula cuadrática
discriminant	p. 202	discriminante
quadratic inequality	p. 209	desigualdad cuadrática

Performance Task Preview

You can use the concepts and skills in the chapter to model the projectile motion of a bottle rocket. Understanding quadratic functions will help you finish the Performance Task at the end of the chapter.

MP In this Performance Task you will:
- make sense of problems and persevere in solving them
- model with mathematics
- attend to precision

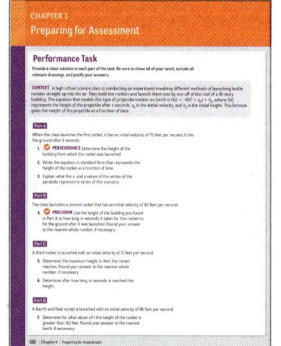

Review Vocabulary

domain *dominio* the set of all *x*-coordinates of the ordered pairs of a relation

function *función* a relation in which each *x*-coordinate is paired with exactly one *y*-coordinate

range *rango* the set of all *y*-coordinates of the ordered pairs of a relation

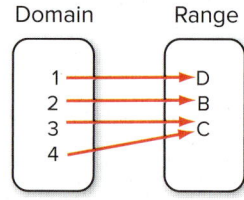

Go Online! for Vocabulary Review Games and key vocabulary in 13 languages.

LESSON 1
Graphing Quadratic Functions

::Then   ::Now   ::Why?

- You identified and manipulated graphs of functions.
- 1. Graph quadratic functions.
  2. Find and interpret the maximum and minimum values of a quadratic function.
- Eddie is organizing a charity tournament. He plans to charge a $20 entry fee for each of the 80 players. He recently decided to raise the entry fee by $5, and 5 fewer players entered with the increase. He used this information to determine how many fee increases will maximize the money raised.

  The quadratic function at the right represents this situation. The tournament prize pool increases when he first increases the fee, but eventually the pool starts to decrease as the fee gets even higher.

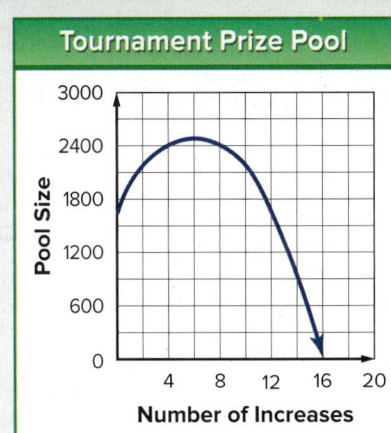

New Vocabulary
quadratic function
quadratic term
linear term
constant term
parabola
axis of symmetry
vertex
maximum value
minimum value

Mathematical Practices
1 Make sense of problems and perservere in solving them.
4 Model with mathematics.

1 Graph Quadratic Functions
In a **quadratic function**, the greatest exponent is 2. These functions can have a **quadratic term**, a **linear term**, and a **constant term**. The general quadratic function is shown below.

$$f(x) = ax^2 + bx + c, \text{ where } a \neq 0$$

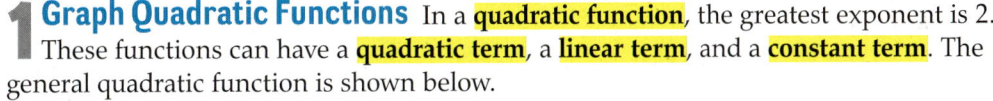

The graph of a quadratic function is called a **parabola**. To graph a quadratic function, graph ordered pairs that satisfy the function.

Example 1   Graph a Quadratic Function by Using a Table

Graph $f(x) = 3x^2 - 12x + 6$ by making a table of values.

Choose integer values for x, and evaluate the function for each value. Graph the resulting coordinate pairs, and connect the points with a smooth curve.

x	$3x^2 - 12x + 6$	$f(x)$	$(x, f(x))$
0	$3(0)^2 - 12(0) + 6$	6	(0, 6)
1	$3(1)^2 - 12(1) + 6$	-3	(1, -3)
2	$3(2)^2 - 12(2) + 6$	-6	(2, -6)
3	$3(3)^2 - 12(3) + 6$	-3	(3, -3)
4	$3(4)^2 - 12(4) + 6$	6	(4, 6)

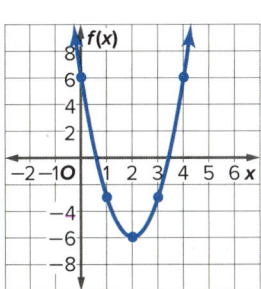

▶ Guided Practice

Graph each function by making a table of values.

1A. $g(x) = -2x^2 + 8x - 3$ **1B.** $h(x) = 4x^2 - 8x + 1$

connectED.mcgraw-hill.com   151

Notice in Example 1 that there seemed to be a pattern in the values for $f(x)$. This is due to the axis of symmetry of parabolas. The **axis of symmetry** is a line through the graph of a parabola that divides the graph into two congruent halves. Each side of the parabola is a reflection of the other side.

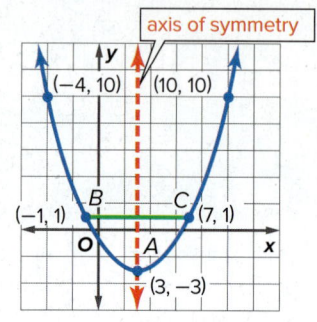

Review Vocabulary

Symmetry When something is symmetrical, its opposite sides are mirror images of each other.

The axis of symmetry will intersect a parabola at only one point, called the **vertex**. The vertex of the graph at the right is $A(3, -3)$.

Notice that the x-coordinates of points B and C are both 4 units away from the x-coordinate of the vertex, and they have the same y-coordinate. This is due to the symmetrical nature of the graph.

Go Online!

How does changing the values of a, b, and c in the equation of a quadratic function affect its graph? Investigate by using the **Graphing Tools** in ConnectED.

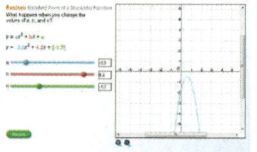

Key Concept Graph of a Quadratic Function—Parabola

Words Consider the graph of $y = ax^2 + bx + c$, where $a \neq 0$.

- The y-intercept is $a(0)^2 + b(0) + c$ or c.
- The equation of the axis of symmetry is $x = -\frac{b}{2a}$.
- The x-coordinate of the vertex is $-\frac{b}{2a}$.

Model

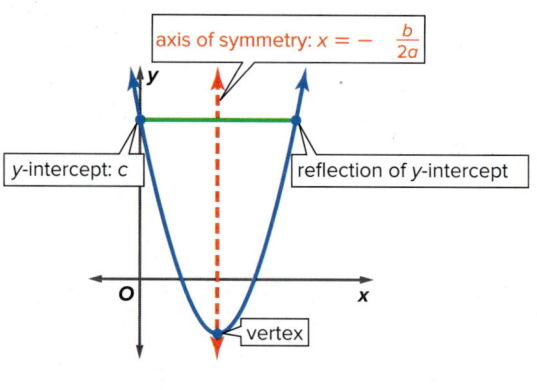

Now you can use the axis of symmetry to help plot points and graph a parabola. For $y = x^2 + 6x - 2$ below, the axis of symmetry is $x = -\frac{b}{2a} = -\frac{6}{2(1)}$ or $x = -3$.

Study Tip

Plotting Reflections The reflection of a point is its mirror image on the other side of the axis of symmetry.

Find the axis of symmetry and the vertex.

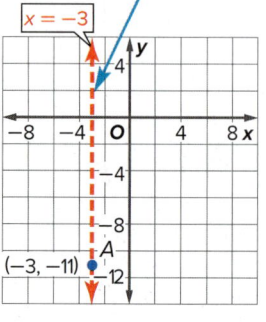

Find the y-intercept and its reflection.

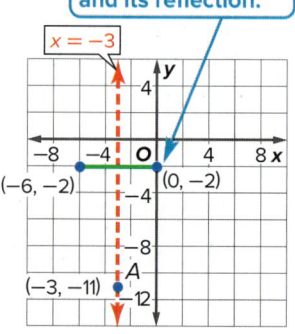

Connect the points with a smooth curve.

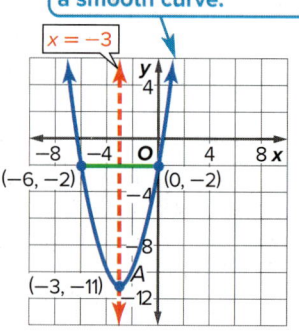

Example 2 Axis of Symmetry, y-intercept, and Vertex

Consider $f(x) = x^2 + 4x - 3$.

a. Find the y-intercept, the equation of the axis of symmetry, and the x-coordinate of the vertex.

The function is of the form $f(x) = ax^2 + bx + c$, so we can identify a, b, and c.

$$f(x) = ax^2 + bx + c$$
$$\downarrow \quad \downarrow \quad \downarrow$$
$$f(x) = 1x^2 + 4x - 3 \quad \rightarrow \quad a = 1, b = 4, \text{ and } c = -3$$

The y-intercept is $c = -3$.

Use a and b to find the equation of the axis of symmetry.

$x = -\dfrac{b}{2a}$ Equation of the axis of symmetry

$ = -\dfrac{4}{2(1)}$ $a = 1$ and $b = 4$

$ = -2$ Simplify.

The equation of the axis of symmetry is $x = -2$. Therefore, the x-coordinate of the vertex is -2.

b. Make a table of values that includes the vertex.

Select five specific points, with the vertex in the middle and two points on either side of the vertex, including the y-intercept and its reflection. Use symmetry to determine the y-values of the reflections.

x	$x^2 + 4x - 3$	f(x)	(x, f(x))
−6	$(-6)^2 + 4(-6) - 3$	9	(−6, 9)
−4	$(-4)^2 + 4(-4) - 3$	−3	(−4, −3)
−2	$(-2)^2 + 4(-2) - 3$	−7	(−2, −7)
0	$(0)^2 + 4(0) - 3$	−3	(0, −3)
2	$(2)^2 + 4(2) - 3$	9	(2, 9)

c. Use this information to graph the function.

Graph the points from the table and connect them with a smooth curve.

Draw the axis of symmetry, $x = -2$, as a dashed line. The graph should be symmetrical about this line.

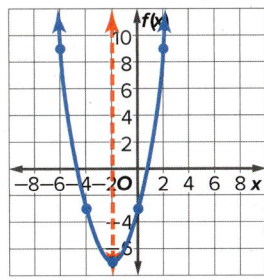

> **Study Tip**
> **Quadratic Form** Make sure the function is in standard quadratic form, $y = ax^2 + bx + c$, before graphing.

> **Study Tip**
> **Fractions** When the x-coordinate of the vertex is a fraction, select the nearest integer for the next point to avoid using fractions and simplify the calculations.

Guided Practice

2. Consider $f(x) = -5x^2 - 10x + 6$.

A. Find the y-intercept, the equation of the axis of symmetry, and the x-coordinate of the vertex.

B. Make a table of values that includes the vertex.

C. Use this information to graph the function.

Watch Out!

Maxima and Minima The terms *minimum point* and *minimum value* are not interchangeable. The minimum point on the graph of a quadratic function is the ordered pair that describes the location of the vertex. The minimum value of a function is the *y*-coordinate of the minimum point. It is the smallest value obtained when $f(x)$ is evaluated for all values of *x*.

2 Maximum and Minimum Values

The *y*-coordinate of the vertex of a quadratic function is the **maximum value** or the **minimum value** of the function. These values represent the greatest or lowest possible value the function can reach.

Key Concept Maximum and Minimum Value

Words   The graph of $f(x) = ax^2 + bx + c$, where $a \neq 0$,
- opens up and has a minimum value when $a > 0$, and
- opens down and has a maximum value when $a < 0$.

Model   *a* is positive.   *a* is negative.

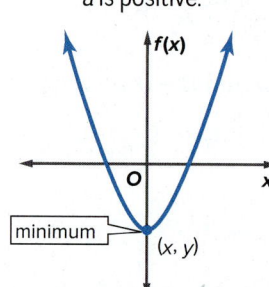
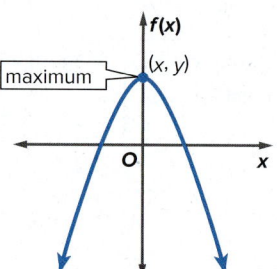

The *y*-coordinate is the minimum value.   The *y*-coordinate is the maximum value.

Example 3   Maximum or Minimum Values

Consider $f(x) = -4x^2 + 12x + 18$.

a. Determine whether the function has a *maximum* or *minimum* value.

For this function, $a = -4$, so the graph opens down and the function has a maximum value.

b. State the maximum or minimum value of the function.

The maximum value of the function is the *y*-coordinate of the vertex.

The *x*-coordinate of the vertex is $-\frac{12}{2(-4)}$ or 1.5.

Find the *y*-coordinate of the vertex by evaluating the function for $x = 1.5$.

$f(x) = -4x^2 + 12x + 18$ Original function

$= -4(1.5)^2 + 12(1.5) + 18$ $x = 1.5$

$= -9 + 18 + 18$ or 27   The maximum value of the function is 27.

c. State the domain and range of the function.

$D = (-\infty, +\infty)$, {all real numbers}, or $\{-\infty < x < +\infty\}$
$R = (-\infty, 27]$, $\{f(x) \mid f(x) \leq 27\}$, or $\{-\infty < x \leq 27\}$

Study Tip

Domain and Range The domain of a quadratic function will always be all real numbers. The range will either be all real numbers less than or equal to the maximum or all real numbers greater than or equal to the minimum.

Guided Practice

3. Consider $f(x) = 4x^2 - 24x + 11$.

A. Determine whether the function has a maximum or minimum value.

B. State the maximum or minimum value of the function.

C. State the domain and range of the function.

Real-World Link

As of 2012, there were approximately 2.3 million nonprofit organizations in the United States.

Source: National Center for Charitable Statistics

Real-World Example 4 Quadratic Equations in the Real World

CHARITY Refer to the beginning of the lesson.

a. How much should Eddie charge in order to maximize charity income?

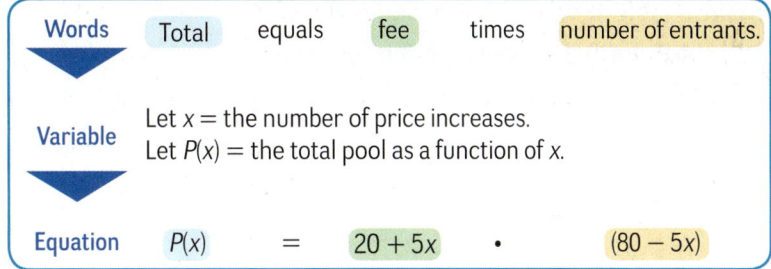

Solve for the x-value of the vertex.

$P(x) = (20 + 5x) \cdot (80 - 5x)$

$\quad\quad = 20(80) + 20(-5x) + 5x(80) + 5x(-5x)$ Distribute.

$\quad\quad = 1600 - 100x + 400x - 25x^2$ Multiply.

$\quad\quad = 1600 + 300x - 25x^2$ Simplify.

$\quad\quad = -25x^2 + 300x + 1600$ $ax^2 + bx + c$ form

Use the formula for the axis of symmetry, $x = -\dfrac{b}{2a}$, to find the x-coordinate.

$x = -\dfrac{300}{2(-25)}$ or 6 $a = -25$ and $b = 300$

Eddie needs to have 6 price increases, so he should charge $20 + 6(5)$ or $50.

b. What will be the maximum value of the pool?

Find the maximum value of the quadratic function $P(x)$ by evaluating $P(6)$.

$P(x) = -25x^2 + 300x + 1600$ Total pool function

$P(6) = -25(6)^2 + 300(6) + 1600$ $x = 6$

$\quad\quad = -900 + 1800 + 1600$ or 2500 Simplify.

Thus, the maximum prize pool is $2500 after 6 price increases.

CHECK Graph the function on a graphing calculator and use the **CALC:maximum** function to confirm the solution.

Select a left bound of 0 and a right bound of 10. The calculator will display the coordinates of the maximum at the bottom of the screen.

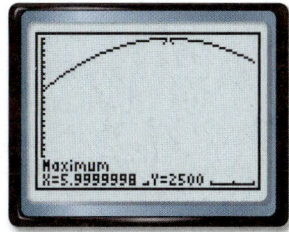

[0, 10] scl: 1 by [0, 2500] scl: 100

The domain is $\{x \mid x \geq 0\}$ because there can be no negative increases in price. The range is $\{y \mid 0 \leq y \leq 2500\}$ because the prize pool cannot have a negative monetary value.

Study Tip

MP Modeling Use logic and the information from the problem to determine the domain and range that are reasonable in the situation.

▶ **Guided Practice**

4. Suppose a different tournament that Eddie organizes has 120 players and the entry fee is $40. Each time he increases the fee by $5, he loses 10 players. Determine what the entry fee should be to maximize the value of the pool.

Check Your Understanding

= Step-by-Step Solutions begin on page R11.

Examples 1–2 Complete parts a–c for each quadratic function.

a. Find the y-intercept, the equation of the axis of symmetry, and the x-coordinate of the vertex.
b. Make a table of values that includes the vertex.
c. Use this information to graph the function.

1. $f(x) = 3x^2$
2. $f(x) = -6x^2$
3. $f(x) = x^2 - 4x$
4. $f(x) = -x^2 - 3x + 4$
5. $f(x) = 4x^2 - 6x - 3$
6. $f(x) = 2x^2 - 8x + 5$

Example 3 Determine whether each function has a *maximum* or *minimum* value, and find that value. Then state the domain and range of the function.

7. $f(x) = -x^2 + 6x - 1$
8. $f(x) = x^2 + 3x - 12$
9. $f(x) = 3x^2 + 8x + 5$
10. $f(x) = -4x^2 + 10x - 6$

Example 4

11. **BUSINESS** Beach Bikes rents 1400 bicycles per week at $22.50 per rental. The owner estimates that she will rent 100 fewer bikes for each $2.50 increase in price. What price will maximize the income of the store?

Practice and Problem Solving

Extra Practice is on page R3.

Examples 1–2 Complete parts a–c for each quadratic function.

a. Find the y-intercept, the equation of the axis of symmetry, and the x-coordinate of the vertex.
b. Make a table of values that includes the vertex.
c. Use this information to graph the function.

12. $f(x) = 4x^2$
13. $f(x) = -2x^2$
14. $f(x) = x^2 - 5$
15. $f(x) = x^2 + 3$
16. $f(x) = 4x^2 - 3$
17. $f(x) = -3x^2 + 5$
18. $f(x) = x^2 - 6x + 8$
19. $f(x) = x^2 - 3x - 10$
20. $f(x) = -x^2 + 4x - 6$
21. $f(x) = -2x^2 + 3x + 9$

Example 3 Determine whether each function has a *maximum* or *minimum* value, and find that value. Then state the domain and range of the function.

22. $f(x) = 5x^2$
23. $f(x) = -x^2 - 12$
24. $f(x) = x^2 - 6x + 9$
25. $f(x) = -x^2 - 7x + 1$
26. $f(x) = 8x - 3x^2 + 2$
27. $f(x) = 5 - 4x - 2x^2$
28. $f(x) = 15 - 5x^2$
29. $f(x) = x^2 + 12x + 27$
30. $f(x) = -x^2 + 10x + 30$
31. $f(x) = 2x^2 - 16x - 42$

Example 4

32. **MODELING** A financial analyst determined that the cost, in thousands of dollars, of producing bicycle frames is $C = 0.000025f^2 - 0.04f + 40$, where f is the number of frames produced.

a. Find the number of frames that minimizes cost.
b. What is the total cost for that number of frames?

Complete parts a–c for each quadratic function.

a. Find the y-intercept, the equation of the axis of symmetry, and the x-coordinate of the vertex.

b. Make a table of values that includes the vertex.

c. Use this information to graph the function.

33. $f(x) = 2x^2 - 6x - 9$

34. $f(x) = -3x^2 - 9x + 2$

35. $f(x) = -4x^2 + 5x$

36. $f(x) = 2x^2 + 11x$

37. $f(x) = 0.25x^2 + 3x + 4$

38. $f(x) = -0.75x^2 + 4x + 6$

39. $f(x) = \frac{3}{2}x^2 + 4x - \frac{5}{2}$

40. $f(x) = \frac{2}{3}x^2 - \frac{7}{3}x + 9$

41. **FINANCIAL LITERACY** A babysitting club sits for 50 different families. They would like to increase their current rate of $9.50 per hour. After surveying the families, the club finds that the number of families will decrease by about 2 for each $0.50 increase in the hourly rate.

   a. Write a quadratic function that models this situation.
   b. State the domain and range of this function as it applies to the situation.
   c. What hourly rate will maximize the club's income? Is this reasonable?
   d. What is the maximum income the club can expect to make?

42. **ACTIVITIES** Last year, 300 people attended the Franklin High School Drama Club's winter play. The ticket price was $8. The advisor estimates that 20 fewer people would attend for each $1 increase in ticket price.

   a. What ticket price would give the greatest income for the Drama Club?
   b. If the Drama Club raised its tickets to this price, how much income should it expect to bring in?

MP TOOLS Use a calculator to find the maximum or minimum of each function. Round to the nearest hundredth if necessary.

43. $f(x) = -9x^2 - 12x + 19$

44. $f(x) = 12x^2 - 21x + 8$

45. $f(x) = -8.3x^2 + 14x - 6$

46. $f(x) = 9.7x^2 - 13x - 9$

47. $f(x) = 28x - 15 - 18x^2$

48. $f(x) = -16 - 14x - 12x^2$

Determine whether each function has a *maximum* or *minimum* value, and find that value. Then state the domain and range of the function.

49. $f(x) = -5x^2 + 4x - 8$

50. $f(x) = -4x^2 - 3x + 2$

51. $f(x) = -9 + 3x + 6x^2$

52. $f(x) = 2x - 5 - 4x^2$

53. $f(x) = \frac{2}{3}x^2 + 6x - 10$

54. $f(x) = -\frac{3}{5}x^2 + 4x - 8$

Determine the function represented by each graph.

55.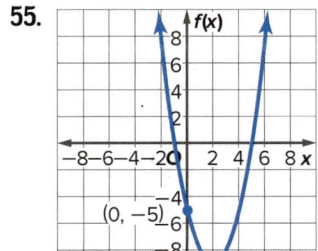

56.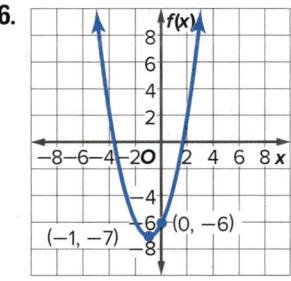

57.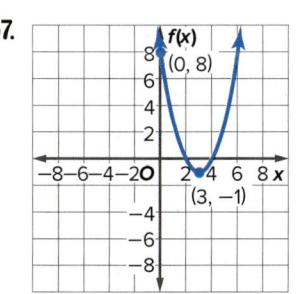

58. **MULTIPLE REPRESENTATIONS** Consider $f(x) = x^2 - 4x + 8$ and $g(x) = 4x^2 - 4x + 8$.
    a. **Tabular** Make a table of values for $f(x)$ and $g(x)$ if $-4 \leq x \leq 4$.
    b. **Graphical** Graph $f(x)$ and $g(x)$.
    c. **Verbal** Explain the difference in the shapes of the graphs of $f(x)$ and $g(x)$. What value was changed to cause this difference?
    d. **Analytical** Predict the appearance of the graph of $h(x) = 0.25x^2 - 4x + 8$. Confirm your prediction by graphing all three functions if $-10 \leq x \leq 10$.

59. **MULTI-STEP** Omar owns a vending machine in a bowling alley. He currently sells 600 bottles of soda per week at $1.50 per bottle. He estimates that he will lose 50 customers for every $0.25 increase in price. He has a fixed cost of $40 for every 100 bottles he buys. He also incurs a variable cost of $10 when he restocks the vending machine. He needs to restock the machine once it falls below 50% of its capacity of 400 bottles.
    a. What is his maximum weekly profit?
    b. Explain your solution process.

60. **BASEBALL** The height h of a baseball in feet after it is thrown straight up t seconds after it is released is given by the function $h(t) = -16t^2 + 30t + 5$.
    a. State the domain and range for this situation.
    b. Find the maximum height the ball will reach.

H.O.T. Problems   Use Higher-Order Thinking Skills

61. **ERROR ANALYSIS** Trent thinks that the function $f(x)$ graphed below, and the function $g(x)$ described next to it have the same maximum. Madison thinks that $g(x)$ has a greater maximum. Is either of them correct? Explain your reasoning.

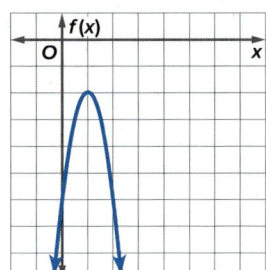

$g(x)$ is a quadratic function with roots of 4 and 2 and a y-intercept of -8.

62. **REASONING** Determine whether the following statement is *sometimes*, *always*, or *never* true. Explain your reasoning.
    In a quadratic function, if two x-coordinates are equidistant from the axis of symmetry, then they will have the same y-coordinate.

63. **CHALLENGE** The table at the right represents some points on the graph of a quadratic function.
    a. Find the values of a, b, c, and d.
    b. What is the x-coordinate of the vertex?
    c. Does the function have a maximum or a minimum?

64. **OPEN-ENDED** Give an example of a quadratic function with a
    a. maximum of 8.   b. minimum of -4.   c. vertex of $(-2, 6)$.

65. **WRITING IN MATH** How can you apply what you have learned about symmetry and maximum and minimum values to the graph of an absolute value function, $f(x) = |x|$?

x	y
-20	-377
c	-13
-5	-2
-1	22
$d - 1$	a
5	$a - 24$
7	$-b$
15	-202
$14 - c$	-377

Preparing for Assessment

66. Consider the function $f(x) = 3x^2 - 11x + 5$. What is the equation of the axis of symmetry? **MP 6**

- A $x = -\frac{11}{3}$
- B $x = -\frac{11}{6}$
- C $x = \frac{11}{6}$
- D $x = \frac{11}{3}$

67. A pizza parlor sells 320 pizzas per night when they charge $9.00 per pizza. The owner believes increasing the price by $0.15 will result in selling four fewer pizzas per night. What price, in dollars, should he charge to generate the maximum total revenue from pizza sales? **MP 1**

68. Consider the function $f(x) = 2x^2 - 16x + 13$. Which statement best describes the function? **MP 7**

- A The function has a minimum value of -19.
- B The function has a minimum value of 4.
- C The function has a maximum value of 13.
- D The function has a maximum value of 17.

69. Consider the function $f(x) = x^2 - 8x - 20$. What is the range of the function? **MP 1**

- A $\{f(x) \mid f(x) \geq -36\}$
- B $\{f(x) \mid f(x) \geq -32\}$
- C $\{f(x) \mid f(x) \geq -20\}$
- D $\{f(x) \mid f(x) \geq 0\}$
- E $\{f(x) \mid f(x) \geq 28\}$

70. Which of the following functions have a graph whose parabola opens up? Select all that apply. **MP 2**

- A $f(x) = 3x^2 + 2x - 5$
- B $f(x) = 3x^2 + x + 1$
- C $f(x) = -x^2 + 3x + 8$
- D $f(x) = x^2 - 7x + 12$
- E $f(x) = -2x^2 - x + 1$
- F $f(x) = -5x^2 - 5x - 5$

71. Which function best represents the graph shown below? **MP 7**

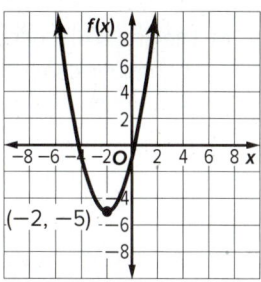

- A $f(x) = x^2 - 4x - 1$
- B $f(x) = x^2 - 2x - 1$
- C $f(x) = x^2 + 2x - 1$
- D $f(x) = x^2 + 4x - 1$

72. MULTI-STEP Consider the function $f(x) = 4x^2 - 24x + 36$. **MP 2**

a. What is the y-intercept of this function?

- A 9
- B -9
- C 36
- D -36

b. Which of the following statements are true?

- A The axis of symmetry of this function is $x = 3$.
- B The axis of symmetry of this function is $y = 3$.
- C A vertical line passed through a graph of this function would cross the line twice.
- D The function is a curved line with the ends pointing up.
- E The function is a curved line with the ends pointing down.

c. Factor to write this function another way.

EXTEND 3-1

Graphing Technology Lab
Modeling Real-World Data

When three points on the graph of a parabola are known, a system of equations can be used to write a quadratic function.

Mathematical Practices

4 Model with mathematics.
5 Use appropriate tools strategically.

Activity 1 Determine Quadratic Model for Real-World Data

WEATHER Andi recorded the outside temperature several times during the day.

a. Write a quadratic function that could be used to model the data. Use your model to predict the temperature at 10 P.M.

Time	Temperature(°F)
8 A.M.	68
12 P.M.	80
4 P.M.	84

Let x represent the number of hours since 12 A.M. and y represent the temperature. So, three points for the quadratic function are (8, 68), (12, 80), and (16, 84).

Substitute each point in $ax^2 + bx + c = y$ to write a system of three equations.

$a(8)^2 + b(8) + c = 68 \rightarrow 64a + 8b + c = 68$ (Equation 1)
$a(12)^2 + b(12) + c = 80 \rightarrow 144a + 12b + c = 80$ (Equation 2)
$a(16)^2 + b(16) + c = 84 \rightarrow 256a + 16b + c = 84$ (Equation 3)

Solve the system.

(Equation 1) − (Equation 2)

$64a + 8b + c = 68$
$(-)144a + 12b + c = 80$
$\overline{{-}80a - 4b \phantom{{}+c} = -12}$

(Equation 2) − (Equation 3)

$144a + 12b + c = 80$
$(-)256a + 16b + c = 84$
$\overline{{-}112a - 4b \phantom{{}+c} = -4}$

Solve the resulting system of two equations.

${-}80a - 4b = -12$
$(-){-}112a - 4b = -4$
$\overline{32a \phantom{{}-4b} = -8}$
$a = -\frac{8}{32} \text{ or } -0.25$

$-80a - 4b = -12$
$-80(-0.25) - 4b = -12$
$20 - 4b = -12$
$-4b = -32$
$b = 8$

Solve for c using one of the initial equations.

$64a + 8b + c = 68$
$64(-0.25) + 8(8) + c = 68$
$-16 + 64 + c = 68$
$c = 20$

So, $a = -0.25$, $b = 8$, and $c = 20$.

A quadratic function that models the situation is $f(x) = -0.25x^2 + 8x + 20$.

CHECK You can use a matrix and a graphing calculator to verify your answer.

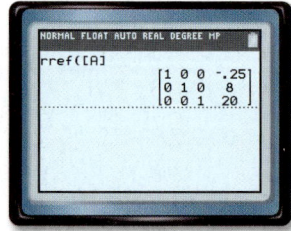

To predict the temperature at 10 P.M., or 22 hours since 12 A.M., substitute 22 for x in the modeling function.

$$f(x) = -0.25x^2 + 8x + 20$$
$$f(22) = -0.25(22)^2 + 8(22) + 20$$
$$= 75$$

According to the modeling function, the temperature at 10 P.M. would be 75°F.

b. Determine whether this quadratic model is suitable for predicting the temperature for the next two days. Explain.

Examine the graph of the function. The parabola opens down. This indicates that as x increases, the value of y decreases. Or, as the number of hours increases, the temperature decreases. The model is not suitable as it would predict temperatures that continually decrease without ever increasing.

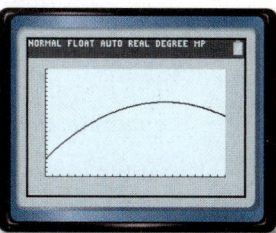

[0, 24] scl: 1 by [0, 120] scl: 5

Previously, you determined a line of regression and its equation for a set of data. Similarly, you can determine a quadratic regression equation for data. Often, you will need to compare the correlation coefficient r for various models to determine which is the most appropriate.

Activity 2 Determine Model for Real-World Data

WATER A container is filled with water. The water is allowed to drain from a hole made near the bottom of the container. The table shows the level of water y measured in centimeters from the bottom of the container after x seconds.

Time (s)	0	20	40	60	80	100	120	140	160	180	200	220
Water level (cm)	42.6	40.7	38.9	37.2	35.8	34.3	33.3	32.3	31.5	30.8	30.4	30.1

a. Determine whether a linear model or a quadratic model best fits the data.

Step 1 Enter the data. Place the values for time in **L1** and the values for the water level in **L2**.

Step 2 Calculate and graph a linear regression equation.

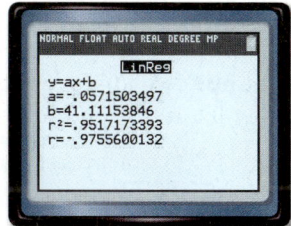 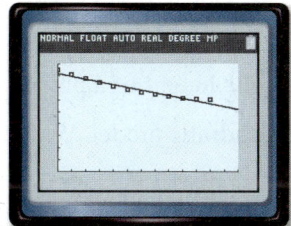

[0, 260] scl: 20 by [0, 45] scl: 5

The r-value, $r \approx -0.976$, indicates that a linear model fits the data well because $|r|$ is close to 1.

(continued on the next page)

EXTEND 3-1

Graphing Technology Lab
Modeling Real-World Data Continued

Step 3 Calculate and graph a quadratic regression equation.

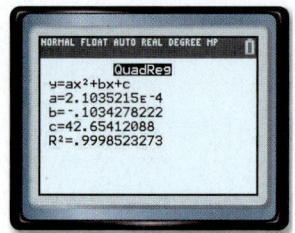 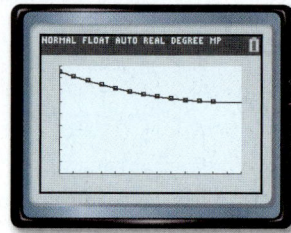

[0, 260] scl: 20 by [0, 45] scl: 5

The R^2-value corresponds to an r-value of about 0.999. So, the quadratic regressions fits the data extremely well.

Recall that the closer $|r|$ is to 1, the better the fit of a regression equation. Because the r-value for the quadratic regression is closer to 1 than that of the linear regression, the quadratic model is more appropriate for the data.

b. Determine which regression equation would make better predictions about the water level for any time after 220 seconds.

The linear regression would be better for making predictions because it is continually decreasing, just as the water level would. The quadratic regression begins to increase indicating that the water level begins to rise, which is not possible.

Exercises

Work cooperatively.

1. **ROCKETS** Ari launched a model rocket straight up. Sensors in the rocket recorded data for 3 seconds before malfunctioning.

   a. Write a quadratic function to represent this situation. Then use the function to determine the altitude of the rocket for each of the next 5 seconds.

   b. Determine whether this quadratic model is suitable for predicting the altitude of the rocket. Explain.

Time (s)	Altitude (ft)
1	160
2	287
3	382

Work cooperatively. The table shows the height of a person jumping straight up.

2. Determine a linear model and a quadratic model. Which best fits the data?

3. Use each regression equation to estimate the height of the player after 1 second and 1.5 seconds.

4. Compare and contrast the estimates you found in Exercise 3.

5. How might choosing a regression equation that does not fit the data well affect predictions made by using the equation?

Height of Player's Feet above Floor

Time (s)	Height (in.)
0.1	3.04
0.2	5.76
0.3	8.16
0.4	10.24
0.5	12
0.6	13.44
0.7	14.56

LESSON 2
Solving Quadratic Equations by Graphing

∵Then | ∵Now | ∵Why?

- You solved systems of equations by graphing.

- **1** Solve quadratic equations by graphing.
- **2** Estimate solutions of quadratic equations by graphing.

- Arielle works in the marketing department of a major retailer. Her job is to set prices for new products sold in the stores. Arielle determined that for a certain product, the function $f(p) = -6p^2 + 192p - 1440$ tells the profit $f(p)$ made at price p.

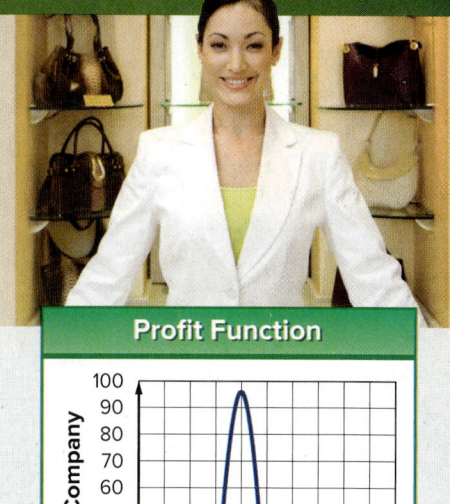

Arielle can determine the price range by finding the prices for which the profit is equal to $0. This can be done by finding the solutions of the related quadratic equation $0 = -6p^2 + 192p - 1440$.

The graph of the function indicates that the profit is zero at 12 and 20, so the profitable price range of the item is between $12 and $20.

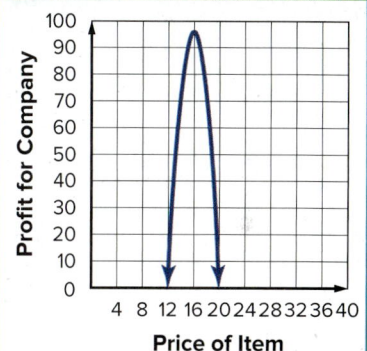

New Vocabulary
quadratic equation
standard form
root
zero

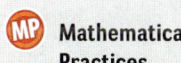 Mathematical Practices
3 Construct viable arguments and critique the reasoning of others.

1 Solve Quadratic Equations **Quadratic equations** are quadratic functions that are set equal to a value. The **standard form** of a quadratic equation is $ax^2 + bx + c = 0$, where $a \neq 0$ and a, b, and c are integers.

The solutions of a quadratic equation are called the **roots** of the equation. One method for finding the roots of a quadratic equation is to find the **zeros** of the related quadratic function.

The zeros of the function are the x-intercepts of its graph.

Quadratic Function

$$f(x) = x^2 - x - 6$$

$$f(-2) = (-2)^2 - (-2) - 6 \text{ or } 0$$
$$f(3) = 3^2 - 3 - 6 \text{ or } 0$$

−2 and 3 are zeros of the function.

Quadratic Equation

$$x^2 - x - 6 = 0$$

$$(-2)^2 - (-2) - 6 \text{ or } 0$$
$$3^2 - 3 - 6 \text{ or } 0$$

−2 and 3 are roots of the equation.

Graph of Function

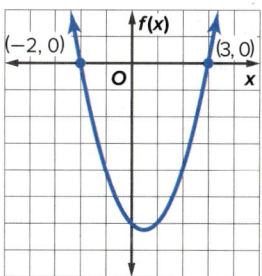

The x-intercepts are −2 and 3.

Example 1 Two Real Solutions

Solve $x^2 - 3x - 4 = 0$ by graphing.

Graph the related function, $f(x) = x^2 - 3x - 4$. The equation of the axis of symmetry is $x = -\frac{-3}{2(1)}$ or 1.5. Make a table using x-values around 1.5. Then graph each point.

x	−1	0	1	1.5	2	3	4
f(x)	0	−4	−6	−6.25	−6	−4	0

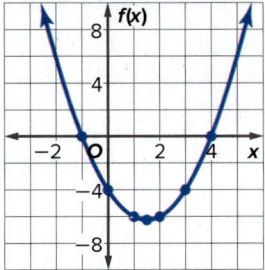

The zeros of the function are −1 and 4. Therefore, the solutions of the equation are −1 and 4 or $\{x \mid x = -1, 4\}$.

> **Study Tip**
> **Set-Builder Notation** In Lesson 1-5, you learned how to express the solution set of an inequality using set-builder notation. The solutions of an equation can also be expressed in set-builder notation. For example, the solutions of $x^2 = 25$ can be expressed as $\{x \mid x = -5, 5\}$.

▶ **Guided Practice**

Solve each equation by graphing.

1A. $x^2 + 2x - 15 = 0$

1B. $x^2 - 8x = -12$

The graph of the related function in Example 1 has two zeros; therefore, the quadratic equation has two real solutions. This is one of the three possible outcomes when solving a quadratic equation.

Key Concept Solutions of a Quadratic Equation

Words A quadratic equation can have one real solution, two real solutions, or no real solutions.

Models

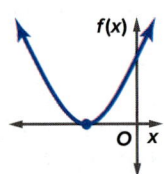
one real solution

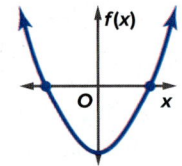
two real solutions

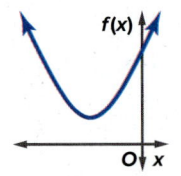
no real solution

Example 2 One Real Solution

Solve $14 - x^2 = -6x + 23$ by graphing.

$14 - x^2 = -6x + 23$ Original equation
$14 = x^2 - 6x + 23$ Add x^2 to each side.
$0 = x^2 - 6x + 9$ Subtract 14.

Graph the related function $f(x) = x^2 - 6x + 9$.

x	1	2	3	4	5
f(x)	4	1	0	1	4

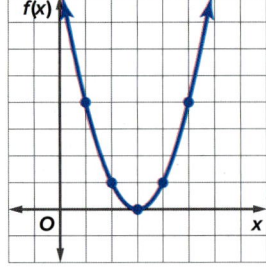

> **Study Tip**
> **Optional Graph**
> $f(x) = -x^2 + 6x - 9$ could also have been graphed for this example. The graph would appear different, but it would have the same solution.

The function has only one zero, 3. Therefore, the solution is 3 or $\{x \mid x = 3\}$.

▶ **Guided Practice**

Solve each equation by graphing.

2A. $x^2 + 5 = -8x - 11$

2B. $12 - x^2 = 48 - 12x$

Example 3 No Real Solution

NUMBER THEORY Use a quadratic equation to find two real numbers with a sum of 15 and a product of 63.

Understand Let x represent one of the numbers. Then $15 - x$ is the other number.

Plan
$x(15 - x) = 63$ The product of the numbers is 63.
$15x - x^2 = 63$ Distributive Property
$-x^2 + 15x - 63 = 0$ Subtract 63.

Solve Graph the related function.

The graph has no x-intercepts. This means the original equation has no real solution.

Check Because the equation has no real solutions, it is not possible for two real numbers to have a sum of 15 and a product of 63.

Try finding the product of several pairs of numbers with sums of 15. Is each product less than 63 as the graph suggests?

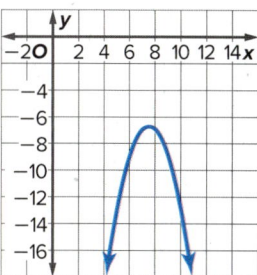

Guided Practice

3. Find two real numbers with a sum of 6 and a product of -55, or show that no such numbers exist.

2 Estimate Solutions

Often exact roots cannot be found by graphing. You can estimate the solutions by stating the integers between which the roots are located.

Watch Out!

Zeros You will see in later chapters that many zeros can appear within small intervals.

When the value of the function is positive for one value and negative for a second value, then there is at least one zero between those two values.

x	−3	−2	−1	0	1	2	3
f(x)	12	3	−6	−2	4	8	14

zero   zero

Go Online!

Watch **Personal Tutor** videos to hear descriptions of problem solving. Try describing how to solve a problem for a partner. Have them ask you questions to help your understanding.

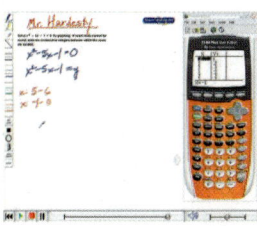

Example 4 Estimate Roots

Solve $x^2 - 6x + 4 = 0$ by graphing. If exact roots cannot be found, state the consecutive integers between which the roots are located.

x	0	1	2	3	4	5	6
f(x)	4	−1	−4	−5	−4	−1	4

The x-intercepts of the graph indicate that one solution is between 0 and 1, and the other solution is between 5 and 6.

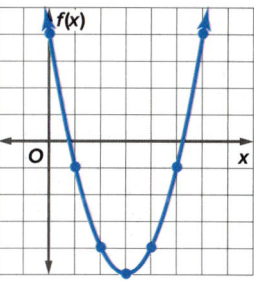

Guided Practice

4. Solve $x^2 - x - 10 = 0$ by graphing. If exact roots cannot be found, state the consecutive integers between which the roots are located.

connectED.mcgraw-hill.com **165**

You can also use tables to solve quadratic equations. After entering the equation in your calculator, scroll through the table to locate the solutions.

Example 5 Solve by Using a Table

Solve $x^2 - 6x + 2 = 0$.

Enter **Y1** $= x^2 - 6x + 2$ in your graphing calculator. Use the **TABLE** window to find where the sign of **Y1** changes. Change △**Tbl** to 0.1 and look again for the sign change. Repeat the process with 0.01 and 0.001 to get a more accurate location of the zero.

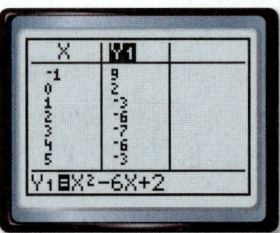 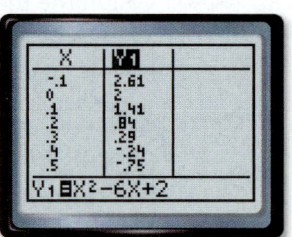 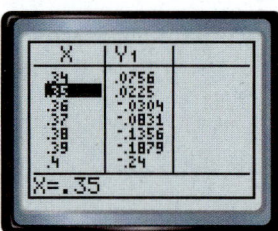

One solution is approximately 0.354.

> **Guided Practice**
>
> **5.** Locate the second zero in the function above to the nearest thousandth.

Quadratic equations can be solved with a calculator as well. After entering the equation, use the **zero** feature in the **CALC** menu.

Real-World Example 6 Solve by Using a Calculator

BALL A golfer hits a ball from an elevated tee box 200 feet above the fairway. The ball's height above the fairway is modeled by $h(t) = -16t^2 + 28t + 200$, where t is the time in seconds after it is hit. How long will it take the ball to reach the fairway?

We need to find t when $h(t)$ is 0. Solve $0 = -16t^2 + 28t + 200$. Then graph the related function $f(t) = -16t^2 + 28t + 200$ on a graphing calculator.

- Use the **zero** feature in the **CALC** menu to find the positive zero of the function, since time cannot be negative.

- Use the arrow keys to select a left bound and press ENTER.

- Locate a right bound and press ENTER twice.

- The positive zero of the function is about 4.52. The ball would take about 4.52 seconds to reach the ground.

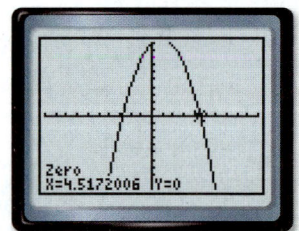

[−10, 10] scl: 1 by [−200, 200] scl: 20

Real-World Link
About 220,000 students participate in high school golf teams in the United States.
Source: Statista

> **Guided Practice**
>
> **6.** How long would it take to reach the ground if the height was modeled by $h(t) = -16t^2 + 48t + 400$?

Check Your Understanding

= Step-by-Step Solutions begin on page R11.

Go Online! for a Self-Check Quiz

Example 1 Use the related graph of each equation to determine its solutions.

1. $x^2 + 2x + 3 = 0$
2. $x^2 - 3x - 10 = 0$
3. $-x^2 - 8x - 16 = 0$

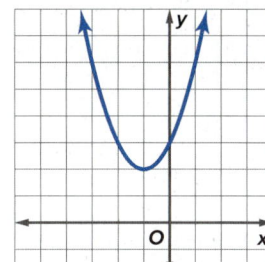 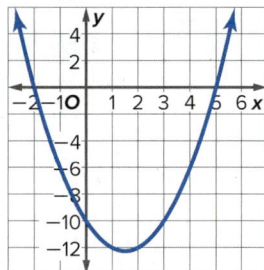 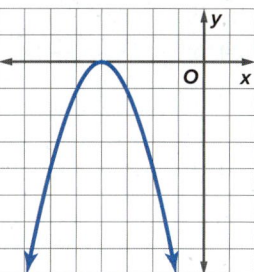

Examples 2–5 **MP PRECISION** Solve each equation. If exact roots cannot be found, state the consecutive integers between which the roots are located.

4. $x^2 + 8x = 0$
5. $x^2 - 3x - 18 = 0$
6. $4x - x^2 + 8 = 0$
7. $-12 - 5x + 3x^2 = 0$
8. $x^2 - 6x + 4 = -8$
9. $9 - x^2 = 12$
10. $5x^2 + 10x - 4 = -6$
11. $x^2 - 20 = 2 + x$

12. **NUMBER THEORY** Use a quadratic equation to find two real numbers with a sum of 2 and a product of -24.

Example 6

13. **PHYSICS** How long will it take an object to fall from the roof of a building 400 feet above the ground? Use the formula $h(x) = -16t^2 + h_0$, where t is the time in seconds and the initial height h_0 is in feet.

Practice and Problem Solving

Extra Practice is on page R3.

Example 1 Use the related graph of each equation to determine its solutions.

14. $x^2 + 4x = 0$
15. $-2x^2 - 4x - 5 = 0$
16. $0.5x^2 - 2x + 2 = 0$

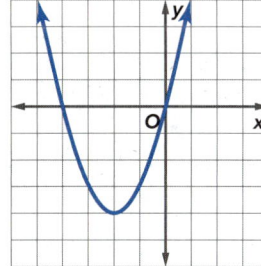 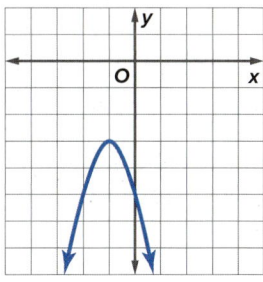 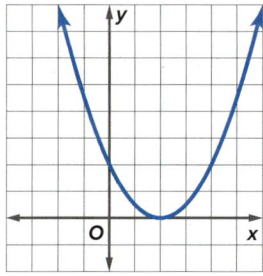

17. $-0.25x^2 - x - 1 = 0$
18. $x^2 - 6x + 11 = 0$
19. $-0.5x^2 + 0.5x + 6 = 0$

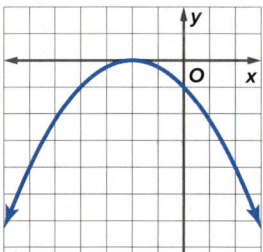 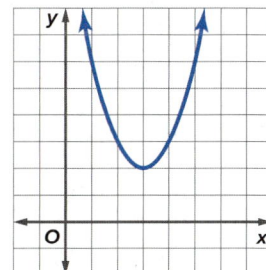 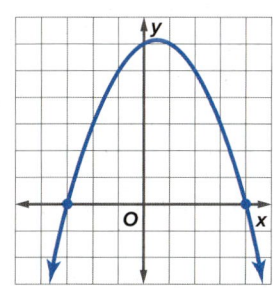

Examples 2–4 Solve each equation. If exact roots cannot be found, state the consecutive integers between which the roots are located.

20. $x^2 = 5x$
21. $-2x^2 - 4x = 0$
22. $x^2 - 5x - 14 = 0$
23. $-x^2 + 2x + 24 = 0$
24. $x^2 - 18x = -81$
25. $2x^2 - 8x = -32$
26. $2x^2 - 3x - 15 = 4$
27. $-3x^2 - 7 + 2x = -11$
28. $-0.5x^2 + 3 = -5x - 2$
29. $-2x + 12 = x^2 + 16$

Example 5 Use the tables to determine the location of the zeros of each quadratic function.

30.
x	−7	−6	−5	−4	−3	−2	−1	0
f(x)	−8	−1	4	4	−1	−8	−22	−48

31.
x	−2	−1	0	1	2	3	4	5
f(x)	32	14	2	−3	−3	2	14	32

32.
x	−6	−3	0	3	6	9	12	15
f(x)	−6	−1	3	5	3	−1	−6	−14

Example 6 **NUMBER THEORY** Use a quadratic equation to find two real numbers that satisfy each situation, or show that no such numbers exist.

33. Their sum is −15, and their product is −54.
34. Their sum is 4, and their product is −117.
35. Their sum is 12, and their product is −84.
36. Their sum is −13, and their product is 42.
37. Their sum is −8 and their product is −209.

MODELING For Exercises 38–40, use the formula $h(t) = v_0 t - 16t^2$, where $h(t)$ is the height of an object in feet, v_0 is the object's initial velocity in feet per second, and t is the time in seconds.

38. **BASEBALL** A baseball is hit with an initial velocity of 80 feet per second. Ignoring the height of the baseball player, how long does it take for the ball to hit the ground?

39. **CANNONS** A cannonball is shot with an initial velocity of 55 feet per second. Ignoring the height of the cannon, how long does it take for the cannonball to hit the ground?

40. **GOLF** A golf ball is hit with an initial velocity of 100 feet per second. How long will it take for it to hit the ground?

Solve each equation. If exact roots cannot be found, state the consecutive integers between which the roots are located.

41. $2x^2 + x = 15$
42. $-5x - 12 = -2x^2$
43. $4x^2 - 15 = -4x$
44. $-35 = -3x - 2x^2$
45. $-3x^2 + 11x + 9 = 1$
46. $13 - 4x^2 = -3x$
47. $-0.5x^2 + 18 = -6x + 33$
48. $0.5x^2 + 0.75 = 0.25x$

49. WATER BALLOONS Tony wants to drop a water balloon so that it splashes on his brother. Use the formula $h(t) = -16t^2 + h_0$, where t is the time in seconds and the initial height h_0 is in feet, to determine how far his brother should be from the target when Tony lets go of the balloon.

50. WATER HOSES A water hose can spray water at an initial velocity of 40 feet per second. Use the formula $h(t) = v_0 t - 16t^2$, where $h(t)$ is the height of the water in feet, v_0 is the initial velocity in feet per second, and t is the time in seconds.

   a. How long will it take the water to hit the nozzle on the way down?

   b. Assuming the nozzle is 5 feet up, what is the maximum height of the water?

51. SKYDIVING In 2003, John Fleming and Dan Rossi became the first two blind skydivers to be in free fall together. They jumped from an altitude of 14,000 feet and free fell to an altitude of 4000 feet before their parachutes opened. Ignoring air resistance and using the formula $h(t) = -16t^2 + h_0$, where t is the time in seconds and the initial height h_0 is in feet, determine how long they were in free fall.

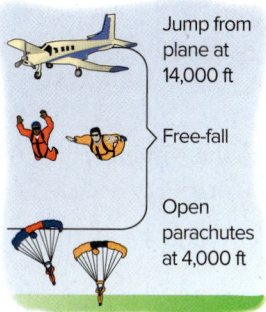

52. MULTI-STEP In economics, the *law of diminishing returns* states that the cost of producing one more unit of a good increases as more of that good is produced. The table below shows the cost associated with the production of canned cheese for Wholesome Dairy.

Units Produced	60	90	120	150	180	210	240	270
Total Cost ($)	1625	2050	2400	2700	3050	3475	4025	4775

   a. The *marginal cost* is the change in the total cost divided by the change in units produced. How many units should the company produce to minimize their marginal cost?

   b. Explain your solution process.

H.O.T. Problems   Use Higher-Order Thinking Skills

53. CHALLENGE Find the value of a positive integer k such that $f(x) = x^2 - 2kx + 55$ has roots at $k + 3$ and $k - 3$.

54. REASONING If a quadratic function has a minimum at $(-6, -14)$ and a root at $x = -17$, what is the other root? Explain your reasoning.

55. OPEN-ENDED Write a quadratic function with a maximum at $(3, 125)$ and roots at -2 and 8.

56. WRITING IN MATH Explain how to solve a quadratic equation by graphing its related quadratic function.

Preparing for Assessment

57. Consider the function graphed below. MP 7

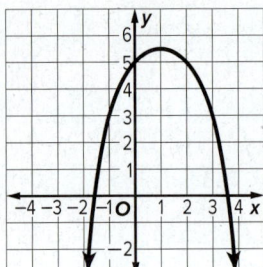

Which best describes the location of a zero?

- A between 0 and 2
- B between 3 and 4
- C between 4 and 5
- D between 5 and 6

58. What are the x-intercepts of the graph of the related function for $x^2 + x - 20 = 2x - 8$? MP 6

- A −7 and 4
- B −5 and 4
- C −3 and 4
- D no x-intercepts

59. Which equation has no real roots? MP 2

- A $x^2 - 16 = 0$
- B $x^2 - 4x + 8 = 0$
- C $x^2 + 6x + 8 = 0$
- D $x^2 + 6x = 0$
- E $x^2 + 8x + 16 = 0$

60. Which statements are always true about the graph of a quadratic function that has only one real zero? MP 2

- A The graph has one x-intercept.
- B The graph will open upward.
- C The x-intercept is the vertex.
- D The equation of the axis of symmetry is $x = 0$.
- E The vertex is on the y-axis.

61. MULTI-PART Consider this equation. MP 2, 6

$x^2 + 3x - 10 = x - 2$

a. When the equation is rearranged to be equal to zero and written in function notation as $f(x) = ax^2 + bx + c$, what are the values of a, b, and c?

- A 1, 2, 8
- B 1, −2, 12
- C 1, 2, −8
- D −1, 2, 12

b. Which of the following statements are true?

- A The graph of $f(x)$ is a parabola that opens down.
- B The graph of $f(x)$ is a parabola that opens up.
- C The vertex of the graph of $f(x)$ is below the x-axis.
- D The vertex of the graph of $f(x)$ is above the x-axis.
- E The graph of $f(x)$ does not intersect the x-axis.

c. What are the zeros of $f(x)$?

d. What is the y-intercept of the graph of $f(x)$?

62. The height of a model rocket launched vertically upward is described by the formula $h(t) = v_0 t - 16t^2$, where $h(t)$ is the height in feet, v_0 is the initial velocity in feet per second, and t is the time in seconds. How long is the model rocket in the air if it is launched with an initial velocity of 96 feet per second? MP 1

EXTEND 3-2
Graphing Technology Lab
Solving Quadratic Equations by Graphing

You can use a TI-83/84 Plus graphing calculator to solve quadratic equations.

Activity 1 Solving Quadratic Equations

Work cooperatively. Solve $x^2 - 8x + 15 = 0$.

Step 1 Let **Y1** $= x^2 - 8x + 15$ and **Y2** $= 0$.

Step 2 Graph **Y1** and **Y2** in the standard viewing window.

KEYSTROKES: Y= X,T,θ,n x^2 (−) 8 X,T,θ,n + 15 ENTER 0 ZOOM 6

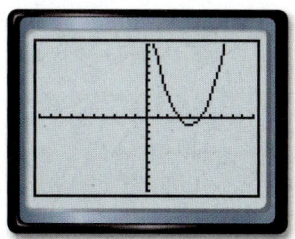

[−10, 10] scl: 1 by [−10, 10] scl: 1

Step 3 To find the x-intercepts, determine the points where **Y1** = **Y2**.

KEYSTROKES: 2nd [CALC] 5

Press ENTER for the first equation. Press ENTER for the second equation.

Move the cursor as close to the left x-intercept and press ENTER.

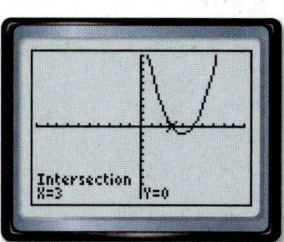

[−10, 10] scl: 1 by [−10, 10] scl: 1

Find the second x-intercept.

KEYSTROKES: 2nd [CALC] 5

Press ENTER for the first equation. Press ENTER for the second equation.

Move the cursor as close to the right x-intercept and press ENTER.

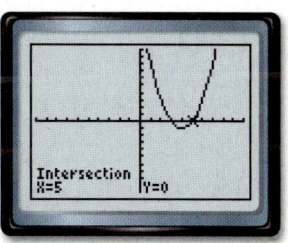

[−10, 10] scl: 1 by [−10, 10] scl: 1

The x-intercepts are 3 and 5, so $x = 3$ and $x = 5$.

Exercises

Work cooperatively. Solve each equation. Round to the nearest tenth if necessary.

1. $x^2 - 7x + 12 = 0$
2. $x^2 + 5x + 6 = 0$
3. $x^2 - 3 = 2x$
4. $x^2 + 5x + 6 = 12$
5. $x^2 + 5x = 0$
6. $x^2 - 4 = 0$
7. $x^2 + 8x + 16 = 0$
8. $x^2 - 10x = -25$
9. $9x^2 + 48x + 64 = 0$
10. $2x^2 + 3x - 1 = 0$
11. $5x^2 - 7x = -2$
12. $6x^2 + 2x + 1 = 0$

LESSON 3
Complex Numbers

Then
- You simplified square roots.

Now
1. Perform operations with pure imaginary numbers.
2. Perform operations with complex numbers.

Why?
Consider the graph of $y = x^2 + 2x + 4$ at the right. Notice how this graph has no x-intercepts and therefore does not have any roots. Does this mean there are no solutions to $0 = x^2 + 2x + 4$?

Use the **Solver** function located in the MATH menu of a graphing calculator. Enter the equation and select $x = 2$ as your guess to a solution.

Press [ALPHA] [ENTER] and the calculator will attempt to solve the equation. The calculator indicates there is no solution with the error message. So there are no real solutions. However, there are imaginary solutions.

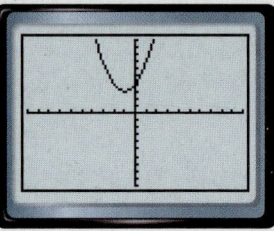

[−10, 10] scl: 1 by [−10, 10] scl: 1

New Vocabulary
imaginary unit
pure imaginary number
complex number
complex conjugates

Mathematical Practices
6 Attend to precision.

1 Pure Imaginary Numbers In your math studies so far, you have worked with real numbers. Equations like the one above led mathematicians to define imaginary numbers. The **imaginary unit** i is defined to be $i^2 = -1$. The number i is the principal square root of -1; that is, $i = \sqrt{-1}$.

Numbers of the form $6i$, $-2i$, and $i\sqrt{3}$ are called **pure imaginary numbers**. Pure imaginary numbers are square roots of negative real numbers. For any positive real number b, $\sqrt{-b^2} = \sqrt{b^2} \cdot \sqrt{-1}$ or bi.

Example 1   Square Roots of Negative Numbers

Simplify.

a. $\sqrt{-27}$

$\sqrt{-27} = \sqrt{-1 \cdot 3^2 \cdot 3}$
$= \sqrt{-1} \cdot \sqrt{3^2} \cdot \sqrt{3}$
$= i \cdot 3 \cdot \sqrt{3}$ or $3i\sqrt{3}$

b. $\sqrt{-216}$

$\sqrt{-216} = \sqrt{-1 \cdot 6^2 \cdot 6}$
$= \sqrt{-1} \cdot \sqrt{6^2} \cdot \sqrt{6}$
$= i \cdot 6 \cdot \sqrt{6}$ or $6i\sqrt{6}$

Guided Practice

1A. $\sqrt{-18}$ **1B.** $\sqrt{-125}$

The Commutative and Associative Properties of Multiplication hold true for pure imaginary numbers. The first few powers of i are shown below.

$i^1 = i$	$i^2 = -1$	$i^3 = i^2 \cdot i$ or $-i$	$i^4 = (i^2)^2$ or 1
$i^5 = i^4 \cdot i$ or i	$i^6 = i^4 \cdot i^2$ or -1	$i^7 = i^4 \cdot i^3$ or $-i$	$i^8 = (i^2)^4$ or 1

Example 2   Products of Pure Imaginary Numbers

Simplify.

a. $-5i \cdot 3i$

$$-5i \cdot 3i = -15i^2 \quad \text{Multiply.}$$
$$= -15(-1) \quad i^2 = -1$$
$$= 15 \quad \text{Simplify.}$$

b. $\sqrt{-6} \cdot \sqrt{-15}$

$$\sqrt{-6} \cdot \sqrt{-15} = i\sqrt{6} \cdot i\sqrt{15} \quad i = \sqrt{-1}$$
$$= i^2\sqrt{90} \quad \text{Multiply.}$$
$$= -1 \cdot \sqrt{9} \cdot \sqrt{10} \quad \text{Simplify.}$$
$$= -3\sqrt{10} \quad \text{Multiply.}$$

▶ **Guided Practice**

2A. $3i \cdot 4i$ **2B.** $\sqrt{-20} \cdot \sqrt{-12}$ **2C.** i^{31}

You can solve some quadratic equations by using the **Square Root Property**. Similar to a difference of squares, the sum of squares can be factored over the complex numbers.

Example 3   Equation with Pure Imaginary Solutions

Solve $x^2 + 64 = 0$.

Method 1   Square Root Property

$$x^2 + 64 = 0$$
$$x^2 = -64$$
$$x = \pm\sqrt{-64}$$
$$x = \pm 8i$$

Method 2   Factoring

$$x^2 + 64 = 0$$
$$x^2 + 8^2 = 0$$
$$x^2 - (-8^2) = 0$$
$$(x + 8i)(x - 8i) = 0$$
$$(x + 8i) = 0 \text{ or } (x - 8i) = 0$$
$$x = -8i \quad\quad x = 8i$$

▶ **Guided Practice**

Solve each equation.

3A. $4x^2 + 100 = 0$ **3B.** $x^2 + 4 = 0$

2 Operations with Complex Numbers

Consider $2 + 3i$. Because 2 is a real number and $3i$ is a pure imaginary number, the terms are not like terms and cannot be combined. This type of expression is called a **complex number**.

Key Concept   Complex Numbers

Words	A complex number is any number that can be written in the form $a + bi$, where a and b are real numbers and i is the imaginary unit. a is called the real part, and b is called the imaginary part.
Examples	$5 + 2i$ $\quad\quad\quad\quad\quad\quad\quad\quad 1 - 3i = 1 + (-3)i$

connectED.mcgraw-hill.com   **173**

The Venn diagram shows the set of complex numbers.

- If $b = 0$, the complex number is a real number.
- If $b \neq 0$, the complex number is imaginary.
- If $a = 0$, the complex number is a pure imaginary number.

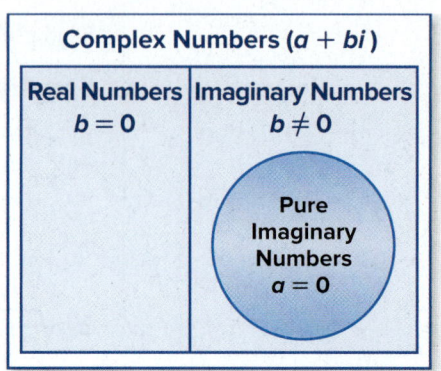

Two complex numbers are equal if and only if their real parts are equal and their imaginary parts are equal. That is, $a + bi = c + di$ if and only if $a = c$ and $b = d$.

Study Tip

MP Precision Whereas all real numbers are also complex, the term *complex number* usually refers to a number that is not real.

Example 4 Equate Complex Numbers

Find the values of x and y that make $3x - 5 + (y - 3)i = 7 + 6i$ true.

Set the real parts equal to each other and the imaginary parts equal to each other.

$3x - 5 = 7$	Real parts	$y - 3 = 6$	Imaginary parts
$3x = 12$	Add 5 to each side.	$y = 9$	Add 3 to each side.
$x = 4$	Divide each side by 3.		

Guided Practice

4. Find the values of x and y that make $5x + 1 + (3 + 2y)i = 2x - 2 + (y - 6)i$ true.

The Commutative, Associative, and Distributive Properties of Multiplication and Addition hold true for complex numbers. To add or subtract complex numbers, combine like terms. That is, combine the real parts, and combine the imaginary parts.

Example 5 Add and Subtract Complex Numbers

Simplify.

a. $(5 - 7i) + (2 + 4i)$

$(5 - 7i) + (2 + 4i) = (5 + 2) + (-7 + 4)i$ Commutative and Associative Properties

$= 7 - 3i$ Simplify.

b. $(4 - 8i) - (3 - 6i)$

$(4 - 8i) - (3 - 6i) = (4 - 3) + [-8 - (-6)]i$ Commutative and Associative Properties

$= 1 - 2i$ Simplify.

Guided Practice

5A. $(-2 + 5i) + (1 - 7i)$

5B. $(4 + 6i) - (-1 + 2i)$

Study Tip

Reading Math Electrical engineers use j as the imaginary unit to avoid confusion with the i for current.

Complex numbers are used with electricity. In these problems, j usually represents the imaginary unit. In a circuit with alternating current, the voltage, current, and impedance, or hindrance to current, can be represented by complex numbers. To multiply these numbers, use the FOIL method.

Real-World Career

Electrical Engineer Electrical engineers design, develop, test, and supervise the making of electrical equipment such as digital music players, electric motors, lighting, and radar and navigation systems. A bachelor's degree in engineering is required for almost all entry-level engineering jobs.

Real-World Example 6 Multiply Complex Numbers

ELECTRICITY In an AC circuit, the voltage V, current C, and impedance I are related by the formula $V = C \cdot I$. Find the voltage in a circuit with current $2 + 4j$ amps and impedance $9 - 3j$ ohms.

$V = C \cdot I$ — Electricity formula
$= (2 + 4j) \cdot (9 - 3j)$ — $C = 2 + 4j$ and $I = 9 - 3j$
$= 2(9) + 2(-3j) + 4j(9) + 4j(-3j)$ — FOIL Method
$= 18 - 6j + 36j - 12j^2$ — Multiply.
$= 18 + 30j - 12(-1)$ — $j^2 = -1$
$= 30 + 30j$ — Add.

The voltage is $30 + 30j$ volts.

Guided Practice

6. Find the voltage in a circuit with current $2 - 4j$ amps and impedance $3 - 2j$ ohms.

Two complex numbers of the form $a + bi$ and $a - bi$ are called **complex conjugates**. The product of complex conjugates is always a real number. You can use this fact to simplify the quotient of two complex numbers.

Example 7 Divide Complex Numbers

Simplify.

a. $\dfrac{2i}{3 + 6i}$

$\dfrac{2i}{3 + 6i} = \dfrac{2i}{3 + 6i} \cdot \dfrac{3 - 6i}{3 - 6i}$ — $3 + 6i$ and $3 - 6i$ are complex conjugates.

$= \dfrac{6i - 12i^2}{9 - 36i^2}$ — Multiply.

$= \dfrac{6i - 12(-1)}{9 - 36(-1)}$ — $i^2 = -1$

$= \dfrac{6i + 12}{45}$ — Simplify.

$= \dfrac{4}{15} + \dfrac{2}{15}i$ — $a + bi$ form

b. $\dfrac{4 + i}{5i}$

$\dfrac{4 + i}{5i} = \dfrac{4 + i}{5i} \cdot \dfrac{i}{i}$ — Multiply by $\dfrac{i}{i}$.

$= \dfrac{4i + i^2}{5i^2}$ — Multiply.

$= \dfrac{4i - 1}{-5}$ — $i^2 = -1$

$= \dfrac{1}{5} - \dfrac{4}{5}i$ — $a + bi$ form

Guided Practice

7A. $\dfrac{-2i}{3 + 5i}$

7B. $\dfrac{2 + i}{1 - i}$

Study Tip

Technology Operations with complex numbers can be preformed with a TI-83/84 Plus graphing calculator. Use the [2nd] [*i*] function to enter the expression. Then press [MATH] [ENTER] [ENTER] to view the answer.

Check Your Understanding

Examples 1–2 Simplify.

1. $\sqrt{-81}$
2. $\sqrt{-32}$
3. $(4i)(-3i)$
4. $3\sqrt{-24} \cdot 2\sqrt{-18}$
5. i^{40}
6. i^{63}

Example 3 Solve each equation.

7. $4x^2 + 32 = 0$
8. $x^2 + 1 = 0$

Example 4 Find the values of a and b that make each equation true.

9. $3a + (4b + 2)i = 9 - 6i$
10. $4b - 5 + (-a - 3)i = 7 - 8i$

Examples 5 and 7 Simplify.

11. $(-1 + 5i) + (-2 - 3i)$
12. $(7 + 4i) - (1 + 2i)$
13. $(6 - 8i)(9 + 2i)$
14. $(3 + 2i)(-2 + 4i)$
15. $\dfrac{3 - i}{4 + 2i}$
16. $\dfrac{2 + i}{5 + 6i}$

Example 6

17. **ELECTRICITY** The current in one part of a series circuit is $5 - 3j$ amps. The current in another part of the circuit is $7 + 9j$ amps. Add these complex numbers to find the total current in the circuit.

Practice and Problem Solving

Extra Practice is on page R3.

Examples 1–2 Simplify.

18. $\sqrt{-121}$
19. $\sqrt{-169}$
20. $\sqrt{-100}$
21. $\sqrt{-81}$
22. $(-3i)(-7i)(2i)$
23. $4i(-6i)^2$
24. i^{11}
25. i^{25}

Examples 5–7

26. $(10 - 7i) + (6 + 9i)$
27. $(-3 + i) + (-4 - i)$
28. $(12 + 5i) - (9 - 2i)$
29. $(11 - 8i) - (2 - 8i)$
30. $(1 + 2i)(1 - 2i)$
31. $(3 + 5i)(5 - 3i)$
32. $(4 - i)(6 - 6i)$
33. $\dfrac{2i}{1 + i}$
34. $\dfrac{5}{2 + 4i}$
35. $\dfrac{5 + i}{3i}$

Example 3 Solve each equation.

36. $4x^2 + 4 = 0$
37. $3x^2 + 48 = 0$
38. $2x^2 + 50 = 0$
39. $2x^2 + 10 = 0$
40. $6x^2 + 108 = 0$
41. $8x^2 + 128 = 0$

Example 4 Find the values of x and y that make each equation true.

42. $9 + 12i = 3x + 4yi$
43. $x + 1 + 2yi = 3 - 6i$
44. $2x + 7 + (3 - y)i = -4 + 6i$
45. $5 + y + (3x - 7)i = 9 - 3i$
46. $a + 3b + (3a - b)i = 6 + 6i$
47. $(2a - 4b)i + a + 5b = 15 + 58i$

Simplify.

48. $\sqrt{-10} \cdot \sqrt{-24}$

49. $4i\left(\frac{1}{2}i\right)^2(-2i)^2$

50. i^{41}

51. $(4 - 6i) + (4 + 6i)$

52. $(8 - 5i) - (7 + i)$

53. $(-6 - i)(3 - 3i)$

54. $\dfrac{(5 + i)^2}{3 - i}$

55. $\dfrac{6 - i}{2 - 3i}$

56. $(-4 + 6i)(2 - i)(3 + 7i)$

57. $(1 + i)(2 + 3i)(4 - 3i)$

58. $\dfrac{4 - i\sqrt{2}}{4 + i\sqrt{2}}$

59. $\dfrac{2 - i\sqrt{3}}{2 + i\sqrt{3}}$

60. **ELECTRICITY** The impedance in one part of a series circuit is $7 + 8j$ ohms, and the impedance in another part of the circuit is $13 - 4j$ ohms. Add these complex numbers to find the total impedance in the circuit.

ELECTRICITY Use the formula $V = C \cdot I$.

61. The current in a circuit is $3 + 6j$ amps, and the impedance is $5 - j$ ohms. What is the voltage?

62. The voltage in a circuit is $20 - 12j$ volts, and the impedance is $6 - 4j$ ohms. What is the current?

63. Find the sum of $ix^2 - (4 + 5i)x + 7$ and $3x^2 + (2 + 6i)x - 8i$.

64. Simplify $[(2 + i)x^2 - ix + 5 + i] - [(-3 + 4i)x^2 + (5 - 5i)x - 6]$.

65. **MULTIPLE REPRESENTATIONS** In this problem, you will explore quadratic equations that have complex roots.

    a. **Algebraic** Write a quadratic equation in standard form with $3i$ and $-3i$ as its roots.

    b. **Graphical** Graph the quadratic equation found in part **a** by graphing its related function.

    c. **Algebraic** Write a quadratic equation in standard form with $2 + i$ and $2 - i$ as its roots.

    d. **Graphical** Graph the quadratic equation found in part **c** by graphing its related function.

    e. **Analytical** How do you know when a quadratic equation will have only complex solutions?

H.O.T. Problems   Use Higher-Order Thinking Skills

66. **ERROR ANALYSIS** Joe and Sue are simplifying $(2i)(3i)(4i)$. Is either of them correct? Explain your reasoning.

    Joe
    $24i^3 = -24$

    Sue
    $24i^3 = -24i$

67. **CHALLENGE** Simplify $(1 + 2i)^3$.

68. **REASONING** Determine whether the following statement is *always*, *sometimes*, or *never* true. Explain your reasoning.

    Every complex number has both a real part and an imaginary part.

69. **OPEN-ENDED** Write two complex numbers with a product of 20.

70. **WRITING IN MATH** Explain how complex numbers are related to quadratic equations.

Preparing for Assessment

71. Which expression is equivalent to $5 - 3(-7 + 4i) - (8 - i)$? MP 6

- A $-22 + 7i$
- B $-22 + 9i$
- C $18 - 11i$
- D $18 + 3i$

72. The formula $V = C \cdot I$ describes the relationship between voltage (V), current (C), and impedance (I). What is the voltage when the current in a circuit is $5 - 7j$ amps and the impedance is $4 + j$ ohms? MP 4

- A $13 - 18j$ volts
- B $13 - 23j$ volts
- C $27 - 18j$ volts
- D $27 - 23j$ volts

73. Which expression is equivalent to $1 - i$? MP 6

- A $i^{30} + i^{31}$
- B $i^{30} - i^{31}$
- C $i^{31} - i^{30}$
- D $(i^{30})(i^{31})$
- E $(i^{30})(-i^{31})$

74. What are the values of x and y when $(9 + 6i) - (x + yi) = -2 + 4i$? MP 6

- A $x = 7, y = 2$
- B $x = 7, y = 10$
- C $x = 11, y = 2$
- D $x = 11, y = 10$

75. If $a > 0$ and $(a - 2i)(a + 2i) = 5$ then what is the value of a? MP 1

76. **MULTI-STEP** Consider the expression $\dfrac{(-5 + 3i)(4 + 2i)}{1 - 2i}$. MP 3

a. Which expressions are equivalent to the numerator? Choose all that apply.

- A $5(4 + 2i) + 3i(4 + 2i)$
- B $-5(4) + (-5)(2i) + (3i)(4) + (3i)(2i)$
- C $-42i + 6i^2$
- D $-26 + 2i$
- E $-20 - 10i + 12i + 6i^2$
- F $-20 - 2i + 6$

b. What is the complex conjugate of the denominator?

c. Which expression is equivalent to the quotient?

- A $6 + 10i$
- B $\dfrac{-22 + 54i}{5}$
- C $-2 + 6i$
- D $-6 - 10i$

77. Which expressions are equal to -1? MP 6

- A i^2
- B i^3
- C $1 + i^2$
- D $(i^5)(i)$
- E $i - 3 - 4i^2$
- F $1 - i^3 + 2i^2 - i$

LESSON 4
Solving Quadratic Equations by Factoring

::**Then**
- You found the greatest common factors of sets of numbers.

::**Now**
1. Write quadratic equations in standard form.
2. Solve quadratic equations by factoring.

::**Why?**
The **factored form** of a quadratic equation is $0 = a(x - p)(x - q)$. In the equation, p and q represent the x-intercepts of the graph of the equation.

The x-intercepts of the graph at the right are 2 and 6. In this lesson, you will learn how to change a quadratic equation in factored form into standard form and vice versa.

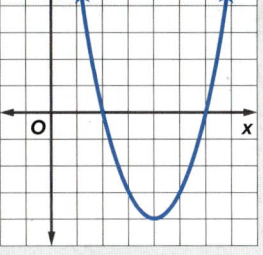

Related Graph
2 and 6 are x-intercepts.

Standard Form	Factored Form
$0 = x^2 - 8x + 12$	$0 = (x - 6)(x - 2)$

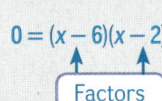
Factors

New Vocabulary
factored form
FOIL method

Mathematical Practices
2 Reason abstractly and quantitatively.

1 Standard Form
You can use the FOIL method to write a quadratic equation that is in factored form in standard form. The **FOIL method** uses the Distributive Property to multiply binomials.

Key Concept FOIL Method for Multiplying Binomials

Words To multiply two binomials, find the sum of the products of **F** the *First terms*, **O** the *Outer terms*, **I** the *Inner terms*, and **L** the *Last terms*.

Examples

	Product of **First** Terms	Product of **Outer** Terms	Product of **Inner** Terms	Product of **Last** Terms
	↓	↓	↓	↓
$(x - 6)(x - 2) =$	$(x)(x)$ +	$(x)(-2)$ +	$(-6)(x)$ +	$(-6)(-2)$

$= x^2 - 2x - 6x + 12$ or $x^2 - 8x + 12$

Example 1 Translate Sentences into Equations

Write a quadratic equation in standard form with $-\frac{1}{3}$ and 6 as its roots.

$(x - p)(x - q) = 0$ Write the pattern.

$\left[x - \left(-\frac{1}{3}\right)\right](x - 6) = 0$ Replace p with $-\frac{1}{3}$ and q with 6.

$\left(x + \frac{1}{3}\right)(x - 6) = 0$ Simplify.

$x^2 - \frac{17}{3}x - 2 = 0$ Multiply.

$3x^2 - 17x - 6 = 0$ Multiply each side by 3 so that b and c are integers.

> **Guided Practice**
>
> 1. Write a quadratic equation in standard form with $\frac{3}{4}$ and -5 as its roots.

connectED.mcgraw-hill.com   179

2 Solve Equations by Factoring
Solving quadratic equations by factoring is an application of the Zero Product Property.

Key Concept Zero Product Property

Words For any real numbers a and b, if $ab = 0$, then either $a = 0$, $b = 0$, or both a and b equal zero.

Example If $(x + 3)(x - 5) = 0$, then $x + 3 = 0$ or $x - 5 = 0$.

Example 2 Factor the GCF

Solve each equation.

a. $16x^2 + 8x = 0$

$16x^2 + 8x = 0$	Original equation
$8x(2x) + 8x(1) = 0$	Factor the GCF.
$8x(2x + 1) = 0$	Distributive Property
$8x = 0$ or $2x + 1 = 0$	Zero Product Property
$x = 0 \quad\quad 2x = -1$	Solve both equations.
$\quad\quad\quad\quad x = -\frac{1}{2}$	

b. $21x^2 - 14x = 0$

$21x^2 - 14x = 0$	Original equation
$7x(3x) - 7x(2) = 0$	Factor the GCF.
$7x(3x - 2) = 0$	Distributive Property
$7x = 0$ or $3x - 2 = 0$	Zero Product Property
$x = 0 \quad\quad x = \frac{2}{3}$	Solve both equations.

▸ **Guided Practice**

2A. $20x^2 + 15x = 0$ **2B.** $4y^2 + 16y = 0$ **2C.** $6a^5 + 18a^4 = 0$

Review Vocabulary

perfect square a number with a positive square root that is a whole number

When factoring, look for patterns like a difference of squares, sum of squares, or a perfect square trinomial. To use the rules for a difference of squares or sum of squares, both terms in the binomial must be perfect squares. To use the rules for perfect square trinomials, the first and last terms of the trinomial must be perfect squares and the middle term must be twice the product of the square roots of the first and last terms.

Factoring Special Products		
Type of Polynomial	Factoring Rule	Example
Difference of squares	$a^2 - b^2 = (a + b)(a - b)$	$x^2 - 49 = (x + 7)(x - 7)$
Sum of squares	$a^2 + b^2 = (a + bi)(a - bi)$	$x^2 + 25 = (x + 5i)(x - 5i)$
Perfect square trinomial	$a^2 - 2ab + b^2 = (a - b)(a - b)$	$x^2 - 18x + 81 = (x - 9)^2$
	$a^2 + 2ab + b^2 = (a + b)(a + b)$	$x^2 + 12x + 36 = (x + 6)^2$

Example 3 Perfect Squares and Differences of Squares

Solve each equation.

a. $x^2 + 16x + 64 = 0$

$x^2 = (x)^2$; $64 = (8)^2$	First and last terms are perfect squares.
$16x = 2(x)(8)$	Middle term equals $2ab$.

$x^2 + 16x + 64$ is a perfect square trinomial.

$x^2 + 16x + 64 = 0$	Original equation
$(x + 8)^2 = 0$	Factor using the pattern.
$x + 8 = 0$	Take the square root of each side.
$x = -8$	Solve.

Study Tip

Square Roots By inspection, notice that the square root of 64 is −8 and 8. Also, for $x^2 = 4$, the solutions would be −2 and 2.

b. $x^2 = 64$

$x^2 = 64$	Original equation
$x^2 - 64 = 0$	Subtract 64 from each side.
$x^2 - (8)^2 = 0$	Write in the form $a^2 - b^2$.
$(x + 8)(x - 8) = 0$	Factor the difference of squares.
$x + 8 = 0$ or $x - 8 = 0$	Zero Product Property
$x = -8$ $x = 8$	Solve.

c. $x^2 + 225 = 0$

$x^2 + 225 = 0$	Original equation
$x^2 + (15)^2 = 0$	Write in the form $a^2 + b^2$.
$(x + 15i)(x - 15i) = 0$	Factor the difference of squares.
$x + 15i = 0$ or $x - 15i = 0$	Zero Product Property
$x = -15i$ $x = 15i$	Solve.

d. $4x^2 - 44x + 121 = 0$

$4x^2 = (2x)^2$; $121 = (11)^2$	First and last terms are perfect squares.
$-44x = 2(2x)(11)$	Middle term is $2ab$.

$x^2 - 44x + 121$ is a perfect square trinomial.

$x^2 - 44x + 121 = 0$	
$(2x - 11)^2 = 0$	Factor using the pattern.
$2x - 11 = 0$	Take the square root of each side.
$x = 5.5$	Solve.

▶ **Guided Practice**

3A. $4x^2 - 12x + 9 = 0$ **3B.** $81x^2 - 9x = 0$ **3C.** $a^2 + 9 = 0$

Study Tip

MP Structure If values for m and p exist, then the trinomial can always be factored.

A special pattern is used when factoring trinomials of the form $ax^2 + bx + c$. First, multiply the values of a and c. Then, find two values, m and p, such that their product equals ac and their sum equals b.

Consider $6x^2 + 13x - 5$: $ac = 6(-5) = -30$.

Factors of −30	Sum	Factors of −30	Sum
1, −30	−29	−1, 30	29
2, −15	−13	**−2, 15**	**13**
3, −10	−7	−3, 10	7
5, −6	−1	−5, 6	1

Now the middle term, $13x$, can be rewritten as $-2x + 15x$.

This polynomial can now be factored by grouping.

$6x^2 + 13x - 5 = 6x^2 + mx + px - 5$ Write the pattern.
$ = 6x^2 - 2x + 15x - 5$ $m = -2$ and $p = 15$
$ = (6x^2 - 2x) + (15x - 5)$ Group terms.
$ = 2x(3x - 1) + 5(3x - 1)$ Factor the GCF.
$ = (2x + 5)(3x - 1)$ Distributive Property

Example 4   Factor Trinomials

Solve each equation.

a. $x^2 + 9x + 20 = 0$

$ac = 20$ $a = 1, c = 20$

Factors of 20	Sum	Factors of 20	Sum
1, 20	21	−1, −20	−21
2, 10	12	−2, −10	−12
4, 5	**9**	−4, −5	−9

Study Tip

Trinomials It does not matter if the values of *m* and *p* are switched when grouping.

$x^2 + 9x + 20 = 0$ Original expression
$x^2 + mx + px + 20 = 0$ Write the pattern.
$x^2 + 4x + 5x + 20 = 0$ $m = 4$ and $p = 5$
$(x^2 + 4x) + (5x + 20) = 0$ Group terms with common factors.
$x(x + 4) + 5(x + 4) = 0$ Factor the GCF from each grouping.
$(x + 5)(x + 4) = 0$ Distributive Property
$x + 5 = 0$ or $x + 4 = 0$ Zero Product Property
$x = -5$ $x = -4$ Solve each equation.

b. $6y^2 - 23y + 20 = 0$

$ac = 120$ $a = 6, c = 20$
$m = -8, p = -15$ $-8(-15) = 120; -8 + (-15) = -23$

$6y^2 - 23y + 20 = 0$ Original equation
$6y^2 + my + py + 20 = 0$ Write the pattern.
$6y^2 - 8y - 15y + 20 = 0$ $m = -8$ and $p = -15$
$(6y^2 - 8y) + (-15y + 20) = 0$ Group terms with common factors.
$2y(3y - 4) - 5(3y - 4) = 0$ Factor the GCF from each grouping.
$(2y - 5)(3y - 4) = 0$ Distributive Property
$2y - 5 = 0$ or $3y - 4 = 0$ Zero Product Property
$2y = 5$ $3y = 4$ Solve both equations.
$y = \dfrac{5}{2}$ $y = \dfrac{4}{3}$

▶ **Guided Practice**

4A. $x^2 - 11x + 30 = 0$ **4B.** $x^2 - 4x - 21 = 0$

4C. $15x^2 - 8x + 1 = 0$ **4D.** $-12x^2 + 8x + 15 = 0$

Real-World Example 5 Solve Equations by Factoring

TRACK AND FIELD The height of a javelin in feet is modeled by $h(t) = -16t^2 + 79t + 5$, where t is the time in seconds after the javelin is thrown. How long is it in the air?

To determine how long the javelin is in the air, we need to find when the height equals 0. We can do this by solving $-16t^2 + 79t + 5 = 0$.

$-16t^2 + 79t + 5 = 0$	Original equation
$m = 80;\ p = -1$	$-16(5) = -80,\ 80 \cdot (-1) = -80,\ 80 + (-1) = 79$
$-16t^2 + 80t - t + 5 = 0$	Write the pattern.
$(-16t^2 + 80t) + (-t + 5) = 0$	Group terms with common factors.
$16t(-t + 5) + 1(-t + 5) = 0$	Factor GCF from each group.
$(16t + 1)(-t + 5) = 0$	Distributive Property
$16t + 1 = 0$ or $-t + 5 = 0$	Zero Product Property
$16t = -1 \qquad -t = -5$	Solve both equations.
$t = -\dfrac{1}{16} \qquad t = 5$	Solve.

CHECK We have two solutions.
- The first solution is negative and since time cannot be negative, this solution can be eliminated.
- The second solution of 5 seconds seems reasonable for the time a javelin spends in the air.
- The answer can be confirmed by substituting back into the original equation.

$$-16t^2 + 79t + 5 = 0$$
$$-16(5)^2 + 79(5) + 5 \stackrel{?}{=} 0$$
$$-400 + 395 + 5 \stackrel{?}{=} 0$$
$$0 = 0 \ \checkmark$$

The javelin is in the air for 5 seconds.

Guided Practice

5. BUNGEE JUMPING Juan recorded his brother bungee jumping from a height of 1100 feet. At the time the cord lifted his brother back up, he was 76 feet above the ground. If Juan started recording as soon as his brother fell, how much time elapsed when the cord snapped back? Use $f(t) = -16t^2 + c$, where c is the height in feet.

Real-World Link

Barbora Spotakova of the Czech Republic broke the javelin world record in 2008 with a distance of 72.28 meters.

Source: IAAF

Go Online!

Solving equations by factoring is an important concept in algebra. Be sure to master it using the **Self-Check Quiz** in ConnectED.

Check Your Understanding

Example 1 Write a quadratic equation in standard form with the given root(s).

1. $-8, 5$
2. $\frac{3}{2}, \frac{1}{4}$
3. $-\frac{2}{3}, \frac{5}{2}$

Examples 2–4 Factor each polynomial.

4. $35x^2 - 15x$
5. $18x^2 - 3x + 24x - 4$
6. $x^2 - 12x + 32$
7. $x^2 - 4x - 21$
8. $x^2 - 22x + 121$
9. $4x^2 - 121$

Example 5 Solve each equation.

10. $x^2 - 36 = 0$
11. $12x^2 - 18x = 0$
12. $12x^2 - 2x - 2 = 0$
13. $x^2 - 9x = 0$
14. $x^2 - 3x - 28 = 0$
15. $2x^2 - 24x = -72$

16. **MP SENSE-MAKING** Tamika wants to double the area of her garden by increasing the length and width by the same amount. What will be the dimensions of her garden then?

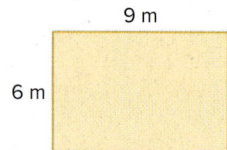

Practice and Problem Solving

Extra Practice is on page R3.

Example 1 Write a quadratic equation in standard form with the given root(s).

17. 7
18. $-5, \frac{1}{2}$
19. $\frac{1}{5}, 6$

Examples 2–4 Factor each polynomial.

20. $40a^2 - 32a$
21. $51c^3 - 34c$
22. $32xy + 40bx - 12ay - 15ab$
23. $3x^2 - 12$
24. $15y^2 - 240$
25. $48cg + 36cf - 4dg - 3df$
26. $x^2 + 13x + 40$
27. $x^2 - 9x - 22$
28. $3x^2 + 12x - 36$
29. $15x^2 + 7x - 2$
30. $4x^2 + 29x + 30$
31. $18x^2 + 15x - 12$
32. $8x^2z^2 - 4xz^2 - 12z^2$
33. $9x^2 + 25$
34. $18x^2y^2 - 24xy^2 + 36y^2$

Example 5 Solve each equation.

35. $15x^2 - 84x - 36 = 0$
36. $12x^2 + 13x - 14 = 0$
37. $12x^2 - 108x = 0$
38. $x^2 + 4x - 45 = 0$
39. $x^2 - 5x - 24 = 0$
40. $x^2 = 121$
41. $x^2 + 17 = 13$
42. $-3x^2 - 10x + 8 = 0$
43. $-8x^2 + 46x - 30 = 0$

44. **GEOMETRY** The hypotenuse of a right triangle is 1 centimeter longer than one side and 4 centimeters longer than three times the other side. Find the dimensions of the triangle.

45. **NUMBER THEORY** Find two consecutive even integers with a product of 624.

GEOMETRY Find x and the dimensions of each rectangle.

46. $A = 96$ ft², $x - 2$ ft, $x + 2$ ft

47. $A = 432$ in², $x - 2$ in., $x + 4$ in.

48. $A = 448$ ft², $3x - 4$ ft, $x + 2$ ft

184 | Lesson 3-4 | Solving Quadratic Equations by Factoring

Solve each equation by factoring.

49. $12x^2 - 4x = 5$

50. $5x^2 = 15x$

51. $16x^2 + 36 = -48x$

52. $75x^2 - 60x = -12$

53. $4x^2 - 144 = 0$

54. $-7x + 6 = 20x^2$

55. MOVIE THEATER A company plans to build a large multiplex theater. The financial analyst told her manager that the profit function for their theater was $P(x) = -x^2 + 48x - 512$, where x is the number of movie screens, and $P(x)$ is the profit earned in thousands of dollars. Determine the range of production of movie screens that will guarantee that the company will not lose money.

Write a quadratic equation in standard form with the given root(s).

56. $-\frac{4}{7}, \frac{3}{8}$

57. $3.4, 0.6$

58. $\frac{2}{11}, \frac{5}{9}$

Solve each equation by factoring.

59. $10x^2 + 25x = 15$

60. $27x^2 + 5 = 48x$

61. $x^2 + 0.25x = 1.25$

62. $48x^2 - 15 = -22x$

63. $3x^2 + 2x = 3.75$

64. $-32x^2 + 56x = 12$

65. DESIGN A square is cut out of the figure at the right. Write an expression for the area of the figure that remains, and then factor the expression.

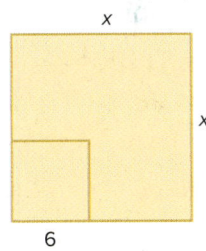

66. MP PERSEVERANCE After analyzing the market, a company that sells websites determined the profitability of their product was modeled by $P(x) = -16x^2 + 368x - 2035$, where x is the price of each website and $P(x)$ is the company's profit. Determine the price range of the websites that will be profitable for the company.

67. PAINTINGS Enola wants to add a border to her painting, distributed evenly, that has the same area as the painting itself. What are the dimensions of the painting with the border included?

68. MULTIPLE REPRESENTATIONS In this problem, you will consider $a(x - p)(x - q) = 0$.

   a. Graphical Graph the related function for $a = 1$, $p = 2$, and $q = -3$.

   b. Analytical What are the solutions of the equation?

   c. Graphical Graph the related functions for $a = 4, -3$, and $\frac{1}{2}$ on the same graph.

   d. Verbal What are the similarities and differences between the graphs?

   e. Verbal What conclusion can you make about the relationship between the factored form of a quadratic equation and its solutions?

69. GEOMETRY The area of the triangle is 26 square centimeters. Find the length of the base.

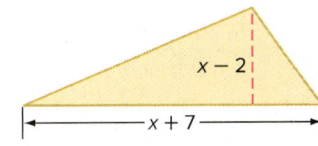

70. SOCCER When a ball is kicked in the air, its height in meters above the ground can be modeled by $h(t) = -4.9t^2 + 14.7t$ and the distance it travels can be modeled by $d(t) = 16t$, where t is the time in seconds.

  a. How long is the ball in the air?

  b. How far does it travel before it hits the ground? (*Hint*: Ignore air resistance.)

  c. What is the maximum height of the ball?

Factor each polynomial.

71. $18a - 24ay + 48b - 64by$

72. $3x^2 + 2xy + 10y + 15x$

73. $6a^2b^2 - 12ab^2 - 18b^3$

74. $12a^2 - 18ab + 30ab^3$

75. $32ax + 12bx - 48ay - 18by$

76. $30ac + 80bd + 40ad + 60bc$

77. $5ax^2 - 2by^2 - 5ay^2 + 2bx^2$

78. $12c^2x + 4d^2y - 3d^2x - 16c^2y$

H.O.T. Problems   Use Higher-Order Thinking Skills

79. ERROR ANALYSIS Gwen and Morgan are solving $-12x^2 + 5x + 2 = 0$. Is either of them correct? Explain your reasoning.

Gwen	Morgan
$-12x^2 + 5x + 2 = 0$	$-12x^2 + 5x + 2 = 0$
$-12x^2 + 8x - 3x + 2 = 0$	$-12x^2 + 8x - 3x + 2 = 0$
$4x(-3x + 2) - (3x + 2) = 0$	$4x(-3x + 2) + (-3x + 2) = 0$
$(4x - 1)(3x + 2) = 0$	$(4x + 1)(-3x + 2) = 0$
$x = \frac{1}{4}$ or $-\frac{2}{3}$	$x = -\frac{1}{4}$ or $\frac{2}{3}$

80. CHALLENGE Solve $3x^6 - 39x^4 + 108x^2 = 0$ by factoring.

81. CHALLENGE The rule for factoring a difference of cubes is shown below. Use this rule to factor $40x^5 - 135x^2y^3$.

$$a^3 - b^3 = (a - b)(a^2 + ab + b^2)$$

82. OPEN-ENDED Choose two integers. Then write an equation in standard form with those roots. How would the equation change if the signs of the two roots were switched?

83. CHALLENGE For a quadratic equation of the form $(x - p)(x - q) = 0$, show that the axis of symmetry of the related quadratic function is located halfway between the x-intercepts p and q.

84. WRITE A QUESTION A classmate is using the guess-and-check strategy to factor trinomials of the form $x^2 + bx + c$. Write a question to help him think of a way to use that strategy for $ax^2 + bx + c$.

85. MP CONSTRUCT ARGUMENTS Determine whether the following statement is *sometimes*, *always*, or *never* true. Explain your reasoning.

In a quadratic equation in standard form where a, b, and c are integers, if b is odd, then the quadratic cannot be a perfect square trinomial.

86. WRITING IN MATH Explain how to factor a trinomial in standard form with $a > 1$.

Preparing for Assessment

87. The height, in feet, of a football is modeled by $h(t) = -16t^2 + 49t$ where t is the time in seconds after the ball is kicked. How long is the ball in the air? **MP 1, 6**

- ○ A $\frac{16}{49}$ second
- ○ B $\frac{4}{7}$ second
- ○ C $1\frac{3}{4}$ seconds
- ○ D $3\frac{1}{16}$ seconds
- ○ E 4 seconds

88. The table represents the quadratic function $f(x)$. What is the negative solution of $f(x) = 0$? **MP 6**

[ ]

x	0	1	2	3
y	−6	−10	−12	−12

89. What are the solutions of $x^2 - x - 42 = 0$? **MP 1**

- ○ A $x = -7$ or $x = -6$
- ○ B $x = -7$ or $x = 6$
- ○ C $x = 7$ or $x = -6$
- ○ D $x = 7$ or $x = 6$

90. Which quadratic equations have solutions of 3 and $\frac{1}{2}$? **MP 2**

- ☐ A $2x^2 - 7x + 3 = 0$
- ☐ B $2x^2 + 6x + 1 = 0$
- ☐ C $-4x^2 + 14x = 6$
- ☐ D $2x^2 + 7x = 3$
- ☐ E $x^2 - 3.5x + 1.5 = 0$

91. What are the solutions to the equation $5x^2 = 20x$? **MP 2**

[ ]

92. MULTI-PART Consider the equation $3x^2 + 4x - 4 = 0$. **MP 3**

a. Derry wrote out the following solution to the equation. Which of the lines have errors in them? Choose all that apply.

- ☐ A $3x^2 + 4x - 4 = 0$
- ☐ B $3x^2 + 6x - 2x - 4 = 0$
- ☐ C $3x^2(x + 2) - 2(x + 2) = 0$
- ☐ D $(3x^2 - 2)(x + 2) = 0$
- ☐ E $3x^2 - 2 = 0$
- ☐ F $x + 2 = 0$

b. What are the solutions of the equation?

- ○ A $x = -2$ or $x = \frac{2}{3}$
- ○ B $x = -2$ or $x = 2$
- ○ C $x = -2$ or $x = 6$
- ○ D $x = -1$ or $x = 4$

93. a. Factor $4x^2 + 64 = 0$. **MP 6, 2**

[ ]

b. What mathematical fact about complex numbers did you use to factor this equation?

[ ]

c. **MP** What mathematical practice did you use to solve this problem?

94. Which are solutions to the equation $x^2 = 144$? **MP 2**

- ☐ A −9
- ☐ B −12
- ☐ C 9
- ☐ D 12
- ☐ E 16
- ☐ F 72

EXTEND 3-4

Algebra Lab
The Complex Plane

A complex number $a + bi$ can be graphed in the **complex plane** by representing it with the point (a, b). Similar to a coordinate plane, the complex plane is comprised of two axes. The real component is plotted on the **real axis**, which is horizontal. The imaginary component is plotted on the **imaginary axis**, which is vertical. The complex plane may also be referred to as the **Argand (ar GON) plane**.

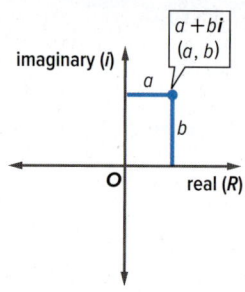

Example 1 Graph in the Complex Plane

Graph $z = 3 + 4i$ in the complex plane.

Step 1 Represent z with the point (a, b).

The real component a of z is 3.

The imaginary component bi of z is $4i$.

z can be represented by the point (a, b) or $(3, 4)$.

Step 2 Graph z in the complex plane.

Construct the complex plane and plot the point $(3, 4)$.

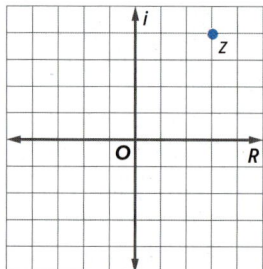

Recall that for a real number, the absolute value is its distance from zero on the number line. Similarly, the **absolute value of a complex number** is its distance from the origin in the complex plane. When $a + bi$ is graphed in the complex plane, the absolute value of $a + bi$ is the distance from (a, b) to the origin. This can be found by using the Distance Formula.

$$\sqrt{(a - 0)^2 + (b - 0)^2} \text{ or } \sqrt{a^2 + b^2}$$

Key Concept Absolute Value of a Complex Number

The absolute value of the complex number $z = a + bi$ is

$|z| = |a + bi| = \sqrt{a^2 + b^2}$.

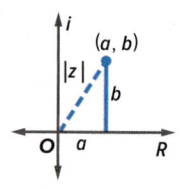

Example 2 Absolute Value of a Complex Number

Find the absolute value of $z = -5 + 12i$.

Step 1 Determine values for a and b.

The real component a of z is -5. The imaginary component bi of z is $12i$.
Thus, $a = -5$ and $b = 12$.

Step 2 Find the absolute value of z.

$|z| = \sqrt{a^2 + b^2}$ Absolute value of a complex number

$ = \sqrt{(-5)^2 + 12^2}$ $a = -5$ and $b = 12$

$ = \sqrt{169}$ or 13 Simplify.

The absolute value of $z = -5 + 12i$ is 13.

Addition and subtraction of complex numbers can be performed graphically.

Example 3 Simplify by Graphing

Simplify $(1 - 2i) - (-2 - 5i)$ by graphing.

Step 1 Write $(1 - 2i) - (-2 - 5i)$ as $(1 - 2i) + (2 + 5i)$.

Step 2 Graph $1 - 2i$ and $2 + 5i$ on the same complex plane. Connect each point with the origin using a dashed segment.

Step 3 Complete the parallelogram that has the two segments as two of its sides. Plot a point where the two additional sides meet. The solution of $(1 - 2i) - (-2 - 5i)$ is $3 + 3i$.

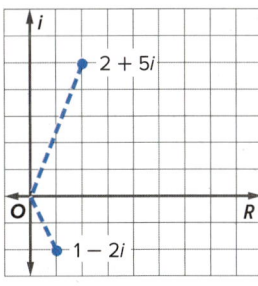
Step 2

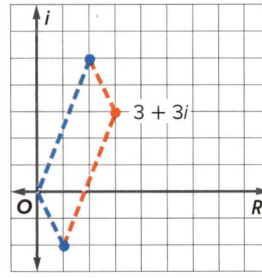
Step 3

Exercises

Graph each number in the complex plane.

1. $z = 3 + i$
2. $z = -4 - 2i$
3. $z = 2 - 2i$

Find the absolute value of each complex number.

4. $z = -4 - 3i$
5. $z = 7 - 2i$
6. $z = -6 - i$

Simplify by graphing.

7. $(6 + 5i) + (-2 - 3i)$
8. $(8 - 2i) - (4 + 7i)$
9. $(5 + 6i) + (-4 + 3i)$

CHAPTER 3
Mid-Chapter Quiz
Lessons 3-1 through 3-4

1. Find the y-intercept, the equation of the axis of symmetry, and the x-coordinate of the vertex for $f(x) = 2x^2 + 8x - 3$. Then graph the function by making a table of values. (Lesson 3-1)

2. **MULTIPLE CHOICE** For which equation is the axis of symmetry $x = 5$? (Lesson 3-1)

   A $f(x) = x^2 - 5x + 3$
   B $f(x) = x^2 - 10x + 7$
   C $f(x) = x^2 + 10x - 3$
   D $f(x) = x^2 + 5x + 2$

3. Determine whether $f(x) = 5 - x^2 + 2x$ has a maximum or a minimum value. Then find this maximum or minimum value and state the domain and range of the function. (Lesson 3-1)

4. **PHYSICAL SCIENCE** From 4 feet above the ground, Maya throws a ball upward with a velocity of 18 feet per second. The height $h(t)$ of the ball t seconds after Maya throws the ball is given by $h(t) = -16t^2 + 18t + 4$. Find the maximum height reached by the ball and the time that this height is reached. (Lesson 3-1)

5. Solve $3x^2 - 17x + 5 = 0$ by graphing. If exact roots cannot be found, state the consecutive integers between which the roots are located. (Lesson 3-2)

Use a quadratic equation to find two real numbers that satisfy each situation, or show that no such numbers exist. (Lesson 3-2)

6. Their sum is 15, and their product is 36.

7. Their sum is 7, and their product is 15.

8. **MULTIPLE CHOICE** Using the graph of the function $f(x) = x^2 + 6x - 7$, what are the solutions to the equation $x^2 + 6x - 7 = 0$? (Lesson 3-2)

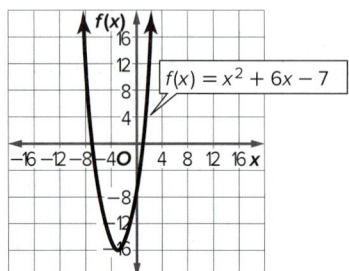

   A −1, 6
   B 1, −6
   C −1, 7
   D 1, −7

9. **BASEBALL** A baseball is hit upward with an initial of 40 feet per second. Ignoring the height of the baseball player, how long does it take for the ball to fall to the ground? Use the formula $h(t) = v_0 t - 16t^2$ where $h(t)$ is the height of an object in feet, v_0 is the object's initial velocity in feet per second, and t is the time in seconds. (Lesson 3-2)

Simplify. (Lesson 3-3)

10. $\sqrt{-81}$
11. $\sqrt{-25x^4 y^5}$
12. $(15 - 3i) - (4 - 12i)$
13. i^{37}
14. $(5 - 3i)(5 + 3i)$
15. $\dfrac{3 - i}{2 + 5i}$

16. The impedance in one part of a series circuit is $3 + 4j$ ohms and the impedance in another part of the circuit is $6 - 7j$ ohms. Add these complex numbers to find the total impedance in the circuit. (Lesson 3-3)

Solve each equation by factoring. (Lesson 3-4)

17. $x^2 - x - 12 = 0$
18. $3x^2 + 7x + 2 = 0$
19. $x^2 - 2x - 15 = 0$
20. $2x^2 + 5x - 3 = 0$
21. $x^2 + 100 = 0$
22. $5x^2 + 180 = 0$

23. **TRIANGLES** Find the dimensions of a triangle if the base is $\dfrac{2}{3}$ the measure of the height and the area is 12 square centimeters. (Lesson 3-4)

24. **PATIO** Eli is putting a cement slab in his backyard. The original slab was going to have dimensions of 8 feet by 6 feet. He decided to make the slab larger by adding x feet to each side. The area of the new slab is 120 square feet. (Lesson 3-4)

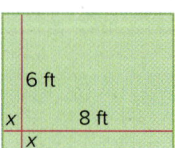

   a. Write a quadratic equation that represents the area of the new slab.

   b. Find the new dimensions of the slab.

   c. **MP** What mathematical practice did you use to solve this problem?

LESSON 5
Solving Quadratic Equations by Completing the Square

Then
- You factored perfect square trinomials.

Now
1. Solve quadratic equations by using the Square Root Property.
2. Solve quadratic equations by completing the square.

Why?
A train moving at 24 meters per second begins to decelerate at a rate of 2 meters per second. The distance it travels after beginning to decelerate is given by the equation $d = -t^2 + 24t$, where t is the number of seconds after the deceleration begins.

Suppose you want to know how long it will take to travel 80 meters. Substitute 80 for d in the equation:

$$80 = -t^2 + 24t$$

You can solve this problem by completing the square and using the Square Root Property.

New Vocabulary
completing the square

Mathematical Practices
7 Look for and make use of structure.

1 Square Root Property
You have solved equations like $x^2 - 25 = 0$ by factoring. You have also used the Square Root Property to solve such equations. This method can be useful with equations like the one above that describes the train's speed.

Example 1 Equation with Rational Roots

Solve $x^2 + 6x + 9 = 36$ by using the Square Root Property.

$x^2 + 6x + 9 = 36$	Original equation
$(x + 3)^2 = 36$	Factor the perfect square trinomial.
$x + 3 = \pm\sqrt{36}$	Square Root Property
$x + 3 = \pm 6$	$\sqrt{36} = 6$
$x = -3 \pm 6$	Subtract 3 from each side.
$x = -3 + 6$ or $x = -3 - 6$	Write as two equations.
$= 3$ $= -9$	Simplify.

The solution set is $\{-9, 3\}$ or $\{x | x = -9, 3\}$.

CHECK Substitute both values into the original equation.

$x^2 + 6x + 9 = 36$ Original equation $x^2 + 6x + 9 = 36$
$3^2 + 6(3) + 9 \stackrel{?}{=} 36$ Substitute 3 and −9. $(-9)^2 + 6(-9) + 9 \stackrel{?}{=} 36$
$9 + 18 + 9 \stackrel{?}{=} 36$ Simplify. $81 - 54 + 9 \stackrel{?}{=} 36$
$36 = 36$ ✓ Both solutions are correct. $36 = 36$ ✓

▶ Guided Practice

Solve each equation by using the Square Root Property.

1A. $x^2 - 12x + 36 = 25$ **1B.** $x^2 - 16x + 64 = 49$

Go Online!

Watch and listen to the **animation** to see how to complete the square with algebra tiles.

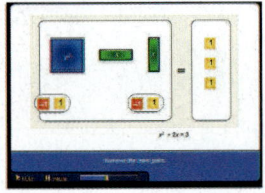

2 Using Completing the Square

All quadratic equations can be solved using the Square Root Property by manipulating the equation until one side is a perfect square. This method is called **completing the square**.

Consider $x^2 + 16x = 9$. Remember to perform each operation on each side of the equation.

$x^2 + 16x + \blacksquare = 9$ What value is needed for the perfect square?

$x^2 + 16x + 64 = 9 + 64$ $\left(\frac{16}{2}\right)^2 = 64$; add 64 to each side.

$x^2 + 16x + 64 = 73$ Simplify.

$(x + 8)^2 = 73$ We can now use the Square Root Property.

Use this pattern of coefficients to complete the square of a quadratic expression.

Key Concept Completing the Square

Words   To complete the square for any quadratic expression of the form $x^2 + bx$, follow the steps below.

> **Step 1** Find one half of b, the coefficient of x.
>
> **Step 2** Square the result in Step 1.
>
> **Step 3** Add the result of Step 2 to $x^2 + bx$.

Symbols $x^2 + bx + \left(\frac{b}{2}\right)^2 = \left(x + \frac{b}{2}\right)^2$

You can solve any quadratic equation by completing the square. Because you are solving an equation, add the value you use to complete the square to each side.

Example 2 Solve an Equation by Completing the Square

Solve $x^2 + 10x - 11 = 0$ by completing the square.

$x^2 + 10x - 11 = 0$ Notice that $x^2 + 10x - 11$ is not a perfect square.

$x^2 + 10x = 11$ Rewrite so the left side is of the form $x^2 + bx$.

$x^2 + 10x + 25 = 11 + 25$ Since $\left(\frac{10}{2}\right)^2 = 25$, add 25 to each side.

$(x + 5)^2 = 36$ Write the left side as a perfect square.

$x + 5 = \pm 6$ Square Root Property

$x = -5 \pm 6$ Subtract 5 from each side.

$x = -5 + 6$ or $x = -5 - 6$ Write as two equations.

$= 1$ $= -11$ Simplify.

The solution set is $\{-11, 1\}$ or $\{x | x = -11, 1\}$. Check the result by using factoring.

> **Watch Out!**
>
> **Each Side** When solving equations by completing the square, don't forget to add $\left(\frac{b}{2}\right)^2$ to *each* side of the equation.

Guided Practice

Solve each equation by completing the square.

2A. $x^2 - 10x + 24 = 0$ **2B.** $x^2 + 10x + 9 = 0$

When the coefficient of the quadratic term is not 1, you must divide the equation by that coefficient before completing the square.

Example 3   Equation with $a \neq 1$

Solve $2x^2 - 7x + 5 = 0$ by completing the square.

$2x^2 - 7x + 5 = 0$	Notice that $2x^2 - 7x + 5$ is not a perfect square.
$x^2 - \frac{7}{2}x + \frac{5}{2} = 0$	Divide by the coefficient of the quadratic term, 2.
$x^2 - \frac{7}{2}x = -\frac{5}{2}$	Subtract $\frac{5}{2}$ from each side.
$x^2 - \frac{7}{2}x + \frac{49}{16} = -\frac{5}{2} + \frac{49}{16}$	Because $\left(-\frac{7}{2} \div 2\right)^2 = \frac{49}{16}$, add $\frac{49}{16}$ to each side.
$\left(x - \frac{7}{4}\right)^2 = \frac{9}{16}$	Write the left side as a perfect square by factoring. Simplify the right side.
$x - \frac{7}{4} = \pm\frac{3}{4}$	Square Root Property
$x = \frac{7}{4} \pm \frac{3}{4}$	Add $\frac{7}{4}$ to each side.
$x = \frac{7}{4} + \frac{3}{4}$ or $x = \frac{7}{4} - \frac{3}{4}$	Write as two equations.
$= \frac{5}{2}$ $= 1$	

The solution set is $\left\{1, \frac{5}{2}\right\}$ or $\left\{x \mid x = 1, \frac{5}{2}\right\}$.

▶ **Guided Practice**

Solve each equation by completing the square.

3A. $3x^2 + 10x - 8 = 0$ **3B.** $3x^2 + 14x - 16 = 0$

Not all solutions of quadratic equations are real numbers. In some cases, the solutions are complex numbers of the form $a + bi$, where $b \neq 0$.

Study Tip

MP Tools A graph of the related function shows that the equation has no real solutions since the graph has no x-intercepts. Imaginary solutions must be checked algebraically by substituting them in the original equation.

Example 4   Equation with Imaginary Solutions

Solve $x^2 + 8x + 22 = 0$ by completing the square.

$x^2 + 8x + 22 = 0$	Notice that $x^2 + 8x + 22$ is not a perfect square.
$x^2 + 8x = -22$	Rewrite so the left side is of the form $x^2 + bx$.
$x^2 + 8x + 16 = -22 + 16$	Because $\left(\frac{8}{2}\right)^2 = 16$, add 16 to each side.
$(x + 4)^2 = -6$	Write the left side as a perfect square.
$x + 4 = \pm\sqrt{-6}$	Square Root Property
$x + 4 = \pm i\sqrt{6}$	$\sqrt{-1} = i$
$x = -4 \pm i\sqrt{6}$	Subtract 4 from each side.

The solution set is $\{-4 + i\sqrt{6}, -4 - i\sqrt{6}\}$ or $\{x \mid x = -4 + i\sqrt{6}, -4 - i\sqrt{6}\}$.

▶ **Guided Practice**

Solve each equation by completing the square.

4A. $x^2 + 2x + 2 = 0$ **4B.** $x^2 - 6x + 25 = 0$

Completing the square can also be used rewrite expressions which contain quadratics. This will often make it easier to determine important characteristics of the function. One such of example of this is to change a quadratic function in standard form to a form known as vertex form.

The vertex form of a quadratic function is $y = a(x - h)^2 + k$, where (h, k) is the vertex of the parabola, $x = h$ is the axis of symmetry, and a determines the shape of the parabola and the direction in which it opens.

When a quadratic equation is in the form $y = ax^2 + bx + c$, you can complete the square to write the function in vertex form. If the coefficient of the quadratic term is not 1, then factor the coefficient from the quadratic and linear terms *before* completing the square. After completing the square and writing the function in vertex form, the value of k indicates a minimum value if $a > 0$ or a maximum value if $a < 0$.

Example 5 Write Functions in Vertex Form

Write each function in vertex form. Identify the vertex.

a. $y = x^2 + 6x - 5$

$y = x^2 + 6x - 5$ Original equation.
$y = (x^2 + 6x + 9) - 5 - 9$ Complete the square.
$y = (x + 3)^2 - 14$ Simplify.

The vertex is $(-3, -14)$.

b. $y = 2x^2 + 8x - 3$

$y = -2x^2 + 8x - 3$ Original equation.
$y = -2(x^2 - 4x) - 3$ Group $ax^2 + bx$ and factor, dividing by a.
$y = -2(x^2 - 4x + 4) - 3 - (-2)(4)$ Complete the square.
$y = -2(x - 2)^2 + 5$ Simplify.

The vertex is $(2, 5)$.

▶ **Guided Practice**

Write each function in vertex form. Identify the vertex.

5A. $y = x^2 + 4x + 6$

5B. $y = 2x^2 - 12x + 17$

Check Your Understanding

= Step-by-Step Solutions begin on page R11.

Example 1 Solve each equation by using the Square Root Property. Round to the nearest hundredth if necessary.

1. $x^2 + 12x + 36 = 6$
2. $x^2 - 8x + 16 = 13$
3. $x^2 + 18x + 81 = 15$
4. $9x^2 + 30x + 25 = 11$

5. **LASER LIGHT SHOW** The area A in square feet of a projected laser light show is given by $A = 0.16d^2$, where d is the distance from the laser to the screen in feet. At what distance will the projected laser light show have an area of 100 square feet?

Examples 2–4 Solve each equation by completing the square.

6. $x^2 + 2x - 8 = 0$
7. $x^2 - 4x + 9 = 0$
8. $2x^2 - 3x - 3 = 0$
9. $2x^2 + 6x - 12 = 0$
10. $x^2 + 4x + 6 = 0$
11. $x^2 + 8x + 10 = 0$

Example 5 Write each function in vertex form. Identify the vertex.

12. $y = x^2 + 3x - 5$
13. $y = -5x^2 + 20x - 51$

Practice and Problem Solving

Extra Practice is on page R3.

Example 1 Solve each equation by using the Square Root Property. Round to the nearest hundredth if necessary.

14. $x^2 + 4x + 4 = 10$
15. $x^2 - 6x + 9 = 20$
16. $x^2 + 8x + 16 = 18$
17. $x^2 + 10x + 25 = 7$
18. $x^2 + 12x + 36 = 5$
19. $x^2 - 2x + 1 = 4$
20. $x^2 - 5x + 6.25 = 4$
21. $x^2 - 15x + 56.25 = 8$
22. $x^2 + 32x + 256 = 1$
23. $x^2 - 3x + \frac{9}{4} = 6$
24. $x^2 + 7x + \frac{49}{4} = 4$
25. $x^2 - 9x + \frac{81}{4} = \frac{1}{4}$

Examples 2–4 Solve each equation by completing the square.

26. $x^2 - 4x + 12 = 0$
27. $x^2 + 2x - 12 = 0$
28. $x^2 + 6x + 8 = 0$
29. $x^2 - 4x + 3 = 0$
30. $2x^2 + x - 3 = 0$
31. $2x^2 - 3x + 5 = 0$
32. $2x^2 + 5x + 7 = 0$
33. $3x^2 - 6x - 9 = 0$
34. $x^2 - 2x + 3 = 0$
35. $x^2 + 4x + 11 = 0$
36. $x^2 - 6x + 18 = 0$
37. $x^2 - 10x + 29 = 0$
38. $3x^2 - 4x = 2$
39. $2x^2 - 7x = -12$
40. $x^2 - 2.4x = 2.2$
41. $x^2 - 5.3x = -8.6$
42. $x^2 - \frac{1}{5}x - \frac{11}{5} = 0$
43. $x^2 - \frac{9}{2}x - \frac{24}{5} = 0$

Example 5 Write each function in vertex form. Identify the vertex.

44. $y = x^2 + 10x + 17$
45. $y = x^2 - 22x + 125$
46. $y = 2x^2 - 4x + 5$
47. $y = -8x^2 - 64x - 146$

48. **MP MODELING** An architect's blueprints call for a dining room measuring 13 feet by 13 feet. The customer would like the dining room to be a square, but with an area of 250 square feet. How much will this add to the dimensions of the room?

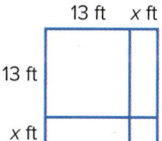

Solve each equation. Round to the nearest hundredth if necessary.

49. $4x^2 - 28x + 49 = 5$
50. $9x^2 + 30x + 25 = 11$
51. $x^2 + x + \frac{1}{3} = \frac{2}{3}$
52. $x^2 + 1.2x + 0.56 = 0.91$

53. FIREWORKS A firework's distance d meters from the ground is given by $d = -1.5t^2 + 25t$, where t is the number of seconds after the firework has been lit.

a. How many seconds have passed since the firework was lit when the firework explodes if it explodes at the maximum height of its path?

b. What is the height of the firework when it explodes?

Write each function in vertex form. Identify the vertex.

54. $y = x^2 - 3x + 9$

55. $y = \frac{1}{2}x^2 - 4x - \frac{21}{2}$

56. $y = 3x^2 - 3x + 1$

57. MULTIPLE REPRESENTATIONS In this problem, you will use quadratic equations to investigate golden rectangles and the golden ratio.

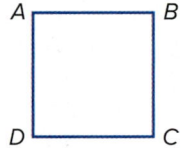

a. **Geometric**
  - Draw square $ABCD$.
  - Locate the midpoint of $\overline{CD}$. Label the midpoint P. Draw $\overline{PB}$.
  - Construct an arc with a radius of $\overline{PB}$ from B clockwise past the bottom of the square.
  - Extend $\overline{CD}$ until it intersects the arc. Label this point Q.
  - Construct rectangle $ARQD$.

b. **Algebraic** Let $AD = x$ and $CQ = 1$. Use completing the square to solve $\dfrac{DQ}{AD} = \dfrac{QR}{CQ}$ for x.

c. **Tabular** Make a table of x and values for $CQ = 2, 3,$ and 4.

d. **Verbal** What do you notice about the x-values? Write an equation you could use to determine x for $CQ = n$, where n is a nonzero real number.

H.O.T. Problems   Use Higher-Order Thinking Skills

58. ERROR ANALYSIS Alonso and Aida are solving $x^2 + 8x - 20 = 0$ by completing the square. Is either of them correct? Explain your reasoning.

Alonso	Aida
$x^2 + 8x - 20 = 0$	$x^2 + 8x - 20 = 0$
$x^2 + 8x = 20$	$x^2 + 8x = 20$
$x^2 + 8x + 16 = 20 + 16$	$x^2 + 8x + 16 = 20$
$(x + 4)^2 = 36$	$(x + 4)^2 = 20$
$x + 4 = \pm 6$	$x + 4 = \pm\sqrt{20}$
$x = -4 \pm 6$	$x = -4 \pm \sqrt{20}$

59. CHALLENGE Solve $x^2 + bx + c = 0$ by completing the square. Your answer will be an expression for x in terms of b and c.

60. CONSTRUCT ARGUMENTS Without solving, determine how many unique solutions there are for each equation. Are they rational, real, or complex? Justify your reasoning.

a. $(x + 2)^2 = 16$
b. $(x - 2)^2 = 16$
c. $-(x - 2)^2 = 16$
d. $36 - (x - 2)^2 = 16$
e. $16(x + 2)^2 = 0$
f. $(x + 4)^2 = (x + 6)^2$

61. OPEN-ENDED Write a perfect square trinomial equation in which the linear coefficient is negative and the constant term is a fraction. Then solve the equation.

62. WRITING IN MATH Explain what it means to complete the square. Describe each step.

Preparing for Assessment

63. The length of a rectangular field is 16 feet less than twice the width. What is the width of the field if its area is 9216 square feet? **MP 7**
- A 60 feet
- B 72 feet
- C 80 feet
- D 96 feet

64. Which value should be added to each side to solve the following equation by completing the square? **MP 7**

$$x^2 - 36x + \underline{\quad} = 30 + \underline{\quad}$$

- A 36
- B 294
- C 324
- D 648
- E 1296

65. What is the solution set for $3x^2 - 7x - 12 = 0$? **MP 7**

66. Solve $x^2 + 8x = 4$ by completing the steps below. **MP 7**

Step 1: $x^2 + 8x + \underline{\quad} = 4 + \underline{\quad}$

Step 2: $(x + \underline{\quad})^2 = \underline{\quad}$

Step 3: $x + \underline{\quad} = \pm \underline{\quad}$

Step 4: $x = -\underline{\quad} \pm \underline{\quad}$

Step 5: $x = -\underline{\quad} \pm 2\underline{\quad}$

67. Solve $x^2 + 6x = -1$ by completing the square. **MP 7**

68. Solve $2x^2 + 15 = 6x$ by completing the square. **MP 7**

69. MULTI-STEP Solve the equation $x^2 + 14x + 9 = -4$ following the steps. **MP 7**

a. What number must be added to each side of the equation to create a perfect square trinomial?

b. Which is the factored form of the equation?
- A $(x + 7)^2 = 14$
- B $(x + 7)^2 = 36$
- C $(x + 14)^2 = 44$
- D $(x + 14)^2 = 49$

c. Which are solutions to the equation?
- A -13
- B -6
- C -1
- D 6
- E 1
- F 13

70. Which are solutions to the equation $x^2 - 10x + 6 = 0$? **MP 7**
- A $\pm\frac{3}{5}$
- B $\pm\frac{19}{5}$
- C $-5 \pm \sqrt{19}$
- D $5 \pm \sqrt{19}$

71. Which equations have solutions with complex solutions? **MP 7**
- A $x^2 + 8x + 20 = 0$
- B $x^2 - 10x - 6 = 0$
- C $x^2 + x - 8 = 0$
- D $x^2 - 6x + 16 = 0$
- E $x^2 + 4x + 5 = 0$
- F $x^2 - 12x + 22 = 0$

EXTEND 3-5
Graphing Technology Lab
Solving Quadratic Equations

You can use TI-Nspire™ CAS technology to solve quadratic equations.

Mathematical Practices
 **5** Use appropriate tools strategically.

Activity Finding Roots

Work cooperatively. Solve each equation.

a. $3x^2 - 4x + 1 = 0$

Step 1 Add a new **Calculator** page.

Step 2 Select the **Solve** tool from the **Algebra** menu.

Step 3 Type $3x^2 - 4x + 1 = 0$ followed by a comma, *x*, and then **enter**.
The solutions are $x = \frac{1}{3}$ or $x = 1$.

b. $6x^2 + 4x - 3 = 0$

Step 1 Select the **Solve** tool from the **Algebra** menu.

Step 2 Type $6x^2 - 4x - 3 = 0$ followed by a comma, *x*, and then **enter**.
The solutions are $x = \frac{-2 \pm \sqrt{22}}{6}$.

c. $x^2 - 6x + 10 = 0.$

Step 1 Select the **Solve** tool from the **Algebra** menu.

Step 2 Type $x^2 - 6x + 10 = 0$ followed by a comma, *x*, and then **enter**.
The calculator returns a value of *false*, meaning that there are no real solutions.

Step 3 Under menu, select **Algebra**, **Complex**, then **Solve**. Reenter the equation.
The solutions are $x = 3 \pm i$.

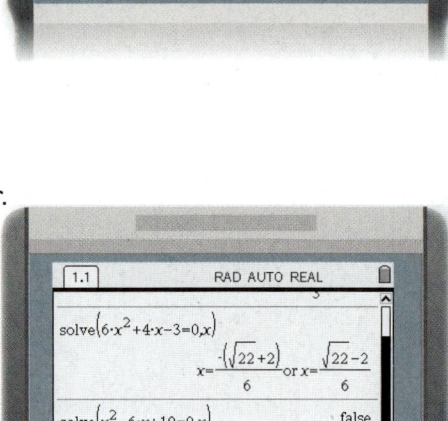

Exercises

Work cooperatively. Solve each equation.

1. $x^2 - 2x - 24 = 0$
2. $-x^2 + 4x - 1 = 0$
3. $0 = -3x^2 - 6x + 9$
4. $x^2 - 2x + 5 = 0$
5. $0 = 4x^2 - 8$
6. $0 = 2x^2 - 4x + 1$
7. $x^2 + 3x + 8 = 5$
8. $25 + 4x^2 = -20x$
9. $x^2 - x = -6$

LESSON 6
The Quadratic Formula and the Discriminant

∷Then
- You solved equations by completing the square.

∷Now
1. Solve quadratic equations by using the Quadratic Formula.
2. Use the discriminant to determine the number and type of roots of a quadratic equation.

∷Why?
- Pumpkin catapult is an event in which a contestant builds a catapult and launches a pumpkin at a target.

  The path of the pumpkin can be modeled by the quadratic function $h = -4.9t^2 + 117t + 42$, where h is the height of the pumpkin and t is the number of seconds.

  To predict when the pumpkin will hit the target, you can solve the equation $0 = -4.9t^2 + 117t + 42$. This equation would be difficult to solve using factoring, graphing, or completing the square.

 New Vocabulary
Quadratic Formula
discriminant

 Mathematical Practices
8 Look for and express regularity in repeated reasoning.

1 Quadratic Formula You have found solutions of some quadratic equations by graphing, by factoring, and by using the Square Root Property. There is also a formula that can be used to solve any quadratic equation. This formula can be derived by solving the standard form of a quadratic equation.

General Case		Specific Case
$ax^2 + bx + c = 0$	Standard quadratic equation	$2x^2 + 8x + 1 = 0$
$x^2 + \frac{b}{a}x + \frac{c}{a} = 0$	Divide each side by a.	$x^2 + 4x + \frac{1}{2} = 0$
$x^2 + \frac{b}{a}x = -\frac{c}{a}$	Subtract $\frac{c}{a}$ from each side.	$x^2 + 4x = -\frac{1}{2}$
$x^2 + \frac{b}{a}x + \frac{b^2}{4a^2} = -\frac{c}{a} + \frac{b^2}{4a^2}$	Complete the square.	$x^2 + 4x + \left(\frac{4}{2}\right)^2 = -\frac{1}{2} + \left(\frac{4}{2}\right)^2$
$\left(x + \frac{b}{2a}\right)^2 = -\frac{c}{a} + \frac{b^2}{4a^2}$	Factor the left side.	$(x + 2)^2 = -\frac{1}{2} + \left(\frac{4}{2}\right)^2$
$\left(x + \frac{b}{2a}\right)^2 = \frac{b^2 - 4ac}{4a^2}$	Simplify the right side.	$(x + 2)^2 = \frac{7}{2}$
$x + \frac{b}{2a} = \pm\frac{\sqrt{b^2 - 4ac}}{2a}$	Square Root Property	$x + 2 = \pm\sqrt{\frac{7}{2}}$
$x = -\frac{b}{2a} \pm \frac{\sqrt{b^2 - 4ac}}{2a}$	Subtract $\frac{b}{2a}$ from each side.	$x = -2 \pm \sqrt{\frac{7}{2}}$
$x = \frac{-b \pm \sqrt{b^2 - 4ac}}{2a}$	Simplify.	$x = \frac{-4 \pm \sqrt{14}}{2}$

The equation $x = \dfrac{-b \pm \sqrt{b^2 - 4ac}}{2a}$ is known as the **Quadratic Formula**.

Go Online!

You will want to reference the Quadratic Formula often. Log into your **eStudent Edition** to bookmark this lesson.

Key Concept Quadratic Formula

Words The solutions of a quadratic equation of the form $ax^2 + bx + c = 0$, where $a \neq 0$, are given by the following formula.

$$x = \frac{-b \pm \sqrt{b^2 - 4ac}}{2a}$$

Example $x^2 + 5x + 6 = 0 \rightarrow x = \dfrac{-5 \pm \sqrt{5^2 - 4(1)(6)}}{2(1)}$

Example 1 Two Rational Roots

Solve $x^2 - 10x = 11$ by using the Quadratic Formula.

First, write the equation in the form $ax^2 + bx + c = 0$ and identify a, b, and c.

$$ax^2 + bx + c = 0$$
$$\downarrow \quad \downarrow \quad \downarrow$$
$$x^2 - 10x = 11 \quad \rightarrow \quad 1x^2 - 10x - 11 = 0$$

Then, substitute these values into the Quadratic Formula.

$x = \dfrac{-b \pm \sqrt{b^2 - 4ac}}{2a}$ Quadratic Formula

$= \dfrac{-(-10) \pm \sqrt{(-10)^2 - 4(1)(-11)}}{2(1)}$ Replace a with 1, b with -10, and c with -11.

$= \dfrac{10 \pm \sqrt{100 + 44}}{2}$ Multiply.

$= \dfrac{10 \pm \sqrt{144}}{2}$ Simplify.

$= \dfrac{10 \pm 12}{2}$ $\sqrt{144} = 12$

$x = \dfrac{10 + 12}{2}$ or $x = \dfrac{10 - 12}{2}$ Write as two equations.

$= 11 \qquad\qquad\qquad = -1$ Simplify.

The solutions are -1 and 11.

CHECK Substitute both values into the original equation.

$$x^2 - 10x = 11 \qquad\qquad x^2 - 10x = 11$$
$$(-1)^2 - 10(-1) \stackrel{?}{=} 11 \qquad (11)^2 - 10(11) \stackrel{?}{=} 11$$
$$1 + 10 \stackrel{?}{=} 11 \qquad\qquad 121 - 110 \stackrel{?}{=} 11$$
$$11 = 11 \checkmark \qquad\qquad\qquad 11 = 11 \checkmark$$

Guided Practice

Solve each equation by using the Quadratic Formula.

1A. $x^2 + 6x = 16$ **1B.** $2x^2 + 25x + 33 = 0$

Review Vocabulary

radicand the value underneath the radical symbol

When the value of the radicand in the Quadratic Formula is 0, the quadratic equation has exactly one rational root.

Math History Link

Brahmagupta (598–668) Indian mathematician Brahmagupta offered the first general solution of the quadratic equation $ax^2 + bx = c$, now known as the Quadratic Formula.

Example 2 One Rational Root

Solve $x^2 + 8x + 16 = 0$ by using the Quadratic Formula.

Identify a, b, and c. Then, substitute these values into the Quadratic Formula.

$x = \dfrac{-b \pm \sqrt{b^2 - 4ac}}{2a}$ Quadratic Formula

$= \dfrac{-(8) \pm \sqrt{(8)^2 - 4(1)(16)}}{2(1)}$ Replace a with 1, b with 8, and c with 16.

$= \dfrac{-8 \pm \sqrt{0}}{2}$ Simplify.

$= \dfrac{-8}{2}$ or -4 $\sqrt{0} = 0$

The solution is -4.

CHECK A graph of the related function shows that there is one solution at $x = -4$.

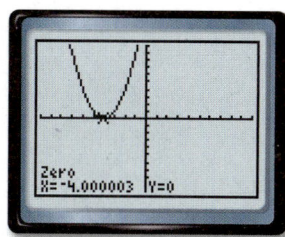

[−10, 10] scl: 1 by [−10, 10] scl: 1

Guided Practice

Solve each equation by using the Quadratic Formula.

2A. $x^2 - 16x + 64 = 0$ **2B.** $x^2 + 34x + 289 = 0$

You can express irrational roots exactly by writing them in radical form.

Example 3 Irrational Roots

Solve $2x^2 + 6x - 7 = 0$ by using the Quadratic Formula.

$x = \dfrac{-b \pm \sqrt{b^2 - 4ac}}{2a}$ Quadratic Formula

$= \dfrac{-(6) \pm \sqrt{(6)^2 - 4(2)(-7)}}{2(2)}$ Replace a with 2, b with 6, and c with -7.

$= \dfrac{-6 \pm \sqrt{92}}{4}$ Simplify.

$= \dfrac{-6 \pm 2\sqrt{23}}{4}$ or $\dfrac{-3 \pm \sqrt{23}}{2}$ $\sqrt{92} = \sqrt{4 \cdot 23}$ or $2\sqrt{23}$

The approximate solutions are -3.9 and 0.9.

CHECK Check these results by graphing the related quadratic function, $y = 2x^2 + 6x - 7$. Using the **zero** function of a graphing calculator, the approximate zeros of the related function are -3.9 and 0.9.

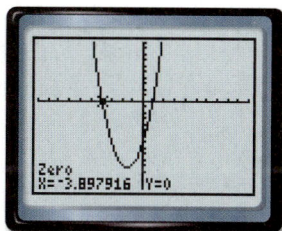

[−10, 10] scl: 1 by [−10, 10] scl: 1

Guided Practice

Solve each equation by using the Quadratic Formula.

3A. $3x^2 + 5x + 1 = 0$ **3B.** $x^2 - 8x + 9 = 0$

connectED.mcgraw-hill.com 201

> **Study Tip**
> **Complex Numbers**
> Remember to write your solutions in the form $a + bi$, sometimes called the *standard form* of a complex number.

When using the Quadratic Formula, if the value of the radicand is negative, the solutions will be complex. Complex solutions always appear in conjugate pairs.

Example 4 Complex Roots

Solve $x^2 - 6x = -10$ by using the Quadratic Formula.

$$x = \frac{-b \pm \sqrt{b^2 - 4ac}}{2a}$$ Quadratic Formula

$$= \frac{-(-6) \pm \sqrt{(-6)^2 - 4(1)(10)}}{2(1)}$$ Replace a with 1, b with -6, and c with 10.

$$= \frac{6 \pm \sqrt{-4}}{2}$$ Simplify.

$$= \frac{6 \pm 2i}{2}$$ $\sqrt{-4} = \sqrt{4 \cdot (-1)}$ or $2i$

$$= 3 \pm i$$ Simplify.

The solutions are the complex numbers $3 + i$ and $3 - i$.

CHECK A graph of the related function shows that the solutions are complex, but it cannot help you find them. To check complex solutions, substitute them into the original equation.

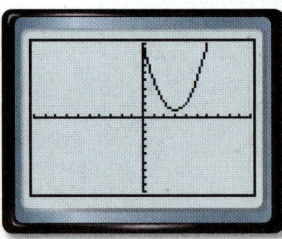
[−10, 10] scl: 1 by [−10, 10] scl: 1

$$x^2 - 6x = -10$$ Original equation
$$(3 + i)^2 - 6(3 + i) \stackrel{?}{=} -10$$ $x = 3 + i$
$$9 + 6i + i^2 - 18 - 6i \stackrel{?}{=} -10$$ Square of a sum; Distributive Property
$$-9 + i^2 \stackrel{?}{=} -10$$ Simplify.
$$-9 - 1 = -10 \checkmark$$ $i^2 = -1$

$$x^2 - 6x = -10$$ Original equation
$$(3 - i)^2 - 6(3 - i) \stackrel{?}{=} -10$$ $x = 3 - i$
$$9 - 6i + i^2 - 18 + 6i \stackrel{?}{=} -10$$ Square of a sum; Distributive Property
$$-9 + i^2 \stackrel{?}{=} -10$$ Simplify.
$$-9 - 1 = -10 \checkmark$$ $i^2 = -1$

▶ **Guided Practice**

Solve each equation by using the Quadratic Formula.

4A. $3x^2 + 5x + 4 = 0$ **4B.** $x^2 - 4x = -13$

2 Roots and the Discriminant

In the previous examples, observe the relationship between the value of the expression under the radical and the roots of the quadratic equation. The expression $b^2 - 4ac$ is called the **discriminant**.

$$x = \frac{-b \pm \sqrt{b^2 - 4ac}}{2a} \leftarrow \text{discriminant}$$

The value of the discriminant can be used to determine the number and type of roots of a quadratic equation. The table on the following page summarizes the possible types of roots.

The discriminant can also be used to confirm the number and type of solutions after you solve the quadratic equation.

202 | Lesson 3-6 | The Quadratic Formula and the Discriminant

> **Study Tip**
>
> **Roots** Remember that the solutions of an equation are called *roots* or *zeros* and are the value(s) where the graph crosses the *x*-axis.

Key Concept Discriminant

Consider $ax^2 + bx + c = 0$, where a, b, and c are rational numbers and $a \neq 0$.

Value of Discriminant	Type and Number of Roots	Example of Graph of Related Function
$b^2 - 4ac > 0$; $b^2 - 4ac$ is a perfect square.	2 real, rational roots	
$b^2 - 4ac > 0$; $b^2 - 4ac$ is *not* a perfect square.	2 real, irrational roots	
$b^2 - 4ac = 0$	1 real rational root	
$b^2 - 4ac < 0$	2 complex roots	

Example 5 Describe Roots

Find the value of the discriminant for each quadratic equation. Then describe the number and type of roots for the equation.

a. $7x^2 - 11x + 5 = 0$

$a = 7, b = -11, c = 5$

$b^2 - 4ac = (-11)^2 - 4(7)(5)$

$ = 121 - 140$

$ = -19$

The discriminant is negative, so there are two complex roots.

b. $x^2 + 22x + 121 = 0$

$a = 1, b = 22, c = 121$

$b^2 - 4ac = (22)^2 - 4(1)(121)$

$ = 484 - 484$

$ = 0$

The discriminant is 0, so there is one rational root.

Guided Practice

5A. $-5x^2 + 8x - 1 = 0$

5B. $-7x + 15x^2 - 4 = 0$

You have studied a variety of methods for solving quadratic equations. The table below summarizes these methods.

Study Tip

Study Notebook You may wish to copy this list of methods to your math notebook or Foldable to keep as a reference as you study.

Concept Summary Solving Quadratic Equations

Method	Can be Used	When to Use
graphing	sometimes	Use only if an exact answer is not required. Best used to check the reasonableness of solutions found algebraically.
factoring	sometimes	Use if the constant term is 0 or if the factors are easily determined. **Example** $x^2 - 7x = 0$
Square Root Property	sometimes	Use for equations in which a perfect square is equal to a constant. **Example** $(x-5)^2 = 18$
completing the square	always	Useful for equations of the form $x^2 + bx + c = 0$, where b is even. **Example** $x^2 + 6x - 14 = 0$
Quadratic Formula	always	Useful when other methods fail or are too tedious. **Example** $2.3x^2 - 1.8x + 9.7 = 0$

Check Your Understanding

Examples 1–4 Solve each equation by using the Quadratic Formula.

1. $x^2 + 12x - 9 = 0$
2. $x^2 + 8x + 5 = 0$
3. $4x^2 - 5x - 2 = 0$
4. $9x^2 + 6x - 4 = 0$
5. $10x^2 - 3 = 13x$
6. $22x = 12x^2 + 6$
7. $-3x^2 + 4x = -8$
8. $x^2 + 3 = -6x + 8$

Examples 3–4 9. **MODELING** An amusement park ride takes riders to the top of a tower and drops them at speeds reaching 80 feet per second. A function that models this ride is $h = -16t^2 - 64t + 60$, where h is the height in feet and t is the time in seconds. About how many seconds does it take for riders to drop from 60 feet to 0 feet?

Example 5 Complete parts a and b for each quadratic equation.
   a. Find the value of the discriminant.
   b. Describe the number and type of roots.

10. $3x^2 + 8x + 2 = 0$
11. $2x^2 - 6x + 9 = 0$
12. $-16x^2 + 8x - 1 = 0$
13. $5x^2 + 2x + 4 = 0$

Practice and Problem Solving

Extra Practice is on page R3.

Examples 1–4 Solve each equation by using the Quadratic Formula.

14. $x^2 + 45x = -200$
15. $4x^2 - 6 = -12x$
16. $3x^2 - 4x - 8 = -6$
17. $4x^2 - 9 = -7x - 4$
18. $5x^2 - 9 = 11x$
19. $12x^2 + 9x - 2 = -17$

20. DIVING Competitors in the 10-meter platform diving competition jump upward and outward before diving into the pool below. The height h of a diver in meters above the pool after t seconds can be approximated by the equation $h = -4.9t^2 + 3t + 10$.

  a. Determine a domain and range for which this function makes sense.

  b. When will the diver hit the water?

Example 5 Complete parts a–c for each quadratic equation.

  a. Find the value of the discriminant.

  b. Describe the number and type of roots.

  c. Find the exact solutions by using the Quadratic Formula.

21. $2x^2 + 3x - 3 = 0$
22. $4x^2 - 6x + 2 = 0$
23. $6x^2 + 5x - 1 = 0$
24. $6x^2 - x - 5 = 0$
25. $3x^2 - 3x + 8 = 0$
26. $2x^2 + 4x + 7 = 0$
27. $-5x^2 + 4x + 1 = 0$
28. $x^2 - 6x = -9$
29. $-3x^2 - 7x + 2 = 6$
30. $-8x^2 + 5 = -4x$
31. $x^2 + 2x - 4 = -9$
32. $-6x^2 + 5 = -4x + 8$

33. PHONES While Darnell is talking to his friend Jack from his balcony, he drops his phone. Darnell stands at his bedroom window, and Jack stands directly below the window. If Jack tosses the phone to Darnell with an initial velocity of 35 feet per second, an equation for the height h in feet of the phone after t seconds is $h = -16t^2 + 35t + 5$.

  a. If the window is 25 feet above the ground, will Darnell have 0, 1, or 2 chances to catch his phone?

  b. If Darnell is unable to catch his phone, when will it hit the ground?

34. **MP SENSE-MAKING** Civil engineers are designing a section of road that is going to dip below sea level. The road's curve can be modeled by the equation $y = 0.00005x^2 - 0.06x$, where x is the horizontal distance in feet between the points where the road is at sea level and y is the elevation. The engineers want to put stop signs at the locations where the elevation of the road is equal to sea level. At what horizontal distances will they place the stop signs?

Complete parts a–c for each quadratic equation.

  a. Find the value of the discriminant.

  b. Describe the number and type of roots.

  c. Find the exact solutions by using the Quadratic Formula.

35. $5x^2 + 8x = 0$
36. $8x^2 = -2x + 1$
37. $4x - 3 = -12x^2$
38. $0.8x^2 + 2.6x = -3.2$
39. $0.6x^2 + 1.4x = 4.8$
40. $-4x^2 + 12 = -6x - 8$

41. SMOKING A decrease in smoking in the United States has resulted in lower death rates caused by lung cancer. The number of deaths per 100,000 people y can be approximated by $y = -0.26x^2 - 0.55x + 91.81$, where x represents the number of years after 2000.

Year	Deaths per 100,000
2000	91.8
2002	89.7
2004	85.5
2010	60.3
2017	?

a. Calculate the number of deaths per 100,000 people for 2017.

b. Use the Quadratic Formula to solve for x when $y = 50$.

c. According to the quadratic function, when will the death rate be 0 per 100,000? Do you think that this prediction is reasonable? Why or why not?

42. NUMBER THEORY The sum S of consecutive integers 1, 2, 3, ..., n is given by the formula $S = \frac{1}{2}n(n+1)$. How many consecutive integers, starting with 1, must be added to get a sum of 666?

H.O.T. Problems Use Higher-Order Thinking Skills

43. ERROR ANALYSIS Tama and Jonathan are determining the number of solutions of $3x^2 - 5x = 7$. Is either of them correct? Explain your reasoning.

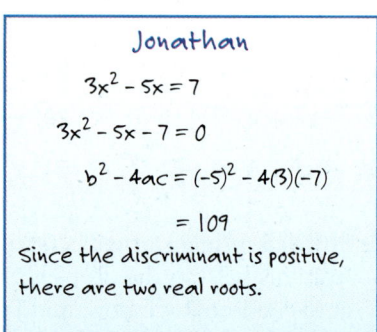

Tama
$3x^2 - 5x = 7$
$b^2 - 4ac = (-5)^2 - 4(3)(7)$
$= -59$
Since the discriminant is negative, there are no real solutions.

Jonathan
$3x^2 - 5x = 7$
$3x^2 - 5x - 7 = 0$
$b^2 - 4ac = (-5)^2 - 4(3)(-7)$
$= 109$
Since the discriminant is positive, there are two real roots.

44. CHALLENGE Find the solutions of $4ix^2 - 4ix + 5i = 0$ by using the Quadratic Formula.

45. REASONING Determine whether each statement is *sometimes*, *always*, or *never* true. Explain your reasoning.

a. In a quadratic equation in standard form, if a and c are different signs, then the solutions will be real.

b. If the discriminant of a quadratic equation is greater than 1, the two roots are real irrational numbers.

46. OPEN-ENDED Sketch the corresponding graph and state the number and type of roots for each of the following.

a. $b^2 - 4ac = 0$

b. A quadratic function in which $f(x)$ never equals zero.

c. A quadratic function in which $f(a) = 0$ and $f(b) = 0$; $a \neq b$.

d. The discriminant is less than zero.

e. a and b are both solutions and can be represented as fractions.

47. CHALLENGE Find the value(s) of m in the quadratic equation $x^2 + x + m + 1 = 0$ such that it has one solution.

48. WRITING IN MATH Describe three different ways to solve $x^2 - 2x - 15 = 0$. Which method do you prefer, and why?

Preparing for Assessment

49. What is the solution set for the equation shown below? **MP** 8

$$x^2 - 10x + 36 = 2$$

- A $3 \pm i\sqrt{11}$
- B $5 \pm 3i$
- C $5 \pm i\sqrt{58}$
- D $10 \pm 6i$

50. Which value describes the discriminant of the equation graphed below? **MP** 8

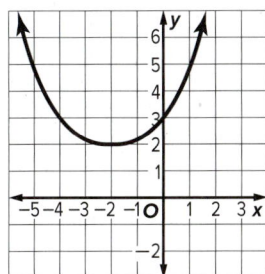

- A -1
- B 0
- C 2
- D 3

51. What is the solution set for the equation $10x^2 - 7x - 3 = 0$? **MP** 8

52. Solve $3x^2 + 8x = 1$ by completing the steps below. **MP** 8

Step 1: $3x^2 + 8x - \square = 0$

Step 2: $x = \dfrac{-\square \pm \sqrt{\square^2 - 4\square}}{2\square}$

Step 3: $x = \dfrac{-\square \pm \sqrt{\square}}{\square}$

Step 4: $x = \dfrac{-\square \pm \sqrt{\square}}{\square}$

53. Find the exact solutions of $x^2 - 10x - 13 = 0$ using the Quadratic Formula. **MP** 8

54. Solve $5x^2 - 7x + 12 = 0$ using the Quadratic Formula. **MP** 8

55. What type of solutions do you expect when you solve $x^2 + 3x + 15 = 0$? **MP** 8

56. MULTI-STEP Solve the equation $2x^2 - 4x + 7 = 0$ following the steps. **MP** 2

a. What does the discriminant tell you about the number and type of solutions?

b. Which is the correct substitution into the quadratic formula?

- A $\dfrac{4 \pm \sqrt{(-4)^2 - 4(2)(7)}}{2(2)}$
- B $\dfrac{-2 \pm \sqrt{(2)^2 - 4(-4)(7)}}{2(-4)}$
- C $\dfrac{-7 \pm \sqrt{(7)^2 - 4(2)(-4)}}{2(2)}$
- D $\dfrac{4 \pm \sqrt{(-4)^2 - 4(7)(2)}}{2(7)}$

c. Which are solutions to the equation?

- A $\dfrac{-4 \pm \sqrt{40}}{4}$
- B $\dfrac{2 \pm i\sqrt{10}}{7}$
- C $\dfrac{4 \pm i\sqrt{10}}{2}$
- D $\dfrac{2 \pm i\sqrt{10}}{2}$

EXTEND 3-6
Algebra Lab
Quadratics and Rate of Change

You have learned that a linear function has a constant rate of change. In this lab, you will investigate the rate of change for quadratic functions.

Mathematical Practices
MP 7 Look for and make use of structure.

Activity Determine Rate of Change

Consider $f(x) = 0.1875x^2 - 3x + 12$.

Step 1 Make a table like the one below. Use values from 0 through 16 for x.

x	0	1	2	3	...	16
y	12	9.1875	6.75			
First-Order Differences						
Second-Order Differences						

Step 2 Find each y-value. For example, when $x = 1$, $y = 0.1875(1)^2 - 3(1) + 12$ or 9.1875.

Step 3 Graph the ordered pairs (x, y). Then connect the points with a smooth curve. Notice that the function *decreases* when $0 < x < 8$ and *increases* when $8 < x < 16$.

Step 4 The rate of change from one point to the next can be found by using the slope formula. From (0, 12) to (1, 9.1875), the slope is $\frac{9.1875 - 12}{1 - 0}$ or -2.8125.

This is the first-order difference at $x = 1$. Complete the table for all the first-order differences. Describe any patterns in the differences.

Step 5 The second-order differences can be found by subtracting consecutive first-order differences. For example, the second-order difference at $x = 2$ is found by subtracting the first order difference at $x = 1$ from the first-order difference at $x = 2$. Describe any patterns in the differences.

Exercises

For each function make a table of values for the given x-values. Graph the function. Then determine the first-order and second-order differences.

1. $y = -x^2 + 2x - 1$ for $x = -3, -2, -1, 0, 1, 2, 3$
2. $y = 0.5x^2 + 2x - 2$ for $x = -5, -4, -3, -2, -1, 0, 1$
3. $y = -3x^2 - 18x - 26$ for $x = -6, -5, -4, -3, -2, -1, 0$
4. **MAKE A CONJECTURE** Repeat the activity for a cubic function. At what order difference would you expect $g(x) = x^4$ to be constant? $h(x) = x^n$?

LESSON 7
Quadratic Inequalities

∷Then
- You solved linear inequalities.

∷Now
1. Graph quadratic inequalities in two variables.
2. Solve quadratic inequalities in one variable.

∷Why?
A water balloon launched from a slingshot can be represented by several different quadratic equations and inequalities.

Suppose the height of a water balloon $h(t)$ in meters above the ground t seconds after being launched is modeled by the quadratic function $h(t) = -4.9t^2 + 32t + 1.2$. You can solve a quadratic inequality to determine how long the balloon will be a certain distance above the ground.

New Vocabulary
quadratic inequality

Mathematical Practices
1 Make sense of problems and persevere in solving them.

1 Graph Quadratic Inequalities
You can graph **quadratic inequalities** in two variables by using the same techniques used to graph linear inequalities in two variables.

Step 1 Graph the related function.

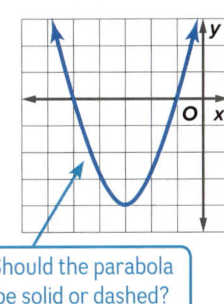

Should the parabola be solid or dashed?

Step 2 Test a point not on the parabola.

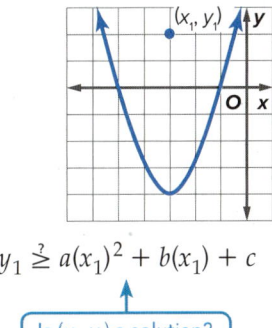

$y_1 \overset{?}{\geq} a(x_1)^2 + b(x_1) + c$

Is (x_1, y_1) a solution?

Step 3 Shade accordingly.

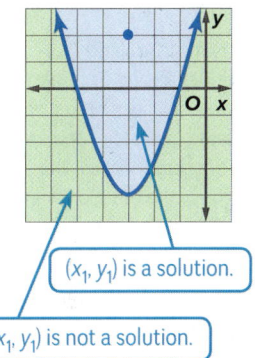

(x_1, y_1) is a solution.

(x_1, y_1) is not a solution.

Example 1 Graph a Quadratic Inequality

Graph $y > x^2 + 2x + 1$.

Step 1 Graph the related function, $y = x^2 + 2x + 1$. The parabola should be dashed.

Step 2 Test a point not on the graph of the parabola.

$y > x^2 + 2x + 1$

$-1 \overset{?}{>} 0^2 + 2(0) + 1$

$-1 \not> 1$ So, $(0, -1)$ is *not* a solution of the inequality.

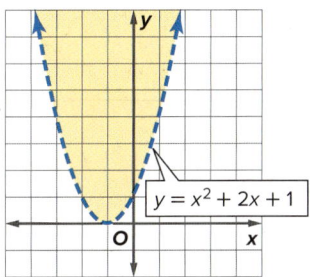

Step 3 Shade the region that does not contain the point $(0, -1)$.

▶ Guided Practice

Graph each inequality.

1A. $y \leq x^2 + 2x + 4$

1B. $y < -2x^2 + 3x + 5$

2 Solve Quadratic Inequalities
Quadratic inequalities in one variable can be solved using the graphs of the related quadratic functions.

$ax^2 + bx + c < 0$

Graph $y = ax^2 + bx + c$ and identify the x-values for which the graph lies *below* the x-axis.

For $\leq$, include the x-intercepts in the solution.

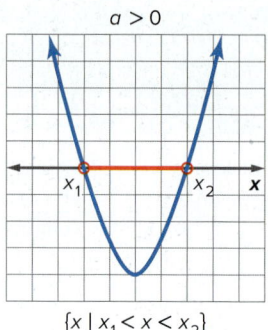
$\{x \mid x_1 < x < x_2\}$

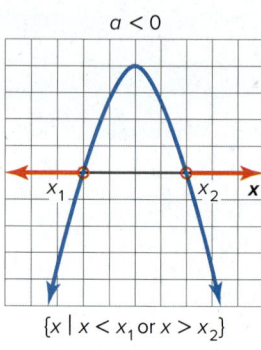
$\{x \mid x < x_1 \text{ or } x > x_2\}$

$ax^2 + bx + c > 0$

Graph $y = ax^2 + bx + c$ and identify the x-values for which the graph lies *above* the x-axis.

For $\geq$, include the x-intercepts in the solution.

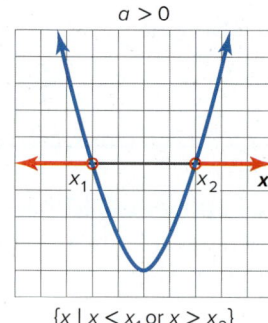
$\{x \mid x < x_1 \text{ or } x > x_2\}$

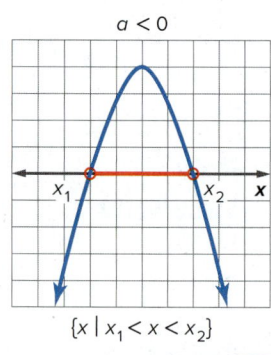
$\{x \mid x_1 < x < x_2\}$

Example 2 Solve $ax^2 + bx + c < 0$ by Graphing

Solve $x^2 + 2x - 8 < 0$ by graphing.

The solution consists of x-values for which the graph of the related function lies *below* the x-axis. Begin by finding the zeros of the related function.

$x^2 + 2x - 8 = 0$ Related equation
$(x - 2)(x + 4) = 0$ Factor.
$x - 2 = 0$ or $x + 4 = 0$ Zero Product Property
$x = 2$ $x = -4$ Solve each equation.

Sketch the graph of a parabola that has x-intercepts at -4 and 2. The graph should open up because $a > 0$.

The graph lies below the x-axis between $x = -4$ and $x = 2$. Thus, the solution set is $\{x \mid -4 < x < 2\}$ or $(-4, 2)$.

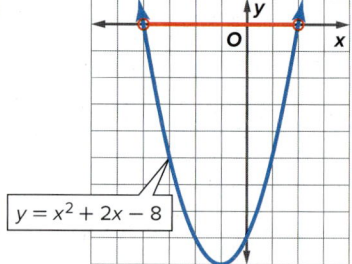

CHECK Test one value of x less than -4, one between -4 and 2, and one greater than 2 in the original inequality.

Test $x = -6$.
$x^2 + 2x - 8 < 0$
$(-6)^2 + 2(-6) - 8 \stackrel{?}{<} 0$
$16 < 0$ ✗

Test $x = 0$.
$x^2 + 2x - 8 < 0$
$0^2 + 2(0) - 8 \stackrel{?}{<} 0$
$-8 < 0$ ✓

Test $x = 5$.
$x^2 + 2x - 8 < 0$
$5^2 + 2(5) - 8 \stackrel{?}{<} 0$
$27 < 0$ ✗

> **Study Tip**
> **Solving Quadratic Inequalities by Graphing** A precise graph of the related quadratic function is not necessary since the zeros of the function were found algebraically.

Guided Practice

Solve each inequality by graphing.

2A. $0 > x^2 + 5x - 6$

2B. $-x^2 + 3x + 10 \leq 0$

A.CED.1

Example 3   Solve $ax^2 + bx + c \geq 0$ by Graphing

Solve $2x^2 + 4x - 5 \geq 0$ by graphing.

The solution consists of x-values for which the graph of the related function lies *on and above* the x-axis. Begin by finding the zeros of the related function.

$2x^2 + 4x - 5 = 0$ Related equation

$x = \dfrac{-b \pm \sqrt{b^2 - 4ac}}{2a}$ Use the Quadratic Formula.

$x = \dfrac{-4 \pm \sqrt{4^2 - 4(2)(-5)}}{2(2)}$ Replace a with 4, b with 2, and c with -5.

$x = \dfrac{-4 + \sqrt{56}}{4}$ or $x = \dfrac{-4 - \sqrt{56}}{4}$ Simplify and write as two equations.

≈ 0.87 ≈ -2.87 Simplify.

Sketch the graph of a parabola with x-intercepts at -2.87 and 0.87. The graph opens up since $a > 0$. The graph lies on and above the x-axis at about $x \leq -2.87$ and $x \geq 0.87$. Therefore, the solution is approximately $\{x \mid x \leq -2.87 \text{ or } x \geq 0.87\}$ or $(-\infty, -2.87] \cup [0.87, \infty)$.

Guided Practice

Solve each inequality by graphing.

3A. $x^2 - 6x + 2 > 0$ **3B.** $-4x^2 + 5x + 7 \geq 0$

Real-world problems can be solved by graphing quadratic inequalities.

A.CED.3

Real-World Example 4   Solve a Quadratic Inequality

WATER BALLOONS Refer to the beginning of the lesson. At what time will a water balloon be within 3 meters of the ground after it has been launched?

The function $h(t) = -4.9t^2 + 32t + 1.2$ describes the height of the water balloon. Therefore, you want to find the values of t for which $h(t) \leq 3$.

$h(t) \leq 3$ Original inequality

$-4.9t^2 + 32t + 1.2 \leq 3$ $h(t) = -4.9t^2 + 32t + 1.2$

$-4.9t^2 + 32t - 1.8 \leq 0$ Subtract 3 from each side.

Graph the related function $y = -4.9x^2 + 32x - 1.8$ using a graphing calculator. The zeros of the function are about 0.06 and 6.47, and the graph lies below the x-axis when $x < 0.06$ and $x > 6.47$.

So, the water balloon is within 3 meters of the ground during the first 0.06 second after being launched and again after about 6.47 seconds until it hits the ground.

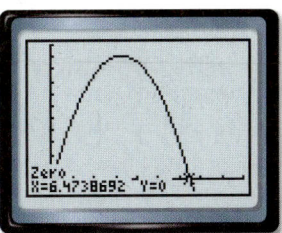

[−1, 9] scl: 1 by [−5, 55] scl: 5

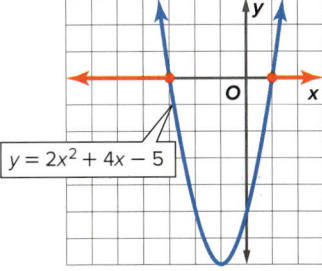

Real-World Career

It takes just milliseconds for a water balloon to break. A high-speed camera can capture the impact on the fluid before gravity makes it fall.
Source: NASA

Guided Practice

4. ROCKETS The height $h(t)$ of a model rocket in feet t seconds after its launch can be represented by the function $h(t) = -16t^2 + 82t + 0.25$. During what interval is the rocket at least 100 feet above the ground?

Go Online!

Watch the **Personal Tutor** to see how to solve a quadratic inequality algebraically. Then try it on your own.

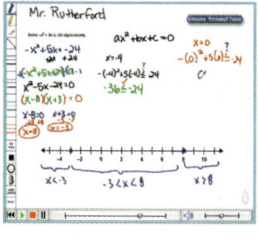

Example 5 Solve a Quadratic Inequality Algebraically

Solve $x^2 - 3x \leq 18$ algebraically.

Step 1 Solve the related quadratic equation $x^2 - 3x = 18$.

$x^2 - 3x = 18$	Related quadratic equation
$x^2 - 3x - 18 = 0$	Subtract 18 from each side.
$(x + 3)(x - 6) = 0$	Factor.
$x + 3 = 0$ or $x - 6 = 0$	Zero Product Property
$x = -3$ $x = 6$	Solve each equation.

Step 2 Plot -3 and 6 on a number line. Use dots since these values are solutions of the original inequality. Notice that the number line is divided into three intervals.

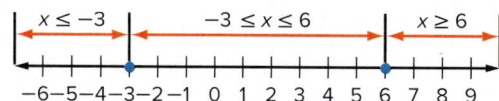

Step 3 Test a value from each interval to see if it satisfies the original inequality.

$x \leq -3$	$-3 \leq x \leq 6$	$x \geq 6$
Test $x = -5$.	Test $x = 0$.	Test $x = 8$.
$x^2 - 3x \leq 18$	$x^2 - 3x \leq 18$	$x^2 - 3x \leq 18$
$(-5)^2 - 3(-5) \stackrel{?}{\leq} 18$	$(0)^2 - 3(0) \stackrel{?}{\leq} 18$	$(8)^2 - 3(8) \stackrel{?}{\leq} 18$
$40 \not\leq 18$	$0 \leq 18$	$40 \not\leq 18$

The solution set is $\{x \mid -3 \leq x \leq 6\}$ or $[-3, 6]$.

▶ **Guided Practice**

Solve each inequality algebraically.

5A. $x^2 + 5x < -6$ **5B.** $x^2 + 11x + 30 \geq 0$

Check Your Understanding

 = Step-by-Step Solutions begin on page R11.

Go Online! for a Self-Check Quiz

Example 1   Graph each inequality.

1. $y \leq x^2 - 8x + 2$ **2.** $y > x^2 + 6x - 2$ **3.** $y \geq -x^2 + 4x + 1$

Examples 2–3   **MP SENSE-MAKING** Solve each inequality by graphing.

4. $0 < x^2 - 5x + 4$ **5.** $x^2 + 8x + 15 < 0$
6. $-2x^2 - 2x + 12 \geq 0$ **7.** $0 \geq 2x^2 - 4x + 1$

Example 4   **8. SOCCER** A midfielder kicks a ball toward the goal during a match. The height of the ball in feet above the ground $h(t)$ at time t can be represented by $h(t) = -0.1t^2 + 2.4t + 1.5$. If the height of the goal is 8 feet, at what time during the kick will the ball be able to enter the goal?

Example 5   Solve each inequality algebraically.

9. $x^2 + 6x - 16 < 0$ **10.** $x^2 - 14x > -49$
(11) $-x^2 + 12x \geq 28$ **12.** $x^2 - 4x \leq 21$

Practice and Problem Solving

Extra Practice is on page R3.

Example 1 Graph each inequality.

13. $y \geq x^2 + 5x + 6$
14. $x^2 - 2x - 8 < y$
15. $y \leq -x^2 - 7x + 8$
16. $-x^2 + 12x - 36 > y$
17. $y > 2x^2 - 2x - 3$
18. $y \geq -4x^2 + 12x - 7$

Examples 2–3 Solve each inequality by graphing.

19. $x^2 - 9x + 9 < 0$
20. $x^2 - 2x - 24 \leq 0$
21. $x^2 + 8x + 16 \geq 0$
22. $x^2 + 6x + 3 > 0$
23. $0 > -x^2 + 7x + 12$
24. $-x^2 + 2x - 15 < 0$
25. $4x^2 + 12x + 10 \leq 0$
26. $-3x^2 - 3x + 9 > 0$
27. $0 > -2x^2 + 4x + 4$
28. $3x^2 + 12x + 36 \leq 0$
29. $0 \leq -4x^2 + 8x + 5$
30. $-2x^2 + 3x + 3 \leq 0$

Example 4

31. **ARCHITECTURE** An arched entry of a room is shaped like a parabola that can be represented by the equation $f(x) = -x^2 + 6x + 1$. How far from the sides of the arch is its height at least 7 feet?

32. **MANUFACTURING** A box is formed by cutting 4-inch squares from each corner of a square piece of cardboard and then folding the sides. If $V(x) = 4x^2 - 64x + 256$ represents the volume of the box, what should the dimensions of the original piece of cardboard be if the volume of the box cannot exceed 750 cubic inches?

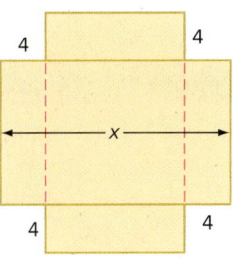

Example 5 Solve each inequality algebraically.

33. $x^2 - 9x < -20$
34. $x^2 + 7x \geq -10$
35. $2 > x^2 - x$
36. $-3 \leq -x^2 - 4x$
37. $-x^2 + 2x \leq -10$
38. $-6 > x^2 + 4x$
39. $2x^2 + 4 \geq 9$
40. $3x^2 + x \geq -3$
41. $-4x^2 + 2x < 3$
42. $-11 \geq -2x^2 - 5x$
43. $-12 < -5x^2 - 10x$
44. $-3x^2 - 10x > -1$

45. **MP PERSEVERANCE** The Sanchez family is adding a deck along two sides of their swimming pool. The deck width will be the same on both sides and the total area of the pool and deck cannot exceed 750 square feet.

   a. Graph the quadratic inequality.
   b. Determine the possible widths of the deck.

Write a quadratic inequality for each graph.

46.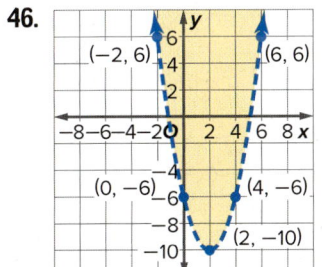
47.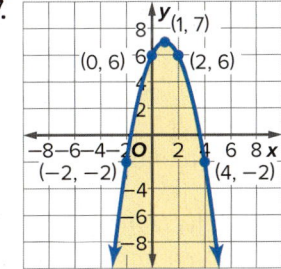
48.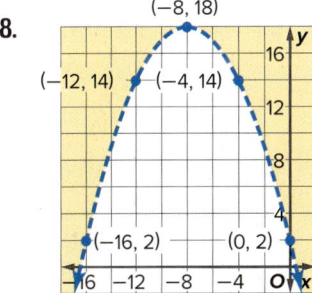

Solve each quadratic inequality by using a graph, a table, or algebraically.

49. $-2x^2 + 12x < -15$
50. $5x^2 + x + 3 \geq 0$
51. $11 \leq 4x^2 + 7x$
52. $x^2 - 4x \leq -7$
53. $-3x^2 + 10x < 5$
54. $-1 \geq -x^2 - 5x$

55. **BUSINESS** An electronics manufacturer uses the function $P(x) = x(-27.5x + 3520) + 20{,}000$ to model their monthly profits when selling x thousand wireless headsets.

   a. Suppose the manufacturer incurs a new monthly expense of $25,000. Explain how this affects the graph of the profit function and what it represents.

   b. Describe how the added expense affects the number of wireless headsets that must be sold in order for the manufacturer to have at least $100,000 in profits. Include details of minimum, maximum, and the range of the number of headsets.

56. **UTILITIES** A contractor is installing drain pipes for a shopping center's parking lot. The outer diameter of the pipe is to be 10 inches. The cross sectional area of the pipe must be at least 35 square inches and should not be more than 42 square inches.

   a. Graph the quadratic inequalities.
   b. What thickness of drain pipe can the contractor use?

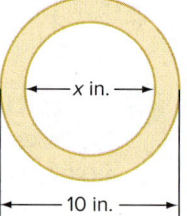

H.O.T. Problems — Use Higher-Order Thinking Skills

57. **OPEN-ENDED** Write a quadratic inequality for each condition.

   a. The solution set is all real numbers.
   b. The solution set is the empty set.

58. **ERROR ANALYSIS** Don and Diego used a graph to solve the quadratic inequality $x^2 - 2x - 8 > 0$. Is either of them correct? Explain.

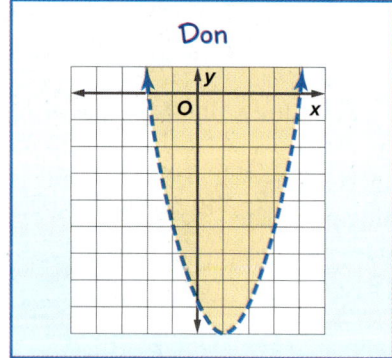
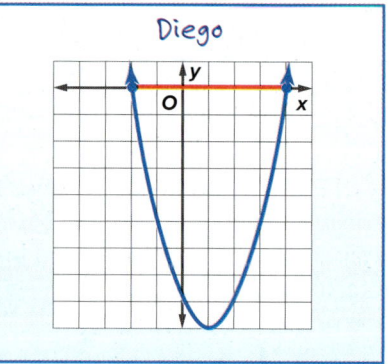

59. **REASONING** Are the boundaries of the solution set of $x^2 + 4x - 12 \leq 0$ twice the value of the boundaries of $\frac{1}{2}x^2 + 2x - 6 \leq 0$? Explain.

60. **REASONING** Determine if the following statement is *sometimes*, *always*, or *never* true. Explain your reasoning.

   The intersection of $y \leq -ax^2 + c$ and $y \geq ax^2 - c$ is the empty set.

61. **CHALLENGE** Graph the intersection of the graphs of $y \leq -x^2 + 4$ and $y \geq x^2 - 4$.

62. **WRITING IN MATH** How are the techniques used when solving quadratic inequalities and quadratic equations similar? different?

Preparing for Assessment

63. What is the solution set of the inequality $4x + 1 > 2 - 5(x + 7)$? 1

- A $\left\{x \mid x > -\frac{37}{9}\right\}$
- B $\left\{x \mid x > -\frac{34}{9}\right\}$
- C $\left\{x \mid x > -\frac{22}{7}\right\}$
- D $\left\{x \mid x > \frac{6}{7}\right\}$
- E $\left\{x \mid x > \frac{8}{9}\right\}$

64. What inequality is represented by the graph? MP 1

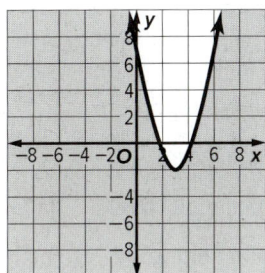

- A $y \geq x^2 - 6x + 7$
- B $y > x^2 - 6x + 7$
- C $y \leq x^2 - 6x + 7$
- D $y < x^2 - 6x + 7$

65. MULTI-STEP Solve $6 < x^2 - x$. MP 2

a. What is the related function of the inequality?

b. Which are the zeros of the related function?

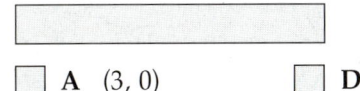

- ☐ A (3, 0)
- ☐ B (−2, 0)
- ☐ C (−3, 0)
- ☐ D (−2, 0)
- ☐ E (4, 4)

c. Which represents the solution set of the original inequality?

- A $\{x \mid x < 1 \text{ or } x > 6\}$
- B $\{x \mid x > -1 \text{ or } x < -6\}$
- C $\{x \mid x < 3 \text{ or } x > 2\}$
- D $\{x \mid x > 3 \text{ or } x < -2\}$

66. Which graph represents the solution of the inequality $y \geq x^2 + 2x - 8$? 1

- A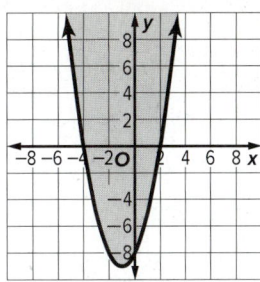
- B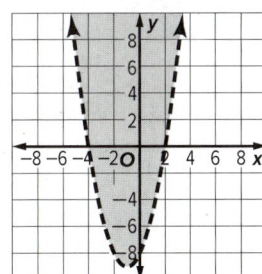
- C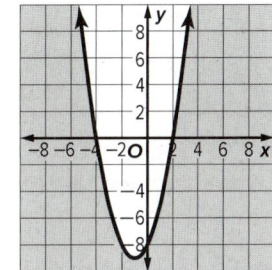
- D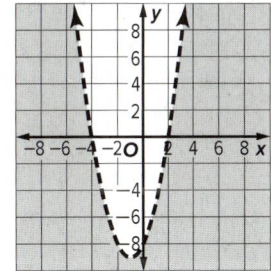

67. Place the correct symbol in the box to create an inequality that contains (2, 4) in the solution set but not (−3, 0). MP 1

$y \; \boxed{} \; x^2 + 2x - 3$

68. Which point is in the solution set of $y < -x^2 + 5x - 3$? MP 1

- A (2.5, 3.25)
- B (0, 0)
- C (2, 0)
- D (0, −2)

CHAPTER 3
Study Guide and Review

Go Online! for Vocabulary Review Games and key vocabulary in 13 languages

Study Guide

Key Concepts

Graphing Quadratic Functions (Lesson 3-1)
- The graph of $y = ax^2 + bx + c$, $a \neq 0$, opens up, and the function has a minimum value when $a > 0$. The graph opens down, and the function has a maximum value when $a < 0$.

Solving Quadratic Equations by Graphing (Lessons 3-2)
- Roots of a quadratic equation are the zeros of the related quadratic function. You can find the zeros of a quadratic function by finding the x-intercepts of the graph.

Complex Numbers (Lesson 3-3)
- i is the imaginary unit; $i^2 = -1$ and $i = \sqrt{-1}$.

Solving Quadratic Equations (Lessons 3-4, 3-5, and 3-6)
- Factoring: **Step 1** Write the equation in standard form. **Step 2** Factor the quadratic expression. **Step 3** Set each factor equal to 0 and solve.
- Completing the square: **Step 1** Find one half of b, the coefficient of x. **Step 2** Square the result in Step 1. **Step 3** Add the result of Step 2 to $x^2 + bx$.
- Quadratic Formula: $x = \dfrac{-b \pm \sqrt{b^2 - 4ac}}{2a}$

Quadratic Inequalities (Lesson 3-7)
- In two variables: Graph the related function, test a point not on the parabola and determine if it is a solution, and shade accordingly.
- In one variable: Graph the related function, and identify the x-values for which the graph lies below the x-axis for $ax^2 + bx + c < 0$ or the x-values for which the graph lies above the x-axis for $ax^2 + bx + c > 0$.

FOLDABLES StudyOrganizer

Use your Foldable to review the chapter. Working with a partner can be helpful. Ask for clarification of concepts as needed.

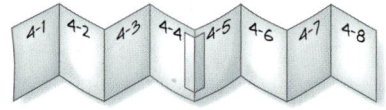

Key Vocabulary

axis of symmetry (p. 152)
completing the square (p. 192)
complex conjugates (p. 175)
complex number (p. 173)
constant term (p. 151)
discriminant (p. 202)
factored form (p. 179)
FOIL method (p. 179)
imaginary unit (p. 172)
linear term (p. 151)
maximum value (p. 154)
minimum value (p. 154)
parabola (p. 151)
pure imaginary number (p. 172)
quadratic equation (p. 163)
Quadratic Formula (p. 199)
quadratic function (p. 151)
quadratic inequality (p. 209)
quadratic term (p. 151)
root (p. 163)
Square Root Property (p. 173)
standard form (p. 163)
vertex (p. 152)
zero (p. 163)

Vocabulary Check

State whether each sentence is *true* or *false*. If *false*, replace the underlined term to make a true sentence.

1. The <u>factored form</u> of a quadratic equation is $ax^2 + bx + c = 0$, where $a \neq 0$ and a, b, and c are integers.
2. The graph of a quadratic function is called a <u>parabola</u>.
3. The axis of symmetry will intersect a parabola in one point called the <u>vertex</u>.
4. A method called <u>FOIL method</u> is used to make a quadratic expression a perfect square in order to solve the related equation.
5. The number $6i$ is called a <u>pure imaginary number</u>.
6. The two numbers $2 + 3i$ and $2 - 3i$ are called <u>complex conjugates</u>.

Concept Check

7. Explain how the value of the discriminant gives the number and type of roots for a quadratic equation.
8. Explain the connection between the solutions of a quadratic equation and the graph of the related quadratic function.

Lesson-by-Lesson Review

3-1 Graphing Quadratic Functions

Complete parts a–c for each quadratic function.
a. Find the y-intercept, the equation of the axis of symmetry, and the x-coordinate of the vertex.
b. Make a table of values that includes the vertex.
c. Use this information to graph the function.

9. $f(x) = x^2 + 5x + 12$
10. $f(x) = x^2 - 7x + 15$
11. $f(x) = -2x^2 + 9x - 5$
12. $f(x) = -3x^2 + 12x - 1$

Determine whether each function has a maximum or minimum value and find the maximum or minimum value. Then state the domain and range of the function.

13. $f(x) = -x^2 + 3x - 1$
14. $f(x) = -3x^2 - 4x + 5$

15. **BUSINESS** Sal's Shirt Store sells 100 T-shirts per week at a rate of $10 per shirt. Sal estimates that he will sell 5 fewer shirts for each $1 increase in price. What price will maximize Sal's T-shirt income?

Example 1

Consider the quadratic function $f(x) = x^2 - 4x + 11$. Find the y-intercept, the equation for the axis of symmetry, and the x-coordinate of the vertex.

In the function, $a = 1$, $b = -4$, and $c = 11$. The y-intercept is $c = 11$.

Use a and b to find the equation of the axis of symmetry.

$x = -\dfrac{b}{2a}$ Equation of the axis of symmetry

$= -\dfrac{-4}{2(1)}$ $a = 1$ and $b = -4$

$= 2$ Simplify.

The equation of the axis of symmetry is $x = 2$. Therefore, the x-coordinate of the vertex is 2.

3-2 Solving Quadratic Equations by Graphing

Solve each equation by graphing. If exact roots cannot be found, state the consecutive integers between which the roots are located.

16. $x^2 - x - 20 = 0$
17. $2x^2 - x - 3 = 0$
18. $4x^2 - 6x - 15 = 0$

19. **BASEBALL** A baseball is hit upward at 120 feet per second. Use the formula $h(t) = v_0 t - 16t^2$, where $h(t)$ is the height of an object in feet, v_0 is the object's initial velocity in feet per second, and t is the time in seconds. Ignoring the height of the ball when it was hit, how long does it take for the ball to hit the ground?

Example 2

Solve $2x^2 - 7x + 3 = 0$ by graphing.

The equation of the axis of symmetry is $-\dfrac{-7}{2(2)}$ or $x = \dfrac{7}{4}$.

x	0	1	$\dfrac{7}{4}$	2	3
f(x)	3	-2	$-2\dfrac{5}{8}$	-3	0

The zeros of the related function are $\dfrac{1}{2}$ and 3. Therefore, the solutions of the equation are $\dfrac{1}{2}$ and 3.

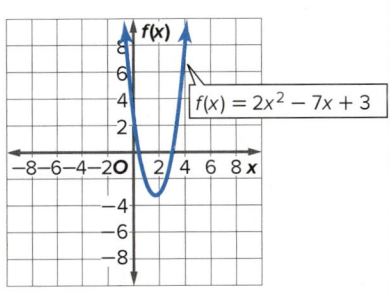

CHAPTER 3
Study Guide and Review Continued

3-3 Complex Numbers

Simplify.

20. $\sqrt{-8}$
21. $(2 - i) + (13 + 4i)$
22. $(6 + 2i) - (4 - 3i)$
23. $(6 + 5i)(3 - 2i)$

24. **ELECTRICITY** The impedance in one part of a series circuit is $3 + 2j$ ohms, and the impedance in the other part of the circuit is $4 - 3j$ ohms. Add these complex numbers to find the total impedance in the circuit.

Solve each equation.

25. $2x^2 + 50 = 0$
26. $4x^2 + 16 = 0$
27. $3x^2 + 15 = 0$
28. $8x^2 + 16 = 0$
29. $4x^2 + 1 = 0$

Example 3
Simplify $(12 + 3i) - (-5 + 2i)$.

$(12 + 3i) - (-5 + 2i)$
$= [12 - (-5)] + (3 - 2)i$ Group the real and imaginary parts.
$= 17 + i$ Simplify.

Example 4
Solve $3x^2 + 12 = 0$.

$3x^2 + 12 = 0$ Original equation
$3x^2 = -12$ Subtract 12 from each side.
$x^2 = -4$ Divide each side by 3.
$x = \pm\sqrt{-4}$ Square Root Property
$x = \pm 2i$ $\sqrt{-4} = \sqrt{4} \cdot \sqrt{-1}$

3-4 Solving Quadratic Equations by Factoring

Write a quadratic equation in standard form with the given roots.

30. $5, 6$
31. $-3, -7$
32. $-4, 2$
33. $-\frac{2}{3}, 1$
34. $\frac{1}{6}, 5$
35. $-\frac{1}{4}, -1$

Solve each equation by factoring.

36. $2x^2 - 2x - 24 = 0$
37. $2x^2 - 5x - 3 = 0$
38. $3x^2 - 16x + 5 = 0$
39. $x^2 + 121 = 0$
40. $4x^2 + 64 = 0$

41. Find x and the dimensions of the rectangle below.

$A = 126$ ft^2, sides $x - 3$ and $x + 2$

Example 5
Write a quadratic equation in standard form with $-\frac{1}{2}$ and 4 as its roots.

$(x - p)(x - q) = 0$ Write the pattern.
$\left[x - \left(-\frac{1}{2}\right)\right](x - 4) = 0$ Replace p with $-\frac{1}{2}$ and q with 4.
$\left(x + \frac{1}{2}\right)(x - 4) = 0$ Simplify.
$x^2 - \frac{7}{2}x - 2 = 0$ Multiply.
$2x^2 - 7x - 4 = 0$ Multiply each side by 2 so that b and c are integers.

Example 6
Solve $2x^2 - 3x - 5 = 0$ by factoring.

$2x^2 - 3x - 5 = 0$ Original equation
$(2x - 5)(x + 1) = 0$ Factor the trinomial.
$2x - 5 = 0$ or $x + 1 = 0$ Zero Product Property
$x = \frac{5}{2}$ $x = -1$

The solution set is $\left\{-1, \frac{5}{2}\right\}$ or $\left\{x \mid x = -1, \frac{5}{2}\right\}$.

3-5 Solving Quadratic Equations by Completing the Square

Find the value of c that makes each trinomial a perfect square. Then write the trinomial as a perfect square.

42. $x^2 + 18x + c$
43. $x^2 - 4x + c$
44. $x^2 - 7x + c$
45. $x^2 + 2.4x + c$
46. $x^2 - \frac{1}{2}x + c$
47. $x^2 + \frac{6}{5}x + c$

Solve each equation by completing the square.

48. $x^2 - 6x - 7 = 0$
49. $x^2 - 2x + 8 = 0$
50. $2x^2 + 4x - 3 = 0$
51. $2x^2 + 3x - 5 = 0$
52. **FLOOR PLAN** Mario's living room has a length 6 feet wider than the width. The area of the living room is 280 square feet. What are the dimensions of his living room?

Example 7

Find the value of c that makes $x^2 + 14x + c$ a perfect square. Then write the trinomial as a perfect square.

Step 1 Find one half of 14.

Step 2 Square the result of Step 1.

Step 3 Add the result of Step 2 to $x^2 + 14x$.

The trinomial $x^2 + 14x + 49$ can be written as $(x + 7)^2$.

Example 8

Solve $x^2 + 12x - 13 = 0$ by completing the square.

$x^2 + 12x - 13 = 0$
$x^2 + 12x = 13$
$x^2 + 12x + 36 = 13 + 36$
$(x + 6)^2 = 49$
$x + 6 = \pm 7$
$x + 6 = 7$ or $x + 6 = -7$
$x = 1$ $\qquad$ $x = -13$

The solution set is $\{-13, 1\}$ or $\{x \mid x = -13, 1\}$.

3-6 The Quadratic Formula and the Discriminant

Complete parts a–c for each quadratic equation.

a. Find the value of the discriminant.

b. Describe the number and type of roots.

c. Find the exact solutions by using the Quadratic Formula.

53. $x^2 - 10x + 25 = 0$
54. $x^2 + 4x - 32 = 0$
55. $2x^2 + 3x - 18 = 0$
56. $2x^2 + 19x - 33 = 0$
57. $x^2 - 2x + 9 = 0$
58. $4x^2 - 4x + 1 = 0$
59. $2x^2 + 5x + 9 = 0$
60. **PHYSICAL SCIENCE** Lauren throws a ball with an initial velocity of 40 feet per second. The equation for the height of the ball is $h = -16t^2 + 40t + 5$, where h represents the height in feet and t represents the time in seconds. When will the ball hit the ground?

Example 9

Solve $x^2 - 4x - 45 = 0$ by using the Quadratic Formula.

In $x^2 - 4x - 45 = 0$, $a = 1$, $b = -4$, and $c = -45$.

$x = \dfrac{-b \pm \sqrt{b^2 - 4ac}}{2a}$ $\qquad$ Quadratic Formula

$= \dfrac{-(-4) \pm \sqrt{(-4)^2 - 4(1)(-45)}}{2(1)}$

$= \dfrac{4 \pm 14}{2}$

Write as two equations.

$x = \dfrac{4 + 14}{2}$ or $x = \dfrac{4 - 14}{2}$

$= 9$ $\qquad\qquad$ $= -5$

The solution set is $\{-5, 9\}$ or $\{x \mid x = -5, 9\}$.

Chapter 3
Study Guide and Review Continued

3-7 Quadratic Inequalities

Graph each quadratic inequality.

61. $y \geq x^2 + 5x + 4$ **62.** $y < -x^2 + 5x - 6$

63. $y > x^2 - 6x + 8$ **64.** $y \leq x^2 + 10x - 4$

65. Solomon wants to put a deck along two sides of his garden. The deck width will be the same on both sides and the total area of the garden and deck cannot exceed 500 square feet. How wide can the deck be?

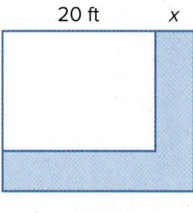

Solve each inequality using a graph or algebraically.

66. $x^2 + 8x + 12 > 0$

67. $6x + x^2 \geq -9$

68. $2x^2 + 3x - 20 > 0$

69. $4x^2 - 3 < -5x$

70. $3x^2 + 4 > 8x$

Example 10

Graph $y > x^2 + 3x + 2$.

Step 1 Graph the related function, $y > x^2 + 3x + 2$. Because the inequality symbol $>$ is used, the parabola should be dashed.

Step 2 Test a point not on the graph of the parabola such as $(0, 0)$.

$$y > x^2 + 3x + 2$$
$$(0) \stackrel{?}{>} (0)^2 + 3(0) + 2$$
$$0 \not> 2$$

So, $(0, 0)$ is not a solution of the inequality.

Step 3 Shade the region that does not contain the point $(0, 0)$.

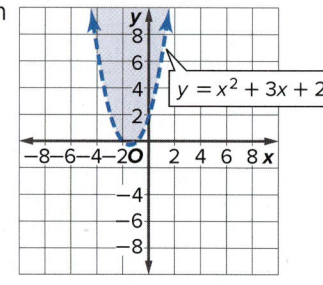

CHAPTER 3
Practice Test

Complete parts a–c for each quadratic function.
a. Find the y-intercept, the equation of the axis of symmetry, and the x-coordinate of the vertex.
b. Make a table of values that includes the vertex.
c. Use this information to graph the function.

1. $f(x) = x^2 + 4x - 7$
2. $f(x) = -2x^2 + 5x$
3. $f(x) = -x^2 - 6x - 9$

Determine whether each function has a maximum or minimum value. State the maximum or minimum value of each function.

4. $f(x) = x^2 + 10x + 25$ 5. $f(x) = -x^2 + 6x$

Simplify.

6. $(3 - 4i) - (9 - 5i)$
7. $\dfrac{4i}{4 - i}$
8. $(4 + 5i)(2 - i)$
9. $(12 - 3i) + (-5 + 8i)$

Solve each equation using the method of your choice. Find exact solutions.

10. $x^2 - 8x - 9 = 0$
11. $-4.8x^2 + 1.6x + 24 = 0$
12. $12x^2 + 15x - 4 = 0$
13. $x^2 - 7x - \dfrac{17}{4} = 0$
14. $4x^2 + x = 3$
15. $-9x^2 + 40x + 84 = 0$
16. $4x^2 + 1 = 0$
17. $2x^2 + x - 1 = -5$

18. **PHYSICAL SCIENCE** Parker throws a ball off the top of a building. The building is 350 feet high and the initial velocity of the ball is 30 feet per second. Find out how long it will take the ball to hit the ground by solving the equation $-16t^2 - 30t + 350 = 0$.

19. **MULTIPLE CHOICE** Which equation below has roots at -6 and $\dfrac{1}{5}$?

   A $0 = 5x^2 - 29x - 6$
   B $0 = 5x^2 + 31x + 6$
   C $0 = 5x^2 + 29x - 6$
   D $0 = 5x^2 - 31x + 6$

20. **PHYSICS** A ball is thrown into the air vertically with a velocity of 112 feet per second. The ball was released 6 feet above the ground. The height above the ground t seconds after release is modeled by $h(t) = -16t^2 + 112t + 6$.
   a. When will the ball reach 130 feet?
   b. Will the ball ever reach 250 feet? Explain.
   c. In how many seconds after its release will the ball hit the ground?

21. The rectangle below has an area of 104 square inches. Find the value of x and the dimensions of the rectangle.

   $A = 104$ in^2
   $x - 1$
   $x + 4$

22. **MULTIPLE CHOICE** Which value of c makes the trinomial $x^2 - 12x + c$ a perfect square trinomial?

   A 6
   B 12
   C 36
   D 144

Complete parts a–c for each quadratic equation.
a. Find the value of the discriminant.
b. Describe the number and type of roots.
c. Find the exact solution by using the Quadratic Formula.

23. $6x^2 + 7x = 0$
24. $5x^2 = -6x + 1$
25. $2x^2 + 5x - 8 = -13$

Solve each inequality by using a graph or algebraically.

26. $x^2 + 6x > -5$
27. $4x^2 - 19x \le -12$
28. $x^2 - 4 < 3x$

CHAPTER 3
Preparing for Assessment

Performance Task

Provide a clear solution to each part of the task. Be sure to show all of your work, include all relevant drawings, and justify your answers.

CONTEXT A high school science class is conducting an experiment involving different methods of launching bottle rockets straight up into the air. They build the rockets and launch them one by one off of the roof of a 10-story building. The equation that models this type of projectile motion on Earth is $h(t) = -16t^2 + v_0 t + h_0$, where $h(t)$ represents the height of the projectile after t seconds, v_0 is the initial velocity, and h_0 is the initial height. This formula gives the height of the projectile as a function of time.

Part A

When the class launches the first rocket, it has an initial velocity of 75 feet per second. It hits the ground after 6 seconds.

1. **MP PERSEVERANCE** Determine the height of the building from which the rocket was launched.

2. Write the equation in standard form that represents the height of the rocket as a function of time.

3. Explain what the x- and y-values of the vertex of the parabola represent in terms of this scenario.

Part B

The class launches a second rocket that has an initial velocity of 60 feet per second.

4. **MP PRECISION** Use the height of the building you found in Part A to how long, in seconds, it takes for this rocket to hit the ground after it was launched. Round your answer to the nearest whole number, if necessary.

Part C

A third rocket is launched with an initial velocity of 72 feet per second.

5. Determine the maximum height, in feet, the rocket reaches. Round your answer to the nearest whole number, if necessary.

6. Determine after how long, in seconds, it reached this height.

Part D

A fourth and final rocket is launched with an initial velocity of 80 feet per second.

7. Determine for what values of t the height of the rocket is greater than 162 feet. Round your answer to the nearest tenth, if necessary.

Test-Taking Strategy

Example

Read the problem. Identify what you need to know. Then use the information in the problem to solve.

The students in Mr. Himebaugh's physics class built a model rocket. The rocket is launched in a large field with an initial upward velocity of 128 feet per second. The function $h(t) = -16t^2 + 128t$ models the height of the rocket above the ground (in feet) t seconds after it is launched. How long will it take for the rocket to reach its maximum height?

A 4 seconds   **C** 6 seconds

B 5 seconds   **D** 8 seconds

Step 1 What are you being asked to solve? What information is given? Could a graph help you solve the problem?
I need to find the amount of time it takes the rocket to reach its maximum height. The problem gives the equation of the parabola. A graph would help because I could use it to find the x-value of the vertex.

Step 2 Would it be better to sketch the graph on a piece of paper, or use your graphing calculator?
Because the graph is a parabola and the problem gives the equation, a graphing calculator would be better.

Step 3 How can you use the graph to solve the problem? What is the correct answer?
I can use the maximum function on the calculator, which will tell me both the x- and y-values of the vertex. Choice A is correct.

Test-Taking Tip
Strategies for Solving Using a Graph
Using a graph can help you solve many different kinds of problems on standardized tests. Graphs can help you solve equations, evaluate functions, and interpret solutions to real-world problems. When you read the problem below, decide whether you should use your calculator to graph the equation representing the rocket's height or to sketch it on a piece of paper.

Apply the Strategy

Read the problem. Identify what you need to know. Then use the information in the problem to solve.

How many times does the graph of $f(x) = 2x^2 - 3x + 2$ cross the x-axis?

A 0
B 1
C 2
D 3

Answer the questions below.

a. What are you being asked to solve? What information is given? Could a graph help you solve the problem?

b. Would it be better to sketch the graph on a piece of paper, or use your graphing calculator?

c. How can you use the graph to solve the problem?

d. What is the correct answer?

CHAPTER 3
Preparing for Assessment
Cumulative Review

Read each question. Then fill in the correct answer on the answer document provided by your teacher or on a sheet of paper.

1. What are the solutions of $x^2 - x - 42 = 0$?

   A $x = -7$ or $x = -6$

   B $x = -7$ or $x = 6$

   C $x = 7$ or $x = -6$

   D $x = 7$ or $x = 6$

2. Consider the equation graphed below.

   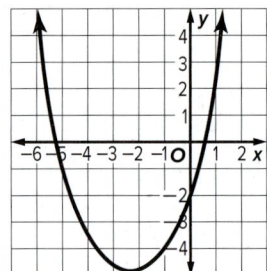

   Which statement describes one of the zeros?

   A between -6 and -5

   B between -5 and -4

   C between -3 and -2

   D between -1 and 0

3. The profit of a telemarketing company as a function of the number of employees it has can be modeled by a quadratic function. If this function is graphed, explain what the x- and y-values of the vertex mean in the context of this scenario.

4. What are the solutions to the inequality $x^2 - 5x \le 36$?

5. What inequality is graphed below?

   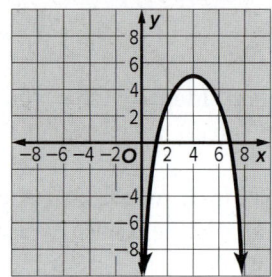

   A $y \ge -\frac{5}{9}(x-4)^2 + 5$

   B $y > -\frac{5}{9}(x-4)^2 + 5$

   C $y \le -\frac{5}{9}(x-4)^2 + 5$

   D $y < -\frac{5}{9}(x-4)^2 + 5$

6. Which expression is equivalent to $(-7 - 6i)(3 + 5i)$?

   A $-51 - 53i$

   B $-51 - 17i$

   C $9 - 53i$

   D $9 - 17i$

7. What is the equation of the function shown in the graph?

   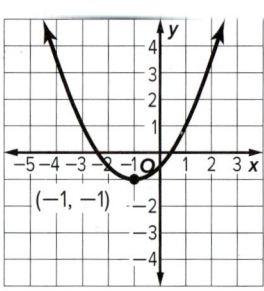

   A $y = 2(x+1)^2 - 1$

   B $y = 2(x-1)^2 - 1$

   C $y = \frac{1}{2}(x+1)^2 - 1$

   D $y = \frac{1}{2}(x-1)^2 - 1$

8. Consider the function $f(x) = x^2 + 16x + 3$. What is the x-coordinate of the vertex of the graph of the function?

9. Which statement describes the solutions of $4x^2 - 12x + 11 = 2$?

   A one real rational root
   B two real, rational roots
   C two real, irrational roots
   D two complex roots

10. Which of the following graphs can be represented by the inequality $16 \geq x^2 + 6x$?

    A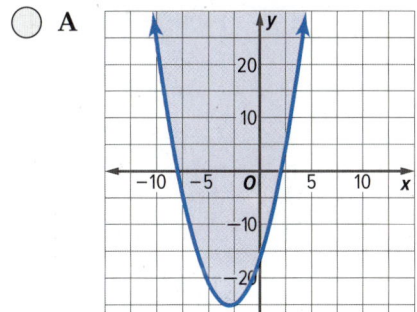

    B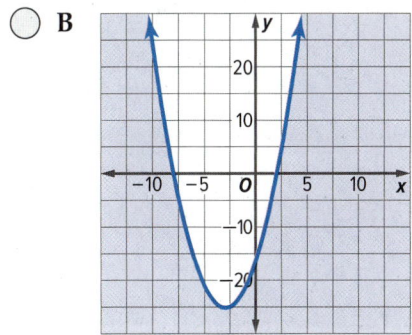

    C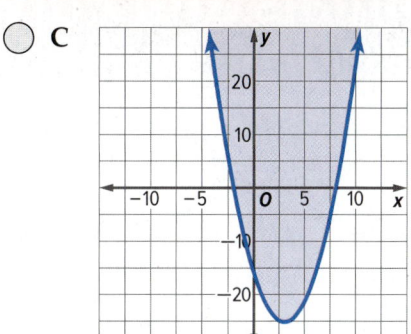

    D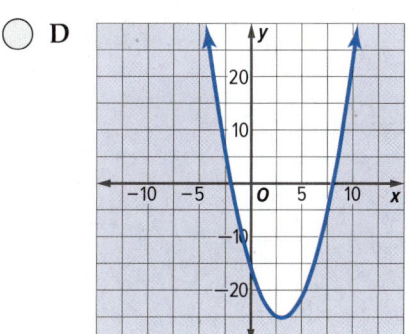

11. What is the solution set for the equation $x^2 - 10x + 8 = 22$?

    A $\{5 + \sqrt{5}, 5 - \sqrt{5}\}$
    B $\{5 + \sqrt{14}, 5 - \sqrt{14}\}$
    C $\{5 + \sqrt{39}, 5 - \sqrt{39}\}$
    D $\{5 + \sqrt{47}, 5 - \sqrt{47}\}$

12. Which equation represents the function graphed below?

    A $y = 2(x - 3)^2 - 1$
    B $y = 2(x + 3)^2 - 1$
    C $y = 2(x - 3)^2 + 1$
    D $y = 2(x + 3)^2 + 1$
    E $y = -2(x + 3)^2 - 1$

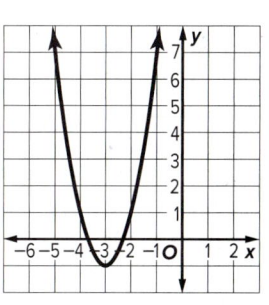

Need Extra Help?											
If you missed Question...	1	2	3	4	5	6	7	8	9	10	11	12
Go to Lesson...	3-4	3-2	3-1	3-7	3-7	3-3	3-5	3-2	3-6	3-7	3-6	3-5

CHAPTER 4

Polynomials and Polynomial Functions

THEN

You graphed quadratic functions and solved quadratic equations.

NOW

You will:

- Add, subtract, multiply, divide, and factor polynomials.
- Analyze and graph polynomial functions.
- Evaluate polynomial functions and solve polynomial equations.
- Find factors and zeros of polynomial functions.

MP WHY

TRANSPORTATION Wherever you live, transportation is vital. If public transportation is not available, you may need to purchase a vehicle.

Use the Mathematical Practices to complete the activity.

1. **Using Tools** Pick a car and use the Internet to find its cost new and in a 10-year-old model. Use the KWL Chart tool to record and organize the information.

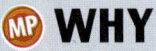

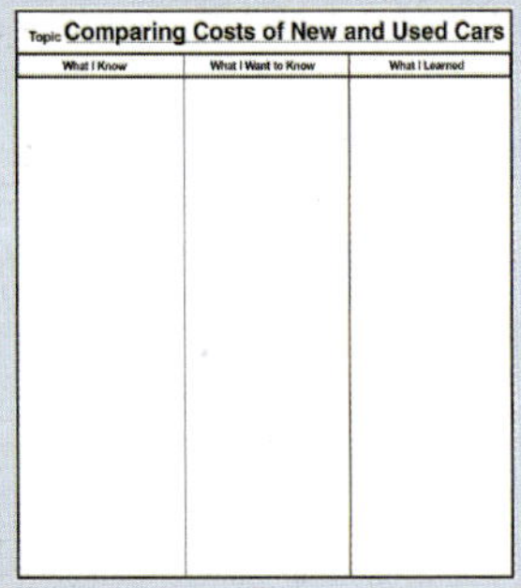

Topic Comparing Costs of New and Used Cars
What I Know
What I Want to Know
What I Learned

2. **Discuss** How will fuel economy impact your annual costs? What is the cost of insurance?
3. **Applying Math** Write equations that use the information you learned to determine the total estimated annual costs of owning the vehicles you selected.

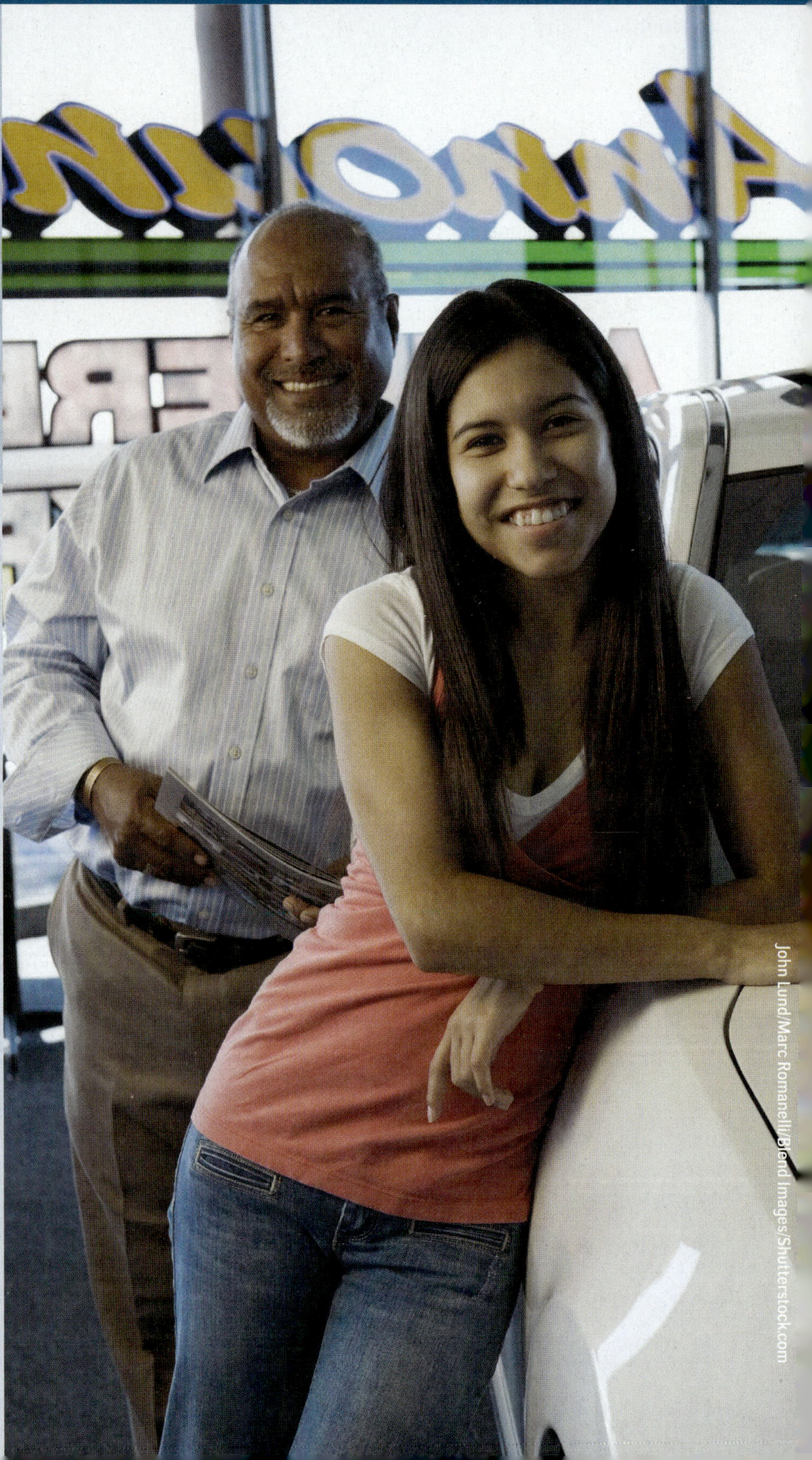

John Lund/Marc Romanelli/Blend Images/Shutterstock.com

# Go Online to Guide Your Learning

Explore & Explain

 Product Mat and Algebra Tiles
Use the **Product Mat** and the **Algebra Tiles** in the eToolkit to enhance your understanding of the division of polynomials in Lesson 4-3.

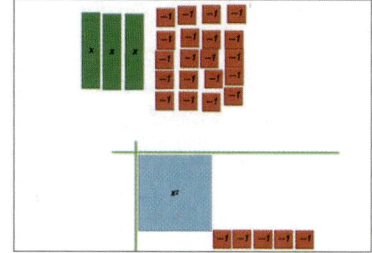

The Geometer's Sketchpad
Use **The Geometer's Sketchpad** to explore using operations with polynomials in Lesson 4-1.

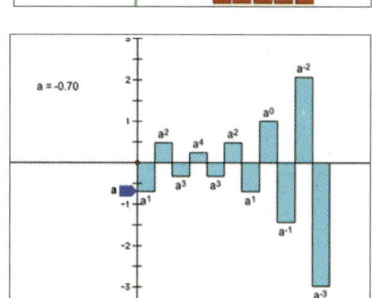

eBook
Interactive Student Guide
Before starting the chapter, answer the **Chapter Focus** preview questions. Check your answers as you complete each lesson. At the end of the chapter, try the **Performance Task**.

Organize

 Foldables
Get organized! Create this **Polynomials and Polynomial Functions Foldable** to help you organize your notes about polynomials and polynomial functions.

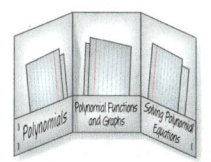

Collaborate

 **Chapter Project**
In the **Here We Grow** project, you will use what you have learned about operations with polynomials and graphs of polynomial functions to complete a project that addresses global awareness.

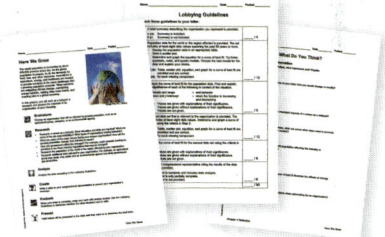

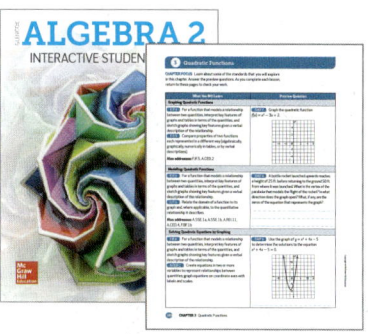

Focus

 LEARNSMART
Need help studying? Complete the **Polynomial, Rational, and Radical Relationships** and **Modeling with Functions** domains in LearnSmart to review for the chapter test.

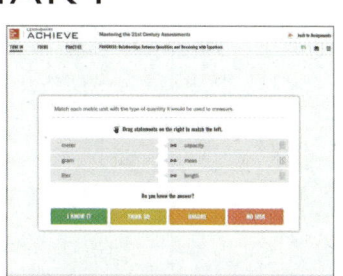

ALEKS
You can use the **Quadratic and Polynomial Functions** and **Exponents and Polynomial Expressions** topics in ALEKS to explore what you know about relations and functions and what you are ready to learn.*

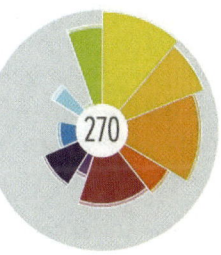

* Ask your teacher if this is part of your program.

Get Ready for the Chapter

Connecting Concepts

Concept Check
Review the concepts used in this chapter by answering the questions below.

1. How would you rewrite $2xy - 3 - z$ as a sum?
2. What property would you use to rewrite the expression $-4(a + 5)$ without parentheses?
3. Rewrite and simplify the expression $-3(a + b - c)$ into an expression that does not contain parentheses.
4. Write the Quadratic Formula.
5. When applying the Quadratic Formula to $2x^2 + 8x + 1$, what are the values for a, b, and c?
6. Given $\dfrac{-2 \pm \sqrt{14}}{2}$, what are the exact solutions?
7. Given $\dfrac{-2 \pm \sqrt{14}}{2}$, what are the approximate solutions?
8. In the coordinate plane shown, in what quadrant is point D?
9. In the coordinate plane shown, is the ordered pair representing point E positive, negative, or a combination?
10. In the coordinate plane shown, in what quadrant is point C?

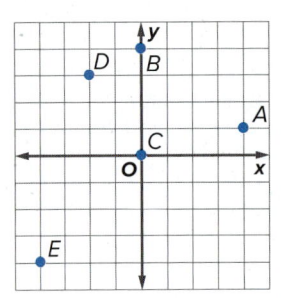

New Vocabulary

English		Español
simplify	p. 229	reducer
degree of a polynomial	p. 231	grado de un polinomio
Pascal's triangle	p. 237	triángulo de Pascal
synthetic division	p. 244	división sintética
Location Principle	p. 262	principio de ubicación
relative maximum	p. 263	máximo relativo
relative minimum	p. 263	mínimo relativo
extrema	p. 263	extrema
turning points	p. 263	momentos cruciales
prime polynomials	p. 274	polinomios primeros
quadratic form	p. 277	forma de ecuación cuadrática
polynomial identity	p. 282	identidad polinómica
synthetic substitution	p. 287	sustitución sintética
depressed polynomial	p. 289	polinomio reducido

Performance Task Preview

You can use the concepts and skills in the chapter to help a container company determine sizes and volumes of various boxes. Understanding polynomials and polynomial functions will help you finish the Performance Task at the end of the chapter.

MP In this Performance Task you will:
- Make sense of problems and persevere in solving them.
- Use tools strategically.

Review Vocabulary

factoring *factorización* to express a polynomial as the product of monomials and polynomials

function *función* a relation in which each element of the domain is paired with exactly one element in the range

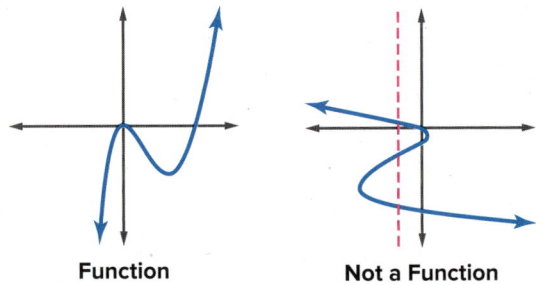

polynomial *polinomio* a monomial or sum of monomials

LESSON 1
Operations with Polynomials

:Then
- You evaluated powers.

:Now
1. Multiply, divide, and simplify monomials and expressions involving powers.
2. Add, subtract, and multiply polynomials.

:Why?
- The light from the Sun takes approximately 8 minutes to reach Earth. So if you are outside right now, you are basking in sunlight that the Sun emitted approximately 8 minutes ago.

  Light travels very fast, at a speed of about 3×10^8 meters per second. How long would it take light to get here from the Andromeda galaxy, which is approximately 2.367×10^{21} meters away?

New Vocabulary
simplify
degree of a polynomial

Mathematical Practices
2 Reason abstractly and quantitatively.

1 Multiply and Divide Monomials To **simplify** an expression containing powers means to rewrite the expression without parentheses or negative exponents. Negative exponents are a way of expressing the multiplicative inverse of a number. The following table summarizes the properties of exponents.

Concept Summary Properties of Exponents

For any real numbers x and y, integers a and b:

Property	Definition	Examples
Product of Powers	$x^a \cdot x^b = x^{a+b}$	$3^2 \cdot 3^4 = 3^{2+4}$ or 3^6 $p^2 \cdot p^9 = p^{2+9}$ or p^{11}
Quotient of Powers	$\dfrac{x^a}{x^b} = x^{a-b}, x \neq 0$	$\dfrac{9^5}{9^2} = 9^{5-2}$ or 9^3 $\dfrac{b^6}{b^4} = b^{6-4}$ or b^2
Negative Exponent	$x^{-a} = \dfrac{1}{x^a}$ and $\dfrac{1}{x^{-a}} = x^a, x \neq 0$	$3^{-5} = \dfrac{1}{3^5}$ $\dfrac{1}{b^{-7}} = b^7$
Power of a Power	$(x^a)^b = x^{ab}$	$(3^3)^2 = 3^{3 \cdot 2}$ or 3^6 $(d^2)^4 = d^{2 \cdot 4}$ or d^8
Power of a Product	$(xy)^a = x^a y^a$	$(2k)^4 = 2^4 k^4$ or $16k^4$ $(ab)^3 = a^3 b^3$
Power of a Quotient	$\left(\dfrac{x}{y}\right)^a = \dfrac{x^a}{y^a}, y \neq 0$, and $\left(\dfrac{x}{y}\right)^{-a} = \left(\dfrac{y}{x}\right)^a$ or $\dfrac{y^a}{x^a}, x \neq 0, y \neq 0$	$\left(\dfrac{x}{y}\right)^2 = \dfrac{x^2}{y^2}$ $\left(\dfrac{a}{b}\right)^{-5} = \dfrac{b^5}{a^5}$
Zero Power	$x^0 = 1, x \neq 0$	$7^0 = 1$

Recall that a *monomial* is a number, a variable, or an expression that is the product of one or more variables with nonnegative integer exponents.

Go Online!

You will want to reference the Properties of Exponents and the rules for simplifying monomials often. Log into your **eStudent Edition** to bookmark this lesson.

When simplifying a monomial, check to be sure that it has been simplified fully.

Key Concept Simplifying Monomials

A monomial expression is in simplified form when:
- there are no powers of powers,
- each base appears exactly once,
- all fractions are in simplest form, and
- there are no negative exponents.

Example 1 Simplify Expressions

Simplify each expression. Assume that no variable equals 0.

a. $(2a^{-2})(3a^3b^2)(c^{-2})$

$(2a^{-2})(3a^3b^2)(c^{-2})$ Original expression

$= 2\left(\dfrac{1}{a^2}\right)(3a^3b^2)\left(\dfrac{1}{c^2}\right)$ Definition of negative exponents

$= \left(\dfrac{2}{a \cdot a}\right)(3 \cdot a \cdot a \cdot a \cdot b \cdot b)\left(\dfrac{1}{c \cdot c}\right)$ Definition of exponents

$= \left(\dfrac{2}{\cancel{a} \cdot \cancel{a}}\right)(3 \cdot \cancel{a} \cdot \cancel{a} \cdot a \cdot b \cdot b)\left(\dfrac{1}{c \cdot c}\right)$ Divide out common factors.

$= \dfrac{6ab^2}{c^2}$ Simplify.

b. $\dfrac{q^2 r^4}{q^7 r^3}$

$\dfrac{q^2 r^4}{q^7 r^3} = q^{2-7} \cdot r^{4-3}$ Quotient of powers

$= q^{-5} r$ Subtract powers.

$= \dfrac{r}{q^5}$ Simplify.

Problem-Solving Tip

Check You can always check your answer using the definition of exponents.

$\dfrac{q^2}{q^7} = \dfrac{q \cdot q}{q \cdot q \cdot q \cdot q \cdot q \cdot q \cdot q}$

$= \dfrac{1}{q^5}$

c. $\left(\dfrac{-2a^4}{b^2}\right)^3$

$\left(\dfrac{-2a^4}{b^2}\right)^3 = \dfrac{(-2a^4)^3}{(b^2)^3}$ Power of a quotient

$= \dfrac{(-2)^3(a^4)^3}{(b^2)^3}$ Power of a product

$= \dfrac{-8a^{12}}{b^6}$ Power of a power

▶ **Guided Practice**

1A. $(2x^{-3}y^3)(-7x^5y^{-6})$ **1B.** $\dfrac{15c^5 d^3}{-3c^2 d^7}$

1C. $\left(\dfrac{a}{4}\right)^{-3}$ **1D.** $(-2x^3 y^2)^5$

Study Tip

MP Precision Remember that a variable with no exponent indicated can be written as a power of 1.

2 Operations With Polynomials
The **degree of a polynomial** is the degree of the monomial with the greatest degree.

Example 2   Degree of a Polynomial

Determine whether each expression is a polynomial. If it is a polynomial, state the degree of the polynomial.

a. $\frac{1}{4}x^4y^3 - 8x^5$

This expression is a polynomial because each term is a monomial. The degree of the first term is 4 + 3 or 7, and the degree of the second term is 5. The degree of the polynomial is 7.

b. $\sqrt{x} + x + 4$

This expression is not a polynomial because $\sqrt{x}$ is not a monomial.

c. $x^{-3} + 2x^{-2} + 6$

This expression is not a polynomial because x^{-3} and x^{-2} are not monomials: $x^{-3} = \frac{1}{x^3}$ and $x^{-2} = \frac{1}{x^2}$. Monomials cannot contain variables in the denominator.

▶ **Guided Practice**

2A. $\frac{x}{y} + 3x^2$ **2B.** $x^5y + 9x^4y^3 - 2xy$

You can simplify a polynomial just like you simplify a monomial. Perform the operations indicated, and combine like terms.

Study Tip

Alternative Methods Notice that Example 3a uses a horizontal method, and Example 3b uses a vertical method to simplify. Either method will yield a correct solution.

Example 3   Simplify Polynomial Expressions

Simplify each expression.

a. $(4x^2 - 5x + 6) - (2x^2 + 3x - 1)$

Remove parentheses, and group like terms together.

$(4x^2 - 5x + 6) - (2x^2 + 3x - 1)$
$= 4x^2 - 5x + 6 - 2x^2 - 3x + 1$ Distribute the -1.
$= (4x^2 - 2x^2) + (-5x - 3x) + (6 + 1)$ Group like terms.
$= 2x^2 - 8x + 7$ Combine like terms.

b. $(6x^2 - 7x + 8) + (-4x^2 + 9x - 5)$

Align like terms vertically and add.

$6x^2 - 7x + 8$
$\underline{(+) -4x^2 + 9x - 5}$
$2x^2 + 2x + 3$

▶ **Guided Practice**

3A. $(-x^2 - 3x + 4) - (x^2 + 2x + 5)$ **3B.** $(3x^2 - 6) + (-x + 1)$

Adding or subtracting integers results in an integer, so the set of integers is closed under addition and subtraction. Similarly, because adding or subtracting polynomials results in a polynomial, the set of polynomials is closed under addition and subtraction.

You can use the Distributive Property to multiply polynomials.

Example 4 Simplify by Using the Distributive Property

Find $3x(2x^2 - 4x + 6)$.

$3x(2x^2 - 4x + 6) = 3x(2x^2) + 3x(-4x) + 3x(6)$ — Distributive Property
$ = 6x^3 - 12x^2 + 18x$ — Multiply the monomials.

▶ **Guided Practice**

Find each product.

4A. $\frac{4}{3}x^2(6x^2 + 9x - 12)$ **4B.** $-2a(-3a^2 - 11a + 20)$

Real-World Example 5 Write a Polynomial Expression

DRIVING The U.S. Department of Transportation limits the time a truck driver can work each day to eleven hours. For the first part of his trip, Tom drives at a speed of 60 miles per hour, and for the second part of the trip, he drives at a speed of 70 miles per hour. Write a polynomial to represent the distance driven.

Words	60 mph for some time and 70 mph for the rest
Variable	Let x = the number of hours he drives at 60 miles per hour.
Expression	60 · x + 70 · (11 − x)

$60x + 70(11 - x)$ — Original expression
$= 60x + 770 - 70x$ — Distributive Property
$= 770 - 10x$ — Combine like terms.

The polynomial is $770 - 10x$.

▶ **Guided Practice**

5. Paul has $900 to invest in a savings account that has an annual interest rate of 1.8%, and a money market account that pays 4.2% per year. Write a polynomial for the interest he will earn in one year if he invests x dollars in the savings account.

Like addition and subtraction, polynomials are closed under multiplication.

Example 6 Multiply Polynomials

Find $(n^2 + 4n - 6)(n + 2)$.

$(n^2 + 4n - 6)(n + 2)$
$= n^2(n + 2) + 4n(n + 2) + (-6)(n + 2)$ — Distributive Property
$= n^2 \cdot n + n^2 \cdot 2 + 4n \cdot n + 4n \cdot 2 + (-6) \cdot n + (-6) \cdot 2$ — Distributive Property
$= n^3 + 2n^2 + 4n^2 + 8n - 6n - 12$ — Multiply monomials.
$= n^3 + 6n^2 + 2n - 12$ — Combine like terms.

▶ **Guided Practice**

Find each product.

6A. $(x^2 + 4x + 16)(x - 4)$ **6B.** $(2x^2 - 4x + 5)(3x - 1)$

Real-World Career

Truck Driver Truck drivers are considered technical professionals because they are required to obtain specialized education and professional licensure. Although state motor vehicle departments administer the Commercial Driver's License program, federal law spells out the requirements to obtain one.

Check Your Understanding

= Step-by-Step Solutions begin on page R11.

Example 1 Simplify. Assume that no variable equals 0.

1. $(2a^3b^{-2})(-4a^2b^4)$
2. $\dfrac{12x^4y^2}{2xy^5}$
3. $\left(\dfrac{2a^2}{3b}\right)^3$
4. $(6g^5h^{-4})^3$

Example 2 Determine whether each expression is a polynomial. If it is a polynomial, state the degree of the polynomial.

5. $3x + 4y$
6. $\dfrac{1}{2}x^2 - 7y$
7. $x^2 + \sqrt{x}$
8. $\dfrac{ab^3 - 1}{az^4 + 3}$

Examples 3–4, and 6 Simplify.

9. $(x^2 - 5x + 2) - (3x^2 + x - 1)$
10. $(3a + 4b) + (6a - 6b)$
11. $2a(4b + 5)$
12. $3x^2(2xy - 3xy^2 + 4x^2y^3)$
13. $(n - 9)(n + 7)$
14. $(a + 4)(a - 6)$

Example 5
15. **EXERCISE** Tara exercises 75 minutes a day. She does cardio, which burns an average of 10 Calories per minute, and weight training, which burns an average of 7.5 Calories per minute. Write a polynomial to represent the amount of Calories Tara burns in one day if she does x minutes of weight training.

Practice and Problem Solving

Extra Practice is on page R4.

Example 1 Simplify. Assume that no variable equals 0.

16. $(5x^3y^{-5})(4xy^3)$
17. $(-2b^3c)(4b^2c^2)$
18. $\dfrac{a^3n^7}{an^4}$
19. $\dfrac{-y^3z^5}{y^2z^3}$
20. $\dfrac{-7x^5y^5z^4}{21x^7y^5z^2}$
21. $\dfrac{9a^7b^5c^5}{18a^5b^9c^3}$
22. $(n^5)^4$
23. $(z^3)^6$

Example 2 Determine whether each expression is a polynomial. If it is a polynomial, state the degree of the polynomial.

24. $2x^2 - 3x + 5$
25. $a^3 - 11$
26. $\dfrac{5np}{n^2} - \dfrac{2g}{h}$
27. $\sqrt{m - 7}$

Examples 3–4, and 6 **MP REGULARITY** Simplify.

28. $(6a^2 + 5a + 10) - (4a^2 + 6a + 12)$
29. $(7b^2 + 6b - 7) - (4b^2 - 2)$
30. $3p(np - z)$
31. $4x(2x^2 + y)$
32. $(x - y)(x^2 + 2xy + y^2)$
33. $(a + b)(a^3 - 3ab - b^2)$
34. $4(a^2 + 5a - 6) - 3(2a^3 + 4a - 5)$
35. $5c(2c^2 - 3c + 4) + 2c(7c - 8)$
36. $5xy(2x - y) + 6y^2(x^2 + 6)$
37. $3ab(4a - 5b) + 4b^2(2a^2 + 1)$
38. $(x - y)(x + y)(2x + y)$
39. $(a + b)(2a + 3b)(2x - y)$

Example 5
40. **PAINTING** Connor has hired two painters to paint his house. The first painter charges $12 per hour and the second painter charges $11 per hour. It will take 15 hours of labor to paint the house.

   a. Write a polynomial to represent the total cost of the job if the first painter does x hours of the labor.

   b. Write a polynomial to represent the total cost of the job if the second painter does y hours of the labor.

Simplify. Assume that no variable equals 0.

41. $\left(\dfrac{8x^2y^3}{24x^3y^2}\right)^4$
42. $\left(\dfrac{12a^3b^5}{4a^6b^3}\right)^3$
43. $\left(\dfrac{4x^{-2}y^3}{xy^{-4}}\right)^{-2}$
44. $\left(\dfrac{5a^{-7}b^2}{ab^{-6}}\right)^{-3}$

45. $(a^2b^3)(ab)^{-2}$
46. $(-3x^3y)^2(4xy^2)$
47. $\dfrac{3c^2d(2c^3d^5)}{15c^4d^2}$

48. $\dfrac{-10g^6h^9(g^2h^3)}{30g^3h^3}$
49. $\dfrac{5x^4y^2(2x^5y^6)}{20x^3y^5}$
50. $\dfrac{-12n^7p^5(n^2p^4)}{36n^6p^7}$

51. ASTRONOMY Refer to the beginning of the lesson.
  a. How long does it take light from Andromeda to reach Earth?
  b. The average distance from the Sun to Mars is approximately 2.28×10^{11} meters. How long does it take light from the Sun to reach Mars?

Simplify.

52. $\dfrac{1}{4}g^2(8g + 12h - 16gh^2)$
53. $\dfrac{1}{3}n^3(6n - 9p + 18np^4)$
54. $x^{-2}(x^4 - 3x^3 + x^{-1})$
55. $a^{-3}b^2(ba^3 + b^{-1}a^2 + b^{-2}a)$
56. $(g^3 - h)(g^3 + h)$
57. $(n^2 - 7)(2n^3 + 4)$
58. $(2x - 2y)^3$
59. $(4n - 5)^3$
60. $(3z - 2)^3$

61. MODELING The polynomials $0.108x^2 - 0.876x + 474.1$ and $0.047x^2 + 9.694x + 361.7$ approximate the number of bachelor's degrees, in thousands, earned by males and females, respectively, where x is the number of years after 1971.
  a. Find the polynomial that represents the total number of bachelor's degrees (in thousands) earned by both men and women.
  b. Find the polynomial that represents the difference between bachelor's degrees earned by men and by women.

62. If $5^{k+7} = 5^{2k-3}$, what is the value of k?

63. What value of k makes $q^{41} = q^{4k} \cdot q^5$ true?

64. MULTIPLE REPRESENTATIONS Use the model at the right that represents the product of $x + 3$ and $x + 4$.

  a. **Geometric** The area of each rectangle is the product of its length and width. Use the model to find the product of $x + 3$ and $x + 4$.
  b. **Algebraic** Use FOIL to find the product of $x + 3$ and $x + 4$.
  c. **Verbal** Explain how each term of the product is represented in the model.

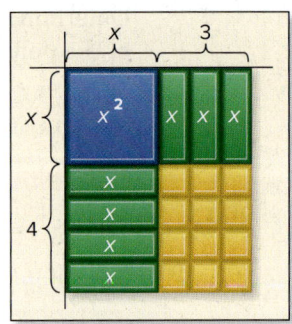

H.O.T. Problems — Use Higher-Order Thinking Skills

65. PROOF Show how the property of negative exponents can be proven using the Quotient of Powers Property and the Zero Power Property.

66. CHALLENGE What happens to the quantity of x^{-y} as y increases, for $y > 0$ and $x > 1$?

67. REASONING Explain why the expression 0^{-2} is undefined.

68. OPEN-ENDED Write three different expressions that are equivalent to x^{12}.

69. WRITING IN MATH Explain why properties of exponents are useful in astronomy. Include an explanation of how to find the amount of time it takes for light from a source to reach a planet.

Preparing for Assessment

70. Which expression is equivalent to $(2y^3 - 5y^2 - 4) - (4y^3 + 11y^2 - 3y + 5)$? MP 6

- A $-6y^3 - 6y^2 + 3y - 1$
- B $-2y^3 - 16y^2 + 3y - 9$
- C $-2y^3 + 6y^2 - 3y + 1$
- D $2y^3 + 16y^2 - 3y + 9$

71. A total of 700 tickets were sold to the Saturday and Sunday performances of the school play. The drama department earns $5.00 from each ticket sold. Which expression represents the dollars the drama department earned from Saturday tickets if x Sunday tickets were sold? MP 1

- A $700 - x$
- B $700 - 5x$
- C $3500 - x$
- D $3500 - 5x$

72. Sammy and Luis each threw a ball into the air at the same time. The height, in feet, of Sammy's ball after t seconds is modeled by the expression $2 + 40t - 16t^2$. The height, in feet, of Luis's ball after t seconds is modeled by the expression $2 + 35t - 16t^2$. Which expression can be evaluated to find how much higher Sammy's ball is than Luis's ball after t seconds? MP 1

- A $5t$
- B $5t + 4$
- C $75t + 4$
- D $-32t^2 + 75t + 4$

73. Which expression is equivalent to $(m^2 - m + 1)(m + 1)$? MP 6

- A $m^3 + 1$
- B $m^3 - m^2 + 1$
- C $m^3 + 2m^2 + 1$
- D $m^3 + m^2 - m + 1$

74. MULTI-STEP Consider the box below. MP 3

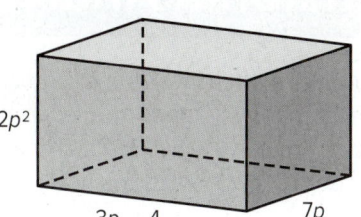

a. Which of the following expressions are expressions for the area of one face of the box? Check all that apply.

- A $2p^2(3p - 4)$
- B $7p(3p - 4)$
- C $2p^2(7p)$
- D $14p^3$
- E $21p^2 - 4p$
- F $6p^3 - 8p$

b. Which expression is correct for the total surface area of the box?

- A $2p^2(3p - 4) + 7p(3p - 4) - 2p^2(7p)$
- B $14p^3 + 27p^2 - 28p$
- C $40p^3 + 26p^2 - 56p$
- D $14p^3 + 7p^3 + 13p^2 - 14p$

c. Georges writes out the following procedure to find an expression for the volume of the box. Identify three errors in the procedure and determine the correct expression for the volume of the box.

Volume $= 7p(3p - 4)(2p^2)$

$= [7p(3p) + 7p(4)](2p^2)$

$= [21p^2 + 28p](2p^2)$

$= 21p^2(2p^2) + 28p(2p^2)$

$= 32p^4 + 56p^2$ units^3

EXTEND 4-1
Algebra Lab
Dimensional Analysis

Real-world problems often involve units of measure. Performing operations with units is called **dimensional analysis** or **unit analysis**. You can use dimensional analysis to convert units or to perform calculations.

Mathematical Practices

1. Make sense of problems and persevere in solving them.
2. Reason abstractly and quantitatively.

Activity Finding Roots

A car is traveling at 65 miles per hour. How fast is the car traveling in meters per second?

You want to find the speed in meters per second, so you need to change the unit of distance from miles to meters and the unit of time from hours to seconds. To make the conversion, use fractions that you can multiply.

Step 1 Change the units of length from miles to meters.
Use the relationships of miles to feet and feet to meters.

$$\frac{65 \text{ miles}}{1 \text{ hour}} \cdot \frac{5280 \text{ feet}}{1 \text{ mile}} \cdot \frac{1 \text{ meter}}{3.3 \text{ feet}}$$

Step 2 Change the units of time from hours to seconds.
Write fractions relating hours to minutes and minutes to seconds.

$$\frac{65 \text{ miles}}{1 \text{ hour}} \cdot \frac{5280 \text{ feet}}{1 \text{ mile}} \cdot \frac{1 \text{ meter}}{3.3 \text{ feet}} \cdot \frac{1 \text{ hour}}{60 \text{ minutes}} \cdot \frac{1 \text{ minute}}{60 \text{ seconds}}$$

Step 3 Simplify and check by canceling the units.

$$\frac{65 \cancel{\text{ miles}}}{1 \cancel{\text{ hour}}} \cdot \frac{5280 \cancel{\text{ feet}}}{1 \cancel{\text{ mile}}} \cdot \frac{1 \text{ meter}}{3.3 \cancel{\text{ feet}}} \cdot \frac{1 \cancel{\text{ hour}}}{60 \cancel{\text{ minutes}}} \cdot \frac{1 \cancel{\text{ minute}}}{60 \text{ seconds}}$$

$$= \frac{65 \cdot 5280}{3.3 \cdot 60 \cdot 60} \text{ m/s} \qquad \text{Simplify.}$$

$$\approx 28.9 \text{ m/s} \qquad \text{Use a calculator.}$$

So, 65 miles per hour is about 28.9 meters per second. This answer is reasonable because the final units are m/s, not m/hr, ft/s, or mi/hr.

Exercises

Work cooperatively. Solve each problem by using dimensional analysis. Include the appropriate units with your answer.

1. A zebra can run 40 miles per hour. How far can a zebra run in 3 minutes?

2. A cyclist traveled 43.2 miles at an average speed of 12 miles per hour. How long did the cyclist ride?

3. If you are driving 50 miles per hour, how many feet per second are you traveling?

4. The equation $d = \frac{1}{2}(9.8 \text{ m/s}^2)(3.5 \text{ s})^2$ represents the distance d that a ball falls 3.5 seconds after it is dropped from a tower. Find the distance.

5. **WRITING IN MATH** Explain how dimensional analysis can be useful in checking the reasonableness of your answer.

LESSON 2
Powers of Binomials

::Then   ::Now   ::Why?

- You worked with combinations.

1. Use Pascal's triangle to expand powers of binomials.
2. Use the Binomial Theorem to expand powers of binomials.

A manager plans to hire 8 new employees. Not wanting to appear biased, the manager wants to hire a combination of males and females that has at least a 10% chance of occurring randomly. If there are an equal number of male and female applicants, is the probability of randomly hiring 6 males and 2 females less than 10%?

New Vocabulary
Pascal's triangle

Mathematical Practices
4 Model with mathematics.

1 Pascal's Triangle In the 13th century, the Chinese discovered a pattern of numbers that would later be referred to as **Pascal's triangle**. This pattern can be used to determine the coefficients of an expanded binomial $(a + b)^n$.

$(a + b)^0$ 1
$(a + b)^1$ 1   1
$(a + b)^2$ 1   2   1
$(a + b)^3$ 1   3   3   1
$(a + b)^4$ 1   4   6   4   1
$(a + b)^5$ **1**   **5**   **10**   **10**   **5**   **1**

For example, the expanded form of
$(a + b)^5 = \mathbf{1}a^5 + \mathbf{5}a^4b + \mathbf{10}a^3b^2 + \mathbf{10}a^2b^3 + \mathbf{5}ab^4 + \mathbf{1}b^5$.

Real-World Example 1   Use Pascal's Triangle

Find the probability of having 6 lambs with white wool and 2 lambs with black wool by expanding $(w + b)^8$.

Write three more rows of Pascal's triangle and use the pattern to write the expansion.

5   1   5   10   10   5   1
6   1   6   15   20   15   6   1
7   1   7   21   35   35   21   7   1
8   1   8   28   56   70   56   28   8   1

$(w + b)^8 = w^8 + 8w^7b + 28w^6b^2 + 56w^5b^3 + 70w^4b^4 + 56w^3b^5 + 28w^2b^6 + 8wb^7 + b^8$

By adding the coefficients of the polynomial, there are 256 combinations of lambs with white and black wool that could be born.

$28w^6b^2$ represents the number of combinations with 6 white-wooled lambs and 2 black-wooled lambs.

Therefore, there is a $\frac{28}{256}$ or about an 11% chance of the sheep giving birth to 6 white-wooled and 2 black-wooled lambs.

▶ **Guided Practice**

1. Expand $(c + d)^9$.

2 The Binomial Theorem Instead of writing out row after row of Pascal's triangle, you can use the **Binomial Theorem** to expand a binomial. Recall that $_nC_r = \frac{n!}{r!(n-r)!}$.

connectED.mcgraw-hill.com   237

Study Tip

Combinations Recall that both $_nC_0$ and $_nC_n$ equal 1.

Key Concept Binomial Theorem

If n is a natural number, then $(a+b)^n =$
$$_nC_0\, a^n b^0 + {_nC_1}\, a^{n-1}b^1 + {_nC_2}\, a^{n-2}b^2 + \cdots + {_nC_n}\, a^0 b^n = \sum_{k=0}^{n} \frac{n!}{k!(n-k)!}\, a^{n-k}b^k.$$

To use the theorem, replace n with the value of the exponent. Notice how the terms will follow the pattern of Pascal's triangle, and the coefficients will be symmetric.

Study Tip

MP Tools You can calculate $_nC_r$ by using a graphing calculator. Press MATH and choose PRB 3.

Example 2   Use the Binomial Theorem

Expand $(a+b)^7$.

Method 1 Use combinations.

Replace n with 7 in the Binomial Theorem.

$(a+b)^7 = {_7C_0}\,a^7 + {_7C_1}\,a^6 b + {_7C_2}\,a^5 b^2 + {_7C_3}\,a^4 b^3 + {_7C_4}\,a^3 b^4 + {_7C_5}\,a^2 b^5 + {_7C_6}\,ab^6 + {_7C_7}\,b^7$

$= a^7 + \dfrac{7!}{6!}a^6 b + \dfrac{7!}{2!5!}a^5 b^2 + \dfrac{7!}{3!4!}a^4 b^3 + \dfrac{7!}{4!3!}a^3 b^4 + \dfrac{7!}{5!2!}a^2 b^5 + \dfrac{7!}{6!}ab^6 + b^7$

$= a^7 + 7a^6 b + 21a^5 b^2 + 35a^4 b^3 + 35a^3 b^4 + 21a^2 b^5 + 7ab^6 + b^7$

Method 2 Use Pascal's triangle.

Use the Binomial Theorem to determine exponents, but instead of finding the coefficients by using combinations, look at the seventh row of Pascal's triangle.

```
6      1    6    15    20    15    6    1
7   1    7    21    35    35    21    7    1
8   1    8    28    56    70    56    28    8    1
```

$(a+b)^7 = a^7 + 7a^6 b + 21a^5 b^2 + 35a^4 b^3 + 35a^3 b^4 + 21a^2 b^5 + 7ab^6 + b^7$

▶ **Guided Practice**

2. Expand $(x+y)^{10}$.

When the binomial to be expanded has coefficients other than 1, the coefficients will no longer be symmetric. In these cases, you may want to use the Binomial Theorem.

Example 3   Coefficients Other Than 1

Expand $(5a - 4b)^4$.

$(5a - 4b)^4$

$= {_4C_0}(5a)^4 + {_4C_1}(5a)^3(-4b) + {_4C_2}(5a)^2(-4b)^2 + {_4C_3}(5a)(-4b)^3 + {_4C_4}(-4b)^4$

$= 625a^4 + \dfrac{4!}{3!}(125a^3)(-4b) + \dfrac{4!}{2!2!}(25a^2)(16b^2) + \dfrac{4!}{3!}(5a)(-64b^3) + 256b^4$

$= 625a^4 - 2000a^3 b + 2400a^2 b^2 - 1280ab^3 + 256b^4$

▶ **Guided Practice**

3. Expand $(3x + 2y)^5$.

Go Online!

Pause and continue to follow along using your graphing calculator as your **Personal Tutor** solves an example.

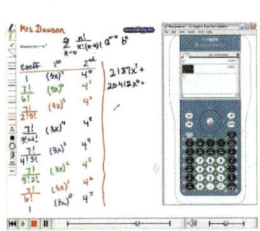

Sometimes you may need to find only one term in a binomial expansion. To do this, you can use the summation formula for the Binomial Theorem, $\displaystyle\sum_{k=0}^{m} \frac{n!}{k!(n-k)!}\, a^{n-k}b^k.$

Example 4 Determine a Single Term

Find the fifth term of $(y + z)^{11}$.

Step 1 Use the Binomial Theorem to write the expansion in sigma notation.

$$(y + z)^{11} = \sum_{k=0}^{11} \frac{11!}{k!(11-k)!} y^{11-k} z^k$$

Step 2 $\frac{11!}{k!(11-k)!} y^{11-k} z^k = \frac{11!}{4!(11-4)!} y^{11-4} z^4$ For the fifth term, $k = 4$.

$= 330 y^7 z^4$ $C(11, 4) = 330$

Guided Practice

4. Find the sixth term of $(c + d)^{10}$.

Concept Summary Binomial Expansion

In a binomial expansion of $(a + b)^n$,
- there are $n + 1$ terms.
- n is the exponent of a in the first term and b in the last term.
- in successive terms, the exponent of a decreases by 1, and the exponent of b increases by 1.
- the sum of the exponents in each term is n.
- the coefficients are symmetric.

Check Your Understanding

 = Step-by-Step Solutions begin on page R11.

Examples 1–3 Expand each binomial.

1. $(c + d)^5$
2. $(g + h)^7$
3. $(x - 4)^6$
4. $(2y - z)^5$
5. $(x + 3)^5$
6. $(y - 4z)^4$

7. **GENETICS** If a woman is equally as likely to have a baby boy or a baby girl, use binomial expansion to determine the probability that 5 of her 6 children are girls. Do not consider identical twins.

Example 4 Find the indicated term of each expression.

8. fourth term of $(b + c)^9$
9. fifth term of $(x + 3y)^8$
10. third term of $(a - 4b)^6$
11. sixth term of $(2c - 3d)^8$
12. last term of $(5x + y)^5$
13. first term of $(3a + 8b)^5$

Practice and Problem Solving

Extra Practice is on page R4.

Examples 1–3 Expand each binomial.

14. $(c - d)^7$
15. $(x + 6)^6$
16. $(y - 5)^7$
17. $(2a + 4b)^4$
18. $(3a - 4b)^5$
19. $(a - b)^6$

20. **BASEBALL** If a pitcher is just as likely to throw a ball as a strike, find the probability that 11 of his first 12 pitches are balls.

21. **COMMITTEES** If an equal number of men and women applied to be on a community planning committee and the committee needs a total of 10 people, find the probability that 7 of the members will be women. Assume that committee members will be chosen randomly.

Example 4 Find the indicated term of each expression.

22. fourth term of $(y - 3x)^6$
23. third term of $(x + 2z)^7$
24. sixth term of $(4x + 5y)^6$
25. seventh term of $(2a - 2b)^8$
26. fourth term of $(c + 6)^8$
27. fifth term of $(x - 4)^9$

28. **MULTI-STEP** In soccer, a tie is decided by 5 penalty kicks taken by each team. The team who scores the most goals wins. Over the course of a season, a team found that for any penalty shot, the kicker has a 13% chance of missing the goal entirely. Otherwise, the goalkeeper has a 10% chance of correctly guessing where the player is aiming. When he guesses correctly, he stops the shot 80.5% of the time.

   a. What is the probability that at least 3 of 5 players score on penalty kicks?
   b. Explain your solution process.

Expand each binomial.

29. $\left(x + \frac{1}{2}\right)^5$
30. $\left(x - \frac{1}{3}\right)^4$
31. $\left(2b + \frac{1}{4}\right)^5$
32. $\left(3c + \frac{1}{3}\right)^5$

33. **SENSE-MAKING** In $\frac{n!}{k!(n-k)!} p^k q^{n-k}$, let p represent the likelihood of a success and q represent the likelihood of a failure.

   a. If a place-kicker makes 70% of his kicks within 40 yards, find the likelihood that he makes 9 of his next 10 attempts from within 40 yards.
   b. If a quarterback completes 60% of his passes, find the likelihood that he completes 8 of his next 10 attempts.
   c. If a team converts 30% of their two-point conversions, find the likelihood that they convert 2 of their next 5 conversions.

H.O.T. Problems   Use Higher-Order Thinking Skills

34. **CHALLENGE** Find the sixth term of the expansion of $(\sqrt{a} + \sqrt{b})^{12}$. Explain your reasoning.

35. **REASONING** Explain how the terms of $(x + y)^n$ and $(x - y)^n$ are the same and how they are different.

36. **REASONING** Determine whether the following statement is *true* or *false*. Explain your reasoning.

    The eighth and twelfth terms of $(x + y)^{20}$ have the same coefficients.

37. **OPEN-ENDED** Write a power of a binomial for which the second term of the expansion is $6x^4y$.

38. **WRITING IN MATH** Explain how to write out the terms of Pascal's triangle.

Preparing for Assessment

39. What are the missing coefficients in this expansion of a binomial raised to a power? **MP** 6

$$a^3 + \underline{} a^2b + \underline{} ab^2 + 8b^3$$

- A 2 and 4
- B 3 and 3
- C 3 and 6
- D 6 and 6
- E 6 and 12

40. Seymour wants to multiply an expression of the form shown below, where m and n are positive integers. **MP** 1

$$x^m(x + 2)^n$$

What will be the coefficient of the x^m term?

- A 2
- B 2^m
- C 2^n
- D 2^{m+n}
- E cannot be determined from the information given

41. What is the missing term in this binomial expansion? **MP** 6

$$x^4 + 12x^3y + \underline{} + 108xy^3 + 81y^4$$

- A $6x^2y^2$
- B $18x^2y^2$
- C $36x^2y^2$
- D $54x^2y^2$
- E $81x^2y^2$

42. A polynomial function is shown below. **MP** 1

$$f(x) = 2 + 3(x - a)^7$$

If $f(0) = -1$, what is the value of a?

- A -1
- B 0
- C 1
- D 3
- E 5

43. Consider this work for the expansion of $(x - y)^5$. **MP** 3

Which line of the work shown has an error? Choose all that apply.

- A $(x - y)^5 = (x + (-y))^5$
- B $_5C_0x^5 + {_5C_1}x^4(-y) + {_5C_2}x^3(-y)^2 + {_5C_3}x^2(-y)^3 + {_5C_4}x^1(-y)^4 + {_5C_5}x(-y)^5$
- C $_5C_0x^5 - {_5C_1}x^4y + {_5C_2}x^3y^2 - {_5C_3}x^2y^3 + {_5C_4}xy^4 - {_5C_5}y^5$
- D $x^5 + 5x^4y + 10x^3y^2 - 10x^2y^3 + 5xy^4 - y^5$

44. MULTI-STEP Consider the expansion of $(x + 3y)^4$ starting with the term x^4. **MP** 1

a. Which option shows the correct fifth term?

- A $108x^4y$
- B $81xy^4$
- C $54x^2y^2$
- D $81y^4$

b. Which of the following statements are true? Choose all that apply.

- A The coefficient of the fourth term is $27 \times \frac{4!}{2!2!}$.
- B The second term includes the variables x^2y^2.
- C The last term is a number with no variables.
- D The first term has only one variable.
- E The middle term has a coefficient of 54.
- F There is only one negative term in this expansion.

c. Write the complete correctly expanded expression.

LESSON 3
Dividing Polynomials

::Then   ::Now   ::Why?

• You divided   • **1** Divide polynomials   • Arianna needed $70x^2 + 30x$ square inches of fabric to make a cover for her tablet that is
  monomials.   using long division. $5x$ inches tall. In figuring the area, she allowed for a front and back flap. If the spine is
                          • **2** Divide polynomials x inches wide, and the front and back are $3x$ inches wide, how wide are the front and
                               using synthetic   back flaps? You can use a quotient of polynomials to help you find the answer.
                               division.

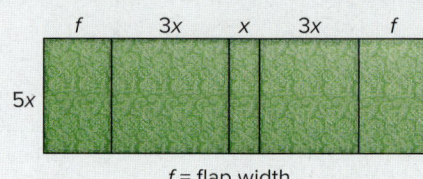

f = flap width

New Vocabulary
synthetic division

Mathematical Practices
6 Attend to precision.

1 Long Division
In Lesson 4-1, you learned how to divide monomials. You can divide a polynomial by a monomial using those same skills.

Example 1 Divide a Polynomial by a Monomial

Simplify $\dfrac{6x^4y^3 + 12x^3y^2 - 18x^2y}{3xy}$.

$\dfrac{6x^4y^3 + 12x^3y^2 - 18x^2y}{3xy} = \dfrac{6x^4y^3}{3xy} + \dfrac{12x^3y^2}{3xy} - \dfrac{18x^2y}{3xy}$ Sum of quotients

$= \dfrac{6}{3} \cdot x^{4-1}y^{3-1} + \dfrac{12}{3} \cdot x^{3-1}y^{2-1} - \dfrac{18}{3} \cdot x^{2-1}y^{1-1}$ Divide.

$= 2x^3y^2 + 4x^2y - 6x$ $y^{1-1} = y^0$ or 1

▶ **Guided Practice**   Simplify.

1A. $(20c^4d^2f - 16cdf^2 + 4cdf) \div (4cdf)$ **1B.** $(18x^2y + 27x^3y^2z)(3xy)^{-1}$

You can use a process similar to long division to divide a polynomial by a polynomial with more than one term. The process is known as the *division algorithm*.

Example 2 Division Algorithm

Use long division to find $(x^2 + 3x - 40) \div (x - 5)$.

```
              x + 8
        ┌─────────────
x − 5 ) x² + 3x − 40
      (−) x² − 5x
        ─────────
              8x − 40
           (−) 8x − 40
              ─────────
                    0
```

Multiply the divisor by x since $\dfrac{x^2}{x} = x$.

Subtract. Bring down the next term.

Multiply the divisor by 8 since $\dfrac{8x}{x} = 8$.

Subtract.

The quotient is $x + 8$. The remainder is 0.

▶ **Guided Practice**   Use long division to find each quotient.

2A. $(x^2 + 7x - 30) \div (x - 3)$ **2B.** $(x^2 - 13x + 12) \div (x - 1)$

Just as with the division of whole numbers, the division of two polynomials may result in a quotient with a remainder. Remember that $11 \div 3 = 3 + R2$, which is often written as $3\frac{2}{3}$. The result of a division of polynomials with a remainder can be written in a similar manner.

Example 3   Divide Polynomials

Which expression is equivalent to $(a^3 + 2a^2 + 4a - 9)(2 - a)^{-1}$?

A $a^2 + 4a + 12 - \dfrac{15}{2-a}$ **C** $-a^2 - 4a - 12 + \dfrac{15}{2-a}$

B $a^2 + 4a + 12 + \dfrac{33}{2-a}$ **D** $-a^2 - 4a - 12 - \dfrac{33}{2-a}$

Read the Item

Because the second factor has an exponent of -1, this is a division problem.
$(a^3 + 2a^2 + 4a - 9)(2-a)^{-1} = \dfrac{a^3 + 2a^2 + 4a - 9}{2-a}$

Solve the Item

$$\begin{array}{r} -a^2 - 4a - 12 \\ -a + 2 \overline{) a^3 + 2a^2 + 4a - 9 } \\ (-)\ \underline{a^3 - 2a^2} \\ 4a^2 + 4a \\ (-)\ \underline{4a^2 - 8a} \\ 12a - 9 \\ (-)\ \underline{12a - 24} \\ 15 \end{array}$$

Rewrite $2 - a$ as $-a + 2$.
$-a^2(-a + 2) = a^3 - 2a^2$
$2a^2 - (-2a^2) = 4a^2$
$-4a(-a + 2) = 4a^2 - 8a$
$4a - (-8a) = 12a$
$-12(-a + 2) = 12a - 24$
$-9 - (-24) = 15$

The quotient is $-a^2 - 4a - 12$, and the remainder is 15.

CHECK Verify your answer by multiplying the divisor by the quotient, then adding the remainder.

$(-a + 2)(-a^2 - 4a - 12) + 15 \stackrel{?}{=} a^3 + 2a^2 + 4a - 9$
$a^3 + 4a^2 + 12a - 2a^2 - 8a - 24 + 15 \stackrel{?}{=} a^3 + 2a^2 + 4a - 9$
$a^3 + 2a^2 + 4a - 9 = a^3 + 2a^2 + 4a - 9$ ✓

So, $(a^3 + 2a^2 + 4a - 9)(2-a)^{-1} = -a^2 - 4a - 12 + \dfrac{15}{2-a}$. The answer is **C**.

Guided Practice

3. Which expression is equal to $(r^3 + 6r^2 - r - 4)(r^2 - 1)^{-1}$?

A $r + 6 - \dfrac{2}{r^2 - 1}$ **C** $r + 5 - \dfrac{1}{r^2 - 1}$

B $r + 6 + \dfrac{2}{r^2 - 1}$ **D** $r + 5 + \dfrac{1}{r^2 - 1}$

Polynomial division can involve divisors with degrees greater than 1. As with real numbers, dividing by larger degree polynomials can result in missing, or zero, terms in the dividend.

Real Numbers

$16{,}384 \div 16 = 1024$

Polynomials

$(x^4 - x^2 - 20) \div (x^2 + 4) = x^2 + 0x - 5$ or $x^2 - 5$

Example 4 Divisor Degree Greater Than 1

Simplify.

a. $(x^3 - 2x^2 - 9x + 18) \div (x^2 - 5x + 6)$

$$
\begin{array}{r}
x + 3 \\
x^2 - 5x + 6 \overline{\smash{\big)} x^3 - 2x^2 - 9x + 18} \\
\underline{(-)\ x^3 - 5x^2 + 6x } \\
3x^2 - 15x + 18 \\
\underline{(-)\ 3x^2 - 15x + 18} \\
0
\end{array}
$$

$x(x^2 - 5x + 6) = x^3 - 5x^2 + 6x$
$-2x^2 - (-5x^2) = 3x^2; -9x - 6x = -15x$
$3(x^2 - 5x + 6) = 3x^2 - 15x + 18$

The quotient is $x + 3$ and the remainder is 0.
So, $(x^3 - 2x^2 - 9x + 18) \div (x^2 - 5x + 6) = x + 3$.

b. $(x^4 + x^3 - 6x^2 - 4x + 14)(x^2 + x - 2)^{-1}$

$$
\begin{array}{r}
x^2 + 0x - 4 \\
x^2 + x - 2 \overline{\smash{\big)} x^4 + x^3 - 6x^2 - 4x + 14} \\
\underline{(-)\ x^4 + x^3 - 2x^2 } \\
0x^3 - 4x^2 - 4x \\
\underline{(-)\ 0x^3 + 0x^2 + 0x } \\
-4x^2 - 4x + 14 \\
\underline{(-)\ -4x^2 - 4x + 8} \\
6
\end{array}
$$

Because there is no x^3-term, the coefficient of x-term in the quotient is 0.

The quotient is $x^2 - 4$ and the remainder is 6.
So, $(x^4 + x^3 - 6x^2 - 4x + 14)(x^2 + x - 2)^{-1} = x^2 - 4 + \dfrac{6}{x^2 + x - 2}$.

Guided Practice

4A. $(x^3 - 3x^2 + x - 3) \div (x^2 + 1)$ **4B.** $(x^3 - 3x^2 + 6x + 4)(x^2 - 4x + 5)^{-1}$

4C. $(x^4 + x^2 - 3x + 5) \div (x^2 + 2)$ **4D.** $(x^4 - x^3 - 8x + 8)(x^2 + 2x + 4)^{-1}$

2 Synthetic Division

Synthetic division is a simpler process for dividing a polynomial by a binomial. Suppose you want to divide $2x^3 - 13x^2 + 26x - 24$ by $x - 4$ using long division. Compare the coefficients in this division with those in Example 4.

$$
\begin{array}{r}
2x^2 - 5x + 6 \\
x - 4 \overline{\smash{\big)} 2x^3 - 13x^2 + 26x - 24} \\
\underline{(-)\ 2x^3 - 8x^2 } \\
-5x^2 + 26x \\
\underline{(-)\ -5x^2 + 20x } \\
6x - 24 \\
\underline{(-)\ 6x - 24} \\
0
\end{array}
$$

When the polynomial in the dividend is missing a term, a zero must be used to represent the missing term. So, with a dividend of $2x^3 - 4x^2 + 6$, a 0 will be used as a placeholder for the x-term.

$\overline{\smash{\big)} 2x^3 - 4x^2 + 0x + 6}$

Key Concept Synthetic Division

Step 1 Write the coefficients of the dividend so that the degrees of the terms are in descending order. Write the constant r of the divisor $x - r$ in the box. Bring the first coefficient down.

Step 2 Multiply the first coefficient by r, and write the product under the second coefficient.

Step 3 Add the product and the second coefficient.

Step 4 Repeat Steps 2 and 3 until you reach a sum in the last column. The numbers along the bottom row are the coefficients of the quotient. The power of the first term is one less than the degree of the dividend. The final number is the remainder.

Example 5 Synthetic Division

Use synthetic division to find $(2x^3 - 13x^2 + 26x - 24) \div (x - 4)$.

Step 1 Write the coefficients of the dividend. Write the constant r in the box. In this case, $r = 4$. Bring the first coefficient, 2, down.

$$\begin{array}{c|cccc} 4 & 2 & -13 & 26 & -24 \\ & \downarrow & & & \\ \hline & 2 & & & | \end{array}$$

Step 2 Multiply the first coefficient by r: $2 \cdot 4 = 8$. Write the product under the second coefficient.

$$\begin{array}{c|cccc} 4 & 2 & -13 & 26 & -24 \\ & & 8 & & \\ \hline & 2 \nearrow & & & | \end{array}$$

Step 3 Add the product and the second coefficient: $-13 + 8 = -5$.

$$\begin{array}{c|cccc} 4 & 2 & -13 & 26 & -24 \\ & & 8 & & \\ \hline & 2 & -5 & & | \end{array}$$

Step 4 Multiply the sum, -5, by r: $-5 \times 4 = -20$. Write the product under the next coefficient, and add: $26 + (-20) = 6$. Multiply the sum, 6, by r: $6 \cdot 4 = 24$. Write the product under the next coefficient and add: $-24 + 24 = 0$.

$$\begin{array}{c|cccc} 4 & 2 & -13 & 26 & -24 \\ & & 8 & -20 & 24 \\ \hline & 2 & -5 \nearrow & 6 \nearrow & | 0 \end{array}$$

Watch Out!
Synthetic Division Remember to *add* terms when performing synthetic division.

CHECK Multiply the quotient by the divisor. The answer should be the dividend.

$$\begin{array}{r} 2x^2 - 5x + 6 \\ (\times) \quad x - 4 \\ \hline -8x^2 + 20x - 24 \\ 2x^3 - 5x^2 + 6x \quad\quad \\ \hline 2x^3 - 13x^2 + 26x - 24 \end{array}$$

The quotient is $2x^2 - 5x + 6$. The remainder is 0.

Guided Practice

Use synthetic division to find each quotient.

5A. $(2x^3 + 3x^2 - 4x + 15) \div (x + 3)$

5B. $(3x^3 - 8x^2 + 11x - 14) \div (x - 2)$

5C. $(4a^4 + 2a^2 - 4a + 12) \div (a + 2)$

5D. $(6b^4 - 8b^3 + 12b - 14) \div (b - 2)$

To use synthetic division, the divisor must be of the form $x - r$. If the coefficient of x in a divisor is not 1, you can rewrite the division expression so that you can use synthetic division.

Watch Out!

MP Precision Remember to divide *all* terms in the numerator and denominator.

Example 6 Divisor with First Coefficient Other than 1

Use synthetic division to find $(3x^4 - 5x^3 + x^2 + 7x) \div (3x + 1)$.

$$\frac{3x^4 - 5x^3 + x^2 + 7x}{3x + 1} = \frac{(3x^4 - 5x^3 + x^2 + 7x) \div 3}{(3x + 1) \div 3}$$ Rewrite the divisor with a leading coefficient of 1. Then divide the numerator and denominator by 3.

$$= \frac{x^4 - \frac{5}{3}x^3 + \frac{1}{3}x^2 + \frac{7}{3}x}{x + \frac{1}{3}}$$ Simplify the numerator and the denominator.

Because the numerator does not have a constant term, use a coefficient of 0 for the constant term.

$x - r = x + \frac{1}{3}$, so $r = -\frac{1}{3}$.

$$\begin{array}{r|rrrrr} -\frac{1}{3} & 1 & -\frac{5}{3} & \frac{1}{3} & \frac{7}{3} & 0 \\ & & -\frac{1}{3} & \frac{2}{3} & -\frac{1}{3} & -\frac{2}{3} \\ \hline & 1 & -2 & 1 & 2 & -\frac{2}{3} \end{array}$$

The result is $x^3 - 2x^2 + x + 2 - \dfrac{\frac{2}{3}}{x + \frac{1}{3}}$. Now simplify the fraction.

$\dfrac{\frac{2}{3}}{x + \frac{1}{3}} = \frac{2}{3} \div \left(x + \frac{1}{3}\right)$ Rewrite as a division expression.

$= \frac{2}{3} \div \frac{3x + 1}{3}$ $x + \frac{1}{3} = \frac{3x}{3} + \frac{1}{3} = \frac{3x + 1}{3}$

$= \frac{2}{3} \cdot \frac{3}{3x + 1}$ Multiply by the reciprocal.

$= \frac{2}{3x + 1}$ Simplify.

The solution is $x^3 - 2x^2 + x + 2 - \dfrac{2}{3x + 1}$.

CHECK Divide using long division.

$$\begin{array}{r} x^3 - 2x^2 + x + 2 \\ 3x + 1 \overline{)3x^4 - 5x^3 + x^2 + 7x} \\ \underline{(-)\ 3x^4 + x^3} \\ -6x^3 + x^2 \\ \underline{(-) -6x^3 - 2x^2} \\ 3x^2 + 7x \\ \underline{(-)\ 3x^2 + x} \\ 6x + 0 \\ \underline{(-)\ 6x + 2} \\ -2 \end{array}$$

The result is $x^3 - 2x^2 + x + 2 - \dfrac{2}{3x + 1}$. ✓

Go Online!

Take the **Self-Check Quiz** with a partner. Take turns describing how to solve each problem using precise terms. Ask for clarification as you need it.

Guided Practice

Use synthetic division to find each quotient.

6A. $(8x^4 - 4x^2 + x + 4) \div (2x + 1)$ **6B.** $(8y^5 - 2y^4 - 16y^2 + 4) \div (4y - 1)$

6C. $(15b^3 + 8b^2 - 21b + 6) \div (5b - 4)$ **6D.** $(6c^3 - 17c^2 + 6c + 8) \div (3c - 4)$

Check Your Understanding

Examples 1, 2, 4, and 5

Simplify.

1. $\dfrac{4xy^2 - 2xy + 2x^2y}{xy}$

2. $(3a^2b - 6ab + 5ab^2)(ab)^{-1}$

3. $(x^2 - 6x - 20) \div (x + 2)$

4. $(2a^2 - 4a - 8) \div (a + 1)$

5. $(3z^3 - 6z^2 - 9z + 3) \div (z^2 - 4z + 3)$

6. $(y^4 - 3y^3 + 2y^2 - 20) \div (y^2 + y - 5)$

Example 3

7. **MULTIPLE CHOICE** Which expression is equal to $(x^2 + 3x - 9)(4 - x)^{-1}$?

   A $-x - 7 + \dfrac{19}{4 - x}$ **B** $-x - 7$ **C** $x + 7 - \dfrac{19}{4 - x}$ **D** $-x - 7 - \dfrac{19}{4 - x}$

Example 6

Simplify.

8. $(10x^2 + 15x + 20) \div (5x + 5)$

9. $(18a^2 + 6a + 9) \div (3a - 2)$

10. $\dfrac{12b^2 + 23b + 15}{3b + 8}$

11. $\dfrac{27y^2 + 27y - 30}{9y - 6}$

Practice and Problem Solving

Extra Practice is on page R4.

Example 1

Simplify.

12. $\dfrac{24a^3b^2 - 16a^2b^3}{8ab}$

13. $\dfrac{5x^2y - 10xy + 15xy^2}{5xy}$

14. $\dfrac{7g^3h^2 + 3g^2h - 2gh^3}{gh}$

15. $\dfrac{4a^3b - 6ab + 2ab^2}{2ab}$

16. $\dfrac{16c^4d^4 - 24c^2d^2}{4c^2d^2}$

17. $\dfrac{9n^3p^3 - 18n^2p^2 + 21n^2p^3}{3n^2p^2}$

18. **ENERGY** Light-emitting diode (LED) bulbs reduce energy waste. The average daily reduction in the energy used expressed as a negative number of watt-hours is estimated by $-b^2 - 600b$, where b is the number of bulbs replaced by CFLs. Divide by b to find the average amount of energy saved per LED bulb.

19. **BAKING** The number of cookies produced in a factory each day can be estimated by $-w^2 + 16w + 1000$, where w is the number of workers. Divide by w to find the average number of cookies produced per worker.

Examples 2, 4, 5, and 6

Simplify.

20. $(a^2 - 8a - 26) \div (a + 2)$

21. $(b^3 - 4b^2 + b - 2) \div (b + 1)$

22. $(z^4 - 3z^3 + 2z^2 - 4z + 4)(z - 1)^{-1}$

23. $(x^5 - 4x^3 + 4x^2) \div (x - 4)$

24. $\dfrac{y^3 + 11y^2 - 10y + 6}{y + 2}$

25. $(g^4 - 3g^2 - 18) \div (g - 2)$

26. $(6a^2 - 3a + 9) \div (3a - 2)$

27. $\dfrac{6x^5 + 5x^4 + x^3 - 3x^2 + x}{3x + 1}$

28. $\dfrac{g^3 - 2g^2 + 3g - 1}{g^2 - 2g}$

29. $(2b^3 - 6b^2 + 8b) \div (2b^2 + b + 1)$

30. $(6z^4 + 3z^2 - 9)(3z^2 - 6)^{-1}$

31. $(25y^4 - 5y^3 + 20y^2 + 2y - 10) \div (5y^2 - y + 5)$

32. **MP REASONING** A rectangular box for a new product is designed in such a way that the three dimensions always have a particular relationship defined by the variable x. The volume of the box can be written as $6x^3 + 31x^2 + 53x + 30$, and the height is always $x + 2$. What are the width and length of the box?

33. **PHYSICS** The voltage V is related to current I and power P by the equation $V = \dfrac{P}{I}$. The power of a generator is modeled by $P(t) = t^3 + 9t^2 + 26t + 24$. If the current of the generator is $I = t + 4$, write an expression that represents the voltage.

34. **ENTERTAINMENT** Your friend says that she has a card trick to show you. She has separated a deck of cards so that the deck contains no face cards and no aces. She asks you to draw a card and then tells you to multiply the number on it by 7. Next she tells you to add the sum of your number and 40 to the product you found. Last she tells you to divide by the sum of your number and 5. She guarantees that she can guess this final quotient. Explain how she is able to do this.

35. **BUSINESS** The number of subscriptions to an online game can be estimated by $n = \dfrac{3500a^2}{a^2 + 100}$, where a is the amount of money the company spent on advertising in hundreds of dollars and n is the number of subscriptions sold.

   a. Perform the division indicated by $\dfrac{3500a^2}{a^2 + 100}$.

   b. About how many subscriptions will be sold if $1500 is spent on advertising?

Simplify.

36. $(x^4 - y^4) \div (x - y)$
37. $(28c^3d^2 - 21cd^2) \div (14cd)$
38. $(a^3b^2 - a^2b + 2b)(-ab)^{-1}$
39. $\dfrac{n^3 + 3n^2 - 5n - 4}{n + 4}$
40. $\dfrac{p^3 + 2p^2 - 7p - 21}{p + 3}$
41. $\dfrac{3z^5 + 5z^4 + z + 5}{z + 2}$

42. **MULTIPLE REPRESENTATIONS** Consider a rectangle with area $2x^2 + 7x + 3$ and length $2x + 1$.

   a. **Concrete** Use algebra tiles to represent this situation. Use the model to find the width.

   b. **Symbolic** Write an expression to represent the model.

   c. **Numerical** Solve this problem algebraically using synthetic or long division. Does your concrete model check with your algebraic model?

H.O.T. Problems Use Higher-Order Thinking Skills

43. **ERROR ANALYSIS** Sharon and Jamal are dividing $2x^3 - 4x^2 + 3x - 1$ by $x - 3$. Sharon claims that the remainder is 26. Jamal argues that the remainder is -100. Is either of them correct? Explain your reasoning.

44. **CHALLENGE** If a polynomial is divided by a binomial and the remainder is 0, what does this tell you about the relationship between the binomial and the polynomial?

45. **MP REASONING** Review any of the division problems in this lesson. What is the relationship between the degrees of the dividend, the divisor, and the quotient?

46. **OPEN-ENDED** Write a quotient of two polynomials for which the remainder is 3.

47. **OPEN-ENDED** Identify the expression that does not belong with the other three. Explain your reasoning.

| $3xy + 6x^2$ | $\dfrac{5}{x^2}$ | $x + 5$ | $5b + 11c - 9ad^2$ |

48. **WRITING IN MATH** Use the information at the beginning of the lesson to write assembly instructions using the division of polynomials to make a fabric cover for your tablet.

Preparing for Assessment

49. Which expression is equivalent to $\dfrac{16m^4n^9 + 36m^8n^6 + 8m^6n^{12}}{4m^2n^3}$? **MP** 6

- A $4m^2n^3 + 9m^4n^2 + 2m^2n^4$
- B $4m^2n^6 + 9m^6n^3 + 2m^4n^9$
- C $12m^2n^3 + 32m^4n^2 + 4m^2n^4$
- D $12m^2n^6 + 32m^6n^3 + 4m^4n^9$

50. What is the remainder when you divide $\dfrac{x^4 + 3x^2 + x + 6}{x + 3}$? **MP** 6

51. Which expression is equivalent to $(m^2 - 3m + 7)(m - 5)^{-1}$? **MP** 6

- A $m - 8 + \dfrac{47}{m - 5}$
- B $m - 1 + \dfrac{2}{m - 5}$
- C $m + 2 + \dfrac{17}{m - 5}$
- D $m + 15 - \dfrac{525}{m - 5}$

52. **MULTI-STEP** Consider the expression $\dfrac{x^3 - 4x + 2}{x + 3}$. **MP** 6, 3

  a. What do you need to do differently when rewriting the numerator $x^3 - 4x + 2$ before you can divide the expression by $x + 3$ using long division?

  - A Change the order of the terms.
  - B Insert $0x^4$.
  - C Insert $0x^2$.
  - D Add a number corresponding to the remainder.

  b. Perform the long division for $\dfrac{x^3 - 4x + 2}{x + 3}$. Show all of your steps.

  c. What is the quotient?

  - A $x^2 + 3x + 13 + \dfrac{23}{x + 3}$
  - B $x^2 - 3x + 13 - \dfrac{37}{x + 3}$
  - C $x^2 - 3x + 5 - \dfrac{13}{x + 3}$
  - D $x^2 + 3x + 5 + \dfrac{17}{x + 3}$

53. Find two errors in the following long division, then state the correct quotient. **MP** 6, 3

$$\begin{array}{r} 8x - 4x \\ x^2 + 1 \overline{\smash{)}8x^3 - 4x^2 + x + 2} \\ \underline{8x^3 + 8x} \\ -4x^2 + 9x \\ \underline{-4x^2 - 9} \\ 9x + 6 \end{array}$$

54. Divide $4x^3$ by $x + 2$. Which terms are part of the quotient? Choose all that apply. **MP** 6

- A $4x^2$
- B $2x^2$
- C $-8x$
- D $4x$
- E $0x$
- F 16
- G 4

55. Perform the following division problems: **MP** 3, 6

  a. $\dfrac{4x^3 - 2x^2 - 3}{2x^2 - 1}$

  b. $\dfrac{12x^3 - 11x^2 + 5x + 18}{4x + 3}$

  c. $\dfrac{3x^3 + 4x + 11}{x^2 - 3x + 2}$

  d. $\dfrac{2x^3 + 4x^2 - 5}{x + 3}$

EXTEND 4-3
Graphing Technology Lab
Dividing Polynomials

In Lesson 4-3, you learned how to divide polynomials by divisors of degree 1 or 2. You can use a computer algebra system (CAS) to divide polynomials with any divisor.

Activity Divide Polynomials

Work cooperatively. Use CAS to find $(4x^5 - 12x^4 - 7x^3 + 32x^2 + 3x + 20) \div (x^2 - 2x + 4)$.

Step 1 Add a new **Calculator** page on the TI-Nspire.

Step 2 From the menu, select **Algebra, Polynomial Tools** and **Quotient of Polynomial**.

Step 3 Type the dividend, a comma, and the divisor.

The CAS indicates that $(4x^5 - 12x^4 - 7x^3 + 32x^2 + 3x + 20) \div (x^2 - 2x + 4)$ is $4x^3 - 4x^2 - 31x + 14$.

Determine whether there is a remainder.

Step 4 Use the **Remainder of a Polynomial** option from the **Algebra, Polynomial Tools** menu to determine the remainder. Then type the dividend, a comma, and the divisor.

The remainder is $99x + 76$.

Therefore, $(4x^5 - 12x^4 - 7x^3 + 32x^2 + 3x + 20) \div (x^2 - 2x + 4) = 4x^3 - 4x^2 - 31x - 14 - \dfrac{99x + 76}{x^2 - 2x + 4}$.

When using a computer algebra system, if the result when calculating the remainder is 0 then there is no remainder.

Exercises

Work cooperatively. Find each quotient.

1. $(2x^4 + x^3 - 8x^2 + 17x - 12) \div (x^2 + 2x - 3)$

2. $(2x^4 + 7x^3 + 8x^2 + x - 12) \div (2x^2 + 3x - 4)$

3. $(9x^5 - 9x^3 - 5x^2 + 5) \div (9x^3 - 5)$

4. $(x^5 - 8x^4 + 10x^3 + 14x^2 + 61x - 30) \div (x^2 - 5x + 3)$

5. $(2x^6 + 2x^5 - 4x^4 - 18x^3 - 16x^2 + 8x + 16) \div (2x^3 + 2x^2 - x - 2)$

6. $(6x^6 - 2x^5 - 14x^4 + 10x^3 - 4x^2 - 28x - 5) \div (3x^3 - x^2 - 7x - 1)$

7. Use synthetic division to find $(x^6 - 7x^5 - 21x^4 + 175x^3 + 56x^2 - 924x + 720) \div (x^3 - 5x^2 - 12x + 36)$.

EXPLORE 4-4

Graphing Technology
Power Functions

A **power function** is any function of the form $f(x) = ax^n$, where *a* and *n* are nonzero constant real numbers. A power function in which *n* is a positive integer is called a **monomial function**.

Activity 1 $y = ax^4$

Work cooperatively. Graph $f(x) = 2x^4$. Analyze the graph.

Step 1 Graph the function.

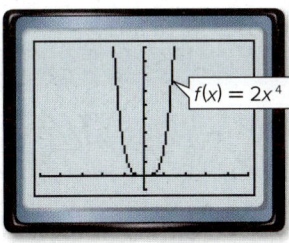

[−5, 5] scl: 1 by [−1, 9] scl: 1

Step 2 Analyze the graph.

The graph resembles the graph of $g(x) = x^2$, but it flattens out more at its vertex.

Analyze the Results

Work cooperatively.

1. What assumptions can you make about the graph of $f(x) = 2x^4$ as *x* becomes more positive or more negative? Use the **TABLE** feature to evaluate the function for values of *x* to confirm this assumption, if necessary.

2. State the domain and range of $f(x) = 2x^4$. Compare these values with the domain and range of $g(x) = x^2$.

3. What other characteristics does the graph of $f(x) = 2x^4$ share with the graph of $g(x) = x^2$?

Activity 2 $y = ax^5$

Work cooperatively. Graph $f(x) = 3x^5$. Analyze the graph.

Step 1 Graph the function.

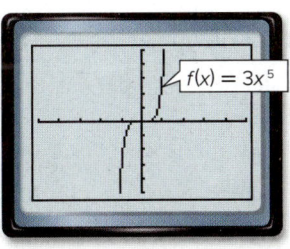

[−5, 5] scl: 1 by [−5, 5] scl: 1

Step 2 Analyze the graph.

The graph resembles the graph of $f(x) = x^3$, but it flattens out more as the graph approaches the origin.

connectED.mcgraw-hill.com 251

EXPLORE 4-4

Graphing Technology
Power Functions Continued

Analyze the Results
Work cooperatively.

4. What assumptions can you make about the graph of $f(x) = 3x^5$ as x becomes more positive or more negative? Use the **TABLE** feature to evaluate the function for values of x to confirm this assumption, if necessary.

5. State the domain and range of $f(x) = 3x^5$. Compare these values with the domain and range of $g(x) = x^3$.

6. What other characteristics does the graph of $f(x) = 3x^5$ share with the graph of $g(x) = x^3$?

7. Compare the characteristics of the graph $h(x) = -3x^5$ with the graph of $f(x) = 3x^5$. What conclusions can you make about $f(x) = ax^5$ when a is positive and when a is negative?

Activity 3 $y = x^n$, where n is even

Work cooperatively. Graph $f(x) = x^2$, $g(x) = x^4$, and $h(x) = x^6$ on the same screen. Analyze the graphs.

Step 1 Graph the function.

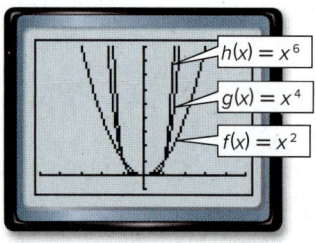 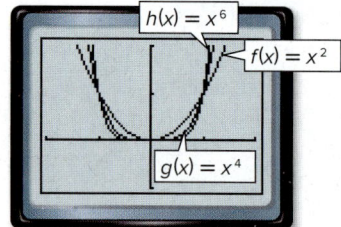

[−5, 5] scl: 1 by [−1, 9] scl: 1   [−2, 2] scl: 1 by [−1, 2] scl: 1

Step 2 The graphs are all U-shaped, but the widths are different. The graphs also differ around the origin. If you zoom in around the origin, you can see this more clearly. You can do this by using the **ZOOM** feature or by adjusting the window manually.

Analyze the Results
Work cooperatively.

8. What assumptions can you make about the graphs of $a(x) = x^8$, $b(x) = x^{10}$, and so on?

9. Identify the common characteristics of the graphs of power functions in which the power is an even number.

10. Graph $f(x) = x^3$, $g(x) = x^5$, and $h(x) = x^7$ on the same graph.

11. What assumptions can you make about the graphs of $a(x) = x^9$, $b(x) = x^{11}$, and so on?

12. Identify the common characteristics of the graphs of power functions in which the power is an odd number.

13. Graph $f(x) = x^4$ and $g(x) = -x^4$ on the same graph.

14. Graph $f(x) = x^5$ and $g(x) = -x^5$ on the same graph.

15. What assumptions can you make about the effects of a negative value of a in $f(x) = ax^n$ when n is even? when n is odd?

LESSON 4
Graphing Polynomial Functions

Then
- You analyzed graphs of quadratic functions.

Now
1. Evaluate polynomial functions.
2. Identify general shapes of graphs of polynomial functions.

Why?
- The volume of air in the lungs during a 5-second respiratory cycle can be modeled by $v(t) = -0.037t^3 + 0.152t^2 + 0.173t$, where v is the volume in liters and t is the time in seconds. This model is an example of a polynomial function.

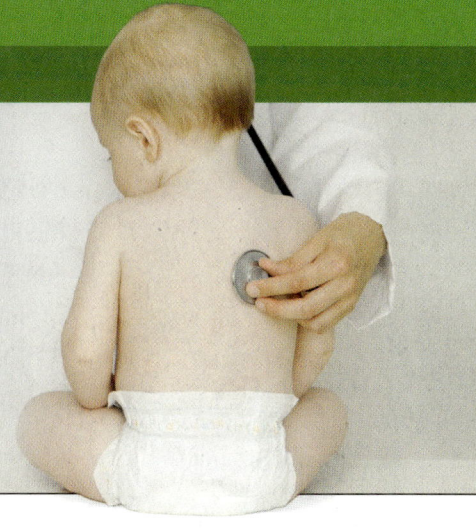

New Vocabulary
polynomial in one variable
leading coefficient
polynomial function
power function
quartic function
quintic function

Mathematical Practices
1 Make sense of problems and persevere in solving them.

1 Polynomial Functions
A **polynomial in one variable** is an expression of the form $a_n x^n + a_{n-1} x^{n-1} + \cdots + a_2 x^2 + a_1 x + a_0$, where $a_n \neq 0$, a_{n-1}, a_2, a_1, and a_0 are real numbers, and n is a nonnegative integer.

The polynomial is written in standard form when the values of the exponents are in descending order. The degree of the polynomial is the value of the greatest exponent. The coefficient of the first term of a polynomial in standard form is called the **leading coefficient**.

Polynomial	Expression	Degree	Leading Coefficient
Constant	12	0	12
Linear	$4x - 9$	1	4
Quadratic	$5x^2 - 6x - 9$	2	5
Cubic	$8x^3 + 12x^2 - 3x + 1$	3	8
General	$a_n x^n + a_{n-1} x^{n-1} + \cdots + a_1 x + a_0$	n	a_n

Example 1 Degrees and Leading Coefficients

State the degree and leading coefficient of each polynomial in one variable. If it is not a polynomial in one variable, explain why.

a. $8x^5 - 4x^3 + 2x^2 - x - 3$

This is a polynomial in one variable. The greatest exponent is 5, so the degree is 5 and the leading coefficient is 8.

b. $12x^2 - 3xy + 8x$

This is not a polynomial in one variable. There are two variables, x and y.

c. $3x^4 + 6x^3 - 4x^8 + 2x$

This is a polynomial in one variable. The greatest exponent is 8, so the degree is 8 and the leading coefficient is -4.

Guided Practice

1A. $5x^3 - 4x^2 - 8x + \dfrac{4}{x}$ **1B.** $5x^6 - 3x^4 + 12x^3 - 14$ **1C.** $8x^4 - 2x^3 - x^6 + 3$

A **polynomial function** is a continuous function that can be described by a polynomial equation in one variable. For example, $f(x) = 3x^3 - 4x + 6$ is a cubic polynomial function. The simplest polynomial functions of the form $f(x) = ax^b$ where a and b are nonzero real numbers are called **power functions**.

If you know an element in the domain of any polynomial function, you can find the corresponding value in the range by inserting the known value and calculating the function.

Real-World Link
Total lung capacity is approximately 6 liters in a healthy young adult.
Source: Family Practice Notebook

Real-World Example 2 Evaluate a Polynomial Function

RESPIRATION Refer to the beginning of the lesson. Find the volume of air in the lungs 2 seconds into the respiratory cycle.

By substituting 2 into the function we can find $v(2)$, the volume of air in the lungs 2 seconds into the respiratory cycle.

$v(t) = -0.037t^3 + 0.152t^2 + 0.173t$ Original function
$v(2) = -0.037(2)^3 + 0.152(2)^2 + 0.173(2)$ Replace t with 2.
$\quad = -0.296 + 0.608 + 0.346$ Simplify.
$\quad = 0.658$ L   Add.

▶ **Guided Practice**

2. Find the volume of air in the lungs 4 seconds into the respiratory cycle.

You can also evaluate functions for variables and algebraic expressions.

Example 3 Function Values of Variables

Find $f(3c - 4) - 5f(c)$ if $f(x) = x^2 + 2x - 3$.

To evaluate $f(3c - 4)$, replace the x in $f(x)$ with $3c - 4$.

$f(x) = x^2 + 2x - 3$ Original function
$f(3c - 4) = (3c - 4)^2 + 2(3c - 4) - 3$ Replace x with $3c - 4$.
$\quad = 9c^2 - 24c + 16 + 6c - 8 - 3$ Multiply.
$\quad = 9c^2 - 18c + 5$ Simplify.

To evaluate $5f(c)$, replace x with c in $f(x)$, then multiply by 5.

$f(x) = x^2 + 2x - 3$ Original function
$5f(c) = 5(c^2 + 2c - 3)$ Replace x with c.
$\quad = 5c^2 + 10c - 15$ Distributive Property

Now evaluate $f(3c - 4) - 5f(c)$.

$f(3c - 4) - 5f(c) = (9c^2 - 18c + 5) - (5c^2 + 10c - 15)$
$\quad = 9c^2 - 18c + 5 - 5c^2 - 10c + 15$ Distribute.
$\quad = 4c^2 - 28c + 20$ Simplify.

▶ **Guided Practice**

3A. Find $g(5a - 2) + 3g(2a)$ if $g(x) = x^2 - 5x + 8$.

3B. Find $h(-4d + 3) - 0.5h(d)$ if $h(x) = 2x^2 + 5x + 3$.

Go Online!

Investigate how changing the values in a power function affects the graph by using the **Graphing Tools** in ConnectED.

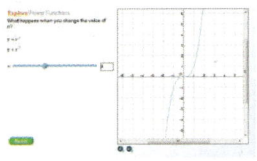

2 Graphs of Polynomial Functions The general shapes of the graphs of several polynomial functions show the *maximum* number of times the graph of each function may intersect the x-axis. This is the same number as the degree of the polynomial.

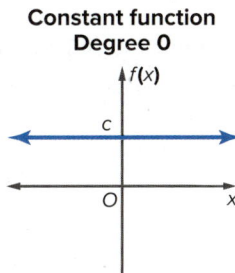
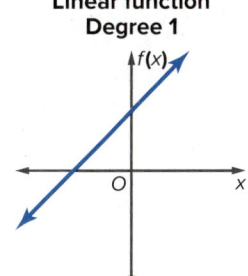
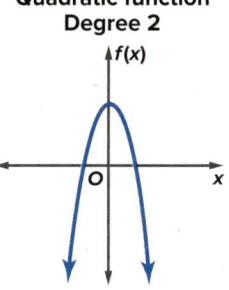
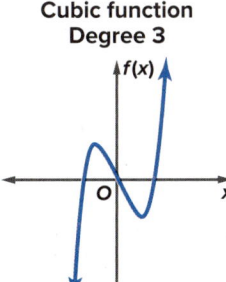
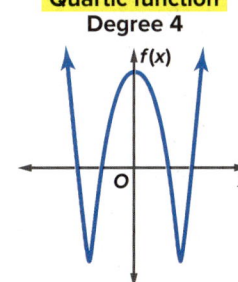
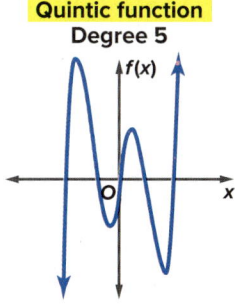

The domain of any polynomial function is all real numbers. The end behavior is the behavior of the graph of $f(x)$ as x approaches positive infinity ($x \to +\infty$) or negative infinity ($x \to -\infty$). The degree and leading coefficient of a polynomial function determine the end behavior of the graph and the range of the function.

Study Tip

MP Sense-Making The leading coefficient and degree are the sole determining factors for the end behavior of a polynomial function. With very large or very small numbers, the rest of the polynomial is insignificant in the appearance of the graph.

Key Concept End Behavior of a Polynomial Function

Degree: even
Leading Coefficient: positive
End Behavior:

$f(x) \to +\infty$
as $x \to -\infty$

$f(x) \to +\infty$
as $x \to +\infty$

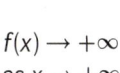

Domain: all real numbers
Range: all real numbers ≥ minimum

Degree: odd
Leading Coefficient: positive
End Behavior:

$f(x) \to -\infty$
as $x \to -\infty$

$f(x) \to +\infty$
as $x \to +\infty$

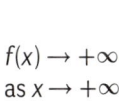

Domain: all real numbers
Range: all real numbers

Degree: even
Leading Coefficient: negative
End Behavior:

$f(x) \to -\infty$
as $x \to -\infty$

$f(x) \to -\infty$
as $x \to +\infty$

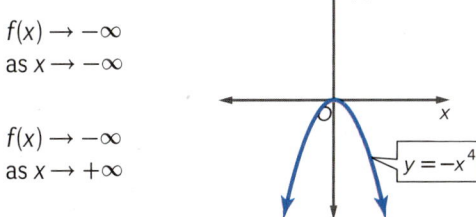

Domain: all real numbers
Range: all real numbers ≤ maximum

Degree: odd
Leading Coefficient: negative
End Behavior:

$f(x) \to +\infty$
as $x \to -\infty$

$f(x) \to -\infty$
as $x \to +\infty$

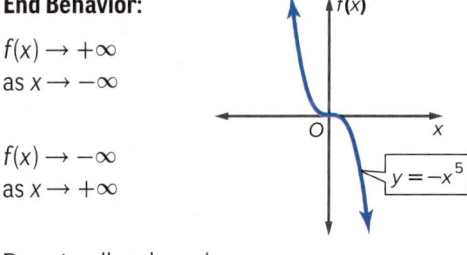

Domain: all real numbers
Range: all real numbers

Review Vocabulary

infinity endless or boundless

connectED.mcgraw-hill.com   255

Review Vocabulary

zero the *x*-coordinate of the point at which a graph intersects the *x*-axis

The number of real zeros of a polynomial function can be determined by examining its graph. Recall that real zeros occur at *x*-intercepts, so the number of times a graph intersects the *x*-axis equals the number of real zeros.

> **Key Concept** Zeros of Even- and Odd-Degree Functions
>
> Odd-degree functions will always have at least one real zero. Even-degree functions may have any number of real zeros or no real zeros at all.

Even-Degree Polynomials

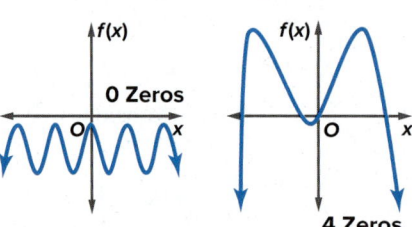

Odd-Degree Polynomials

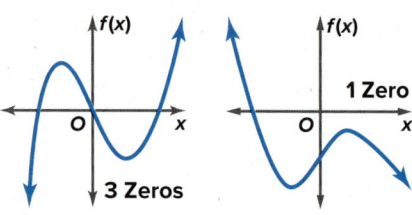

Study Tip

Double zero When a graph is tangent to the *x*-axis, the *x*-intercept at the point of tangency is called a double zero.

Example 4 Graphs of Polynomial Functions

For each graph,
- describe the end behavior,
- determine whether it represents an odd-degree or an even-degree polynomial function, and
- state the number of real zeros.

a.

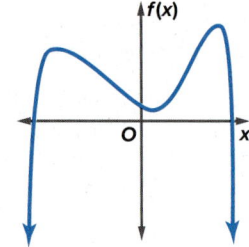

$f(x) \to -\infty$ as $x \to -\infty$.
$f(x) \to -\infty$ as $x \to +\infty$.

Because the end behavior is in the same direction, it is an even-degree function. The graph intersects the *x*-axis at two points, so there are two real zeros.

b.

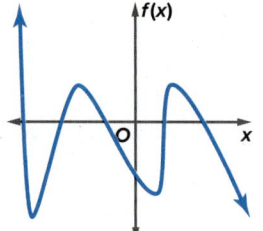

$f(x) \to +\infty$ as $x \to -\infty$.
$f(x) \to -\infty$ as $x \to +\infty$.

Because the end behavior is in opposite directions, it is an odd-degree function. The graph intersects the *x*-axis at five points, so there are five real zeros.

▶ **Guided Practice**

4A.

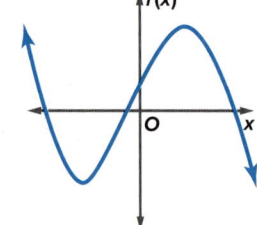

4B.

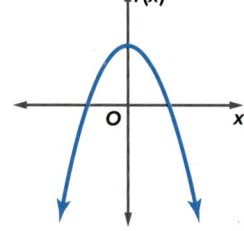

To graph a polynomial function, make a table of values to find several points. Then connect the points to make a smooth, continuous curve. Use what you know about the end behavior or zeros to help sketch the graph.

Study Tip

MP Precision For linear functions, two points are enough. For higher order polynomial functions, you may need many more points to locate zeros and determine the shape of the graph.

Example 5 Graph Polynomial Functions Using a Table of Values

Graph $f(x) = -x^4 + x^3 + 3x^2 + 2x$ by making a table of values.

Make a table of values.

x	f(x)	x	f(x)
−2.5	≈ −41	0.5	≈ 1.8
−2.0	−16	1.0	5.0
−1.5	≈ −4.7	1.5	≈ 8.1
−1.0	−1.0	2.0	8.0
−0.5	≈ −0.4	2.5	≈ 0.3
0.0	0.0	3.0	−21

Choose values for x and substitute in the function rule to find the corresponding values for y.

Choose negative and positive values for x as well as 0. Choose decimal values to help make a more accurate sketch.

Plot the points and sketch the graph.

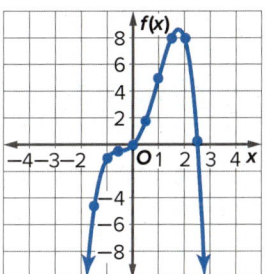

Start with the zeros. There is one zero at $x = 0$.

The function changes from positive to negative between 2.5 and 3, so there is a zero between $x = 2.5$ and $x = 3$.

Plot the known points and connect them with a smooth curve.

Then use what you know about the end behaviour to help you check your graph.

This is an even-degree polynomial with a negative leading coefficient, so $f(x) \to -\infty$ as $x \to -\infty$ and $f(x) \to -\infty$ as $x \to +\infty$. Notice that the graph intersects the x-axis at two points, indicating there are two zeros for this function.

▶ **Guided Practice**

5. Graph $f(x) = x^4 - x^3 - 2x^2 + 4x - 6$ by making a table of values.

Check Your Understanding   = Step-by-Step Solutions begin on page R11.

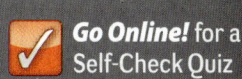
Go Online! for a Self-Check Quiz

Example 1 State the degree and leading coefficient of each polynomial in one variable. If it is not a polynomial in one variable, explain why.

1. $11x^6 - 5x^5 + 4x^2$
2. $-10x^7 - 5x^3 + 4x - 22$
3. $14x^4 - 9x^3 + 3x - 4y$
4. $8x^5 - 3x^2 + 4xy - 5$

Example 2 Find $w(5)$ and $w(-4)$ for each function.

5. $w(x) = -2x^3 + 3x - 12$
6. $w(x) = 2x^4 - 5x^3 + 3x^2 - 2x + 8$

connectED.mcgraw-hill.com 257

Example 3 If $c(x) = 4x^3 - 5x^2 + 2$ and $d(x) = 3x^2 + 6x - 10$, find each value.

7. $c(y^3)$
8. $-4[d(3z)]$
9. $6c(4a) + 2d(3a - 5)$
10. $-3c(2b) + 6d(4b - 3)$

Example 4 For each graph,
a. describe the end behavior,
b. determine whether it represents an odd-degree or an even-degree function, and
c. state the number of real zeros.

11.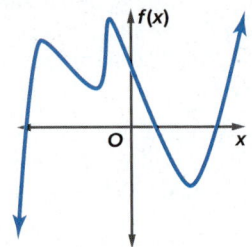

12.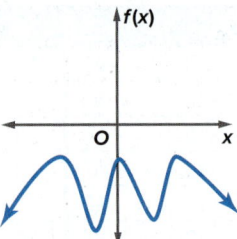

Example 5 Graph the following functions by making a table of values.

13. $f(x) = x^5 - 3x^4 + 9x^3 - x^2 + 2$
14. $f(x) = -9x^4 + 2x^3 + x + 1$

Practice and Problem Solving

Extra Practice is on page R4.

Example 1 **PERSEVERANCE** State the degree and leading coefficient of each polynomial in one variable. If it is not a polynomial in one variable, explain why.

15. $8x^5 - 12x^6 + 14x^3 - 9$
16. $-12 - 8x^2 + 5x - 21x^7$
17. $15x - 4x^3 + 3x^2 - 5x^4$
18. $13b^3 - 9b + 3b^5 - 18$
19. $(d + 5)(3d - 4)$
20. $(5 - 2y)(4 + 3y)$
21. $6x^5 - 5x^4 + 2x^9 - 3x^2$
22. $7x^4 + 3x^7 - 2x^8 + 7$

Example 2 Find $p(-6)$ and $p(3)$ for each function.

23. $p(x) = x^4 - 2x^2 + 3$
24. $p(x) = -3x^3 - 2x^2 + 4x - 6$
25. $p(x) = 2x^3 + 6x^2 - 10x$
26. $p(x) = x^4 - 4x^3 + 3x^2 - 5x + 24$
27. $p(x) = -x^3 + 3x^2 - 5$
28. $p(x) = 2x^4 + x^3 - 4x^2$

Example 3 If $c(x) = 2x^2 - 4x + 3$ and $d(x) = -x^3 + x + 1$, find each value.

29. $c(3a)$
30. $5d(2a)$
31. $c(b^2)$
32. $d(4a^2)$
33. $d(4y - 3)$
34. $c(y^2 - 1)$

Example 4 For each graph,
a. describe the end behavior,
b. determine whether it represents an odd-degree or an even-degree function, and
c. state the number of real zeros.

35.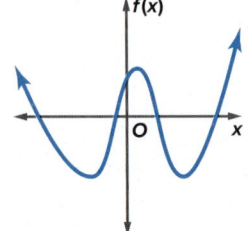

36.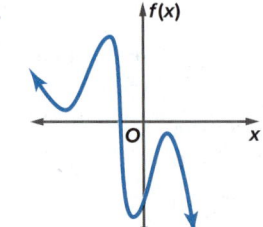

37.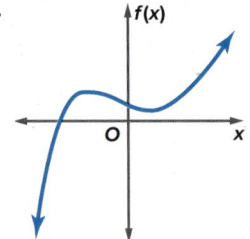

258 | Lesson 4-4 | Graphing Polynomial Functions

38. 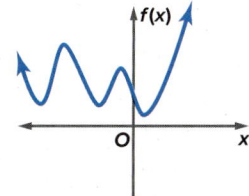   **39.** 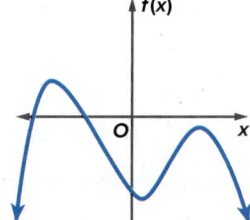   **40.**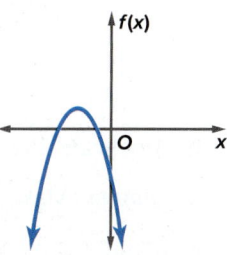

41. PHYSICS In physics, the formula for the amount of energy stored in an inductor is $W = 0.5LI^2$, where W represents the energy stored in joules (J), L represents the inductance in henries (H), and I represents the current in amperes (A). Find the energy stored in an inductor of 24 henries with a current of 8 amperes.

42. MODELING A microwave manufacturing firm has determined that their profit function is $P(x) = -0.0014x^3 + 0.3x^2 + 6x - 355$, where x is the number of microwaves sold annually.

    a. Graph the profit function using a calculator.

    b. Determine a reasonable viewing window for the function.

    c. Approximate all of the zeros of the function using the **CALC** menu.

    d. What must be the range of microwaves sold in order for the firm to have a profit?

Find $p(-2)$ and $p(8)$ for each function.

43. $p(x) = \frac{1}{4}x^4 + \frac{1}{2}x^3 - 4x^2$ **44.** $p(x) = \frac{1}{8}x^4 - \frac{3}{2}x^3 + 12x - 18$

45. $p(x) = \frac{3}{4}x^4 - \frac{1}{8}x^2 + 6x$ **46.** $p(x) = \frac{5}{8}x^3 - \frac{1}{2}x^2 + \frac{3}{4}x + 10$

Use the degree and end behavior to match each polynomial to its graph.

A 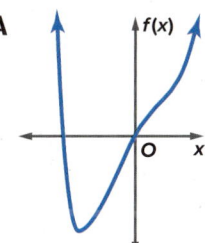   **B** 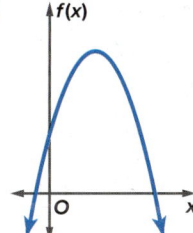   **C** 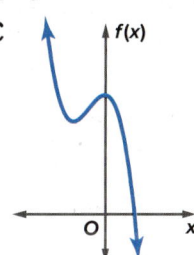   **D**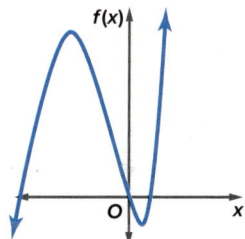

47. $f(x) = x^3 + 3x^2 - 4x$ **48.** $f(x) = -2x^2 + 8x + 5$

49. $f(x) = x^4 - 3x^2 + 6x$ **50.** $f(x) = -4x^3 - 4x^2 + 8$

If $c(x) = x^3 - 2x$ and $d(x) = 4x^2 - 6x + 8$, find each value.

51. $3c(a - 4) + 3d(a + 5)$ **52.** $-2d(2a + 3) - 4c(a^2 + 1)$

53. $5c(a^2) - 8d(6 - 3a)$ **54.** $-7d(a^3) + 6c(a^4 + 1)$

55. BUSINESS A clothing manufacturer's profitability can be modeled by $p(x) = -x4 + 40x2 - 144$, where x is the number of items sold in thousands and $p(x)$ is the company's profit in thousands of dollars.

    a. Use a table of values to sketch the function.

    b. Determine the zeros of the function.

    c. Between what two values should the company sell in order to be profitable?

    d. Explain why only two of the zeros are considered in part **c**.

56. **MULTIPLE REPRESENTATIONS** Consider $g(x) = (x - 2)(x + 1)(x - 3)(x + 4)$.

   a. **Analytical** Determine the *x*- and *y*-intercepts, roots, degree, and end behavior of $g(x)$.

   b. **Algebraic** Write the function in standard form

   c. **Tabular** Make a table of values for the function.

   d. **Graphical** Sketch a graph of the function by plotting points and connecting them with a smooth curve. Be sure to choose an appropriate scale on each axis and show the end behavior of the function in your graph.

Describe the end behavior of the graph of each function.

57. $f(x) = -5x^4 + 3x^2$

58. $g(x) = 2x^5 + 6x^4$

59. $h(x) = -4x^7 + 8x^6 - 4x$

60. $f(x) = 6x - 7x^2$

61. $g(x) = 8x^4 + 5x^5$

62. $h(x) = 9x^6 - 5x^7 + 3x^2$

H.O.T. Problems   Use Higher-Order Thinking Skills

63. **ERROR ANALYSIS** Shenequa and Virginia are determining the number of real zeros of the graph at the right. Is either of them correct? Explain your reasoning.

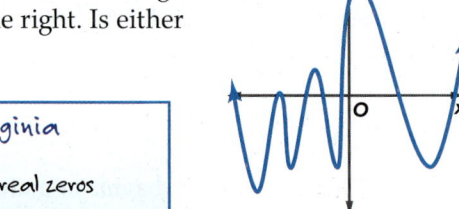

Shenequa
There are 7 real zeros because the graph intersects the x-axis 7 times.

Virginia
There are 8 real zeros because the graph intersects the x-axis 7 times, and there is a double zero.

64. **CHALLENGE** Suppose $f(x)$ is a polynomial function. What is the degree of that polynomial function?

x	−24	−18	−12	−6	0	6	12	18	24
f(x)	−8	−1	3	−2	4	7	−1	−8	5

$g(x) = x^4 + x^3 - 13x^2 + x + 4$

65. **CHALLENGE** If $f(x)$ has a degree of 5 and a positive leading coefficient and $g(x)$ has a degree of 3 and a positive leading coefficient, determine the end behavior of $\frac{f(x)}{g(x)}$. Explain your reasoning.

66. **OPEN-ENDED** Sketch the graph of an even-degree polynomial with 7 real zeros, one of them a double zero.

67. **REASONING** Determine whether the following statement is *always*, *sometimes*, or *never* true. Explain.

   A polynomial function that has four real zeros is a fourth-degree polynomial.

68. **WRITING IN MATH** Describe what the end behavior of a polynomial function is and how to determine it.

Preparing for Assessment

69. What is $4h(2y^2)$ if $h(x) = -2 - x^2 + 5x$? **MP 6**

- A $-8 - 4y^4 + 20y^2$
- B $-16 - 8y^4 + 40y^2$
- C $-8 - 16y^4 + 40y^2$
- D $-16y - 8y^4 + 40y^3$
- E $-16y - 16y^4 + 40y^3$

70. The value of which function approaches ∞ as x approaches $-\infty$ and ∞? **MP 2**

- A $f(x) = -5x^4 + x^2$
- B $g(x) = 2x^5 + 6x^4$
- C $h(x) = 4x^5 - 8x + 5$
- D $j(x) = 7x^4 - 2x^3 + 5x^2$

71. Which function has the same end behavior as $g(x)$ pictured below? **MP 2**

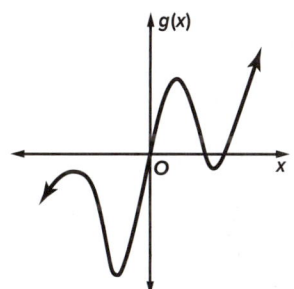

- A $f(x) = 2x^3 + x$
- B $f(x) = 5x^4 + 2x^3 + x$
- C $f(x) = -8x^5 + 5x^4 + 2x^3 + x$
- D $f(x) = -10x^6 - 8x^5 + 5x^4 + 2x^3 + x$

72. Describe the end behavior of the following functions: **MP 2**

a. $f(x) = 3x^7 - 5x + 3$

b. $f(x) = -3x^7 - 5x + 3$

73. MULTI-STEP Consider the function $g(x) = 2 - 3x^3$. **MP 2, 6**

a. How is the function different from the parent function, $g(x) = x^3$? Choose all answers that apply.

- ☐ A It is flipped upside down.
- ☐ B It is flipped right to left.
- ☐ C It is moved two units up.
- ☐ D It is moved two units down.
- ☐ E It is three times wider.
- ☐ F It is three times narrower.

b. Which of the following options describes the end behavior of $g(x) = 2 - 3x^3$?

- A As $x \to \infty$, $g(x) \to \infty$, and as $x \to -\infty$, $g(x) \to \infty$.
- B As $x \to \infty$, $g(x) \to -\infty$, and as $x \to -\infty$, $g(x) \to \infty$.
- C As $x \to \infty$, $g(x) \to \infty$, and as $x \to -\infty$, $g(x) \to -\infty$.
- D As $x \to \infty$, $g(x) \to -\infty$, and as $x \to -\infty$, $g(x) \to -\infty$.

c. What is the y-intercept of this function?

d. What is the value of the function when $x = i$?

74. Which of the following graphs fits the end behavior described below: **MP 2**

As $x \to \infty$, $f(x) \to \infty$ and as $x \to -\infty$, $f(x) \to \infty$

A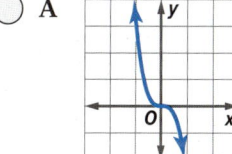

C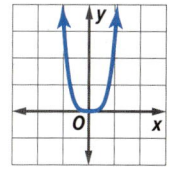

B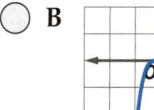

D

LESSON 5
Analyzing Graphs of Polynomial Functions

∷Then
- You used maxima and minima and graphs of polynomials.

∷Now
1. Graph polynomial functions and locate their zeros.
2. Find the relative maxima and minima of polynomial functions.

∷Why?
- Annual attendance at the movies has fluctuated since the first movie theater opened in 1906. Overall movie attendance peaked during the 1920s, and it was at its lowest during the 1970s. A graph of the annual attendance to the movies can be represented by a polynomial function.

 New Vocabulary
Location Principle

 Mathematical Practices
3 Construct viable arguments and critique the reasoning of others.

1 Location Principle
Knowing where to expect zeros is essential when graphing a polynomial. **The Location Principle** states that if the value of $f(x)$ changes signs from one value of x to the next, then there is a zero between those two x-values.

In this graph, one of the zeros occurs at $x = 0$. Another zero occurs between $x = 2.5$ and $x = 3.0$. Because $f(x)$ is positive for $x = 2.5$ and negative for $x = 3.0$ and all polynomial functions are continuous, we know there is a zero between these two values.

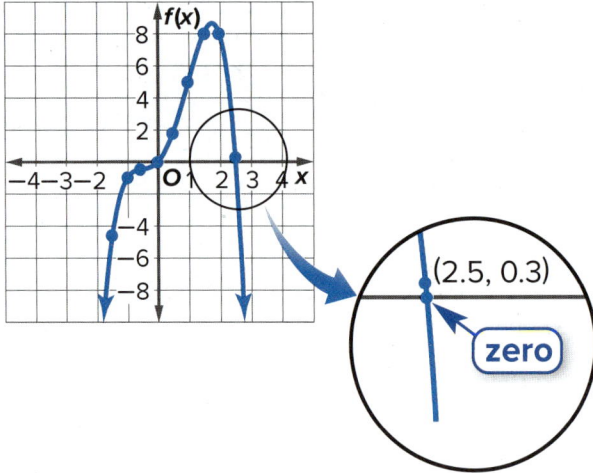

Key Concept Location Principle

Words Suppose $y = f(x)$ represents a polynomial function and a and b are two real numbers such that $f(a) < 0$ and $f(b) > 0$. Then the function has at least one real zero between a and b.

Model

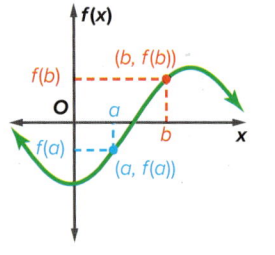

262 | Lesson 4-5

Go Online!

The Location Principle guides you as you find the zeros of functions. Watch an **animation** of the Location Principle in ConnectED.

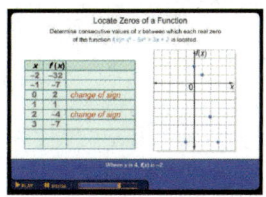

Example 1   Locate Zeros of a Function

Determine consecutive integer values of x between which each real zero of $f(x) = x^3 - 4x^2 + 3x + 1$ is located. Then draw the graph.

Make a table of values. Because $f(x)$ is a third-degree polynomial function, it will have either 3 or 1 real zeros. Look at the values of $f(x)$ to locate the zeros. Then use the points to sketch a graph of the function.

x	$f(x)$
-2	-29
-1	-7
0	1
1	1
2	-1
3	1
4	13

← change in sign (between 0 and −1)
← change in sign (between 2 and 1)
← change in sign (between 3 and 2)

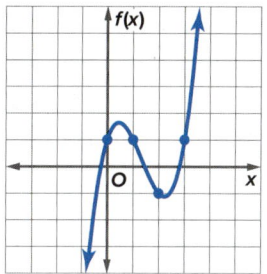

The changes in sign indicate that there are zeros between $x = -1$ and $x = 0$, between $x = 1$ and $x = 2$, and between $x = 2$ and $x = 3$.

Guided Practice

1. Determine consecutive integer values of x between which each real zero of the function $f(x) = x^4 - 3x^3 - 2x^2 + x + 1$ is located. Then draw the graph.

Study Tip

Degree Recall that the degree of the function is also the maximum number of zeros the function can have.

2 Maximum and Minimum Points

The graph below shows the general shape of a third-degree polynomial function.

Point A on the graph is a relative maximum of the function since no other nearby points have a greater y-coordinate. The graph is increasing as it approaches A and decreasing as it moves from A.

Likewise, point B is a relative minimum because no other nearby points have a lesser y-coordinate. The graph is decreasing as it approaches B and increasing as it moves from B. The maximum and minimum values of a function are called the extrema.

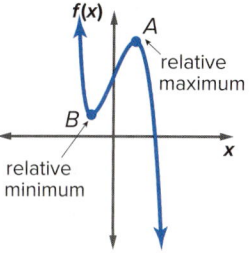

These points are often referred to as turning points. The graph of a polynomial function of degree n has at most $n - 1$ turning points.

Study Tip

Odd Functions Some odd functions, like $f(x) = x^3$, have no turning points.

Example 2   Maximum and Minimum Points

Graph $f(x) = x^3 - 4x^2 - 2x + 3$. Estimate the x-coordinates at which the relative maxima and relative minima occur.

Make a table of values and graph the function.

x	$f(x)$
-2	-17
-1	0
0	3
1	-2
2	-9
3	-12
4	-4
5	18

← zero
← indicates a relative maximum
← indicates a relative minimum
← zero between 4 and 5

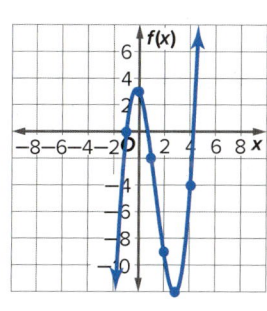

Study Tip

Maximum and Minimum A polynomial with a degree greater than 3 may have more than one relative maximum or relative minimum.

Study Tip

Rational Values Zeros and turning points will not always occur at integral values of x.

Look at the table of values and the graph.

The value of $f(x)$ changes signs between $x = 4$ and $x = 5$, indicating a zero of the function.

The value of $f(x)$ at $x = 0$ is greater than the surrounding points, so there must be a relative maximum *near* $x = 0$.

The value of $f(x)$ at $x = 3$ is less than the surrounding points, so there must be a relative minimum *near* $x = 3$.

CHECK You can use a graphing calculator to find the relative maximum and relative minimum of a function and confirm your estimates.

Enter $y = x^3 - 4x^2 - 2x + 3$ in the **Y=** list and graph the function.

Use the **CALC** menu to find each maximum and minimum.

When selecting the left bound, move the cursor to the left of the maximum or minimum. When selecting the right bound, move the cursor to the right of the maximum or minimum.

Press ENTER twice.

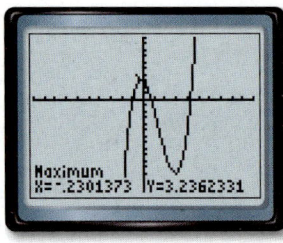
[−10, 10] scl: 1 by [−15, 10] scl: 1

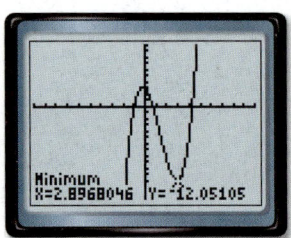
[−10, 10] scl: 1 by [−15, 10] scl: 1

The estimates for a relative maximum near $x = 0$ and a relative minimum near $x = 3$ are accurate.

▶ **Guided Practice**

2. Graph $f(x) = 2x^3 + x^2 - 4x - 2$. Estimate the *x*-coordinates at which the relative maxima and relative minima occur.

The graph of a polynomial function can reveal trends in real-world data. It is often helpful to note when the graph is increasing or decreasing.

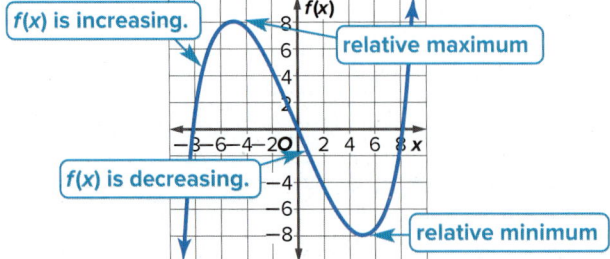

264 | Lesson 4-5 | Analyzing Graphs of Polynomial Functions

Real World Example 3 Graph a Polynomial Model

MOVIES Annual admissions to movies in the United States can be modeled by the function $f(x) = 0.006x^4 - 0.0896x^3 + 0.37x^2 - 0.746x + 8.95$, where x is the number of years since 2005 and $f(x)$ is the annual revenue in billions.

a. Graph the function.

Make a table of values for the years 2005–2013. Plot the points and connect with a smooth curve. Finding and plotting the points for every year gives a good approximation of the graph.

x	f(x)
0	8.95
1	8.50
2	8.33
3	8.13
4	7.71
5	7.05
6	6.25
7	5.57
8	5.41
9	6.31

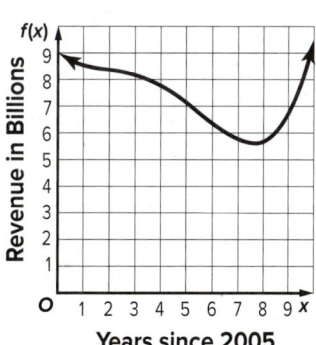

Real-World Link

Moviegoers in the United States bought a record $10.8 billion in movie tickets in 2012, with the number of tickets sold rising for the first time in three years.

Source: CNN

b. Describe the turning points of the graph and its end behavior.

There are relative maxima near 2008 and 2013 and a relative minimum between 2012 and 2013. $f(x) \to \infty$ as $x \to -\infty$ and $f(x) \to \infty$ as $x \to \infty$.

c. What trends in revenue does the graph suggest?

Movie attendance peaked around 2008 and declined until about 2012. It then increased.

d. Is it reasonable that the trend will continue indefinitely?

This trend may continue for a couple of years, but the graph will soon become unreasonable as it predicts positive attendance for the future.

▶ Guided Practice

3. COLLEGE The percent of high school graduates who enroll in college can be modeled by $f(x) = 0.015x^4 - 0.270x^3 + 1.432x^2 - 1.429x + 62.853$, where x is the number of years since 2000 and $f(x)$ is the annual percent of graduates who enroll in college.

A Graph the function.
B Describe the turning points of the graph and its end behavior.
C What trends does the graph suggest?
D Is it reasonable that the trend will continue indefinitely?

Check Your Understanding

 = Step-by-Step Solutions begin on page R11.

Example 1 Determine the consecutive integer values of x between which each real zero of each function is located. Then draw the graph.

1. $f(x) = x^3 - 2x^2 + 5$
2. $f(x) = -x^4 + x^3 + 2x^2 + x + 1$
3. $f(x) = -3x^4 + 5x^3 + 4x^2 + 4x - 8$
4. $f(x) = 2x^4 - x^3 - 3x^2 + 2x - 4$

Example 2 Graph each polynomial function. Estimate the x-coordinates at which the relative maxima and relative minima occur. State the domain and range for each function.

5. $f(x) = x^3 + x^2 - 6x - 3$
6. $f(x) = 3x^3 - 6x^2 - 2x + 2$
7. $f(x) = -x^3 + 4x^2 - 2x - 1$
8. $f(x) = -x^3 + 2x^2 - 3x + 4$

Example 3

9. **SENSE-MAKING** Annual compact disc sales can be modeled by the quartic function $f(x) = 0.48x^4 - 9.6x^3 + 53x^2 - 49x + 599$, where x is the number of years after 1995 and $f(x)$ is annual sales in millions.

   a. Graph the function for $0 \leq x \leq 10$.

   b. Describe the turning points of the graph, its end behavior, and the intervals on which the graph is increasing or decreasing.

   c. Continue the graph for $x = 11$ and $x = 12$. What trends in compact disc sales does the graph suggest?

   d. Is it reasonable that the trend will continue indefinitely? Explain.

Practice and Problem Solving

Extra Practice is on page R4.

Examples 1–2 Complete each of the following.

a. Graph each function by making a table of values.

b. Determine the consecutive integer values of x between which each real zero is located.

c. Estimate the x-coordinates at which the relative maxima and minima occur.

10. $f(x) = x^3 + 3x^2$
11. $f(x) = -x^3 + 2x^2 - 4$
12. $f(x) = x^3 + 4x^2 - 5x$
13. $f(x) = x^3 - 5x^2 + 3x + 1$
14. $f(x) = -2x^3 + 12x^2 - 8x$
15. $f(x) = 2x^3 - 4x^2 - 3x + 4$
16. $f(x) = x^4 + 2x - 1$
17. $f(x) = x^4 + 8x^2 - 12$

Example 3

18. **FINANCIAL LITERACY** The average annual price of crude oil can be modeled by the cubic function $f(x) = 0.2728x^3 - 2.612x^2 + 7.02x + 89.18$, where x is the number of years after Gena began driving and $f(x)$ is the price in dollars.

   a. Graph the function for $0 \leq x \leq 10$.

   b. Describe the turning points of the graph and its end behavior.

   c. What trends in crude oil prices does the graph suggest?

   d. Is it reasonable that the trend will continue indefinitely? Explain.

Use a graphing calculator to estimate the x-coordinates at which the maxima and minima of each function occur. Round to the nearest hundredth.

19. $f(x) = x^3 + 3x^2 - 6x - 6$
20. $f(x) = -2x^3 + 4x^2 - 5x + 8$
21. $f(x) = -2x^4 + 5x^3 - 4x^2 + 3x - 7$
22. $f(x) = x^5 - 4x^3 + 3x^2 - 8x - 6$

Sketch the graph of polynomial functions with the following characteristics.

23. an odd function with zeros at $-5, -3, 0, 2$ and 4

24. an even function with zeros at $-2, 1, 3$, and 5

25. a 4-degree function with a zero at -3, maximum at $x = 2$, and minimum at $x = -1$

26. a 5-degree function with zeros at $-4, -1$, and 3, maximum at $x = -2$

27. an odd function with zeros at $-1, 2$, and 5 and a negative leading coefficient

28. an even function with a minimum at $x = 3$ and a positive leading coefficient

29. DIVING The deflection d of a 10-foot-long diving board can be calculated using the function $d(x) = 0.015x^2 - 0.0005x^3$, where x is the distance between the diver and the stationary end of the board in feet.

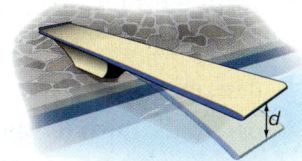

a. Complete a table of values for the function $d(x)$.

x	d(x)
0	
2	
4	
6	
8	
10	

b. Graph the function.

c. What does the end behavior of the graph suggest as x increases?

d. Will this trend continue indefinitely? Explain your reasoning.

Complete each of the following.

a. Estimate the x-coordinate of every turning point and determine if those coordinates are relative maxima or relative minima.

b. Estimate the x-coordinate of every zero.

c. Determine the smallest possible degree of the function.

d. Determine the domain and range of the function.

30.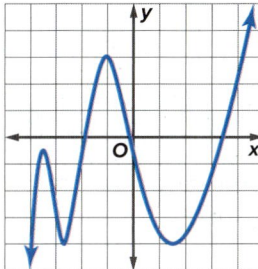

31.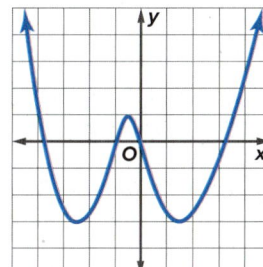

32.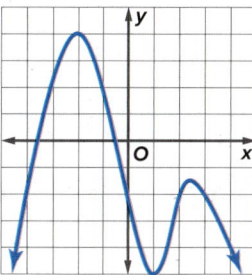

33.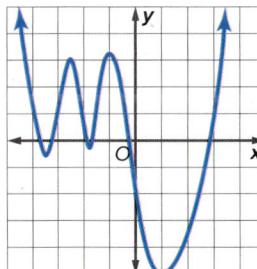

34.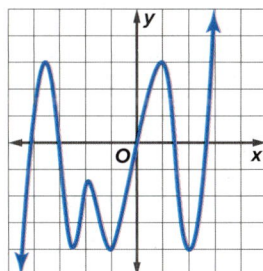

35.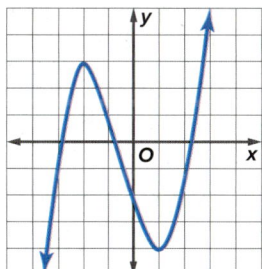

36. REASONING The number of subscribers to a website can be modeled by $f(x) = 0.015x^4 - 0.44x^3 + 3.46x^2 - 2.7x + 9.68$, where x is the number of years after 2000 and $f(x)$ is the number of subscribers in millions.

a. Graph the function.

b. Describe the end behavior of the graph.

c. What does the end behavior suggest about the number of subscribers?

d. Will this trend continue indefinitely? Explain your reasoning.

37. PRICING Jin's vending machines currently sell an average of 3500 beverages per week at a rate of $1.50 per bottle. She is considering increasing the price. Her weekly earnings can be represented by $f(x) = -5x^2 + 100x + 2625$, where x is the number of $0.05 increases. Graph the function and determine the most profitable price for Jin.

For each function,

a. determine the zeros, *x*- and *y*-intercepts, and turning points,

b. determine the axis of symmetry, and

c. determine the intervals for which it is increasing, decreasing, or constant.

38. $y = x^3 + 6x^2 - 32$

39. $y = -x^4 + 3x^2$

40. $y = x^4 - 8x^2 + 16$

41. $y = x^5 - 3x^3 + 2x - 4$

42. $y = -2x^4 + 4x^3 - 5x$

43. $y = \begin{cases} x^2 & \text{if } x \leq -4 \\ 5 & \text{if } -4 < x \leq 0 \\ x^3 & \text{if } x > 0 \end{cases}$

44. MULTIPLE REPRESENTATIONS Consider the following function.

$$f(x) = x^4 - 8.65x^3 + 27.34x^2 - 37.2285x + 18.27$$

a. Analytical What are the degree, leading coefficient, and end behavior?

b. Tabular Make a table of integer values $f(x)$ if $-4 \leq x \leq 4$. How many zeros does the function appear to have from the table?

c. Graphical Graph the function by using a graphing calculator.

d. Graphical Change the viewing window to [0, 4] scl: 1 by [−0.4, 0.4] scl: 0.2. What conclusions can you make from this new view of the graph?

H.O.T. Problems   Use Higher-Order Thinking Skills

45. MP REASONING Explain why the leading coefficient and the degree are the only determining factors in the end behavior of a polynomial function.

46. MP REASONING The table below shows the values of $g(x)$, a cubic function. Could there be a zero between $x = 2$ and $x = 3$? Explain your reasoning.

x	−2	−1	0	1	2	3
g(x)	4	−2	−1	1	−2	−2

47. OPEN-ENDED Sketch the graph of an odd-degree polynomial function with 6 turning points and 2 double roots.

48. MP CONSTRUCT ARGUMENTS Determine whether the following statement is *sometimes*, *always*, or *never* true. Explain your reasoning.

For any continuous polynomial function, the y-coordinate of a turning point is also either a relative maximum or relative minimum.

49. MP REASONING A function is said to be *even* if for every *x* in the domain of *f*, $f(x) = f(-x)$. Is every even-degree polynomial function also an even function? Explain.

50. MP REASONING A function is said to be *odd* if for every *x* in the domain, $-f(x) = f(-x)$. Is every odd-degree polynomial function also an odd function? Explain.

51. WRITING IN MATH How can you use the characteristics of a polynomial function to sketch its graph?

Preparing for Assessment

52. Which graph represents an odd-degree function with a negative leading coefficient and zeros at $x = -2$ and $x = 3$? **MP** 3

A 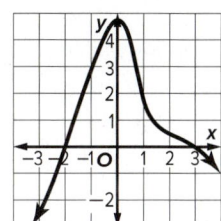   B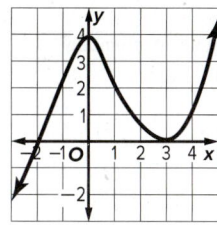

C 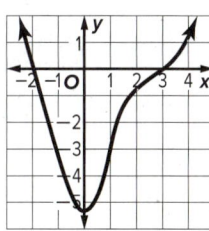   D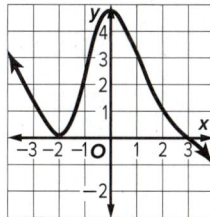

53. The table below describes the values of $g(x)$. Which statement describes the relative maxima of $g(x)$? **MP** 3

x	−7	−6	−5	−4	−3	−2	−1
g(x)	15	−4	−9	−3	−1	−2	7

- A A relative maximum occurs between $x = -7$ and $x = -6$.
- B A relative maximum occurs between $x = -6$ and $x = -4$.
- C A relative maximum occurs between $x = -4$ and $x = -2$.
- D A relative maximum occurs between $x = -2$ and $x = -1$.

54. Which polynomial is equivalent to $(x - 2)^2 + 3$? **MP** 7

- A $x^2 - 1$
- B $x^2 + 7$
- C $x^2 - 2x + 1$
- D $x^2 - 4x - 1$
- E $x^2 - 4x + 7$

55. MULTI-STEP Consider the function in the graph shown. **MP** 4

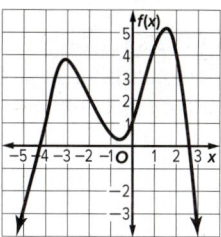

a. Which statement describes the relative minima of $f(x)$?

- A A relative minimum occurs between $x = -5$ and $x = -4$.
- B A relative minimum occurs between $x = -4$ and $x = -3$.
- C A relative minimum occurs between $x = -1$ and $x = 0$.
- D A relative minimum occurs between $x = 0$ and $x = 1$.

b. Which of the following statements are true?

- A The value of the function when $x = 1$ is close to 4.
- B The value of the function when $x = -1$ is zero.
- C A vertical line passed through a graph of this function would not cross the graph twice.
- D The graph is a line.
- E The graph is a curve with the ends pointing up.
- F The graph is a curve with the ends pointing down.

c. If this function is known as $g(x)$, which statement describes the end behavior of this function?

- A As $x \to \infty$, $g(x) \to \infty$, and as $x \to -\infty$, $g(x) \to \infty$.
- B As $x \to \infty$, $g(x) \to -\infty$, and as $x \to -\infty$, $g(x) \to \infty$.
- C As $x \to \infty$, $g(x) \to \infty$, and as $x \to -\infty$, $g(x) \to -\infty$.
- D As $x \to \infty$, $g(x) \to -\infty$, and as $x \to -\infty$, $g(x) \to -\infty$.

EXTEND 4-5

Graphing Technology Lab
Modeling Data Using Polynomial Functions

You can use a TI-83/84 Plus graphing calculator to model data points when a curve of best fit is a polynomial function.

Mathematical Practices
MP 5 Use appropriate tools strategically.

Example Determine and Use a Model

The table shows the distance a seismic wave produced by an earthquake travels from the epicenter. Work cooperatively to draw a scatter plot and a curve of best fit to show how the distance is related to time. Then determine approximately how far away from the epicenter a seismic wave will be felt 8.5 minutes after an earthquake occurs.

Travel Time (min)	1	2	5	7	10	12	13
Distance (km)	400	800	2500	3900	6250	8400	10,000

Source: University of Arizona

Step 1 Enter time in **L1** and distance in **L2**.

KEYSTROKES: STAT 1 1 ENTER 2 ENTER 5 ENTER 7 ENTER 10 ENTER 12 ENTER 13 ENTER ▶ 400 ENTER 800 ENTER 2500 ENTER 3900 ENTER 6250 ENTER 8400 ENTER 10000 ENTER

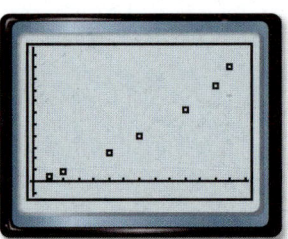

[−0.2, 14.2] scl: 1 by [−1232, 11632] scl: 1000

Step 2 Graph the scatter plot.

KEYSTROKES: 2nd [STAT PLOT] 1 ENTER ▼ ENTER ZOOM 9

Step 3 Determine and graph the equation for a curve of best fit. Use a quartic regression for the data.

KEYSTROKES: STAT ▶ 7 ENTER Y= VARS 5 ▶ ▶ 1 GRAPH

The equation is shown in the Y= screen. If rounded, the regression equation shown on the calculator can be written as the algebraic equation $y = 0.7x^4 - 17x^3 + 161x^2 - 21x + 293$.

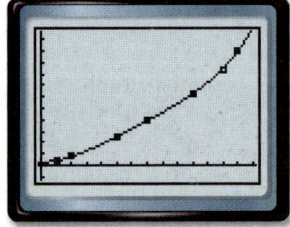

Step 4 Use the [CALC] feature to find the value of the function for $x = 8.5$.

KEYSTROKES: 2nd [CALC] 1 8.5 ENTER

After 8.5 minutes, the wave could be expected to be felt approximately 4980 kilometers from the epicenter.

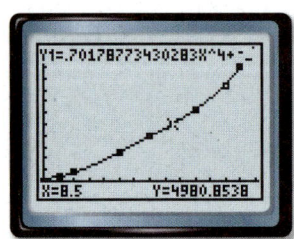

MENTAL CHECK The table gives the distance for 7 minutes as 3900 and the distance for 10 minutes as 6250. Because 8.5 is halfway between 7 and 10 a reasonable estimate for the distance is halfway between 3900 and 6250. ✓

(continued on the next page)

Exercises

The table shows how many minutes out of each eight-hour work day are used to pay one day's worth of taxes.

1. Draw a scatter plot of the data. Then graph several curves of best fit that relate the number of minutes to the number of years. Try LinReg, QuadReg, and CubicReg.

2. Write the equation for the curve that best fits the data.

3. Based on this equation, how many minutes should you expect to work each day in the year 2020 to pay one day's taxes? Use mental math to check the reasonableness of your estimate.

Year	Minutes
2003	136
2004	137
2005	145
2006	150
2007	149
2008	139
2009	128
2010	129
2011	133

Source: Tax Foundation

The table shows the estimated number of alternative-fueled vehicles in use in the United States per year from 1998 to 2007.

4. Draw a scatter plot of the data. Then graph several curves of best fit that relate the number of vehicles to the year.

5. Which curve best fits the data? Is that curve best for predicting future values?

6. Use the best-fit equation you think will give the most accurate prediction for how many alternative-fuel vehicles will be in use in 2025. Use mental math to check the reasonableness of your estimate.

Year	Number of Vehicles	Year	Number of Vehicles
2001	425,457	2006	634,562
2002	471,098	2007	695,766
2003	533,999	2008	775,667
2004	565,492	2009	826,318
2005	592,125	2010	938,643

Source: U.S. Department of Energy

The table shows the average distance from the Sun to Earth during each month of the year.

7. Draw a scatter plot of the data. Then graph several curves of best fit that relate the distance to the month.

8. Write the equation for the curve that best fits the data.

9. Based on your regression equation, what is the distance from the Sun to Earth halfway through September?

10. Would you use this model to find the distance from the Sun to Earth in subsequent years? Explain your reasoning.

Extension

11. Write a question that could be answered by examining data. For example, you might estimate the number of people living in your town 5 years from now or predict the future cost of a car.

12. Collect and organize the data you need to answer the question you wrote. You may need to research your topic on the Internet or conduct a survey to collect the data you need.

13. Make a scatter plot and find a regression equation for your data. Then use the regression equation to answer the question.

Month	Distance (astronomical units)
January	0.9840
February	0.9888
March	0.9962
April	1.0050
May	1.0122
June	1.0163
July	1.0161
August	1.0116
September	1.0039
October	0.9954
November	0.9878
December	0.9837

Source: The Astronomy Cafe

CHAPTER 4
Mid-Chapter Quiz
Lessons 4-1 through 4-4

Simplify. Assume that no variable equals 0. (Lesson 4-1)

1. $(3x^2y^{-3})(-2x^3y^5)$
2. $4t(3rt - r)$
3. $\dfrac{3a^4b^3c}{6a^2b^5c^3}$
4. $\left(\dfrac{p^2r^3}{pr^4}\right)^2$
5. $(4m^2 - 6m + 5) - (6m^2 + 3m - 1)$
6. $(x + y)(x^2 + 2xy - y^2)$

7. **MULTIPLE CHOICE** The volume of the rectangular prism is $6x^3 + 19x^2 + 2x - 3$. Which polynomial expression represents the area of the base? (Lesson 4-1)

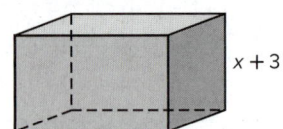

A $6x^4 + 37x^3 + 59x^2 + 3x - 9$

B $6x^2 + x + 1$

C $6x^2 + x - 1$

D $6x + 1$

Simplify. (Lesson 4-3)

8. $(4r^3 - 8r^2 - 13r + 20) \div (2r - 5)$
9. $\dfrac{3x^3 - 16x^2 + 9x - 24}{x - 5}$

10. Expand the binomials.
    a. $(x - 7y)^3$
    b. $(a + 2b)^5$

11. **MULTIPLE CHOICE** Find $p(-3)$ if $p(x) = \frac{2}{3}x^3 + \frac{1}{3}x^2 - 5x$. (Lesson 4-3)

    A 0
    B 11
    C 30
    D 36

12. **PENDULUMS** The formula $L(t) = \dfrac{8t^2}{\pi^2}$ can be used to find the length of a pendulum in feet when it swings back and forth in t seconds. Find the length of a pendulum that makes one complete swing in 4 seconds. (Lesson 4-3)

13. **MULTIPLE CHOICE** Find $3f(a - 4) - 2h(a)$ if $f(x) = x^2 + 3x$ and $h(x) = 2x^2 - 3x + 5$. (Lesson 4-3)

    A $-a^2 + 15a - 74$
    B $-a^2 - 2a - 1$
    C $a^2 + 9a - 2$
    D $-a^2 - 9a + 2$

14. **ENERGY** The power generated by a windmill is a function of the speed of the wind. The approximate power is given by the function $P(s) = \dfrac{s^3}{1000}$, where s represents the speed of the wind in kilometers per hour. Find the units of power $P(s)$ generated by a windmill when the wind speed is 18 kilometers per hour. (Lesson 4-3)

Use $f(x) = x^3 - 2x^2 - 3x$ for Exercises 15–17. (Lesson 4-4)

15. Graph the function.

16. Estimate the x-coordinates at which the relative maxima and relative minima occur.

17. State the domain and range of the function.

18. Determine the consecutive integer values of x between which each real zero is located for $f(x) = 3x^2 - 3x - 1$. (Lesson 4-4)

Refer to the graph below for Exercises 19–21. (Lesson 4-4)

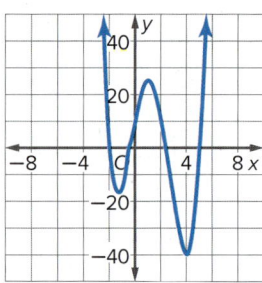

19. Estimate the x-coordinate of every turning point, and determine if those points are relative maxima or relative minima.

20. Estimate the x-coordinate of every zero.

21. What is the least possible degree of the function?

EXPLORE 4-6
Graphing Technology Lab
Solving Polynomial Equations by Graphing

You can use a TI-83/84 Plus graphing calculator to solve polynomial equations.

Mathematical Practices
MP 5 Use appropriate tools strategically.

Activity Solve Polynomial Equations

Work cooperatively. Solve $x^4 + 2x^3 = 7$.

Method 1

Step 1 Graph each side of the equation separately in a standard viewing window.

Let $Y1 = x^4 + 2x^3$ and $Y2 = 7$.

Step 2 Find the points of intersection.

Step 3 Examine the graphs.

Determine where the graph of $y = x^4 + 2x^3$ intersects $y = 7$.

The intersections of the graphs of **Y1** and **Y2** are approximately -2.47 and 1.29, so the solution is approximately -2.47 and 1.29.

Method 2

Step 1 Rewrite the related equation so it is equal to 0.

$$x^4 + 2x^3 = 7$$
$$x^4 + 2x^3 - 7 = 0$$

Let $Y1 = x^4 + 2x^3 - 7$ and $Y2 = 0$.

Step 2 Because $Y2 = 0$, to find the intersection points of **Y1** and **Y2**, find where **Y1** crosses the x-axis.

Step 3 Examine the graphs.

Determine where the graph of $y = x^4 + 2x^3 - 7$ crosses the x-axis.

The intersections of the graphs of **Y1** and **Y2** are approximately -2.47 and 1.29, so the solution is approximately -2.47 and 1.29.

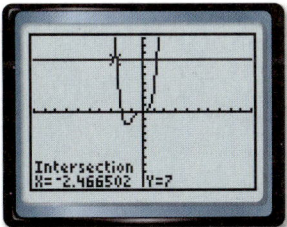
[−10, 10] scl: 1 by [−10, 10] scl: 1

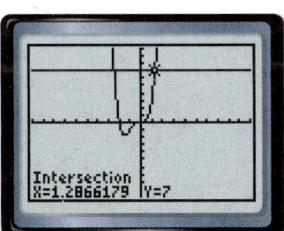
[−10, 10] scl: 1 by [−10, 10] scl: 1

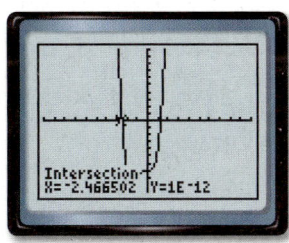
[−10, 10] scl: 1 by [−10, 10] scl: 1

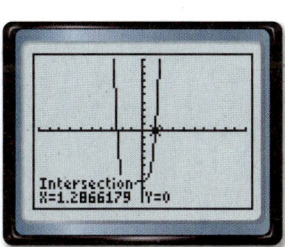
[−10, 10] scl: 1 by [−10, 10] scl: 1

Exercises

Work cooperatively. Solve each equation. Round to the nearest hundredth.

1. $\frac{2}{3}x^3 + x^2 - 5x = -9$
2. $x^3 - 9x^2 + 27x = 20$
3. $x^3 + 1 = 4x^2$
4. $x^6 - 15 = 5x^4 - x^2$
5. $\frac{1}{2}x^5 = \frac{1}{5}x^2 - 2$
6. $x^8 = -x^7 + 3$
7. $x^4 - 15x^2 = -24$
8. $x^3 - 6x^2 + 4x = -6$
9. $x^4 - 15x^2 + x + 65 = 0$

LESSON 6
Solving Polynomial Equations

∷Then
- You solved quadratic functions by factoring.

∷Now
1. Factor polynomials.
2. Solve polynomial equations by factoring.

∷Why?
- A small cube is cut out of a larger cube. The volume of the remaining figure is given and the dimensions of each cube need to be determined.

  This can be accomplished by factoring the cubic polynomial $x^3 - y^3$.

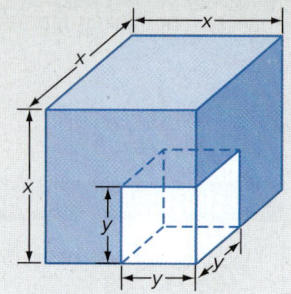

New Vocabulary
prime polynomials
quadratic form

Mathematical Practices
4 Model with mathematics.

1 Factor Polynomials
In Lesson 4-3, you learned that quadratics can be factored just like whole numbers. Their factors, however, are other polynomials. Like quadratics, some cubic polynomials can also be factored with special rules.

Key Concept — Sum and Difference of Cubes

Factoring Technique	General Case
Sum of Two Cubes	$a^3 + b^3 = (a + b)(a^2 - ab + b^2)$
Difference of Two Cubes	$a^3 - b^3 = (a - b)(a^2 + ab + b^2)$

Polynomials that cannot be factored are called **prime polynomials**.

Example 1 — Sum and Difference of Cubes

Factor each polynomial. If the polynomial cannot be factored, write *prime*.

a. $16x^4 + 54xy^3$

$16x^4 + 54xy^3 = 2x(8x^3 + 27y^3)$ Factor out the GCF.

$8x^3$ and $27y^3$ are both perfect cubes, so we can factor the sum of two cubes.

$8x^3 + 27y^3 = (2x)^3 + (3y)^3$ $(2x)^3 = 8x^3; (3y)^3 = 27y^3$

$\quad\quad\quad\quad\quad = (2x + 3y)[(2x)^2 - (2x)(3y) + (3y)^2]$ Sum of two cubes

$\quad\quad\quad\quad\quad = (2x + 3y)(4x^2 - 6xy + 9y^2)$ Simplify.

$16x^4 + 54xy^3 = 2x(2x + 3y)(4x^2 - 6xy + 9y^2)$ Replace the GCF.

b. $8y^3 + 5x^3$

The first term is a perfect cube, but the second term is not. So, the polynomial cannot be factored using the sum of two cubes pattern. The polynomial also cannot be factored using quadratic methods or the GCF. Therefore, it is a prime polynomial.

Guided Practice

1A. $5y^4 - 320yz^3$ **1B.** $-54w^4 - 250wz^3$

The table below summarizes the most common factoring techniques used with polynomials.

Key Concept Factoring Techniques

Number of Terms	Factoring Technique	General Case
any number	Greatest Common Factor (GCF)	$4a^3b^2 - 8ab = 4ab(a^2b - 2)$
two	Difference of Two Squares	$a^2 - b^2 = (a + b)(a - b)$
two	Sum of Two Cubes	$a^3 + b^3 = (a + b)(a^2 - ab + b^2)$
two	Difference of Two Cubes	$a^3 - b^3 = (a - b)(a^2 + ab + b^2)$
three	Perfect Square Trinomials	$a^2 + 2ab + b^2 = (a + b)^2$ $a^2 - 2ab + b^2 = (a - b)^2$
three	General Trinomials	$acx^2 + (ad + bc)x + bd$ $= (ax + b)(cx + d)$
four or more	Grouping	$ax + bx + ay + by$ $= x(a + b) + y(a + b)$ $= (a + b)(x + y)$

Math History Link

Sophie Germain (1776–1831)
Sophie Germain taught herself mathematics with books from her father's library during the French Revolution when she was confined for safety. Germain discovered the identity $x^4 + 4y^4 = (x^2 + 2y^2 + 2xy)(x^2 + 2y^2 - 2xy)$, which is named for her.

Example 2   Factoring by Grouping

Factor each polynomial. If the polynomial cannot be factored, write *prime*.

a. $x^4 + 7x^3 - 4x^2 + 10x^2 - 28x - 40$

$\quad x^4 + 7x^3 - 4x^2 + 10x^2 - 28x - 40$ Original expression
$\quad = x^4 + 7x^3 + 10x^2 - 4x^2 - 28x - 40$ Group to find a GCF.
$\quad = x^2(x^2 + 7x + 10) - 4(x^2 + 7x + 10)$ Factor the GCF.
$\quad = (x^2 - 4)(x^2 + 7x + 10)$ Distributive Property
$\quad = (x + 2)(x - 2)(x + 2)(x + 5)$ Identify linear factors.

b. $2x^4 - 3x^3 + 12x^2 - 18x - 5x^2 - 30$

$\quad 2x^4 - 3x^3 + 12x^2 - 18x - 5x^2 - 30$ Original expression
$\quad = 2x^4 - 3x^3 - 5x^2 + 12x^2 - 18x - 30$ Group to find a GCF.
$\quad = x^2(2x^2 - 3x - 5) + 6(2x^2 - 3x - 5)$ Factor the GCF.
$\quad = (x^2 + 6)(2x^2 - 3x - 5)$ Distributive Property
$\quad = (x^2 + 6)(x + 1)(2x - 5)$ Identify linear factors.

Study Tip

MP Precision Multiply the factors to check your answer.

Guided Practice

2A. $x^3 - 8x^2 - 64x + 8x^2 + 12x - 96$

2B. $x^4 - 5x^3 - 4x^2 - 5x^2 + 20x + 16$

Factoring by grouping is the only method that can be used to factor polynomials with four or more terms. For polynomials with two or three terms, it may be possible to factor according to one of the patterns listed above.

When factoring two terms in which the exponents are 6 or greater, look to factor perfect squares before factoring perfect cubes.

Example 3 Combine Cubes and Squares

Factor each polynomial. If the polynomial cannot be factored, write *prime*.

a. $x^6 - y^6$

This polynomial could be considered the difference of two squares or the difference of two cubes. The difference of two squares should always be done before the difference of two cubes for easier factoring.

$$x^6 - y^6 = (x^3 + y^3)(x^3 - y^3) \qquad \text{Difference of two squares}$$
$$= (x + y)(x^2 - xy + y^2)(x - y)(x^2 + xy + y^2) \qquad \text{Sum and difference of two cubes}$$

b. $a^3x^2 - 6a^3x + 9a^3 - b^3x^2 + 6b^3x - 9b^3$

With six terms, factor by grouping first.

$$a^3x^2 - 6a^3x + 9a^3 - b^3x^2 + 6b^3x - 9b^3$$
$$= (a^3x^2 - 6a^3x + 9a^3) + (-b^3x^2 + 6b^3x - 9b^3) \qquad \text{Group to find a GCF.}$$
$$= a^3(x^2 - 6x + 9) - b^3(x^2 - 6x + 9) \qquad \text{Factor the GCF.}$$
$$= (a^3 - b^3)(x^2 - 6x + 9) \qquad \text{Distributive Property}$$
$$= (a - b)(a^2 + ab + b^2)(x^2 - 6x + 9) \qquad \text{Difference of cubes}$$
$$= (a - b)(a^2 + ab + b^2)(x - 3)^2 \qquad \text{Perfect squares}$$

> **Study Tip**
> **Grouping 6 or more terms** Group the terms that have the *most* common values.

Guided Practice

3A. $a^6 + b^6$

3B. $x^5 + 4x^4 + 4x^3 + x^2y^3 + 4xy^3 + 4y^3$

2 Solve Polynomial Equations
In Chapter 3, you learned to solve quadratic equations by factoring and using the Zero Product Property. You can extend these techniques to solve higher-degree polynomial equations.

Real-World Example 4 Solve Polynomial Functions by Factoring

GEOMETRY Refer to the beginning of the lesson. If the small cube is half the length of the larger cube and the figure is 7000 cubic centimeters, what should be the dimensions of the cubes?

Because the length of the smaller cube is half the length of the larger cube, their lengths can be represented by x and $2x$, respectively. The volume of the object equals the volume of the larger cube minus the volume of the smaller cube.

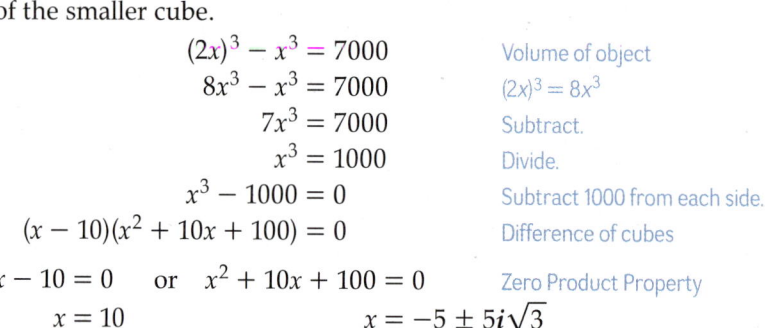

$$(2x)^3 - x^3 = 7000 \qquad \text{Volume of object}$$
$$8x^3 - x^3 = 7000 \qquad (2x)^3 = 8x^3$$
$$7x^3 = 7000 \qquad \text{Subtract.}$$
$$x^3 = 1000 \qquad \text{Divide.}$$
$$x^3 - 1000 = 0 \qquad \text{Subtract 1000 from each side.}$$
$$(x - 10)(x^2 + 10x + 100) = 0 \qquad \text{Difference of cubes}$$
$$x - 10 = 0 \quad \text{or} \quad x^2 + 10x + 100 = 0 \qquad \text{Zero Product Property}$$
$$x = 10 \qquad\qquad x = -5 \pm 5i\sqrt{3}$$

Because 10 is the only real solution, the lengths of the cubes are 10 cm and 20 cm.

> **Go Online!**
> Ask your teacher to assign the **Graphing Calculator Easy File**™ 5-Minute Check to you in ConnectED to take a quick review with your graphing calculator.

Guided Practice

4. Determine the dimensions of the cubes if the length of the smaller cube is one third of the length of the larger cube, and the volume of the object is 3250 cubic centimeters.

In some cases, you can rewrite a polynomial in x in the form $au^2 + bu + c$. For example, by letting $u = x^2$, the expression $x^4 + 12x^2 + 32$ can be written as $(x^2)^2 + 12(x^2) + 32$ or $u^2 + 12u + 32$. This new, but equivalent, expression is said to be in **quadratic form**.

> ### Key Concept Quadratic Form
>
> **Words** An expression that is in quadratic form can be written as $au^2 + bu + c$ for any numbers a, b, and c, $a \neq 0$, where u is some expression in x. The expression $au^2 + bu + c$ is called the quadratic form of the original expression.
>
> **Example** $12x^6 + 4x^3 + 1 = 3(2x^3)^2 + 2(2x^3) + 1$

Study Tip

Quadratic Form When writing a polynomial in quadratic form, choose the expression equal to u by examining the terms with variables. Pay special attention to the exponents in those terms. Not every polynomial can be written in quadratic form.

Example 5 Quadratic Form

Write each expression in quadratic form, if possible.

a. $150n^8 + 40n^4 - 15$

$150n^8 + 40n^4 - 15 = 6(5n^4)^2 + 8(5n^4) - 15$ $(5n^4)^2 = 25n^8$

b. $y^8 + 12y^3 + 8$

This cannot be written in quadratic form since $y^8 \neq (y^3)^2$.

▶ **Guided Practice**

5A. $x^4 + 5x + 6$

5B. $8x^4 + 12x^2 + 18$

You can use quadratic form to solve equations with larger degrees.

Example 6 Solve Equations in Quadratic Form

Solve $18x^4 - 21x^2 + 3 = 0$.

$18x^4 - 21x^2 + 3 = 0$	Original equation
$2(3x^2)^2 - 7(3x^2) + 3 = 0$	$2(3x^2)^2 = 18x^4$
$2u^2 - 7u + 3 = 0$	Let $u = 3x^2$.
$(2u - 1)(u - 3) = 0$	Factor.
$u = \frac{1}{2}$ or $u = 3$	Zero Product Property
$3x^2 = \frac{1}{2}$ $3x^2 = 3$	Replace u with $3x^2$.
$x^2 = \frac{1}{6}$ $x^2 = 1$	Divide by 3.
$x = \pm\frac{\sqrt{6}}{6}$ $x = \pm 1$	Take the square root.

The solutions of the equation are 1, -1, 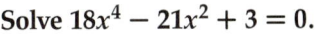 $\frac{\sqrt{6}}{6}$, and $-\frac{\sqrt{6}}{6}$.

▶ **Guided Practice**

6A. $4x^4 - 8x^2 + 3 = 0$

6B. $8x^4 + 10x^2 - 12 = 0$

Check Your Understanding

= Step-by-Step Solutions begin on page R11.

Go Online! for a Self-Check Quiz

Examples 1–3 Factor completely. If the polynomial is not factorable, write *prime*.

1. $3ax + 2ay - az + 3bx + 2by - bz$
2. $2kx + 4mx - 2nx - 3ky - 6my + 3ny$
3. $2x^3 + 5y^3$
4. $16g^3 + 2h^3$
5. $12qw^3 - 12q^4$
6. $x^3 - 8x^2 + 7x^2 + 16x - 56x + 112$
7. $a^6x^2 - b^6x^2$
8. $x^4 + 2x^3 - 3x^2 + x^2 - 6x - 3$
9. $8c^3 - 125d^3$
10. $x^4 + 6x^3 + 5x^2 - 54x - 9x^2 - 45$

Example 4 Solve each equation.

11. $x^4 - 19x^2 + 48 = 0$
12. $x^3 - 64 = 0$
13. $x^3 + 27 = 0$
14. $x^4 - 33x^2 + 200 = 0$

15. **ZOO** The zoo is looking to put a sidewalk around the alligator enclosure. The alligator enclosure is 90 feet wide and 35 feet long. The combined area of the alligator enclosure and sidewalk is 3800 square feet. What is the width of the sidewalk?

Example 5 Write each expression in quadratic form, if possible.

16. $4x^6 - 2x^3 + 8$
17. $25y^6 - 5y^2 + 20$

Example 6 Solve each equation.

18. $x^4 - 6x^2 + 8 = 0$
19. $y^4 - 18y^2 + 72 = 0$

Practice and Problem Solving

Extra Practice is on page R4.

Examples 1–3 Factor completely. If the polynomial is not factorable, write *prime*.

20. $8c^3 - 27d^3$
21. $64x^4 + xy^3$
22. $a^8 - a^2b^6$
23. $x^6y^3 + y^9$
24. $18x^6 + 5y^6$
25. $w^3 - 2y^3$
26. $gx^2 - 3hx^2 - 6fy^2 - gy^2 + 6fx^2 + 3hy^2$
27. $12ax^2 - 20cy^2 - 18bx^2 - 10ay^2 + 15by^2 + 24cx^2$
28. $a^3x^2 - 16a^3x + 64a^3 - b^3x^2 + 16b^3x - 64b^3$
29. $8x^5 - 25y^3 + 80x^4 - x^2y^3 + 200x^3 - 10xy^3$

Example 4 Solve each equation.

30. $x^4 + x^2 - 90 = 0$
31. $x^4 - 16x^2 - 720 = 0$
32. $x^4 - 7x^2 - 44 = 0$
33. $x^4 + 6x^2 - 91 = 0$
34. $x^3 + 216 = 0$
35. $64x^3 + 1 = 0$

Example 5 Write each expression in quadratic form, if possible.

36. $x^4 + 12x^2 - 8$
37. $-15x^4 + 18x^2 - 4$
38. $8x^6 + 6x^3 + 7$
39. $5x^6 - 2x^2 + 8$
40. $9x^8 - 21x^4 + 12$
41. $16x^{10} + 2x^5 + 6$

Example 6 Solve each equation.

42. $x^4 + 6x^2 + 5 = 0$
43. $x^4 - 3x^2 - 10 = 0$
44. $4x^4 - 14x^2 + 12 = 0$
45. $9x^4 - 27x^2 + 20 = 0$
46. $4x^4 - 5x^2 - 6 = 0$
47. $24x^4 + 14x^2 - 3 = 0$

48. ZOOLOGY A species of animal is introduced to a small island. Suppose the population of the species is represented by $P(t) = -t^4 + 9t^2 + 400$, where t is the time in years. Determine when the population becomes zero.

Factor completely. If the polynomial is not factorable, write *prime*.

49. $x^4 - 625$ **50.** $x^6 - 64$ **51.** $x^5 - 16x$ **52.** $8x^5y^2 - 27x^2y^5$

53. $15ax - 10bx + 5cx + 12ay - 8by + 4cy + 15az - 10bz + 5cz$

54. $6a^2x^2 - 24b^2x^2 + 18c^2x^2 - 5a^2y^3 + 20b^2y^3 - 15c^2y^3 + 2a^2z^2 - 8b^2z^2 + 6c^2z^2$

55. $6x^5 - 11x^4 - 10x^3 - 54x^3 + 99x^2 + 90x$

56. $20x^6 - 7x^5 - 6x^4 - 500x^4 + 175x^3 + 150x^2$

57. GEOMETRY The volume of the figure at the right is 440 cubic centimeters. Find the value of x and the length, height, and width.

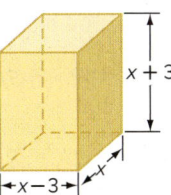

Solve each equation.

58. $8x^4 + 10x^2 - 3 = 0$ **59.** $6x^4 - 5x^2 - 4 = 0$

60. $20x^4 - 53x^2 + 18 = 0$ **61.** $18x^4 + 43x^2 - 5 = 0$

62. $8x^4 - 18x^2 + 4 = 0$ **63.** $3x^4 - 22x^2 - 45 = 0$

64. $x^6 + 7x^3 - 8 = 0$ **65.** $x^6 - 26x^3 - 27 = 0$

66. $8x^6 + 999x^3 = 125$ **67.** $4x^4 - 4x^2 - x^2 + 1 = 0$

68. $x^6 - 9x^4 - x^2 + 9 = 0$ **69.** $x^4 + 8x^2 + 15 = 0$

70. SENSE-MAKING A rectangular prism with dimensions $x - 2$, $x - 4$, and $x - 6$ has a volume equal to $40x$ cubic units.

  a. Write a polynomial equation using the formula for volume.

  b. Use factoring to solve for x.

  c. Are any values for x unreasonable? Explain.

  d. What are the dimensions of the prism?

71. MULTI-STEP Karen is starting a job that pays an annual salary of $50,000, 2% of which goes to health coverage and 5% to her retirement plan. After those deductions, 28% goes to taxes. She pays $850 a month in rent and has a $225 monthly car payment. She plans to allocate $1400 per month for all her other expenses and invest the rest at the end of each year. She wants to invest in CDs, which pay out a percentage of the investment after one year. She will reinvest the principal, interest, and her additional deposit for the year in a new CD. She plans to eventually use her investment to place a 10% down payment on a $200,000 house.

  a. How much can she invest at the end of each year? Explain your reasoning.

  b. Let r be the rate of growth and x be her yearly investment. At the end of the first year, she invests x. After year 2, this amount is rx to account for the interest earned. After year 3, it is r^2x, and so on. What interest rate does the CD need to be in order for her to have enough money invested after 5 years?

  c. Describe your solution process.

  d. What assumptions did you make?

72. **BIOLOGY** During an experiment, the number of cells of a virus can be modeled by $P(t) = -0.012t^3 - 0.24t^2 + 6.3t + 8000$, where t is the time in hours and P is the number of cells. Jack wants to determine the times at which there are 8000 cells.

   a. Solve for t by factoring.

   b. What method did you use to factor?

   c. Which values for t are reasonable and which are unreasonable? Explain.

   d. Graph the function for $0 \leq t \leq 20$ using your calculator.

73. **HOME BUILDING** Alicia's parents want their basement home theater designed according to the diagram at the right.

   a. Write a function in terms of x for the area of the basement.

   b. If the basement is to be 1366 square feet, what is x?

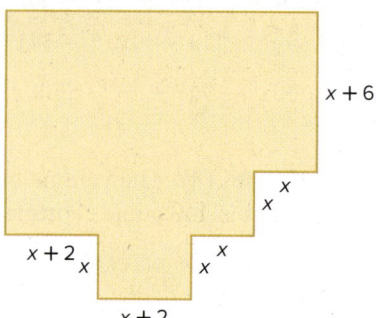

74. **BIOLOGY** A population of parasites in an experiment can be modeled by $f(t) = t^3 + 5t^2 - 4t - 20$, where t is the time in days.

   a. Use factoring by grouping to determine the values of t for which $f(t) = 0$.

   b. At what times does the population reach zero?

   c. Are any of the values of t unreasonable? Explain.

Factor completely. If the polynomial is not factorable, write *prime*.

75. $x^6 - 4x^4 - 8x^4 + 32x^2 + 16x^2 - 64$

76. $y^9 - y^6 - 2y^6 + 2y^3 + y^3 - 1$

77. $x^6 - 3x^4y^2 + 3x^2y^4 - y^6$

78. **SENSE-MAKING** Fredo's corral, an enclosure for livestock, is currently 32 feet by 40 feet. He wants to enlarge the area to 4.5 times its current area by increasing the length and width by the same amount.

   a. Draw a diagram to represent the situation.

   b. Write a polynomial equation for the area of the new corral. Then solve the equation by factoring.

   c. Graph the function.

   d. Which solution is irrelevant? Explain.

H.O.T. Problems — Use Higher-Order Thinking Skills

79. **CHALLENGE** Factor $36x^{2n} + 12x^n + 1$.

80. **CHALLENGE** Solve $6x - 11\sqrt{3x} + 12 = 0$.

81. **MP REASONING** Find a counterexample to the statement $a^2 + b^2 = (a + b)^2$.

82. **OPEN-ENDED** The cubic form of an equation is $ax^3 + bx^2 + cx + d = 0$. Write an equation with degree 6 that can be written in *cubic* form.

83. **WRITING IN MATH** Explain how the graph of a polynomial function can help you factor the polynomial.

Preparing for Assessment

84. What is the complete solution set of $x^4 - 13x^2 + 36 = 0$?
- A $\{-\sqrt{3}, -\sqrt{2}, \sqrt{2}, \sqrt{3}\}$
- B $\{-3, -2, 2, 3\}$
- C $\{3i, 2i, 2, 3\}$
- D $\{-9, -4, 4, 9\}$

85. Let $_c a_b = (a)(a-b)(a-c)$. What is the value $_m(m-1)_1$?
- A 0
- B $2m - 2$
- C $m^2 - 3m + 2$
- D $-2 + 3m - m^2$
- E $2m^2 - 2m$

86. Which expression is prime?
- A $25x^4 - 16$
- B $x^2 - 16x + 1$
- C $x^5 + 8x^3 - 2x^2 - 16$
- D $x^6 - x^3 - 20$

87. A small box is placed inside a larger box. The dimensions of the smaller box are x, $x + 6$, and $x - 1$. The dimensions of the larger box are $x + 12$, $2x$, and $x + 10$. If the volume of the space inside the larger box but outside the smaller box is 46,150 cubic inches, use a graphing calculator to find the volume of the smaller box in cubic inches.

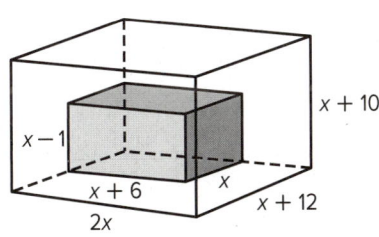

Volume = _____ cubic inches

88. The figure to the right is formed by removing an x-by-x-by-$2x$ rectangular prism from a $3x$-by-$2x$-by-$7x$ rectangular prism. If the volume of the figure is 20,480 cubic inches, what is the value, in inches, of x?

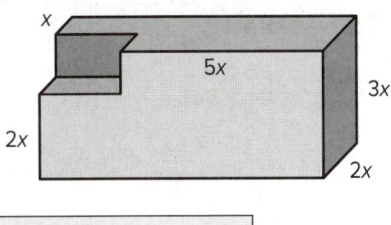

$x =$ _____

89. Solve $-x^3 - 18x^2 = 4x$ by completing the steps below.

Step 1: $-8x^3 - 18x^2 - \square = 4x - \square$.

Step 2: $\square(4x^2 + \square + 2) = \square$

Step 3: $\square(4x + \square)(x + \square) = \square$

Step 4: $0 = \square$; $0 = 4x + \square$; $0 = x + \square$

Step 5: $\{\quad\quad\}$

90. Solve $a^4 - 16 = 0$.

91. Solve the following equations:
a. $81x^4 - 16 = 0$
b. $x^5 + 13x^3 + 36x = 0$
c. $3x^7 = 81x^4$

92. Assuming dry road conditions and average reaction times, the safe stopping distance in feet is given by $d(x) = \frac{1}{20}x^2 + x$, where x represents the speed of the car in miles per hour. Determine the safe speed of the car if you expect to stop in 40 feet.

_____

93. Factor the polynomial, $x^4 - 23x^2 - 50$.

LESSON 7
Proving Polynomial Identities

∴Then
- You solved polynomial equations.

∴Now
1. Prove polynomial identities.

∴Why?
- Polynomials can represent situations like the area of a rectangle. Some polynomial identities have recognizable patterns, such as the sums and differences of powers.

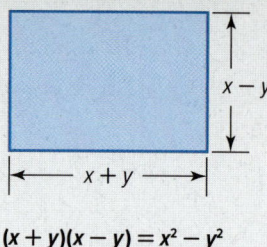

$(x + y)(x - y) = x^2 - y^2$

 New Vocabulary
polynomial identity

 Mathematical Practices
7 Look for and make use of structure.
8 Look for and express regularity in repeated reasoning.

1 Prove Polynomial Identities
An identity is an equation that is satisfied by any number that replaces the variables, without just one specific value or set of values as a solution. Thus, a **polynomial identity** is a polynomial equation that is true for any values that are substituted for the variables, without only one specific solution.

To prove that an equation is an identity, show that it is true for all values of the variables.

Example 1   Transform One Side

Prove that $(x + y)^2 = x^2 + 2xy + y^2$ is an identity.

$(x + y)^2 = x^2 + 2xy + y^2$ Original equation
$(x + y)(x + y) = x^2 + 2xy + y^2$ Expand $(x + y)^2$.
$x^2 + xy + xy + y^2 = x^2 + 2xy + y^2$ FOIL Method
$x^2 + 2xy + y^2 = x^2 + 2xy + y^2$ Simplify.

The identity has been proved because the expressions on both sides of the equation are identical.

Guided Practice

1. Prove that $(x^2 + y^2)^2 = (x^2 - y^2)^2 + (2xy)^2$ is an identity.

Some polynomial equations need to be transformed on both sides to prove they are identities.

Example 2   Transform Each Side

Prove that $(x + 1)^2(x^3 - 6x^2 + 12x - 8) = (x^2 + 2x + 1)(x - 2)^3$ is an identity.

$(x + 1)^2(x^3 - 6x^2 + 12x - 8) = (x^2 + 2x + 1)(x - 2)^3$ Original equation
$(x + 1)(x + 1)(x^3 - 6x^2 + 12x - 8) = (x^2 + 2x + 1)(x - 2)(x - 2)(x - 2)$ Expand.
$(x^2 + x + x + 1)(x^3 - 6x^2 + 12x - 8) = (x^2 + 2x + 1)(x^2 - 4x + 4)(x - 2)$ FOIL
$(x^2 + 2x + 1)(x^3 - 6x^2 + 12x - 8) = (x^2 + 2x + 1)(x^3 - 6x^2 + 12x - 8)$ Simplify.

Guided Practice

2. Prove that $(x + 1)(x - 2)^2 = (x^2 - x - 2)(x - 2)$ is an identity.

You can use a table to determine whether a polynomial equation may be an identity. Substitute values for the variables and then simplify both sides of the equation. If both sides of the equation are equal for several values of the variables, it is likely that the equation is a polynomial identity. This is not a formal proof, so it cannot be determined definitively.

Example 3 Use Technology

Use a table to determine whether $x^3 - y^3 = (x - y)(x^2 + xy + y^2)$ may be an identity.

Step 1 Use spreadsheet software. Put values for x in column A and values for y in column B. Label Column A with x. Label Column B with y.

Column C will show computed values for the left side of the equation, and column D will show computed values for the right side of the equation. Label Column C with x^3-y^3. Label Column D with (x-y)(x^2+xy+y^2).

Step 2 In cell C2, enter the formula = A2^3 − B2^3. In cell D2, enter the formula = (A2-B2)*(A2^2+A2*B2+B2^2). Copy these formulas into other rows in columns C and D.

Step 3 Enter values for x and y in columns A and B.

	A	B	C	D
	x	y	x^3 – y^3	(x – y)(x^2 + xy + y^2)
1				
2	1	10	-999	-999
3	2	15	-3367	-3367
4	3	20	-7973	-7973
5	4	25	-15561	-15561
6	5	30	-26875	-26875

> **Study Tip**
> **Reasoning** Remember to think of order of operations as you input expressions into a table.

Step 4 Examine the values in columns C and D.

The values in columns C and D are equal when different values are substituted for x and y. The equation is likely to be a polynomial identity.

Guided Practice

3. Use a table to determine whether $p^4 - q^4 = (p - q^2p + q^2p^2 + q^2)$ may be an identity.

Check Your Understanding

= Step-by-Step Solutions begin on page R11.

Go Online! for a Self-Check Quiz

Example 1 Transform one side of the equation to determine if it is a polynomial identity.

1. $a^2 - b^2 = (a + b)(a - b)$

2. $a^2 + b^2 = (a + b)(a + b)$

Example 2 Transform each side of the equation to determine if it is an identity.

3. $(x + 1)^2(x - 4)^3 = (x^2 - 3x - 4)(x^3 - 7x^2 + 8x + 16)$

Example 3 Use technology to determine whether the equation may be an identity.

4. $x^3 + y^3 = (x + y)(x^2 - xy + y^2)$

5. $g^6 + h^6 = (g^2 + h^2)(g^4 - g^2h^2 + h^4)$

Practice and Problem Solving

Extra Practice is found on page R4.

Example 1 Transform one side of the equation to determine if it is an identity.

6. $(a + b)(c + d) = a(c + d) + b(c + d)$

7. $(x + 3)^2 = x^2 + 27$

8. $x^4 + x^2 - 20 = (x^2 + 5)(x^2 - 4)$

9. $5(x - 10)^2 = 5x^2 - 100x + 500$

10. $2(2x - 1)(x + 3) = 4x^2 + 10x - 6$

11. $(x + 4)(x - 4) = x^2 - 16$

Example 2 Transform each side of the equation to determine if it is an identity.

12. $(x^2 + 4)(x - 2)(x + 2) = (x^3 + 2x^2 - 8)(x + 2)$
13. $(6g + 2h)^2 = 9g^2 + h(6g + h)$
14. $(x + 3)(x + 4)(x + 5) = x^2(x + 12x) + (47x + 60)$

Example 3 Use technology to determine whether the equation may be an identity.

15. $a^5 - b^5 = (a - b)(a^4 + a^3b + a^2b^2 + ab^3 + b^4)$
16. $a^5 + b^5 = (a + b)(a^4 - a^3b + a^2b^2 - ab^3 + b^4)$

Determine whether the equation may be an identity using transformation or technology.

17. $u^6 - w^6 = (u + w)(u - w)(u^2 + uw + w^2)(u^2 - uw + w^2)$

18. **USE TECHNOLOGY** In Example 1, $(x + y)^2 = x^2 + 2xy + y^2$ was proven to be a polynomial identity. Use technology to verify this proof.

19. **USE TECHNOLOGY** In Example 2, $(x + 1)^2(x^3 - 6x^2 + 12x - 8) = (x^2 + 2x + 1)(x - 2)^3$ was proven to be a polynomial identity. Use technology to verify this proof.

20. **DIFFERENCE OF CUBES** Prove the polynomial $(a - b)^3 = a^3 - 3a^2b + 3ab^2 - b^3$ is an identity.

21. **CRITIQUE ARGUMENTS** DeMarco wanted to investigate whether the equation $(x - 1)(x - 2)(x - 3)(x - 4) = 2x^4 - 20x^3 + 70x^2 - 100x + 48$ may be an identity. He used spreadsheet software and entered values for x as shown. He concluded that the equation is an identify. Do you agree? Explain why or why not.

	A	B	C
	x	(x − 1)(x − 2)(x − 3)(x − 4)	2x^4 − 20x^3 + 70x^2 − 100x + 48
1			
2	1	0	0
3	2	0	0
4	3	0	0
5	4	0	0

22. Use the figure to write and justify an identity as follows.
    a. Write an expression for the length of each side of the large square. Then write an expression that represents the area of the large square.
    b. Write an expression for the area of each of the small rectangles within the large square. Then use these expressions to write an expression that represents the area of the large square.
    c. What identity is justified by your work in parts **a** and **b**?

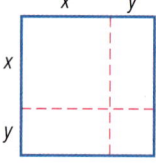

23. Use the figure to write and justify an identity as follows.
    a. Write an expression for the length of each side of the red rectangle. Then write an expression that represents the area of the red rectangle.
    b. Explain how the area of the red rectangle is related to the area of the blue square. Use this relationship to write an expression that represents the area of the red rectangle.
    c. What identity is justified by your work in parts **a** and **b**?

24. **TOOLS** Explain how you can use graphing software to verify that $(x + 1)^2(x - 1)^2 = x^4 - 2x^2 + 1$ is an identity.

25. **TOOLS** Explain how you can use graphing software to show that $(x + 1)^3 = x^3 + x^2 + x + 1$ is not an identity.

Each of the following equations is an identity. The letters x and y represent variables, and all other letters represent constants. Find the value of the constant or constants in each identity.

26. $(x - 2y)^2 = ax^2 + bxy + cy^2$
27. $(x + y)^4 = ax^4 + bx^3y + cx^2y^2 + dxy^3 + ey^4$
28. $x^6 + y^6 = (x^2 + y^2)(x^4 + kx^2y^2 + y^4)$

Transform one side of the equation to determine if it is a polynomial identity.

29. $(x + y + z)^2 = x^2 + y^2 + z^2 + 2xy + 2xz + 2yz$
30. $(x + y + z)^3 = x^3 + y^3 + z^3 + 3x^2y + 3xz^2 + 3y^2z$
31. $(x + y + z)(x - y - z) = x^2 - y^2 - 2yz - z^2$
32. $(x + y)(x + z)(y + z) = x^2y + x^2z + xy^2 + y^2z + xz^2 + yz^2$

33. **PYTHAGOREAN TRIPLE** Pythagorean triples are three values or terms that are the sides of a right triangle. These values or terms satisfy the Pythagorean Theorem $a^2 + b^2 = c^2$, where a and b are the legs of the right triangle and c is the hypotenuse.
    a. Simplify $(x^2 + y^2)^2$.
    b. Find the sum of $(x^2 - y^2)^2$ and $(2xy)^2$.
    c. Identify expressions that would represent the length of the legs of a right triangle from parts **a** and **b**.

34. Sara believes that if she picks any two integers and looks at the sum of their squares, the difference of their squares, and twice the product of the two integers, that the three integers are sides of a right triangle. Write a polynomial equation to represent Sara's idea.

H.O.T. Problems   Use Higher-Order Thinking Skills

35. Elaine is creating a garden in the shape of a right triangle. The length of one leg can be represented by the expression $x^2 - 16$. Elaine decided that the length of the hypotenuse side of her garden should be the same as the given leg increased by 32. What is the length of the third side of Elaine's garden?

36. **ERROR ANALYSIS** Ethan determined that the equation $(a + b)^3 = (a^2 + 2ab + b^2)(a + b)$ was not a polynomial identity. What did he do wrong?

    $(a + b)^3 = (a^2 + 2ab + b^2)(a + b)$
    $\dfrac{(a + b)^3}{(a + b)} = \dfrac{(a^2 + 2ab + b^2)(a + b)}{(a + b)}$
    $(a + b)^2 = (a^2 + 2ab + b^2)$
    $a^2 + b^2 \neq a^2 + 2ab + b^2$
    This is not an identity.

37. **CHALLENGE** Explain whether $x^3 + y^3 = \left(\dfrac{x^4 + 2xy^3}{x^3 - y^3}\right)^3 + \left(\dfrac{y^4 + 2x^3y}{y^3 - x^3}\right)^3$ is an identity.

38. **OPEN-ENDED** Determine whether a polynomial equation is an identity using technology and substitution as a method to check for equivalence on both sides of the equation. Create an example of a polynomial equation that appears to be an identity by using substitution, but is not actually a polynomial identity.

Preparing for Assessment

39. When transforming the left side of a polynomial equation to determine if it is an identity, which would be the first step? **MP** 7, 8

$$(a + b)^3 = a^3 + 3a^2b - 3ab^2 + b^3$$

- ◯ **A** Subtract b^3 from both sides of the equation.
- ◯ **B** Divide both sides of the equation by $(a + b)$.
- ◯ **C** Expand $(a + b)^3$.
- ◯ **D** Factor $a^3 + 3a^2b$.

40. What is the missing term on the right side of the equation for this Pythagorean triple? **MP** 7, 8

$$(x^2 + 4)^2 = (4x)^2 + \underline{\hspace{2cm}}$$

- ◯ **A** $(x^2 + 4)^2$
- ◯ **B** $(x^2 + 2)^2$
- ◯ **C** $(x^2 - 2)^2$
- ◯ **D** $(x^2 - 4)^2$

41. Which of the following transformations would be unnecessary for determining whether the equation below is a polynomial identity? **MP** 7, 8

$$(x - 5)(x^2 - 2x - 15) = (x + 3)(x - 5)^2$$

- ☐ **A** $x^3 - 2x^2 - 15x - 5x^2 + 10x + 75 = (x + 3)(x - 5)^2$
- ☐ **B** $(x - 5)(x^2 - 2x - 15) = (x + 3)(x - 5)(x - 5)$
- ☐ **C** $(x - 5)(x^2 - 2x - 15) = (x + 3)(x^2 - 10x + 25)$
- ☐ **D** $\dfrac{(x - 5)(x^2 - 2x - 15)}{x - 5} = \dfrac{(x + 3)(x - 5)^2}{x - 5}$

42. Determine whether the equation below is a polynomial identity. Show your steps. **MP** 7, 8

$$(x - y)^2 = x^2 + y^2$$

43. MULTI-STEP Use the equation $(x + y)^4 = x^4 + 4x^3y + 6x^2y^2 + 4xy^3 + y^4$. **MP** 7, 8

a. Copy the table. Use technology to complete the columns for the left side and right side of the equation.

x	y	Left side	Right side
0	2		
1	3		
2	4		
3	5		
4	6		

b. What conclusion can you make about the equation, based on the results in your table? Explain.

44. Determine whether the equation below is a polynomial identity. Show your steps. **MP** 7, 8

$$(x + 2)^3 = (x^2 + 4x + 4)(x + 2)$$

45. Choose all equations that are polynomial identities. **MP** 7, 8

- ☐ **A** $x^2 + y^2 = x^2 + 2xy + y^2$
- ☐ **B** $x^3 + y^3 = (x + y)(x^2 - xy + y^2)$
- ☐ **C** $x^2 - y^2 = (x + y)(x - y)$
- ☐ **D** $x^3 - y^3 = (x - y)(x^3 + xy + y^2)$
- ☐ **E** $(x + y)^2 = x^2 + y^2$
- ☐ **F** $(x - y)^2 = x^2 - 2xy + y^2$

46. Write out the expansions using polynomial identities: **MP** 7, 8

- ☐ **A** $(a + b)^3$
- ☐ **B** $(a - b)^3$
- ☐ **C** $(a + b)^2$
- ☐ **D** $(a - b)^2$
- ☐ **E** $(a - b)(a + b)$

LESSON 8
The Remainder and Factor Theorem

::Then   ::Now   ::Why?

- You used the Distributive Property and factoring to simplify algebraic expressions.
- 1 Evaluate functions by using synthetic substitution.
- 2 Determine whether a binomial is a factor of a polynomial by using synthetic substitution.
- The number of college students from the United States who study abroad can be modeled by the function $S(x) = 0.02x^4 - 0.52x^3 + 4.03x^2 + 0.09x + 77.54$, where x is the number of years since 1993 and $S(x)$ is the number of students in thousands.

  You can use this function to estimate the number of U.S. college students studying abroad in 2020 by evaluating the function for $x = 27$. Another method you can use is *synthetic substitution*.

New Vocabulary
synthetic substitution
depressed polynomial

MP Mathematical Practices
7 Look for and make use of structure.

1 Synthetic Substitution Synthetic division can be used to find the value of a function. Consider the polynomial function $f(x) = -3x^2 + 5x + 4$. Divide the polynomial by $x - 3$.

Method 1 Long Division

$$\begin{array}{r} -3x - 4 \\ x-3 \overline{\smash{)}-3x^2 + 5x + 4} \\ \underline{-3x^2 + 9x} \\ -4x + 4 \\ \underline{-4x + 12} \\ -8 \end{array}$$

Method 2 Synthetic Division

$$\begin{array}{r|rrr} 3 & -3 & 5 & 4 \\ & & -9 & -12 \\ \hline & -3 & -4 & -8 \end{array}$$

Compare the remainder of -8 to $f(3)$.

$f(3) = -3(3)^2 + 5(3) + 4$ Replace x with 3.
$ = -27 + 15 + 4$ Multiply.
$ = -8$ Simplify.

Notice that the value of $f(3)$ is the same as the remainder when the polynomial is divided by $x - 3$. This illustrates the **Remainder Theorem**.

> **Key Concept** Remainder Theorem
>
> **Words** If a polynomial $P(x)$ is divided by $x - r$, the remainder is a constant $P(r)$, and
>
>   Dividend equals quotient times divisor plus remainder.
> $P(x) = Q(x) \cdot (x - r) + P(r)$,
>
> where $Q(x)$ is a polynomial with degree one less than $P(x)$.
>
> **Example** $x^2 + 6x + 2 = (x - 4) \cdot (x + 10) + 42$

Applying the Remainder Theorem using synthetic division to evaluate a function is called **synthetic substitution**. It is a convenient way to find the value of a function, especially when the degree of the polynomial is greater than 2.

connectED.mcgraw-hill.com   287

Example 1   Synthetic Substitution

If $f(x) = 3x^4 - 2x^3 + 5x + 2$, find $f(4)$.

Method 1   Synthetic Substitution

By the Remainder Theorem, $f(4)$ should be the remainder when the polynomial is divided by $x - 4$.

```
4 | 3   -2    0    5     2
  |      12   40   160   660
  ----------------------------
    3    10   40   165 | 662
```

Because there is no x^2 term, a zero is placed in this position as a placeholder.

The remainder is 662. Therefore, by using synthetic substitution, $f(4) = 662$.

Method 2   Direct Substitution

Replace x with 4.

$f(x) = 3x^4 - 2x^3 + 5x + 2$ Original function
$f(4) = 3(4)^4 - 2(4)^3 + 5(4) + 2$ Replace x with 4.
$\quad\quad = 768 - 128 + 20 + 2$ or 662 Simplify.

By using direct substitution, $f(4) = 662$. Both methods give the same result.

▶ **Guided Practice**

1A. If $f(x) = 3x^3 - 6x^2 + x - 11$, find $f(3)$.

1B. If $g(x) = 4x^5 + 2x^3 + x^2 - 1$, find $f(-1)$.

Synthetic substitution can be used in situations in which direct substitution would involve cumbersome calculations.

Real-World Link
Some benefits of studying abroad include learning a new language, becoming more independent, and improving communication skills. Studying abroad also gives students a chance to experience different cultures and customs as well as try new foods.

Source: StudyAbroad

Real-World Example 2   Find Function Values

COLLEGE Refer to the beginning of the lesson. How many U.S. college students will study abroad in 2018?

Use synthetic substitution to divide $0.02x^4 - 0.52x^3 + 4.03x^2 + 0.09x + 77.54$ by $x - 25$.

```
25 | 0.02   -0.52    4.03    0.09      77.54
   |         0.5    -0.5    88.25    2208.5
   ---------------------------------------------
     0.02   -0.02    3.53   88.34  |  2286.04
```

In 2018, there will be about 2,286,040 U.S. college students studying abroad.

▶ **Guided Practice**

2. COLLEGE The function $C(x) = 2.46x^3 - 22.37x^2 + 53.81x + 548.24$ can be used to approximate the number, in thousands, of international college students studying in the United States x years since 2010. How many international college students can be expected to study in the U.S. in 2025?

Watch Out!

MP Precision Remember that synthetic substitution is used to divide a polynomial by $(x - a)$. If the binomial is $(x - a)$, use a. If the binomial is $(x + a)$, use $-a$.

2 Factors of Polynomials
The synthetic division below shows that the quotient of $2x^3 - 3x^2 - 17x + 30$ and $x + 3$ is $2x^2 - 9x + 10$.

$$\begin{array}{r|rrrr} -3 & 2 & -3 & -17 & 30 \\ & & -6 & 27 & -30 \\ \hline & 2 & -9 & 10 & 0 \end{array}$$

When you divide a polynomial by one of its binomial factors, the quotient is called a depressed polynomial. A **depressed polynomial** has a degree that is one less than the original polynomial. From the results of the division, and by using the Remainder Theorem, we can make the following statement.

$$\underbrace{2x^3 - 3x^2 - 17x + 30}_{\text{Dividend}} \underbrace{=}_{\text{equals}} \underbrace{(2x^2 - 9x + 10)}_{\text{quotient}} \underbrace{\cdot}_{\text{times}} \underbrace{(x + 3)}_{\text{divisor}} \underbrace{+}_{\text{plus}} \underbrace{0}_{\text{remainder.}}$$

Since the remainder is 0, $f(-3) = 0$. This means that $x + 3$ is a factor of $2x^3 - 3x^2 - 17x + 30$. This illustrates the **Factor Theorem**, which is a special case of the Remainder Theorem.

> **Key Concept** Factor Theorem
>
> The binomial $x - r$ is a factor of the polynomial $P(x)$ if and only if $P(r) = 0$.

The Factor Theorem can be used to determine whether a binomial is a factor of a polynomial. It can also be used to determine all of the factors of a polynomial.

Example 3 Use the Factor Theorem

Determine whether $x - 5$ is a factor of $x^3 - 7x^2 + 7x + 15$. Then find the remaining factors of the polynomial.

The binomial $x - 5$ is a factor of the polynomial if 5 is a zero of the related polynomial function. Use the Factor Theorem and synthetic division.

$$\begin{array}{r|rrrr} 5 & 1 & -7 & 7 & 15 \\ & & 5 & -10 & -15 \\ \hline & 1 & -2 & -3 & 0 \end{array}$$

Study Tip

Factoring The factors of a polynomial do not have to be binomials. For example, the factors of $x^3 + x^2 - x + 15$ are $x + 3$ and $x^2 - 2x + 5$.

Because the remainder is 0, $x - 5$ is a factor of the polynomial. The polynomial $x^3 - 7x^2 + 7x + 15$ can be factored as $(x - 5)(x^2 - 2x - 3)$. The polynomial $x^2 - 2x - 3$ is the depressed polynomial. Check to see if this polynomial can be factored.

$x^2 - 2x - 3 = (x + 1)(x - 3)$ Factor the trinomial.

So, $x^3 - 7x^2 + 7x + 15 = (x - 5)(x + 1)(x - 3)$.

You can check your answer by multiplying out the factors and seeing if you come up with the initial polynomial.

▶ **Guided Practice**

3A. Show that $x - 2$ is a factor of $x^3 - 7x^2 + 4x + 12$. Then find the remaining factors of the polynomial.

3B. Show that $x + 4$ and $x + 3$ are factors of $x^4 + 9x^3 + 23x^2 + 3x - 36$. Then find the remaining factors of the polynomial.

Check Your Understanding

= Step-by-Step Solutions begin on page R11.

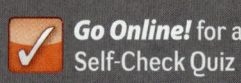

Example 1   Use synthetic substitution to find $f(4)$ and $f(-2)$ for each function.

1. $f(x) = 2x^3 - 5x^2 - x + 14$
2. $f(x) = x^4 + 8x^3 + x^2 - 4x - 10$

Example 2

3. **NATURE** The approximate number of bald eagle nesting pairs in the United States can be modeled by the function $P(x) = -0.16x^3 + 15.83x^2 - 154.15x + 1147.97$, where x is the number of years since 1970. About how many nesting pairs of bald eagles can be expected in 2018?

Example 3   Given a polynomial and one of its factors, find the remaining factors of the polynomial.

4. $x^3 - 6x^2 + 11x - 6;\ x - 1$
5. $x^3 + x^2 - 16x - 16;\ x + 1$
6. $3x^3 + 10x^2 - x - 12;\ x - 1$
7. $2x^4 - 3x^3 - 33x^2 - 13x + 15;\ x + 3$ and $x + 1$

Practice and Problem Solving

Extra Practice is on page R4.

Example 1   Use synthetic substitution to find $f(-5)$ and $f(2)$ for each function.

8. $f(x) = x^3 + 2x^2 - 3x + 1$
9. $f(x) = x^2 - 8x + 6$
10. $f(x) = 3x^4 + x^3 - 2x^2 + x + 12$
11. $f(x) = 2x^3 - 8x^2 - 2x + 5$
12. $f(x) = x^3 - 5x + 2$
13. $f(x) = x^5 + 8x^3 + 2x - 15$
14. $f(x) = x^6 - 4x^4 + 3x^2 - 10$
15. $f(x) = x^4 - 6x - 8$

Example 2

16. **FINANCIAL LITERACY** A specific car's fuel economy in miles per gallon can be approximated by $f(x) = 0.00000056x^4 - 0.000018x^3 - 0.016x^2 + 1.38x - 0.38$, where x represents the car's speed in miles per hour. Determine the fuel economy when the car is traveling 40, 50, and 60 miles per hour.

Example 3   Given a polynomial and one of its factors, find the remaining factors of the polynomial.

17. $x^3 - 3x + 2;\ x + 2$
18. $x^4 + 2x^3 - 8x - 16;\ x + 2$
19. $x^3 - x^2 - 10x - 8;\ x + 2$
20. $x^3 - x^2 - 5x - 3;\ x - 3$
21. $2x^3 + 17x^2 + 23x - 42;\ x - 1$
22. $2x^3 + 7x^2 - 53x - 28;\ x - 4$
23. $x^4 + 2x^3 + 2x^2 - 2x - 3;\ x - 1$
24. $x^4 + 6x^3 + 7x^2 - 6x - 8;\ x + 2$ and $x + 4$
25. $6x^3 - 25x^2 + 2x + 8;\ 2x + 1$
26. $4x^4 - 16x^3 + 7x^2 + 36x - 36;\ (x - 2)^2$

27. **PERSONAL WATERCRAFT** A personal watercraft traveling against the waves on Lake Tahoe accelerates from a resting position. Suppose the speed of the watercraft in feet per second is given by the function $f(t) = -0.001t^4 + 0.059t^3 - 1.51t^2 + 18.94t + 0.027$ where t is the time in seconds.
    a. Find the speed of the watercraft at 1, 2, and 3 seconds.
    b. It takes 8 seconds for the watercraft to travel between two buoys while it is accelerating. Use synthetic substitution to evaluate $f(8)$ and explain its meaning.

28. **REASONING** A company's sales, in millions of dollars, of consumer electronics can be modeled by $S(x) = -1.2x^3 + 18x^2 + 26.4x + 678$, where x is the number of years since 2010.
    a. Use synthetic substitution to estimate the sales for 2022 and 2025.
    b. Do you think this model is useful in estimating future sales? Explain.

Use the graph to find all of the factors for each polynomial function.

29.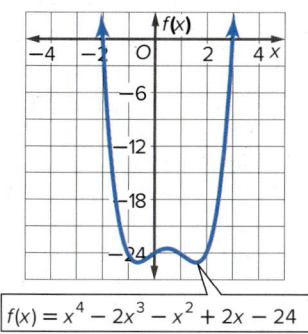
$f(x) = x^4 - 2x^3 - x^2 + 2x - 24$

30.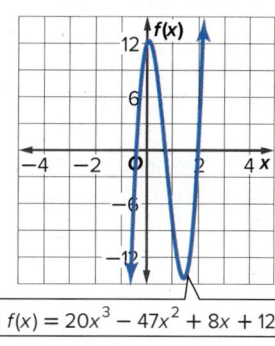
$f(x) = 20x^3 - 47x^2 + 8x + 12$

31. **MULTIPLE REPRESENTATIONS** In this problem, you will consider the function $f(x) = -9x^5 + 104x^4 - 249x^3 - 456x^2 + 828x + 432$.

   a. **Algebraic** If $x - 6$ is a factor of the function, find the depressed polynomial.

   b. **Tabular** Make a table of values for $-5 \leq x \leq 6$ for the depressed polynomial.

   c. **Analytical** What conclusions can you make about the locations of the other zeros based on the table? Explain your reasoning.

   d. **Graphical** Graph the original function to confirm your conclusions.

MP PERSEVERANCE Find values of k so that each remainder is 3.

32. $(x^2 - x + k) \div (x - 1)$

33. $(x^2 + kx - 17) \div (x - 2)$

34. $(x^2 + 5x + 7) \div (x + k)$

35. $(x^3 + 4x^2 + x + k) \div (x + 2)$

H.O.T. Problems   Use Higher-Order Thinking Skills

36. **OPEN-ENDED** Write a polynomial function that has a double zero of 1 and a double zero of −5. Graph the function.

CHALLENGE Find the solutions of each polynomial function.

37. $(x^2 - 4)^2 - (x^2 - 4) - 2 = 0$

38. $(x^2 + 3)^2 - 7(x^2 + 3) + 12 = 0$

39. **MP REASONING** Polynomial $f(x)$ is divided by $x - c$. What can you conclude if:

   a. the remainder is 0?
   b. the remainder is 1?
   c. the quotient is 1, and the remainder is 0?

40. **CHALLENGE** Review the definition for the Factor Theorem. Provide a proof of the theorem.

41. **OPEN-ENDED** Write a cubic function that has a remainder of 8 for $f(2)$ and a remainder of −5 for $f(3)$.

42. **CHALLENGE** Show that the quartic function $f(x) = ax^4 + bx^3 + cx^2 + dx + e$ will always have a rational zero when the numbers 1, −2, 3, 4, and −6 are randomly assigned to replace a through e, and all of the numbers are used.

43. **WRITING IN MATH** Explain how the zeros of a function can be located by using the Remainder Theorem and making a table of values for different input values and then comparing the remainders.

Preparing for Assessment

44. Which binomial is a factor of $x^3 - 6x^2 + 3x + 10$? **MP 7**

- A $x + 1$
- B $x + 2$
- C $x + 5$
- D $x + 10$

45. For what value of n is the polynomial $x^3 - 9x^2 + nx - 12$ divisible by $x - 3$? **MP 7**

- A -40
- B 22
- C 1
- D 6
- E 32

46. Which binomial is a factor of $6x^3 - 35x^2 + 34x + 40$? **MP 7**

- A $2x - 3$
- B $2x - 5$
- C $3x - 2$
- D $5x - 2$

47. Which binomial is a factor of $x^3 + x^2 - 24x + 36$? **MP 7**

- A $x + 2$
- B $x + 3$
- C $x + 6$
- D $x + 9$

48. What is the remainder of $(-3x^3 - 4x^2 + 10x - 7) \div (x - 3)$? What is the quotient? **MP 7**

49. What is the remainder when $a^3 - 4$ is divided by $a + 2$? **MP 7**

- A -12
- B -6
- C -2
- D 0
- E 4

50. Which binomials are factors of $P(x) = 2x^4 - 11x^3 + 15x^2 + 4x - 12$? Explain. Select all the apply. **MP 7**

- A $x - 2$
- B $x - 3$
- C $x - 4$
- D $x - 6$
- E $x - 12$

51. For what value of c is $p(x) = 2x^4 - 5x^2 + cx - 1$ divisible by $x - 1$? **MP 7**

52. Which of the following is NOT a factor of the polynomial, $x^3 + 4x^2 - 25x = 100$? **MP 7**

- A $x - 4$
- B $x - 5$
- C $x + 4$
- D $x + 5$

53. Find the value of c and of d such that $(x + 1)$ and $(x - 1)$ are factors of $x^3 - 3x^2 + cx + d$. **MP 7**

54. Show that $x - 4$ is a factor of $x^3 - 8x^2 + 4x + 48$. **MP 2**

LESSON 9
Roots and Zeros

::Then
- You used complex numbers to describe solutions of quadratic equations.

::Now
1. Determine the number and type of roots for a polynomial equation.
2. Find the zeros of a polynomial function.

::Why?
The function $g(x) = 0.002x^4 - 0.049x^3 + 0.404x^2 - 1.00x + 1.979$ can be used to model the average price of a gallon of gasoline in a given year if x is the number of years since 2000. To find the average price of gasoline in a specific year, you can use the roots of the related polynomial equation.

MP Mathematical Practices
6 Attend to precision.

1 Types of Roots Previously, you learned that a zero of a function $f(x)$ is any value c such that $f(c) = 0$. When the function is graphed, the real zeros of the function are the x-intercepts of the graph.

Concept Summary Zeros, Factors, Roots, and Intercepts

Words Let $P(x) = a_n x^n + \cdots + a_1 x + a_0$ be a polynomial function. Then the following statements are equivalent.
- c is a zero of $P(x)$.
- c is a root or solution of $P(x) = 0$.
- $x - c$ is a factor of $a_n x^n + \cdots + a_1 x + a_0$.
- If c is a real number, then $(c, 0)$ is an x-intercept of the graph of $P(x)$.

Example Consider the polynomial function $P(x) = x^4 + 2x^3 - 7x^2 - 8x + 12$.

The zeros of $P(x) = x^4 + 2x^3 - 7x^2 - 8x + 12$ are $-3, -2, 1,$ and 2.

The roots of $x^4 + 2x^3 - 7x^2 - 8x + 12 = 0$ are $-3, -2, 1,$ and 2.

The factors of $x^4 + 2x^3 - 7x^2 - 8x + 12$ are $(x + 3), (x + 2), (x - 1),$ and $(x - 2)$.

The x-intercepts of the graph of $P(x) = x^4 + 2x^3 - 7x^2 - 8x + 12$ are $(-3, 0), (-2, 0), (1, 0),$ and $(2, 0)$.

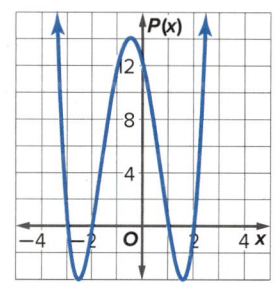

When solving a polynomial equation with degree greater than zero, there may be one or more real roots or no real roots (the roots are imaginary numbers). Since real numbers and imaginary numbers both belong to the set of complex numbers, all polynomial equations with degree greater than zero will have at least one root in the set of complex numbers. This is the **Fundamental Theorem of Algebra**.

Key Concept Fundamental Theorem of Algebra

Every polynomial equation with degree greater than zero has at least one root in the set of complex numbers.

connectED.mcgraw-hill.com   293

Example 1 Determine Number and Type of Roots

Solve each equation. State the number and type of roots.

a. $x^2 + 6x + 9 = 0$

$$x^2 + 6x + 9 = 0 \quad \text{Original equation}$$
$$(x + 3)^2 = 0 \quad \text{Factor.}$$
$$x + 3 = 0 \quad \text{Take the root of each side.}$$
$$x = -3 \quad \text{Solve for } x.$$

Because $(x + 3)$ is twice a factor of $x^2 + 6x + 9$, -3 is a double root. Thus, the equation has one real repeated root, -3.

CHECK The graph of the equation touches the x-axis at $x = -3$. Because -3 is a double root, the graph does not cross the axis. ✓

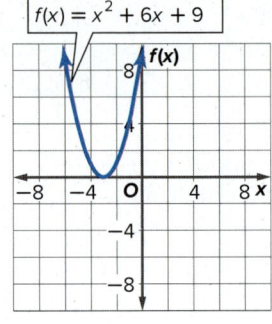

> **Reading Math**
>
> **Repeated Roots** Polynomial equations can have double roots, triple roots, quadruple roots, and so on. In general, these are referred to as *multiple roots*.

b. $x^3 + 25x = 0$

$$x^3 + 25x = 0 \quad \text{Original equation}$$
$$x(x^2 + 25) = 0 \quad \text{Factor.}$$
$$x = 0 \text{ or } x^2 + 25 = 0$$
$$x^2 = -25$$
$$x = \pm\sqrt{-25} \text{ or } \pm 5i$$

This equation has one real root, 0, and two imaginary roots, $5i$ and $-5i$.

CHECK The graph of this equation crosses the x-axis at only one place, $x = 0$. ✓

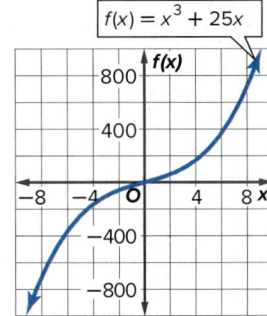

Guided Practice

1A. $x^3 + 2x = 0$ **1B.** $x^4 - 16 = 0$

1C. $x^3 + 4x^2 - 7x - 10 = 0$ **1D.** $3x^3 - x^2 + 9x - 3 = 0$

Examine the solutions for each equation in Example 1. Notice that the number of solutions for each equation is the same as the degree of each polynomial. The following corollary to the Fundamental Theorem of Algebra describes this relationship between the degree and the number of roots of a polynomial equation.

Key Concept Corollary to the Fundamental Theorem of Algebra

Words	A polynomial equation of degree n has exactly n roots in the set of complex numbers, including repeated roots.
Example	$x^3 + 2x^2 + 6 = 0$ $4x^4 - 3x^3 + 5x - 6 = 0$ $-2x^5 - 3x^2 + 8 = 0$
	3 roots 4 roots 5 roots

Similarly, an *n*th degree polynomial function has exactly *n* zeros.

Additionally, French mathematician René Descartes discovered a relationship between the signs of the coefficients of a polynomial function and the number of positive and negative real zeros.

Study Tip

Zero at the Origin If a zero of a function is at the origin, the sum of the number of positive real zeros, negative real zeros, and imaginary zeros is reduced by how many times 0 is a zero of the function.

Key Concept Descartes' Rule of Signs

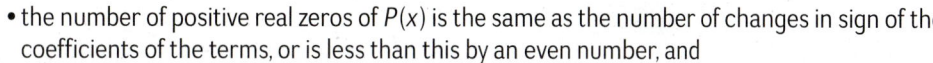

Let $P(x) = a_n x^n + \cdots + a_1 x + a_0$ be a polynomial function with real coefficients. Then

- the number of positive real zeros of $P(x)$ is the same as the number of changes in sign of the coefficients of the terms, or is less than this by an even number, and
- the number of negative real zeros of $P(x)$ is the same as the number of changes in sign of the coefficients of the terms of $P(-x)$, or is less than this by an even number.

Example 2 Find Numbers of Positive and Negative Zeros

State the possible number of positive real zeros, negative real zeros, and imaginary zeros of $f(x) = x^6 + 3x^5 - 4x^4 - 6x^3 + x^2 - 8x + 5$.

Because $f(x)$ has degree 6, it has six zeros, either real or imaginary. Use Descartes' Rule of Signs to determine the possible number and type of *real* zeros.

Count the number of changes in sign for the coefficients of $f(x)$.

$$f(x) = x^6 \underset{\text{+ to +}}{\overset{\text{no}}{+}} 3x^5 \underset{\text{+ to -}}{\overset{\text{yes}}{-}} 4x^4 \underset{\text{- to -}}{\overset{\text{no}}{-}} 6x^3 \underset{\text{- to +}}{\overset{\text{yes}}{+}} x^2 \underset{\text{+ to -}}{\overset{\text{yes}}{-}} 8x \underset{\text{- to +}}{\overset{\text{yes}}{+}} 5$$

There are 4 sign changes, so there are 4, 2, or 0 positive real zeros.

Count the number of changes in sign for the coefficients of $f(-x)$.

$$f(-x) = (-x)^6 + 3(-x)^5 - 4(-x)^4 - 6(-x)^3 + (-x)^2 - 8(-x) + 5$$
$$= x^6 \underset{\text{+ to -}}{\overset{\text{yes}}{-}} 3x^5 \underset{\text{- to -}}{\overset{\text{no}}{-}} 4x^4 \underset{\text{- to +}}{\overset{\text{yes}}{+}} 6x^3 \underset{\text{+ to +}}{\overset{\text{no}}{+}} x^2 \underset{\text{+ to +}}{\overset{\text{no}}{+}} 8x \underset{\text{+ to +}}{\overset{\text{no}}{+}} 5$$

There are 2 sign changes, so there are 2 or 0 negative real zeros.
Make a chart of the possible combinations of real and imaginary zeros.

Number of Positive Real Zeros	Number of Negative Real Zeros	Number of Imaginary Zeros	Total Number of Zeros
4	2	0	$4 + 2 + 0 = 6$
4	0	2	$4 + 0 + 2 = 6$
2	2	2	$2 + 2 + 2 = 6$
2	0	4	$2 + 0 + 4 = 6$
0	2	4	$0 + 2 + 4 = 6$
0	0	6	$0 + 0 + 6 = 6$

Go Online!

Watch and listen to the **animation** to learn about finding the number of positive and negative zeros. Summarize what you hear for a partner, and have them ask you questions to help your understanding.

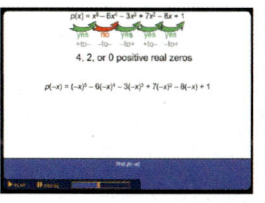

Guided Practice

2. State the possible number of positive real zeros, negative real zeros, and imaginary zeros of $h(x) = 2x^5 + x^4 + 3x^3 - 4x^2 - x + 9$.

2 Find Zeros
You can use the various strategies and theorems you have learned to find all of the zeros of a function.

Example 3   Use Synthetic Substitution to Find Zeros

Find all of the zeros of $f(x) = x^4 - 18x^2 + 12x + 80$.

Step 1 Determine the total number of zeros.

Because $f(x)$ has degree 4, the function has 4 zeros.

Step 2 Determine the type of zeros.

Examine the number of sign changes for $f(x)$ and $f(-x)$.

$f(x) = x^4 - 18x^2 + 12x + 80$ $f(-x) = x^4 - 18x^2 - 12x + 80$

       yes   yes   no   yes   no   yes

Because there are 2 sign changes for the coefficients of $f(x)$, the function has 2 or 0 positive real zeros. Because there are 2 sign changes for the coefficients of $f(-x)$, $f(x)$ has 2 or 0 negative real zeros. Thus, $f(x)$ has 4 real zeros, 2 real zeros, and 2 imaginary zeros, or 4 imaginary zeros.

> **Study Tip**
>
> **Testing for Zeros** If a value is not a zero for a polynomial, then it will not be a zero for the depressed polynomial either, so it does not need to be checked again.

Step 3 Determine the real zeros. List some possible values, and then use synthetic substitution to evaluate $f(x)$ for real values of x.

x	1	0	-18	12	80
-3	1	-3	-9	39	-37
-2	1	-2	-14	40	0
-1	1	-1	-17	29	51
0	1	0	-18	12	80
1	1	1	-17	-5	75
2	1	2	-14	-2	76

Each row shows the coefficients of the depressed polynomial and the remainder.

From the table, we can see that one zero occurs at $x = -2$. Because there are 2 negative real zeros, use synthetic substitution with the depressed polynomial function $f(x) = x^3 - 2x^2 - 14x + 40$ to find a second negative zero.

A second negative zero is at $x = -4$. Because the depressed polynomial $x^2 - 6x + 10$ is quadratic, use the Quadratic Formula to find the remaining zeros of $f(x) = x^2 - 6x + 10$.

x	1	-2	-14	40
-4	1	-6	10	0
-5	1	-7	21	-65
-6	1	-8	34	-164

$x = \dfrac{-b \pm \sqrt{b^2 - 4ac}}{2a}$ Quadratic Formula

$= \dfrac{-(-6) \pm \sqrt{(-6)^2 - 4(1)(10)}}{2(1)}$ Replace a with 1, b with -6, and c with 10.

$= 3 \pm i$ Simplify.

The function has zeros at -4, -2, $3 + i$, and $3 - i$.

> **Study Tip**
>
> **MP Tools** Refer to Lesson 4-2 on how to use the **CALC** menu to locate a zero on your calculator.

CHECK Graph the function on a graphing calculator. The graph crosses the x-axis two times, so there are two real zeros. Use the zero function under the **CALC** menu to locate each zero. The two real zeros are -4 and -2.

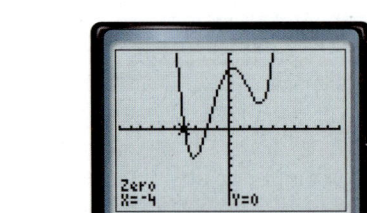

[−10, 10] scl: 1 by [−100, 100] scl: 10

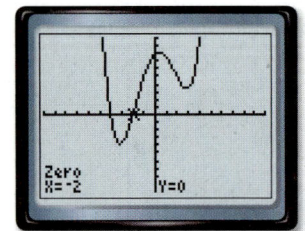

[−10, 10] scl: 1 by [−100, 100] scl: 10

▶ **Guided Practice**

3. Find all of the zeros of $h(x) = x^3 + 2x^2 + 9x + 18$.

Review Vocabulary

complex conjugates two complex numbers of the form $a + bi$ and $a - bi$

In Chapter 4, you learned that the product of complex conjugates is always a real number and that complex roots always come in conjugate pairs. For example, if one root of $x^2 - 8x + 52 = 0$ is $4 + 6i$, then the other root is $4 - 6i$.

This applies to the zeros of polynomial functions as well. For any polynomial function with real coefficients, if an imaginary number is a zero of that function, its conjugate is also a zero. This is called the **Complex Conjugates Theorem**.

> **Key Concept** Complex Conjugates Theorem
>
> **Words** Let a and b be real numbers, and $b \neq 0$. If $a + bi$ is a zero of a polynomial function with real coefficients, then $a - bi$ is also a zero of the function.
>
> **Example** If $3 + 4i$ is a zero of $f(x) = x^3 - 4x^2 + 13x + 50$, then $3 - 4i$ is also a zero of the function.

When you are given all of the zeros of a polynomial function and are asked to determine the function, convert the zeros to factors and then multiply all of the factors together. The result is the polynomial function.

Example 4 Use Zeros to Write a Polynomial Function

Write a polynomial function of least degree with integral unconscious, the zeros of which include -1 and $5 - i$.

Understand If $5 - i$ is a zero, then $5 + i$ is also a zero according to the Complex Conjugates Theorem. So, $x + 1$, $x - (5 - i)$, and $x - (5 + i)$ are factors of the polynomial.

Plan Write the polynomial function as a product of its factors.
$P(x) = (x + 1)[x - (5 - i)][x - (5 + i)]$

Solve Multiply the factors to find the polynomial function.

$P(x) = (x + 1)[x - (5 - i)][x - (5 + i)]$ Write the equation.
$= (x + 1)[(x - 5) + i][(x - 5) - i]$ Regroup terms.
$= (x + 1)[(x - 5)^2 - i^2]$ Difference of squares
$= (x + 1)[(x^2 - 10x + 25) - (-1)]$ Square terms.
$= (x + 1)(x^2 - 10x + 26)$ Simplify.
$= x^3 - 10x^2 + 26x + x^2 - 10x + 26$ Multiply.
$= x^3 - 9x^2 + 16x + 26$ Combine like terms.

Check Substitute $x = -1$, $x = 5 - i$, and $x = 5 + i$ into $P(x)$ to verify that these are the zeros of the polynomial function.

Because there are 3 zeros, the degree of the polynomial function must be 3, so $P(x) = x^3 - 9x^2 + 16x + 26$ is a reasonable solution for a polynomial function of least degree.

▶ **Guided Practice**

4. Write a polynomial function of least degree with integral coefficients having zeros that include -1 and $1 + 2i$.

Check Your Understanding

= Step-by-Step Solutions begin on page R11.

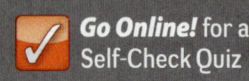

Example 1 Solve each equation. State the number and type of roots.

1. $x^2 - 3x - 10 = 0$
2. $x^3 + 12x^2 + 32x = 0$
3. $16x^4 - 81 = 0$
4. $0 = x^3 - 8$

Example 2 State the possible number of positive real zeros, negative real zeros, and imaginary zeros of each function.

5. $f(x) = x^3 - 2x^2 + 2x - 6$
6. $f(x) = 6x^4 + 4x^3 - x^2 - 5x - 7$
7. $f(x) = 3x^5 - 8x^3 + 2x - 4$
8. $f(x) = -2x^4 - 3x^3 - 2x - 5$

Example 3 Find all of the zeros of each function.

9. $f(x) = x^3 + 9x^2 + 6x - 16$
10. $f(x) = x^3 + 7x^2 + 4x + 28$
11. $f(x) = x^4 - 2x^3 - 8x^2 - 32x - 384$
12. $f(x) = x^4 - 6x^3 + 9x^2 + 6x - 10$

Example 4 Write a polynomial function of least degree with integral coefficients that have the given zeros.

13. $4, -1, 6$
14. $3, -1, 1, 2$
15. $-2, 5, -3i$
16. $-4, 4 + i$

Practice and Problem Solving

Extra Practice is on page R4.

Example 1 Solve each equation. State the number and type of roots.

17. $2x^2 + x - 6 = 0$
18. $4x^2 + 1 = 0$
19. $x^3 + 1 = 0$
20. $2x^2 - 5x + 14 = 0$
21. $-3x^2 - 5x + 8 = 0$
22. $8x^3 - 27 = 0$
23. $16x^4 - 625 = 0$
24. $x^3 - 6x^2 + 7x = 0$
25. $x^5 - 8x^3 + 16x = 0$
26. $x^5 + 2x^3 + x = 0$

Example 2 State the possible number of positive real zeros, negative real zeros, and imaginary zeros of each function.

27. $f(x) = x^4 - 5x^3 + 2x^2 + 5x + 7$
28. $f(x) = 2x^3 - 7x^2 - 2x + 12$
29. $f(x) = -3x^5 + 5x^4 + 4x^2 - 8$
30. $f(x) = x^4 - 2x^2 - 5x + 19$
31. $f(x) = 4x^6 - 5x^4 - x^2 + 24$
32. $f(x) = -x^5 + 14x^3 + 18x - 36$

Example 3 Find all of the zeros of each function.

33. $f(x) = x^3 + 7x^2 + 4x - 12$
34. $f(x) = x^3 + x^2 - 17x + 15$
35. $f(x) = x^4 - 3x^3 - 3x^2 - 75x - 700$
36. $f(x) = x^4 + 6x^3 + 73x^2 + 384x + 576$
37. $f(x) = x^4 - 8x^3 + 20x^2 - 32x + 64$
38. $f(x) = x^5 - 8x^3 - 9x$

Example 4 Write a polynomial function of least degree with integral coefficients that have the given zeros.

39. $5, -2, -1$
40. $-4, -3, 5$
41. $-1, -1, 2i$
42. $-3, 1, -3i$
43. $0, -5, 3 + i$
44. $-2, -3, 4 - 3i$

45. **MP REASONING** A frozen yogurt manufacturer determines that the profit for producing x batches of frozen yogurt per day is $P(x) = -0.006x^4 + 0.15x^3 - 0.05x^2 - 1.8x$.

   a. How many positive real zeros, negative real zeros, and imaginary zeros exist?

   b. What is the meaning of the zeros in this situation?

Sketch the graph of each function using its zeros.

46. $f(x) = x^3 - 5x^2 - 2x + 24$

47. $f(x) = 4x^3 + 2x^2 - 4x - 2$

48. $f(x) = x^4 - 6x^3 + 7x^2 + 6x - 8$

49. $f(x) = x^4 - 6x^3 + 9x^2 + 4x - 12$

Match each graph to the given zeros.

a. $-3, 4, i, -i$ b. $-4, 3$ c. $-4, 3, i, -i$

50.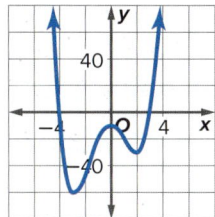

51.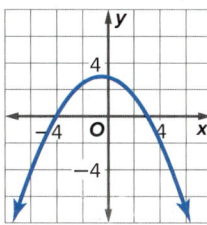

52.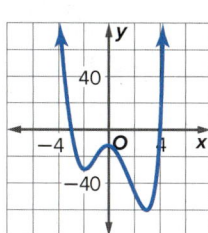

53. **MULTI-STEP** Morgan paid $5000 for new furniture last month with a credit card. She hopes to pay off the card in 6 months at $900 per month.
    a. What monthly interest rate would she need to get in order to pay off the card in 6 months?
    b. Explain your solution process.

Determine the number of positive real zeros, negative real zeros, and imaginary zeros for each function. Explain your reasoning.

54.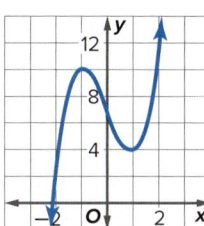

degree: 3

55.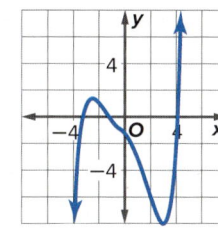

degree: 5

H.O.T. Problems Use Higher-Order Thinking Skills

56. **OPEN-ENDED** Sketch the graph of a polynomial function with:
    a. 3 real, 2 imaginary zeros b. 4 real zeros c. 2 imaginary zeros

57. **CHALLENGE** Write an equation in factored form of a polynomial function of degree 5 with 2 imaginary zeros, 1 nonintegral zero, and 2 irrational zeros. Explain.

58. **MP ARGUMENTS** Determine which equation is not like the others. Explain.

    $r^4 + 1 = 0$ $r^3 + 1 = 0$ $r^2 - 1 = 0$ $r^3 - 8 = 0$

59. **MP REASONING** Provide a counterexample for each statement.
    a. All polynomial functions of degree greater than 2 have at least 1 negative real zero.
    b. All polynomial functions of degree greater than 2 have at least 1 positive real zero.

60. **WRITING IN MATH** Explain to a friend how you would use Descartes' Rule of Signs to determine the number of possible positive real zeros and the number of possible negative zeros of the polynomial function $f(x) = x^4 - 2x^3 + 6x^2 + 5x - 12$.

Preparing for Assessment

61. Which statement describes the roots of $x^6 - 8x^4 - 21x^2 + 108$? **MP 6**

- A 1 real root, 5 imaginary
- B 3 real roots, 3 imaginary
- C 4 real roots, 2 imaginary
- D 5 real roots, 1 imaginary
- E 6 real roots, 0 imaginary

62. Let $4 + 2i$ be a zero of the function $f(x)$. Which value must also be a zero of the function? **MP 6**

- A $-4 - 2i$
- B $-4 + 2i$
- C $2 - 4i$
- D $2 + 4i$
- E $4 - 2i$

63. Let $6i$ be a zero of the function $g(x)$. Which value must also be a zero of the function? **MP 6**

- A -6
- B 6
- C $-6i$
- D $6 - i$
- E $-6 - i$

64. Let $f(x)$ be a function with 2 real zeros and 4 imaginary zeros. What is the number of real zeros of $g(x) = x \cdot f(x)$? **MP 6**

- A 2
- B 3
- C 4
- D 5
- E 7

65. Let $-3i$ be a zero of $p(x)$. Which value must also be a zero of the function? **MP 6**

66. Let $3 - 5i$ be a zero of function $q(x)$. Which value must also be a zero of the function? **MP 6**

67. Which polynomial function is the function of least degree whose zeros include -3, 2, and $5 + 2i$? **MP 6**

- A $x^3 - 2x^2 - 9x + 18$
- B $x^4 + x^3 + 15x^2 + 21x - 126$
- C $x^4 - 9x^3 + 10x^2 + 86x - 126$
- D $x^4 + x^3 + 23x^2 + 29x - 174$
- E $x^4 - 9x^3 + 13x^2 + 89x - 174$

68. Which statement describes the roots of $x^5 - 12x^3 + 32x$? **MP 6**

- A 5 real root
- B 3 real root, 2 imaginary
- C 1 real root, 4 imaginary
- D 5 imaginary

69. MULTI-STEP Consider the function, $f(x) = 2x^5 + 4x^4 + 9x^3 + 18x^2 - 35x - 70$. **MP 6**

a. What is the number of possible positive real zeros?

b. What is the number of possible negative real zeros?

c. Show that $x = -2$ is a real zero of the polynomial.

70. Write a polynomial function that has 0 possible positive real zeros and 5, 3, or 1 possible negative real zero. **MP 6**

71. Write the number of possible positive real zeros of each function: **MP 6**

a. $x^6 - 64$

b. $x^7 - 64x$

c. $16x^8 - 153x^4 + 81$

72. Write the number of possible positive complex zeros of each function: **MP 6**

a. $x^6 - 64$

b. $x^7 - 64x$

c. $16x^8 - 153x^4 + 81$

EXTEND 4-9

Graphing Technology Lab
Analyzing Polynomial Functions

You can use graphing technology to help you identify real zeros, maximum and minimum points, number and type of zeros, *y*-intercepts, and symmetry of polynomial functions.

Mathematical Practices
 5 Use appropriate tools strategically.

Activity Identify Polynomial Characteristics

Work cooperatively. Graph each function. Identify the real zeros, maximum and minimum points, number and type of zeros, *y*-intercepts, and symmetry.

a. $g(x) = 3x^4 - 15x^3 + 87x^2 - 375x + 300$

Step 1 Graph the equation.

Step 2 Use 2nd [CALC] **zero** to find the real zeros at $x = 1$ and $x = 4$.

Step 3 Use 2nd [CALC] **minimum** to find the relative minimum at $(2.68, -214.11)$. There is no relative maximum point.

Step 4 $g(x)$ has degree 4 and can have at most 4 zeros. Two real zeros were found through graphing. The other two zeros are either multiple zeros or imaginary zeros.

Step 5 Use 2nd [CALC] **value** 0 to find the *y*-intercept, 300.

Step 6 The line of symmetry passes through the vertex. Its equation is $x = 2.68$.

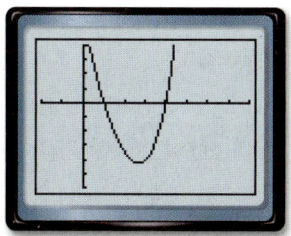
[−2, 8] scl: 1 by [−300, 200] scl: 50

b. $f(x) = 2x^5 - 5x^4 - 3x^3 + 8x^2 + 4x$

Step 1 Graph the equation.

Step 2 Locate the real zeros at $x = -1$, $x = -\frac{1}{2}$, $x = 0$ and $x = 2$.

Step 3 Find the relative maxima at $(-0.81, 0.75)$ and $(1.04, 6.02)$ and the relative minima at $(-0.24, -0.48)$ and $(2, 0)$.

Step 4 $f(x)$ has degree 5 and can have at most 5 zeros. Four real zeros were found through graphing. The other zero is either a multiple zero or an imaginary zero. In this case, there is a double zero at $x = 2$.

Step 5 The *y*-intercept is 0 because the graph goes through the origin.

Step 6 There is no symmetry.

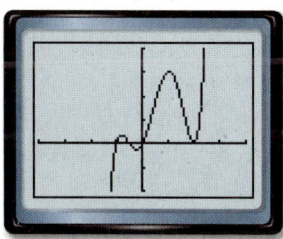
[−4, 4] scl: 1 by [−4, 8] scl: 2

Exercises

Work cooperatively. Graph each function. Identify the real zeros, maximum and minimum points, number and type of zeros, *y*-intercepts, and symmetry.

1. $f(x) = x^3 - 5x^2 + 6x$
2. $g(x) = x^4 - 3x^2 - 4$
3. $k(x) = -x^4 - x^3 + 2x^2$
4. $f(x) = -2x^3 - 4x^2 + 16x$
5. $g(x) = 3x^5 - 18x^4 + 27x^3$
6. $k(x) = x^4 - 8x^2 + 15$
7. $f(x) = -x^3 + 2x^2 + 8x$
8. $g(x) = x^5 + 3x^4 - 10x^2$

CHAPTER 4
Study Guide and Review

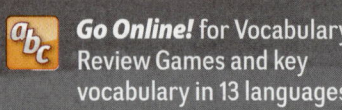
Go Online! for Vocabulary Review Games and key vocabulary in 13 languages

Study Guide

Key Concepts

Operations with Polynomials (Lessons 4-1 and 4-3)
- To add or subtract: Combine like terms.
- To multiply: Use the Distributive Property.
- To divide: Use long division or synthetic division.

Powers of Binomials (Lesson 4-2)
- The Binomial Theorem:
$$(a+b)^n = \sum_{k=0}^{n} \frac{n!}{(n-k)!\,k!} a^{n-k} b^k$$

Polynomial Functions and Graphs (Lessons 4-4 and 4-5)
- Turning points of a function are called *relative maxima* and *relative minima*.

Solving Polynomial Equations (Lesson 4-6)
- You can factor polynomials by using the GCF, grouping, or quadratic techniques.

Proving Polynomial Identities (Lesson 4-7)
- To prove a polynomial identity, simplify one side of the equation until the two sides of the equation are the same.

The Remainder and Factor Theorems (Lesson 4-8)
- Factor Theorem: The binomial $x - a$ is a factor of the polynomial $f(x)$ if and only if $f(a) = 0$.

Roots and Zeros (Lesson 4-9)
- Fundamental Theorem of Algebra: Every polynomial equation with degree greater than zero has at least one root in the set of complex numbers.
- Descartes' Rule of Signs gives the possible number of positive and negative real zeros.
- Complex Conjugates Theorem: If $a + bi$ is a zero of a function, then $a - bi$ is also a zero.

 **Study Organizer**

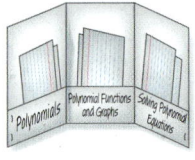

Use your Foldable to review the chapter. Working with a partner can be helpful. Ask for clarification of concepts as needed.

Key Vocabulary

Binomial Theorem (p. 237)
degree of a polynomial (p. 231)
depressed polynomial (p. 289)
end behavior (p. 255)
extrema (p. 263)
identity (p. 282)
leading coefficient (p. 253)
Location principle (p. 262)
Pascal's triangle (p. 237)
polynomial identity (p. 282)
polynomial function (p. 254)
polynomial in one variable (p. 253)
power function (p. 254)
prime polynomials (p. 274)
quadratic form (p. 277)
relative maximum (p. 263)
relative minimum (p. 263)
simplify (p. 229)
synthetic division (p. 244)
synthetic substitution (p. 287)
turning points (p. 263)

Vocabulary Check

State whether each sentence is *true* or *false*. If *false*, replace the underlined term to make a true sentence.

1. The coefficient of the first term of a polynomial in standard form is called the <u>leading coefficient</u>.
2. Polynomials that cannot be factored are called <u>polynomials in one variable</u>.
3. A <u>prime polynomial</u> has a degree that is one less than the original polynomial.
4. A point on the graph of a function where no other nearby point has a greater *y*-coordinate is called a <u>relative maximum</u>.
5. A <u>polynomial function</u> is a continuous function that can be described by a polynomial equation in one variable.
6. To <u>simplify</u> an expression containing powers means to rewrite the expression without parentheses or negative exponents.
7. <u>Synthetic division</u> is a shortcut method for dividing a polynomial by a binomial.
8. Using the <u>Binomial Theorem</u>, $(x-2)^4$ can be expanded to $x^4 - 8x^3 + 24x^2 - 32x + 16$.

Concept Check

9. Explain how Descartes' Rule of Signs gives the possible number of positive and negative real zeros.
10. Explain how to prove a polynomial identity.

Lesson-by-Lesson Review

4-1 Operations with Polynomials

Simplify. Assume that no variable equals 0.

11. $\dfrac{14x^4 y}{2x^3 y^5}$

12. $3t(tn - 5)$

13. $(4r^2 + 3r - 1) - (3r^2 - 5r + 4)$

14. $(x^4)^3$

15. $(m + p)(m^2 - 2mp + p^2)$

16. $3b(2b - 1) + 2b(b + 3)$

Example 1

Simplify each expression.

a. $(-4a^3 b^5)(5ab^3)$

$(-4a^3 b^5)(5ab^3) = (-4)(5)(a^3 \cdot a)(b^5 \cdot b^3)$ Product of Powers

$ = -20a^4 b^8$ Simplify.

b. $(2x^2 + 3x - 8) + (3x^2 - 5x - 7)$

$(2x^2 + 3x - 8) + (3x^2 - 5x - 7)$

$= (2x^2 + 3x^2) + (3x - 5x) + [-8 + (-7)]$

$= 5x^2 - 2x - 15$

4-2 Powers of Binomials

Expand each binomial.

17. $(a + b)^3$

18. $(y - 3)^7$

19. $(3 - 2z)^5$

20. $(4a - 3b)^4$

21. $\left(x - \dfrac{1}{4}\right)^5$

Find the indicated term of each expression.

22. third term of $(a + 2b)^8$

23. sixth term of $(3x + 4y)^7$

24. second term of $(4x - 5)^{10}$

Example 2

Expand $(x - 3y)^4$.

$(x - 3y)^4$

$= x^4 + {}_4C_1 x^3(-3y) + {}_4C_2 x^2(-3y)^2 + {}_4C_3(-3y)^4 + {}_4C_4(-3y)^4$

$= x^4 + \dfrac{4!}{3!}x^3(-3y) + \dfrac{4!}{2!2!}x^2(9y^2) + \dfrac{4!}{3!}x(-27y^3) + 81y^4$

$= x^4 + -12x^3 y + 54x^2 y^2 + -108xy^3 + 81y^4$

Example 3

Find the fourth term of $(x + y)^8$.

Use the Binomial Theorem to write the expansion in sigma notation.

$(x + y)^8 = \sum_{k=0}^{8} \dfrac{8!}{k!(8-k)!} x^{8-k} y^k$

For the fourth term, $k = 3$.

$\dfrac{8!}{k!(8-k)!} x^{8-k} y^k = \dfrac{8!}{3!(8-3)!} x^{8-3} y^3$

$= 56x^5 y^3$

connectED.mcgraw-hill.com   303

CHAPTER 4
Study Guide and Review Continued

4-3 Dividing Polynomials

Simplify.

25. $\dfrac{12x^4y^5 + 8x^3y^7 - 16x^2y^6}{4xy^5}$

26. $(6y^3 + 13y^2 - 10y - 24) \div (y + 2)$

27. $(a^4 + 5a^3 + 2a^2 - 6a + 4)(a + 2)^{-1}$

28. $(4a^6 - 5a^4 + 3a^2 - a) \div (2a + 1)$

29. **GEOMETRY** The volume of the rectangular prism is $3x^3 + 11x^2 - 114x - 80$ cubic units. What is the area of the base?

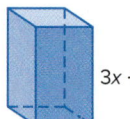
$3x + 2$

Example 4

Simplify $(6x^3 - 31x^2 - 34x + 22) \div (2x - 1)$.

$$
\begin{array}{r}
3x^2 - 14x - 24 \\
2x - 1 \overline{\smash{)}6x^3 - 31x^2 - 34x + 22} \\
\underline{(-)\; 6x^3 - 3x^2} \\
-28x^2 - 34x \\
\underline{(-)\; -28x^2 + 14x} \\
-48x + 22 \\
\underline{(-)\; -48x + 24} \\
-2
\end{array}
$$

The result is $3x^2 - 14x - 24 - \dfrac{2}{2x - 1}$.

4-4 Graphing Polynomial Functions

State the degree and leading coefficient of each polynomial in one variable. If it is not a polynomial in one variable, explain why.

30. $5x^6 - 3x^4 + x^3 - 9x^2 + 1$

31. $6xy^2 - xy + y^2$

32. $12x^3 - 5x^4 + 6x^8 - 3x - 3$

Find $p(-2)$ and $p(x + h)$ for each function.

33. $p(x) = x^2 + 2x - 3$

34. $p(x) = 3x^2 - x$

35. $p(x) = 3 - 5x^2 + x^3$

36. Identify the type of polynomial function based on the general shape of the graph.

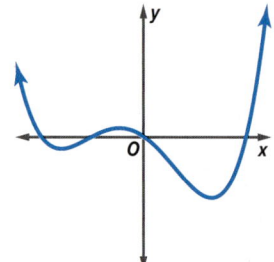

Example 5

What are the degree and leading coefficient of $4x^3 + 3x^2 - 7x^7 + 4x - 1$?

The greatest exponent is 7, so the degree is 7. The leading coefficient is -7.

Example 6

Find $p(a - 2)$ if $p(x) = 3x + 2x^2 - x^3$.

$p(a - 2) = 3(a - 2) + 2(a - 2)^2 - (a - 2)^3$

$= 3a - 6 + 2a^2 - 8a + 8 - (a^3 - 6a^2 + 12a - 8)$

$= -a^3 + 8a^2 - 17a + 10$

Example 7

Identify the type of polynomial function based on the general shape of the graph.

The graph represents a function of degree 5, or a quintic function.

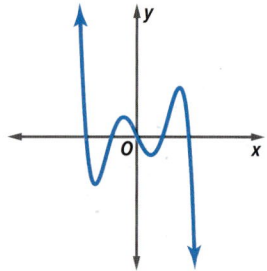

4-5 Analyzing Graphs of Polynomial Functions

Complete each of the following.

a. Graph each function by making a table of values.
b. Determine the consecutive integer values of x between which each real zero is located.
c. Estimate the x-coordinates at which the relative maxima and minima occur.

37. $h(x) = x^3 - 4x^2 - 7x + 10$
38. $g(x) = 4x^4 - 21x^2 + 5$
39. $f(x) = x^3 - 3x^2 - 4x + 12$
40. $h(x) = 4x^3 - 6x^2 + 1$
41. $p(x) = x^5 - x^4 + 1$

42. **BUSINESS** Milo tracked the monthly profits for his sports store business for the first six months of the year. They can be modeled by using the following six points: (1, 675), (2, 950), (3, 550), (4, 250), (5, 600), and (6, 400). How many turning points would the graph of a polynomial function through these points have? Describe them.

Example 8

Graph $f(x) = x^3 + 3x^2 - 4$ by making a table of values.

Make a table of values for several values of x.

x	−3	−2	−1	0	1	2
$f(x)$	−4	0	−2	−4	0	16

Plot the points and connect the points with a smooth curve.

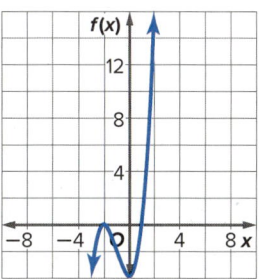

4-6 Solving Polynomial Equations

Factor completely. If the polynomial is not factorable, write *prime*.

43. $a^4 - 16$
44. $x^3 + 6y^3$
45. $54x^3y - 16y^4$
46. $6ay + 4by - 2cy + 3az + 2bz - cz$

Solve each equation.

47. $x^3 + 2x^2 - 35x = 0$
48. $8x^4 - 10x^2 + 3 = 0$

49. **GEOMETRY** The volume of the prism is 315 cubic inches. Find the value of x and the length, height, and width.

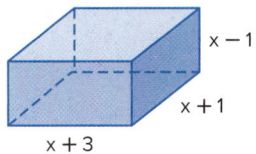

Example 9

Factor $r^7 + 64r$.

$r^7 + 64r = r(r^6 + 64)$ Factor by GCF.
$ = r[(r^2)^3 + 4^3]$ Write as cubes.
$ = r(r^2 + 4)(r^4 - 4r^2 + 16)$

Example 10

Solve $4x^4 - 25x^2 + 36 = 0$.

$(x^2 - 4)(4x^2 - 9) = 0$

$x^2 - 4 = 0$ or $4x^2 - 9 = 0$
$x^2 = 4$ $x^2 = \dfrac{9}{4}$
$x = \pm 2$ $x = \pm\dfrac{3}{2}$

The solutions are $-2, 2, -\dfrac{3}{2}$, and $\dfrac{3}{2}$.

CHAPTER 4
Study Guide and Review Continued

4-7 Proving Polynomial Identities

Prove each polynomial identity.

50. $(x + 2)^2 - 4 = x^2 + 4x$
51. $(a + b)^3 - (a^3 + b^3) = 3ab(a + b)$
52. $1 - p^4 = (1 - p)(p + 1)(p^2 + 1)$
53. $w(w + 2) + 3(w - 2) = (w + 6)(w - 1)$

Example 11

Prove the polynomial identity below.
$(x + 1)(4x + y) - y = x(4x + y + 4)$

Simplify the left side of the equation until the two sides of the equation are the same.

$(x + 1)(4x + y) - y = 4x^2 + xy + 4x + y - y$ Distribute.
$\qquad\qquad\qquad\qquad = 4x^2 + xy + 4x$ Combine like terms.
$\qquad\qquad\qquad\qquad = x(4x + y + 4)$ Factor out x.

4-8 The Remainder and Factor Theorems

Use synthetic substitution to find $f(-2)$ and $f(4)$ for each function.

54. $f(x) = x^2 - 3$
55. $f(x) = x^2 - 5x + 4$
56. $f(x) = x^3 + 4x^2 - 3x + 2$
57. $f(x) = 2x^4 - 3x^3 + 1$

Given a polynomial and one of its factors, find the remaining factors of the polynomial.

58. $3x^3 + 20x^2 + 23x - 10$; $x + 5$
59. $2x^3 + 11x^2 + 17x + 5$; $2x + 5$
60. $x^3 + 2x^2 - 23x - 60$; $x - 5$

Example 12

Determine whether $x - 6$ is a factor of $x^3 - 2x^2 - 21x - 18$.

```
6 | 1   -2   -21   -18
  |      6    24    18
  ---------------------
    1    4     3  |  0
```

$x - 6$ is a factor because $r = 0$.
$x^3 - 2x^2 - 21x - 18 = (x - 6)(x^2 + 4x + 3)$

4-9 Roots and Zeros

State the possible number of positive real zeros, negative real zeros, and imaginary zeros of each function.

61. $f(x) = -2x^3 + 11x^2 - 3x + 2$
62. $f(x) = -4x^4 - 2x^3 - 12x^2 - x - 23$
63. $f(x) = x^6 - 5x^3 + x^2 + x - 6$
64. $f(x) = -2x^5 + 4x^4 + x^2 - 3$
65. $f(x) = -2x^6 + 4x^4 + x^2 - 3x - 3$

Example 13

State the possible number of positive real zeros, negative real zeros, and imaginary zeros of $f(x) = 3x^4 + 2x^3 - 2x^2 - 26x - 48$.

$f(x)$ has one sign change, so there is 1 positive real zero.

$f(-x)$ has 3 sign changes, so there are 3 or 1 negative real zeros.

There are 0 or 2 imaginary zeros.

CHAPTER 4
Practice Test

Go Online! for another Chapter Test

Simplify.

1. $(3a)^2(7b)^4$

2. $(7x - 2)(2x + 5)$

3. $(2x^2 + 3x - 4) - (4x^2 - 7x + 1)$

4. $(4x^3 - x^2 + 5x - 4) + (5x - 10)$

5. $(x^4 + 5x^3 + 3x^2 - 8x + 3) \div (x + 3)$

6. $(3x^3 - 5x^2 - 23x + 24) \div (x - 3)$

7. Expand $(2a - 3b)^4$.

8. What is the coefficient of the fifth term of $(m + 3n)^6$?

9. Find the fourth term of the expansion of $(c + d)^9$.

10. **MULTIPLE CHOICE** How many unique real zeros does the graph have?

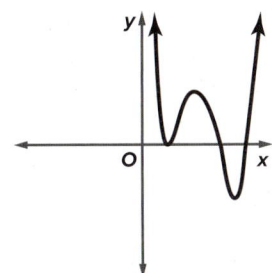

A 0
B 2
C 3
D 5

11. If $c(x) = 3x^3 + 5x^2 - 4$, what is the value of $4c(3b)$?

Complete each of the following.
a. Graph each function by making a table of values.
b. Determine consecutive integer values of x between which each real zero is located.
c. Estimate the x-coordinates at which the relative maxima and relative minima occur.

12. $g(x) = x^3 + 4x^2 - 3x + 1$

13. $h(x) = x^4 - 4x^3 - 3x^2 + 6x + 2$

Factor completely. If the polynomial is not factorable, write *prime*.

14. $8y^4 + x^3y$

15. $2x^2 + 2x + 1$

16. $a^2x + 3ax + 2x - a^2y - 3ay - 2y$

Solve each equation.

17. $8x^3 + 1 = 0$

18. $x^4 - 11x^2 + 28 = 0$

19. **FRAMING** The area of the picture and frame shown below is 168 square inches. What is the width of the frame?

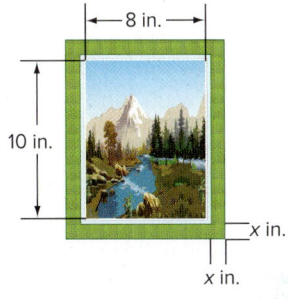

Prove each polynomial identity.

20. $(g + h)^2 - 2gh = g^2 + h^2$

21. $(2x + 1)(3x + 1) + (x - 1) = 6x(x + 1)$

22. **MULTIPLE CHOICE** Let $f(x) = x^4 - 3x^3 + 5x - 3$. Use synthetic substitution to find $f(-2)$.

A 37
B 27
C -21
D -33

Given a polynomial and one of its factors, find the remaining factors of the polynomial.

23. $2x^3 + 15x^2 + 22x - 15;\ x + 5$

24. $x^3 - 4x^2 + 10x - 12;\ x - 2$

State the possible number of positive real zeros, negative real zeros, and imaginary zeros of each function.

25. $p(x) = x^3 - x^2 - x - 3$

26. $p(x) = 2x^6 + 5x^4 - x^3 - 5x - 1$

Find all zeros of each function.

27. $p(x) = x^3 - 4x^2 + x + 6$

28. $p(x) = x^3 + 2x^2 + 4x + 8$

CHAPTER 4
Preparing for Assessment

Performance Task

Provide a clear solution to each part of the task. Be sure to show all of your work, include all relevant drawings, and justify your answers.

APPLY MATH A container company makes containers of all different shapes and sizes.

Part A
One of their most popular items, a plastic rectangular box, has a width w. The height is 2 more than 3 times the width, and the length is 10 less than twice the square of the width.

1. Write individual expressions that represent the width, height, and length of the box.
2. Write an expression in simplest form that represents the volume of the box.
3. **Tools** Determine the highest degree and leading coefficient of the polynomial expression you found above.

Part B
The company also makes cardboard moving boxes. They have a large box whose volume can be represented by the polynomial $4x^3 + 2x^2 - 8x + 10$. A medium box has a volume that can be represented by the polynomial $2x^3 - 4x^2 + 10x + 1$.

4. Determine the volume of both boxes combined.
5. **Sense-Making** Determine how much greater the volume of the large box is than the medium box.

Part C
Another plastic box has a volume that can be represented by the polynomial $3w^3 + 29w^2 - 10w$.

6. If the width of the polynomial is w, determine the height and length of the box.

Part D
The company makes a small cube box for packing fragile items. These each have a side length s. A larger rectangular box has a width that is 4 times the side length of the cube, a height that is 2 more than 6 times the side length of the cube, and a length that is 8 less than 10 times the side length of the cube.

7. Determine how many whole cubes will fit into the larger rectangular box.

Part E
The company also makes a large cube box, whose side length is 4 times the side length of the smaller cube. After 1 of the smaller cubes is put into the larger cube, the remaining volume in the larger cube is 1701 cubic units.

8. Determine the side lengths of the smaller and larger cubes.

Test-Taking Strategy

Example

Read the problem. Identify what you need to know. Then use the information in the problem to solve.

Mr. Nolan has a rectangular swimming pool that measures 25 feet by 14 feet. He wants to have a cement walkway installed around the perimeter of the pool. The combined area of the pool and walkway will be 672 square feet. What will be the width of the walkway?

- A 2.75 ft
- B 3 ft
- C 3.25 ft
- D 3.5 ft

Test-Taking Tip

Strategies for Drawing a Picture Drawing a picture can be a helpful way for you to visualize how to solve a problem. Sketch your picture of the pool and surrounding walkway on scrap paper. Do not make any marks on your answer sheet other than your answers.

Step 1 What are you being asked to solve? What information is given? Could a drawing help you solve the problem?

The width of the walkway is required. The problem gives the dimensions of the pool and the total area of the pool and walkway. A drawing would help because it could be used to visualize the swimming pool and walkway and label the dimensions given.

Step 2 How can you make sure your drawing is as helpful as possible?

Label it with all of the dimensions given in the problem.

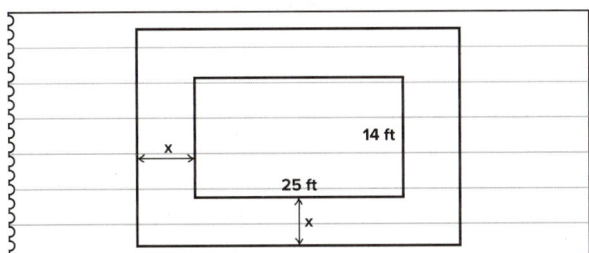

Step 3 How can you use the drawing to solve the problem? What is the correct answer?

I can write expressions representing the side lengths and solve the resulting polynomial expression. The answer is D.

Apply the Strategy

Read the problem. Identify what you need to know. Then use the information in the problem to solve.

A farmer has 240 feet of fencing that he wants to use to enclose a rectangular area for his chickens. He plans to build the enclosure using the wall of his barn as one of the walls. What is the maximum amount of area he can enclose?

- A 7200 ft²
- B 4960 ft²
- C 3600 ft²
- D 3280 ft²

Answer the questions below.

a. What are you being asked to solve? What information is given? Could a drawing help you solve the problem?
b. How can you make sure your drawing is as helpful as possible?
c. What is the correct answer?

CHAPTER 4
Preparing for Assessment
Cumulative Review

Read each question. Then fill in the correct answer on the answer document provided by your teacher or on a sheet of paper.

1. Which expression is equivalent to $\frac{x^3 - 6x^2 + 8}{x + 2}$?

   ◯ A $x^2 - 3x + \frac{8}{x+2}$

   ◯ B $x^2 - 4x - 8 - \frac{8}{x+2}$

   ◯ C $x^2 - 6x + 4 + \frac{4x}{x+2}$

   ◯ D $x^2 - 8x + 16 - \frac{24}{x+2}$

2. Create Pascal's triangle to expand $(x + y)^5$.

3. Use the Binomial Theorem to find the 8th term in the expansion of $(3x - 2)^{10}$.

4. What is the solution set of the equation?
   $$2x^3 + 6x^2 - 50x - 150 = 0?$$

   ◯ A $\{5i, 3i, 2, 5\}$

   ◯ B $\{-5, 3i, 5\}$

   ◯ C $\{-5, -3, 2, 5\}$

   ◉ D $\{-5, -3, 5\}$

5. What is the remainder when $3x^3 - 5x^2 - 4x + 1$ is divided by $x + 3$?

   Test-Taking Tip

   Question 4 Some four-term polynomials can be factored by grouping. Use the distributive property to factor the GCF from the first two terms and the GCF from the last two terms of the polynomial. If the binomials left in the parentheses are the same, the polynomial is factorable.

6. The function $p(x)$ is a polynomial function. As $x \to \infty$, $p(x) \to \infty$, and as $x \to -\infty$, $p(x) \to \infty$. If $p(x)$ has no real zeros, which of the following could be the graph of $p(x)$?

   ◯ A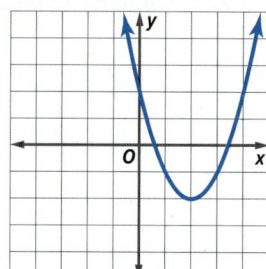

   ◯ B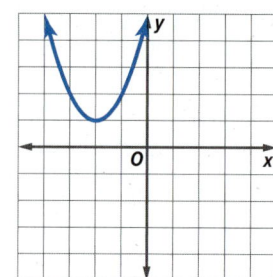

   ◯ C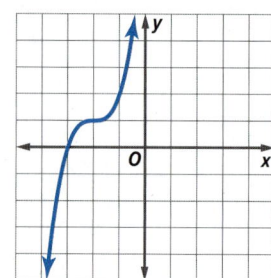

   ◯ D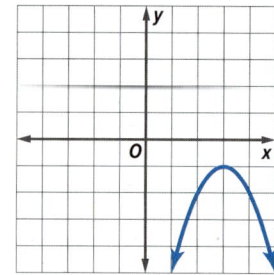

7. Which binomial is a factor of $20x^3 + 29x^2 - 9x - 18$?

   ◯ A $3x - 4$

   ◯ B $4x - 3$

   ◯ C $5x - 6$

   ◯ D $6x - 5$

Go Online! for Standardized Test Practice

8. The volume of a sphere is $\frac{4}{3}(\pi r^3)$, where r is the radius of the sphere. Harry has a globe with a radius 3 centimeters greater than a globe belonging to Ryan. The volume of Harry's globe is 684 cubic centimeters greater than the volume of Ryan's globe. What is the radius of Harry's globe?

   A 0.67 cm
   B 2.67 cm
   C 5.67 cm
   D 8.67 cm

9. How many real zeros exist for the function $f(x)$?

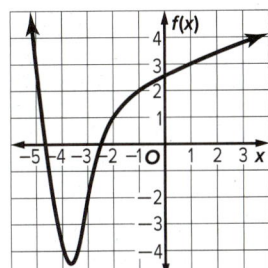

10. How many negative real zeros does the function $f(x)$ have if $f(x) = 2x^4 + 8x^3 - 16x^2 - 40x + 30$?

    A 0
    B 1
    C 2 or 0
    D 3 or 1
    E 4, 2, or 0

11. Which expression is a polynomial of degree 3?

    A $4x^2 + 5x$
    B $7x - 5x^3$
    C $9x - 2x^5 + 6x^2$
    D $7x^5 + 2x - 9x^2 + 4$
    E $3x^4 - x^6 - 8x + 10x^3$

12. Let $3 - 10i$ be a zero of the function $f(x)$. Which value must also be a zero of the function?

    A $-3 - 10i$
    B $-3 + 10i$
    C $3 + 10i$
    D $10 - 3i$
    E $10 + 3i$

13. The area of a rectangle is the product of its length and width. Write an expression in standard form that represents the area of the shaded region shown below.

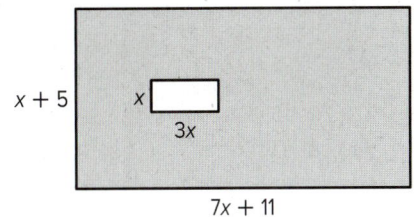

14. Complete the polynomial identities below:

    a. $a^2 - b^2 = $
    b. $a^2 + 2ab + b^2 = $
    c. $a^2 - 2ab + b^2 = $
    d. $a^3 + b^3 = $
    e. $a^3 - b^3 = $

15. What are the complex roots of $x^3 + 1 = 0$?

Need Extra Help?

If you missed Question…	1	2	3	4	5	6	7	8	9	10	11	12	13	14	15
Go to Lesson…	4-3	4-2	4-2	4-6	4-3	4-5	4-8	4-6	4-5	4-9	4-4	4-9	4-1	4-7	4-7

connectED.mcgraw-hill.com

CHAPTER 5

Inverses and Radical Functions

THEN

You simplified polynomial expressions.

NOW

You will:

- Find compositions and inverses of functions.
- Graph and analyze square root functions and inequalities.
- Simplify and solve equations involving roots, radicals, and rational exponents.

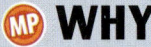

MP WHY

FINANCES Connecting finances to mathematics is a skill that, once mastered, you will use your entire life. Learning to manage your finances includes anticipating costs, such as college, and preparing for them.

Use the Mathematical Practices to complete the activity.

1. **Using Tools** Use the Internet to find the annual tuition for the college you plan to attend. Then, use the Internet to find the tuition costs over the last four years.
2. **Applying Math** Calculate the average annual rate of increase in tuition. Then, create a Line Graph that plots the tuition costs over the last four years and projects the tuition costs through the end of your four years in college.

Peathegee Inc/Blend Images/Getty Images

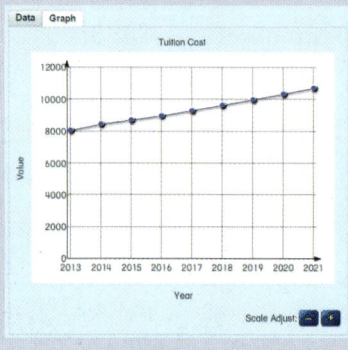

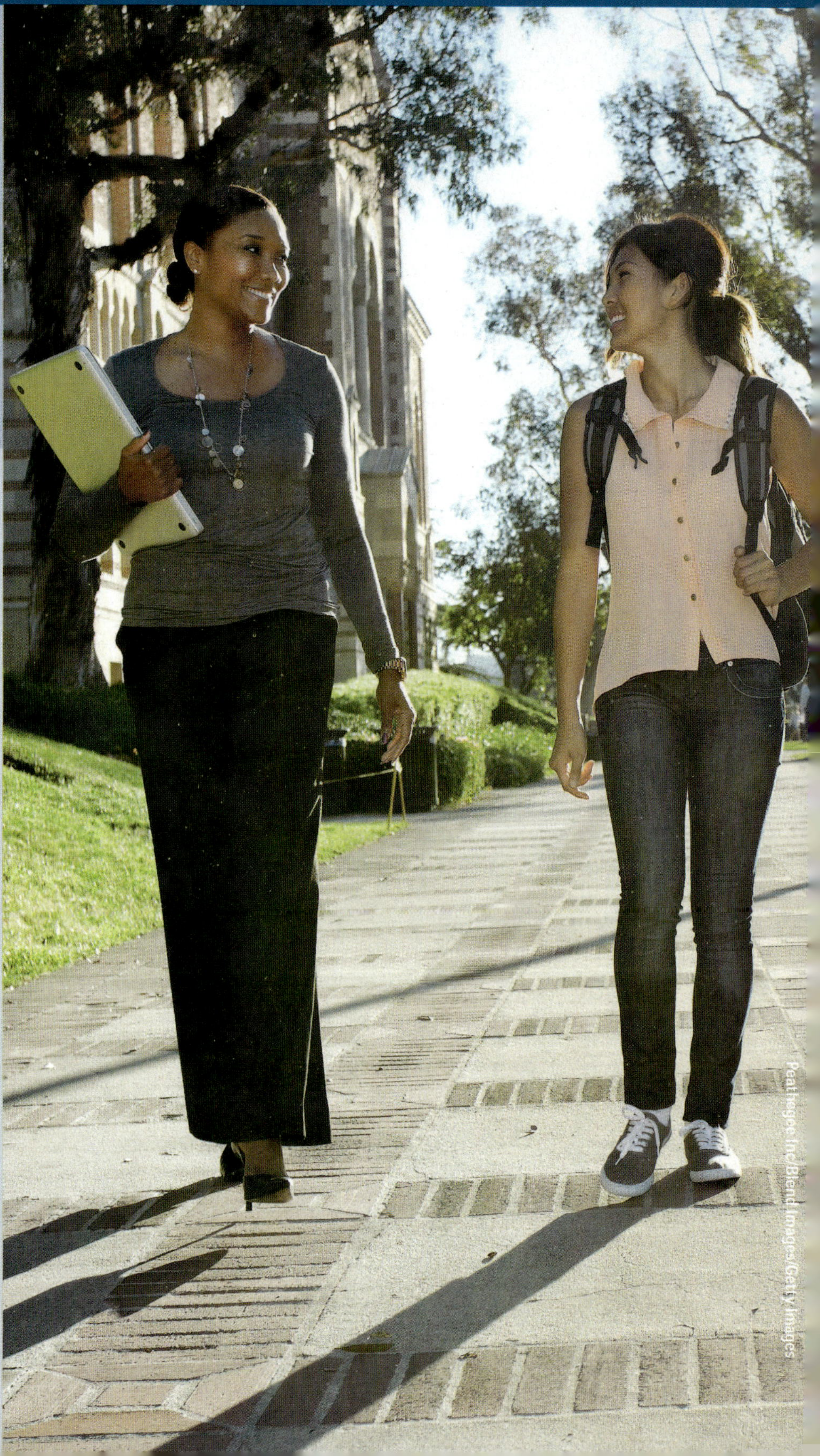

Go Online to Guide Your Learning

Explore & Explain

 Product Mat and Algebra Tiles
Use the **Product Mat** and the **Algebra Tiles** to practice multiplying polynomials.

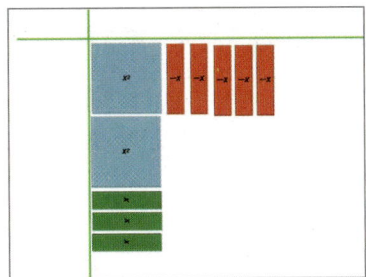

 The Geometer's Sketchpad
Use **The Geometer's Sketchpad** to perform operations with functions in Lesson 5-1 and Lesson 5-2, to illustrate inverses of relations and functions in Lesson 5-3, and to explore square root functions in Lesson 5-4.

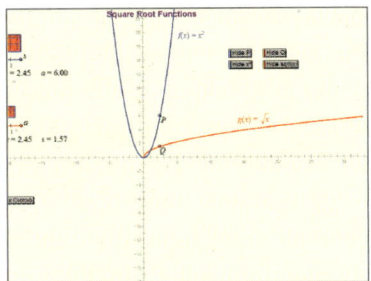

eBook
Interactive Student Guide
Before starting the chapter, answer the **Chapter Focus** preview questions. Check your answers as you complete each lesson. At the end of the chapter, try the **Performance Task**.

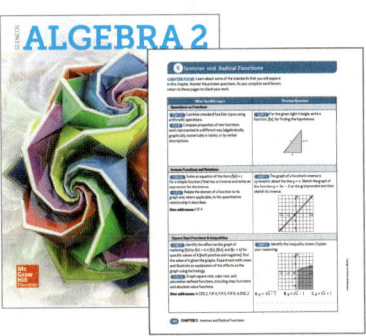

Organize

 Foldables
Get organized! Create an **Inverses and Radical Functions Foldable** before you start the chapter to arrange your notes about radical equations and inequalities.

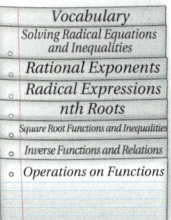

Collaborate

 Chapter Project
In the **Driving in Style** project, you will use what you have learned about inverse functions to complete a project that addresses financial literacy.

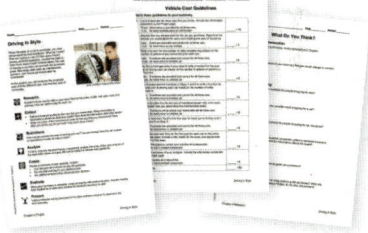

Focus

 LEARNSMART
Need help studying? Complete the **Modeling with Functions** domain in LearnSmart to review for the chapter test.

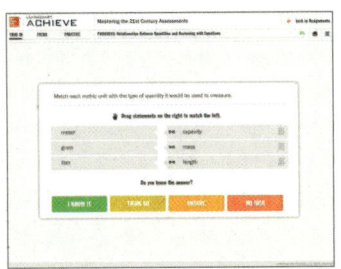

ALEKS
You can use the **Radicals and Advanced Functions** topic in ALEKS to explore what you know about relations and functions and what you are ready to learn.*

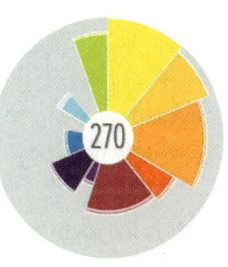

* Ask your teacher if this is part of your program.

connectED.mcgraw-hill.com

Get Ready for the Chapter

Connecting Concepts

Concept Check
Review the concepts used in this chapter by answering the questions below.

1. In a graph such as the one shown, how do you define the "roots" of the graph?

2. If the exact roots of a graph cannot be found, what is typically stated?

3. What are the roots of the graph shown?

4. What type of division would you use to simplify $(5x^2 - 22x - 15) \div (x - 5)$?

5. What property would you use to rewrite $-\frac{1}{2}(2m - 5)$ without parentheses?

6. Given $0 = 3x^2 + 3x - 5x - 5$, what property can you apply to begin to simplify the equation?

7. Given $0 = 3x(x + 1) - 5(x + 1)$, what property can you apply to simplify the equation?

8. How would $-1(3b^2 + 2b - 1)$ be written so that it does not contain parentheses?

9. Given $4x^2 + 7x + 3$, what are the values for a, b, and c when applying the Quadratic Formula?

10. Given $\frac{3m + 5n}{p}$, where $m = -5$, $n = 2$, and $p = -1$, would you have a positive or negative number?

Performance Task Preview
You can use the concepts and skills in the chapter to perform various calculations on data collected in a university lab. Understanding inverses and radical functions will help you finish the Performance Task at the end of the chapter.

MP In this Performance Task you will:
- make sense of problems and persevere in solving them
- reason abstractly and quantitatively
- attend to precision

New Vocabulary

English		Español
composition of functions	p. 322	composición de funciones
inverse relation	p. 329	relaciones inversas
inverse function	p. 329	función inversa
square root function	p. 338	función raíz cuadrada
radical function	p. 338	función radical
cube root function	p. 345	función raíz cúbica
inflection point	p. 345	punto de inflexión
radical equation	p. 352	ecuación radical
extraneous solution	p. 352	solución extraña
radical inequality	p. 354	desigualdad radical

Review Vocabulary

absolute value valor absoluto a number's distance from zero on the number line, represented by $|x|$

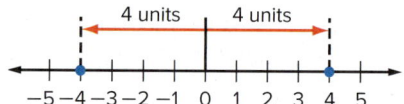

rational number número racional any number $\frac{m}{n}$, where m and n are integers and n is not zero; the decimal form is either a terminating or repeating decimal.

relation relación a set of ordered pairs

LESSON 1
Operations with Functions

::Then::
- You performed operations on polynomials.

::Now::
1. Perform arithmetic operations with functions.
2. Apply arithmetic operations with functions.

::Why?::
- The graphs model the income for the Brooks family since 2005, where $m(x)$ represents Mr. Brooks' income and $f(x)$ represents Mrs. Brooks' income.

  The total household income for the Brooks household can be represented by $f(x) + m(x)$.

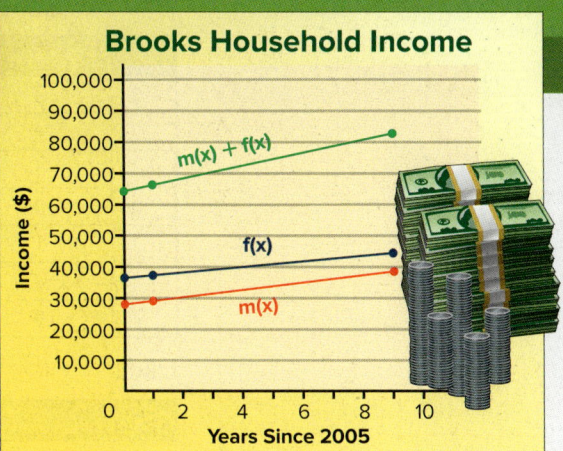

New Vocabulary
composition of functions

Mathematical Practices
4 Model with mathematics.
7 Look for and make use of structure.

1 Perform Operations with Functions You have performed arithmetic operations with polynomials. You can also use addition, subtraction, multiplication, and division with functions.

You can perform arithmetic operations according to the following rules.

Key Concept Operations on Functions

Operation	Definition	Example Let $f(x) = 2x$ and $g(x) = -x + 5$.
Addition	$(f + g)(x) = f(x) + g(x)$	$2x + (-x + 5) = x + 5$
Subtraction	$(f - g)(x) = f(x) - g(x)$	$2x - (-x + 5) = 3x - 5$
Multiplication	$(f \cdot g)(x) = f(x) \cdot g(x)$	$2x(-x + 5) = -2x^2 + 10x$
Division	$\left(\dfrac{f}{g}\right)(x) = \dfrac{f(x)}{g(x)}, g(x) \neq 0$	$\dfrac{2x}{-x + 5}, x \neq 5$

Example 1 Add and Subtract Functions

Given $f(x) = x^2 - 4$ and $g(x) = 2x + 1$, find each function.

a. $(f + g)(x)$

$(f + g)(x) = f(x) + g(x)$ Addition of functions
$= (x^2 - 4) + (2x + 1)$ $f(x) = x^2 - 4$ and $g(x) = 2x + 1$
$= x^2 + 2x - 3$ Simplify.

b. $(f - g)(x)$

$(f - g)(x) = f(x) - g(x)$ Subtraction of functions
$= (x^2 - 4) - (2x + 1)$ $f(x) = x^2 - 4$ and $g(x) = 2x + 1$
$= x^2 - 2x - 5$ Simplify.

▶ **Guided Practice**

Given $f(x) = x^2 + 5x - 2$ and $g(x) = 3x - 2$, find each function.

1A. $(f + g)(x)$ **1B.** $(f - g)(x)$

You can graph sum and difference functions by graphing each function involved separately, then adding their corresponding functional values. Let $f(x) = x^2$ and $g(x) = x$. Examine the graphs of $f(x)$, $g(x)$, and their sum and difference.

Find $(f + g)(x)$.

x	$f(x) = x^2$	$g(x) = x$	$(f + g)(x) = x^2 + x$
−3	9	−3	9 + (−3) = 6
−2	4	−2	4 + (−2) = 2
−1	1	−1	1 + (−1) = 0
0	0	0	0 + 0 = 0
1	1	1	1 + 1 = 2
2	4	2	4 + 2 = 6
3	9	3	9 + 3 = 12

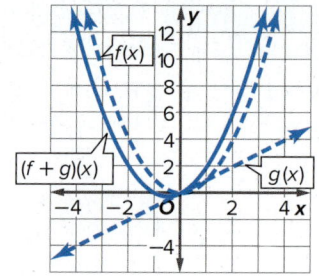

Find $(f - g)(x)$.

x	$f(x) = x^2$	$g(x) = x$	$(f - g)(x) = x^2 - x$
−3	9	−3	9 − (−3) = 12
−2	4	−2	4 − (−2) = 6
−1	1	−1	1 − (−1) = 2
0	0	0	0 − 0 = 0
1	1	1	1 − 1 = 0
2	4	2	4 − 2 = 2
3	9	3	9 − 3 = 6

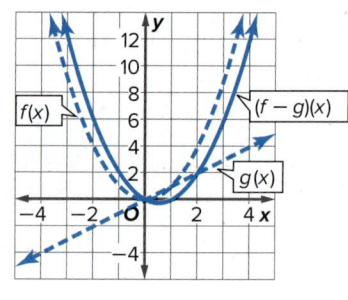

Reading Math Tip

intersection Everyday use—the intersection of two roads is where the two roads meet; Math meaning—the intersection of two sets is the set of elements common to them.

In Example 1, the functions $f(x)$ and $g(x)$ have the same domain of all real numbers. The functions $(f + g)(x)$ and $(f - g)(x)$ also have domains that include all real numbers. For each new function, the domain consists of the intersection of the domains of $f(x)$ and $g(x)$. Under division, the domain of the new function is restricted by excluded values that cause the denominator to equal zero.

Example 2   Multiply and Divide Functions

Given $f(x) = x^2 + 7x + 12$ and $g(x) = 3x - 4$, find each function. Indicate any restrictions in the domain.

a. $(f \cdot g)(x)$

$(f \cdot g)(x) = f(x) \cdot g(x)$ Multiplication of functions

$ = (x^2 + 7x + 12)(3x - 4)$ Substitution

$ = 3x^3 + 21x^2 + 36x - 4x^2 - 28x - 48$ Distributive Property

$ = 3x^3 + 17x^2 + 8x - 48$ Simplify.

b. $\left(\dfrac{f}{g}\right)(x)$

$\left(\dfrac{f}{g}\right)(x) = \dfrac{f(x)}{g(x)}$ Division of functions

$\phantom{\left(\dfrac{f}{g}\right)(x)} = \dfrac{x^2 + 7x + 12}{3x - 4}, x \neq \dfrac{4}{3}$ Substitution

Because $x = \dfrac{4}{3}$ makes the denominator $3x - 4 = 0$, $\dfrac{4}{3}$ is excluded from the domain of $\left(\dfrac{f}{g}\right)(x)$.

▶ **Guided Practice** Given $f(x) = x^2 - 7x + 2$ and $g(x) = x + 4$, find each function.

2A. $(f \cdot g)(x)$ **2B.** $\left(\dfrac{f}{g}\right)(x)$

Real-World Link

In a robotics competition, teams compete to build robots that can accomplish a given task. In 2016, about 29,000 students from 40 countries participated in a robotics competition in St. Louis.

Source: U.S. News & World Report

2 Apply Operations with Functions In many real-world modeling situations you write multiple functions that each represent one aspect of the problem. You can combine the functions to build a new function that models a different aspect of the problem.

Real-World Example 3   Build a New Function

ROBOTICS A robotics competition team is preparing for a tournament. The team's captain is ordering custom T-shirts for each of the members. The T-shirts cost $8 each, plus a one-time set-up fee of $25. Sales tax on the order is 8%. The team decides that the members will split the cost of the T-shirts equally, and that the team's captain and vice-captain will not have to pay for their shirts.

a. Write a function $C(x)$ that represents the total cost of the T-shirts, where x is the number of team members.

$8x + 25$ represents the cost of T-shirts before sales tax. Multiply by 1.08 to find the total cost after the 8% sales tax is applied.

$C(x) = (1.08)(8x + 25)$ Multiply $8x + 25$ by 1.08.
$ = (1.08)(8x) + (1.08)(25)$ Distributive Property
$ = 8.64x + 27$ Simplify.

So, $C(x) = 8.64x + 27$ represents the total cost of the T-shirts for x team members.

b. Write a function $N(x)$ to represent the number of team members who pay for the T-shirts.

All of the team members except the captain and vice-captain pay for T-shirts.

So, $N(x) = x - 2$.

c. Find $\left(\dfrac{C}{N}\right)(x)$ and explain what this function represents.

$\left(\dfrac{C}{N}\right)(x) = \dfrac{C(x)}{N(x)}$ Division of functions

$\phantom{\left(\dfrac{C}{N}\right)(x)} = \dfrac{8.64x + 27}{x - 2}$ Substitution

So, $\left(\dfrac{C}{N}\right)(x) = \dfrac{8.64x + 27}{x - 2}$, $x \neq 2$. This function represents the dollar amount that each paying team member will contribute to the cost of the T-shirts.

d. If the team has 15 members, how much does each paying team member contribute to the cost of the T-shirts?

$\left(\dfrac{C}{N}\right)(15) = \dfrac{8.64(15) + 27}{15 - 2}$ Evaluate the function for $x = 15$.

$\phantom{\left(\dfrac{C}{N}\right)(15)} \approx 12.05$ Simplify.

Each paying member of the team contributes $12.05.

Watch Out!

Evaluating the Function Remember that x is the total number of team members, so evaluate the function for $x = 15$, not $x = 13$.

▶ **Guided Practice**

3. **CHEMISTRY** A chemist has 500 grams of a 15% saline solution. She adds x grams of salt to the solution. Write a function $S(x)$ that represents the number of grams of salt in the new solution, a function $T(x)$ that represents the total number of grams of the new solution, and then find $\left(\dfrac{S}{T}\right)(x)$ and explain what this function represents.

Check Your Understanding

= Step-by-Step Solutions begin on page R11.

Go Online! for a Self-Check Quiz

Examples 1–2 Find $(f + g)(x)$, $(f - g)(x)$, $(f \cdot g)(x)$, and $\left(\dfrac{f}{g}\right)(x)$ for each $f(x)$ and $g(x)$. Indicate any restrictions in the domain.

1. $f(x) = x + 2$
$g(x) = 3x - 1$

2. $f(x) = x^2 - 5$
$g(x) = -x + 8$

Example 3

3. **PHOTOGRAPHY** A group of photographers is planning an exhibit of their work. Each photographer will contribute 5 prints to the exhibit and the cost of framing each print is $7.85. There is also a flat fee of $200 to rent the room for the exhibit. The photographers plan to split the cost of the exhibit equally.

   a. Write a function $C(x)$ that represents the total cost of the exhibit, where x is the number of photographers.

   b. In addition to the photographers, 3 family members offer to participate in sharing the cost of the exhibit. Write a function $P(x)$ to represent the number of people who pay for the exhibit.

   c. Find $\left(\dfrac{C}{P}\right)(x)$ and explain what this function represents.

   d. If there are 8 photographers, how much does each photographer contribute to the cost of the exhibit?

Practice and Problem Solving

Extra Practice is on page R5.

Examples 1–2 Find $(f + g)(x)$, $(f - g)(x)$, $(f \cdot g)(x)$, and $\left(\dfrac{f}{g}\right)(x)$ for each $f(x)$ and $g(x)$. Indicate any restrictions in the domain.

4. $f(x) = 2x$
$g(x) = -4x + 5$

5. $f(x) = x - 1$
$g(x) = 5x - 2$

6. $f(x) = x^2$
$g(x) = -x + 1$

7. $f(x) = 3x$
$g(x) = -2x + 6$

8. $f(x) = x - 2$
$g(x) = 2x - 7$

9. $f(x) = x^2$
$g(x) = x - 5$

10. $f(x) = -x^2 + 6$
$g(x) = 2x^2 + 3x - 5$

11. $f(x) = 3x^2 - 4$
$g(x) = x^2 - 8x + 4$

Examples 3

12. **BASKETBALL** During a practice session, Dimitri makes 20 free throw attempts and makes 60% of the free throws. Then, on his next attempt, he begins a streak in which he makes x free throws in a row.

    a. What do the functions $f(x) = 12 + x$ and $g(x) = 20 + x$ represent?

    b. Find $\left(\dfrac{f}{g}\right)(x)$ and explain what this function represents.

    b. Find $\left(\dfrac{f}{g}\right)(7)$ and explain what this value represents.

13. **POPULATION** In a particular county, the population of the two largest cities can be modeled by $f(x) = 200x + 25$ and $g(x) = 175x - 15$, where x is the number of years since 2010 and the population is in thousands.

    a. What is the population of the two cities combined after any number of years?

    b. What is the difference in the populations of the two cities?

Perform each operation if $f(x) = x^2 + x - 12$ and $g(x) = x - 3$. State the domain of the resulting function.

14. $(f - g)(x)$ **15.** $2(g \cdot f)(x)$ **16.** $\left(\dfrac{f}{g}\right)(x)$

17. MULTIPLE REPRESENTATIONS Let $f(x) = x^2$ and $g(x) = x$.

a. Tabular Make a table showing values for $f(x)$, $g(x)$, $(f + g)(x)$, and $(f - g)(x)$.

b. Graphical Graph $f(x)$, $g(x)$, and $(f + g)(x)$ on the same coordinate grid.

c. Graphical Graph $f(x)$, $g(x)$, and $(f - g)(x)$ on the same coordinate grid.

d. Verbal Describe the relationship among the graphs of $f(x)$, $g(x)$, $(f + g)(x)$, and $(f - g)(x)$.

Use the table to find each value.

18. $(f + g)(-5)$ **19.** $(g - f)(-1)$

20. $(f \cdot g)(3)$ **21.** $(h \cdot f)(0)$

22. $\left(\dfrac{f}{g}\right)(-1)$ **23.** $\left(\dfrac{h}{g}\right)(0)$

24. $\left(\dfrac{g}{f}\right)(4)$ **25.** $\left(\dfrac{g}{h}\right)(-5)$

x	f(x)	g(x)	h(x)
−5	−8	8	2
−2	4	5	−10
−1	−2	−4	0
0	3	−5	−5
3	2	0	8
4	0	−1	7

Use the graph of $f(x)$ and $g(x)$ to find each value.

26. $(f - g)(1)$ **27.** $(f \cdot g)(0)$

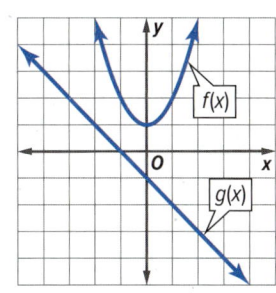 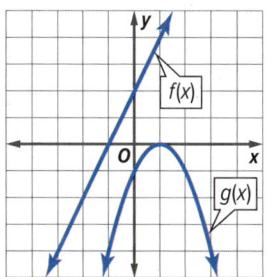

28. $\left(\dfrac{g}{f}\right)(-3)$ **29.** $\left(\dfrac{f}{g}\right)(-2)$

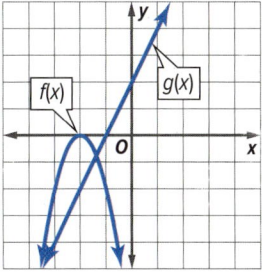 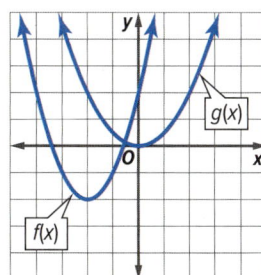

If $f(x) = -x + 1$, $g(x) = 4x + 2$, and $h(x) = x^2 - 1$, find each value.

30. $(2f + g)(1)$ **31.** $(3f + 2h)(0)$ **32.** $(-f + 2g)(3)$

33. $(5f \cdot h)(-1)$ **34.** $\left(\dfrac{3f}{g}\right)(2)$ **35.** $\left(\dfrac{g}{2h}\right)(0)$

36. $(h - 2f)(5)$ **37.** $(-f - h)(1)$ **38.** $(5h - 0.1g)(2)$

39. EMPLOYMENT The number of women and men age 16 and over employed each year in the United States can be modeled by the following equations, where x is the number of years since 2000 and y is the number of people in thousands.

women: $y = 548.6x + 66{,}527$ men: $y = 2090.7x + 62{,}243$

a. Write a function that models the total number of men and women employed in the United States during this time.

b. If f is the function for the number of men, and g is the function for the number of women, what does $(f - g)(x)$ represent?

If $f(x) = x + 2$, $g(x) = -4x + 3$, and $h(x) = x^2 - 2x + 1$, find each value.

40. $(f \cdot g \cdot h)(3)$ **41.** $[(f + g) \cdot h](1)$ **42.** $\left(\dfrac{h}{fg}\right)(-6)$

43. MULTIPLE REPRESENTATIONS You will explore $(f \cdot g)(x)$, and $\left(\dfrac{f}{g}\right)(x)$, if $f(x) = x^2 + 1$ and $g(x) = x - 3$.

a. **Tabular** Make a table showing values for $(f \cdot g)(x)$ and $\left(\dfrac{f}{g}\right)(x)$.

b. **Graphical** Use a graphing calculator to graph $(f \cdot g)(x)$ and $\left(\dfrac{f}{g}\right)(x)$ on the same coordinate plane.

c. **Verbal** Explain the relationship between $(f \cdot g)(x)$ and $\left(\dfrac{f}{g}\right)(x)$.

44. MULTI-STEP Ice cream cones are one of many treats sold at Sam's Desserts. They sell 60 scoops for every gallon of ice cream. They pay $6 per gallon of ice cream, $2 for every box of 24 cones, and allocate a fixed monthly cost of $400 to ice cream. Their sales reports for the past 6 months are shown.

Month	January	February	March	April	May	June
Price	$3.50	$3.70	$3.90	$3.75	$3.55	$3.80
Scoops Sold	224	208	188	205	219	199

a. What is their maximum monthly profit from ice cream sales?

b. Describe your solution process.

H.O.T. Problems   Use Higher-Order Thinking Skills

45. OPEN-ENDED Write two functions $f(x)$ and $g(x)$ such that $(f \cdot g)(x) = 2x^2 - 2$.

46. CHALLENGE Given that $(f + g)(4) = 8$ and $(f - g)(4) = -6$, find $f(4)$ and $g(4)$.

47. REASONING State whether each statement is *sometimes*, *always*, or *never* true. Explain.

a. If $f(x)$ and $g(x)$ are linear functions, then there is one value that is excluded from the domain of $(f + g)(x)$.

b. If $f(x)$ and $g(x)$ are linear functions, then there is one value that is excluded from the domain of $\left(\dfrac{f}{g}\right)(x)$.

48. STRUCTURE Suppose $f(x) = ax^2 + bx + c$ and $g(x) = mx^2 + nx + p$, for constants a, b, c, m, n, and p, with $a \neq 0$ and $m \neq 0$. What can you conclude about the constants if the domain of $\left(\dfrac{f}{g}\right)(x)$ is all real numbers? Explain.

49. WRITING IN MATH If $f(x)$ and $g(x)$ are polynomials, what can you say about the domains of $(f + g)(x)$, $(f - g)(x)$, $(f \cdot g)(x)$, and $\left(\dfrac{f}{g}\right)(x)$?

Preparing for Assessment

50. Let $f(x) = x^2 - 4$ and $g(x) = x^2 - 1$. What is the domain of the function $\left(\dfrac{f}{g}\right)(x)$? **MP** 7

- A all real numbers
- B all real numbers except $x = 0$
- C all real numbers except $x = \pm 1$
- D all real numbers except $x = \pm 2$

51. Let $f(m) = p$ where m and p are both nonzero integers. Which statement(s) must be true? **MP** 7

  I. $f\left(\dfrac{m}{2}\right) = \dfrac{p}{2}$
  II. $2f(m) = 2p$
  III. $f(2m) = 2p$

- A I only
- B II only
- C III only
- D I and II only

52. The graphs of $f(x)$ and $g(x)$ are shown. **MP** 7

What is $\left(\dfrac{f}{g}\right)(-2)$?

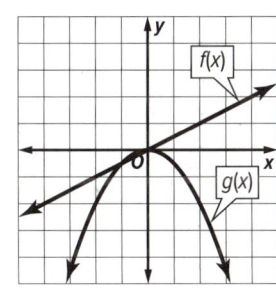

- A -2
- B $-\dfrac{1}{2}$
- C $\dfrac{1}{2}$
- D 2

53. If $f(x) = 2x - 10$ and $g(x) = x^2 + 3x + 1$, what is $(f \cdot g)(3)$? **MP** 6

$(f \cdot g)(3) = $ _____

54. Find $(f + g)(x)$ for the following functions. **MP** 1, 6

$f(x) = -5x^2 + 4x - 7$
$g(x) = 6x^2 - 4x + 12$

55. Find $(f \cdot g)(-2)$ for the following functions. **MP** 1, 6

$f(x) = -x^2 + 2x - 2$
$g(x) = 2x + 3$

- A -11
- B -9
- C 2
- D 10

56. For which pair(s) of functions is the domain of $\left(\dfrac{f}{g}\right)(x)$ all real numbers? **MP** 7

- A $f(x) = x$ and $g(x) = x^2 + 4$
- B $f(x) = x$ and $g(x) = x^2 - 4$
- C $f(x) = x^2 - 4$ and $g(x) = 4$
- D $f(x) = x + 4$ and $g(x) = x - 4$
- E $f(x) = 4$ and $g(x) = x^2 + 4$
- F $f(x) = x^2 + 4$ and $g(x) = x^2 - 4$

57. If $f(-2) = a$ and $(f \cdot g)(-2) = 2a^2$, which of the following is $g(-2)$? **MP** 2

- A $2a$
- B $-4a^2$
- C $-2a$
- D $2a^3$

58. MULTI-STEP Jordan is ordering books online for the members of his book club. Each member of the club will receive a copy of the book and each book costs $8.95. Because Jordan is ordering a large quantity of books, one of them is free. The shipping fee for the order is a flat rate of $4.50 and there is no sales tax. **MP** 1, 4

  a. Let x represent the number of members of the club. Write a function $B(x)$ that represents the total cost of the books.

  b. The club members decide to split the cost evenly and they decide that Jordan should not have to pay anything since he placed the order. Write a function $N(x)$ that represents the number of club members who pay for books.

  c. Find $\left(\dfrac{B}{N}\right)(x)$ and explain what it represents.

  d. The book club has 12 members. How much does each member pay?

LESSON 2
Composition of Functions

::Then::
- You performed arithmetic operations on functions.

::Now::
1. Perform compositions of functions.
2. Apply compositions of functions.

::Why?::
Submersibles can descend several miles below the surface of the ocean. You can write a function $d(t)$ that gives the depth of the submersible after t minutes and a function $p(d)$ that gives the pressure at depth d. The composition of the functions $p[d(t)]$ gives the pressure on the submersible after t minutes.

New Vocabulary
composition of functions

Mathematical Practices
3 Construct viable arguments and critique the reasoning of others.
4 Model with mathematics.
7 Look for and make use of structure.

1 Perform Compositions of Functions
You have already combined functions with arithmetic operations. Another method used to combine functions is a composition of functions. In a **composition of functions**, the results of one function are used to evaluate a second function.

Key Concept Composition of Functions

Words Suppose f and g are functions such that the range of g is a subset of the domain of f. Then the composition function $f \circ g$ can be described by

$$[f \circ g](x) = f[g(x)].$$

Model

domain of g | range of g domain of f | range of f

$x \rightarrow g(x) \rightarrow f[g(x)]$

$[f \circ g](x)$

Example 1 Evaluate Compositions of Functions

Given $f(x) = x^2 - 4$ and $g(x) = 2x + 1$, find each value.

a. $[f \circ g](3)$

$[f \circ g](3) = f[g(3)]$ — Composition of functions
$\quad = f(7)$ — $g(3) = 2(3) + 1 = 7$
$\quad = 45$ — $f(7) = 7^2 - 4 = 45$

So, $[f \circ g](3) = 45$.

b. $[g \circ f](3)$

$[g \circ f](3) = g[f(3)]$ — Composition of functions
$\quad = g(5)$ — $f(3) = 3^2 - 4 = 5$
$\quad = 11$ — $g(5) = 2(5) + 1 = 11$

So, $[g \circ f](3) = 11$.

Guided Practice

Given $f(x) = 3x - 6$ and $g(x) = x^3 + 1$, find each value.

1A. $[f \circ g](5)$ **1B.** $[g \circ f](5)$

Go Online!
The composition of f and g, denoted by $f \circ g$ or $f[g(x)]$, is read f of g. To hear more pronunciations of expressions, log into your eStudent Edition. Ask your teacher or a partner for clarification as you need it.

If $f(x) = x + 2$, $g(x) = -4x + 3$, and $h(x) = x^2 - 2x + 1$, find each value.

59. $[f \circ (g \circ h)](2)$ **60.** $[g \circ (h \circ f)](-4)$ **61.** $[h \circ (f \circ g)](5)$

62. MULTIPLE REPRESENTATIONS You will explore $[f \circ g](x)$ and $[g \circ f](x)$ if $f(x) = x^2 + 1$ and $g(x) = x - 3$.

  a. Tabular Make a table showing values for $[f \circ g](x)$ and $[g \circ f](x)$.

  b. Graphical Use a graphing calculator to graph $[f \circ g](x)$, and $[g \circ f](x)$ on the same coordinate plane.

  c. Verbal Explain the relationship between $[f \circ g](x)$, and $[g \circ f](x)$.

63. REASONING Copy and complete the table. Use the following clues and logical reasoning to help you.

- $f(x)$ and $g(x)$ are linear functions.
- $[f \circ g](2) = 6$
- $[g \circ f](3) = 10$

x	f(x)	g(x)
1		
2		
3		
4	6	10
5		

Given that $f(x) = mx + d$ and $g(x) = ax^2 + bx + c$, find each composition.

64. $(f \circ g)(x)$ **65.** $(g \circ f)(x)$ **66.** $(f \circ f)(x)$

67. Suppose $f(x) = x^p$ and $g(x) = x^q$, where p and q are positive integers. What can you say about the power of $(f \circ g)(x)$ and $(g \circ f)(x)$? Explain.

H.O.T. Problems   Use Higher-Order Thinking Skills

68. OPEN-ENDED Write two functions $f(x)$ and $g(x)$ such that $(f \circ g)(4) = 0$.

69. CRITIQUE REASONING Denise and Keiko were asked to find $[f \circ g](x)$ given that $f(x) = 6x + 5$ and $g(x) = 2x - 1$. Is either of them correct? Explain your reasoning. If neither student is correct, provide the correct answer.

Denise	Keiko
$[f \circ g](x) = (6x + 5)(2x - 1)$	$[f \circ g](x) = f[g(x)]$
$= (6x)(2x) + 6x(-1) + 5(2x) + 5(-1)$	$= f(2x - 1)$
$= 12x^2 - 6x + 10x - 5$	$= 6(2x - 1) + 5$
$= 12x^2 + 4x - 5$	$= 12x - 1 + 5$
	$= 12x + 4$

70. CHALLENGE Given that $f(x) = 3x + 4$, find $[f \circ f \circ f](2)$.

71. REASONING State whether each statement is *sometimes*, *always*, or *never* true. Explain.

  a. The domain of two functions $f(x)$ and $g(x)$ that are composed $g[f(x)]$ is restricted by the domain of $f(x)$.

  b. The domain of two functions $f(x)$ and $g(x)$ that are composed $g[f(x)]$ is restricted by the domain of $g(x)$.

72. WRITING IN MATH In the real world, why would you ever perform a composition of functions?

Preparing for Assessment

73. What is the value of $f[g(5)]$ if $f(x) = \dfrac{x+2}{2}$ and $g(x) = x^2 - 2$? MP 1
- A $3\tfrac{1}{2}$
- B $10\tfrac{1}{4}$
- C $12\tfrac{1}{2}$
- D 23

74. Let $f(x) = x^2 + 1$. What is the value of $f(f(3))$? MP 1
- A 2
- B 10
- C 82
- D 101

75. For which pair(s) of functions is $[f \circ g](x)$ a quadratic function? MP 7
- A $f(x) = x^2$ and $g(x) = 3x + 7$
- B $f(x) = x - 4$ and $g(x) = x^2 - 1$
- C $f(x) = x + 9$ and $g(x) = 2x - 3$
- D $f(x) = x$ and $g(x) = 6x$
- E $f(x) = x^2 - 1$ and $g(x) = x^2 + 1$
- F $f(x) = x^2 - 4$ and $g(x) = 3x$

76. If $f(x) = 2x + 1$ and $g(x) = 4x - 5$, which of the following is $[f \circ g](x)$? MP 1
- A $[f \circ g](x) = 8x - 9$
- B $[f \circ g](x) = 8x - 4$
- C $[f \circ g](x) = 8x - 1$
- D $[f \circ g](x) = 8x - 10$

77. Find $(f \circ g)(4)$ for the following functions. MP 1, 6

$f(x) = 2x + 5$

$g(x) = x^2 - 2x + 3$

78. Find $(g \circ f)(-2)$ for the following functions. MP 1, 6

$f(x) = -x^2 + 6x - 1$

$g(x) = 2x^2 - 3x$

- A -238
- B -113
- C 77
- D 629

79. The graphs of $f(x)$ and $g(x)$ are shown.

What is $[f \circ g](2)$? MP 7

- A -8
- B -3
- C 2
- D 4

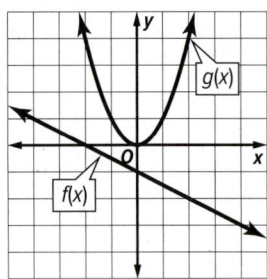

80. If $f(x) = x^2 - 8$ and $g(x) = 3x + 1$, what is the value of $[g \circ f](-4)$? MP 6

$[g \circ f](-4) = $ _____

81. MULTI-STEP Amelia is shopping at an online store. Shipping for each order costs a flat fee of $6.75. Sales tax is 8%. MP 1, 4

a. Let x represent the cost of an order before shipping or sales tax. Write a function $s(x)$ that represents the cost of an order with shipping, and a function $t(x)$ that represents the cost of an order with sales tax.

b. Find $[s \circ t](x)$ and explain what this function represents.

c. Find $[t \circ s](x)$ and explain what this function represents.

d. Will Amelia get a better deal if the shipping fee is applied to the order before sales tax, or after sales tax? Explain.

LESSON 3
Inverse Functions and Relations

∴Then
- You transformed and solved equations for a specific variable.

∴Now
1. Find the inverse of a function or relation.
2. Determine whether two functions or relations are inverses.

∴Why?
The table shows the value of $1 (U.S.) compared to Canadian dollars and Mexican pesos.

The equation $p = 12.45d$ represents the number of pesos p you can receive for every U.S. dollar d. To determine how many U.S. dollars you can receive for one Mexican peso, solve the equation $p = 12.45d$ for d. The result, $d \approx 0.08p$, is the inverse function.

	U.S.	Canada	Mexico
U.S.		1.05	12.45
Canada	0.95		11.97
Mexico	0.08	0.08	

New Vocabulary
inverse relation
inverse function

Mathematical Practices
7 Look for and make use of structure.
8 Look for and express regularity in repeated reasoning.

1 Find Inverses
Recall that a relation is a set of ordered pairs. The **inverse relation** is the set of ordered pairs obtained by exchanging the coordinates of each ordered pair. The domain of a relation becomes the range of its inverse, and the range of the relation becomes the domain of its inverse.

Key Concept Inverse Relations

Words Two relations are inverse relations if and only if whenever one relation contains the element (a, b), the other relation contains the element (b, a).

Example A and B are inverse relations.

$A = \{(1, 5), (2, 6), (3, 7)\}$ $B = \{(5, 1), (6, 2), (7, 3)\}$

Example 1 Find an Inverse Relation

GEOMETRY The vertices of $\triangle ABC$ can be represented by the relation $\{(1, -2), (2, 5), (4, -1)\}$. Find the inverse of this relation. Describe the graph of the inverse.

Graph the relation. To find the inverse, exchange the coordinates of the ordered pairs. The inverse of the relation is $\{(-2, 1), (5, 2), (-1, 4)\}$.

Plotting these points shows that the ordered pairs describe the vertices of $\triangle A'B'C'$ as a reflection of $\triangle ABC$ in the line $y = x$.

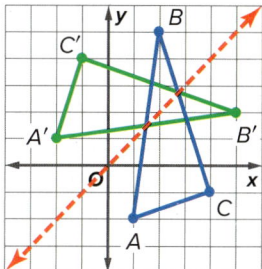

▶ **Guided Practice**

1. **GEOMETRY** The ordered pairs of the relation $\{(-8, -3), (-8, -6), (-3, -6)\}$ are the coordinates of the vertices of a right triangle. Find the inverse of this relation. Describe the graph of the inverse.

As with relations, the ordered pairs of **inverse functions** are also related. We can write the inverse of the function $f(x)$ as $f^{-1}(x)$.

connectED.mcgraw-hill.com

Reading Math

Sense-Making f^{-1} is read *f inverse* or *the inverse of f*. Note that -1 is *not* an exponent.

Key Concept Property of Inverses

Words If f and f^{-1} are inverses, then $f(a) = b$ if and only if $f^{-1}(b) = a$.

Example Let $f(x) = x - 4$ and represent its inverse as $f^{-1}(x) = x + 4$.

Evaluate $f(6)$.
$f(x) = x - 4$
$f(6) = 6 - 4$ or 2

Evaluate $f^{-1}(2)$.
$f^{-1}(x) = x + 4$
$f^{-1}(2) = 2 + 4$ or 6

Because $f(x)$ and $f^{-1}(x)$ are inverses, $f(6) = 2$ and $f^{-1}(2) = 6$.

The inverse of a function can be found by exchanging the domain and the range.

Example 2 Find and Graph an Inverse

Find the inverse of each function. Then graph the function and its inverse.

a. $f(x) = 2x - 5$

Step 1 Rewrite the function as an equation relating x and y.
$f(x) = 2x - 5 \rightarrow y = 2x - 5$

Step 2 Exchange x and y in the equation. $x = 2y - 5$

Step 3 Solve the equation for y.
$x = 2y - 5$ Inverse of $y = 2x - 5$
$x + 5 = 2y$ Add 5 to each side.
$\dfrac{x + 5}{2} = y$ Divide each side by 2.

Step 4 Replace y with $f^{-1}(x)$.
$y = \dfrac{x + 5}{2} \rightarrow f^{-1}(x) = \dfrac{x + 5}{2}$

The inverse of $f(x) = 2x - 5$ is $f^{-1}(x) = \dfrac{x + 5}{2}$.
The graph of $f^{-1}(x) = \dfrac{x + 5}{2}$ is the reflection of the graph of $f(x) = 2x - 5$ in the line $y = x$.

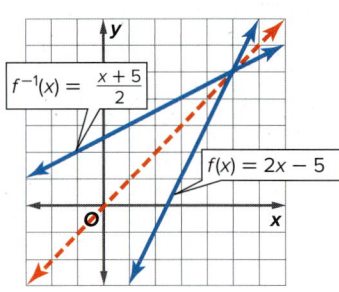

Go Online!

Explore the inverses of several functions and figure out when inverse functions exist and when they don't using **The Geometer's Sketchpad®** activity in ConnectED.

b. $f(x) = x^2 + 1$

Step 1 $f(x) = x^2 + 1 \rightarrow y = x^2 + 1$

Step 2 $x = y^2 + 1$

Step 3
$x = y^2 + 1$
$x - 1 = y^2$
$\pm\sqrt{x - 1} = y$ Take the square root of each side.

Step 4 $y = \pm\sqrt{x - 1}$

Graph $y = \pm\sqrt{x - 1}$ by reflecting the graph of $f(x) = x^2 + 1$ in the line $y = x$.

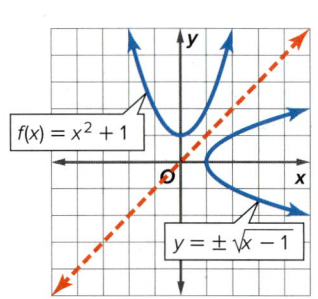

Study Tip

Functions The inverse of the function in part **b** is not a function since it does not pass the vertical line test.

Guided Practice

Find the inverse of each function. Then graph the function and its inverse.

2A. $f(x) = \dfrac{x - 3}{5}$

2B. $f(x) = 3x^2$

Not all functions have an inverse function. The graph of the initial relation in Example 2b is a function because it passes the vertical line test. However, its inverse relation fails this test, so it is not a function. The reflection relationship between the graph of a function and its inverse relation leads to the horizontal line test for determining whether an inverse of a function is itself a function.

Key Concept Horizontal Line Test

Words A function f has an inverse function f^{-1} if and only if each horizontal line intersects the graph of the function in at most one point.

Example Because no horizontal line intersects the graph of f more than once, the inverse function f^{-1} exists.

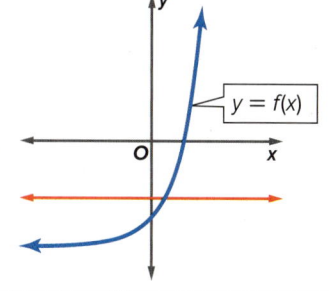

Sometimes it is necessary to restrict the domain of a function in order for its inverse to be a function.

Example 3 Inverses with Restricted Domains

Find the inverse of $f(x) = x^2 - 6x + 8$. Then graph the function and its inverse. If necessary, restrict the domain of $f(x)$ so that the inverse is a function.

Step 1 Use a graph to determine whether $f(x)$ and $f^{-1}(x)$ are functions.

$f(x)$ is a function because it passes the vertical line test. However, $f(x)$ does not pass the horizontal line test, which indicates that $f^{-1}(x)$ is not a function.

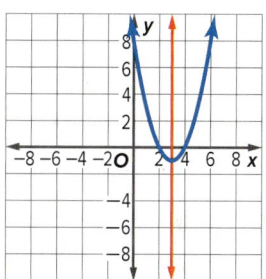

Step 2 Identify the axis of symmetry.

The axis of symmetry is $x = 3$.

Step 3 Find $f^{-1}(x)$.

$f(x) = x^2 - 6x + 8$ — Original function

$y = x^2 - 6x + 8$ — Replace $f(x)$ with y.

$x = y^2 - 6y + 8$ — Exchange x and y.

$x - 8 + 9 = y^2 - 6y + 9$ — Complete the square.

$x + 1 = (y - 3)^2$ — Simplify.

$\pm\sqrt{x + 1} = y - 3$ — Take the square root of each side.

$3 \pm \sqrt{x + 1} = y$ — Add 3 to each side.

$f^{-1}(x) = 3 \pm \sqrt{x + 1}$ — Replace y with $f^{-1}(x)$.

Step 4 Find a restricted domain of $f(x)$ so that $f^{-1}(x)$ will be a function.

If the domain is restricted to $(-\infty, 3]$, then the inverse is $f^{-1}(x) = 3 - \sqrt{x + 1}$.

If the domain is restricted to $[3, \infty)$, then the inverse is $f^{-1}(x) = 3 + \sqrt{x + 1}$.

Step 5 Graph.

Notice that in each case, the range of $f^{-1}(x)$ is restricted so that the graph passes the vertical line test.

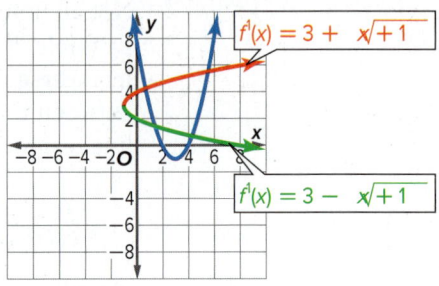

Guided Practice

3. Find the inverse of $f(x) = x^2 + 7x + 12$. Then graph the function and its inverse. If necessary, restrict the domain of $f(x)$ so that the inverse is a function.

2 Verifying Inverses

You can determine whether two functions are inverses by finding both of their compositions. If both compositions equal the identity function $I(x) = x$, then the functions are inverse functions.

Key Concept Inverse Functions

Words Two functions f and g are inverse functions if and only if both of their compositions are the identity function.

Symbols $f(x)$ and $g(x)$ are inverses if and only if $[f \circ g](x) = x$ and $[g \circ f](x) = x$.

Example 4 Verify that Two Functions are Inverses

Determine whether each pair of functions are inverse functions. Explain your reasoning.

a. $f(x) = 3x + 9$ and $g(x) = \frac{1}{3}x - 3$

Verify that the compositions of $f(x)$ and $g(x)$ are identity functions.

$[f \circ g](x) = f[g(x)]$
$\quad = f\left(\frac{1}{3}x - 3\right)$
$\quad = 3\left(\frac{1}{3}x - 3\right) + 9$
$\quad = x - 9 + 9 \text{ or } x$

$[g \circ f](x) = g[f(x)]$
$\quad = g(3x + 9)$
$\quad = \frac{1}{3}(3x + 9) - 3$
$\quad = x + 3 - 3 \text{ or } x$

The functions are inverses because $[f \circ g](x) = [g \circ f](x) = x$.

b. $f(x) = 4x^2$ and $g(x) = 2\sqrt{x}$

$[f \circ g](x) = f(2\sqrt{x})$
$\quad = 4(2\sqrt{x})^2$
$\quad = 4(4x) \text{ or } 16x$

Because $[f \circ g](x) \neq x$, $f(x)$ and $g(x)$ are not inverses.

> **Watch Out!**
>
> **Inverse Functions** Be sure to check both $[f \circ g](x)$ and $[g \circ f](x)$ to verify that functions are inverses. By definition, both compositions must be the identity function.

Guided Practice

4A. $f(x) = 3x - 3$, $g(x) = \frac{1}{3}x + 4$

4B. $f(x) = 2x^2 - 1$, $g(x) = \sqrt{\frac{x+1}{2}}$

Check Your Understanding

= Step-by-Step Solutions begin on page R11.

Go Online! for a Self-Check Quiz

Example 1 Find the inverse of each relation.

1. $\{(-9, 10), (1, -3), (8, -5)\}$
2. $\{(-2, 9), (4, -1), (-7, 9), (7, 0)\}$

Example 2 Find the inverse of each function. Then graph the function and its inverse.

3. $f(x) = -3x$
4. $g(x) = 4x - 6$
5. $h(x) = x^2 - 3$

Example 3 Find the inverse for each function. Then graph the function and its inverse. If necessary, restrict the domain of $f(x)$ so that the inverse is a function.

6. $f(x) = x^2 + 4x - 5$
7. $f(x) = x^2 - 16x + 63$

Example 4 Determine whether each pair of functions are inverse functions. Write *yes* or *no*.

8. $f(x) = x - 7$
   $g(x) = x + 7$
9. $f(x) = \frac{1}{2}x + \frac{3}{4}$
   $g(x) = 2x - \frac{4}{3}$
10. $f(x) = 2x^3$
    $g(x) = \frac{1}{3}\sqrt{x}$

Practice and Problem Solving

Extra Practice is on page R5.

Example 1 Find the inverse of each relation.

11. $\{(-8, 6), (6, -2), (7, -3)\}$
12. $\{(7, 7), (4, 9), (3, -7)\}$
13. $\{(8, -1), (-8, -1), (-2, -8), (2, 8)\}$
14. $\{(4, 3), (-4, -4), (-3, -5), (5, 2)\}$

Example 2 **MP SENSE-MAKING** Find the inverse of each function. Then graph the function and its inverse.

15. $f(x) = x + 2$
16. $g(x) = 5x$
17. $f(x) = -2x + 1$
18. $h(x) = \frac{x - 4}{3}$
19. $f(x) = -\frac{5}{3}x - 8$
20. $g(x) = x + 4$
21. $h(x) = x^2 + 4$
22. $f(x) = \frac{1}{2}x^2 - 1$
23. $f(x) = (x + 1)^2 + 3$

Example 3 Find the inverse for each function. Then graph the function and its inverse. If necessary, restrict the domain of $f(x)$ so that the inverse is a function.

24. $f(x) = 4x$
25. $f(x) = -8x + 9$
26. $f(x) = 5x^2$
27. $f(x) = x^2 + 12x + 32$
28. $f(x) = x^2 - 22x + 120$
29. $f(x) = x^2 - 36x - 160$

Example 4 Determine whether each pair of functions are inverse functions. Write *yes* or *no*.

30. $f(x) = 2x + 3$
    $g(x) = 2x - 3$
31. $f(x) = 4x + 6$
    $g(x) = \frac{x - 6}{4}$
32. $f(x) = -\frac{1}{3}x + 3$
    $g(x) = -3x + 9$
33. $f(x) = -6x$
    $g(x) = \frac{1}{6}x$
34. $f(x) = \frac{1}{2}x + 5$
    $g(x) = 2x - 10$
35. $f(x) = \frac{x + 10}{8}$
    $g(x) = 8x - 10$
36. $f(x) = \frac{2}{3}x^3$
    $g(x) = \sqrt{\frac{2}{3}x}$
37. $f(x) = (x + 6)^2$
    $g(x) = \sqrt{x} - 6$
38. $f(x) = 2\sqrt{x - 5}$
    $g(x) = \frac{1}{4}x^2 - 5$

39. **FUEL** The average miles traveled for every gallon g of gas consumed by Leroy's car is represented by the function $m(g) = 28g$.

    a. Find a function $c(g)$ to represent the cost per gallon of gasoline.

    b. Use inverses to determine the function used to represent the cost per mile traveled in Leroy's car.

40. MULTI-STEP Carlos is looking to trade in his old car for a new one. He has 6 payments remaining on his old car. The dealer is offering 0% financing and a $4000 trade-in with the purchase of a new car. Carlos plans to take out a 5-year loan on the new car.

　a. If his current monthly payment is $280 and he doesn't want to pay more than $300 per month on a car, what is the most expensive new car that he can afford?

　b. Describe your solution process.

41 GEOMETRY The formula for the area of a circle is $A = \pi r^2$.

　a. Find the inverse of the function.

　b. Use the inverse to find the radius of a circle with an area of 36 square centimeters.

Use the horizontal line test to determine whether the inverse of each function is also a function.

42. $f(x) = 2x^2$　　　**43.** $f(x) = x^3 - 8$　　　**44.** $g(x) = x^4 - 6x^2 + 1$

45. $h(x) = -2x^4 - x - 2$　　　**46.** $g(x) = x^5 + x^2 - 4x$　　　**47.** $h(x) = x^3 + x^2 - 6x + 12$

48. SHOPPING Felipe bought a used car. The sales tax rate was 7.25% of the selling price, and he paid $350 in processing and registration fees. Find the selling price if Felipe paid a total of $8395.75.

49. TEMPERATURE A formula for converting degrees Celsius to Fahrenheit is $F(x) = \frac{9}{5}x + 32$.

　a. Find the inverse $F^{-1}(x)$. Show that $F(x)$ and $F^{-1}(x)$ are inverses.

　b. Explain what purpose $F^{-1}(x)$ serves.

50. MEASUREMENT There are approximately 1.852 kilometers in a nautical mile.

　a. Write a function that converts nautical miles to kilometers.

　b. Find the inverse of the function that converts kilometers back to nautical miles.

　c. Using composition of functions, verify that these two functions are inverses.

51. MULTIPLE REPRESENTATIONS Consider the functions $y = x^n$ for $n = 0, 1, 2, \ldots$.

　a. Graphing Use a graphing calculator to graph $y = x^n$ for $n = 0, 1, 2, 3,$ and 4.

　b. Tabular For which values of n is the inverse a function? Record your results in a table.

　c. Analytical Make a conjecture about the values of n for which the inverse of $f(x) = x^n$ is a function. Assume that n is a whole number.

H.O.T. Problems　　Use Higher-Order Thinking Skills

52. REASONING If a relation is *not* a function, then its inverse is *sometimes*, *always*, or *never* a function. Explain your reasoning.

53. OPEN-ENDED Give an example of a function and its inverse. Verify that the two functions are inverses.

54. CHALLENGE Give an example of a function that is its own inverse.

55. CONSTRUCT ARGUMENTS Show that the inverse of a linear function $y = mx + b$, where $m \neq 0$ and $x \neq b$, is also a linear function.

56. WRITING IN MATH Suppose you have a composition of two functions that are inverses. When you put in a value of 5 for x, why is the result always 5?

Preparing for Assessment

57. The inverse relation of which function is *not* a function? **MP 3**

- A $f(x) = x^2 - 2$
- B $f(x) = 2x^3 + 3$
- C $f(x) = \sqrt{x - 3}$
- D $f(x) = \sqrt[3]{2x + 1}$

58. Find the inverse of $f(x) = \dfrac{1}{x^2}$. **MP 1, 6**

59. Let $f(n) = 2n$ where $n \neq 0$. If $n = m$, which statement must be true? **MP 3**

- A $f^{-1}(m) = m$
- B $f^{-1}(m) = 2m$
- C $f^{-1}\left(\dfrac{1}{2}m\right) = m$
- D $f^{-1}(2m) = m$
- E $f^{-1}(2m) = 4m$

60. MULTI-STEP Consider the function $f(x) = \dfrac{x + 2}{2}$.

a. Work is shown below to find the inverse function $f^{-1}(x)$. Which lines contain errors if the intention is to find the inverse function? Choose all that apply. **MP 1**

- A $f(x) = \dfrac{x + 2}{2}$
- B $y = \dfrac{x + 2}{2}$
- C $x = \dfrac{y + 2}{2}$
- D $2y = y + 2$
- E $2x - 2 = y$
- F $f^{-1}(x) = 2y - 2$

b. Write the correct inverse function for $f(x) = \dfrac{x + 2}{2}$.

c. What results when $f(f^{-1}(x))$ is calculated?

d. What results when $f^{-1}(f(x))$ is calculated?

61. Which pairs of functions below are inverse functions? Choose all that apply. **MP 1**

- A $f(x) = \dfrac{x + 2}{2}$ and $g(x) = \dfrac{x - 2}{2}$
- B $f(x) = 3x + 9$ and $g(x) = \dfrac{1}{3}x - 3$
- C $f(x) = 2x^2$ and $g(x) = \dfrac{\sqrt{x}}{2}$
- D $f(x) = 4x^2$ and $g(x) = \dfrac{\sqrt{x}}{2}$
- E $f(x) = 2x - 3$ and $g(x) = \dfrac{x + 3}{2}$
- F $f(x) = 2x - 3$ and $g(x) = \dfrac{x - 3}{2}$

62. Which of the following functions have an inverse that passes the vertical line test. Choose all for which the answer is yes. **MP 3**

- A $f(x) = \dfrac{x + 2}{2}$
- B $f(x) = 3\sqrt{x} + 9$
- C $f(x) = \sqrt{\dfrac{x}{2}}$
- D $f(x) = 4x^2$
- E $f(x) = \dfrac{x + 3}{3}$

63. Find the inverse of $f(x) = \dfrac{\sqrt{x}}{2}$ and then graph the function and its inverse. **MP 1, 4, 6**

64. Given that $f(2) = 4$, $f(4) = 6$, and $f(6) = 8$, what is the value of $f^{-1}(f^{-1}(6))$? **MP 2**

- A 2
- B 4
- C 6
- D 8

EXTEND 5-3

Graphing Technology Lab
Inverse Functions and Relations

You can use a TI-83/84 Plus graphing calculator to compare a function and its inverse using tables and graphs. Note that before you enter any values in the calculator, you should clear all lists.

Activity 1 Graph Inverses with Ordered Pairs

Graph $f(x) = \{(1, 2), (2, 4), (3, 6), (4, 8), (5, 10), (6, 12)\}$ and its inverse.

Step 1 Enter the *x*-values in **L1** and the *y*-values in **L2**. Then graph the function.

KEYSTROKES: STAT ENTER 1 ENTER 2 ENTER 3 ENTER 4 ENTER 5 ENTER 6 ENTER ▶ 2 ENTER 4 ENTER 6 ENTER 8 ENTER 10 ENTER 12 ENTER 2nd [STAT PLOT] ENTER ENTER GRAPH

Adjust the window to reflect the domain and range.

Step 2:

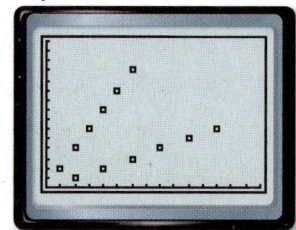

[0, 14] scl: 1 by [0, 14] scl: 1

Step 2 Define the inverse function by setting **Xlist** to **L2** and **Ylist** to **L1**. Then graph the inverse function.

KEYSTROKES: 2nd [STAT PLOT] ▼ ENTER ENTER ▼ ▼ 2nd [L2] ▼ 2nd [L1] GRAPH

Step 3:

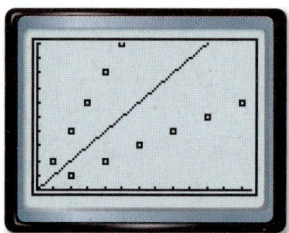

Step 3 Graph the line $y = x$.

KEYSTROKES: Y= X,T,θ,n GRAPH

[0, 14] scl: 1 by [0, 14] scl: 1

Activity 2 Graph Inverses with Function Notation

Graph $f(x) = 3x$ and its inverse $g(x) = \frac{x}{3}$.

Step 1 Clear the data from Activity 1.

KEYSTROKES: 2nd [STAT PLOT] ENTER ▶ ENTER ▲ ▶ ENTER ▶ ENTER 2nd [QUIT]

Step 2 Enter $f(x)$ as **Y1**, $g(x)$ as **Y2**, and $y = x$ as **Y3**. Then graph.

KEYSTROKES: Y= 3 X,T,θ,n ENTER X,T,θ,n ÷ 3 ENTER X,T,θ,n GRAPH

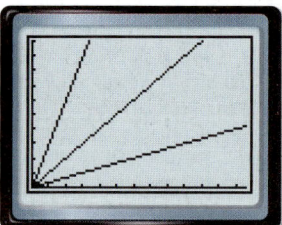

[0, 14] scl: 1 by [0, 14] scl: 1

Exercises

Graph each function $f(x)$ and its inverse $g(x)$. Then graph $(f \circ g)(x)$.

1. $f(x) = 5x$
2. $f(x) = x - 3$
3. $f(x) = 2x + 1$
4. $f(x) = \frac{1}{2}x + 3$
5. $f(x) = x^2$
6. $f(x) = x^2 - 3$

7. What is the relationship between the graphs of a function and its inverse?

8. **MAKE A CONJECTURE** For any function $f(x)$ and its inverse $g(x)$, what is $(f \circ g)(x)$?

CHAPTER 5
Mid-Chapter Quiz
Lessons 5-1 through 5-3

Given $f(x) = 2x^2 + 4x - 3$ and $g(x) = 5x - 2$, find each function. (Lesson 5-1)

1. $(f + g)(x)$

2. $(f - g)(x)$

3. $(f \cdot g)(x)$

4. $\left(\dfrac{f}{g}\right)(x)$

5. **FINANCE** A small company is producing a new product. The revenue $r(x)$ from the sale of x units of the new product is expected to be $r(x) = 10x$. The cost of manufacturing x units is $c(x) = 2.25x + 2000$. (Lesson 5-1)

   a. Write the profit function.
   b. Find the profit on 1000 units of the product.
   c. **MP** What mathematical practice did you use to solve this problem?

Given $f(x) = 2x^2 + 4x - 3$ and $g(x) = 5x - 2$, find each function. (Lesson 5-2)

6. $[f \circ g](x)$

7. $[g \circ f](x)$

8. **PRODUCTION** The cost in dollars of producing p cell phones in a factory is represented by $C(p) = 5p + 60$. The number of cell phones produced in h hours is represented by $P(h) = 40h$. (Lesson 5-2)

   a. Find the composition function.
   b. Determine the cost of producing cell phones for 8 hours.

Find $[f \circ g](x)$ and $[g \circ f](x)$, if they exist. State the domain and range for each composed function. (Lesson 5-2)

9. $f(x) = 4x$
   $g(x) = x - 8$

10. $f(x) = 3x - 1$
    $g(x) = 5x + 1$

11. $f(x) = -2x$
    $g(x) = x^2 - 8$

12. **SHOPPING** Mrs. Ross is shopping for her children's school clothes. She has a coupon for 25% off her total. The sales tax of 6% is added to the total after the coupon is applied. (Lesson 5-2)

    a. Express the total price after the discount and the total price after the tax using function notation. Let x represent the price of the clothing, $p(x)$ represent the price after the 25% discount, and $g(x)$ represent the price after the tax is added.

    b. Which composition of functions represents the final price, $p[g(x)]$ or $g[p(x)]$? Explain your reasoning.

Determine whether each pair of functions are inverse functions. Write *yes* or *no*. (Lesson 5-3)

13. $f(x) = 2x + 16$
    $g(x) = \dfrac{1}{2}x - 8$

14. $g(x) = 4x + 15$
    $h(x) = \dfrac{1}{4}x - 15$

15. $f(x) = x^2 - 5$
    $g(x) = 5 + x^{-2}$

16. $g(x) = -6x + 8$
    $h(x) = \dfrac{8 - x}{6}$

Find the inverse of each function, if it exists. (Lesson 5-3)

17. $h(x) = \dfrac{2}{5}x + 8$

18. $f(x) = \dfrac{4}{9}(x - 3)$

19. $h(x) = -\dfrac{10}{3}(x + 5)$

20. $f(x) = \dfrac{x + 12}{7}$

Use the horizontal line test to determine whether the inverse of each function is also a function. (Lesson 5-3)

21. $f(x) = 4x - 1$

22. $f(x) = 10x^2$

23. $f(x) = 3x^3 - 8$

24. **JOBS** Louise runs a lawn care service. She charges $25 for supplies plus $15 per hour. The function $f(h) = 15h + 25$ gives the cost $f(h)$ for h hours of work. (Lesson 5-3)

    a. Find $f^{-1}(h)$. What is the significance of $f^{-1}(h)$?

    b. If Louise charges a customer $85, how many hours did she work?

LESSON 4
Graphing Square Root Functions

::Then
- You simplified expressions with square roots.

::Now
1. Graph square root functions.
2. Analyze square root functions.

::Why?
With guitars, pitch is dependent on string length and string tension. The longer the string is, the higher the tension needs to be to produce a desired pitch. Likewise, the heavier the string is, the higher the tension needs to be to reach a desired pitch.

This can be modeled by the square root function $f = \frac{1}{2L}\sqrt{\frac{T}{P}}$, where T is the tension, P is the mass of the string, L is the length of the string, and f is the pitch.

New Vocabulary
square root function
radical function

Mathematical Practices
1 Make sense of problems and persevere in solving them.
2 Reason abstractly and quantitatively.
4 Model with mathematics.

1 Square Root Functions
If a function contains the square root of a variable, it is called a **square root function**. The square root function is a type of **radical function**.

Key Concept Parent Function of Square Root Functions

Parent function: $f(x) = \sqrt{x}$
Domain: $\{x \mid x \geq 0\}$ or $[0, +\infty)$
Range: $\{f(x) \mid f(x) \geq 0\}$ or $[0, +\infty)$
Intercepts: $x = 0, f(x) = 0$
Symmetry: none
Not defined: $x < 0$
End behavior: $x \to 0, f(x) \to 0; x \to +\infty, f(x) \to +\infty$
Extrema: minimum at $(0, 0)$

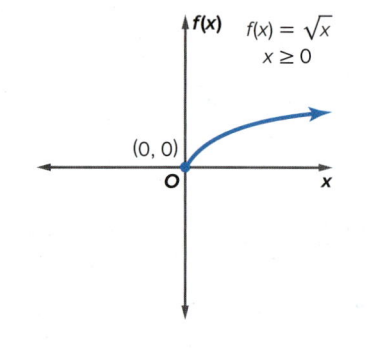

The domain of a square root function is limited to values for which the square root function is defined.

Example 1 Identify Domain and Range

Identify the domain and range of $f(x) = \sqrt{x + 4}$.

The domain only includes values for which the radicand is nonnegative.

$x + 4 \geq 0$ Write an inequality.
$x \geq -4$ Subtract 4 from each side.

$D = [-4, +\infty), \{x \mid x \geq -4\}$, or $\{-4 \leq x < \infty\}$.

Find $f(-4)$ to determine the lower limit of the range.

$f(-4) = \sqrt{-4 + 4}$ or 0

$R = [0, +\infty), \{f(x) \mid f(x) \geq 0\}$, or $\{0 \leq x < \infty\}$

▶ **Guided Practice**

Identify the domain and range of each function.

1A. $f(x) = \sqrt{x - 3}$ **1B.** $f(x) = \sqrt{x + 6} + 2$

The same techniques used to transform the graph of other functions you have studied can be applied to the graphs of square root functions.

> **Key Concept** Transformations of Square Root Functions
>
> $$f(x) = a\sqrt{x - h} + k$$
>
> | h—Horizontal Translation | k—Vertical Translation |
> |---|---|
> | h units right if h is positive | k units up if k is positive |
> | $\|h\|$ units left if h is negative | $\|k\|$ units down if k is negative |
> | The domain is $\{x \mid x \geq h\}$. | If $a > 0$, then the range is $\{f(x) \mid f(x) \geq k\}$. |
> | | If $a < 0$, then the range is $\{f(x) \mid f(x) \leq k\}$. |
>
> **a—Orientation and Shape**
> - If $a < 0$, the graph is reflected across the x-axis.
> - If $\|a\| > 1$, the graph is stretched vertically.
> - If $0 < \|a\| < 1$, the graph is compressed vertically.

Study Tip

Domain and Range The limits on the domain and range also represent the initial point of the graph of a square root function.

Example 2 Graph Square Root Functions

Graph each function. State the domain and range.

a. $y = \sqrt{x - 2} + 5$

The minimum point is at $(h, k) = (2, 5)$. Make a table of values for $x \geq 2$, and graph the function. The graph is the same shape as $f(x) = \sqrt{x}$, but is translated 2 units right and 5 units up. Notice the end behavior. As x increases, y increases.

$D = [2, +\infty), \{x \mid x \geq 2\}$, or $\{2 \leq x < +\infty\}$
$R = [5, +\infty), \{y \mid y \geq 5\}$, or $\{5 \leq x < +\infty\}$

x	y
2	5
3	6
4	6.4
5	6.7
6	7
7	7.2
8	7.4

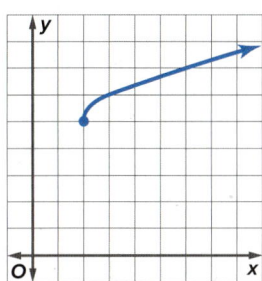

Go Online!

Follow along with your graphing calculator as you watch a **Personal Tutor** graph a square root function.

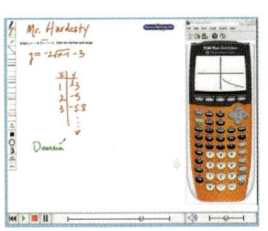

b. $y = -2\sqrt{x + 3} - 1$

The minimum domain value is at h or -3. Make a table of values for $x \geq -3$, and graph the function. Because a is negative, the graph is similar to the graph of $f(x) = \sqrt{x}$, but is reflected in the x-axis. Because $|a| > 1$, the graph is vertically stretched. It is also translated 3 units left and 1 unit down.

x	y
−3	−1
−2	−3
−1	−3.8
0	−4.5
1	−5
2	−5.5
3	−5.9

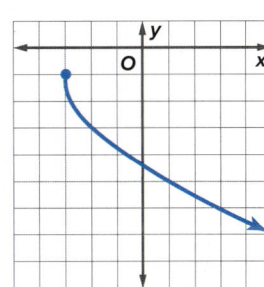

$D = [-3, +\infty), \{x \mid x \geq -3\}$,
or $\{-3 \leq x < +\infty\}$
$R = (-\infty, -1], \{y \mid y \leq -1\}$,
or $\{-\infty < x \leq -1\}$

Guided Practice

2A. $f(x) = 2\sqrt{x + 4}$

2B. $f(x) = \frac{1}{4}\sqrt{x - 5} + 3$

2 Analyze Square Root Functions

Previously you learned that an inverse relation interchanges the x- and y-coordinates of the original relation. For the power function $f(x) = x^2$, if the domain of x is restricted to nonnegative values, then the inverse of f is the function $f^{-1}(x) = \sqrt{x}$, $x \geq 0$.

Real-World Link
On every string, the guitar player has an option of decreasing the length of the string in about 24 different ways. This will produce 24 different frequencies on each string.
Source: *Guitar World*

Real-World Example 3   Use Graphs to Analyze Square Root Functions

MUSIC Refer to the application at the beginning of the lesson. The pitch, or frequency, measured in hertz (Hz) of a certain string can be modeled by $f(T) = \dfrac{1}{1.28}\sqrt{\dfrac{T}{0.0000708}}$, where T is tension in kilograms.

a. Graph the function for tension in the domain $\{T \mid 0 \leq T \leq 10\}$.

Make a table of values for $0 \leq T \leq 10$ and graph.

T	y(T)
0	0
1	92.8
2	131.3
3	160.8
4	185.7
5	207.6

T	f(T)
6	227.4
7	245.7
8	262.6
9	278.5
10	293.6

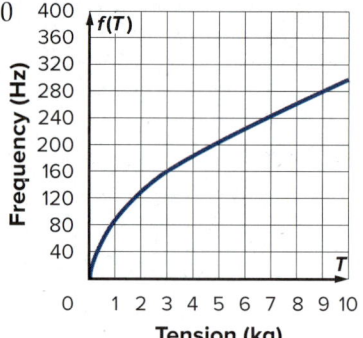

b. How much tension is needed for a pitch of over 200 Hz?

According to the graph and the table, more than 4.5 kilograms of tension is needed for a pitch of more than 200 hertz.

Problem-Solving Tip

MP Modeling Making a table is a good way to organize ordered pairs in order to see the general behavior of a graph.

▶ Guided Practice

3. MUSIC The frequency of vibrations for a certain guitar string when it is plucked can be determined by $F = 200\sqrt{T}$, where F is the number of vibrations per second and T is the tension measured in pounds. Graph the function for $0 \leq T \leq 10$. Then determine the frequency for $T = 3, 6,$ and 9 pounds.

Problem-Solving Tip

MP Reasoning Point out that if the domain of $f(x) = x^2$ is restricted to $x \leq 0$, its inverse would be the function $f^{-1}(x) = -\sqrt{x}$.

Example 4   Find the Inverse of Power Function $f(x) = x^2$

Find the inverse of $f(x) = x^2$, $x \geq 0$. Graph $f(x)$ and $f^{-1}(x)$ on the same coordinate plane.

$f(x) = x^2$ Write the original function.
$y = x^2$ Replace $f(x)$ with y.
$x = y^2$ Interchange x and y.
$\pm\sqrt{x} = y$ Take the square root of both sides.

Because the domain of f is restricted to nonnegative values of x, the range of f^{-1} must also be restricted to nonnegative values. So, the inverse of f is $f^{-1}(x) = \sqrt{x}$, $x \geq 0$.

The graph of $f^{-1}(x) = \sqrt{x}$ is a reflection of the graph of $f(x) = x^2$, $x \geq 0$, in the line $y = x$, shown as a dashed line on the graph.

▶ Guided Practice

4. Find the inverse of $f(x) = x^2 + 1$, $x \geq 0$ and graph $f(x)$ and $f^{-1}(x)$ on the same coordinate plane.

Check Your Understanding

= Step-by-Step Solutions begin on page R11.

Go Online! for a Self-Check Quiz

Example 1 Identify the domain and range of each function.

1. $f(x) = \sqrt{4x}$
2. $f(x) = \sqrt{x-5}$
3. $f(x) = \sqrt{x+8} - 2$

Example 2 Graph each function. State the domain and range.

4. $f(x) = \sqrt{x} - 2$
5. $f(x) = 3\sqrt{x-1}$
6. $f(x) = \frac{1}{2}\sqrt{x+4} - 1$
7. $f(x) = -\sqrt{3x-5} + 5$

Example 3

8. **OCEAN** The speed that a tsunami, or tidal wave, can travel is modeled by the equation $v = 356\sqrt{d}$, where v is the speed in kilometers per hour and d is the average depth of the water in kilometers. A tsunami in the ocean is found to be traveling at 145 kilometers per hour. What is the average depth of the water rounded to the nearest hundredth of a kilometer?

Example 4 Find the inverse of each function. Then graph the function and its inverse on the same coordinate plane.

9. $f(x) = 3x^2, x \geq 0$
10. $f(x) = x^2 + 2, x \geq 0$
11. $f(x) = -6x^2, x \geq 0$
12. $f(x) = \frac{1}{2}x^2, x \geq 0$

Practice and Problem Solving

Extra Practice is on page R5.

Example 1 Identify the domain and range of each function.

13. $f(x) = -\sqrt{2x} + 2$
14. $f(x) = \sqrt{x} - 6$
15. $f(x) = 4\sqrt{x-2} - 8$
16. $f(x) = \sqrt{x+2} + 5$
17. $f(x) = \sqrt{x-4} - 6$
18. $f(x) = -\sqrt{x-6} + 5$

Example 2 Graph each function. State the domain and range.

19. $f(x) = \sqrt{6x}$
20. $f(x) = -\sqrt{5x}$
21. $f(x) = \sqrt{x-8}$
22. $f(x) = \sqrt{x+1}$
23. $f(x) = \sqrt{x+3} + 2$
24. $f(x) = \sqrt{x-4} - 10$
25. $f(x) = 2\sqrt{x-5} - 6$
26. $f(x) = \frac{3}{4}\sqrt{x+12} + 3$
27. $f(x) = -\frac{1}{5}\sqrt{x-1} - 4$
28. $f(x) = -3\sqrt{x+7} + 9$

Example 3

29. **SKYDIVING** The approximate time t in seconds that it takes an object to fall a distance of d feet is given by $t = \sqrt{\frac{d}{16}}$. Suppose a parachutist falls 11 seconds before the parachute opens. How far does the parachutist fall during this time?

30. **MODELING** The velocity of a roller coaster as it moves down a hill is $V = \sqrt{v^2 + 64h}$, where v is the initial velocity in feet per second and h is the vertical drop in feet. The designer wants the coaster to have a velocity of 90 feet per second when it reaches the bottom of the hill.

   a. If the initial velocity of the coaster at the top of the hill is 10 feet per second, write an equation that models the situation.

   b. How high should the designer make the hill?

Example 4 Find the inverse of the function. Then graph the function and its inverse on the same coordinate plane.

31. $f(x) = 2x^2, x \geq 0$
32. $f(x) = x^2 + 1, x \geq 0$
33. $f(x) = -4x^2, x \geq 0$
34. $f(x) = \frac{1}{4}x^2, x \geq 0$
35. $f(x) = -\frac{1}{2}x^2, x \geq 0$
36. $f(x) = 4x^2 + 2, x \geq 0$
37. $f(x) = 9x^2 - 4, x \geq 0$
38. $f(x) = \frac{3}{4}x^2 + 8, x \geq 0$

39. **PHYSICS** The kinetic energy of an object is the energy produced due to its motion and mass. The formula for kinetic energy, measured in joules j, is $E = 0.5mv^2$, where m is the mass in kilograms and v is the velocity of the object in meters per second.

   a. Solve the above formula for v.

   b. If a 1500-kilogram vehicle is generating 1 million joules of kinetic energy, how fast is it traveling?

   c. *Escape velocity* is the minimum velocity at which an object must travel to escape the gravitational field of a planet or other object. Suppose a ship that weighs 100,000 kilograms must have a kinetic energy of 3.624×10^{14} joules to escape the gravitational field of Jupiter. Estimate the escape velocity of Jupiter.

40. **REASONING** After an accident, police can determine how fast a car was traveling before the driver put on his or her brakes by using the equation $v = \sqrt{30fd}$. In this equation, v represents the speed in miles per hour, f represents the coefficient of friction, and d represents the length of the skid marks in feet. The coefficient of friction varies depending on road conditions. Assume that $f = 0.6$.

   a. Find the speed of a car that skids 25 feet.

   b. If your car is going 35 miles per hour, how many feet would it take you to stop?

   c. If the speed of a car is doubled, will the skid be twice as long? Explain.

Write the square root function represented by each graph.

41.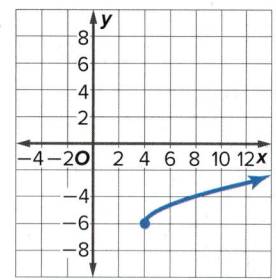
42.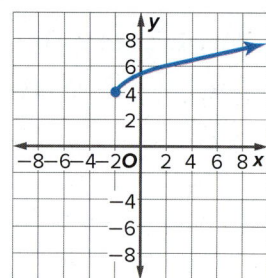
43.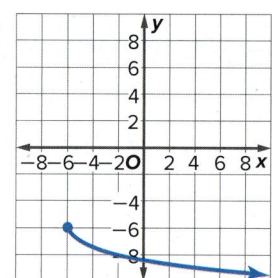

44. **REASONING** In this problem, you will use the following functions to investigate transformations of square root functions.

   $f(x) = 4\sqrt{x - 6} + 3$ $g(x) = \sqrt{16x + 1} - 6$ $h(x) = \sqrt{x + 3} + 2$

   a. **Graphical** Graph each function on the same set of axes.

   b. **Analytical** Identify the transformation on the graph of the parent function. What values caused each transformation?

   c. **Analytical** Which functions appear to be stretched or compressed vertically? Explain your reasoning.

   d. **Verbal** The two functions that are stretched appear to be stretched by the same magnitude. How is this possible?

   e. **Tabular** Make a table of the rate of change for all three functions between 8 and 12 as compared to 12 and 16. What generalization about rate of change in square root functions can be made as a result of your findings?

45. PENDULUMS The period of a pendulum can be represented by $T = 2\pi\sqrt{\frac{L}{g}}$, where T is the time in seconds, L is the length in feet, and g is gravity, 32 feet per second squared.

a. Graph the function for $0 \leq L \leq 10$.

b. What is the period for lengths of 2, 5, and 8 feet?

46. PHYSICS Using the function $m = \dfrac{m_0}{\sqrt{1 - \left(\frac{v^2}{c^2}\right)}}$, Einstein's theory of relativity states that the apparent mass m of a particle depends on its velocity v. An object that is traveling extremely fast, close to the speed of light c, will appear to have more mass compared to its mass at rest, m_0.

a. Use a graphing calculator to graph the function for a ship that weighs 10,000 kilograms for the domain $0 \leq v \leq 300{,}000{,}000$. Use 300 million meters per second for the speed of light.

b. What viewing window did you use to view the graph?

c. Determine the apparent mass m of the ship for speeds of 100 million, 200 million, and 299 million meters per second.

H.O.T. Problems   Use Higher-Order Thinking Skills

47. CHALLENGE Write an equation for a square root function with a domain of $\{x \mid x \geq -4\}$, a range of $\{y \mid y \leq 6\}$, and that passes through (5, 3).

48. MP REASONING For what positive values of a are the domain and range of $f(x) = \sqrt[a]{x}$ the set of real numbers?

49. OPEN-ENDED Write a square root function for which the domain is $\{x \mid x \geq 8\}$ and the range is $\{y \mid y \leq 14\}$.

50. WRITING IN MATH Explain why there are limitations on the domain and range of functions that have inverses that are functions.

51. ERROR ANALYSIS Cleveland thinks that the graph and the equation represent the same function. Molly disagrees. Who is correct? Explain your reasoning.

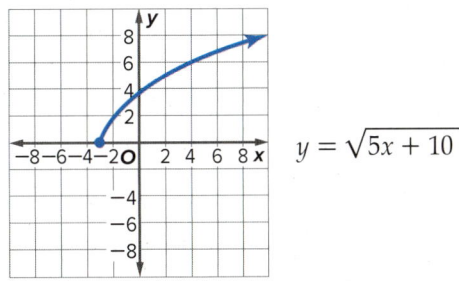

$y = \sqrt{5x + 10}$

52. WRITING IN MATH Explain why $y = \pm\sqrt{x}$ is not a function.

53. OPEN-ENDED Write an equation of a relation that contains a radical and its inverse such that

a. the original relation is a function, and its inverse is not a function.

b. the original relation is not a function, and its inverse is a function.

Preparing for Assessment

54. The graph shown is a transformation of the graph of $f(x) = \sqrt{x}$ MP 1, 2

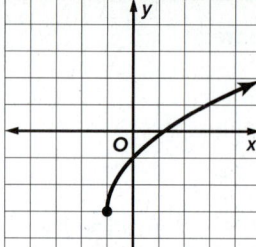

a. Which function is represented by the graph?

- A $g(x) = 2\sqrt{x+1} + 3$
- B $g(x) = 2\sqrt{x-1} + 3$
- C $g(x) = 2\sqrt{x+1} - 3$
- D $g(x) = 2\sqrt{x-1} - 3$

b. Which of the following statements about $g(x)$ are true? Select all true solutions.

- A An x-intercept is 1.
- B The domain is all real numbers.
- C The range is all real numbers.
- D The graph is of the inverse of $f(x) = \dfrac{(x-1)^2}{4} + 3$.
- E The graph has a y-intercept of -1.

55. What is the domain of the function? MP 2

$$f(x) = 3\sqrt{x-6} - 2$$

- A $\{x \mid x \geq -6\}$
- B $\{x \mid x \geq -2\}$
- C $\{x \mid x \geq 2\}$
- D $\{x \mid x \geq 6\}$

56. What is the range of the function? MP 2

$$f(x) = -4\sqrt{x-5} + 3$$

- A $\{f(x) \mid f(x) \leq -12\}$
- B $\{f(x) \mid f(x) \geq -12\}$
- C $\{f(x) \mid f(x) \leq 3\}$
- D $\{f(x) \mid f(x) \geq 3\}$

57. Let $f(x) = \sqrt{x-3} + 1$ and $g(x) = 4f(x)$. Which of the following statements are true? MP 2

- A The functions have the same domain and range.
- B The functions have the same domain, but the graph of $f(x)$ has lesser y-values for each value of x than $g(x)$ does.
- C The graph of $g(x)$ is a vertical compression of $f(x)$.
- D The graph of $g(x)$ is a vertical stretch of $f(x)$.
- E The graph of $g(x)$ is a translation of the graph of $f(x)$.

58. Which of the following is the graph of $f(x) = \dfrac{1}{2}x^2 - 4$ and its inverse? MP 1

A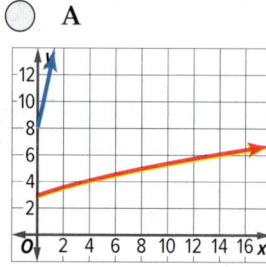

B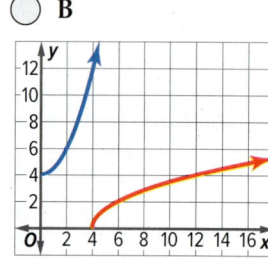

C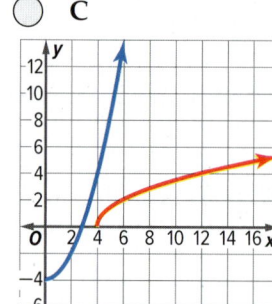

D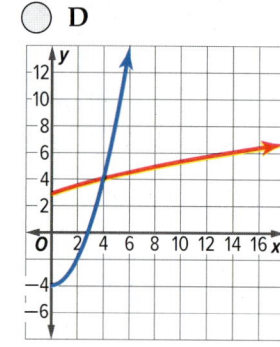

59. For which value(s) of x would $f(x) = \sqrt{2x-4}$ be undefined? Select all solutions. MP 2

- A $x = -4$
- B $x = 0$
- C $x = 2$
- D $x = 4$

60. Write the equation of a square root function that has been vertically stretched by a factor of 5, translated 3 units to the left, and translated 2 units up from the parent function, $f(x) = \sqrt{x}$. MP 2

LESSON 5
Graphing Cube Root Functions

Then
- You graphed square root functions.

Now
1. Graph cube root functions.
2. Analyze cube root functions.

Why?
A giraffe is a mammal that has an average life span of 25 years in the wild. It can range in height from 4 to 6 meters, and in weight from 794 to 1270 kilograms.

The height h in meters of some small giraffes weighing x kilograms can be approximated by the cube root function $h(x) = 0.45\sqrt[3]{x}$.

New Vocabulary
cube root function
inflection point

Mathematical Practices
1 Make sense of problems.
2 Reason abstractly and quantitatively.
4 Model with mathematics.
7 Make use of structure.

1 Graph Cube Root Functions
A radical function that contains the cube root of a variable is called a **cube root function**. The domain and range of a cube root function are both all real numbers, and the graph of a cube root function has an **inflection point**, a point on the curve where the curvature changes direction.

Key Concept Parent Function of Cube Root Functions

Parent function:	$f(x) = \sqrt[3]{x}$
Domain:	$\{x \mid -\infty < x < +\infty\}$ or $(-\infty, +\infty)$
Range:	$\{f(x) \mid -\infty < f(x) < +\infty\}$ or $(-\infty, +\infty)$
Intercepts:	$x = 0, y = 0$
Symmetry:	Point symmetry about the origin
End behavior:	$f(x) \to +\infty$ as $x \to +\infty$, $f(x) \to -\infty$ as $x \to -\infty$
Extrema:	None
Inflection point:	$(0, 0)$

Example 1 Identify Key Characteristics of Cube Root Functions

Describe the key characteristics of the graph of $f(x) = -\sqrt[3]{x - 2}$.

$D = \{x \mid -\infty < x < +\infty\}$ or $(-\infty, +\infty)$

$R = \{x \mid -\infty < f(x) < +\infty\}$ or $(-\infty, +\infty)$

End behavior: $f(x) \to -\infty$ as $x \to +\infty$
$f(x) \to +\infty$ as $x \to -\infty$

Inflection point: $(2, 0)$

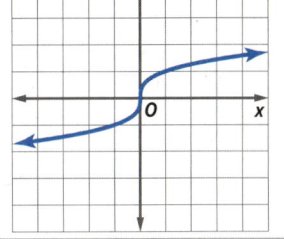

Guided Practice

1. Describe the key characteristics of the graph of $f(x) = -\sqrt[3]{x + 3}$.

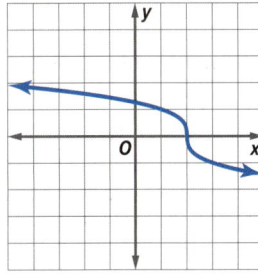

The same techniques used to transform the graph of other functions you have studied can be applied to the graphs of cube root functions.

Key Concept Transformations of Cube Root Functions

$f(x) = a\sqrt[3]{x - h} + k$	Inflection Point (h, k)
h—Horizontal Translation h units right if h is positive $\|h\|$ units left if h is negative The domain is all real numbers.	**k—Vertical Translation** k units up if k is positive $\|k\|$ units down if k is negative The range is all real numbers.

a—Orientation and Shape
- If $a < 0$, the graph is reflected across the x-axis.
- If $|a| > 1$, the graph is stretched vertically.
- If $0 < |a| < 1$, the graph is compressed vertically.

Example 2 Graph Cube Root Functions

Graph each function. Describe the key characteristics.

a. $y = \sqrt[3]{x + 3} - 4$

Make a table of values and graph the function. The graph is the same shape as $f(x) = \sqrt[3]{x}$, but is translated 3 units to the left and 4 units down.

$D = \{x \mid -\infty < x < +\infty\}$ or $(-\infty, +\infty)$
$R = \{x \mid -\infty < f(x) < +\infty\}$ or $(-\infty, +\infty)$
End behavior: $f(x) \to +\infty$ as $x \to +\infty$
$f(x) \to -\infty$ as $x \to -\infty$
Inflection point: $(-3, 4)$

x	y
−4	−5
−3	−4
−2	−3
0	−2.6
2	−2.3
5	−2

b. $y = -2\sqrt[3]{x - 1} + 3$

Make a table of values and graph the function. The graph of $f(x) = \sqrt[3]{x}$ is stretched by a factor of 2, translated 1 unit to the right, 3 units up, and reflected in the line $x = 1$.

Domain and Range:

$D = \{x \mid -\infty < x < +\infty\}$ or $(-\infty, +\infty)$
$R = \{x \mid -\infty < f(x) < +\infty\}$ or $(-\infty, +\infty)$
End behavior: $f(x) \to -\infty$ as $x \to +\infty$
$f(x) \to +\infty$ as $x \to -\infty$
Inflection point: $(1, 3)$

x	y
−7	7
−1	5.5
0	5
1	3
2	1
3	0.5

Guided Practice

2A. $y = \sqrt[3]{x - 3} + 2$

2B. $y = -3\sqrt[3]{x + 1} - 2$

2 Analyze Cube Root Functions

Previously you learned that an inverse relation interchanges the x- and y-coordinates of the original relation. For the power function $f(x) = x^3$, the inverse of f is the function $f^{-1}(x) = \sqrt[3]{x}$ for all real numbers.

Real-World Example 3   Use Graphs to Analyze Cube Root Functions

NATURE Refer to the application at the beginning of the lesson. A zookeeper found that the height of some small giraffes weighing x kilograms can be modeled by the cube root function $h(x) = 0.45\sqrt[3]{x}$, where h is the height in meters.

a. Make a table and graph the function in the domain $\{x \mid 0 \le x \le 1500\}$.

Round output values to the nearest tenth.

x	h
0	0
200	2.6
300	3.0
400	3.3
500	3.6

x	h
600	3.8
700	4.0
800	4.2
900	4.3
1000	4.5

x	h
1100	4.6
1200	4.8
1300	4.9
1400	5.0
1500	5.2

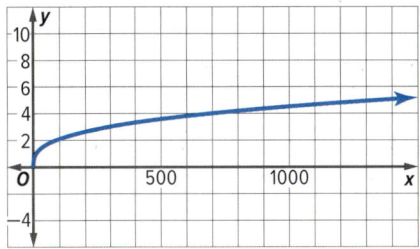

b. Analyze the graph. How should the domain be restricted?

The domain should be restricted to be between about 700 to 1300 kilograms, because that is the approximate weight range for giraffes.

c. What are key features of the graph?

The graph is in the first quadrant only. The values of y increase as the values of x increase.

d. What is the approximate height of a giraffe that weighs 850 kilograms?

According to the graph and the table, the height is about 4.26 meters.

Guided Practice

3. The height of some larger giraffes weighing x kilograms is better modeled by the cube root function $h(x) = 0.55\sqrt[3]{x}$, where h is the height in meters. Graph the function in the domain $\{x \mid 0 \le x \le 1300\}$ and analyze the graph. Then determine the approximate height of a giraffe that weighs 1150 kilograms.

Real-World Link

As the world's tallest mammals, giraffes use their height for reaching the leaves of tall trees, and their long legs to help them run as fast as 35 miles per hour for short distances. For longer distances, they run 10 miles per hour.

Source: http://animals.nationalgeographic.com/animals/mammals/giraffe/

Problem-Solving Tip

MP Modeling Making a table is a good way to organize ordered pairs in order to see the general behavior of a graph.

Example 4   Find the Inverse of Power Function $f(x) = x^3$

Find the inverse of $f(x) = x^3$. Then graph $f(x)$ and $f^{-1}(x)$ on the same coordinate plane.

$f(x) = x^3$ Write the original function.

$y = x^3$ Replace $f(x)$ with y.

$x = y^3$ Interchange x and y.

$\sqrt[3]{x} = y$ Take the cube root of both sides.

Because the domain of f is all real numbers, the range of f^{-1} is also all real numbers. So, the inverse of f is $f^{-1}(x) = \sqrt[3]{x}$ with domain and range all real numbers.

The graph of $f^{-1}(x) = \sqrt[3]{x}$ is a reflection of the graph of $f(x) = x^3$ in the line $y = x$, shown as a dashed line on the graph.

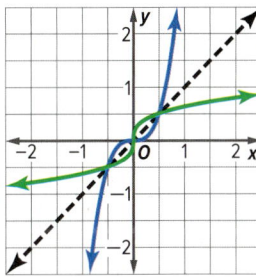

Guided Practice

4. Find the inverse of $f(x) = x^3 - 1$. Then graph $f(x)$ and $f^{-1}(x)$ on the same coordinate plane.

Check Your Understanding

= Step-by-Step Solutions begin on page R11.

Example 1 Describe the key characteristics of each function.

1. $f(x) = \sqrt[3]{x-4}$
2. $f(x) = \sqrt[3]{x+2} - 3$
3. $f(x) = 2\sqrt[3]{x} - 6$

Example 2 Graph each function. Describe the key characteristics.

4. $f(x) = \sqrt[3]{x} + 2$
5. $f(x) = 2\sqrt[3]{x-1}$
6. $f(x) = \frac{1}{2}\sqrt[3]{x+2} - 3$
7. $f(x) = -3\sqrt[3]{x-4} + 4$

Example 3

8. **GEOMETRY** The radius r of a sphere with volume V can be found using the formula $r = \sqrt[3]{\frac{3V}{4\pi}}$.

   a. Graph the function.

   b. Use the graph to determine the approximate radius for volumes of 1000 cubic meters, 8000 cubic meters, and 64,000 cubic meters.

   c. How does the volume of the sphere change if the radius is doubled? Explain.

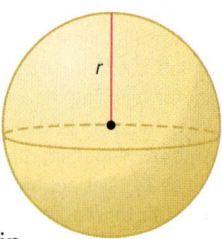

Example 4 Find the inverse of the function. Then graph the function and its inverse on the same coordinate plane.

9. $f(x) = 2x^3$
10. $f(x) = x^3 + 1$
11. $f(x) = -4x^3$
12. $f(x) = \frac{1}{2}x^3$

Practice and Problem Solving

Extra Practice is on page R5.

Example 1 Describe the key characteristics of each function.

13. $f(x) = 2\sqrt[3]{x+4}$
14. $f(x) = -\frac{1}{4}\sqrt[3]{x} + 1$
15. $f(x) = \frac{1}{3}\sqrt[3]{x-2} + 1$

Example 2 Graph each function. Describe the key characteristics.

16. $f(x) = \sqrt[3]{x} - 3$
17. $f(x) = 3\sqrt[3]{x+1}$
18. $f(x) = \frac{1}{4}\sqrt[3]{x-1} - 4$
19. $f(x) = -\sqrt[3]{x+1} - 2$
20. $f(x) = 2\sqrt[3]{x} + 3$
21. $f(x) = 3\sqrt[3]{x+2} - 4$
22. $f(x) = \frac{1}{2}\sqrt[3]{x-3} + 1$
23. $f(x) = -2\sqrt[3]{x-4} - 2$
24. $f(x) = \frac{1}{3}\sqrt[3]{x} - 1$
25. $f(x) = \frac{1}{2}\sqrt[3]{x-2} + 2$

26. **PERSERVERANCE** The radius r of the orbit of a communications satellite is given by $r = \sqrt[3]{\frac{GMt^2}{4\pi^2}}$, where G is the universal gravitational constant, M is the mass of Earth, and t is the time it takes the satellite to complete one orbit. Find the radius of the satellite's orbit if G is 6.67×10^{-11} N · m²/kg², M is 5.98×10^{24} kilograms, and t is 2.6×10^6 seconds.

27. **MODELING** A fruit grower has found that the circumference C, in inches, of most of his fruit crop can be modeled by the cube root equation $C = \sqrt[3]{\frac{W}{0.009}}$, where W is the weight in ounces.

    a. Graph the function.

    b. What is the approximate circumference of a piece of fruit that weighs 8.5 ounces?

    c. What is the approximate weight of a piece of fruit with a circumference of 8 inches?

    d. Explain why it is difficult to give a function that will work on all fruit.

28. **PHYSICS** Johannes Kepler developed the formula $d = \sqrt[3]{6t^2}$, where d is the distance of a planet to the Sun in millions of miles and t is the number of Earth days that it takes for the planet to orbit the Sun.

   a. Use a graphing calculator to graph the function. Explain why the graph of the function $f(t) = \sqrt[3]{6t^2}$ is not the graph of a standard cube root function.

   b. If the length of a year on Venus is 224.7 Earth days, how far is the Sun from Venus?

29. **MODELING** The surface area S of a sphere can be determined from the volume of the sphere using the formula $S = \sqrt[3]{36\pi V^2}$, where V is the volume.

   a. Use a graphing calculator to graph the function. Explain why the graph of $S = \sqrt[3]{36\pi V^2}$ is not the graph of a standard cube root function.

   b. Determine the surface area of a sphere with a volume of 200 cubic inches.

Write the cube root function represented by each graph. Points with integer coordinates are shown, with one point the inflection point.

30.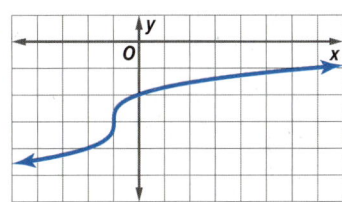

31.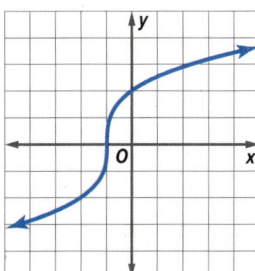

32.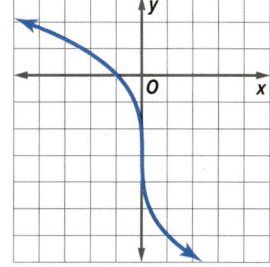

33. **GEOMETRY** The side length of a cube is determined by $r = \sqrt[3]{V}$, where V is the volume in cubic units.

   a. Graph the function.

   b. What is the domain and range of the function? Justify your reasoning.

   c. What is the side length of a cube with a volume of 512 cubic centimeters?

H.O.T. Problems   Use Higher-Order Thinking Skills

34. **REASONING** Explain how to find an equation for a cube root function that passes through the points $(-8, -1)$ if the inflection point is $(0, 0)$.

35. **REASONING** Describe any symmetry in the graph of $f(x) = \sqrt[3]{x}$.

36. **WRITING IN MATH** Explain why there are no limitations on the domain and range of a cube root function.

37. **ERROR ANALYSIS** Lin thinks that the graph and the equation $g(x) = \sqrt[3]{x}$ represent the same function. Milo disagrees. Who is correct? Explain your reasoning.

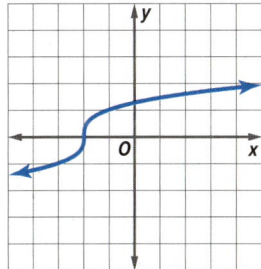

38. **CHALLENGE** Use a graphing calculator to explore the graphs of $f(x) = \sqrt{x}$, $g(x) = \sqrt[3]{x}$, $h(x) = \sqrt[4]{x}$, and $j(x) = \sqrt[5]{x}$. Compare the symmetry of the graphs of $s(x) = \sqrt[n]{x}$ when n is even or odd.

Preparing for Assessment

39. The graph shown is a transformation of the graph of $y = \sqrt[3]{x}$. MP 1, 2

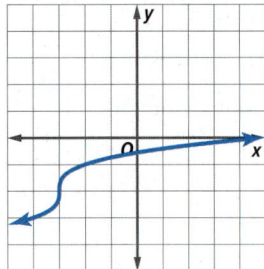

a. Which function is represented by the graph?

- A $g(x) = \sqrt[3]{x - 3} - 2$
- B $g(x) = \sqrt[3]{x - 3} + 2$
- C $g(x) = \sqrt[3]{x + 3} - 2$
- D $g(x) = \sqrt[3]{x + 2} - 3$

b. Which of the following statements about g(x) are true? Select all true solutions. MP 2

- A An x-intercept is 5.
- B The domain is all real numbers.
- C A point of inflection is $(-3, -2)$.
- D The graph is a dilation of the graph of f(x).
- E The graph has a y-intercept of -1.

40. Which of the following will have a graph that is stretched by a factor of 2, translated 4 units right and 1 unit down? MP 2

- A $y = 2\sqrt[3]{x + 4} - 1$
- B $y = 2\sqrt[3]{x - 4} - 1$
- C $y = \sqrt[3]{2x - 4} - 1$
- D $y = 2\sqrt[3]{x - 1} + 4$

41. The formula $d = \sqrt[3]{6t^2}$ represents d, the distance in millions of miles a planet is from the Sun, in terms of t, the number of Earth days it takes for the planet to orbit the Sun. It takes Mercury 88 Earth days to complete one orbit. To the nearest hundredth, how many millions of miles is Mercury from the Sun? MP 2

42. What is the equation of the inverse of $f(x) = \frac{x^3}{3} + 6$? MP 1

43. Let $f(x) = \sqrt[3]{x}$ and $g(x) = 3\sqrt[3]{x + 2} - 5$. How do the domain and range of f(x) and g(x) compare? MP 2

- A They have the same domain and range.
- B They have the same domain, but the graph of f(x) has higher y-values for each value of x.
- C They have the same domain, but the graph of g(x) has higher y-values for each value of x.
- D They have the same range, but the domain of f(x) has lower x-values for each value of y.

44. Which of the following is the graph of the inverse of $f(x) = \frac{1}{2}x^3 + 4$? MP 1

A
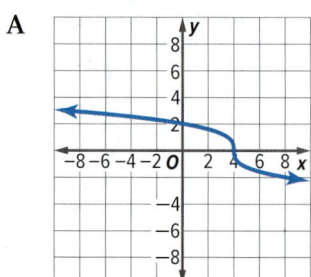

B
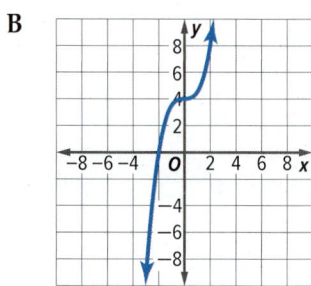

C
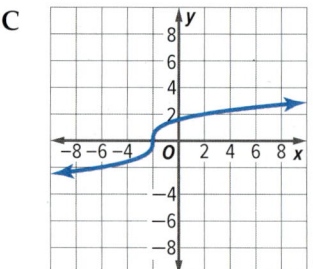

D
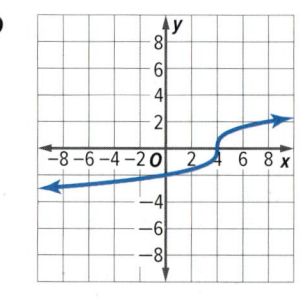

EXTEND 5

Graphing Technology Lab
Graphing *n*th Root Functions

In general, the graph of a cube root function is an s-shaped curve. The graphs are not symmetric with the *x*- or *y*-axis or the line $y = -x$ there are no asymptotes. The parent function $f(x) = \sqrt[3]{x}$ can be transformed by using $af(x)$, $f(x) + d$, $f(bx)$, and $f(x - c)$ for positive and negative values of *a*, *b*, *c*, and *d*.

Mathematical Practices
MP **4** Model with Mathematics

Activity 1 $af(x)$, $-af(x)$, $f(bx)$, and $f(-bx)$

Work cooperatively. Graph the set of functions on the same screen. Describe any similarities and differences among the graphs.

$f(x) = \sqrt[3]{x}$, $g(x) = 2\sqrt[3]{x}$, $h(x) = \sqrt[3]{2x}$,
$j(x) = -2\sqrt[3]{x}$, $k(x) = \sqrt[3]{-2x}$

The graphs have the same general shape. The functions $f(x)$, $g(x)$, and $h(x)$ lie in the first and third quadrants. The graphs $j(x)$ and $k(x)$ are reflections and lie in the second and fourth quadrants. The graphs of $g(x)$ and $j(x)$ are more vertically stretched than $f(x)$. The graphs of $h(x)$ and $k(x)$ are more horizontally compressed than $f(x)$ but less than $g(x)$ and $j(x)$.

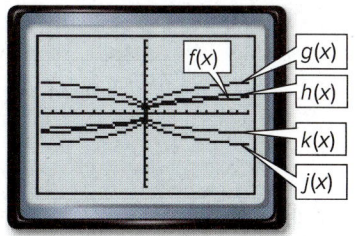

[−10, 10] scl: 1 by [−10, 10] scl: 1

Some transformations cause the graph of a function to change location but do not change the shape from the parent function.

Activity 2 $f(x - c)$, $f(x + c)$, $f(x) + d$, and $f(x) - d$

Work cooperatively. Graph the set of functions on the same screen. Describe any similarities and differences among the graphs.

$f(x) = \sqrt[3]{x}$, $g(x) = \sqrt[3]{x + 2}$, $h(x) = \sqrt[3]{x - 2}$,
$j(x) = \sqrt[3]{x} + 2$, $k(x) = \sqrt[3]{x} - 2$

The graphs have the same shape, but have been translated. The graph of $g(x)$ is shifted 2 units left and $h(x)$ is shifted 2 units right while the graph of $j(x)$ is shifted 2 units up and $k(x)$ shifted 2 units down.

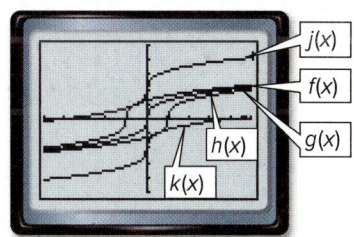

[−10, 10] scl: 2 by [−5, 5] scl: 1

Exercises

Work cooperatively. Examine each set of functions and predict the similarities and differences in their graphs. Use a graphing calculator to confirm your prediction. Write a sentence or two comparing the graphs.

1. $f(x) = \sqrt[3]{x}$, $g(x) = 3\sqrt[3]{x}$, $h(x) = -3\sqrt[3]{x}$
2. $f(x) = \sqrt[3]{x}$, $g(x) = \sqrt[3]{x} + 1$, $h(x) = \sqrt[3]{x} - 1$
3. $f(x) = \sqrt[3]{x}$, $g(x) = \sqrt[3]{8x}$, $h(x) = \sqrt[3]{-8x}$
4. $f(x) = \sqrt[3]{x}$, $g(x) = \sqrt[3]{x - 8}$, $h(x) = \sqrt[3]{x + 8}$

LESSON 6
Solving Radical Equations

::Then::
- You solved polynomial equations.

::Now::
1. Solve equations containing radicals.
2. Solve inequalities containing radicals.

::Why?::
When you jump, the time that you are in the air is your hang time. Hang time can be calculated in seconds t if you know the height h of the jump in feet. The formula for hang time is $t = 0.5\sqrt{h}$.

A volleyball player had a hang time of about 0.67 second. How would you calculate the height of her jump?

New Vocabulary
radical equation
extraneous solution
radical inequality

Mathematical Practices
4 Model with mathematics.
6 Attend to precision.

1 Solve Radical Equations Radical equations include radical expressions. You can solve a radical equation by raising each side of the equation to a power.

Key Concept Solving Radical Equations

Step 1 Isolate the radical on one side of the equation.

Step 2 Raise each side of the equation to a power equal to the index of the radical to eliminate the radical.

Step 3 Solve the resulting polynomial equation. Check your results.

When solving radical equations, the result may be a number that does not satisfy the original equation. Such a number is called an **extraneous solution**.

Example 1 Solve Radical Equations

Solve each equation.

a. $\sqrt{x + 2} + 4 = 7$

$\sqrt{x + 2} + 4 = 7$	Original equation
$\sqrt{x + 2} = 3$	Subtract 4 from each side to isolate the radical.
$(\sqrt{x + 2})^2 = 3^2$	Square each side to eliminate the radical.
$x + 2 = 9$	Find the squares.
$x = 7$	Subtract 2 from each side.

CHECK
$\sqrt{x + 2} + 4 = 7$	Original equation
$\sqrt{7 + 2} + 4 \stackrel{?}{=} 7$	Replace x with 7.
$7 = 7$ ✓	Simplify.

b. $\sqrt{x - 12} = 2 - \sqrt{x}$

$\sqrt{x - 12} = 2 - \sqrt{x}$	Original equation
$(\sqrt{x - 12})^2 = (2 - \sqrt{x})^2$	Square each side.
$x - 12 = 4 - 4\sqrt{x} + x$	Find the squares.
$-16 = -4\sqrt{x}$	Isolate the radical.
$4 = \sqrt{x}$	Divide each side by -4.
$16 = x$	Square each side.

Study Tip

Tools You can use a graphing calculator to check solutions of equations. Graph each side of the original equation, and examine the intersection.

CHECK $\sqrt{x-12} = 2 - \sqrt{x}$

$\sqrt{16-12} \stackrel{?}{=} 2 - \sqrt{16}$

$\sqrt{4} \stackrel{?}{=} 2 - 4$

$2 \neq -2$ ✗

The solution does not check, so the equation has an extraneous solution. The graphs of $y = \sqrt{x-12}$ and $y = 2 - \sqrt{x}$ do not intersect, which confirms that there is no real solution.

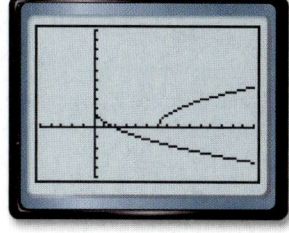

[−10, 30] scl: 2 by [−5, 10] scl: 1

▶ **Guided Practice**

1A. $5 = \sqrt{x-2} - 1$

1B. $\sqrt{x+15} = 5 + \sqrt{x}$

To undo a square root, you square the expression. To undo a cube root, you must raise the expression to the third power.

Example 2   Solve a Cube Root Equation

Solve $2(6x-3)^{\frac{1}{3}} - 4 = 0$.

In order to remove the $\frac{1}{3}$ power, or cube root, you must first isolate it and then raise each side of the equation to the third power.

$2(6x-3)^{\frac{1}{3}} - 4 = 0$ Original equation

$2(6x-3)^{\frac{1}{3}} = 4$ Add 4 to each side.

$(6x-3)^{\frac{1}{3}} = 2$ Divide each side by 2.

$\left[(6x-3)^{\frac{1}{3}}\right]^3 = 2^3$ Cube each side.

$6x - 3 = 8$ Evaluate the cubes.

$6x = 11$ Add 3 to each side.

$x = \frac{11}{6}$ Divide each side by 6.

CHECK $2(6x-3)^{\frac{1}{3}} - 4 = 0$ Original equation

$2\left(6 \cdot \frac{11}{6} - 3\right)^{\frac{1}{3}} - 4 \stackrel{?}{=} 0$ Replace x with $\frac{11}{6}$.

$2(8)^{\frac{1}{3}} - 4 \stackrel{?}{=} 0$ Simplify.

$2(2) - 4 \stackrel{?}{=} 0$ The cube root of 8 is 2.

$0 = 0$ ✓   Subtract.

▶ **Guided Practice**

Solve each equation.

2A. $(3n+2)^{\frac{1}{3}} + 1 = 0$

2B. $3(5y-1)^{\frac{1}{3}} - 2 = 0$

You can apply the methods used to solve square and cube root equations to solving equations with roots of any index. To undo an nth root, raise to the nth power.

Example 3 Solve a Radical Equation

What value of n is a solution to $3(\sqrt[4]{2n+6}) - 6 = 0$?

A -1 **B** 1 **C** 5 **D** 11

$3(\sqrt[4]{2n+6}) - 6 = 0$	Original equation
$3(\sqrt[4]{2n+6}) = 6$	Add 6 to each side.
$\sqrt[4]{2n+6} = 2$	Divide each side by 3.
$(\sqrt[4]{2n+6})^4 = 2^4$	Raise each side to the fourth power.
$2n + 6 = 16$	Evaluate each side.
$2n = 10$	Subtract 6 from each side.
$n = 5$	The answer is C.

Study Tip

Substitute Values You could also solve the multiple-choice question by substituting each answer for n in the equation to see if the solution is correct.

Guided Practice

3. What value of x is a solution of $4(3x + 6)^{\frac{1}{4}} - 12 = 0$?

A $x = 7$ **B** $x = 25$ **C** $x = 29$ **D** $x = 37$

2 Solve Radical Inequalities
A **radical inequality** has a variable in the radicand. To solve radical inequalities, complete the following steps.

Key Concept Solving Radical Inequalities

Step 1 If the index of the root is even, identify the values of the variable for which the radicand is nonnegative.

Step 2 Solve the inequality algebraically.

Step 3 Test values to check your solution.

Example 4 Solve Radical Inequalities

Solve $3 + \sqrt{5x - 10} \leq 8$.

Step 1 Because the radicand of a square root must be greater than or equal to zero, first solve $5x - 10 \geq 0$ to identify the values of x for which the left side of the inequality is defined.

$5x - 10 \geq 0$	Set the radicand ≥ 0.
$5x \geq 10$	Add 10 to each side.
$x \geq 2$	Divide each side by 5.

Step 2 Solve $3 + \sqrt{5x - 10} \leq 8$.

$3 + \sqrt{5x - 10} \leq 8$	Original inequality
$\sqrt{5x - 10} \leq 5$	Isolate the radical.
$5x - 10 \leq 25$	Eliminate the radical.
$5x \leq 35$	Add 10 to each side.
$x \leq 7$	Divide each side by 5.

Study Tip

Radical Inequalities Because a principal square root is never negative, inequalities that simplify to the form $\sqrt{ax + b} \leq c$, where c is a negative number, have no solutions.

Go Online!

Check your mastery of solving radical inequalities by using the **Self-Check Quiz** in ConnectED.

Step 3 It appears that $2 \leq x \leq 7$. You can test some x-values to confirm the solution. Use three test values: one less than 2, one between 2 and 7, and one greater than 7. Organize the test values in a table.

$x = 0$	$x = 4$	$x = 9$
$3 + \sqrt{5(0) - 10} \stackrel{?}{\leq} 8$ $3 + \sqrt{-10} \leq 8$ ✗	$3 + \sqrt{5(4) - 10} \stackrel{?}{\leq} 8$ $6.16 \leq 8$ ✓	$3 + \sqrt{5(9) - 10} \stackrel{?}{\leq} 8$ $8.92 \leq 8$ ✗
Because $\sqrt{-10}$ is not a real number, the inequality is not satisfied.	Because $6.16 \leq 8$, the inequality is satisfied.	Because $8.92 \not\leq 8$, the inequality is not satisfied.

The solution checks. Only values in the interval $2 \leq x \leq 7$ satisfy the inequality. You can summarize the solution with a number line.

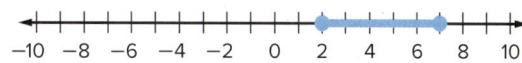

Guided Practice

Solve each inequality.

4A. $\sqrt{2x + 2} + 1 \geq 5$ **4B.** $\sqrt{4x - 4} - 2 < 4$

Check Your Understanding

 = Step-by-Step Solutions begin on page R11.

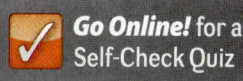 **Go Online!** for a Self-Check Quiz

Examples 1–2 Solve each equation.

1. $\sqrt{x - 4} + 6 = 10$
2. $\sqrt{x + 13} - 8 = -2$
3. $8 - \sqrt{x + 12} = 3$
4. $\sqrt{x - 8} + 5 = 7$
5. $\sqrt[3]{x - 2} = 3$
6. $(x - 5)^{\frac{1}{3}} - 4 = -2$
7. $(4y)^{\frac{1}{3}} + 3 = 5$
8. $\sqrt[3]{n + 8} - 6 = -3$
9. $\sqrt{y} - 7 = 0$
10. $2 + 4z^{\frac{1}{2}} = 0$
11. $5 + \sqrt{4y - 5} = 12$
12. $\sqrt{2t - 7} = \sqrt{t + 2}$

13. **REASONING** The time T in seconds that it takes a pendulum to make a complete swing back and forth is given by the formula $T = 2\pi\sqrt{\dfrac{L}{g}}$, where L is the length of the pendulum in feet and g is the acceleration due to gravity, 32 feet per second squared.

   a. In Tokyo, Japan, a huge pendulum in the Shinjuku building measures 73 feet 9.75 inches. How long does it take for the pendulum to make a complete swing?

   b. A clockmaker wants to build a pendulum that takes 20 seconds to swing back and forth. How long should the pendulum be?

Example 3

14. **MULTIPLE CHOICE** Solve $(2y + 6)^{\frac{1}{4}} - 2 = 0$.

   A $y = 1$ **B** $y = 5$ **C** $y = 11$ **D** $y = 15$

Example 4 Solve each inequality.

15. $\sqrt{3x + 4} - 5 \leq 4$
16. $\sqrt{b - 7} + 6 \leq 12$
17. $2 + \sqrt{4y - 4} \leq 6$
18. $\sqrt{3a + 3} - 1 \leq 2$
19. $1 + \sqrt{7x - 3} > 3$
20. $\sqrt{3x + 6} + 2 \leq 5$
21. $-2 + \sqrt{9 - 5x} \geq 6$
22. $6 - \sqrt{2y + 1} < 3$

Practice and Problem Solving

Extra Practice is on page R5.

Example 1 Solve each equation. Confirm by using a graphing calculator.

23. $\sqrt{2x+5} - 4 = 3$
24. $6 + \sqrt{3x+1} = 11$
25. $\sqrt{x+6} = 5 - \sqrt{x+1}$
26. $\sqrt{x-3} = \sqrt{x+4} - 1$
27. $\sqrt{x-15} = 3 - \sqrt{x}$
28. $\sqrt{x-10} = 1 - \sqrt{x}$
29. $6 + \sqrt{4x+8} = 9$
30. $2 + \sqrt{3y-5} = 10$
31. $\sqrt{x-4} = \sqrt{2x-13}$
32. $\sqrt{7a-2} = \sqrt{a+3}$
33. $\sqrt{x-5} - \sqrt{x} = -2$
34. $\sqrt{b-6} + \sqrt{b} = 3$

35. **MP SENSE-MAKING** Isabel accidentally dropped her keys from the top of a Ferris wheel. The formula $t = \frac{1}{4}\sqrt{d-h}$ describes the time t in seconds at which the keys are h meters above the ground and Isabel is d meters above the ground. If Isabel was 65 meters high when she dropped the keys, how many meters above the ground will the keys be after 2 seconds?

Example 2 Solve each equation.

36. $(5n-6)^{\frac{1}{3}} + 3 = 4$
37. $(5p-7)^{\frac{1}{3}} + 3 = 5$
38. $(6q+1)^{\frac{1}{4}} + 2 = 5$
39. $(3x+7)^{\frac{1}{4}} - 3 = 1$
40. $(3y-2)^{\frac{1}{5}} + 5 = 6$
41. $(4z-1)^{\frac{1}{5}} - 1 = 2$
42. $2(x-10)^{\frac{1}{3}} + 4 = 0$
43. $3(x+5)^{\frac{1}{3}} - 6 = 0$
44. $\sqrt[3]{5x+10} - 5 = 0$
45. $\sqrt[3]{4n-8} - 4 = 0$
46. $\frac{1}{7}(14a)^{\frac{1}{3}} = 1$
47. $\frac{1}{4}(32b)^{\frac{1}{3}} = 1$

Example 3

48. **MULTIPLE CHOICE** Solve $\sqrt[4]{y+2} + 9 = 14$.
    - A 23
    - B 53
    - C 123
    - D 623

49. **MULTIPLE CHOICE** Solve $(2x-1)^{\frac{1}{4}} - 2 = 1$.
    - F 41
    - G 28
    - H 13
    - J 1

Example 4 Solve each inequality.

50. $1 + \sqrt{5x-2} > 4$
51. $\sqrt{2x+14} - 6 \geq 4$
52. $10 - \sqrt{2x+7} \leq 3$
53. $6 + \sqrt{3y+4} < 6$
54. $\sqrt{2x+5} - \sqrt{9+x} > 0$
55. $\sqrt{d+3} + \sqrt{d+7} > 4$
56. $\sqrt{3x+9} - 2 < 7$
57. $\sqrt{2y+5} + 3 \leq 6$
58. $-2 + \sqrt{8-4z} \geq 8$
59. $-3 + \sqrt{6a+1} > 4$
60. $\sqrt{2} - \sqrt{b+6} \leq -\sqrt{b}$
61. $\sqrt{c+9} - \sqrt{c} > \sqrt{3}$

62. **PENDULUMS** The formula $s = 2\pi\sqrt{\frac{\ell}{32}}$ represents the swing of a pendulum, where s is the time in seconds to swing back and forth and ℓ is the length of the pendulum in feet. Find the length of a pendulum that makes one swing in 1.5 seconds.

63. **TURTLES** The relationship between the length and weight of certain turtles can be approximated by the equation $L = 0.55\sqrt[3]{W}$, where L is the length in feet and W is the weight in pounds. Solve this equation for W.

64. **HANG TIME** Refer to the information at the beginning of the lesson regarding hang time. Describe how the height of a jump is related to the amount of time in the air. Write a step-by-step explanation of how to determine the height of the volleyball player's 0.98-second jump.

65. **CONCERTS** The organizers of a concert are preparing for the arrival of 50,000 people in the open field where the concert will take place. Each person is allotted 5 square feet of space, so the organizers rope off a circular area of 250,000 square feet. Using the formula $A = \pi r^2$, where A represents the area of the circular region and r represents the radius of the region, find the radius of this region.

66. **WEIGHTLIFTING** The formula $M = 512 - 146{,}230B^{-\frac{8}{5}}$ can be used to estimate the maximum total mass that a weightlifter of mass B kilograms can lift using the snatch and the clean and jerk. According to the formula, how much does a person weigh who can lift at most 470 kilograms?

H.O.T. Problems   Use Higher-Order Thinking Skills

67. **ERROR ANALYSIS** Which equation does not have a solution?

$\sqrt{x-1} + 3 = 4$	$\sqrt{x+1} + 3 = 4$
$\sqrt{x-2} + 7 = 10$	$\sqrt{x+2} - 7 = -10$

68. **CHALLENGE** Lola is working to solve $(x+5)^{\frac{1}{4}} = -4$. She said that she could tell there was no real solution without even working the problem. Is Lola correct? Explain your reasoning.

69. **REASONING** Determine whether $\dfrac{\sqrt{(x^2)^2}}{-x} = x$ is *sometimes*, *always*, or *never* true when x is a real number. Explain your reasoning.

70. **OPEN-ENDED** Select a whole number. Now work backward to write two radical equations that have that whole number as solutions. Write one square root equation and one cube root equation. You may need to experiment until you find a whole number you can easily use.

71. **WRITING IN MATH** Explain the relationship between the index of the root of a variable in an equation and the power to which you raise each side of the equation to solve the equation.

72. **OPEN-ENDED** Write an equation that can be solved by raising each side of the equation to the given power.
    a. $\frac{3}{2}$ power
    b. $\frac{5}{4}$ power
    c. $\frac{7}{8}$ power

73. **CHALLENGE** Solve $7^{3x-1} = 49^{x+1}$ for x. (Hint: $b^x = b^y$ if and only if $x = y$.)

REASONING Determine whether the following statements are *sometimes*, *always*, or *never* true for $x^{\frac{1}{n}} = a$. Explain your reasoning.

74. If n is odd, there will be extraneous solutions.

75. If n is even, there will be extraneous solutions.

Preparing for Assessment

76. For what value of x is $\sqrt{x-8} = 1 - \sqrt{x}$? MP 1

- A 1.5
- B 4.5
- C 20.25
- D no real solution

77. The equation $v = \sqrt{64h}$ approximates the velocity v of a roller coaster car, in feet per second, at the bottom of a hill h feet high. If the car travels at a velocity of 55 feet per second at the bottom of a hill, how high is the hill? Round to the nearest tenth of a foot. MP 1

78. If $(2y + 3)^{\frac{1}{3}} + 2 = 5$, what is the value of y? MP 6

- A 6
- B 12
- C 30
- D 170

79. If $\sqrt{p-1} = \frac{2}{w} + 3$, what is the value of p in terms of w? MP 5

- A $p = \frac{4}{w^2} + 16$
- B $p = \frac{4}{w^2} + 10$
- C $p = \frac{4}{w^2} + \frac{6}{w} + 10$
- D $p = \frac{4}{w^2} + \frac{12}{w} + 10$
- E $p = \frac{4}{w^2} + \frac{12}{w} + 16$

80. Solve for x. MP 6

$3 + \sqrt{3x + 2} = 9$

- A $\frac{68}{5}$
- B $\frac{34}{3}$
- C $\frac{22}{9}$
- D 12

81. Solve for x. MP 6

$\sqrt{9x + 18} \geq x - 2$

82. MULTI-PART Consider the following equation.

$\sqrt{x - 4} + 1 = 5$ MP 1, 3, 6,

a. Which lines of the work shown correctly belong in the solution of the equation above? Choose all that apply.

- A $\sqrt{x - 4} = 4$
- B $(\sqrt{x - 4})^2 + 1^2 = 4$
- C $(\sqrt{x - 4})^2 = 4$
- D $x - 4 = 16$
- E $(\sqrt{x - 4})^2 = 24$
- F $x^2 = 20$

b. What is the solution to $\sqrt{x - 4} + 1 = 5$?

- A $\sqrt{20}$
- B $\sqrt{20}$
- C 4
- D 20

c. How can you check the answer?

d. Without doing the arithmetic, rearrange the equation to show the solution. (Write the equation as $x =$ ... with an arrangement of 4, 1, and 5 on the other side.)

83. a. Why does the solution to $11 + \sqrt{11x - 123} \leq 14$ have two parts? MP 1

b. Solve for both parts of the solution of $11 + \sqrt{11x - 123} \leq 14$.

c. Does the inequality have a real solution?

84. Find the height of a right triangle whose area is 27 square centimeters with a base of 3 cm and a height of $\sqrt{3x - 5}$ centimeters. MP 1, 6

- A $\frac{329}{3}$
- B $\frac{34}{3}$
- C 18
- D 21

358 | Lesson 5-6 | Solving Radical Equations

EXTEND 6
Graphing Technology Lab
Solving Radical Equations

You can use a TI-83/84 Plus graphing calculator to solve radical equations and inequalities. One way to do this is to rewrite the equation or inequality so that one side is 0. Then use the zero feature on the calculator.

Mathematical Practices
MP 4 Model with Mathematics

Example 1 Radical Equation

Solve $\sqrt{x} + \sqrt{x+2} = 3$.

Step 1 Rewrite the equation.

- Subtract 3 from each side of the equation to get $\sqrt{x} + \sqrt{x+2} - 3 = 0$.
- Enter the function $y = \sqrt{x} + \sqrt{x+2} - 3$ in the Y= list.

  KEYSTROKES: [Y=] [2nd] [√] [X,T,θ,n] [)] [+] [2nd] [√] [X,T,θ,n] [+] 2 [)] [−] 3 [ENTER]

Step 2 Use a table.

- You can use the **TABLE** function to locate intervals where the solution(s) lie. First, enter the starting value and the interval for the table.

  KEYSTROKES: [2nd] [TBLSET] 0 [ENTER] 1 [ENTER]

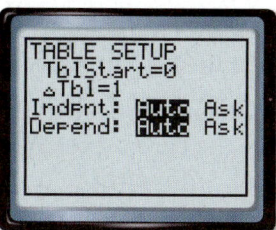

Step 3 Estimate the solution.

- Complete the table and estimate the solution(s).

  KEYSTROKES: [2nd] [TABLE]

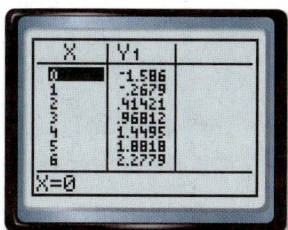

Because the function changes sign from negative to positive between $x = 1$ and $x = 2$, there is a solution between 1 and 2.

Step 4 Use the **ZERO** feature.

- Graph the function in the standard viewing window; then select **ZERO** from the **CALC** menu.

  KEYSTROKES: [ZOOM] 6 [2nd] [CALC] 2

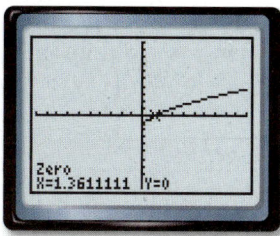

[−10, 10] scl: 1 by [−10, 10] scl: 1

Place the cursor on a point at which $y < 0$ and press [ENTER] for the **LEFT BOUND**. Then place the cursor on a point at which $y > 0$ and press [ENTER] for the **RIGHT BOUND**. You can use the same point for the **GUESS**. The solution is about 1.36. This is consistent with the estimate made by using the **TABLE**.

(continued on the next page)

EXTEND 6

Graphing Technology Lab
Solving Radical Equations *Continued*

Example 2 Radical Inequality

Solve $2\sqrt{x} > \sqrt{x+2} + 1$.

Step 1 Graph each side of the inequality and use the **TRACE** feature.

- In the **Y=** list, enter $y_1 = 2\sqrt{x}$ and $y_2 = \sqrt{x+2} + 1$. Then press [GRAPH].

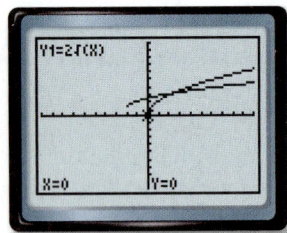

[−10, 10] scl: 1 by [−10, 10] scl: 1

- Press [TRACE]. You can use ▲ or ▼ to switch the cursor between the two curves.

The calculator screen above shows that, for points to the left of where the curves cross, **Y1 < Y2** or $2\sqrt{x} < \sqrt{x+2} + 1$. To solve the original inequality, you must find points for which **Y1 > Y2**. These are the points to the right of where the curves cross.

Step 2 Use the **intersect** feature.

- You can use the **intersect** feature on the **CALC** menu to approximate the *x*-coordinate of the point at which the curves cross.

  KEYSTROKES: [2nd] [CALC] 5

- Press [ENTER] for each of **FIRST CURVE?**, **SECOND CURVE?**, and **GUESS?**.

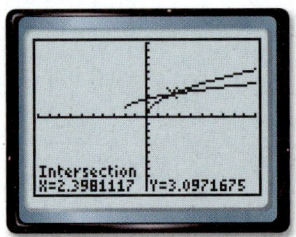

[−10, 10] scl: 1 by [−10, 10] scl: 1

The calculator screen shows that the *x*-coordinate of the point at which the curves cross is about 2.40. Therefore, the solution of the inequality is about $x > 2.40$. Use the symbol > in the solution because the symbol in the original inequality is >.

Step 3 Use the **TABLE** feature to check your solution.

- Start the table at 2 and show *x*-values in increments of 0.1. Scroll through the table.

  KEYSTROKES: [2nd] [TBLSET] 2 [ENTER] .1 [ENTER] [2nd] [TABLE]

Notice that when *x* is less than or equal to 2.4, **Y1 < Y2**. This verifies the solution $\{x \mid x > 2.40\}$.

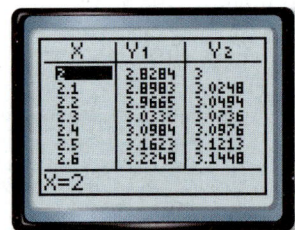

Exercises

Use a graphical method to solve each equation or inequality.

1. $\sqrt{x+4} = 3$
2. $\sqrt{3x-5} = 1$
3. $\sqrt{x+5} = \sqrt{3x+4}$
4. $\sqrt{x+3} + \sqrt{x-2} = 4$
5. $\sqrt{3x-7} = \sqrt{2x-2} - 1$
6. $\sqrt{x+8} - 1 = \sqrt{x+2}$
7. $\sqrt{x-3} \geq 2$
8. $\sqrt{x+3} > 2\sqrt{x}$
9. $\sqrt{x} + \sqrt{x-1} < 4$

10. **WRITING IN MATH** Explain how you could apply the technique in the first example to solving an inequality.

CHAPTER 5
Study Guide and Review

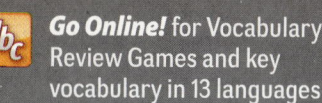 **Go Online!** for Vocabulary Review Games and key vocabulary in 13 languages

Study Guide

Key Concepts

Operations on Functions (Lessons 5-1 and 5-2)

Operation	Definition
Sum	$(f + g)(x) = f(x) + g(x)$
Difference	$(f - g)(x) = f(x) - g(x)$
Product	$(f \cdot g)(x) = f(x) \cdot g(x)$
Quotient	$\left(\dfrac{f}{g}\right)(x) = \dfrac{f(x)}{g(x)}, g(x) \neq 0$
Composition	$[f \circ g](x) = f[g(x)]$

Inverse Functions and Relations (Lesson 5-3)

- Two functions are inverses if and only if both their compositions are the identity function.
- The inverse relation is the set of ordered pairs obtained by exchanging the coordinates of each ordered par.
- The inverse of a function can be found by exchanging the independent and dependent variables (x and y, respectively) and then solving for the dependent variable (y).

Square Root and Cube Root Functions (Lessons 5-4 and 5-5)

- The domain of a square root function is limited to values for which the function is defined.
- The same techniques used to transform the graph of other functions can be applied to the graphs of square root and cube root functions.

Solving Radical Equations (Lesson 5-6)

- To solve a radical equation: **Step 1** Isolate the radical on one side of the equation. **Step 2** Raise each side of the equation to a power equal to the index. **Step 3** Solve the resulting polynomial equation.

 **Study Organizer**

Use your Foldable to review the chapter. Working with a partner can be helpful. Ask for clarification of concepts as needed.

Key Vocabulary

composition of functions (p. 322)
cube root function (p. 345)
extraneous solution (p. 352)
inverse function (p. 329)
inverse relation (p. 329)
inflection point (p. 345)
radical equation (p. 352)
radical function (p. 338)
radical inequality (p. 354)
square root function (p. 338)

Vocabulary Check

Choose a word or term that best completes each statement.

1. If both compositions result in the _____, then the functions are inverse functions.

2. In a(n) _____, the results of one function are used to evaluate a second function.

3. Equations with radicals that have variables in the radicands are called _____.

4. Two relations are _____ if and only if one relation contains the element (b, a) when the other relation contains the element (a, b).

5. When solving a radical equation, sometimes you will obtain a number that does not satisfy the original equation. Such a number is called a(n) _____.

6. The square root function is a type of _____.

Concept Check

7. Explain how to algebraically determine the inverse of a function.

8. Explain how to determine the domain of a square root function.

CHAPTER 5
Study Guide and Review Continued

Lesson-by-Lesson Review

5-1 Operations with Functions

Given $f(x) = 2x + 9$ and $g(x) = x^2 + 2x + 1$, find each function.

9. $(f + g)(x)$
10. $(f - g)(x)$
11. $(f \cdot g)(x)$
12. $\left(\dfrac{f}{g}\right)(x)$

Given $f(x) = 10x$ and $g(x) = x^3 - 8$, find each function.

13. $(f + g)(x)$
14. $(f - g)(x)$
15. $(f \cdot g)(x)$
16. $\left(\dfrac{f}{g}\right)(x)$

Example 2

Given $f(x) = 3x + 1$ and $g(x) = x^3 + 1$, find each function.

a. $(f - g)(x)$

$$\begin{aligned}(x - g)(x) &= f(x) - g(x) && \text{Subtraction of functions}\\ &= (3x + 1) - (x^3 + 1) && \text{Substitution}\\ &= 3x - x^3 && \text{Simplify.}\end{aligned}$$

b. $\left(\dfrac{f}{g}\right)(x)$

$$\begin{aligned}\left(\dfrac{f}{g}\right)(x) &= \dfrac{f(x)}{g(x)} && \text{Division of functions}\\ &= \dfrac{(3x + 1)}{(x^3 + 1)} && \text{Substitution}\end{aligned}$$

5-2 Composition of Functions

Find $[f \circ g](x)$ and $[g \circ f](x)$.

17. $f(x) = 2x + 1$
    $g(x) = 4x - 5$
18. $f(x) = x^2 + 1$
    $g(x) = x - 7$
19. $f(x) = x^2 + 4$
    $g(x) = -2x + 1$
20. $f(x) = 4x$
    $g(x) = 5x - 1$
21. $f(x) = x^3$
    $g(x) = x - 1$
22. $f(x) = x^2 + 2x - 3$
    $g(x) = x + 1$

23. **MEASUREMENT** The formula $f = 3y$ converts yards y to feet f and $f = \dfrac{n}{12}$ converts inches n to feet f. Write a composition of functions that converts yards to inches.

Example 2

If $f(x) = x^2 + 3$ and $g(x) = 3x - 2$, find $g[f(x)]$ and $f[g(x)]$.

$$\begin{aligned}g[f(x)] &= 3(x^2 + 3) - 2 && \text{Replace } f(x) \text{ with } x^2 + 3.\\ &= 3x^2 + 9 - 2 && \text{Distributive Property}\\ &= 3x^2 + 7 && \text{Simplify.}\end{aligned}$$

$$\begin{aligned}f[g(x)] &= (3x - 2)^2 + 3 && \text{Replace } g(x) \text{ with } 3x - 2.\\ &= 9x^2 - 12x + 4 + 3 && \text{Multiply.}\\ &= 9x^2 - 12x + 7 && \text{Simplify.}\end{aligned}$$

5-3 Inverse Functions and Relations

Find the inverse of each function. Then graph the function and its inverse.

24. $f(x) = 5x - 6$
25. $f(x) = -3x - 5$
26. $f(x) = \dfrac{1}{2}x + 3$
27. $f(x) = \dfrac{4x + 1}{5}$
28. $f(x) = x^2$
29. $f(x) = (2x + 1)^2$

30. **SHOPPING** Samuel bought a computer. The sales tax rate was 6% of the sale price, and he paid $50 for shipping. Find the sale price if Samuel paid a total of $1322.

Example 3

Find the inverse of $f(x) = -2x + 7$.

Rewrite $f(x)$ as $y = -2x + 7$. Then interchange the variables and solve for y.

$$\begin{aligned}x &= -2y + 7 && \text{Interchange the variables.}\\ 2y &= -x + 7 && \text{Solve for } y.\\ y &= \dfrac{-x + 7}{2} && \text{Divide each side by 2.}\\ f^{-1}(x) &= \dfrac{-x + 7}{2} && \text{Rewrite using function notation.}\end{aligned}$$

362 | Chapter 5 | Study Guide and Review

5-4 Graphing Square Root Functions

Graph each function. State the domain and range.

31. $f(x) = \sqrt{3x}$
32. $f(x) = -\sqrt{6x}$
33. $f(x) = \sqrt{x-7}$
34. $f(x) = \sqrt{x+5} - 3$
35. $f(x) = \frac{3}{4}\sqrt{x-1} + 5$
36. $f(x) = -\frac{1}{3}\sqrt{x+4} - 1$

37. **GEOMETRY** The area of a circle is given by the formula $A = \pi r^2$. What is the radius of a circle with an area of 300 square inches?

Example 4

Graph $f(x) = \sqrt{x+1} - 2$. State the domain and range.

Identify the domain.

$x + 1 \geq 0$ Write the radicand as greater than or equal to 0.

$x \geq -1$ Subtract 1 from each side.

Make a table of values for $x \geq -1$ and graph the function.

x	f(x)
−1	−2
0	−1
1	−0.59
2	−0.27
3	0
4	0.24
5	0.45

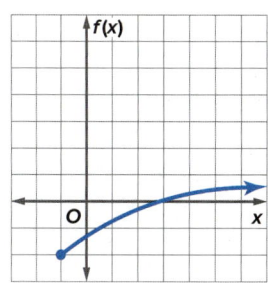

The domain is $\{x \mid x \geq -1\}$, and the range is $\{f(x) \mid f(x) \geq -2\}$.

5-5 Graphing Cube Root Equations

Graph each function.

38. $f(x) = \sqrt[3]{2x}$
39. $f(x) = \sqrt[3]{-4x}$
40. $f(x) = \sqrt[3]{1-x}$
41. $f(x) = 2 - \sqrt[3]{x}$
42. $f(x) = \sqrt[3]{x+4} + 1$
43. $f(x) = 2\sqrt[3]{x-1} + 2$

44. **GEOMETRY** The volume of a sphere is given by the formula $V = \frac{4}{3}\pi r^3$. What is the radius of a sphere with a volume of 400 cubic inches?

Example 5

Graph the function $f(x) = \sqrt[3]{x-2} + 1$.

Make a table of values and graph the function.

x	f(x)
−5	−0.91
−4	−0.82
−3	−0.71
−2	−0.59
−1	−0.44
0	−0.26
1	0
2	1
3	2
4	2.26
5	2.44

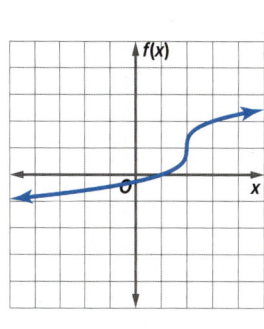

CHAPTER 5
Study Guide and Review Continued

5-6 Solving Radical Equations

Solve each equation.

45. $\sqrt{x-3} + 5 = 15$
46. $-\sqrt{x-11} = 3 - \sqrt{x}$
47. $4 + \sqrt{3x-1} = 8$
48. $\sqrt{m+3} = \sqrt{2m+1}$
49. $\sqrt{2x+3} = 3$
50. $(x+1)^{\frac{1}{4}} = -3$
51. $a^{\frac{1}{3}} - 4 = 0$
52. $3(3x-1)^{\frac{1}{3}} - 6 = 0$

53. **PHYSICS** The formula $t = 2\pi\sqrt{\dfrac{\ell}{32}}$ represents the swing of a pendulum, where t is the time in seconds for the pendulum to swing back and forth and ℓ is the length of the pendulum in feet. Find the length of a pendulum that makes one swing in 2.75 seconds.

Solve each inequality.

54. $2 + \sqrt{3x-1} < 5$
55. $\sqrt{3x+13} - 5 \geq 5$
56. $6 - \sqrt{3x+5} \leq 3$
57. $\sqrt{-3x+4} - 5 \geq 3$
58. $5 + \sqrt{2y-7} < 5$
59. $3 + \sqrt{2x-3} \geq 3$
60. $\sqrt{3x+1} - \sqrt{6+x} > 0$

Example 6
Solve $\sqrt{2x+9} - 2 = 5$.

$\sqrt{2x+9} - 2 = 5$ Original equation
$\sqrt{2x+9} = 7$ Add 2 to each side.
$(\sqrt{2x+9})^2 = 7^2$ Square each side.
$2x + 9 = 49$ Evaluate the squares.
$2x = 40$ Subtract 9 from each side.
$x = 20$ Divide each side by 2.

Example 7
Solve $\sqrt{2x-5} + 2 > 5$.

$\sqrt{2x-5} \geq 0$ Radicand must be ≥ 0.
$2x - 5 \geq 0$ Square each side.
$2x \geq 5$ Add 5 to each side.
$x \geq 2.5$ Divide each side by 2.

The solution must be greater than or equal to 2.5 to satisfy the domain restriction.

$\sqrt{2x-5} + 2 > 5$ Original inequality
$\sqrt{2x-5} > 3$ Subtract 2 from each side.
$(\sqrt{2x-5})^2 > 3^2$ Square each side.
$2x - 5 > 9$ Evaluate the squares.
$2x > 14$ Add 5 to each side.
$x > 7$ Divide each side by 2.

Because $x \geq 2.5$ contains $x > 7$, the solution of the inequality is $x > 7$.

CHAPTER 5
Practice Test

Go Online! for another Chapter Test

Determine whether each pair of functions are inverse functions. Write *yes* or *no*. Explain your reasoning.

1. $f(x) = 3x + 8$, $g(x) = \frac{x-8}{3}$
2. $f(x) = \frac{1}{3}x + 5$, $g(x) = 3x - 15$
3. $f(x) = x + 7$, $g(x) = x - 7$
4. $g(x) = 3x - 2$, $f(x) = \frac{x-2}{3}$

5. **MULTIPLE CHOICE** Which equation represents the graph below?

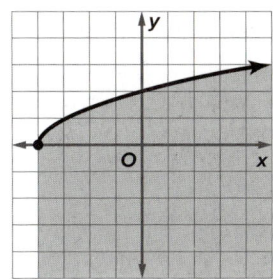

A $y = \sqrt{x-4}$
B $y = \sqrt{x+4}$
C $y = \sqrt[3]{x-4}$
D $y = \sqrt[3]{x+4}$

If $f(x) = 3x + 2$ and $g(x) = x^2 - 2x + 1$, find each function.

6. $(f + g)(x)$
7. $(f \cdot g)(x)$
8. $(f - g)(x)$
9. $\left(\frac{f}{g}\right)(x)$

Solve each equation.

10. $\sqrt{a + 12} = \sqrt{5a - 4}$
11. $\sqrt{3x} = \sqrt{x - 2}$
12. $4\left(\sqrt[4]{3x + 1}\right) - 8 = 0$
13. $\sqrt[3]{5m + 6} + 15 = 21$
14. $\sqrt{3x + 21} = \sqrt{5x + 27}$
15. $1 + \sqrt{x + 11} = \sqrt{2x + 15}$
16. $\sqrt{x - 5} = \sqrt{2x - 4}$
17. $\sqrt{x - 6} - \sqrt{x} = 3$

If $f(x) = 3x + 2$ and $g(x) = x^2 - 2x + 1$, find each function.

18. $[f \circ g](x)$
19. $[g \circ f](x)$

Graph each function. State the domain and range of each function.

20. $y = 2 + \sqrt{x}$
21. $y = \sqrt{x + 4} - 1$

22. **MULTIPLE CHOICE** What is the domain of $f(x) = \sqrt{2x + 5}$?

A $\left\{x \mid x > \frac{5}{2}\right\}$
B $\left\{x \mid x > -\frac{5}{2}\right\}$
C $\left\{x \mid x \geq \frac{5}{2}\right\}$
D $\left\{x \mid x \geq -\frac{5}{2}\right\}$

Graph each function. State the domain and range of each function.

23. $f(x) = 4 - \sqrt[3]{2x}$
24. $f(x) = 2\sqrt[3]{x + 2} - 4$

25. **MULTIPLE CHOICE** The radius of the cylinder below is equal to the height of the cylinder. The radius r can be found using the formula $r = \sqrt[3]{\frac{V}{\pi}}$. Find the radius of the cylinder if the volume is 500 cubic inches.

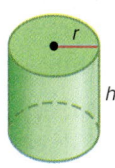

A 2.53 inches
B 5.42 inches
C 7.94 inches
D 24.92 inches

Solve each inequality.

26. $\sqrt{4x - 3} < 5$
27. $-2 + \sqrt{3m - 1} < 4$
28. $2 + \sqrt{4x - 4} \leq 6$
29. $\sqrt{2x + 3} - 4 \leq 5$
30. $\sqrt{b + 12} - \sqrt{b} > 2$
31. $\sqrt{y - 7} + 5 \geq 10$
32. $\sqrt{a - 5} - \sqrt{a + 7} \leq 4$
33. $\sqrt{c + 5} + \sqrt{c + 10} > 2$

Find the inverse of each function, if it exists.

34. $f(x) = -2x + 8$
35. $f(x) = \frac{x + 1}{3}$

Use the horizontal line test to determine whether the inverse of each function is also a function.

36. $f(x) = -4x^2$
37. $f(x) = x^3 - 1$

CHAPTER 5
Preparing for Assessment

Performance Task

Provide a clear solution to each part of the task. Be sure to show all of your work, include all relevant drawings, and justify your answers.

APPLY MATH Two students at a university are each research assistants. Johanna is a physics research assistant and Ravi is a biology research assistant. They are both performing various calculations on data collected in their labs.

Part A After studying production of two different varieties of tomatoes grown in two identical gardens, Ravi comes up with the following equations to represent the tomatoes' production rates:

Variety A: $t = 2d^2 + 4d - 2$ Variety B: $t = d^2 + d + 1$

In Ravi's models, t represents the total number of tomatoes produced in each garden and d represents the number of whole days since the plants first started producing tomatoes.

1. Write an expression that models the total production rates of both varieties.
2. Determine which variety is more productive. Explain your reasoning.
3. Write an expression that models how much more productive one variety is than the other.

Part B While studying the decline of a certain kind of cell in children with certain illnesses, Ravi calculates the following functions to represent how the decrease in the numbers of cells c as a function of the number of days d the child has the illness:

Illness A: $c_1(d) = 1218e^{-0.05d}$ Illness B: $c_2(d) = 420e^{-0.125d}$

Ravi also finds that if a child has *both* illnesses, because of the increased trauma to the immune system, the total decrease in cells is the product of the two functions, rather than the sum.

4. Write a function that models the decrease in the number of cells if a child had *both* illnesses.

Part C Ravi is working with data collected by researchers all over the world concerning bacteria growth in different bodies of water with different temperatures. The data collected is in Kelvin, Celsius, and Fahrenheit. The formula to convert Kelvin to Celsius is $T_{(C)} = T_{(K)} - 273.15$, and the formula to convert Celsius to Fahrenheit is $T_{(F)} = \frac{9}{5}T_{(C)} + 32$.

5. Write a composition function that can be used to directly convert Kelvin to Fahrenheit.

Part D Johanna is working on a project involving asteroids stuck in the orbit of a planet. The formula for escape velocity is $v \geq \sqrt{\frac{2GM}{r}}$, where v is the necessary velocity needed to escape orbit, G is the gravitational constant, M is the mass of the planet, and r is the radius of the planet. The gravitational constant is 6.673×10^{-11}.

6. If the mass of the Earth is 5.98×10^{24}, in kilograms, and the escape velocity needed for an asteroid to leave Earth's orbit is 11,184 meters per second, determine the radius of the Earth. (Note: do not worry about units.)

Part E

Johanna next turns her focus to a project involving springs. The period of oscillation of a spring can be found using the formula $T = 2\pi\sqrt{\frac{m}{k}}$, where T is the period, m is the mass of the oscillating body, and k is a constant.

7. Write an equation for the mass of the oscillating body, in terms of the period.

Test-Taking Strategy

Example

Read the problem. Identify what you need to know. Then use the information in the problem to solve.

Given the equation $\sqrt{\frac{2x}{3y}} = 8$, which of the following represents y in terms of x?

A $y = \frac{x}{32}$

C $y = \frac{x}{96}$

B $y = \frac{\sqrt{8x}}{3}$

D $y = \frac{\sqrt{2x}}{3}$

Step 1 What information is given? Are any of the following involved: earlier quantities, answer choices that are easy to test, or inverse operations?
An equation is given. Because I need to solve for a variable, I will use inverse operations.

Step 2 How will working backward help solve this problem?
I'll use inverse operations to "undo" operations and isolate the variable y.

Step 3 How can you check your answer? What is the correct answer?
I can "work forward" by beginning with my answer and seeing if I arrive at the same result given in the problem. Choice C is correct.

Test-Taking Tip

Strategies for Working Backward
In certain math problems, you are given information about an end result, but you are asked to find out something that happened earlier. In other problems, it can be faster to work from the answer choices, instead of from the problem. Finally, you'll sometimes need to use inverse operations, which is a specific way of working backward. You can work backward to solve all of these kinds of problems.

Apply the Strategy

Read the problem. Identify what you need to know. Then use the information in the problem to solve.

An object is shot straight upward into the air with an initial speed of 800 feet per second. The height h that the object will be after t seconds is given by the equation $h = -16t^2 + 800t$. When will the object reach a height of 10,000 feet?

A 10 seconds

C 100 seconds

B 25 seconds

D 625 seconds

Answer the questions below.

a. What information is given? Are any of the following involved: earlier quantities, answer choices that are easy to test, or inverse operations?

b. How will working backward help solve this problem?

c. How can you check your answer?

d. What is the correct answer?

connectED.mcgraw-hill.com 367

CHAPTER 5
Preparing for Assessment
Cumulative Review

Read each question. Then fill in the correct answer on the answer document provided by your teacher or on a sheet of paper.

1. What is the inverse of function $f(x)$ graphed below?

   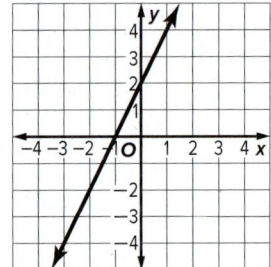

   - A $f^{-1}(x) = \dfrac{2}{x-2}$
   - B $f^{-1}(x) = \dfrac{x-2}{2}$
   - C $f^{-1}(x) = \dfrac{x}{2} - 2$
   - D $f^{-1}(x) = -2x - 2$

2. For what value of x is $\sqrt{x-3} - 9 = 6$?

3. For what value of x is $9 + \sqrt{5x+5} = 24$?
   - A -24
   - B 18
   - C 30
   - D 44

4. The formula $d = \sqrt[3]{6t^2}$ represents d, the distance in millions of miles a planet is from the Sun, in terms of t, the number of Earth days it takes for the planet to orbit the Sun. It takes Venus 224.7 Earth days to complete one orbit. To the nearest tenth, how many millions of miles is Venus from the Sun?
   - A 37.0
   - B 67.2
   - C 550.4
   - D 100,980.2
   - E 302,940.5

5. Given the functions $f(x) = x^2 - 9$ and $g(x) = x + 1$, find each of the following functions:

   $(f + g)(x)$ ⬜

   $(f - g)(x)$ ⬜

6. What is $(f \cdot g)(2)$ and $\left(\dfrac{f}{g}\right)(-2)$ for the functions $f(x) = x^2 - 36$ and $g(x) = x + 4$?
   - A $-24; -14$
   - B $-128; -24$
   - C $-192; -16$
   - D $212; 32$

7. Determine whether each pair of functions are inverse functions.

   $f(x) = x + 1$ $g(x) = -x - 1$

   $f(x) = (x + 4)^2$ $g(x) = \sqrt{x} - 4$

   $f(x) = -\dfrac{1}{2}x - 8$ $g(x) = -2x - 16$

   $f(x) = -\dfrac{1}{16}x^2 - 1$ $g(x) = 4\sqrt{x} + 1$

8. Which of the following functions is graphed below?

   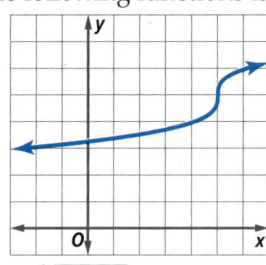

   - A $f(x) = \sqrt[3]{x-5} - 5$
   - B $f(x) = \sqrt[3]{x-5} + 5$
   - C $f(x) = \sqrt[3]{x+5} - 5$
   - D $f(x) = \sqrt[3]{x+5} + 5$

Test-Taking Tip

Question 7 Remember, two functions are inverse functions if both of their compositions are the identity function.

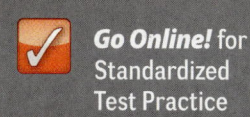

Go Online! for Standardized Test Practice

9. Solve the equation.

$\sqrt[3]{2x+5} + 8 = 11$

☐

10. What is the value of x for the equation $4 + \sqrt[3]{9x + 81} = 10$?
    - A -37
    - B 2
    - C 15
    - D 244

11. Andrew works at a suit store where he earns $360 per forty-hour work week, plus a 10% commission on sales over $500 each week, which can be represented by the functions $f(x) = 0.1x$ and $g(x) = x - 500$, where x is the amount Andrew makes in sales in a week. Assuming Andrew made enough to earn a commission, which composition would give his commission: $(f \circ g)(x)$ or $(g \circ f)(x)$? Explain your reasoning.

12. Which of the following belong to the solution set of $10 - \sqrt{4x + 20} > 2$?
    - ☐ A -3
    - ☐ B 6
    - ☐ C 11
    - ☐ D 16
    - ☐ E 21

13. If $a = 10$ and $b = 13$, what is the value of $(6\sqrt{a} + 2\sqrt{b})(6\sqrt{a} - 2\sqrt{b})$?

14. What is $(f \circ g)(-2)$ and $(g \circ f)(-2)$ for the functions $f(x) = x^2 - 100$ and $g(x) = x + 10$?
    - A $-134; -28$
    - B $-36; -86$
    - C $192; 28$
    - D $242; 64$

15. Let $f(x) = x^2 - 1$. What is the value of $f[f(2)]$?

☐

16. Which of the following functions is graphed below?

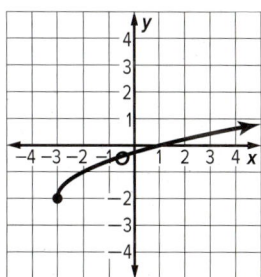

- A $y = \sqrt{x - 2} - 3$
- B $y = \sqrt{x + 2} - 3$
- C $y = \sqrt{x - 3} - 2$
- D $y = \sqrt{x + 3} - 2$

Need Extra Help?

If you missed Question...	1	2	3	4	5	6	7	8	9	10	11	12	13	14	15	16
Go to Lesson...	5-3	5-6	5-6	5-5	5-1	5-1	5-3	5-5	5-6	5-6	5-2	5-6	5-4	5-4	5-1	5-4

connectED.mcgraw-hill.com

CHAPTER 6

Exponential and Logarithmic Functions

THEN

You graphed functions and transformations of functions.

NOW

You will:

- Graph exponential and logarithmic functions.
- Solve exponential and logarithmic equations and inequalities.
- Solve problems involving exponential growth and decay.

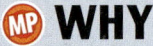

MP WHY

SCIENCE Scientists often use aspects of mathematics to quantify their studies and to illustrate their observations.

Use the Mathematical Practices to complete the activity.

1. **Using Tools** Use the Internet to learn about bacterial growth.
2. **Applying Math** Use the Coordinate Plane tool to create a graph that reflects a bacterial growth curve, with the x-axis set as the time interval and the y-axis as the bacterial population. Use the line segment tool to connect the points.

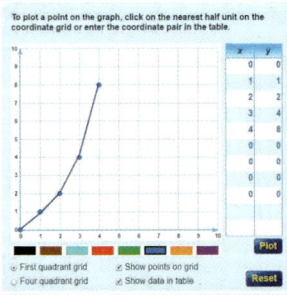

3. **Discuss** Is this the kind of growth typically seen in nature or do other factors impact the population? How might a scientist use mathematics to account for other factors that affect populations?

Erik Isakson/Blend Images LLC

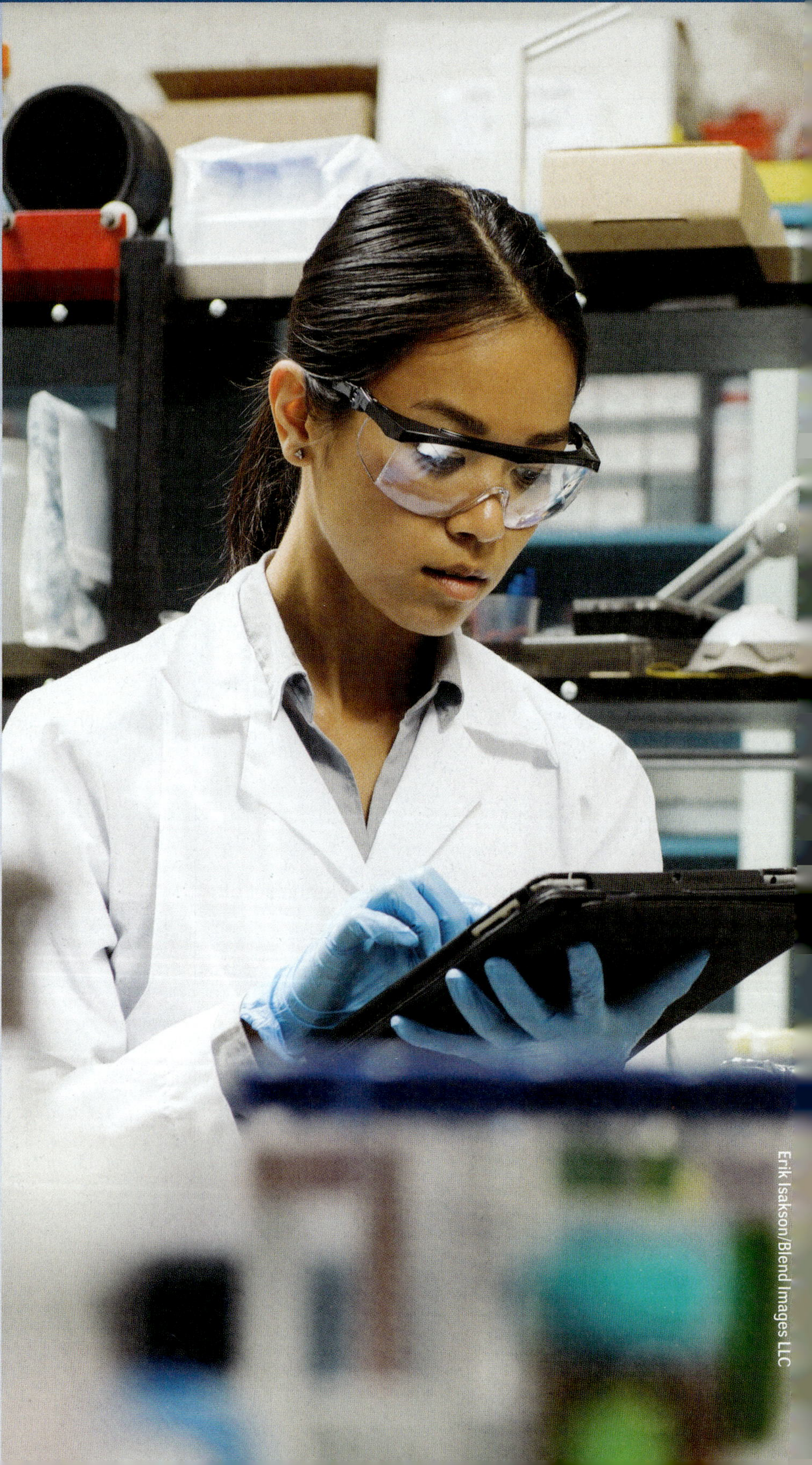

 *Go Online* to Guide Your Learning

Explore & Explain

 The Geometer's Sketchpad

Use **The Geometer's Sketchpad** to illustrate how to graph exponential functions in Lesson 6-1 and to explore geometric sequences in Lesson 6-3.

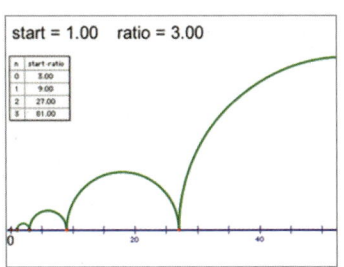

Graphing Tools: Logarithmic Functions

Use **Graphing Tools: Logarithmic Functions** to explore logarithmic functions and to enhance your understanding of logarithmic equations and inequalities in Lesson 6-9.

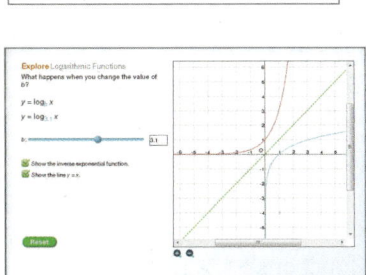

eBook
Interactive Student Guide

Before starting the chapter, answer the **Chapter Focus** preview questions. Check your answers as you complete each lesson. At the end of the chapter, try the **Performance Task**.

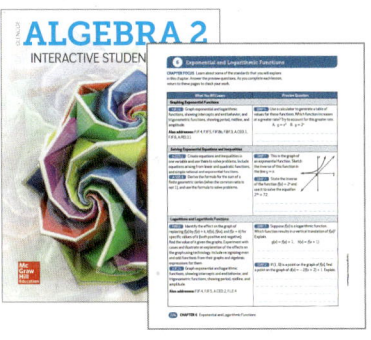

Organize

 Foldables

Get organized! Create an **Exponential and Logarithmic Functions Foldable** before you start the chapter to arrange your notes.

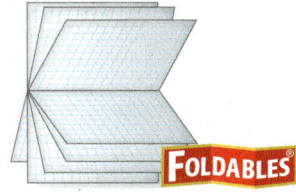

Collaborate

 Chapter Project

In the **The Population Puzzle** project, you will use what you have learned about exponential and logarithmic functions to complete a project that addresses environmental literacy.

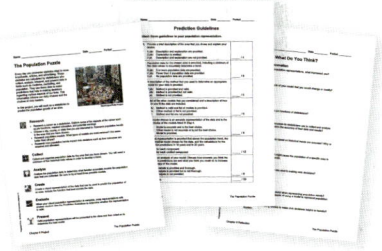

Focus

 LEARNSMART

Need help studying? Complete the **Modeling with Functions** domain in LearnSmart to review for the chapter test.

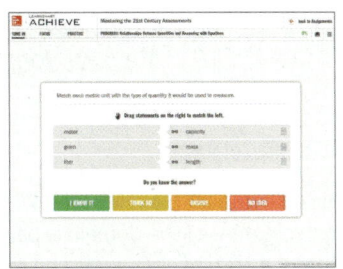

ALEKS

You can use the **Exponential and LogarithmicFunctions** topic in ALEKS to explore what you know about relations and functions and what you are ready to learn.*

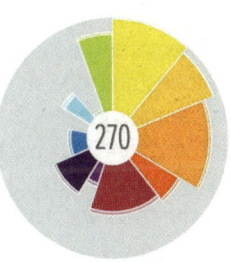

* Ask your teacher if this is part of your program.

connectED.mcgraw-hill.com 371

Get Ready for the Chapter

Connecting Concepts

Concept Check
Review the concepts used in this chapter by answering the questions below.

1. Given $a^4 a^3 a^5$, where a does not equal zero, what would you do to simplify the expression?

2. Given $\dfrac{(a^3 bc^2)^2}{a^4 a^2 b^2 bc^5 c^3}$, where no variable equals zero, what rule can be applied to simplify the numerator?

3. Given $\dfrac{(a^3 bc^2)^2}{a^4 a^2 b^2 bc^5 c^3}$, where no variable equals zero, what rule can be applied to simplify the denominator?

4. Given $\dfrac{a^6 b^2 c^4}{a^6 b^3 c^8}$, where no variable equals zero, what rule can be applied to simplify this expression?

5. Are $f(x) = 2x + 5$ and $g(x) = 2x - 5$ inverse functions?

6. Are $f(x) = x - 6$ and $g(x) = x + 6$ inverse functions?

7. In the graph shown, one line is $f(x) = \dfrac{x-1}{2}$ and the other is the inverse of that function. Which line is the function and which is its inverse?

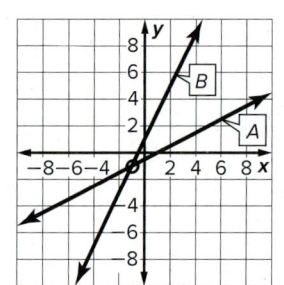

8. What are the steps to find the inverse of a function?

Performance Task Preview

You can use the concepts and skills in the chapter to perform data analysis for a United Nations committee to determine the health of various nations' economies, populations, and food supplies. Understanding exponential and logarithmic functions will help you finish the Performance Task at the end of the chapter.

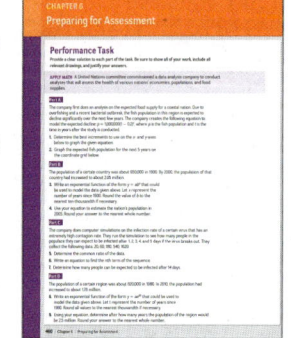

MP In this Performance Task you will:
- make sense of problems and persevere in solving them
- model with mathematics
- attend to precision

New Vocabulary

English		Español
exponential growth	p. 373	crecimiento exponencial
asymptote	p. 373	asíntota
growth factor	p. 375	factor de crecimiento
exponential decay	p. 375	desintegración exponencial
decay factor	p. 376	factor de desintegración
exponential equation	p. 383	ecuación exponencial
compound interest	p. 384	interés compuesto
exponential inequality	p. 385	desigualdad exponencial
geometric mean	p. 391	media geométrica
geometric series	p. 392	serie geométrica
logarithm	p. 397	logaritmo
logarithmic function	p. 398	función logarítmica
regression line	p. 408	reca de regresión
correlation coefficient	p. 409	coeficiente de correlación
common logarithm	p. 423	logaritmos communes
Change of Base Formula	p. 425	fórmula del cambio de base
natural base, e	p. 430	e base natural
natural logarithm	p. 430	logaritmo natural
logarithmic equation	p. 437	ecuación logarítmica
logarithmic inequality	p. 438	desigualdad logarítmica
rate of continuous growth	p. 445	el índice del crecimiento continuo
rate of continuous decay	p. 445	índice de desintegración continúa
logistic growth model	p. 448	modelo logístico del crecimiento

Review Vocabulary

domain dominio the set of all x-coordinates of the ordered pairs of a relation

function función a relation in which each element of the domain is paired with exactly one element in the range

range rango the set of all y-coordinates of the ordered pairs of a relation

$\{(-3, 1), (0, 2), (2, 4)\}$

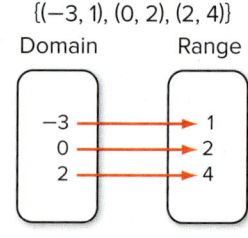

LESSON 1
Graphing Exponential Functions

::Then   ::Now   ::Why?

- You graphed polynomial functions.
- 1 Graph exponential growth functions.
- 2 Graph exponential decay functions.
- If you see a funny video online and share it with 100 of your friends, and each of your friends shares the video with 100 of their friends, the number of people who see the video increases exponentially until you can say that the video has gone viral. The equation $y = 100^x$ can be used to represent this situation where x is the number of rounds that the video has been shared.

New Vocabulary
exponential function
exponential growth
asymptote
growth factor
exponential decay
decay factor

Mathematical Practices
2 Reason abstractly and quantitatively.

1 Exponential Growth
A function like $y = 100^x$, where the base is a constant and the exponent is the independent variable, is an **exponential function**. An **exponential growth** function is a function of the form $f(x) = b^x$, where $b > 1$. The graph of an exponential function has an **asymptote**, which is a line that the graph approaches.

Key Concept Parent Function of Exponential Growth Functions

Parent Functions: $f(x) = b^x, b > 1$

Type of graph: continuous, one-to-one, and increasing

Domain: $(-\infty, \infty)$, {all real numbers},
                 or $\{-\infty < x < \infty\}$

Range: $(0, \infty)$, $\{f(x) \mid f(x) > 0\}$, or $\{0 < x < \infty\}$

Asymptote: x-axis

Intercept: $(0, 1)$

Symmetry: none   Extrema: none

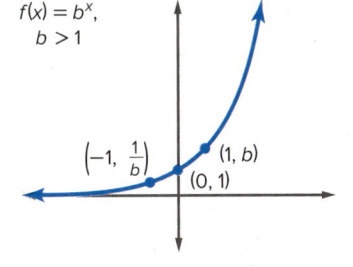

Example 1   Graph Exponential Growth Functions

Graph $y = 3^x$. State the domain and range.

Make a table of values. Then plot the points and sketch the graph.

x	-3	-2	$-\frac{1}{2}$	0
$y = 3^x$	$3^{-3} = \frac{1}{27}$	$3^{-2} = \frac{1}{9}$	$3^{-\frac{1}{2}} = \frac{\sqrt{3}}{3}$	$3^0 = 1$
x	1	$\frac{3}{2}$	2	
$y = 3^x$	$3^1 = 3$	$3^{\frac{3}{2}} = \sqrt{27}$	$3^2 = 9$	

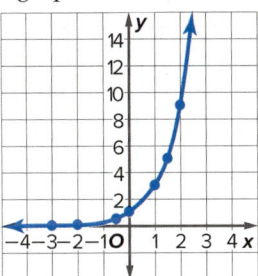

The domain is all real numbers, and the range is all positive real numbers.

▶ **Guided Practice**

1. Graph $y = 4^x$. State the domain and range.

connectED.mcgraw-hill.com 373

The graph of $f(x) = b^x$ represents a parent graph of the exponential functions. The same techniques used to transform the graphs of other functions you have studied can be applied to the graphs of exponential functions.

Key Concept Transformations of Exponential Functions

$$f(x) = ab^{x-h} + k$$

h — Horizontal Translation	k — Vertical Translation
h units right if h is positive	k units up if k is positive
$\|h\|$ units left if h is negative	$\|k\|$ units down if k is negative

a — Orientation and Shape	
If $a < 0$, the graph is reflected in the x-axis.	If $\|a\| > 1$, the graph is stretched vertically. If $0 < \|a\| < 1$, the graph is compressed vertically.

Example 2 Graph Transformations

Graph each function. State the domain and range.

a. $y = 2^x + 1$

The equation represents a translation of the graph of $y = 2^x$ one unit up.

x	$y = 2^x + 1$
-3	$2^{-3} + 1 = 1.125$
-2	$2^{-2} + 1 = 1.25$
-1	$2^{-1} + 1 = 1.5$
0	$2^0 + 1 = 2$
1	$2^1 + 1 = 3$
2	$2^2 + 1 = 5$
3	$2^3 + 1 = 9$

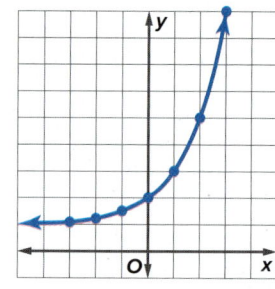

$D = (-\infty, \infty)$, {all real numbers}, or $\{-\infty < x < \infty\}$;
$R = (1, \infty)$, $\{f(x) \mid f(x) > 1\}$, or $\{1 < x < \infty\}$

b. $y = -\frac{1}{2} \cdot 5^{x-2}$

The equation represents a transformation of the graph of $y = 5^x$.

Graph $y = 5^x$ and transform the graph.

- $a = -\frac{1}{2}$: The graph is reflected in the x-axis and compressed vertically.
- $h = 2$: The graph is translated 2 units right.
- $k = 0$: The graph is not translated vertically.

$D = (-\infty, \infty)$, {all real numbers}, or $\{-\infty < x < \infty\}$;
$R = (-\infty, 0)$ or $\{f(x) \mid f(x) < 0\}$, or $\{-\infty < x < 0\}$;

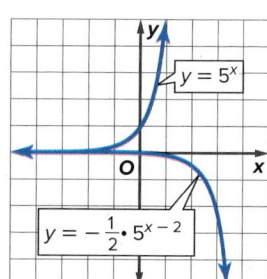

Guided Practice

2A. $y = 2^{x+3} - 5$

2B. $y = 0.1(6)^x - 3$

Study Tip

MP Precision Remember that end behavior is the action of the graph as x approaches positive infinity or negative infinity. In Example 2a, as x approaches infinity, y approaches infinity. In Example 2b, as x approaches infinity, y approaches negative infinity.

Go Online!

Investigate how changing the values in the equation of an exponential function affects the graph by using the **Graphing Tools** in ConnectED.

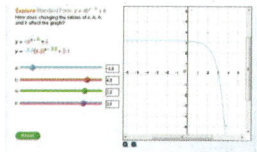

You can model exponential growth with a constant percent increase over specific time periods using the following function.

$$A(t) = a(1 + r)^t$$

The function can be used to find the amount $A(t)$ after t time periods, where a is the initial amount and r is the percent of increase per time period. Note that the base of the exponential expression, $1 + r$, is called the **growth factor**.

The exponential growth function is often used to model population growth.

Real-World Link
The U.S. Census Bureau's American Community Survey is mailed to approximately 3.5 million randomly selected households.

Source: Census Bureau

Real-World Example 3   Graph Exponential Growth Functions

CENSUS The first U.S. census was conducted in 1790. At that time, the population was 3,929,214. Since then, the U.S. population has grown by approximately 2.03% annually. Draw a graph showing the population growth of the United States since 1790.

First, write an equation using $a = 3,929,214$, and $r = 0.0203$.

$$y = 3,929,214(1.0203)^t$$

Then graph the equation.

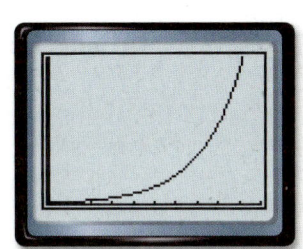

[0, 250] scl: 25 by [0, 400,000,000] scl: 40,000,000

Study Tip

Interest The formula for simple interest, $i = prt$, illustrates linear growth over time. However, the formula for compound interest, $A(t) = a(1 + r)^t$, illustrates exponential growth over time. This is why investments with compound interest make more money.

Guided Practice

3. FINANCIAL LITERACY Teen spending is expected to grow 3.5% annually from $208.7 billion in 2012. Draw a graph to show the spending growth.

2 Exponential Decay
The second type of exponential function is **exponential decay**.

Key Concept   Parent Function of Exponential Decay Functions

Parent Functions: $f(x) = b^x, 0 < b < 1$

Type of graph: continuous, one-to-one, and decreasing

Domain: $(-\infty, \infty)$, {all real numbers}, or $\{-\infty < x < \infty\}$

Range: $(0, \infty)$, $\{f(x) \mid f(x) > 0\}$, or $\{0 < x < \infty\}$

Asymptote: x-axis

Intercept: (0, 1)

Model

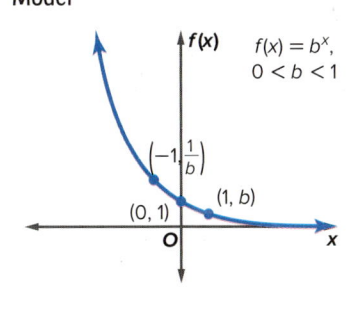

The graphs of exponential decay functions can be transformed in the same manner as those of exponential growth.

Study Tip

Exponential Decay Be sure not to confuse a dilation in which $|a| < 1$ with exponential decay in which $0 < b < 1$.

Example 4 Graph Exponential Decay Functions

Graph each function. State the domain and range.

a. $y = \left(\dfrac{1}{3}\right)^x$

x	$y = \left(\dfrac{1}{3}\right)^x$
-3	$\left(\dfrac{1}{3}\right)^{-3} = 27$
-2	$\left(\dfrac{1}{3}\right)^{-2} = 9$
$-\dfrac{1}{2}$	$\left(\dfrac{1}{3}\right)^{-\frac{1}{2}} = \sqrt{3}$
0	$\left(\dfrac{1}{3}\right)^0 = 1$
1	$\left(\dfrac{1}{3}\right)^1 = \dfrac{1}{3}$
$\dfrac{3}{2}$	$\left(\dfrac{1}{3}\right)^{\frac{3}{2}} = \sqrt{\dfrac{1}{27}}$
2	$\left(\dfrac{1}{3}\right)^2 = \dfrac{1}{9}$

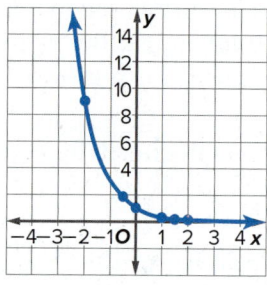

$D = (-\infty, \infty)$, {all real numbers}, or $\{-\infty < x < \infty\}$;
$R = (0, \infty)$, $\{f(x) \mid f(x) > 0\}$, or $\{0 < x < \infty\}$

b. $y = 2\left(\dfrac{1}{4}\right)^{x+2} - 3$

The equation represents a transformation of the graph of $y = \left(\dfrac{1}{4}\right)^x$.

Examine each parameter.

- $a = 2$: The graph is stretched vertically.
- $h = -2$: The graph is translated 2 units left.
- $k = -3$: The graph is translated 3 units down.

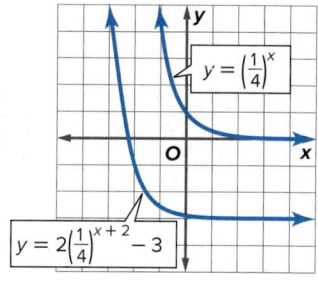

$D = (-\infty, \infty)$, {all real numbers}, or $\{-\infty < x < \infty\}$;
$R = (-3, \infty)$, $\{f(x) \mid f(x) > -3\}$, or $\{-3 < x < \infty\}$

▶ **Guided Practice**

4A. $y = -3\left(\dfrac{2}{5}\right)^{x-4} + 2$ **4B.** $y = \dfrac{3}{8}\left(\dfrac{5}{6}\right)^{x-1} + 1$

Similar to exponential growth, you can model exponential decay with a constant percent of decrease over specific time periods using the following function.

$$A(t) = a(1 - r)^t$$

The base of the exponential expression, $1 - r$, is called the **decay factor**.

Lesson 6-1 | Graphing Exponential Functions

Real-World Example 5   Graph Exponential Decay Functions

TEA A cup of green tea contains 35 milligrams of caffeine. The average teen can eliminate approximately 12.5% of the caffeine from their system per hour.

a. Draw a graph to represent the amount of caffeine remaining after drinking a cup of green tea.

$$y = a(1-r)^t$$
$$= 35(1-0.125)^t$$
$$= 35(0.875)^t$$

Graph the equation.

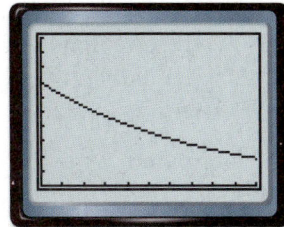

[0, 10] scl: 1 by [0, 50] scl: 5

Real-World Link
After water, tea is the most consumed beverage in the United States. It can be found in almost all American households. Just over half the American population drinks tea daily.

Source: The Tea Association of the USA

b. Estimate the amount of caffeine in a teenager's body 3 hours after drinking a cup of green tea.

$y = 35(0.875)^t$ Equation from part a
$= 35(0.875)^3$ Replace t with 3.
≈ 23.45 Use a calculator.

The caffeine in a teenager will be about 23.45 milligrams after 3 hours.

Guided Practice

5. A cup of black tea contains about 68 milligrams of caffeine. Draw a graph to represent the amount of caffeine remaining in the body of an average teen after drinking a cup of black tea. Estimate the amount of caffeine in the body 2 hours after drinking a cup of black tea.

Check Your Understanding

 = Step-by-Step Solutions begin on page R11.

Examples 1–2   Graph each function. State the domain and range.

1. $f(x) = 2^x$ **2.** $f(x) = 5^x$ **3.** $f(x) = 3^{x-2} + 4$

4. $f(x) = 2^{x+1} + 3$ **5.** $f(x) = 0.25(4)^x - 6$ **6.** $f(x) = 3(2)^x + 8$

Example 3

7. **SENSE-MAKING** A virus spreads through a network of computers such that each minute, 25% more computers are infected. If the virus began at only one computer, graph the function for the first hour of the spread of the virus.

Example 4   Graph each function. State the domain and range.

8. $f(x) = 2\left(\dfrac{2}{3}\right)^{x-3} - 4$ **9.** $f(x) = -\dfrac{1}{2}\left(\dfrac{3}{4}\right)^{x+1} + 5$

10. $f(x) = -\dfrac{1}{3}\left(\dfrac{4}{5}\right)^{x-4} + 3$ **11.** $f(x) = \dfrac{1}{8}\left(\dfrac{1}{4}\right)^{x+6} + 7$

Example 5

12. FINANCIAL LITERACY A new SUV depreciates in value each year by a factor of 15%. Draw a graph of the SUV's value for the first 20 years after the initial purchase.

Practice and Problem Solving

Extra Practice is on page R6.

Examples 1–2 Graph each function. State the domain and range.

13. $f(x) = 2(3)^x$
14. $f(x) = -2(4)^x$
15. $f(x) = 4^{x+1} - 5$
16. $f(x) = 3^{2x} + 1$
17. $f(x) = -0.4(3)^{x+2} + 4$
18. $f(x) = 1.5(2)^x + 6$

Example 3

19. **SCIENCE** The population of a colony of beetles grows 30% each week for 10 weeks. If the initial population is 65 beetles, graph the function that represents the situation.

Example 4 Graph each function. State the domain and range.

20. $f(x) = -4\left(\dfrac{3}{5}\right)^{x+4} + 3$
21. $f(x) = 3\left(\dfrac{2}{5}\right)^{x-3} - 6$
22. $f(x) = \dfrac{1}{2}\left(\dfrac{1}{5}\right)^{x+5} + 8$
23. $f(x) = \dfrac{3}{4}\left(\dfrac{2}{3}\right)^{x+4} - 2$
24. $f(x) = -\dfrac{1}{2}\left(\dfrac{3}{8}\right)^{x+2} + 9$
25. $f(x) = -\dfrac{5}{4}\left(\dfrac{4}{5}\right)^{x+4} + 2$

Example 5

26. **ATTENDANCE** The attendance for a basketball team declined at a rate of 5% per game throughout a losing season. Graph the function modeling the attendance if 15 home games were played and 23,500 people were at the first game.

27. **SKATING** The function $P(x) = 3.9(0.9^x)$ can be used to model the number of inline skaters in millions x years since 2004.
    a. Classify the function as either exponential *growth* or *decay*, and identify the growth or decay factor. Then graph the function.
    b. Explain what the $P(x)$-intercept and the asymptote represent in this situation.

28. **HEALTH** Each day, 10% of a certain drug dissipates from the system.
    a. Classify the function representing this situation as either exponential *growth* or *decay*, and identify the growth or decay factor. Then graph the function.
    b. How much of the original amount remains in the system after 9 days?
    c. If a second dose should not be taken if more than 50% of the original amount is in the system, when should the label say it is safe to redose? Design the label and explain your reasoning.

29. **REASONING** A sequence of numbers follows a pattern in which the next number is 125% of the previous number. The first number in the pattern is 18.
    a. Write the function that represents the situation.
    b. Classify the function as either exponential *growth* or *decay*, and identify the growth or decay factor. Then graph the function for the first 10 numbers.
    c. What is the value of the tenth number? Round to the nearest whole number.

For each graph, $f(x)$ is the parent function and $g(x)$ is a transformation of $f(x)$. Use the graph to determine the equation of $g(x)$.

30. $f(x) = 3^x$
31. $f(x) = 2^x$
32. $f(x) = 4^x$

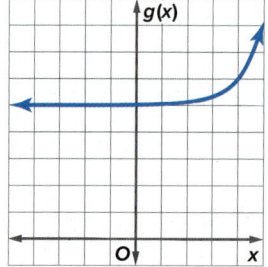
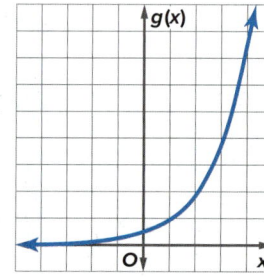
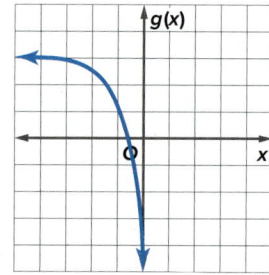

33. **MULTIPLE REPRESENTATIONS** In this problem, you will use the tables below for exponential functions $f(x)$, $g(x)$, and $h(x)$.

x	−1	0	1	2	3	4	5
f(x)	2.5	2	1	−1	−5	−13	−29

x	−1	0	1	2	3	4	5
g(x)	5	11	23	47	95	191	383

x	−1	0	1	2	3	4	5
h(x)	3	2.5	2.25	2.125	2.0625	2.0313	2.0156

a. **Graphical** Graph the functions for $-1 \leq x \leq 5$ on separate graphs.

b. **Verbal** List any function with a negative coefficient. Explain your reasoning.

c. **Analytical** List any function with a graph that is translated to the left.

d. **Analytical** Determine which functions are growth models and which are decay models.

H.O.T. Problems   Use Higher-Order Thinking Skills

34. **REASONING** Determine whether each statement is *sometimes*, *always*, or *never* true. Explain your reasoning.

    a. An exponential function of the form $y = ab^{x-h} + k$ has a y-intercept.

    b. An exponential function of the form $y = ab^{x-h} + k$ has an x-intercept.

    c. The function $f(x) = |b|^x$ is an exponential growth function if b is an integer.

35. **ERROR ANALYSIS** Vince and Grady were asked to graph the following functions. Vince thinks they are the same, but Grady disagrees. Who is correct? Explain your reasoning.

x	y
0	2
1	1
2	0.5
3	0.25
4	0.125
5	0.0625
6	0.03125

an exponential function with rate of decay of $\frac{1}{2}$ and an initial amount of 2

36. **CHALLENGE** A substance decays 35% each day. After 8 days, there are 8 milligrams of the substance remaining. How many milligrams were there initially?

37. **OPEN-ENDED** Give an example of a value of b for which $f(x) = \left(\frac{8}{b}\right)^x$ represents exponential decay.

38. **WRITING IN MATH** Write the procedure for transforming the graph of $g(x) = b^x$ to the graph of $f(x) = ab^{x-h} + k$. Justify each step.

Preparing for Assessment

39. Which is the best model for the data shown? **MP 2**

x	−1	0	1	2	3
y	1.6	4	16	76	376

- A $y = 3(5)^x + 1$
- B $y = 3(5)^x - 1$
- C $y = 5(3^x) + 1$
- D $y = 5(3^x) - 1$

40. The value of Louise's comic book collection has been increasing by 11% every year. Her collection began with a value of $150.00. About how long did it take her collection to double in value? **MP 1**

- A between 5 and 6 years
- B between 6 and 7 years
- C between 7 and 8 years
- D between 8 and 9 years
- E between 9 and 10 years

41. Which is the best model for the data shown? **MP 2**

x	−3	−2	−1	0	1	2	3
y	64	32	16	8	4	2	1

- A $y = 8(2)^x$
- B $y = 8\left(\frac{1}{2}\right)^x$
- C $y = 2(8^x)$
- D $y = 2\left(\frac{1}{8}\right)^x$

42. A bacteria grows exponentially so that it doubles every 12 days. There are 500 bacteria initially. **MP 2**

a. Write a function that models the number of bacteria after t days.

b. Find the number of bacteria after 60 days.

43. MULTI-STEP The function $g(x)$ is transformed from the function $f(x)$.

$$g(x) = 1 - \left(\frac{1}{3}\right)^{x-2}$$ **MP 6**

a. Describe the transformations from $f(x) = \left(\frac{1}{3}\right)^x$.

b. Is the function $g(x)$ increasing or decreasing?

c. Identify any asymptotes.

d. What is the y-intercept?

e. Sketch its graph.

44. MULTI-STEP Suppose you accidentally open a canister of plutonium in your living room and 200 units of radiation leak out. Every year, there is half as much radiation as there was the year before. **MP 6**

a. Write a function that represents the amount of radiation in your living room after t years.

b. How many units of radiation will there be after 4 years?

c. Will your living room ever be free of radiation? Explain.

45. Your grandfather would like you to invest in his company. He guarantees that if you invest in the stock of his company, you will earn 15% on your money every year. If you invest $100, and your grandfather is right, how much money will you have after 20 years? Round your answer to the nearest cent. **MP 6**

EXPLORE 6-2
Graphing Technology Lab
Solving Exponential Equations and Inequalities

You can use a TI-83/84 Plus graphing calculator to solve exponential equations by graphing or by using the table feature. To do this, you will write the equations as systems of equations.

Mathematical Practices
 5 Use appropriate tools strategically.

Activity 1

Solve $3^{x-4} = \frac{1}{9}$.

Step 1 Graph each side of the equation as a separate function. Enter 3^{x-4} as **Y1**. Be sure to include parentheses around the exponent.

Enter $\frac{1}{9}$ as **Y2**. Then graph the two equations.

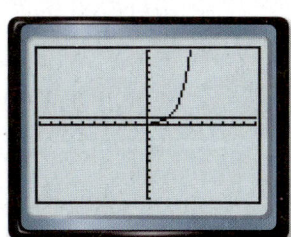
[−10, 10] scl: 1 by [−1, 1] scl: 0.1

Step 2 Use the intersect feature.

You can use the **intersect** feature on the **CALC** menu to approximate the ordered pair of the point at which the graphs cross.

The calculator screen shows that the x-coordinate of the point at which the curves cross is 2. Therefore, the solution of the equation is 2.

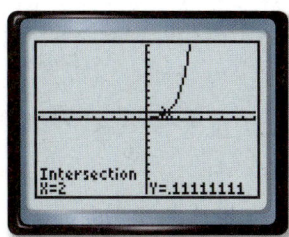
[−10, 10] scl: 1 by [−1, 1] scl: 0.1

Step 3 Use the **TABLE** feature.

You can also use the **TABLE** feature to locate the point at which the curves intersect.

The table displays x-values and corresponding y-values for each graph. Examine the table to find the x-value for which the y-values of the graphs are equal.

At $x = 2$, both functions have a y-value of $0.\overline{1}$ or $\frac{1}{9}$. Thus, the solution of the equation is 2.

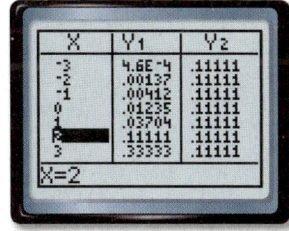

CHECK Substitute 2 for x in the original equation.

$3^{x-4} \stackrel{?}{=} \frac{1}{9}$ Original equation

$3^{2-4} \stackrel{?}{=} \frac{1}{9}$ Substitute 2 for x.

$3^{-2} \stackrel{?}{=} \frac{1}{9}$ Simplify.

$\frac{1}{9} = \frac{1}{9}$ ✓   The solution checks.

A similar procedure can be used to solve exponential inequalities.

(continued on the next page)

EXPLORE 6-2

Graphing Technology Lab
Solving Exponential Equations and Inequalities Continued

Activity 2

Solve $2^{x-2} \geq 0.5^{x-3}$.

Step 1 Enter the related inequalities.

Rewrite the problem as a system of inequalities.

The first inequality is $2^{x-2} \geq y$ or $y \leq 2^{x-2}$. Because this inequality includes the *less than or equal to* symbol, shade below the curve.

First enter the boundary, and then use the arrow and ENTER keys to choose the shade below icon, ▜.

The second inequality is $y \geq 0.5^{x-3}$. Shade above the curve since this inequality contains *greater than or equal to*.

KEYSTROKES: Y= ◀ ◀ ENTER ENTER ENTER ▶ ▶ 2 ∧ (
X,T,θ,n − 2) ENTER ◀ ◀ ENTER ENTER ▶
▶ .5 ∧ (X,T,θ,n − 3)

Step 2 Graph the system.

keystrokes: GRAPH

The *x*-values of the points in the region where the shadings overlap is the solution set of the original inequality. Using the **intersect** feature, you can conclude that the solution set is $\{x \mid x \geq 2.5\}$.

[−10, 10] scl: 1 by [−10, 10] scl: 1

Step 3 Use the **TABLE** feature.

Verify using the **TABLE** feature. Set up the table to show *x*-values in increments of 0.5.

KEYSTROKES: 2nd [TBLSET] 0 ENTER .5 ENTER 2nd [TABLE]

Notice that for *x*-values greater than $x = 2.5$, **Y1 > Y2**. This confirms that the solution of the inequality is $\{x \mid x \geq 2.5\}$.

Exercises

Solve each equation or inequality.

1. $9^{x-1} = \dfrac{1}{81}$
2. $4^{x+3} = 25x$
3. $5^{x-1} = 2x$
4. $3.5^{x+2} = 1.75^{x+3}$
5. $-3^{x+4} = -0.5^{2x+3}$
6. $6^{2-x} - 4 < -0.25^{x-2.5}$
7. $16^{x-1} > 2^{2x+2}$
8. $3^{x-4} \leq 5^{\frac{x}{2}}$
9. $5^{x+3} \leq 2^{x+4}$

10. **WRITING IN MATH** Explain why this technique of graphing a system of equations or inequalities works to solve exponential equations and inequalities.

LESSON 2
Solving Exponential Equations and Inequalities

::Then::
- You graphed exponential functions.

::Now::
1. Solve exponential equations.
2. Solve exponential inequalities.

::Why?::
Feral hogs are a highly adaptive species that causes significant ecological and economic damage. Suppose the population of feral hogs is growing exponentially and can be modeled by $y = 2.5(1.45)^x$, where x is the number of years since 2016 and y is the number of hogs in millions.

You can use $y = 2.5(1.45)^x$ to determine how many hogs there will be in a given year or to determine the year in which the hog population is at a certain level.

 New Vocabulary
exponential equation
compound interest
exponential inequality

 Mathematical Practices
2 Reason abstractly and quantitatively.

1 Solve Exponential Equations
In an **exponential equation**, variables occur as exponents.

> **Key Concept** Property of Equality for Exponential Functions
>
> **Words** Let $b > 0$ and $b \neq 1$. Then $b^x = b^y$ if and only if $x = y$.
>
> **Example** If $3^x = 3^5$, then $x = 5$. If $x = 5$, then $3^x = 3^5$.

The Property of Equality can be used to solve exponential equations.

Example 1 Solve Exponential Equations

Solve each equation.

a. $2^x = 8^3$

$$\begin{aligned}
2^x &= 8^3 & &\text{Original equation} \\
2^x &= (2^3)^3 & &\text{Rewrite 8 as } 2^3. \\
2^x &= 2^9 & &\text{Power of a Power} \\
x &= 9 & &\text{Property of Equality for Exponential Functions}
\end{aligned}$$

b. $9^{2x-1} = 3^{6x}$

$$\begin{aligned}
9^{2x-1} &= 3^{6x} & &\text{Original equation} \\
(3^2)^{2x-1} &= 3^{6x} & &\text{Rewrite 9 as } 3^2. \\
3^{4x-2} &= 3^{6x} & &\text{Power of a Power} \\
4x - 2 &= 6x & &\text{Property of Equality for Exponential Functions} \\
-2 &= 2x & &\text{Subtract } 4x \text{ from each side.} \\
-1 &= x & &\text{Divide each side by 2.}
\end{aligned}$$

▶ **Guided Practice**

1A. $4^{2n-1} = 64$ **1B.** $5^{5x} = 125^{x+2}$

You can use information about growth or decay to write the equation of an exponential function.

Real-World Example 2   Write an Exponential Function

SCIENCE Kristin starts an experiment with 7500 bacteria cells. After 4 hours, there are 23,000 cells.

a. **Write an exponential function that could be used to model the number of bacteria after *x* hours if the number of bacteria changes at the same rate.**

At the beginning of the experiment, the time is 0 hours and there are 7500 bacteria cells. Thus, the *y*-intercept, and the value of *a*, is 7500.

When $x = 4$, the number of bacteria cells is 23,000. Substitute these values into an exponential function to determine the value of *b*.

$y = ab^x$ Exponential function
$23{,}000 = 7500 \cdot b^4$ Replace *x* with 4, *y* with 23,000, and *a* with 7500.
$3.067 \approx b^4$ Divide each side by 7500.
$\sqrt[4]{3.067} \approx b$ Take the 4th root of each side.
$1.323 \approx b$ Use a calculator.

An equation that models the number of bacteria is $y \approx 7500(1.323)^x$.

b. **How many bacteria cells can be expected in the sample after 12 hours?**

$y \approx 7500(1.323)^x$ Modeling equation
$\approx 7500(1.323)^{12}$ Replace *x* with 12.
$\approx 215{,}665$ Use a calculator.

There will be approximately 215,665 bacteria cells after 12 hours.

Guided Practice

2. **RECYCLING** A manufacturer distributed 3.2 million cans in 2010 made from recycled aluminum.

   A. In 2015, the manufacturer distributed 420,000 cans made from recycled aluminum. Assuming that the recycling rate continues, write an equation to model the distribution each year of cans that are made from recycled aluminum.

   B. How many cans made from recycled aluminum can be expected in the year 2055?

Real-World Link
In 2011, the U.S. recycling rate for metals of 34% prevented the release of approximately 20 million metric tons of carbon into the air—roughly the amount emitted annually by 4 million cars.

Source: Environmental Protection Agency

Exponential functions are used in situations involving compound interest. **Compound interest** is interest paid on the principal of an investment and any previously earned interest.

Key Concept   Compound Interest

You can calculate compound interest using the following formula.

$$A = P\left(1 + \frac{r}{n}\right)^{nt},$$

where *A* is the amount in the account after *t* years, *P* is the principal amount invested, *r* is the annual interest rate, and *n* is the number of compounding periods each year.

Go Online!

Watch the **Personal Tutor** describe how to solve a compound interest problem with a partner. Then try describing how to solve a problem for them. Have them ask questions to help your understanding.

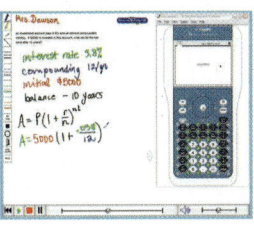

Example 3   Compound Interest

An investment account pays 4.2% annual interest compounded monthly. If $2500 is invested in this account, what will be the balance after 15 years?

Analyze   Find the total amount in the account after 15 years.

Formulate   Use the compound interest formula.
$$P = 2500, r = 0.042, n = 12, \text{ and } t = 15$$

Determine
$$A = P\left(1 + \frac{r}{n}\right)^{nt}$$ Compound Interest Formula

$$= 2500\left(1 + \frac{0.042}{12}\right)^{12 \cdot 15}$$ $P = 2500, r = 0.042, n = 12, t = 15$

$$\approx 4688.87$$ Use a calculator.

Justify   Graph the corresponding equation $y = 2500(1.0035)^{12t}$. Use **CALC: value** to find y when $x = 15$.

The y-value 4688.8662 is very close to 4688.87, so the answer is reasonable.

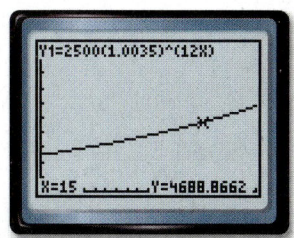

[0, 20] scl: 1 by [0, 10,000] scl: 1000

Evaluate   The compound interest formula is a model used to solve problems in the real world.

Watch Out!

Percents Remember to convert all percents to decimal form; 4.2% is 0.042.

▶ Guided Practice

3. Find the account balance after 20 years if $100 is placed in an account that pays 1.2% interest compounded twice a month.

2 Solve Exponential Inequalities

An **exponential inequality** is an inequality involving exponential functions.

🔑 Key Concept   Property of Inequality for Exponential Functions

Words   Let $b > 1$. Then $b^x > b^y$ if and only if $x > y$, and $b^x < b^y$ if and only if $x < y$.

Example   If $2^x > 2^6$, then $x > 6$. If $x > 6$, then $2^x > 2^6$.

This property also holds true for $\leq$ and $\geq$.

Example 4   Solve Exponential Inequalities

Solve $16^{2x-3} < 8$.

$16^{2x-3} < 8$ Original inequality

$(2^4)^{2x-3} < 2^3$ Rewrite 16 as 2^4 and 8 as 2^3.

$2^{8x-12} < 2^3$ Power of a Power

$8x - 12 < 3$ Property of Inequality for Exponential Functions

$8x < 15$ Add 12 to each side.

$x < \frac{15}{8}$ Divide each side by 8.

▶ Guided Practice

Solve each inequality.

4A. $3^{2x-1} \geq \frac{1}{243}$ **4B.** $2^{x+2} > \frac{1}{32}$

Check Your Understanding

= Step-by-Step Solutions begin on page R11.

Example 1 Solve each equation.

1. $3^{5x} = 27^{2x-4}$
2. $16^{2y-3} = 4^{y+1}$
3. $2^{6x} = 32^{x-2}$
4. $49^{x+5} = 7^{8x-6}$

Example 2

5. **SCIENCE** Mitosis is a process in which one cell divides into two. The *Escherichia coli* is one of the fastest growing bacteria. It can reproduce itself in 15 minutes.

   a. Write an exponential function to represent the number of cells c after t minutes.
   b. If you begin with one *Escherichia coli* cell, how many cells will there be in one hour?

Example 3

6. A certificate of deposit (CD) pays 2.25% annual interest compounded biweekly. If you deposit $500 into this CD, what will the balance be after 6 years?

Example 4 Solve each inequality.

7. $4^{2x+6} \le 64^{2x-4}$
8. $25^{y-3} \le \left(\dfrac{1}{125}\right)^{y+2}$

Practice and Problem Solving

Extra Practice is on page R6.

Example 1 Solve each equation.

9. $8^{4x+2} = 64$
10. $5^{x-6} = 125$
11. $81^{a+2} = 3^{3a+1}$
12. $256^{b+2} = 4^{2-2b}$
13. $9^{3c+1} = 27^{3c-1}$
14. $8^{2y+4} = 16^{y+1}$

Example 2

15. **MODELING** In 2015, My-Lien received $10,000 from her grandmother. Her parents invested all of the money, and by 2027, the amount will have grown to $16,960.

    a. Write an exponential function that could be used to model the money y. Write the function in terms of x, the number of years since 2015.
    b. Assume that the amount of money continues to grow at the same rate. What would be the balance in the account in 2037?

Write an exponential function for the graph that passes through the given points.

16. (0, 6.4) and (3, 100)
17. (0, 256) and (4, 81)
18. (0, 128) and (5, 371,293)
19. (0, 144), and (4, 21,609)

Example 3

20. Find the balance of an account after 7 years if $700 is deposited into an account paying 4.3% interest compounded monthly.
21. Determine how much is in a retirement account after 20 years if $5000 was invested at 6.05% interest compounded weekly.
22. A savings account offers 1.1% interest compounded bimonthly. If $110 is deposited in this account, what will the balance be after 15 years?
23. A college savings account pays 8.2% annual interest compounded semiannually. What is the balance of an account after 12 years if $21,000 was initially deposited?

Example 4 Solve each inequality.

24. $625 \ge 5^{a+8}$
25. $10^{5b+2} > 1000$
26. $\left(\dfrac{1}{64}\right)^{c-2} < 32^{2c}$
27. $\left(\dfrac{1}{27}\right)^{2d-2} \le 81^{d+4}$
28. $\left(\dfrac{1}{9}\right)^{3t+5} \ge \left(\dfrac{1}{243}\right)^{t-6}$
29. $\left(\dfrac{1}{36}\right)^{w+2} < \left(\dfrac{1}{216}\right)^{4w}$

386 | Lesson 6-2 | Solving Exponential Equations and Inequalities

30. **MULTI-STEP** George is leaving money in a college savings plan that earns 4% interest for his newborn granddaughter, Jenny. Currently, 4-year tuition at his alma mater costs $15,000 per year and has been increasing at an annual rate of 5%.

   a. How much money would he need to deposit now in order to cover her tuition expenses at his alma mater?

   b. Describe your solution process.

   c. What assumptions did you make?

31. **ANIMALS** Studies show that an animal will defend a territory, with area in square yards, that is directly proportional to the 1.31 power of the animal's weight in pounds.

   a. If a 45-pound beaver will defend 170 square yards, write an equation for the area a defended by a beaver weighing w pounds.

   b. Scientists believe that thousands of years ago, the beaver's ancestors were 11 feet long and weighed 430 pounds. Use your equation to determine the area defended by these animals.

Solve each equation.

32. $\left(\dfrac{1}{2}\right)^{4x+1} = 8^{2x+1}$

33. $\left(\dfrac{1}{5}\right)^{x-5} = 25^{3x+2}$

34. $216 = \left(\dfrac{1}{6}\right)^{x+3}$

35. $\left(\dfrac{1}{8}\right)^{3x+4} = \left(\dfrac{1}{4}\right)^{-2x+4}$

36. $\left(\dfrac{2}{3}\right)^{5x+1} = \left(\dfrac{27}{8}\right)^{x-4}$

37. $\left(\dfrac{25}{81}\right)^{2x+1} = \left(\dfrac{729}{125}\right)^{-3x+1}$

38. **MODELING** In 1950, the world population was about 2.556 billion. By 2000, it had increased to about 6.08 billion.

   a. Write an exponential function of the form $y = ab^x$ that could be used to model the world population y in billions for 1950 to 2000. Write the equation in terms of x, the number of years since 1950. (Round the value of b to the nearest ten-thousandth.)

   b. Suppose the population continued to grow at that rate. Estimate the population in 2013.

   c. In 2013, the population of the world was about 7.164 billion. Compare your estimate to the actual population.

   d. Use the equation you wrote in part **a** to estimate the world population in the year 2040. How accurate do you think the estimate is? Explain your reasoning.

39. **TREES** The diameter of the base of a tree trunk in centimeters varies directly with the $\dfrac{3}{2}$ power of its height in meters.

   a. A young sequoia tree is 6 meters tall, and the diameter of its base is 19.1 centimeters. Use this information to write an equation for the diameter d of the base of a sequoia tree if its height is h meters high.

   b. The General Sherman Tree in Sequoia National Park, California, is approximately 84 meters tall. Find the diameter of the General Sherman Tree at its base.

40. **FINANCIAL LITERACY** Mrs. Jackson has two different retirement investment plans from which to choose.

   Option A: 6.5% annual rate compounded quarterly; minimum deposit $5000

   Option B: 4.2% annual rate compounded monthly; minimum deposit $5000
   PLUS 2.3% annual rate compounded weekly; minimum deposit $5000

   a. Write equations for Option A and Option B given the minimum deposits.

   b. Draw a graph to show the balances for each investment option after t years.

   c. Explain whether Option A or Option B is the better investment choice.

41. **MULTIPLE REPRESENTATIONS** In this problem, you will explore the rapid increase of an exponential function. A large sheet of paper is cut in half, and one of the resulting pieces is placed on top of the other. Then the pieces in the stack are cut in half and placed on top of each other. Suppose this procedure is repeated several times.

   a. **Concrete** Perform this activity and count the number of sheets in the stack after the first cut. How many pieces will there be after the second cut? How many pieces after the third cut? How many pieces after the fourth cut?

   b. **Tabular** Record your results in a table.

   c. **Symbolic** Use the pattern in the table to write an equation for the number of pieces in the stack after x cuts.

   d. **Analytical** The thickness of ordinary paper is about 0.003 inch. Write an equation for the thickness of the stack of paper after x cuts.

   e. **Analytical** How thick will the stack of paper be after 30 cuts?

H.O.T. Problems   Use Higher-Order Thinking Skills

42. **WRITING IN MATH** In a problem about compound interest, describe what happens as the compounding period becomes more frequent while the principal and overall time remain the same.

43. **ERROR ANALYSIS** Beth and Liz are solving $6^{x-3} > 36^{-x-1}$. Is either of them correct? Explain your reasoning.

Beth	Liz
$6^{x-3} > 36^{-x-1}$	$6^{x-3} > 36^{-x-1}$
$6^{x-3} > (6^2)^{-x-1}$	$6^{x-3} > (6^2)^{-x-1}$
$6^{x-3} > 6^{-2x-2}$	$6^{x-3} > 6^{-x+1}$
$x - 3 > -2x - 2$	$x - 3 > -x + 1$
$3x > 1$	$2x > 4$
$x > \frac{1}{3}$	$x > 2$

44. **CHALLENGE** Solve for x: $16^{18} + 16^{18} + 16^{18} + 16^{18} + 16^{18} = 4^x$.

45. **OPEN-ENDED** What would be a more beneficial change to a 5-year loan at 8% interest compounded monthly: reducing the term to 4 years or reducing the interest rate to 6.5%?

46. **CONSTRUCT ARGUMENTS** Determine whether the following statements are *sometimes*, *always*, or *never* true. Explain your reasoning.
    a. $2^x > -8^{20x}$ for all values of x.
    b. The graph of an exponential growth equation is increasing.
    c. The graph of an exponential decay equation is increasing.

47. **OPEN-ENDED** Write an exponential inequality with a solution of $x \leq 2$.

48. **PROOF** Show that $27^{2x} \cdot 81^{x+1} = 3^{2x+2} \cdot 9^{4x+1}$.

49. **WRITING IN MATH** If you were given the initial and final amounts of a radioactive substance and the amount of time that passes, how would you determine the rate at which the amount was increasing or decreasing in order to write an equation?

Preparing for Assessment

50. Carmen invested $3000 in an account that pays 3.6% annual interest compounded monthly. To the nearest whole dollar, what will be the balance in her account after 6 years? **MP 6**

- A $3223
- B $3648
- C $3709
- D $3722
- E $38,285

51. Let $p \# q = p^q - q^p$ for all nonzero values of p and q. What is the value of $-2 \# 4$? **MP 2**

- A $-16\frac{1}{16}$
- B 0
- C $15\frac{15}{16}$
- D $16\frac{1}{16}$
- E 32

52. What is the value of x in the equation $8^{12-4x} = 4^{x+18}$? **MP 1**

- A $-5\frac{1}{7}$
- B -4
- C 0
- D $1\frac{1}{5}$
- E $3\frac{3}{5}$

53. Let m^x be greater than 1 for all negative values of x. What are the possible values of m? **MP 2**

- A $-\infty < m < -1$
- B $-\infty < m < 0$
- C $-1 < m < 0$
- D $-1 < m < 1$
- E $0 < m < 1$

54. What is the value of x in $\left(\frac{1}{9}\right)^{3x-4} = 27^{2x+1}$? **MP 1**

- A -1
- B -0.8
- C 0.42
- D 5

55. Solve the following equations for x. **MP 2**

a. $9^{2x} = \left(\frac{1}{27}\right)^{x+1}$

b. $25(5^{3x}) = \left(\frac{1}{5}\right)^{-x}$

56. MULTI-STEP Suppose the amount of a radioactive substance $a(t)$, in grams, after t years is given by the function $a(t) = 900(2)^{\frac{-t}{10}}$. **MP 1**

a. How much of the substance remains after 3 years? Round your answer to the nearest gram.

The half-life of the substance is how many years it takes for half of the substance to decay.

b. Write down an equation whose solution gives the half-life of the substance.

c. What is the half-life of the substance?

57. Solve the equation and inequalities: **MP 1**

a. $\frac{2}{5^x} = 0.08$

b. $4^{\frac{x}{2}} > 8$

c. $\frac{3}{2^x} \leq 0.375$

LESSON 3
Geometric Sequences and Series

::Then   ::Now   ::Why?

- You determined whether a sequence was geometric.
- 1 Use geometric sequences.
  2 Find sums of geometric series.
- Julian sees a band at a concert. He posts a link for the band's website on his social network page. Five of his friends repost the link, then five of each of their friends repost the link, and so on. If this pattern continues, how many people will post the link on the eighth round?

New Vocabulary
geometric means
geometric series

Mathematical Practices
8 Look for and express regularity in repeated reasoning.

1 Geometric Sequences As with arithmetic sequences, there is a formula for the nth term of a geometric sequence.

Key Concept nth Term of a Geometric Sequence

The nth term a_n of a geometric sequence in which the first term is a_1 and the common ratio is r is given by the following formula, where n is any natural number.

$$a_n = a_1 r^{n-1}$$

You will prove this formula in Exercise 68.

Real-World Example 1   Find the nth Term

MUSIC If the pattern continues, how many people will post the link on the eighth round?

Understand You need to determine how many people posted the link on the eighth round. Five people posted the link on the first round. Each of those people had five people repost the link on the second round.

Plan This is a geometric sequence, and the common ratio is 5. Use the formula for the nth term of a geometric sequence.

Solve $a_n = a_1 r^{n-1}$ nth term of a geometric sequence

$a_8 = 5(5)^{8-1}$ $a_1 = 5, r = 5,$ and $n = 8$

$a_8 = 5(78{,}125)$ or $390{,}625$ $5^7 = 78{,}125$

Check There will be 390,625 posts on the 8th round. Write out the first eight terms by multiplying by the common ratio.

5, 25, 125, 625, 3125, 15,625, 78,125, 390,625

The formula for the nth term of a geometric sequence is more efficient than calculating each term using the common ratio.

Guided Practice

1. **SOCIAL NETWORKING** Shira posts a joke on her social network page. Four of her friends repost the joke on their pages. Four of each of their friends also repost the joke, and so on. How many people will post the joke on the ninth round?

If you are given some of the terms of a geometric sequence, you can determine an equation for finding the nth term of the sequence.

Example 2 Write an Equation for the nth Term

Write an equation for the nth term of each geometric sequence.

a. 0.5, 2, 8, 32, ...

$r = 8 \div 2$ or 4; 0.5 is the first term.
$a_n = a_1 r^{n-1}$ nth term of a geometric sequence
$a_n = 0.5(4)^{n-1}$ $a_1 = 0.5$ and $r = 4$

b. $a_4 = 5$ and $r = 6$

Step 1 Find a_1.
$a_n = a_1 r^{n-1}$ nth term of a geometric sequence
$5 = a_1(6^{4-1})$ $a_n = 5, r = 6$, and $n = 4$
$5 = a_1(216)$ Evaluate the power.
$\frac{5}{216} = a_1$ Divide each side by 216.

Step 2 Write the equation.
$a_n = a_1 r^{n-1}$ nth term of a geometric sequence
$a_n = \frac{5}{216}(6)^{n-1}$ $a_1 = \frac{5}{216}$ and $r = 6$

▶ **Guided Practice**

Write an equation for the nth term of each geometric sequence.

2A. $-0.25, 2, -16, 128, \ldots$ **2B.** $a_3 = 16, r = 4$

> **Math History Link**
> **Archytas** (428–347 B.C.) Geometric sequences, or geometric progressions, were first studied by the Greek mathematician Archytas. His studies of these sequences came from his interest in music and octaves.

Like arithmetic means, **geometric means** are the terms between two nonconsecutive terms of a geometric sequence. The common ratio r can be used to find the geometric means.

Example 3 Find Geometric Means

Find three geometric means between 2 and 1250.

Step 1 Because there are three terms between the first and last term, there are $3 + 2$ or 5 total terms, so $n = 5$.

Step 2 Find r.
$a_n = a_1 r^{n-1}$ nth term of a geometric sequence
$1250 = 2r^{5-1}$ $a_n = 1250, a_1 = 2$, and $n = 5$
$625 = r^4$ Divide each side by 2.
$\pm 5 = r$ Take the 4th root of each side.

Step 3 Use r to find the four arithmetic means.

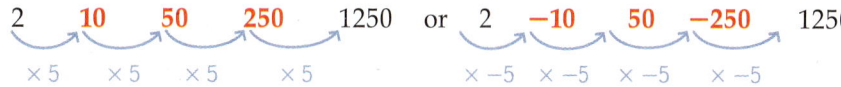

The geometric means are 10, 50, and 250 or -10, 50, and -250.

> **Reading Math**
> **Geometric Means** A geometric mean can also be represented geometrically. In the figure below, h is the geometric mean between x and y.
>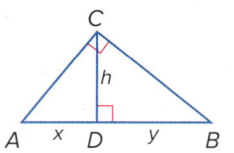

▶ **Guided Practice**

3. Find four geometric means between 0.5 and 512.

2 Geometric Series

A **geometric series** is the indicated sum of the terms of a geometric sequence. The sum of the first n terms of a series is denoted S_n. You can use either of the following formulas to find the partial sum S_n of the first n terms of a geometric series.

Key Concept Partial Sum of a Geometric Series

Given	The sum S_n of the first n terms is:
a_1 and n	$S_n = \dfrac{a_1 - a_1 r^n}{1 - r}, r \neq 1$
a_1 and a_n	$S_n = \dfrac{a_1 - a_n r}{1 - r}, r \neq 1$

Go Online!

Watch **Personal Tutor** videos to hear descriptions of problem solving. Try describing how to solve a problem for a partner. Have them ask you questions to help your understanding.

Real-World Example 4 Find the Sum of a Geometric Series

MUSIC Refer to the beginning of the lesson. If the pattern continues, how many total people posted the link after the eighth round?

Five people reposted the link in the first round and there are 8 rounds of reposts. So, $a_1 = 5$, $r = 5$ and $n = 8$.

$S_n = \dfrac{a_1 - a_1 r^n}{1 - r}$ Sum formula

$S_8 = \dfrac{5 - 5 \cdot 5^8}{1 - 5}$ $a_1 = 5, r = 5,$ and $n = 8$

$S_8 = \dfrac{-1{,}953{,}120}{-4}$ Simplify the numerator and denominator.

$S_8 = 488{,}280$ Divide.

There will be 488,280 total people who reposted the link after 8 rounds.

Guided Practice

Find the sum of each geometric series.

4A. $a_1 = 2, n = 10, r = 3$

4B. $a_1 = 2000, a_n = 125, r = \dfrac{1}{2}$

As with arithmetic series, sigma notation can also be used to represent geometric series.

Example 5 Sum in Sigma Notation

Watch Out!

Sigma Notation Notice in Example 5 that you are being asked to evaluate the sum from the 3rd term to the 10th term.

Find $\displaystyle\sum_{k=3}^{10} 4(2)^{k-1}$.

Find a_1, r, and n. In the first term, $k = 3$ and $a_1 = 4 \cdot 2^{3-1}$ or 16. The base of the exponential function is r, so $r = 2$. There are $10 - 3 + 1$ or 8 terms, so $n = 8$.

$S_n = \dfrac{a_1 - a_1 r^n}{1 - r}$ Sum formula

$= \dfrac{16 - 16(2)^8}{1 - 2}$ $a_1 = 16, r = 2,$ and $n = 8$

$= 4080$ Use a calculator.

Guided Practice

Find each sum.

5A. $\displaystyle\sum_{k=4}^{12} \dfrac{1}{4} \cdot 3^{k-1}$

5B. $\displaystyle\sum_{k=2}^{9} \dfrac{2}{3} \cdot 4^{k-1}$

You can use the formula for the sum of a geometric series to help find a particular term of the series.

> **Example 6**   **Find the First Term of a Series**
>
> Find a_1 in a geometric series for which $S_n = 13{,}116$, $n = 7$, and $r = 3$.
>
> $S_n = \dfrac{a_1 - a_1 r^n}{1 - r}$ Sum formula
>
> $13{,}116 = \dfrac{a_1 - a_1(3^7)}{1 - 3}$ $S_n = 13{,}116$, $r = 3$, and $n = 7$
>
> $13{,}116 = \dfrac{a_1(1 - 3^7)}{1 - 3}$ Distributive Property
>
> $13{,}116 = \dfrac{-2186 a_1}{-2}$ Subtract.
>
> $13{,}116 = 1093 a_1$ Simplify.
>
> $12 = a_1$ Divide each side by 1093.
>
> ▶ **Guided Practice**
>
> **6.** Find a_1 in a geometric series for which $S_n = -26{,}240$, $n = 8$, and $r = -3$.

Check Your Understanding

= Step-by-Step Solutions begin on page R11.

Example 1

1. REGULARITY Dean is making a family tree for his grandfather. He was able to trace many generations. If Dean could trace his family back 10 generations, how many ancestors are in the 10th generation back?

Example 2   Write an equation for the nth term of each geometric sequence.

2. 2, 4, 8, …
3. 18, 6, 2, …
4. −4, 16, −64, …

5 $a_2 = 4$, $r = 3$
6. $a_6 = \dfrac{1}{8}$, $r = \dfrac{3}{4}$
7. $a_2 = -96$, $r = -8$

Example 3   Find the geometric means of each sequence.

8. 0.25, _?_, _?_, _?_, 64
9. 0.20, _?_, _?_, _?_, 125

Example 4

10. GAMES Miranda arranges some rows of dominoes so that after she knocks over the first one, each domino knocks over two more dominoes when it falls. If there are ten rows, how many dominoes does Miranda use?

Example 5   Find the sum of each geometric series.

11. $\displaystyle\sum_{k=1}^{6} 3(4)^{k-1}$
12. $\displaystyle\sum_{k=1}^{8} 4\left(\dfrac{1}{2}\right)^{k-1}$

Example 6   Find a_1 for each geometric series described.

13. $S_n = 85\dfrac{5}{16}$, $r = 4$, $n = 6$
14. $S_n = 91\dfrac{1}{12}$, $r = 3$, $n = 7$

15. $S_n = 1020$, $a_n = 4$, $r = \dfrac{1}{2}$
16. $S_n = 121\dfrac{1}{3}$, $a_n = \dfrac{1}{3}$, $r = \dfrac{1}{3}$

Practice and Problem Solving

Extra Practice is on page R6.

Example 1

17. WEATHER Heavy rain in Brieanne's town caused the Pecos River to rise. The river rose three inches the first day, and each day after rose twice as much as the previous day. How much did the river rise in five days?

Find a_n for each geometric sequence.

18. $a_1 = 2400, r = \frac{1}{4}, n = 7$

19. $a_1 = 800, r = \frac{1}{2}, n = 6$

20. $a_1 = \frac{2}{9}, r = 3, n = 7$

21. $a_1 = -4, r = -2, n = 8$

22. BIOLOGY A certain bacteria grows at a rate of 3 cells every 2 minutes. If there were 260 cells initially, how many are there after 21 minutes?

Example 2

Write an equation for the nth term of each geometric sequence.

23. $-3, 6, -12, \ldots$

24. $288, -96, 32, \ldots$

25. $-1, 1, -1, \ldots$

26. $\frac{1}{3}, \frac{2}{9}, \frac{4}{27}, \ldots$

27. $8, 2, \frac{1}{2}, \ldots$

28. $12, -16, \frac{64}{3}, \ldots$

29. $a_3 = 28, r = 2$

30. $a_4 = -8, r = 0.5$

31. $a_6 = 0.5, r = 6$

32. $a_3 = 8, r = \frac{1}{2}$

33. $a_4 = 24, r = \frac{1}{3}$

34. $a_4 = 80, r = 4$

Example 3

Find the geometric means of each sequence.

35. $810, \underline{?}, \underline{?}, \underline{?}, 10$

36. $640, \underline{?}, \underline{?}, \underline{?}, 2.5$

37. $\frac{7}{2}, \underline{?}, \underline{?}, \underline{?}, \frac{56}{81}$

38. $\frac{729}{64}, \underline{?}, \underline{?}, \underline{?}, \frac{324}{9}$

39. Find two geometric means between 3 and 375.

40. Find two geometric means between 16 and -2.

Example 4

41. **MP PERSEVERANCE** A certain water filtration system can remove 70% of the contaminants each time a sample of water is passed through it. If the same water is passed through the system four times, what percent of the original contaminants will be removed from the water sample?

Find the sum of each geometric series.

42. $a_1 = 36, r = \frac{1}{3}, n = 8$

43. $a_1 = 16, r = \frac{1}{2}, n = 9$

44. $a_1 = 240, r = \frac{3}{4}, n = 7$

45. $a_1 = 360, r = \frac{4}{3}, n = 8$

46. VACUUMS A vacuum claims to pick up 80% of the dirt every time it is run over the carpet. Assuming this is true, what percent of the original amount of dirt is picked up after the seventh time the vacuum is run over the carpet?

Example 5

Find the sum of each geometric series.

47. $\sum_{k=1}^{7} 4(-3)^{k-1}$

48. $\sum_{k=1}^{8} (-3)(-2)^{k-1}$

49. $\sum_{k=1}^{9} (-1)(4)^{k-1}$

50. $\sum_{k=1}^{10} 5(-1)^{k-1}$

Example 6

Find a_1 for each geometric series described.

51. $S_n = -2912, r = 3, n = 6$

52. $S_n = -10{,}922, r = 4, n = 7$

53. $S_n = 1330, a_n = 486, r = \frac{3}{2}$

54. $S_n = 4118, a_n = 128, r = \frac{2}{3}$

55. $a_n = 1024, r = 8, n = 5$

56. $a_n = 1875, r = 5, n = 7$

57. SCIENCE One minute after it is released, a gas-filled balloon has risen 100 feet. In each succeeding minute, the balloon rises only 50% as far as it rose in the previous minute. How far will it rise in 5 minutes?

58. CHEMISTRY Radon has a half-life of about 4 days. This means that about every 4 days, half of the mass of radon decays into another element. How many grams of radon remain from an initial 60 grams after 4 weeks?

59. REASONING A virus goes through a computer, infecting the files. If one file was infected initially and the total number of files infected doubles every minute, how many files will be infected in 20 minutes?

60. GEOMETRY In the figure, the sides of each equilateral triangle are twice the size of the sides of its inscribed triangle. If the pattern continues, find the sum of the perimeters of the first eight triangles.

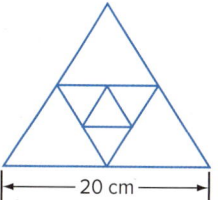

61. PENDULUMS The first swing of a pendulum travels 30 centimeters. If each subsequent swing travels 95% as far as the previous swing, find the total distance traveled by the pendulum after the 30th swing.

62. PHONE CHAINS A school established a phone chain in which every staff member calls two other staff members to notify them when the school closes due to weather. The first round of calls begins with the superintendent calling both principals. If there are 94 total staff members and employees at the school, how many rounds of calls are there?

63. TABLETS High Tech Electronics advertises a monthly installment plan for the purchase of a popular tablet. A buyer pays $40 at the end of the first month, $44 at the end of the second month, $48.40 at the end of the third month, and so on for one year.

  a. What will the payments be at the end of the 4th, 6th, and 10th months?
  b. Find the total cost of the tablet.
  c. Why is the cost found in part **b** not entirely accurate?

H.O.T. Problems   Use Higher-Order Thinking Skills

64. PROOF Derive the General Sum Formula using the Alternate Sum Formula.

65. PROOF Derive a sum formula that does not include a_1.

66. OPEN-ENDED Write a geometric series for which $r = \frac{3}{4}$ and $n = 6$.

67. REASONING Explain how $\sum_{k=1}^{10} 3(2)^{k-1}$ needs to be altered to refer to the same series if $k = 1$ changes to $k = 0$. Explain your reasoning.

68. PROOF Prove the formula for the nth term of a geometric sequence.

69. CHALLENGE The fifth term of a geometric sequence is $\frac{1}{27}$th of the eighth term. If the ninth term is 702, what is the eighth term?

70. CHALLENGE Use the fact that h is the geometric mean between x and y in the figure at the right to find h^4 in terms of x and y.

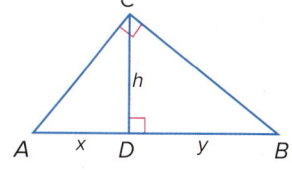

71. OPEN-ENDED Write a geometric series with 6 terms and a sum of 252.

72. WRITING IN MATH How can you classify a sequence? Explain your reasoning.

Preparing for Assessment

73. Janelle makes jewelry. She e-mails a link for her website to four potential customers. They each forward the link to three friends. If the link is forwarded again following the same pattern, how many people will receive the link on the tenth round of e-mails? **MP 4**

74. A certain bacteria doubles in number every 30 minutes. If there were 140 cells initially, how many cells are there after 4 hours? **MP 8**

- A 1120
- B 3840
- C 17,920
- D 35,840

75. Radon has a half-life of 4 days. This means that about every 4 days, half of the mass of radon decays into another element. How many grams of radon remain from an initial 120 grams after 28 days? **MP 8**

76. Refer to question 73. After the tenth round, what is the total number of people who have received e-mails containing a link to Janelle's website? **MP 1**

- A 29,522.5
- B 39,364
- C 78,731
- D 118,096
- E 1,048,576

77. Find $\sum_{k=3}^{7} \left(\frac{1}{2} \cdot 4^{k-1}\right)$. **MP 2**

- A 170.5
- B 680
- C 2728
- D 5456
- E 43,688

78. The curve below could be part of the graph of which function? **MP 2**

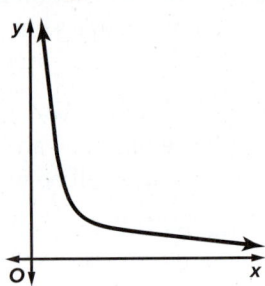

- A $y = \sqrt{x}$
- B $y = x^2 - 5x + 4$
- C $y = -x + 20$
- D $y = \log x$
- E $xy = 4$

79. A bacteria grows exponentially so that it doubles every 12 days. If there are 500 bacteria initially, find the number of bacteria after: **MP 2**

a. 12 days

b. 60 days

80. MULTI-STEP Given the geometric sequence $a_n = \dfrac{5}{2^n}$: **MP 6**

a. Find the first term.

b. Find the common ratio.

c. Find the ninth term.

d. Show that the fifth term is the geometric mean between the first term and the ninth term.

e. Find the sum of the first nine terms.

396 | Lesson 6-3 | Geometric Sequences and Series

LESSON 4
Logarithms and Logarithmic Functions

∷Then
- You found the inverse of a function.

∷Now
1. Evaluate logarithmic expressions.
2. Graph logarithmic functions.

∷Why?
Many scientists believe the extinction of the dinosaurs was caused by an asteroid striking Earth. Astronomers use the Palermo scale to classify objects near Earth based on the likelihood of impact. To make comparing several objects easier, the scale was developed using *logarithms*. The Palermo scale value of any object can be found using the equation $PS = \log_{10} R$, where R is the relative risk posed by the object.

 New Vocabulary
logarithm
logarithmic function

Mathematical Practices
4 Model with mathematics.
6 Attend to precision.
7 Look for and make use of structure.

1 Logarithmic Functions and Expressions
Consider the exponential function $f(x) = 2^x$ and its inverse. Recall that you can graph an inverse function by interchanging the *x*- and *y*-values in the ordered pairs of the function. Notice that the domain and range of the functions are switched.

$y = 2^x$	
x	y
−3	$\frac{1}{8}$
−2	$\frac{1}{4}$
−1	$\frac{1}{2}$
0	1
1	2
2	4
3	8

$x = 2^y$	
x	y
$\frac{1}{8}$	−3
$\frac{1}{4}$	−2
$\frac{1}{2}$	−1
1	0
2	1
4	2
8	3

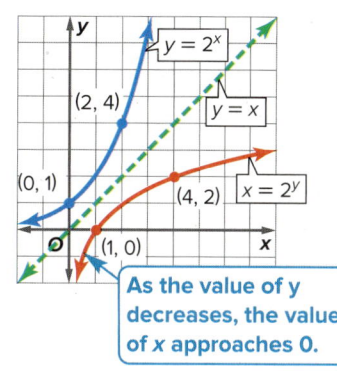

As the value of *y* decreases, the value of *x* approaches 0.

The inverse of $y = 2^x$ can be defined as $x = 2^y$. In general, the inverse of $y = b^x$ is $x = b^y$. In $x = b^y$, the variable *y* is called the **logarithm** of *x*. This is usually written as $y = \log_b x$, which is read *y equals log base b of x*.

🔑 Key Concept Logarithm with Base *b*

Words Let *b* and *x* be positive numbers, $b \neq 1$. The *logarithm of x with base b* is denoted $\log_b x$ and is defined as the exponent *y* that makes the equation $b^y = x$ true.

Symbols Suppose $b > 0$ and $b \neq 1$. For $x > 0$, there is a number *y* such that
$$\log_b x = y \text{ if and only if } b^y = x.$$

Example If $\log_3 27 = y$, then $3^y = 27$.

The definition of logarithms can be used to express logarithms in exponential form.

Example 1 Logarithmic to Exponential Form

Write each equation in exponential form.
a. $\log_2 8 = 3$

$\log_2 8 = 3 \rightarrow 8 = 2^3$

b. $\log_4 \frac{1}{256} = -4$

$\log_4 \frac{1}{256} = -4 \rightarrow \frac{1}{256} = 4^{-4}$

▶ **Guided Practice**
1A. $\log_4 16 = 2$
1B. $\log_3 729 = 6$

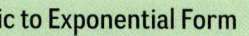 connectED.mcgraw-hill.com

The definition of logarithms can also be used to write exponential equations in logarithmic form.

> **Example 2**   **Exponential to Logarithmic Form**
>
> **Write each equation in logarithmic form.**
>
> a. $15^3 = 3375$ b. $4^{\frac{1}{2}} = 2$
>
> $15^3 = 3375 \rightarrow \log_{15} 3375 = 3$ $4^{\frac{1}{2}} = 2 \rightarrow \log_4 2 = \frac{1}{2}$
>
> ▶ **Guided Practice**
>
> **2A.** $4^3 = 64$ **2B.** $125^{\frac{1}{3}} = 5$

You can use the definition of a logarithm to evaluate a logarithmic expression.

Watch Out!

Logarithmic Base It is easy to get confused about which number is the base and which is the exponent in logarithmic equations. Consider highlighting each number as you solve to help organize your calculations.

> **Example 3**   **Evaluate Logarithmic Expressions**
>
> **Evaluate $\log_{16} 4$.**
>
> $\log_{16} 4 = y$ Let the logarithm equal y.
> $4 = 16^y$ Definition of logarithm
> $4^1 = (4^2)^y$ $16 = 4^2$
> $4^1 = 4^{2y}$ Power of a Power
> $1 = 2y$ Property of Equality for Exponential Functions
> $\frac{1}{2} = y$ Divide each side by 2.
>
> Thus, $\log_{16} 4 = \frac{1}{2}$.
>
> ▶ **Guided Practice**
>
> **Evaluate each expression.**
>
> **3A.** $\log_3 81$ **3B.** $\log_{\frac{1}{2}} 256$

2 Graphing Logarithmic Functions

$y = \log_b x$, where $b \neq 1$, is a **logarithmic function**. The graph of $f(x) = \log_b x$ represents a parent graph of the logarithmic functions.

> **Key Concept**   **Parent Function of Logarithmic Functions**
>
> | | | | | |
> |---|---|---|---|---|
> | **Parent function:** | $f(x) = \log_b x$ | | **Type of graph:** | continuous, one-to-one |
> | **Domain:** | $(0, \infty)$ or $\{x \mid x > 0\}$ | | **Range:** | {all real numbers} |
> | **Asymptote:** | $f(x)$-axis | | **Intercept:** | $(1, 0)$ |
> | **Symmetry:** | none | | **Extrema:** | none |
>
> 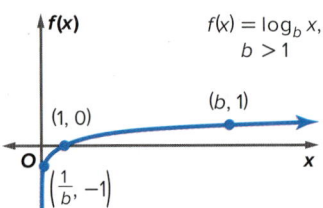   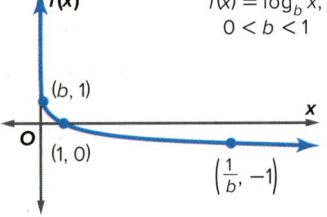

Example 4 Graph Logarithmic Functions

Graph each function.

a. $f(x) = \log_5 x$

Step 1 Identify the base.
$b = 5$

Study Tip

Zero Exponent Recall that for any $b \neq 0$, $b^0 = 1$. Therefore, $\log_2 0$ is undefined because $2^x \neq 0$ for any x-value.

Step 2 Determine points on the graph.
Because $5 > 1$, use the points $\left(\frac{1}{b}, -1\right)$, $(1, 0)$, and $(b, 1)$.

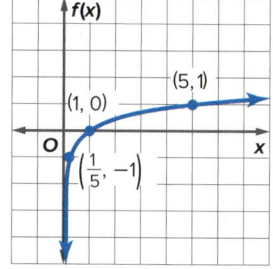

Step 3 Plot the points and sketch the graph.
$\left(\frac{1}{b}, -1\right) \rightarrow \left(\frac{1}{5}, -1\right)$
$(1, 0)$
$(b, 1) \rightarrow (5, 1)$

b. $f(x) = \log_{\frac{1}{3}} x$

Step 1 $b = \frac{1}{3}$

Step 2 $0 < \frac{1}{3} < 1$, so use the points $\left(\frac{1}{3}, 1\right)$, $(1, 0)$ and $(3, -1)$.

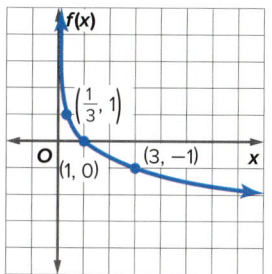

Step 3 Sketch the graph.

▶ **Guided Practice**

4A. $f(x) = \log_2 x$ **4B.** $f(x) = \log_{\frac{1}{8}} x$

The same techniques used to transform the graphs of other functions you have studied can be applied to the graphs of logarithmic functions.

Go Online!

How does changing the values in the equation of a logarithmic function affect its graph? Investigate by using the **Graphing Tools** in ConnectED.

Key Concept Transformations of Logarithmic Functions

$$f(x) = a \log_b (x - h) + k$$

h – Horizontal Translation	k – Vertical Translation
h units right if h is positive |h| units left if h is negative	k units up if k is positive |k| units down if k is negative

a – Orientation and Shape	
If $a < 0$, the graph is reflected across the x-axis.	If $\|a\| > 1$, the graph is stretched vertically. If $0 < \|a\| < 1$, the graph is compressed vertically.

connectED.mcgraw-hill.com **399**

Example 5 Graph Logarithmic Functions

Graph each function.

Study Tip

End Behavior In Example 5a, as x approaches infinity, $f(x)$ approaches infinity.

a. $f(x) = 3 \log_{10} x + 1$

This represents a transformation of the graph of $f(x) = \log_{10} x$.

- $|a| = 3$: The graph stretches vertically
- $h = 0$: There is no horizontal shift.
- $k = 1$: The graph is translated 1 unit up.

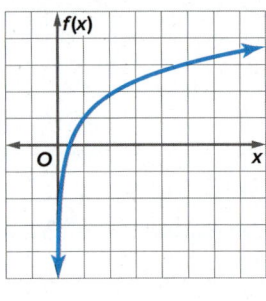

b. $f(x) = \frac{1}{2} \log_{\frac{1}{4}} (x - 3)$

This is a transformation of the graph of $f(x) = \log_{\frac{1}{4}} x$.

- $|a| = \frac{1}{2}$: The graph is compressed vertically.
- $h = 3$: The graph is translated 3 units to the right.
- $k = 0$: There is no vertical shift.

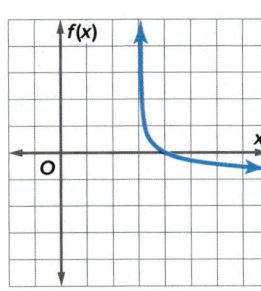

Guided Practice

Graph each function.

5A. $f(x) = 2 \log_3 (x - 2)$

5B. $f(x) = \frac{1}{4} \log_{\frac{1}{2}} (x + 1) - 5$

Real-World Example 6 Find Inverses of Exponential Functions

EARTHQUAKES The Richter scale measures earthquake intensity. The increase in intensity between each number is 10 times. For example, an earthquake with a rating of 7 is 10 times more intense than one measuring 6. The intensity of an earthquake can be modeled by $y = 10^{x-1}$, where x is the Richter scale rating.

Real-World Link

The largest recorded earthquake in the United States was a magnitude 9.2 that struck Prince William Sound, Alaska, on Good Friday, March 28, 1964.

Source: United States Geological Survey

a. Use the information at the left to find the intensity of the strongest recorded earthquake in the United States.

$y = 10^{x-1}$ Original equation
$ = 10^{9.2 - 1}$ Substitute 9.2 for x.
$ = 10^{8.2}$ Simplify.
$ = 158{,}489{,}319.2$ Use a calculator.

b. Write an equation of the form $y = \log_{10} x + c$ for the inverse of the function.

$y = 10^{x-1}$ Original equation
$x = 10^{y-1}$ Replace x with y, replace y with x, and solve for y.
$y - 1 = \log_{10} x$ Definition of logarithm
$y = \log_{10} x + 1$ Add 1 to each side.

Guided Practice

6. Write an equation for the inverse of the function $y = 0.5^x$.

Check Your Understanding

= Step-by-Step Solutions begin on page R11.

 Go Online! for a Self-Check Quiz

Example 1 Write each equation in exponential form.
1. $\log_8 512 = 3$
2. $\log_5 625 = 4$

Example 2 Write each equation in logarithmic form.
3. $11^3 = 1331$
4. $16^{\frac{3}{4}} = 8$

Example 3 Evaluate each expression.
5. $\log_{13} 169$
6. $\log_2 \frac{1}{128}$
7. $\log_6 1$

Examples 4–5 Graph each function.
8. $f(x) = \log_3 x$
9. $f(x) = \log_{\frac{1}{6}} x$
10. $f(x) = 4 \log_4 (x - 6)$
11. $f(x) = 2 \log_{\frac{1}{10}} x - 5$

Example 6
12. **SCIENCE** Use the information at the beginning of the lesson. The Palermo scale value of any object can be found using the equation $PS = \log_{10} R$, where R is the relative risk posed by the object. Write an equation in exponential form for the inverse of the function.

Practice and Problem Solving

Extra Practice is on page R6.

Example 1 Write each equation in exponential form.
13. $\log_2 16 = 4$
14. $\log_7 343 = 3$
15. $\log_9 \frac{1}{81} = -2$
16. $\log_3 \frac{1}{27} = -3$
17. $\log_{12} 144 = 2$
18. $\log_9 1 = 0$

Example 2 Write each equation in logarithmic form.
19. $9^{-1} = \frac{1}{9}$
20. $6^{-3} = \frac{1}{216}$
21. $2^8 = 256$
22. $4^6 = 4096$
23. $27^{\frac{2}{3}} = 9$
24. $25^{\frac{3}{2}} = 125$

Example 3 Evaluate each expression.
25. $\log_3 \frac{1}{9}$
26. $\log_4 \frac{1}{64}$
27. $\log_8 512$
28. $\log_6 216$
29. $\log_{27} 3$
30. $\log_{32} 2$
31. $\log_9 3$
32. $\log_{121} 11$
33. $\log_{\frac{1}{5}} 3125$
34. $\log_{\frac{1}{8}} 512$
35. $\log_{\frac{1}{3}} \frac{1}{81}$
36. $\log_{\frac{1}{6}} \frac{1}{216}$

Examples 4–5 **PRECISION** Graph each function.
37. $f(x) = \log_6 x$
38. $f(x) = \log_{\frac{1}{5}} x$
39. $f(x) = 4 \log_2 x + 6$
40. $f(x) = \log_{\frac{1}{9}} x$
41. $f(x) = \log_{10} x$
42. $f(x) = -3 \log_{\frac{1}{12}} x + 2$
43. $f(x) = 6 \log_{\frac{1}{8}} (x + 2)$
44. $f(x) = -8 \log_3 (x - 4)$
45. $f(x) = \log_{\frac{1}{4}} (x + 1) - 9$
46. $f(x) = \log_5 (x - 4) - 5$
47. $f(x) = -\frac{1}{6} \log_8 (x - 3) + 4$
48. $f(x) = -\frac{1}{3} \log_{\frac{1}{6}} (x + 2) - 5$

Example 6

49. PHOTOGRAPHY The formula $n = \log_2 \frac{1}{p}$ represents the change in the f-stop setting n to use in less light where p is the fraction of sunlight.

  a. Benito's camera is set up to take pictures in direct sunlight, but it is a cloudy day. If the amount of sunlight on a cloudy day is $\frac{1}{4}$ as bright as direct sunlight, how many f-stop settings should he move to accommodate less light?

  b. Graph the function.

  c. Use the graph in part b to predict what fraction of daylight Benito is accommodating if he moves down 3 f-stop settings. Is he allowing more or less light into the camera?

50. EDUCATION To measure a student's retention of knowledge, the student is tested after a given amount of time. A student's score on an Algebra 2 test t months after the school year is over can be approximated by $y(t) = 85 - 6 \log_2 (t + 1)$, where $y(t)$ is the student's score as a percent.

  a. What was the student's score at the time the school year ended ($t = 0$)?

  b. What was the student's score after 3 months?

  c. What was the student's score after 15 months?

Graph each function.

51. $f(x) = 4 \log_2 (2x - 4) + 6$

52. $f(x) = -3 \log_{12} (4x + 3) + 2$

53. $f(x) = 15 \log_{14} (x + 1) - 9$

54. $f(x) = 10 \log_5 (x - 4) - 5$

55. $f(x) = -\frac{1}{6} \log_8 (x - 3) + 4$

56. $f(x) = -\frac{1}{3} \log_6 (6x + 2) - 5$

57. MODELING In general, the more money a company spends on advertising, the higher the sales. The amount of money in sales for a company, in thousands, can be modeled by the equation $S(a) = 10 + 20 \log_4(a + 1)$, where a is the amount of money spent on advertising in thousands, when $a \geq 0$.

  a. The value of $S(0) = 10$, which means that if no money is spent on advertising, $10,000 is returned in sales. Find the values of $S(3)$, $S(15)$, and $S(63)$.

  b. Interpret the meaning of each function value in the context of the problem.

  c. Graph the function.

  d. Use the graph in part c and your answers from part a to explain why the money spent in advertising becomes less "efficient" as it is used in larger amounts.

58. BIOLOGY The generation time for bacteria is the time that it takes for the population to double. The generation time G for a specific type of bacteria can be found using experimental data and the formula $G = \frac{t}{3.3 \log_b f}$, where t is the time period, b is the number of bacteria at the beginning of the experiment, and f is the number of bacteria at the end of the experiment.

  a. The generation time for mycobacterium tuberculosis is 16 hours. How long will it take four of these bacteria to multiply into 1024 bacteria?

  b. An experiment involving rats that had been exposed to salmonella showed that the generation time for the salmonella was 5 hours. After how long would 20 of these bacteria multiply into 8000?

  c. E. coli are fast growing bacteria. If 6 E. coli can grow to 1296 in 4.4 hours, what is the generation time of E. coli?

59. FINANCIAL LITERACY Jacy has charged $2000 on a credit card. The credit card company charges 24% interest, compounded monthly. The credit card company uses $\log_{\left(1+\frac{0.24}{12}\right)} \frac{A}{2000} = 12t$ to determine how much time it will be until Jacy's debt reaches a certain amount, if A is the amount of debt after a period of time, and t is time in years.

  a. Graph the function for Jacy's debt.
  b. Approximately how long will it take Jacy's debt to double?
  c. Approximately how long will it be until Jacy's debt triples?

60. GRAPHING CALCULATOR Graph $f(x) = \log_2 x$ and the transformation graph, $g(x)$. Determine the effects on each of the key attributes of the graph of $f(x)$.

  a. $g(x) = 3f(x)$ **b.** $g(x) = -2f(x)$ **c.** $g(x) = f(x) + 4$
  d. $g(x) = f(x) - 6$ **e.** $g(x) = f(x+5)$ **f.** $g(x) = f(x-8)$

H.O.T. Problems   Use Higher-Order Thinking Skills

61. CRITIQUE ARGUMENTS Consider $y = \log_b x$ in which b, x, and y are real numbers. Zero can be in the domain *sometimes*, *always* or *never*. Justify your answer.

62. ERROR ANALYSIS Betsy says that the graphs of all logarithmic functions cross the y-axis at (0, 1) because any number to the zero power equals 1. Tyrone disagrees. Is either of them correct? Explain your reasoning.

63. REASONING Without using a calculator, compare $\log_7 51$, $\log_8 61$, and $\log_9 71$. Which of these is the greatest? Explain your reasoning.

64. OPEN-ENDED Write a logarithmic equation of the form $y = \log_b x$ for each of the following conditions.

  a. y is equal to 25.   **b.** y is negative.
  c. y is between 0 and 1.   **d.** x is 1.
  e. x is 0.

65. ERROR ANALYSIS Elisa and Matthew are evaluating $\log_{\frac{1}{7}} 49$. Is either of them correct? Explain your reasoning.

Elisa	Matthew
$\log_{\frac{1}{7}} 49 = y$	$\log_{\frac{1}{7}} 49 = y$
$\frac{1}{7}^y = 49$	$49^y = \frac{1}{7}$
$(7^{-1})^y = 7^2$	$(7^2)^y = (7)^{-1}$
$(7)^{-y} = 7^2$	$7^{2y} = (7)^{-1}$
$y = 2$	$2y = -1$
	$y = -\frac{1}{2}$

66. WRITING IN MATH A transformation of $\log_{10} x$ is $g(x) = a \log_{10}(x - h) + k$. Explain the process of graphing this transformation.

Preparing for Assessment

67. Which graph is the function $f(x) = -\log(x - 2)$?
MP 4

○ A
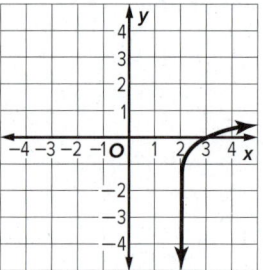

○ B
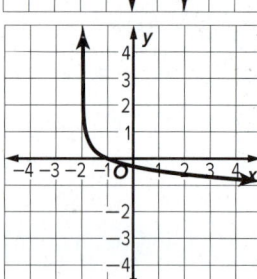

○ C
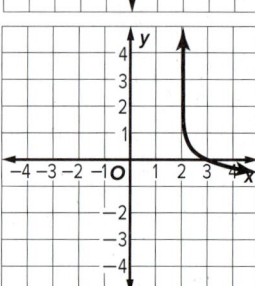

○ D
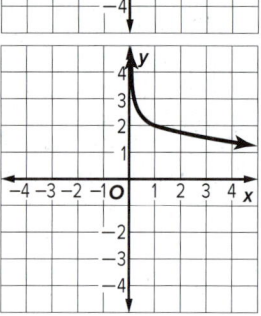

68. Let $n = \log_m p$. If $p > 1 > m > 0$, which statement must be true? MP 7

○ A $p > n > 1$

○ B $1 > n > m$

○ C $m > n > 0$

○ D $0 > n$

69. Which of the following is an asymptote for the graph of $y = \log_4 x + 1$? MP 6

○ A $x = -1$

○ B $x = 0$

○ C $y = 1$

○ D $y = 4$

70. If $-3 = \log_3 x$, what is the value of x?
MP 7

○ A -27 ○ C $\frac{1}{27}$

○ B $-\frac{1}{27}$ ○ D 27

71. Select all of the transformations for the function $f(x) = -\log_5(x + 4) - 5$ based on the parent function $f(x) = \log_5 x$. MP 7

☐ A   translated up 5 units

☐ B   translated down 5 units

☐ C   translated left 4 units

☐ D   translated right 4 units

☐ E   reflected in the y-axis

☐ F   reflected in the x-axis

72. MULTI-STEP The function $f(x) = 10 - \log(x + 10)$ represents the home attendance of a football team in ten thousands, when $f(x)$ is based on the number of home games x. MP 4, 6, 7

a. Sketch a graph of the function.

b. What is the domain of the function?

c. What is the range of the function?

d. What is the attendance for the second home game?

e. What is the attendance for the eighth home game?

f. Explain the general trend in attendance based on the graph of the function.

73. A logarithmic function, $f(x) = \log_a(x - b)$ is increasing with a vertical asymptote of $x = 3$. The point $(5, 1)$ lies on its graph. Find the value of a and of b. MP 6

LESSON 5
Modeling Data

:Then
- You learned to understand the relationship between an independent and dependent variable and how that relationship is represented in a function.

:Now
1. Find equations of best fit for data modeled by exponential and logarithmic functions.
2. Choose the best model for a data set.

:Why?
- Leila made a study card showing the steps needed to use her graphing calculator to make a scatter plot for life expectancy based on year of birth.

> Make a scatter plot.
> - Enter the years of birth in L1 and the ages in L2.
> **KEYSTROKES:** STAT ENTER 1980 ENTER 1983 ENTER 1990 ENTER ...
> - Set the viewing window to fit the data.
> **KEYSTROKES:** WINDOW 1975 ENTER 2010 ENTER 5 ENTER 70 ENTER 90 ENTER 2
> - Use STAT PLOT to graph the scatter plot.
> **KEYSTROKES:** 2nd [STAT PLOT] ENTER ENTER GRAPH

New Vocabulary
correlation coefficient
regression line
regression curve

Mathematical Practices
3 Construct viable arguments and critique the reasoning of others.
4 Model with mathematics.
6 Attend to precision.

1 Equations of Best Fit
An equation of best fit helps to analyze data patterns between two variables. A scatter plot is one of the tools used to display data and provides information on the relationship between the variables. This relationship is between the independent, or the predictor, variable on the x-axis and the dependent, or response, variable on the y-axis. When you describe a data set displayed on a scatterplot, the relationship between the variables is described by the direction, form, and strength.

Direction (positive or negative)
If y increases as x increases, there is a positive slope. If y decreases as x increases, there is a negative slope.

Form (linear or nonlinear)
If all the data lies on the same line, then it is linear. If the data is scattered and the line has to curve, then it is nonlinear.

Strength (weak, strong, no correlation)
If the points cluster around the line of best fit, there is a strong correlation. If it is hard to see a line that the data clusters around, there may be a weak or no correlation.

Example 1 Analyze Patterns using a Scatterplot

POPULATION The population per square mile of a country has changed dramatically over a period of years. The table shows the number of people per square mile for several years. Graph the data and describe its direction, form, and strength.

U.S. Population Density			
Year	People per square mile	Year	People per square mile
1790	4.5	1900	21.5
1800	6.1	1910	26.0
1810	4.3	1920	29.9
1820	5.5	1930	34.7
1830	7.4	1940	37.2
1840	9.8	1950	42.6
1850	7.9	1960	50.6
1860	10.6	1970	57.5
1870	10.9	1980	64.0
1880	14.2	1990	70.3
1890	17.8	2000	80.0

Source: Northeast-Midwest Institute

Use a graphing calculator to enter the data. Then draw a scatter plot that shows how the number of people per square mile is related to year.

Step 1 Enter the years in L1. These are the x-values in the domain. Enter the people per square mile in L2. These are the y-values in the range.

Step 2 Determine the scale for the graph. The years are from 1790 to 2000, so use a scale of $1790 \leq x \leq 2000$ in increments of 10. The population per square mile ranges from 4.5 to 80, so use a scale of $0 \leq y \leq 115$ in increments of 5.

connectED.mcgraw-hill.com   405

Study Tip

Structure Organizing the domain and range values from least to greatest can help you determine the best scale.

Step 3 Graph the scatterplot.

Step 4 Describe the scatterplot.

The scatterplot is nonlinear, with a positive direction and a strong correlation.

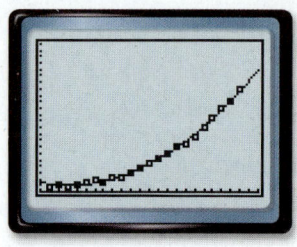

[1780, 2020] scl: 10 by [0, 115] scl: 5

Guided Practice

1. The table shows the cost of a smartphone in different years. Graph the data and describe its direction, form, and strength.

Year	Cost
2012	325.50
2013	406.88
2014	508.60
2015	635.75
2016	794.69

Sometimes a graphing calculator is not available. Understanding how to model data without a calculator is an essential skill.

Example 2 Graph and Analyze a Data Set

Sahara is selling cups of lemonade as a fundraiser. She records the grams of lemon powder she uses to make a cup of lemonade and the time it takes to sell the lemonade. Her data set is: (0, 0.05), (1, 0.1), (2, 0.2), (3, 0.4), (4, 0.8), (5, 1.6). Graph the data and describe its direction, form, and strength.

Create a table to organize the data.

x	y
0	0.05
1	0.1
2	0.2
3	0.4
4	0.8
5	1.6

Graph the data on a coordinate grid.

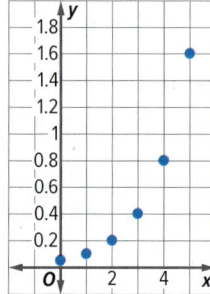

Describe the scatterplot:

Positive direction, nonlinear form, with a strong correlation

If this was a linear scatterplot, we could use a ruler to draw a line that represents the data set and then write a linear equation, but this data set curves.

Study the relationship between the variables and the graph.

The y values are increasing by a constant rate of 2. This data set appears to be exponential.

Study Tip

MP Reasoning The line of best fit does not have to intersect all or even most of the points. Most of the points just have to cluster around the line.

Guided Practice

Use a graph to describe the direction, form, and strength of the data set. If the scatterplot is linear, then write the linear equation for the line of best fit.

2A. $(-3, 0), (-2, 2), (-2.5, 1), (-1, 4)$

2B. $(0, 6), (1, 12), (2, 24), (3, 48)$

2C. $(2, 150), (10, 141), (15, 132.54), (20, 124.56)$

When data is nonlinear, a linear equation, or line of best fit, will not accurately represent the data. Nonlinear data may be modeled by a curve of fit based on a quadratic, exponential, or logarithmic function. For data that fits an exponential curve, use the relationship between exponents and logarithms to find an equation for the curve of fit.

Exponential Function

$f(x) = a^x, a > 0$

When $a = 2$:

x	y
0	1
1	2
2	4
3	8

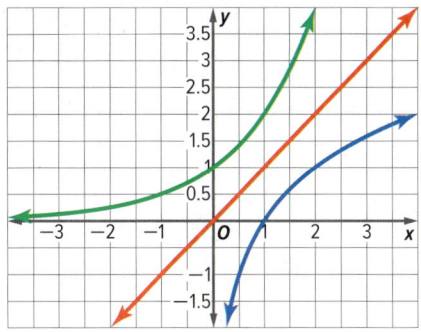

Logarithmic Function

$f(x) = \log_a x, a > 0$ and $a \neq 1$

When $a = 2$:

x	y
1	0
2	1
4	2
8	3

The exponential data are the inverse of the logarithmic data. One graph is the reflection of the other in the line $y = x$. Using the idea of inverse operations, an exponential expression can be written as a logarithmic expression. For example, the exponential expression $4^3 = 64$ can be rewritten as $\log_4 (64) = 3$.

The following formulas are helpful when working with exponential and logarithmic functions.

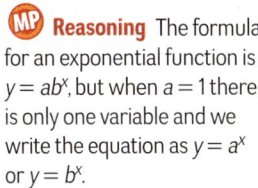

Study Tip

MP Reasoning The formula for an exponential function is $y = ab^x$, but when $a = 1$ there is only one variable and we write the equation as $y = a^x$ or $y = b^x$.

Key Concepts Important Formulas

Exponential growth or decay:
$A = Pe^{rt}$

Natural exponential function:
$f(x) = e^x$

Natural logarithmic function:
$f(x) = \log_e(x)$ or $f(x) = \ln(x)$

Standard form of an exponential function:
$a^x = b$

Standard form of a logarithmic function:
$\log_a(b) = x$

To use a graphing calculator to find a curve of fit, you may need to change the base of a logarithm.

Change the base of a logarithm from b to a: $\log_b x = \dfrac{\log_a x}{\log_a b}$

Example 3 Convert an Exponential Function to a Logarithmic Function

VIRUS In 1991, the Internet was introduced and the first computer virus threatened 500,000 computers. Computer engineers calculated that one person's computer would have infected 5 other computers every minute. They created an exponential function to calculate how long it would take for the virus to spread to all 500,000 computers.

How many minutes would it take for the computer virus to destroy 500,000 computers?

Convert the exponential function $500{,}000 = 5^x$ to a logarithmic function to solve for x.

Step 1 Rewrite the exponential function as a logarithmic function.

$500{,}000 = 5^x$ becomes $\log_5 500{,}000 = x$

Step 2 Study the logarithm. If you know the power of 5, that would equal 500,000 then solve for x. If not, then change the base to a base of 10 and use a calculator.

Step 3 Change the base 5 to base 10.

$$\log_5 500{,}000 = \frac{\log_{10} 500{,}000}{\log_{10} 5}$$

5 is the old base
10 is the new base

Step 4 Input the equation into the calculator and solve for x. Round to the nearest hundredth.

The 500,000 computers would be destroyed in 8.15 minutes.

> **Study Tip**
> **Precision** Take the time to study the formula and walk through the steps a few times to increase understanding and accuracy. Use simple exponential formulas that you can solve mentally, then use a calculator to change the base and check your work.

▶ **Guided Practice**

Use a scientific or graphing calculator to solve for x.

3A. $4^x = 262{,}144$

3B. $27^x = 19{,}683$

3C. $7^x = 2401$

---

2 Best Model for a Data Set A <mark>regression line</mark> or <mark>regression curve</mark> is an equation of a function that represents a line or curve of fit. Choosing the best model for a data set allows you to find the regression line or curve with the strongest correlation; an increased *correlation coefficient* means an increase in the precision of a prediction using the equation of the regression line or curve.

🔑 Key Concepts Types of Regression Equations and Models

Linear	Quadratic
$y = ax + b$	$y = ax^2 + bx + c$

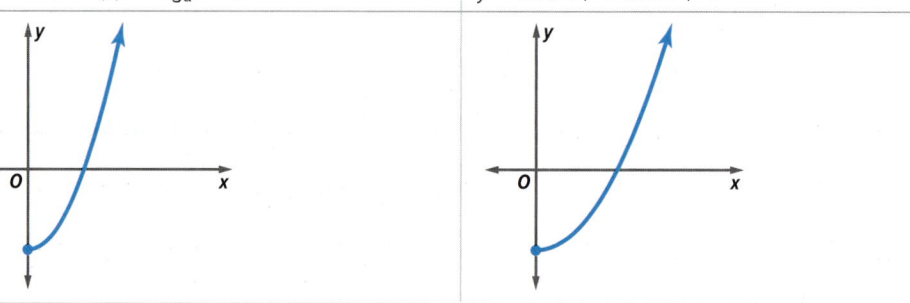

Exponential and Logarithmic	Power
$y = ab^x$ and $f(x) = \log_a x = b$	$y = ax^b$; $a \neq 0$ and $b \neq 0$

> **Study Tip**
> **Modeling** Knowing the general shape of the regression models will help you choose the best model for the data set.

408 | Lesson 6-5 | Modeling Data

The graphing calculator is a good tool to find the line of best fit for nonlinear scatter plots, and will automatically calculate the ==correlation coefficient== r. The correlation coefficient is a measure that shows how well a data set is modeled by a given equation. It is similar to analyzing the relationship between the variables, but it analyzes the relationship between the data set and the line of best fit, represented by an equation.

- Positive correlation: when x increases, y increases.
- Negative correlation: when x increases, y decreases.
- No correlation: the data are randomly dispersed (r is close to 0).
- Perfect correlation: the data lie exactly on the regression line ($r = \pm 1$).
- Strong correlation: $|r| \geq 0.8$
- Weak correlation: $|r| \leq 0.5$

Example 4 Use Regression Equation and Models

Use the data set from Example 1. What is the equation for the line or curve of best fit?

Find a regression equation.

To find an equation that best fits the data, use the regression feature of the calculator. Examine various regressions to determine the best model.

Recall that the calculator returns the correlation coefficient r, which is used to indicate how well the model fits the data. The closer r is to 1 or -1, the better the fit.

Linear regression

KEYSTROKES: STAT ▶ 4 ENTER

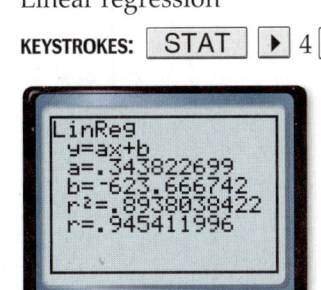

Quadratic regression

KEYSTROKES: STAT ▶ 5 ENTER

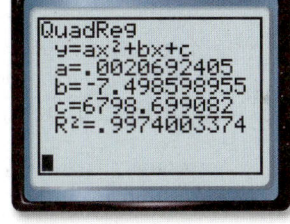

$r^2 = 0.9974003374$
$r = \sqrt{0.9974003374}$
$r \approx 0.9986993228$

Exponential regression

KEYSTROKES: STAT ▶ 0 ENTER

Power regression

KEYSTROKES: STAT ▶ ALPHA [A] ENTER

Compare the r-values.

Linear: 0.945411996   Quadratic: 0.9986993228

Exponential: 0.991887235   Power: 0.9917543535

The r-value of the quadratic regression is closest to 1, so the equation of the curve of best fit is about $y = 0.0005x^2 - 0.0041x + 1.6292$. You can examine the equation visually by graphing the regression equation with the scatter plot.

KEYSTROKES: STAT ▶ 5 ENTER Y= VARS 5 ▶ ▶ 1 GRAPH

Guided Practice

Identify the type of regression for each of the following equations.

4A. $y = 2(3)^x$

4B. $y = 2x^3$

4C. $y = x^2 + 3x + 7$

The statistical value r^2 represents the *correlation of determination*. It can be used to increase the accuracy of the equation used to model data. It considers variation in the data (outliers) and represents the accuracy of the correlation coefficient to the data.

Example 5   Making a Prediction

Use the data set from Example 1 and a graphing calculator to predict the population per square mile in 2020.

Enter the data and make a scatterplot.

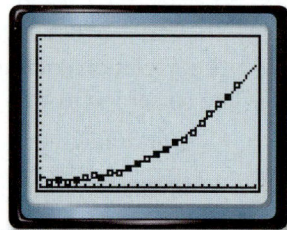

[1780, 2020] scl: 10 by [0, 115] scl: !

To determine the population per square mile in 2020, find the value of y when $x = 20$.

KEYSTROKES: 2nd [CALC] ENTER 2020

U.S. Population Density			
Year	People per square mile	Year	People per square mile
1790	4.5	1900	21.5
1800	6.1	1910	26.0
1810	4.3	1920	29.9
1820	5.5	1930	34.7
1830	7.4	1940	37.2
1840	9.8	1950	42.6
1850	7.9	1960	50.6
1860	10.6	1970	57.5
1870	10.9	1980	64.0
1880	14.2	1990	70.3
1890	17.8	2000	80.0

Source: Northeast-Midwest Institute

If the trend continues, there will be approximately 94.9 people per square mile in 2020. To check your work, you can substitute the values into the equation.

Guided Practice

Fredrick has 20 weeks to train for a 15K race. He tracks his progress by running a practice race every Wednesday and Thursday. The table shows the distance of the race he is able to run each week. At the end of 7 weeks he stops keeping track of his progress, but continues to train for the race.

Weeks	Distance (km)
1	1.5
2	1.75
3	1.3
4	1.8
5	2.1
6	2.45
7	2.5

5A. Will Fredrick be able to run the 15K in 20 weeks? Approximately how far will he be able to run?

5B. How many weeks will it take for him to confidently run the 15K?

5C. Will Fredrick be able to run a 20K race in 6 months?

Check Your Understanding

= Step-by-Step Solutions begin on page R11.

Go Online! for a Self-Check Quiz

Example 1 Describe the direction, form, and strength of each scatterplot.

1.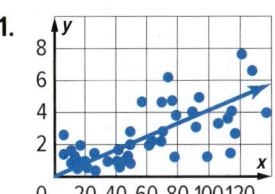

2.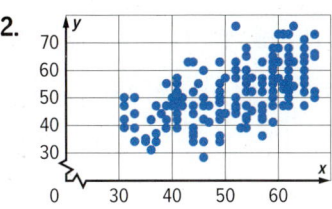

3.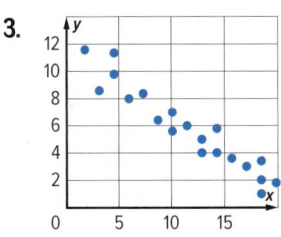

4.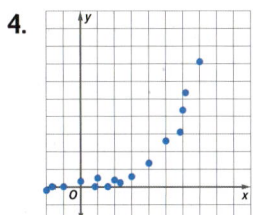

5. Why is the x variable sometimes referred to as the predictor variable?

6. What values are input into L1 and L2?

Example 2 Use a graph to describe the direction, form, and strength of the data set. If the scatterplot is linear, then write the linear equation for the line of best fit.

7. (0, 3), (−1, 2.5), (−2, 2), (1, 3.5), (2, 4), (3, 4.5)

8. (0, 6), (1, 18), (2, 54), (3, 162)

9. (3, 10), (7, 16), (4, 20), (8, 35)

Example 3 Solve for x.

10. $3^x = 27$

11. $99^x = 1$

12. $12^x = 288$

13. $11^x = 121$

14. $45^x = 91{,}125$

15. $52^x = 7{,}311{,}616$

Examples 1, 4, 5 A teacher asked eight random students how many hours they spent each week studying. She then matched the students' answers with their first quarter report card.

16. Make a graph of the data. What type of regression model works best with the data?

17. What is the minimum number of hours that you should spend studying if you want to get a 90 in this class?

Hours	Grade
10	100
8	83
5	92
2.5	57
4	74
9	95
6	75
6.5	85

Practice and Problem Solving

Extra Practice is found on page R6.

Example 1

18. Describe the direction, form, and strength of the data set.

x	0	−2.1	−3	−2.1	−3	−3.25	5	9.4	11.5
y	1	4	7	7	7	10.6	0.3	0.2	0

19. Describe the direction, form, and strength of the data set.

x	0	5	5	3	0.5	16	4	7	4
y	7	7	12	12	8	8	6	14	9

Example 2 Use a graph to describe the direction, form, and strength of the data set. If the scatterplot is linear, then write the linear equation for the line of best fit.

20. (0, 1), (3, 27), (6, 729), (5, 243), (4, 81), (2, 9), (1, 3)

21. (3, 8.5), (1, 5.5), (5, 11.5), (12, 22), (20, 34), (17, 29.5), (2, 7)

22. (1, 6), (2, 12), (3, 24), (4, 48), (5, 96)

Example 3 Solve for x.

23. $9^x + 5 = 6566$

24. $22^x = 484$

25. $36^x + 2 = 14739$

26. $5^x = 20$

27. $e^x = 1$

28. $10^x = 1778.28$

Examples 1, 4, 5 Each equation represents a data set. Choose the best model for each.

29. $y = \frac{1}{4}x^3$

30. $y = 4x^2 + 2x - 3$

31. $y = 4x + 7$

32. $y = 24^x$

Marcy studied a poster at her dentist's office. It showed the average number of grams of sugar a person consumes per day and the number of decayed teeth that person had.

Sugar (g)	140	25	100	85	95	10	40	60	75	130	135	110
Teeth	10	1	5	6	7	0	3	5	8	9	7	6

33. What is the regression model and line?

34. Estimate approximately how many of Marcy's teeth will have some decay if she eats 55 grams of sugar a day.

Describe the correlation for each value of r.

35. $r = 0.5$

36. $r = 0.82$

37. $r = 1$

38. $r = 0$

39. How do you visually determine the direction of data when describing a scatter plot?

40. Graph the data in the table.

    a. Describe the scatterplot.

    b. Give an example of a regression model that does not accurately describe the set. Explain.

x	y
0	0
1	1
2	4
3	9
4	16

41. **REASONING** In the mid-1990s, several natural disasters plagued a large city, so its inhabitants left. The surrounding cities witnessed an influx of population growth. Starting in 1994, one city had a population of 502,000 and began to grow at a constant rate of 3.2% every year.

    a. Let x represent years since 1994. Make a table for $x = 0, 1, 2, 3,$ and 4. Round to the nearest whole number.

    b. Graph the data.

    c. Find the equation that models the given situation.

    d. What is the best model?

    e. What will the population be in 2018?

    f. In how many years will the population be 1.5 million?

42. **PRECISION** Francine is calculating bacterial growth in a culture in her science class. She put 50 bacteria on a slide under a microscope and documented the population growth of the bacteria every hour.

Hours (x)	0	1	2	3	4
Bacteria population (y)	50	100	200	400	800

   a. Graph the data.
   b. What is the relationship between the variables?
   c. What is the best model?
   d. What is the regression line?
   e. When will the bacteria exceed a million?

43. How does describing the scatter plot help you determine the best regression model?

44. **MULTI-STEP** Jewel deposits $50 into a savings account and does not make any deposits or withdrawals. The table shows the balance in the account after 12 years.

Time	Balance
0	$50
2	$55.80
4	$64.80
6	$83.09
8	$101.40
10	$123.14
12	$162.67

   a. Make a scatterplot of the data.
   b. Calculate and graph a curve of fit using an exponential regression.
   c. Write the equation of best fit.
   d. Based on the model, what will the account balance be after 25 years?
   e. Is an exponential model the best fit for the data? Explain.
   f. What does the variable r represent and why is it important?

45. Paleontologists think they discovered a new dinosaur that is half the size of a T-rex. They measured the length and diameter, in centimeters, of the arms of a miniature T-rex.

Diameter	Length
1.76	15.99
2.6	20.69
3.19	23.68
4.58	30.06
5.12	32.36
5.81	35.17
6.47	37.76
6.67	38.41
8.08	43.72
8.29	44.47

   a. Graph the data.
   b. What is the scale?
   c. What is the best regression model?
   d. Is there a strong correlation? Explain.
   e. Predict the diameter if the length is 30.57 centimeters.

H.O.T. Problems   Use Higher-Order Thinking Skills

46. **CRITIQUE ARGUMENTS** Malcolm changed the quadratic function $7^x = 343$ into the logarithmic function $\log_7 343 = x$ and entered the function into his calculator. The answer it gave was $x = 2.54$ (rounded to the nearest hundredth). Is he correct? Explain your answer.

47. **OPEN-ENDED** Explain why we have to choose the best model before making a prediction.

48. **CHALLENGE** Write a word problem that would lead to a data set for one of the regression models. Graph the data set using the regression model you chose, and find the regression line and correlation coefficient.

49. **PRECISION** For each type of regression model, explain whether you would get a more accurate result by using a graphing calculator or by sketching the graph and using a linear equation for the line of best fit.
   a. linear regression
   b. quadratic regression
   c. power regression
   d. exponential regression

50. **WRITING IN MATH** Describe the difference between the process of how to make a prediction using a graphing calculator and using a sketch.

Preparing for Assessment

51. Describe the scatter plot. **MP** 2

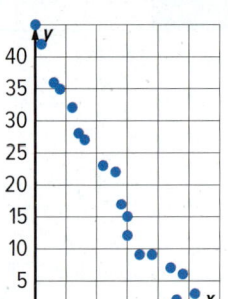

- A nonlinear, positive direction, weak correlation
- B linear, negative direction, strong correlation
- C nonlinear, negative direction, weak correlation
- D linear, positive direction, strong correlation

52. Choose the data set that represents an exponential model. **MP** 2

- A $y = 3x^2 + x - 17$
- B $y = 0.5x - 9$
- C $y = 2(1.67)^x$
- D $y = 2x^2$

53. Convert the exponential function $4(20)^x = 3{,}200{,}000$ into a logarithmic function to solve for x. What is the value of x? Round to the nearest hundredth. **MP** 6

- A 4.54
- B 5.00
- C 10.80
- D 8.64

54. Choose the best model for the data set. **MP** 2, 6

linear: $r = .945421996$

quadratic: $r = .9984103974$

exponential: $r = .994897438$

power: $r = .9917643588$

- A linear
- B quadratic
- C exponential
- D power

55. MULTI-STEP Every year the local golf club sponsors a tournament. The tournament starts with 512 participants. During each round, one half of the players are eliminated. **MP** 2, 6

a. What values go into L1?

b. What values go into L2?

c. Describe the scatter plot.

d. Choose the best model. Explain.

e. How many players remain after 5 rounds?

f. How many players remain after 8 rounds?

g. How many rounds are in the tournament?

56. The fish population in pond grows by 10% every year. At the beginning of the observation, there are 100 fish in this pond. **MP** 1

a. Write an exponential function to model the fish population in the pond.

b. How many fish will be in the pond after 2 years?

c. After how many years will the fish population in the pond reach 146? Round your answer to the nearest whole number.

CHAPTER 6
Mid-Chapter Quiz
Lessons 6-1 through 6-5

Graph each function. State the domain and range. (Lesson 6-1)

1. $f(x) = 3(4)^x$

2. $f(x) = -(2)^x + 5$

3. $f(x) = -0.5(3)^{x+2} + 4$

4. $f(x) = -3\left(\dfrac{2}{3}\right)^{x-1} + 8$

5. **SCIENCE** You are studying a bacteria population. The population originally started with 6000 bacteria cells. After 2 hours, there were 28,000 bacteria cells. (Lesson 6-1)

   a. Write an exponential function that could be used to model the number of bacteria after x hours if the number of bacteria changes at the same rate.

   b. How many bacteria cells can be expected after 4 hours?

   c. **MP** What mathematical practice did you use to solve this problem?

6. **MULTIPLE CHOICE** Which exponential function has a graph that passes through the points at (0, 125) and (3, 1000)? (Lesson 6-1)

   A $f(x) = 125(3)^x$ C $f(x) = 125(1000)^x$

   B $f(x) = 1000(3)^x$ D $f(x) = 125(2)^x$

7. **POPULATION** In 2002, a certain city had a population of 45,000. It increased to 68,000 by 2014. (Lesson 6-2)

   a. What is an exponential function that could be used to model the population of this city x years after 2002?

   b. Use your model to estimate the population in 2027.

Solve each equation or inequality. Check your solution. (Lesson 6-2)

8. $4^{3x-1} = 16^x$

9. $\dfrac{1}{9} = 243^{2x+1}$

10. $16^{2x+3} < 64$

11. $\left(\dfrac{1}{32}\right)^{x+3} \geq 16^{3x}$

Find the indicated term for each geometric sequence. (Lesson 6-3)

12. $a_2 = 8, r = 2, a_8 = ?$

13. $a_3 = 0.5, r = 8, a_{10} = ?$

14. Find the geometric means of the sequence below. (Lesson 6-3)

    0.5, _____, _____, _____, 2048

Evaluate the sum of each geometric series. (Lesson 6-3)

15. $\displaystyle\sum_{k=1}^{8} 3 \cdot 2^{k-1}$

16. $\displaystyle\sum_{k=1}^{9} 4 \cdot (-1)^{k-1}$

17. **INCOME** Peter works for a house building company for 4 months per year. He starts out making $3000 per month. At the end of each month, his salary increases by 5%. How much money will he make in those 4 months? (Lesson 6-3)

18. **MULTIPLE CHOICE** Find the value of x for $\log_3 (x^2 + 2x) = \log_3 (x + 2)$. (Lesson 6-4)

    A $x = -2, 1$ C $x = 1$

    B $x = -2$ D no solution

Graph each function. (Lesson 6-4)

19. $f(x) = 3 \log_2 (x - 1)$

20. $f(x) = -4 \log_3 (x - 2) + 5$

Evaluate each expression. (Lesson 6-4)

21. $\log_4 32$

22. $\log_5 5^{12}$

23. $\log_{16} 4$

24. Write $\log_9 729 = 3$ in exponential form. (Lesson 6-4)

25. Determine the type of function that would best model the data in the table. Then determine the regression equation. (Lesson 6-5)

x	1	2	3	4	5
y	3	9	20	45	100

LESSON 6
Properties of Logarithms

Then
You evaluated logarithmic expressions and solved logarithmic equations.

Now
1. Simplify and evaluate expressions using the properties of logarithms.
2. Solve logarithmic equations using the properties of logarithms.

Why?
The level of acidity in food is important to some consumers with sensitive stomachs. Most of the foods that we consume are more acidic than basic. The pH scale measures acidity; a low pH indicates an acidic solution, and a high pH indicates a basic solution. It is another example of a logarithmic scale based on powers of ten. Black coffee has a pH of 5, while neutral water has a pH of 7. Black coffee is one hundred times as acidic as neutral water, because $10^{7-5} = 10^2$ or 100.

Product	pH Level
lemon juice	2.1
hot sauce	3.5
tomatoes	4.2
black coffee	5.0
milk	6.4
pure water	7.0
eggs	7.8
milk of magnesia	10.0

Mathematical Practices
8 Look for and express regularity in repeated reasoning.

1. Properties of Logarithms
Because logarithms are exponents, the properties of logarithms can be derived from the properties of exponents. The Product Property of Logarithms can be derived from the Product of Powers Property of Exponents.

Key Concept Product Property of Logarithms

Words	The logarithm of a product is the sum of the logarithms of its factors.
Symbols	For all positive numbers a, b, and x, where $x \neq 1$, $\log_x ab = \log_x a + \log_x b$.
Example	$\log_2 [(5)(6)] = \log_2 5 + \log_2 6$

To show that this property is true, let $b^x = a$ and $b^y = c$. Then, using the definition of logarithm, $x = \log_b a$ and $y = \log_b c$.

$b^x b^y = ac$	Substitution
$b^{x+y} = ac$	Product of Powers
$\log_b b^{x+y} = \log_b ac$	Property of Equality for Logarithmic Functions
$x + y = \log_b ac$	Inverse Property of Exponents and Logarithms
$\log_b a + \log_b c = \log_b ac$	Replace x with $\log_b a$ and y with $\log_b c$.

You can use the Product Property of Logarithms to approximate logarithmic expressions.

Example 1 Use the Product Property

Use $\log_4 3 \approx 0.7925$ to approximate the value of $\log_4 192$.

$\log_4 192 = \log_4 (4^3 \cdot 3)$	Replace 192 with $64 \cdot 3$ or $4^3 \cdot 3$.
$= \log_4 4^3 + \log_4 3$	Product Property
$= 3 + \log_4 3$	Inverse Property of Exponents and Logarithms
$\approx 3 + 0.7925$ or 3.7925	Replace $\log_4 3$ with 0.7925.

▶ Guided Practice

1. Use $\log_4 2 = 0.5$ to approximate the value of $\log_4 32$.

Recall that the quotient of powers is found by subtracting exponents. The property for the logarithm of a quotient is similar. Let $b^x = a$ and $b^y = c$. Then $\log_b a = x$ and $\log_b c = y$.

$$\frac{b^x}{b^y} = \frac{a}{c}$$

$$b^{x-y} = \frac{a}{c} \qquad \text{Quotient Property}$$

$$\log_b b^{x-y} = \log_b \frac{a}{c} \qquad \text{Property of Equality for Logarithmic Equations}$$

$$x - y = \log_b \frac{a}{c} \qquad \text{Inverse Property of Exponents and Logarithms}$$

$$\log_b a - \log_b c = \log_b \frac{a}{c} \qquad \text{Replace } x \text{ with } \log_b a \text{ and } y \text{ with } \log_b c.$$

Key Concept Quotient Property of Logarithms

Words The logarithm of a quotient is the difference of the logarithms of the numerator and the denominator.

Symbols For all positive numbers a, b, and x, where $x \neq 1$,
$$\log_x \frac{a}{b} = \log_x a - \log_x b.$$

Example $\log_2 \frac{5}{6} = \log_2 5 - \log_2 6$

Real-World Example 2 Quotient Property

SCIENCE The pH of a substance is defined as the concentration of hydrogen ions $[H^+]$ in moles. It is given by the formula $\text{pH} = \log_{10} \frac{1}{H^+}$. Find the amount of hydrogen in a liter of acid rain that has a pH of 4.2.

Understand The formula for finding pH and the pH of the rain is given.

Plan Write the equation. Then, solve for $[H^+]$.

Solve
$\text{pH} = \log_{10} \frac{1}{H^+}$ Original equation

$4.2 = \log_{10} \frac{1}{H^+}$ Substitute 4.2 for pH.

$4.2 = \log_{10} 1 - \log_{10} H^+$ Quotient Property

$4.2 = 0 - \log_{10} H^+$ $\log_{10} 1 = 0$

$4.2 = -\log_{10} H^+$ Simplify.

$-4.2 = \log_{10} H^+$ Multiply each side by -1.

$10^{-4.2} = H^+$ Definition of logarithm

There are $10^{-4.2}$, or about 0.000063, mole of hydrogen in a liter of this rain.

Check
$4.2 = \log_{10} \frac{1}{10^{-4.2}}$ pH = 4.2, $H^+ = 10^{-4.2}$

$4.2 \stackrel{?}{=} \log_{10} 1 - \log_{10} 10^{-4.2}$ Quotient Property

$4.2 \stackrel{?}{=} 0 - (-4.2)$ Simplify.

$4.2 = 4.2$ ✓

Using the Quotient Property, the equation is simplified to two terms that can easily be evaluated.

Real-World Link

Acid rain is more acidic than normal rain. Smoke and fumes from burning fossil fuels rise into the atmosphere and combine with the moisture in the air to form acid rain. Acid rain can be responsible for the erosion of statues, as in the photo above.

Guided Practice

2. SOUND The loudness L of a sound, measured in decibels, is given by $L = 10 \log_{10} R$, where R is the sound's relative intensity. Suppose one person talks with a relative intensity of 10^6 or 60 decibels. How much louder would 100 people be, talking at the same intensity?

Recall that the power of a power is found by multiplying exponents. The property for the logarithm of a power is similar.

> **Key Concept** Power Property of Logarithms
>
> **Words** The logarithm of a power is the product of the logarithm and the exponent.
>
> **Symbols** For any real number p, and positive numbers m and b, where $b \neq 1$, $\log_b m^p = p \log_b m$.
>
> **Example** $\log_2 6^5 = 5 \log_2 6$

Study Tip

MP Tools You can check this answer by evaluating $2^{4.6438}$ on a calculator. The calculator should give a result of about 25, because $\log_2 25 \approx 4.6438$ means $2^{4.6438} \approx 25$.

Example 3 Power Property of Logarithms

Given $\log_2 5 \approx 2.3219$, approximate the value of $\log_2 25$.

$\log_2 25 = \log_2 5^2$ Replace 25 with 5^2.

$ = 2 \log_2 5$ Power Property

$ \approx 2(2.3219)$ or 4.6438 Replace $\log_2 5$ with 2.3219.

▶ **Guided Practice**

3. Given $\log_3 7 \approx 1.7712$, approximate the value of $\log_3 49$.

2 Solve Logarithmic Equations
You can use the properties of logarithms to solve equations involving logarithms.

Example 4 Solve Equations Using Properties of Logarithms

Solve $\log_6 x + \log_6 (x - 9) = 2$.

$\log_6 x + \log_6 (x - 9) = 2$ Original equation

$\log_6 x(x - 9) = 2$ Product Property

$x(x - 9) = 6^2$ Definition of logarithm

$x^2 - 9x - 36 = 0$ Subtract 36 from each side.

$(x - 12)(x + 3) = 0$ Factor.

$x - 12 = 0$ or $x + 3 = 0$ Zero Product Property

$x = 12 x = -3$ Solve each equation.

CHECK $\log_6 x + \log_6 (x - 9) = 2$ $\log_6 x + \log_6 (x - 9) = 2$

$\log_6 \mathbf{12} + \log_6 (\mathbf{12} - 9) \stackrel{?}{=} 2$ $\log_6 (\mathbf{-3}) + \log_6 (\mathbf{-3} - 9) \stackrel{?}{=} 2$

$\log_6 12 + \log_6 3 \stackrel{?}{=} 2$ $\log_6 (-3) + \log_6 (-12) \stackrel{?}{=} 2$

$\log_6 (12 \cdot 3) \stackrel{?}{=} 2$

$\log_6 36 \stackrel{?}{=} 2$ Because $\log_6 (-3)$ and $\log_6 (-12)$ are undefined, -3 is an extraneous

$2 = 2 \checkmark$ solution.

The solution is $x = 12$.

Go Online!

You will want to reference the Properties of Logarithms often. Log into your **eStudent Edition** to bookmark this lesson.

▶ **Guided Practice**

4A. $2 \log_7 x = \log_7 27 + \log_7 3$ **4B.** $\log_6 x + \log_6 (x + 5) = 2$

418 | Lesson 6-6 | Properties of Logarithms

Check Your Understanding

= Step-by-Step Solutions begin on page R11.   Go Online! for a Self-Check Quiz

Example 1 Use $\log_4 3 \approx 0.7925$ and $\log_4 5 \approx 1.1610$ to approximate the value of each expression.

1. $\log_4 18$
2. $\log_4 15$
3. $\log_4 \frac{5}{3}$
4. $\log_4 \frac{3}{4}$

Example 2

5. **MOUNTAIN CLIMBING** As elevation increases, the atmospheric air pressure decreases. The formula for pressure based on elevation is $a = 15{,}500(5 - \log_{10} P)$, where a is the altitude in meters and P is the pressure in pascals (1 psi ≈ 6900 pascals). What is the air pressure at the summit in pascals (Pa) for each mountain listed in the table at the right?

Mountain	County	Height (m)
Guadalupe Peak	Culberson	2667
Emory Peak	Brewster	2385
Anthony's Nose	El Paso	2111
Panther Peak	Hudspeth	1936
Buck Mountain	Jeff Davis	1789

Example 3 Given $\log_3 5 \approx 1.465$ and $\log_5 7 \approx 1.2091$, approximate the value of each expression.

6. $\log_3 25$
7. $\log_5 49$

Example 4 Solve each equation. Check your solutions.

8. $\log_4 48 - \log_4 n = \log_4 6$
9. $\log_3 2x + \log_3 7 = \log_3 28$
10. $3 \log_2 x = \log_2 8$
11. $\log_{10} a + \log_{10} (a - 6) = 2$

Practice and Problem Solving

Extra Practice is on page R6.

Example 1 Use $\log_4 2 = 0.5$, $\log_4 3 \approx 0.7925$, and $\log_4 5 \approx 1.1610$ to approximate the value of each expression.

12. $\log_4 30$
13. $\log_4 20$
14. $\log_4 \frac{2}{3}$
15. $\log_4 \frac{4}{3}$
16. $\log_4 9$
17. $\log_4 8$

Example 2

18. **SCIENCE** The magnitude M of an earthquake is measured on the Richter scale using the formula $M = \log 10x$, where x is the intensity of the seismic wave causing the ground motion. In 2007, an earthquake near San Francisco registered approximately 5.6 on the Richter scale. The famous San Francisco earthquake of 1906 measured 8.3 in magnitude.

    a. How many times more intense, was the 1906 earthquake than the 2007 earthquake?

    b. Richter himself classified the 1906 earthquake as having a magnitude of 8.3. More recent research indicates it was most likely a 7.9. How many times greater in intensity was Richter's measure of the earthquake?

Example 3 Given $\log_6 8 \approx 1.1606$ and $\log_7 9 \approx 1.1292$, approximate the value of each expression.

19. $\log_6 48$
20. $\log_7 81$
21. $\log_6 512$
22. $\log_7 729$

Example 4 **PERSEVERANCE** Solve each equation. Check your solutions.

23. $\log_3 56 - \log_3 n = \log_3 7$
24. $\log_2 (4x) + \log_2 5 = \log_2 40$
25. $5 \log_2 x = \log_2 32$
26. $\log_{10} a + \log_{10} (a + 21) = 2$

27. PROBABILITY In the 1930s, Dr. Frank Benford demonstrated a way to determine whether a set of numbers has been randomly chosen or manually chosen. If the sets of numbers were not randomly chosen, then the Benford formula, $P = \log_{10}\left(1 + \frac{1}{d}\right)$, predicts the probability of a digit d being the first digit of the set. For example, there is a 4.6% probability that the first digit is 9.

    a. Rewrite the formula to solve for the digit if given the probability.

    b. Find the digit that has a 9.7% probability of being selected.

    c. Find the probability that the first digit is 1 ($\log_{10} 2 \approx 0.30103$).

Use $\log_5 3 \approx 0.6826$ and $\log_5 4 \approx 0.8614$ to approximate the value of each expression.

28. $\log_5 40$ **29.** $\log_5 30$

30. $\log_5 \frac{3}{4}$ **31.** $\log_5 \frac{4}{3}$

32. $\log_5 9$ **33.** $\log_5 16$

34. $\log_5 12$ **35.** $\log_5 27$

Solve each equation. Check your solutions.

36. $\log_3 6 + \log_3 x = \log_3 12$ **37.** $\log_4 a + \log_4 8 = \log_4 24$

38. $\log_{10} 18 - \log_{10} 3x = \log_{10} 2$ **39.** $\log_7 100 - \log_7 (y + 5) = \log_7 10$

40. $\log_2 n = \frac{1}{3} \log_2 27 + \log_2 36$ **41.** $3 \log_{10} 8 - \frac{1}{2} \log_{10} 36 = \log_{10} x$

Solve for n.

42. $\log_a 6n - 3 \log_a x = \log_a x$ **43.** $2 \log_b 16 + 6 \log_b n = \log_b (x - 2)$

Solve each equation. Check your solutions.

44. $\log_{10} z + \log_{10} (z + 9) = 1$ **45.** $\log_3 (a^2 + 3) + \log_3 3 = 3$

46. $\log_2 (15b - 15) - \log_2 (-b^2 + 1) = 1$ **47.** $\log_4 (2y + 2) - \log_4 (y - 2) = 1$

48. $\log_6 0.1 + 2 \log_6 x = \log_6 2 + \log_6 5$ **49.** $\log_7 64 - \log_7 \frac{8}{3} + \log_7 2 = \log_7 4p$

50. REASONING Suppose there are 5000 humpback whales in existence today, and the population decreases at a rate of 4% per year.

    a. Write a logarithmic function for the time in years based upon population.

    b. After how many years will the population drop below 1000? Is this reasonable?

State whether each identity is *true* or *false*.

51. $\log_8 (x - 3) = \log_8 x - \log_8 3$ **52.** $\log_5 22x = \log_5 22 + \log_5 x$

53. $\log_{10} 19k = 19 \log_{10} k$ **54.** $\log_2 y^5 = 5 \log_2 y$

55. $\log_7 \frac{x}{3} = \log_7 x - \log_7 3$ **56.** $\log_4 (z + 2) = \log_4 z + \log_4 2$

57. $\log_8 p^4 = (\log_8 p)^4$ **58.** $\log_9 \frac{x^2 y^3}{z^4} = 2 \log_9 x + 3 \log_9 y - 4 \log_9 z$

59. MULTI-STEP Teresa's retirement account balance is currently $320,000 and it is increasing about 15% per year from deposits and interest. When she retires, she plans to invest this money in a CD with 5% annual interest. Each year, she will only spend the interest and reinvest the principal in a new CD. She also expects to earn $1050 per month from Social Security when she retires. Her goal is to have enough saved in order to earn $50,000 per year before taxes.

    a. In how many years will she be able to retire?

    b. Describe and evaluate your solution process.

    c. What assumptions did you make?

60. **FINANCIAL LITERACY** The average American carries a credit card debt of approximately $8600 with an annual percentage rate (APR) of 18.3%. The formula $m = \dfrac{b\left(\dfrac{r}{n}\right)}{1 - \left(1 + \dfrac{r}{n}\right)^{-nt}}$ can be used to compute the monthly payment m that is necessary to pay off a credit card balance b in a given number of years t, where r is the annual percentage rate and n is the number of payments per year.

   a. What monthly payment should be made in order to pay off the debt in exactly three years? What is the total amount paid?

   b. The equation $t = \dfrac{\log\left(1 - \dfrac{br}{mn}\right)}{-n \log\left(1 + \dfrac{r}{n}\right)}$ can be used to calculate the number of years necessary for a given payment schedule. Copy and complete the table.

Payment (m)	Years (t)
$50	
$100	
$150	
$200	
$250	
$300	

   c. Graph the information in the table from part **b**.

   d. If you could only afford to pay $100 a month, will you be able to pay off the debt? If so, how long will it take? If not, why not?

   e. What is the minimum monthly payment that will work toward paying off the debt?

H.O.T. Problems   Use Higher-Order Thinking Skills

61. **OPEN-ENDED** Write a logarithmic expression for each condition. Then write the expanded expression.

    a. a product and a quotient
    b. a product and a power
    c. a product, a quotient, and a power

62. **MP CONSTRUCT ARGUMENTS** Use the properties of exponents to prove the Power Property of Logarithms.

63. **WRITING IN MATH** Explain why the following are true.

    a. $\log_b 1 = 0$ b. $\log_b b = 1$ c. $\log_b b^x = x$

64. **CHALLENGE** Simplify $\log_{\sqrt{a}}(a^2)$ to find an exact numerical value.

65. **WHICH ONE DOESN'T BELONG?** Find the expression that does not belong. Explain.

    $\log_b 24 = \log_b 2 + \log_b 12$

    $\log_b 24 = \log_b 20 + \log_b 4$

    $\log_b 24 = \log_b 8 + \log_b 3$

    $\log_b 24 = \log_b 4 + \log_b 6$

66. **MP REASONING** Use the properties of logarithms to prove that $\log_a \dfrac{1}{x} = -\log_a x$.

67. **CHALLENGE** Simplify $x^{3 \log_x 2 - \log_x 5}$ to find an exact numerical value.

68. **WRITING IN MATH** Explain how the properties of exponents and logarithms are related. Include examples like the one shown at the beginning of the lesson illustrating the Product Property, but with the Quotient Property and Power Property of Logarithms.

Preparing for Assessment

69. What is the solution set in $\log_4 x + \log_4 (x - 12) = 3$? MP 8

- A $\{-4, 16\}$
- B $\{-4\}$
- C $\{16\}$
- D $\emptyset$

70. What is the value of x in $\log_4 (3x + 2) - 2 = 7$? MP 8

- A $800\frac{1}{3}$
- B 805
- C $5461\frac{1}{3}$
- D $87{,}380\frac{2}{3}$

71. Let $\log_q m = p$ and $\log_q n = r$. What is the value of $\log_q \dfrac{m^3}{n}$? MP 8

- A $\dfrac{3p}{r}$
- B $\dfrac{3p}{r}$
- C $3p - r$
- D $3 + p - r$

72. What is the solution set of $\log_3 (9x^2) + 1 = \log_3 (4)$? MP 8

- A $\left\{-\dfrac{2\sqrt{3}}{9}, \dfrac{2\sqrt{3}}{9}\right\}$
- B $\left\{-\dfrac{1}{3}, \dfrac{1}{3}\right\}$
- C $\left\{\dfrac{2\sqrt{3}}{9}\right\}$
- D $\emptyset$

73. If $\log_2 y = 0.451$, what is the value of $\log_2 16y^2$? MP 1

74. What is the extraneous solution to $\log_2 x + \log_2 (x - 2) = 3$? MP 8

- A $\{4, -2\}$
- B $\{-2\}$
- C $\{4\}$
- D $\emptyset$

75. **MULTI-STEP** Charles Richter defined the magnitude of an earthquake to be the equation

$$M = \log_{10}\left(\frac{I}{S}\right)$$

where I is the intensity of the earthquake (measured by the amplitude in centimeters) and S is the intensity of a standard earthquake (which is a value of 10^{-4} cm). MP 8

a. What is the replacement set for I?

b. What are the possible values of M?

c. If one earthquake is 20,000 times as intense as a standard earthquake, then the value of $\dfrac{I}{S} = 20{,}000$. What is the magnitude of the earthquake?

d. If one earthquake is 200,000 times as intense as a standard earthquake, then what is its magnitude?

e. If one earthquake is 2,000,000 times as intense as a standard earthquake, then what is its magnitude?

f. Explain the pattern from the responses in **c**, **d**, and **e**.

76. The pH of an aqueous solution is given by the formula, pH $= -\log [H^+]$, where H^+ is the hydrogen ion concentration. MP 8

a. What is the pH of an aqueous solution in which $H^+ = 2.7 \times 10^{-3}$ M?

b. If the pH of an aqueous solution is 6.52, what is the hydrogen ion concentration?

LESSON 7
Common Logarithms

:·Then
- You simplified expressions and solved equations using properties of logarithms.

:·Now
1. Solve exponential equations and inequalities using common logarithms.
2. Evaluate logarithmic expressions using the Change of Base Formula.

:·Why?
- Seismologists use the Richter scale to measure the strength or magnitude of earthquakes. The magnitude of an earthquake is determined using the logarithm of the amplitude of waves recorded by seismographs.

  The logarithmic scale used by the Richter scale is based on the powers of 10. For example, a magnitude 6.4 earthquake can be represented by $6.4 = \log_{10} x$.

Richter Number	Intensity
1	10^1 micro
2	10^2 minor
3	10^3 minor
4	10^4 light
5	10^5 moderate
6	10^6 strong
7	10^7 major
8	10^8 great

New Vocabulary
common logarithm
Change of Base Formula

Mathematical Practices
4 Model with mathematics.

1 Common Logarithms You have seen that the base 10 logarithm function, $y = \log_{10} x$, is used in many applications. Base 10 logarithms are called **common logarithms**. Common logarithms are usually written without the subscript 10.

$$\log_{10} x = \log x, \quad x > 0$$

Most scientific calculators have a [LOG] key for evaluating common logarithms. The graph of the common log is shown.

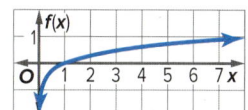

Example 1 Find Common Logarithms

Use a calculator to evaluate each expression to the nearest ten-thousandth.

a. log 5
KEYSTROKES: [LOG] 5 [ENTER]
.6989700043   log 5 ≈ 0.6990

b. log 0.3
KEYSTROKES: [LOG] 0.3 [ENTER]
−.5228787453   log 0.3 ≈ −0.5229

▶ **Guided Practice**

1A. log 7

1B. log 0.5

The common logarithms of numbers that differ by integral powers of ten are closely related. Remember that a logarithm is an exponent. For example, in the equation $y = \log x$, y is the power to which 10 is raised to obtain the value of x.

$\log x = y$ → means → $10^y = x$
$\log 1 = 0$ because $10^0 = 1$
$\log 10 = 1$ because $10^1 = 10$
$\log 10^m = m$ because $10^m = 10^m$

connectED.mcgraw-hill.com   423

Common logarithms are used in the measurement of sound. Soft recorded music is about 36 decibels (dB).

Real-World Example 2 Solve Logarithmic Equations

ROCK CONCERT The loudness L, in decibels, of a sound is $L = 10 \log \frac{I}{m}$, where I is the intensity of the sound and m is the minimum intensity of sound detectable by the human ear. Residents living several miles from a concert venue can hear the music at an intensity of 66.6 decibels. How many times the minimum intensity of sound detectable by the human ear was this sound, if m is defined to be 1?

$L = 10 \log \frac{I}{m}$ Original equation

$66.6 = 10 \log \frac{I}{1}$ Replace L with 66.6 and m with 1.

$6.66 = \log I$ Divide each side by 10 and simplify.

$I = 10^{6.66}$ Exponential form

$I \approx 4{,}570{,}882$ Use a calculator.

The sound heard by the residents was approximately 4,570,000 times the minimum intensity of sound detectable by the human ear.

Guided Practice

2. EARTHQUAKES The amount of energy E in ergs that an earthquake releases is related to its Richter scale magnitude M by the equation $\log E = 11.8 + 1.5M$. Use the equation to find the amount of energy released by the 2004 Sumatran earthquake, which measured 9.0 on the Richter scale and led to a tsunami.

Real-World Link

Acoustical Engineer
Acoustical engineers are concerned with reducing unwanted sounds, noise control, and making useful sounds. Examples of useful sounds are ultrasound, sonar, and sound reproduction. Employment in this field requires a minimum of a bachelor's degree.

If both sides of an exponential equation cannot easily be written as powers of the same base, you can solve by taking the logarithm of each side.

Example 3 Solve Exponential Equations Using Logarithms

Solve $4^x = 19$. Round to the nearest ten-thousandth.

$4^x = 19$ Original equation

$\log 4^x = \log 19$ Property of Equality for Logarithmic Functions

$x \log 4 = \log 19$ Power Property of Logarithms

$x = \dfrac{\log 19}{\log 4}$ Divide each side by log 4.

$x \approx 2.1240$ Use a calculator.

The solution is approximately 2.1240.

CHECK You can check this answer graphically by using a graphing calculator. Graph the line $y = 4^x$ and the line $y = 19$. Then use the **CALC** menu to find the intersection of the two graphs. The intersection is very close to the answer that was obtained algebraically. ✓

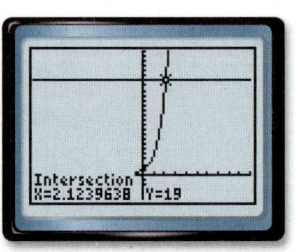

[−10, 10] scl: 1 by [−5, 25] scl: 1

Guided Practice

3A. $3^x = 15$ **3B.** $6^x = 42$

The same strategies that are used to solve exponential equations can be used to solve exponential inequalities.

> **Study Tip**
>
> **MP Reasoning** Remember that the direction of an inequality must be switched if each side is multiplied or divided by a negative number. Because $5 \log 3 - \log 7 > 0$, the inequality does not change.

Example 4 Solve Exponential Inequalities Using Logarithms

Solve $3^{5y} < 7^{y-2}$. Round to the nearest ten-thousandth.

$3^{5y} < 7^{y-2}$	Original inequality
$\log 3^{5y} < \log 7^{y-2}$	Property of Inequality for Logarithmic Functions
$5y \log 3 < (y-2) \log 7$	Power Property of Logarithms
$5y \log 3 < y \log 7 - 2 \log 7$	Distributive Property
$5y \log 3 - y \log 7 < -2 \log 7$	Subtract $y \log 7$ from each side.
$y(5 \log 3 - \log 7) < -2 \log 7$	Distributive Property
$y < \dfrac{-2 \log 7}{5 \log 3 - \log 7}$	Divide each side by $5 \log 3 - \log 7$.
$\{y \mid y < -1.0972\}$	Use a calculator.

CHECK Test $y = -2$.

$3^{5y} < 7^{y-2}$ Original inequality

$3^{5(-2)} \stackrel{?}{<} 7^{(-2)-2}$ Replace y with -2.

$3^{-10} \stackrel{?}{<} 7^{-4}$ Simplify.

$\dfrac{1}{59,049} < \dfrac{1}{2401}$ ✓ Negative Exponent Property

▶ **Guided Practice**

Solve each inequality. Round to the nearest ten-thousandth.

4A. $3^{2x} \geq 6^{x+1}$ **4B.** $4^y < 5^{2y+1}$

2 Change of Base Formula

The **Change of Base Formula** allows you to write equivalent logarithmic expressions that have different bases.

Key Concept Change of Base Formula

Symbols	For all positive numbers a, b, and n, where $a \neq 1$ and $b \neq 1$, $\log_a n = \dfrac{\log_b n}{\log_b a}$. ← log base b of original number ← log base b of old base
Example	$\log_3 11 = \dfrac{\log_{10} 11}{\log_{10} 3}$

> **Math History Link**
>
> **John Napier** (1550–1617)
> John Napier was a Scottish mathematician and theologian who began the use of logarithms to aid in calculations. He is also known for popularizing the use of the decimal point.

To prove this formula, let $\log_a n = x$.

$a^x = n$	Definition of logarithm
$\log_b a^x = \log_b n$	Property of Equality for Logarithmic Functions
$x \log_b a = \log_b n$	Power Property of Logarithms
$x = \dfrac{\log_b n}{\log_b a}$	Divide each side by $\log_b a$.
$\log_a n = \dfrac{\log_b n}{\log_b a}$	Replace x with $\log_a n$.

The Change of Base Formula makes it possible to evaluate a logarithmic expression of any base by translating the expression into one that involves common logarithms.

Example 5 Change of Base Formula

Express $\log_3 20$ in terms of common logarithms. Then round to the nearest ten-thousandth.

$\log_3 20 = \dfrac{\log_{10} 20}{\log_{10} 3}$ Change of Base Formula

$ \approx 2.7268$ Use a calculator.

> **Guided Practice**
>
> 5. Express $\log_6 8$ in terms of common logarithms. Then round to the nearest ten-thousandth.

Check Your Understanding

 = Step-by-Step Solutions begin on page R11.

Example 1 Use a calculator to evaluate each expression to the nearest ten-thousandth.

1. $\log 5$
2. $\log 21$
3. $\log 0.4$
4. $\log 0.7$

Example 2

5. **SCIENCE** The amount of energy E in ergs that an earthquake releases is related to its Richter scale magnitude M by the equation $\log E = 11.8 + 1.5M$. Use the equation to find the amount of energy released by the 1960 Chilean earthquake, which measured 8.5 on the Richter scale.

Example 3 Solve each equation. Round to the nearest ten-thousandth.

6. $6^x = 40$
7. $2.1^{a+2} = 8.25$
8. $7^{x^2} = 20.42$
9. $11^{b-3} = 5^b$

Example 4 Solve each inequality. Round to the nearest ten-thousandth.

10. $5^{4n} > 33$
11. $6^{p-1} \leq 4^p$

Example 5 Express each logarithm in terms of common logarithms. Then approximate its value to the nearest ten-thousandth.

12. $\log_3 7$
13. $\log_4 23$
14. $\log_9 13$
15. $\log_2 5$

Practice and Problem Solving

Extra Practice is on page R6.

Example 1 Use a calculator to evaluate each expression to the nearest ten-thousandth.

16. $\log 3$
17. $\log 11$
18. $\log 3.2$
19. $\log 8.2$
20. $\log 0.9$
21. $\log 0.04$

Example 2

22. **SENSE-MAKING** Loretta had a new muffler installed on her car. The noise level of the engine dropped from 85 decibels to 73 decibels.

    a. How many times the minimum intensity of sound detectable by the human ear was the car with the old muffler, if m is defined to be 1?

    b. How many times the minimum intensity of sound detectable by the human ear is the car with the new muffler? Find the percent of decrease of the intensity of the sound with the new muffler.

Example 3 Solve each equation. Round to the nearest ten-thousandth.

23. $8^x = 40$
24. $5^x = 55$
25. $2.9^{a-4} = 8.1$
26. $9^{b-1} = 7^b$
27. $13^{x^2} = 33.3$
28. $15^{x^2} = 110$

Example 4 Solve each inequality. Round to the nearest ten-thousandth.

29. $6^{3n} > 36$
30. $2^{4x} \leq 20$
31. $3^{y-1} \leq 4^y$
32. $5^{p-2} \geq 2^p$

Example 5 Express each logarithm in terms of common logarithms. Then approximate its value to the nearest ten-thousandth.

33. $\log_7 18$
34. $\log_5 31$
35. $\log_2 16$
36. $\log_4 9$
37. $\log_3 11$
38. $\log_6 33$

39. **PETS** The number n of pet owners in thousands after t years can be modeled by $n = 35[\log_4 (t + 2)]$. Let $t = 0$ represent 2000. Use the Change of Base Formula to answer the following questions.

   a. How many pet owners were there in 2010?

   b. In what year are there 80,000 pet owners? Does your answer seem reasonable?

40. **MP PRECISION** Five years ago the grizzly bear population in a certain national park was 325. Today it is 450. Studies show that the park can support a population of 750.

   a. What is the average annual rate of growth in the population if the grizzly bears reproduce once a year?

   b. How many more years will it take to reach the maximum population if the population growth continues at the same average rate? Does your answer seem reasonable?

Solve each equation or inequality. Round to the nearest ten-thousandth.

41. $3^x = 40$
42. $5^{3p} = 15$
43. $4^{n+2} = 14.5$
44. $8^{z-4} = 6.3$
45. $7.4^{n-3} = 32.5$
46. $3.1^{y-5} = 9.2$
47. $5^x \geq 42$
48. $9^{2a} < 120$
49. $3^{4x} \leq 72$
50. $7^{2n} > 52^{4n+3}$
51. $6^p \leq 13^{5-p}$
52. $2^{y+3} \geq 8^{3y}$

Express each logarithm in terms of common logarithms. Then approximate its value to the nearest ten-thousandth.

53. $\log_4 12$
54. $\log_3 21$
55. $\log_5 (2.7)^2$
56. $\log_7 \sqrt{5}$

57. **MUSIC** A musical cent is a unit in a logarithmic scale of relative pitch or intervals. One octave is equal to 1200 cents. The formula $n = 1200\left(\log_2 \frac{a}{b}\right)$ can be used to determine the difference in cents between two notes with frequencies a and b. Find the interval in cents when the frequency changes from 443 Hertz (Hz) to 415 Hz.

GRAPHING CALCULATOR Graph $f(x) = \log x$ and the transformation graph, $g(x)$. Determine the effects on each of the key attributes of the graph of $f(x)$.

58. $g(x) = 2f(x)$
59. $g(x) = -4f(x)$
60. $g(x) = f(x) + 3$
61. $g(x) = f(x) - 5$
62. $g(x) = f(x + 2)$
63. $g(x) = f(x - 6)$

Solve each equation. Round to the nearest ten-thousandth.

64. $10^{x^2} = 60$

65. $4^{x^2-3} = 16$

66. $9^{6y-2} = 3^{3y+1}$

67. $8^{2x-4} = 4^{x+1}$

68. $16^x = \sqrt{4^{x+3}}$

69. $2^y = \sqrt{3^{y-1}}$

70. ENVIRONMENTAL SCIENCE An environmental engineer is testing drinking water wells in coastal communities for pollution, specifically unsafe levels of arsenic. The safe standard for arsenic is 0.025 parts per million (ppm). Also, the pH of the arsenic level should be less than 9.5. The formula for hydrogen ion concentration is pH = $-\log H$. (*Hint*: 1 kilogram of water occupies approximately 1 liter. 1 ppm = 1 mg/kg.)

  a. Suppose the hydrogen ion concentration of a well is 1.25×10^{-11}. Should the environmental engineer be worried about too high an arsenic content?

  b. The environmental engineer finds 1 milligram of arsenic in a 3-liter sample, is the well safe?

  c. What is the hydrogen ion concentration that meets the troublesome pH level of 9.5?

71. MULTIPLE REPRESENTATIONS In this problem, you will solve the exponential equation $4^x = 13$.

  a. Tabular Enter the function $y = 4^x$ into a graphing calculator, create a table of values for the function, and scroll through the table to find x when $y = 13$.

  b. Graphical Graph $y = 4^x$ and $y = 13$ on the same screen. Use the **intersect** feature to find the point of intersection.

  c. Numerical Solve the equation algebraically. Do all of the methods produce the same result? Explain why or why not.

H.O.T. Problems   Use Higher-Order Thinking Skills

72. CRITIQUE ARGUMENTS Sam and Rosamaria are solving $4^{3p} = 10$. Is either of them correct? Explain your reasoning.

Sam
$4^{3p} = 10$
$\log 4^{3p} = \log 10$
$p \log 4 = \log 10$
$p = \dfrac{\log 10}{\log 4}$

Rosamaria
$4^{3p} = 10$
$\log 4^{3p} = \log 10$
$3p \log 4 = \log 10$
$p = \dfrac{\log 10}{3 \log 4}$

73. CHALLENGE Solve $\log_{\sqrt{a}} 3 = \log_a x$ for x and explain each step.

74. REASONING Write $\dfrac{\log_5 9}{\log_5 3}$ as a single logarithm.

75. PROOF Find the values of $\log_3 27$ and $\log_{27} 3$. Make and prove a conjecture about the relationship between $\log_a b$ and $\log_b a$.

76. WRITING IN MATH Explain how exponents and logarithms are related. Include examples like how to solve a logarithmic equation using exponents and how to solve an exponential equation using logarithms.

Preparing for Assessment

77. Let $p^{2x} = q^{16x}$ for all nonzero values of x. What is the value of p in terms of q? **MP 4**

- A q^4
- B q^8
- C q^{14}
- D q^{32}

78. Let $\log_k m = 3.6$ and $\log_k n = 0.9$. What is the value of $\log_m n$? **MP 4**

- A 0.25
- B 2.7
- C 3.24
- D 4

79. What is the value of x if $5^{2-x} = 20$? **MP 4**

- A -1.560
- B 0.139
- C 0.500
- D 0.602

80. What is the solution set of $6^{4-x} > 11$? **MP 4**

- A $\{x \mid x < -2.6617\}$
- B $\{x \mid x > -2.6617\}$
- C $\{x \mid x < 2.6617\}$
- D $\{x \mid x > 2.6617\}$

81. If $5^{x+2} = 9$, what is the value of x? **MP 4**

- A -1.26751
- B -0.63479
- C 0.63479
- D 1.26751

82. If $7^{2x+2} = 9^x$, what is the value of x to the nearest thousandth? **MP 4**

83. MULTI-STEP The population of a town $f(t)$ in thousands after t years can be represented by the function $f(t) = 5[\log_6(t + 5)]$. Let $t = 0$ represent 2010. **MP 4**

a. What was the population of the town in the year 2010?

b. What was the population of the town in the year 2025?

c. In what year does the town have 12,000 people? Does this year seem reasonable? Explain.

d. What is the domain of the function? Explain your reasoning.

e. What is the range of the function? Explain your reasoning.

f. What was the population of the town in the year 2007?

84. Solve each equation or inequality. Round to the nearest ten-thousandths. **MP 1**

a. $2^x = 30$

b. $4^{\frac{x}{2}} > 7$

c. $\dfrac{5}{12^x} = 100$

d. $\dfrac{2^x}{3} = \dfrac{5^x}{4}$

LESSON 8
Natural Logarithms

∵Then
- You worked with common logarithms.

∵Now
1. Evaluate expressions involving the natural base and natural logarithm.
2. Solve exponential equations and inequalities using natural logarithms.

∵Why?
The St. Louis Gateway Arch in Missouri is in the form of an inverted catenary curve. A catenary curve directs the force of its weight along itself, so that:
- if a rope or chain is hanging, it is pulled into that shape, and,
- if a catenary is standing upright, it can support itself.

The equation for the catenary curve involves e, a special number that appears throughout mathematics and science.

New Vocabulary
natural base, e
natural base exponential function
natural logarithm

Mathematical Practices
4 Model with mathematics.

1 Base e and Natural Logarithms
Like π and $\sqrt{2}$, the number e is an irrational number. The value of e is 2.71828... . It is referred to as the **natural base, e**. An exponential function with base e is called a **natural base exponential function**.

Key Concept Natural Base Functions

The function $f(x) = e^x$ is used to model continuous exponential growth.
The function $f(x) = e^{-x}$ is used to model continuous exponential decay.

The inverse of a natural base exponential function is called the **natural logarithm**. This logarithm can be written as $\log_e x$, but is more often abbreviated as $\ln x$.

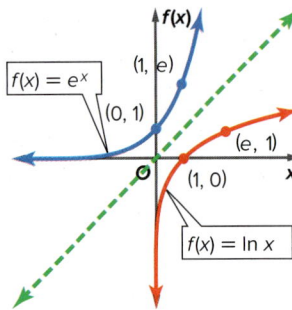 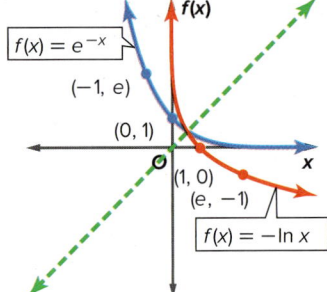

Exponential Growth   Exponential Decay

You can write an equivalent base e exponential equation for a natural logarithmic equation by using the fact that $\ln x = \log_e x$.

$$\ln 4 = x \quad \rightarrow \quad \log_e 4 = x \quad \rightarrow \quad e^x = 4$$

Example 1 Write Equivalent Expressions

Write each exponential equation in logarithmic form.

a. $e^x = 8$
$e^x = 8 \rightarrow \log_e 8 = x$
$\ln 8 = x$

b. $e^5 = x$
$e^5 = x \rightarrow \log_e x = 5$
$\ln x = 5$

▶ **Guided Practice**
1A. $e^x = 9$ **1B.** $e^7 = x$

430 | Lesson 6-8

You can also write an equivalent natural logarithm equation for a natural base e exponential equation.

$$e^x = 12 \quad \rightarrow \quad \log_e 12 = x \quad \rightarrow \quad \ln 12 = x$$

Example 2 Write Equivalent Expressions

Write each logarithmic equation in exponential form.

a. $\ln x \approx 0.7741$

$\ln x \approx 0.7741 \quad \rightarrow \quad \log_e x = 0.7741$
$$x \approx e^{0.7741}$$

b. $\ln 10 = x$

$\ln 10 = x \quad \rightarrow \quad \log_e 10 = x$
$$10 = e^x$$

▶ **Guided Practice**

2A. $\ln x \approx 2.1438$ **2B.** $\ln 18 = x$

The properties of logarithms you learned in Lesson 6-6 also apply to the natural logarithms. The logarithmic expressions below can be simplified into a single logarithmic term.

Study Tip

Simplifying When you simplify logarithmic expressions, verify that the logarithm contains no operations and no powers.

Example 3 Simplify Expressions with e and the Natural Log

Write each expression as a single logarithm.

a. $3 \ln 10 - \ln 8$

$\begin{aligned}
3 \ln 10 - \ln 8 &= \ln 10^3 - \ln 8 &&\text{Power Property of Logarithms} \\
&= \ln \frac{10^3}{8} &&\text{Quotient Property of Logarithms} \\
&= \ln 125 &&\text{Simplify.} \\
&= \ln 5^3 &&\text{$5^3 = 125$} \\
&= 3 \ln 5 &&\text{Power Property of Logarithms}
\end{aligned}$

CHECK Use a calculator to verify the solution.

KEYSTROKES: 3 [LN] 10) − [LN] 8) [ENTER] 4.828313737

KEYSTROKES: 3 [LN] 5) [ENTER] 4.828313737 ✓

b. $\ln 40 + 2 \ln \frac{1}{2} + \ln x$

$\begin{aligned}
\ln 40 + 2 \ln \tfrac{1}{2} + \ln x &= \ln 40 + \ln \tfrac{1}{4} + \ln x &&\text{Power Property of Logarithms} \\
&= \ln \left(40 \cdot \tfrac{1}{4} \cdot x\right) &&\text{Product Property of Logarithms} \\
&= \ln 10x &&\text{Simplify.}
\end{aligned}$

▶ **Guided Practice**

3A. $6 \ln 8 - 2 \ln 4$ **3B.** $2 \ln 5 + 4 \ln 2 + \ln 5y$

Because the natural base and natural log are inverse functions, they can be used to *undo* or eliminate each other.

$$e^{\ln x} = x \qquad\qquad \ln e^x = x$$

2 Equations and Inequalities with *e* and ln
Equations and inequalities involving base e are easier to solve by using natural logarithms rather than by using common logarithms, because $\ln e = 1$.

Example 4 Solve Base *e* Equations

Solve $4e^{-2x} - 5 = 3$. Round to the nearest ten-thousandth.

$4e^{-2x} - 5 = 3$	Original equation
$4e^{-2x} = 8$	Add 5 to each side.
$e^{-2x} = 2$	Divide each side by 4.
$\ln e^{-2x} = \ln 2$	Property of Equality for Logarithms
$-2x = \ln 2$	$\ln e^x = x$
$x = \dfrac{\ln 2}{-2}$	Divide each side by -2.
$x \approx -0.3466$	Use a calculator.

KEYSTROKES: LN 2) ÷ −2 ENTER −.34657359

Study Tip

MP Tools Most calculators have an e^x and LN key for evaluating natural base and natural log expressions.

Guided Practice

Solve each equation. Round to the nearest ten-thousandth.

4A. $3e^{4x} - 12 = 15$ **4B.** $4e^{-x} + 8 = 17$

Just like the natural logarithm can be used to eliminate e^x, the natural base exponential function can eliminate $\ln x$.

Example 5 Solve Natural Log Equations and Inequalities

Solve each equation or inequality. Round to the nearest ten-thousandth.

a. $3 \ln 4x = 24$

$3 \ln 4x = 24$	Original equation
$\ln 4x = 8$	Divide each side by 3.
$e^{\ln 4x} = e^8$	Property of Equality for Exponential Functions
$4x = e^8$	$e^{\ln x} = x$
$x = \dfrac{e^8}{4}$	Divide each side by 4.
$x \approx 745.2395$	Use a calculator.

b. $\ln (x - 8)^4 < 4$

$\ln (x - 8)^4 < 4$	Original equation
$e^{\ln (x-8)^4} < e^4$	Write each side using exponents and base e.
$(x - 8)^4 < e^4$	$e^{\ln x} = x$
$x - 8 < e$	Property of Equality for Exponential Functions
$x < e + 8$	Add 8 to each side.
$x < 10.7183$	Use a calculator.

Go Online!

Natural logarithms are widely used in problems arising in everyday life, society, and the workplace. Got a question about logarithms? Send a message to your teacher in ConnectED.

Guided Practice

Solve each equation or inequality. Round to the nearest ten-thousandth.

5A. $5 \ln 6x = 8$ **5B.** $\ln (2x - 3)^3 > 6$

Interest that is compounded continuously can be found using e or the natural logarithm.

> **Key Concept** Continuously Compounded Interest
>
> The formula for continuously compounded interest can be presented in exponential or logarithmic form, where A is the amount in the account after t years, P is the principal amount invested, and r is the annual interest rate.
>
> **Exponential** $A = Pe^{rt}$
>
> **Logarithmic** $\ln \frac{A}{P} = rt$

Real-World Example 6 Solve Base e Equations and Inequalities

FINANCIAL LITERACY When Angelina was born, her grandparents deposited $3000 into a college savings account paying 4% interest compounded continuously.

Real-World Link
The average cost of tuition, room, and board at four-year public colleges in Texas is about $16,232 per year.

Source: College for Texans

a. Assuming there are no deposits or withdrawals from the account, what will the balance be after 10 years?

$A = Pe^{rt}$ Continuous Compounding Exponential Formula
$= 3000e^{(0.04)(10)}$ $P = 3000, r = 0.04,$ and $t = 10$
$= 3000e^{0.4}$ Simplify.
≈ 4475.47 Use a calculator.

The balance will be $4475.47.

b. How long will it take the balance to reach at least $10,000?

$\ln \frac{A}{P} < rt$ Continuous Compounding Logarithmic Formula

$\ln \frac{10,000}{3000} < 0.04t$ $P = 3000, r = 0.04,$ and $A = 10,000$

$\ln \frac{10}{3} < 0.04t$ Simplify fraction.

$\frac{\ln \frac{10}{3}}{0.04} < t$ Divide each side by 0.04.

$30.099 < t$ Use a calculator.

It will take about 30 years to reach at least $10,000.

c. If her grandparents want Angelina to have $10,000 after 18 years, how much would they need to invest?

$10,000 = Pe^{(0.04)18}$ $A = 10,000, r = 0.04,$ and $t = 18$

$\frac{10,000}{e^{0.72}} = P$ Divide each side by $e^{0.72}$.

$4867.52 \approx P$ Use a calculator.

They need to invest $4867.52.

Study Tip
Rounding In order to avoid any errors due to rounding, do not round until the very end of your calculations.

Guided Practice

6. Use the information in Example 6 to answer the following.

a. If they invested $8000 at 3.75% interest compounded continuously, how much money would be in the account in 30 years?

b. If they invested $10,000 at 6% interest compounded continuously, then how long would it take the balance to reach at least $30,000?

c. If Angelina's grandparents found an account that paid 5% compounded continuously and wanted her to have $30,000 after 18 years, how much would they need to deposit?

Check Your Understanding

Examples 1–2 Write an equivalent exponential or logarithmic function.

1. $e^x = 30$
2. $\ln x = 42$
3. $e^3 = x$
4. $\ln 18 = x$

Example 3 Write each as a single logarithm.

5. $3 \ln 2 + 2 \ln 4$
6. $5 \ln 3 - 2 \ln 9$
7. $3 \ln 6 + 2 \ln 9$

Example 4 Solve each equation. Round to the nearest ten-thousandth.

8. $-3e^x + 9 = 4$
9. $3e^{-3x} + 4 = 6$
10. $2e^{-x} - 3 = 8$

Example 5 Solve each equation or inequality. Round to the nearest ten-thousandth.

11. $\ln 3x = 8$
12. $-4 \ln 2x = -26$
13. $\ln (x + 5)^2 < 6$
14. $\ln (x - 2)^3 > 15$
15. $e^x > 29$
16. $5 + e^{-x} > 14$

Example 6

17. **TECHNOLOGY** A virus is spreading through a computer network according to the formula $v(t) = 30e^{0.1t}$, where v is the number of computers infected and t is the time in minutes.
    a. How many computers will be infected after 1.5 hours?
    b. How long will it take the virus to infect 10,000 computers?
    c. How many computers would need to be infected initially for the virus to spread to 1 million computers in 2 hours?

Practice and Problem Solving

Extra Practice is on page R6.

Examples 1–2 Write an equivalent exponential or logarithmic function.

18. $e^{-x} = 8$
19. $e^{-5x} = 0.1$
20. $\ln 0.25 = x$
21. $\ln 5.4 = x$
22. $e^{x-3} = 2$
23. $\ln (x + 4) = 36$
24. $e^{-2} = x^6$
25. $\ln e^x = 7$

Example 3 Write each as a single logarithm.

26. $\ln 125 - 2 \ln 5$
27. $3 \ln 10 + 2 \ln 100$
28. $4 \ln \frac{1}{3} - 6 \ln \frac{1}{9}$
29. $7 \ln \frac{1}{2} + 5 \ln 2$
30. $8 \ln x - 4 \ln 5$
31. $3 \ln x^2 + 4 \ln 3$

Example 4 Solve each equation. Round to the nearest ten-thousandth.

32. $6e^x - 3 = 35$
33. $4e^x + 2 = 180$
34. $3e^{2x} - 5 = -4$
35. $-2e^{3x} + 19 = 3$
36. $6e^{4x} + 7 = 4$
37. $-4e^{-x} + 9 = 2$

Examples 5–6

38. **SENSE-MAKING** Due to depreciation, the value of a car after t years is given by $v(t) = v_0 e^{-0.186t}$, where v_0 is the cost of the car new.
    a. What will the value of a car that cost $28,500 new be in 18 months?
    b. When will the value of a car be half of its cost new?
    c. If a car were worth $30,000 after 1 year, then how much did it cost new?

Solve each inequality. Round to the nearest ten-thousandth.

39. $e^x \le 8.7$
40. $e^x \ge 42.1$
41. $\ln (3x + 4)^3 > 10$
42. $4 \ln x^2 < 72$
43. $\ln (8x^4) > 24$
44. $-2 [\ln (x - 6)^{-1}] \le 6$

GRAPHING CALCULATOR Graph $f(x) = \ln x$ and the transformation graph, $g(x)$. Determine the effects on each of the key attributes of the graph of $f(x)$.

45. $g(x) = 0.5f(x)$
46. $g(x) = -0.25f(x)$
47. $g(x) = f(x) + 8$
48. $g(x) = f(x) - 9$
49. $g(x) = f(x + 5)$
50. $g(x) = f(x - 4)$

51. FINANCIAL LITERACY Use the logarithmic and exponential formulas for continuously compounded interest.

    a. If you deposited $800 in an account paying 4.5% interest compounded continuously, how much money would be in the account in 5 years?

    b. How long would it take you to double your money?

    c. If you want to double your money in 9 years, what rate would you need?

    d. If you want to open an account that pays 4.75% interest compounded continuously and have $10,000 in the account 12 years after your deposit, how much would you need to deposit?

Write the expression as a sum or difference of logarithms or multiples of logarithms.

52. $\ln 12x^2$ **53.** $\ln \frac{16}{125}$ **54.** $\ln \sqrt[5]{x^3}$ **55.** $\ln xy^4 z^{-3}$

Use the natural logarithm to solve each equation.

56. $8^x = 24$ **57.** $3^x = 0.4$ **58.** $2^{3x} = 18$ **59.** $5^{2x} = 38$

60. MODELING Newton's Law of Cooling, which can be used to determine how fast an object will cool in given surroundings, is represented by $T(t) = T_s + (T_0 - T_s)e^{-kt}$, where T_0 is the initial temperature of the object, T_s is the temperature of the surroundings, t is the time in minutes, and k is a constant value that depends on the type of object.

    a. If a cup of coffee with an initial temperature of 180° is placed in a room with a temperature of 70° and the coffee cools to 140° after 10 minutes, find k.

    b. Use this value of k to determine the temperature of the coffee after 20 minutes.

    c. When will the temperature of the coffee reach 75°?

61. MULTIPLE REPRESENTATIONS In this problem, you will use $f(x) = e^x$ and $g(x) = \ln x$.

    a. Graphical Graph both functions and their axis of symmetry, $y = x$, for $-5 \leq x \leq 5$. Then graph $a(x) = e^{-x}$ on the same graph.

    b. Analytical The graphs of $a(x)$ and $f(x)$ are reflections in which axis? What function would be a reflection of $f(x)$ in the other axis?

    c. Graphical Determine the two functions that are reflections of $g(x)$. Graph these new functions.

    d. Verbal We know that $f(x)$ and $g(x)$ are inverses. Are any of the other functions that we have graphed inverses as well? Explain your reasoning.

H.O.T. Problems   Use Higher-Order Thinking Skills

62. CHALLENGE Solve $4^x - 2^{x+1} = 15$ for x.

63. PROOF Prove $\ln ab = \ln a + \ln b$ for natural logarithms.

64. REASONING Determine whether $x > \ln x$ is *sometimes*, *always*, or *never* true. Explain your reasoning.

65. OPEN-ENDED Express the value 3 using e^x and the natural log.

66. WRITING IN MATH Explain how the natural log can be used to solve a natural base exponential function.

Preparing for Assessment

67. MULTI-STEP Population growth can be modeled by exponential growth, using the formula $P(t) = P_0 e^{rt}$, where P_0 is the initial population size, $P(t)$ is the population at time t (measured in years) after the population size is measured, and r is the rate of growth of the population. **MP** 2, 4

a. The human population growth has an estimated average growth rate of 0.013 (1.3%). If the population in 2016 was measured to be 7.4 billion people, estimate the population we can expect in 2022.

b. Estimate the human population of 2000.

c. How long ago was the human population half of its value in 2016?

d. Estimate the year in which the population was half of the projected population in 2022.

68. Simplify the following expression. **MP** 1, 8

$$\ln 3x^2 z^4 \sqrt[3]{y^5}$$

69. Solve the inequality: $6 \ln(x^3) > 14$. **MP** 1

70. Which account would have more money in it, given a 5% interest rate with continuously compounded interest: an initial investment of $1000 for 30 years, or an investment of $3000 for 8 years? **MP** 4, 6

71. What is the approximate value of x in this equation? **MP** 1

$$\ln(5x - 1) + 6 = 15$$

- A 7.39
- B 1620.82
- C 4405.29
- D 653,802.47

72. Let $m \lozenge n = \ln m - \ln n + \ln(mn)$. What is the value of $e^x \lozenge e^x$? **MP** 2, 5

- A 0
- B e^{2x}
- C e^{x^2}
- D $2x$
- E x^2

73. What is the value of x, to the nearest thousandth, if $-5^{3x} + 15 = 8$? **MP** 1, 6

74. Which equation is equivalent to $e^{2x} = 14$? **MP** 1

- A $x = \ln 7$
- B $x = \ln 28$
- C $x = \frac{1}{2} \ln 14$
- D $x = 2 \ln 14$

75. Which function generates this table of values? **MP** 1, 2

x	−2	−1	0	1	2
f(x)	1.25	1.5	2	3	5

- A $f(x) = \left(\frac{1}{2}\right)^x + 1$
- B $f(x) = 2^x + 1$
- C $f(x) = 3^x - 1$
- D $f(x) = 3^x$
- E $f(x) = 3^x + 1$

76. Given that $\ln 5 = x$ and $\ln 3 = y$, express in terms of x and y: **MP** 1

a. $\ln 75$

b. $\ln 45$

c. $\ln 15$

d. $\ln 0.6$

e. $\ln 0.36$

LESSON 9
Solving Logarithmic Equations and Inequalities

Then
- You evaluated logarithmic expressions.

Now
 Solve logarithmic equations.
 Solve logarithmic inequalities.

Why?
Each year the National Weather Service documents about 1000 tornado touchdowns in the United States. The intensity of a tornado is measured on the Enhanced Fujita scale. Tornados are divided into six categories according to their wind speed, path length, path width, and damage caused.

EF-Scale	Wind Speed (mph)	Type of Damage
EF-0	65-85	chimneys, branches
EF-1	86-110	mobile homes overturned
EF-2	111-135	roof torn off
EF-3	136-165	tree uprooted
EF-4	166-200	homes leveled, cars thrown
EF-5	201+	homes thrown

New Vocabulary
logarithmic equation
logarithmic inequality

Mathematical Practices
4 Model with mathematics.

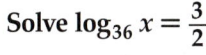

1 Solve Logarithmic Equations A **logarithmic equation** contains one or more logarithms. You can use the definition of a logarithm to help you solve logarithmic equations.

Example 1   Solve a Logarithmic Equation

Solve $\log_{36} x = \frac{3}{2}$.

$\log_{36} x = \frac{3}{2}$ Original equation

$x = 36^{\frac{3}{2}}$ Definition of logarithm

$x = (6^2)^{\frac{3}{2}}$ $36 = 6^2$

$x = 6^3$ or 216   Power of a Power

$x^{\frac{3}{2}}$ is $x \cdot x^{\frac{1}{2}}$, so $36^{\frac{3}{2}} = 36 \cdot 6$ or 216. The answer is reasonable.

▶ **Guided Practice**

Solve each equation.

1A. $\log_9 x = \frac{3}{2}$ **1B.** $\log_{16} x = \frac{5}{2}$

🔑 Key Concept   Property of Equality for Logarithmic Functions

Symbols   If b is a positive number other than 1, then $\log_b x = \log_b y$ if and only if $x = y$.

Example   If $\log_5 x = \log_5 8$, then $x = 8$. If $x = 8$, then $\log_5 x = \log_5 8$.

connectED.mcgraw-hill.com   437

Go Online!

You can use TI-Nspire® technology to solve logarithmic equations. Learn how with the **TI-Nspire Worksheet** in ConnectED.

Example 2 Solve a Logarithmic Equation

Solve $\log_2 (x^2 - 4) = \log_2 3x$.

A -2 **B** -1 **C** 2 **D** 4

Read the Item

You need to find x for the logarithmic equation.

Solve the Item

$\log_2 (x^2 - 4) = \log_2 3x$	Original equation
$x^2 - 4 = 3x$	Property of Equality for Logarithmic Functions
$x^2 - 3x - 4 = 0$	Subtract $3x$ from each side.
$(x - 4)(x + 1) = 0$	Factor.
$x - 4 = 0$ or $x + 1 = 0$	Zero Product Property
$x = 4 \qquad x = -1$	Solve each equation.

CHECK Substitute each value into the original equation.

$x = 4$
$\log_2 (4^2 - 4) \stackrel{?}{=} \log_2 3(4)$
$\log_2 12 = \log_2 12$ ✓

$x = -1$
$\log_2 [(-1)^2 - 4] \stackrel{?}{=} \log_2 3(-1)$
$\log_2 (-3) \stackrel{?}{=} \log_2 (-3)$ ✗

The domain of a logarithmic function is $(0, \infty)$, so $\log_2 (-3)$ is undefined and -1 is an extraneous solution. The answer is D.

Guided Practice

2. Solve $\log_3 (x^2 - 15) = \log_3 2x$.

A -3 **B** -1 **C** 5 **D** 15

2 Solve Logarithmic Inequalities

A **logarithmic inequality** is an inequality that involves logarithms. The following property can be used to solve logarithmic inequalities.

> **Key Concept** Property of Inequality for Logarithmic Functions
>
> If $b > 1$, $x > 0$, and $\log_b x > y$, then $x > b^y$.
>
> If $b > 1$, $x > 0$, and $\log_b x < y$, then $0 < x < b^y$.
>
> This property also holds true for $\leq$ and $\geq$.

Example 3 Solve a Logarithmic Inequality

Solve $\log_3 x > 4$.

$\log_3 x > 4$	Original inequality
$x > 3^4$	Property of Inequality for Logarithmic Functions
$x > 81$	Simplify.

Guided Practice

Solve each inequality.

3A. $\log_4 x \geq 3$ **3B.** $\log_2 x < 4$

Math History Link

Zhang Heng (A.D. 78–139) The earliest known seismograph was invented by Zhang Heng in China in 132 B.C. It was a large brass vessel with a heavy pendulum and several arms that tripped when an earthquake tremor was felt. This helped determine the direction of the quake.

The following property can be used to solve logarithmic inequalities that have logarithms with the same base on each side. Exclude from your solution set values that would result in taking the logarithm of a number less than or equal to zero in the original inequality.

> **Key Concept** Property of Inequality for Logarithmic Functions
>
> **Symbols** If $b > 1$, then $\log_b x > \log_b y$ if and only if $x > y$, and $\log_b x < \log_b y$ if and only if $x < y$.
>
> **Example** If $\log_6 x > \log_6 35$, then $x > 35$

This property also holds true for $\leq$ and $\geq$.

Example 4 Solve Inequalities with Logarithms on Each Side

Solve $\log_4 (x + 3) > \log_4 (2x + 1)$.

$\log_4 (x + 3) > \log_4 (2x + 1)$ Original inequality

$x + 3 > 2x + 1$ Property of Inequality for Logarithmic Functions

$2 > x$ Subtract $x + 1$ from each side.

Exclude all values of x for which $x + 3 \leq 0$ or $2x + 1 \leq 0$. So, $x > -3$, $x > -\frac{1}{2}$, and $x < 2$. The solution set is $\{x \mid -\frac{1}{2} < x < 2\}$ or $(-\frac{1}{2}, 2)$.

▶ **Guided Practice**

4. Solve $\log_5 (2x + 1) \leq \log_5 (x + 4)$. Check your solution.

Check Your Understanding

= Step-by-Step Solutions begin on page R11.

Go Online! for a Self-Check Quiz

Example 1 Solve each equation.

1. $\log_8 x = \frac{4}{3}$

2. $\log_{16} x = \frac{3}{4}$

Example 2 **3. MULTIPLE CHOICE** Solve $\log_5 (x^2 - 10) = \log_5 3x$.

A 10   B 2   C 5   D 2, 5

Example 3 Solve each inequality.

4. $\log_5 x > 3$

5. $\log_8 x \leq -2$

6. $\log_4 (2x + 5) \leq \log_4 (4x - 3)$

7. $\log_8 (2x) > \log_8 (6x - 8)$

Practice and Problem Solving

Extra Practice is on page R6.

Examples 1–2 **STRUCTURE** Solve each equation.

8. $\log_{81} x = \frac{3}{4}$

9. $\log_{25} x = \frac{5}{2}$

10. $\log_8 \frac{1}{2} = x$

11. $\log_6 \frac{1}{36} = x$

12. $\log_x 32 = \frac{5}{2}$

13. $\log_x 27 = \frac{3}{2}$

14. $\log_3 (3x + 8) = \log_3 (x^2 + x)$

15. $\log_{12} (x^2 - 7) = \log_{12} (x + 5)$

16. $\log_6 (x^2 - 6x) = \log_6 (-8)$

17. $\log_9 (x^2 - 4x) = \log_9 (3x - 10)$

18. $\log_4 (2x^2 + 1) = \log_4 (10x - 7)$

19. $\log_7 (x^2 - 4) = \log_7 (-x + 2)$

SCIENCE The equation for wind speed w, in miles per hour, near the center of a tornado is $w = 93 \log_{10} d + 65$, where d is the distance in miles that the tornado travels.

20. Write this equation in exponential form.

21. In May of 1999, a tornado devastated Oklahoma City with the fastest wind speed ever recorded. If the tornado traveled 525 miles, estimate the wind speed near the center of the tornado. Is your answer reasonable?

Solve each inequality.

Examples 3–4

22. $\log_6 x < -3$

23. $\log_4 x \geq 4$

24. $\log_3 x \geq -4$

25. $\log_2 x \leq -2$

26. $\log_5 x > 2$

27. $\log_7 x < -1$

28. $\log_2 (4x - 6) > \log_2 (2x + 8)$

29. $\log_7 (x + 2) \geq \log_7 (6x - 3)$

30. $\log_3 (7x - 6) < \log_3 (4x + 9)$

31. $\log_5 (12x + 5) \leq \log_5 (8x + 9)$

32. $\log_{11} (3x - 24) \geq \log_{11} (-5x - 8)$

33. $\log_9 (9x + 4) \leq \log_9 (11x - 12)$

34. **MODELING** The magnitude of an earthquake is measured on a logarithmic scale called the Richter scale. The magnitude M is given by $M = \log_{10} x$, where x represents the amplitude of the seismic wave causing ground motion.

a. How many times as great is the amplitude caused by an earthquake with a Richter scale rating of 8 as an aftershock with a Richter scale rating of 5?

b. In 1906, San Francisco was almost completely destroyed by a 7.8 magnitude earthquake. In 1911, an earthquake estimated at magnitude 8.1 occurred along the New Madrid fault in the Mississippi River Valley. How many times greater was the New Madrid earthquake than the San Francisco earthquake?

35. **MUSIC** The first key on a piano keyboard corresponds to a pitch with a frequency of 27.5 cycles per second. With every successive key, going up the black and white keys, the pitch multiplies by a constant. The formula for the frequency of the pitch sounded when the nth note up the keyboard is played is given by $n = 1 + 12 \log_2 \frac{f}{27.5}$.

a. A note has a frequency of 220 cycles per second. How many notes up the piano keyboard is this?

b. Another pitch on the keyboard has a frequency of 880 cycles per second. After how many notes up the keyboard will this be found?

36. **MULTIPLE REPRESENTATIONS** In this problem, you will explore the graphs shown: $y = \log_4 x$ and $y = \log_{\frac{1}{4}} x$.

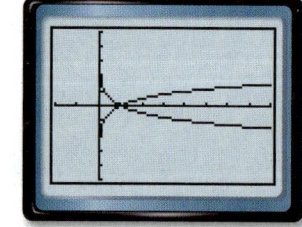

[−2, 8] scl: 1 by [−5, 5] scl: 1

a. Analytical How do the shapes of the graphs compare? How do the asymptotes and the x-intercepts of the graphs compare?

b. Verbal Describe the relationship between the graphs.

c. Graphical Use what you know about transformations of graphs to compare and contrast the graph of each function and the graph of $y = \log_4 x$.

1. $y = \log_4 x + 2$ **2.** $y = \log_4 (x + 2)$ **3.** $y = 3 \log_4 x$

d. Analytical Describe the relationship between $y = \log_4 x$ and $y = -1(\log_4 x)$. What are a reasonable domain and range for each function?

e. Analytical Write an equation for a function for which the graph is the graph of $y = \log_3 x$ translated 4 units left and 1 unit up.

37. SOUND The relationship between the intensity of sound I and the number of decibels β is $\beta = 10 \log_{10}\left(\dfrac{I}{10^{-12}}\right)$, where I is the intensity of sound in watts per square meter.

  a. Find the number of decibels of a sound with an intensity of 1 watt per square meter.

  b. Find the number of decibels of sound with an intensity of 10^{-2} watts per square meter.

  c. The intensity of the sound of 1 watt per square meter is 100 times as much as the intensity of 10^{-2} watts per square meter. Why are the decibels of sound not 100 times as great?

H.O.T. Problems   Use Higher-Order Thinking Skills

38. MP CRITIQUE ARGUMENTS Ryan and Heather are solving $\log_3 x \geq -3$. Is either of them correct? Explain your reasoning.

Ryan	Heather
$\log_3 x \geq -3$	$\log_3 x \geq -3$
$x \geq 3^{-3}$	$x \geq 3^{-3}$
$x \geq \dfrac{1}{27}$	$0 < x \leq \dfrac{1}{27}$

39. CHALLENGE Find $\log_3 27 + \log_9 27 + \log_{27} 27 + \log_{81} 27 + \log_{243} 27$.

40. MP REASONING The Property of Inequality for Logarithmic Functions states that when $b > 1$, $\log_b x > \log_b y$ if and only if $x > y$. What is the case for when $0 < b < 1$? Explain your reasoning.

41. WRITING IN MATH Explain how the domain and range of logarithmic functions are related to the domain and range of exponential functions.

42. OPEN-ENDED Give an example of a logarithmic equation that has no solution.

43. MP REASONING Choose the appropriate term. Explain your reasoning. All logarithmic equations are of the form $y = \log_b x$.

  a. If the base of a logarithmic equation is greater than 1 and the value of x is between 0 and 1, then the value for y is (*less than, greater than, equal to*) 0.

  b. If the base of a logarithmic equation is between 0 and 1 and the value of x is greater than 1, then the value of y is (*less than, greater than, equal to*) 0.

  c. There is/are (*no, one, infinitely many*) solution(s) for b in the equation $y = \log_b 0$.

  d. There is/are (*no, one, infinitely many*) solution(s) for b in the equation $y = \log_b 1$.

44. WRITING IN MATH Explain why any logarithmic function of the form $y = \log_b x$ has an x-intercept of $(1, 0)$ and no y-intercept.

Preparing for Assessment

45. MULTI-STEP An aftershock is a smaller earthquake that happens in the same area after a large earthquake. A typical aftershock has an amplitude approximately one-thirtieth that of the initial earthquake. **MP 1, 4**

The Richter scale reports the magnitude M of an earthquake using the equation $M = \log_{10} x$, where x is the amplitude of the earthquake.

   a. Determine the amplitude of an earthquake that measured 7.3 on the Richter scale.

   b. Determine the amplitude of a typical aftershock of an initial earthquake with magnitude 7.3.

   c. Determine the magnitude of an aftershock with the amplitude from part b.

46. How many solutions are there to the following problem? **MP 2**

$$\log_4 (x^2) = \log_4 (x)$$

- A 0
- B 1
- C 2
- D 3

47. If a noise has $ß = 30$ decibels, what is the intensity I of the sound in watts per square meter, where $ß = 10 \log_{10}\left(\dfrac{I}{10^{-12}}\right)$? **MP 2, 4**

- A 10^4
- B 10^{-3}
- C 30
- D 10^{-9}

48. What is the value of x? **MP 2**

$$\log_4 1 = x$$

- A 0
- B 1
- C 4
- D $\dfrac{1}{4}$

49. What is the value of x in $\log_{19}(x + 8) = \log_{19}(3x + 4)$? **MP 2, 4**

- A -8
- B 2
- C 4
- D $-\dfrac{9}{4}$

50. What is the solution set of this equation?

$$\log_3 (x^2 - 16) = \log_3 (6x)$$

MP 1, 2

- A $\{-2, 8\}$
- B $\{-2\}$
- C $\{8\}$
- D $\varnothing$

51. The intensity I of normal conversation is about 1×10^{-6} watts per square meter. How many decibels $ß$ is this sound if $ß = 10 \log_{10}\left(\dfrac{I}{10^{-12}}\right)$? **MP 2, 4**

- A -6
- B 6
- C 60
- D 600

52. Let $\log_{\frac{1}{64}} m = \dfrac{3}{2}$. What is the value of m? **MP 1, 2**

- A $\dfrac{1}{512}$
- B $\dfrac{1}{96}$
- C $\dfrac{3}{128}$
- D $\dfrac{1}{16}$
- E $\dfrac{1}{8}$

53. What is the solution set for $\log_7 (2x - 1) < \log_7 (5 - 3x)$? **MP 1, 7**

- A $\left\{x \mid \dfrac{1}{2} < x < \dfrac{4}{5}\right\}$
- B $\left\{x \mid \dfrac{1}{2} < x < \dfrac{6}{5}\right\}$
- C $\left\{x \mid \dfrac{1}{2} < x < \dfrac{5}{3}\right\}$
- D $\left\{x \mid \dfrac{4}{5} < x < \dfrac{5}{3}\right\}$
- E $\left\{x \mid \dfrac{6}{5} < x < \dfrac{5}{3}\right\}$

EXTEND 6-9
Graphing Technology Lab
Solving Logarithmic Equations and Inequalities

You have solved logarithmic equations algebraically. You can also solve logarithmic equations by graphing or by using a table. The TI-83/84 Plus has $y = \log_{10} x$ as a built-in function. Enter [Y=] [LOG] [X,T,θ,n] [GRAPH] to view this graph. To graph logarithmic functions with bases other than 10, you must use the Change of Base Formula, $\log_a n = \dfrac{\log_b n}{\log_b a}$.

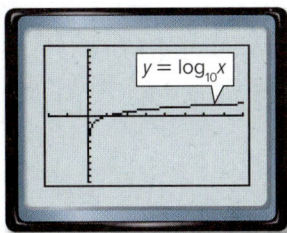

[−2, 8] scl: 1 by [−10, 10] scl: 1

Activity 1

Solve $\log_2 (6x - 8) = \log_3 (20x + 1)$.

Step 1 Graph each side of the equation.

Graph each side of the equation as a separate function. Enter $\log_2 (6x - 8)$ as **Y1** and $\log_3 (20x + 1)$ as **Y2**. Then graph the two equations.

KEYSTROKES: [Y=] [LOG] 6 [X,T,θ,n] [−] 8 [)] [÷] [LOG] 2 [)]
[ENTER] [LOG] 20 [X,T,θ,n] [+] 1 [)] [÷] [LOG] 3 [)] [GRAPH]

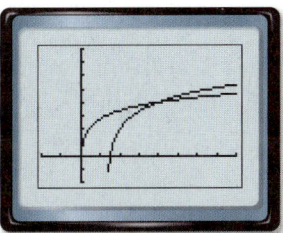

[−2, 8] scl: 1 by [−2, 8] scl: 1

Step 2 Use the **intersect** feature.

Use the **intersect** feature on the **CALC** menu to approximate the ordered pair of the point at which the curves intersect.

The calculator screen shows that the x-coordinate of the point at which the curves intersect is 4. Therefore, the solution of the equation is 4.

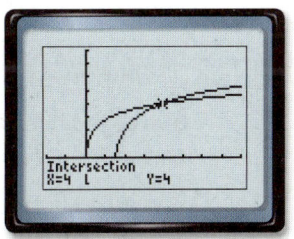

[−2, 8] scl: 1 by [−2, 8] scl: 1

Step 3 Use the **TABLE** feature.

Examine the table to find the x-value for which the y-values for the graphs are equal. At $x = 4$, both functions have a y-value of 4. Thus, the solution of the equation is 4.

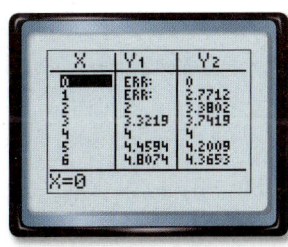

You can use a similar procedure to solve logarithmic inequalities using a graphing calculator.

(continued on the next page)

EXTEND 6-9

Graphing Technology Lab
Solving Logarithmic Equations and Inequalities *Continued*

Activity 2

Solve $\log_4 (10x + 1) < \log_5 (16 + 6x)$.

Step 1 Enter the inequalities.

Rewrite the problem as a system of inequalities.

The first inequality is $\log_4 (10x + 1) < y$ or $y > \log_4 (10x + 1)$. Because this inequality includes the *greater than* symbol, shade above the curve.

First enter the boundary and then use the arrow and ENTER keys to choose the shade above icon, .

The second inequality is $y < \log_5 (16 + 6x)$. Shade below the curve because this inequality contains *less than*.

KEYSTROKES: Y= ◄ ◄ ENTER ENTER ► ► LOG 10 X,T,θ,n + 1) ÷ LOG 4) ENTER ◄ ◄ ENTER ENTER ENTER ► ► LOG 16 + 6 X,T,θ,n) ÷ LOG 5)

Step 2 Graph the system.

KEYSTROKES: GRAPH

The left boundary of the solution set is where the first inequality is undefined. It is undefined for $10x + 1 \leq 0$.

$10x + 1 \leq 0$

$10x \leq -1$

$x \leq -\dfrac{1}{10}$

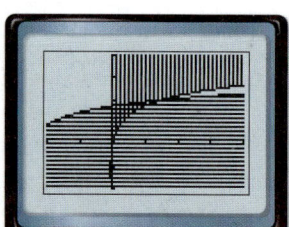

[−2, 4] scl: 1 by [−2, 4] scl: 1

Use the calculator's **intersect** feature to find the right boundary. You can conclude that the solution set is $\{x \mid -0.1 < x < 1.5\}$.

Step 3 Use the **TABLE** feature to check your solution.

Start the table at -0.1 and show x-values in increments of 0.1. Scroll through the table.

KEYSTROKES: 2nd [TBLSET] −0.1 ENTER 0.1 ENTER 2nd [TABLE]

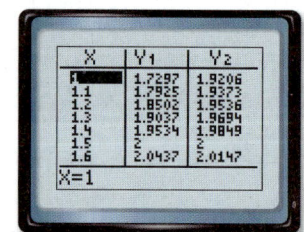

The table confirms the solution of the inequality is $\{x \mid -0.1 < x < 1.5\}$.

Exercises

Solve each equation or inequality. Check your solution.

1. $\log_2 (3x + 2) = \log_3 (12x + 3)$
2. $\log_6 (7x + 1) = \log_4 (4x - 4)$
3. $\log_2 3x = \log_3 (2x + 2)$
4. $\log_{10} (1 - x) = \log_5 (2x + 5)$
5. $\log_4 (9x + 1) > \log_3 (18x - 1)$
6. $\log_3 (3x - 5) \geq \log_3 (x + 7)$
7. $\log_5 (2x + 1) < \log_4 (3x - 2)$
8. $\log_2 2x \leq \log_4 (x + 3)$

LESSON 10
Using Logarithms to Solve Exponential Problems

::Then:: · You used exponential growth and decay formulas.

::Now::
1. Use logarithms to solve problems involving exponential growth and decay.
2. Use logarithms to solve problems involving logistic growth.

::Why?:: · Scientists can use carbon dating to determine the age of an artifact that contained organic material. A living thing exchanges carbon with the environment. When it dies, this carbon exchange stops and the amount of Carbon-14 begins to decrease. This can be modeled using an exponential decay function.

 New Vocabulary
rate of continuous growth
rate of continuous decay
logistic growth model

 Mathematical Practices
1 Make sense of problems and persevere in solving them.

1 Exponential Growth and Decay
Scientists and researchers frequently use alternate forms of the growth and decay formulas.

Key Concept Exponential Growth and Decay

Exponential Growth	Exponential Decay
Exponential growth can be modeled by the function $$f(t) = ae^{kt},$$ where a is the initial value, t is time in years, and k is a constant representing the **rate of continuous growth**.	Exponential decay can be modeled by the function $$f(t) = ae^{-kt},$$ where a is the initial value, t is time in years, and k is a constant representing the **rate of continuous decay**.

Real-World Example 1   Exponential Decay

SCIENCE The half-life of a radioactive substance is the time it takes for half of the atoms of the substance to disintegrate. The half-life of Carbon-14 is 5730 years. Determine the value of k and the equation of decay for Carbon-14.

If a is the initial amount of the substance, then the amount y that remains after 5730 years can be represented by $\frac{1}{2}a$ or $0.5a$.

$y = ae^{-kt}$ Exponential Decay Formula

$0.5a = ae^{-k(5730)}$ $y = 0.5a$ and $t = 5730$

$0.5 = e^{-5730k}$ Divide each side by a.

$\ln 0.5 = \ln e^{-5730k}$ Property of Equality for Logarithmic Functions

$\ln 0.5 = -5730k$ $\ln e^x = x$

$\dfrac{\ln 0.5}{-5730} = k$ Divide each side by -5730.

$0.00012 \approx k$ Use a calculator.

Thus, the equation for the decay of Carbon-14 is $y = ae^{-0.00012t}$.

▶ **Guided Practice**

1. The half-life of Plutonium-239 is 24,000 years. Determine the value of k.

connectED.mcgraw-hill.com   445

Now that the value of k for Carbon-14 is known, it can be used to date fossils.

Real-World Example 2 Carbon Dating

SCIENCE A paleontologist examining the bones of a prehistoric animal estimates that they contain 2% as much Carbon-14 as they would have contained when the animal was alive.

a. How long ago did the animal live?

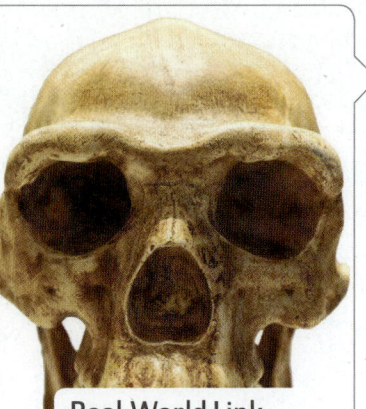

Real-World Link
The oldest modern human fossil, found in Ethiopia, is approximately 160,000 years old.
Source: National Public Radio

Understand The formula for the decay of Carbon-14 is $y = ae^{-0.00012t}$. You want to find out how long ago the animal lived.

Plan Let a be the initial amount of Carbon-14 in the animal's body. The amount y that remains after t years is 2% of a or $0.02a$.

Solve

$y = ae^{-0.00012t}$	Formula for the decay of Carbon-14
$0.02a = ae^{-0.00012t}$	$y = 0.02a$
$0.02 = e^{-0.00012t}$	Divide each side by a.
$\ln 0.02 = \ln e^{-0.00012t}$	Property of Equality for Logarithmic Functions
$\ln 0.02 = -0.00012t$	$\ln e^x = x$
$\dfrac{\ln 0.02}{-0.00012} = t$	Divide each side by -0.00012.
$32{,}600 \approx t$	Use a calculator.

The animal lived about 32,600 years ago.

Check Use the formula to find the amount of a sample remaining after 32,600 years. Use an original amount of 1.

$y = ae^{-0.00012t}$ Original equation
$ = 1e^{-0.00012(32{,}600)}$ $a = 1$ and $t = 32{,}600$
$ \approx 0.02$ or 2% ✓   Use a calculator.

Study Tip
Carbon Dating When given a percent or fraction of decay, use an original amount of 1 for a.

The half-life of Carbon-14 is 5730 years. 32,600 years is a reasonable answer given the level of Carbon-14 that remains in the bones.

b. If prior research points to the animal being around 20,000 years old, how much Carbon-14 should be in the animal?

$y = ae^{-0.00012t}$ Formula for the decay of Carbon-14
$ = 1e^{-0.00012(20{,}000)}$ $a = 1$ and $t = 20{,}000$
$ = e^{-2.4}$ Simplify.
$ = 0.09$ or 9%   Use a calculator.

Guided Practice

2. A specimen that originally contained 42 milligrams of Carbon-14 now contains 8 milligrams. How old is the fossil?

Go Online!
Investigate how changing the values in the equation of an exponential function with base e affects the graph by using the **eToolkit** in ConnectED.

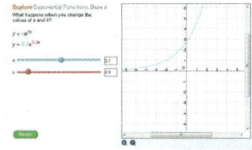

The exponential growth equation $y = ae^{kt}$ is identical to the continuously compounded interest formula.

Continuous Compounding

$$A = Pe^{rt}$$

P = initial amount
A = amount at time t
r = interest rate

Population Growth

$$y = ae^{kt}$$

a = initial population
y = population at time t
k = rate of continuous growth

446 | Lesson 6-10 | Using Logarithms to Solve Exponential Problems

Real-World Example 3 Continuous Exponential Growth

POPULATION In 2015, the population of the state of Georgia was 10.2 million people. In 2010, it was 9.7 million.

a. Determine the value of k, Georgia's relative rate of growth.

$y = ae^{kt}$	Formula for continuous exponential growth
$10.2 = 9.7e^{k(5)}$	$y = 10.2$, $a = 9.7$, and $t = 2015 - 2010$ or 5
$\dfrac{10.2}{9.7} = e^{5k}$	Divide each side by 9.7.
$\ln \dfrac{10.2}{9.7} = \ln e^{5k}$	Property of Equality for Logarithmic Functions
$\ln \dfrac{10.2}{9.7} = 5k$	$\ln e^x = x$
$\dfrac{\ln \frac{10.2}{9.7}}{5} = k$	Divide each side by 5.
$0.01005 = k$	Use a calculator.

Georgia's relative rate of growth is about 0.01005 or about 1%.

b. When will Georgia's population reach 12 million people?

$y = ae^{kt}$	Formula for continuous exponential growth
$12 = 9.7e^{0.01005t}$	$y = 12$, $a = 9.7$, and $k = 0.01005$
$1.237 = e^{0.01005t}$	Divide each side by 9.36.
$\ln 1.237 = \ln e^{0.01005t}$	Property of Equality for Logarithmic Functions
$\ln 1.237 = 0.01005t$	$\ln e^x = x$
$\dfrac{\ln 1.237}{0.01005} = t$	Divide each side by 0.01005.
$21.16 \approx t$	Use a calculator.

Georgia's population will reach 12 million people about 21 years after 2010, or in 2031.

c. Ohio's population in 2010 was 11.6 million and can be modeled by $y = 11.6e^{0.004t}$. Determine when Georgia's population will surpass Ohio's.

$9.7e^{0.01005t} > 11.6e^{0.004t}$	Formula for exponential growth
$\ln(9.7e^{0.01005t}) > \ln(11.6e^{0.004t})$	Property of Inequality for Logarithms
$\ln 9.7 + \ln e^{0.01005t} > \ln 11.6 + \ln e^{0.004t}$	Product Property of Logarithms
$\ln 9.7 + 0.01005t > \ln 11.6 + 0.004t$	$\ln e^x = x$
$0.00605t > \ln 11.6 - \ln 9.7$	Subtract (0.004t + ln 9.7) from each side.
$t > \dfrac{\ln 11.6 - \ln 9.7}{0.00605}$	Divide each side by 0.00605.
$t > 29.57$	Use a calculator.

Georgia's population will surpass Ohio's about 29 years after 2010, or in 2039.

> **Problem-Solving Tip**
>
> **Use a Formula** When dealing with population, it is almost always necessary to use an exponential growth or decay formula.

Guided Practice

3. BIOLOGY A type of bacteria is growing exponentially according to the model $y = 1000e^{kt}$, where t is the time in minutes.

A. If there are 1000 cells initially and 1650 cells after 40 minutes, find the value of k for the bacteria.

B. Suppose a second type of bacteria is growing exponentially according to the model $y = 50e^{0.0432t}$. Determine how long it will be before the number of cells of this bacteria exceed the number of cells in the other bacteria.

2 Logistic Growth

Refer to the equation representing Georgia's population in Example 3. According to the graph at the right, Georgia's population will be about one billion by the year 2130. Does this seem logical?

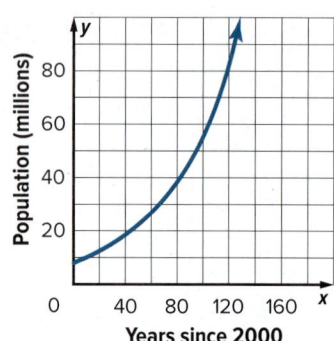

Populations cannot grow infinitely large. There are limitations, such as food supplies, war, living space, diseases, available resources, and so on.

Exponential growth is unrestricted, meaning it will increase without bound. A **logistic growth model**, however, represents growth that has a limiting factor. Logistic models are the most accurate models for representing population growth.

Key Concept Logistic Growth Function

Let a, b, and c be positive constants where $b < 1$. The logistic growth function is represented by $f(t) = \dfrac{c}{1 + ae^{-bt}}$, where t represents time.

Real-World Example 4 Logistic Growth

The population of Phoenix, Arizona, in millions can be modeled by the logistic function $f(t) = \dfrac{2.0666}{1 + 1.66e^{-0.048t}}$, where t is the number of years after 1980.

a. Graph the function for $0 \le t \le 500$.

b. What is the horizontal asymptote?

The horizontal asymptote is at $y = 2.0666$.

c. Will the population of Phoenix increase indefinitely? If not, what will be their maximum population?

No. The population will reach a maximum of a little less than 2.0666 million people.

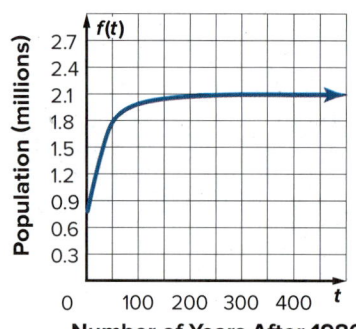

d. According to the function, when will the population of Phoenix reach 1.8 million people?

The graph indicates the population will reach 1.8 million people at $t \approx 50$. Replacing $f(t)$ with 1.8 and solving for t in the equation yields $t = 50.35$ years. So, the population of Phoenix will reach 1.8 million people by 2031.

Real-World Link
Phoenix is the fifth largest city in the country and has a population of 1.5 million.

Study Tip

Tools To determine where the graph intersects 1.8 on the calculator, graph $y = 1.8$ on the same graph and select *intersection* in the CALC menu.

Guided Practice

4. The population of a certain species of fish in a lake after t years can be modeled by the function $P(t) = \dfrac{1880}{1 + 1.42e^{-0.037t}}$, where $t \ge 0$.

A. Graph the function for $0 \le t \le 500$.

B. What is the horizontal asymptote?

C. What is the maximum population of the fish in the lake?

D. When will the population reach 1875?

Check Your Understanding

= Step-by-Step Solutions begin on page R11.

Go Online! for a Self-Check Quiz

Examples 1–2

1. **PALEONTOLOGY** The half-life of Potassium-40 is about 1.25 billion years.
   a. Determine the value of k and the equation of decay for Potassium-40.
   b. A specimen currently contains 36 milligrams of Potassium-40. How long will it take the specimen to decay to only 15 milligrams of Potassium-40?
   c. How many milligrams of Potassium-40 will be left after 300 million years?
   d. How long will it take Potassium-40 to decay to one eighth of its original amount?

Example 3

2. **SCIENCE** A certain food is dropped on the floor and is growing bacteria exponentially according to the model $y = 2e^{kt}$, where t is the time in seconds.
   a. If there are 2 cells initially and 8 cells after 20 seconds, find the value of k for the bacteria.
   b. The "5-second rule" says that if a person who drops food on the floor eats it within 5 seconds, there will be no harm. How much bacteria is on the food after 5 seconds?
   c. Would you eat food that had been on the floor for 5 seconds? Why or why not? Do you think that the information you obtained in this exercise is reasonable? Explain.

Example 4

3. **ZOOLOGY** Suppose the red fox population in a restricted habitat follows the function $P(t) = \dfrac{16{,}500}{1 + 18e^{-0.085t}}$, where t represents the time in years.
   a. Graph the function for $0 \leq t \leq 200$.
   b. What is the horizontal asymptote?
   c. What is the maximum population?
   d. When does the population reach 16,450?

Practice and Problem Solving

Extra Practice is on page R6.

Examples 1–2

4. **MP PERSEVERANCE** The half-life of Rubidium-87 is about 48.8 billion years.
   a. Determine the value of k and the equation of decay for Rubidium-87.
   b. A specimen currently contains 50 milligrams of Rubidium-87. How long will it take the specimen to decay to only 18 milligrams of Rubidium-87?
   c. How many milligrams of Rubidium-87 will be left after 800 million years?
   d. How long will it take Rubidium-87 to decay to one-sixteenth its original amount?

Example 3

5. **BIOLOGY** A certain bacteria is growing exponentially according to the model $y = 80e^{kt}$, where t is the time in minutes.
   a. If there are 80 cells initially and 675 cells after 30 minutes, find the value of k for the bacteria.
   b. When will the bacteria reach a population of 6000 cells?
   c. If a second type of bacteria is growing exponentially according to the model $y = 35e^{0.0978t}$, determine how long it will be before the number of cells of this bacteria exceed the number of cells in the other bacteria.

Example 4

6. **FORESTRY** The population of trees in a certain forest follows the function $f(t) = \dfrac{18{,}000}{1 + 16e^{-0.084t}}$, where t is the time in years.
   a. Graph the function for $0 \leq t \leq 100$.
   b. When does the population reach 17,500 trees?

7. **PALEONTOLOGY** A paleontologist finds a human bone and determines that the Carbon-14 found in the bone is 85% of that found in living bone tissue. How old is the bone?

8. **ANTHROPOLOGY** An anthropologist has determined that a newly discovered human bone is 8000 years old. How much of the original amount of Carbon-14 is in the bone?

9. **RADIOACTIVE DECAY** 100 milligrams of Uranium-238 are stored in a container. If Uranium-238 has a half-life of about 4.47 billion years, after how many years will only 10 milligrams be present?

10. **POPULATION GROWTH** The population of the state of Oregon has grown from 3.4 million in 2000 to 3.9 million in 2012.
    a. Write an exponential growth equation of the form $y = ae^{kt}$ for Oregon, where t is the number of years after 2000.
    b. Use your equation to predict the population of Oregon in 2025.
    c. According to the equation, when will Oregon reach 6 million people?

11. **HALF-LIFE** A substance decays 99.9% of its total mass after 200 years. Determine the half-life of the substance.

12. **LOGISTIC GROWTH** The population in millions of the state of Ohio after 1900 can be modeled by $P(t) = \dfrac{7.85}{1 + 12.19e^{-kt}} + 3.83$, where t is the number of years after 1900 and k is a constant.
    a. If Ohio had a population of 10.7 million in 1970, find the value of k.
    b. According to the equation, will the population of Ohio reach 12 million? What is the maximum population?

13. **MULTIPLE REPRESENTATIONS** In this problem, you will explore population growth. The population growth of a country follows the exponential function $f(t) = 8e^{0.075t}$ or the logistic function $g(t) = \dfrac{400}{1 + 16e^{-0.025t}}$. The population is measured in millions and t is time in years.
    a. **Graphical** Graph both functions for $0 \leq t \leq 100$.
    b. **Analytical** Determine the intersection of the graphs. What is the significance of this intersection?
    c. **Analytical** Which function is a more accurate estimate of the country's population 100 years from now? Explain your reasoning.

H.O.T. Problems   Use Higher-Order Thinking Skills

14. **OPEN-ENDED** Give an example of a quantity that grows or decays at a fixed rate. Write a real-world problem involving the rate and solve by using logarithms.

15. **CHALLENGE** Solve $\dfrac{120{,}000}{1 + 48e^{-0.015t}} = 24e^{0.055t}$ for t.

16. **CONSTRUCT ARGUMENTS** Explain mathematically why $f(t) = \dfrac{c}{1 + 60e^{-0.5t}}$ approaches, but never reaches the value of c as $t \to +\infty$.

17. **OPEN-ENDED** Give an example of a quantity that grows logistically and has limitations to growth. Explain why the quantity grows in this manner.

18. **WRITING IN MATH** How are exponential, continuous exponential, and logistic functions used to model different real-world situations?

Preparing for Assessment

19. Over the last six years, the population of dolphins in Ocean Bay has increased from 308 to 353. What is the annual relative rate of growth, to the nearest hundredth of a percent? **MP** 1, 4

20. During its exponential phase, E. coli bacteria in a culture increase in number at a rate proportional to the current population. If the growth rate is 2.9% per minute and the current population is 292 million, what will the population be 5.6 minutes from now? **MP** 1, 4

- A 292 million
- B 343.5 million
- C 47.4 million
- D 33,087 million

21. Solve for $f(3)$ if **MP** 1, 4
$$f(t) = \frac{5692}{1 + 9e^{-2t}}$$

- A 5,568
- B 5,692
- C 255,146.3

22. Jeremy has $2500 that he deposits in the bank. The interest rate is 3.2% with continuous compounding. How much money does he have after 10 years? **MP** 1, 4

- A $2581.29
- B $61,331.33
- C $3442.82

23. Jada bought a used car for $6000. The value of the car is expected to depreciate at a uniform rate of 30% per year. What will be the approximate value of the car in 3 years? **MP** 1, 4

- A $600
- B $735
- C $2060
- D $2440

24. Let $p = \log_q r$. What is the value of $\log_q r^2$ in terms of p? **MP** 2

- A $p + 2$
- B $2p$
- C p^2
- D $\frac{p}{2}$
- E $\frac{2}{p}$

25. The formula for decay of Carbon-14 is $y = ae^{-0.00012t}$. After how many years will 3% of the original amount of carbon be left in a sample? **MP** 1, 4

- A approximately 10,000 years
- B approximately 25,000 years
- C approximately 30,000 years
- D approximately 36,000 years

26. The population of a certain species of squirrel in a forest after t years is modeled by the function $P(t) = \frac{1540}{1 + 1.35e^{-0.07t}}$. What is the maximum population of the squirrels in the forest? **MP** 1, 4

27. **MULTI STEP** The temperature of an object, T (in degrees Celsius), after t minutes is given by the model $T(t) = 25 + 75e^{kt}$. **MP** 1

a. What is the initial temperature of the object?

b. If the temperature cools to 50 degrees Celsius after 10 minutes, what is the value of k?

c. How long will it take for the temperature of the object to reach 30 degrees Celsius? Round your answer to the nearest tenth of a minute.

EXTEND 6-10
Graphing Technology Lab
Cooling

In this lab, you will explore the type of equation that models the change in the temperature of water as it cools under various conditions.

Set Up the Lab

- Collect a variety of containers, such as a foam cup, a ceramic coffee mug, and an insulated cup.
- Boil water or collect hot water from a tap.
- Choose a container to test and fill with hot water. Place the temperature probe in the cup.
- Connect the temperature probe to your data collection device.

Activity

Step 1 Program the device to collect 20 or more samples in 1 minute intervals.

Step 2 Wait a few seconds for the probe to warm to the temperature of the water.

Step 3 Press the button to begin collecting data.

Analyze the Results

1. When the data collection is complete, graph the data in a scatter plot. Use time as the independent variable and temperature as the dependent variable. Write a sentence that describes the points on the graph.

2. Use the **STAT** menu to find an equation to model the data you collected. Try linear, quadratic, and exponential models. Which model appears to fit the data best? Explain.

3. Would you expect the temperature of the water to drop below the temperature of the room? Explain your reasoning.

4. Use the data collection device to find the temperature of the air in the room. Graph the function $y = t$, where t is the temperature of the room, along with the scatter plot and the model equation. Describe the relationship among the graphs. What is the meaning of the relationship in the context of the experiment?

Make a Conjecture

5. Do you think the results of the experiment would change if you used an insulated container for the water? What part of the function will change, the constant or the rate of decay? Repeat the experiment to verify your conjecture.

6. How might the results of the experiment change if you added ice to the water? What part of the function will change, the constant or the rate of decay? Repeat the experiment to verify your conjecture.

CHAPTER 6
Study Guide and Review

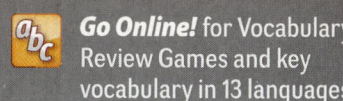

Go Online! for Vocabulary Review Games and key vocabulary in 13 languages

Study Guide

Key Concepts

Exponential Functions (Lessons 6-1 and 6-2)
- An exponential function is in the form $y = ab^x$, where $a \neq 0$, $b > 0$ and $b \neq 1$.
- Property of Equality for Exponential Functions: If b is a positive number other than 1, then $b^x = b^y$ if and only if $x = y$.

Geometric Sequences and Series (Lesson 6-3)
- The nth term a_n of a geometric sequence with first term a_1 and common ratio r is given by $a_n = a_1 \cdot r^{n-1}$, where n is any positive integer.
- The sum S_n of the first n terms of a geometric series is given by $S_n = \frac{a_1(1-r^n)}{1-r}$ or $S_n = \frac{a_1 - a_1 r^n}{1-r}$, where $r \neq 1$.

Modeling Data (Lesson 6-5)
- The closer the correlation coefficient r is to 1 or -1, the better the fit.

Logarithms and Logarithmic Functions
(Lessons 6-4, 6-6, 6-7, and 6-9)
- Suppose $b > 0$ and $b \neq 1$. For $x > 0$, there is a number y such that $\log_b x = y$ if and only if $b^y = x$.
- Product Property of Logarithms: $\log_x ab = \log_x a + \log_x b$
- Quotient Property of Logarithms: $\log_x \frac{a}{b} = \log_x a - \log_x b$
- Power Property of Logarithms: $\log_b m^p = p \log_b m$
- The Change of Base Formula: $\log_a n = \frac{\log_b n}{\log_b a}$

Natural Logarithms (Lesson 6-8)
- Because the natural base function and the natural logarithmic function are inverses, these two can be used to "undo" each other.

Using Exponential and Logarithmic Functions (Lesson 6-10)
- Exponential growth can be modeled by the function $f(x) = ae^{kt}$, where k is a constant representing the rate of continuous growth.
- Exponential decay can be modeled by the function $f(x) = ae^{-kt}$, where k is a constant representing the rate of continuous decay.

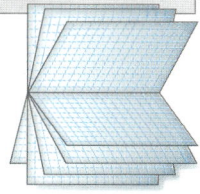 **Study Organizer**

Use your Foldable to review the chapter. Working with a partner can be helpful. Ask for clarification of concepts as needed.

Key Vocabulary

asymptote (p. 373)
Change of Base Formula (p. 425)
common logarithm (p. 423)
compound interest (p. 384)
correlation coefficient (p. 409)
decay factor (p. 376)
exponential decay (p. 375)
exponential equation (p. 383)
exponential function (p. 373)
exponential growth (p. 373)
exponential inequality (p. 385)
geometric means (p. 391)
geometric sequences (p. 390)
geometric series (p. 392)
growth factor (p. 375)
logarithm (p. 397)
logarithmic equation (p. 437)
logarithmic function (p. 398)
logarithmic inequality (p. 438)
logistic growth model (p. 448)
natural base, e (p. 430)
natural base exponential function (p. 430)
natural logarithm (p. 430)
rate of continuous decay (p. 445)
rate of continuous growth (p. 445)

Vocabulary Check

Choose a word or term from the list above that best completes each statement or phrase.

1. In $x = b^y$, the variable y is called the _____ of x.

2. A(n) _____ is an equation in which variables occur as exponents.

3. The _____ allows you to write equivalent logarithmic expressions that have different bases.

4. The base of the exponential function, $A(t) = a(1-r)^t$, $1-r$ is called the _____.

5. The function $y = \log_b x$, where $b > 0$ and $b \neq 1$, is called a(n) _____.

6. An exponential function with base e is called the _____.

Concept Check

7. Explain the difference between common logarithms and natural logarithms.

8. Explain how to determine whether an exponential function of the form $f(x) = b^x$ represents growth or decay.

CHAPTER 6
Study Guide and Review Continued

Lesson-by-Lesson Review

6-1 Graphing Exponential Functions

Graph each function. State the domain and range.

9. $f(x) = 3^x$
10. $f(x) = -5(2)^x$
11. $f(x) = 3(4)^x - 6$
12. $f(x) = 3^{2x} + 5$
13. $f(x) = 3\left(\frac{1}{4}\right)^{x+3} - 1$
14. $f(x) = \frac{3}{5}\left(\frac{2}{3}\right)^{x-2} + 3$

15. **POPULATION** A city with a population of 120,000 decreases at a rate of 3% annually.

   a. Write the function that represents this situation.

   b. What will the population be in 10 years?

Example 1

Graph $f(x) = -2(3)^x + 1$. State the domain and range.

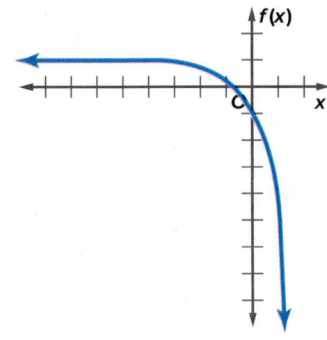

The domain is all real numbers or $D = (-\infty, \infty)$, and the range is all real numbers less than 1 or $R = (-\infty, 1)$.

6-2 Solving Exponential Equations and Inequalities

Solve each equation or inequality.

16. $16^x = \frac{1}{64}$
17. $3^{4x} = 9^{3x+7}$
18. $64^{3n} = 8^{2n-3}$
19. $8^{3-3y} = 256^{4y}$
20. $5^{1-x} = 125^x$
21. $4^{2x} = \left(\frac{1}{16}\right)^{x+4}$
22. $9^{x-2} > \left(\frac{1}{81}\right)^{x+2}$
23. $27^{3x} \leq 9^{2x-1}$
24. $6^x < 36^{x-2}$
25. $49^{2x-1} \geq 343^{x-1}$

26. **BACTERIA** A bacteria population started with 5000 bacteria. After 8 hours there were 28,000 in the sample.

   a. Write an exponential function that could be used to model the number of bacteria after x hours if the number of bacteria changes at the same rate.

   b. How many bacteria can be expected in the sample after 32 hours?

Example 2

Solve $4^{3x} = 32^{x-1}$ for x.

$4^{3x} = 32^{x-1}$	Original equation
$(2^2)^{3x} = (2^5)^{x-1}$	Rewrite so each side has the same base.
$2^{6x} = 2^{5x-5}$	Power of a Power
$6x = 5x - 5$	Property of Equality for Exponential Functions
$x = -5$	Subtract $5x$ from each side.

The solution is -5.

6-3 Geometric Sequences and Series

Find the indicated term for each geometric sequence.

27. $a_1 = 5, r = 2, n = 7$
28. $a_1 = 11, r = 3, n = 3$
29. $a_1 = 128, r = -\frac{1}{2}, n = 5$
30. a_8 for $\frac{1}{8}, \frac{3}{8}, \frac{9}{8}, \ldots$

Find the geometric means in each sequence.

31. 6, __, __, 162
32. 8, __, __, __, 648
33. −4, __, __, 108

34. **SAVINGS** Nolan has a savings account with a current balance of $1500. What would be Nolan's account balance after 4 years if he receives 5% interest annually?

Find S_n for each geometric series.

35. $a_1 = 15, r = 2, n = 4$
36. $a_1 = 9, r = 4, n = 6$
37. $.5 - 10 + 20 - \cdots$ to 7 terms
38. $243 + 81 + 27 + \cdots$ to 5 terms

Evaluate the sum of each geometric series.

39. $\sum_{k=1}^{7} 3 \cdot (-2)^{k-1}$
40. $\sum_{k=1}^{8} -1\left(\frac{2}{3}\right)^{k-1}$

Example 3

Find the sixth term of a geometric sequence for which $a_1 = 9$ and $r = 4$.

$a_n = a_1 \cdot r^{n-1}$ Formula for the nth term
$a_6 = 9 \cdot 4^{6-1}$ $n = 6, a_1 = 9, r = 4$
$a_6 = 9216$

The sixth term is 9216.

Example 4

Find two geometric means between 1 and 27.

$a_n = a_1 \cdot r^{n-1}$ Formula for the nth term
$a_4 = 1 \cdot r^{4-1}$ $n = 4$ and $a_1 = 1$
$27 = r^3$ $a_4 = 27$
$3 = r$ Simplify.

The geometric means are 1(3) or 3 and 3(3) or 9.

Example 5

Find the sum of a geometric series for which $a_1 = 3, r = 5$, and $n = 11$.

$S_n = \dfrac{a_1 - a_1 r^n}{1 - r}$ Sum formula
$S_{11} = \dfrac{3 - 3 \cdot 5^{11}}{1 - 5}$ $n = 11, a_1 = 3, r = 5$
$S_{11} = 36,621,093$ Use a calculator.

6-4 Logarithms and Logarithmic Functions

41. Write $\log_2 \frac{1}{16} = -4$ in exponential form.
42. Write $10^2 = 100$ in logarithmic form.

Evaluate each expression.

43. $\log_4 256$
44. $\log_2 \frac{1}{8}$

Graph each function.

45. $f(x) = 2 \log_{10} x + 4$
46. $f(x) = \frac{1}{6} \log_{\frac{1}{3}} (x - 2)$

Example 6

Evaluate $\log_2 64$.

$\log_2 64 = y$ Let the logarithm equal y.
$64 = 2^y$ Definition of logarithm
$2^6 = 2^y$ $64 = 2^6$
$6 = y$ Property of Equality for Exponential Functions

CHAPTER 6
Study Guide and Review Continued

6-5 Modeling Data

Determine the type of function that would best model the data in the table. Then determine the regression equation.

47.
x	2	3	6	10	15
y	6	20	35	50	60

48.
x	1	2	3	7	8
y	5	20	25	30	50

49.
x	1	5	8	10	15
y	1	20	50	200	2000

Example 7

Determine the type of function that would best model the data in the table. Then determine the regression equation.

x	1	2	3	4	5	6
y	2.4	5	7.8	12.3	20.1	50.8

Use a graphing calculator to determine the correlation coefficient for each regression model.

Linear: $r \approx 0.8682$ Power: $r \approx 0.9550$
Quadratic: $r \approx 0.9438$ Logarithmic: $r \approx 0.7486$
Exponential: $r \approx 0.9918$

The exponential model has the best fit as the value of r is closest to 1. Therefore, the regression equation is $y = 1.4055(1.7653)^x$.

6-6 Properties of Logarithms

Use $\log_5 16 \approx 1.7227$ and $\log_5 2 \approx 0.4307$ to approximate the value of each expression.

50. $\log_5 8$
51. $\log_5 64$
52. $\log_5 4$
53. $\log_5 \frac{1}{8}$
54. $\log_5 \frac{1}{2}$

Solve each equation. Check your solution.

55. $\log_5 x - \log_5 2 = \log_5 15$
56. $3 \log_4 a = \log_4 27$
57. $2 \log_3 x + \log_3 3 = \log_3 36$
58. $\log_4 n + \log_4 (n - 4) = \log_4 5$
59. $2 \log_5 x + 3 \log_5 2 = \log_5 10$
60. $\log_6 8 + \log_6 (n - 4) = \log_6 (n + 11) - \log_6 2$
61. **SOUND** Use the formula $L = 10 \log_{10} R$, where L is the loudness of a sound and R is the sound's relative intensity, to find out how many times louder 20 people talking would be than one person talking. Suppose one person talks with a loudness of 80 decibels.

Example 8

Use $\log_5 16 \approx 1.7227$ and $\log_5 2 \approx 0.4307$ to approximate $\log_5 32$.

$\log_5 32 = \log_5 (16 \cdot 2)$ Replace 32 with 16.
$ = \log_5 16 + \log_5 2$ Product Property
$ \approx 1.7227 + 0.4307$ Use a calculator.
$ \approx 2.1534$

Example 9

Solve $\log_3 3x + \log_3 4 = \log_3 36$.

$\log_3 3x + \log_3 4 = \log_3 36$ Original equation
$\log_3 3x(4) = \log_3 36$ Product Property
$3x(4) = 36$ Definition of logarithm
$12x = 36$ Multiply.
$x = 3$ Divide each side by 12.

456 | Chapter 6 | Study Guide and Review

6-7 Common Logarithms

Solve each equation or inequality. Round to the nearest ten-thousandth.

62. $3^x = 15$
63. $6^{x^2} = 28$
64. $8^{m+1} = 30$
65. $12^{r-1} = 7r$
66. $3^{5n} > 24$
67. $5^{x+2} \leq 3^x$
68. $2^{3x} < 5^{x-1}$
69. $6^{2w-5} \geq 23$

70. **SAVINGS** You deposited $1000 into an account that pays an annual interest rate r of 5% compounded quarterly. Use $A = P\left(1 + \dfrac{r}{n}\right)^{nt}$.

   a. How long will it take until you have $1500 in your account?

   b. How long it will take for your money to double?

Express each logarithm in terms of common logarithms. Then approximate to the nearest ten-thousandth.

71. $\log_8 61$
72. $\log_3 42$
73. $\log_5 97$
74. $\log_9 150$
75. $\log_7 128$
76. $\log_6 295$

Example 10

Solve $5^{3x} > 7^{x+1}$.

$5^{3x} > 7^{x+1}$	Original inequality
$\log 5^{3x} > \log 7^{x+1}$	Property of Inequality
$3x \log 5 > (x + 1) \log 7$	Power Property
$3x \log 5 > x \log 7 + \log 7$	Distributive Property
$3x \log 5 - x \log 7 > \log 7$	Subtract $x \log 7$.
$x(3 \log 5 - \log 7) > \log 7$	Distributive Property
$x > \dfrac{\log 7}{3 \log 5 - \log 7}$	Divide by $3 \log 5 - \log 7$.
$x > 0.6751$	Use a calculator.

The solution set is $\{x \mid x > 0.6751\}$.

Example 11

Express $\log_4 15$ in terms of common logarithms. Then round to the nearest ten-thousandth.

$\log_4 15 = \dfrac{\log_{10} 15}{\log_{10} 4}$	Change of Base formula
≈ 1.9534	Use a calculator

6-8 Natural Logarithms

Solve each equation or inequality. Round to the nearest ten-thousandth.

77. $4e^x - 11 = 17$
78. $2e^{-x} + 1 = 15$
79. $\ln 2x = 6$
80. $\ln(4x - 1) = 5$
81. $\ln(x + 3)^5 < 5$
82. $e^{-x} > 18$
83. $2 + e^x < 9$
84. $\ln(x + 1) > 2$

85. **SAVINGS** If you deposit $2000 in an account paying 6.4% interest compounded continuously, how long will it take for your money to triple? Use $A = Pe^{rt}$.

Example 12

Solve $3e^{5x} + 1 = 10$. Round to the nearest ten-thousandth.

$3e^{5x} + 1 = 10$	Original equation
$3e^{5x} = 9$	Subtract 1 from each side.
$e^{5x} = 3$	Divide each side by 3.
$\ln e^{5x} = \ln 3$	Property of Equality
$5x = \ln 3$	$\ln e^x = x$
$x = \dfrac{\ln 3}{5}$	Divide each side by 5.
$x \approx 0.2197$	Use a calculator.

CHAPTER 6
Study Guide and Review Continued

6-9 Solving Logarithmic Equations and Inequalities

Solve each equation or inequality.

86. $\log_4 x = \frac{3}{2}$
87. $\log_2 \frac{1}{64} = x$
88. $\log_4 x < 3$
89. $\log_5 x < -3$
90. $\log_9 (3x - 1) = \log_9 (4x)$
91. $\log_2 (x^2 - 18) = \log_2 (-3x)$
92. $\log_3 (3x + 4) \leq \log_3 (x - 2)$

93. **EARTHQUAKE** The magnitude of an earthquake is measured on a logarithmic scale called the Richter scale. The magnitude M is given by $M = \log_{10} x$, where x represents the amplitude of the seismic wave causing ground motion. How many times as great is the amplitude caused by an earthquake with a Richter scale rating of 10 as an aftershock with a Richter scale rating of 7?

Example 13
Solve $\log_{27} x < \frac{2}{3}$.

$\log_{27} x < \frac{2}{3}$	Original inequality
$x < 27^{\frac{2}{3}}$	Logarithmic to Exponential Inequality
$x < 9$	Simplify.

Example 14
Solve $\log_5 (p^2 - 2) = \log_5 p$.

$\log_5 (p^2 - 2) = \log_5 p$	Original equation
$p^2 - 2 = p$	Property of Equality
$p^2 - p - 2 = 0$	Subtract p from each side.
$(p - 2)(p + 1) = 0$	Factor.
$p - 2 = 0$ or $p + 1 = 0$	Zero Product Property
$p = 2$ $p = -1$	Solve each equation.

The solution is $p = 2$, since $\log_5 p$ is undefined for $p = -1$.

6-10 Using Logarithms to Solve Exponential Problems

94. **CARS** Abe bought a used car for $2500. It is expected to depreciate at a rate of 25% per year. What will be the value of the car in 3 years?

95. **BIOLOGY** For a certain strain of bacteria, k is 0.728 when t is measured in days. Using the formula $y = ae^{kt}$, how long will it take 10 bacteria to increase to 675 bacteria?

96. **POPULATION** The population of a city 20 years ago was 24,330. Since then, the population has increased at a steady rate each year. If the population is currently 55,250, find the annual rate of growth for this city.

Example 15
A certain culture of bacteria will grow from 250 to 2000 bacteria in 1.5 hours. Find the constant k for the growth formula. Use $y = ae^{kt}$.

$y = ae^{kt}$	Exponential Growth Formula
$2000 = 250e^{k(1.5)}$	Replace y with 2000, a with 250, and t with 1.5.
$8 = e^{1.5k}$	Divide each side by 250.
$\ln 8 = \ln e^{1.5k}$	Property of Equality
$\ln 8 = 1.5k$	Inverse Property
$\frac{\ln 8}{1.5} = k$	Divide each side by 1.5.
$1.3863 \approx k$	Use a calculator.

CHAPTER 6
Practice Test

Graph each function. State the domain and range.

1. $f(x) = 3^{x-3} + 2$
2. $f(x) = 2\left(\frac{3}{4}\right)^{x+1} - 3$

Solve each equation or inequality. Round to the nearest ten-thousandth if necessary.

3. $8^{c+1} = 16^{2c+3}$
4. $9^{x-2} > \left(\frac{1}{27}\right)^x$
5. $2^{a+3} = 3^{2a-1}$
6. $\log_2(x^2 - 7) = \log_2 6x$
7. $\log_5 x > 2$
8. $\log_3 x + \log_3(x-3) = \log_3 4$
9. $6^{n-1} \leq 11^n$
10. $4e^{2x} - 1 = 5$
11. $\ln(x+2)^2 > 2$

Use $\log_5 11 \approx 1.4899$ and $\log_5 2 \approx 0.4307$ to approximate the value of each expression.

12. $\log_5 44$
13. $\log_5 \frac{11}{2}$

14. **POPULATION** The population of a city 10 years ago was 150,000. Since then, the population has increased at a steady rate each year. The population is currently 185,000.

    a. Write an exponential function that could be used to model the population after x years if the population changes at the same rate.

    b. What will the population be in 25 years?

15. Write $\log_9 27 = \frac{3}{2}$ in exponential form.

16. **AGRICULTURE** An equation that models the declining number of U.S. farms is $y = 3{,}962{,}520(0.98)^x$, where x is the number of years since 1960 and y is the number of farms.

    a. How can you tell that the number is declining?

    b. By what annual rate is the number declining?

    c. Predict when the number of farms will be less than 1 million.

17. **MULTIPLE CHOICE** What is the value of $\log_4 \frac{1}{64}$?

    A -3
    B $-\frac{1}{3}$
    C $\frac{1}{3}$
    D 3

18. **MULTIPLE CHOICE** What is the next term in the geometric sequence below?

    $$10, \frac{5}{2}, \frac{5}{8}, \frac{5}{32}, ...$$

    A $\frac{5}{8}$
    B $\frac{5}{32}$
    C $\frac{5}{128}$
    D $\frac{5}{256}$

19. Find the three geometric means between 6 and 1536.

20. Find the sum of the geometric series for which $a_1 = 15$, $r = \frac{2}{3}$, and $n = 5$.

21. Determine the type of function that would best model the data in the table. Then determine the regression equation.

x	1	5	15	25	60
y	5	50	80	100	120

22. **MULTIPLE CHOICE** What is the solution of $\log_4 16 - \log_4 x = \log_4 8$?

    A $\frac{1}{2}$
    B 2
    C 4
    D 8

23. **MULTIPLE CHOICE** Which function is graphed below?

    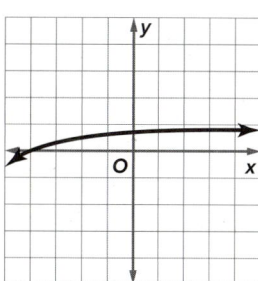

    A $y = \log_{10}(x - 5)$
    B $y = 5\log_{10} x$
    C $y = \log_{10}(x + 5)$
    D $y = -5\log_{10} x$

24. Write $2\ln 6 + 3\ln 4 - 5\ln\left(\frac{1}{3}\right)$ as a single logarithm.

CHAPTER 6
Preparing for Assessment

Performance Task

Provide a clear solution to each part of the task. Be sure to show all of your work, include all relevant drawings, and justify your answers.

APPLY MATH A United Nations committee commissioned a data analysis company to conduct analyses that will assess the health of various nations' economies, populations, and food supplies.

Part A

The company first does an analysis on the expected food supply for a coastal nation. Due to overfishing and a recent bacterial outbreak, the fish population in this region is expected to decline significantly over the next few years. The company creates the following equation to model the expected decline: $p = 1,000,000(1 - 0.2)^t$, where p is the fish population and t is the time in years after the study is conducted.

1. Determine the best increments to use on the x- and y-axes to graph the given equation.
2. Graph the expected fish population for the next 5 years on a coordinate grid.

Part B

The population of a certain country was about 650,000 in 1900. By 2000, the population of that country had increased to about 2.05 million.

3. Write an exponential function of the form $y = ab^x$ that could be used to model the data given above. Let x represent the number of years since 1900. Round the value of b to the nearest ten-thousandth if necessary.
4. Use your equation to estimate the nation's population in 2005. Round your answer to the nearest whole number.

Part C

The company does computer simulations on the infection rate of a certain virus that has an extremely high contagion rate. They run the simulation to see how many people in the populace they can expect to be infected after 1, 2, 3, 4, and 5 days if the virus breaks out. They collect the following data: 20, 60, 180, 540, 1620.

5. Determine the common ratio of the data.
6. Write an equation to find the nth term of the sequence.
7. Determine how many people can be expected to be infected after 14 days.

Part D

The population of a certain region was about 820,000 in 1980. In 2010, the population had increased to about 1.78 million.

8. Write an exponential function of the form $y = ae^{kt}$ that could be used to model the data given above. Let t represent the number of years since 1980. Round all values to the nearest thousandth if necessary.
9. Using your equation, determine after how many years the population of the region would be 2.5 million. Round your answer to the nearest whole number.

Test-Taking Strategy

Example

Read the problem. Identify what you need to know. Then use the information in the problem to solve.

A certain can of soda contains 60 milligrams of caffeine. The caffeine is eliminated from the body at a rate of 15% per hour. What is the half-life of the caffeine? That is, how many hours does it take for half of the caffeine to be eliminated from the body?

A 4 hours
B 4.25 hours
C 4.5 hours
D 4.75 hours

Step 1 How would you normally solve a problem like this?
I would create an exponential decay function and solve it using logarithms.

Step 2 Is a calculator necessary? Would using a calculator be faster? Is there any part that would be quicker to do by hand?
A calculator is necessary to evaluate the log functions.

Step 3 What is the correct answer?
The answer is B.

Test-Taking Tip

Strategies for Using Technology
Your calculator can be a useful tool in taking tests. Some problems that you encounter might have steps or computations that require the use of a calculator. A calculator may also help you solve a problem more quickly, especially in cases involving decimals, large numbers, or percents, as in the problem below.

Apply the Strategy

Read the problem. Identify what you need to know. Then use the information in the problem to solve.

Jason recently purchased a new truck for $34,750. The value of the truck decreases by 12% each year. What will the approximate value of the truck be 7 years after Jason purchased it?

A $13,775
B $13,890
C $14,125
D $14,200

Answer the questions below.

a. How would you normally solve a problem like this?

b. Is a calculator necessary? Would using a calculator be faster? Is there any part that would be quicker to do by hand?

c. What is the correct answer?

connectED.mcgraw-hill.com

CHAPTER 6
Preparing for Assessment
Cumulative Review

Read each question. Then fill in the correct answer on the answer document provided by your teacher or on a sheet of paper.

1. What is the value of x in this equation?

   $$16^{x+2} = 8^{x+3}$$

   ○ A -6
   ○ B -4
   ○ C -1
   ○ D 1

2. Rivka shares a funny meme on a social media site. If six of her friends share the meme, then six of each of their friends share the meme, and the pattern continues, how many times will the picture have been shared, including the time Rivka posted it, after eight rounds of sharing?

   ☐

3. Which expression is equivalent to $\log_7 9$?

   ○ A $(\log_{10} 9)(\log_{10} 7)$
   ○ B $(\log_{10} 9) + (\log_{10} 7)$
   ○ C $(\log_{10} 7)^{(\log_{10} 9)}$
   ○ D $(\log_{10} 9)^{(\log_{10} 7)}$
   ○ E $\dfrac{\log_{10} 9}{\log_{10} 7}$

4. Which are solutions to the inequality? Select all that apply.

   $$27^{x-1} \leq 3^{2x}$$

   ☐ A -2
   ☐ B 0
   ☐ C 1
   ☐ D 3
   ☐ E 9

5. What are the domain and range of the function $y = 2^{x-4} + 3$?

   ○ A $D = \{\text{all real numbers}\}, R = \{y \mid y > -4\}$
   ○ B $D = \{\text{all real numbers}\}, R = \{y \mid y > 0\}$
   ○ C $D = \{\text{all real numbers}\}, R = \{y \mid y > 3\}$
   ○ D $D = \{x \mid x > -4\}, R = \{\text{all real numbers}\}$
   ○ E $D = \{x \mid x > 3\}, R = \{\text{all real numbers}\}$

6.

x	f(x)	x	f(x)
1	6245	6	6
2	1560	7	1.5
3	390	8	0.25
4	100	9	0.1
5	25	10	0.02

   Given the data in the table above, determine whether a linear, quadratic, exponential, or power regression would be most appropriate. Justify your answer.

   ☐

7. What is the solution set of this equation?

   $$\log_5 4x^2 + 2 = \log_5 (41)$$

   ○ A $\{-2, 2\}$
   ○ B $\left\{-\dfrac{\sqrt{41}}{10}, \dfrac{\sqrt{41}}{10}\right\}$
   ○ C $\left\{\dfrac{\sqrt{41}}{10}\right\}$
   ○ D $\emptyset$

8. Write an equation for the nth term of the geometric sequence below.

   $$a_6 = 243;\ r = -3$$

   ☐

9. Let $\log_5 p = x$ and $\log_5 q = y$. What is the value of $\log_5 \frac{pq}{5}$?

  A xy

  B $xy - 1$

  C $xy - 5$

  D $x + y - 1$

  E $x + y - 5$

10. Leslie invested some money in an account that paid 2% annual interest compounded continuously. After 40 years, her investment was worth $6231.51. To the nearest whole dollar, what was the value of Leslie's original investment?

11. Let $6 - 7i$ be a zero of the function $f(x)$. Which value must also be a zero of the function?

  A $-6 - 7i$

  B $-6 + 7i$

  C $6 + 7i$

  D $7 - 6i$

  E $7 + 6i$

12. What is the solution set for the inequality $\log_6 (x - 2) < \log_6 (18 - 3x)$?

  A $\{x \mid -8 < x < 2\}$

  B $\{x \mid -8 < x < 6\}$

  C $\{x \mid 2 < x < 5\}$

  D $\{x \mid 2 < x < 6\}$

  E $\{x \mid 5 < x < 6\}$

13. Which of the following graphs shows the equation $f(x) = \log_8 x - 4$?

  A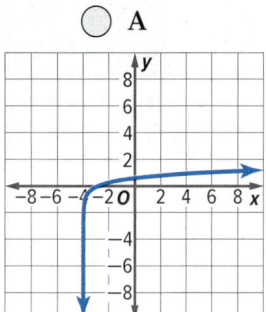

  B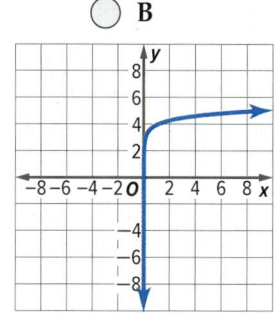

  C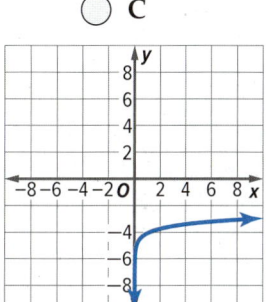

  D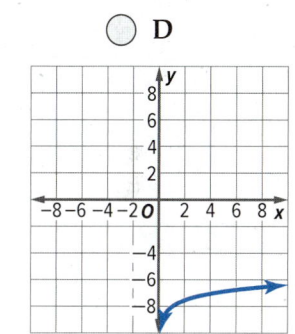

14. Find four geometric means between 0.0625 and 2048.

15. Which equation represents the inverse of the function $y = 0.2^x$?

  A $y = -0.2^x$

  B $y = 2^x$

  C $y = 5^x$

  D $y = \log_{\frac{1}{5}} x$

  E $y = \log_{\frac{1}{2}} x$

Need Extra Help?

If you missed Question…	1	2	3	4	5	6	7	8	9	10	11	12	13	14	15
Go to Lesson…	6-2	6-3	6-7	6-2	6-1	6-5	6-6	6-3	6-8	6-10	3-3	6-9	6-4	6-3	6-4

CHAPTER 7

Rational Functions

THEN

You used factoring to solve quadratic equations and you graphed quadratic equations.

NOW

You will:

- Simplify rational expressions.
- Graph rational functions.
- Solve direct, joint, and inverse variation problems.
- Solve rational equations and inequalities.

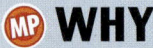

MP WHY

TRAVEL Whatever way you travel, mathematical functions can be used to find distance traveled, time spent traveling, and speed.

Use the Mathematical Practices to complete the activity.

1. **Using Tools** Use the Internet to find the National Park closest to your home. Use a mapping tool to determine the distance to that park. How long will it take to get there? Record the information you learn in a KWL chart.

Topic: Distance and Time to National Park

What I Know	What I Want to Know	What I Learned

2. **Applying Math** Set up an equation to find the time it takes to reach the park if your speed is 50 mph.
3. **Discuss** Compare your answer to the one provided by the mapping tool. If there are differences, why do you think that is?

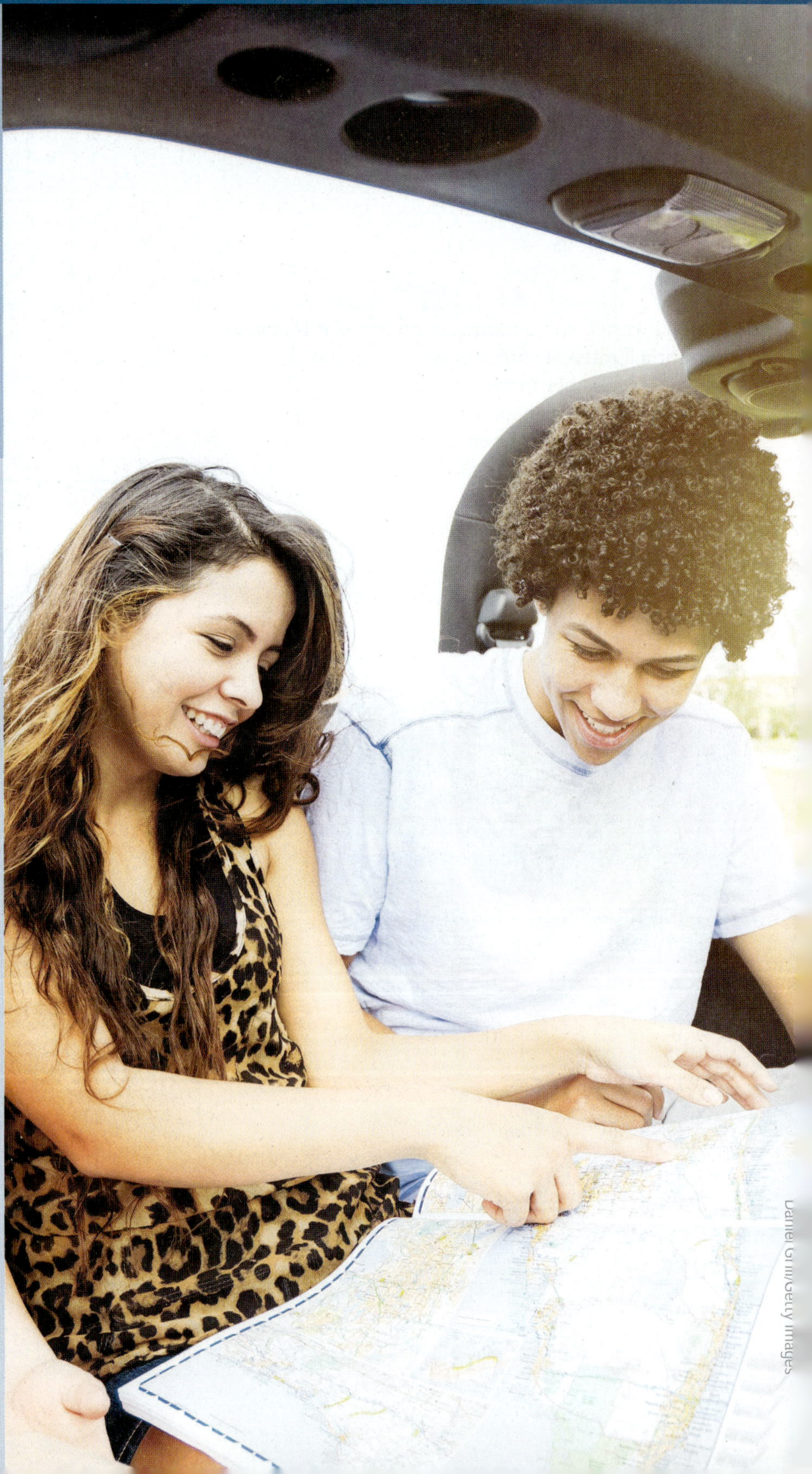

Daniel Grill/Getty Images

 # *Go Online* to Guide Your Learning

Explore & Explain

Graphing Tools: Reciprocal Functions
Use the **Graphing Tools: Reciprocal Functions** to explore the graphs of reciprocal functions in Lesson 7-3.

 ### The Geometer's Sketchpad
Use **The Geometer's Sketchpad** to explore graphs of reciprocal and rational functions and to learn more about variation functions in Lesson 7-5.

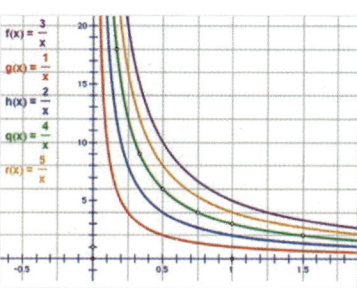

eBook
Interactive Student Guide
Before starting the chapter, answer the **Chapter Focus** preview questions. Check your answers as you complete each lesson. At the end of the chapter, try the **Performance Task**.

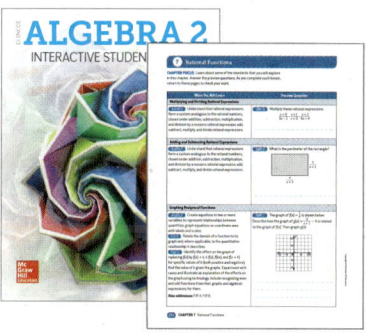

Organize

Foldables
Get organized! Create this **Rational Functions Foldable** before you start the chapter to help you organize your notes about rational functions.

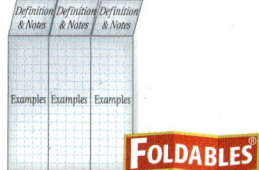

Collaborate

Chapter Project
In the **Stop and Smell the Roses** project, you will use what you have learned about rational functions to complete a project that addresses financial literacy.

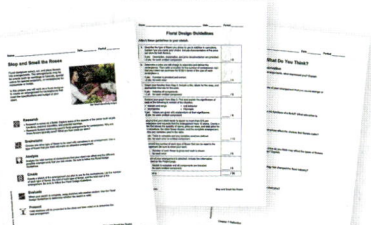

Focus

LEARNSMART
Need help studying? Complete the **Polynomial, Rational, and Radical Relationships** and the **Modeling with Functions** domains in LearnSmart to review for the chapter test.

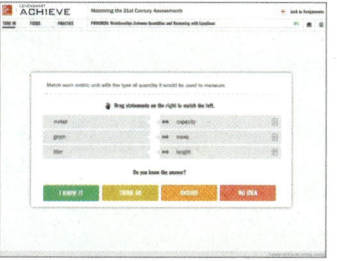

ALEKS
You can use the **Rational Expressions with Functions** topic in ALEKS to explore what you know about relations and functions and what you are ready to learn.*

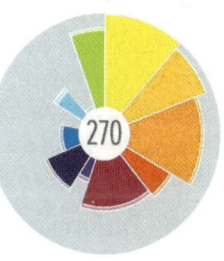

* Ask your teacher if this is part of your program.

connectED.mcgraw-hill.com 465

Get Ready for the Chapter

Connecting Concepts

Concept Check
Review the concepts used in this chapter by answering the questions below.

1. What step would you take first to solve $\frac{9}{11} = \frac{7}{8}r$ for r?

2. What step would you take first to solve $\frac{72}{11} = 7r$ for r?

3. How do you know that $\frac{72}{77} = r$ is in simplest form?

4. What do you need to determine in order to simplify the expression $\frac{1}{3} + \frac{3}{4} + \frac{5}{6}$?

5. When applying the Quadratic Formula to $-4x^2 + 16x - 1$, what are the values for a, b, and c?

6. What property would you use to rewrite the expression $\frac{1}{4(4b+6)}$ without parentheses?

7. Given $0 = 7x^2 + 7x - 9x - 9$, what property can you apply to begin to simplify the equation?

8. In the graph shown, one line is $f(x) = -4x$ and the other is the inverse of that function. Which line is the function and which is its inverse?

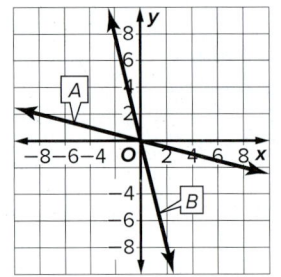

9. How could you use cross products to solve the proportion $\frac{5}{8} = \frac{p}{11}$ for p?

Performance Task Preview

You can use the concepts and skills in this chapter to help a car company evaluate the safety and efficiency of its vehicles. Understanding rational functions will help you finish the Performance Task at the end of the chapter.

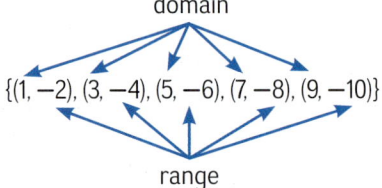

MP In this Performance Task you will:
- make sense of problems and persevere in solving them
- model with mathematics
- attend to precision

New Vocabulary

English		Español
rational expression	p. 467	expresión racional
complex fraction	p. 470	fracción compleja
reciprocal function	p. 483	función recíproco
hyperbola	p. 483	hipérbola
rational function	p. 491	función racional
vertical asymptote	p. 491	asíntota vertical
horizontal asymptote	p. 491	asíntota horizontal
oblique asymptote	p. 493	asíntota oblicua
point discontinuity	p. 494	discontinuidad evitable
direct variation	p. 500	variación directa
constant of variation	p. 500	constante de variación
joint variation	p. 501	variación conjunta
inverse variation	p. 502	variación inversa
combined variation	p. 503	variación combinada
rational equation	p. 508	ecuación racional
weighted average	p. 510	media ponderada
rational inequality	p. 513	desigualdad racional

Review Vocabulary

function función a relation in which each element of the domain is paired with exactly one element of the range

domain
$\{(1, -2), (3, -4), (5, -6), (7, -8), (9, -10)\}$
range

least common multiple mínimo común múltiplo the least number that is a common multiple of two or more numbers

rational number número racional a number expressed in the form $\frac{a}{b}$, where a and b are integers and $b \neq 0$

LESSON 1
Multiplying and Dividing Rational Expressions

Then
- You factored polynomials.

Now
1. Simplify rational expressions.
2. Simplify complex fractions.

Why?
- If a scuba diver goes to depths greater than 33 feet, the rational function $T(d) = \dfrac{1700}{d-33}$ gives the maximum time a diver can remain at those depths and still surface at a steady rate with no stops. $T(d)$ represents the dive time in minutes and d represents the depth in feet.

New Vocabulary
rational expression
complex fraction

Mathematical Practices
8 Look for and express regularity in repeated reasoning.

1 Simplify Rational Expressions
A ratio of two polynomial expressions such as $\dfrac{1700}{d-33}$ is called a **rational expression**.

Because variables in algebra often represent real numbers, operations with rational numbers and rational expressions are similar. Just as with reducing fractions, to simplify a rational expression, you divide the numerator and denominator by their greatest common factor (GCF).

$$\frac{8}{12} = \frac{2 \cdot \cancel{4}}{3 \cdot \cancel{4}} = \frac{2}{3} \qquad \frac{x^2 - 4x + 3}{x^2 - 6x + 5} = \frac{(x-3)\cancel{(x-1)}}{(x-5)\cancel{(x-1)}} = \frac{(x-3)}{(x-5)}$$

GCF = 4   GCF = $(x-1)$

Example 1 Simplify a Rational Expression

a. Simplify $\dfrac{5x(x^2+4x+3)}{(x-6)(x^2-9)}$.

$\dfrac{5x(x^2+4x+3)}{(x-6)(x^2-9)} = \dfrac{5x(x+3)(x+1)}{(x-6)(x+3)(x-3)}$ Factor numerator and denominator.

$= \dfrac{5x(x+1)}{(x-6)(x-3)} \cdot \dfrac{\cancel{(x+3)}}{\cancel{(x+3)}}$ Eliminate common factors.

$= \dfrac{5x(x+1)}{(x-6)(x-3)}$ Simplify.

b. Under what conditions is this expression undefined?

The original factored denominator is $(x-6)(x+3)(x-3)$.
Determine the values that would make the denominator equal to 0.
These values are 6, −3, or 3, so the expression is undefined when $x = 6, -3$, or 3.

Guided Practice

Simplify each expression. Under what conditions is the expression undefined?

1A. $\dfrac{4y(y-3)(y+4)}{y(y^2-y-6)}$

1B. $\dfrac{2z(z+5)(z^2+2z-8)}{(z-1)(z+5)(z-2)}$

connectED.mcgraw-hill.com   467

Example 2   Determine Undefined Values

For what value(s) is $\dfrac{x^2(x^2 - 5x - 14)}{4x(x^2 + 6x + 8)}$ undefined?

A $-2, -4$ **C** $0, -2, -4$

B $-2, 7$ **D** $0, -2, -4, 7$

Read the Item

You want to determine which values of x make the denominator equal to 0.

Study Tip

Eliminating Choices Sometimes you can save time by looking at the possible answers and eliminating choices.

Solve the Item

With $4x$ in the denominator, x cannot equal 0. So, choices A and B can be eliminated. Next, factor the denominator.

$x^2 + 6x + 8 = (x + 2)(x + 4)$, so the denominator is $4x(x + 2)(x + 4)$.

Because the denominator equals 0 when $x = 0, -2,$ and -4, the answer is C.

▶ **Guided Practice**

2. For what value(s) of x is $\dfrac{x(x^2 + 8x + 12)}{-6(x^2 - 3x - 10)}$ undefined?

A $0, 5, -2$ **B** $5, -2$ **C** $0, -2, -6$ **D** $5, -2, -6$

Sometimes you can factor out -1 in the numerator or denominator to help simplify a rational expression.

Example 3   Simplify Using -1

Simplify $\dfrac{(4w^2 - 3wy)(w + y)}{(3y - 4w)(5w + y)}$.

$\dfrac{(4w^2 - 3wy)(w + y)}{(3y - 4w)(5w + y)} = \dfrac{w(4w - 3y)(w + y)}{(3y - 4w)(5w + y)}$ Factor.

$= \dfrac{w(-1)(3y - 4w)(w + y)}{(3y - 4w)(5w + y)}$ $4w - 3y = -1(3y - 4w)$

$= \dfrac{(-w)(w + y)}{5w + y}$ Simplify.

▶ **Guided Practice**

Simplify each expression.

3A. $\dfrac{(xz - 4z)}{z^2(4 - x)}$ **3B.** $\dfrac{ab^2 - 5ab}{(5 + b)(5 - b)}$

The method for multiplying and dividing fractions also works with rational expressions. Remember that to multiply two fractions, you multiply the numerators and multiply the denominators. To divide two fractions, you multiply by the multiplicative inverse, or the reciprocal, of the divisor.

Multiplication

$\dfrac{2}{9} \cdot \dfrac{15}{4} = \dfrac{2 \cdot 3 \cdot 5}{3 \cdot 3 \cdot 2 \cdot 2} = \dfrac{5}{3 \cdot 2} = \dfrac{5}{6}$

Division

$\dfrac{3}{5} \div \dfrac{6}{35} = \dfrac{3}{5} \cdot \dfrac{35}{6} = \dfrac{3 \cdot 5 \cdot 7}{5 \cdot 2 \cdot 3} = \dfrac{7}{2}$

The following table summarizes the rules for multiplying and dividing rational expressions.

🔑 Key Concept

Multiplying Rational Expressions

Words	To multiply rational expressions, multiply the numerators and multiply the denominators.
Symbols	For all rational expressions $\frac{a}{b}$ and $\frac{c}{d}$ with $b \neq 0$ and $d \neq 0$, $\frac{a}{b} \cdot \frac{c}{d} = \frac{ac}{bd}$.

Dividing Rational Expressions

Words	To divide rational expressions, multiply by the reciprocal of the divisor.
Symbols	For all rational expressions $\frac{a}{b}$ and $\frac{c}{d}$ with $b \neq 0$, $c \neq 0$, and $d \neq 0$, $\frac{a}{b} \div \frac{c}{d} = \frac{a}{b} \cdot \frac{d}{c} = \frac{ad}{bc}$.

Study Tip

Eliminating Common Factors Be sure to eliminate factors from both the numerator and denominator.

Example 4 Multiply and Divide Rational Expressions

Simplify each expression.

a. $\dfrac{6c}{5d} \cdot \dfrac{15cd^2}{8a}$

$\dfrac{6c}{5d} \cdot \dfrac{15cd^2}{8a} = \dfrac{2 \cdot 3 \cdot c \cdot 5 \cdot 3 \cdot c \cdot d \cdot d}{5 \cdot d \cdot 2 \cdot 2 \cdot 2 \cdot a}$ Factor.

$= \dfrac{\cancel{2} \cdot 3 \cdot c \cdot \cancel{5} \cdot 3 \cdot c \cdot \cancel{d} \cdot d}{\cancel{5} \cdot \cancel{d} \cdot \cancel{2} \cdot 2 \cdot 2 \cdot a}$ Eliminate common factors.

$= \dfrac{3 \cdot 3 \cdot c \cdot c \cdot d}{2 \cdot 2 \cdot a}$ Simplify.

$= \dfrac{9c^2d}{4a}$ Simplify.

b. $\dfrac{18xy^3}{7a^2b^2} \div \dfrac{12x^2y}{35a^2b}$

$\dfrac{18xy^3}{7a^2b^2} \div \dfrac{12x^2y}{35a^2b} = \dfrac{18xy^3}{7a^2b^2} \cdot \dfrac{35a^2b}{12x^2y}$ Multiply by reciprocal of the divisor.

$= \dfrac{2 \cdot 3 \cdot 3 \cdot x \cdot y \cdot y \cdot y \cdot 5 \cdot 7 \cdot a \cdot a \cdot b}{7 \cdot a \cdot a \cdot b \cdot b \cdot 2 \cdot 2 \cdot 3 \cdot x \cdot x \cdot y}$ Factor.

$= \dfrac{\cancel{2} \cdot \cancel{3} \cdot 3 \cdot \cancel{x} \cdot \cancel{y} \cdot y \cdot y \cdot 5 \cdot \cancel{7} \cdot \cancel{a} \cdot \cancel{a} \cdot \cancel{b}}{\cancel{7} \cdot \cancel{a} \cdot \cancel{a} \cdot \cancel{b} \cdot b \cdot \cancel{2} \cdot 2 \cdot \cancel{3} \cdot x \cdot x \cdot \cancel{y}}$ Eliminate common factors.

$= \dfrac{3 \cdot 5 \cdot y \cdot y}{2 \cdot b \cdot x}$ Simplify.

$= \dfrac{15y^2}{2bx}$ Simplify.

▶ **Guided Practice**

4A. $\dfrac{12d^2}{21b} \cdot \dfrac{14b}{8c^2}$

4B. $\dfrac{6y}{15a} \cdot \dfrac{21a}{18y}$

4C. $\dfrac{16m}{21a} \div \dfrac{24m}{7a}$

4D. $\dfrac{12x^4y^2}{40a^4b^4} \div \dfrac{6x^2y^4}{16a^2x}$

Sometimes you must factor the numerator and/or the denominator first before you can simplify a product or a quotient of rational expressions.

> **Study Tip**
>
> **MP Regularity** When simplifying rational expressions, factors in one polynomial will often reappear in other polynomials. In Example 5a, $x - 8$ appears four times. Use this as a guide when factoring challenging polynomials.

Example 5 Polynomials in the Numerator and Denominator

Simplify each expression.

a. $\dfrac{x^2 - 6x - 16}{x^2 - 16x + 64} \cdot \dfrac{x - 8}{x^2 + 5x + 6}$

$\dfrac{x^2 - 6x - 16}{x^2 - 16x + 64} \cdot \dfrac{x - 8}{x^2 + 5x + 6} = \dfrac{(x - 8)(x + 2)}{(x - 8)(x - 8)} \cdot \dfrac{x - 8}{(x + 3)(x + 2)}$ Factor.

$= \dfrac{\cancel{(x - 8)}\cancel{(x + 2)}}{\cancel{(x - 8)}\cancel{(x - 8)}} \cdot \dfrac{\cancel{x - 8}}{(x + 3)\cancel{(x + 2)}}$ Eliminate common factors.

$= \dfrac{1}{x + 3}$ Simplify.

b. $\dfrac{x^2 - 16}{12y + 36} \div \dfrac{x^2 - 12x + 32}{y^2 - 3y - 18}$

$\dfrac{x^2 - 16}{12y + 36} \div \dfrac{x^2 - 12x + 32}{y^2 - 3y - 18} = \dfrac{x^2 - 16}{12y + 36} \cdot \dfrac{y^2 - 3y - 18}{x^2 - 12x + 32}$ Multiply by reciprocal.

$= \dfrac{(x + 4)(x - 4)}{12(y + 3)} \cdot \dfrac{(y - 6)(y + 3)}{(x - 4)(x - 8)}$ Factor.

$= \dfrac{(x + 4)\cancel{(x - 4)}}{12\cancel{(y + 3)}} \cdot \dfrac{(y - 6)\cancel{(y + 3)}}{\cancel{(x - 4)}(x - 8)}$ Eliminate common factors.

$= \dfrac{(x + 4)(y - 6)}{12(x - 8)}$ Simplify.

▶ **Guided Practice**

5A. $\dfrac{8x^2 - 4x - 40}{x^2 + 2x - 35} \cdot \dfrac{x^2 - 7x + 10}{4x^2 - 16}$

5B. $\dfrac{x^2 - 9x + 20}{x^2 + 10x + 21} \div \dfrac{x^2 - x - 12}{6x^2 + 60x + 126}$

2 Simplify Complex Fractions
A **complex fraction** is a rational expression with a numerator and/or denominator that is also a rational expression. The following expressions are complex fractions.

$$\dfrac{\dfrac{c}{6}}{5d} \qquad \dfrac{\dfrac{8}{x}}{x - 2} \qquad \dfrac{\dfrac{x - 3}{8}}{\dfrac{x - 2}{x + 4}} \qquad \dfrac{\dfrac{4}{a} + 6}{\dfrac{12}{a} - 3}$$

To simplify a complex fraction, first rewrite it as a division expression.

Example 6 Simplify Complex Fractions

Simplify each expression.

a. $\dfrac{\dfrac{a + b}{4}}{\dfrac{a^2 + b^2}{4}}$

$\dfrac{\dfrac{a + b}{4}}{\dfrac{a^2 + b^2}{4}} = \dfrac{a + b}{4} \div \dfrac{a^2 + b^2}{4}$ Express as a division expression.

$= \dfrac{a + b}{4} \cdot \dfrac{4}{a^2 + b^2}$ Multiply by the reciprocal.

$= \dfrac{a + b}{\cancel{4}} \cdot \dfrac{\cancel{4}}{a^2 + b^2}$ or $\dfrac{a + b}{a^2 + b^2}$ Simplify.

470 | Lesson 7-1 | Multiplying and Dividing Rational Expressions

Go Online!

Watch and listen to the **animation** to learn about simplifying complex fractions. Summarize what you hear for a partner, and have them ask you questions to help your understanding.

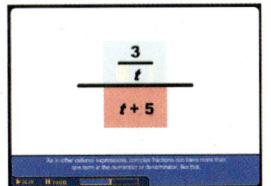

b. $\dfrac{\dfrac{x^2}{x^2-y^2}}{\dfrac{4x}{y-x}}$

$\dfrac{\dfrac{x^2}{x^2-y^2}}{\dfrac{4x}{y-x}} = \dfrac{x^2}{x^2-y^2} \div \dfrac{4x}{y-x}$ Express as a division expression.

$= \dfrac{x^2}{x^2-y^2} \cdot \dfrac{y-x}{4x}$ Multiply by the reciprocal.

$= \dfrac{x \cdot x}{(x+y)(x-y)} \cdot \dfrac{(-1)(x-y)}{4x}$ Factor.

$= \dfrac{x \cdot \overset{1}{\cancel{x}}}{(x+y)\cancel{(x-y)}} \cdot \dfrac{(-1)\overset{1}{\cancel{(x-y)}}}{\underset{1}{\cancel{4x}}}$ Eliminate Factors.

$= \dfrac{-x}{4(x+y)}$ Simplify.

Guided Practice

Simplify each expression.

6A. $\dfrac{\dfrac{(x-2)^2}{2(x^2-5x+4)}}{\dfrac{x^2-4}{4x-10}}$

6B. $\dfrac{\dfrac{x^2-y^2}{y^2-49}}{\dfrac{y-x}{y+7}}$

Check Your Understanding

 = Step-by-Step Solutions begin on page R11.

 Go Online! for a Self-Check Quiz

Example 1 Simplify each expression. Under what conditions is the expression undefined?

1. $\dfrac{x^2 - 5x - 24}{x^2 - 64}$

2. $\dfrac{c+d}{3c^2 - 3d^2}$

Example 2 3. **MULTIPLE CHOICE** Identify all values of x for which $\dfrac{x+7}{x^2 - 3x - 28}$ is undefined.

A $-7, 4$ B $7, 4$ C $4, -7, 7$ D $-4, 7$

Examples 3–6 Simplify each expression.

4. $\dfrac{y^2 + 3y - 40}{25 - y^2}$

5. $\dfrac{a^2 x - b^2 x}{by - ay}$

6. $\dfrac{27x}{16y} \cdot \dfrac{8z}{9x}$

7. $\dfrac{12y}{13a} \div \dfrac{36x}{26b}$

8. $\dfrac{x^2 - 4x - 21}{x^2 - 6x + 8} \cdot \dfrac{x-4}{x^2 - 2x - 35}$

9. $\dfrac{a^2 - b^2}{3a^2 - 6a + 3} \div \dfrac{4a + 4b}{a^2 - 1}$

10. $\dfrac{\dfrac{a^3 b^3}{xy^4}}{\dfrac{a^2 b}{x^2 y}}$

11. $\dfrac{\dfrac{4x}{x+6}}{\dfrac{x^2 - 3x}{x^2 + 3x - 18}}$

12. **MP SENSE-MAKING** The volume of a shipping container in the shape of a rectangular prism can be represented by the polynomial $6x^3 + 11x^2 + 4x$, where the height is x.

   a. Find the length and width of the container if each dimension is a linear polynomial in x.

   b. Find the ratio of the three dimensions of the container when $x = 2$.

   c. Will the ratio of the three dimensions be the same for all values of x?

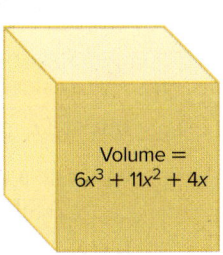

Volume = $6x^3 + 11x^2 + 4x$

Practice and Problem Solving

Extra Practice is on page R7.

Example 1 Simplify each expression. Under what conditions is this expression undefined?

13. $\dfrac{x(x-3)(x+6)}{x^2+x-12}$

14. $\dfrac{y^2(y^2+3y+2)}{2y(y-4)(y+2)}$

15. $\dfrac{(x^2-9)(x^2-z^2)}{4(x+z)(x-3)}$

16. $\dfrac{(x^2-16x+64)(x+2)}{(x^2-64)(x^2-6x-16)}$

17. $\dfrac{x^2(x+2)(x-4)}{6x(x^2+x-20)}$

18. $\dfrac{3y(y-8)(y^2+2y-24)}{15y^2(y^2-12y+32)}$

Example 2

19. **MULTIPLE CHOICE** Identify all values of x for which $\dfrac{(x-3)(x+6)}{(x^2-7x+12)(x^2-36)}$ is undefined.

    A 3, −6
    B 4, 6
    C −6, 6
    D −6, 3, 4, 6

Example 3 Simplify each expression.

20. $\dfrac{x^2-5x-14}{28+3x-x^2}$

21. $\dfrac{x^3-9x^2}{x^2-3x-54}$

22. $\dfrac{(x-4)(x^2+2x-48)}{(36-x^2)(x^2+4x-32)}$

23. $\dfrac{16-c^2}{c^2+c-20}$

24. **GEOMETRY** The cylinder at the right has a volume of $(x+3)(x^2-3x-18)\pi$ cubic centimeters. Find the height of the cylinder.

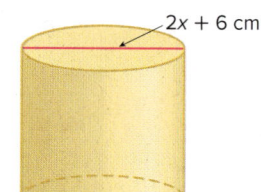
2x + 6 cm

Examples 4–6 Simplify each expression.

25. $\dfrac{3ac^3f^3}{8a^2bcf^4} \cdot \dfrac{12ab^2c}{18ab^3c^2f}$

26. $\dfrac{14xy^2z^3}{21w^4x^2yz} \cdot \dfrac{7wxyz}{12w^2y^3z}$

27. $\dfrac{64a^2b^5}{35b^2c^3f^4} \div \dfrac{12a^4b^3c}{70abcf^2}$

28. $\dfrac{9x^2yz}{5z^4} \div \dfrac{12x^4y^2}{50xy^4z^2}$

29. $\dfrac{15a^2b^2}{21ac} \cdot \dfrac{14a^4c^2}{6ab^3}$

30. $\dfrac{14c^2f^5}{9a^2} \div \dfrac{35cf^4}{18ab^3}$

31. $\dfrac{y^2+8y+15}{y-6} \cdot \dfrac{y^2-9y+18}{y^2-9}$

32. $\dfrac{c^2-6c-16}{c^2-d^2} \div \dfrac{c^2-8c}{c+d}$

33. $\dfrac{x^2+9x+20}{8x+16} \cdot \dfrac{4x^2+16x+16}{x^2-25}$

34. $\dfrac{3a^2+6a+3}{a^2-3a-10} \div \dfrac{12a^2-12}{a^2-4}$

35. $\dfrac{\dfrac{x^2-9}{6x-12}}{\dfrac{x^2+10x+21}{x^2-x-2}}$

36. $\dfrac{\dfrac{y-x}{z^3}}{\dfrac{x-y}{6z^2}}$

37. $\dfrac{\dfrac{a^2-b^2}{b^3}}{\dfrac{b^2-ab}{a^2}}$

38. $\dfrac{\dfrac{x-y}{a+b}}{\dfrac{x^2-y^2}{b^2-a^2}}$

39. **MP REASONING** At the end of her high school soccer career, Ashley had made 33 goals out of 121 attempts.

   a. Write a ratio to represent the ratio of the number of goals made to goals attempted by Ashley at the end of her high school career.

   b. Suppose Ashley attempted a goals and made m goals during her first year at college. Write a rational expression to represent the ratio of the number of goals made to the number of goals attempted from her first year in college and high school career combined.

40. GEOMETRY Parallelogram F has an area of $12x^2 - 10x - 42$ square meters and a height of $4x + 6$ meters. Parallelogram G has an area of $24x^2 + 22x - 10$ square meters and a height of $6x - 2$ meters. Find the area of right triangle H.

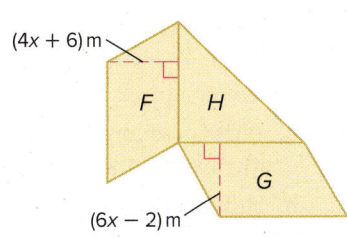

41. POLLUTION The thickness of an oil spill from a ruptured pipe on a rig is modeled by the function $T(x) = \dfrac{0.4(x^2 - 2x)}{x^3 + x^2 - 6x}$, where T is the thickness of the oil slick in meters and x is the distance from the rupture in meters.

a. Simplify the function.

b. How thick is the slick 100 meters from the rupture?

Simplify each expression.

42. $\dfrac{x^2 - 16}{3x^3 + 18x^2 + 24x} \cdot \dfrac{x^3 - 4x}{2x^2 - 7x - 4}$

43. $\dfrac{3x^2 - 17x - 6}{4x^2 - 20x - 24} \div \dfrac{6x^2 - 7x - 3}{2x^2 - x - 3}$

44. $\dfrac{9 - x^2}{x^2 - 4x - 21} \cdot \left(\dfrac{2x^2 + 7x + 3}{2x^2 - 15x + 7}\right)^{-1}$

45. $\left(\dfrac{2x^2 + 2x - 12}{x^2 + 4x - 5}\right)^{-1} \cdot \dfrac{2x^3 - 8x}{x^2 - 2x - 35}$

46. $\left(\dfrac{3xy^3z}{2a^2bc^2}\right)^3 \cdot \dfrac{16a^4b^3c^5}{15x^7yz^3}$

47. $\dfrac{20x^2y^6z^{-2}}{3a^3c^2} \cdot \left(\dfrac{16x^3y^3}{9acz}\right)^{-1}$

48. $\left(\dfrac{2xy^3}{3abc}\right)^{-2} \div \dfrac{6a^2b}{x^2y^4}$

49. $\dfrac{\dfrac{8x^2 - 10x - 3}{10x^2 + 35x - 20}}{\dfrac{2x^2 + x - 6}{4x^2 + 18x + 8}}$

50. $\dfrac{\dfrac{2x^2 + 7x - 30}{-6x^2 + 13x + 5}}{\dfrac{4x^2 + 12x - 72}{3x^2 - 11x - 4}}$

51. $\dfrac{\dfrac{4x^2 - 1}{3x^3 - 6x^2 - 24x}}{\dfrac{12x^2 + 12x - 9}{-2x^2 + 5x + 12}}$

52. GEOMETRY The area of the base of the rectangular prism at the right is 20 square centimeters.

a. Find the length of $\overline{BC}$ in terms of x.

b. If $DC = 3BC$, determine the area of the shaded region in terms of x.

c. Determine the volume of the prism in terms of x.

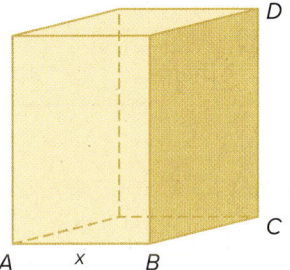

Simplify each expression.

53. $\dfrac{x^2 + 4x - 32}{2x^2 + 9x - 5} \cdot \dfrac{3x^2 - 75}{3x^2 - 11x - 4} \div \dfrac{6x^2 - 18x - 60}{x^3 - 4x}$

54. $\dfrac{8x^2 + 10x - 3}{3x^2 - 12x - 36} \div \dfrac{2x^2 - 5x - 12}{3x^2 - 17x - 6} \cdot \dfrac{4x^2 + 3x - 1}{4x^2 - 40x + 24}$

55. $\dfrac{4x^2 - 9x - 9}{3x^2 + 6x - 18} \div \dfrac{-2x^2 + 5x + 3}{x^2 - 4x - 32} \div \dfrac{8x^2 + 10x + 3}{6x^2 - 6x - 12}$

56. MULTI-STEP Mike and Scott rode their bikes home from the park. As soon as Mike arrived, he sent Scott a text saying, "Home." Scott responded with, "Wow! I've been home for ten minutes." Shocked, Mike replies, "How did you do that? You live a mile farther away than I do!" Scott retorts, "Apparently, I ride my bike twice as fast as you!"

a. Write a rational expression representing Scott's average speed in miles per hour in terms of Mike's distance d and time t.

b. Find a rational expression describing the ratio of Scott's speed to Mike's speed.

c. Assuming Scott's statement is true, write a rational expression for Mike's distance in terms of his time.

d. If it took Mike 24 minutes to get home, how far did he ride?

57. TRAINS Trying to get into a train yard one evening, all of the trains are backed up for 2 miles along a system of tracks. Assume that each car occupies an average of 75 feet of space on a track and that the train yard has 5 tracks.

a. Write an expression that could be used to determine the number of train cars involved in the backup.

b. How many train cars are involved in the backup?

c. Suppose that there are 8 attendants doing safety checks on each car, and it takes each vehicle an average of 45 seconds for each check. Approximately how many hours will it take for all the vehicles in the backup to exit?

58. MULTIPLE REPRESENTATIONS In this problem, you will investigate the graph of a rational function.

a. **Algebraic** Simplify $\frac{x^2 - 5x + 4}{x - 4}$.

b. **Tabular** Let $f(x) = \frac{x^2 - 5x + 4}{x - 4}$. Use the expression you wrote in part **a** to write the related function $g(x)$. Use a graphing calculator to make a table for both functions for $0 \leq x \leq 10$.

c. **Analytical** What are $f(4)$ and $g(4)$? Explain the significance of these values.

d. **Graphical** Graph the functions on the graphing calculator. Use the **TRACE** function to investigate each graph, using the ▲ and ▼ keys to switch from one graph to the other. Compare and contrast the graphs.

e. **Verbal** What conclusions can you draw about the expressions and the functions?

H.O.T. Problems   Use Higher-Order Thinking Skills

59. MP REASONING Compare and contrast $\frac{(x-6)(x+2)(x+3)}{x+3}$ and $(x-6)(x+2)$.

60. MP CRITIQUE ARGUMENTS Troy and Beverly are simplifying $\frac{x+y}{x-y} \div \frac{4}{y-x}$. Is either of them correct? Explain your reasoning.

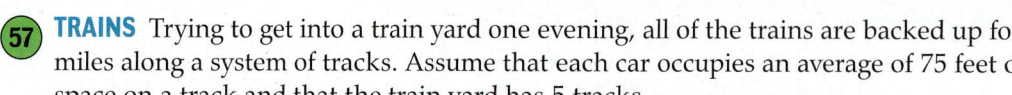

61. CHALLENGE Find the expression that makes the following statement true.
$$\frac{x-6}{x+3} \cdot \frac{?}{x-6} = x - 2$$

62. WHICH ONE DOESN'T BELONG? Identify the expression that does not belong with the other three. Explain your reasoning.

$\frac{1}{x-1}$ $\frac{x^2 + 3x + 2}{x - 5}$ $\frac{x+1}{\sqrt{x+3}}$ $\frac{x^2 + 1}{3}$

63. MP REASONING Determine whether the following statement is *sometimes*, *always*, or *never* true. Explain your reasoning.

A rational function that has a variable in the denominator is defined for all real values of x.

64. OPEN-ENDED Write a rational expression that simplifies to $\frac{x-1}{x+4}$.

65. WRITING IN MATH The rational expression $\frac{x^2 + 3x}{4x}$ is simplified to $\frac{x+3}{4}$. Explain why this new expression is not defined for all values of x.

Preparing for Assessment

66. The surface area of the cylindrical oatmeal container shown is $(54x^2 - 48x + 8)\pi$ square inches. What is the height of the box h in terms of x? **MP** 4

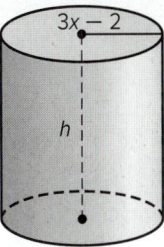

- A 6
- B $6x$
- C $6x^2$
- D $12\pi x$

67. For what values of x is the quotient undefined? **MP** 2

$$\frac{2x^2 - 3x - 9}{x(2x^2 - 5x - 3)} \div \frac{4x^2 + 12x + 9}{x^2(4x^2 + 4x + 1)}$$

- A 0
- B $-\frac{3}{2}$
- C $0, -\frac{1}{2}$
- D $0, -\frac{1}{2}, 3$
- E $0, -\frac{3}{2}, -\frac{1}{2}, 3$

68. Divide $14x^3 - 15x^2 - 98x + 15$ by $x - 5$. Which option correctly shows the factored dividend? **MP** 2

- A $(7x - 1)(2x + 3)(x - 5)$
- B $(7x + 1)(2x - 3)(x - 5)$
- C $(2x - 1)(7x + 3)(x - 5)$
- D $(2x + 1)(7x - 3)(x - 5)$

69. Divide $3x^3 + 37x^2 + 82x - 72$ by $x + 4$. Which option correctly shows the factored quotient? **MP** 2

- A $(3x + 2)(x + 9)$
- B $(3x - 2)(x - 9)$
- C $(3x + 2)(x - 9)$
- D $(3x - 2)(x + 9)$

70. MULTI-PART Consider the figure of the tissue box below. The volume of this box is $2x^3 - 3x^2 - 2x$ cubic inches. **MP** 1, 2, 4

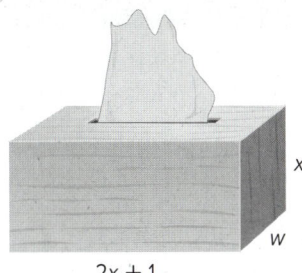

a. Which rearranged formula for the volume of a rectangular prism would you use to find the width of the box in terms of x?

- A $w = \dfrac{V}{\ell h}$
- B $w = \dfrac{\ell h}{V}$
- C $w = \dfrac{V\ell}{h}$
- D $w = \dfrac{Vh}{\ell}$

b. Complete the work to divide the volume of the box by the length and height to find the width. Which of the expressions below are correct expressions that could be included in this work? Choose all that apply.

- ☐ A $\dfrac{2x^3 - 3x^2 - 2x}{(2x + 1)x}$
- ☐ B $\dfrac{x(2x^2 - 3x - 2)}{(2x + 1)}$
- ☐ C $\dfrac{x(2x + 1)(x - 2)}{(2x + 1)x}$
- ☐ D $\dfrac{(2x + 1)(x - 2)}{(2x + 1)x}$
- ☐ E $\dfrac{x(x - 2)}{(2x + 1)}$
- ☐ F $x(x - 2)$

c. What is the correct expression for the width of this tissue box?

- A $x - 2$
- B x
- C $2x + 1$
- D $x + 2$

LESSON 2
Adding and Subtracting Rational Expressions

∷Then
- You added and subtracted polynomial expressions.

∷Now
1. Determine the LCM of polynomials.
2. Add and subtract rational expressions.

∷Why?
As a fire engine moves toward a person, the pitch of the siren sounds higher to that person than it would if the fire engine were at rest. This is because the sound waves are compressed closer together, referred to as the *Doppler effect*. The Doppler effect can be represented by the rational expression $P_0\left(\dfrac{s_0}{s_0 - v}\right)$, where P_0 is the actual pitch of the siren, v is the speed of the fire truck, and s_0 is the speed of sound in air.

 Mathematical Practices
3 Construct viable arguments and critique the reasoning of others.

1 LCM of Polynomials
Just as with rational numbers in fractional form, to add or subtract two rational expressions that have unlike denominators, you must first find the least common denominator (LCD). The LCD is the least common multiple (LCM) of the denominators.

To find the LCM of two or more numbers or polynomials, factor them. The LCM contains each factor the greatest number of times it appears as a factor.

Numbers	Polynomials
$\dfrac{5}{6} + \dfrac{4}{9}$	$\dfrac{3}{x^2 - 3x + 2} + \dfrac{5}{2x^2 - 2}$
LCM of 6 and 9	LCM of $x^2 - 3x + 2$ and $2x^2 - 2$
$6 = 2 \cdot 3$	$x^2 - 3x + 2 = (x - 1)(x - 2)$
$9 = 3 \cdot 3$	$2x^2 - 2 = 2 \cdot (x - 1)(x + 1)$
LCM = $2 \cdot 3 \cdot 3$ or 18	LCM = $2(x - 1)(x - 2)(x + 1)$

Example 1 LCM of Monomials and Polynomials

Find the LCM of each set of polynomials.

a. $6xy$, $15x^2$, and $9xy^4$

$6xy = 2 \cdot 3 \cdot x \cdot y$ Factor the first monomial.
$15x^2 = 3 \cdot 5 \cdot x^2$ Factor the second monomial.
$9xy^4 = 3 \cdot 3 \cdot x \cdot y^4$ Factor the third monomial.
LCM $= 2 \cdot 3 \cdot 3 \cdot 5 \cdot x^2 \cdot y^4$ Use each factor the greatest number of times it appears.
$= 90x^2y^4$ Then simplify.

b. $y^4 + 8y^3 + 15y^2$ and $y^2 - 3y - 40$

$y^4 + 8y^3 + 15y^2 = y^2(y + 5)(y + 3)$ Factor the first polynomial.
$y^2 - 3y - 40 = (y + 5)(y - 8)$ Factor the second polynomial.
LCM $= y^2(y + 5)(y + 3)(y - 8)$ Use each factor the greatest number of times it appears as a factor.

Guided Practice

1A. $12a^2b$, $15abc$, $8b^3c^4$

1B. $4a^2 - 12a - 16$ and $a^3 - 9a^2 + 20a$

2 Add and Subtract Rational Expressions
As with fractions, rational expressions must have common denominators in order to be added or subtracted.

Key Concept

Adding Rational Expressions

Words: To add rational expressions, find the least common denominator (LCD). Rewrite each expression with the LCD. Then add.

Symbols: For all $\frac{a}{b}$ and $\frac{c}{d}$, with $b \neq 0$ and $d \neq 0$, $\frac{a}{b} + \frac{c}{d} = \frac{ad}{bd} + \frac{bc}{bd} = \frac{ad + bc}{bd}$.

Subtracting Rational Expressions

Words: To subtract rational expressions, find the least common denominator (LCD). Rewrite each expression with the LCD. Then subtract.

Symbols: For all $\frac{a}{b}$ and $\frac{c}{d}$, with $b \neq 0$ and $d \neq 0$, $\frac{a}{b} - \frac{c}{d} = \frac{ad}{bd} - \frac{bc}{bd} = \frac{ad - bc}{bd}$.

Example 2 Monomial Denominators

Simplify each expression.

a. $\frac{3y}{2x} + \frac{2z}{5x}$

$\frac{3y}{2x} + \frac{2z}{5x} = \frac{3y}{2x} \cdot \frac{5}{5} + \frac{2z}{5x} \cdot \frac{2}{2}$

$= \frac{15y}{10x} + \frac{4z}{10x}$

$= \frac{15y + 4z}{10x}$

b. $\frac{4}{x^2 - 4x + 3} + \frac{x}{x^2 + 5x - 6}$

$\frac{4}{x^2 - 4x + 3} + \frac{x}{x^2 + 5x - 6}$

$= \frac{4}{(x-3)(x-1)} + \frac{x}{(x-1)(x+6)}$

$= \frac{4}{(x-3)(x-1)} \cdot \frac{x+6}{x+6} + \frac{x}{(x-1)(x+6)} \cdot \frac{x-3}{x-3}$

$= \frac{4x + 24}{(x-3)(x-1)(x+6)} + \frac{x^2 - 3x}{(x-3)(x-1)(x+6)}$

$= \frac{x^2 + x + 24}{(x-3)(x-1)(x+6)}$

Guided Practice

2A. $\frac{6}{7y} - \frac{2c}{3y}$

2B. $\frac{a}{a^2 + 7a + 12} - \frac{3}{a^2 + 9a + 20}$

The LCD is also used to combine rational expressions with polynomial denominators.

Example 3 Polynomial Denominators

Simplify $\frac{5}{6x - 18} - \frac{x - 1}{4x^2 - 14x + 6}$.

$\frac{5}{6x - 18} - \frac{x - 1}{4x^2 - 14x + 6} = \frac{5}{6(x - 3)} - \frac{x - 1}{2(2x - 1)(x - 3)}$ Factor denominators.

$= \frac{5(2x - 1)}{6(x - 3)(2x - 1)} - \frac{(x - 1)(3)}{2(2x - 1)(x - 3)(3)}$ Multiply by missing factors.

$= \frac{10x - 5 - 3x + 3}{6(x - 3)(2x - 1)}$ Subtract numerators.

$= \frac{7x - 2}{6(x - 3)(2x - 1)}$ Simplify.

Study Tip

MP Precision After you add or subtract rational expressions, it is possible the resulting expression can be further simplified.

Guided Practice

Simplify each expression.

3A. $\frac{x - 1}{x^2 - x - 6} - \frac{4}{5x + 10}$

3B. $\frac{x - 8}{4x^2 + 21x + 5} + \frac{6}{12x + 3}$

One way to simplify a complex fraction is to simplify the numerator and the denominator separately, and then simplify the resulting expressions.

Example 4 — Complex Fractions with Different LCDs

Simplify $\dfrac{1 + \dfrac{1}{x}}{1 - \dfrac{x}{y}}$.

$\dfrac{1 + \dfrac{1}{x}}{1 - \dfrac{x}{y}} = \dfrac{\dfrac{x}{x} + \dfrac{1}{x}}{\dfrac{y}{y} - \dfrac{x}{y}}$ The LCD of the numerator is x.
The LCD of the denominator is y.

$= \dfrac{\dfrac{x+1}{x}}{\dfrac{y-x}{y}}$ Simplify the numerator and denominator.

$= \dfrac{x+1}{x} \div \dfrac{y-x}{y}$ Write as a division expression.

$= \dfrac{x+1}{x} \cdot \dfrac{y}{y-x}$ Multiply by the reciprocal of the divisor.

$= \dfrac{xy + y}{xy - x^2}$ Simplify.

▶ Guided Practice

Simplify each expression.

4A. $\dfrac{1 - \dfrac{y}{x}}{\dfrac{1}{y} + \dfrac{1}{x}}$

4B. $\dfrac{\dfrac{c}{d} - \dfrac{d}{c}}{\dfrac{d}{c} + 2}$

Another method of simplifying complex fractions is to find the LCD of all of the denominators. Then, the denominators are all eliminated by multiplying by the LCD.

Example 5 — Complex Fractions with Same LCDs

Simplify $\dfrac{1 + \dfrac{1}{x}}{1 - \dfrac{x}{y}}$.

$\dfrac{1 + \dfrac{1}{x}}{1 - \dfrac{x}{y}} = \dfrac{\left(1 + \dfrac{1}{x}\right)}{\left(1 - \dfrac{x}{y}\right)} \cdot \dfrac{xy}{xy}$ The LCD of all of the denominators is xy.
Multiply by $\dfrac{xy}{xy}$.

$= \dfrac{xy + y}{xy - x^2}$ Distribute xy.

Notice that the same problem is solved in Examples 4 and 5 using different methods, but both produce the same answer. So, how you solve problems similar to these is left up to your own discretion.

▶ Guided Practice

Simplify each expression.

5A. $\dfrac{1 + \dfrac{2}{x}}{\dfrac{3}{y} - \dfrac{4}{x}}$

5B. $\dfrac{\dfrac{1}{d} - \dfrac{d}{c}}{\dfrac{1}{c} + 6}$

5C. $\dfrac{\dfrac{1}{y} + \dfrac{1}{x}}{\dfrac{1}{y} - \dfrac{1}{x}}$

5D. $\dfrac{\dfrac{a}{b} + 1}{1 - \dfrac{b}{a}}$

> **Study Tip**
> **Undefined Terms** Remember that there are restrictions on variables in the denominator.

Check Your Understanding

Example 1 Find the LCM of each set of polynomials.

1. $16x$, $8x^2y^3$, $5x^3y$
2. $7a^2$, $9ab^3$, $21abc^4$
3. $3y^2 - 9y$, $y^2 - 8y + 15$
4. $x^3 - 6x^2 - 16x$, $x^2 - 4$

Examples 2–3 Simplify each expression.

5. $\dfrac{12y}{5x} + \dfrac{5x}{4y^3}$
6. $\dfrac{5}{6b} + \dfrac{3b^2}{14a}$
7. $\dfrac{7b}{12a} - \dfrac{1}{18a}$
8. $\dfrac{y^2}{8c^2d^2} - \dfrac{3x}{14c^4d}$
9. $\dfrac{4x}{x^2 + 9x + 18} + \dfrac{5}{x + 6}$
10. $\dfrac{8}{y - 3} + \dfrac{2y - 5}{y^2 - 12y + 27}$
11. $\dfrac{4}{3x + 6} - \dfrac{x + 1}{x^2 - 4}$
12. $\dfrac{3a + 2}{a^2 - 16} - \dfrac{7}{6a + 24}$

13. **GEOMETRY** Find the perimeter of the rectangle.

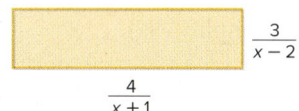

Examples 4–5 Simplify each expression.

14. $\dfrac{4 + \frac{2}{x}}{3 - \frac{2}{x}}$
15. $\dfrac{6 + \frac{4}{y}}{2 + \frac{6}{y}}$
16. $\dfrac{\frac{3}{x} + \frac{2}{y}}{1 + \frac{4}{y}}$
17. $\dfrac{\frac{2}{b} + \frac{5}{a}}{\frac{3}{a} - \frac{8}{b}}$

Practice and Problem Solving

Extra Practice is on page R7.

Example 1 Find the LCM of each set of polynomials.

18. $24cd$, $40a^2c^3d^4$, $15abd^3$
19. $4x^2y^3$, $18xy^4$, $10xz^2$
20. $x^2 - 9x + 20$, $x^2 + x - 30$
21. $6x^2 + 21x - 12$, $4x^2 + 22x + 24$

Examples 2–3 **MP PERSEVERANCE** Simplify each expression.

22. $\dfrac{5a}{24cf^4} + \dfrac{a}{36bc^4f^3}$
23. $\dfrac{4b}{15x^3y^2} - \dfrac{3b}{35x^2y^4z}$
24. $\dfrac{5b}{6a} + \dfrac{3b}{10a^2} + \dfrac{2}{ab^2}$
25. $\dfrac{4}{3x} + \dfrac{8}{x^3} + \dfrac{2}{5xy}$
26. $\dfrac{8}{3y} + \dfrac{2}{9} - \dfrac{3}{10y^2}$
27. $\dfrac{1}{16a} + \dfrac{5}{12b} - \dfrac{9}{10b^3}$
28. $\dfrac{8}{x^2 - 6x - 16} + \dfrac{9}{x^2 - 3x - 40}$
29. $\dfrac{6}{y^2 - 2y - 35} + \dfrac{4}{y^2 + 9y + 20}$
30. $\dfrac{12}{3y^2 - 10y - 8} - \dfrac{3}{y^2 - 6y + 8}$
31. $\dfrac{6}{2x^2 + 11x - 6} - \dfrac{8}{x^2 + 3x - 18}$
32. $\dfrac{2x}{4x^2 + 9x + 2} + \dfrac{3}{2x^2 - 8x - 24}$
33. $\dfrac{4x}{3x^2 + 3x - 18} - \dfrac{2x}{2x^2 + 11x + 15}$

34. **BIOLOGY** After a person eats something, the pH or acid level A of his or her mouth can be determined by the formula $A = \dfrac{20.4t}{t^2 + 36} + 6.5$, where t is the number of minutes that have elapsed since the food was eaten.

   a. Simplify the equation.
   b. What would the acid level be after 30 minutes?

35. **GEOMETRY** Both triangles in the figure at the right are equilateral. If the area of the smaller triangle is 200 square centimeters and the area of the larger triangle is 300 square centimeters, find the minimum distance from A to B in terms of x and y and simplify.

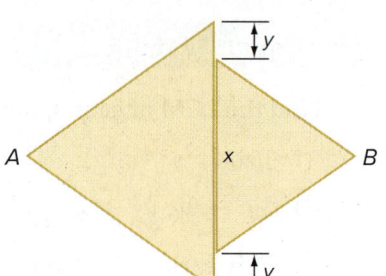

Examples 4–5 Simplify each expression.

36. $\dfrac{\dfrac{2}{x-3}+\dfrac{3x}{x^2-9}}{\dfrac{3}{x+3}-\dfrac{4x}{x^2-9}}$

37. $\dfrac{\dfrac{4}{x+5}+\dfrac{9}{x-6}}{\dfrac{5}{x-6}-\dfrac{8}{x+5}}$

38. $\dfrac{\dfrac{5}{x+6}-\dfrac{2x}{2x-1}}{\dfrac{x}{2x-1}+\dfrac{4}{x+6}}$

39. $\dfrac{\dfrac{8}{x-9}-\dfrac{x}{3x+2}}{\dfrac{3}{3x+2}+\dfrac{4x}{x-9}}$

40. **OIL PRODUCTION** Managers of an oil company have estimated that oil will be pumped from a certain well at a rate based on the function $R(x) = \dfrac{20}{x} + \dfrac{200x}{3x^2 + 20}$, where $R(x)$ is the rate of production in thousands of barrels per year x years after pumping begins.

   a. Simplify $R(x)$.

   b. At what rate will oil be pumping from the well in 50 years?

Find the LCM of each set of polynomials.

41. $12xy^4$, $14x^4y^2$, $5xyz^3$, $15x^5y^3$

42. $-6abc^2$, $18a^2b^2$, $15a^4c$, $8b^3$

43. $x^2 - 3x - 28$, $2x^2 + 9x + 4$, $x^2 - 16$

44. $x^2 - 5x - 24$, $x^2 - 9$, $3x^2 + 8x - 3$

Simplify each expression.

45. $\dfrac{1}{12a} + 6 - \dfrac{3}{5a^2}$

46. $\dfrac{5}{16y^2} - 4 - \dfrac{8}{3x^2y}$

47. $\dfrac{5}{6x^2 + 46x - 16} + \dfrac{2}{6x^2 + 57x + 72}$

48. $\dfrac{1}{8x^2 - 20x - 12} + \dfrac{4}{6x^2 + 27x + 12}$

49. $\dfrac{x^2 + y^2}{x^2 - y^2} + \dfrac{y}{x + y} - \dfrac{x}{x - y}$

50. $\dfrac{x^2 + x}{x^2 - 9x + 8} + \dfrac{4}{x - 1} - \dfrac{3}{x - 8}$

51. $\dfrac{\dfrac{2}{a-1}+\dfrac{3}{a-4}}{\dfrac{6}{a^2 - 5a + 4}}$

52. $\dfrac{\dfrac{1}{x}+\dfrac{1}{y}}{\left(\dfrac{1}{x}-\dfrac{1}{y}\right)(x+y)}$

53. **GEOMETRY** The length of one rectangular room is $\dfrac{x^2 - 16}{x - 5}$. The length of a similar rectangular room is expressed as $\dfrac{x + 4}{x^2 - 25}$. What is the scale factor of the lengths of the two rooms? Write in simplest form.

54. **MODELING** Cameron is taking a 20-mile kayaking trip. He travels half the distance at one rate. The rest of the distance he travels 2 miles per hour slower.

   a. If x represents the faster pace in miles per hour, write an expression that represents the time spent at that pace.

   b. Write an expression for the amount of time spent at the slower pace.

   c. Write an expression for the amount of time Cameron needed to complete the trip.

Find the slope of the line that passes through each pair of points.

55. $A\left(\dfrac{2}{p}, \dfrac{1}{2}\right)$ and $B\left(\dfrac{1}{3}, \dfrac{3}{p}\right)$

56. $C\left(\dfrac{1}{4}, \dfrac{4}{q}\right)$ and $D\left(\dfrac{5}{q}, \dfrac{1}{5}\right)$

57. $E\left(\dfrac{7}{w}, \dfrac{1}{7}\right)$ and $F\left(\dfrac{1}{7}, \dfrac{7}{w}\right)$

58. $G\left(\dfrac{6}{n}, \dfrac{1}{6}\right)$ and $H\left(\dfrac{1}{6}, \dfrac{6}{n}\right)$

59. PHOTOGRAPHY The focal length of a lens establishes the field of view of the camera. The shorter the focal length is, the larger the field of view. For a camera with a fixed focal length of 70 millimeters to focus on an object x millimeters from the lens, the film must be placed a distance y from the lens. This is represented by $\frac{1}{x} + \frac{1}{y} = \frac{1}{70}$.

   a. Express y as a function of x.

   b. What happens to the focusing distance when the object is 70 millimeters away?

60. PHARMACOLOGY Two drugs are administered to a patient. The concentrations in the bloodstream of each are given by $f(t) = \frac{2t}{3t^2 + 9t + 6}$ and $g(t) = \frac{3t}{2t^2 + 6t + 4}$ where t is the time, in hours, after the drugs are administered.

   a. Add the two functions together to determine a function for the total concentration of drugs in the patient's bloodstream.

   b. What is the concentration of drugs after 8 hours?

61 DOPPLER EFFECT Refer to the application at the beginning of the lesson. George is equidistant from two fire engines traveling toward him from opposite directions.

   a. Let x be the speed of the faster fire engine and y be the speed of the slower fire engine. Write and simplify a rational expression representing the difference in pitch between the two sirens according to George.

   b. If one is traveling at 45 meters per second and the other is traveling at 70 meters per second, what is the difference in their pitches according to George? The speed of sound in air is 332 meters per second, and both engines have a siren with a pitch of 500 Hz.

62. MULTIPLE REPRESENTATIONS Consider the following sets of rational functions.

$af(x) = \frac{a}{x}$ for $a = \{-2, -1, -0.5, 0.5, 2, 4\}$ $g(bx) = \frac{1}{bx}$ for $b = \{-2, -1, -0.5, 0.5, 2, 4\}$

$h(x - c) = \frac{1}{x - c}$ for $c = \{-4, -2, -0.5, 0.5, 2, 4\}$ $k(x) + d = \frac{1}{x} + d$ for $d = \{-4, -2, -0.5, 0.5, 2, 4\}$

   a. Graphical Graph each set of functions on a graphing calculator.

   b. Verbal Compare and contrast the graphs of the functions in each set.

   c. Tabular Choose two functions from any set. Find the slope between consecutive points on the graphs.

   d. Verbal Describe how the slopes of the two sections of a rational function graph are related.

H.O.T. Problems   Use Higher-Order Thinking Skills

63. CHALLENGE Simplify $\dfrac{5x^{-2} - \frac{x+1}{x}}{\frac{4}{3 - x^{-1}} + 6x^{-1}}$.

64. CONSTRUCT ARGUMENTS The sum of any two rational numbers is always a rational number. So, the set of rational numbers is said to be closed under addition. Determine whether the set of rational expressions is closed under addition, subtraction, multiplication, and division by a nonzero rational expression. Justify your reasoning.

65. OPEN-ENDED Write three monomials with an LCM of $180a^4b^6c$.

66. WRITING IN MATH Write a how-to manual for adding rational expressions that have unlike denominators. How does this compare to adding rational numbers?

Preparing for Assessment

67. Kim drove to work at an average speed of 48 miles per hour. On the ride home, her average speed was 36 miles per hour because of traffic. Which is closest to her average speed for the entire commute? **MP 4**

- A 36 mph
- B 38 mph
- C 41 mph
- D 42 mph

68. What is the domain of the function shown? **MP 6**

$$f(x) = 3 + \frac{2}{x^2}$$

- A $\{x \mid x = 0\}$
- B $\{x \mid x \neq 0\}$
- C $\{x \mid x > 0\}$
- D $\{x \mid x < 0\}$
- E all real numbers

69. What is the perimeter of the triangle? **MP 1**

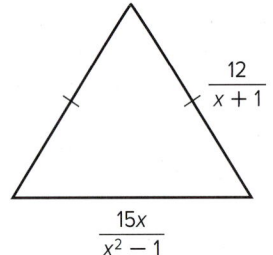

- A $\frac{27x - 12}{x^2 - 1}$
- B $\frac{39x - 24}{x^2 - 1}$
- C $\frac{15x + 24}{x^2 - 1}$
- D $\frac{27x - 2}{x^2 - 1}$
- E $\frac{24x - 24}{x^2 - 1}$

70. What is the sum of $\frac{6x - 2}{3x^2 + 2x - 1} + \frac{7}{x + 1} + \frac{2}{x}$? **MP 1**

- A $\frac{7}{x + 1} + \frac{1}{2x}$
- B $\frac{18x}{x + 1}$
- C $\frac{11x + 2}{x^2 + x}$
- D $\frac{11}{2x + 1}$

71. MULTI-STEP Jason hiked up a mountain trail at an average rate of 2 miles per hour. He hiked down the same trail at a rate of 3 miles per hour. **MP 4**

a. Let d represent the distance from the bottom of the trail to the top of the mountain (the end of the trail). Which expression shows the total time Jason spent hiking the trail?

- A d
- B $\frac{d}{2 + 3}$
- C $\frac{d}{2} + \frac{d}{3}$
- D $2d + 3d$

b. Let d represent the distance from the bottom of the trail to the top of the mountain (the end of the trail). Which expression shows Jason's average rate for the whole hike?

- A dt
- B $\frac{12d}{2d + 4d}$
- C $\frac{2d}{\frac{d}{2} + \frac{d}{3}}$
- D $\frac{5d}{12d}$

c. What was Jason's average rate for the whole hike?

- A 1 mph
- B 2.4 mph
- C 2.5 mph
- D 5 mph

482 | Lesson 7-2 | Adding and Subtracting Rational Expressions

LESSON 3
Graphing Reciprocal Functions

::Then:: ::Now:: ::Why?::

- You graphed polynomial functions.

1. Determine properties of reciprocal functions.
2. Graph transformations of reciprocal functions.

- The sophomore class is renting an indoor trampoline park for a class party. The cost of renting the facility is $900, which is shared equally among all students who attend. If c represents the cost to each student and n represents the number of students, then $c = \frac{900}{n}$.

New Vocabulary
reciprocal function
hyperbola

Mathematical Practices
2 Reason abstractly and quantitatively.

1 Vertical and Horizontal Asymptotes
The function $c = \frac{5000}{n}$ is a reciprocal function. A **reciprocal function** has an equation of the form $f(x) = \frac{1}{a(x)}$, where $a(x)$ is a linear function and $a(x) \neq 0$.

Key Concept Parent Function of Reciprocal Functions

Parent function:	$f(x) = \frac{1}{x}$
Type of graph:	**hyperbola**
Domain and range:	all nonzero real numbers, $(-\infty, 0) \cup (0, +\infty)$
Asymptotes:	$x = 0$ and $f(x) = 0$
Intercepts:	none
Not defined:	$x = 0$

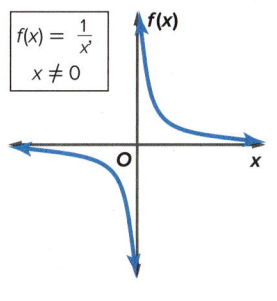

The domain of a reciprocal function is limited to values for which the denominator is nonzero.

Functions: $\quad f(x) = \frac{-3}{x+2} \qquad g(x) = \frac{4}{x-5} \qquad h(x) = \frac{3}{x}$

Not defined at: $\quad x = -2 \qquad\qquad x = 5 \qquad\qquad x = 0$

Example 1 Limitations on Domain

Determine the value of x for which $f(x) = \frac{3}{2x+5}$ is not defined.

Find the value for which the denominator of the expression equals 0.

$\frac{3}{2x+5} \quad \rightarrow \quad 2x + 5 = 0$

$\qquad\qquad\qquad x = -\frac{5}{2}$ The function is undefined for $x = -\frac{5}{2}$.

▶ Guided Practice

Determine the value of x for which each function is not defined.

1A. $f(x) = \frac{2}{x-1}$ \qquad\qquad **1B.** $f(x) = \frac{7}{3x+2}$

connectED.mcgraw-hill.com

The graphs of reciprocal functions have breaks in continuity for excluded values. They have asymptotes which are lines that the graph of the function approaches.

> **Study Tip**
>
> **Structure** Vertical asymptotes show where a function is undefined, while horizontal asymptotes show the end behavior of a graph.

> **Go Online!**
>
> Investigate how changing the values in the equation of a reciprocal function affects the graph by using the **eToolkit** in ConnectED.
>
>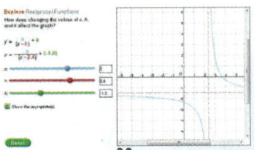

Example 2 Determine Properties of Reciprocal Functions

Identify the asymptotes, domain, and range of each function.

a.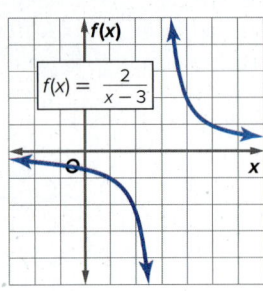

Identify x-values for which $f(x)$ is undefined.
$$x - 3 = 0$$
$$x = 3$$

$f(x)$ is not defined when $x = 3$. So there is an asymptote at $x = 3$.

From $x = 3$, as x-values decrease, $f(x)$-values approach 0, and as x-values increase, $f(x)$-values approach 0. So there is an asymptote at $f(x) = 0$.

The domain is all real numbers not equal to 3 and the range is all real numbers not equal to 0. This is represented as D = $\{x \mid x \neq 3\}$ or $(-\infty, 3) \cup (3, +\infty)$ and R = $\{x \mid x \neq 0\}$ or $(-\infty, 0) \cup (0, +\infty)$.

b.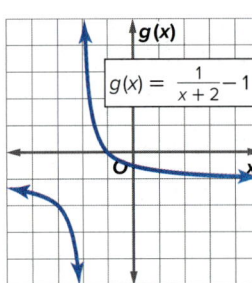

Identify x-values for which $g(x)$ is undefined.
$$x + 2 = 0$$
$$x = -2$$

$g(x)$ is not defined when $x = -2$. So there is an asymptote at $x = -2$.

From $x = -2$, as x-values decrease, $g(x)$-values approach -1, and as x-values increase, $g(x)$-values approach -1. So there is an asymptote at $g(x) = -1$.

The domain is all real numbers not equal to -2. The range is all real numbers not equal to -1. This is represented as D = $\{x \mid x \neq -2\}$, $(-\infty, -2) \cup (-2, +\infty)$, or $\{-\infty < -2 < +\infty\}$ and R = $\{x \mid x \neq -1\}$, $(-\infty, -1) \cup (-1, +\infty)$, or $\{-\infty < -1 < +\infty\}$.

▶ **Guided Practice**

2A.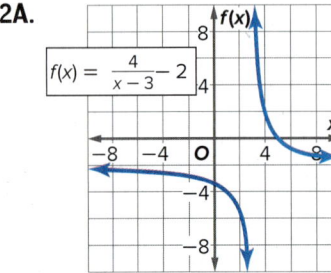

2B.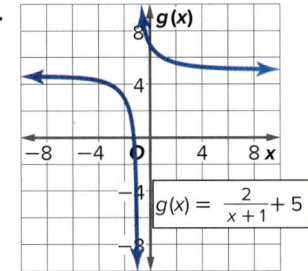

2 Transformations of Reciprocal Functions
The same techniques used to transform the graphs of other functions you have studied can be applied to the graphs of reciprocal functions.

Key Concept Transformations of Reciprocal Functions

$$f(x) = \frac{a}{b(x-h)} + k$$

h—Horizontal Translation	**k—Vertical Translation**
h units right if h is positive	k units up if k is positive
$\|h\|$ units left if h is negative	$\|k\|$ units down if k is negative
The *vertical* asymptote is at $x = h$.	The *horizontal* asymptote is at $f(x) = k$.
a—Orientation and Shape	**b—Orientation and Shape**
If $a < 0$, the graph is reflected in the x-axis.	If $b < 0$, the graph is reflected in the y-axis.
If $\|a\| > 1$, the graph is stretched vertically.	If $\|b\| > 1$, the graph is compressed horizontally.
If $0 < \|a\| < 1$, the graph is compressed vertically.	If $0 < \|b\| < 1$, the graph is stretched horizontally.

Study Tip

Asymptotes The asymptotes of a reciprocal function move with the graph of the function and intersect at (h, k).

Example 3 Graph Transformations

Graph each function. State the domain and range.

a. $f(x) = \dfrac{1}{\frac{1}{2}x - 2} + 2$

This represents a transformation of the graph of $f(x) = \frac{1}{x}$.

$b = \frac{1}{2}$: The graph is stretched horizontally.

$h = 4$: The graph is translated 4 units right because $\frac{1}{2}x - 2 = \frac{1}{2}(x - 4)$. There is an asymptote at $x = 4$.

$k = 2$: The graph is translated 2 units up. There is an asymptote at $f(x) = 2$.

Domain: $\{x \mid x \neq 4\}$ or $(-\infty, 4) \cup (4, +\infty)$

Range: $\{f(x) \mid f(x) \neq 2\}$ or $(-\infty, 2) \cup (2, +\infty)$

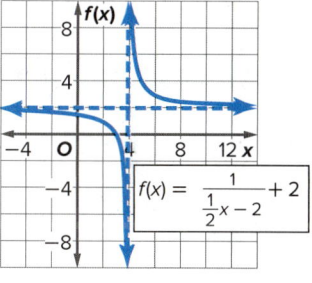

b. $f(x) = \dfrac{-3}{x + 1} - 4$

This represents a transformation of the graph of $f(x) = \frac{1}{x}$.

$a = -3$: The graph is stretched vertically and reflected across the x-axis.

$h = -1$: The graph is translated 1 unit left. There is an asymptote at $x = -1$.

$k = -4$: The graph is translated 4 units down. There is an asymptote at $f(x) = -4$.

Domain: $\{x \mid x \neq -1\}$ or $(-\infty, -1) \cup (-1, +\infty)$

Range: $\{f(x) \mid f(x) \neq -4\}$ or $(-\infty, -4) \cup (-4, +\infty)$

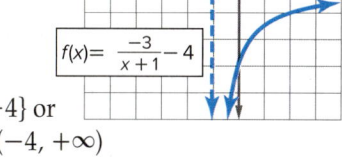

▶ Guided Practice

3A. $f(x) = \dfrac{-2}{x + 4} + 1$

3B. $g(x) = \dfrac{1}{-3x + 3} - 2$

Real-World Career

Pilot Pilots are responsible for transporting many people through the air. Pilots are required to complete many hours of flight training and theoretical study. The first step, in the United States, is acquiring a Private Pilot License at the age of at least 16 years old. Then, an aspiring pilot must get either an instrument rating or multi-engine rating endorsement. Finally, to obtain a professional career, a commercial pilot license endorsement is also required.

Real-World Example 4   Write Equations

TRAVEL Philip is taking a nonstop flight to Hyderabad, India, for a business trip. A one-way trip is about 9000 miles.

a. Write an equation to represent the travel time to Hyderabad as a function of flight speed. Then graph the equation.

Solve $rt = d$ for t.

$rt = d$ Original formula

$t = \dfrac{d}{r}$ Divide each side by r.

$t = \dfrac{9000}{r}$ $d = 9000$

Graph $t = \dfrac{9000}{r}$.

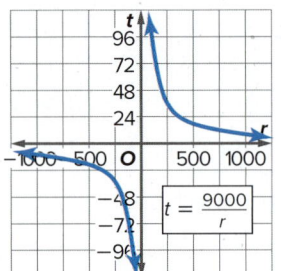

b. Explain any limitations to the range or domain in this situation.

In this situation, the range and domain are limited to all real numbers greater than zero because negative values do not make sense. There will be further restrictions to the domain because the aircraft has minimum and maximum speeds at which it can travel.

Guided Practice

4. HOMECOMING DANCE The junior and senior class officers are sponsoring a homecoming dance. The total cost for the facilities and catering is $45 per person plus a $2500 deposit. Write and graph an equation to represent the average cost per person. Then explain any limitations to the domain and range.

Check Your Understanding

○ = Step-by-Step Solutions begin on page R11.

Go Online! for a Self-Check Quiz

Examples 1–2 Identify the asymptotes, domain, and range of each function.

1.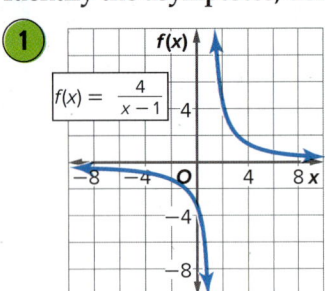

2.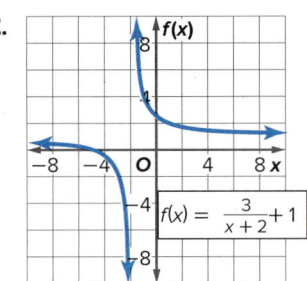

Example 3   Graph each function. State the domain and range.

3. $f(x) = \dfrac{5}{x}$

4. $f(x) = \dfrac{2}{x+3}$

5. $f(x) = \dfrac{-1}{x-2} + 4$

Example 4

6. **SENSE-MAKING** A group of friends plans to get their youth group leader a gift certificate for a day at a spa. The certificate costs $150.

   a. If c represents the cost for each friend and f represents the number of friends, write an equation to represent the cost to each friend as a function of how many friends give.

   b. Graph the function.

   c. Explain any limitations to the range or domain in this situation.

Practice and Problem Solving

Extra Practice is on page R7.

Examples 1–2 Identify the asymptotes, domain, and range of each function.

7.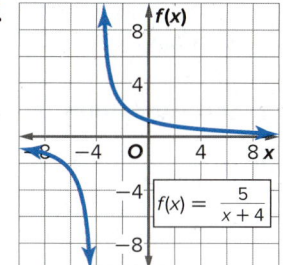

8.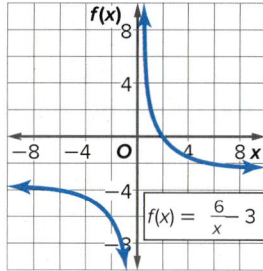

9.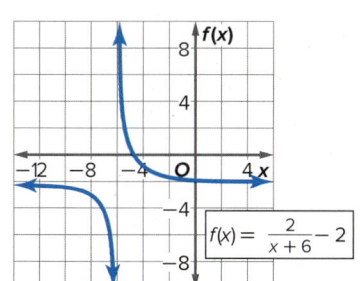

10.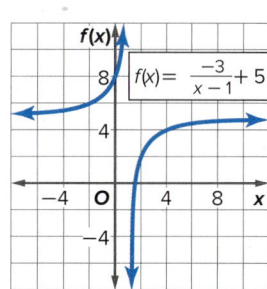

Example 3 Graph each function. State the domain and range.

11. $f(x) = \dfrac{3}{x}$

12. $f(x) = \dfrac{-4}{x+2}$

13. $f(x) = \dfrac{2}{x-6}$

14. $f(x) = \dfrac{6}{x} - 5$

15. $f(x) = \dfrac{2}{x} + 3$

16. $f(x) = \dfrac{8}{x}$

17. $f(x) = \dfrac{-2}{x-5}$

18. $f(x) = \dfrac{3}{x-7} - 8$

19. $f(x) = \dfrac{9}{x+3} + 6$

20. $f(x) = \dfrac{8}{-x+3}$

21. $f(x) = \dfrac{-6}{x+4} - 2$

22. $f(x) = \dfrac{-5}{\frac{1}{4}x - 2} + 2$

Example 4

23. **CYCLING** Marina's New Year's resolution is to ride her bike 5000 miles.

   a. If m represents the mileage Marina rides each day and d represents the number of days, write an equation to represent the mileage each day as a function of the number of days that she rides.

   b. Graph the function.

   c. If she rides her bike every day of the year, how many miles should she ride each day to meet her goal?

24. **MODELING** Parker has 200 grams of an unknown liquid. Knowing the density will help him discover what type of liquid this is.

   a. Density of a liquid is found by dividing the mass by the volume. Write an equation to represent the density of this unknown as a function of volume.

   b. Graph the function.

   c. From the graph, identify the asymptotes, domain, and range of the function.

Graph each function. State the domain and range.

25. $f(x) = \dfrac{3}{2x-4}$

26. $f(x) = \dfrac{5}{3x}$

27. $f(x) = \dfrac{2}{4x+1}$

28. $f(x) = \dfrac{1}{2x+3}$

29. BASEBALL The distance from the pitcher's mound to home plate is 60.5 feet.
  a. If r represents the speed of the pitch and t represents the time it takes the ball to get to the plate, write an equation to represent the speed as a function of time.
  b. Graph the function.
  c. If a two-seam fastball reaches the plate in 0.48 second, what was its speed?

Graph each function. State the domain and range, and identify the asymptotes.

30. $f(x) = \dfrac{-3}{x+7} - 1$
31. $f(x) = \dfrac{-4}{x+2} - 5$
32. $f(x) = \dfrac{6}{x-1} + 2$

33. $f(x) = \dfrac{2}{x-4} + 3$
34. $f(x) = \dfrac{-7}{x-8} - 9$
35. $f(x) = \dfrac{-6}{x-7} - 8$

36. FINANCIAL LITERACY Lawanda's car went 440 miles on one tank of gas.
  a. If g represents the number of miles to the gallon that the car gets and t represents the size of the gas tank, write an equation to represent the miles to the gallon as a function of tank size.
  b. Graph the function.
  c. How many miles does the car get per gallon if it has a 15-gallon tank?

37. MULTIPLE REPRESENTATIONS In this problem you will investigate the similarities and differences between power functions with positive and negative exponents.
  a. **TABULAR** Make a table of values for $a(x) = x^2$, $b(x) = x^{-2}$, $c(x) = x^3$, and $d(x) = x^{-3}$.
  b. **GRAPHICAL** Graph $a(x)$ and $b(x)$ on the same coordinate plane.
  c. **VERBAL** Compare the domain, range, end behavior, and behavior at $x = 0$ for $a(x)$ and $b(x)$.
  d. **GRAPHICAL** Graph $c(x)$ and $d(x)$ on the same coordinate plane.
  e. **VERBAL** Compare the domain, range, end behavior, and behavior at $x = 0$ for $c(x)$ and $d(x)$.
  f. **ANALYTICAL** What conclusions can you make about the similarities and differences between power functions with positive and negative exponents?

H.O.T. Problems   Use Higher-Order Thinking Skills

38. OPEN-ENDED Write a reciprocal function for which the graph has a vertical asymptote at $x = -4$ and a horizontal asymptote at $f(x) = 6$.

39. REASONING Compare and contrast the graphs of each pair of equations.
  a. $y = \dfrac{1}{x}$ and $y - 7 = \dfrac{1}{x}$
  b. $y = \dfrac{1}{x}$ and $y = 4\left(\dfrac{1}{x}\right)$
  c. $y = \dfrac{1}{x}$ and $y = \dfrac{1}{x+5}$
  d. Without making a table of values, use what you observed in parts **a–c** to sketch a graph of $y - 7 = 4\left(\dfrac{1}{x+5}\right)$.

40. CONSTRUCT ARGUMENTS Find the function that does not belong. Explain.

$f(x) = \dfrac{3}{x+1}$ $g(x) = \dfrac{x+2}{x^2+1}$ $h(x) = \dfrac{5}{x^2+2x+1}$ $j(x) = \dfrac{20}{x-7}$

41. CHALLENGE Write two different reciprocal functions with graphs having the same vertical and horizontal asymptotes. Then graph the functions.

42. WRITING IN MATH Refer to the beginning of the lesson. Explain how rational functions can be used to represent shared costs. Explain why only part of the graph is meaningful in the context of the problem.

Preparing for Assessment

43. What is the domain of the function?

$$f(x) = \frac{-3}{x-1} - 2$$

- A $D = \{x \mid x \neq 0\}$
- B $D = \{x \mid x \neq -1\}$
- C $D = \{x \mid x \neq -2\}$
- D $D = \{x \mid x \neq 1\}$

44. For which reciprocal function are the x- and y-values for the asymptotes the same?

- A $f(x) = \frac{-4}{x-4}$
- B $f(x) = \frac{-4}{x-3} - 4$
- C $f(x) = \frac{4}{x} + 4$
- D $f(x) = \frac{1}{x+4} - 4$
- E $f(x) = \frac{1}{x+4} + 4$

45. **MULTI-STEP** Consider the graph of the function $f(x) = \frac{-9}{x+9}$.

a. Which statements about this function are true? Choose all that apply.

- A It has an asymptote at $x = 9$.
- B It has an asymptote at $f(x) = 0$.
- C It has an asymptote at $x = -9$.
- D Its transformation from $f(x) = \frac{1}{x}$ includes a translation 9 units down.
- E Its transformation from $f(x) = \frac{1}{x}$ includes a translation 9 units up.
- F Its transformation from $f(x) = \frac{1}{x}$ includes a translation 9 units left.

b. What is the value of the function at $x = -1$?

- A undefined
- B negative
- C positive
- D infinity

46. **MULTI-STEP** Consider the graph of the function $f(x) = \frac{-1}{x-1} + 1$.

a. Which statements about this function are true? Choose all that apply.

- A It has an asymptote at $x = -1$.
- B It has an asymptote at $x = 1$.
- C It has an asymptote at $f(x) = 1$.
- D Its transformation from $f(x) = \frac{1}{x}$ includes a translation 1 unit down.
- E Its transformation from $f(x) = \frac{1}{x}$ includes a translation 1 unit up.
- F $f(x) = \frac{-8}{x-1} + 1$ has the same asymptotes but is stretched taller.

b. What is the value of the function at $x = -1$?

- A undefined
- B infinity
- C -1
- D 1.5

c. Which is the graph of the function?

○ A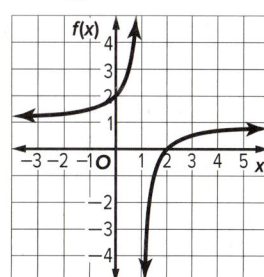

○ B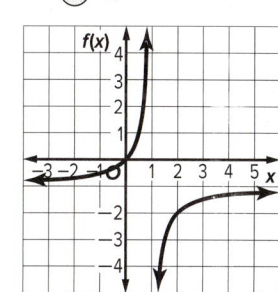

○ C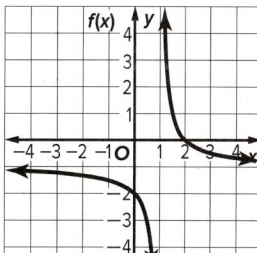

○ D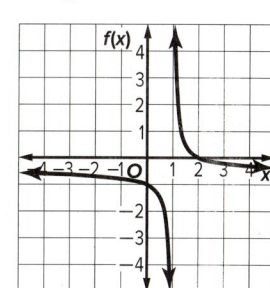

CHAPTER 7
Mid-Chapter Quiz
Lessons 7-1 through 7-3

Simplify each expression. (Lesson 7-1)

1. $\dfrac{2x^2y^5}{7x^3yz} \cdot \dfrac{14xyz^2}{18x^4y}$

2. $\dfrac{24a^4b^6}{35ab^3} \div \dfrac{12abc}{7a^2c}$

3. $\dfrac{3x-3}{x^2+x-2} \cdot \dfrac{4x+8}{6x+18}$

4. $\dfrac{m^2+3m+2}{9} \div \dfrac{m+1}{3m+15}$

5. $\dfrac{\frac{r^2+3r}{r+1}}{\frac{3r}{3r+3}}$

6. $\dfrac{\frac{2y}{y^2-4}}{\frac{3}{y^2-4y+4}}$

7. **MULTIPLE CHOICE** For all $r \neq \pm 2$, $\dfrac{r^2+6r+8}{r^2-4} = $ ___. (Lesson 7-1)

   A $\dfrac{r-2}{r+4}$ C $\dfrac{r+2}{r-4}$

   B $\dfrac{r+4}{r-2}$ D $\dfrac{r+4}{r+2}$

8. **MULTIPLE CHOICE** Identify all values of x for which $\dfrac{x^2-16}{(x^2-6x-27)(x+1)}$ is undefined. (Lesson 7-1)

   A $-3, -1$ C $-3, -1, 9$

   B $3, 1, -9$ D -1

9. What is the LCM of $x^2 - x$ and $3 - 3x$? (Lesson 7-2)

Simplify each expression. (Lesson 7-2)

10. $\dfrac{2x}{4x^2y} + \dfrac{x}{3xy^3}$

11. $\dfrac{3}{4m} + \dfrac{2}{3mn^2} - \dfrac{4}{n}$

12. $\dfrac{6}{r^2-3r-18} - \dfrac{1}{r^2+r-6}$

13. $\dfrac{3x+6}{x+y} + \dfrac{6}{-x-y}$

14. $\dfrac{x-4}{x^2-3x-4} + \dfrac{x+1}{2x-8}$

15. Determine the perimeter of the rectangle. (Lesson 7-2)

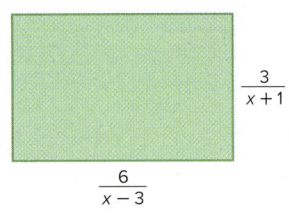

16. **TRAVEL** Lucita is going to a beach 100 miles away. She travels half the distance at one rate. The rest of the distance, she travels 15 miles per hour slower. (Lesson 7-2)

   a. If x represents the faster pace in miles per hour, write an expression that represents the time spent at that pace.

   b. Write an expression for the amount of time spent at the slower pace.

   c. Write an expression for the amount of time Lucita needs to complete the trip.

Identify the asymptotes, domain, and range of each function. (Lesson 7-3)

17.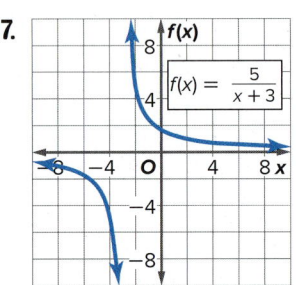

18.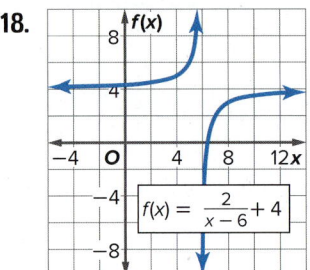

Graph each reciprocal function. State the domain and range. (Lesson 7-3)

19. $f(x) = \dfrac{4}{x}$

20. $f(x) = \dfrac{1}{3x}$

21. $f(x) = \dfrac{6}{x-1}$

22. $f(x) = \dfrac{-2}{x} + 4$

23. $f(x) = \dfrac{3}{x+2} - 5$

24. $f(x) = -\dfrac{1}{x-3} + 2$

25. **SANDWICHES** A group makes 45 sandwiches to take on a picnic. The number of sandwiches a person can eat depends on how many people go on the trip. (Lesson 7-3)

   a. Write a function to represent the numbers of sandwiches that can be taken on the picnic.

   b. Graph the function.

   c. **MP** What mathematical practice did you use to solve this problem?

LESSON 4
Graphing Rational Functions

::Then
- You graphed reciprocal functions.

::Now
1. Graph rational functions with vertical and horizontal asymptotes.
2. Graph rational functions with oblique asymptotes and point discontinuity.

::Why?
Regina bought a season pass to a water park for $146. She plans on paying for one meal in the park every time she visits. The park claims that meals on average cost $12.49. The rational function $W(m) = \frac{12.49m + 146}{m}$ can be used to determine the average cost $W(m)$ for visiting the park m times.

New Vocabulary
rational function
vertical asymptote
horizontal asymptote
oblique asymptote
point discontinuity

Mathematical Practices
7 Look for and make use of structure.

1 Vertical and Horizontal Asymptotes A **rational function** has an equation of the form $f(x) = \frac{a(x)}{b(x)}$, where $a(x)$ and $b(x)$ are polynomial functions and $b(x) \neq 0$.

In order to graph a rational function, it is helpful to locate the zeros and asymptotes. A zero of a rational function $f(x) = \frac{a(x)}{b(x)}$ occurs at every value of x for which $a(x) = 0$.

Key Concept Vertical and Horizontal Asymptotes

Words If $f(x) = \frac{a(x)}{b(x)}$, $a(x)$ and $b(x)$ are polynomial functions with no common factors other than 1, and $b(x) \neq 0$, then:

- $f(x)$ has a **vertical asymptote** whenever $b(x) = 0$.
- $f(x)$ has at most one **horizontal asymptote**.
  - If the degree of $a(x)$ is greater than the degree of $b(x)$, there is no horizontal asymptote.
  - If the degree of $a(x)$ is less than the degree of $b(x)$, the horizontal asymptote is the line $y = 0$.
  - If the degree of $a(x)$ equals the degree of $b(x)$, the horizontal asymptote is the line $y = \frac{\text{leading coefficient of } a(x)}{\text{leading coefficient of } b(x)}$.

Examples

No horizontal asymptote

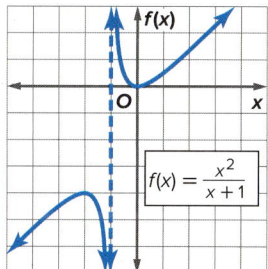

Vertical asymptote:
$x = -1$

One horizontal asymptote

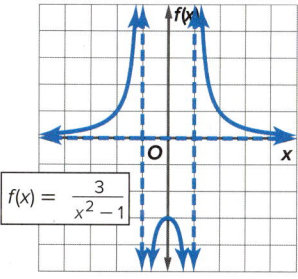

Vertical asymptotes:
$x = -1, x = 1$
Horizontal asymptote:
$f(x) = 0$

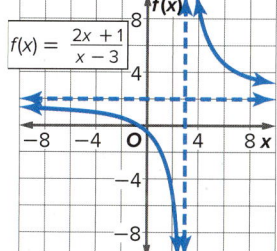

Vertical asymptote:
$x = 3$
Horizontal asymptote:
$f(x) = 2$

connectED.mcgraw-hill.com 491

Watch Out!

Zeros vs. Vertical Asymptotes Zeros of rational functions occur at the values that make the numerator equal to zero. Vertical asymptotes occur at the values that make the denominator equal to zero.

The asymptotes of a rational function can be used to draw the graph of the function. Additionally, the asymptotes can be used to divide a graph into regions to find ordered pairs on the graph.

Example 1 Graph with No Horizontal Asymptote

Graph $f(x) = \dfrac{x^3}{x-1}$.

Step 1 Find the zeros.

$x^3 = 0$ Set $a(x) = 0$.

$x = 0$ Take the cube root of each side.

There is a zero at $x = 0$.

Step 2 Draw the asymptotes.

Find the vertical asymptote.

$x - 1 = 0$ Set $b(x) = 0$.

$x = 1$ Add 1 to each side.

There is a vertical asymptote at $x = 1$.

The degree of the numerator is greater than the degree of the denominator. So, there is no horizontal asymptote.
$D = \{x \mid x \neq 1\}$ and $R = \{\text{all real numbers}\}$.

Study Tip

MP Tools The TABLE feature of a graphing calculator can be used to calculate decimal values for x and y.

Step 3 Draw the graph.

Use a table to find ordered pairs on the graph. Then connect the points.

x	f(x)
−3	6.75
−2	2.67
−1	0.5
0	0
0.5	−0.25
1.5	6.75
2	8
3	13.5

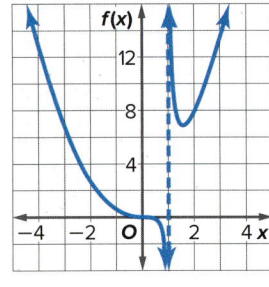

▶ **Guided Practice**

Graph each function.

1A. $f(x) = \dfrac{x^2 - x - 6}{x + 1}$

1B. $f(x) = \dfrac{(x+1)^3}{(x+2)^2}$

In the real world, sometimes values on the graph of a rational function are not meaningful. In the graph at the right, x-values such as time, distance, and number of people cannot be negative in the context of the problem. So, you do not even need to consider that portion of the graph.

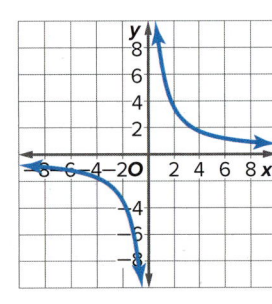

492 | Lesson 7-4 | Graphing Rational Functions

Real-World Career

U.S. Coast Guard Boatswain's Mate
The most versatile member of the U.S. Coast Guard's operational team is the boatswain's mate. BMs are capable of performing almost any task. Training for BMs is accomplished through 12 weeks of intensive training.

Real-World Example 2 Use Graphs of Rational Functions

AVERAGE SPEED A boat traveled upstream at r_1 miles per hour. During the return trip to its original starting point, the boat traveled at r_2 miles per hour. The average speed for the entire trip R is given by the formula
$$R = \frac{2r_1 r_2}{r_1 + r_2}.$$

a. Let r_1 be the independent variable, and let R be the dependent variable. Draw the graph if $r_2 = 10$ miles per hour.

The function is $R = \dfrac{2r_1(10)}{r_1 + (10)}$ or $R = \dfrac{20r_1}{r_1 + 10}$.

The vertical asymptote is $r_1 = -10$.
Graph the vertical asymptote and the function.
Notice that the horizontal asymptote is $R = 20$.

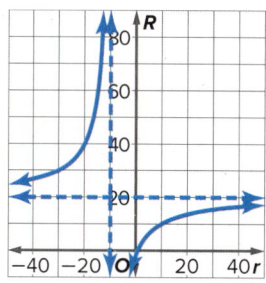

b. What is the R-intercept of the graph?

The R-intercept is 0.

c. What domain and range values are meaningful in the context of the problem?

In the problem context, speeds are nonnegative values. Therefore, only values of r_1 greater than or equal to 0 and values of R between 0 and 20 are meaningful.

▶ **Guided Practice**

2. SALARIES A company uses the formula $S(x) = \dfrac{45x + 25}{x + 1}$ to determine the salary in thousands of dollars of an employee during his xth year. Graph $S(x)$. What domain and range values are meaningful in the context of the problem? What is the meaning of the horizontal asymptote for the graph?

2 Oblique Asymptotes and Point Discontinuity
An **oblique asymptote**, sometimes called a *slant asymptote*, is an asymptote that is neither horizontal nor vertical.

Key Concept Oblique Asymptotes

Words If $f(x) = \dfrac{a(x)}{b(x)}$, $a(x)$ and $b(x)$ are polynomial functions with no common factors other than 1 and $b(x) \neq 0$, then $f(x)$ has an oblique asymptote if the degree of $a(x)$ minus the degree of $b(x)$ equals 1. The equation of the asymptote is $f(x) = \dfrac{a(x)}{b(x)}$ with no remainder.

Example $f(x) = \dfrac{x^4 + 3x^3}{x^3 - 1}$

Vertical asymptote: $x = 1$
Oblique asymptote: $f(x) = x + 3$

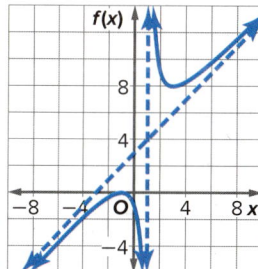

> **Study Tip**
> **Oblique Asymptotes**
> Oblique asymptotes occur for rational functions that have a numerator polynomial that is one degree higher than the denominator polynomial.

Example 3   Determine Oblique Asymptotes

Graph $f(x) = \dfrac{x^2 + 4x + 4}{2x - 1}$.

Step 1 Find the zeros.

$x^2 + 4x + 4 = 0$ Set $a(x) = 0$.

$(x + 2)^2 = 0$ Factor.

$x + 2 = 0$ Take the square root of each side.

$x = -2$ Subtract 2 from each side.

There is a zero at $x = -2$.

Step 2 Find the asymptotes.

$2x - 1 = 0$ Set $b(x) = 0$.

$2x = 1$ Add 1 to each side.

$x = \dfrac{1}{2}$ Divide each side by 2.

There is a vertical asymptote at $x = \dfrac{1}{2}$.

The degree of the numerator is greater than the degree of the denominator, so there is no horizontal asymptote.

The difference between the degree of the numerator and the degree of the denominator is 1, so there is an oblique asymptote.

Divide the numerator by the denominator to determine the equation of the oblique asymptote.

The equation of the asymptote is the quotient excluding any remainder.

$$\begin{array}{r}\frac{1}{2}x + \frac{9}{4}\\2x-1\overline{)x^2 + 4x + 4}\\(-)x^2 - \frac{1}{2}x\\\hline\frac{9}{2}x + 4\\(-)\frac{9}{2}x - \frac{9}{4}\\\hline\frac{25}{4}\end{array}$$

Thus, the oblique asymptote is the line $f(x) = \dfrac{1}{2}x + \dfrac{9}{4}$.

Step 3 Draw the asymptotes, and then use a table of values to graph the function.

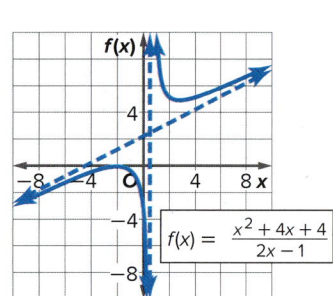

Guided Practice

Graph each function.

3A. $f(x) = \dfrac{x^2}{x - 2}$

3B. $f(x) = \dfrac{x^3 - 1}{x^2 - 4}$

In some cases, graphs of rational functions may have **point discontinuity**, which looks like a hole in the graph. This is because the function is undefined at that point.

> **Go Online!**
> Explore the graphs of rational functions using **The Geometer's Sketchpad®** activity in ConnectED.
>
>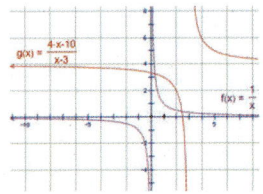

Key Concept Point Discontinuity

Words If $f(x) = \dfrac{a(x)}{b(x)}$, $b(x) \neq 0$, and $x - c$ is a factor of both $a(x)$ and $b(x)$, then there is a point discontinuity at $x = c$.

Example $f(x) = \dfrac{(x+2)(x+1)}{x+1}$
$= x + 2; x \neq -1$

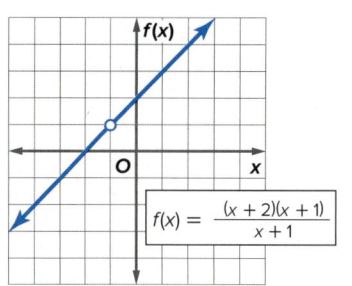

Example 4 Graph with Point Discontinuity

Graph $f(x) = \dfrac{x^2 - 16}{x - 4}$.

Notice that $\dfrac{x^2 - 16}{x - 4} = \dfrac{(x+4)(x-4)}{x-4}$ or $x + 4$.

Therefore, the graph of $f(x) = \dfrac{x^2 - 16}{x - 4}$ is the graph of $f(x) = x + 4$ with a point discontinuity at $x = 4$.

$D = \{x \mid x \neq 4\}$ and $R = \{f(x) \mid f(x) \neq 8\}$

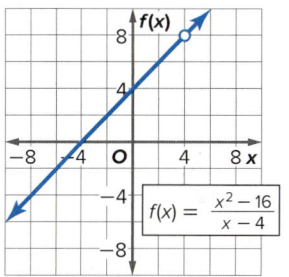

> **Watch Out!**
> **Holes** Remember that a common factor in the numerator and denominator can signal a point discontinuity.

Guided Practice

Graph each function.

4A. $f(x) = \dfrac{x^2 + 4x - 5}{x + 5}$

4B. $f(x) = \dfrac{x^3 + 2x^2 - 9x - 18}{x^2 - 9}$

Check Your Understanding

 = Step-by-Step Solutions begin on page R11.

Example 1 Graph each function.

1. $f(x) = \dfrac{x^4 - 2}{x^2 - 1}$

2. $f(x) = \dfrac{x^3}{x + 2}$

Example 2

3. **REASONING** Eduardo is a kicker for his high school football team. So far this season, he has made 7 out of 11 field goals. He would like to improve his field goal percentage. If he can make x consecutive field goals, his field goal percentage can be determined using the function $P(x) = \dfrac{7 + x}{11 + x}$.
   a. Graph the function.
   b. What part of the graph is meaningful in the context of this problem?
   c. Describe the meaning of the intercept of the vertical axis.
   d. What is the equation of the horizontal asymptote? Explain its meaning with respect to Eduardo's field goal percentage.

Examples 3–4 Graph each function.

4. $f(x) = \dfrac{6x^2 - 3x + 2}{x}$

5. $f(x) = \dfrac{x^2 + 8x + 20}{x + 2}$

6. $f(x) = \dfrac{x^2 - 4x - 5}{x + 1}$

7. $f(x) = \dfrac{x^2 + x - 12}{x + 4}$

Practice and Problem Solving

Extra Practice is on page R7.

Example 1 Graph each function.

8. $f(x) = \dfrac{x^4}{6x+12}$

9. $f(x) = \dfrac{x^3}{8x-4}$

10. $f(x) = \dfrac{x^4-16}{x^2-1}$

11. $f(x) = \dfrac{x^3+64}{16x-24}$

Example 2

12. **SCHOOL SPIRIT** As president of Student Council, Brandy is getting T-shirts made for a pep rally. Each T-shirt costs $9.50, and there is a set-up fee of $75. The student council plans to sell the shirts, but each of the 15 council members will get one for free.

   a. Write a function for the average cost of a T-shirt to be sold. Graph the function.
   b. What is the average cost if 200 shirts are ordered? if 500 shirts are ordered?
   c. How many T-shirts must be ordered to bring the average cost under $9.75?

Examples 3–4 Graph each function.

13. $f(x) = \dfrac{x}{x+2}$

14. $f(x) = \dfrac{5}{(x-1)(x+4)}$

15. $f(x) = \dfrac{4}{(x-2)^2}$

16. $f(x) = \dfrac{x-3}{x+1}$

17. $f(x) = \dfrac{1}{(x+4)^2}$

18. $f(x) = \dfrac{2x}{(x+2)(x-5)}$

19. $f(x) = \dfrac{(x-4)^2}{x+2}$

20. $f(x) = \dfrac{(x+3)^2}{x-5}$

21. $f(x) = \dfrac{x^3+1}{x^2-4}$

22. $f(x) = \dfrac{4x^3}{2x^2+x-1}$

23. $f(x) = \dfrac{3x^2+8}{2x-1}$

24. $f(x) = \dfrac{2x^2+5}{3x+4}$

25. $f(x) = \dfrac{x^4-2x^2+1}{x^3+2}$

26. $f(x) = \dfrac{x^4-x^2-12}{x^3-6}$

27. **MP PERSEVERANCE** The graph of a certain rational function has 2 branches, passes through the point (0, 0), has no horizontal asymptotes, and a vertical asymptote of $x = -1$.

   a. Find an equation of the rational function.
   b. How did you develop a plan for solving this problem?
   c. What assumptions did you make in your solution process?

Example 4 Graph each function.

28. $f(x) = \dfrac{x^2-2x-8}{x-4}$

29. $f(x) = \dfrac{x^2+4x-12}{x-2}$

30. $f(x) = \dfrac{x^2-25}{x+5}$

31. $f(x) = \dfrac{x^2-64}{x-8}$

32. $f(x) = \dfrac{(x-4)(x^2-4)}{x^2-6x+8}$

33. $f(x) = \dfrac{(x+5)(x^2+2x-3)}{x^2+8x+15}$

34. $f(x) = \dfrac{3x^4+6x^3+3x^2}{x^2+2x+1}$

35. $f(x) = \dfrac{2x^4+10x^3+12x^2}{x^2+5x+6}$

496 | Lesson 7-4 | Graphing Rational Functions

36. **BUSINESS** Liam purchased a riding lawn mower for $4500 and mows the lawns of local businesses. Each time he mows a lawn, he incurs a cost of $50 for gas and maintenance.

   a. Write and graph the rational function representing his average cost per customer as a function of the number of lawns.
   b. What are the asymptotes of the graph?
   c. Why is the first quadrant in the graph the only relevant quadrant?
   d. How many total lawns does Liam need to mow for his average cost per lawn to be less than $80?

37. **FINANCIAL LITERACY** Kristina bought a new smartphone with a data plan. The phone cost $150, and her monthly usage charge is $30 plus $10 for the data plan.

   a. Write and graph the rational function representing her average monthly cost as a function of the number of months Kristina uses the phone.
   b. What are the asymptotes of the graph?
   c. Why is the first quadrant in the graph the only relevant quadrant?
   d. After how many months will the average monthly charge be $45?

38. **SENSE-MAKING** Alana plays softball for Centerville High School. So far this season she has gotten a hit 4 out of 12 times at bat. She is determined to improve her batting average. If she can get x consecutive hits, her batting average can be determined using $B(x) = \frac{4 + x}{12 + x}$.

   a. Graph the function.
   b. What part of the graph is meaningful in the context of the problem?
   c. Describe the meaning of the intercept of the vertical axis.
   d. What is the equation of the horizontal asymptote? Explain its meaning with respect to Alana's batting average.

Graph each function.

39. $f(x) = \frac{x + 1}{x^2 + 6x + 5}$ 40. $f(x) = \frac{x^2 - 10x - 24}{x + 2}$ 41. $f(x) = \frac{6x^2 + 4x + 2}{x + 2}$

H.O.T. Problems Use Higher-Order Thinking Skills

42. **OPEN-ENDED** Sketch the graph of a rational function with a horizontal asymptote $y = 1$ and a vertical asymptote $x = -2$.

43. **CHALLENGE** Compare and contrast $g(x) = \frac{x^2 - 1}{x(x^2 - 2)}$ and $f(x)$ shown at the right.

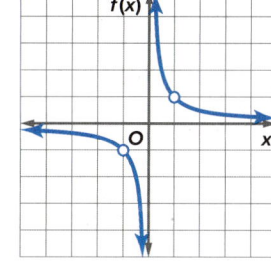

44. **REASONING** What is the difference between the graphs of $f(x) = x - 2$ and $g(x) = \frac{(x + 3)(x - 2)}{x + 3}$?

45. **PROOF** A rational function has an equation of the form $f(x) = \frac{a(x)}{b(x)}$, where $a(x)$ and $b(x)$ are polynomial functions and $b(x) \neq 0$. Show that $f(x) = \frac{x}{a - b} + c$ is a rational function.

46. **WRITING IN MATH** How can factoring be used to determine the vertical asymptotes or point discontinuity of a rational function?

Preparing for Assessment

47. Which function has a graph with an oblique asymptote? **MP 7**

- A $f(x) = \dfrac{(x-1)^3}{x+4}$
- B $f(x) = \dfrac{(x-1)^3(x-2)^2}{(x-4)^3}$
- C $f(x) = \dfrac{(2x-1)^2}{x^2+4}$
- D $f(x) = \dfrac{x(x+3)^2}{x^2+2}$

48. Which function is graphed below? **MP 7**

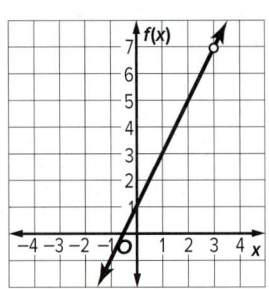

- A $f(x) = \dfrac{2x^2 - 5x - 3}{x - 3}$
- B $f(x) = 2x + 1$
- C $f(x) = -2x + 1$
- D $f(x) = \dfrac{2x^2 - 5x - 3}{x + 3}$
- E $f(x) = \dfrac{2x - 5}{1}$
- F $f(x) = -2x + 5$

49. A group of friends are attending a ball game in another state. Each person spends $25 for a ticket and they share the transportation costs of $150 equally. Which function represents the total cost for each person? **MP 7**

- A $f(x) = 150x + 25$
- B $f(x) = 25x + 150$
- C $f(x) = \dfrac{150}{x} + 25$
- D $f(x) = \dfrac{150}{x} + \dfrac{25}{x}$
- E $f(x) = \dfrac{150}{x} - 25$
- F $f(x) = \dfrac{25}{x} + 150$

50. Which of the following functions is its own inverse? **MP 7**

- A $f(x) = 2x^2 - 7x - 15$
- B $f(x) = \dfrac{7}{x}$
- C $f(x) = 7x^2 + 2$
- D $f(x) = \dfrac{x}{7}$

51. Consider the function $f(x) = \dfrac{1}{6x^2 - 12x - 18}$. **MP 7**

a. Where are the discontinuities?
- A $x = -3, x = -1$
- B $x = -3, x = 1$
- C $x = 3, x = -1$
- D $x = 3, x = 1$

b. How many horizontal asymptotes does the function have?
- A 1
- B 2
- C 3
- D none

c. What are the vertical asymptotes?
- A $x = 3, x = 1$
- B $x = 3, x = -1$
- C $x = -3, x = -1$
- D $x = -3, x = 1$

d. How many x-intercepts are there?
- A 1
- B 2
- C 3
- D none

498 | Lesson 7-4 | Graphing Rational Functions

EXTEND 7-4

Graphing Technology Lab
Graphing Rational Functions

A TI-83/84 Plus graphing calculator can be used to explore graphs of rational functions. These graphs have some features that never appear in the graphs of polynomial functions.

Activity 1 Graph with Asymptotes

Work cooperatively. Graph $y = \dfrac{8x - 5}{2x}$ in the standard viewing window. Find the equations of any asymptotes. State the domain and range of the function.

Step 1 Enter the equation in the **Y=** list, and then graph.

Step 2 Examine the graph.

By looking at the equation, we can determine that if $x = 0$, the function is undefined. The equation of the vertical asymptote is $x = 0$. Notice what happens to the y-values as x grows larger and as x gets smaller. The y-values approach 4. So, the equation for the horizontal asymptote is $y = 4$. The domain is $\{x \mid x \neq 0\}$, and the range is all real numbers.

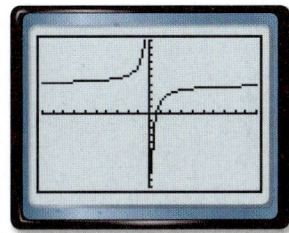

[−10, 10] scl: 1 by [−10, 10] scl: 1

Activity 2 Graph with Point Discontinuity

Work cooperatively. Graph $y = \dfrac{x^2 - 16}{x + 4}$ in the window [−5, 4.4] by [−10, 2] with scale factors of 1.

Step 1 Because the function is not continuous, put the calculator in dot mode.

Step 2 Examine the graph.

This graph looks like a line. Because the denominator has a factor of $x + 4$ and there is no vertical asymptote at $x = -4$, the graph must have a break in continuity at $x = -4$. Therefore, the function is undefined and has point discontinuity at $x = -4$.

If you **TRACE** along the graph, when you come to $x = -4$, you will see that there is no corresponding y-value.

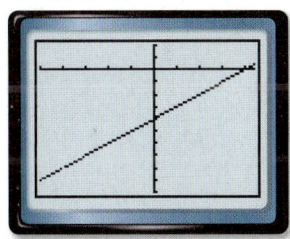

[−5, 4.4] scl: 1 by [−10, 2] scl: 1

Exercises

Work cooperatively. Use a graphing calculator to graph each function. Write the x-coordinates of any points of discontinuity and/or the equations of any asymptotes. State the domain and range.

1. $f(x) = \dfrac{1}{x}$

2. $f(x) = \dfrac{x}{x + 2}$

3. $f(x) = \dfrac{2}{x - 4}$

4. $f(x) = \dfrac{2x}{3x - 6}$

5. $f(x) = \dfrac{4x + 2}{x - 1}$

6. $f(x) = \dfrac{x^2 - 9}{x + 3}$

LESSON 5

Variation Functions

::Then
- You wrote and graphed linear equations.

::Now
 Recognize and solve direct and joint variation problems.

 Recognize and solve inverse and combined variation problems.

::Why?
- While building skateboard ramps, Yu determined that the best ramps were the ones in which the length of the top of the ramp was 1.5 times as long as the height of the ramp.

As shown in the table, the length of the top of the ramp depends on the height of a ramp. The length increases as the height increases, but the ratio remains the same, or is *constant*.

The equation $\frac{\ell}{h} = 1.5$ can be written as $\ell = 1.5h$.

The length *varies directly* with the height of the ramp.

Length (ℓ)	Height (h)	Ratio $\frac{\ell}{h}$
3	2	1.5
6	4	1.5
9	6	1.5
12	8	1.5

New Vocabulary
direct variation
constant of variation
joint variation
inverse variation
combined variation

MP Mathematical Practices
1 Make sense of problems and persevere in solving them.
2 Reason abstractly and quantitatively.
4 Model with mathematics.

Direct Variation and Joint Variation The relationship given by $\ell = 1.5h$ is an example of direct variation. A **direct variation** can be expressed in the form $y = kx$. In this equation, k is called the **constant of variation**.

Notice that the graph of $\ell = 1.5h$ is a straight line through the origin. A direct variation is a special case of an equation written in slope-intercept form, $y = mx + b$. When $m = k$ and $b = 0$, $y = mx + b$ becomes $y = kx$. So the slope of a direct variation equation is its constant of variation.

To express a direct variation, we say that y varies directly as x. In other words, as x increases, y increases or decreases at a constant rate.

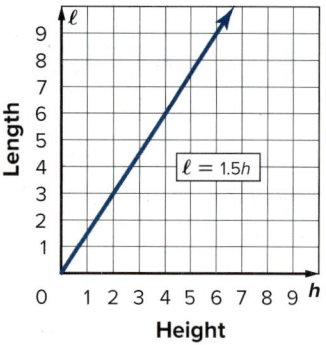

Key Concept Direct Variation

Words y varies directly as x if there is some nonzero constant k such that $y = kx$. k is called the *constant of variation*.

Example   If $y = 3x$ and $x = 7$, then $y = 3(7)$ or 21.

If you know that y varies directly as x and one set of values, you can use a proportion to find the other set of corresponding values.

$$y_1 = kx_1 \qquad \text{and} \qquad y_2 = kx_2$$
$$\frac{y_1}{x_1} = k \qquad\qquad\qquad \frac{y_2}{x_2} = k \qquad \text{Therefore, } \frac{y_1}{x_1} = \frac{y_2}{x_2}.$$

Using the properties of equality, you can find many other proportions that relate these same x- and y-values.

Example 1   Direct Variation

If y varies directly as x and $y = 15$ when $x = -5$, find y when $x = 7$.

Use a proportion that relates the values.

$\dfrac{y_1}{x_1} = \dfrac{y_2}{x_2}$ Direct variation

$\dfrac{15}{-5} = \dfrac{y_2}{7}$ $y_1 = 15$, $x_1 = -5$, and $x_2 = 7$

$15(7) = -5(y_2)$ Cross multiply.

$105 = -5y_2$ Simplify.

$-21 = y_2$ Divide each side by -5.

Guided Practice

1. If r varies directly as t and $r = -20$ when $t = 4$, find r when $t = -6$.

Another type of variation is joint variation. **Joint variation** occurs when one quantity varies directly as the product of two or more other quantities.

> **StudyTip**
>
> **Joint Variation** Some mathematicians consider joint variation a special type of combined variation.

Key Concept   Joint Variation

Words y varies jointly as x and z if there is some nonzero constant k such that $y = kxz$.

Example   If $y = 5xz$, $x = 6$, and $z = -2$, then $y = 5(6)(-2)$ or -60.

If you know that y varies jointly as x and z and one set of values, you can use a proportion to find the other set of corresponding values.

$y_1 = kx_1z_1$ and $y_2 = kx_2z_2$

$\dfrac{y_1}{x_1z_1} = k$ $\dfrac{y_2}{x_2z_2} = k$ Therefore, $\dfrac{y_1}{x_1z_1} = \dfrac{y_2}{x_2z_2}$.

Example 2   Joint Variation

Suppose y varies jointly as x and z. Find y when $x = 9$ and $z = 2$, if $y = 20$ when $z = 3$ and $x = 5$.

Use a proportion that relates the values.

$\dfrac{y_1}{x_1z_1} = \dfrac{y_2}{x_2z_2}$ Joint variation

$\dfrac{20}{5(3)} = \dfrac{y_2}{9(2)}$ $y_1 = 20$, $x_1 = 5$, $z_1 = 3$, $x_2 = 9$, and $z_2 = 2$

$20(9)(2) = 5(3)(y_2)$ Cross multiply.

$360 = 15y_2$ Simplify.

$24 = y_2$ Divide each side by 15.

Guided Practice

2. Suppose r varies jointly as v and t. Find r when $v = 2$ and $t = 8$, if $r = 70$ when $v = 10$ and $t = 4$.

connectED.mcgraw-hill.com

2 Inverse Variation and Combined Variation

Another type of variation is inverse variation. If two quantities x and y show **inverse variation**, their product is equal to a constant k.

Inverse variation is often described as one quantity increasing while the other quantity is decreasing. For example, speed and time for a fixed distance vary inversely with each other; the faster you go, the less time it takes you to get there.

> **Key Concept** Inverse Variation
>
> **Words** y varies inversely as x if there is some nonzero constant k such that $xy = k$ or $y = \frac{k}{x}$, where $x \neq 0$ and $y \neq 0$.
>
> **Example**   If $xy = 2$, and $x = 6$, then $y = \frac{2}{6}$ or $\frac{1}{3}$.

Study Tip

MP Reasoning You can identify the type of variation by looking at a table of values for x and y. If the quotient $\frac{y}{x}$ has a constant value, y varies directly as x. If the product xy has a constant value, y varies inversely as x.

Suppose y varies inversely as x such that $xy = 6$ or $y = \frac{6}{x}$. The graph of this equation is shown at the right. Because k is a positive value, as the values of x increase, the values of y decrease.

Notice that the graph of an inverse variation is a reciprocal function.

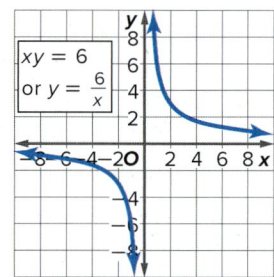

A proportion can be used with inverse variation to solve problems in which some quantities are known. The following proportion is only one of several that can be formed.

$$x_1 y_1 = k \text{ and } x_2 y_2 = k$$

$$x_1 y_1 = x_2 y_2 \qquad \text{Substitution Property of Equality}$$

$$\frac{x_1}{y_2} = \frac{x_2}{y_1} \qquad \text{Divide each side by } y_1 y_2.$$

Example 3   Inverse Variation

If a varies inversely as b and $a = 28$ when $b = -2$, find a when $b = -10$.

Use a proportion that relates the values.

$$\frac{a_1}{b_2} = \frac{a_2}{b_1} \qquad \text{Inverse variation}$$

$$\frac{28}{-10} = \frac{a_2}{-2} \qquad a_1 = 28, b_1 = -2, \text{ and } b_2 = -10$$

$$28(-2) = -10(a_2) \qquad \text{Cross multiply.}$$

$$-56 = -10(a_2) \qquad \text{Simplify.}$$

$$5\frac{3}{5} = a_2 \qquad \text{Divide each side by } -10.$$

Go Online!

Investigate how variation functions can be used to model astronauts' weights on other planets in a **Spreadsheet Activity** from the Resources in ConnectED.

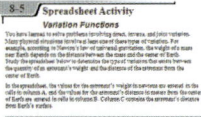
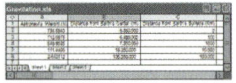

▶ **Guided Practice**

3. If x varies inversely as y and $x = 24$ when $y = 4$, find x when $y = 12$.

Inverse variation is often used in real-world situations.

Real-World Link

When you pluck a string, it vibrates back and forth. This causes mechanical energy to travel through the air in waves. The number of times per second these waves hit our ear is called the *frequency*. The more waves per second, the higher the pitch.

Study Tip

Combined Variation
Quantities that vary directly appear in the numerator. Quantities that vary inversely appear in the denominator.

Real-World Example 4 Write and Solve an Inverse Variation

MUSIC The length of a violin string varies inversely as the frequency of its vibrations. A violin string 10 inches long vibrates at a frequency of 512 cycles per second. Find the frequency of an 8-inch violin string.

Let $v_1 = 10$, $f_1 = 512$, and $v_2 = 8$. Solve for f_2.

$v_1 f_1 = v_2 f_2$ Original equation
$10 \cdot 512 = 8 \cdot f_2$ $v_1 = 10$, $f_1 = 512$, and $v_2 = 8$
$\frac{5120}{8} = f_2$ Divide each side by 8.
$640 = f_2$ Simplify.

The 8-inch violin string vibrates at a frequency of 640 cycles per second.

▶ **Guided Practice**

4. The apparent length of an object is inversely proportional to one's distance from the object. Earth is about 93 million miles from the Sun. Jupiter is about 483.6 million miles from the Sun. Find how many times as large the diameter of the Sun would appear on Earth as on Jupiter.

Another type of variation is combined variation. **Combined variation** occurs when one quantity varies directly and/or inversely as two or more other quantities.

If you know that y varies directly as x, y varies inversely as z, and one set of values, you can use a proportion to find the other set of corresponding values.

$y_1 = \frac{kx_1}{z_1}$ and $y_2 = \frac{kx_2}{z_2}$
$\frac{y_1 z_1}{x_1} = k$ $\frac{y_2 z_2}{x_2} = k$ Therefore, $\frac{y_1 z_1}{x_1} = \frac{y_2 z_2}{x_2}$.

Example 5 Combined Variation

Suppose f varies directly as g, and f varies inversely as h. Find g when $f = 18$ and $h = -3$, if $g = 24$ when $h = 2$ and $f = 6$.

First set up a correct proportion for the information given.

$f_1 = \frac{kg_1}{h_1}$ and $f_2 = \frac{kg_2}{h_2}$ g varies directly as f, so g goes in the numerator. h varies inversely as f, so h goes in the denominator.

$k = \frac{f_1 h_1}{g_1}$ and $k = \frac{f_2 h_2}{g_2}$ Solve for k.

$\frac{f_1 h_1}{g_1} = \frac{f_2 h_2}{g_2}$ Set the two proportions equal to each other.

$\frac{6(2)}{24} = \frac{18(-3)}{g_2}$ $f_1 = 6$, $g_1 = 24$, $h_1 = 2$, $f_2 = 18$, and $h_2 = -3$

$24(18)(-3) = 6(2)(g_2)$ Cross multiply.
$-1296 = 12g_2$ Simplify.
$-108 = g_2$ Divide each side by 12.

When $f = 18$ and $h = -3$, the value of g is -108.

▶ **Guided Practice**

5. Suppose p varies directly as r, and p varies inversely as t. Find t when $r = 10$ and $p = -5$, if $t = 20$ when $p = 4$ and $r = 2$.

Check Your Understanding

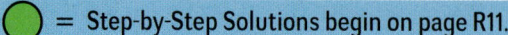

Examples 1–3

1. If y varies directly as x and $y = 12$ when $x = 8$, find y when $x = 14$.

2. Suppose y varies jointly as x and z. Find y when $x = 9$ and $z = -3$, if $y = -50$ when z is 5 and x is -10.

3. If y varies inversely as x and $y = -18$ when $x = 16$, find x when $y = 9$.

Example 4

4. **TRAVEL** A map of Texas is scaled so that 2 inches represents 30 miles. How far apart are El Paso and Odessa if they are 20 inches apart on the map?

Example 5

5. Suppose a varies directly as b, and a varies inversely as c. Find b when $a = 8$ and $c = -3$, if $b = 16$ when $c = 2$ and $a = 4$.

6. Suppose d varies directly as f, and d varies inversely as g. Find g when $d = 6$ and $f = -7$, if $g = 12$ when $d = 9$ and $f = 3$.

Practice and Problem Solving

Extra Practice is on page R7.

Example 1 If x varies directly as y, find x when $y = 8$.

7. $x = 6$ when $y = 32$
8. $x = 11$ when $y = -3$
9. $x = 14$ when $y = -2$
10. $x = -4$ when $y = 10$

11. **MOON** Astronaut Neil Armstrong, the first man on the Moon, weighed 360 pounds on Earth with all his equipment on, but weighed only 60 pounds on the Moon. Write an equation that relates weight on the Moon m with weight on Earth w.

Example 2 Suppose a varies jointly as b and c. Find a when $b = 4$ and $c = -3$.

12. $a = -96$ when $b = 3$ and $c = -8$
13. $a = -60$ when $b = -5$ and $c = 4$
14. $a = -108$ when $b = 2$ and $c = 9$
15. $a = 24$ when $b = 8$ and $c = 12$

16. **MODELING** According to the A.C. Nielsen Company, the average American watches about 5 hours of television per day.

   a. Write an equation to represent the average number of hours spent watching television by m household members during a period of d days.

   b. Assume that members of your household watch the same amount of television each day as the average American. How many hours of television would the members of your household watch in a week?

Example 3 If f varies inversely as g, find f when $g = -6$.

17. $f = 15$ when $g = 9$
18. $f = 4$ when $g = 28$
19. $f = -12$ when $g = 19$
20. $f = 0.6$ when $g = -21$

21. **COMMUNITY SERVICE** Every year students at West High School collect canned goods for a local food pantry. They plan to distribute flyers to homes in the community asking for donations. Last year, 12 students were able to distribute 1000 flyers in four hours.

   a. Write an equation that relates the number of students s to the amount of time t it takes to distribute 1000 flyers.

   b. How long would it take 15 students to hand out the same number of flyers this year?

Example 4

22. **BIRDS** When a group of snow geese migrate, the distance that they fly varies directly with the amount of time they are in the air.

    a. A group of snow geese migrated 375 miles in 7.5 hours. Write a direct variation equation that represents this situation.

    b. Every year, geese migrate 3000 miles from their winter home in the southwest United States to their summer home in the Canadian Arctic. Estimate the number of hours of flying time that it takes for the geese to migrate.

Example 5

23. Suppose a varies directly as b, and a varies inversely as c. Find b when $a = 5$ and $c = -4$, if $b = 12$ when $c = 3$ and $a = 8$.

24. Suppose x varies directly as y, and x varies inversely as z. Find z when $x = 10$ and $y = -7$, if $z = 20$ when $x = 6$ and $y = 14$.

Determine whether each relation shows *direct* or *inverse* variation, or *neither*.

25.
x	y
4	12
8	24
16	48
32	96

26.
x	y
8	2
4	4
−2	−8
−8	−2

27.
x	y
2	4
3	9
4	16
5	25

28. If y varies inversely as x and $y = 6$ when $x = 19$, find y when $x = 2$.

29. If x varies inversely as y and $x = 16$ when $y = 5$, find x when $y = 20$.

30. Suppose a varies directly as b, and a varies inversely as c. Find b when $a = 7$ and $c = -8$, if $b = 15$ when $c = 2$ and $a = 4$.

31. Suppose x varies directly as y, and x varies inversely as z. Find z when $x = 8$ and $y = -6$, if $z = 26$ when $x = 8$ and $y = 13$.

State whether each equation represents a *direct*, *joint*, *inverse*, or *combined* variation. Then name the constant of variation.

32. $\frac{x}{y} = 2.75$ 33. $fg = -2$ 34. $a = 3bc$ 35. $10 = \frac{xy^2}{z}$

36. $y = -11x$ 37. $\frac{n}{p} = 4$ 38. $9n = pr$ 39. $-2y = z$

40. $a = 27b$ 41. $c = \frac{7}{d}$ 42. $-10 = gh$ 43. $m = 20cd$

44. **PRECISION** The volume of a gas v varies inversely as the pressure p and directly as the temperature t.

    a. Write an equation to represent the volume of a gas in terms of pressure and temperature. Is your equation a *direct*, *joint*, *inverse*, or *combined* variation?

    b. A certain gas has a volume of 8 liters, a temperature of 275 Kelvin, and a pressure of 1.25 atmospheres. If the gas is compressed to a volume of 6 liters and is heated to 300 Kelvin, what will the new pressure be?

    c. If the volume stays the same, but the pressure drops by half, then what must have happened to the temperature?

45. **VACATION** The time it takes the Levensteins to reach Lake Tahoe varies inversely with their average rate of speed.

    a. If they are 800 miles away, write and graph an equation relating their travel time to their average rate of speed.

    b. What minimum average speed will allow them to arrive within 18 hours?

46. VIDEO The maximum number of videos that a smartphone can store depends on the lengths of the videos and the quality of the files. A movie will take up more space on the phone than a television show.

  a. If a certain phone has 64 gigabytes of storage space, write a function that represents the number of files the player can hold as a function of the average size of the files.

  b. Is your function a *direct*, *joint*, *inverse*, or *combined* variation?

  c. Suppose the average file size for a movie is 0.75 gigabyte and the average size for a television show is 0.25 gigabyte. Determine how many more files the phone can hold if they are television shows than if they are movies.

47. GRAVITY According to the Law of Universal Gravitation, the attractive force F in newtons between any two bodies in the universe is directly proportional to the product of the masses m_1 and m_2 in kilograms of the two bodies and inversely proportional to the square of the distance d in meters between the bodies. That is, $F = \dfrac{Gm_1m_2}{d^2}$. G is the universal gravitational constant. Its value is 6.67×10^{-11} Nm²/kg².

  a. The distance between Earth and the Moon is about 3.84×10^8 meters. The mass of the Moon is 7.36×10^{22} kilograms. The mass of Earth is 5.97×10^{24} kilograms. What is the gravitational force that the Moon and Earth exert upon each other?

  b. The distance between Earth and the Sun is about 1.5×10^{11} meters. The mass of the Sun is about 1.99×10^{30} kilograms. What is the gravitational force that the Sun and Earth exert upon each other?

  c. Find the gravitational force exerted on each other by two 1000-kilogram iron balls at a distance of 0.1 meter apart.

H.O.T. Problems   Use Higher-Order Thinking Skills

48. CRITIQUE ARGUMENTS Jamil and Savannah are setting up a proportion to begin solving the combined variation in which z varies directly as x and z varies inversely as y. Who has set up the correct proportion? Explain your reasoning.

Jamil
$$z_1 = \frac{kx_1}{y_1} \text{ and } z_2 = \frac{kx_2}{y_2}$$
$$k = \frac{z_1 y_1}{x_1} \text{ and } k = \frac{z_2 y_2}{x_2}$$
$$\frac{z_1 y_1}{x_1} = \frac{z_2 y_2}{x_2}$$

Savannah
$$z_1 = \frac{kx_1}{y_1} \text{ and } z_2 = \frac{kx_2}{y_2}$$
$$k = \frac{z_1 x_1}{y_1} \text{ and } k = \frac{z_2 x_2}{y_2}$$
$$\frac{z_1 x_1}{y_1} = \frac{z_2 x_2}{y_2}$$

49. CHALLENGE If a varies inversely as b, c varies jointly as b and f, and f varies directly as g, how are a and g related?

50. REASONING Explain why some mathematicians consider every joint variation a combined variation, but not every combined variation a joint variation.

51. OPEN-ENDED Describe three real-life quantities that vary jointly with each other.

52. WRITING IN MATH Determine the type(s) of variation(s) for which 0 cannot be one of the values. Explain your reasoning.

Preparing for Assessment

53. MULTI-STEP The time t it takes to paint a house varies inversely as the number of people p painting it. **MP** 4

a. Which of the following equations accurately represents the relationship between t and p?

- A $t_1 p_1 = t_2 p_2$
- B $\dfrac{t_1}{p_1} = \dfrac{t_2}{p_2}$
- C $\dfrac{t_1}{t_2} = \dfrac{p_1}{p_2}$
- D $t_1 p_2 = p_1 t_2$

b. If 5 people can paint the house in 7.5 hours, how many people are needed to get the house painted in no more than 4.5 hours? Round up to the nearest whole number.

- A 3 people
- B 5 people
- C 7 people
- D 9 people

c. Finish the chart to show the change in time t as the number of painters p increases.

Number of Painters (p)	Time to Paint the House (t)
10	
15	
20	
25	

d. Draw a graph to illustrate the time to paint the house in terms of the number of painters.

54. Julia earns d total dollars for the number of hours h that she works. What type of variation represents the relationship between d and h? **MP** 4

- A combined
- B direct
- C inverse
- D joint

55. Which relation shows an inverse variation? **MP** 4

- A

x	y
3	12
3.5	14
7.5	30
8.5	34

- B

x	y
3	7.5
2.5	6.25
9	18
12.2	24.4

- C

x	y
2	3
3	2
1	12
4	12

- D

x	y
4.5	8
12	3
18	2
6	6

56. The variable x varies directly as y, and x varies inversely as z. What is the value of y when $x = 20$ and $z = -2$, if $y = 30$ when $z = 3$ and $x = 5$? 4

- A -180
- B -80
- C -5
- D 5

57. The total number of hours h that a construction job will take to reach completion is related to the average daily number of workers w that are on the job site throughout the whole project. What type of variation represents the relationship between h and w? **MP** 2

- A combined
- B direct
- C inverse
- D joint

LESSON 6
Solving Rational Equations and Inequalities

::Then::
- You simplified rational expressions.

::Now::
1. Solve rational equations.
2. Solve rational inequalities.

::Why?::
- A gaming club charges $20 per month for membership. Members also have to pay $5 each time they visit the club. If a member visits the club x times in one month, then the charge for that month will be $20 + 5x$. The actual cost per visit will be $\frac{20 + 5x}{x}$. To determine how many visits are needed for the cost per visit to be $6, you would need to solve the equation $\frac{20 + 5x}{x} = 6$.

New Vocabulary
rational equation
weighted average
rational inequality

Mathematical Practices
6 Attend to precision.

1 Solve Rational Equations
Equations that contain one or more rational expressions are called **rational equations**. These equations are often easier to solve once the fractions are eliminated. You can eliminate the fractions by multiplying each side by the least common denominator (LCD).

Example 1 Solve a Rational Equation

Solve $\frac{4}{x+3} + \frac{5}{6} = \frac{23}{18}$. Check your solution.

First estimate. Then, you can later determine if the answer is reasonable. $\frac{23}{18}$ is about $1\frac{1}{3}$, so $\frac{4}{x+3}$ is about $1\frac{1}{3} - \frac{5}{6}$ or $\frac{1}{2}$. Cross multiply to solve for x. So, x is about 5.

The LCD for the terms is $18(x + 3)$.

$$\frac{4}{x+3} + \frac{5}{6} = \frac{23}{18}$$ Original equation

$$18(x+3)\left(\frac{4}{x+3}\right) + 18(x+3)\left(\frac{5}{6}\right) = 18(x+3)\frac{23}{18}$$ Multiply by LCD.

$$\cancel{18(x+3)}\left(\frac{4}{\cancel{x+3}}\right) + \cancel{18}(x+3)\left(\frac{5}{\cancel{6}}\right) = \cancel{18}(x+3)\left(\frac{23}{\cancel{18}}\right)$$ Divide common factors.

$$72 + 15x + 45 = 23x + 69$$ Multiply.

$$15x + 117 = 23x + 69$$ Simplify.

$$48 = 8x$$ Subtract 15x and 69.

$$6 = x$$ Divide.

This answer is close to the estimate, so the answer is reasonable.

Guided Practice
Solve each equation. Check your solution.

1A. $\frac{2}{x+3} + \frac{3}{2} = \frac{19}{10}$

1B. $\frac{7}{12} + \frac{9}{x-4} = \frac{55}{48}$

508 | Lesson 7-6

Multiplying each side of an equation by the LCD of rational expressions can yield results that are not solutions of the original equation. These are extraneous solutions.

Example 2 Solve a Rational Equation

Solve $\dfrac{2x}{x+5} - \dfrac{x^2 - x - 10}{x^2 + 8x + 15} = \dfrac{3}{x+3}$. Check your solution.

The LCD for the terms is $(x+3)(x+5)$.

$\dfrac{2x}{x+5} - \dfrac{x^2 - x - 10}{x^2 + 8x + 15} = \dfrac{3}{x+3}$ Original equation

$\dfrac{(x+3)(x+5)(2x)}{x+5} - \dfrac{(x+3)(x+5)(x^2 - x - 10)}{x^2 + 8x + 15} = \dfrac{(x+3)(x+5)3}{x+3}$ Multiply by LCD.

$\dfrac{(x+3)(x+5)(2x)}{x+5} - \dfrac{(x+3)(x+5)(x^2-x-10)}{x^2+8x+15} = \dfrac{(x+5)(x+3)3}{x+3}$ Divide common factors.

$(x+3)(2x) - (x^2 - x - 10) = 3(x+5)$ Simplify.

$2x^2 + 6x - x^2 + x + 10 = 3x + 15$ Distribute.

$x^2 + 7x + 10 = 3x + 15$ Simplify.

$x^2 + 4x - 5 = 0$ Subtract $3x + 15$.

$(x+5)(x-1) = 0$ Factor.

$x + 5 = 0$ or $x - 1 = 0$ Zero Product Property

$x = -5 \qquad\qquad x = 1$

CHECK Try $x = -5$.

$\dfrac{2x}{x+5} - \dfrac{x^2 - x - 10}{x^2 + 8x + 15} = \dfrac{3}{x+3}$

$\dfrac{2(-5)}{-5+5} - \dfrac{(-5)^2 - (-5) - 10}{(-5)^2 + 8(-5) + 15} \stackrel{?}{=} \dfrac{3}{-5+3}$

$\dfrac{-10}{0} - \dfrac{25 + 5 - 10}{25 - 40 + 15} \neq -\dfrac{3}{2}$ ✗

Try $x = 1$.

$\dfrac{2x}{x+5} - \dfrac{x^2 - x - 10}{x^2 + 8x + 15} = \dfrac{3}{x+3}$

$\dfrac{2(1)}{1+5} - \dfrac{1^2 - 1 - 10}{1^2 + 8(1) + 15} \stackrel{?}{=} \dfrac{3}{1+3}$

$\dfrac{2}{6} - \dfrac{-10}{24} \stackrel{?}{=} \dfrac{3}{4}$

$\dfrac{8}{24} + \dfrac{10}{24} \stackrel{?}{=} \dfrac{3}{4}$

$\dfrac{3}{4} = \dfrac{3}{4}$ ✓

When solving a rational equation, any possible solution that results in a zero in the denominator must be excluded from your list of solutions.

Because $x = -5$ results in a zero in the denominator, it is extraneous. Eliminate -5 from the list of solutions. The solution is 1.

> **Math History Link**
> **Brook Taylor** (1685–1731) English mathematician Taylor developed a theorem used in calculus known as Taylor's Theorem that relies on the remainders after computations with rational expressions.

> **Reading Math Tip**
> **Extraneous** Everyday use—something extraneous is not essential, it is unnecessary; Math meaning—extraneous solutions arise in solving an equation, but do not satisfy the original equation.

Guided Practice

2A. $\dfrac{5}{y-2} + 2 = \dfrac{17}{6}$

2B. $\dfrac{2}{z+1} - \dfrac{1}{z-1} = \dfrac{-2}{z^2 - 1}$

2C. $\dfrac{7n}{3n+3} - \dfrac{5}{4n-4} = \dfrac{3n}{2n+2}$

2D. $\dfrac{1}{p-2} = \dfrac{2p+1}{p^2 + 2p - 8} + \dfrac{2}{p+4}$

connectED.mcgraw-hill.com

The **weighted average** is a method for finding the mean of a set of numbers in which some elements of the set carry more importance, or weight, than others. Many real-world problems involving weighted averages can be solved by using rational equations.

Real-World Example 3 — Mixture Problem

CHEMISTRY How much of a 70% acid solution should Mia add to 12 milliliters of a solution that is 15% acid to create a solution that is 60% acid?

Understand Mia needs to know how much of a solution needs to be added to an original solution to create a new solution.

Plan Each solution has a certain percentage that is acid. The percentage of acid in the final solution must equal the amount of acid divided by the total solution. Estimate the amount of 70% solution by seeing that the second solution is about five times as acidic. So, add five times the amount of the first solution, or 60 mL.

	Original	Added	New
Amount of Acid	0.15(12)	0.7(x)	0.15(12) + 0.7x
Total Solution	12	x	12 + x

Study Tip

 **Modeling** Tables like the one in Example 3 are useful in organizing and solving mixture, work, weighted average, and distance problems.

Solve

$\dfrac{\text{percent}}{100} = \dfrac{\text{amount of acid}}{\text{total solution}}$ Write a proportion.

$\dfrac{60}{100} = \dfrac{0.15(12) + 0.7x}{12 + x}$ Substitute.

$\dfrac{60}{100} = \dfrac{1.8 + 0.7x}{12 + x}$ Simplify numerator.

$100(12 + x)\dfrac{60}{100} = 100(12 + x)\dfrac{1.8 + 0.7x}{12 + x}$ LCD is $100(12 + x)$. Multiply by LCD.

$\cancel{100}(12 + x)\dfrac{60}{\cancel{100}} = 100\cancel{(12 + x)}\dfrac{1.8 + 0.7x}{\cancel{12 + x}}$ Divide common factors.

$(12 + x)60 = 100(1.8 + 0.7x)$ Simplify.

$720 + 60x = 180 + 70x$ Distribute.

$540 = 10x$ Subtract $60x$ and 180.

$54 = x$ Divide by 10.

Go Online!

Many students find mixture, distance, and work problems challenging. Watch the **Personal Tutor** describe how to solve these problems with a partner. Then try describing how to solve a problem for them. Have them ask questions to help your understanding.

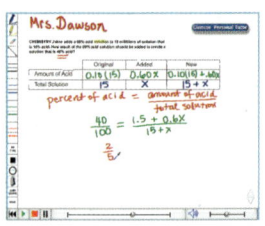

Check

$\dfrac{60}{100} = \dfrac{0.15(12) + 0.7x}{12 + x}$ Original equation

$\dfrac{60}{100} \stackrel{?}{=} \dfrac{0.15(12) + 0.7(54)}{12 + 54}$ $x = 54$

$\dfrac{60}{100} \stackrel{?}{=} 39.6$ Simplify.

$0.6 = 0.6$ ✓ Simplify.

Mia needs to add 54 milliliters of the 70% acid solution.

The acidity of each solution is the "weight" that must be considered in solving this problem. The answer is close to our estimate and seems reasonable.

▶ Guided Practice

3. How much of a 65% fruit juice solution must Justin add to 15 milliliters of a drink that is 10% fruit juice to create a fruit punch that is 35% fruit juice?

The formula relating distance, rate, and time, $d = rt$, can be used to solve rational equations. However, it can also be represented by $r = \frac{d}{t}$ and $t = \frac{d}{r}$.

Real-World Example 4 Distance Problem

ROWING Sandra's canoeing rate in still water is 6 miles per hour. It takes Sandra 3 hours to travel 10 miles round trip. Assuming a constant rate of speed, determine the rate of the current.

Study Tip

Distance Problems When distances involve round trips, the distance in one direction usually equals the distance in the other direction.

Understand You know her speed in still water and the time it takes her to travel 5 miles with the current and 5 miles against it. You need to determine the speed of the current.

Plan The formula that relates distance, rate, and time is $d = rt$, or $t = \frac{d}{r}$. If the current is 6 miles per hour or more, Sandra makes no progress against the current. The current must be greater than 0 miles per hour, so $0 \leq r \leq 6$.

Time with the Current	Time Against the Current	Total Time
$\frac{5}{6+r}$	$\frac{5}{6-r}$	3 hours

Solve

$\frac{5}{6+r} + \frac{5}{6-r} = 3$ Write the equation.

$(6+r)(6-r)\frac{5}{6+r} + (6+r)(6-r)\frac{5}{6-r} = (6+r)(6-r)3$ LCD $= (6+r)(6-r)$. Multiply by LCD.

$(6+r)(6-r)\frac{5}{6+r} + (6+r)(6-r)\frac{5}{6-r} = (6+r)(6-r)3$ Divide common factors.

$(6-r)5 + (6+r)5 = (36 - r^2)3$ Simplify.

$30 - 5r + 30 + 5r = 108 - 3r^2$ Distribute.

$60 = 108 - 3r^2$ Simplify.

$0 = -3r^2 + 48$ Subtract 10r.

$0 = -3(r+4)(r-4)$ Factor.

$0 = (r+4)(r-4)$ Divide each side by -3.

$r = 4$ or -4 Zero Product Property

Because speed cannot be negative, the speed of the current is 4 miles per hour.

Check $\frac{5}{6+r} + \frac{5}{6-r} = 3$ Original equation

$\frac{5}{6+4} + \frac{5}{6-4} \stackrel{?}{=} 3$ $r = 4$

$\frac{5}{10} + \frac{5}{2} \stackrel{?}{=} 3$ Simplify.

$\frac{1}{2} + \frac{5}{2} = \frac{6}{2}$ ✓   Simplify.

It takes Sandra one half hour to row with the current. It take her two and one half hours to row against the current. The answer is reasonable.

Guided Practice

4. FLYING The speed of the wind is 20 miles per hour. If it takes a plane 7 hours to fly 2368 miles round trip, determine the plane's speed in still air.

Real-world problems that involve work can often be solved using rational equations.

Real-World Example 5 Work Problems

COMMUNITY SERVICE Every year, the junior and senior classes at Hillcrest High School build a house for the community. If it takes the senior class 24 days to complete a house and 18 days if they work with the junior class, how long would it take the junior class to complete a house if they worked alone?

Understand We are given how long it takes the senior class working alone and when the classes work together. We need to determine how long it would take the junior class by themselves.

Plan The senior class can complete 1 house in 24 days, so their rate is $\frac{1}{24}$ of a house per day.

The rate for the junior class is $\frac{1}{j}$.

The combined rate for both classes is $\frac{1}{18}$.

Senior Rate	Junior Rate	Combined Rate
$\frac{1}{24}$	$\frac{1}{j}$	$\frac{1}{18}$

Solve

$\frac{1}{24} + \frac{1}{j} = \frac{1}{18}$ Write the equation.

$72j\frac{1}{24} + 72j\frac{1}{j} = 72j\frac{1}{18}$ LCD = 72j
Multiply by LCD.

$\overset{3}{72}j\frac{1}{\underset{1}{24}} + 72\overset{1}{j}\frac{1}{\underset{1}{j}} = \overset{4}{72}j\frac{1}{\underset{1}{18}}$ Divide common factors.

$3j + 72 = 4j$ Distribute.

$72 = j$ Subtract 3j.

Check Two methods are possible.

Method 1 Substitute values.

$\frac{1}{24} + \frac{1}{j} = \frac{1}{18}$ Original equation

$\frac{1}{24} + \frac{1}{72} \stackrel{?}{=} \frac{1}{18}$ j = 72

$\frac{3}{72} + \frac{1}{72} \stackrel{?}{=} \frac{4}{72}$ LCD = 72

$\frac{4}{72} = \frac{4}{72}$ ✓   Simplify.

Method 2 Use a calculator.

It would take the junior class 72 days to complete the house by themselves.

Guided Practice

5A. Working together, it took Anthony and Travis 6 hours to mow the lawns of all their clients together last week. The previous week it took Travis 10 hours to do it alone. How long will it take Anthony if he mows by himself this week?

5B. Noah and Owen build birdhouses together. If Noah can build a particular house in 6 days and Owen can build the same house in 5 days, how long would it take the two of them if they work together?

2 Solve Rational Inequalities

To solve **rational inequalities**, which are inequalities that contain one or more rational expressions, follow these steps.

> **Key Concept** Solving Rational Inequalities
>
> **Step 1** State the excluded values. These are the values for which the denominator is 0.
>
> **Step 2** Solve the related equation.
>
> **Step 3** Use the values determined from the previous steps to divide a number line into intervals.
>
> **Step 4** Test a value in each interval to determine which intervals contain values that satisfy the inequality.

Example 6 Solve a Rational Inequality

Solve $\dfrac{x}{3} - \dfrac{1}{x-2} < \dfrac{x+1}{4}$.

Step 1 The excluded value for this inequality is 2.

Step 2 Solve the related equation.

$$\dfrac{x}{3} - \dfrac{1}{x-2} = \dfrac{x+1}{4} \qquad \text{Related equation}$$

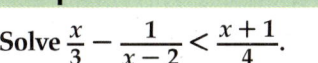

$$12(x-2)\dfrac{x}{3} - 12(x-2)\dfrac{1}{x-2} = 12(x-2)\dfrac{x+1}{4} \qquad \text{LCD is } 12(x-2). \text{ Multiply by LCD.}$$

$$4x^2 - 8x - 12 = 3x^2 - 3x - 6 \qquad \text{Distribute.}$$

$$x^2 - 5x - 6 = 0 \qquad \text{Subtract } 3x^2 - 3x - 6.$$

$$(x - 6)(x + 1) = 0 \qquad \text{Factor.}$$

$$x = 6 \text{ or } -1 \qquad \text{Zero Product Property}$$

Step 3 Draw vertical lines at the excluded value and at the solutions to separate the number line into intervals.

Study Tip
Rational Inequalities It is possible that none or all of the intervals will produce a true statement.

Step 4 Now test a sample value in each interval to determine whether the values in the interval satisfy the inequality.

Test $x = -3$.

$$\dfrac{-3}{3} - \dfrac{1}{-3-2} \stackrel{?}{<} \dfrac{-3+1}{4}$$
$$-1 + \dfrac{1}{5} \stackrel{?}{<} -\dfrac{2}{4}$$
$$-\dfrac{4}{5} < -\dfrac{1}{2} \checkmark$$

Test $x = 0$.

$$\dfrac{0}{3} - \dfrac{1}{0-2} \stackrel{?}{<} \dfrac{0+1}{4}$$
$$0 + \dfrac{1}{2} \stackrel{?}{<} \dfrac{1}{4}$$
$$\dfrac{1}{2} \not< \dfrac{1}{4}$$

Test $x = 4$.

$$\dfrac{4}{3} - \dfrac{1}{4-2} \stackrel{?}{<} \dfrac{4+1}{4}$$
$$\dfrac{4}{3} - \dfrac{1}{2} \stackrel{?}{<} \dfrac{5}{4}$$
$$\dfrac{5}{6} < \dfrac{5}{4} \checkmark$$

Test $x = 8$.

$$\dfrac{8}{3} - \dfrac{1}{8-2} \stackrel{?}{<} \dfrac{8+1}{4}$$
$$\dfrac{32}{12} - \dfrac{2}{12} \stackrel{?}{<} \dfrac{27}{12}$$
$$\dfrac{30}{12} \not< \dfrac{27}{12}$$

The statement is true for $x = -3$ and $x = 4$. Therefore, the solution is $x < -1$ or $2 < x < 6$.

▶ **Guided Practice** Solve each inequality.

6A. $\dfrac{5}{x} + \dfrac{6}{5x} > \dfrac{2}{3}$

6B. $\dfrac{4}{3x} + \dfrac{7}{x} < \dfrac{5}{9}$

Check Your Understanding

= Step-by-Step Solutions begin on page R11.

Examples 1–2 Solve each equation. Check your solution.

1. $\dfrac{4}{7} + \dfrac{3}{x-3} = \dfrac{53}{56}$

2. $\dfrac{7}{3} - \dfrac{3}{x-5} = \dfrac{19}{12}$

3. $\dfrac{10}{2x+1} + \dfrac{4}{3} = 2$

4. $\dfrac{11}{4} - \dfrac{5}{y+3} = \dfrac{23}{12}$

5. $\dfrac{8}{x-5} - \dfrac{9}{x-4} = \dfrac{5}{x^2 - 9x + 20}$

6. $\dfrac{14}{x+3} + \dfrac{10}{x-2} = \dfrac{122}{x^2 + x - 6}$

7. $\dfrac{14}{x-8} - \dfrac{5}{x-6} = \dfrac{82}{x^2 - 14x + 48}$

8. $\dfrac{5}{x+2} - \dfrac{3}{x-2} = \dfrac{12}{x^2 - 4}$

Example 3

9. **MP STRUCTURE** Sara has 10 pounds of dried fruit selling for $6.25 per pound. She wants to know how many pounds of mixed nuts selling for $4.50 per pound she needs to make a trail mix selling for $5 per pound.

   a. Let m = the number of pounds of mixed nuts. Complete the following table and estimate an answer.

	Pounds	Price per Pound	Total Price
Dried Fruit	10	$6.25	6.25(10)
Mixed Nuts			
Trail Mix			

   b. Write a rational equation using the last column of the table.

   c. Solve the equation to determine how many pounds of mixed nuts are needed. Is your answer reasonable? Why or why not?

Example 4

10. **DISTANCE** Alicia's average speed riding her bike is 11.5 miles per hour. She takes a round trip of 40 miles. It takes her 1 hour and 20 minutes with the wind and 2 hours and 30 minutes against the wind.

    a. Write an expression for Alicia's time with the wind.
    b. Write an expression for Alicia's time against the wind.
    c. How long does it take to complete the trip?
    d. Write and solve the rational equation to determine the speed of the wind.

Example 5

11. **WORK** Kendal and Chandi wax cars. Kendal can wax a particular car in 60 minutes and Chandi can wax the same car in 80 minutes. They plan on waxing the same car together and want to know how long it will take. How much will Kendal and Chandi complete individually in

    a. 1 minute?
    b. x minutes?
    c. Write a rational equation representing Kendal and Chandi working together on the car and estimate an answer.
    d. Solve the equation to determine how long it will take them to finish the car. Is your answer reasonable? Why or why not?

Example 6 Solve each inequality. Check your solutions.

12. $\dfrac{3}{5x} + \dfrac{1}{6x} > \dfrac{2}{3}$

13. $\dfrac{1}{4c} + \dfrac{1}{9c} < \dfrac{1}{2}$

14. $\dfrac{4}{3y} + \dfrac{2}{5y} < \dfrac{3}{2}$

15. $\dfrac{1}{3b} + \dfrac{1}{4b} < \dfrac{1}{5}$

Practice and Problem Solving

Extra Practice is on page R7.

Examples 1–2 Solve each equation. Check your solutions.

16. $\dfrac{9}{x-7} - \dfrac{7}{x-6} = \dfrac{13}{x^2-13x+42}$

17. $\dfrac{13}{y+3} - \dfrac{12}{y+4} = \dfrac{18}{y^2+7y+12}$

18. $\dfrac{14}{x-2} - \dfrac{18}{x+1} = \dfrac{22}{x^2-x-2}$

19. $\dfrac{11}{a+2} - \dfrac{10}{a+5} = \dfrac{36}{a^2+7a+10}$

20. $\dfrac{x}{2x-1} + \dfrac{3}{x+4} = \dfrac{21}{2x^2+7x-4}$

21. $\dfrac{2}{y-5} + \dfrac{y-1}{2y+1} = \dfrac{2}{2y^2-9y-5}$

Examples 3–5

22. **CHEMISTRY** How many milliliters of a 20% acid solution must be added to 40 milliliters of a 75% acid solution to create a 30% acid solution?

23. **GROCERIES** Ellen bought 3 pounds of bananas for $0.90 per pound. How many pounds of apples costing $1.25 per pound must she purchase so that the total cost for fruit is $1 per pound?

24. **BUILDING** Bryan's volunteer group can build a garage in 12 hours. Sequoia's group can build it in 16 hours. How long would it take them if they worked together?

Example 6 Solve each inequality. Check your solutions.

25. $3 - \dfrac{4}{x} > \dfrac{5}{4x}$

26. $\dfrac{5}{3a} - \dfrac{3}{4a} > \dfrac{5}{6}$

27. $\dfrac{x-2}{x+2} + \dfrac{1}{x-2} > \dfrac{x-4}{x-2}$

28. $\dfrac{3}{4} - \dfrac{1}{x-3} > \dfrac{x}{x+4}$

29. $\dfrac{x}{5} + \dfrac{2}{3} < \dfrac{3}{x-4}$

30. $\dfrac{x}{x+2} + \dfrac{1}{x-1} < \dfrac{3}{2}$

31. **AIR TRAVEL** It takes a plane 20 hours to fly to its destination against the wind. The return trip takes 16 hours. If the plane's average speed in still air is 500 miles per hour, what is the average speed of the wind during the flight?

32. **FINANCIAL LITERACY** Judie wants to invest $10,000 in two different accounts. The risky account could earn 9% interest, while the other account earns 5% interest. She wants to earn $750 interest for the year. Of tables, graphs, or equations, choose the best representation needed and determine how much should be invested in each account.

33. **MULTIPLE REPRESENTATIONS** Consider $\dfrac{2}{x-3} + \dfrac{1}{x} = \dfrac{x-1}{x-3}$.

   a. **Algebraic** Solve the equation for x. Were any values of x extraneous?
   b. **Graphical** Graph $y_1 = \dfrac{2}{x-3} + \dfrac{1}{x}$ and $y_2 = \dfrac{x-1}{x-3}$ on the same graph for $0 < x < 5$.
   c. **Analytical** For what value(s) of x do they intersect? Do they intersect where x is extraneous for the original equation?
   d. **Verbal** Use this knowledge to describe how you can use a graph to determine whether an apparent solution of a rational equation is extraneous.

Solve each equation. Check your solutions.

34. $\dfrac{2}{y+3} - \dfrac{3}{4-y} = \dfrac{2y-2}{y^2-y-12}$

35. $\dfrac{2}{y+2} - \dfrac{y}{2-y} = \dfrac{y^2+4}{y^2-4}$

H.O.T. Problems   Use Higher-Order Thinking Skills

36. **OPEN-ENDED** Give an example of a rational equation that can be solved by multiplying each side of the equation by $4(x+3)(x-4)$.

37. **CHALLENGE** Solve $\dfrac{1 + \dfrac{9}{x} + \dfrac{20}{x^2}}{1 - \dfrac{25}{x^2}} = \dfrac{x+4}{x-5}$.

38. **TOOLS** While using the table feature on the graphing calculator to explore $f(x) = \dfrac{1}{x^2-x-6}$, the values -2 and 3 say "**ERROR.**" Explain its meaning.

39. **WRITING IN MATH** Why should you check solutions of rational equations and inequalities?

Preparing for Assessment

40. Jason can water all the plants at the botanical garden in 32 minutes. Celia can water them in 25 minutes. If they work together, about how long will it take for them to water the plants? **MP** 1, 6

- A 7 min
- B 14 min
- C 16 min
- D 29 min

41. Thirty-two ounces of trail mix containing 25% raisins were mixed with 18 ounces of trail mix containing 15% raisins. About what percent of the new mixture is *not* raisins? **MP** 1, 6

- A 18%
- B 21%
- C 40%
- D 79%

42. John is preparing to paint his garage. If it would take 3 workers 8 hours to apply 2 coats of paint, how long would it take 4 workers to apply 1 coat of paint? **MP** 1, 6

- A 3 hours
- B 4 hours
- C 6 hours
- D 7 hours

43. A team of runners is planning to do a 27-mile relay as part of their ongoing training. The first runner will do 8 miles, the second runner will do 9 miles, and the third runner will do 10 miles. If the first runner averages 10 mph, the second runner averages 7 mph, and the third runner averages 9 mph, how long will the relay take in all? **MP** 1, 6

- A 87.145 minutes
- B 136.97 minutes
- C 168.55 minutes
- D 191.81 minutes

44. MULTI-STEP Manuel's engineering team is surveying a triangular parcel of land. The team measures its perimeter using a measuring wheel. Each of his three team members uses the wheel to measure a side and then passes it to the next member. Each person's average walking speed and the measure of their side is shown in the table.

April	3 mph	1567 ft
Matt	3.5 mph	1125 ft
Henry	3.2 mph	920 ft

Manuel walks at an average speed of 4 mph. How much faster would the measurements have been taken if Manuel had taken them all himself? **MP** 1, 6

- A 108 seconds faster
- B 155 seconds faster
- C 615 seconds faster
- D 771 seconds faster

45. MULTI-STEP Dara is paddling a kayak on a river. It takes her a total of 6 hours to paddle 4 miles upstream and four miles downstream. When the water is still, Dara can paddle at an average speed of 2 miles per hour. Let r represent the average rate of the current. **MP** 1, 6

a. Fill in the table below to organize the equation:

	Distance (mi)	Avg Speed (mph)	Time (h)
With Current			
Against Current			

b. Write an equation to represent the average rate of the current.

c. What LCD can you use to simplify the equation?

d. To the nearest hundredth, what is the average rate of the current?

- A About 0.76 miles per hour
- B About 1.15 miles per hour
- C About 1.78 miles per hour
- D About 2.21 miles per hour

EXTEND 7-6
Graphing Technology Lab
Solving Rational Equations and Inequalities

You can use a TI-83/84 Plus graphing calculator to solve rational equations by graphing or by using the table feature. Graph both sides of the equation, and locate the point(s) of intersection.

Mathematical Practices
5 Use appropriate tools strategically.

Activity 1 Rational Equation

Work cooperatively. Solve $\dfrac{4}{x+1} = \dfrac{3}{2}$.

Step 1 Graph each side of the equation.

Graph each side of the equation as a separate function. Enter $\dfrac{4}{x+1}$ as **Y1** and $\dfrac{3}{2}$ as **Y2**. Then graph the two equations in the standard viewing window.

KEYSTROKES: Y= 4 ÷ (X,T,θ,n + 1)
ENTER 3 ÷ 2 ZOOM 6

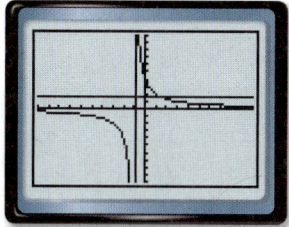

[−10, 10] scl: 1 by [−10, 10] scl: 1

Because the calculator is in connected mode, a vertical line may appear connecting the two branches of the hyperbola. This line is not part of the graph.

Step 2 Use the **intersect** feature.

The **intersect** feature on the **CALC** menu allows you to approximate the ordered pair of the point at which the graphs cross.

KEYSTROKES: 2nd [CALC] 5

Select one graph and press ENTER. Select the other graph, press ENTER, and press ENTER again.

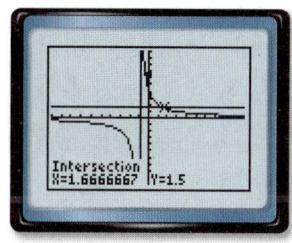

[−10, 10] scl: 1 by [−10, 10] scl: 1

The solution is $1\dfrac{2}{3}$.

Step 3 Use the **TABLE** feature.

Verify the solution using the **TABLE** feature. Set up the table to show x-values in increments of $\dfrac{1}{3}$.

KEYSTROKES: 2nd [TBLSET] 0 ENTER 1 ÷ 3 ENTER 2nd [TABLE]

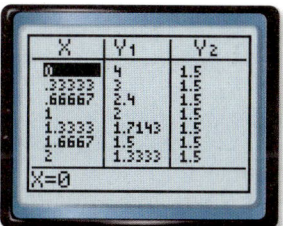

The table displays x-values and corresponding y-values for each graph. At $x = 1\dfrac{2}{3}$, both functions have a y-value of 1.5. Thus, the solution of the equation is $1\dfrac{2}{3}$.

(continued on the next page)

EXTEND 7-6

Graphing Technology Lab
Solving Rational Equations and Inequalities *Continued*

You can use a similar procedure to solve rational inequalities using a graphing calculator.

Activity 2 Rational Inequality

Work cooperatively. Solve $\frac{3}{x} + \frac{7}{x} > 9$.

Step 1 Enter the inequalities.

Rewrite the problem as a system of inequalities.

The first inequality is $\frac{3}{x} + \frac{7}{x} > y$ or $y < \frac{3}{x} + \frac{7}{x}$. Because this inequality includes the *less than* symbol, shade below the curve. First enter the boundary and then use the arrow and ENTER keys to choose the shade below icon, ▐▄.

The second inequality is $y > 9$. Shade above the curve since this inequality contains *greater than*.

KEYSTROKES: Y= ◄ ◄ ENTER ENTER ENTER ► ► 3 ÷ X,T,θ,n + 7 ÷ X,T,θ,n ENTER ◄ ◄ ENTER ENTER ► ► 9

Step 2 Graph the system.

KEYSTROKES: GRAPH

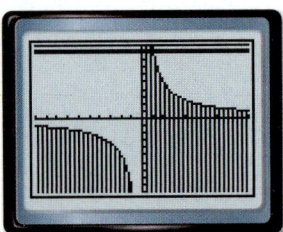

[−10, 10] scl: 1 by [−10, 10] scl: 1

The solution set of the original inequality is the set of *x*-values of the points in the region where the shadings overlap. Using the calculator's **intersect** feature, you can conclude that the solution set is $\left\{x \mid 0 < x < 1\frac{1}{9}\right\}$.

Step 3 Use the **TABLE** feature.

Verify using the **TABLE** feature. Set up the table to show *x*-values in increments of $\frac{1}{9}$.

KEYSTROKES: 2nd [TBLSET] 0 ENTER 1 ÷ 9 ENTER
2nd [TABLE]

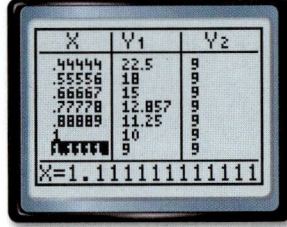

Scroll through the table. Notice that for *x*-values greater than 0 and less than $1\frac{1}{9}$, Y1 > Y2. This confirms that the solution of the inequality is $\left\{x \mid 0 < x < 1\frac{1}{9}\right\}$.

Exercises

Work cooperatively. Solve each equation or inequality.

1. $\frac{1}{x} + \frac{1}{2} = \frac{2}{x}$

2. $\frac{1}{x-4} = \frac{2}{x-2}$

3. $\frac{4}{x} = \frac{6}{x^2}$

4. $\frac{1}{1-x} = 1 - \frac{x}{x-1}$

5. $\frac{1}{x+4} = \frac{2}{x^2+3x-4} - \frac{1}{1-x}$

6. $\frac{1}{x} + \frac{1}{2x} > 5$

7. $\frac{1}{x-1} + \frac{2}{x} < 0$

8. $1 + \frac{5}{x-1} \leq 0$

9. $2 + \frac{1}{x-1} \geq 0$

CHAPTER 7
Study Guide and Review

Go Online! for Vocabulary Review Games and key vocabulary in 13 languages

Study Guide

Key Concepts

Rational Expressions (Lessons 7-1 and 7-2)
- Multiplying and dividing rational expressions is similar to multiplying and dividing fractions.
- To simplify complex fractions, simplify the numerator and the denominator separately, and then simplify the resulting expression.

Reciprocal and Rational Functions (Lessons 7-3 and 7-4)
- A reciprocal function is of the form $f(x) = \frac{1}{a(x)}$, where $a(x)$ is a linear function and $a(x) \neq 0$.
- A rational function is of the form $\frac{a(x)}{b(x)}$, where $a(x)$ and $b(x)$ are polynomial functions and $b(x) \neq 0$.

Direct, Joint, and Inverse Variation (Lesson 7-5)
- Direct variation: There is a nonzero number k such that $y = kx$.
- Joint variation: There is a nonzero number k such that $y = kxz$.
- Inverse variation: There is a nonzero constant k such that $xy = k$ or $y = \frac{k}{x}$, where $x \neq 0$ and $y \neq 0$.

Rational Equations and Inequalities (Lesson 7-6)
- Eliminate fractions in rational equations by multiplying each side of the equation by the LCD.
- Possible solutions of a rational equation must exclude values that result in zero in the denominator.

 **Study Organizer**

Use your Foldable to review the chapter. Working with a partner can be helpful. Ask for clarification of concepts as needed.

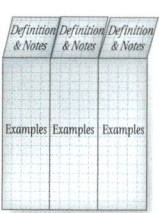

Key Vocabulary

combined variation (p. 503)
complex fraction (p. 470)
constant of variation (p. 500)
direct variation (p. 500)
horizontal asymptote (p. 491)
hyperbola (p. 483)
inverse variation (p. 502)
joint variation (p. 501)
oblique asymptote (p. 493)
point discontinuity (p. 494)
rational equation (p. 508)
rational expression (p. 467)
rational function (p. 491)
rational inequality (p. 513)
reciprocal function (p. 483)
vertical asymptote (p. 491)
weighted average (p. 510)

Vocabulary Check

Choose a term from the list above that best completes each statement or phrase.

1. A(n) _____ is a rational expression whose numerator and/or denominator contains a rational expression.

2. A(n) _____ asymptote is a linear asymptote that is neither horizontal nor vertical.

3. Equations that contain one or more rational expressions are called _____.

4. The graph of $y = \frac{x}{x+2}$ has a(n) _____ at $x = -2$.

5. _____ occurs when one quantity varies directly as the product of two or more other quantities.

6. A ratio of two polynomial expressions is called a(n) _____.

7. _____ looks like a hole in a graph because the graph is undefined at that point.

8. _____ occurs when one quantity varies directly and/or inversely as two or more other quantities.

Concept Check

9. Explain how to algebraically determine the horizontal asymptotes of the graph of a rational function $f(x) = \frac{a(x)}{b(x)}$.

10. Explain the difference between direct and inverse variation.

CHAPTER 7
Study Guide and Review Continued

Lesson-by-Lesson Review

7-1 Multiplying and Dividing Rational Expressions

Simplify each expression.

11. $\dfrac{-16xy}{27z} \cdot \dfrac{15z^3}{8x^2}$

12. $\dfrac{x^2 - 2x - 8}{x^2 + x - 12} \cdot \dfrac{x^2 + 2x - 15}{x^2 + 7x + 10}$

13. $\dfrac{x^2 - 1}{x^2 - 4} \cdot \dfrac{x^2 - 5x - 14}{x^2 - 6x - 7}$

14. $\dfrac{x + y}{15x} \div \dfrac{x^2 - y^2}{3x^2}$

15. $\dfrac{\frac{x^2 + 3x - 18}{x + 4}}{\frac{x^2 + 7x + 6}{x + 4}}$

16. **GEOMETRY** A triangle has an area of $3x^2 + 9x - 54$ square centimeters. If the height of the triangle is $x + 6$ centimeters, find the length of the base.

Example 1

Simplify $\dfrac{4a}{3b} \cdot \dfrac{9b^4}{2a^2}$.

$\dfrac{4a}{3b} \cdot \dfrac{9b^4}{2a^2} = \dfrac{2 \cdot 2 \cdot a \cdot 3 \cdot 3 \cdot b \cdot b \cdot b \cdot b}{3 \cdot b \cdot 2 \cdot a \cdot a}$

$= \dfrac{6b^3}{a}$

Example 2

Simplify $\dfrac{r^2 + 5r}{2r} \div \dfrac{r^2 - 25}{6r - 12}$.

$\dfrac{r^2 + 5r}{2r} \div \dfrac{r^2 - 25}{6r - 12} = \dfrac{r^2 + 5r}{2r} \cdot \dfrac{6r - 12}{r^2 - 25}$

$= \dfrac{r(r + 5)}{2r} \cdot \dfrac{6(r - 2)}{(r + 5)(r - 5)}$

$= \dfrac{3(r - 2)}{r - 5}$

7-2 Adding and Subtracting Rational Expressions

Simplify each expression.

17. $\dfrac{9}{4ab} + \dfrac{5a}{6b^2}$

18. $\dfrac{3}{4x - 8} - \dfrac{x - 1}{x^2 - 4}$

19. $\dfrac{y}{2x} + \dfrac{4y}{3x^2} - \dfrac{5}{6xy^2}$

20. $\dfrac{2}{x^2 - 3x - 10} - \dfrac{6}{x^2 - 8x + 15}$

21. $\dfrac{3}{3x^2 + 2x - 8} + \dfrac{4x}{2x^2 + 6x + 4}$

22. $\dfrac{\frac{3}{2x+3} - \frac{x}{x+1}}{\frac{2x}{x+1} + \frac{5}{2x+3}}$

23. **GEOMETRY** What is the perimeter of the rectangle?

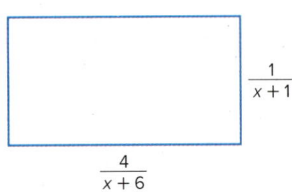

Example 3

Simplify $\dfrac{3a}{a^2 - 4} - \dfrac{2}{a - 2}$.

$\dfrac{3a}{a^2 - 4} - \dfrac{2}{a - 2} = \dfrac{3a}{(a - 2)(a + 2)} - \dfrac{2}{a - 2}$

$= \dfrac{3a}{(a - 2)(a + 2)} - \dfrac{2(a + 2)}{(a - 2)(a + 2)}$

$= \dfrac{3a - 2(a + 2)}{(a - 2)(a + 2)}$ Subtract numerators.

$= \dfrac{3a - 2a - 4}{(a - 2)(a + 2)}$ Distributive Property

$= \dfrac{a - 4}{(a - 2)(a + 2)}$ Simplify.

7-3 Graphing Reciprocal Functions

Graph each function. State the domain and range.

24. $f(x) = \dfrac{10}{x}$
25. $f(x) = -\dfrac{12}{x} + 2$
26. $f(x) = \dfrac{3}{x+5}$
27. $f(x) = \dfrac{6}{x-9}$
28. $f(x) = \dfrac{7}{x-2} + 3$
29. $f(x) = -\dfrac{4}{x+4} - 8$

30. **CONSERVATION** The student council is planting 28 trees for a service project. The number of trees each person plants depends on the number of student council members.

    a. Write a function to represent this situation.

    b. Graph the function.

Example 4

Graph $f(x) = \dfrac{3}{x+2} - 1$. State the domain and range.

$a = 3$: The graph is stretched vertically.
$h = -2$: The graph is translated 2 units left.
There is an asymptote at $x = -2$.
$k := -1$: The graph is translated 1 unit down.
There is an asymptote at $f(x) = -1$.

Domain: $\{x \mid x \neq -2\}$,
Range: $\{f(x) \mid f(x) \neq -1\}$

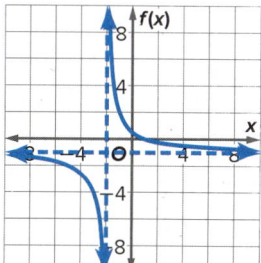

7-4 Graphing Rational Functions

Determine the equations of any vertical asymptotes and the values of x for any holes in the graph of each rational function.

31. $f(x) = \dfrac{3}{x^2 + 4x}$
32. $f(x) = \dfrac{x+2}{x^2 + 6x + 8}$
33. $f(x) = \dfrac{x^2 - 9}{x^2 - 5x - 24}$

Graph each rational function.

34. $f(x) = \dfrac{x+2}{(x+5)^2}$
35. $f(x) = \dfrac{x}{x+1}$
36. $f(x) = \dfrac{x^2 + 4x + 4}{x+2}$
37. $f(x) = \dfrac{x-1}{x^2 + 5x + 6}$

38. **FUNDRAISERS** Adelle is selling cookies for a fundraiser. Out of the first 15 houses, she sold cookies to 10 of them. Suppose Adelle goes to x more houses and sells cookies to all of them. The percentage of houses that she sold to out of the total houses can be determined using $P(x) = \dfrac{10 + x}{15 + x}$.

    a. Graph the function.

    b. What domain and range values are meaningful in the context of the problem?

Example 5

Determine the equation of any vertical asymptotes and the values of x for any holes in the graph of $f(x) = \dfrac{x^2 - 1}{x^2 + 2x - 3}$.

$\dfrac{x^2 - 1}{x^2 + 2x - 3} = \dfrac{(x-1)(x+1)}{(x-1)(x+3)}$

The function is undefined for $x = 1$ and $x = -3$.
Because $\dfrac{(x-1)(x+1)}{(x-1)(x+3)} = \dfrac{x+1}{x+3}$, $x = -3$ is a vertical asymptote, and $x = 1$ represents a hole in the graph.

Example 6

Graph $f(x) = \dfrac{1}{6x(x-1)}$.

The function is undefined for $x = 0$ and $x = 1$. Because $\dfrac{1}{6x(x-1)}$ is in simplest form, $x = 0$ and $x = 1$ are vertical asymptotes. Draw the two asymptotes and sketch the graph.

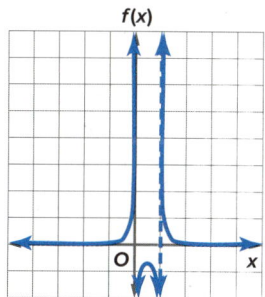

CHAPTER 7
Study Guide and Review Continued

7-5 Variation Functions

39. If a varies directly as b and $b = 18$ when $a = 27$, find a when $b = 10$.

40. If y varies inversely as x and $y = 15$ when $x = 3.5$, find y when $x = -5$.

41. If y varies inversely as x and $y = -3$ when $x = 9$, find y when $x = 81$.

42. If y varies jointly as x and z, and $x = 8$ and $z = 3$ when $y = 72$, find y when $x = -2$ and $z = -5$.

43. If y varies jointly as x and z, and $y = 18$ when $x = 6$ and $z = 15$, find y when $x = 12$ and $z = 4$.

44. **JOBS** Lisa's earnings vary directly with how many hours she babysits. If she earns $68 for 8 hours of babysitting, find her earnings after 5 hours of babysitting.

Example 7

If y varies inversely as x and $x = 24$ when $y = -8$, find x when $y = 15$.

$\dfrac{x_1}{y_2} = \dfrac{x_2}{y_1}$ Inverse variation

$\dfrac{24}{15} = \dfrac{x_2}{-8}$ $x_1 = 24, y_1 = -8, y_2 = 15$

$24(-8) = 15(x_2)$ Cross multiply.

$-192 = 15x_2$ Simplify.

$-12\dfrac{4}{5} = x_2$ Divide each side by 15.

When $y = 15$, the value of x is $-12\dfrac{4}{5}$.

7-6 Solving Rational Equations and Inequalities

Solve each equation or inequality. Check your solutions.

45. $\dfrac{1}{3} + \dfrac{4}{x-2} = 6$

46. $\dfrac{6}{x+5} - \dfrac{3}{x-3} = \dfrac{6}{x^2 + 2x - 15}$

47. $\dfrac{2}{x^2 - 9} = \dfrac{3}{x^2 - 2x - 3}$

48. $\dfrac{4}{2x-3} + \dfrac{x}{x+1} = \dfrac{-8x}{2x^2 - x - 3}$

49. $\dfrac{x}{x+4} - \dfrac{28}{x^2 + x - 12} = \dfrac{1}{x-3}$

50. $\dfrac{x}{2} + \dfrac{1}{x-1} < \dfrac{x}{4}$

51. $\dfrac{1}{2x} - \dfrac{4}{5x} > \dfrac{1}{3}$

52. **YARD WORK** Lana can plant a garden in 3 hours. Milo can plant the same garden in 4 hours. How long will it take them if they work together?

Example 8

Solve $\dfrac{3}{x+2} + \dfrac{1}{x} = 0$.

The LCD is $x(x+2)$.

$\dfrac{3}{x+2} + \dfrac{1}{x} = 0$

$x(x+2)\left(\dfrac{3}{x+2} + \dfrac{1}{x}\right) = x(x+2)(0)$

$x(x+2)\left(\dfrac{3}{x+2}\right) + x(x+2)\left(\dfrac{1}{x}\right) = 0$

$3(x) + 1(x+2) = 0$

$3x + x + 2 = 0$

$4x + 2 = 0$

$4x = -2$

$x = -\dfrac{1}{2}$

CHAPTER 7
Practice Test

 Go Online! for another Chapter Test

Simplify each expression.

1. $\dfrac{r^2 + rt}{2r} \div \dfrac{r + t}{16r^2}$

2. $\dfrac{m^2 - 4}{3m^2} \cdot \dfrac{6m}{2 - m}$

3. $\dfrac{m^2 + m - 6}{n^2 - 9} \div \dfrac{m - 2}{n + 3}$

4. $\dfrac{\dfrac{x^2 + 4x + 3}{x^2 - 2x - 15}}{\dfrac{x^2 - 1}{x^2 - x - 20}}$

5. $\dfrac{x + 4}{6x + 3} + \dfrac{1}{2x + 1}$

6. $\dfrac{x}{x^2 - 1} - \dfrac{3}{2x + 2}$

7. $\dfrac{1}{y} + \dfrac{2}{7} - \dfrac{3}{2y^2}$

8. $\dfrac{2 + \dfrac{1}{x}}{5 - \dfrac{1}{x}}$

9. Identify the asymptotes, domain, and range of the function graphed.

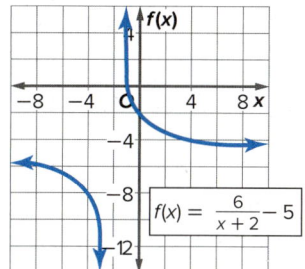

$f(x) = \dfrac{6}{x + 2} - 5$

10. **MULTIPLE CHOICE** What is the equation for the vertical asymptote of the rational function $f(x) = \dfrac{x + 1}{x^2 + 3x + 2}$?

    A $x = -2$
    B $x = -1$
    C $x = 1$
    D $x = 2$

Graph each function.

11. $f(x) = -\dfrac{8}{x} - 9$

12. $f(x) = \dfrac{2}{x + 4}$

13. $f(x) = \dfrac{3}{x - 1} + 8$

14. $f(x) = \dfrac{5x}{x + 1}$

15. $f(x) = \dfrac{x}{x - 5}$

16. $f(x) = \dfrac{x^2 + 5x - 6}{x - 1}$

17. Determine the equations of any vertical asymptotes and the values of x for any holes in the graph of the function $f(x) = \dfrac{x + 5}{x^2 - 2x - 35}$.

18. Determine the equations of any oblique asymptotes in the graph of the function $f(x) = \dfrac{x^2 + x - 5}{x + 3}$.

Solve each equation or inequality.

19. $\dfrac{-1}{x + 4} = 6 - \dfrac{x}{x + 4}$

20. $\dfrac{1}{3} = \dfrac{5}{m + 3} + \dfrac{8}{21}$

21. $7 + \dfrac{2}{x} < -\dfrac{5}{x}$

22. $r + \dfrac{6}{r} - 5 = 0$

23. $\dfrac{6}{7} - \dfrac{3m}{2m - 1} = \dfrac{11}{7}$

24. $\dfrac{r + 2}{3r} = \dfrac{r + 4}{r - 2} - \dfrac{2}{3}$

25. If y varies inversely as x and $y = 18$ when $x = -\dfrac{1}{2}$, find x when $y = -10$.

26. If m varies directly as n and $m = 24$ when $n = -3$, find n when $m = 30$.

27. Suppose r varies jointly as s and t. If $s = 20$ when $r = 140$ and $t = -5$, find s when $r = 7$ and $t = 2.5$.

28. **BICYCLING** When Susan rides her bike, the distance that she travels varies directly with the amount of time she is biking. Suppose she bikes 50 miles in 2.5 hours. At this rate, how many hours would it take her to bike 80 miles?

29. **PAINTING** Peter can paint a house in 10 hours. Melanie can paint the same house in 9 hours. How long would it take if they worked together?

30. **MULTIPLE CHOICE** How many liters of a 25% acid solution must be added to 30 liters of an 80% acid solution to create a 50% acid solution?

    A 18
    B 30
    C 36
    D 66

31. What is the volume of the rectangular prism?

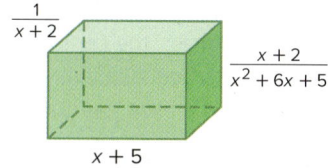

CHAPTER 7

Preparing for Assessment

Performance Task

Provide a clear solution to each part of the task. Be sure to show all of your work, include all relevant drawings, and justify your answers.

MANUFACTURING A car company evaluates the safety and efficiency of its cars to ensure it meets government standards and consumer expectations.

Part A

EFFICIENCY One of the company's SUVs gets an average of 320 miles out of a single tank of gas.

1. Write a function for y in terms of x, where y represents the number of miles per gallon the SUV gets and x represents the number of gallons of gas in one tank.

2. **Model** Make a graph of the function you wrote.

3. Determine the number of miles per gallon the SUV gets if the tank holds 20 gallons of gas.

Part B

SAFETY The company is dealing with a defective airbag issue. When conducting crash tests, the rear passenger side airbags deploy every 5 out of 8 times. This is called the crash test success ratio. After the company resolves an issue, the air bags deploy in x consecutive crash tests.

4. **Reasoning** Write a function for the car company's crash test success ratio y after x successful crash tests.

5. Make a graph of the function you wrote.

6. **Sense-Making** Explain which parts of the graph are meaningful in this situation.

7. Explain the meaning of the y-intercept in this context.

8. Write the equation of the horizontal asymptote and explain the meaning of the horizontal asymptote in this situation.

Part C

FINDINGS The company researches gas mileage for one of their vehicles. Determine whether each finding represents direct, inverse, joint, or combined variation.

9. The amount of gas consumed by a car varies based on the age of the vehicle. As a car ages, it consumes a greater amount of gas.

10. The amount of gas consumed is affected by the weight of the load in the car and grade of the road on which the car is driving. As the load weight and the absolute value of the grade increases, so does the amount of gas consumed.

11. Average miles per gallon is affected by the ingredients in the fuel used. As ethanol content increases, average miles per gallon decreases. As the purity of the petrol used increases, the average miles per gallon increases.

12. For the first 5000 miles, the car's engine experiences a "break-in period" in which the car's engine has not reached maximum efficiency. During this period, the more miles that are driven, the less gas is used per mile.

Test-Taking Strategy

Example

Read the problem. Identify what you need to know. Then use the information in the problem to solve.

Solve: $\dfrac{2}{x-3} - \dfrac{4}{x+3} = \dfrac{8}{x^2-9}$.

A -1
B 1
C 5
D 7

Step 1 Are there any answer choices you can eliminate because they are unreasonable?

No, the answer must be a real number, but all of the answer choices are real numbers.

Step 2 Which answer choice should you start with?

Either B or C, because they are in the middle. I'll start with B.

Step 3 Is the answer you chose correct? If not, in which direction do you need to go to find the correct answer?

Choice B was not correct. The result was too small, so I need to go up to find the correct answer. Choice C results in a true statement.

> **Test-Taking Tip**
>
> **Strategies for Guessing and Checking**
> It is very important to pace yourself and keep track of how much time you have when taking a standardized test. If time is running short, or if you are unsure how to solve a problem, the guess-and-check strategy may help you determine the correct answer quickly, especially in a problem where it would be fast and easy to check your answer, such as in the problem below.

Apply the Strategy

Read the problem. Identify what you need to know. Then use the information in the problem to solve.

Solve: $\dfrac{2}{5x} - \dfrac{1}{2x} = -\dfrac{1}{2}$.

A $\dfrac{1}{10}$
B $\dfrac{1}{5}$
C $\dfrac{1}{4}$
D $\dfrac{1}{2}$

Answer the questions below.

a. Are there any answer choices you can eliminate because they are unreasonable?

b. Which answer choice should you start with?

c. Is the answer you chose correct? If not, in which direction do you need to go to find the correct answer?

connectED.mcgraw-hill.com

CHAPTER 7
Preparing for Assessment
Cumulative Review

Read each question. Then fill in the correct answer on the answer document provided by your teacher or on a sheet of paper.

1. The height of each cylinder is half of the height of the cylinder that it sits upon. The height of the bottom cylinder is 2 inches. How tall is this stack of five cylinders?

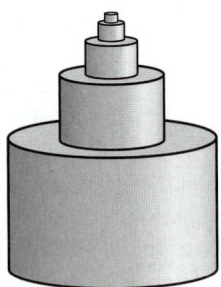

- A 4 in.
- B $3\frac{15}{16}$ in.
- C $3\frac{7}{8}$ in.
- D $3\frac{3}{4}$ in.
- E $3\frac{1}{2}$ in.

2. What is the domain of the function?

$$f(x) = \frac{2x}{x^2 - 1} + \frac{1}{2}$$

- A $D = \{x \mid x \neq 1\}$
- B $D = \{x \mid x \neq -1, 1\}$
- C $D = \{x \mid x \neq -1\}$
- D $D = \{x \mid x \neq -2\}$
- E $D = \{x \mid x \neq 0\}$

3. **MULTI-STEP** Pilar can build a fence in 12.5 hours. Jackson can build the same fence in 15 hours.

   a. If they work together, about how long will it take for them to build the fence? Round to the nearest tenth.

   b. **MP** What mathematical practice did you use to solve this problem?

4. Which is an equation of a line perpendicular to the line passing through the points $(-1, 8)$ and $(0, 12)$?
- A $y = -\frac{1}{4}x - 12$
- B $y = -x + 4$
- C $y = \frac{1}{4}x - 48$
- D $y = -4x - 8$
- E $y = 4x - 48$

5. Which function has a graph with no horizontal asymptote?
- A $f(x) = \frac{1}{x}$
- B $f(x) = \frac{x^2 - 1}{x(x + 2)}$
- C $f(x) = \frac{(x + 3)(x - 1)}{2x}$
- D $f(x) = \frac{x^3 - 1}{(x^2 - 1)(x^2 + 1)}$

> **Test-Taking Tip**
> **Question 6** Substitute each of the answer choices for the numerator of the first term. Evaluate the expression on the left to see if it evaluates to $2x + 6$.

6. What expression makes the following a true statement?

$$\frac{?}{x + 1} \cdot \frac{2(x + 1)}{x - 5} = 2x + 6$$

- A $x - 5$
- B $x^2 + 8x + 15$
- C $x^2 - 2x - 15$
- D $x^2 - 4x - 5$

7. Jeremy bought a painting in 2005 for $350. It is estimated that the value of the painting will increase by 12% each year. In what year does the value of the painting exceed $1000?
- A 2011
- B 2013
- C 2015
- D 2017

526 | Chapter 7 | Preparing for Assessment

Go Online! for Standardized Test Practice

8. If x varies inversely as y and $x = 15$ when $y = 3$, what is the value of x when $y = 8$?

9. Which function has a graph with a vertical asymptote at $x = 7$ and a horizontal asymptote at $f(x) = -3$?
   - A $f(x) = \frac{1}{x+7} + 3$
   - B $f(x) = \frac{1}{x-3} - 7$
   - C $f(x) = \frac{1}{x+3} + 7$
   - D $f(x) = \frac{1}{x-7} - 3$
   - E $f(x) = \frac{1}{x+7} - 3$

10. For what value of x is the function $f(x) = \frac{2x^2 - 7x - 4}{x - 4}$ discontinuous?

11. The function below is a transformation of the function $f(x) = \frac{1}{x}$.
    $$g(x) = \frac{1}{x-2} + 3$$
    Which correctly describes the transformation?
    - A a horizontal translation 2 units right and a vertical translation 3 units up
    - B a horizontal translation 2 units right and a vertical translation 3 units down
    - C a horizontal translation 2 units left and a vertical translation 3 units down
    - D a horizontal translation 3 units right and a vertical translation 2 units up
    - E A horizontal translation 3 units left and a vertical translation 2 units down

12. Select all true statements about the function graphed below.

    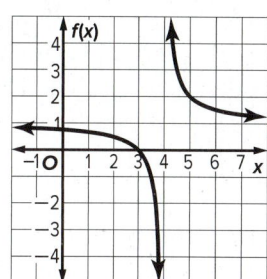

    - A The domain of the function is all real numbers except 4.
    - B The range of the function is all real numbers.
    - C The vertical asymptote is $x = 1$.
    - D The horizontal asymptote is $f(x) = 1$.
    - E The graph represents a transformation of the function $f(x) = \frac{1}{x}$.

13. Janice spent $3.85 for 2 pounds of apples and 3 pounds of pears. Lisa paid $4.94 for 4 pounds of apples and 2 pounds of pears. How much does a pound of apples cost?
    - A $0.69
    - B $0.77
    - C $0.82
    - D $0.89
    - E $1.58

14. What are the zeros of the function?
    $$f(x) = \frac{x^3 + 2x^2 - 35x}{x - 3}$$
    - A 3
    - B 0 and 3
    - C 5 and -7
    - D 0, 5, and -7
    - E 3, 5, and -7

Need Extra Help?

If you missed Question...	1	2	3	4	5	6	7	8	9	10	11	12	13	14
Go to Lesson...	6-3	7-2	7-6	1-4	7-4	7-1	6-1	7-5	7-3	7-4	7-4	7-3	1-6	7-4

CHAPTER 8
Statistics and Probability

THEN
You calculated weighted averages.

NOW
You will:
- Evaluate surveys, studies, and experiments.
- Create and use graphs of probability distributions.
- Compare sample statistics and population statistics.

 WHY

EDUCATION Probability and statistics are used in all facets of education, including determining grades or when teachers weight their grades.

Use the Mathematical Practices to complete the activity.

1. **Using Tools** Use what you already know, your Algebra 1 text, or the Internet to refresh your memory about distributions of data.

2. **Sense Making** This graph represents exam grades. What does this graph tell you? What grade did most students receive?

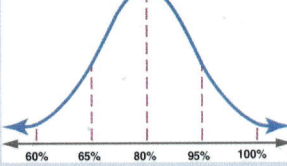

3. **Applying Math** Is the data symmetrically distributed or skewed? Can the mean and standard deviation be used to represent the data set or should the five-number summary be used?

4. **Discuss** What might positively skewed grades say about the exam or the teaching methods used? How can a positively skewed set of data be altered to become symmetrically distributed?

 # Go Online to Guide Your Learning

Explore & Explain

 ### Coin Toss
Use the **Coin Toss** tool, the **Number Cube**, or the **Spinner** tool to enhance your understanding of statistical experiments and to run simulations in Lesson 8-2.

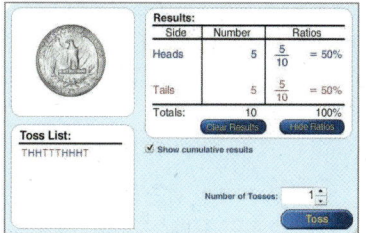

The Geometer's Sketchpad
Use **The Geometer's Sketchpad** to illustrate the normal distribution of data in Lesson 8-6.

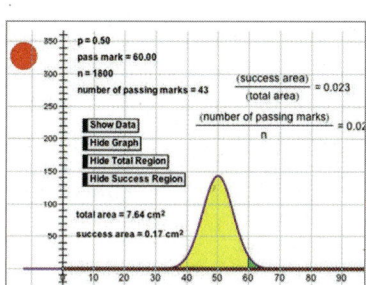

eBook
Interactive Student Guide
Before starting the chapter, answer the **Chapter Focus** preview questions. Check your answers as you complete each lesson. At the end of the chapter, try the **Performance Task**.

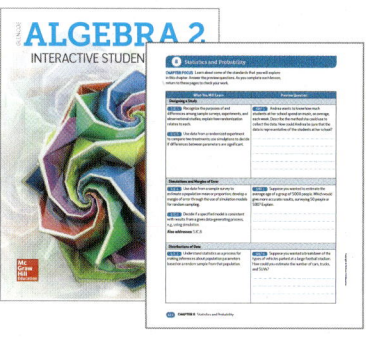

Organize

 ### Foldables
Get organized! Create this **Statistics and Probability Foldable** before you begin this chapter to help you organize your notes about statistics and probability.

Collaborate

 ### Chapter Project
In the **Please Complete This Survey** project, you will use what you have learned about statistics and probability to complete a project that addresses business literacy.

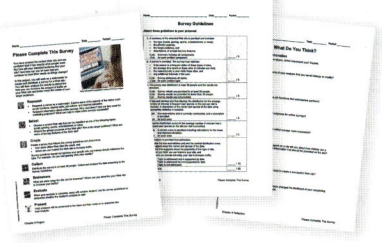

Focus

 LEARNSMART
Need help studying? Complete the **Modeling with Functions** domain in LearnSmart to review for the chapter test.

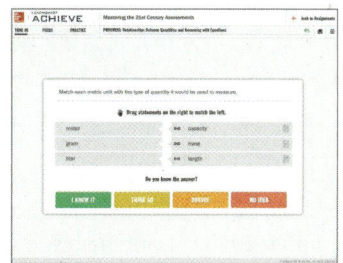

ALEKS
You can use the **Sequences and Probability** topic in ALEKS to explore what you know about statistics and probability and what you are ready to learn.*

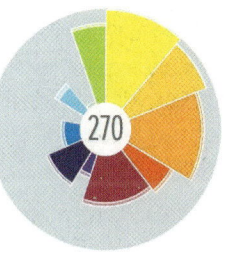

*Ask your teacher if this is part of your program.

Get Ready for the Chapter

Go Online! for Vocabulary Review Games and key vocabulary in 13 languages.

Connecting Concepts

Concept Check
Review the concepts used in this chapter by answering the questions below.

1. What is the mean of a set of data? How is it determined?
2. What is the median of a set of data?
3. What is the mode of a set of data?
4. How is the median of a set of data determined?
5. If a number cube is rolled and a coin is tossed, how would you determine the probability that the number cube would show a 1 and the coin would land tails up?
6. In the box-and-whisker plot shown, is the data negatively skewed, symmetric, or positively skewed?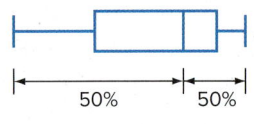
7. When plotting data in a distribution chart, how will an outlier affect the shape of the distribution?
8. When a distribution is symmetric, what can be said about the relationship between the mean and the median?

New Vocabulary

English		Español
statistic	p. 531	estadística
parameter	p. 531	parámetro
survey	p. 531	exámenes
experiment	p. 531	experimento
observational study	p. 531	estudio de observación
random sample	p. 531	muestra aleatoria
bias	p. 531	sesgo
experiment	p. 531	experimento
simulation	p. 531	simulación
probability model	p. 538	modelo de probabilidad
theoretical probability	p. 538	probabilidad teórica
experimental probability	p. 538	probabilidad experimental
relative frequency	p. 538	frecuencia relativa
margin of error	p. 546	margen de error muestral
distribution	p. 551	distribución
normal distribution	p. 566	distribución normal
z-value	p. 568	valor de z
standard normal distribution	p. 568	distribución normal estándar

Performance Task Preview

You can use the concepts and skills in this chapter to perform statistical analysis on data for a company that specializes in this service. Understanding statistics and probability will help you finish the Performance Task at the end of the chapter.

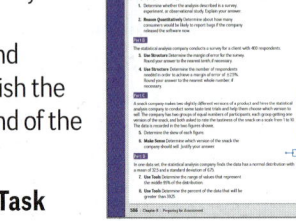

In this Performance Task you will:
- make sense of problems and persevere in solving them
- model with mathematics
- attend to precision
- look for and express regularity in repeated reasoning

Review Vocabulary

combination combinación an arrangement or selection of objects in which order is not important

permutation permutación a group of objects or people arranged in a certain order

random arbitrario Unpredictable, or not based on any predetermined characteristics of the population; when a number cube is tossed, a coin is flipped, or a spinner is spun, the outcome is a random event.

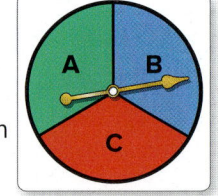

LESSON 1
Random Sampling

:Then
- You summarized data using measures of center and measures of variation.

:Now
1. Distinguish among sample surveys, experiments, and observational studies.
2. Make inferences about population parameters based on random samples of the population.

:Why?
- According to a recent study, 88% of teen cell phone users in the United States send text messages, and one in three teens sends more than 100 texts per day.

New Vocabulary
parameter
statistic
bias
random sample
survey
experiment
observational study

Mathematical Practices
1 Make sense of problems and persevere in solving them.
4 Model with mathematics.

1 Classifying Studies

Statistics is the collection, analysis, interpretation, and organization of data. You have used *descriptive statistics* to summarize data using measures like the mean. Due to time and money constraints, it may not be possible to collect data from each member of a population to calculate a population characteristic or **parameter**. In *inferential statistics*, data is collected to make inferences about a population. In these studies, a sample of the population is taken, and a measure called a **statistic** is calculated using the data. The sample statistic, such as the sample mean or sample standard deviation, is then used to make inferences about the population parameter.

The steps in a typical statistical study are shown below.

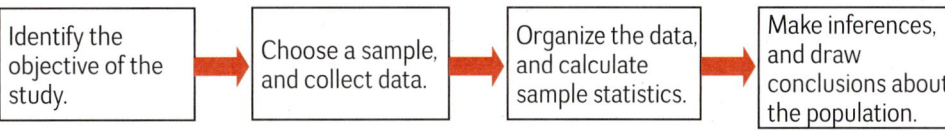

| Identify the objective of the study. | → | Choose a sample, and collect data. | → | Organize the data, and calculate sample statistics. | → | Make inferences, and draw conclusions about the population. |

To obtain good information and draw accurate conclusions about a population, it is important to select an *unbiased* sample. A **bias** is an error that results in a misrepresentation of members of a population. A poorly chosen sample can cause biased results. To reduce the possibility of selecting a biased sample, a **random sample** can be taken, in which members of the population are selected entirely by chance.

The following study types can be used to collect sample information.

Key Concept Study Types

Definition	Example
In a **survey**, data are collected from responses given by members of a population regarding their characteristics, behaviors, or opinions.	To determine whether the student body likes the new cafeteria menu, the student council asks a random sample of students for their opinion.
In an **experiment**, the sample is divided into two groups: • an *experimental group* that undergoes a change, and • a *control group* that does not undergo the change. The effect on the experimental group is then compared to the control group.	A restaurant is considering creating meals with chicken instead of beef. They randomly give half of a group of participants meals with chicken and the other half meals with beef. Then they ask how they like the meals.
In an **observational study**, members of a sample are measured or observed without being affected by the study.	Researchers at an electronics company observe a group of teenagers using different laptops and note their reactions.

connectED.mcgraw-hill.com   531

Example 1 Classify Study Types

Determine whether each situation describes a *survey*, an *experiment*, or an *observational study*. Then identify the sample, and suggest a population from which it may have been selected.

a. **MUSIC** A band wants to test three designs for an album cover. They randomly select 50 members of their fan club to view the covers while they watch and record their reactions.

This is an observational study, because the band is going to observe the fans without them being affected by the study. The sample is the 50 fans selected, and the population is all potential purchasers of this album.

b. **RECYCLING** The city council wants to start a recycling program. They send out a questionnaire to 200 random citizens asking what items they would recycle.

This is a survey, because the data are collected from participants' responses in the questionnaire. The sample is the 200 people who received the questionnaire, and the population is all of the citizens of the city.

> **Study Tip**
> **Census** A census is a survey in which each member of a population is questioned. Therefore, when a census is conducted, there is no sample.

▶ **Guided Practice**

1A. RESEARCH Scientists study the behavior of one group of dogs given a new heartworm treatment and another group of dogs given a false treatment or *placebo*.

1B. YEARBOOKS The yearbook committee conducts a study to determine whether students would prefer to have a print yearbook or both print and digital yearbooks.

To determine when to use a survey, experiment, or observational study, think about how the data will be obtained and whether or not the participants will be affected by the study.

Example 2 Choose a Study Type

Determine whether each situation calls for a *survey*, an *experiment*, or an *observational study*. Explain your reasoning.

a. **MEDICINE** A pharmaceutical company wants to test whether a new medicine is effective.

The treatment will need to be tested on a sample group, which means that the members of the sample will be affected by the study. Therefore, this situation calls for an experiment.

b. **ELECTIONS** A news organization wants to randomly call citizens to gauge opinions on a presidential election.

This situation calls for a survey because members of the sample population are asked for their opinion.

▶ **Guided Practice**

2A. RESEARCH A research company wants to study smokers and nonsmokers to determine whether 10 years of smoking affects lung capacity.

2B. PETS A national pet chain wants to know whether customers would pay a small annual fee to participate in a rewards program. They randomly select 200 customers and send them questionnaires.

2 Make Inferences
Once data have been collected using a random sample, you can analyze the data to calculate a sample statistic and make inferences about the population.

Real-World Example 3   Make an Inference About a Population

TECHNOLOGY A random sample of the 684 students at Sanchez High School were surveyed and asked to name the type of technology they would most like to use for learning a foreign language. Based on the results in the bar graph, what is the most reasonable inference about the number of students at Sanchez High School who would like to use a tablet for learning a foreign language?

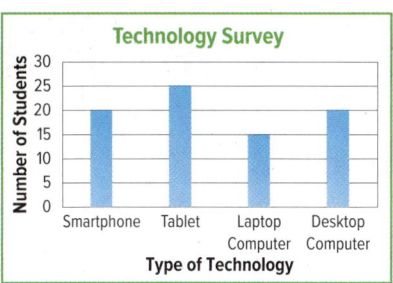

Step 1 Determine the number of students in the random sample.

$20 + 25 + 15 + 20 = 80$

So, there were 80 students in the random sample.

Step 2 Calculate the sample proportion.

25 students in the random sample chose a tablet as their preferred technology.

$\frac{25}{80} = 0.3125$

The sample proportion is 0.3125.

Step 3 Use the sample proportion to make an inference about the population.

$0.3125 \cdot 684 = 213.75$

It is reasonable to infer that approximately 214 students at the school would most like to use a tablet for learning a foreign language.

> **Reading Math**
>
> **MP Reasoning** To make an inference is to make a logical conclusion from given evidence. When you make an inference, be sure each step of the process can be justified.

Guided Practice

3. What is the most reasonable inference about the number of students who would like to use a laptop computer for learning a foreign language?

Your inferences might not reflect your population if there is bias in the survey or sample.

Real-World Example 4   Identify Bias

CLUBS Jamar surveyed a random sample of his school's film club. He found that 30 of the 40 members that he surveyed also play video games. He uses this survey to infer that, at his school of 800 students, approximately 600 students at the school play video games. Identify and explain any bias that might affect the validity of Jamar's inference.

Step 1 Identify the sample and population.

The sample is students in the film club at Jamar's school. The population is all of the students at Jamar's school, because that is the group he makes an inference about.

Step 2 Identify and explain potential bias.

Jamar takes a random sample of his school's film club, but makes an inference about all of the students at the school. Because students with a specific interest in film are in the club, they do not accurately reflect the student population, and thus Jamar's inference might be affected by the bias in his survey.

connectED.mcgraw-hill.com   533

> **Guided Practice**
>
> 4. **CARS** Sumitra wants to estimate how many cars in a mall parking lot have bumper stickers. She drives around the lot and takes random sample of the cars and finds that 30 of the 52 cars she sampled have bumper stickers, and thus determines that approximately 346 of the 600 cars in the whole lot have bumper stickers. Identify and explain any bias that might affect the validity of Sumitra's inference.

Check Your Understanding

 = Step-by-Step Solutions begin on page R11.

Example 1 Determine whether each situation describes a *survey*, an *experiment*, or an *observational study*. Then identify the sample, and suggest a population from which it may have been selected.

1. **SCHOOL** A group of high school students is randomly selected and asked to complete the form shown.

2. **DESIGN** An advertising company wants to test a new logo design. They randomly select 20 participants and watch them discuss the logo.

Do you agree with the new lunch rules?
☐ agree
☐ disagree
☐ don't care

Example 2 **CONSTRUCT ARGUMENTS** Determine whether each situation calls for a *survey*, an *experiment*, or an *observational study*. Explain your reasoning.

3. **LITERACY** A literacy group wants to determine whether high school students that participated in a recent national reading program had higher standardized test scores than high school students that did not participate in the program.

4. **RETAIL** The research department of a retail company plans to conduct a study to determine whether a dye used on a new T-shirt will begin fading before 50 washes.

Example 3

5. **EXERCISE** A random sample of the 4240 gym members at Work It! were surveyed about their favorite way to exercise outdoors. Based on the results in the graph, what is the most reasonable inference about the number of gym members whose favorite way to exercise outdoors is cycling?

Example 4

6. **EGGS** An employee at a supermarket wants to see how many eggs in their latest shipment are broken. He checks the top carton in each stack of cartons and sees that 4 of the 72 eggs are broken. From this, he infers that approximately 24 of the 432 total eggs are broken. Identify any bias that might affect the validity of the employee's inference.

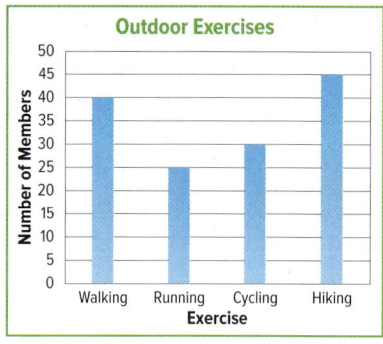

Practice and Problem Solving

Extra Practice is on page R8.

Example 1 Determine whether each situation describes a *survey*, an *experiment*, or an *observational study*. Then identify the sample, and suggest a population from which it may have been selected.

7. **GRADES** A research group randomly selects 80 college students, half of whom took a physics course in high school, and compares their grades in a college physics course.

8. **FOOD** A grocery store conducts an online study in which customers are randomly selected and asked to provide feedback on their shopping experience.

Example 2 Determine whether each situation calls for a *survey*, an *experiment*, or an *observational study*. Explain your reasoning.

9. **FASHION** A fashion magazine plans to poll 100 people in the United States to determine whether they would be more likely to buy a subscription if given a free issue.

10. **TRAVEL** A travel agency randomly calls 250 U.S. citizens and asks them what their favorite vacation destination is.

11. **FOOD** Chee wants to examine the eating habits of 100 random students at lunch to determine how many students eat in the cafeteria.

12. **ENGINEERING** An engineer is planning to test 50 metal samples to determine whether a new titanium alloy has a higher strength than a different alloy.

Example 3 **TOOTHPASTE** A random sample of a dentist's 847 patients were surveyed about their favorite flavor of toothpaste. Use the results in the graph to determine the most reasonable inference about each of the following.

13. the number of patients whose favorite flavor is spearmint

14. the number of patients whose favorite flavor is cinnamon

15. the number of patients whose favorite flavor is peppermint or spearmint

16. the number of patients whose favorite flavor is not spearmint

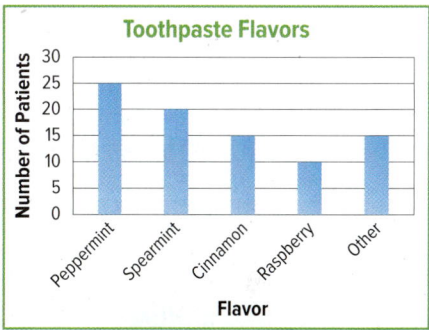

WINTER SPORTS A random sample of the 5358 subscribers to a sports blog were asked to name their favorite winter sport. Use the results in the table to determine the most reasonable inference about each of the following.

17. the number of subscribers whose favorite winter sport is ice skating

18. the number of subscribers whose favorite winter sport is not skiing

19. the number of subscribers whose favorite winter sport is hockey or snowboarding

20. the number of subscribers whose favorite winter sport is neither skiing nor hockey

Favorite Winter Sports	
Sport	Frequency
skiing	23
hockey	14
snowboarding	51
ice skating	40

Example 4 21. **CUSTOMERS** Gabrielle owns a café. One weekend, she hands out an anonymous survey at random to her customers asking how satisfied they were with their service. Of the 35 respondents, 18 stated they were extremely satisfied. So, she infers that approximately 257 of her average weekend crowd of 500 are extremely satisfied with their experiences. Identify any bias that might affect the validity of the employee's inference.

22. **ART** Miguel chose a random sample of 32 paintings in the modern art section and found that 12 were watercolors. Thus, he infers that approximately 171 of the museum's 455 paintings are watercolors. Identify any bias that might affect the validity of the employee's inference.

23. **REPORTS** The graph shown is from a report on the average number of minutes 8- to 18-year-olds in the United States spend on cell phones each day.

  a. Describe the sample and suggest a population.
  b. What type of sample statistic do you think was calculated for this report?
  c. Describe the results of the study for each age group.
  d. Who do you think would be interested in this type of report? Explain your reasoning.

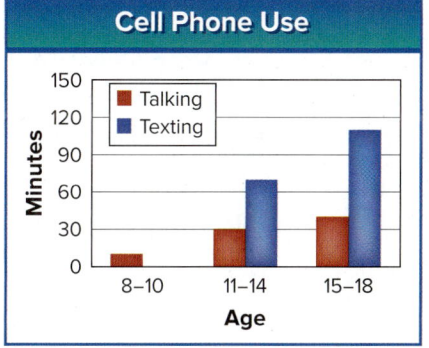

24. **MP MODELING** In 1936, the *Literary Digest* reported the results of a statistical study used to predict whether Alf Landon or Franklin D. Roosevelt would win the presidential election that year. The sample consisted of 2.4 million Americans, including subscribers to the magazine, registered automobile owners, and telephone users. The results concluded that Landon would win 57% of the popular vote. The actual election results are shown.

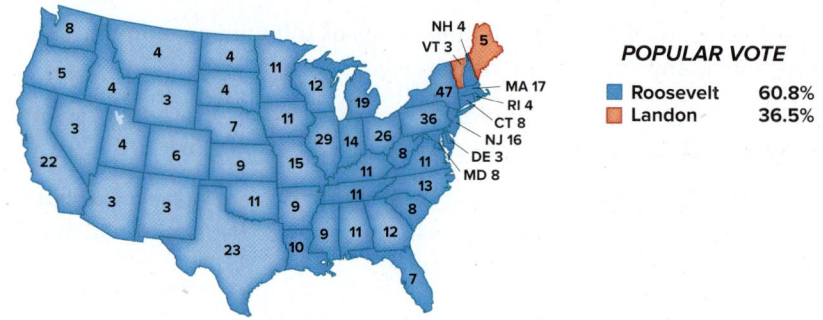

a. Describe the type of study performed, the sample taken, and the population.
b. How do the predicted and actual results compare?
c. Do you think that the survey was biased? Explain your reasoning.

25. **MULTIPLE REPRESENTATIONS** The results of two experiments concluded that Product A is 70% effective and Product B is 80% effective.

   a. **NUMERICAL** To simulate the experiment for Product A, use the random number generator on a graphing calculator to generate 30 integers between 0 and 9. Let 0–6 represent an effective outcome and 7–9 represent an ineffective outcome.

   b. **TABULAR** Copy and complete the frequency table shown using the results from part **a**. Then use the data to calculate the probability that Product A was effective. Repeat to find the probability for Product B.

   c. **ANALYTICAL** Compare the probabilities that you found in part **b**. Do you think that the difference in the effectiveness of each product is significant enough to justify selecting one product over the other? Explain.

   d. **LOGICAL** Suppose Product B costs twice as much as Product A. Do you think the probability of the product's effectiveness justifies the price difference to a consumer? Explain.

Product A	
Number	Frequency
0–6	
7–9	

H.O.T. Problems Use Higher-Order Thinking Skills

MP CONSTRUCT ARGUMENTS Determine whether each statement is *true* or *false*. If false, explain.

26. To save time and money, population parameters are used to estimate sample statistics.

27. Observational studies and experiments can both be used to study cause-and-effect relationships.

28. **OPEN-ENDED** Design an observational study. Identify the objective of the study, define the population and sample, collect and organize the data, and calculate a sample statistic.

29. **CHALLENGE** What factors should be considered when determining whether a given statistical study is reliable?

30. **WRITING IN MATH** Research each of the following sampling methods. Then describe each method and discuss whether using the method could result in bias.

   a. convenience sample
   b. self-selected sample
   c. stratified sample
   d. systematic sample

Preparing for Assessment

31. The principal of a school sends out a questionnaire to a random sample of 60 parents of students at the school and asks them to choose a date for back-to-school night. Which of the following is the best description for this study? MP 6

- A census
- B experiment
- C observational study
- D survey

32. Amani takes a random sample of the nuts in his snack mix and finds that 7 of the 11 nuts he chose are cashews. Because the bag has 50 nuts, he infers that approximately 32 of the nuts are cashews. Identify any bias that might affect the validity of the employee's inference. MP 1

- A There is no bias because the sample accurately represents the population.
- B There is bias because the sample is not random.
- C There is bias because the sample does not accurately reflect the population, which is all snack mixes.
- D There is bias because he incorrectly calculated the expected number of cashews in his snack mix.

33. Which of these situations call for an observational study? MP 6

- A A biologist wants to know how long penguins sleep each day.
- B A scientist at a skin-care company wants to know if a new sunscreen is more effective than the current version.
- C The editor of a magazine wants to know if subscribers like the new format of the magazine.
- D A cafeteria employee wants to know if students finish everything on their trays at lunchtime.
- E A veterinarian wants to know if flea powder X works better than flea powder Y.

34. Rich surveyed a random sample of 50 of the 695 employees at his company and found that 32 of them take a bus or train to get to work. Which is the most reasonable inference about the number of employees who take a bus or train to work? MP 1, 4

- A 222
- B 348
- C 445
- D 1086

35. Morgan surveyed a random sample s of registered voters in her town. She asked them for which candidate for mayor they planned to vote. The results are shown in the table. There are P registered voters in Morgan's town. MP 1, 4

Candidate	Frequency
Fernandez	f
Harrison	h
Taylor	t
Wong	w

Which expression can Morgan use to make an inference about the number of registered voters who plan to vote for Fernandez or Harrison?

- A $\dfrac{f+h}{2} \cdot P$
- B $\dfrac{f+h}{s}$
- C $\dfrac{s}{f+h} \cdot P$
- D $\dfrac{f+h}{s} \cdot P$

36. MULTI-STEP A random sample of the 952 subscribers to a travel blog were surveyed about the country they would most like to visit. The results are shown in the graph. MP 2, 4

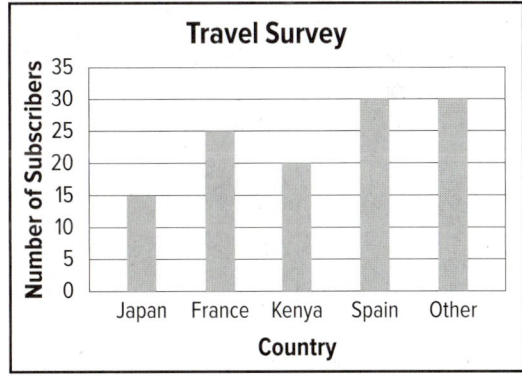

a. What is the sample proportion of subscribers who did not choose France?

b. What is the most reasonable inference about the number of subscribers who would choose Spain as the country they would most like to visit?

c. Suppose the number of subscribers to the blog doubles but these survey results remain unchanged. Would your answers to parts **a** and **b** change? If so, explain how. If not, explain why not.

LESSON 2
Using Statistical Experiments

Then
- You calculated simple probability.

Now
1. Collect and analyze data by conducting simulations of real-life situations.
2. Use data to compare theoretical and experimental probabilities.

Why?
Alex has been practicing his penalty kicks. Based on practice, he predicts that he scores on at least 66% of his kicks. To test this, he takes 50 penalty kicks, of which he scores on 35.

 New Vocabulary
simulation
probability model
theoretical probability
experimental probability
relative frequency

 Mathematical Practices
1 Make sense of problems.
2 Reason abstractly and quantitatively.
3 Construct viable arguments.
4 Model with mathematics.

1 Simulations
An experiment that would be difficult or impractical to perform can be modeled by a **simulation**. In a simulation, a **probability model** to represent the **theoretical probability** is used to recreate a situation in which the **experimental probability** or **relative frequency** of an outcome can be found.

Example Event	Theoretical Probability	Experimental Probability (Relative Frequency)
	Ratio of number of favorable outcomes to total number of outcomes	Ratio of number of outcomes in an experiment to total number of trials
Coin toss lands heads up	$\frac{1}{2}$, or 50%	3 heads land up out of 10 tosses $\frac{3}{10}$, or 30%

Real-World Example 1 — Design a Simulation

QUALITY CONTROL Eloy inspects bike frames as they come through the assembly line. From previous observations, he expects to find a weld defect in one out of every 10 frames and a design defect in one out of every 20 frames that he inspects. Design a simulation using Eloy's expectation of defects. Assume that a frame can only have one of the defects.

Step 1 Determine each possible outcome and its theoretical probability. There are three possible outcomes: weld defect, design defect, and no defects. Use Eloy's expectation of defects to calculate the theoretical probability of each outcome.

Possible Outcomes	Theoretical Probability
weld defect	10%
design defect	5%
no defects	85%

Step 2 Describe an appropriate probability model for the situation that accurately represents the theoretical probability of each outcome. We can use the random number generator on a graphing calculator. Assign the integers 0–19 to accurately represent the probability data.

Outcome	Represented by
weld defect	0, 1
design defect	2
no defects	3–19

Step 3 Define what a trial is for the situation, and state the number of trials to be conducted. A trial will represent selecting a frame at random. The simulation can consist of any number of trials. We will use 40.

Guided Practice

1. A survey asked students which method they would choose to travel to school each morning. The results are shown in the circle graph. Design a simulation that can be used to estimate the probability that a randomly chosen student will choose each of the four transportation methods.

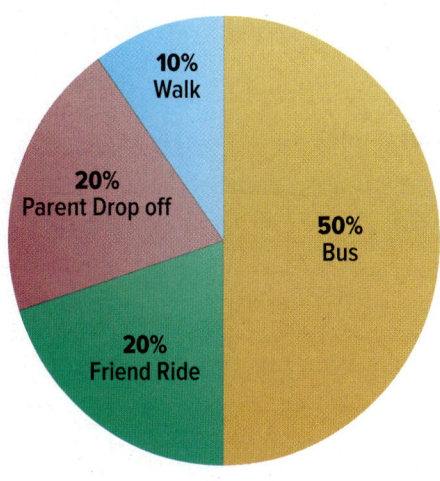

Problem-Solving Tip

Use a Simulation
Simulations often provide a safe and efficient problem-solving strategy in situations that otherwise may be costly, dangerous, or impossible to solve using theoretical techniques. Simulations should involve data that are easier to obtain than the actual data you are modeling.

Simulations can also be conducted using number cubes, coin tosses, random number tables, and random number generators, such as those available on graphing calculators.

Example 2 Design a Simulation by Using Random Numbers

EYE COLOR A survey of East High School students found that 40% had brown eyes, 30% had hazel eyes, 20% had blue eyes, and 10% had green eyes. Design a simulation that can be used to estimate the probability that a randomly chosen student will have one of these eye colors. Assume that the student's eye color will fall into one of these categories.

Step 1 Possible Outcomes   Theoretical Proability
Possible Outcomes		Theoretical Proability
brown eyes	→	40%
hazel eyes	→	30%
blue eyes	→	20%
green eyes	→	10%

Step 2 Describe an appropriate probability model for the situation.

Use the random number generator on your calculator. Assign the integers 0–9 to accurately represent the probability data. The actual numbers chosen to represent the outcomes do not matter.

Outcome	Represented by
brown eyes	0, 1, 2, 3
hazel eyes	4, 5, 6
blue eyes	7, 8
green eyes	9

Step 3 A trial will represent selecting a student at random and recording his or her eye color. The simulation will consist of 20 trials.

Study Tip

Random Number Generator
To generate a set of random integers on a graphing calculator, press MATH and select randInt(under the PRB menu. Then enter the beginning and ending integer values for your range and the number of integers you want in each trial.

Guided Practice

2. **SOCCER** Last season, Yao made 18% of his free kicks. Design a simulation using a random number generator that can be used to estimate the probability that he will make his next free kick.

After designing a simulation, you will need to conduct the simulation and report the results. Include both numerical and graphical summaries of the simulation data, as well as an estimate of the probability of the desired outcome.

Example 3 Conduct and Evaluate a Simulation

QUALITY CONTROL Refer to the simulation in Example 1. Conduct the simulation and report the results.

Press MATH ◄ and select [randInt (]. Then press 0 , 19 , 40) ENTER. Use the left and right arrow buttons to view the results. Make a frequency table and record the results.

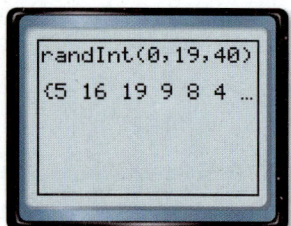

Outcome	Tally	Frequency
weld defect	|||	3
design defect	||	2
no defects	|||| |||| |||| |||| |||| |||| ||||	35
Total		40

Calculate the experimental probability of finding each type of defect.

Weld defect $= \dfrac{\text{frequency}}{\text{total}} = \dfrac{3}{40}$ or 0.075

Design defect $= \dfrac{\text{frequency}}{\text{total}} = \dfrac{2}{40}$ or 0.05

No defects $= \dfrac{\text{frequency}}{\text{total}} = \dfrac{35}{40}$ or 0.875

The experimental probabilities that a frame will have a weld defect, a design defect, or no defects in this case are 7.5%, 5%, and 87.5%, respectively.

Make a bar graph of these results.

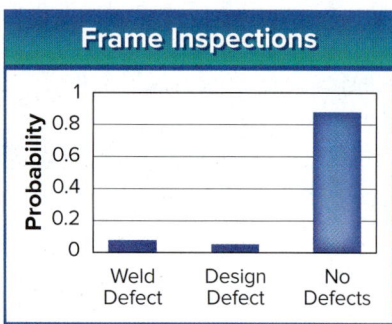

Technology Tip

Random Number Generator After running the random number generator, you can store the results as a list by pressing STO▶ and entering **L1**.

Real-World Career

Assemblers Assemblers may work as part of a team, where all members are capable of performing each task. In the automobile manufacturing industry, the median wage for a team assembler in 2013 was $30,554.
Source: Pay Scale

▶ **Guided Practice**

3. Conduct the simulation in Guided Practice 1. Then report the results.

Concept Summary — Designing, Conducting, and Reporting a Simulation

Designing a Simulation

Step 1 Determine each possible outcome and its theoretical probability.

Step 2 Describe an appropriate probability model for the situation.

Step 3 Define what a trial is for the situation and state the number of trials to be conducted.

Conducting a Simulation

Step 1 Conduct the simulation.

Step 2 Report the results using appropriate numerical and graphical summaries, including frequency tables and bar graphs.

Step 3 Compare the experimental probability results with the theoretical probability.

Example 4   Conduct and Summarize Data from a Simulation

Refer to the simulation in Example 2. Conduct the simulation and report the results using appropriate numerical and graphical summaries.

Make a frequency table after using a graphing calculator to conduct the simulation 20 times.

Outcome	Tally	Frequency
brown eyes	𝍬 IIII	9
hazel eyes	𝍬	5
blue eyes	IIII	4
green eyes	II	2

Based on the simulation data, calculate the probability that a randomly selected student will have brown eyes.

$$\frac{\text{number of students with brown eyes}}{\text{total number of students}} = \frac{9}{20} \text{ or } 45\%$$

The experimental probability that a randomly selected student will have brown eyes is 45%. Notice that this is close to the theoretical probability, 40%. So, the experimental probability of not having brown eyes is 1 − 0.45, or 55%.

Make a bar graph of the results.

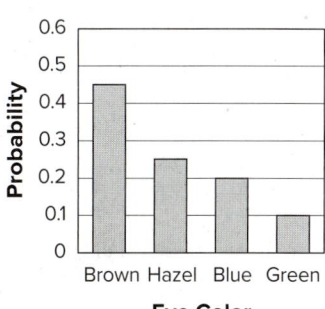

Guided Practice

4. Use a graphing calculator to conduct the simulation in Example 4 for 30 trials. Report the results using appropriate numerical and graphical summaries. Compare the experimental and theoretical probabilities.

The *Law of Large Numbers* states that as the number of trials of a random process increases, the experimental probability will approach the theoretical probability. The table at the right shows the result of running the simulation again for 40, 80, 150, and 200 trials. Notice the probability that a randomly chosen student having brown eyes is $\frac{81}{200}$, or 40.5%. This experimental probability is very close to the theoretical value of 40%.

Number of Trials	Frequency of Brown Eyes
40	20
80	30
150	58
200	81

Check Your Understanding   ⬤ = Step-by-Step Solutions begin on page R11.

Go Online! for a Self-Check Quiz

Examples 1, 3 **1. GRADES** Clara got an A on 80% of her first semester Biology quizzes. Design and conduct a simulation using a geometric model to estimate the probability that she will get an A on a second semester Biology quiz. Report the results using appropriate numerical and graphical summaries.

Examples 2–3 **2. FOOTBALL** Rico is a pitcher on the baseball team. Last season, 72% of the pitches he threw were strikes.

    a. Design a simulation that can be used to estimate the probability that Rico will throw a strike for his next pitch.

    b. Conduct the simulation, and report the results.

Age	Frequency
0–7	13
8–12	28
13–17	48
18–23	42
24–31	27
32–45	40
46–64	33
65+	23

3. **FITNESS** The table shows the percent of members participating in four classes offered at a gym. Design and conduct a simulation to estimate the probability that a new gym member will take each class. Report the results using appropriate numerical and graphical summaries.

Class	Sign-Up %
tae kwon do	45%
yoga	30%
swimming	15%
kick-boxing	10%

Examples 3–4

4. Bridget is a member of the bowling club at her school. Last season, she bowled a strike 60% of the time.

   a. Design a simulation to estimate the probability that she will get a strike in the next frame of bowling.

   b. Conduct the simulation and report the results using appropriate numerical and graphical summaries.

   c. Compare the experimental probability to the theoretical probability.

Practice and Problem Solving

Extra Practice is on page R8.

Examples 1, 3 Design and conduct a simulation using a probability model. Then report the results.

5. **VIDEO GAMES** Ian works at a video game store. Last year he sold 95% of the new-release video games.

6. **MUSIC** Kadisha is listening to music from a playlist set to random. There are 10 songs in the playlist.

Examples 2–4 **MP MODELING** Design and conduct a simulation using a random number generator. Then report the results.

7. **MOVIES** A movie theater reviewed sales from the previous year to determine which genre of movie sold the most tickets. The results are shown at the right.

Genre	Ticket %
drama	40%
mystery	30%
comedy	25%
action	5%

8. **BASEBALL** According to a baseball player's on-base percentages, he gets a single 60% of the time, a double 25% of the time, a triple 10% of the time, and a home run 5% of the time.

9. **VACATION** According to a survey done by a travel agency, 45% of their clients went on vacation to Europe, 25% went to Asia, 15% went to South America, 10% went to Africa, and 5% went to Australia.

10. **TRANSPORTATION** A car dealership's analysis indicated that 35% of the customers purchased a blue car, 30% purchased a red car, 15% purchased a white car, 15% purchased a black car, and 5% purchased any other color.

11. **BATTING AVERAGE** In a computer baseball game, a player has a batting average of .300. That is, he gets a hit 300 out of 1000, or 30%, of the times he is at bat.

    a. Design a simulation that can be used to estimate the probability that the player will get a hit at his next at bat.

    b. Conduct the simulation, and report the results.

12. **JEANS** Julie examines the stitching on pairs of jeans that are produced at a manufacturing plant. She expects to find defects in 1 out of every 20 pairs.

    a. Design a simulation that can be used to estimate the probability that the next pair of jeans that Julie examines has a defect.

    b. Conduct the simulation, and report the results.

13. **FOOD** For a promotion, the concession stands at a football stadium are giving away free items. For every tenth customer, a wheel is spun to choose the customer's prize. Each prize is equally likely.

    a. Design a simulation that can be used to estimate the probability that the next spin is one of the five prizes.

    b. Conduct the simulation, and report the results.

14. **TEST** Jack forgot to study for his multiple-choice science quiz and is going to guess for each question. There are 20 questions, each with 4 possible answers.

    a. Design a simulation that can be used to estimate the number of questions that Jack answers correctly.

    b. Conduct the experiment from part **a** five times, and complete the table.

Simulation	Number of Correct Answers
1	
2	
3	
4	
5	

15. **MULTIPLE REPRESENTATIONS** In this problem, you will investigate expected value.

    a. **Concrete** Roll two dot cubes 20 times and record the sum of each roll.

    b. **Numerical** Use the random number generator on a calculator to generate 20 pairs of integers between 1 and 6. Record the sum of each pair.

    c. **Tabular** Copy and complete the table below using your results from parts **a** and **b**.

Trial	Sum of Roll	Sum of Output from Random Number Generator
1		
2		
...		
20		

    d. **Graphical** Use a bar graph to graph the number of times each possible sum occurred in the first 5 rolls. Repeat the process for the first 10 rolls and then all 20 outcomes.

    e. **Verbal** How does the shape of the bar graph change with each additional trial?

    f. **Graphical** Graph the number of times each possible sum occurred with the random number generator as a bar graph.

    g. **Verbal** How do the graphs of the cube trial and the random number trial compare?

H.O.T. Problems   Use Higher-Order Thinking Skills

16. **OPEN-ENDED** Describe a situation at your school that could be represented by a simulation. Then design the simulation.

17. **JUSTIFY ARGUMENTS** An experiment has three equally likely outcomes A, B, and C. Is it possible to use the spinner shown in a simulation to predict the probability of outcome C? Explain your reasoning.

    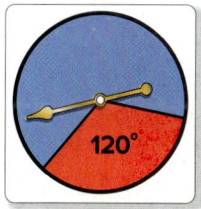

18. **JUSTIFY ARGUMENTS** Can tossing a coin *sometimes*, *always*, or *never* be used to simulate an experiment with two possible outcomes? Explain.

19. **WRITING IN MATH** What should you consider when using the results of a simulation to make a prediction?

20. **WRITING IN MATH** How is designing a simulation like solving a word problem?

Preparing for Assessment

21. Pilar is playing a board game with eight different categories, each with questions that must be answered correctly to win. **MP 4**

  a. Design a simulation that can be used to estimate the probabilities of landing on the eight categories.

  b. Conduct the simulation and report the results using appropriate numerical and graphical summaries.

  c. Use the data to compare the theoretical and experimental probabilities. Which statement is true for the results in part **b**?

   ○ A The experimental and theoretical probabilities are close in value.

   ○ B The experimental and theoretical probabilities are not close in value.

   ○ C The experimental and theoretical probabilities are the same in value.

22. MULTI-STEP A basketball player made 62% of her free throws. **MP 4**

  a. Design a simulation using 40 trials to estimate the probability that she will make her next free throw.

  b. Which result(s) shown below may be a result of the simulation in part **a**?

   ☐ A

Outcome	Tally	Frequency
make free throw	||||| ||||| ||||| ||||| |||||		26
miss free throw	||||| ||||| ||||	14
Total		40

   ☐ B

Outcome	Tally	Frequency
1																											25
2																						20
Total		45

   ☐ C
   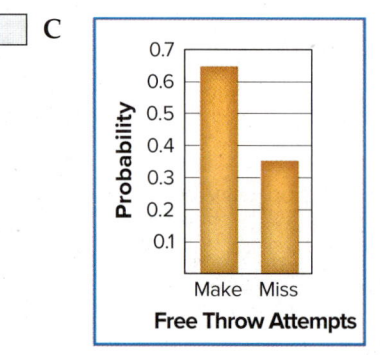
   Free Throw Attempts

  c. What is the experimental probability that she will make her next free throw? **MP 4**

   ○ A 70%   ○ C 35%
   ○ B 65%   ○ D 30%

  d. Use the data to compare the theoretical and experimental probabilities.

  e. Explain how to change the simulation so that the experimental probability may be closer to the theoretical probability.

23. A national survey showed that 70% of Americans aged 12 and older attended a movie in 2013. Which of the following are the best steps that could be used for a simulation to estimate the probability that the next attendee at a movie theater will be aged 12 or over? Select all that apply. **MP 4**

  ☐ A Possible outcomes: Moviegoer is 12 or older—70% Moviegoer is under 12—30%

  ☐ B Possible outcomes: Moviegoer is 12 years old—70% Moviegoer is under 12—30%

  ☐ C Possible outcomes: Moviegoer is 12 or older—30% Moviegoer is under 12—70%

  ☐ D Have theater employees ask attendees their age as they exit the theater. Make a frequency table with categories "12 and over" and "under 12."

  ☐ E Send a survey through the mail about the age of movie attendees and ask people to return the survey.

  ☐ F Conduct 20 trials.

  ☐ G Conduct 200 trials.

24. You roll a dot cube to see how many times you can roll a 1. You generate the following table. **MP 4**

Trials	10	20	50	100
Frequency	3	4	12	18
Experimental Probability	0.3	0.2	0.24	0.18

  a. Describe the simulation that may have generated this table.

  b. As the number of trials increase, what is happening to the experimental probability?

  c. Evaluate the simulation in part **a**. How would you describe the experimental probability of rolling a 1?

LESSON 3
Population Parameters

::Then
- You defined and calculated random samples of a population.

::Now
1. Use data from sample surveys to estimate population means or proportions.
2. Develop margins of error by using simulation models.

::Why?
A company wants to test out their new product before it launches. To save on cost, they conduct a survey on a sample of people who best represent the point of view of the general population.

New Vocabulary
confidence level
population mean
population parameters
population proportion
margin of error

Mathematical Practices
1 Make sense of problems and persevere in solving them.
4 Model with mathematics.
5 Use appropriate tools strategically.

1 Estimate Population Means or Proportions
A population parameter is a number that represents the whole population. These are usually unknown values and must be estimated from given data. The **population mean** is a population parameter that calculates the mean, or average, of the entire population. The population mean can be represented by μ. Another parameter is the **population proportion** p, which is the number of members in the population sharing a particular attribute divided by the number of members in the population.

Key Concept — Population Mean and Population Proportion

Five students with test scores 100, 100, 91, 84, and 75

Population Mean
average of a sample survey

Formula
$$\mu = \frac{\Sigma x_i}{N}$$
$$= \frac{x_1 + x_2 + x_3 + \ldots + x_N}{N}$$

Σ means "the sum of."
x = all the individual items in the group
N = the number of items in the group

Total sum of test scores divided by total number of test scores:

$$\mu = \frac{(100 + 100 + 91 + 84 + 75)}{5}$$
$$\mu = \frac{450}{5} = 90$$

Population Proportion
fraction or percentage of the sample survey that possess a particular trait

Proportion of test scores higher than 90:

$$p = \frac{\text{(test scores higher)}}{\text{(number of test)}}$$
$$p = \frac{3}{5}$$
$$p = 0.60$$

Real-World Example 1 Use Data to Estimate Population Mean

COLLEGE All 51 residents in a dorm were surveyed to see how many classes they are taking this semester. The results are shown in the table at the right.

What is the population mean for the number of classes taken this semester?

Step 1 Sum the total number of classes taken by the residents. This represents the numerator of the formula Σx_i.

Number of Classes	Number of Residents
1	6
2	5
3	26
4	11
5	3

Note that each term x_i represents the number of classes taken by resident i, so there should be 51 values that are summed.

$\Sigma x_i = (1 \cdot 6) + (2 \cdot 5) + (3 \cdot 26) + (4 \cdot 11) + (5 \cdot 3) = 153$

Step 2 Divide the answer to Step 1 by the number of items in the data set. This is the N part of the population mean formula.

There are 51 residents, so:
$$\frac{153}{51} = 3$$

The population mean is 3 classes per resident.

> **Study Tip**
>
> **Precision** Recall that a decimal can be converted to a percentage by multiplying the decimal value by 100.

Guided Practice

What is the population mean for the number of hours worked per day?

Number of Hours	Number of Employees
3	4
4	12
5	21

1. Step 1: $(4 \cdot 3) + (12 \cdot 4) + (21 \cdot 5) = 165$

Step 2: $\frac{165}{37} \approx 4.459459$

Example 2 Use Data to Estimate Population Proportion

In a population of 250 people, 90 said they prefer running over hiking.

  a. What is the population proportion of people who prefer running over hiking?

  b. What percent of the population prefers running to hiking?

Part a: Divide the number of people who prefer running by the total number of people in the data set. There are 250 people, so:
$$p = \frac{90}{250} = 0.36$$

Part b: A population proportion of 0.36 is equivalent to 36% of the population. Therefore, 36% of the population prefers running to hiking.

Guided Practice

Find the following population proportions.

2A. 105 people prefer running over hiking.

2B. 55 people prefer hiking over running.

2C. 140 people prefer hiking over running.

---

2 Margin of Error

A survey of a random sample is a valuable tool for generalizing information about a larger population. The accuracy of a sample survey is based on two key factors—the **margin of error** and the confidence level. The margin of error tells you the interval in which the result will fall relative to the real population value. Statisticians have found that for large populations, the margin of error for a random sample of size n can be approximated by the formula, margin of error $= \pm \frac{1}{\sqrt{n}} (100)$.

> **Study Tip**
>
> **Sense-Making** The margin of error determines how reliable a survey is. The larger the sample size, the smaller the margin of error.

You can use the following formula to calculate the margin of error.

> **Key Concept** Margin of Error Formula
>
> $$\text{margin of error} = \pm \frac{1}{\sqrt{n}}(100)$$
>
> where n is the sample size

Example 3   Calculate Margin of Error

A survey is conducted to determine how people will vote for the school presidential candidacy. The result of the 500 students surveyed showed that 62% will vote for Candidate B. Find the margin of error.

$$\frac{1}{\sqrt{n}} \cdot 100 = \frac{1}{\sqrt{500}} \cdot 100 \approx 4.47 \text{ or about } \pm 4.5\%$$

The real proportion is in the range 57.5% to 66.5%.

Guided Practice

3. A survey is conducted to determine the most liked color of the rainbow. The results of the 200 people surveyed showed that 44% of people like the color blue the most. Find the margin of error.

Real-World Example 4   Use Margin of Error to Find Sample Size

RESEARCH You are a member of a research team and are going to run a simulation. The simulation needs to result in an adequate sample size and margin of error for each trial.

a. Find the sample size you should use in your simulation for a margin of error of ±3%.

Substitute ±3% for the margin of error and solve for n in the margin of error formula.

$$\pm 3\% = \pm \frac{1}{\sqrt{n}}(100) \quad \text{Margin of error formula}$$
$$0.03\sqrt{n} = 1 \quad \text{Multiply by } \frac{\sqrt{n}}{100}.$$
$$\sqrt{n} = 33.333 \quad \text{Divide.}$$
$$n = 1111.11 \quad \text{Square each side.}$$

A sample size of about 1111 would have a margin of error of ±3%.

b. Determine the sample size that has a margin of error of ±10%.

Substitute ±10% for the margin of error and solve for n in the margin of error formula.

$$\pm 10\% = \pm \frac{1}{\sqrt{n}}(100) \quad \text{Margin of error formula}$$
$$0.1\sqrt{n} = 1 \quad \text{Multiply by } \frac{\sqrt{n}}{100}.$$
$$n = 10 \quad \text{Divide.}$$
$$n = 100 \quad \text{Square each side.}$$

A sample size of 100 would have a margin of error of ±10%.

Guided Practice

4. The finance director decides to conduct a survey with a margin of error of ±4%. Find the appropriate sample size.

Check Your Understanding

= Step-by-Step Solutions begin on page R11.

Example 1
1. **MODELING** Jesse does community service keeping the local park clean. He noticed people who come to the park on a regular basis to play chess for several hours. Jesse surveyed some of the players to find the total number of hours they play each week. Use the table to calculate the population mean.

Number of Players	Hours per Week
3	10
5	12
4	15
2	20
2	25

Example 2
2. A survey is conducted to find the most-watched TV network during the times of 5:00 P.M. to 10:00 P.M. Out of a sample of 175 people, 59 viewers are watching the news channel during this time frame. What is the population proportion of people who are watching the news channel?

Example 3
3. For a sample size of 100 people, what is the margin of error that 53% of consumers will vote "yes" on a new product line from Company A to Z?

Example 4
4. A university needs to perform an annual survey and obtain a margin of error of ±5%. What would be the best sample size to reach this goal?

Practice and Problem Solving

Extra Practice is on page R8.

Example 1
5. **MANUFACTURING** A clothing manufacturer wants to begin selling their own products. They decide to run a small survey to determine how many items per color they currently manufacture at the warehouse. Find the population mean.

Color	Number of items
Yellow	13
Red	76
Blue	23
Black	98

Example 2 For a sample survey of 500 employees, find the following population proportions.

6. 75 prefer the weekend shift
7. 270 prefer to stay on site for lunch
8. 165 prefer the morning shift
9. 300 prefer a flexible schedule

Example 3 Compute the margin of error for each sample size.

10. 100
11. 300
12. 500
13. 755
14. 1000
15. 2500

Example 4 Find the sample size for each margin of error.

16. ±2%
17. ±3%
18. ±6%
19. ±7%

20. **CONSTRUCT ARGUMENTS** The Pew Research Center recently conducted a survey of a random sample of 200 teens and concluded that 43% of all teens who take their cell phones to school text in class on a daily basis. How accurately did their random sample represent all teens?

21. **CRITIQUE ARGUMENTS** Fitness center A claims that more of their members burn more calories per day than at the competing fitness center B. A random sample of 10 people from each center results in the following calories burned.

Fitness center A: 391 326 322 297 326 289 293 264 327 331

Fitness center B: 311 304 321 302 307 297 311 313 336 299

Estimate the population mean of calories burned by the members at each fitness center. Does the claim appear to be legitimate?

22. **CEREAL** A company is conducting a small survey about a new cereal that they want to release next summer. Ten customers produced the following ratings for the cereal after a sample.

Rating	1	2	3	4
Number of Customers	1	3	4	2

a. Estimate the population mean for these ratings.

b. What is the margin of error?

c. What does the margin of error say about the poll?

d. What sample size will produce a margin of 4%?

23. **APPLES** A survey of 250 kids was conducted to determine their favorite type of apple between Granny Smith and Golden Delicious apples. It showed that 43% prefer Granny Smith apples. Calculate the margin of error.

24. A survey of 1600 customers at a travel agency showed that 66% would prefer to live in a cooler climate. What is the true percent range that those surveyed will choose to live in a cooler climate?

25. The Department of Education surveyed 20,000 students to determine the graduation rate over a course of five years at a local high school. The results showed that 86% of seniors graduated. Determine the real percentage range of graduated students.

H.O.T. Problems Use Higher-Order Thinking Skills

26. What happens to the margin of error if the sample size increases?

27. Describe how to calculate the population mean rating for 100 candy bars.

28. The Warriors cross-country team is preparing for a national run. Each team member runs laps around a track and records their time in order to monitor their progress.

Team Members	Running Time (in secs)
Runner 1	180
Runner 2	172
Runner 3	185
Runner 4	168
Runner 5	149

a. Calculate the mean of the runners' running time.

b. What proportion of the team runs 170 seconds or faster?

29. **DOWNTOWN** The Mountain Hill High School student council conducted a survey of 300 students and reported that 250 of them hang out in the downtown area on Friday and Saturday nights. Estimate the population proportion of the students who hang out downtown.

30. A survey showed that 250 out of 500 adults prefer coffee over tea. Determine the population proportion of the adults who prefer coffee and estimate the margin of error.

31. Every attendee at the carnival filled out a short survey before leaving. Out of 1750 people, 75% said they will refer the carnival to a friend next year. What is the population proportion of attendees that will refer the carnival for the following year?

Preparing for Assessment

32. In a fifth grade class, 26 students were asked how many children their parents have. The results are shown in the frequency table. Estimate the mean number of children. **MP 4**

Children	Frequency
1	5
2	9
3	8
4	4

$\mu =$ _____

33. In Nomax County, 3000 residents are registered to vote. Out of this population of voters, 40% are registered for Party A. What is the population proportion of registered voters who belong to Party A? **MP 1**

- A $p = 0.3$
- B $p = 0.4$
- C $p = 0.6$
- D $p = 0.12$
- E $p = 0.16$
- F $p = 0.20$

34. A random sample of middle school students revealed that 15% of children were experiencing reading difficulties. **MP 1**

a. Find the margin of error if the sample size was 325.

b. Find the margin of error if the sample size was 500.

35. Every student taking Geology 101 was asked how they felt about the last exam. Twenty percent of the 300 students surveyed found it to be difficult. Which of the following statements below is true about the value 0.20? **MP 1**

- A It is the population proportion.
- B It is the population mean.
- C It is the margin of error.
- D Is is the confidence interval.
- E More information is needed.

36. At a day care, there are 17 children under the age of 4. The ages, in months, are: 19 28 25 24 26 21 20 27 36 35 30 22 23 33 13 34 12. What is the population mean age of children at the day care? **MP 1**

- A 22
- B 25.2
- C 26.5
- D 30

37. 30 children were given a random amount of marbles. The table represents the number of marbles each child possesses. What information can be gathered from the data? Select all that apply. **MP 1**

Marbles	Children
2	5
3	19
4	8

- A proportion of children with 3 marbles
- B proportion of children under the age of 10 with marbles
- C mean number of marbles
- D mean number of children without marbles

38. Explain how to estimate the population mean of the number of times that each player attempts to beat the enemy in Level 5 of a computer game. **MP 5**

39. MULTI-STEP Cassy wants to conduct a survey of the amount of water consumed per day for homeowners in her area. Because of the large population, she wants to determine the best sample size with a low margin of error. **MP 1**

a. What is the margin of error if she surveys 150 homeowners?

b. What is the sample size if she wants to accomplish a margin of error of ±4.6%? Round to the nearest whole number.

c. Use the answers above to determine the better option for Cassy. Explain.

LESSON 4
Distributions of Data

∴Then
You calculated measures of central tendency and variation.

∴Now
1. Use the shapes of distributions to select appropriate statistics.
2. Use the shapes of distributions to compare data.

∴Why?
After four games as a reserve player, Craig joined the starting lineup and averaged 18 points per game over the *remaining* games. Craig's scoring average for the *entire* season was less than 18 points per game as a result of the lack of playing time in the first four games.

 New Vocabulary
distribution
negatively skewed distribution
symmetric distribution
positively skewed distribution

Mathematical Practices
1 Make sense of problems and perservere in solving them.
7 Look for and make use of structure.

1 Analyzing Distributions
A **distribution** of data shows the observed or theoretical frequency of each possible data value. In Lesson 0-9, you described distributions of sample data using statistics. You used the mean or median to describe a distribution's center and standard deviation or quartiles to describe its spread. Analyzing the shape of a distribution can help you decide which measure of center or spread best describes a set of data.

The shape of the distribution for a set of data can be seen by drawing a curve over its histogram.

Key Concept Symmetric and Skewed Distributions

Negatively Skewed Distribution	**Symmetric Distribution**	**Positively Skewed Distribution**
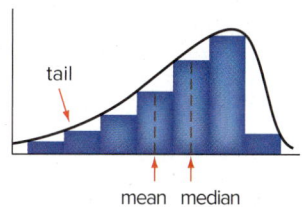	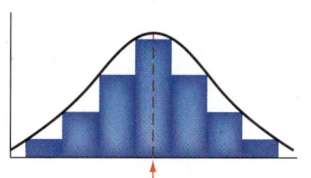	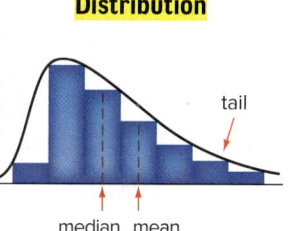
• The mean is less than the median. • The majority of the data are on the right of the mean.	• The mean and median are approximately equal. • The data are evenly distributed on both sides of the mean.	• The mean is greater than the median. • The majority of the data are on the left of the mean.

When a distribution is symmetric, the mean and standard deviation accurately reflect the center and spread of the data. However, when a distribution is skewed, these statistics are not as reliable. Recall that outliers have a strong effect on the mean of a data set, while the median is less affected. Similarly, when a distribution is skewed, the mean lies away from the majority of the data toward the tail. The median is less affected, so it stays near the majority of the data.

When choosing appropriate statistics to represent a set of data, first determine the skewness of the distribution.

- If the distribution is relatively symmetric, the mean and standard deviation can be used.
- If the distribution is skewed or has outliers, use the five-number summary to describe the center and spread of the data.

connectED.mcgraw-hill.com   **551**

Real-World Example 1 Describe a Distribution Using a Histogram

LAPTOPS The prices for a random sample of laptops are shown.

Price (dollars)							
723	605	847	410	440	386	572	523
374	915	734	472	420	508	613	659
706	463	470	752	671	618	538	425
811	502	490	552	390	512	389	621

Real-World Link
The first portable computer, the Osborne I, was available for sale in 1981 for $1795. The computer weighed 24 pounds and included a 5-inch display. Laptops can now be purchased for as little as $199 and can weigh as little as 2.3 pounds.
Source: Computer History Museum

a. **Use a graphing calculator to create a histogram. Then describe the shape of the distribution.**

First, press STAT ENTER and enter each data value. Then, press 2nd [STAT PLOT] ENTER ENTER and choose 📊. Finally, adjust the window to the dimensions shown.

The majority of the laptops cost between $400 and $700. Some of the laptops are priced significantly higher, forming a tail for the distribution on the right. Therefore, the distribution is positively skewed.

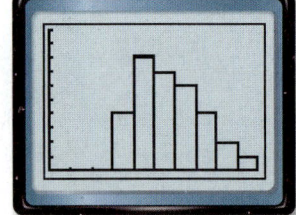

[0, 1000] scl: 100 by [0, 10] scl: 1

b. **Describe the center and spread of the data using either the mean and standard deviation or the five-number summary. Justify your choice.**

The distribution is skewed, so use the five-number summary to describe the center and spread. Press STAT ▶ ENTER ENTER and scroll down to view the five-number summary.

The prices for this sample range from $374 to $915. The median price is $530.50, and half of the laptops are priced between $451.50 and $665.

▶ **Guided Practice**

1. **RAINFALL** The annual rainfall for a region over a 24-year period is shown below.

   A. Use a graphing calculator to create a histogram. Then describe the shape of the distribution.

   B. Describe the center and spread of the data using either the mean and standard deviation or the five-number summary. Justify your choice.

Annual Rainfall (in.)					
27.2	30.2	35.8	26.1	39.3	20.6
28.9	23.0	32.7	26.8	22.7	25.4
29.6	36.8	33.4	28.4	21.9	20.8
24.7	30.6	27.7	31.4	34.9	37.1

A box-and-whisker plot can also be used to identify the shape of a distribution. The position of the line representing the median indicates the center of the data. The "whiskers" show the spread of the data. If one whisker is considerably longer than the other and the median is closer to the shorter whisker, then the distribution is skewed.

Go Online!

You can use the box-and-whisker plot in the **eToolkit** to create box-and-whisker plots from given data.

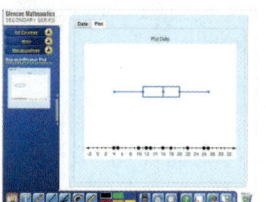

Key Concept Box-and-Whisker Plots as Distributions

Negatively Skewed

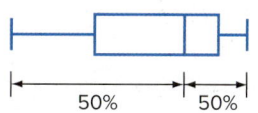

The data to the left of the median are distributed over a wider range than the data to the right. The data have a tail to the left.

Symmetric

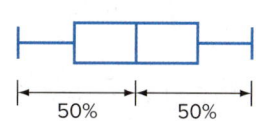

The data are equally distributed to the left and right of the median.

Positively Skewed

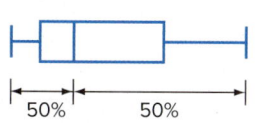

The data to the right of the median are distributed over a wider range than the data to the left. The data have a tail to the right.

Example 2   Describe a Distribution Using a Box-and-Whisker Plot

HOMEWORK The students in Mr. Fejis' language arts class found the average number of minutes that they each spent on homework each night.

Minutes per Night					
62	53	46	66	38	45
52	46	73	39	42	56
64	54	48	59	70	60
49	54	48	57	70	33

a. Use a graphing calculator to create a box-and-whisker plot. Then describe the shape of the distribution.

Enter the data as **L1**. Press [2nd] [STAT PLOT] [ENTER] [ENTER] and choose ⊡⋯. Adjust the window to the dimensions shown.

The lengths of the whiskers are approximately equal, and the median is in the middle of the data. This indicates that the data are equally distributed to the left and right of the median. Thus, the distribution is symmetric.

[30, 75] scl: 5 by [0, 5] scl: 1

b. Describe the center and spread of the data using either the mean and standard deviation or the five-number summary. Justify your choice.

The distribution is symmetric, so use the mean and standard deviation to describe the center and spread. The average number of minutes that a student spent on homework each night was 53.5 with standard deviation of about 10.5.

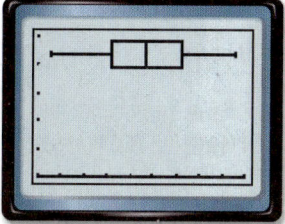

> **Watch Out!**
>
> **Standard Deviation** Recall from Lesson 0-9 that the formulas for standard deviation for a population σ and for a sample s are slightly different. In Example 2, times for all of the students in Mr. Fejis' class are being analyzed, so use the population standard deviation.

Guided Practice

2. CELL PHONE Janet reviewed the number of minutes she spent talking to her friends each month for the last two years.

A. Use a graphing calculator to create a box-and-whisker plot. Then describe the shape of the distribution.

B. Describe the center and spread of the data using either the mean and standard deviation or the five-number summary. Justify your choice.

Minutes Used per Month			
582	608	670	620
667	598	671	613
537	511	674	627
638	661	642	641
668	673	680	695
658	653	670	688

connectED.mcgraw-hill.com   **553**

2 Comparing Distributions

To compare two sets of data, first analyze the shape of each distribution. Use the mean and standard deviation to compare two symmetric distributions. Use the five-number summaries to compare two skewed distributions or a symmetric distribution and a skewed distribution.

Example 3 Compare Data Using Histograms

TEST SCORES Test scores from Mrs. Morash's class are shown below.

Chapter 3 Test Scores
81, 81, 92, 99, 61, 67, 86, 82, 76, 73, 62, 97, 97, 72, 72, 84, 77, 88, 92, 93, 76, 74, 66, 78, 76, 69, 84, 87, 83, 87, 92, 87, 82

Chapter 4 Test Scores
87, 73, 69, 83, 74, 86, 74, 69, 79, 84, 79, 74, 83, 74, 86, 69, 91, 73, 79, 83, 69, 79, 83, 74, 86, 79, 79, 78, 83, 79, 86, 79, 84

a. Use a graphing calculator to create a histogram for each data set. Then describe the shape of each distribution.

Chapter 3 Test Scores

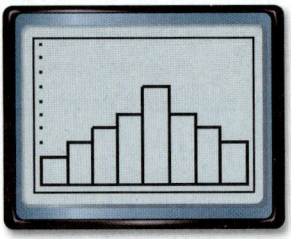

[60, 100] scl: 5 by [0, 10] scl: 1

Chapter 4 Test Scores

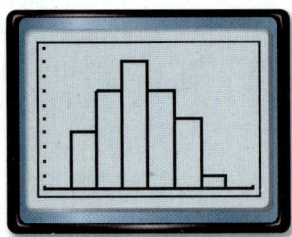

[60, 100] scl: 5 by [0, 10] scl: 1

Both distributions are symmetric.

b. Compare the distributions using either the means and standard deviations or the five-number summaries. Justify your choice.

The distributions are symmetric, so use the means and standard deviations.

Chapter 3 Test Scores

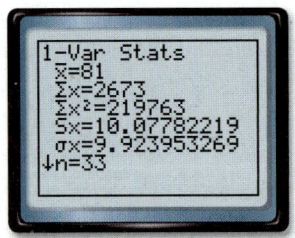

Chapter 4 Test Scores

The Chapter 4 test scores, while lower in average, have a much smaller standard deviation, indicating that the scores are more closely grouped about the mean. Therefore, the mean for the Chapter 4 test scores is a better representation of the data than the mean for the Chapter 3 test scores.

> **Study Tip**
>
>  **Tools** To compare two sets of data, enter one set as **L1** and the other as **L2**. In order to calculate statistics for a set of data in **L2**, press STAT ▶ ENTER 2nd [L2] ENTER.

Guided Practice

3. TYPING The typing speeds of the students in two classes are shown below.

A. Use a graphing calculator to create a histogram for each data set. Then describe the shape of each distribution.

B. Compare the distributions using either the means and standard deviations or the five-number summaries. Justify your choice.

3rd Period (wpm)
23, 38, 27, 28, 40, 45, 32, 33, 34, 27, 40, 22, 26, 34, 29, 31, 35, 33, 37, 38, 28, 29, 39, 42

6th Period (wpm)
38, 26, 43, 46, 23, 24, 27, 36, 22, 21, 26, 27, 31, 32, 27, 25, 23, 22, 28, 29, 28, 33, 23, 24

Box-and-whisker plots can be displayed alongside one another, making them useful for side-by-side comparisons of data.

Example 4   Compare Data Using Box-and-Whisker Plots

POINTS The points scored per game by a football team for the 2013 and 2014 football seasons are shown.

2013							
7	51	24	27	17	35	27	33
28	30	27	21	24	30	14	20

2014							
20	9	3	10	6	14	3	10
3	37	7	21	13	41	20	23

a. Use a graphing calculator to create a box-and-whisker plot for each data set. Then describe the shape of each distribution.

Enter the 2013 scores as **L1**. Graph these data as **Plot1** by pressing [2nd] [STAT PLOT] [ENTER] [ENTER] and choosing ⊡⋯. Enter the 2014 scores as **L2**. Graph these data as **Plot2** by pressing [2nd] [STAT PLOT] [▼] [ENTER] [ENTER] and choosing ⊡⋯. For **Xlist**, enter **L2**. Adjust the window to the dimensions shown.

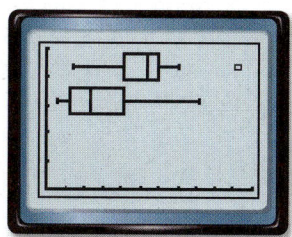

[0, 55] scl: 5 by [0, 5] scl: 1

For the 2013 scores, the left whisker is longer than the right and the median is closer to the right whisker. The distribution is negatively skewed.

For the 2014 scores, the right whisker is longer than the left and the median is closer to the left whisker. The distribution is positively skewed.

> **Study Tip**
>
> **Outliers** Recall from Lesson 0-9 that outliers are data that are more than 1.5 times the interquartile range beyond the upper or lower quartile. All outliers should be plotted, but the whiskers should be drawn to the least and greatest values that are not outliers.

b. Compare the distributions using either the means and standard deviations or the five-number summaries. Justify your choice.

The distributions are skewed, so use the five-number summaries to compare the data.

The lower quartile for the 2013 season and the upper quartile for the 2014 season are both 20.5. This means that 75% of the scores from the 2013 season were greater than 20.5 and 75% of the scores from the 2014 season were less than 20.5.

The minimum of the 2013 season is approximately equal to the lower quartile for the 2014 season. This means that 25% of the scores from the 2014 season are lower than any score achieved in the 2013 season. Therefore, we can conclude that the team scored a significantly higher amount of points during the 2013 season than the 2014 season.

▶ **Guided Practice**

4. GOLF Robert recorded his golf scores for his sophomore and junior seasons.

   A. Use a graphing calculator to create a box-and-whisker plot for each data set. Then describe the shape of each distribution.

   B. Compare the distributions using either the means and standard deviations or the five-number summaries. Justify your choice.

Sophomore Season
42, 47, 43, 46, 50, 47, 52, 45, 53, 55, 48, 39, 40, 49, 47, 50

Junior Season
44, 38, 46, 48, 42, 41, 42, 46, 43, 40, 43, 43, 44, 45, 39, 44

connectED.mcgraw-hill.com   555

Check Your Understanding

= Step-by-Step Solutions begin on page R11.

Go Online! for a Self-Check Quiz

Example 1

1. **EXERCISE** The amount of time that James ran on a treadmill for the first 24 days of his workout is shown.

Time (minutes)											
23	10	18	24	13	27	19	7	25	30	15	22
10	28	23	16	29	26	26	22	12	23	16	27

 a. Use a graphing calculator to create a histogram. Then describe the shape of the distribution.

 b. Describe the center and spread of the data using either the mean and standard deviation or the five-number summary. Justify your choice.

Example 2

2. **RESTAURANTS** The total number of times that 20 random people either ate at a restaurant or bought fast food in a month are shown.

Restaurants or Fast Food										
4	7	5	13	3	22	13	6	5	10
7	18	4	16	8	5	15	3	12	6

 a. Use a graphing calculator to create a box-and-whisker plot. Then describe the shape of the distribution.

 b. Describe the center and spread of the data using either the mean and standard deviation or the five-number summary. Justify your choice.

Example 3

3. **TOOLS** The total fundraiser sales for the students in two classes are shown.

Mrs. Johnson's Class (dollars)					
6	14	17	12	38	15
11	12	23	6	14	28
16	13	27	34	25	32
21	24	21	17	16	

Mr. Edmunds' Class (dollars)					
29	38	21	28	24	33
14	19	28	15	30	6
31	23	33	12	38	28
18	34	26	34	24	37

 a. Use a graphing calculator to create a histogram for each data set. Then describe the shape of each distribution.

 b. Compare the distributions using either the means and standard deviations or the five-number summaries. Justify your choice.

Example 4

4. **RECYCLING** The weekly totals of recycled paper for the junior and senior classes are shown.

Junior Class (pounds)						
14	24	8	26	19	38	
12	15	12	18	9	24	
12	21	9	15	13	28	

Senior Class (pounds)						
25	31	35	20	37	27	
22	32	24	28	18	32	
25	32	22	29	26	35	

 a. Use a graphing calculator to create a box-and-whisker plot for each data set. Then describe the shape of each distribution.

 b. Compare the distributions using either the means and standard deviations or the five-number summaries. Justify your choice.

Practice and Problem Solving

Extra Practice is on page R8.

Examples 1–2 For Exercises 5 and 6, complete each step.
   a. Use a graphing calculator to create a histogram and a box-and-whisker plot. Then describe the shape of the distribution.
   b. Describe the center and spread of the data using either the mean and standard deviation or the five-number summary. Justify your choice.

5. **FANTASY** The weekly total points of Kevin's fantasy football team are shown.

Total Points							
165	140	88	158	101	137	112	127
53	151	120	156	142	179	162	79

6. **MOVIES** The students in one of Mr. Peterson's classes recorded the number of movies they saw over the past month.

Movies Seen										
14	11	17	9	6	11	7	8	12	13	10	9
5	11	7	13	9	12	10	9	15	11	13	15

Example 3 **MODELING** For Exercises 7 and 8, complete each step.
   a. Use a graphing calculator to create a histogram for each data set. Then describe the shape of each distribution.
   b. Compare the distributions using either the means and standard deviations or the five-number summaries. Justify your choice.

7. **SAT** A group of students took the SAT their sophomore year and again their junior year. Their scores are shown.

Sophomore Year Scores					
1327	1663	1708	1583	1406	1563
1637	1521	1282	1752	1628	1453
1368	1681	1506	1843	1472	1560

Junior Year Scores					
1728	1523	1857	1789	1668	1913
1834	1769	1655	1432	1885	1955
1569	1704	1833	2093	1608	1753

8. **INCOME** The total incomes for 18 households in two neighboring cities are shown.

Yorkshire (thousands of dollars)					
68	59	61	78	58	66
56	72	86	58	63	53
68	58	74	60	103	64

Applewood (thousands of dollars)					
52	55	60	61	55	65
65	60	45	37	41	71
50	61	65	66	87	55

Example 4

9. **TUITION** The annual tuitions for a sample of public and private colleges are shown. Complete each step.
   a. Use a graphing calculator to create a box-and-whisker plot for each data set. Then describe the shape of each distribution.
   b. Compare the distributions using either the means and standard deviations or the five-number summaries. Justify your choice.

Public Colleges (dollars)					
5760	7304	8230	6248	9064	9794
7155	8736	7344	6640	6960	6869
6283	5978	5760	8480	9211	6207
7630	7328	6664	7462	9152	6558

Private Colleges (dollars)					
26,770	32,665	11,664	10,804	12,297	15,835
14,250	4200	18,000	17,400	20,910	20,670
12,240	6000	16,360	23,600	13,120	14,976
6800	30,586	9108	9600	9000	21,450

10. **DANCE** The total amount of money that a random sample of seniors spent on prom is shown. Complete each step.

    a. Use a graphing calculator to create a box-and-whisker plot for each data set. Then describe the shape of each distribution.
    b. Compare the distributions using either the means and standard deviations or the five-number summaries. Justify your choice.

Boys (dollars)					
253	288	304	283	348	276
322	368	247	404	450	341
291	260	394	302	297	272

Girls (dollars)					
682	533	602	504	635	541
489	703	453	521	472	368
562	426	382	668	352	587

11. **BASKETBALL** Refer to the beginning of the lesson. The points that Craig scored in the remaining games are shown.

    a. Use a graphing calculator to create a box-and-whisker plot. Describe the center and spread of the data.
    b. Craig scored 0, 2, 1, and 0 points in the first four games. Use a graphing calculator to create a box-and-whisker plot that includes the new data. Then find the mean and median of the new data set.
    c. What effect does adding the scores from the first four games have on the shape of the distribution and on how you should describe the center and spread?

Points Scored			
18	10	18	21
9	25	13	17
17	12	24	19
20	17	27	21

12. **SCORES** Allison's quiz scores are shown.

    a. Use a graphing calculator to create a box-and-whisker plot. Describe the center and spread.
    b. Allison's teacher allows students to drop their two lowest quiz scores. Use a graphing calculator to create a box-and-whisker plot that reflects this change. Then describe the center and spread of the new data set.

Math Quiz Scores					
83	76	86	82	84	57
86	62	90	96	76	89
76	88	86	86	92	94

H.O.T. Problems Use Higher-Order Thinking Skills

13. **CHALLENGE** Approximate the mean and median for each distribution of data.

    a.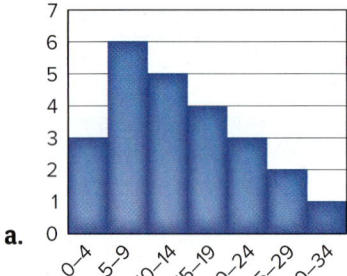
    b.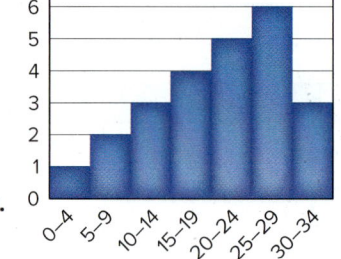
    c.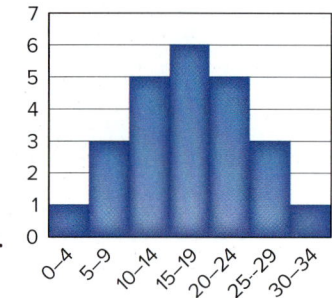

14. **MP CONSTRUCT ARGUMENTS** Distributions of data are not always symmetric or skewed. If a distribution has a gap in the middle, like the one shown, two separate clusters of data may result, forming a *bimodal distribution*. How can the center and spread of a bimodal distribution be described?

    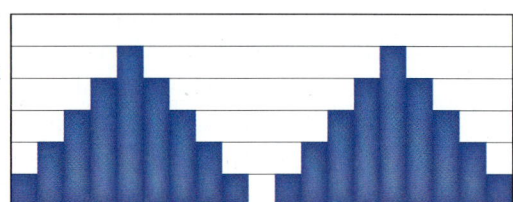

15. **OPEN-ENDED** Find a real-world data set that appears to represent a symmetric distribution and one that does not. Describe each distribution. Create a visual representation of each set of data.

16. **WRITING IN MATH** Explain the difference between positively skewed, negatively skewed, and symmetric sets of data, and give an example of each.

Preparing for Assessment

17. A histogram is used to show the annual snowfall, in inches, for a city over a 20-year period. The center and spread of the data are accurately described as having a mean snowfall of 7.5 inches with a standard deviation of 1.1 inches. Which statement about the data can be assumed from this information? **MP** 2

- A The data distribution is flat.
- B The data distribution is negatively skewed.
- C The data distribution is symmetric.
- D The data distribution is skewed to the left.
- E The mean of the data is much greater than the median.

18. Jane displays attendance data using a box-and-whisker plot. One whisker is much longer than the other. The median is closest to the shorter whisker. Which statement can reasonably be deduced from this information? **MP** 2

- A The mean and the median are equal.
- B The distribution of data is skewed.
- C The median is less than the mean.
- D The median is greater than the mean.
- E The distribution of data is symmetrical.

19. The distributions of exam scores for two high school biology classes are shown in the histograms. Which of the following statements about the histograms is true? **MP** 1

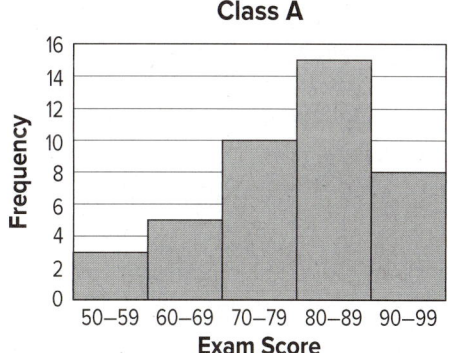

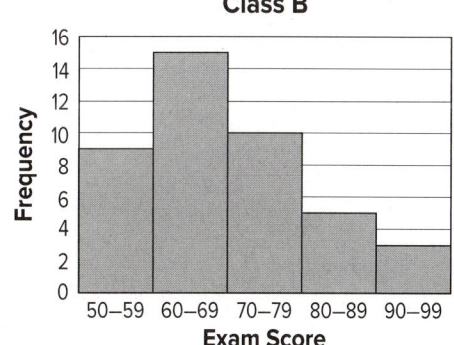

- A Most students in class A and in class B scored higher than the mean score.
- B The data are negatively distributed in class A and symmetrically skewed in class B.
- C The data are symmetrically distributed in class A and positively skewed in class B.
- D The mean exam score in class A and in class B is about the same as the corresponding median exam score.
- E Most students in class A scored higher than the mean score, and most students in class B scored lower than the mean score.

20. A set of data is given as: **MP** 2

18, 18, 19, 19, 20, 22, 22, 23, 27, 28, 28, 31, 34, 34, 36.

The box and whisker plot for this data is shown below.

Write down the values of A, B, C, D, and E.

A = ☐   D = ☐
B = ☐   E = ☐
C = ☐

21. MULTI-STEP The Spanish exam results of classes 9A and 9B are given in the box-and-whisker plots below. **MP** 2

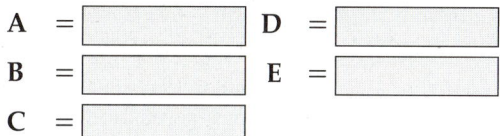

a. Find the maximum and minimum scores in each class. ☐

b. About what percent of the students in 9B scored in the same interval as the students in 9A? ☐

c. About what percent of the students in 9A scored between 65 and 70? ☐

d. About what percent of the students in 9B scored between 45 and 55? ☐

22. The distribution of a set of data is given in the box-and-whisker plot below. **MP** 1

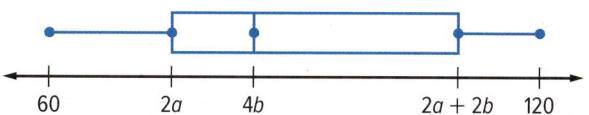

The difference between Q1 and Q3 is 40. Find the median. ☐

CHAPTER 8
Mid-Chapter Quiz
Lessons 8-1 through 8-4

Determine whether each situation describes a *survey*, an *experiment*, or an *observational study*. Then identify the sample, and suggest a population from which it may have been selected. (Lesson 8-1)

1. A high school principal wants to test five ideas for a new school mascot. He randomly selects 15 high school students to view pictures of the ideas while he watches and records their reactions.

2. Half of the employees of a grocery store are randomly chosen for an extra hour of lunch break. The managers then compare their attitudes with those of their coworkers.

3. Students want to create a school yearbook. They send out a questionnaire to 100 students asking what they would like to showcase in the yearbook.

4. The producers of a sitcom want to determine if a new character that they are planning to introduce will be well received. They show a clip of the show with the new character to 50 randomly chosen participants and then record the participants' reactions.

5. A random sample of 85 out of 390 students were surveyed and found that 26 of them own a laptop computer. Which is the most reasonable inference about the total number of students who own a laptop computer? (Lesson 8-1)

6. Julia chose a random sample of 25% of the green peppers in crate from a farm and found that 4 of them had a crack in them. Which is the most reasonable inference about the total number of peppers with a crack in them in the crate? (Lesson 8-1)
   - A 4 peppers
   - B 12 peppers
   - C 16 peppers
   - D 20 peppers

7. A statistician would like to obtain a margin of error of ±4% in his research study. What is the sample size needed to achieve that margin of error? (Lesson 8-3)
   - A 25
   - B 100
   - C 400
   - D 625

8. **TRAINING** Aiden and Mark's training times for the 40-meter dash are shown. (Lesson 8-4)

Aiden's 40-Meter Dash Times (seconds)					
4.84	4.94	4.87	4.78	5.04	4.98
4.83	5.03	4.74	5.15	4.82	4.91
4.62	4.83	4.76	4.93	4.85	4.82
4.76	4.98	4.94	5.05	4.94	5.04
4.86	4.85	4.71	4.66	4.91	4.82

Mark's 40-Meter Dash Times (seconds)					
5.03	4.76	4.69	4.52	4.81	4.78
4.65	4.66	4.83	4.95	4.64	4.76
4.43	4.64	4.50	4.58	4.68	4.65
4.83	4.78	4.71	4.81	4.76	4.84
4.61	4.63	4.33	4.46	4.74	4.63

   a. Use a graphing calculator to create a histogram for each data set. Then describe the shape of each distribution.

   b. Compare the distributions using either the means and standard deviations or the five-number summaries. Justify your choice.

9. **MULTI-STEP** Miguel is taking a 10-question true-false quiz. He is wondering what his chances of passing the test are if he guesses on each question. (Lesson 8-2)

   a. Design a simulation that can be used to estimate the number of correct answers on the quiz.

   b. Conduct the simulation in part **a** five times and report the results using an appropriate summary.

   c. Use the data to compare the theoretical and experimental probabilities. Which statement is true for the results in part **b**? Explain your choice.
      - A The experimental and theoretical probabilities are close in value.
      - B The experimental and theoretical probabilities are not close in value.
      - C The experimental and theoretical probabilities are the same in value.

   d. **MP** What mathematical practice did you use to solve this problem?

LESSON 5
Evaluating Published Data

::Then
- You solved problems about random sampling and statistical experiments.

::Now
1. Evaluate reports based on data.
2. Identify and explain misleading uses of data.

::Why?
- Over the past decade, açai berries have become a popular addition to smoothies and snacks. Some news reports have presented data to promote the health benefits of the berries, while others have pointed out the lack of scientifically controlled studies.

 Mathematical Practices
4 Model with mathematics.
6 Attend to precision.

1 Evaluate Reports Reports published on news sites, in magazines, and in advertising may present data from surveys or studies. It is important to read the reports critically to determine whether the conclusions presented in the reports are valid. Consider the following.

- Was the survey based on a random sample?
- What was the size of the sample?
- Could bias have been introduced in conducting the survey?
- Was the study based on a controlled experiment with individuals randomly assigned to a control group and to a treatment group?
- Can a cause-and-effect relationship be deduced from the data?

Real-World Example 1   Evaluate a Report

BLOOD PRESSURE Explain whether the conclusion presented in the following report is valid.

> **Eating Fruit Lowers Blood Pressure**
>
> The cafeteria at Midville Hospital offers a bowl of fresh fruit and other items for dessert. Researchers tracked 200 hospital employees who eat lunch at the cafeteria each day. They divided the employees into two groups: those who chose fruit for dessert and those who chose other items for dessert. At the end of six months, the researchers measured the blood pressure of the employees and found that those in the group that chose fruit had a lower average blood pressure than those who chose other desserts.

Was the study based on a controlled experiment with individuals randomly assigned to a control group and to a treatment group? The report is based on the results of an observational study rather than a controlled experiment. This means that it may not be valid to draw a cause-and-effect conclusion from the study.

Can a cause-and-effect relationship be deduced from the data? There may be other reasons why employees in the group that chose fruit had a lower average blood pressure. For example, employees who chose fruit may be more likely to exercise or make other healthy lifestyle choices that affect their blood pressure.

connectED.mcgraw-hill.com   561

▶ **Guided Practice**

1. **TEXTING** Explain whether the conclusion presented in the following report is valid.

   > **U.S. Students Text Less Often in Class Than Originally Thought**
   >
   > A survey asked 450 students at Madison High School how often they sent one or more text messages while in class. Of the students surveyed, 86% said they never send a text in class. The results of the survey show that texting in class is not as serious a problem as some may have feared.

2 Identify Misleading Uses of Data

Data from a survey or study can be presented in a variety of ways. In some cases, the data may be presented in a graph so that someone might draw an incorrect conclusion from the data. Consider the following.

- Do the scales on the axes begin at zero? If not, is a break shown on the axes?
- Do the scales on the axes use consistent intervals between tick marks?
- Do the scales on the axes distort differences in the data values?
- Does a histogram group data values appropriately?
- Could the person or organization presenting the data have a reason to create a misleading graph?

Real-World Example 2   Identify Misleading Uses of Data

MASCOTS Explain whether the data presented in the following report is misleading.

> **Hawks: The Overwhelming Choice**
>
> Students at Redwood High School are choosing a new mascot for the school's teams. A random sample of 56 students at the school were surveyed and given a choice of four mascots. The results, shown in the bar graph, indicate an overwhelming preference for Hawks as the new name for the school's teams.

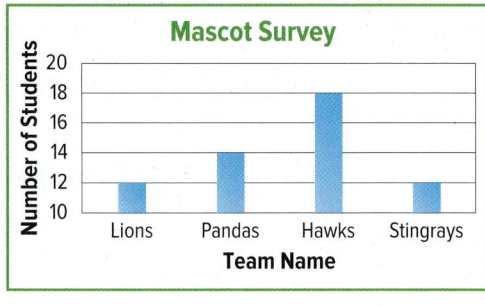

Study Tip

Sense-Making
It can be helpful to compare percentages rather than raw data values. For example, $\frac{18}{56} \approx 32\%$ chose Hawks, but $\frac{14}{56} = 25\%$ chose Pandas, which shows that the difference in popularity was not as great as described in the report.

Although Hawks was the most popular team name, the vertical axis of the graph does not begin at zero. This distorts the relative lengths of the bars and makes Hawks appear much more popular than the other choices.

▶ **Guided Practice**

2. **CHARITY** Explain whether the data presented are misleading.

   > **Donations Are Looking Good!**
   >
   > A random sample of 10 donations received this week at the Gray Foundation show that most donations to the charity are in the $41 to $60 range. The histogram displays the data. The dollar values of the donations in the random sample are given below.
   > 42, 21, 35, 41, 43, 20, 43, 25, 41, 42

562 | Lesson 8-5 | Evaluating Published Data

Check Your Understanding

= Step-by-Step Solutions begin on page R11.

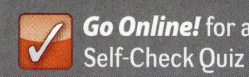 Go Online! for a Self-Check Quiz

Example 1

1. **NEWSPAPERS** Explain whether the conclusion presented in the report is valid.

   > **Reading a Newspaper Increases Test Scores**
   >
   > A group of 150 students were divided into two groups: those that read a newspaper at least once a week and those that do not. Test scores in all subject areas were monitored over the course of a school year. At the end of the year, the average test score of students who read a newspaper at least once a week was 6% greater than the average score of students who did not.

Example 2 Explain whether the data presented in each graph is misleading.

2.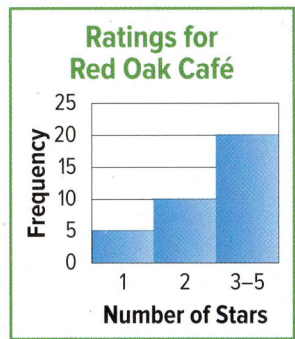

3.

Practice and Problem Solving

Extra Practice is found on page R8.

Example 1

4. **SPORTS** Explain whether the conclusion presented in the following report is valid.

   > **Soccer Is Town's Favorite Sport**
   >
   > A survey of 15 randomly-selected customers at an Elmwood athletic shoe store were asked whether they prefer to watch soccer or another sport. Twenty percent of those surveyed named soccer as their favorite sport to watch. No other sport was named by 20% or more of the participants, so it is safe to conclude that soccer is Elmwood's favorite sport.

5. **MEDICINE** Explain whether the conclusion presented in the following report is valid.

   > **New Drug Helps Headaches**
   >
   > A random sample of 820 patients who get headaches was randomly split into two equal groups. One group tried the new drug every time they felt a headache starting. The other group was given a placebo to use in the same way. After six months, members of the group that tried the new drug reported 35% fewer headaches than members of the group that used the placebo.

Example 2 Explain whether the data presented in each graph is misleading.

6.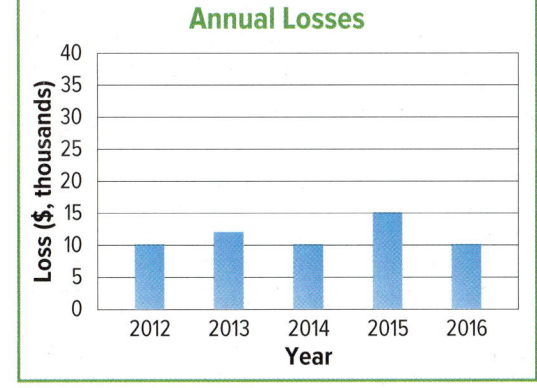

7.

connectED.mcgraw-hill.com 563

8. **MP SENSE-MAKING** The table shows the number of frozen yogurts sold each month at YoZone.

Month	Number Sold
April	1250
May	1360
June	1420
July	1480
August	1390

    a. Make a bar graph of the data that YoZone's manager might use to impress the owner of the shop.
    b. Make a bar graph of the data that a rival shop owner might use to convince customers that sales are low at YoZone.

9. **TRENDS** Explain whether the conclusion presented in the report is valid.

    > **Millions of Americans Know Viral Dance Moves**
    >
    > A recent study was designed to determine the number of Americans who have learned dance moves from videos that have spread around the internet. Researchers surveyed a random sample of 5235 Americans and found that 6 of them knew dance moves from videos that have gone viral. The researchers concluded that millions of Americans are doing viral dance moves.

10. The graphs show the average attendance at games for two college basketball teams.

    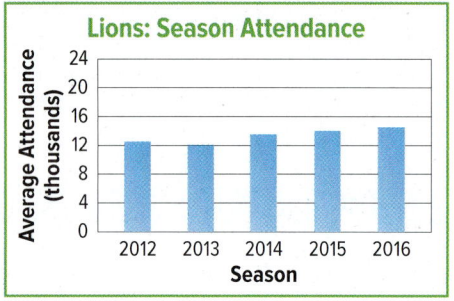 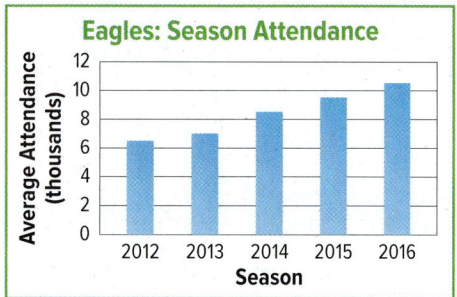

    a. Explain why the data presented in the graphs may be misleading.
    b. Describe a better way to display the data.

11. **MP PRECISION** Explain whether the conclusion presented in the report is valid.

    > **Mayoral Election Is a Toss-Up**
    >
    > A random sample of 2050 registered voters were asked who they intend to vote for in the upcoming mayoral election. The survey showed that 53% support Rodriguez and 47% support Baird. According to Baird's campaign manager, the results are within the survey's margin of error and the election is essentially a tie at this point.

H.O.T. Problems   Use Higher-Order Thinking Skills

12. **WRITING IN MATH** Explain how choosing a random sample for a survey could introduce bias. Provide a specific example.

13. **MP CRITIQUE ARGUMENTS** Mayumi said that it is never possible to draw a cause-and-effect conclusion from the results of a study. Do you agree or disagree? Explain.

14. **MP SENSE-MAKING** A news site reported that 56% of registered voters in Ferndale support a proposition to build a new civic center. The survey involved calling a random sample of 1130 registered voters on their cell phones. On election day, the proposition failed to pass. Explain why the conclusion of the report may not have been valid.

15. **CHALLENGE** An employee of a science museum surveyed a random sample of visitors to the museum to find out their age. The results were 22, 30, 11, 17, 12, 50, 69, 24, 8, 16, 27, 51, 26, 60, and 42. Make a histogram of the data that makes it appear as though the museum is equally popular with visitors of all ages.

Preparing for Assessment

16. The line graph shows the population of Linfield over several years.

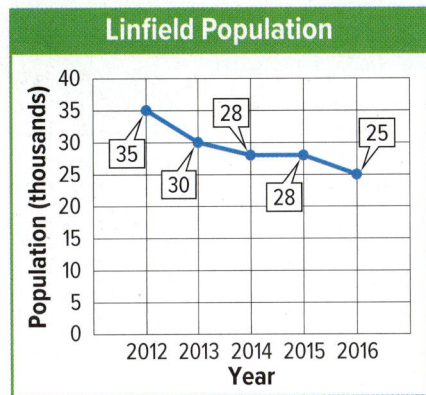

Which scale on the vertical axis will make it appear as though the population declined more slowly? Assume the graph is drawn at the same overall size. **MP** 2

- A 1 to 60 in increments of 5
- B 1 to 20 in increments of 2
- C 20 to 40 in increments of 5
- D 10 to 40 in increments of 2

17. A group of scientists conducted a study. At the end of the study they were able to make a cause-and-effect conclusion based on the data. Which of the following was most likely *not* a part of their research? **MP** 6

- A control group
- B observational study
- C random sample
- D treatment group

18. MULTI-STEP Alana surveyed a random sample of seniors at her high school to determine which city the class should visit for the annual senior trip. The data are shown in the table. **MP** 4

City	Number of Votes
Boston	8
New York	12
Philadelphia	10
Washington	16

  a. Make a bar graph of the data that might cause someone to think that Washington received many more votes than the other cities.

  b. Make a bar graph of the data that might cause someone to think that the four cities all received about the same number of votes.

19. Consider the report below.

> **Car Washing Improves Performance**
>
> A random sample of 1400 drivers was divided into two groups: those who wash their cars at least once a week and those who do not. After nine months, researchers found that drivers who washed their cars at least once a week had fewer car problems and fewer repairs. So, wash your car to improve its performance.

Which statements about the report are true? **MP** 1

- ☐ A The conclusion of the report is valid.
- ☐ B The report is based on an observational study.
- ☐ C The report is based on a controlled experiment.
- ☐ D The sample is likely to be biased.
- ☐ E The report makes an incorrect cause-and-effect conclusion.
- ☐ F The sample is too small to be representative of the population.

20. The table shows the number of tablets sold over four days by a salesperson at a computer store.

Day	Number Sold
Monday	12
Tuesday	15
Wednesday	14
Thursday	12

Which scale on the *y*-axis should the salesperson use to make a bar graph of the data if she wants to make her sales figures appear as impressive as possible? **MP** 4

- A 1 to 20 in increments of 5
- B 1 to 16 in increments of 2
- C 10 to 20 in increments of 2
- D 5 to 20 in increments of 5

21. Ming is making a histogram of the data set {2, 2, 3, 4, 5, 6, 7, 8, 8}. Which intervals along the horizontal axis will make it appear that the data are evenly distributed? **MP** 6

- A 2–4, 5–7, 8–10
- B 1–4, 5–8
- C 2–5, 6–9
- D 1–3, 4–6, 7–9

LESSON 6
Normal Distributions

::Then
- You constructed and analyzed discrete probability distributions.

::Now
1. Use the Empirical Rule to analyze normally distributed variables.
2. Apply the standard normal distribution and z-values.

::Why?
- Extensive observations of Swiss cherry trees found that the mean flowering date is April 21 with a standard deviation of about 10 days. Therefore, 95% of the time, a Swiss cherry tree will have a flowering date between April 1 and May 11.

 New Vocabulary
normal distribution
Empirical Rule
z-value
standard normal distribution

 Mathematical Practices
6 Attend to precision.
8 Look for and express regularity in repeated reasoning.

1 The Normal Distribution Distributions of mileages of different sample sizes of cars are shown below. As the sample size increases, the distributions become more and more symmetrical and resemble the curve at the right, due to the Law of Large Numbers.

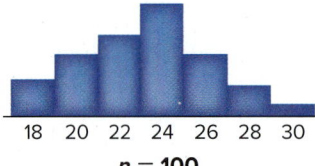
n = 100

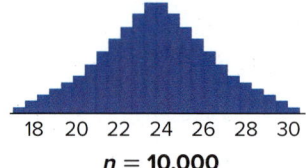
n = 10,000

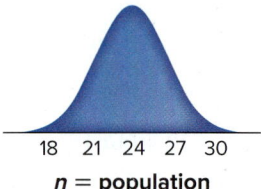
n = population

The curve at the right is a **normal distribution**, a continuous, symmetric, bell-shaped distribution of a random variable. It is the most common *continuous probability distribution*. The characteristics of the normal distribution are as follows.

Key Concept The Normal Distribution

- The graph of the curve is continuous, bell-shaped, and symmetric with respect to the mean.
- The mean, median, and mode are equal and located at the center.
- The curve approaches, but never touches, the x-axis.
- The total area under the curve is equal to 1 or 100%.

The area under the normal curve represents the amount of data within a certain interval or the probability that a random data value falls within that interval. The **Empirical Rule** can be used to determine the area under the normal curve at specific intervals.

Key Concept The Empirical Rule

In a normal distribution with mean μ and standard deviation σ,

- approximately 68% of the data fall within 1σ of the mean,
- approximately 95% of the data fall within 2σ of the mean, and
- approximately 99.7% of the data fall within 3σ of the mean.

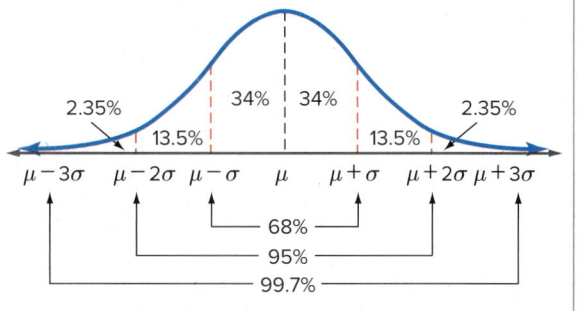

Study Tip

Normal Distributions In all of these cases, the number of data values must be large for the distribution to be approximately normal.

Example 1 Use the Empirical Rule to Analyze Data

A normal distribution has a mean of 21 and a standard deviation of 4.

a. Find the range of values that represent the middle 68% of the distribution.

The middle 68% of data in a normal distribution is the range from $\mu - \sigma$ to $\mu + \sigma$. Therefore, the range of values in the middle 68% is $17 < X < 25$.

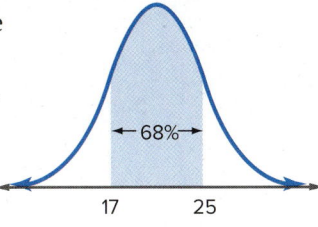

b. What percent of the data will be greater than 29?

29 is 2σ more than μ. 95% of the data fall between $\mu - 2\sigma$ and $\mu + 2\sigma$, so the remaining data values represented by the two tails covers 5% of the distribution. We are only concerned with the upper tail, so 2.5% of the data will be greater than 29.

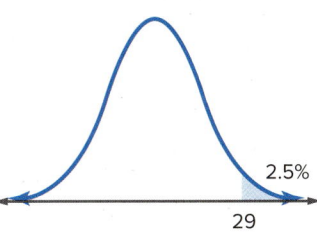

Guided Practice

1. A normal distribution has a mean of 8.2 and a standard deviation of 1.3.

   A. Find the range of values that represent the middle 95% of the distribution.

   B. What percent of the data will be less than 4.3?

Real-World Example 2 Use the Empirical Rule to Analyze a Distribution

HEIGHTS The heights of 1800 adults are normally distributed with a mean of 70 inches and a standard deviation of 2 inches.

a. About how many adults are between 66 and 74 inches?

66 and 74 are 2σ away from the mean. Therefore, about 95% of the data are between 66 and 74.

Because $1800 \times 95\% = 1710$, we know that about 1710 of the adults are between 66 and 74 inches tall.

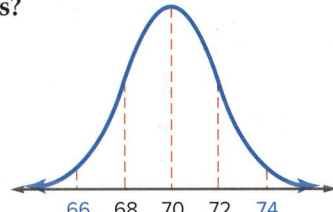

Real-World Link

While the average adult American male is 5 feet 10 inches, the average height of adult males in the Netherlands is the highest worldwide, at almost 6 feet 1 inch.

Source: Eurostats Statistical Yearbook

b. What is the probability that a random adult is more than 72 inches tall?

From the curve, values greater than 72 are more than 1σ from the mean. 13.5% are between 1σ and 2σ, 2.35% are between 2σ and 3σ, and 0.15% are greater than 3σ.

So, the probability that an adult selected at random has a height greater than 72 inches is $13.5 + 2.35 + 0.15$ or 16%.

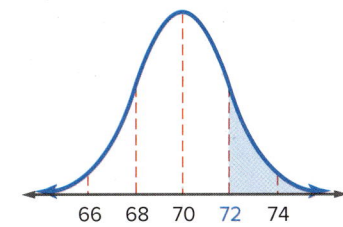

Guided Practice

2. **NETWORKING SITES** The number of friends per member in a sample of 820 members is normally distributed with a mean of 38 and a standard deviation of 12.

   A. About how many members have between 26 and 50 friends?

   B. What is the probability that a random member will have more than 14 friends?

Go Online!

Explore the normal distribution using **The Geometer's Sketchpad®** activity in ConnectED.

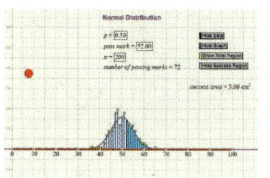

2 Standard Normal Distribution

The Empirical Rule is only useful for evaluating specific values, such as μ and σ. Once the data set is *standardized*, however, any data value can be evaluated. Data are standardized by converting them to z-values, also known as z-scores. The **z-value** represents the number of standard deviations that a given data value is from the mean. Therefore, z-values can be used to determine the position of any data value within a set of data.

Key Concept   Formula for z-Values

The z-value for a data value X in a set of normally distributed data is given by $z = \dfrac{X - \mu}{\sigma}$, where μ is the mean and σ is the standard deviation.

Example 3   Use z-Values to Locate Position

Find z if $X = 18$, $\mu = 22$, and $\sigma = 3.1$.
Indicate the position of X in the distribution.

$z = \dfrac{X - \mu}{\sigma}$ Formula for z-values

$= \dfrac{18 - 22}{3.1}$ $X = 18, \mu = 22, \sigma = 3.1$

≈ -1.29 Simplify.

The z-value that corresponds to $X = 18$ is approximately -1.29. Therefore, 18 is about 1.29 standard deviations less than the mean of the distribution.

Study Tip

Symmetry The normal distribution is symmetrical, so when you are asked for the middle or outside set of data, the z-values will be opposites.

Guided Practice

3. Find X if $\mu = 39$, $\sigma = 8.2$, and $z = 0.73$. Indicate the position of X in the distribution.

Any combination of mean and standard deviation is possible for a normally distributed set of data. As a result, there are infinitely many normal probability distributions. This makes comparing two individual distributions difficult. Different distributions *can* be compared, however, once they are standardized using z-values. The **standard normal distribution** is a normal distribution with a mean of 0 and a standard deviation of 1.

Study Tip

Standard Normal Distribution The standard normal distribution is the set of all z-values.

Key Concept   Characteristics of the Standard Normal Distribution

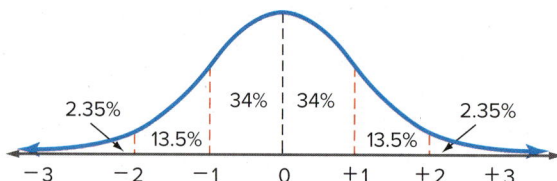

- The total area under the curve is equal to 1 or 100%.
- Almost all of the area is between $z = -3$ and $z = 3$.
- The distribution is symmetric.
- The mean is 0, and the standard deviation is 1.

The standard normal distribution allows us to assign actual areas to the intervals created by z-values. The area under the normal curve corresponds to the proportion of data values in an interval as well as the probability of a random data value falling within the interval. For example, the area between $z = 0$ and $z = 1$ is 0.34. Therefore, the probability of a z-value being in this interval is 34%.

Real-World Example 4 Find Probabilities

VIDEOS The number of videos uploaded daily to a video sharing site is normally distributed with $\mu = 181{,}099$ videos and $\sigma = 35{,}644$ videos. Find each probability. Then use a graphing calculator to sketch the corresponding area under the curve.

a. $P(180{,}000 < X < 200{,}000)$

The question is asking for the percent of days when between 180,000 and 200,000 videos are uploaded. First, find the corresponding z-values for $X = 180{,}000$ and $X = 200{,}000$.

$$z = \frac{X - \mu}{\sigma}$$ Formula for z-values

$$= \frac{180{,}000 - 181{,}099}{35{,}644} \text{ or about } -0.03$$ $X = 180{,}000, \mu = 181{,}099,$ and $\sigma = 35{,}644$

Use 200,000 to find the other z-value.

$$z = \frac{X - \mu}{\sigma}$$ Formula for z-values

$$= \frac{200{,}000 - 181{,}099}{35{,}644} \text{ or about } 0.53$$ $X = 200{,}000, \mu = 181{,}099,$ and $\sigma = 35{,}644$

The range of z-values that corresponds to $180{,}000 < X < 200{,}000$ is $-0.03 < z < 0.53$. Find the area under the normal curve within this interval.

You can use a graphing calculator to display the area that corresponds to any z-value by selecting [2nd] [DISTR]. Then, under the **DRAW** menu, select **ShadeNorm**(lower z value, upper z value). The area between $z = -0.03$ and $z = 0.53$ is about 0.21 as shown in the graph.

Therefore, about 21% of the time, there will be between 180,000 and 200,000 video uploads on a given day.

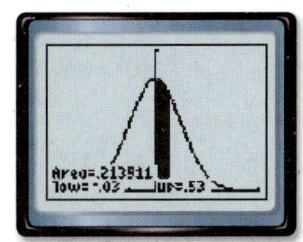

[−4, 4] scl: 1 by [0, 0.5] scl: 0.125

b. $P(X > 250{,}000)$

$$z = \frac{X - \mu}{\sigma}$$ Formula for z-values

$$= \frac{250{,}000 - 181{,}099}{35{,}644} \text{ or about } 1.93$$ $X = 250{,}000, \mu = 181{,}099,$ and $\sigma = 35{,}644$

Using a graphing calculator, you can find the area between $z = 1.93$ and $z = 4$ to be about 0.027.

Therefore, the probability that more than 250,000 videos will be uploaded is about 2.7%.

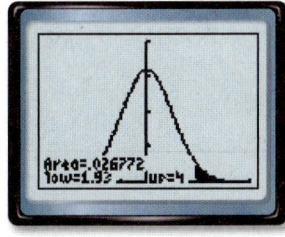

[−4, 4] scl: 1 by [0, 0.5] scl: 0.125

Real-World Link

Video Uploading According to a recent study, 37% of Internet-using students aged 12–17 participate in video chats. 27% record and upload video to the Internet, and 13% stream video to the Internet.

Source: Pew Research Center

Study Tip

Range of z-values The majority of data values are within $\pm 4\sigma$ of the mean, so setting the maximum value of z equal to 4 is sufficient in part **b**. Use the window [−4, 4] by [0, 0.5] when using **ShadeNorm**.

▸ **Guided Practice**

4. TIRES The life spans of a certain tread of tire are normally distributed with $\mu = 31{,}066$ miles and $\sigma = 1644$ miles. Find each probability. Then use a graphing calculator to sketch the corresponding area under the curve.

A. $P(30{,}000 < X < 32{,}000)$ **B.** $P(X > 35{,}000)$

Another method for calculating the area between two z-values is [2nd] [DISTR] **normalcdf**(*lower z value, upper z value*).

Check Your Understanding

= Step-by-Step Solutions begin on page R11.

Go Online! for a Self-Check Quiz

Example 1 A normal distribution has a mean of 416 and a standard deviation of 55.

1. Find the range of values that represent the middle 99.7% of the distribution.
2. What percent of the data will be less than 361?

Example 2 3. **MP TOOLS** The number of texts sent per day by a sample of 811 teens is normally distributed with a mean of 38 and a standard deviation of 7.
   a. About how many teens sent between 24 and 38 texts?
   b. What is the probability that a teen selected at random sent less than 45 texts?

Example 3 Find the missing variable. Indicate the position of X in the distribution.

4. z if $\mu = 89$, $X = 81$, and $\sigma = 11.5$
5. z if $\mu = 13.3$, $X = 17.2$, and $\sigma = 1.9$
6. X if $z = -1.38$, $\mu = 68.9$, and $\sigma = 6.6$
7. σ if $\mu = 21.1$, $X = 13.7$, and $z = -2.40$

Example 4 8. **CONCERTS** The number of concerts attended per year by a sample of 925 teens is normally distributed with a mean of 1.8 and a standard deviation of 0.5. Find each probability. Then use a graphing calculator to sketch the area under each curve.
   a. $P(X < 2)$
   b. $P(1 < X < 3)$

Practice and Problem Solving

Extra Practice is on page R8.

Example 1 A normal distribution has a mean of 29.3 and a standard deviation of 6.7.

9. Find the range of values that represent the outside 5% of the distribution.
10. What percent of the data will be between 22.6 and 42.7?

Example 2 11. **GYMS** The number of visits to a gym per year by a sample of 522 members is normally distributed with a mean of 88 and a standard deviation of 19.
   a. About how many members went to the gym at least 50 times?
   b. What is the probability that a member selected at random went to the gym more than 145 times?

Example 3 Find the missing variable. Indicate the position of X in the distribution.

12. z if $\mu = 3.3$, $X = 3.8$, and $\sigma = 0.2$
13. z if $\mu = 19.9$, $X = 18.7$, and $\sigma = 0.9$
14. μ if $z = -0.92$, $X = 44.2$, and $\sigma = 8.3$
15. X if $\mu = 138.8$, $\sigma = 22.5$, and $z = 1.73$

Example 4 16. **VENDING** A vending machine dispenses about 8.2 ounces of coffee. The amount varies and is normally distributed with a standard deviation of 0.3 ounce. Find each probability. Then use a graphing calculator to sketch the corresponding area under the curve.
   a. $P(X < 8)$
   b. $P(X > 7.5)$

17. **CAR BATTERIES** The useful life of a certain car battery is normally distributed with a mean of 113,627 miles and a standard deviation of 14,266 miles. The company makes 20,000 batteries a month.
   a. About how many batteries will last between 90,000 and 110,000 miles?
   b. About how many batteries will last more than 125,000 miles?
   c. What is the probability that if you buy a car battery at random, it will last less than 100,000 miles?

18. **FOOD** The shelf life of a particular snack chip is normally distributed with a mean of 173.3 days and a standard deviation of 23.6 days.
   a. About what percent of the product lasts between 150 and 200 days?
   b. About what percent of the product lasts more than 225 days?
   c. What range of values represents the outside 5% of the distribution?

19. FINANCIAL LITERACY The insurance industry uses various factors including age, type of car driven, and driving record to determine an individual's insurance rate. Suppose insurance rates for a sample population are normally distributed.

   a. If the mean annual cost per person is $829 and the standard deviation is $115, what is the range of rates you would expect the middle 68% of the population to pay annually?

   b. If 900 people were sampled, how many would you expect to pay more than $1000 annually?

   c. Where on the distribution would you expect a person with several traffic citations to lie? Explain your reasoning.

   d. How do you think auto insurance companies use each factor to calculate an individual's insurance rate?

20. STANDARDIZED TESTS Nikki took three national standardized tests and scored an 86 on all three. The table shows the mean and standard deviation of each test.

	Math	Science	Social Studies
μ	76	81	72
σ	9.7	6.2	11.6

   a. Calculate the z-values that correspond to her score on each test.

   b. What is the probability of a student scoring an 86 or *lower* on each test?

   c. On which test was Nikki's standardized score the highest? Explain your reasoning.

H.O.T. Problems   Use **H**igher-**O**rder **T**hinking Skills

21. ERROR ANALYSIS A set of normally distributed tree diameters have mean 11.5 centimeters, standard deviation 2.5, and range from 3.6 to 19.8. Monica and Hiroko are to find the range that represents the middle 68% of the data. Is either of them correct? Explain.

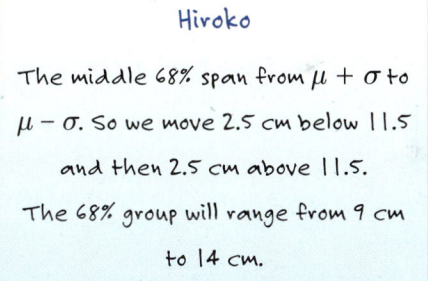

Monica
The data span 16.2 cm. 68% of 16.2 is about 11 cm. Center this 11-cm range around the mean of 11.5 cm. This 68% group will range from about 6 cm to about 17 cm.

Hiroko
The middle 68% span from $\mu + \sigma$ to $\mu - \sigma$. So we move 2.5 cm below 11.5 and then 2.5 cm above 11.5. The 68% group will range from 9 cm to 14 cm.

22. CHALLENGE A selection of tablets has an average battery life of 8.2 hours with a standard deviation of 0.7 hour. Eight of the tablets have a battery life greater than 9.3 hours. If the sample is normally distributed, how many tablets are in the selection?

23. REASONING The term *six sigma process* comes from the notion that if one has six standard deviations between the mean of a process and the nearest specification limit, there will be practically no items that fail to meet the specifications. Is this a true assumption? Explain.

24. REASONING *True* or *false*: According to the Empirical Rule, in a normal distribution, most of the data will fall within one standard deviation of the mean. Explain.

25. OPEN-ENDED Find a set of real-world data that appears to be normally distributed. Calculate the range of values that represent the middle 68%, the middle 95%, and the middle 99.7% of the distribution.

26. WRITING IN MATH Describe the relationship between the z-value, the position of an interval of X in the normal distribution, the area under the normal curve, and the probability of the interval occurring. Use an example to explain your reasoning.

Preparing for Assessment

27. The production costs of 600 candles are normally distributed. The mean cost is $80 and the standard deviation is $12. About how many candles have a production cost between $68 and $92? **MP 1**

- A 81
- B 204
- C 408
- D 570
- E 598

28. A normal distribution has a mean of 18 and a standard deviation of 2.5. What is the range of values that represent the outside 5% of the distribution? **MP 6**

- A $X < 13$ or $X > 23$
- B $X < 15.5$ or $X > 20.5$
- C $X < 10.5$ or $X > 25.5$
- D $X > 13$ or $X < 23$
- E $X > 15.5$ or $X < 20.5$

29. The heights of 1000 trees are normally distributed. The mean height is 30 feet and the standard deviation is 5 feet. How many trees are between 25 and 35 feet? **MP 1**

- A 47
- B 135
- C 500
- D 680
- E 950

30. MULTI-STEP The time taken for a student to complete a task is normally distributed with a mean of 20 minutes and a standard deviation of 2.2 minutes. **MP 6**

  a. A student is selected at random. Find the probability that the student completes the task in less than 21.8 minutes.

  b. The probability that a student takes between k and 21.8 minutes is 0.3. Find the value of k.

The hourly wages of 2400 construction workers are normally distributed with a mean of $36 per hour and a standard deviation of $12 per hour. Use this information for Questions 31–32.

31. About how many construction workers earn between $24 per hour and $48 per hour? **MP 1**

- A 480
- B 800
- C 1632
- D 2280
- E 2393

32. What is the probability that a construction worker selected at random earns an hourly wage of less than $24 per hour? **MP 6**

- A 0.15%
- B 2.35%
- C 13.5%
- D 16.0%
- E 34.0%

33. A random variable has the normal distribution with mean 82.0 and standard deviation 4.8. Find the probabilities that it will take on a value. **MP 6**

  a. less than 89.2
  b. greater than 78.4
  c. between 83.2 and 88.0
  d. between 73.6 and 90.4

34. Find z if: **MP 6**

  a. The standard-normal-curve area between 0 and z is 0.4484.
  b. The standard-normal-curve area to the left of z is 0.9868.

LESSON 7
Using Probability to Make Decisions

∴Then
- You calculated probabilities of dependent and independent events.

∴Now
1. Use probability to make fair decisions.
2. Analyze decisions by using probability concepts.

∴Why?
- A veterinarian knows that 1% of all dogs are infected with a new virus that has spread around the country. A test for the virus is 98% accurate. Is it a good decision to treat every dog that tests positive for the virus? You can use probability to analyze this situation.

 New Vocabulary
fair

 Mathematical Practices
4 Model with mathematics.
5 Use appropriate tools strategically.

1 Make Fair Decisions
A probability experiment is considered fair if all outcomes are equally likely. For example, flipping a coin is a fair experiment. This assumes that the coin is equally likely to land heads up or tails up. Such a coin is called a fair coin. A decision is **fair** if all of the possible options have the same probability of being chosen.

Real-World Example 1 Make a Fair Decision

CLUBS Kalil, Amy, Diego, and Jenna are the officers of their school's video game club. They need to choose one club officer to attend a meeting of representatives from all of the school's clubs. All four officers would like to attend the meeting. Kalil proposes that they flip a coin three times, record the number of heads that appear, and use the table to determine who attends the meeting.

Result	Officer
No heads	Kalil
1 head	Amy
2 heads	Diego
3 heads	Jenna

a. Does Kalil's method result in a fair decision? Why or why not?

Make a table to show all of the possible outcomes. For each outcome, determine which officer would attend the meeting. Then determine the probability that each officer attends the meeting.

Because the four options have different probabilities, this is not a fair decision.

Outcome	Officer	Probability
T T T	Kalil	$P(\text{Kalil}) = \frac{1}{8}$
H T T		
T H T	Amy	$P(\text{Amy}) = \frac{3}{8}$
T T H		
H H T		
H T H	Diego	$P(\text{Diego}) = \frac{3}{8}$
T H H		
H H H	Jenna	$P(\text{Jenna}) = \frac{1}{8}$

b. Is it possible to flip a coin three times to make a fair decision in this situation? Explain.

One possibility is that Kalil attends the meeting if there are no heads, Amy attends if the result is heads-tails-tails, Diego attends if the result is heads-heads-tails, and Jenna attends if the result is three heads. In case of any other result, the coin is flipped three more times until an officer is chosen.

This is fair because the probability of each officer attending the meeting at the end of three coin flips is $\frac{1}{8}$.

connectED.mcgraw-hill.com   573

▶ **Guided Practice**

1. **BOARD GAMES** Yoshio and Kaitlyn are playing a board game that uses two spinners. Each spinner has three equal sections that are numbered 1, 2, and 3. To determine who goes first in the game, they decide to spin both spinners. If the sum of the numbers shown is even, Yoshio goes first. If the sum is odd, Kaitlyn goes first. Does this method result in a fair decision? Explain. If it is not fair, explain how they can use the two spinners to make a fair decision.

Real-World Example 2   Make a Fair Decision

MANUFACTURING The owner of a candle-making business wants to chose among three suppliers of wax. All three suppliers offer the same wax at the same price. The owner has a random-number app on her phone that chooses a random whole number from 1 to 1000, inclusive. How can she use the app to make a fair decision?

Assign numbers to each supplier so that each supplier has the same probability of being chosen. One way to do this is shown in the table.

If the random-number generator returns a number from 901 to 1000, choose a new number until a supplier is chosen.

Supplier	Outcomes	Probability
A	1 through 300	$P(A) = \frac{300}{1000}$ or $\frac{3}{10}$
B	301 through 600	$P(B) = \frac{300}{1000}$ or $\frac{3}{10}$
C	601 through 900	$P(C) = \frac{300}{1000}$ or $\frac{3}{10}$

> **Study Tip**
>
> **MP Tools** Most problems that ask how to use a tool to make a fair decision will have many possible correct answers. Check that your answer is correct by making sure each possible option has the same probability.

▶ **Guided Practice**

2. The owner of the candle-making business also needs to choose among seven suppliers of wicks. All of these suppliers offer the same wicks at the same price. How can the owner use the random-number app to make a fair decision?

2 Analyze Decisions You can use what you know about the probability of dependent and independent events and conditional probability to analyze decisions and strategies.

Real-World Example 3   Analyze a Decision

PRIZES Every bottle of Citro orange juice has a cap with a number of points printed inside the cap. The probability of getting a cap with various numbers of points is shown in the table. Caps worth a total of 100 points can be redeemed for a prize valued at $7.95. Dalton decides to pay $54 to buy a case of 24 bottles of the juice. He plans to redeem all of the points he gets for as many prizes as possible. Considering only the financial aspect of the strategy, is this a good decision? Explain.

Points	Probability
10	60%
25	30%
40	9%
75	1%

Step 1 Find the expected or average number of points on each bottle cap.

Multiply each possible number of points by its probability and add.

$10(0.6) + 25(0.3) + 40(0.09) + 75(0.01) = 17.85$

Dalton can expect that each bottle cap will be worth an average of 17.85 points.

Step 2 Find the number of points Dalton can expect from a case of juice.

$24(17.85) = 428.4$

Step 3 Find the value of the prizes Dalton can expect to win from a case of juice.

Dalton can redeem 428 points for 4 prizes.

The value of the prizes is 4($7.95) = $31.80

Step 4 Analyze the decision.

Dalton paid $54 for the case of juice, which is greater than the value of the prizes, so this was not a good decision.

▶ **Guided Practice**

3. **CARNIVALS** A wheel at a carnival has 20 equal sections. One section is worth 50 prize tickets, 5 sections are each worth 10 tickets, and the remaining sections are each worth 1 ticket. Each spin costs $1, and tickets worth a total of 25 points can be redeemed for a prize valued at $3.25. Jaycee plans to spin the wheel 5 times to win as many prizes as possible. Considering only the financial aspect of the strategy, is this a good decision? Explain.

Real-World Example 4 **Analyze a Decision**

VETERINARY MEDICINE A veterinarian knows that 1% of the dogs in the country are infected with a virus. There is a test for the virus that is 98% accurate. This means that it returns a positive result 98% of the time if a dog is infected with the virus and a negative result 98% of the time if a dog is not infected. The vet decides to treat any dog that tests positive for the virus. Do you think this a good decision? Explain.

Make a two-way frequency table to show the data. You can start by assuming a large population of 100,000 dogs.

Enter the total population of 100,000 dogs in the cell in bottom right corner.

Fill in the *Totals* column using the fact that 1% of the dogs have the virus.

Fill in the interior cells using the fact that the test is 98% accurate.

Add to fill in the *Totals* row.

> **Watch Out!**
>
> **Two-Way Tables** When you make a two-way frequency table, the values in the *Totals* row and the values in the *Totals* column must add up to the value in the cell in the bottom right corner.

	Tests Positive	Tests Negative	Totals
Has Virus	980	20	1000
Does Not Have Virus	1980	97,020	99,000
Totals	2960	97,040	100,000

Find the conditional probability that a dog that tests positive for the virus actually has the virus. Use the values in the *Tests Positive* column.

$P(\text{has virus} \mid \text{tests positive}) = \frac{980}{2960} \approx 33.1\%$

The probability that a dog that tests positive for the virus is actually infected with the virus is only about 33%. Therefore, it may not be an efficient use of resources to treat every dog that tests positive for the virus and the vet's decision may not be a good one.

▶ **Guided Practice**

4. **PRODUCT TESTING** A factory produces and packages hamburger patties. It is known that 0.5% of the patties are not the correct weight. A scale at the factory is 99% accurate in identifying whether or not a patty is the correct weight. A manager at the factory decides to throw out any patty that is identified as not the correct weight. Do you think this is a good decision? Explain.

Check Your Understanding

= Step-by-Step Solutions begin on page R11.

Go Online! for a Self-Check Quiz

Example 1

1. **SOFTBALL** Alexa, Gayle, and Mei are trying to decide which one of them will be captain of their softball team for the coming season. They decide to roll two number cubes and use the sum of the numbers rolled to determine the captain. If the sum is 2 through 5, Alexa is the captain; if the sum is 6 through 8, Gayle is the captain; and if the sum is 9 through 12, Mei is the captain.

   a. Explain why this method does not result in a fair decision.

   b. Explain how they can use the sum of the numbers rolled to make a fair decision.

Example 2

2. **GARDENING** There are eight types of seeds that Derek wants to plant in his garden, but he only has enough space to plant three different types. He wants to choose the three types of seeds to plant by using a random-number generator that chooses a whole number from 1 to 50, inclusive. Explain how he can use the random-number generator to make a fair decision.

Example 3

3. **ARCADE GAMES** An arcade game consists of rolling a ball onto a platform with several holes. The player wins tokens based on which hole the ball lands in. The game has been set up to have the probabilities shown in the table. Each roll costs $0.25 and players can redeem 15 tickets for a prize worth $4.95. Latrell decides to play the game 10 times and redeem his tickets for as many prizes as possible. Considering only the financial aspect of the strategy, is this a good decision? Explain.

Tokens	Probability
20	2%
10	10%
5	18%
2	30%
1	40%

Example 4

4. **TRAFFIC ACCIDENTS** All of the taxis in a town are yellow, except for 5% that are blue. After a hit-and-run traffic accident, a witness says that the accident was caused by a blue taxi. A detective knows that witnesses in the town are 90% accurate in identifying the color of vehicles involved in accidents. The detective decides to restrict the search to blue taxis. Do you think this is a good decision? Explain.

Practice and Problem Solving

Extra Practice is found on page R8.

Example 1

Micah and Carrie want to decide who gets the last granola bar in a box. They spin two spinners at the same time and find the sum of the resulting numbers. One spinner has three equal sections labeled 1–3 and the other has four equal sections labeled 1–4. Determine whether each method results in a fair decision. Explain.

5. Micah wins if the sum is odd; Carrie wins if the sum is even.

6. Micah wins if the sum is prime; Carrie wins if the sum is composite.

7. Micah wins if the sum is a multiple of 3; Carrie wins if the sum is not a multiple of 3.

Example 2

8. **DRAWING LOTS** Drawing lots is a method of using sticks, straws, or other objects to make a fair decision. Sara has 20 straws that are all the same length. She wants to use the straws to decide which two members of the drama club will serve as copresidents. Assuming there are 15 members in the club, describe how Sara can cut some of the straws to make them shorter and then have members choose straws to make a fair decision.

Example 3

9. **CEREAL** Each box of Rice Crunchies cereal is printed so that the inside of the box top shows a number from 1 to 5. The numbers appear with the probabilities shown in the table. Customers can collect the box tops and redeem them for prizes based on the sum of the numbers. A prize worth $18.95 requires a sum of 10 or greater. Ricardo decides to buy 8 boxes of cereal and redeem the box tops for as many prizes as possible. Each box of cereal costs $3.65. Considering only the financial aspect of the strategy, is this a good decision? Explain.

Number	Probability
1	35%
2	29%
3	22%
4	13%
5	1%

Example 4

10. QUALITY CONTROL It is known that 0.1% of the smartphones produced at a factory are defective. A quality-control engineer has a quick way of testing the phones that is 97% accurate. The engineer tests some phones and discards any that the test shows to be defective. Do you think this is a good decision? Explain.

MP TOOLS Latoya and David are biologists. They need to decide which one of them will present their work at a conference. Explain how they can use each of the following tools to make a fair decision.

11. two number cubes that each have faces numbered 1 through 6

12. a spinner with 5 sections labeled 1–5 and a spinner with 2 sections labeled 1 and 2

13. a random-number generator that chooses a whole number from 5 to 23, inclusive

14. Visitors to a school fair can spin a wheel to win prize tickets. The sections of the wheel each show a number of tickets according to the probabilities in the table, where the variables represent a probability written as a decimal between 0 and 1.

Tickets	Probability
5	a
10	b
15	c
20	d

   a. Mollie spins the wheel x times. Write an expression for the total number of tickets she can expect to win.

   b. Mollie determines that each ticket has a value of t dollars. She decides she will only spin the wheel if the expected dollar value of the tickets after x spins is greater than $12. Write an inequality to represent this decision.

15. MP SENSE-MAKING A company makes yogurt pops in four flavors: berry, peach, lemon, and cherry. The pops come in boxes of six pops, and, according to the company, the four flavors are produced in equal numbers and are chosen at random for the boxes.

   a. Brian bought a box of the yogurt pops and was surprised to find that it did not contain any berry pops. He decided that this was very unlikely to happen by chance and wrote an angry email to the company. Do you agree with this decision? Use probability to justify your answer.

   b. Suppose Brian had purchased a case of five boxes of the pops and did not get any berry pops. Would you agree with his decision to write the email in this case? Explain.

H.O.T. Problems Use **H**igher-**O**rder **T**hinking Skills

16. OPEN-ENDED Ellie, Mitchell, and Rosa all want to attend a concert, but they only have two tickets. They have a set of 20 cards that are numbered 1 to 20. Describe one way they can use the cards to make a fair decision about which two of the three friends will get the tickets.

17. MP CRITIQUE ARGUMENTS A spinner has 7 equal sections that numbered 1 to 7. Dylan said that it is not possible to make a fair decision between two options using this spinner since there is an odd number of outcomes for each spin. Do you agree or disagree? Explain.

18. MP REASONING A carnival wheel has a 65% chance of landing on "2 tickets" and a 35% chance of landing on "5 tickets." Margo only wants to spin the wheel if the expected value of the prize tickets after one spin is greater than $2. What should be true about the value of each ticket in order for Margo to decide to spin the wheel? Explain.

19. ERROR ANALYSIS Ray said he could use a standard deck of cards to make a fair decision when choosing between three options by assigning one option to the red cards, one option to the clubs, and one option to the spades and choosing a card at random. Explain his error and describe a correct way to use the cards to make a fair decision.

20. WRITING IN MATH Explain how making a fair decision is related to the idea of a fair probability experiment.

Preparing for Assessment

21. Chitra has a spinner with eight equal sections that are numbered 1 through 8. She wants to use the spinner to make a decision between two options, X and Y. Which methods of assigning outcomes to the options result in a fair decision? **MP 1**

☐ A option X: multiple of 3; option Y: not a multiple of 3

☐ B option X: prime number; option Y: not a prime number

☐ C option X: odd number; option Y: even number

☐ D option X: less than 5; option Y: greater than 5

☐ E option X: divisible by 2; option Y: not divisible by 2

☐ F option X: 1, 2, 7, or 8; option Y: any other number

22. A street fair has four different wheels that visitors can spin to win prize tickets. For each wheel, the probability of winning different numbers of tickets is shown in the table. Mario wants to spin the wheel for which he can expect to win the greatest number of tickets on a single spin. Which wheel should he decide to spin? **MP 1**

Wheel A	2 tickets: 80% 4 tickets: 20%
Wheel B	1 ticket: 75% 3 tickets: 15% 5 tickets: 10%
Wheel C	2 tickets: 50% 3 tickets: 35% 4 tickets: 15%
Wheel D	1 ticket: 70% 6 tickets: 30%

○ A wheel A ○ C wheel C

○ B wheel B ○ D wheel D

23. Tyra and Connor need to decide which one of them will mow the lawn. They decide to use a random-number generator that chooses a whole number between 1 and 99. Which method of assigning numbers results in a fair decision? **MP 5**

○ A Tyra: odd numbers; Connor: even numbers

○ B Tyra: less than 50; Connor: greater than 50

○ C Tyra: 1 through 50; Connor: 51 through 99

○ D Tyra: prime numbers; Connor: composite numbers

24. Danielle wants to use the two spinners shown here to decide whether she goes to a movie or goes to the mall. She plans to spin the spinners at the same time and find the sum of the resulting numbers. Which method of assigning outcomes to the options results in a fair decision? **MP 1**

○ A movie: sum is less than or equal to 7; mall: sum is greater than 7

○ B movie: sum is even; mall: sum is odd

○ C movie: sum is odd; mall: sum is even

○ D movie: sum is 5, 8, or 9; mall: sum is 6 or 7

○ E movie: sum is less than 7; mall: sum is greater than 7; spin again if the sum equals 7

25. Every can of Toby's tomato sauce has a number (6, 8, or 10) printed on the inside of the lid. The numbers appear with the following probabilities. 6: 70%, 8: 20%, and 10: 10%. When a customer has lids that total 30 or more, they can be redeemed for a prize worth $5.20. The cans of sauce cost $1.69 each. Miguel decides to buy 10 cans and redeem the lids for as many prizes as possible. Is this a good or bad decision? Explain. **MP 4**

26. MULTI-STEP There is a 1.5% probability that a cat coming into a shelter will have a particular type of eye infection. A vet has a test that is 92% accurate in detecting the infection. The vet decides to treat every cat at the shelter that tests positive for the infection. **MP 4**

a. Make a two-way frequency table to represent this situation, assuming a total population of 1,000,000 cats.

b. Find the probability of an eye infection, given that the test result is positive.

c. Do you think it was a good decision to treat every cat that tests positive? Explain.

CHAPTER 8
Study Guide and Review

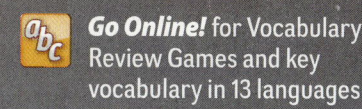
Go Online! for Vocabulary Review Games and key vocabulary in 13 languages

Study Guide

Key Concepts

Random Sampling (Lesson 8-1)
- A survey, an experiment, or an observational study can be used to collect information.
- Samples can be used to make inferences about a population.

Using Statistical Experiments (Lesson 8-2)
- A simulation can be used to predict outcomes of an experiment.
- Data can be used to compare theoretical and experimental probabilities.

Population Parameters (Lesson 8-3)
- The population mean and population proportion can be predicted using samples.
- The margin of error depends on the sample size.

Distributions of Data (Lesson 8-4)
- Use the mean and standard deviation to describe a symmetric distribution.
- Use the five-number summary to describe a skewed distribution.

Evaluating Published Data (Lesson 8-5)
- Analyzing reports is important to discern if the conclusions are valid.

Normal Distributions (Lesson 8-6)
- The graph of a normal distribution is bell-shaped.
- The z-value represents the number of standard deviations that a given data value is from the mean.

Using Probability to Make Decisions (Lesson 8-7)
- Probability experiments can be used to make fair decisions.

FOLDABLES Study Organizer

Use your Foldable to review the chapter. Working with a partner can be helpful. Ask for clarification of concepts as needed.

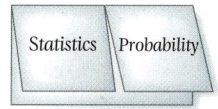

Key Vocabulary

bias (p. 531)
☆ confidence level (p. 546)
distribution (p. 551)
Empirical Rule (p. 566)
experiment (p. 531)
experimental probability (p. 538)
fair (p. 573)
margin of error (p. 546)
negatively skewed distribution (p. 551)
normal distribution (p. 566)
observational study (p. 531)
parameter (p. 531)
population mean (p. 545)
population proportion (p. 545)
positively skewed distribution (p. 551)
probability model (p. 538)
random sample (p. 531)
relative frequency (p. 538)
simulation (p. 538)
standard normal distribution (p. 568)
statistic (p. 531)
statistics (p. 531)
survey (p. 531)
symmetric distribution (p. 551)
theoretical probability (p. 538)
☆ z-value (p. 568)

Vocabulary Check

Choose a term from the list above that best completes each statement.

1. A(n) _____ is an error that results in a misrepresentation of members of a population.

2. In a statistical study, data are collected and used to answer questions about a population characteristic or _____.

3. The _____ can be used to determine the area under the normal curve at specific intervals.

4. In a(n) _____, members of a sample are measured or observed without being affected by the study.

Concept Check

5. Explain the difference between a negatively skewed and a positively skewed distribution.

6. Explain how the sample size affects the margin of error.

CHAPTER 8
Study Guide and Review Continued

Lesson-by-Lesson Review

8-1 Random Sampling

Determine whether each situation describes a *survey*, an *experiment*, or an *observational study*. Then identify the sample, and suggest a population from which it may have been selected.

7. **SHOPPING** Every tenth shopper coming out of a store is asked questions about his or her satisfaction with the store.

8. **MILK SHAKE** A fast food restaurant gives 25 of their customers a sample of a new milk shake and employees monitor their reactions as they taste it.

9. **SCHOOL** Every fifth person coming out of a high school is asked what their favorite class is.

Example 1

A dealership wants to test different promotions. They randomly select 100 customers and ask them which promotion they prefer. Does this situation describe a *survey*, an *experiment*, or an *observational study*? Identify the sample, and suggest a population from which it may have been selected.

This is a survey, because the data are collected from participants' responses. The sample is the 100 customers that were selected, and the population is all potential customers.

8-2 Using Statistical Experiments

10. **BASEBALL** A baseball player gets on base 30% of the time that he is up to bat.
    a. Design a simulation that can be used to estimate the probability that the player will get a on base his next bat.
    b. Conduct the simulation and report the results.

11. **SCHOOL BUS** Christy has determined that the school bus is late 60% of the time.
    a. Design a simulation that can be used to estimate the probability that the school bus is late today.
    b. Conduct the simulation, and report the results.

Example 2

GROUPS Before the random drawing of groups, Dawn has determined that she is 20% likely to get placed in the same group as Sherry. Design a simulation that can be used to estimate the probability of Dawn and Sherry being in the same group.

Step 1 There are two possible outcomes.

Possible Outcomes	Theoretical Probability
grouped together	20%
not grouped together	80%

Step 2 We can use the random number generator on a graphing calculator. Assign the integers 1–5 to accurately represent the probability data.

Outcome	Represented by
grouped together	1
not grouped together	2–5

Step 3 A trial will represent one drawing of groups. The simulation can consist of any number of trials. We will use 20.

8-3 Population Parameters

In an eleventh grade class, there are 140 total students.

12. In a survey, 35 of the eleventh grade students reported that they share a bedroom with a sibling. What is the population proportion of students who do not share a bedroom?

13. In the same survey, the students were asked how many pets they have. The results are shown in the table. What is the population mean number of pets owned by the students?

Number of Pets	Number of Students
0	26
1	39
2	56
3	17
4	2

14. A sample of 30 of the eleventh grade students is conducted to determine the mean pulse rate of all eleventh graders. What is the margin of error?

Example 3

HOUSING In a new development, 40 new houses are being built. The design plans were analyzed to determine how many bathrooms are in each house. The results are shown in the table. What is the population mean number of bathrooms in the houses?

Number of Bathrooms	Number of Houses
1	3
2	17
3	13
4	6

Step 1 Sum the total number of bathrooms.

$(1 \times 3) + (2 \times 17) + (3 \times 13) + (4 \times 6) = 100$

Step 2 Divide the answer in Step 1 by the total number of houses.

$\frac{100}{40} = 2.5$

The mean number of bathrooms in the population is 2.5.

CHAPTER 8
Study Guide and Review Continued

8-4 Distributions of Data

15. DOGSLED The Iditarod is a race across Alaska. The table shows the winning times, in days, for recent years.

Iditarod Winning Times
9.1, 9.4, 10.3, 9.3, 9.6, 8.7, 9.5, 9.4, 9.2, 17.3, 15.4,
15.5, 14.2, 12.0, 16.6, 13.5, 13.0, 18.1, 12.4, 11.6,
11.5, 11.3, 11.3, 13.1, 11.2, 11.6, 11.6, 9.7

a. Use a graphing calculator to create a histogram. Then describe the shape of the distribution.

b. Describe the center and spread of the data using either the mean and standard deviation or the five-number summary. Justify your choice.

16. SWIMMING Kelly's practice times in the 400-meter individual medley are shown in the table.

Times in Seconds
301, 311, 320, 308, 312, 307, 303, 305, 309, 308,
304, 302, 311, 313, 313, 316, 314, 306, 329, 326,
319, 310, 306, 309, 320, 318, 315, 318, 314, 309

a. Use a graphing calculator to create a box-and-whisker plot. Then describe the shape of the distribution.

b. Describe the center and spread of the data using either the mean and standard deviation or the five-number summary. Justify your choice.

Example 4

Data collected from a group of sixth graders is shown.

Number of Years Playing an Instrument
2.5, 2.4, 3.1, 2.9, 4.2, 1.3, 2.6, 2.4, 3.3, 1.9, 3.4, 4.8,
2.3, 1.7, 3.2, 2.3, 3.5, 2.2, 3.6, 1.2, 4.4, 2.1, 3.4, 4.5,
1.9, 1.5, 1.4, 0.7, 1.2, 2.5, 1.9, 2.0, 2.4, 2.5, 3.4

a. Use a graphing calculator to create a histogram. Then describe the shape of the distribution.

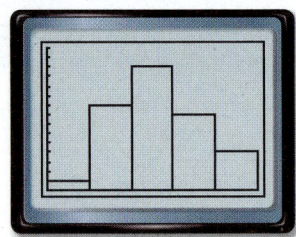

[0, 5] scl: 1 by [0, 15] scl: 1

The distribution is symmetric.

b. Describe the center and spread of the data using either the mean and standard deviation or the five-number summary. Justify your choice.

The distribution is symmetric, so use the mean and standard deviation. The mean number of years is about 2.6 with standard deviation of about 1 year.

8-5 Evaluating Published Data

17. TEST SCORES Explain whether the conclusion in the following report is valid.

> **Sleeping 8 hours a Night Causes Higher Grades**
> A group of students were surveyed. If they slept 8 hours or more per night, they were put into one group, and everyone else in another group. Then, each student's grades were analyzed and it was determined that the group who slept more than 8 hours per night got better grades.

Example 5

TEST SCORES Explain whether the conclusion in the following report is valid.

> **We Have the Highest Test Scores**
> The first 10 students to arrive at school this morning were surveyed to determine the average test score on the latest history test. The survey showed that the average grade was a 98%.

The conclusion may not be valid. The sample size is too small and the sample was not obtained using a simple random sampling method. Also, students may not be truthful when reporting their grade.

8-6 Normal Distributions

18. RUNNING TIMES The times in the 40-meter dash for a select group of professional football players are normally distributed with a mean of 4.74 seconds and a standard deviation of 0.13 second.

  a. About what percent of players have times between 4.6 and 4.8 seconds?

  b. About how many of a sample of 800 players will have times below 4.5 seconds?

19. ATTENDANCE The number of tickets sold at high school basketball games in a particular conference are normally distributed with a mean of 68.7 and a standard deviation of 13.1.

  a. About what percent of the games sell fewer than 75 tickets?

  b. About how many of a sample of 200 games will sell more than 100 tickets?

20. COMMUTING The number of minutes it takes Phil to commute to work each day are normally distributed with a mean of 18.6 and a standard deviation of 3.5.

  a. About what percent of the time will it take Phil more than 20 minutes to commute to work?

  b. About how many of a sample of 50 days will it take Phil less than 15 minutes to commute to work?

Example 6

TEST SCORES The midterm test scores for the students in Mrs. Hendrix's classes are normally distributed with a mean of 73.2 and standard deviation of 7.8. About how many test scores are between 70 and 80?

Find the corresponding z-values for $X = 70$ and 80.

$$z = \frac{X - \mu}{\sigma}$$
$$= \frac{70 - 73.2}{7.8}$$
$$\approx -0.41$$

$$z = \frac{X - \mu}{\sigma}$$
$$= \frac{80 - 73.2}{7.8}$$
$$\approx 0.87$$

Using a graphing calculator, you can find the area between $-0.41 < z < 0.87$ to be about 0.47.

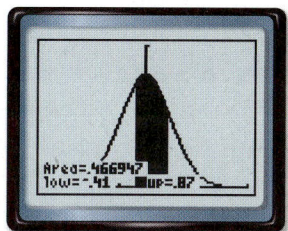

[−4, 4] scl: 1 by [0, 0.5] scl: 0.125

8-7 Using Probability to Make Decisions

21. SPEECH Of the five members of student council, one must make a speech. To choose who will make the speech, the advisor decided to roll a number cube. He assigns the president numbers 1 and 2 and each of the other positions the other numbers.

  a. Explain why this does not result in a fair decision.

  b. Explain how they can use the results to make a fair decision.

22. MEDICAL TESTS A certain medical test reports errors 0.4% of the time. 8% of people actually have the disease. The treatment for the disease is very risky. The doctor decides to treat everyone who tests positive. Is this a good decision?

Example 7
MATH COMPETITION A principal must choose one student to represent the school at a mathematics competition. There are 10 eligible students who are all equally qualified. The principal has a random number generator that generates numbers from 1 to 500, inclusive. To determine which student to choose, how can the principal use the random number generator to make a fair decision?

Assign an equal amount of numbers to each of the ten students. 1–50, 51–100, etc. The probability is $\frac{50}{500}$ or $\frac{1}{10}$ for each student.

Example 8
BATTERIES A company determines that 0.8% of their batteries are defective. The machine that tests the batteries is accurate 98% of the time. A quality control specialist decides to destroy all of the batteries that are tested and marked defective. Is this a good decision?

No, the probability that the battery is defective if it tests defective is 28%. Therefore, they would be destroying many batteries that were actually working.

CHAPTER 8
Practice Test

1. **BUTTERFLIES** Students in a biology class are learning about the monarch butterfly's life cycle. Each student is given a caterpillar. When a caterpillar turns into chrysalis, it is placed in a glass enclosure with food and a heat lamp and examined.

   a. Determine whether the situation describes a *survey*, an *experiment* or an *observational study*.

   b. Identify the sample, and suggest a population from which it was selected.

2. **HEIGHTS** The heights of Ms. Joy's dance students are shown.

Height (Inches)				
60	64	62	69	64
63	65	64	66	73
74	63	62	65	64
68	70	66	63	61

   a. Use a graphing calculator to create a box-and-whisker plot. Then describe the shape of the distribution.

   b. Describe the center and spread of the data using either the mean and standard deviation or the five-number summary. Justify your choice.

3. A binomial distribution has a 65% rate of success. There are 15 trials.

   a. What is the probability that there will be exactly 12 successes?

   b. What is the probability that there will be at least 10 successes?

4. **AIRPLANE** A certain airline's planes take off on time 75% of the time.

   a. Design a simulation that can be used to estimate the probability that the next plane will take off on time.

   b. Conduct the simulation and report the results.

5. **MULTIPLE CHOICE** A sample of 200 nurses is conducted to determine the mean number of hours nurses typically work in one week. Which represents the margin of error?

   A 3.5%
   B 7%
   C 10.5%
   D 14%

6. A survey was done of every adult in a small town. 8574 people stated that they are satisfied with the condition of the streets. There are 24,503 people in the town. Which can be calculated with this data?

   A sample mean
   B sample proportion
   C population mean
   D population proportion

7. When evaluating a report to determine its validity, which are things to consider? Select all that apply.

   A Was there bias in survey questions?
   B Was the sample chosen randomly?
   C Was the report published?
   D Was the sample size large enough?
   E Was it an experiment or an observational study?

8. **WEIGHTS** The weights of 1500 bodybuilders are normally distributed with a mean of 190.6 pounds and a standard deviation of 5.8 pounds.

   a. About how many bodybuilders are between 180 and 190 pounds?

   b. What is the probability that a bodybuilder selected at random has a weight greater than 195 pounds?

A normal distribution has a mean of 16.4 and a standard deviation of 2.6.

9. Find the range of values that represent the middle 95% of the distribution.

10. What percent of the data will be less than 19?

11. **PROMOTION** A restaurant serves lunches to 80 local small businesses employees each month. To say thank you, the restaurant wants to choose one employee to receive a gift card. The restaurant has a random number generator that chooses whole numbers between 1 and 1000, inclusive. To determine which person to choose, how can the restaurant use the random number generator to make a fair decision?

CHAPTER 8
Preparing for Assessment

Performance Task

Provide a clear solution to each part of the task. Be sure to show all of your work, include all relevant drawings, and justify your answers.

TECHNOLOGY A company specializes in statistical analysis. They collect and analyze various types of data.

Part A

A software company wants to get feedback on a new program, so they release a beta version and hire the statistical analysis company to collect data from the beta testers. There are 100 beta testers and of these, 62 report a bug in the software. Based on past sales, the software company expects to sell approximately 1.4 million licenses for the software once they release it.

1. Determine whether the analysis described is a survey, experiment, or observational study. Explain your answer.

2. **Reason Quantitatively** Determine about how many consumers would be likely to report bugs if the company released the software now.

Part B

The statistical analysis company conducts a survey for a client with 400 respondents.

3. **Use Structure** Determine the margin of error for the survey. Round your answer to the nearest tenth, if necessary.

4. **Use Structure** Determine the number of respondents needed in order to achieve a margin of error of ±2.5%. Round your answer to the nearest whole number, if necessary.

Part C

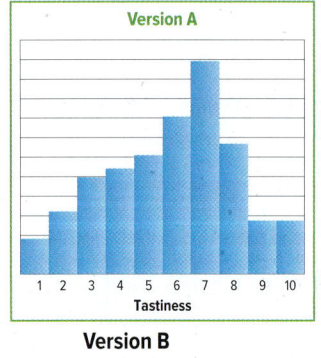

A snack company makes two slightly different versions of a product and hires the statistical analysis company to conduct some taste test trials and help them choose which version to sell. The company has two groups of equal numbers of participants, each group getting one version of the snack, and both asked to rate the tastiness of the snack on a scale from 1 to 10. The data is recorded in the two figures shown.

5. Determine the skew of each figure.

6. **Make Sense** Determine which version of the snack the company should sell. Justify your answer.

Part D

In one data set, the statistical analysis company finds the data has a normal distribution with a mean of 32.5 and a standard deviation of 6.75.

7. **Use Tools** Determine the range of values that represent the middle 95% of the distribution.

8. **Use Tools** Determine the percent of the data that will be greater than 39.25.

Test-Taking Strategy

Example

Read the problem. Identify what you need to know. Then use the information in the problem to solve.

There are 15 boys and 12 girls in Mrs. Lawrence's homeroom. Suppose a committee is to be made up of 6 randomly selected students. What is the probability that the committee will contain 3 boys and 3 girls? Round your answer to the nearest tenth of a percent.

A 27.2%   C 31.5%

B 29.6%   D 33.8%

Step 1 What are you being asked to solve? What information is given and what do you still need?
The problem asks for the probability of the committee containing 3 boys and 3 girls. I have the number of boys and number of girls. I need to know the probability of 3 boys being chosen and the probability of 3 girls being chosen.

Step 2 What intermediate steps will you take to solve the problem?
I will find the number of possible successes and the number of possible outcomes. Then, I will use them to compute the probability.

Step 3 What is the correct answer?
The answer is D.

> **Test-Taking Tip**
>
> **Strategies for Solving Multi-Step Problems**
> Some problems that you will encounter on standardized tests require you to solve multiple parts in order to come up with the final solution. The question below is such a problem.

Apply the Strategy

Read the problem. Identify what you need to know. Then use the information in the problem to solve.

There are 52 cards in a standard deck. Of these, 4 of the cards are aces. What is the probability of a randomly dealt five-card hand containing exactly one pair of aces? Round your answer to the nearest whole percent.

A 4%

B 5%

C 6%

D 7%

Answer the questions below.

a. What are you being asked to solve? What information is given and what do you still need?

b. What intermediate steps will you take to solve the problem?

c. What is the correct answer?

CHAPTER 8
Preparing for Assessment
Cumulative Review

Read each question. Then fill in the correct answer on the answer document provided by your teacher or on a sheet of paper.

1. What is the value of the function $f(x) = 4\left(\frac{2}{3}\right)^x$ when $x = -4$?
   - A $-\frac{32}{3}$
   - B $-\frac{4}{3}$
   - C $-\frac{2}{3}$
   - D $-16\frac{2}{3}$
   - E $20\frac{1}{4}$

Test-Taking Tip

Question 1 To evaluate an expression that contains a fraction with a negative exponent, take the reciprocal of the fraction and change the sign on the exponent to positive.

2. What is the solution to the equation?
   $$x^2 + 6x = -10$$
   - A $x = -3 \pm i$
   - B $x = \pm 3i$
   - C $x = 2$ or 4
   - D $x = -2$ or -4
   - E $x = -1$ or -3

3. What is the solution of the linear system shown below?
   $$x + y - 3z = 6$$
   $$2x + y = 5$$
   $$y - 2z = 5$$
   - A $\left(\frac{3}{2}, \frac{3}{2}, -1\right)$
   - B $(1, 3, 0)$
   - C $\left(3, 2, -\frac{1}{3}\right)$
   - D $\left(\frac{1}{2}, 4, -\frac{1}{2}\right)$
   - E $(0, 3, -1)$

4. Which of the following is a characteristic of a set of data in which the mean and median are approximately equal?
   - A The data have a symmetrical distribution.
   - B The data have a positively skewed distribution.
   - C The data have a negatively skewed distribution.
   - D The majority of the data is on the left of the mean.
   - E The majority of the data is on the right of the median.

5. Which box-and-whisker plot is the best representation of a positvely skewed data distribution?
   - A
   - B
   - C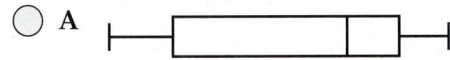
   - D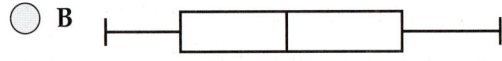
   - E

6. Select all situations that call for an experiment.
   - A A teacher wants to examine handwriting skills of her students.
   - B A drug company wants to test whether a new vaccine is effective.
   - C A farmer wants to know which of two fertilizers will make plants grow faster.
   - D A store wants to know whether shoppers would be interested in receiving weekly ads online.
   - E A school district superintendent wants to determine what percentage of residents support the school budget.

Questions 7–8 refer to the following table, which shows the number of goals scored by the varsity soccer team at Golden Valley High School.

Varsity Soccer Scoring	
Game	Number of Goals
1	5
2	0
3	2
4	1
5	3
6	2
7	0

7. What is the median number of goals scored by the varsity soccer team?

    ☐ goals

8. The coach wants the soccer team to have an average of 2 goals per game after 8 games played. How many goals does the team need to score in game 8 to achieve this average?

    ☐ goals

9. What is the domain of the function?
$$f(x) = \frac{3}{x-1} + \frac{1}{2} - \frac{2}{(x-1)^2}$$
   - A $\{x \mid x = 1\}$
   - B $\{x \mid x \neq 0\}$
   - C $\{x \mid x = \pm 1\}$
   - D $\{x \mid x \neq 1\}$
   - E all real numbers

10. Select all phrases that describe the graph shown.

    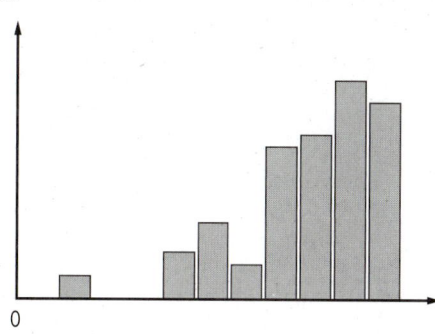

   - ☐ A not symmetric
   - ☐ B no correlation
   - ☐ C negatively skewed
   - ☐ D positively skewed
   - ☐ E normally distributed

11. The test scores of 500 students are normally distributed. The mean score is 74 points, and the standard deviation is 6 points.

    a. How many more students scored between 74 and 80 points than between 80 and 86 points?

    ☐ students

    b. **MP** What mathematical practice did you use to solve this problem?

12. What is the vertex of the function $y = 4 - x^2$?
   - A $(0, -4)$
   - B $(-4, -4)$
   - C $(4, 0)$
   - D $(0, 4)$
   - E $(4, 4)$

Need Extra Help?

If you missed Question...	1	2	3	4	5	6	7	8	9	10	11	12
Go to Lesson...	6-2	3-6	1-9	8-4	8-4	8-1	1-1	1-4	7-2	8-4	8-6	3-1

CHAPTER 9
Trigonometric Functions

THEN
You have graphed and analyzed functions.

NOW
You will:
- Find values of trigonometric functions.
- Solve problems by using right triangle trigonometry.
- Graph trigonometric functions.

WHY

WATER SPORTS Knowing trigonometric functions has practical applications in water sports. You can use right triangle trigonometry to find the distance a kayak has traveled when paddling against the current.

Use the Mathematical Practices to complete the activity.

1. **Sense-Making** You want to reach a point 100 meters up river, but know the current is too strong to paddle there directly. You can paddle straight out for 50 meters. How would you determine how far you have to paddle diagonally upriver to reach the desired point?

2. **Model with Math** Use the Triangle Special Segments tool to model the problem.

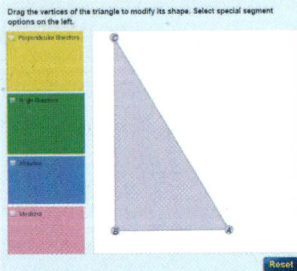

3. **Apply Math** Determine both the distance traveled diagonally upriver as well as the total distance traveled in the kayak.

 # Go Online to Guide Your Learning

Explore & Explain	Organize

 ### Graphing Tools

Use the **Explore: Trigonometric Functions** tool to enhance your understanding of graphing trigonometric functions in Lesson 9-5.

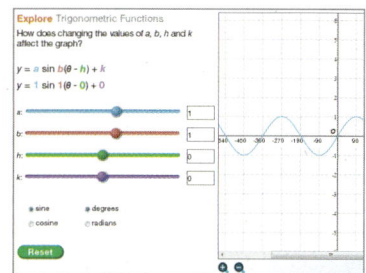

 ### Foldables

Get organized! Before you begin this chapter, create this **Trigonometric Functions Foldable** to help you organize your notes about trigonometric functions.

The Geometer's Sketchpad

The Geometer's Sketchpad can be used throughout this chapter to illustrate trigonometric ratios in right angles, to graph and analyze trigonometric functions, and to explore angles, angle measures, transformations of trigonometric graphs, and periodic functions.

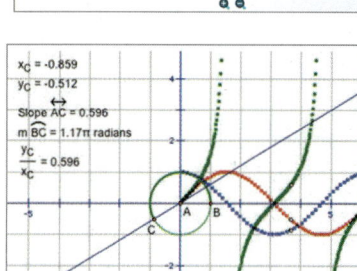

Collaborate

 ### Chapter Project

In the **To Bee or Not to Bee** project, you will use what you have learned about trigonometry to complete a project that addresses environmental awareness.

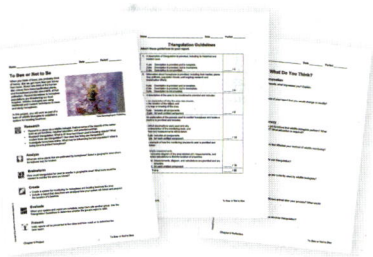

eBook
Interactive Student Guide

Before starting the chapter, answer the **Chapter Focus** preview questions. Check your answers as you complete each lesson. At the end of the chapter, try the **Performance Task**.

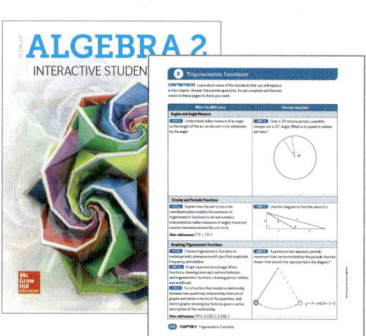

Focus

 ### LEARNSMART

Need help studying? Complete the **Trigonometric Functions** domain in LearnSmart to review for the chapter test.

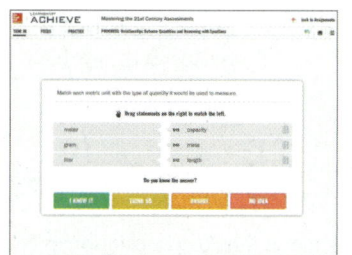

ALEKS

You can use the **Trigonometry** topic in ALEKS to explore what you know about statistics and probability and what you are ready to learn.*

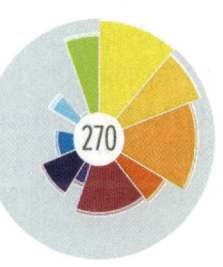

* Ask your teacher if this is part of your program.

connectED.mcgraw-hill.com

Get Ready for the Chapter

Connecting Concepts

Concept Check
Review the concepts used in this chapter by answering the questions below.

1. How is the Pythagorean Theorem stated mathematically?
2. What type of triangle does the Pythagorean Theorem apply to?
3. In the triangle shown, can you determine the measures of the two angles that are not the right angle? If so, what are they?

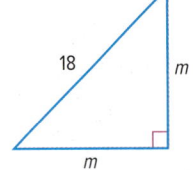

4. If you applied the Pythagorean Theorem to the triangle shown, what would the equation be?
5. In the triangle shown, what is the value of x?
6. In the triangle shown, what is the value of y?

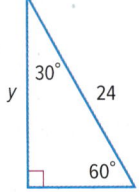

7. Given a 45°–45°–90° special right triangle with a hypotenuse of x, what will the values of the other two sides be?
8. Given a 30°–60°–90° special right triangle with the shortest side measuring x, what will the values of the longer side and of the hypotenuse be?

Performance Task Preview
You can use the concepts and skills in the chapter to perform calculations for a safety inspector. Understanding trigonometry and other principles of mathematics will help you finish the Performance Task at the end of the chapter.

MP In this Performance Task you will:
- make sense of problems and persevere in solving them
- reason abstractly and quantitatively
- construct viable arguments and critique the reasoning of others
- model with mathematics
- attend to precision

New Vocabulary

English		Español
trigonometry	p. 594	trigonometría
trigonometric ratio	p. 594	rázon trigonometric
sine	p. 594	seno
cosine	p. 594	coseno
tangent	p. 594	tangente
cosecant	p. 594	cosecante
secant	p. 594	secante
cotangent	p. 594	cotangente
reciprocal functions	p. 595	funciones recíprocas
angle of elevation	p. 598	ángulo de depresión
angle of depression	p. 598	ángulo de elevación
standard position	p. 604	posición estándar
initial side	p. 604	lado inicial de un ángulo
terminal side	p. 604	lado terminal
coterminal angles	p. 605	ángulos coterminales
radian	p. 606	radián
quadrantal angle	p. 613	ángulo de cuadrante
reference angle	p. 613	ángulo de referencia
unit circle	p. 620	círculo unitario
circular function	p. 620	funciones circulares
periodic function	p. 621	función periódica
cycle	p. 621	ciclo
period	p. 621	período
amplitude	p. 627	amplitud
frequency	p. 628	frecuencia
phase shift	p. 635	desplazamiento de fase
vertical shift	p. 636	cambio vertical
midline	p. 636	linea media

Review Vocabulary

function función a relation in which each element of the domain is paired with exactly one element in the range

inverse function función inversa two functions f and g are inverse functions if and only if both of their compositions are the identity function

EXPLORE 9-1

Spreadsheet Lab
Investigating Special Right Triangles

You can use a spreadsheet to investigate side measures of special right triangles.

Mathematical Practices
 8 Look for and express regularity in repeated reasoning.

Activity 45°-45°-90° Triangle

The legs of a 45°-45°-90° triangle, a and b, are equal in measure. What patterns do you observe in the ratios of the side measures of these triangles?

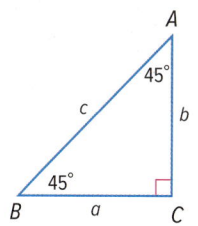

Step 1 Enter the indicated formulas in the spreadsheet. The formula uses the Pythagorean Theorem in the form $c = \sqrt{a^2 + b^2}$.

=SQRT(A2^2+B2^2)   =B2/A2   =B2/C2   =A2/C2

45-45-90 triangles

	A	B	C	D	E	F
1	a	b	c	b/a	b/c	a/c
2	1	1	1.414213562	1	0.707106781	0.707106781
3	2	2	2.828427125	1	0.707106781	0.707106781
4	3	3	4.242640687	1	0.707106781	0.707106781
5	4	4	5.656854249	1	0.707106781	0.707106781

Sheet 1 / Sheet 2 / Sheet 3

Step 2 Examine the results. Because 45°-45°-90° triangles share the same angle measures, these triangles are all similar. The ratios of the sides of these triangles are all the same. The ratios of side b to side a are 1. The ratios of side b to side c and of side a to side c are approximately 0.71.

Model and Analyze

Work cooperatively. Use the spreadsheet below for 30°-60°-90° triangles.

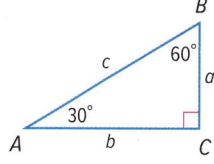

30-60-90 triangles

	A	B	C	D	E	F
1	a	b	c	b/a	b/c	a/c
2	1		2			
3	2		4			
4	3		6			
5	4		8			

Sheet 1 / Sheet 2 / Sheet 3

1. Copy and complete the spreadsheet above.

2. Describe the relationship among the 30°-60°-90° triangles with the dimensions given.

3. What patterns do you observe in the ratios of the side measures of these triangles?

LESSON 1
Trigonometric Functions in Right Triangles

Then · **Now** · **Why?**

- You used the Pythagorean Theorem to find side lengths of right triangles.

1. Find values of trigonometric functions for acute angles.
2. Use trigonometric functions to find side lengths and angle measures of right triangles.

- The altitude of a person parasailing depends on the length of the tow rope ℓ and the angle the rope makes with the horizontal $x°$. If you know these two values, you can use a ratio to find the altitude of the person parasailing.

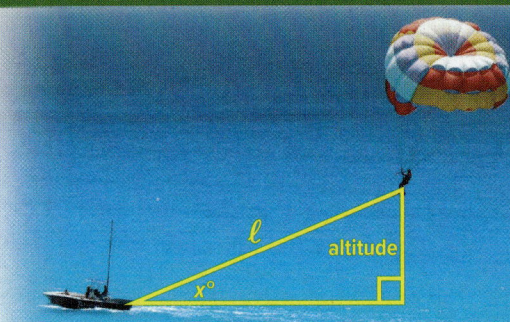

New Vocabulary
trigonometry
trigonometric ratio
trigonometric function
sine
cosine
tangent
cosecant
secant
cotangent
reciprocal functions
inverse sine
inverse cosine
inverse tangent
angle of elevation
angle of depression

Mathematical Practices
6 Attend to precision.

1 Trigonometric Functions for Acute Angles
Trigonometry is the study of relationships among the angles and sides of a right triangle. A **trigonometric ratio** compares the side lengths of a right triangle. A **trigonometric function** has a rule given by a trigonometric ratio.

The Greek letter *theta* θ is often used to represent the measure of an acute angle in a right triangle. The *hypotenuse*, the *leg opposite θ*, and the *leg adjacent to θ* are used to define the six trigonometric functions.

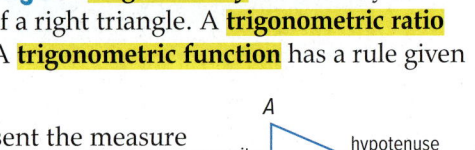

Key Concept — Trigonometric Functions in Right Triangles

Words If θ is the measure of an acute angle of a right triangle, then the following trigonometric functions involving the opposite side *opp*, the adjacent side *adj*, and the hypotenuse *hyp* are true.

Symbols
$\sin \theta = \dfrac{opp}{hyp}$ $\csc \theta = \dfrac{hyp}{opp}$

$\cos \theta = \dfrac{adj}{hyp}$ $\sec \theta = \dfrac{hyp}{adj}$

$\tan \theta = \dfrac{opp}{adj}$ $\cot \theta = \dfrac{adj}{opp}$

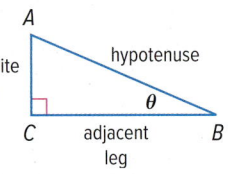

Examples
$\sin \theta = \dfrac{4}{5}$ $\cos \theta = \dfrac{3}{5}$ $\tan \theta = \dfrac{4}{3}$

$\csc \theta = \dfrac{5}{4}$ $\sec \theta = \dfrac{5}{3}$ $\cot \theta = \dfrac{3}{4}$

Example 1 Evaluate Trigonometric Functions

Find the values of the six trigonometric functions for angle θ.

leg opposite θ: $BC = 8$ leg adjacent θ: $AC = 15$ hypotenuse: $AB = 17$

$\sin \theta = \dfrac{opp}{hyp} = \dfrac{8}{17}$ $\cos \theta = \dfrac{adj}{hyp} = \dfrac{15}{17}$ $\tan \theta = \dfrac{opp}{adj} = \dfrac{8}{15}$

$\csc \theta = \dfrac{hyp}{opp} = \dfrac{17}{8}$ $\sec \theta = \dfrac{hyp}{adj} = \dfrac{17}{15}$ $\cot \theta = \dfrac{adj}{opp} = \dfrac{15}{8}$

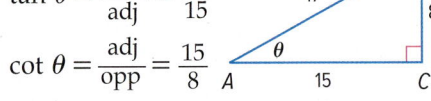

Guided Practice

1. Find the values of the six trigonometric functions for angle B.

> **StudyTip**
>
> **Tools** SOH-CAH-TOA is a mnemonic device for remembering the first letter of each word in the ratios for sine, cosine, and tangent.
>
> $\sin \theta = \dfrac{\text{opp}}{\text{hyp}}$
>
> $\cos \theta = \dfrac{\text{adj}}{\text{hyp}}$
>
> $\tan \theta = \dfrac{\text{opp}}{\text{adj}}$

Notice that the cosecant, secant, and cotangent ratios are reciprocals of the sine, cosine, and tangent ratios, respectively. These are called the **reciprocal functions**.

$$\csc \theta = \dfrac{1}{\sin \theta} \qquad \sec \theta = \dfrac{1}{\cos \theta} \qquad \cot \theta = \dfrac{1}{\tan \theta}$$

The domain of any trigonometric function is the set of all acute angles θ of a right triangle. So, trigonometric functions depend only on the measures of the acute angles, not on the side lengths of a right triangle.

Example 2 Find Trigonometric Ratios

If $\sin B = \dfrac{5}{8}$, find the exact values of the five remaining trigonometric functions for B.

Step 1 Draw a right triangle and label one acute angle B. Label the opposite side 5 and the hypotenuse 8.

Step 2 Use the Pythagorean Theorem to find a.

$a^2 + b^2 = c^2$ Pythagorean Theorem

$a^2 + 5^2 = 8^2$ $b = 5$ and $c = 8$

$a^2 + 25 = 64$ Simplify.

$a^2 = 39$ Subtract 25 from each side.

$a = \pm\sqrt{39}$ Take the square root of each side.

$a = \sqrt{39}$ Length cannot be negative.

Step 3 Find the other values.

Because $\sin B = \dfrac{5}{8}$, $\csc B = \dfrac{\text{hyp}}{\text{opp}}$ or $\dfrac{8}{5}$.

$\cos B = \dfrac{\text{adj}}{\text{hyp}} = \dfrac{\sqrt{39}}{8}$ $\sec B = \dfrac{\text{hyp}}{\text{adj}} = \dfrac{8}{\sqrt{39}}$ or $\dfrac{8\sqrt{39}}{39}$

$\tan B = \dfrac{\text{opp}}{\text{adj}} = \dfrac{5}{\sqrt{39}}$ or $\dfrac{5\sqrt{39}}{39}$ $\cot B = \dfrac{\text{adj}}{\text{opp}} = \dfrac{\sqrt{39}}{5}$

Guided Practice

2. If $\tan B = \dfrac{3}{7}$, find exact values of the remaining trigonometric fuctions for B.

> **Reading Math**
>
> **Labeling Triangles** Throughout this chapter, a capital letter is used to represent both a vertex of a triangle and the measure of the angle at that vertex. The same letter in lowercase is used to represent both the side opposite that angle and the length of the side.

Angles that measure 30°, 45°, and 60° occur frequently in trigonometry.

Key Concept Trigonometric Values for Special Angles

30°-60°-90°

$\sin 30° = \dfrac{1}{2}$ $\cos 30° = \dfrac{\sqrt{3}}{2}$ $\tan 30° = \dfrac{\sqrt{3}}{3}$

$\sin 60° = \dfrac{\sqrt{3}}{2}$ $\cos 60° = \dfrac{1}{2}$ $\tan 60° = \sqrt{3}$

45°-45°-90°

$\sin 45° = \dfrac{\sqrt{2}}{2}$ $\cos 45° = \dfrac{\sqrt{2}}{2}$ $\tan 45° = 1$

2 Use Trigonometric Functions
You can use trigonometric functions to find missing side lengths and missing angle measures of right triangles.

Example 3 Find a Missing Side Length

Use a trigonometric function to find the value of x. Round to the nearest tenth if necessary.

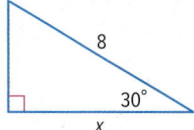

The length of the hypotenuse is 8. The missing measure is for the side adjacent to the 30° angle. Use the cosine function.

$\cos \theta = \dfrac{\text{adj}}{\text{hyp}}$ Cosine function

$\cos 30° = \dfrac{x}{8}$ Replace θ with 30°, adj with x, and hyp with 8.

$\dfrac{\sqrt{3}}{2} = \dfrac{x}{8}$ $\cos 30° = \dfrac{\sqrt{3}}{2}$

$\dfrac{8\sqrt{3}}{2} = x$ Multiply each side by 8.

$6.9 \approx x$ Use a calculator.

Study Tip

Choose a Function If the length of the hypotenuse is unknown, then either the sine or cosine function must be used to find the missing measure.

Guided Practice

Use a trigonometric function to find the value of x. Round to the nearest tenth if necessary.

3A.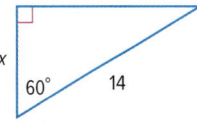

3B.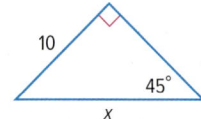

You can use a calculator to find the missing side lengths of triangles.

Real-World Example 4 Find a Missing Side Length

BUILDINGS To calculate the height of a building, Joel walked 200 feet from the base of the building and used an inclinometer to measure the angle from his eye to the top of the building. If his eye level is at 6 feet, how tall is the building?

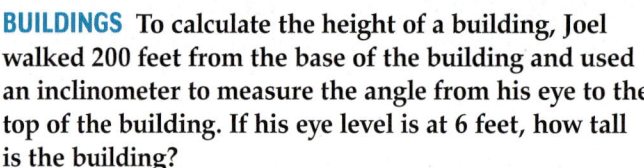

The measured angle is 76°. The side adjacent to the angle is 200 feet. The missing measure is the side opposite the angle. Use the tangent function to find d.

$\tan \theta = \dfrac{\text{opp}}{\text{adj}}$ Tangent function

$\tan 76° = \dfrac{d}{200}$ Replace θ with 76°, opp with d, and adj with 200.

$200 \tan 76° = d$ Multiply each side by 200.

$802 \approx d$ Use a calculator to simplify: 200 TAN 76 ENTER.

Because the inclinometer was 6 feet above the ground, the height of the building is approximately 808 feet.

Real-World Link

Inclinometers measure the angle of Earth's magnetic field as well as the pitch and roll of vehicles, sailboats, and airplanes. They are also used for monitoring volcanoes and well drilling.

Source: Science Magazine

Guided Practice

4. Use a trigonometric function to find the value of x. Round to the nearest tenth if necessary.

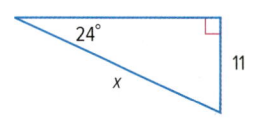

When solving equations like $3x = -27$, you use the inverse of multiplication to find x. You also can find angle measures by using the inverse of sine, cosine, or tangent.

Reading Math

Inverse Trigonometric Ratios The expression $\sin^{-1} x$ is read *the inverse sine of x* and is interpreted as *the angle whose sine is x*. Be careful not to confuse this notation with the notation for negative exponents; $\sin^{-1} x \neq \dfrac{1}{\sin x}$. Instead, this notation is similar to the notation for an inverse function, $f^{-1}(x)$.

Key Concept Inverse Trigonometric Ratios

Words	If $\angle A$ is an acute angle and the sine of A is x, then the **inverse sine** of x is the measure of $\angle A$.
Symbols	If $\sin A = x$, then $\sin^{-1} x = m\angle A$.
Example	$\sin A = \dfrac{1}{2} \rightarrow \sin^{-1}\dfrac{1}{2} = m\angle A \rightarrow m\angle A = 30°$
Words	If $\angle A$ is an acute angle and the cosine of A is x, then the **inverse cosine** of x is the measure of $\angle A$.
Symbols	If $\cos A = x$, then $\cos^{-1} x = m\angle A$.
Example	$\cos A = \dfrac{\sqrt{2}}{2} \rightarrow \cos^{-1}\dfrac{\sqrt{2}}{2} = m\angle A \rightarrow m\angle A = 45°$
Words	If $\angle A$ is an acute angle and the tangent of A is x, then the **inverse tangent** of x is the measure of $\angle A$.
Symbols	If $\tan A = x$, then $\tan^{-1} x = m\angle A$.
Example	$\tan A = \sqrt{3} \rightarrow \tan^{-1}\sqrt{3} = m\angle A \rightarrow m\angle A = 60°$

Go Online!

Trigonometric functions in right triangles are an important concept as you continue to study advanced math. **Search** *trigonometry* in ConnectED to find resources to review.

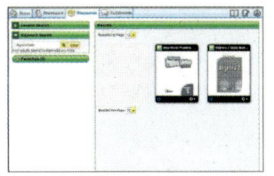

If you know the sine, cosine, or tangent of an acute angle, you can use a calculator to find the measure of the angle, which is the inverse of the trigonometric ratio.

Example 5 Find a Missing Angle Measure

Find the measure of each angle. Round to the nearest tenth if necessary.

a. $\angle N$

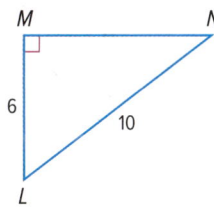

You know the measure of the side opposite $\angle N$ and the measure of the hypotenuse. Use the sine function.

$\sin N = \dfrac{6}{10}$ $\quad \sin \theta = \dfrac{\text{opp}}{\text{hyp}}$

$\sin^{-1}\dfrac{6}{10} = m\angle N$ $\quad$ Inverse sine

$36.9° \approx m\angle N$ $\quad$ Use a calculator.

b. $\angle B$

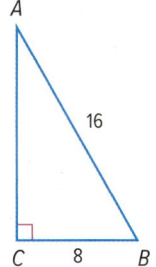

Use the cosine function.

$\cos B = \dfrac{8}{16}$ $\quad \cos \theta = \dfrac{\text{adj}}{\text{hyp}}$

$\cos^{-1}\dfrac{8}{16} = m\angle B$ $\quad$ Inverse cosine

$60° = m\angle B$ $\quad$ Use a calculator.

▶ **Guided Practice** Find x. Round to the nearest tenth if necessary.

5A.

(triangle with sides 17, 15, and angle $x°$)

5B.

(triangle with sides 18, 27, and angle $x°$)

Study Tip

Angles of Elevation and Depression The angle of elevation and the angle of depression are congruent since they are alternate interior angles of parallel lines.

In the figure at the right, the angle formed by the line of sight from the swimmer and a line parallel to the horizon is called the **angle of elevation**. The angle formed by the line of sight from the lifeguard and a line parallel to the horizon is called the **angle of depression**.

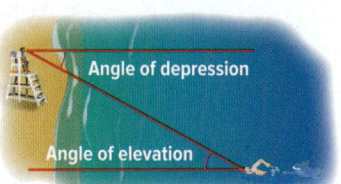

Real-World Example 6 Use Angles of Elevation and Depression

a. GOLF A golfer is standing at the tee, looking up to the green on a hill. If the tee is 36 feet lower than the green and the angle of elevation from the tee to the hole is 12°, find the distance from the tee to the hole.

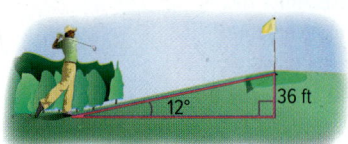

Write an equation using a trigonometric function that involves the ratio of the vertical rise (side opposite the 12° angle) and the distance from the tee to the hole (hypotenuse).

$\sin 12° = \dfrac{36}{x}$ $\sin \theta = \dfrac{\text{opp}}{\text{hyp}}$

$x \sin 12° = 36$ Multiply each side by x.

$x = \dfrac{36}{\sin 12°}$ Divide each side by sin 12°.

$x \approx 173.2$ Use a calculator.

So, the distance from the tee to the hole is about 173.2 feet.

b. ROLLER COASTER The hill of the roller coaster has an *angle of descent*, or an angle of depression, of 60°. Its vertical drop is 195 feet. Estimate the length of the hill.

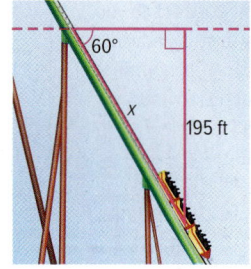

Write an equation using a trigonometric function that involves the ratio of the vertical drop (side opposite the 60° angle) and the length of the hill (hypotenuse).

$\sin 60° = \dfrac{195}{x}$ $\sin \theta = \dfrac{\text{opp}}{\text{hyp}}$

$x \sin 60° = 195$ Multiply each side by x.

$x = \dfrac{195}{\sin 60°}$ Divide each side by sin 60°.

$x \approx 225.2$ Use a calculator.

So, the length of the hill is about 225.2 feet.

Real-World Link

The steepest roller coasters in the world have angles of descent that are close to 90°.

Source: Ultimate Roller Coaster

Guided Practice

6A. MOVING A ramp for unloading a moving truck has an angle of elevation of 32°. If the top of the ramp is 4 feet above the ground, estimate the length of the ramp.

6B. LADDERS A 14-foot long ladder is placed against a house at an angle of elevation of 72°. How high above the ground is the top of the ladder?

Check Your Understanding

= Step-by-Step Solutions begin on page R11.

Go Online! for a Self-Check Quiz

Example 1 Find the values of the six trigonometric functions for angle θ.

1.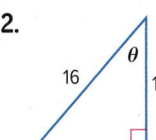
2.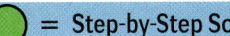

Example 2 In a right triangle, ∠A is acute. Find the values of the five remaining trigonometric funtions.

3. $\cos A = \frac{4}{7}$
4. $\tan A = \frac{20}{21}$

Examples 3–4 Use a trigonometric function to find the value of x. Round to the nearest tenth.

5.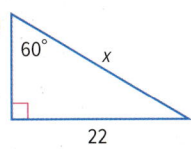
6.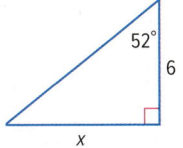
7.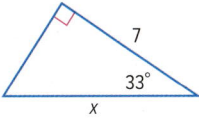

Example 5 Find the value of x. Round to the nearest tenth.

8.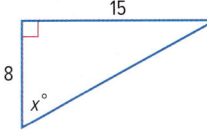
9.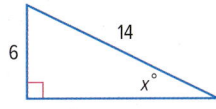
10.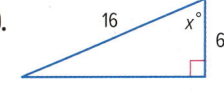

Example 6

11. **MP SENSE-MAKING** Christian found two trees directly across from each other in a canyon. When he moved 100 feet from the tree on his side (parallel to the edge of the canyon), the angle formed by the tree on his side and the tree on the other side was 70°. Find the distance across the canyon.

12. **LADDERS** The recommended angle of elevation for a ladder used in fire fighting is 75°. At what height on a building does a 21-foot ladder reach if the recommended angle of elevation is used? Round to the nearest tenth.

Practice and Problem Solving

Extra Practice is on page R9.

Example 1 Find the values of the six trigonometric functions for angle θ.

13.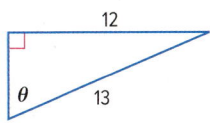
14.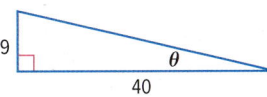
15.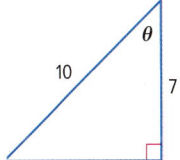
16.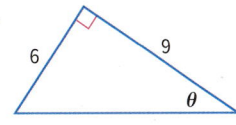

Example 2 In a right triangle, ∠A and ∠B are acute. Find the values of the five remaining trigonometric funtions.

17. $\tan A = \frac{8}{15}$
18. $\cos A = \frac{3}{10}$
19. $\tan B = 3$
20. $\sin B = \frac{4}{9}$

Examples 3–4 Use a trigonometric function to find each value of x. Round to the nearest tenth.

21.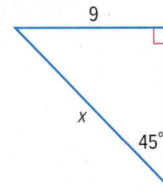
22.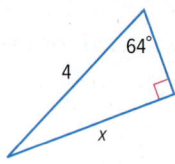
23.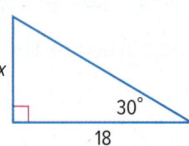

24.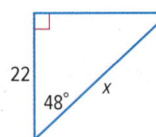
25.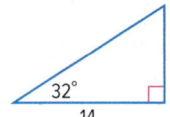
26.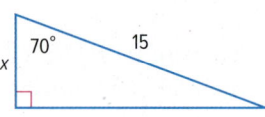

27. **PARASAILING** Refer to the beginning of the lesson and the figure at the right. Find a, the altitude of a person parasailing, if the tow rope is 250 feet long and the angle formed is 32°. Round to the nearest tenth.

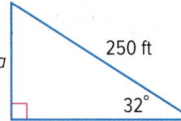

28. **MODELING** Devon wants to build a rope bridge between his treehouse and Cheng's treehouse. Suppose Devon's treehouse is directly behind Cheng's treehouse. At a distance of 20 meters to the left of Devon's treehouse, an angle of 52° is measured between the two treehouses. Find the length of the rope.

Example 5 Find the value of x. Round to the nearest tenth.

29.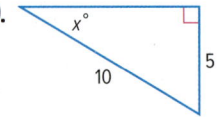
30.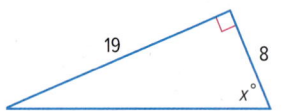
31.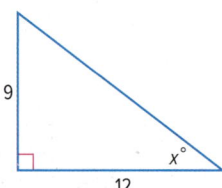

32.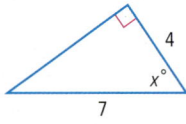
33.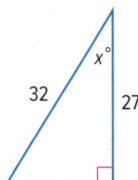
34.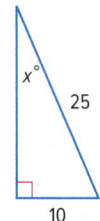

Example 6

35. **SQUIRRELS** Adult flying squirrels can make glides of up to 160 feet. If a flying squirrel glides a horizontal distance of 160 feet and the angle of descent is 9°, find its change in height.

36. **HANG GLIDING** A hang glider climbs at a 20° angle of elevation. Find the change in altitude of the hang glider when it has flown a horizontal distance of 60 feet.

Use trigonometric functions to find the values of x and y. Round to the nearest tenth.

37.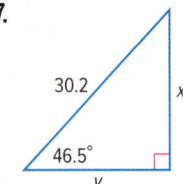
38.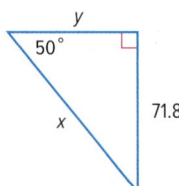
39.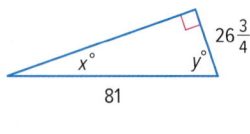

Solve each equation.

40. $\cos A = \dfrac{3}{19}$
41. $\sin N = \dfrac{9}{11}$
42. $\tan X = 15$
43. $\sin T = 0.35$
44. $\tan G = 0.125$
45. $\cos Z = 0.98$

46. MONUMENTS The San Jacinto Monument casts a shadow 370 feet long. The angle of elevation from the end of the shadow to the top of the monument is 57°.

  a. Draw and label a right triangle to represent this situation.

  b. Write a trigonometric function that can be used to find the height of the monument.

  c. Find the value of the function to determine the height of the monument to the nearest tenth.

47. NESTS Tabitha's eyes are 5 feet above the ground as she looks up to a bird's nest in a tree. If the angle of elevation is 74.5° and she is standing 12 feet from the tree's base, what is the height of the bird's nest? Round to the nearest tenth of a foot.

48. RAMPS Two bicycle ramps each cover a horizontal distance of 8 feet. One ramp has a 20° angle of elevation, and the other ramp has a 35° angle of elevation, as shown at the right.

  a. How much taller is the second ramp than the first? Round to the nearest tenth.

  b. How much longer is the second ramp than the first? Round to the nearest tenth.

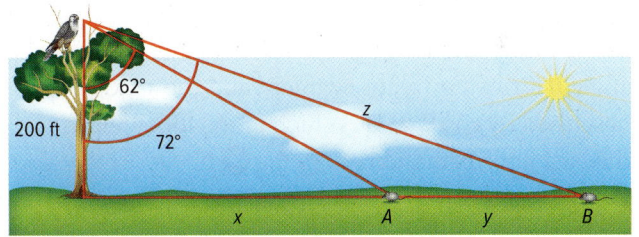

49. FALCONS A falcon at a height of 200 feet sees two mice A and B, as shown in the diagram.

  a. What is the approximate distance z between the falcon and mouse B?

  b. How far apart are the two mice?

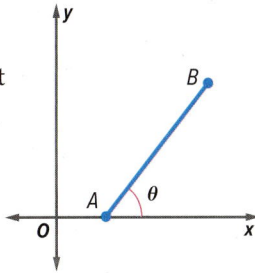

In △ABC, ∠C is a right angle. Use the given measurements to find the missing side lengths and missing angle measures of △ABC. Round to the nearest tenth if necessary.

50. $m\angle A = 36°$, $a = 12$

51. $m\angle B = 31°$, $b = 19$

52. $a = 8$, $c = 17$

53. $\tan A = \frac{4}{5}$, $a = 6$

H.O.T. Problems   Use Higher-Order Thinking Skills

54. CHALLENGE A line segment has endpoints A(2, 0) and B(6, 5), as shown in the figure at the right. What is the measure of the acute angle θ formed by the line segment and the x-axis? Explain how you found the measure.

55. CONSTRUCT ARGUMENTS Determine whether the following statement is *true* or *false*. Explain your reasoning.

  For any acute angle, the sine function will never have a negative value.

56. OPEN-ENDED In right triangle ABC, sin A = sin C. What can you conclude about △ABC? Justify your reasoning.

57. WRITING IN MATH A roof has a slope of $\frac{2}{3}$. Describe the connection between the slope and the angle of elevation θ that the roof makes with the horizontal. Then use an inverse trigonometric function to find θ.

Preparing for Assessment

58. The angle of elevation of a ramp is 17°. The length of the ramp is 5 meters. Which of the following expressions gives the height of the top of the ramp from the ground? **MP 6**

- A $5 \sin 17°$
- B $\dfrac{5}{\sec 17°}$
- C $5 \tan 17°$
- D $\dfrac{5}{\cos 17°}$
- E $\dfrac{5}{\sin 17°}$

59. In the first quadrant of a coordinate plane, the angle θ formed by a line segment and the x-axis is described by $\cos \theta = \dfrac{12}{13}$. Which of the following is the best approximation of $\tan \theta$? **MP 6**

- A -0.9231
- B 0.3846
- C 0.4167
- D 1.0833
- E 2.4000

60. The measure of an acute angle of a right triangle is given by θ. If $\sin \theta = \dfrac{8}{17}$, which of the following is the best approximation of $\tan \theta$? **MP 6**

- A 0.4705
- B 0.5333
- C 0.8824
- D 2.125
- E 1.875

61. MULTI-STEP A designer is laying out a section for a mosaic in a garden wall. The measurements of the section are shown on the figure. The width of the base of the triangle is 15 feet and the angle at the top of the section is 32°. Use the image below. **MP 6**

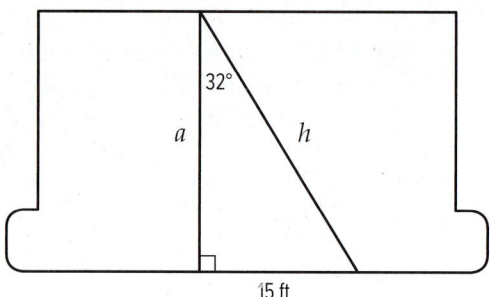

a. Which expression can be used to find the hypotenuse, h?

- A $h = \dfrac{\sin 32°}{15}$
- B $15h = \sin 32°$
- C $h = 15 \sin 32°$
- D $h = \dfrac{15}{\sin 32°}$

b. Which expression can be used to find the triangle's height, a?

- A $a = \dfrac{\tan 32°}{15}$
- B $15a = \tan 32°$
- C $a = 15 \tan 32°$
- D $a = \dfrac{15}{\tan 32°}$

c. What is the length of the hypotenuse of the triangle?

- A 18.0
- B 20.8
- C 24.0
- D 28.3
- E 49.2

d. What is the height of the triangle?

- A 18.0
- B 20.8
- C 24.0
- D 28.3
- E 49.2

e. Use the Pythagorean Theorem to show that your answers are correct.

EXTEND 9-1
Geometry Lab
Regular Polygons

You can use central angles of circles to investigate characteristics of regular polygons inscribed in a circle. Recall that a regular polygon is inscribed in a circle if each of its vertices lies on the circle.

Mathematical Practices
 5 Use appropriate tools strategically.

Activity Collect the Data

Step 1 Use a compass to draw a circle with a radius of one inch.

Step 2 Inscribe an equilateral triangle inside the circle. To do this, use a protractor to measure three angles of 120° at the center of the circle, because $\frac{360°}{3} = 120°$. Then connect the points where the sides of the angles intersect the circle using a straightedge.

Step 3 The **apothem** of a regular polygon is a segment that is drawn from the center of the polygon perpendicular to a side of the polygon. Use the cosine of angle θ to find the length of an apothem, labeled a in the diagram.

Model and Analyze
Work cooperatively.

1. Make a table like the one shown below and record the length of the apothem of the equilateral triangle. Inscribe each regular polygon named in the table in a circle with radius one inch. Complete the table.

Number of Sides, n	θ	a	Number of Sides, n	θ	a
3	60		7		
4	45		8		
5			9		
6			10		

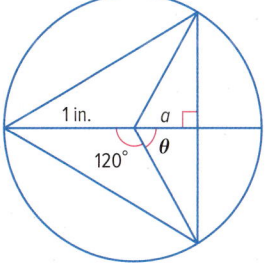

2. What do you notice about the measure of θ as the number of sides of the inscribed polygon increases?

3. What do you notice about the value of a?

4. **MAKE A CONJECTURE** Suppose you inscribe a 30-sided regular polygon inside a circle. Find the measure of angle θ.

5. Write a formula that gives the measure of angle θ for a polygon with n sides.

6. Write a formula that gives the length of the apothem of a regular polygon inscribed in a circle with radius one inch.

7. How would the formula you wrote in Exercise 6 change if the radius was not one inch?

LESSON 2
Angles and Angle Measure

::Then
- You used angles with degree measures.

::Now
1. Draw and find angles in standard position.
2. Convert between degree measures and radian measures.

::Why?
- A sundial is an instrument that indicates the time of day by the shadow that it casts on a surface marked to show hours or fractions of hours. The shadow moves around the dial 15° every hour.

 New Vocabulary
standard position
initial side
terminal side
coterminal angles
radian
central angle
arc length

 Mathematical Practices
2 Reason abstractly and quantitatively.

1 Angles in Standard Position
An angle on the coordinate plane is in **standard position** if the vertex is at the origin and one ray is on the positive x-axis.

- The ray on the x-axis is called the **initial side** of the angle.
- The ray that rotates about the center is called the **terminal side**.

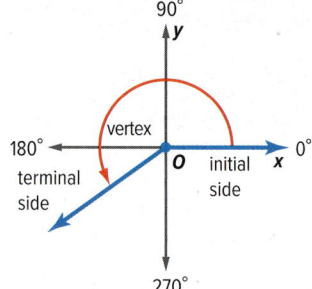

Key Concept   Angle Measures

If the measure of an angle is positive, the terminal side is rotated counterclockwise.

If the measure of an angle is negative, the terminal side is rotated clockwise.

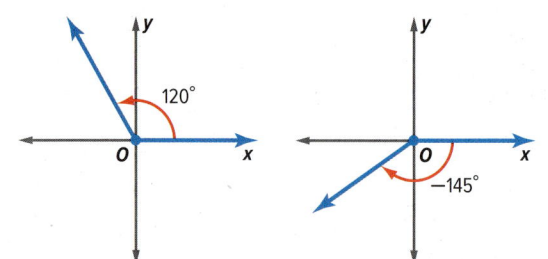

Example 1   Draw an Angle in Standard Position

Draw an angle with the given measure in standard position.

a. 215°   215° = 180° + 35°

Draw the terminal side of the angle 35° counterclockwise past the negative x-axis.

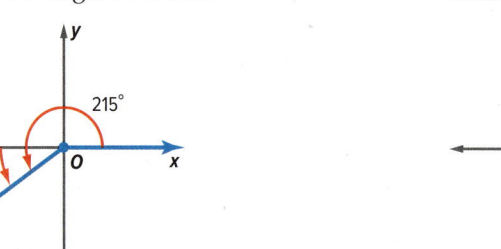

b. −40°

The angle is negative. Draw the terminal side of the angle 40° clockwise from the positive x-axis.

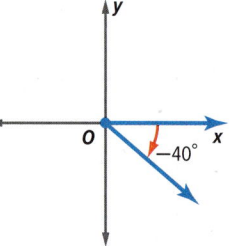

▶ **Guided Practice**

1A. 80°

1B. −105°

604 | Lesson 9-2

The terminal side of an angle can make more than one complete rotation. For example, a complete rotation of 360° plus a rotation of 120° forms an angle that measures 360° + 120° or 480°.

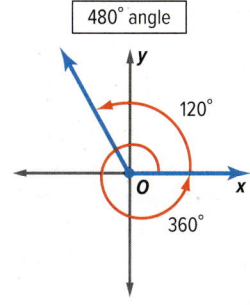

Real-World Example 2 Draw an Angle in Standard Position

WAKEBOARDING *Wakeboarding* is a combination of surfing, skateboarding, snowboarding, and water skiing. One maneuver involves a 540-degree rotation in the air. Draw an angle in standard position that measures 540°.

540° = 360° + 180°

Draw the terminal side of the angle 180° past the positive *x*-axis.

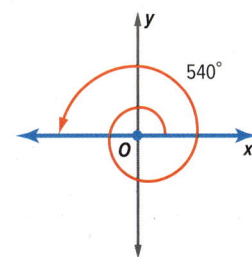

Guided Practice

2. Draw an angle in standard position that measures 600°.

Two or more angles in standard position with the same terminal side are called **coterminal angles**. For example, angles that measure 60°, 420°, and −300° are coterminal, as shown in the figure at the right.

An angle that is coterminal with another angle can be found by adding or subtracting a multiple of 360°.

- 60° + 360° = 420°
- 60° − 360° = −300°

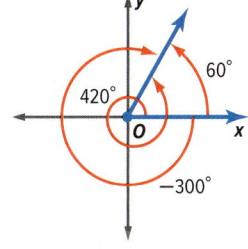

Example 3 Find Coterminal Angles

Find an angle with a positive measure and an angle with a negative measure that are coterminal with each angle.

a. 130°

positive angle: 130° + 360° = 490°   Add 360°.
negative angle: 130° − 360° = −230°   Subtract 360°.

b. −200°

positive angle: −200° + 360° = 160°   Add 360°.
negative angle: −200° − 360° = −560°   Subtract 360°.

Guided Practice

3A. 15°   **3B.** −45°

Real-World Link
Wakeboarding is one of the fastest-growing water sports in the United States. Participation has increased more than 100% in recent years.
Source: King of Wake

Reading Math
Angle of Rotation
In trigonometry, an angle is sometimes referred to as an *angle of rotation*.

Study Tip

Radians As with degrees, radians measure the amount of rotation from the initial side to the terminal side.
- The measure of an angle in radians is positive if its rotation is counterclockwise.
- The measure is negative if the rotation is clockwise.

2 Convert Between Degrees and Radians

Angles can also be measured in units that are based on arc length. One **radian** is the measure of an angle θ in standard position with a terminal side that intercepts an arc with the same length as the radius of the circle.

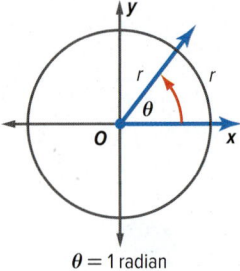

$\theta = 1$ radian

The circumference of a circle is $2\pi r$. So, one complete revolution around a circle equals 2π radians. Because 2π radians $= 360°$, degree measure and radian measure are related by the following equations.

$$2\pi \text{ radians} = 360° \qquad \pi \text{ radians} = 180°$$

Key Concept Convert Between Degrees and Radians

Degrees to Radians	Radians to Degrees
To convert from degrees to radians, multiply the number of degrees by $\dfrac{\pi \text{ radians}}{180°}$.	To convert from radians to degrees, multiply the number of radians by $\dfrac{180°}{\pi \text{ radians}}$.

Reading Math

Radian Measures The word *radian* is usually omitted when angles are expressed in radian measure. Thus, when no units are given for an angle measure, radian measure is implied.

Example 4 Convert Between Degrees and Radians

Rewrite the degree measure in radians and the radian measure in degrees.

a. $-30°$

$-30° = -30° \cdot \dfrac{\pi \text{ radians}}{180°}$

$= \dfrac{-30\pi}{180}$ or $-\dfrac{\pi}{6}$ radians

b. $\dfrac{5\pi}{2}$

$\dfrac{5\pi}{2} = \dfrac{5\pi}{2} \text{ radians} \cdot \dfrac{180°}{\pi \text{ radians}}$

$= \dfrac{900°}{2}$ or $450°$

Guided Practice

4A. $120°$

4B. $-\dfrac{3\pi}{8}$

Concept Summary Degrees and Radians

The diagram shows equivalent degree and radian measures for special angles.

You may find it helpful to memorize the following equivalent degree and radian measures. The other special angles are multiples of these angles.

$30° = \dfrac{\pi}{6} \qquad 45° = \dfrac{\pi}{4}$

$60° = \dfrac{\pi}{3} \qquad 90° = \dfrac{\pi}{2}$

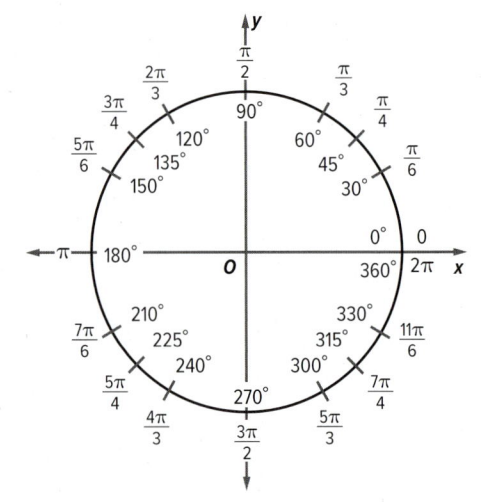

A **central angle** of a circle is an angle with a vertex at the center of the circle. If you know the measure of a central angle and the radius of the circle, you can find the length of the arc that is intercepted by the angle.

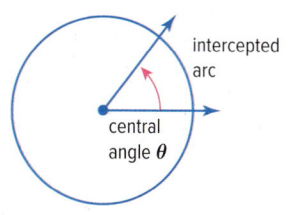

Key Concept Arc Length

| Words | For a circle with radius r and central angle θ (in radians), the **arc length** s equals the product of r and θ. | Model |

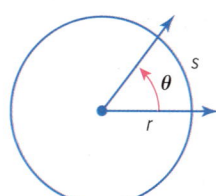

Symbols $s = r\theta$

Real-World Example 5 Find Arc Length

TRUCKS Monster truck tires have a radius of 33 inches. How far does a monster truck travel in feet after just three fourths of a tire rotation?

Step 1 Find the central angle in radians.

$\theta = \dfrac{3}{4} \cdot 2\pi$ or $\dfrac{3\pi}{2}$ The angle is $\dfrac{3}{4}$ of a complete rotation.

Step 2 Use the radius and central angle to find the arc length.

$s = r\theta$ Write the formula for arc length.

$ = 33 \cdot \dfrac{3\pi}{2}$ Replace r with 33 and θ with $\dfrac{3\pi}{2}$.

$ \approx 155.5$ in.   Use a calculator to simplify.

$ \approx 13.0$ ft   Divide by 12 to convert to feet.

So, the truck travels about 13 feet after three fourths of a tire rotation.

> **Watch Out!**
> **Arc Length** Remember to write the angle measure in radians, not degrees, when finding arc length. Also, recall that the number of radians in a complete rotation is 2π.

Guided Practice

5. A circle has a diameter of 9 centimeters. Find the arc length if the central angle is 60°. Round to the nearest tenth.

Check Your Understanding

 = Step-by-Step Solutions begin on page R11.

Examples 1–2 Draw an angle with the given measure in standard position.

    1. 140°   **2.** −60°   **3.** 390°

Example 3 Find an angle with a positive measure and an angle with a negative measure that are coterminal with each angle.

    4. 25°   **(5)** 175°   **6.** −100°

Example 4 Rewrite each degree measure in radians and each radian measure in degrees.

    7. $\dfrac{\pi}{4}$ **8.** 225°   **9.** −40°

Example 5 **10.** **REASONING** A tennis player's swing moves along the path of an arc. If the radius of the arc's circle is 4 feet and the angle of rotation is 100°, what is the length of the arc? Round to the nearest tenth.

Practice and Problem Solving

Extra Practice is on page R9.

Examples 1–2 Draw an angle with the given measure in standard position.

11. 75° 12. 160° 13. −90°
14. −120° 15. 295° 16. 510°

17. **GYMNASTICS** A gymnast on the uneven bars swings to make a 240° angle of rotation.

18. **FOOD** The lid on a jar of pasta sauce is turned 420° before it comes off.

Example 3 Find an angle with a positive measure and an angle with a negative measure that are coterminal with each angle.

19. 50° 20. 95° 21. 205°
22. 350° 23. −80° 24. −195°

Example 4 Rewrite each degree measure in radians and each radian measure in degrees.

25. 330° 26. $\frac{5\pi}{6}$ 27. $-\frac{\pi}{3}$
28. −50° 29. 190° 30. $-\frac{7\pi}{3}$

Example 5

31. **SKATEBOARDING** The skateboard ramp at the right is called a *quarter pipe*. The curved surface is determined by the radius of a circle. Find the length of the curved part of the ramp.

32. **RIVERBOATS** The paddlewheel of a riverboat has a diameter of 24 feet. Find the arc length of the circle made when the paddlewheel rotates 300°.

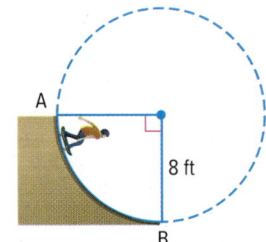

Find the length of each arc. Round to the nearest tenth.

33.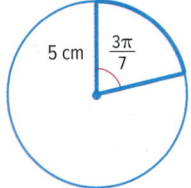

34.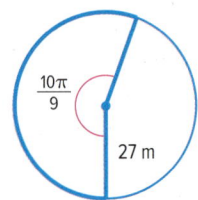

35. **CLOCKS** How long does it take for the minute hand on a clock to pass through 2.5π radians?

36. **MP PERSEVERANCE** Refer to the beginning of the lesson. A shadow moves around a sundial 15° every hour.
    a. After how many hours is the angle of rotation of the shadow $\frac{8\pi}{5}$ radians?
    b. What is the angle of rotation in radians after 5 hours?
    c. A sundial has a radius of 8 inches. What is the arc formed by a shadow after 14 hours? Round to the nearest tenth.

Find an angle with a positive measure and an angle with a negative measure that are coterminal with each angle.

37. 620° 38. −400° 39. $-\frac{3\pi}{4}$ 40. $\frac{19\pi}{6}$

41. SWINGS A swing has a 165° angle of rotation.
  a. Draw the angle in standard position.
  b. Write the angle measure in radians.
  c. If the chains of the swing are $6\frac{1}{2}$ feet long, what is the length of the arc that the swing makes? Round to the nearest tenth.
  d. Describe how the arc length would change if the chain lengths were doubled.

42. MULTIPLE REPRESENTATIONS Consider $A(-4, 0)$, $B(-4, 6)$, $C(6, 0)$, and $D(6, 8)$.
  a. **Geometric** Draw $\triangle EAB$ and $\triangle ECD$ with E at the origin.
  b. **Algebraic** Find the values of the tangent of $\angle BEA$ and the tangent of $\angle DEC$.
  c. **Algebraic** Find the slope of $\overline{BE}$ and $\overline{ED}$.
  d. **Verbal** What conclusions can you make about the relationship between slope and tangent?

Rewrite each degree measure in radians and each radian measure in degrees.

43. $\frac{21\pi}{8}$ **44.** 124° **45.** −200° **46.** 5

47. CAROUSELS At its peak, the carousel of the Sky Screamer at Six Flags Over Texas makes 0.5 revolution per minute. The circle formed by riders sitting in the outside has a radius of 49 feet. The circle formed by riders sitting in the inside has a radius of 47 feet.

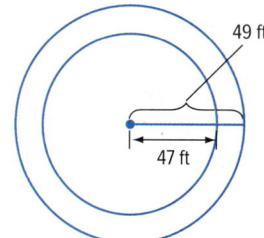

  a. Find the angle θ, in degrees, through which the carousel rotates in one second.
  b. In one second, what is the difference in arc lengths between the riders sitting in the outside row and the riders sitting in the inside row?

H.O.T. Problems Use Higher-Order Thinking Skills

48. ERROR ANALYSIS Tarshia and Alan are writing an expression for the measure of an angle coterminal with the angle shown at the right. Is either of them correct? Explain your reasoning.

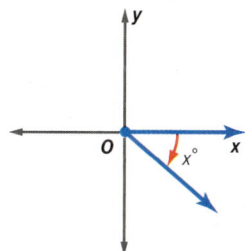

Tarshia	Alan
The measure of a coterminal angle is $(x - 360)°$.	The measure of a coterminal angle is $(360 - x)°$.

49. CHALLENGE A line makes an angle of $\frac{\pi}{2}$ radians with the positive x-axis at the point (2, 0). Find an equation for this line.

50. REASONING Express $\frac{1}{8}$ of a revolution in degrees and in radians. Explain your reasoning.

51. OPEN-ENDED Draw and label an acute angle in standard position. Find two angles, one positive and one negative, that are coterminal with the angle.

52. REASONING Justify the formula for the length of an arc.

53. WRITING IN MATH Use a circle with radius r to describe what one degree and one radian represent. Then explain how to convert between the measures.

Preparing for Assessment

54. Given two angle measures in standard position, which of the following pairs of angles are coterminal? **MP 2**

- A $-360°$ and $\frac{\pi}{2}$
- B $-90°$ and 4π
- C $120°$ and $\frac{4\pi}{3}$
- D $180°$ and 4π
- E $270°$ and $\frac{7\pi}{2}$

55. Where on a coordinate plane does the terminal side of the angle $\frac{6\pi}{5}$ appear? **MP 2**

- A Quadrant I
- B Quadrant II
- C Quadrant III
- D Quadrant IV
- E lies on y-axis

56. The shortest distance along circle O from point R to point Q is 800 yards. The radius of circle O is 1200 yards. What is the approximate measure of $\angle ROQ$? **MP 6**

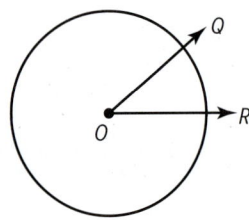

- A less than 20°
- B between 20° and 30°
- C between 30° and 40°
- D between 40° and 50°
- E greater than 50°

57. MULTI-STEP

a. Which of the following angle measures is coterminal with the angle $\frac{5\pi}{12}$? **MP 2**

- A 25°
- B 180°
- C 150°
- D 435°
- E 510°

b. Some students are creating a school tile mosaic to cover an industrial pipe spool's circumference. What would the arc length be for a section of tiles that will cover $\frac{5\pi}{12}$ radians of the spool if the diameter of the spool is 8 feet?

- A 5.2 ft
- B 10.4 ft
- C 3.3 ft
- D 6.7 ft

c. Two other groups of students will split the remaining part of the spool circumference evenly. Calculate the degree measure to which each group is entitled.

d. Calculate how many feet of the spool that will be.

58. a. What is the measure of an angle in degrees if the angle measures 2 radians? **MP 2**

b. What is the measure of an angle in radians if the angle measures π degrees?

59. Aisha runs 120 yards around a circular racetrack. The radius of the racetrack is 40 yards. **MP 1**

a. Find the central angle of Aisha's rotation, as she runs the 120 yards. Express your answer in radians.

b. Find the central angle of Aisha's rotation, as she runs the 120 yards. Express your answer in degrees.

EXTEND 9-2

Geometry Lab
Areas of Parallelograms

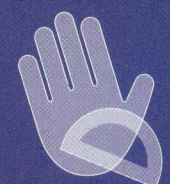

The area of any triangle can be found using the sine ratios in the triangle. A similar process can be used to find the area of a parallelogram.

Activity

Work cooperatively. Find the area of parallelogram *ABCD*.

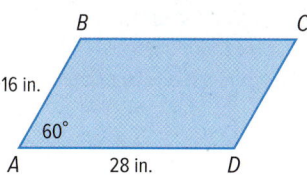

Step 1 Draw diagonal $\overline{BD}$.

$\overline{BD}$ divides the parallelogram into two congruent triangles, $\triangle ABD$ and $\triangle CDB$.

Step 2 Find the area of $\triangle ABD$.

Use the sine ratio to determine the height h from B to $\overline{AD}$.

$\sin \theta = \dfrac{\text{opp}}{\text{hyp}}$	Definition of sine
$\sin \theta = \dfrac{h}{AB}$	$h = $ opp, $AB = $ hyp
$AB \sin \theta = h$	Solve for h.

So, $h = AB \sin \theta$.

Area $= \dfrac{1}{2}bh$	Area of a triangle
$= \dfrac{1}{2}(AD)(AB) \sin A$	$b = AD$, $h = AB \sin A$
$= \dfrac{1}{2}(28)(16) \sin 60°$	$AD = 28$, $AB = 16$, $A = 60°$
$= 224 \left[\dfrac{\sqrt{3}}{2}\right]$	Multiply and evaluate sin 60°.
$= 112\sqrt{3}$	Simplify.

Step 3 Find the area of $\square ABCD$.

The area of $\square ABCD$ is equal to the sum of the areas of $\triangle ABD$ and $\triangle CDB$. Because $\triangle ABD \cong \triangle CDB$, the areas of $\triangle ABD$ and $\triangle CDB$ are equal. So, the area of $\square ABCD$ equals twice the area of $\triangle ABD$.

$2 \cdot 112\sqrt{3} = 224\sqrt{3}$ or about 387.98 square inches.

Exercises

Work cooperatively. For each of the following:

a. Find the area of each parallelogram.

b. Find the area of each parallelogram when the included angle is half the given measure.

c. Find the area of each parallelogram when the included angle is twice the given measure.

1.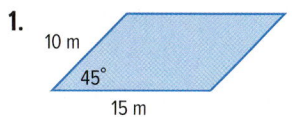

2.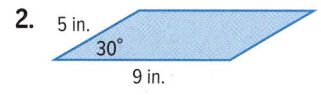

3.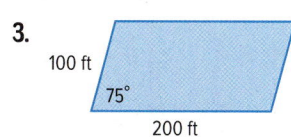

LESSON 3
Trigonometric Functions of General Angles

::Then
- You found values of trigonometric functions for acute angles.

::Now
1. Find values of trigonometric functions for general angles.
2. Find values of trigonometric functions by using reference angles.

::Why?
- In the ride at the right, the cars rotate back and forth about a central point. The positions of the arms supporting the cars can be described using trigonometric angles in standard position, with the central point of the ride at the origin of a coordinate plane.

 New Vocabulary
quadrantal angle
reference angle

 Mathematical Practices
6 Attend to precision.

1 Trigonometric Functions for General Angles
You can find values of trigonometric functions for angles greater than 90° or less than 0°.

Key Concept Trigonometric Functions of General Angles

Let θ be an angle in standard position and let $P(x, y)$ be a point on its terminal side. Using the Pythagorean Theorem, $r = \sqrt{x^2 + y^2}$. The six trigonometric functions of θ are defined below.

$\sin \theta = \dfrac{y}{r}$ $\qquad \cos \theta = \dfrac{x}{r} \qquad \tan \theta = \dfrac{y}{x}, x \neq 0$

$\csc \theta = \dfrac{r}{y}, y \neq 0 \qquad \sec \theta = \dfrac{r}{x}, x \neq 0 \qquad \cot \theta = \dfrac{x}{y}, y \neq 0$

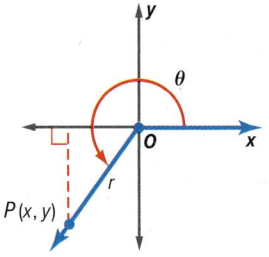

Example 1 Evaluate Trigonometric Functions Given a Point

The terminal side of θ in standard position contains the point at $(-3, -4)$. Find the exact values of the six trigonometric functions of θ.

Step 1 Draw the angle, and find the value of r.

$r = \sqrt{x^2 + y^2}$
$= \sqrt{(-3)^2 + (-4)^2}$
$= \sqrt{25}$ or 5

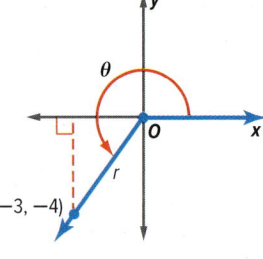

Step 2 Use $x = -3$, $y = -4$, and $r = 5$ to write the six trigonometric ratios.

$\sin \theta = \dfrac{y}{r} = \dfrac{-4}{5}$ or $-\dfrac{4}{5} \qquad \cos \theta = \dfrac{x}{r} = \dfrac{-3}{5}$ or $-\dfrac{3}{5} \qquad \tan \theta = \dfrac{y}{x} = \dfrac{-4}{-3}$ or $\dfrac{4}{3}$

$\csc \theta = \dfrac{r}{y} = \dfrac{5}{-4}$ or $-\dfrac{5}{4} \qquad \sec \theta = \dfrac{r}{x} = \dfrac{5}{-3}$ or $-\dfrac{5}{3} \qquad \cot \theta = \dfrac{x}{y} = \dfrac{-3}{-4}$ or $\dfrac{3}{4}$

Guided Practice

1. The terminal side of θ in standard position contains the point at $(-6, 2)$. Find the exact values of the six trigonometric functions of θ.

If the terminal side of angle θ in standard position lies on the x- or y-axis, the angle is called a **quadrantal angle**.

Study Tip

Quadrantal Angles The measure of a quadrantal angle is a multiple of $90°$ or $\frac{\pi}{2}$.

Key Concept — Quadrantal Angles

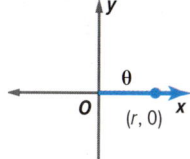
$\theta = 0°$ or 0 radians

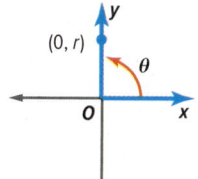
$\theta = 90°$ or $\frac{\pi}{2}$ radians

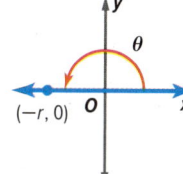
$\theta = 180°$ or π radians

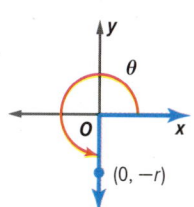
$\theta = 270°$ or $\frac{3\pi}{2}$ radians

Example 2 — Quadrantal Angles

The terminal side of θ in standard position contains the point at $(0, 6)$. Find the values of the six trigonometric functions of θ.

The point at $(0, 6)$ lies on the positive y-axis, so the quadrantal angle θ is $90°$. Use $x = 0$, $y = 6$, and $r = 6$ to write the trigonometric functions.

$\sin \theta = \frac{y}{r} = \frac{6}{6}$ or 1 $\qquad \cos \theta = \frac{x}{r} = \frac{0}{6}$ or 0 $\qquad \tan \theta = \frac{y}{x} = \frac{6}{0}$ undefined

$\csc \theta = \frac{r}{y} = \frac{6}{6}$ or 1 $\qquad \sec \theta = \frac{r}{x} = \frac{6}{0}$ undefined $\qquad \cot \theta = \frac{x}{y} = \frac{0}{6}$ or 0

Guided Practice

2. The terminal side of θ in standard position contains the point at $(-2, 0)$. Find the values of the six trigonometric functions of θ.

2 Trigonometric Functions with Reference Angles

If θ is a nonquadrantal angle in standard position, its **reference angle** θ' is the acute angle formed by the terminal side of θ and the x-axis. The rules for finding the measures of reference angles for $0° < \theta < 360°$ or $0° < \theta < 2\pi$ are shown below.

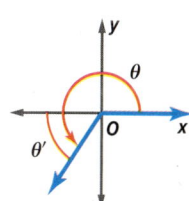

Go Online!

θ' is read *theta prime*. To hear more pronunciations of expressions, log into your **eStudent Edition**. Ask your teacher or a partner for clarification as you need it.

Key Concept — Reference Angles

Quadrant I	Quadrant II	Quadrant III	Quadrant IV
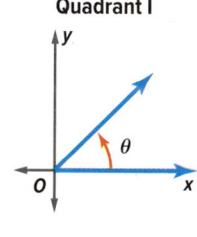		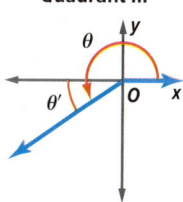	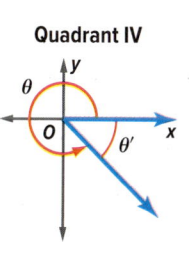
$\theta' = \theta$	$\theta' = 180° - \theta$ $\theta' = \pi - \theta$	$\theta' = \theta - 180°$ $\theta' = \theta - \pi$	$\theta' = 360° - \theta$ $\theta' = 2\pi - \theta$

If the measure of θ is greater than 360° or less than 0°, then use a coterminal angle with a positive measure between 0° and 360° to find the reference angle.

> **Study Tip**
>
> **Modeling** You can refer to the diagram in the Lesson 9-2 Concept Summary to help you sketch angles.

Example 3   Find Reference Angles

Sketch each angle. Then find its reference angle.

a. 210°

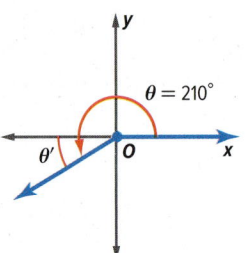

The terminal side of 210° lies in Quadrant III.
$\theta' = \theta - 180°$
$= 210° - 180°$ or $30°$

b. $-\dfrac{5\pi}{4}$

coterminal angle: $-\dfrac{5\pi}{4} + 2\pi = \dfrac{3\pi}{4}$

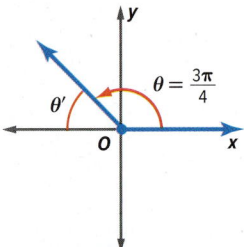

The terminal side of $\dfrac{3\pi}{4}$ lies in Quadrant III.
$\theta' = \pi - \theta$
$= \pi - \dfrac{3\pi}{4}$ or $\dfrac{\pi}{4}$

▶ **Guided Practice**

3A. $-110°$

3B. $\dfrac{2\pi}{3}$

You can use reference angles to evaluate trigonometric functions for any angle θ. The sign of a function is determined by the quadrant in which the terminal side of θ lies. Use these steps to evaluate a trigonometric function for any angle θ.

Key Concept   Evaluate Trigonometric Functions

Step 1 Find the measure of the reference angle θ′.

Step 2 Evaluate the trigonometric function for θ′.

Step 3 Determine the sign of the trigonometric function value. Use the quadrant in which the terminal side of θ lies.

Quadrant II	Quadrant I
sin θ, csc θ: +	sin θ, csc θ: +
cos θ, sec θ: −	cos θ, sec θ: +
tan θ, cot θ: −	tan θ, cot θ: +
Quadrant III	**Quadrant IV**
sin θ, csc θ: −	sin θ, csc θ: −
cos θ, sec θ: −	cos θ, sec θ: +
tan θ, cot θ: +	tan θ, cot θ: −

You can use the trigonometric values of angles measuring 30°, 45°, and 60° that you learned in Lesson 9-1.

Trigonometric Values for Special Angles					
Sine	Cosine	Tangent	Cosecant	Secant	Cotangent
$\sin 30° = \dfrac{1}{2}$	$\cos 30° = \dfrac{\sqrt{3}}{2}$	$\tan 30° = \dfrac{\sqrt{3}}{3}$	$\csc 30° = 2$	$\sec 30° = \dfrac{2\sqrt{3}}{3}$	$\cot 30° = \sqrt{3}$
$\sin 45° = \dfrac{\sqrt{2}}{2}$	$\cos 45° = \dfrac{\sqrt{2}}{2}$	$\tan 45° = 1$	$\csc 45° = \sqrt{2}$	$\sec 45° = \sqrt{2}$	$\cot 45° = 1$
$\sin 60° = \dfrac{\sqrt{3}}{2}$	$\cos 60° = \dfrac{1}{2}$	$\tan 60° = \sqrt{3}$	$\csc 60° = \dfrac{2\sqrt{3}}{3}$	$\sec 60° = 2$	$\cot 60° = \dfrac{\sqrt{3}}{3}$

Example 4 Use a Reference Angle to Find a Trigonometric Value

Find the exact value of each trigonometric function.

a. cos 240°

The terminal side of 240° lies in Quadrant III.

$\theta' = \theta - 180°$ Find the measure of the reference angle.
$ = 240° - 180°$ or $60°$ $\theta = 240°$

$\cos 240° = -\cos 60°$ or $-\dfrac{1}{2}$ The cosine function is negative in Quadrant III.

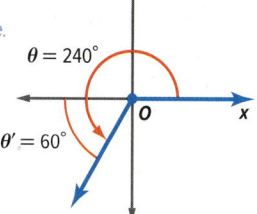

b. $\csc \dfrac{5\pi}{6}$

The terminal side of $\dfrac{5\pi}{6}$ lies in Quadrant II.

$\theta' = \pi - \theta$ Find the measure of the reference angle.
$ = \pi - \dfrac{5\pi}{6}$ or $\dfrac{\pi}{6}$ $\theta = \dfrac{5\pi}{6}$

$\csc \dfrac{5\pi}{6} = \csc \dfrac{\pi}{6}$ The cosecant function is positive in Quadrant II.
$\phantom{\csc \dfrac{5\pi}{6}} = \csc 30°$ $\dfrac{\pi}{6}$ radians = 30°
$\phantom{\csc \dfrac{5\pi}{6}} = 2$ $\csc 30° = \dfrac{1}{\sin 30}$

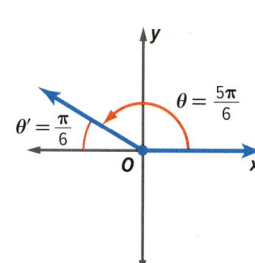

Guided Practice

4A. cos 135°   **4B.** $\tan \dfrac{5\pi}{6}$

Real-World Link
On a swing ride, riders experience weightlessness just like the drop side of a roller coaster. The ride lasts one minute and reaches speeds of 60 miles per hour in both directions.
Source: Cedar Point

Real-World Example 5 Use Trigonometric Functions

RIDES The swing arms of the ride at the right are 84 feet long and the height of the axis from which the arms swing is 97 feet. What is the total height of the ride at the peak of the arc?

coterminal angle: $-200° + 360° = 160°$

reference angle: $180° - 160° = 20°$

$\sin \theta = \dfrac{y}{r}$ Sine function
$\sin 20° = \dfrac{y}{84}$ $\theta = 20°$ and $r = 84$
$84 \sin 20° = y$ Multiply each side by 84.
$28.7 \approx y$ Use a calculator to solve for y.

Because y is approximately 28.7 feet, the total height of the ride at its peak is $28.7 + 97$ or about 125.7 feet.

Guided Practice

5. RIDES A similar ride that is smaller has swing arms that are 72 feet long. The height of the axis from which the arms swing is 88 feet, and the angle of rotation from the standard position is $-195°$. What is the total height of the ride at the peak of the arc?

Check Your Understanding

= Step-by-Step Solutions begin on page R11.

Examples 1–2 The terminal side of θ in standard position contains each point. Find the exact values of the six trigonometric functions of θ.

1. (1, 2)
2. (−8, −15)
3. (0, −4)

Example 3 Sketch each angle. Then find its reference angle.

4. 300°
5. 115°
6. $-\dfrac{3\pi}{4}$

Example 4 Find the exact value of each trigonometric function.

7. $\sin \dfrac{3\pi}{4}$
8. $\tan \dfrac{5\pi}{3}$
9. sec 120°
10. sin 300°

Example 5

11. **ENTERTAINMENT** Alejandra opens her portable DVD player so that it forms a 125° angle. The screen is $5\dfrac{1}{2}$ inches long.

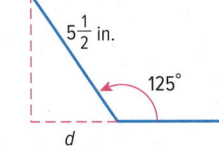

a. Redraw the diagram so that the angle is in standard position on the coordinate plane.

b. Find the reference angle. Then write a trigonometric function that can be used to find the distance to the wall d that she can place the DVD player.

c. Use the function to find the distance. Round to the nearest tenth.

Practice and Problem Solving

Extra Practice is on page R9.

Examples 1–2 The terminal side of θ in standard position contains each point. Find the exact values of the six trigonometric functions of θ.

12. (5, 12)
13. (−6, 8)
14. (3, 0)
15. (0, −7)
16. (4, −2)
17. (−9, −3)

Example 3 Sketch each angle. Then find its reference angle.

18. 195°
19. 285°
20. −250°
21. $\dfrac{7\pi}{4}$
22. $-\dfrac{\pi}{4}$
23. 400°

Example 4 Find the exact value of each trigonometric function.

24. sin 210°
25. tan 315°
26. cos 150°
27. csc 225°
28. $\sin \dfrac{4\pi}{3}$
29. $\cos \dfrac{5\pi}{3}$
30. $\cot \dfrac{5\pi}{4}$
31. $\sec \dfrac{11\pi}{6}$

Example 5

32. **REASONING** A soccer player x feet from the goalie kicks the ball toward the goal, as shown in the figure. The goalie jumps up and catches the ball 7 feet in the air.

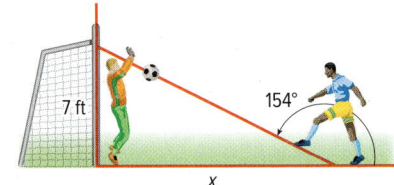

a. Find the reference angle. Then write a trigonometric function that can be used to find how far from the goalie the soccer player was when he kicked the ball.

b. About how far away from the goalie was the soccer player?

616 | Lesson 9-3 | Trigonometric Functions of General Angles

33. SPRINKLER A sprinkler rotating back and forth shoots water out a distance of 10 feet. From the horizontal position, it rotates 145° before reversing its direction. At a 145° angle, about how far to the left of the sprinkler does the water reach?

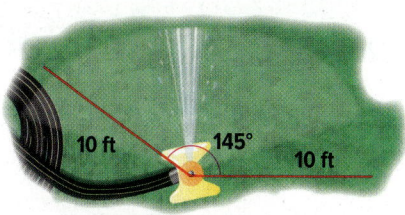

34. MULTI-STEP Seven hikers are stranded at the edge of a cliff. They have determined that the most direct way out is to create a rope long enough to reach their companion in the trees below. The companion estimates the horizontal distance from the cliff to the trees to be about 90 feet and the angle from the safest high point in the tree-line up to the cliff to be between 30 and 40 degrees.

  a. What is the minimum amount of rope needed to reach the tree line?

  b. What other factors do you think can cause them to need more rope? Use these factors to determine a new minimum amount of rope needed. Explain your reasoning.

  c. What assumptions did you make?

35. PHYSICS A rock is shot off the edge of a ravine with a slingshot at an angle of 65° and with an initial velocity of 6 meters per second. The equation that represents the horizontal distance of the rock x is $x = v_0 (\cos \theta) t$, where v_0 is the initial velocity, θ is the angle at which it is shot, and t is the time in seconds. About how far does the rock travel after 4 seconds?

36. FERRIS WHEELS Geri is in line to ride a Ferris wheel that is 212 feet tall and has a radius of about 98 feet. After a person gets on the bottom car, the Ferris wheel rotates 202.5° counterclockwise before stopping. How high above the ground is this car when it has stopped?

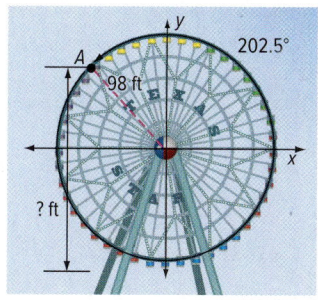

Suppose θ is an angle in standard position whose terminal side is in the given quadrant. For each function, find the exact values of the remaining five trigonometric functions of θ.

37. $\sin \theta = \frac{4}{5}$, Quadrant II

38. $\tan \theta = -\frac{2}{3}$, Quadrant IV

39. $\cos \theta = -\frac{8}{17}$, Quadrant III

40. $\cot \theta = -\frac{12}{5}$, Quadrant IV

Find the exact value of each trigonometric function.

41. $\cot 270°$

42. $\csc 180°$

43. $\sin 570°$

44. $\tan \left(-\frac{7\pi}{6}\right)$

45. $\cos \left(-\frac{11\pi}{6}\right)$

46. $\cot \frac{9\pi}{4}$

H.O.T. Problems   Use Higher-Order Thinking Skills

47. CHALLENGE For an angle θ in standard position, $\sin \theta = \frac{\sqrt{2}}{2}$ and $\tan \theta = -1$. Can the value of θ be 225°? Justify your reasoning.

48. MP CONSTRUCT ARGUMENTS Determine whether $3 \sin 60° = \sin 180°$ is *true* or *false*. Explain your reasoning.

49. MP REASONING Use the sine and cosine functions to explain why $\cot 180°$ is undefined.

50. OPEN-ENDED Identify a negative angle θ for which $\sin \theta > 0$ and $\cos \theta < 0$.

51. WRITING IN MATH Describe the steps for evaluating a trigonometric function for an angle θ that is greater than 90°. Include a description of a reference angle.

Preparing for Assessment

52. Which pair of expressions is equal to $\frac{-\sqrt{2}}{2}$? **MP 2**
- A $\sin \frac{3\pi}{4}$ and $\cos \frac{3\pi}{4}$
- B $\sin \frac{5\pi}{4}$ and $\sin \frac{7\pi}{4}$
- C $\cos \frac{3\pi}{4}$ and $\sin \frac{9\pi}{4}$
- D $\cos \frac{5\pi}{4}$ and $\sin \frac{3\pi}{4}$
- E $\cos \frac{\pi}{4}$ and $\cos \frac{3\pi}{4}$

53. The terminal side of θ in standard position contains the point at $(-2, 1)$. What is the exact measure of $(\tan \theta \cdot \cot \theta)$? **MP 6**
- A -2
- B -1
- C $-\frac{1}{2}$
- D $\frac{1}{2}$
- E 1

54. Which angle has a tangent and a sine that are both positive? **MP 2**
- A 25°
- B 110°
- C 150°
- D 225°
- E 330°

55. MULTI-STEP

a. When is a trigonometric function undefined? Choose all that apply. **MP 2**
- A when it does not exist
- B when it is equal to zero
- C when it is equal to one
- D when it is a cosecant function for an angle with a terminal arm along the x-axis
- E when it is a secant function for an angle with a terminal arm along the y-axis
- F when it is divided by one
- G when it is divided by zero

b. Look at the diagram. The terminal side of the angle in standard position contains the point $(0, -4)$.

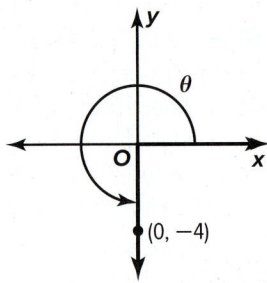

Which of the following trigonometric functions is undefined?
- A $\sin \theta$
- B $\cos \theta$
- C $\csc \theta$
- D $\cot \theta$
- E $\sec \theta$

c. Which angles are coterminal with the angle shown? Choose all that apply.
- A 90°
- B 730°
- C 270°
- D $-330°$
- E $-270°$
- F $-540°$
- G $-90°$

56. The terminal side of an angle in standard position contains the point $(-8, 0)$. Which functions are undefined? Choose all that apply. **MP 2**
- A $\sin \theta$
- B $\cos \theta$
- C $\csc \theta$
- D $\cot \theta$
- E $\sec \theta$
- F $\tan \theta$

CHAPTER 9
Mid-Chapter Quiz
Lessons 9-1 through 9-3

Solve △XYZ by using the given measurements. Round measures of sides to the nearest tenth and measures of angles to the nearest degree. (Lesson 9-1)

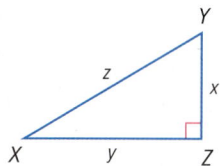

1. $Y = 65°$, $x = 16$
2. $X = 25°$, $x = 8$

3. Find the values of the six trigonometric functions for angle θ. (Lesson 9-1)

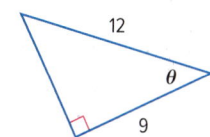

4. Draw an angle measuring $-80°$ in standard position. (Lesson 9-2)

Rewrite each degree measure in radians and each radian measure in degrees. (Lesson 9-2)

5. $215°$
6. $-350°$
7. $\dfrac{8\pi}{5}$
8. $\dfrac{9\pi}{2}$

9. **MULTIPLE CHOICE** What is the length of the arc below rounded to the nearest tenth? (Lesson 9-2)

   A 4.2 cm
   B 17.1 cm
   C 53.9 cm
   D 2638.9 cm

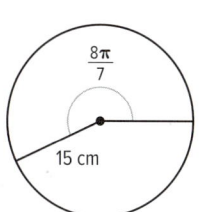

Find the exact value of each trigonometric function. (Lesson 9-3)

10. $\tan \pi$
11. $\cos \dfrac{3\pi}{4}$

The terminal side of θ in standard position contains each point. Find the exact values of the six trigonometric functions of θ. (Lesson 9-3)

12. $(0, -5)$
13. $(6, 8)$

14. **MULTIPLE CHOICE** Suppose θ is an angle in standard position with $\cos \theta > 0$. In which quadrant(s) does the terminal side of θ lie? (Lesson 9-3)

   A I
   B II
   C III
   D I and IV

15. **MULTI-PART** The angle of elevation of a ramp is 11°. The length of the ramp is 14 meters.

   a. Which of the following expressions gives the height of the top of the ramp from the ground?

      A 14 sin 11°
      B 14 cos 11°
      C 14 sec 11°
      D 14 csc 11°

   b. Which of the following expressions gives the distance on the ground that the ramp covers?

      A 14 sin 11°
      B 14 cos 11°
      C 14 sec 11°
      D 14 csc 11°

16. If both angles are in standard position, which of the following pairs are coterminal? Choose all that apply.

   A 180° and $\dfrac{\pi}{2}$
   B 150° and $\dfrac{5\pi}{6}$
   C 210° and $\dfrac{3\pi}{2}$
   D 240° and $\dfrac{4\pi}{3}$
   E 60° and $\dfrac{2\pi}{3}$
   F 315° and $\dfrac{-\pi}{4}$

17. Which of the following angles has a cosine value of $\dfrac{-\sqrt{2}}{2}$?

   A 45°
   B 135°
   C 180°
   D 315°

LESSON 4
Circular and Periodic Functions

:Then
- You evaluated trigonometric functions using reference angles..

:Now
1. Find values of trigonometric functions based on the unit circle.
2. Use the properties of periodic functions to evaluate trigonometric functions.

:Why?
- The pedals on a bicycle rotate as the bike is being ridden. The height of a pedal is a function of time, as shown in the figure at the right.

  Notice that the pedal makes one complete rotation every two seconds.

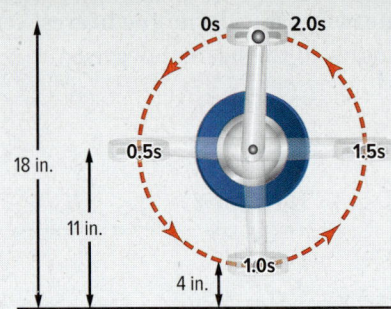

 New Vocabulary
unit circle
circular function
periodic function
cycle
period

MP Mathematical Practices
3 Construct viable arguments and critique the reasoning of others.
7 Look for and make use of structure.

1 Circular Functions
A **unit circle** is a circle with a radius of 1 unit centered at the origin on the coordinate plane. You can use a point P on the unit circle to generalize sine and cosine functions.

$$\sin \theta = \frac{y}{r} = \frac{y}{1} \text{ or } y \qquad \cos \theta = \frac{x}{r} = \frac{x}{1} \text{ or } x$$

So, the values of $\sin \theta$ and $\cos \theta$ are the y-coordinate and x-coordinate, respectively, of the point where the terminal side of θ intersects the unit circle.

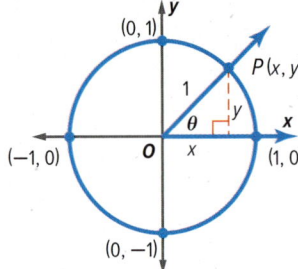

🔑 Key Concept Functions on a Unit Circle

Words	If the terminal side of an angle θ in standard position intersects the unit circle at $P(x, y)$, then $\cos \theta = x$ and $\sin \theta = y$.
Symbols	$P(x, y) = P(\cos \theta, \sin \theta)$
Example	If $\theta = 120°$, $P(x, y) = P(\cos 120°, \sin 120°)$.

Model

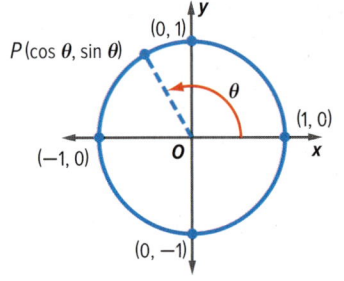

Both $\cos \theta = x$ and $\sin \theta = y$ are functions of θ. Because they are defined using a unit circle, they are called **circular functions**.

Example 1 Find Sine and Cosine Given a Point on the Unit Circle

The terminal side of angle θ in standard position intersects the unit circle at $P\left(\frac{1}{2}, \frac{\sqrt{3}}{2}\right)$. Find $\cos \theta$ and $\sin \theta$.

$P\left(\frac{1}{2}, \frac{\sqrt{3}}{2}\right) = P(\cos \theta, \sin \theta)$

$\cos \theta = \frac{1}{2} \qquad \sin \theta = \frac{\sqrt{3}}{2}$

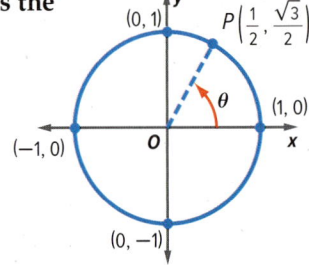

▶ **Guided Practice**

1. The terminal side of angle θ in standard position intersects the unit circle at $P\left(\frac{3}{5}, -\frac{4}{5}\right)$. Find $\cos \theta$ and $\sin \theta$.

620 | Lesson 9-4

Study Tip

Cycles A cycle can begin at any point on the graph of a periodic function. In Example 2, if the beginning of the cycle is at $\frac{\pi}{2}$, then the pattern repeats at $\frac{3\pi}{2}$. The period is $\frac{3\pi}{2} - \frac{\pi}{2}$ or π.

2 Periodic Functions A **periodic function** has y-values that repeat at regular intervals. One complete pattern is a **cycle**, and the horizontal length of one cycle is a **period**.

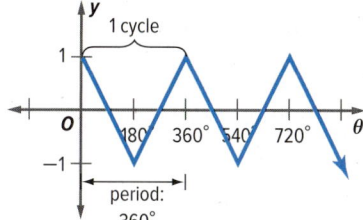

θ	y
0°	1
180°	−1
360°	1
540°	−1
720°	1

The cycle repeats every 360°.

Example 2 Identify the Period

Determine the period of the function.

The pattern repeats at π, 2π, and so on. So, the period is π.

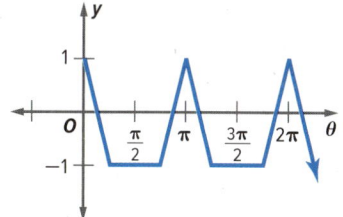

Guided Practice

2. Graph a function with a period of 4.

The rotations of wheels, pedals, carousels, and objects in space are all periodic.

Real-World Example 3 Use Trigonometric Functions

CYCLING Refer to the beginning of the lesson. The height of a bicycle pedal varies periodically as a function of time, as shown in the figure.

a. Make a table showing the height of a bicycle pedal at 0, 0.5, 1.0, 1.5, 2.0, 2.5, and 3.0 seconds.

At 0 seconds, the pedal is 18 inches high. At 0.5 second, the pedal is 11 inches high. At 1.0 second, the pedal is 4 inches high, and so on.

Time (s)	Height (in.)
0	18
0.5	11
1.0	4
1.5	11
2.0	18
2.5	11
3.0	4

b. Identify the period of the function.

The period is the time it takes to complete one rotation. So, the period is 2 seconds.

c. Graph the function. Let the horizontal axis represent the time t and the vertical axis represent the height h in inches that the pedal is from the ground.

The maximum height of the pedal is 18 inches, and the minimum height is 4 inches. Because the period of the function is 2 seconds, the pattern of the graph repeats in intervals of 2 seconds.

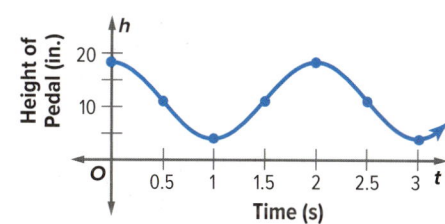

Real-World Link

Most competitive cyclists pedal at rates of more than 200 rotations per minute. Most other people pedal at between 90 and 120 rotations per minute.

Source: SpringerLink

Guided Practice

3. CYCLING Another cyclist pedals the same bike at a rate of 1 revolution per second.

A. Make a table showing the height of a bicycle pedal at times 0, 0.5, 1.0, 1.5, 2.0, 2.5, and 3.0 seconds.

B. Identify the period and graph the function.

Study Tip

MP Modeling To help you remember that for (x, y) on a unit circle, x = cos θ and y = sin θ, notice that alphabetically *x* comes before *y* and *cosine* comes before *sine*.

The exact values of cos θ and sin θ for special angles are shown on the unit circle at the right. The cosine values are the *x*-coordinates of the points on the unit circle, and the sine values are the *y*-coordinates.

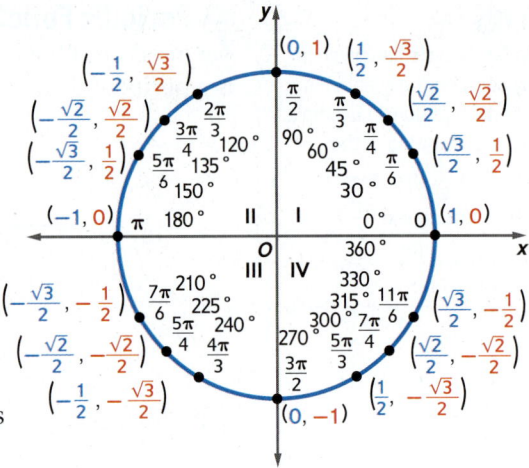

You can use this information to graph the sine and cosine functions. Let the horizontal axis represent the values of θ and the vertical axis represent the values of sin θ or cos θ.

The cycles of the sine and cosine functions repeat every 360°. So, they are periodic functions. The period of each function is 360° or 2π.

Consider the points on the unit circle for θ = 45°, θ = 150°, and θ = 270°.

$(\cos 45°, \sin 45°) = \left(\dfrac{\sqrt{2}}{2}, \dfrac{\sqrt{2}}{2}\right)$

$(\cos 150°, \sin 150°) = \left(-\dfrac{\sqrt{3}}{2}, \dfrac{1}{2}\right)$

$(\cos 270°, \sin 270°) = (0, -1)$

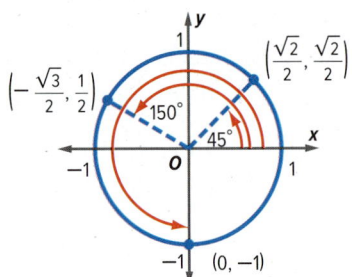

Study Tip

Radians The sine and cosine functions can also be graphed using radians as the units on the θ-axis.

These points can also be shown on the graphs of the sine and cosine functions.

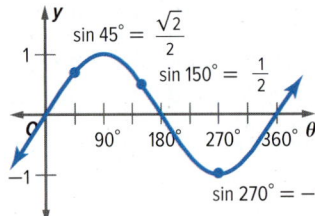 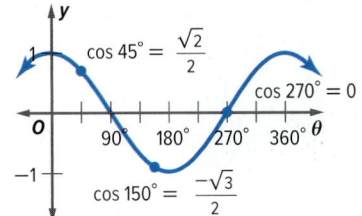

Because the period of the sine and cosine functions is 360°, the values repeat every 360°. So, sin (x + 360°) = sin x, and cos (x + 360°) = cos x.

Example 4   Evaluate Trigonometric Expressions

Find the exact value of each expression.

a. cos 480°

$\cos 480° = \cos (120° + 360°)$

$= \cos 120°$

$= -\dfrac{1}{2}$

b. $\sin \dfrac{11\pi}{4}$

$\sin \dfrac{11\pi}{4} = \sin \left(\dfrac{3\pi}{4} + \dfrac{8\pi}{4}\right)$

$= \sin \dfrac{3\pi}{4}$

$= \dfrac{\sqrt{2}}{2}$

▶ **Guided Practice**

4A. $\cos \left(-\dfrac{3\pi}{4}\right)$

4B. sin 420°

Check Your Understanding

= Step-by-Step Solutions begin on page R11.

Go Online! for a Self-Check Quiz

Example 1 **STRUCTURE** The terminal side of angle θ in standard position intersects the unit circle at each point P. Find $\cos \theta$ and $\sin \theta$.

1. $P\left(\dfrac{15}{17}, \dfrac{8}{17}\right)$

2. $P\left(-\dfrac{\sqrt{2}}{2}, \dfrac{\sqrt{2}}{2}\right)$

Example 2 Determine the period of each function.

3.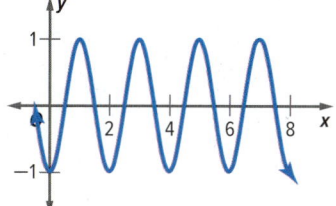

4.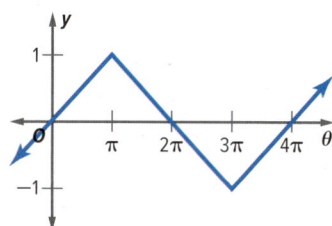

Example 3

5. **SWINGS** The height of a swing varies periodically as the function of time. The swing goes forward and reaches its high point of 6 feet. It then goes backward and reaches 6 feet again. Its lowest point is 2 feet. The time it takes to swing from its high point to its low point is 1 second.

   a. How long does it take for the swing to go forward and back one time?

   b. Graph the height of the swing h as a function of time t.

Example 4 Find the exact value of each expression.

6. $\sin \dfrac{13\pi}{6}$

7. $\sin(-60°)$

8. $\cos 540°$

Practice and Problem Solving

Extra Practice is on page R9.

Example 1 The terminal side of angle θ in standard position intersects the unit circle at each point P. Find $\cos \theta$ and $\sin \theta$.

9. $P\left(\dfrac{6}{10}, -\dfrac{8}{10}\right)$

10. $P\left(-\dfrac{10}{26}, -\dfrac{24}{26}\right)$

11. $P\left(\dfrac{\sqrt{3}}{2}, \dfrac{1}{2}\right)$

12. $P\left(\dfrac{\sqrt{6}}{5}, \dfrac{\sqrt{19}}{5}\right)$

Example 2 Determine the period of each function.

13.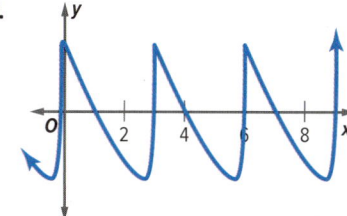

14.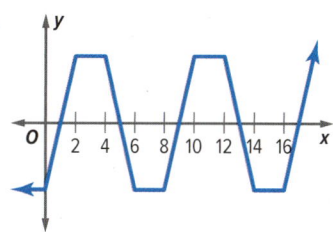

15.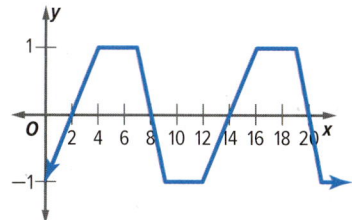

16.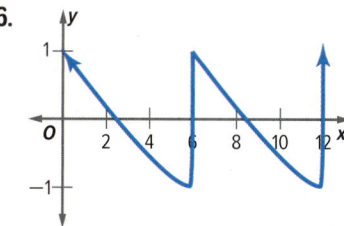

connectED.mcgraw-hill.com 623

Determine the period of each function.

17.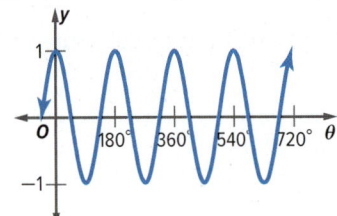

18.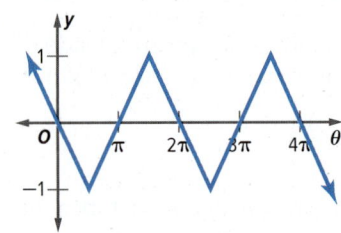

Example 3

19. **WEATHER** In a city, the average high temperature for each month is shown in the table.

    a. Sketch a graph of the function representing this situation.

    b. Describe the period of the function.

Average High Temperatures			
Month	Temperature (°F)	Month	Temperature (°F)
Jan	35	July	79
Feb.	39	Aug.	77
Mar.	47	Sept.	69
Apr.	57	Oct.	59
May	65	Nov.	46
Jun.	74	Dec.	37

Source: The Weather Channel

Example 4

Find the exact value of each expression.

20. $\sin \frac{7\pi}{3}$

21. $\cos(-60°)$

22. $\cos 450°$

23. $\sin \frac{11\pi}{4}$

24. $\sin(-45°)$

25. $\cos 570°$

26. **SENSE-MAKING** In the engine at the right, the distance d from the piston to the center of the circle, called the *crankshaft*, is a function of the speed of the piston rod. Point R on the piston rod rotates 150 times per second.

    a. Identify the period of the function as a fraction of a second.

    b. The shortest distance d is 0.5 inch, and the longest distance is 3.5 inches. Sketch a graph of the function. Let the horizontal axis represent the time t. Let the vertical axis represent the distance d.

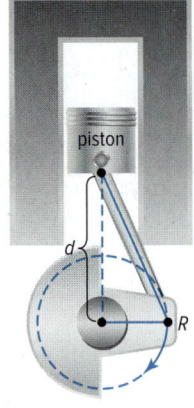
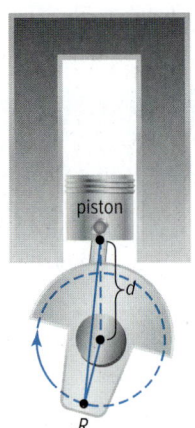

27. **TORNADOES** A tornado siren makes 2.5 rotations per minute and the beam of sound has a radius of 1 mile. Ms. Miller's house is 1 mile from the siren. The distance of the sound beam from her house varies periodically as a function of time.

    a. Identify the period of the function in seconds.

    b. Sketch a graph of the function. Let the horizontal axis represent the time t from 0 seconds to 60 seconds. Let the vertical axis represent the distance d the sound beam is from Ms. Miller's house at time t.

28. **FERRIS WHEEL** A Ferris wheel in China has a diameter of approximately 520 feet. The height of a compartment h is a function of time t. It takes about 30 seconds to make one complete revolution. Let the height at the center of the wheel represent the height at time 0. Sketch a graph of the function.

29. MULTIPLE REPRESENTATIONS The terminal side of an angle in standard position intersects the unit circle at *P*, as shown in the figure.

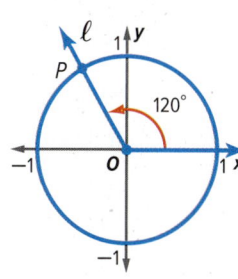

  a. **Geometric** Copy the figure. Draw lines representing 30°, 60°, 150°, 210°, and 315°.

  b. **Tabular** Use a table of values to show the slope of each line to the nearest tenth.

  c. **Analytical** What conclusions can you make about the relationship between the terminal side of the angle and the slope? Explain your reasoning.

30. POGO STICK A person is jumping up and down on a pogo stick at a constant rate. The difference between his highest and lowest points is 2 feet. He jumps 50 times per minute.

  a. Describe the independent variable and dependent variable of the periodic function that represents this situation. Then state the period of the function in seconds.

  b. Sketch a graph of the jumper's change in height in relation to his starting point. Assume that his starting point is halfway between his highest and lowest points. Let the horizontal axis represent the time *t* in seconds. Let the vertical axis represent the height *h*.

Find the exact value of each expression.

31. $\cos 45° - \cos 30°$

32. $6(\sin 30°)(\sin 60°)$

33. $2 \sin \frac{4\pi}{3} - 3 \cos \frac{11\pi}{6}$

34. $\cos\left(-\frac{2\pi}{3}\right) + \frac{1}{3} \sin 3\pi$

35. $(\sin 45°)^2 + (\cos 45°)^2$

36. $\frac{(\cos 30°)(\cos 150°)}{\sin 315°}$

H.O.T. Problems Use Higher-Order Thinking Skills

37. **CRITIQUE ARGUMENTS** Francis and Benita are finding the exact value of $\cos \frac{-\pi}{3}$. Is either of them correct? Explain your reasoning.

Francis	Benita
$\cos \frac{-\pi}{3} = -\cos \frac{\pi}{3}$ $= -0.5$	$\cos \frac{-\pi}{3} = \cos\left(-\frac{\pi}{3} + 2\pi\right)$ $= \cos \frac{5\pi}{3}$ $= 0.5$

38. CHALLENGE A ray has its endpoint at the origin of the coordinate plane, and point $P\left(\frac{1}{2}, -\frac{\sqrt{3}}{2}\right)$ lies on the ray. Find the angle θ formed by the positive *x*-axis and the ray.

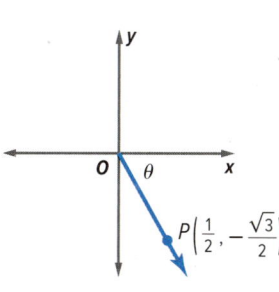

39. **REASONING** Is the period of a sine curve *sometimes*, *always*, or *never* a multiple of π? Justify your reasoning.

40. OPEN-ENDED Draw the graph of a periodic function that has a maximum value of 10 and a minimum value of −10. Describe the period of the function.

41. WRITING IN MATH Explain how to determine the period of a periodic function from its graph. Include a description of a cycle.

Preparing for Assessment

42. The terminal side of angle θ in standard position intersects the unit circle at $P\left(\frac{\sqrt{3}}{2}, \frac{1}{2}\right)$, as shown in the figure. **MP 6**

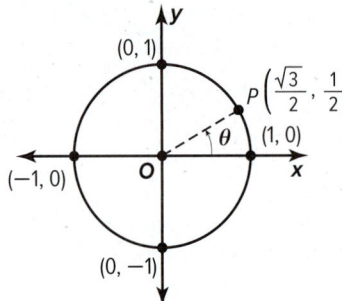

What is two times the sum of $\cos \theta$ and $\sin \theta$?

- A $\frac{\sqrt{3}}{2}$
- B 1
- C $\frac{\sqrt{3}+1}{2}$
- D $\sqrt{3}$
- E $\sqrt{3}+1$

43. What is the exact value of $2 \cos \frac{4\pi}{3} - 4 \cos \frac{\pi}{3}$? **MP 6**

- A $-3\sqrt{3}$
- B -4
- C -3
- D -1
- E 1

44. What is the period of the function shown in the graph? **MP 2, 6**

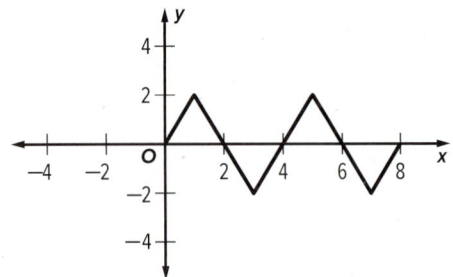

- A 1
- B 2
- C 4
- D 6
- E 8

45. MULTI-STEP Consider $\cos \frac{31\pi}{6}$.

a. Which of the following values are equal to $\cos \frac{31\pi}{6}$? Choose all that apply. **MP 2**

- ☐ A $\cos\left(\frac{\pi}{6} + \frac{24\pi}{6}\right)$
- ☐ B $\cos\left(\frac{7\pi}{6} + \frac{24\pi}{6}\right)$
- ☐ C $\cos\left(\frac{3\pi}{6} + 5\pi\right)$
- ☐ D $\cos \frac{9\pi}{6}$
- ☐ E $\cos \frac{7\pi}{6}$
- ☐ F $\cos\left(\frac{7\pi}{6} + 4\pi\right)$

b. What is the exact value of $\cos \frac{31\pi}{6}$?

- A $-\frac{\sqrt{3}}{2}$
- B $-\frac{1}{2}$
- C $\frac{-\sqrt{3}+1}{2}$
- D $\frac{1}{2}$
- E $\frac{2\sqrt{3}}{2}$

c. Which facts did you need to use to solve this trigonometry problem? Choose all that apply.

- ☐ A The sine function repeats every 2π radians.
- ☐ B The cosine function repeats every 2π radians.
- ☐ C The sine function values are shifted 0.5π to the right of the cosine function values.
- ☐ D $\cos \theta = \cos(\theta + 2\pi)$
- ☐ E $\sin \theta = \sin(\theta + 2\pi)$
- ☐ F The standard amplitude of the sine function is 1.
- ☐ G The standard amplitude of the cosine function is 1.

46. The period of a function $f(x)$ is 3 and the period of a function $g(x)$ is 4. What is the fundamental period of the function $h(x) = f(x) + g(x)$? **MP 1**

LESSON 5
Graphing Trigonometric Functions

:'Then
- You examined periodic functions.

:'Now
1. Describe and graph the sine, cosine, and tangent functions.
2. Describe and graph other trigonometric functions.

:'Why?
- Visible light waves have different wavelengths or periods. Red has the longest wavelength and violet has the shortest wavelength.

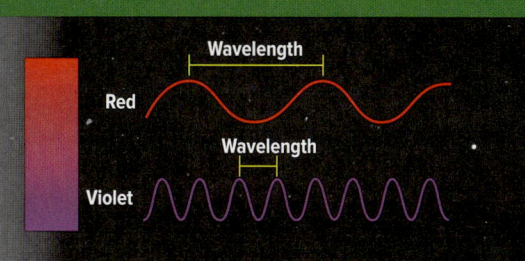

 New Vocabulary
amplitude
frequency

 Mathematical Practices
1 Make sense of problems and persevere in solving them.

1 Sine, Cosine, and Tangent Functions Trigonometric functions can also be graphed on the coordinate plane. Recall that graphs of periodic functions have repeating patterns, or *cycles*. The horizontal length of each cycle is the *period*. The **amplitude** of the graph of a sine or cosine function equals half the difference between the maximum and minimum values of the function.

Key Concept Sine and Cosine Functions

Parent Function	$y = \sin \theta$	$y = \cos \theta$
Graph		
Domain	{all real numbers}	{all real numbers}
Range	$\{y \mid -1 \leq y \leq 1\}$	$\{y \mid -1 \leq y \leq 1\}$
Amplitude	1	1
Period	360°	360°

As with other functions, trigonometric functions can be transformed. For the graphs of $y = a \sin b\theta$ and $y = a \cos b\theta$, the amplitude $= |a|$ and the period $= \dfrac{360°}{|b|}$.

Example 1   Find Amplitude and Period

Find the amplitude and period of
$y = 4 \cos 3\theta$.

amplitude: $|a| = |4|$ or 4

period: $\dfrac{360°}{|b|} = \dfrac{360°}{|3|}$ or 120°

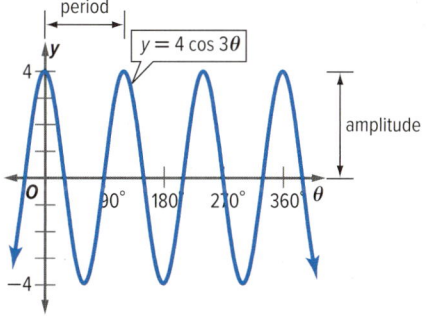

▶ **Guided Practice**

Find the amplitude and period of each function.

1A. $y = \cos \dfrac{1}{2}\theta$

1B. $y = 3 \sin 5\theta$

Study Tip

Periods In $y = a \sin b\theta$ and $y = a \cos b\theta$, b represents the number of cycles in 360°. In Example 1, the 3 in $y = 4 \cos 3\theta$ indicates that there are three cycles in 360°. So, there is one cycle in 120°.

Use the graphs of the parent functions to graph $y = a \sin b\theta$ and $y = a \cos b\theta$. Then use the amplitude and period to draw the appropriate sine and cosine curves. You can also use θ-intercepts to help you graph the functions.

The θ-intercepts of $y = a \sin b\theta$ and $y = a \cos b\theta$ in one cycle are as follows.

$y = a \sin b\theta$	$y = a \cos b\theta$
$(0, 0), \left(\frac{1}{2} \cdot \frac{360°}{b}, 0\right), \left(\frac{360°}{b}, 0\right)$	$\left(\frac{1}{4} \cdot \frac{360°}{b}, 0\right), \left(\frac{3}{4} \cdot \frac{360°}{b}, 0\right)$

Example 2 Graph Sine and Cosine Functions

Graph each function.

a. $y = 2 \sin \theta$

Find the amplitude, the period, and the x-intercepts: $a = 2$ and $b = 1$.

amplitude: $|a| = |2|$ or 2 → The graph is stretched vertically so that the maximum value is 2 and the minimum value is -2.

period: $\dfrac{360°}{|b|} = \dfrac{360°}{|1|}$ or 360° → One cycle has a length of 360°.

x-intercepts: $(0, 0)$

$\left(\dfrac{1}{2} \cdot \dfrac{360°}{b}, 0\right) = (180°, 0)$

$\left(\dfrac{360°}{b}, 0\right) = (360°, 0)$

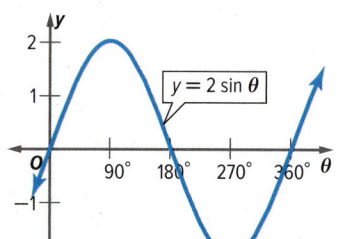

Study Tip

Amplitude The graphs of $y = a \sin b\theta$ and $y = a \cos b\theta$ with amplitude of $|a|$ have maxima at $y = a$ and minima at $y = -a$.

b. $y = \cos 4\theta$

amplitude: $|a| = |1|$ or 1

period: $\dfrac{360°}{|b|} = \dfrac{360°}{|4|}$ or 90°

x-intercepts: $\left(\dfrac{1}{4} \cdot \dfrac{360°}{b}, 0\right) = (22.5°, 0)$

$\left(\dfrac{3}{4} \cdot \dfrac{360°}{b}, 0\right) = (67.5°, 0)$

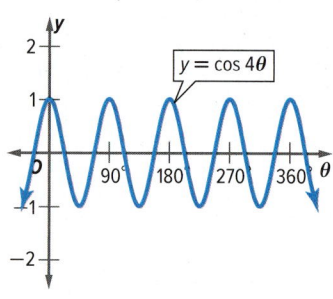

Go Online!

Investigate how changing the values in the equation of a trigonometric function affects the graph by using the **eToolkit** in ConnectED.

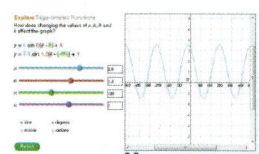

▶ **Guided Practice**

2A. $y = 3 \cos \theta$ **2B.** $y = \dfrac{1}{2} \sin 2\theta$

Trigonometric functions are useful for modeling real-world periodic motion such as electromagnetic waves or sound waves. Often these waves are described using *frequency*. **Frequency** is the number of cycles in a given unit of time.

The frequency of the graph of a function is the reciprocal of the period of the function. So, if the period of a function is $\dfrac{1}{100}$ second, then the frequency is 100 cycles per second.

Real-World Link
Elephants are able to hear sound coming from up to 5 miles away. Humans can hear sounds with frequencies between 20 Hz and 20,000 Hz.
Source: School for Champions

Study Tip
MP Reasoning Note that the amplitude affects the graph along the vertical axis, and the period affects it along the horizontal axis.

Real-World Example 3 Model Periodic Situations

SOUND Sound that has a frequency below the human range is known as *infrasound*. Elephants can hear sounds in the infrasound range, with frequencies as low as 5 hertz (Hz), or 5 cycles per second.

a. Find the period of the function that models the sound waves.

There are 5 cycles per second, and the period is the time it takes for one cycle. So, the period is $\frac{1}{5}$ or 0.2 second.

b. Let the amplitude equal 1 unit. Write a sine equation to represent the sound wave y as a function of time t. Then graph the equation.

$\text{period} = \dfrac{2\pi}{|b|}$ Write the relationship between the period and b.

$0.2 = \dfrac{2\pi}{|b|}$ Substitution

$0.2|b| = 2\pi$ Multiply each side by $|b|$.

$b = 10\pi$ Multiply each side by 5; b is positive.

$y = a \sin b\theta$ Write the general equation for the sine function.

$y = 1 \sin 10\pi t$ $a = 1$, $b = 10\pi$, and $\theta = t$

$y = \sin 10\pi t$ Simplify.

Guided Practice

3. SOUND Humans can hear sounds with frequencies as low as 20 hertz.

A. Find the period of the function.

B. Let the amplitude equal 1 unit. Write a cosine equation to model the sound waves. Then graph the equation.

The tangent function is one of the trigonometric functions whose graphs have asymptotes.

Key Concept Tangent Functions

Parent Function	$y = \tan \theta$
Domain	$\{\theta \mid \theta \neq 90 + 180n,$ n is an integer$\}$
Range	{all real numbers}
Amplitude	no amplitude
Period	180°
θ intercepts in one cycle	$180n$

Graph

For the graph of $y = a \tan b\theta$, the period is $\dfrac{180°}{|b|}$, there is no amplitude, and the asymptotes are odd multiples of $\dfrac{180°}{2|b|}$.

Study Tip

Tangent The tangent function does not have an amplitude because it has no maximum or minimum values.

Example 4 Graph Tangent Functions

Find the period of $y = \tan 2\theta$. Then graph the function.

period: $\dfrac{180°}{|b|} = \dfrac{180°}{|2|}$ or $90°$

asymptotes: $\dfrac{180°}{2|b|} = \dfrac{180°}{2|2|}$ or $45°$

Sketch asymptotes at $-1 \cdot 45°$ or $-45°$, $1 \cdot 45°$ or $45°$, $3 \cdot 45°$ or $135°$, and so on.

Use $y = \tan \theta$, but draw one cycle every $90°$.

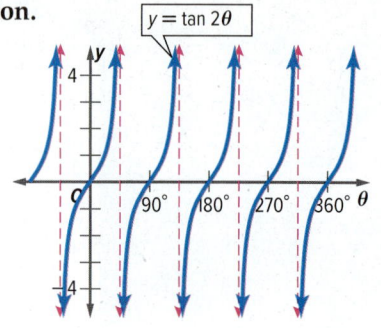

▶ **Guided Practice**

4. Find the period of $y = \dfrac{1}{2} \tan \theta$. Then graph the function.

2 Graphs of Other Trigonometric Functions
The graphs of the cosecant, secant, and cotangent functions are related to the graphs of the sine, cosine, and tangent functions.

Key Concept Cosecant, Secant, and Cotangent Functions

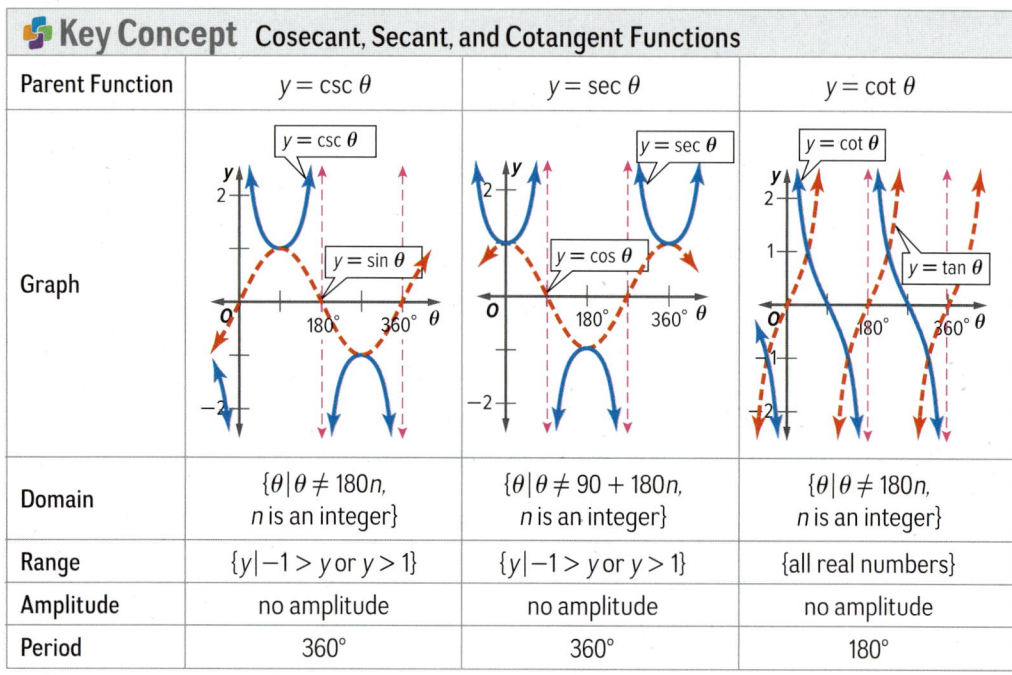

Parent Function	$y = \csc \theta$	$y = \sec \theta$	$y = \cot \theta$
Graph			
Domain	$\{\theta \mid \theta \neq 180n,$ n is an integer$\}$	$\{\theta \mid \theta \neq 90 + 180n,$ n is an integer$\}$	$\{\theta \mid \theta \neq 180n,$ n is an integer$\}$
Range	$\{y \mid -1 > y$ or $y > 1\}$	$\{y \mid -1 > y$ or $y > 1\}$	{all real numbers}
Amplitude	no amplitude	no amplitude	no amplitude
Period	$360°$	$360°$	$180°$

Study Tip

Reciprocal Functions You can use the graphs of $y = \sin \theta$, $y = \cos \theta$, and $y = \tan \theta$ to graph the reciprocal functions, but these graphs are not part of the graphs of the cosecant, secant, and cotangent functions.

Example 5 Graph Other Trigonometric Functions

Find the period of $y = 2 \sec \theta$. Then graph the function.

Because $2 \sec \theta$ is a reciprocal of $2 \cos \theta$, the graphs have the same period, $360°$. The vertical asymptotes occur at the points where $2 \cos \theta = 0$. So, the asymptotes are at $\theta = 90°$ and $\theta = 270°$.

Sketch $y = 2 \cos \theta$ and use it to graph $y = 2 \sec \theta$.

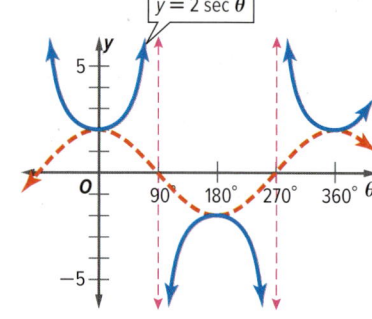

▶ **Guided Practice**

5. Find the period of $y = \csc 2\theta$. Then graph the function.

Check Your Understanding

= Step-by-Step Solutions begin on page R11.

Go Online! for a Self-Check Quiz

Examples 1–2 Find the amplitude and period of each function. Then graph the function.

1. $y = 4 \sin \theta$
2. $y = \sin 3\theta$
3. $y = \cos 2\theta$
4. $y = \frac{1}{2} \cos 3\theta$

Example 3

5. **SPIDERS** When an insect gets caught in a spider web, the web vibrates with a frequency of 14 hertz.

   a. Find the period of the function.

   b. Let the amplitude equal 1 unit. Write a sine equation to represent the vibration of the web y as a function of time t. Then graph the equation.

Examples 4–5 Find the period of each function. Then graph the function.

6. $y = 3 \tan \theta$
7. $y = 2 \csc \theta$
8. $y = \cot 2\theta$

Practice and Problem Solving

Extra Practice is on page R9.

Examples 1–2 Find the amplitude and period of each function. Then graph the function.

9. $y = 2 \cos \theta$
10. $y = 3 \sin \theta$
11. $y = \sin 2\theta$
12. $y = \cos 3\theta$
13. $y = \cos \frac{1}{2}\theta$
14. $y = \sin 4\theta$
15. $y = \frac{3}{4} \cos \theta$
16. $y = \frac{3}{2} \sin \theta$
17. $y = \frac{1}{2} \sin 2\theta$
18. $y = 4 \cos 2\theta$
19. $y = 3 \cos 2\theta$
20. $y = 5 \sin \frac{2}{3}\theta$

Example 3

21. **REASONING** A boat on Lake Klondike bobs up and down with the waves. The difference between the lowest and highest points of the boat is 8 inches. The boat is at *equilibrium* when it is halfway between the lowest and highest points. Each cycle of the periodic motion lasts 3 seconds.

    a. Write an equation for the motion of the boat. Let h represent the height in inches and let t represent the time in seconds. Assume that the boat is at equilibrium at $t = 0$ seconds.

    b. Draw a graph showing the height of the boat as a function of time.

22. **ELECTRICITY** The voltage supplied by an electrical outlet is a periodic function that *oscillates*, or goes up and down, between −165 volts and 165 volts with a frequency of 50 cycles per second.

    a. Write an equation for the voltage V as a function of time t. Assume that at $t = 0$ seconds, the current is 165 volts.

    b. Graph the function.

Examples 4–5 Find the period of each function. Then graph the function.

23. $y = \tan \frac{1}{2}\theta$
24. $y = 3 \sec \theta$
25. $y = 2 \cot \theta$
26. $y = \csc \frac{1}{2}\theta$
27. $y = 2 \tan \theta$
28. $y = \sec \frac{1}{3}\theta$

29. **EARTHQUAKES** A seismic station detects an earthquake wave that has a frequency of 0.5 hertz and an amplitude of 1 meter.

   a. Write an equation involving sine to represent the height of the wave h as a function of time t. Assume that the equilibrium point of the wave, $h = 0$, is halfway between the lowest and highest points.

   b. Graph the function. Then determine the height of the wave after 20.5 seconds.

30. **PERSEVERANCE** An object is attached to a spring as shown at the right. It oscillates according to the equation $y = 20 \cos \pi t$, where y is the distance in centimeters from its equilibrium position at time t.

   a. Describe the motion of the object by finding the following: the amplitude in centimeters, the frequency in vibrations per second, and the period in seconds.

   b. Find the distance of the object from its equilibrium to its position at $t = \frac{1}{4}$ second.

   c. The equation $v = (-20 \text{ cm})(\pi \text{ rad/s}) \cdot \sin(\pi \text{ rad/s} \cdot t)$ represents the velocity v of the object at time t. Find the velocity at $t = \frac{1}{4}$ second.

31. **PIANOS** A piano string vibrates at a frequency of 130 hertz.

   a. Write and graph an equation using cosine to model the vibration of the string y as a function of time t. Let the amplitude equal 1 unit.

   b. Suppose the frequency of the vibration doubles. Do the amplitude and period increase, decrease, or remain the same? Explain.

Find the amplitude, if it exists, and period of each function. Then graph the function.

32. $y = 3 \sin \frac{2}{3}\theta$

33. $y = \frac{1}{2} \cos \frac{3}{4}\theta$

34. $y = 2 \tan \frac{1}{2}\theta$

35. $y = 2 \sec \frac{4}{5}\theta$

36. $y = 5 \csc 3\theta$

37. $y = 2 \cot 6\theta$

Identify the period of the graph and write an equation for each function.

38.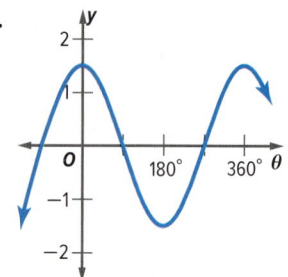

39.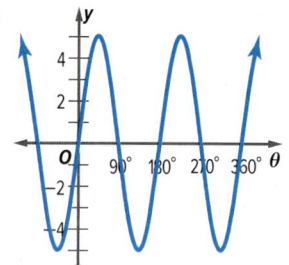

40.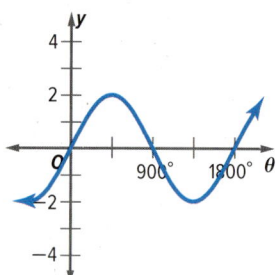

H.O.T. Problems   Use Higher-Order Thinking Skills

41. **CHALLENGE** Describe the domain and range of $y = a \cos \theta$ and $y = a \sec \theta$, where a is any positive real number.

42. **REASONING** Compare and contrast the graphs of $y = \frac{1}{2} \sin \theta$ and $y = \sin \frac{1}{2}\theta$.

43. **OPEN-ENDED** Write a trigonometric function that has an amplitude of 3 and a period of 180°. Then graph the function.

44. **WRITING IN MATH** How can you use the characteristics of a trigonometric function to sketch its graph?

Preparing for Assessment

45. Which of the following equations is best represented by the graph shown below? **MP 2, 7**

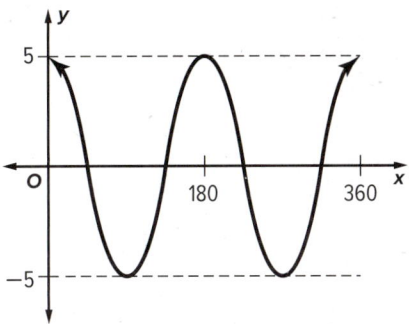

- A $y = 2 \cos 5\theta$
- B $y = -5 \cos \theta$
- C $y = -\cos 2\theta$
- D $y = 5 \cos 2\theta$
- E $y = 10 \cos \frac{1}{2}\theta$

46. For the graph of the trigonometric function $y = 4 \tan 4\theta$, which of the following is (are) true? **MP 2, 7**

  I. The amplitude is 4.

  II. The period is 45°.

  III. There are no asymptotes.

- A I only
- B II only
- C I and II only
- D I, II, and III
- E II and III only

47. What is the amplitude and period of $y = 6 \cos 2\theta$? **MP 2, 7**

- A amplitude 6; period 720°
- B amplitude 6; period 360°
- C amplitude 6; period 180°

48. MULTI-STEP Use this graph of a trigonometric function. **MP 1, 2, 8**

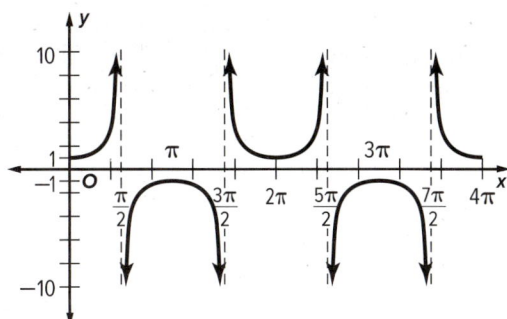

a. At which values of x does the function have a vertical asymptote? Choose all that apply.
- A π
- B $\frac{\pi}{2}$
- C 2π
- D $\frac{3\pi}{2}$
- E 3π
- F $\frac{5\pi}{2}$

b. What is the period of this function?
- A $\frac{5\pi}{2}$
- B 2π
- C $\frac{\pi}{2}$
- D π

c. Which function does this graph represent?
- A $y = \sin \theta$
- B $y = \cos \theta$
- C $y = \cot \theta$
- D $y = \tan \theta$
- E $y = \sec \theta$

d. What is the reciprocal function of this graph?
- A $y = \sin \theta$
- B $y = \cos \theta$
- C $y = \cot \theta$
- D $y = \tan \theta$
- E $y = \sec \theta$

EXTEND 9-5

Graphing Technology Lab
Trigonometric Graphs

You can use a TI-83/84 Plus graphing calculator to explore transformations of the graphs of trigonometric functions.

Mathematical Practices
 5 Use appropriate tools strategically.

Activity 1 k in $y = \sin \theta + k$

Work cooperatively. Graph $y = \sin \theta$, $y = \sin \theta + 2$, and $y = \sin \theta - 3$ on the same coordinate plane. Describe any similarities and differences among the graphs.

Set the viewing window to match the window shown at the right. Let Y1 = sin θ, Y2 = sin θ + 2, and Y3 = sin θ − 3.

KEYSTROKES: Y= SIN X,T,θ,n) ENTER
SIN X,T,θ,n) + 2 ENTER
SIN X,T,θ,n) − 3 GRAPH

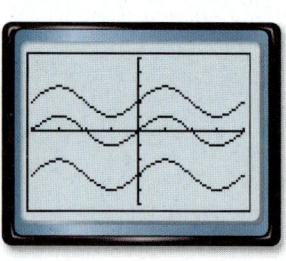

[−360, 360] scl: 90 by [−5, 5] scl: 1

The graphs have the same shape, but different vertical positions.

Activity 2 h in $y = \sin(\theta - h)$

Work cooperatively. Graph $y = \sin \theta$, $y = \sin(\theta + 45°)$, and $y = \sin(\theta - 90°)$ on the same coordinate plane. Describe any similarities and differences among the graphs.

Let Y1 = sin θ, Y2 = sin (θ + 45), and Y3 = sin (θ − 90).
Be sure to clear the entries from Activity 1.

KEYSTROKES: Y= SIN X,T,θ,n) ENTER
SIN X,T,θ,n + 45) ENTER
SIN X,T,θ,n − 90) GRAPH

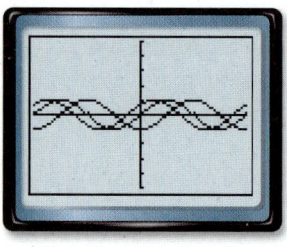

[−360, 360] scl: 90 by [−5, 5] scl: 1

The graphs have the same shape, but different horizontal positions.

Model and Analyze

Work cooperatively. Repeat the activities for the cosine and tangent functions.

1. What are the domain and range of the functions in Activities 1 and 2?
2. What is the effect of adding a constant to a trigonometric function?
3. What is the effect of adding a constant to θ in a trigonometric function?

Repeat the activities for each of the following. Describe the relationship between each pair of graphs.

4. $y = \sin \theta + 4$
   $y = \sin(2\theta) + 4$

5. $y = \cos\left(\frac{1}{2}\theta\right)$
   $y = \cos \frac{1}{2}(\theta + 45°)$

6. $y = 2 \sin \theta$
   $y = 2 \sin \theta - 1$

7. $y = \cos \theta - 3$
   $y = \cos(\theta - 90°) - 3$

8. Write a general equation for the sine, cosine, and tangent functions after changes in amplitude a, period b, horizontal position h, and vertical position k.

LESSON 6
Translations of Trigonometric Graphs

::Then
- You translated exponential functions.

::Now
1. Graph horizontal translations of trigonometric graphs and find phase shifts.
2. Graph vertical translations of trigonometric graphs.

::Why?
- The graphs at the right represent the waves in a bay during high and low tides. Notice that the shape of the waves does not change.

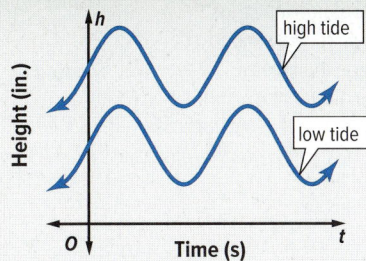

 New Vocabulary
phase shift
vertical shift
midline

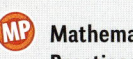 **Mathematical Practices**
4 Model with mathematics.

1 Horizontal Translations Recall that a *translation* occurs when a figure is moved from one location to another on the coordinate plane without changing its orientation. A horizontal translation of a periodic function is called a **phase shift**.

Key Concept Phase Shift

Words The phase shift of the functions $y = a \sin b(\theta - h)$, $y = a \cos b(\theta - h)$, and $y = a \tan b(\theta - h)$ is h, where $b > 0$.

Models

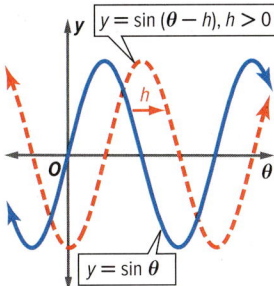

If $h > 0$, the shift is h units to the right.

If $h < 0$, the shift is $|h|$ units to the left.

Examples $y = \cos(\theta - 90°)$ The phase shift is 90° to the right.
$y = \tan(\theta + 30°)$ The phase shift is 30° to the left.

The secant, cosecant, and cotangent can be graphed using the same rules.

Example 1 Graph Phase Shifts

State the amplitude, period, and phase shift for $y = \sin(\theta - 90°)$. Then graph the function.

amplitude: $a = 1$

period: $\dfrac{360°}{|b|} = \dfrac{360°}{1}$ or $360°$

phase shift: $h = 90°$

Graph $y = \sin \theta$ shifted 90° to the right.

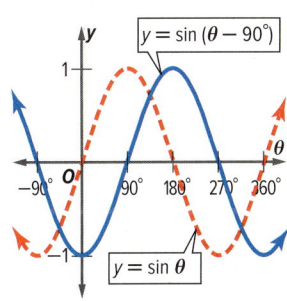

Guided Practice

1. State the amplitude, period, and phase shift for $y = 2 \cos(\theta + 45°)$. Then graph the function.

connectED.mcgraw-hill.com   635

2 Vertical Translations

Recall that the graph of $y = x^2 + 5$ is the graph of the parent function $y = x^2$ shifted up 5 units. Similarly, graphs of trigonometric functions can be translated vertically through a **vertical shift**.

Study Tip

Notation Note that $\sin(\theta + x) \neq \sin\theta + x$. The first expression indicates a phase shift. The second expression indicates a vertical shift.

Key Concept · Vertical Shift

Words The vertical shift of the functions $y = a\sin b\theta + k$, $y = a\cos b\theta + k$, and $y = a\tan b\theta + k$ is k.

Models

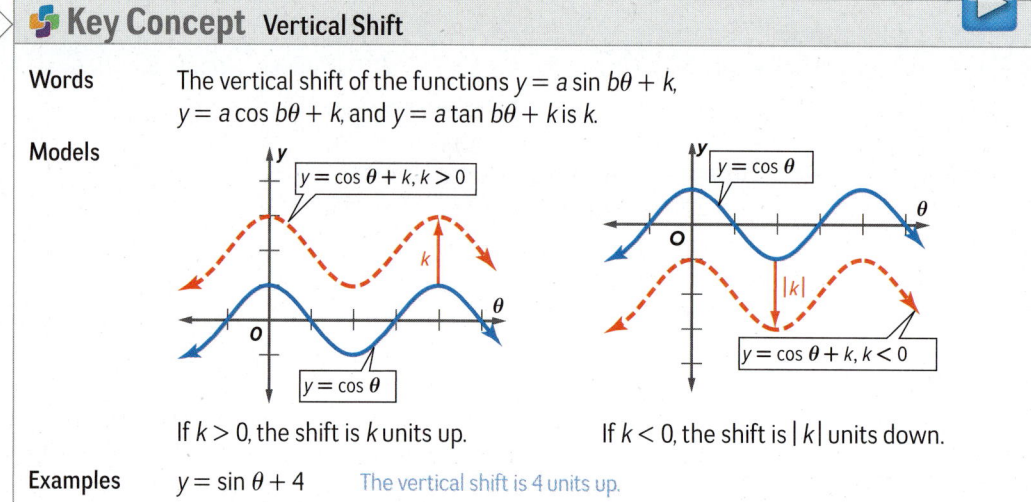

If $k > 0$, the shift is k units up. If $k < 0$, the shift is $|k|$ units down.

Examples $y = \sin\theta + 4$ The vertical shift is 4 units up.
$y = \tan\theta - 3$ The vertical shift is 3 units down.

The secant, cosecant, and cotangent can be graphed using the same rules.

When a trigonometric function is shifted vertically k units, the line $y = k$ is the new horizontal axis about which the graph oscillates. This line is called the **midline**, and it can be used to help draw vertical translations.

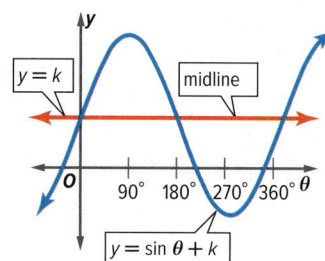

Study Tip

MP Tools It may be helpful to first graph the parent function in one color. Next, apply the vertical shift and graph the function in another color. Then apply the change in amplitude and graph the function in the final color.

Example 2   Graph Vertical Translations

State the amplitude, period, vertical shift, and equation of the midline for $y = \frac{1}{2}\cos\theta - 2$. Then graph the function.

amplitude: $|a| = \frac{1}{2}$

period: $\frac{2\pi}{|b|} = \frac{2\pi}{|1|}$ or 2π

vertical shift: $k = -2$

midline: $y = -2$

To graph $y = \frac{1}{2}\cos\theta - 2$, first draw the midline. Then use it to graph $y = \frac{1}{2}\cos\theta$ shifted 2 units down.

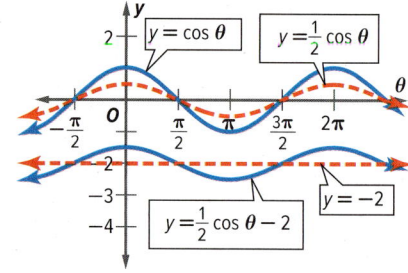

Guided Practice

2. State the amplitude, period, vertical shift, and equation of the midline for $y = \tan\theta + 3$. Then graph the function.

You can use the following steps to graph trigonometric functions involving phase shifts and vertical shifts.

> **Key Concept** Graph Trigonometric Functions
>
> $$y = a \sin b(\theta - h) + k$$
>
> where a is the amplitude, b relates to the period, h is the phase shift, and k is the vertical shift.
>
> **Step 1** Determine the vertical shift, and graph the midline.
>
> **Step 2** Determine the amplitude, if it exists. Use dashed lines to indicate the maximum and minimum values of the function.
>
> **Step 3** Determine the period of the function, and graph the appropriate function.
>
> **Step 4** Determine the phase shift, and translate the graph accordingly.

Go Online!

Follow along with your graphing calculator as you watch a **Personal Tutor** graph a translated trigonometric function.

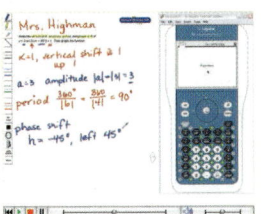

Example 3 Graph Transformations

State the amplitude, period, phase shift, and vertical shift for $y = 3 \sin \frac{2}{3}(\theta - \pi) + 4$. Then graph the function.

amplitude: $|a| = 3$

period: $\frac{2\pi}{|b|} = \frac{2\pi}{\left|\frac{2}{3}\right|}$ or 3π The period indicates that the graph will be stretched.

phase shift: $h = \pi$ The graph will shift π to the right.
vertical shift: $k = 4$ The graph will shift 4 units up.
midline: $y = 4$ The graph will oscillate around the line $y = 4$.

Step 1 Graph the midline.

Step 2 Because the amplitude is 3, draw dashed lines 3 units above and 3 units below the midline.

Step 3 Graph $y = 3 \sin \frac{2}{3}\theta + 4$ using the midline as a reference.

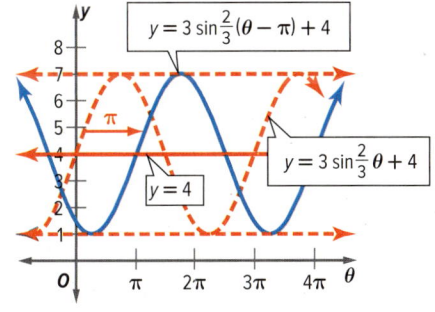

Step 4 Shift the graph π units to the right.

CHECK You can check the accuracy of your transformation by evaluating the function for various values of θ and confirming their location on the graph.

Study Tip

Verifying a Graph After drawing the graph of a trigonometric function, select values of θ and evaluate them in the equation to verify your graph.

▶ **Guided Practice**

3. State the amplitude, period, phase shift, and vertical shift for $y = 2 \cos \frac{1}{2}\left(\theta + \frac{\pi}{2}\right) - 2$. Then graph the function.

The sine wave occurs often in physics, signal processing, music, electrical engineering, and many other fields.

Real-World Example 4 Represent Periodic Functions

WAVE POOL The height of water in a wave pool oscillates between a maximum of 13 feet and a minimum of 5 feet. The wave generator pumps 6 waves per minute. Write a sine function that represents the height of the water at time *t* seconds. Then graph the function.

Step 1 Write the equation for the midline, and determine the vertical shift.

$y = \dfrac{13 + 5}{2}$ or 9   The midline lies halfway between the maximum and minimum values.

Because the midline is $y = 9$, the vertical shift is $k = 9$.

Step 2 Find the amplitude.

$|a| = |13 - 9|$ or 4   Find the difference between the midline value and the maximum value.

So, $a = 4$.

Step 3 Find the period.

Because there are 6 waves per minute, there is 1 wave every 10 seconds. So, the period is 10 seconds.

$10 = \dfrac{2\pi}{|b|}$ Period $= \dfrac{2\pi}{|b|}$

$|b| = \dfrac{2\pi}{10}$ Solve for $|b|$.

$b = \pm\dfrac{\pi}{5}$ Simplify.

Step 4 Write an equation for the function.

$h = a \sin b(t - h) + k$ Write the equation for sine relating height *h* and time *t*.

$= 4 \sin \dfrac{\pi}{5}(t - 0) + 9$ Substitution: $a = 4, b = \dfrac{\pi}{5}, h = 0, k = 9$

$= 4 \sin \dfrac{\pi}{5}t + 9$ Simplify.

Then graph the function.

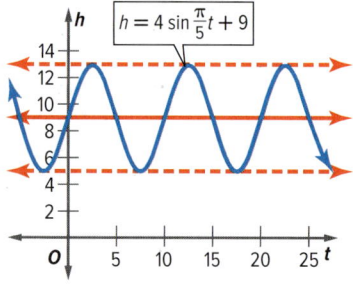

Real-World Link

Waves at the Surf and Swim in Garland, Texas reach an apex of 3 feet.

Source: Surf and Swim

Watch Out

Parent Functions Often the graph of a trigonometric function can be represented by more than one equation. For example, the graphs of $y = \cos \theta$ and $y = \sin (\theta + 90°)$ are the same.

▶ **Guided Practice**

4. **WAVE POOL** The height of water in a wave pool oscillates between a maximum of 14 feet and a minimum of 6 feet. The wave generator pumps 5 waves per minute. Write a cosine function that represents the height of water at time *t* seconds. Then graph the function.

Check Your Understanding

= Step-by-Step Solutions begin on page R11.

Example 1 State the amplitude, period, and phase shift for each function. Then graph the function.

1. $y = \sin(\theta - 180°)$
2. $y = \tan\left(\theta - \frac{\pi}{4}\right)$
3. $y = \sin\left(\theta - \frac{\pi}{2}\right)$
4. $y = \frac{1}{2}\cos(\theta + 90°)$

Example 2 State the amplitude, period, vertical shift, and equation of the midline for each function. Then graph the function.

5. $y = \cos\theta + 4$
6. $y = \sin\theta - 2$
7. $y = \frac{1}{2}\tan\theta + 1$
8. $y = \sec\theta - 5$

Example 3 **MP REGULARITY** State the amplitude, period, phase shift, and vertical shift for each function. Then graph the function.

9. $y = 2\sin(\theta + 45°) + 1$
10. $y = \cos 3(\theta - \pi) - 4$
11. $y = \frac{1}{4}\tan 2(\theta + 30°) + 3$
12. $y = 4\sin\frac{1}{2}\left(\theta - \frac{\pi}{2}\right) + 5$

Example 4

13. **EXERCISE** While doing some moderate physical activity, a person's blood pressure oscillates between a maximum of 130 and a minimum of 90. The person's heart rate is 90 beats per minute. Write a sine function that represents the person's blood pressure P at time t seconds. Then graph the function.

Practice and Problem Solving

Extra Practice is on page R9.

Example 1 State the amplitude, period, and phase shift for each function. Then graph the function.

14. $y = \cos(\theta + 180°)$
15. $y = \tan(\theta - 90°)$
16. $y = \sin(\theta + \pi)$
17. $y = 2\sin\left(\theta + \frac{\pi}{2}\right)$
18. $y = \tan\frac{1}{2}(\theta + 30°)$
19. $y = 3\cos\left(\theta - \frac{\pi}{3}\right)$

Example 2 State the amplitude, period, vertical shift, and equation of the midline for each function. Then graph the function.

20. $y = \cos\theta + 3$
21. $y = \tan\theta - 1$
22. $y = \tan\theta + \frac{1}{2}$
23. $y = 2\cos\theta - 5$
24. $y = 2\sin\theta - 4$
25. $y = \frac{1}{3}\sin\theta + 7$

Example 3 State the amplitude, period, phase shift, and vertical shift for each function. Then graph the function.

26. $y = 4\sin(\theta - 60°) - 1$
27. $y = \cos\frac{1}{2}(\theta - 90°) + 2$
28. $y = \tan(\theta + 30°) - 2$
29. $y = 2\tan 2\left(\theta + \frac{\pi}{4}\right) - 5$
30. $y = \frac{1}{2}\sin\left(\theta - \frac{\pi}{2}\right) + 4$
31. $y = \cos 3(\theta - 45°) + \frac{1}{2}$
32. $y = 3 + 5\sin 2(\theta - \pi)$
33. $y = -2 + 3\sin\frac{1}{3}\left(\theta - \frac{\pi}{2}\right)$

Example 4

34. **TIDES** The height of the water in a harbor rose to a maximum height of 15 feet at 6:00 P.M. and then dropped to a minimum level of 3 feet by 3:00 A.M. The water level can be modeled by the sine function. Write an equation that represents the height h of the water t hours after noon on the first day.

35. LAKES A buoy marking the swimming area in a lake oscillates each time a speed boat goes by. Its distance d in feet from the bottom of the lake is given by $d = 1.8 \sin \frac{3\pi}{4} t + 12$, where t is the time in seconds. Graph the function. Describe the minimum and maximum distances of the buoy from the bottom of the lake when a boat passes by.

36. FERRIS WHEEL Suppose a Ferris wheel has a diameter of approximately 100 feet and makes one revolution every 24 seconds. Write an equation for the height of a car h as a function of time t in seconds, assuming at $t = 0$, the car is at its lowest point, 5 feet above ground.

Write an equation for each translation.

37. $y = \sin x$, 4 units to the right and 3 units up

38. $y = \cos x$, 5 units to the left and 2 units down

39. $y = \tan x$, π units to the right and 2.5 units up

40. JUMP ROPE The graph at the right approximates the height of a jump rope h in inches as a function of time t in seconds. A maximum point on the graph is (1.25, 68), and a minimum point is (2.75, 2).

  a. Describe what the maximum and minimum points mean in the context of the situation.

  b. What is the equation for the midline, the amplitude, and the period of the function?

  c. Write an equation for the function.

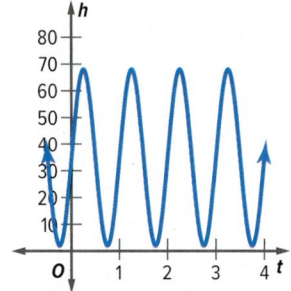

41. MULTI-STEP You are designing a carousel with 32 horses. You want only 4 horses to be at their maximum point at any particular time. You want the carousel to do a complete revolution every 20 seconds and the range of motion of each horse to be 5 feet.

  a. Use trigonometry to model the movement of the carousel horses.

  b. Describe each part of your trigonometric function(s).

  c. What assumptions did you make in designing your carousel?

  d. Suppose you also want each horse to be at a different height at any particular point for each rotation so parents will be able to take pictures of their children at different heights without having to move. How do you accomplish this?

42. MP REASONING During one month, the outside temperature fluctuates between 40°F and 50°F. A cosine curve approximates the change in temperature, with a high of 50°F being reached every four days.

  a. Describe the amplitude, period, and midline of the function that approximates the temperature y on day d.

  b. Write a cosine function to estimate the temperature y on day d.

  c. Sketch a graph of the function.

  d. Estimate the temperature on the 7th day of the month.

Find a coordinate that represents a maximum for each graph.

43. $y = -2 \cos \left(x - \frac{\pi}{2}\right)$

44. $y = 4 \sin \left(x + \frac{\pi}{3}\right)$

45. $y = 3 \tan \left(x + \frac{\pi}{2}\right) + 2$

46. $y = -3 \sin \left(x - \frac{\pi}{4}\right) - 4$

Compare each pair of graphs.

47. $y = -\cos 3\theta$ and $y = \sin 3(\theta - 90°)$

48. $y = 2 + 0.5 \tan \theta$ and $y = 2 + 0.5 \tan(\theta + \pi)$

49. $y = 2 \sin\left(\theta - \frac{\pi}{6}\right)$ and $y = -2 \sin\left(\theta + \frac{5\pi}{6}\right)$

Identify the period of each function. Then write an equation for the graph using the given trigonometric function.

50. sine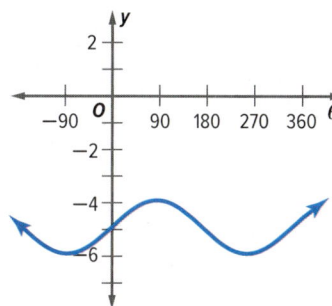

51. cosine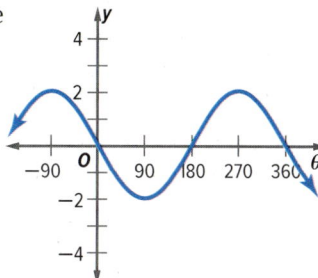

52. cosine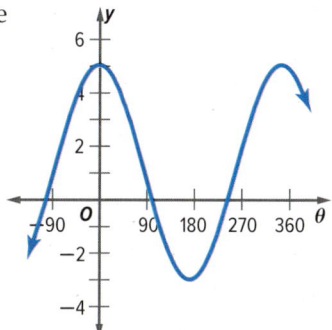

53. sine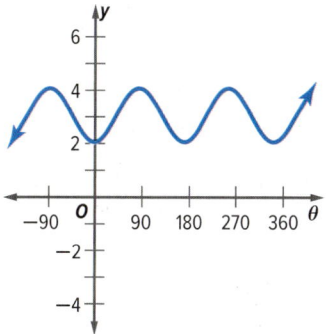

State the period, phase shift, and vertical shift. Then graph the function.

54. $y = \csc(\theta + \pi)$

55. $y = \cot \theta + 6$

56. $y = \cot\left(\theta - \frac{\pi}{6}\right) - 2$

57. $y = \frac{1}{2} \csc 3(\theta - 45°) + 1$

58. $y = 2 \sec \frac{1}{2}(\theta - 90°)$

59. $y = 4 \sec 2\left(\theta + \frac{\pi}{2}\right) - 3$

H.O.T. Problems   Use Higher-Order Thinking Skills

60. **CONSTRUCT ARGUMENTS** If you are given the amplitude and period of a cosine function, is it *sometimes*, *always*, or *never* possible to find the maximum and minimum values of the function? Explain your reasoning.

61. **REASONING** Describe how the graph of $y = 3 \sin 2\theta + 1$ is different from $y = \sin \theta$.

62. **WRITING IN MATH** Describe two different phase shifts that will translate the sine curve onto the cosine curve shown at the right. Then write an equation for the new sine curve using each phase shift.

63. **OPEN-ENDED** Write a periodic function that has an amplitude of 2 and midline at $y = -3$. Then graph the function.

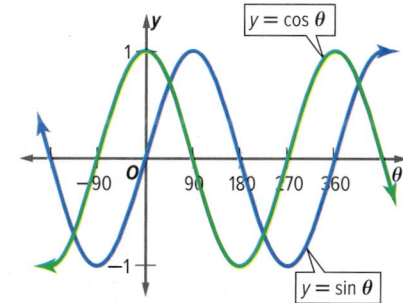

64. **REASONING** How many different sine graphs pass through the origin and all points of the form $(n\pi, 0)$? Explain.

Preparing for Assessment

65. The graph of $y = \cos \theta$ was transformed to produce the graph shown below. **MP 2**

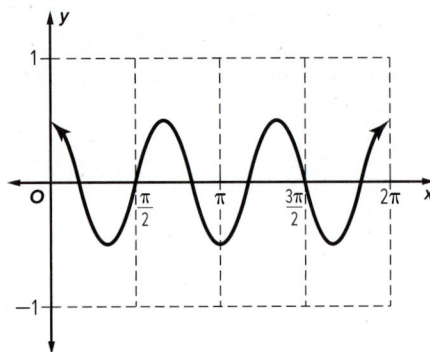

What equation best represents the graph?

- A $y = \frac{1}{2} \cos 3x$
- B $y = \cos 3x$
- C $y = -\cos 3x$
- D $y = 2 \cos 3x$
- E $y = 2 \cos x$

66. Which of the following explains how the graph of $y = \sin \theta$ can be used to produce the graph of $y = \sin 2(\theta - \pi)$? **MP 2**

- A Shift right π units and double amplitude.
- B Translate left π units and shift up 2 units.
- C Move up 2 units and double period.
- D Shift right π units and divide period by 2.
- E Reflect across x-axis and double amplitude.

67. The period of the graph of $y = \cos \theta$ is reduced by 50%, and the amplitude is doubled. Which function represents this change? **MP 2**

- A $y = 0.5 \cos \frac{\theta}{2}$
- B $y = 0.5 \cos 2\theta$
- C $y = 2 \cos 2\theta$
- D $y = 2 \cos \theta + 0.5$
- E $y = 2 \cos \frac{\theta}{2}$

68. MULTI-STEP The graph of $y = \cos \theta$ is to be transformed to produce the graph of $y = 4 \cos \theta + 2$.

a. Which of the following transformation(s) of $y = \cos \theta$ need to occur to produce this new function? **MP 2**

    I. The amplitude will increase to 4.
    II. The graph will be stretched horizontally.
    III. The graph will shift 2 units up.

- A I only
- B II only
- C I and II only
- D I and III only
- E II and III only

b. Which of the transformations would change the period?

c. Give an example of a function which uses all three transformations on the sine function.

69. Look at the list of functions. Write the letters of all the functions that match each description. **MP 2**

A $y = 0.5 \cos(0.5\theta) + 2$

B $y = 2 \cos\theta - 2$

C $y = 0.5 \cos(0.5\theta + 0.5\pi) + 2$

D $y = 2 \cos 2\theta - 2$

E $y = 0.5 \cos(0.5\theta - \pi)$

a. amplitude doubled
b. a period of 2π
c. shifted π to the left
d. shifted two units up
e. a midline on the x-axis

70. The graph of $y = \sin x$ is transformed so that its amplitude becomes 6, its period 3π and its maximum value 4. Find an expression for this new function. **MP 1**

CHAPTER 9
Study Guide and Review

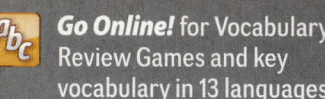
Go Online! for Vocabulary Review Games and key vocabulary in 13 languages

Study Guide

Key Concepts

Right Triangle Trigonometry (Lesson 9-1)

- $\sin\theta = \dfrac{opp}{hyp}$, $\cos\theta = \dfrac{adj}{hyp}$, $\tan\theta = \dfrac{opp}{adj}$,
  $\csc\theta = \dfrac{hyp}{opp}$, $\sec\theta = \dfrac{hyp}{adj}$, $\cot\theta = \dfrac{adj}{opp}$

Angle Measures and Trigonometric Functions of General Angles (Lessons 9-2 and 9-3)

- The measure of an angle is determined by the amount of rotation from the initial side to the terminal side.
- You can find the exact values of the six trigonometric functions of θ, given the coordinates of a point $P(x, y)$ on the terminal side of the angle.

Circular and Periodic Functions (Lesson 9-4)

- If the terminal side of an angle θ in standard position intersects the unit circle at $P(x, y)$, then $\cos\theta = x$ and $\sin\theta = y$.

Graphing Trigonometric Functions (Lesson 9-5)

- For trigonometric functions of the form $y = a\sin b\theta$ and $y = a\cos b\theta$, the amplitude is $|a|$, and the period is $\dfrac{360°}{|b|}$ or $\dfrac{2\pi}{b}$.
- The period of $y = a\tan b\theta$ is $\dfrac{180°}{|b|}$ or $\dfrac{\pi}{|b|}$.

Translations of Trigonometric Graphs (Lesson 9-6)

- The same techniques used to transform the graphs of other functions can be applied to the graphs of trigonometric functions
- A horizontal translation of a trigonometric function is called a phase shift.
- The midline can be used to help draw vertical translations of trigonometric functions.

FOLDABLES Study Organizer

Use your Foldable to review the chapter. Working with a partner can be helpful. Ask for clarification of concepts as needed.

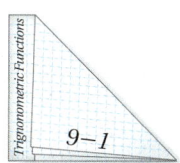

Key Vocabulary

amplitude (p. 627)
angle of depression (p. 598)
angle of elevation (p. 598)
central angle (p. 607)
circular function (p. 620)
cosecant (p. 594)
cosine (p. 594)
cotangent (p. 594)
coterminal angles (p. 605)
cycle (p. 621)
frequency (p. 628)
initial side (p. 604)
midline (p. 636)
period (p. 621)
periodic function (p. 621)

phase shift (p. 635)
quadrantal angle (p. 613)
radian (p. 606)
reference angle (p. 613)
secant (p. 594)
sine (p. 594)
standard position (p. 604)
tangent (p. 594)
terminal side (p. 604)
trigonometric function (p. 594)
trigonometric ratio (p. 594)
trigonometry (p. 794)
unit circle (p. 620)
vertical shift (p. 636)

Vocabulary Check

State whether each sentence is *true* or *false*. If *false*, replace the underlined term to make a true sentence.

1. An angle on the coordinate plane is in standard position if the vertex is at the origin and one ray is on the positive x-axis.

2. Coterminal angles are angles in standard position that have the same terminal side.

3. A horizontal translation of a periodic function is called a phase shift.

4. The cycle of the graph of a sine or cosine function equals half the difference between the maximum and minimum values of the function.

Concept Check

5. When given one trigonometric ratio, explain how to find the other five trigonometric ratios.

6. Explain how points on the unit circle are related to the values of sine and cosine.

CHAPTER 9
Study Guide and Review *Continued*

Lesson-by-Lesson Review

9-1 Trigonometric Functions in Right Triangles

Solve △ABC by using the given measurements. Round measures of sides to the nearest tenth and measures of angles to the nearest degree.

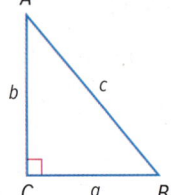

7. $c = 12$, $b = 5$
8. $a = 10$, $B = 55°$
9. $B = 75°$, $b = 15$
10. $B = 45°$, $c = 16$
11. $A = 35°$, $c = 22$
12. $\sin A = \frac{2}{3}$, $a = 6$

13. **TRUCK** The back of a moving truck is 3 feet off of the ground. What length does a ramp off the back of the truck need to be in order for the angle of elevation of the ramp to be 20°?

Example 1

Solve △ABC by using the given measurements. Round measures of sides to the nearest tenth and measures of angles to the nearest degree.

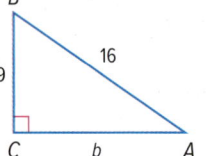

Find b.
$$a^2 + b^2 = c^2$$
$$9^2 + b^2 = 16^2$$
$$b = \sqrt{16^2 - 9^2}$$
$$b \approx 13.2$$

Find A. $\sin A = \frac{9}{16}$

Use a calculator.
To the nearest degree, $A = 34°$.

Find B. $34° + B \approx 90°$
$B \approx 56°$

Therefore, $b \approx 13.2$, $A \approx 34°$, and $B \approx 56°$.

9-2 Angles and Angle Measure

Rewrite each degree measure in radians and each radian measure in degrees.

14. $215°$
15. $\frac{5\pi}{2}$
16. -3π
17. $-315°$

Find one angle with positive measure and one angle with negative measure coterminal with each angle.

18. $265°$
19. $-65°$
20. $\frac{7\pi}{2}$

21. **BICYCLE** A bicycle tire makes 8 revolutions in one second. The tire has a radius of 15 inches. Find the angle θ in radians through which the tire rotates in one second.

Example 2

Rewrite 160° in radians.
$$160° = 160°\left(\frac{\pi \text{ radians}}{180°}\right)$$
$$= \frac{160\pi}{180} \text{ radians or } \frac{8\pi}{9}$$

Example 3

Find one angle with positive measure and one angle with negative measure coterminal with 150°.

positive angle:
$150° + 360° = 510°$ Add 360°.

negative angle:
$150° - 360° = -210°$ Subtract 360°.

9-3 Trigonometric Functions of General Angles

Find the exact value of each trigonometric function.

22. cos 135°
23. tan 150°
24. sin 2π
25. $\cos \frac{3\pi}{2}$

The terminal side of θ in standard position contains each point. Find the exact values of the six trigonometric functions of θ.

26. $P(-4, 3)$
27. $P(5, 12)$
28. $P(16, -12)$

29. **BALL** A ball is thrown off the edge of a building at an angle of 70° and with an initial velocity of 5 meters per second. The equation that represents the horizontal distance of the ball x is $x = v_0 (\cos \theta) t$, where v_0 is the initial velocity, θ is the angle at which it is thrown, and t is the time in seconds. About how far will the ball travel in 10 seconds?

Example 4
Find the exact value of sin 120°.

Because the terminal side of 120° lies in Quadrant II, the reference angle θ′ is 180° − 120° or 60°. The sine function is positive in Quadrant II, so sin 120° = sin 60° or $\frac{\sqrt{3}}{2}$.

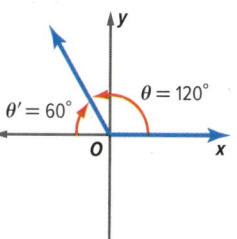

Example 5
The terminal side of θ in standard position contains the point (6, 5). Find the exact values of the six trigonometric functions of θ.

$\sin \theta = \frac{y}{r}$ or $\frac{5\sqrt{61}}{61}$ $\csc \theta = \frac{r}{y}$ or $\frac{\sqrt{61}}{5}$

$\cos \theta = \frac{x}{r}$ or $\frac{6\sqrt{61}}{61}$ $\sec \theta = \frac{r}{x}$ or $\frac{\sqrt{61}}{6}$

$\tan \theta = \frac{y}{x}$ or $\frac{5}{6}$ $\cot \theta = \frac{x}{y}$ or $\frac{6}{5}$

9-4 Circular and Periodic Functions

Find the exact value of each function.

30. cos (−210°)
31. (cos 45°)(cos 210°)
32. $\sin \left(-\frac{7\pi}{4}\right)$
33. $\left(\cos \frac{\pi}{2}\right)\left(\sin \frac{\pi}{2}\right)$

34. Determine the period of the function.

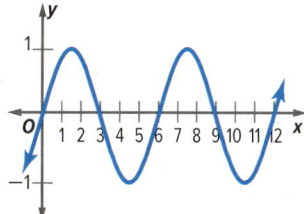

35. A wheel with a diameter of 18 inches completes 4 revolutions in 1 minute. What is the period of the function that describes the height of one spot on the outside edge of the wheel as a function of time?

Example 6
Find the exact value of sin 510°.

sin 510° = sin (360° + 150°)
= sin 150°
= $\frac{1}{2}$

Example 7
Determine the period of the function below.

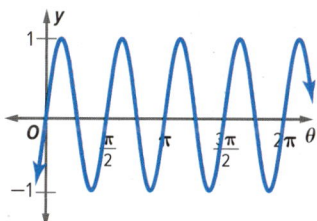

The pattern repeats itself at $\frac{\pi}{2}$, π, and so on. So, the period is $\frac{\pi}{2}$.

CHAPTER 9
Study Guide and Review Continued

9-5 Graphing Trigonometric Functions

Find the amplitude, if it exists, and period of each function. Then graph the function.

36. $y = 4 \sin 2\theta$
37. $y = \cos \frac{1}{2}\theta$
38. $y = 3 \csc \theta$
39. $y = 3 \sec \theta$
40. $y = \tan 2\theta$
41. $y = 2 \csc \frac{1}{2}\theta$

42. When Lauren jumps on a trampoline it vibrates with a frequency of 10 hertz. Let the amplitude equal 5 feet. Write a sine equation to represent the vibration of the trampoline y as a function of time t.

Example 8

Find the amplitude and period of $y = 2 \cos 4\theta$. Then graph the function.

amplitude: $|a| = |2|$ or 2. The graph is stretched vertically so that the maximum value is 2 and the minimum value is −2.

period:
$$\frac{360°}{|b|} = \frac{360°}{|4|} \text{ or } 90°$$

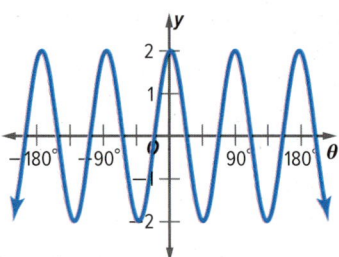

9-6 Translations of Trigonometric Graphs

State the vertical shift, amplitude, period, and phase shift of each function. Then graph the function.

43. $y = 3 \sin [2(\theta - 90°)] + 1$
44. $y = \frac{1}{2} \tan [2(\theta - 30°)] - 3$
45. $y = 2 \sec \left[3\left(\theta - \frac{\pi}{2}\right)\right] + 2$
46. $y = \frac{1}{2} \cos \left[\frac{1}{4}\left(\theta + \frac{\pi}{4}\right)\right] - 1$
47. $y = \frac{1}{3} \sin \left[\frac{1}{3}(\theta - 90°)\right] + 2$

48. The graph below approximates the height y of a rope that two people are twirling as a function of time t in seconds. Write an equation for the function.

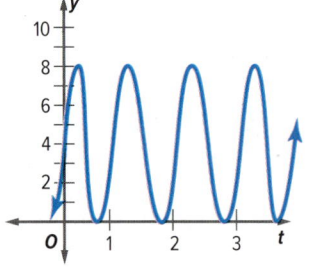

Example 9

State the vertical shift, amplitude, period, and phase shift of $y = 2 \sin \left[3\left(\theta + \frac{\pi}{2}\right)\right] + 4$. Then graph the function.

Identify the values of k, a, b, and h.

$k = 4$, so the vertical shift is 4.

$a = 2$, so the amplitude is 2.

$b = 3$, so the period is $\frac{2\pi}{|3|}$ or $\frac{2\pi}{3}$.

$h = -\frac{\pi}{2}$, so the phase shift is $\frac{\pi}{2}$ to the left.

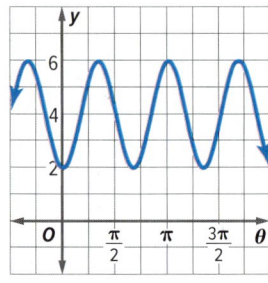

CHAPTER 9
Practice Test

Solve △ABC by using the given measurements. Round measures of sides to the nearest tenth and measures of angles to the nearest degree.

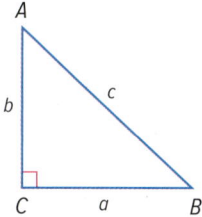

1. $A = 36°, c = 9$
2. $a = 12, A = 58°$
3. $B = 85°, b = 8$
4. $a = 9, c = 12$

Rewrite each degree measure in radians and each radian measure in degrees.

5. $325°$
6. $-175°$
7. $\dfrac{9\pi}{4}$
8. $-\dfrac{5\pi}{6}$

Find the exact value of each function. Write angle measures in degrees.

9. $\cos(-90°)$
10. $\sin 585°$
11. $\cot \dfrac{4\pi}{3}$
12. $\sec\left(-\dfrac{9\pi}{4}\right)$

13. The terminal side of angle θ in standard position intersects the unit circle at point $P\left(\dfrac{1}{2}, \dfrac{\sqrt{3}}{2}\right)$. Find $\cos \theta$ and $\sin \theta$.

14. **MULTIPLE CHOICE** What angle has a tangent and sine that are both negative?

    A 65°
    B 120°
    C 265°
    D 310°

15. **NAVIGATION** Airplanes and ships measure distance in nautical miles. The formula 1 nautical mile = $6077 - 31 \cos 2\theta$ feet, where θ is the latitude in degrees, can be used to find the approximate length of a nautical mile at a certain latitude. Find the length of a nautical mile when the latitude is 120°.

Find the amplitude and period of each function. Then graph the function.

16. $y = 2 \sin 3\theta$
17. $y = \dfrac{1}{2} \cos 2\theta$

18. **MULTIPLE CHOICE** What is the period of the function $y = 3 \cot \theta$?

    A 120°
    B 180°
    C 360°
    D 1080°

Write an equation for each translation.

19. $y = \sin x$, 6 units to the left and 4 units down
20. $y = \tan x$, $\dfrac{\pi}{2}$ units to the right and 1 unit up
21. Write an equation for the graph using the given trigonometric function.

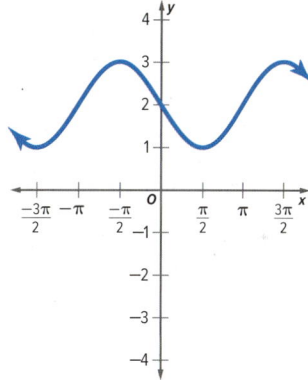

State the amplitude, period, and phase shift for each function. Then graph the function.

22. $y = \cos(\theta + 180)$
23. $y = \dfrac{1}{2} \tan\left(\theta - \dfrac{\pi}{2}\right)$

24. **WHEELS** A water wheel has a diameter of 20 feet. It makes one complete revolution in 45 seconds. Let the height at the top of the wheel represent the height at time 0. Write an equation for the height of point h in the diagram below as a function of time t. Then graph the function.

connectED.mcgraw-hill.com   647

CHAPTER 9
Preparing for Assessment

Performance Task

Provide a clear solution to each part of the task. Be sure to show all of your work, include all relevant drawings, and justify your answers.

SAFETY INSPECTOR An inspector is often hired to take detailed measurements for government safety commissions, architects, and construction companies. They frequently rely on trigonometry to make their calculations.

Part A

A telephone company needs to perform routine maintenance on some telephone poles. An inspector is taking measurements so they can order materials. The telephone pole is perpendicular to the ground, and a wire runs from the top of the pole diagonally to the ground. The angle formed by the wire and the pole is 20°. The distance from the base of the pole to the point where the wire meets the ground is 12 feet.

1. Find the length of the wire. Round your answer to the nearest tenth, if necessary.
2. Find the height of the pole. Round your answer to the nearest tenth, if necessary.

Part B

A safety inspector is evaluating plans for a spiral staircase to make sure it complies with local safety regulations. The inspector needs to draw several angles and make some calculations for the report.

3. Draw a 280° angle in standard position.
4. Draw a 630° angle in standard position.
5. Find a positive and negative coterminal angle measure for a 110° angle.
6. Convert 105° to radians.
7. Convert $\frac{7\pi}{4}$ to degrees.

Part C

An inspector is reviewing plans for a circular building that will have a domed roof. There are eight evenly-spaced support beams for the dome that form a design similar to the spokes of a wheel. The inspector makes a sketch of the beams where the center is at (0, 0) and four support beams have endpoints at (−36, 0), (36, 0), (0, −36), and (0, 36).

8. Use the unit circle to find the exact locations of the end points of the other four support beams.

Part D

Regulations require that the tornado alarm on a large university's campus be audible at every point on the campus within 8 seconds of the alarm being activated. The alarm makes 3 rotations per minute and has a period of 20 seconds. The length of the radius of the sound beam is exactly as long as the farthest place on campus from the alarm. The distance of the sound beam from the edge of campus varies periodically as a function of time.

9. Explain whether the alarm is in compliance with safety regulations.

Test-Taking Strategy

Example

Read the problem. Identify what you need to know. Then use the information in the problem to solve.

When Molly stands at a distance of 18 feet from the base of a tree, she forms an angle of 57° with the top of the tree. What is the height of the tree to the nearest tenth?

A 27.7 ft
B 28.5 ft
C 29.2 ft
D 30.1 ft

Step 1 Does the problem involve any concepts that may require a calculator?
Yes, it involves a trigonometric function.

Step 2 What steps can you do by hand? How will you use a calculator?
Make a sketch by hand and then set up a trigonometric ratio. Use the calculator to evaluate the trigonometric function.

Step 3 What is the correct answer?
Choice A is correct.

Test-Taking Tip

Using a Scientific Calculator Problems involving nth roots, logarithms, and exponential or trigonometric functions may require a scientific calculator for one or more steps.

Apply the Strategy

Read the problem. Identify what you need to know. Then use the information in the problem to solve.

What is the angle of the bike ramp below?

A 26.3°
B 28.5°
C 30.4°
D 33.6°

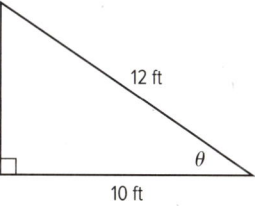

Answer the questions below.

a. Does the problem involve any concepts that may require a calculator?

b. What steps can you do by hand? How will you use a calculator?

c. What is the correct answer?

CHAPTER 9
Preparing for Assessment
Cumulative Review

Read each question. Then fill in the correct answer on the answer document provided by your teacher or on a sheet of paper.

1. A line segment extends from the origin of the coordinate plane $O(0, 0)$ to $B(30, 40)$, as shown in the figure below. Which equation represents the measure of the acute angle θ formed by $\overline{OB}$ and the x-axis?

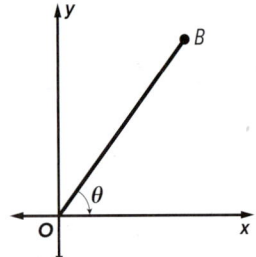

- A $\sin \theta = \frac{4}{5}$
- B $\cos \theta = \frac{5}{4}$
- C $\tan \theta = \frac{4}{5}$
- D $\sin \theta = \frac{4}{3}$
- E $\cos \theta = \frac{3}{4}$

2. Determine the exact value of csc 210°.

3. The two acute angles of a right triangle have the same measure. The hypotenuse measures 32 inches. What is the sum of the measures of the two legs of the triangle?
- A $\sqrt{2}$ in.
- B 32 in.
- C $16\sqrt{2}$ in.
- D $32\sqrt{2}$ in.
- E 2048 in.

4. The radius of a children's bicycle tire is 12 inches. What is the arc length of the circle made when the tire rotates 270°?
- A 4.5π in.
- B 9π in.
- C 18π in.
- D 24π in.
- E 108π in.

Test-Taking Tip

Question 3 Pay attention to the measures given in the question. If diameter is used in a formula, you need to multiply the radius by 2.

5. Which of the following angle measures are ordered from least to greatest?
- A $\frac{3\pi}{4}$, 180°, $\frac{2\pi}{3}$, 225°
- B $\frac{2\pi}{3}$, $\frac{3\pi}{4}$, 180°, 225°
- C 180°, 225°, $\frac{2\pi}{3}$, $\frac{3\pi}{4}$
- D $\frac{3\pi}{4}$, $\frac{2\pi}{3}$, 180°, 225°
- E 225°, 180°, $\frac{3\pi}{4}$, $\frac{2\pi}{3}$

6. A line with slope of $\frac{3}{4}$ passes through $(-3, -4)$. Which is the equation of the line?
- A $y = \frac{3}{4}x - 4$
- B $-3x + 4y = -7$
- C $y = \frac{3}{4}x + 2$
- D $3x - 4y = -7$
- E $y = \frac{3}{4}x - 3$

7. Find the exact value of sin 75°.

650 | Chapter 9 | Preparing for Assessment

Go Online! for Standardized Test Practice

8. What are the missing coefficients, from left to right, in this binomial expansion of $(a + 2b)^5$?

$$a^5 + 10a^4b + \_\_ a^3b^2 + \_\_ a^2b^3 + 80ab^4 + 32b^5$$

- ○ A 10 and 10
- ○ B 10 and 20
- ○ C 20 and 10
- ○ D 40 and 80
- ○ E 80 and 40

9. What is the solution to the system of equations below?

$$x - y = 5$$
$$3x + 5y = 7$$

- ○ A (4, −1)
- ○ D (5, 7)
- ○ B (5, 0)
- ○ E (−1, 2)
- ○ C (3, 5)

10. For which of the trigonometric functions is the period and amplitude the same as $y = 2\pi \sin x$?

- ○ A $y = -2\pi \cos x$
- ○ D $y = 4 \sin x$
- ○ B $y = -2 \sin 2\pi x$
- ○ E $y = \cos 2x$
- ○ C $y = -2 \cos x$

11. For what value(s) of W will the function $y = R \cos (W \cdot x)$ result in a horizontal stretch of $y = \cos x$?

- ○ A $W = \pi$
- ○ D $0 < W < 1$
- ○ B $W > 1$
- ○ E $W =$ any real number
- ○ C $W < 0$

12. Select all true characteristics of the trigonometric function graphed below.

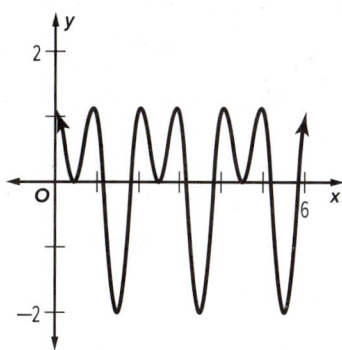

- ☐ A The domain of the function is all real numbers.
- ☐ B The range of the function is all real numbers between −2 and 1.
- ☐ C The minimum value of the function is −2.
- ☐ D The function cycles 6 times.
- ☐ E The period of the function is 6.

13. The exponential function $f(x) = a(2)^x$ passes through the point (−1, 6). What is the value of a?

[ ]

14. A certain medication has an effective biological half-life of 2 hours. This means that after every 2 hours, the amount of medication remaining in the body is reduced by half.

   a. What fraction of an initial dose of the medication has been eliminated after 8 hours? Express your answer as a decimal.

   [ ]

   b. **MP** What mathematical practice did you use to solve this problem?

Need Extra Help?

If you missed Question...	1	2	3	4	5	6	7	8	9	10	11	12	13	14
Go to Lesson...	9-1	9-3	9-1	9-2	9-2	1-4	10-3	4-2	1-6	9-5	9-4	9-6	6-1	6-3

connectED.mcgraw-hill.com 651

CHAPTER 10

Trigonometric Identities and Equations

Hero Images/Getty Images

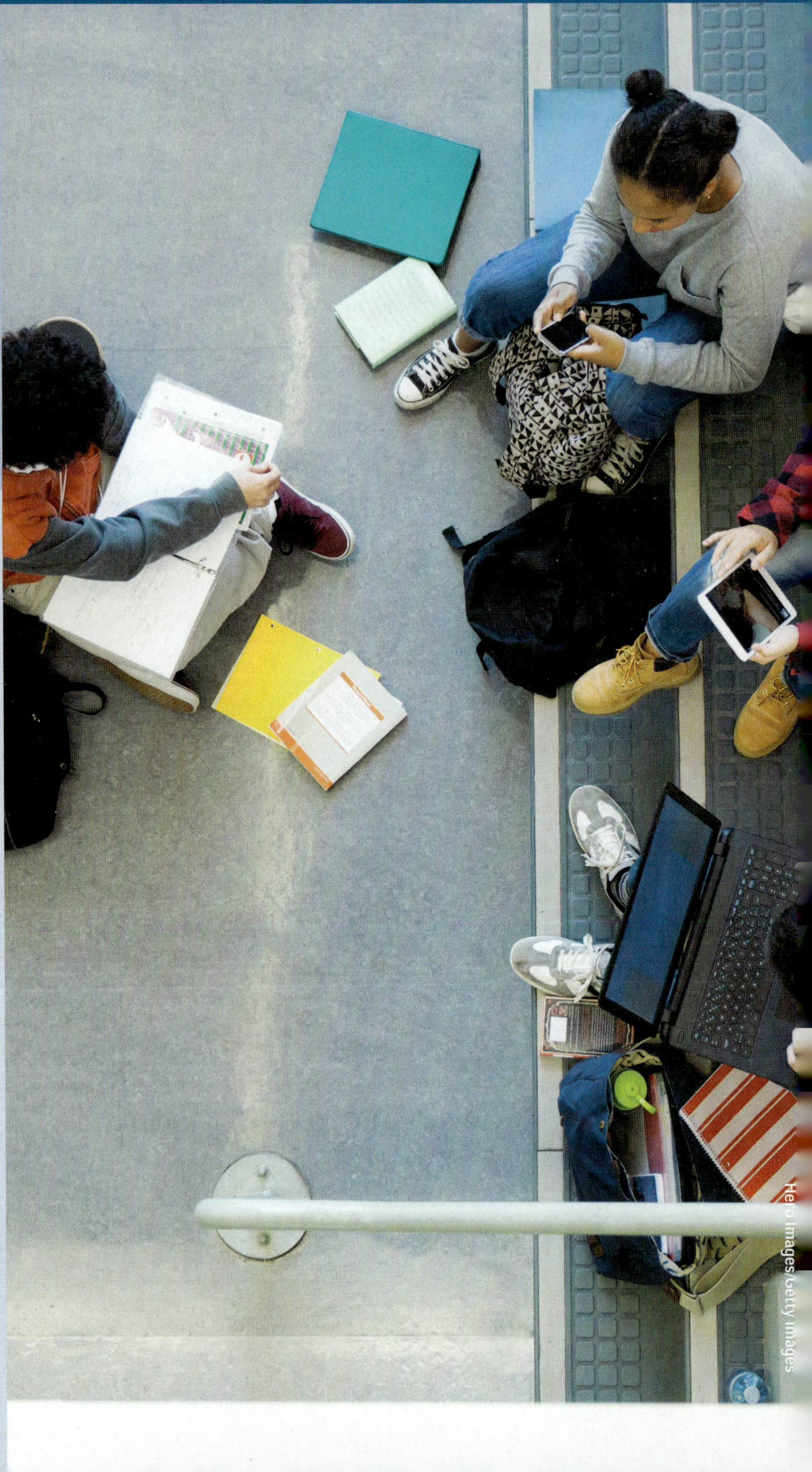

THEN

You graphed trigonometric functions and determined the period, amplitude, phase shifts, and vertical shifts.

NOW

You will:

- Use and verify trigonometric identities.
- Use the sum and difference of angles identities.
- Use the double- and half-angle identities.
- Solve trigonometric equations.

MP WHY

ELECTRONICS Radios, smartphones, and wireless Internet all use waves that can be modeled by trigonometric functions.

Use the Mathematical Practices to complete the activity.

1. **Use Tools** Use the Internet to learn about the ways in which waves are used for Wi-Fi signals, cell phones, or other wireless devices.
2. **Model with Math** Use the Pen tool on the Coordinate Grid mat to model the wave you researched. Use the Text tool to label the amplitude, frequency, and so on.

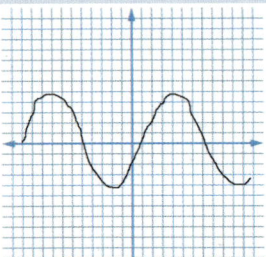

3. **Discuss** How does the cyclical or periodic behavior of the wave enable it to transmit information to and from our wireless devices?

 # Go Online to Guide Your Learning

Explore & Explain

 The Geometer's Sketchpad

The Geometer's Sketchpad can be used to enhance your understanding of trigonometric identities in Lesson 10-1 and throughout this chapter.

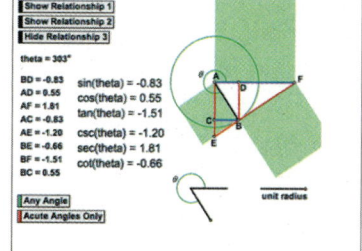

 Graphing Tools

Use the **Explore: Trigonometric Functions** tool to enhance your understanding of graphing trigonometric functions.

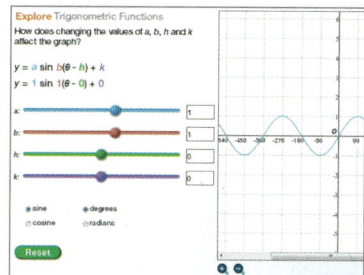

eBook

Interactive Student Guide

Before starting the chapter, answer the **Chapter Focus** preview questions. Check your answers as you complete each lesson. At the end of the chapter, try the **Performance Task**.

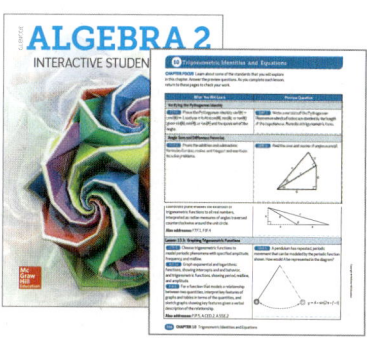

Organize

 Foldables

Get organized! Before you begin this chapter, create this **Trigonometric Identities and Equations Foldable** to help you organize your notes about trigonometric identities and equations.

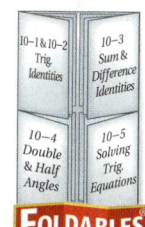

Collaborate

 Chapter Project

In the **Spin the Wheel** project, you will use what you have learned about trigonometric identities to complete a project that addresses business literacy.

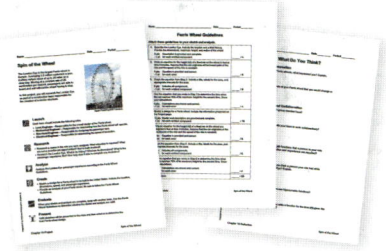

Focus

 LEARNSMART

Need help studying? Complete the **Trigonometric Functions** domain in LearnSmart to review for the chapter test.

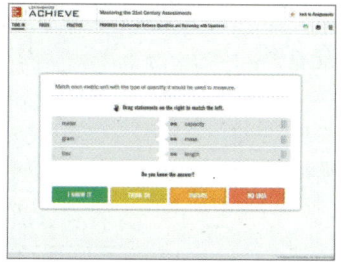

 ALEKS

You can use the **Trigonometry** topic in ALEKS to explore what you know about trigonometry and what you are ready to learn.*

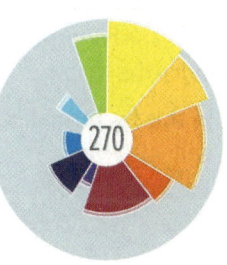

* Ask your teacher if this is part of your program.

connectED.mcgraw-hill.com

Get Ready for the Chapter

Connecting Concepts

Concept Check
Review the concepts used in this chapter by answering the questions below.

1. Is the polynomial $-16a^2 + 4a$ factorable or is it prime? If it is factorable, write the expression in its completely factored form.

2. Is the polynomial $x(x^2 + 2x - 24)$ completely factored? If not, write its completely factored form.

3. You are designing a flower bed that has an area of 42 square feet, with the dimensions shown. Write the equation, in polynomial form, to determine the value of x, and hence the length of each side.

x ft

$x + 1$ ft

4. Solve the polynomial equation in question 3. How do you know which value of x is a reasonable solution?

5. To find the cos 135°, you need to know its reference angle. How do you determine the reference angle?

6. What is the exact value of cos 45°?

Performance Task Preview
You can use the concepts and skills in the chapter to perform error analysis for a teacher. Understanding trigonometric identities and equations will help you finish the Performance Task at the end of the chapter.

MP In this Performance Task you will:
- make sense of problems and persevere in solving them
- reason abstractly and quantitatively
- construct viable arguments and critique the reasoning of others
- attend to precision
- look for and make use of structure

New Vocabulary

English		Español
trigonometric identity	p. 655	identidad trigonométrica
trigonometric equation	p. 683	ecuación trigonométrica

Review Vocabulary

formula fórmula a mathematical sentence that expresses the relationship between certain quantities

identity identidad an equality that remains true regardless of the values of any variables that are in it

trigonometric functions funciones rigonométricas
For any angle, with measure θ, a point $P(x, y)$ on its terminal side, $r = \sqrt{x^2 + y^2}$, the trigonometric functions of θ are as follows.

$\sin \theta = \dfrac{y}{r}$ $\cos \theta = \dfrac{x}{r}$ $\tan \theta = \dfrac{y}{x}$

$\csc \theta = \dfrac{r}{y}$ $\sec \theta = \dfrac{r}{x}$ $\cot \theta = \dfrac{x}{y}$

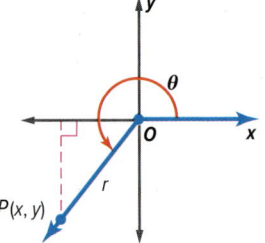

LESSON 1
Trigonometric Identities

Then
- You evaluated trigonometric functions.

Now
1. Use trigonometric identities to find trigonometric values.
2. Use trigonometric identities to simplify expressions.

Why?
The amount of light that a source provides to a surface is called the *illuminance*. The illuminance E in foot candles on a surface is related to the distance R in feet from the light source. The formula $\sec \theta = \dfrac{I}{ER^2}$, where I is the intensity of the light source measured in candles and θ is the angle between the light beam and a line perpendicular to the surface, can be used in situations in which lighting is important, as in photography.

New Vocabulary
trigonometric identity

Mathematical Practices
2 Reason abstractly and quantitatively.
7 Look for and make use of structure.

1 Find Trigonometric Values The equation above can also be written as $E = \dfrac{I \cos \theta}{R^2}$. This is an example of a trigonometric identity. A **trigonometric identity** is an equation involving trigonometric functions that is true for all values for which every expression in the equation is defined.

If you can show that a specific value of the variable in an equation makes the equation false, then you have produced a *counterexample*. It only takes one counterexample to prove that an equation is not an identity.

Key Concept Basic Trigonometric Identities

Quotient Identities

$\tan \theta = \dfrac{\sin \theta}{\cos \theta}$, $\cos \theta \neq 0$ $\cot \theta = \dfrac{\cos \theta}{\sin \theta}$, $\sin \theta \neq 0$

Reciprocal Identities

$\sin \theta = \dfrac{1}{\csc \theta}$, $\csc \theta \neq 0$ $\csc \theta = \dfrac{1}{\sin \theta}$, $\sin \theta \neq 0$

$\cos \theta = \dfrac{1}{\sec \theta}$, $\sec \theta \neq 0$ $\sec \theta = \dfrac{1}{\cos \theta}$, $\cos \theta \neq 0$

$\tan \theta = \dfrac{1}{\cot \theta}$, $\cot \theta \neq 0$ $\cot \theta = \dfrac{1}{\tan \theta}$, $\tan \theta \neq 0$

Pythagorean Identities

$\cos^2 \theta + \sin^2 \theta = 1$ $\tan^2 \theta + 1 = \sec^2 \theta$ $\cot^2 \theta + 1 = \csc^2 \theta$

Cofunction Identities

$\sin\left(\dfrac{\pi}{2} - \theta\right) = \cos \theta$ $\cos\left(\dfrac{\pi}{2} - \theta\right) = \sin \theta$ $\tan\left(\dfrac{\pi}{2} - \theta\right) = \cot \theta$

Negative Angle Identities

$\sin(-\theta) = -\sin \theta$ $\cos(-\theta) = \cos \theta$ $\tan(-\theta) = -\tan \theta$

The negative angle identities are sometimes called *odd-even* identities.

The identity $\tan \theta = \dfrac{\sin \theta}{\cos \theta}$ is true except for angle measures such as 90°, 270°, …, $90° + k180°$, where k is an integer. The cosine of each of these angle measures is 0, so $\tan \theta$ is not defined when $\cos \theta = 0$. An identity similar to this is $\cot \theta = \dfrac{\cos \theta}{\sin \theta}$.

You can use trigonometric identities to find exact values of trigonometric functions. You can find approximate values by using a graphing calculator.

Example 1   Use Trigonometric Identities

a. Find the exact value of $\cos \theta$ if $\sin \theta = \frac{1}{4}$ and $90° < \theta < 180°$.

$\cos^2 \theta + \sin^2 \theta = 1$ Pythagorean identity

$\cos^2 \theta = 1 - \sin^2 \theta$ Subtract $\sin^2 \theta$ from each side.

$\cos^2 \theta = 1 - \left(\frac{1}{4}\right)^2$ Substitute $\frac{1}{4}$ for $\sin \theta$.

$\cos^2 \theta = 1 - \frac{1}{16}$ Square $\frac{1}{4}$.

$\cos^2 \theta = \frac{15}{16}$ Subtract: $\frac{16}{16} - \frac{1}{16} = \frac{15}{16}$.

$\cos \theta = \pm \frac{\sqrt{15}}{4}$ Take the square root of each side.

Because θ is in the second quadrant, $\cos \theta$ is negative. Thus, $\cos \theta = -\frac{\sqrt{15}}{4}$.

CHECK Use a calculator to find an approximate answer.

Step 1 Find Arcsin $\frac{1}{4}$.

$\sin^{-1} \frac{1}{4} \approx 14.48°$ Use a calculator.

Because $90° < \theta < 180°$, $\theta \approx 180° - 14.48°$ or about $165.52°$.

Step 2 Find $\cos \theta$.

Replace θ with $165.52°$.

$\cos 165.52° \approx -0.97$

Step 3 Compare with the exact value.

$-\frac{\sqrt{15}}{4} \stackrel{?}{\approx} 0.97$

$-0.968 \approx 0.97$ ✓

b. Find the exact value of $\csc \theta$ if $\cot \theta = -\frac{3}{5}$ and $270° < \theta < 360°$.

$\cot^2 \theta + 1 = \csc^2 \theta$ Pythagorean identity

$\left(-\frac{3}{5}\right)^2 + 1 = \csc^2 \theta$ Substitute $-\frac{3}{5}$ for $\cot \theta$.

$\frac{9}{25} + 1 = \csc^2 \theta$ Square $-\frac{3}{5}$.

$\frac{34}{25} = \csc^2 \theta$ Add: $\frac{9}{25} + \frac{25}{25} = \frac{34}{25}$.

$\pm \frac{\sqrt{34}}{5} = \csc \theta$ Take the square root of each side.

Because θ is in the fourth quadrant, $\csc \theta$ is negative. Thus, $\csc \theta = -\frac{\sqrt{34}}{5}$.

Guided Practice

1A. Find $\sin \theta$ if $\cos \theta = \frac{1}{3}$ and $270° < \theta < 360°$.

1B. Find $\sec \theta$ if $\sin \theta = -\frac{2}{7}$ and $180° < \theta < 270°$.

Study Tip

Quadrants Here is a table to help you remember which ratios are positive and which are negative in each quadrant.

Function	+	−
$\sin \theta$	1, 2	3, 4
$\cos \theta$	1, 4	2, 3
$\tan \theta$	1, 3	2, 4
$\csc \theta$	1, 2	3, 4
$\sec \theta$	1, 4	2, 3
$\cot \theta$	1, 3	2, 4

Go Online!

You will want to reference the Basic Trigonometric Identities as you study this chapter. Log into your eStudent Edition to bookmark this lesson.

2 Simplify Expressions

Simplifying an expression that contains trigonometric functions means that the expression is written as a numerical value or in terms of a single trigonometric function, if possible.

Study Tip

Simplifying It is often easiest to write all expressions in terms of sine and/or cosine.

Example 2 Simplify an Expression

Simplify $\dfrac{\sin \theta \csc \theta}{\cot \theta}$.

$$\dfrac{\sin \theta \csc \theta}{\cot \theta} = \dfrac{\sin \theta \left(\dfrac{1}{\sin \theta}\right)}{\dfrac{1}{\tan \theta}} \qquad \csc \theta = \dfrac{1}{\sin \theta} \text{ and } \cot \theta = \dfrac{1}{\tan \theta}$$

$$= \dfrac{1}{\dfrac{1}{\tan \theta}} \qquad \dfrac{\sin \theta}{\sin \theta} = 1$$

$$= \dfrac{1}{1} \cdot \dfrac{\tan \theta}{1} \text{ or } \tan \theta \qquad \dfrac{a}{b} \div \dfrac{c}{d} = \dfrac{a}{b} \cdot \dfrac{d}{c}$$

▶ **Guided Practice**

Simplify each expression.

2A. $\dfrac{\tan^2 \theta \csc^2 \theta - 1}{\sec^2 \theta}$

2B. $\dfrac{\sec \theta}{\sin \theta}(1 - \cos^2 \theta)$

Simplifying trigonometric expressions can be helpful when solving real-world problems.

Real-World Example 3 Simplify and Use an Expression

LIGHTING Refer to the beginning of the lesson.

a. Solve the formula in terms of E.

$$\sec \theta = \dfrac{I}{ER^2} \qquad \text{Original equation}$$

$$ER^2 \sec \theta = I \qquad \text{Multiply each side by } ER^2.$$

$$ER^2 \dfrac{1}{\cos \theta} = I \qquad \dfrac{1}{\cos \theta} = \sec \theta$$

$$\dfrac{E}{\cos \theta} = \dfrac{I}{R^2} \qquad \text{Divide each side by } R^2.$$

$$E = \dfrac{I \cos \theta}{R^2} \qquad \text{Multiply each side by } \cos \theta.$$

b. Is the equation in part a equivalent to $R^2 = \dfrac{I \tan \theta \cos \theta}{E}$? Explain.

$$R^2 = \dfrac{I \tan \theta \cos \theta}{E} \qquad \text{Original equation}$$

$$ER^2 = I \tan \theta \cos \theta \qquad \text{Multiply each side by } E.$$

$$E = \dfrac{I \tan \theta \cos \theta}{R^2} \qquad \text{Divide each side by } R^2.$$

$$E = \dfrac{I \left(\dfrac{\sin \theta}{\cos \theta}\right) \cos \theta}{R^2} \qquad \tan \theta = \dfrac{\sin \theta}{\cos \theta}$$

$$E = \dfrac{I \sin \theta}{R^2} \qquad \text{Simplify.}$$

No; the equations are not equivalent. $R^2 = \dfrac{I \tan \theta \cos \theta}{E}$ simplifies to $E = \dfrac{I \sin \theta}{R^2}$.

▶ **Guided Practice**

3. Rewrite $\cot^2 \theta - \tan^2 \theta$ in terms of $\sin \theta$.

Math History Link

Aryabhatta (476–550 A.D.) Among Indian mathematicians, Aryabhatta is probably the most famous. His name is closely associated with trigonometry. He was the first to introduce inverse trigonometric functions and spherical trigonometry. Aryabhatta also calculated approximations for pi and trigonometric functions.

Check Your Understanding

= Step-by-Step Solutions begin on page R11.

Example 1 Find the exact value of each expression if $0° < \theta < 90°$.

1. If $\cot \theta = 2$, find $\tan \theta$.
2. If $\sin \theta = \frac{4}{5}$, find $\cos \theta$.
3. If $\cos \theta = \frac{2}{3}$, find $\sin \theta$.
4. If $\cos \theta = \frac{2}{3}$, find $\csc \theta$.

Example 2 Simplify each expression.

5. $\tan \theta \cos^2 \theta$
6. $\csc^2 \theta - \cot^2 \theta$
7. $\dfrac{\cos \theta \csc \theta}{\tan \theta}$

Example 3

8. **PERSEVERANCE** When unpolarized light passes through polarized sunglass lenses, the intensity of the light is cut in half. If the light then passes through another polarized lens with its axis at an angle of θ to the first, the intensity of the light is again diminished. The intensity of the emerging light can be found by using the formula $I = I_0 - \dfrac{I_0}{\csc^2 \theta}$, where I_0 is the intensity of the light incoming to the second polarized lens, I is the intensity of the emerging light, and θ is the angle between the axes of polarization.

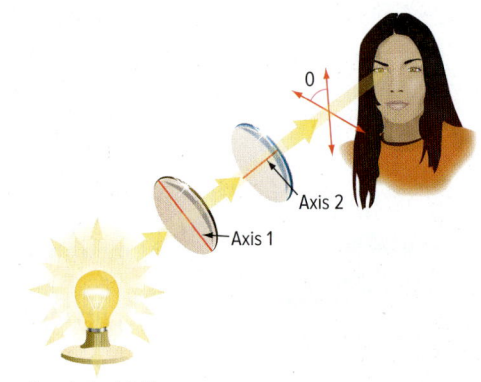
Unpolarized light

  a. Simplify the formula in terms of $\cos \theta$.
  b. Use the simplified formula to determine the intensity of light that passes through a second polarizing lens with axis at 30° to the original.

Practice and Problem Solving

Extra Practice is on page R10.

Example 1 Find the exact value of each expression if $0° < \theta < 90°$.

9. If $\cos \theta = \frac{3}{5}$, find $\csc \theta$.
10. If $\sin \theta = \frac{1}{2}$, find $\tan \theta$.
11. If $\sin \theta = \frac{3}{5}$, find $\cos \theta$.
12. If $\tan \theta = 2$, find $\sec \theta$.

Find the exact value of each expression if $180° < \theta < 270°$.

13. If $\cos \theta = -\frac{3}{5}$, find $\csc \theta$.
14. If $\sec \theta = -3$, find $\tan \theta$.
15. If $\cot \theta = \frac{1}{4}$, find $\csc \theta$.
16. If $\sin \theta = -\frac{1}{2}$, find $\cos \theta$.

Find the exact value of each expression if $270° < \theta < 360°$.

17. If $\cos \theta = \frac{5}{13}$, find $\sin \theta$.
18. If $\tan \theta = -1$, find $\sec \theta$.
19. If $\sec \theta = \frac{5}{3}$, find $\cos \theta$.
20. If $\csc \theta = -\frac{5}{3}$, find $\cos \theta$.

Example 2 Simplify each expression.

21. $\sec \theta \tan^2 \theta + \sec \theta$
22. $\cos\left(\dfrac{\pi}{2} - \theta\right)\cot \theta$
23. $\cot \theta \sec \theta$
24. $\sin \theta (1 + \cot^2 \theta)$
25. $\sin\left(\dfrac{\pi}{2} - \theta\right)\sec \theta$
26. $\dfrac{\cos(-\theta)}{\sin(-\theta)}$

658 | Lesson 10-1 | Trigonometric Identities

Example 3

27. ELECTRONICS When there is a current in a wire in a magnetic field, such as in a hairdryer, a force acts on the wire. The strength of the magnetic field can be determined using the formula $B = \frac{F \csc \theta}{I\ell}$, where F is the force on the wire, I is the current in the wire, ℓ is the length of the wire, and θ is the angle the wire makes with the magnetic field. Rewrite the equation in terms of $\sin \theta$. (*Hint:* Solve for F.)

Simplify each expression.

28. $\frac{1 - \sin^2 \theta}{\sin^2 \theta}$

29. $\tan \theta \csc \theta$

30. $\frac{1}{\sin^2 \theta} - \frac{\cos^2 \theta}{\sin^2 \theta}$

31. $2(\csc^2 \theta - \cot^2 \theta)$

32. $(1 + \sin \theta)(1 - \sin \theta)$

33. $2 - 2 \sin^2 \theta$

34. SUN The ability of an object to absorb energy is related to a factor called the emissivity e of the object. The emissivity can be calculated by using the formula $e = \frac{W \sec \theta}{AS}$, where W is the rate at which a person's skin absorbs energy from the Sun, S is the energy from the Sun in watts per square meter, A is the surface area exposed to the Sun, and θ is the angle between the Sun's rays and a line perpendicular to the body.

  a. Solve the equation for W. Write your answer using only $\sin \theta$ or $\cos \theta$.

  b. Find W if $e = 0.80$, $\theta = 40°$, $A = 0.75$ m^2, and $S = 1000$ W/m^2. Round to the nearest hundredth.

35. MODELING The map shows some of the buildings in Maria's neighborhood that she visits on a regular basis. The sine of the angle θ formed by the roads connecting the dance studio, the school, and Maria's house is $\frac{4}{9}$.

  a. What is the cosine of the angle?

  b. What is the tangent of the angle?

  c. What are the sine, cosine, and tangent of the angle formed by the roads connecting the piano teacher's house, the school, and Maria's house?

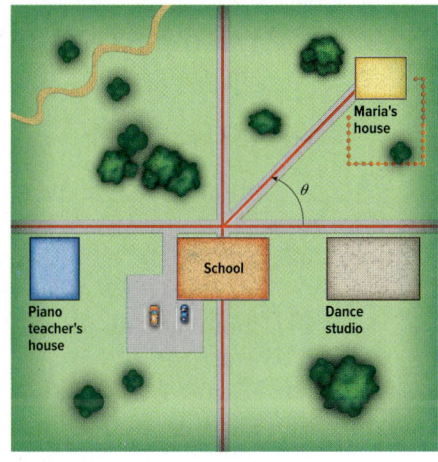

36. MULTIPLE REPRESENTATIONS In this problem, you will use a graphing calculator to determine whether an equation may be a trigonometric identity. Consider the trigonometric identity $\tan^2 \theta - \sin^2 \theta = \tan^2 \theta \sin^2 \theta$.

  a. Tabular Copy and complete the table below.

θ	0°	30°	45°	60°
$\tan^2 \theta - \sin^2 \theta$				
$\tan^2 \theta \sin^2 \theta$				

  b. Graphical Use a graphing calculator to graph $\tan^2 \theta - \sin^2 \theta = \tan^2 \theta \sin^2 \theta$ as two separate functions. Sketch the graph.

  c. Analytical If the graphs of the two functions do not match, then the equation is not an identity. Do the graphs coincide?

  d. Analytical Use a graphing calculator to determine whether the equation $\sec^2 x - 1 = \sin^2 x \sec^2 x$ may be an identity. (Be sure your calculator is in degree mode.)

37. **SKIING** A skier of mass m descends a θ-degree hill at a constant speed. When Newton's laws are applied to the situation, the following system of equations is produced: $F_n - mg \cos \theta = 0$ and $mg \sin \theta - \mu_k F_n = 0$, where g is the acceleration due to gravity, F_n is the normal force exerted on the skier, and μ_k is the coefficient of friction. Use the system to define μ_k as a function of θ.

Simplify each expression.

38. $\dfrac{\tan\left(\dfrac{\pi}{2} - \theta\right)\sec \theta}{1 - \csc^2 \theta}$

39. $\dfrac{\cos\left(\dfrac{\pi}{2} - \theta\right) - 1}{1 + \sin(-\theta)}$

40. $\dfrac{\sec \theta \sin \theta + \cos\left(\dfrac{\pi}{2} - \theta\right)}{1 + \sec \theta}$

41. $\dfrac{\cot \theta \cos \theta}{\tan(-\theta) \sin\left(\dfrac{\pi}{2} - \theta\right)}$

H.O.T. Problems   Use Higher-Order Thinking Skills

42. **MP CRITIQUE ARGUMENTS** Clyde and Rosalina are debating whether an equation from their homework assignment is an identity. Clyde says that since he has tried ten specific values for the variable and all of them worked, it must be an identity. Rosalina argues that specific values could only be used as counterexamples to prove that an equation is not an identity. Is either of them correct? Explain your reasoning.

43. **CHALLENGE** Find a counterexample to show that $1 - \sin x = \cos x$ is *not* an identity.

44. **MP REASONING** Demonstrate how the formula about illuminance from the beginning of the lesson can be rewritten to show that $\cos \theta = \dfrac{ER^2}{I}$.

45. **WRITING IN MATH** Pythagoras is most famous for the Pythagorean Theorem. The identity $\cos^2 \theta + \sin^2 \theta = 1$ is an example of a Pythagorean identity. Why do you think that this identity is classified in this way?

46. **PROOF** Prove that $\tan(-a) = -\tan a$ by using the quotient and negative angle identities.

47. **OPEN-ENDED** Write two expressions that are equivalent to $\tan \theta \sin \theta$.

48. **MP REASONING** Explain how you can use division to rewrite $\sin^2 \theta + \cos^2 \theta = 1$ as $1 + \cot^2 \theta = \csc^2 \theta$.

49. **CHALLENGE** Find $\cot \theta$ if $\sin \theta = \dfrac{3}{5}$ and $90° \leq \theta < 180°$.

50. **ERROR ANALYSIS** Jordan and Ebony are simplifying $\dfrac{\sin^2 \theta}{\cos^2 \theta + \sin^2 \theta}$. Is either of them correct? Explain your reasoning.

Jordan

$\dfrac{\sin^2 \theta}{\cos^2 \theta + \sin^2 \theta} = \dfrac{\sin^2 \theta}{\cos^2 \theta} + \dfrac{\sin^2 \theta}{\sin^2 \theta}$

$= \tan^2 \theta + 1$

$= \sec^2 \theta$

Ebony

$\dfrac{\sin^2 \theta}{\cos^2 \theta + \sin^2 \theta} = \dfrac{\sin^2 \theta}{1}$

$= \sin^2 \theta$

Preparing for Assessment

51. Of the following expressions, which is/are equivalent to $\cos \theta$?

    I $\sin(90° - \theta)$
    II $\dfrac{1}{\sec \theta}$
    III $\cos(-\theta)$

- A I only
- B III only
- C I and II only
- D II and III only
- E I, II, and III

52. Which expression, when defined, can be used in place of $\cos\left(\dfrac{\pi}{2} - x\right) \sec x$?

- A -1
- B $-\tan x$
- C 1
- D $\tan x$
- E $\sin^2 x$

53. Suppose $90° < \theta < 180°$ and $\cos \theta = -\dfrac{4}{5}$. What is $\sin^2 \theta$?

- A $-\dfrac{5}{4}$
- B $\dfrac{3}{5}$
- C $-\dfrac{9}{25}$
- D $\dfrac{9}{25}$
- E $\dfrac{16}{25}$

54. For $0° < \theta < 90°$, which expression is **NOT** valid?

- A $\cos(-\theta) = -\cos \theta$
- B $\sin(-\theta) = -\sin \theta$
- C $\tan(-\theta) = -\tan \theta$
- D $\cos(-\theta) = \cos \theta$
- E $\tan(-\theta) = -\dfrac{\sin \theta}{\cos \theta}$

55. MULTI-STEP

  a. Which of the following trigonometric expressions is equivalent to $\dfrac{\sin(90° - \theta)}{\cos(\theta - 90°)}$?

- A $\sin \theta$
- B $\cos \theta$
- C $\tan \theta$
- D $\cot \theta$
- E $\sec \theta$

  b. Use the trigonometric identities to simplify $\dfrac{\sin(90° - \theta)}{\cos(\theta - 90°)}$.

56. To what is $\dfrac{\tan \theta}{\sec \theta}$ equivalent?

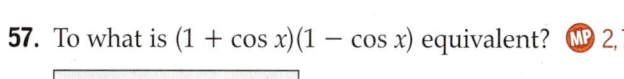

57. To what is $(1 + \cos x)(1 - \cos x)$ equivalent?

58. If $\sec \theta = \dfrac{3}{5}$, what is $\cos \theta$ for $0° < \theta < 90°$?

- A $\dfrac{3}{4}$
- B $\dfrac{5}{3}$
- C $\dfrac{4}{3}$
- D $\dfrac{5}{4}$
- E $\dfrac{4}{5}$

59. Which is **NOT** a Pythagorean identity? Choose all that apply.

- A $\cos^2 \theta = 1 - \sin^2 \theta$
- B $\tan^2 \theta + 1 = \sec^2 \theta$
- C $\sin^2 \theta + 1 = \cos^2 \theta$
- D $\tan^2 \theta + \cot^2 \theta = 1$
- E $\cot^2 \theta = \csc^2 \theta - 1$

LESSON 2
Verifying Trigonometric Identities

::Then
- You used identities to find trigonometric values and simplify expressions.

::Now
1. Verify trigonometric identities by transforming one side of an equation into the form of the other side.
2. Verify trigonometric identities by transforming each side of the equation into the same form.

::Why?
- While running on a circular track, Lamont notices that his body is not perpendicular to the ground. Instead, it leans away from a vertical position. The nonnegative acute angle θ that Lamont's body makes with the vertical is called the *angle of incline* and is described by the equation $\tan \theta = \dfrac{v^2}{gR}$.

This is not the only equation that describes the angle of incline in terms of trigonometric functions. Another such equation is $\sin \theta = \cos \dfrac{v^2}{gR}\theta$, where $0 \le \theta \le 90°$.

Are these two equations completely independent of one another or are they merely different versions of the same relationship?

 Mathematical Practices
1. Make sense of problems and persevere in solving them.
7. Look for and express regularity in repeated reasoning.

1 Transform One Side of an Equation You can use the basic trigonometric identities along with the definitions of the trigonometric functions to verify identities. If you wish to show an identity, you need to show that it is true for all values of θ.

> **Key Concept** Verifying Identities by Transforming One Side
>
> **Step 1** Simplify one side of an equation until the two sides of the equation are the same. It is often easier to work with the more complicated side of the equation.
>
> **Step 2** Transform that expression into the form of the simpler side.

Example 1 Transform One Side of an Equation

Verify that $\dfrac{\sin^2 \theta}{1 - \cos \theta} = 1 + \cos \theta$ is an identity.

$\dfrac{\sin^2 \theta}{1 - \cos \theta} \stackrel{?}{=} 1 + \cos \theta$ Original equation

$\dfrac{1 + \cos \theta}{1 + \cos \theta} \cdot \dfrac{\sin^2 \theta}{1 - \cos \theta} \stackrel{?}{=} 1 + \cos \theta$ Multiply the numerator and denominator by $1 + \cos \theta$.

$\dfrac{\sin^2 \theta (1 + \cos \theta)}{1 - \cos^2 \theta} \stackrel{?}{=} 1 + \cos \theta$ $(1 + \cos \theta)(1 - \cos \theta) = 1 - \cos^2 \theta$

$\dfrac{\sin^2 \theta (1 + \cos \theta)}{\sin^2 \theta} \stackrel{?}{=} 1 + \cos \theta$ $\sin^2 \theta = 1 - \cos^2 \theta$

$1 + \cos \theta = 1 + \cos \theta$ ✓   Divide the numerator and denominator by $\sin^2 \theta$.

▶ **Guided Practice**

1. Verify that $\cot^2 \theta - \cos^2 \theta = \cot^2 \theta \cos^2 \theta$ is an identity.

When verifying trigonometric identities, it is often helpful to rewrite an expression in terms of sine and cosine before simplifying.

> **Example 2**   **Rewrite Expressions**
>
> Verify that $\dfrac{1 - \cos \theta}{1 + \cos \theta} = (\csc \theta - \cot \theta)^2$.
>
> Rewrite the expression on the right side of the equation in terms of $\sin \theta$ and $\cos \theta$. Then simplify.
>
> $\dfrac{1 - \cos \theta}{1 + \cos \theta} \stackrel{?}{=} (\csc \theta - \cot \theta)^2$ Original equation
>
> $\dfrac{1 - \cos \theta}{1 + \cos \theta} \stackrel{?}{=} \csc^2 \theta - 2 \cot \theta \csc \theta + \cot^2 \theta$ Multiply.
>
> $\dfrac{1 - \cos \theta}{1 + \cos \theta} \stackrel{?}{=} \dfrac{1}{\sin^2 \theta} - 2 \cdot \dfrac{\cos \theta}{\sin \theta} \cdot \dfrac{1}{\sin \theta} + \dfrac{\cos^2 \theta}{\sin^2 \theta}$ $\csc \theta = \dfrac{1}{\sin \theta}, \cot \theta = \dfrac{\cos \theta}{\sin \theta}$
>
> $\dfrac{1 - \cos \theta}{1 + \cos \theta} \stackrel{?}{=} \dfrac{1}{\sin^2 \theta} - \dfrac{2 \cos \theta}{\sin^2 \theta} + \dfrac{\cos^2 \theta}{\sin^2 \theta}$ Simplify.
>
> $\dfrac{1 - \cos \theta}{1 + \cos \theta} \stackrel{?}{=} \dfrac{1 - 2\cos\theta + \cos^2 \theta}{\sin^2 \theta}$ Add and subtract.
>
> $\dfrac{1 - \cos \theta}{1 + \cos \theta} \stackrel{?}{=} \dfrac{(1 - \cos \theta)(1 - \cos \theta)}{1 - \cos^2 \theta}$ Factor the numerator; $\sin^2 \theta = 1 - \cos^2 \theta$.
>
> $\dfrac{1 - \cos \theta}{1 + \cos \theta} \stackrel{?}{=} \dfrac{(1 - \cos \theta)(1 - \cos \theta)}{(1 - \cos \theta)(1 + \cos \theta)}$ Factor the denominator.
>
> $\dfrac{1 - \cos \theta}{1 + \cos \theta} = \dfrac{1 - \cos \theta}{1 + \cos \theta}$ ✓   Simplify.
>
> ▶ **Guided Practice**
>
> **Verify that each equation is an identity.**
>
> **2A.** $\tan^2 \theta (\cot^2 \theta - \cos^2 \theta) = \cos^2 \theta$ **2B.** $\cos^2 \theta \sec \theta \csc \theta = \cot \theta$

Watch Out!

Verify Identities Verifying an identity is like checking the solution of an equation. You must simplify one or both sides separately until they are the same.

2 Transform Each Side of an Equation
Sometimes it is easier to transform each side of an equation separately into a common form. The following suggestions may be helpful as you verify trigonometric identities.

> **Key Concept**   Suggestions for Verifying Identities
>
> • Substitute one or more basic trigonometric identities to simplify the expression.
>
> • Factor or multiply as necessary. You may have to multiply both the numerator and denominator by the same trigonometric expression.
>
> • Write each side of the identity in terms of sine and cosine only. Then simplify each side as much as possible.
>
> • The properties of equality do not apply to identities as with equations. Do not perform operations to the quantities on each side of an unverified identity.

connectED.mcgraw-hill.com

Go Online!

Watch **Personal Tutor** videos to hear descriptions of how to verify trigonometric identities. Try describing how to verify an identity for a partner.

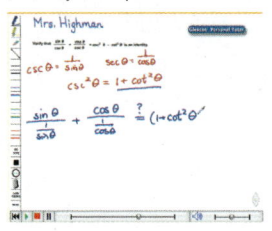

Example 3   Verify by Transforming Each Side

Verify that $1 - \tan^4 \theta = 2\sec^2 \theta - \sec^4 \theta$ is an identity.

$1 - \tan^4 \theta \stackrel{?}{=} 2\sec^2 \theta - \sec^4 \theta$	Original equation
$(1 - \tan^2 \theta)(1 + \tan^2 \theta) \stackrel{?}{=} \sec^2 \theta (2 - \sec^2 \theta)$	Factor each side.
$[1 - (\sec^2 \theta - 1)]\sec^2 \theta \stackrel{?}{=} (2 - \sec^2 \theta)\sec^2 \theta$	$1 + \tan^2 \theta = \sec^2 \theta$
$(2 - \sec^2 \theta)\sec^2 \theta = (2 - \sec^2 \theta)\sec^2 \theta \checkmark$	Simplify.

▶ Guided Practice

3. Verify that $\csc^2 \theta - \cot^2 \theta = \cot \theta \tan \theta$ is an identity.

Go Online! for a Self-Check Quiz

Check Your Understanding   ⬤ = Step-by-Step Solutions begin on page R11.

Examples 1–3   **MP PRECISION** Verify that each equation is an identity.

1. $\cot \theta + \tan \theta = \dfrac{\sec^2 \theta}{\tan \theta}$

2. $\cos^2 \theta = (1 + \sin \theta)(1 - \sin \theta)$

3. $\sin \theta = \dfrac{\sec \theta}{\tan \theta + \cot \theta}$

4. $\tan^2 \theta = \dfrac{1 - \cos^2 \theta}{\cos^2 \theta}$

5. $\tan^2 \theta \csc^2 \theta = 1 + \tan^2 \theta$

6. $\tan^2 \theta = (\sec \theta + 1)(\sec \theta - 1)$

7. $\dfrac{\tan^2 \theta + 1}{\tan^2 \theta} = \csc^2 \theta$

8. $\cot \theta + \sec \theta = \dfrac{\cos^2 \theta + \sin \theta}{\sin \theta \cos \theta}$

9 $\sin^2 \theta + \tan^2 \theta = (1 - \cos^2 \theta) + \dfrac{\sec^2 \theta}{\csc^2 \theta}$

Practice and Problem Solving

Extra Practice is on page R10.

Example 1   Verify that each equation is an identity.

10. $\cos^2 \theta + \tan^2 \theta \cos^2 \theta = 1$

11. $\cot \theta (\cot \theta + \tan \theta) = \csc^2 \theta$

12. $1 + \sec^2 \theta \sin^2 \theta = \sec^2 \theta$

13. $\sin \theta \sec \theta \cot \theta = 1$

14. $\dfrac{1 - \cos \theta}{1 + \cos \theta} = (\csc \theta - \cot \theta)^2$

15. $\dfrac{1 - 2\cos^2 \theta}{\sin \theta \cos \theta} = \tan \theta - \cot \theta$

16. $\tan \theta = \dfrac{\sec \theta}{\csc \theta}$

17. $\cos \theta = \sin \theta \cot \theta$

18. $(\sin \theta - 1)(\tan \theta + \sec \theta) = -\cos \theta$

19. $\cos \theta \cos(-\theta) - \sin \theta \sin(-\theta) = 1$

Example 2   **20. LADDER** Some students derived an expression for the length of a ladder that, when carried flat, could fit around a corner from a 5-foot-wide hallway into a 7-foot-wide hallway, as shown. They determined that the maximum length ℓ of a ladder that would fit was given by $\ell(\theta) = \dfrac{7\sin \theta + 5\cos \theta}{\sin \theta \cos \theta}$. When their teacher worked the problem, she concluded that $\ell(\theta) = 7\sec \theta + 5\csc \theta$. Are the two expressions equivalent?

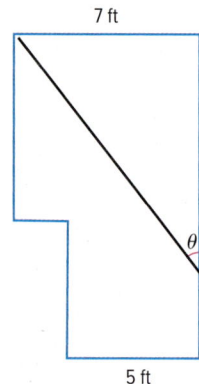

Example 3 Verify that each equation is an identity.

21. $\sec\theta - \tan\theta = \dfrac{1-\sin\theta}{\cos\theta}$

22. $\dfrac{1+\tan\theta}{\sin\theta+\cos\theta} = \sec\theta$

23. $\sec\theta\csc\theta = \tan\theta + \cot\theta$

24. $\sin\theta + \cos\theta = \dfrac{2\sin^2\theta - 1}{\sin\theta - \cos\theta}$

25. $(\sin\theta + \cos\theta)^2 = \dfrac{2+\sec\theta\csc\theta}{\sec\theta\csc\theta}$

26. $\dfrac{\cos\theta}{1-\sin\theta} = \dfrac{1+\sin\theta}{\cos\theta}$

27. $\csc\theta - 1 = \dfrac{\cot^2\theta}{\csc\theta + 1}$

28. $\cos\theta\cot\theta = \csc\theta - \sin\theta$

29. $\sin\theta\cos\theta\tan\theta + \cos^2\theta = 1$

30. $(\csc\theta - \cot\theta)^2 = \dfrac{1-\cos\theta}{1+\cos\theta}$

31. $\csc^2\theta = \cot^2\theta + \sin\theta\csc\theta$

32. $\dfrac{\sec\theta - \csc\theta}{\csc\theta\sec\theta} = \sin\theta - \cos\theta$

33. $\sin^2\theta + \cos^2\theta = \sec^2\theta - \tan^2\theta$

34. $\sec\theta - \cos\theta = \tan\theta\sin\theta$

35. **SENSE-MAKING** The diagram at the right represents a game of tetherball. As the ball rotates around the pole, a conical surface is swept out by the line segment $\overline{SP}$. A formula for the relationship between the length L of the string and the angle θ that the string makes with the pole is given by the equation $L = \dfrac{g\sec\theta}{\omega^2}$. Is $L = \dfrac{g\tan\theta}{\omega^2\sin\theta}$ also an equation for the relationship between L and θ?

36. **RUNNING** A portion of a racetrack has the shape of a circular arc with a radius of 16.7 meters. As a runner races along the arc, the sine of her angle of incline θ is found to be $\dfrac{1}{4}$. Find the speed of the runner. Use the Angle of Incline Formula given at the beginning of the lesson, $\tan\theta = \dfrac{v^2}{gR}$, where $g = 9.8$ and R is the radius. (*Hint*: Find $\cos\theta$ first.)

When simplified, would the expression be equal to 1 or -1?

37. $\cot(-\theta)\tan(-\theta)$

38. $\sin\theta\csc(-\theta)$

39. $\sin^2(-\theta) + \cos^2(-\theta)$

40. $\sec(-\theta)\cos(-\theta)$

41. $\sec^2(-\theta) - \tan^2(-\theta)$

42. $\cot(-\theta)\cot\left(\dfrac{\pi}{2} - \theta\right)$

Simplify the expression to either a constant or a basic trigonometric function.

43. $\dfrac{\tan\left(\dfrac{\pi}{2} - \theta\right)\csc\theta}{\csc^2\theta}$

44. $\dfrac{1+\tan\theta}{1+\cot\theta}$

45. $(\sec^2\theta + \csc^2\theta) - (\tan^2\theta + \cot^2\theta)$

46. $\dfrac{\sec^2\theta - \tan^2\theta}{\cos^2 x + \sin^2 x}$

47. $\tan\theta\cos\theta$

48. $\cot\theta\tan\theta$

49. $\sec\theta\sin\left(\dfrac{\pi}{2} - \theta\right)$

50. $\dfrac{1+\tan^2\theta}{\csc^2\theta}$

51. **PHYSICS** When a firework is fired from the ground, its height y and horizontal displacement x are related by the equation $y = \dfrac{-gx^2}{2v_0^2\cos^2\theta} + \dfrac{x\sin\theta}{\cos\theta}$, where v_0 is the initial velocity of the projectile, θ is the angle at which it was fired, and g is the acceleration due to gravity. Rewrite this equation so that $\tan\theta$ is the only trigonometric function that appears in the equation.

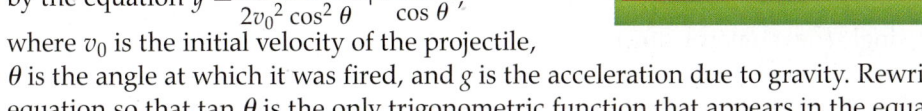

52. **ELECTRONICS** When an alternating current of frequency f and peak current I_0 passes through a resistance R, the power delivered to the resistance at time t seconds is $P = I_0^2 R \sin^2 2\pi ft$.

   a. Write an expression for the power in terms of $\cos^2 2\pi ft$.

   b. Write an expression for the power in terms of $\csc^2 2\pi ft$.

53. **THROWING A BALL** In this problem, you will investigate the path of a ball represented by the equation $h = \dfrac{v_0^2 \sin^2 \theta}{2g}$, where θ is the measure of the angle between the ground and the path of the ball, v_0 is its initial velocity in meters per second, and g is the acceleration due to gravity. The value of g is 9.8 m/s².

   a. If the initial velocity of the ball is 47 meters per second, find the height of the ball at 30°, 45°, 60°, and 90°. Round to the nearest tenth.

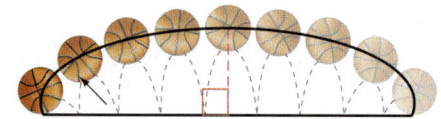

   b. Graph the equation on a graphing calculator.

   c. Show that the formula $h = \dfrac{v_0^2 \tan^2 \theta}{2g \sec^2 \theta}$ is equivalent to the one given above.

H.O.T. Problems — Use Higher-Order Thinking Skills

54. **WHICH ONE DOESN'T BELONG?** Identify the equation that does not belong with the other three. Explain your reasoning.

$\sin^2 \theta + \cos^2 \theta = 1$	$1 + \cot^2 \theta = \csc^2 \theta$
$\sin^2 \theta - \cos^2 \theta = 2 \sin^2 \theta$	$\tan^2 \theta + 1 = \sec^2 \theta$

55. **CHALLENGE** Transform the right side of $\tan^2 \theta = \dfrac{\sin^2 \theta}{\cos^2 \theta}$ to show that $\tan^2 \theta = \sec^2 \theta - 1$.

56. **WRITING IN MATH** Explain why you cannot square each side of an equation when verifying a trigonometric identity.

57. **REASONING** Explain why $\sin^2 \theta + \cos^2 \theta = 1$ is an identity, but $\sin \theta = \sqrt{1 - \cos \theta}$ is not.

58. **WRITE A QUESTION** A classmate is having trouble trying to verify a trigonometric identity involving multiple trigonometric functions to multiple degrees. Write a question to help her work through the problem.

59. **WRITING IN MATH** Why do you think expressions in trigonometric identities are often rewritten in terms of sine and cosine?

60. **CHALLENGE** Let $x = \dfrac{1}{2} \tan \theta$, where $-\dfrac{\pi}{2} < \theta < \dfrac{\pi}{2}$. Write $f(x) = \dfrac{x}{\sqrt{1 + 4x^2}}$ in terms of a single trigonometric function of θ.

61. **REASONING** Justify the three basic Pythagorean identities.

Preparing for Assessment

62. If $\tan(-\theta)\cos(-\theta) = Y$, then which of the following expressions could be Y? **MP** 1, 8

- A $\dfrac{\cos^2 \theta}{\sin \theta}$
- B 1
- C −1
- D $-\csc \theta$
- E $-\sin \theta$

63. Which trigonometric function when defined is equivalent to $\dfrac{\cos^2 \alpha}{\sin \alpha \cot \alpha}$? **MP** 1, 8

- A $\cos \alpha$
- B $\dfrac{\cos \alpha}{\sin^2 \alpha}$
- C $\dfrac{1}{\cos \alpha}$
- D $\cot^2 \alpha$
- E $\dfrac{\cos^3 \alpha}{\sin^2 \alpha}$

64. For which value(s) for h is $h \sin \theta = \sin(h\theta)$ an identity? **MP** 1, 8

    I $h = -1$
    II $h = 0$
    III $h = 1$

- A I and II only
- B II only
- C III only
- D II and III only
- E I, II, and III

65. MULTI-STEP

  a. For all $\theta \neq \dfrac{n\pi}{2}$, the identity shown below is complete for which of the following expressions? **MP** 1, 8

$$\dfrac{\sin^2 \theta + \cos^2 \theta}{\tan \theta \cos \theta} = \underline{\qquad}$$

- A 1
- B $\dfrac{1}{\cos \theta}$
- C $\dfrac{1}{\sin \theta}$
- D $\sin \theta + \cos \theta$
- E $\sin \theta + \dfrac{\cos^2 \theta}{\sin \theta}$

  b. Use trigonometric identities to simplify the expression in part **a**.

66. Prove that $\dfrac{\tan \theta}{\sin \theta} = \sec \theta$. **MP** 1, 8

67. Prove that $\sin x \cos x \tan x = 1 - \cos^2 x$. **MP** 1, 8

68. When simplified, which expressions would be equal to −1? Check all that apply. **MP** 1, 8

- ☐ A $\csc \theta \sin(-\theta)$
- ☐ B $\cos \theta \sec(-\theta)$
- ☐ C $\sin^2 \theta + \cos^2 \theta$
- ☐ D $\cot(-\theta) \tan \theta$
- ☐ E $\sec^2(-\theta) - \tan^2(-\theta)$

69. What is the simplified form of $\cot \theta \sec \theta$? **MP** 1, 8

- A 1
- B −1
- C $\sin \theta$
- D $\csc \theta$
- E $\tan \theta$

LESSON 3

Sum and Difference Identities

::Then::
- You found values of trigonometric functions for general angles.

::Now::
1. Find values of sine and cosine by using sum and difference identities.
2. Verify trigonometric identities by using sum and difference identities.

::Why?::
Have you ever been using a wireless Internet provider and temporarily lost the signal? Waves that pass through the same place at the same time cause interference. Interference occurs when two waves combine to have a greater, or smaller, amplitude than either of the component waves.

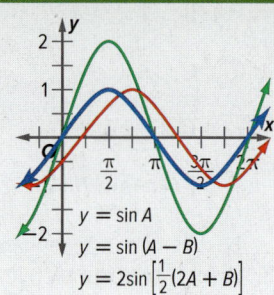

MP Mathematical Practices
3 Construct viable arguments and critique the reasoning of others.
6 Attend to precision.

1 Sum and Difference Identities
Notice that the third equation shown above involves the sum of A and B. It is often helpful to use formulas for the trigonometric values of the difference or sum of two angles. For example, you could find the exact value of sin 15° by evaluating sin (60° − 45°). Formulas exist that can be used to evaluate expressions like sin $(A − B)$ or cos $(A + B)$.

Key Concept Sum and Difference Identities

Sum Identities
- $\sin(A + B) = \sin A \cos B + \cos A \sin B$
- $\cos(A + B) = \cos A \cos B - \sin A \sin B$
- $\tan(A + B) = \dfrac{\tan A + \tan B}{1 - \tan A \tan B}$

Difference Identities
- $\sin(A - B) = \sin A \cos B - \cos A \sin B$
- $\cos(A - B) = \cos A \cos B + \sin A \sin B$
- $\tan(A - B) = \dfrac{\tan A - \tan B}{1 + \tan A \tan B}$

Example 1 Find Trigonometric Values

Find the exact value of each expression.

a. sin 105°

Use the identity $\sin(A + B) = \sin A \cos B + \cos A \sin B$.

$\sin 105° = \sin(60° + 45°)$ *A = 60° and B = 45°*

$= \sin 60° \cos 45° + \cos 60° \sin 45°$ *Sum identity*

$= \left(\dfrac{\sqrt{3}}{2} \cdot \dfrac{\sqrt{2}}{2}\right) + \left(\dfrac{1}{2} \cdot \dfrac{\sqrt{2}}{2}\right)$ *Evaluate each expression.*

$= \dfrac{\sqrt{6}}{4} + \dfrac{\sqrt{2}}{4}$ or $\dfrac{\sqrt{6} + \sqrt{2}}{4}$ *Multiply.*

b. cos (−120°)

Use the identity $\cos(A - B) = \cos A \cos B + \sin A \sin B$.

$\cos(-120) = \cos(60° - 180°)$ *A = 60° and B = 180°*

$= \cos 60° \cos 180° + \sin 60° \sin 180°$ *Difference identity*

$= \dfrac{1}{2} \cdot (-1) + \dfrac{\sqrt{3}}{2} \cdot 0$ *Evaluate each expression.*

$= -\dfrac{1}{2}$ *Multiply.*

▶ Guided Practice

1A. sin 15°

1B. cos (−15°)

Real-World Example 2 Sum and Difference of Angles Identities

SURVEYING A surveyor measures the angle between one side of a rectangular lot and the line from her position to the opposite corner of the lot as 30°. She then measures the angle between that line and the west bank of a creek as 45°. She is standing 100 yards from the opposite corner of the property. How long is the west bank of the creek within the lot?

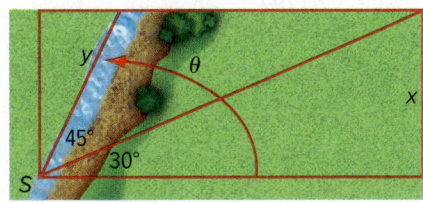

Understand The question asks for the length of the creek within the lot, or y.

Plan Draw a picture that labels all the things that you know from the information given.

Solve Solve for x.

$\sin 30° = \dfrac{x}{100}$ Definition of sine

$x = 100 \sin 30°$

$x = 50$ Since the lot is rectangular, opposite sides are equal.

Now look at the triangle on the far left and solve for y.

$\cos 15° = \dfrac{50}{y}$ Definition of cosine

$\cos (45° - 30°) = \dfrac{50}{y}$ $15 = 45 - 30$

$\cos 45° \cos 30° + \sin 45° \sin 30° = \dfrac{50}{y}$ Difference identity

$\dfrac{\sqrt{2}}{2} \cdot \dfrac{\sqrt{3}}{2} + \dfrac{\sqrt{2}}{2} \cdot \dfrac{1}{2} = \dfrac{50}{y}$ Evaluate.

$\dfrac{\sqrt{6} + \sqrt{2}}{4} = \dfrac{50}{y}$ Simplify.

$(\sqrt{6} + \sqrt{2})y = 200$ Cross products

$y = \dfrac{200}{(\sqrt{6} + \sqrt{2})} \cdot \dfrac{(\sqrt{6} - \sqrt{2})}{(\sqrt{6} - \sqrt{2})}$

$y = 50(\sqrt{6} - \sqrt{2})$

$y = 50\sqrt{6} - 50\sqrt{2}$ or about 51.8

The length of the creek within the lot is about 51.8 yards.

Check Use a calculator to find Arccos $\dfrac{50}{51.8} \approx 15°$. ✓

> **Problem-Solving Tip**
>
> **Make a Model** Make a model to visualize a problem situation. A model can be a drawing or a figure made of different objects, such as algebra tiles or folded paper.

Guided Practice

2. The harmonic motion of an object can be described by $x = 4 \cos \left(2\pi t - \dfrac{\pi}{4}\right)$, where x is the distance from the equilibrium point in inches and t is time in minutes. Find the exact distance from the equilibrium point at 45 seconds.

Study Tip

MP Sense-Making Make a list of the trigonometric values for the angles between 0° and 360° for which the sum and difference identities can be easily used. Use your list as a reference.

2 Verify Trigonometric Identities
You can also use the sum and difference identities to verify identities.

Example 3 Verify Trigonometric Identities

Verify that each equation is an identity.

a. $\cos(90° - \theta) = \sin\theta$

$\cos(90° - \theta) \stackrel{?}{=} \sin\theta$ — Original equation

$\cos 90° \cos\theta + \sin 90° \sin\theta \stackrel{?}{=} \sin\theta$ — Sum identity

$0 \cdot \cos\theta + 1 \cdot \sin\theta \stackrel{?}{=} \sin\theta$ — Evaluate each expression.

$\sin\theta = \sin\theta$ ✓ — Simplify.

b. $\sin\left(\theta + \dfrac{\pi}{2}\right) = \cos\theta$

$\sin\left(\theta + \dfrac{\pi}{2}\right) \stackrel{?}{=} \cos\theta$ — Original equation

$\sin\theta \cos\dfrac{\pi}{2} + \cos\theta \sin\dfrac{\pi}{2} \stackrel{?}{=} \cos\theta$ — Sum identity

$\sin\theta \cdot 0 + \cos\theta \cdot 1 \stackrel{?}{=} \cos\theta$ — Evaluate each expression.

$\cos\theta = \cos\theta$ ✓ — Simplify.

Guided Practice

3A. $\sin(90° - \theta) = \cos\theta$

3B. $\cos(90° + \theta) = -\sin\theta$

Check Your Understanding

◯ = Step-by-Step Solutions begin on page R11.

✓ *Go Online!* for a Self-Check Quiz

Example 1 Find the exact value of each expression.

1. $\cos 165°$
2. $\cos 105°$
3. $\cos 75°$
4. $\sin(-30°)$
5. $\sin 135°$
6. $\sin(-210°)$

Example 2

7. **MP MODELING** Refer to the beginning of the lesson. *Constructive interference* occurs when two waves combine to have a greater amplitude than either of the component waves. *Destructive interference* occurs when the component waves combine to have a smaller amplitude. The first signal can be modeled by the equation $y = 20\sin(3\theta + 45°)$. The second signal can be modeled by the equation $y = 20\sin(3\theta + 225°)$.

   a. Find the sum of the two functions.

   b. What type of interference results when signals modeled by the two equations are combined?

Example 3 Verify that each equation is an identity.

8. $\sin(90° + \theta) = \cos\theta$
9. $\cos\left(\dfrac{3\pi}{2} - \theta\right) = -\sin\theta$
10. $\tan\left(\theta + \dfrac{\pi}{2}\right) = -\cot\theta$
11. $\sin(\theta + \pi) = -\sin\theta$

670 | Lesson 10-3 | Sum and Difference Identities

Practice and Problem Solving

Extra Practice is on page R10.

Example 1 Find the exact value of each expression.

12. $\sin 165°$
13. $\cos 135°$
14. $\cos \frac{7\pi}{12}$
15. $\sin \frac{\pi}{12}$
16. $\tan 195°$
17. $\cos \left(-\frac{\pi}{12}\right)$

Example 2

18. **ELECTRONICS** In a certain circuit carrying alternating current, the formula $c = 2 \sin (120t)$ can be used to find the current c in amperes after t seconds.

    a. Rewrite the formula using the sum of two angles.

    b. Use the sum of angles formula to find the exact current at $t = 1$ second.

Example 3 Verify that each equation is an identity.

19. $\cos \left(\frac{\pi}{2} + \theta\right) = -\sin \theta$
20. $\cos (60° + \theta) = \sin (30° - \theta)$
21. $\cos (180° + \theta) = -\cos \theta$
22. $\tan (\theta + 45°) = \frac{1 + \tan \theta}{1 - \tan \theta}$

23. **REASONING** The monthly high temperatures for Minneapolis, Minnesota, can be modeled by the equation $y = 31.65 \sin \left(\frac{\pi}{6}x - 2.09\right) + 52.35$, where the months x are represented by January = 1, February = 2, and so on. The monthly low temperatures for Minneapolis can be modeled by the equation $y = 30.15 \sin \left(\frac{\pi}{6}x - 2.09\right) + 32.95$.

    a. Write a new function by adding the expressions on the right side of each equation and dividing the result by 2.

    b. What is the meaning of the function you wrote in part **a**?

Find the exact value of each expression.

24. $\tan 165°$
25. $\sec 1275°$
26. $\sin 735°$
27. $\tan \frac{23\pi}{12}$
28. $\csc \frac{5\pi}{12}$
29. $\cot \frac{113\pi}{12}$

30. **FORCE** In the figure at the right, the effort F necessary to hold a safe in position on a ramp is given by $F = \frac{W(\sin A + \mu \cos A)}{\cos A - \mu \sin A}$, where W is the weight of the safe and $\mu = \tan \theta$. Show that $F = W \tan (A + \theta)$.

31. **QUILTING** As part of a quilt that is being made, the quilter places two right triangular swatches together to make a new triangular piece. One swatch has sides 6 inches, 8 inches, and 10 inches long. The other swatch has sides 8 inches, $8\sqrt{3}$ inches, and 16 inches long. The pieces are placed with the sides of eight inches against each other, as shown in the figure, to form triangle ABC.

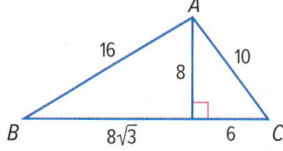

    a. What is the exact value of the sine of angle BAC?

    b. What is the exact value of the cosine of angle BAC?

    c. What is the measure of angle BAC?

    d. Is the new triangle formed from the two triangles also a right triangle?

32. **OPTICS** When light passes symmetrically through a prism, the index of refraction n of the glass with respect to air is
$$n = \frac{\sin\left[\frac{1}{2}(a+b)\right]}{\sin\frac{b}{2}},$$
where a is the measure of the deviation angle and b is the measure of the prism apex angle.

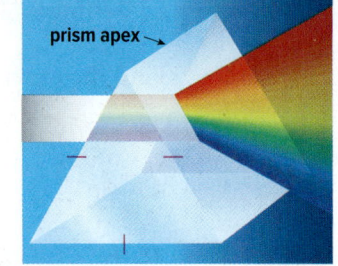

a. Show that for the prism shown, $n = \sqrt{3}\sin\frac{a}{2} + \cos\frac{a}{2}$.

b. Find n for the prism shown.

33. **MULTIPLE REPRESENTATIONS** In this problem, you will disprove the hypothesis that $\sin(A + B) = \sin A + \sin B$.

    a. **Tabular** Copy and complete the table.

    b. **Graphical** Assume that B is always 15° less than A. Use a graphing calculator to graph $y = \sin(x + x - 15)$ and $y = \sin x + \sin(x - 15)$ on the same screen.

    c. **Analytical** Determine whether $\cos(A + B) = \cos A + \cos B$ is an identity. Explain your reasoning.

A	B	sin A	sin B	sin (A + B)	sin A + sin B
30°	90°				
45°	60°				
60°	45°				
90°	30°				

Verify that each equation is an identity.

34. $\sin(A + B) = \dfrac{\tan A + \tan B}{\sec A \sec B}$

35. $\cos(A + B) = \dfrac{1 - \tan A \tan B}{\sec A \sec B}$

36. $\sec(A - B) = \dfrac{\sec A \sec B}{1 + \tan A \tan B}$

37. $\sin(A + B)\sin(A - B) = \sin^2 A - \sin^2 B$

H.O.T. Problems   Use Higher-Order Thinking Skills

38. **REASONING** Simplify the following expression without expanding any of the sums or differences.
$$\sin\left(\frac{\pi}{3} - \theta\right)\cos\left(\frac{\pi}{3} + \theta\right) - \cos\left(\frac{\pi}{3} - \theta\right)\sin\left(\frac{\pi}{3} + \theta\right)$$

39. **WRITING IN MATH** Use the information at the beginning of the lesson and in Exercise 7 to explain how the sum and difference identities are used to describe wireless Internet interference. Include an explanation of the difference between constructive and destructive interference.

40. **CHALLENGE** Derive an identity for $\cot(A + B)$ in terms of $\cot A$ and $\cot B$.

41. **REASONING** The figure shows two angles A and B in standard position on the unit circle. Use the Distance Formula to find d, where $(x_1, y_1) = (\cos B, \sin B)$ and $(x_2, y_2) = (\cos A, \sin A)$.

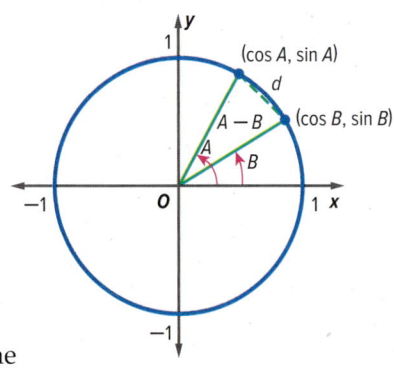

42. **OPEN-ENDED** Consider the following theorem. *If A, B, and C are the angles of an oblique triangle, then $\tan A + \tan B + \tan C = \tan A \tan B \tan C$.* Choose values for A, B, and C. Verify that the conclusion is true for your specific values.

Preparing for Assessment

43. The expression sin 45° cos 30° + cos 45° sin 30° can be used to evaluate which of the following measures? **MP 1**
- A sin 15°
- B cos 15°
- C tan 75°
- D sin 75°
- E cos 75°

44. What is the exact measure of tan 15°? **MP 6**
- A $\dfrac{1-\sqrt{3}}{1+\sqrt{3}}$
- B $1 - \dfrac{\sqrt{3}}{3}$
- C 1
- D $2 + \sqrt{3}$
- E $2 - \sqrt{3}$

45. If cos 60° = $\dfrac{1}{2}$, which of the following expressions is equal to 1? **MP 6**
- A sin² 30° − cos² 30°
- B sin² 30° + cos² 30°
- C sin² 60° + cos² 0°
- D −sin 60° + cos 0°
- E −sin 30° + cos 30°

46. What is the exact value of tan $\dfrac{5\pi}{12}$? **MP 6**

☐

47. Which expression can be used to find cos 10°? **MP 1**
- A sin 60° cos 50° − cos 60° sin 50°
- B cos 60° cos 50° − sin 60° sin 50°
- C sin 60° cos 50° + cos 60° sin 50°
- D cos 60° cos 50° + sin 60° sin 50°
- E cos 60° cos 50° − cos 60° cos 50°

48. MULTI-STEP
  a. Which of the following expressions has the same value as cos (π + θ)? **MP 2**
   - A −cos θ
   - B −sin θ
   - C $\dfrac{1}{\sin \theta}$
   - D $\dfrac{1}{\cos \theta}$
   - E sin θ − cos θ
  b. Use the sum identity to determine the value of cos (π + θ).

49. Which expressions have a value of $\dfrac{1}{\sqrt{2}}$? Check all that apply. **MP 6, 8**
- ☐ A sin (−45°)
- ☐ B cos (−45°)
- ☐ C sin 45°
- ☐ D sin 135°
- ☐ E cos (−135°)

50. Which expression can be evaluated using the Angle Difference Identity? **MP 2**
- A $\dfrac{\tan 30° + \tan 45°}{1 - \tan 30° \tan 45°}$
- B cos 30° cos 45° − sin 30° sin 45°
- C sin 30° cos 45° + cos 30° sin 45°
- D cos 30° cos 45° + sin 30° sin 45°

CHAPTER 10
Mid-Chapter Quiz
Lessons 10-1 through 10-3

Simplify each expression. (Lesson 10-1)

1. $\cot\theta \sec\theta$

2. $\dfrac{1 - \cos^2\theta}{\sin^2\theta}$

3. $\dfrac{1}{\cos\theta} - \dfrac{\sin^2\theta}{\cos\theta}$

4. $\cos\left(\dfrac{\pi}{2} - \theta\right)\csc\theta$

5. **HISTORY** In 1861, the United States 34-star flag was adopted. For this flag, $\tan\theta = \dfrac{31.5}{51}$. Find $\sin\theta$.

Find the value of each expression. (Lesson 10-1)

6. $\sin\theta$, if $\cos\theta = \dfrac{3}{5}$; $0° < \theta < 90°$

7. $\csc\theta$, if $\cot\theta = \dfrac{1}{2}$; $270° < \theta < 360°$

8. $\tan\theta$, if $\sec\theta = \dfrac{4}{3}$; $0° < \theta < 90°$

9. **MULTIPLE CHOICE** Which of the following is equivalent to $\dfrac{\cos\theta}{1 - \sin^2\theta}$? (Lesson 10-1)

   A $\cos\theta$
   B $\csc\theta$
   C $\tan\theta$
   D $\sec\theta$

10. **AMUSEMENT PARKS** Suppose a child on a merry-go-round is seated on an outside horse. The diameter of the merry-go-round is 16 meters. The angle of inclination is represented by the equation $\tan\theta = \dfrac{v^2}{gR}$, where R is the radius of the circular path, v is the speed in meters per second, and g is 9.8 meters per second squared. (Lesson 10-1)

    a. If the sine of the angle of inclination of the child is $\dfrac{1}{5}$, what is the angle of inclination made by the child?

    b. What is the velocity of the merry-go-round?

    c. **MP** What mathematical practice did you use to solve this problem?

Verify that each of the following is an identity. (Lesson 10-2)

11. $\cot^2\theta + 1 = \dfrac{\cot\theta}{\cos\theta \cdot \sin\theta}$

12. $\dfrac{\cos\theta \csc\theta}{\cot\theta} = 1$

13. $\dfrac{\sin\theta \tan\theta}{1 - \cos\theta} = (1 + \cos\theta)\sec\theta$

14. $\tan\theta(1 - \sin\theta) = \dfrac{\cos\theta \sin\theta}{1 + \sin\theta}$

15. **COMPUTER** The front of a computer monitor is usually measured along the diagonal of the screen as shown below. (Lesson 10-2)

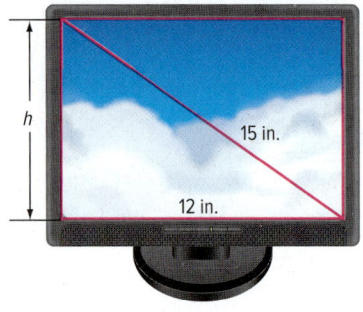

   a. Find h.

   b. Using the diagram shown, show that $\cot\theta = \dfrac{\cos\theta}{\sin\theta}$.

Verify that each of the following is an identity. (Lesson 10-2)

16. $\tan^2\theta + 1 = \dfrac{\tan\theta}{\cos\theta \cdot \sin\theta}$

17. $\dfrac{\sin\theta \cdot \sec\theta}{\sec\theta - 1} = (\sec\theta + 1)\cot\theta$

18. $\sin^2\theta \cdot \tan^2\theta = \tan^2\theta - \sin^2\theta$

19. $\cot\theta(1 - \cos\theta) = \dfrac{\cos\theta \cdot \sin\theta}{1 + \cos\theta}$

Find the exact value of each expression. (Lesson 10-3)

20. $\cos 105°$

21. $\sin(-135°)$

22. $\tan 15°$

23. $\cot 75°$

24. **MULTIPLE CHOICE** What is the exact value of $\cos\dfrac{5\pi}{12}$? (Lesson 10-3)

    A $\sqrt{2}$
    B $\dfrac{\sqrt{6} + \sqrt{2}}{2}$
    C $\dfrac{\sqrt{6} - \sqrt{2}}{4}$
    D $\dfrac{\sqrt{6} + \sqrt{2}}{4}$

25. Verify that $\cos 30° \cos\theta + \sin 30° \sin\theta = \sin 60° \cos\theta + \cos 60° \sin\theta$ is an identity. (Lesson 10-3)

LESSON 4
Double-Angle and Half-Angle Identities

∴Then
- You found values of sine and cosine by using sum and difference identities.

∴Now
1. Find values of sine and cosine by using double-angle identities.
2. Find values of sine and cosine by using half-angle identities.

∴Why?
- Chicago's Buckingham Fountain contains jets placed at specific angles to create arcs. When a stream of water shoots into the air with velocity v at an angle of θ with the horizontal, the model predicts that the water will travel a horizontal distance of $D = \frac{v^2}{g} \sin 2\theta$ and reach a maximum height of $H = \frac{v^2}{2g} \sin^2 \theta$. The ratio of H to D helps determine the total height and width of the fountain. Express $\frac{H}{D}$ as a function of θ.

MP Mathematical Practices
3 Construct viable arguments and critique the reasoning of others.
6 Attend to precision.

1 Double-Angle Identities
It is sometimes useful to have identities to find the value of a function of twice an angle or half an angle.

Key Concept Double-Angle Identities

The following identities hold true for all values of θ.

$$\sin 2\theta = 2 \sin \theta \cos \theta$$

$$\cos 2\theta = \cos^2 \theta - \sin^2 \theta$$
$$\cos 2\theta = 2 \cos^2 \theta - 1$$
$$\cos 2\theta = 1 - 2 \sin^2 \theta$$

$$\tan 2\theta = \frac{2 \tan \theta}{1 - \tan^2 \theta}$$

Example 1 Double-Angle Identities

Find the exact value of $\sin 2\theta$ if $\sin \theta = \frac{2}{3}$ and θ is between $0°$ and $90°$.

Step 1 Use the identity $\sin 2\theta = 2 \sin \theta \cos \theta$ to find the value of $\cos \theta$.

$\cos^2 \theta = 1 - \sin^2 \theta$ $\cos^2 \theta + \sin^2 \theta = 1$

$\cos^2 \theta = 1 - \left(\frac{2}{3}\right)^2$ $\sin \theta = \frac{2}{3}$

$\cos^2 \theta = \frac{5}{9}$ Subtract.

$\cos \theta = \pm \frac{\sqrt{5}}{3}$ Take the square root of each side.

Because θ is in the first quadrant, cosine is positive. Thus, $\cos \theta = \frac{\sqrt{5}}{3}$.

Step 2 Find $\sin 2\theta$.

$\sin 2\theta = 2 \sin \theta \cos \theta$ Double-angle identity

$= 2\left(\frac{2}{3}\right)\left(\frac{\sqrt{5}}{3}\right)$ $\sin \theta = \frac{2}{3}$ and $\cos \theta = \frac{\sqrt{5}}{3}$

$= \frac{4\sqrt{5}}{9}$ Multiply.

Guided Practice

1. Find the exact value of $\sin 2\theta$ if $\cos \theta = -\frac{1}{3}$ and $90° < \theta < 180°$.

Example 2 Double-Angle Identities

Find the exact value of each expression if $\sin \theta = \frac{2}{3}$ and θ is between 0° and 90°.

a. $\cos 2\theta$

Because we know the values of $\cos \theta$ and $\sin \theta$, we can use any of the double-angle identities for cosine. We will use the identity $\cos 2\theta = 1 - 2\sin^2 \theta$.

$\cos 2\theta = 1 - 2\sin^2 \theta$ — Double-angle identity

$= 1 - 2\left(\frac{2}{3}\right)^2$ or $\frac{1}{9}$ — $\sin \theta = \frac{2}{3}$

b. $\tan 2\theta$

Step 1 Find $\tan \theta$ to use the double-angle identity for $\tan 2\theta$.

$\tan \theta = \dfrac{\sin \theta}{\cos \theta}$ — Definition of tangent

$= \dfrac{\frac{2}{3}}{\frac{\sqrt{5}}{3}}$ — $\sin \theta = \frac{2}{3}$ and $\cos \theta = \frac{\sqrt{5}}{3}$

$= \dfrac{2}{\sqrt{5}}$ or $\dfrac{2\sqrt{5}}{5}$ — Rationalize the denominator.

Step 2 Find $\tan 2\theta$.

$\tan 2\theta = \dfrac{2\tan \theta}{1 - \tan^2 \theta}$ — Double-angle identity

$= \dfrac{2\left(\frac{2\sqrt{5}}{5}\right)}{1 - \left(\frac{2\sqrt{5}}{5}\right)^2}$ — $\tan \theta = \frac{2\sqrt{5}}{5}$

$= \dfrac{2\left(\frac{2\sqrt{5}}{5}\right)}{\frac{25}{25} - \frac{20}{25}}$ — Square the denominator.

$= \dfrac{\frac{4\sqrt{5}}{5}}{\frac{1}{5}}$ — Simplify.

$= \dfrac{4\sqrt{5}}{5} \cdot \dfrac{5}{1}$ or $4\sqrt{5}$ — $\dfrac{a}{b} \div \dfrac{c}{d} = \dfrac{a}{b} \cdot \dfrac{d}{c}$

Guided Practice

Find the exact value of each expression if $\cos \theta = -\frac{1}{3}$ and $90° < \theta < 180°$.

2A. $\cos 2\theta$ **2B.** $\tan 2\theta$

> **Study Tip**
> **Deriving Formulas** You can use the identity for $\sin(A + B)$ to find the sine of twice an angle θ, $\sin 2\theta$, and the identity for $\cos(A + B)$ to find the cosine of twice an angle θ, $\cos 2\theta$.

Real-World Career

Electrician An electrician specializes in the wiring of electrical components. Electricians serve an apprenticeship lasting 3–5 years. Schooling in electrical theory and building codes is required. Certification requires work experience and a passing score on a written test.

2 Half-Angle Identities
It is sometimes useful to have identities to find the value of a function of half an angle.

> **Key Concept** Half-Angle Identities
>
> The following identities hold true for all values of θ.
>
> $\sin \dfrac{\theta}{2} = \pm\sqrt{\dfrac{1 - \cos \theta}{2}}$ $\cos \dfrac{\theta}{2} = \pm\sqrt{\dfrac{1 + \cos \theta}{2}}$ $\tan \dfrac{\theta}{2} = \pm\sqrt{\dfrac{1 - \cos \theta}{1 + \cos \theta}}$, $\cos \theta \neq -1$

Study Tip

Choosing the Sign In the first step of the solution, you may want to determine the quadrant in which the terminal side of $\frac{\theta}{2}$ will lie. Then you can use the correct sign from that point on.

Example 3 Half-Angle Identities

a. Find the exact value of $\cos \frac{\theta}{2}$ if $\sin \theta = -\frac{4}{5}$ and θ is in the third quadrant.

$\cos^2 \theta = 1 - \sin^2 \theta$ Use a Pythagorean identity to find $\cos \theta$.

$\cos^2 \theta = 1 - \left(-\frac{4}{5}\right)^2$ $\sin \theta = -\frac{4}{5}$

$\cos^2 \theta = 1 - \frac{16}{25}$ Evaluate exponent.

$\cos^2 \theta = \frac{9}{25}$ Subtract.

$\cos \theta = \pm \frac{3}{5}$ Take the square root of each side.

Because θ is in the third quadrant, $\cos \theta = -\frac{3}{5}$.

$\cos \frac{\theta}{2} = \pm \sqrt{\frac{1 + \cos \theta}{2}}$ Half-angle identity

$= \pm \sqrt{\frac{1 - \frac{3}{5}}{2}}$ $\cos \theta = -\frac{3}{5}$

$= \pm \sqrt{\frac{1}{5}}$ Simplify.

$= \pm \frac{1}{\sqrt{5}} \cdot \frac{\sqrt{5}}{\sqrt{5}}$ or $\pm \frac{\sqrt{5}}{5}$ Rationalize the denominator.

If θ is between 180° and 270°, $\frac{\theta}{2}$ is between 90° and 135°. So, $\cos \frac{\theta}{2}$ is $-\frac{\sqrt{5}}{5}$.

Reading Math

Precision The first sign of the half-angle identity is read *plus or minus*. Unlike with the double-angle identities, you must determine the sign.

b. Find the exact value of $\cos 67.5°$.

$\cos 67.5° = \cos \frac{135°}{2}$ $67.5° = \frac{135°}{2}$

$= \sqrt{\frac{1 + \cos 135°}{2}}$ $\cos \frac{\theta}{2} = \pm \sqrt{\frac{1 + \cos \theta}{2}}$

$= \sqrt{\frac{1 - \frac{\sqrt{2}}{2}}{2}}$ 67.5° is in Quadrant I; the value is positive.

$= \sqrt{\frac{\frac{2}{2} - \frac{\sqrt{2}}{2}}{2}}$ $1 = \frac{2}{2}$

$= \sqrt{\frac{\frac{2 - \sqrt{2}}{2}}{2}}$ Subtract fractions.

$= \sqrt{\frac{2 - \sqrt{2}}{2} \cdot \frac{1}{2}}$ $\frac{a}{b} \div \frac{c}{d} = \frac{a}{b} \cdot \frac{d}{c}$

$= \sqrt{\frac{2 - \sqrt{2}}{4}}$ Multiply.

$= \frac{\sqrt{2 - \sqrt{2}}}{\sqrt{4}}$ $\sqrt{\frac{a}{b}} = \frac{\sqrt{a}}{\sqrt{b}}$

$= \frac{\sqrt{2 - \sqrt{2}}}{2}$ Simplify.

▶ **Guided Practice**

3. Find the exact value of $\sin \frac{\theta}{2}$ if $\sin \theta = \frac{2}{3}$ and θ is in the second quadrant.

Real-World Link
The City Hall Park Fountain in New York City is located in the heart of Manhattan in front of City Hall.
Source: Fodor's

Real-World Example 4 Simplify Using Double-Angle Identities

FOUNTAIN Refer to the beginning of the lesson. Find $\frac{H}{D}$.

$$\frac{H}{D} = \frac{\frac{v^2}{2g}\sin^2\theta}{\frac{v^2}{g}\sin 2\theta} \qquad \text{Original equation}$$

$$= \frac{\frac{v^2 \sin^2\theta}{2g}}{\frac{v^2 \sin 2\theta}{g}} \qquad \text{Simplify the numerator and denominator.}$$

$$= \frac{v^2 \sin^2\theta}{2g} \cdot \frac{g}{v^2 \sin 2\theta} \qquad \frac{a}{b} \div \frac{c}{d} = \frac{a}{b} \cdot \frac{d}{c}$$

$$= \frac{\sin^2\theta}{2\sin 2\theta} \qquad \text{Simplify.}$$

$$= \frac{\sin^2\theta}{4\sin\theta \cos\theta} \qquad \sin 2\theta = 2\sin\theta\cos\theta$$

$$= \frac{1}{4} \cdot \frac{\sin\theta}{\cos\theta} \qquad \text{Simplify.}$$

$$= \frac{1}{4}\tan\theta \qquad \frac{\sin\theta}{\cos\theta} = \tan\theta$$

▶ **Guided Practice**

Find each value.

4A. $\sin 135°$ 　　　　　　**4B.** $\cos \frac{7\pi}{8}$

Recall that you can use the sum and difference identities to verify identities. Double- and half-angle identities can also be used to verify identities.

Example 5 Verify Identities

Verify that $\dfrac{\cos 2\theta}{1 + \sin 2\theta} = \dfrac{\cot\theta - 1}{\cot\theta + 1}$ **is an identity.**

$$\frac{\cos 2\theta}{1 + \sin 2\theta} \stackrel{?}{=} \frac{\cot\theta - 1}{\cot\theta + 1} \qquad \text{Original equation}$$

$$\frac{\cos 2\theta}{1 + \sin 2\theta} \stackrel{?}{=} \frac{\frac{\cos\theta}{\sin\theta} - 1}{\frac{\cos\theta}{\sin\theta} + 1} \qquad \cot\theta = \frac{\cos\theta}{\sin\theta}$$

$$\frac{\cos 2\theta}{1 + \sin 2\theta} \stackrel{?}{=} \frac{\cos\theta - \sin\theta}{\cos\theta + \sin\theta} \qquad \text{Multiply numerator and denominator by } \sin\theta.$$

$$\frac{\cos 2\theta}{1 + \sin 2\theta} \stackrel{?}{=} \frac{\cos\theta - \sin\theta}{\cos\theta + \sin\theta} \cdot \frac{\cos\theta + \sin\theta}{\cos\theta + \sin\theta} \qquad \text{Multiply the right side by 1.}$$

$$\frac{\cos 2\theta}{1 + \sin 2\theta} \stackrel{?}{=} \frac{\cos^2\theta - \sin^2\theta}{\cos^2\theta + 2\cos\theta\sin\theta + \sin^2\theta} \qquad \text{Multiply.}$$

$$\frac{\cos 2\theta}{1 + \sin 2\theta} \stackrel{?}{=} \frac{\cos^2\theta - \sin^2\theta}{1 + 2\cos\theta\sin\theta} \qquad \text{Simplify.}$$

$$\frac{\cos 2\theta}{1 + \sin 2\theta} = \frac{\cos 2\theta}{1 + \sin 2\theta} \checkmark \qquad \cos^2\theta - \sin^2\theta = \cos 2\theta;\ 2\cos\theta\sin\theta = \sin 2\theta$$

▶ **Guided Practice**

5. Verify that $4\cos^2 x - \sin^2 2x = 4\cos^4 x$.

Go Online!

Check your mastery of double-angle and half-angle identities by using the **Self-Check Quiz** in ConnectED.

Check Your Understanding

= Step-by-Step Solutions begin on page R11.

Go Online! for a Self-Check Quiz

Examples 1–3 **PRECISION** Find the exact values of $\sin 2\theta$, $\cos 2\theta$, $\sin \frac{\theta}{2}$, and $\cos \frac{\theta}{2}$.

1. $\sin \theta = \frac{1}{4}$; $0° < \theta < 90°$
2. $\sin \theta = \frac{4}{5}$; $90° < \theta < 180°$
3. $\cos \theta = -\frac{5}{13}$; $\frac{\pi}{2} < \theta < \pi$
4. $\cos \theta = \frac{3}{5}$; $270° < \theta < 360°$
5. $\tan \theta = -\frac{8}{15}$; $90° < \theta < 180°$
6. $\tan \theta = \frac{5}{12}$; $\pi < \theta < \frac{3\pi}{2}$

Find the exact value of each expression.

7. $\sin \frac{\pi}{8}$
8. $\cos 15°$

Example 4

9. **SOCCER** A soccer player kicks a ball at an angle of 37° with the ground with an initial velocity of 52 feet per second. The distance d that the ball will go in the air if it is not blocked is given by $d = \frac{2v^2 \sin \theta \cos \theta}{g}$. In this formula, g is the acceleration due to gravity and is equal to 32 feet per second squared, and v is the initial velocity.

  a. Simplify this formula by using a double-angle identity.
  b. Using the simplified formula, how far will this ball go?

Example 5 Verify that each equation is an identity.

10. $\tan \theta = \frac{1 - \cos 2\theta}{\sin 2\theta}$
11. $(\sin \theta + \cos \theta)^2 = 1 + 2 \sin \theta \cos \theta$

Practice and Problem Solving

Extra Practice is on page R10.

Examples 1–3 Find the exact values of $\sin 2\theta$, $\cos 2\theta$, $\sin \frac{\theta}{2}$, and $\cos \frac{\theta}{2}$.

12. $\sin \theta = \frac{2}{3}$; $90° < \theta < 180°$
13. $\sin \theta = -\frac{15}{17}$; $\pi < \theta < \frac{3\pi}{2}$
14. $\cos \theta = \frac{3}{5}$; $\frac{3\pi}{2} < \theta < 2\pi$
15. $\cos \theta = \frac{1}{5}$; $270° < \theta < 360°$
16. $\tan \theta = \frac{4}{3}$; $180° < \theta < 270°$
17. $\tan \theta = -2$; $\frac{\pi}{2} < \theta < \pi$

Find the exact value of each expression.

18. $\sin 75°$
19. $\sin \frac{3\pi}{8}$
20. $\cos \frac{7\pi}{12}$
21. $\tan 165°$
22. $\tan \frac{5\pi}{12}$
23. $\tan 22.5°$

24. **GEOGRAPHY** The Mercator projection of the globe is a projection on which the distance between the lines of latitude increases with their distance from the equator. The calculation of the location of a point on this projection involves the expression $\tan\left(45° + \frac{L}{2}\right)$, where L is the latitude of the point.

  a. Write this expression in terms of a trigonometric function of L.
  b. The latitude of Tallahassee, Florida, is 30° north. Find the value of the expression if $L = 30°$.

Example 4

25. ELECTRONICS Consider an AC circuit consisting of a power supply and a resistor. If the current I_0 in the circuit at time t is $I_0 \sin t\theta$, then the power delivered to the resistor is $P = I_0^2 R \sin^2 t\theta$, where R is the resistance. Express the power in terms of $\cos 2t\theta$.

Example 5 Verify that each equation is an identity.

26. $\tan 2\theta = \dfrac{2}{\cot \theta - \tan \theta}$

27. $1 + \dfrac{1}{2}\sin 2\theta = \dfrac{\sec \theta + \sin \theta}{\sec \theta}$

28. $\sin \dfrac{\theta}{2} \cos \dfrac{\theta}{2} = \dfrac{\sin \theta}{2}$

29. $\tan \dfrac{\theta}{2} = \dfrac{\sin \theta}{1 + \cos \theta}$

30. FOOTBALL Suppose a place kicker consistently kicks a football with an initial velocity of 95 feet per second. Prove that the horizontal distance the ball travels in the air will be the same for $\theta = 45° + A$ as for $\theta = 45° - A$. Use the formula given in Exercise 9.

Find the exact values of $\sin 2\theta$, $\cos 2\theta$, and $\tan 2\theta$.

31. $\cos \theta = \dfrac{4}{5}; 0° < \theta < 90°$

32. $\sin \theta = \dfrac{1}{3}; 0 < \theta < \dfrac{\pi}{2}$

33. $\tan \theta = -3; 90° < \theta < 180°$

34. $\sec \theta = -\dfrac{4}{3}; 90° < \theta < 180°$

35. $\csc \theta = -\dfrac{5}{2}; \dfrac{3\pi}{2} < \theta < 2\pi$

36. $\cot \theta = \dfrac{3}{2}; 180° < \theta < 270°$

H.O.T. Problems Use Higher-Order Thinking Skills

37. ERROR ANALYSIS Teresa and Nathan are calculating the exact value of $\sin 15°$. Is either of them correct? Explain your reasoning.

Teresa	Nathan
$\sin(A - B) = \sin A \cos B - \cos A \sin B$ $\sin(45 - 30) = \sin 45 \cos 30 - \cos 45 \sin 30$ $= \dfrac{\sqrt{2}}{2} \cdot \dfrac{\sqrt{3}}{2} - \dfrac{\sqrt{2}}{2} \cdot \dfrac{1}{2}$ $= \dfrac{\sqrt{4}}{4}$	$\sin \dfrac{A}{2} = \pm\sqrt{\dfrac{1 - \cos A}{2}}$ $\sin \dfrac{30}{2} = \pm\sqrt{\dfrac{1 - \frac{1}{2}}{2}}$ $= 0.5$

38. CONSTRUCT ARGUMENTS Circle O is a unit circle. Use the figure to prove that $\tan \dfrac{1}{2}\theta = \dfrac{\sin \theta}{1 + \cos \theta}$.

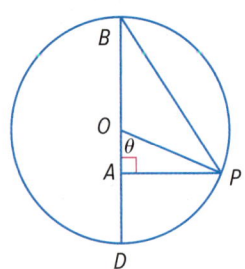

39. WRITING IN MATH Write a short paragraph about the conditions under which you would use each of the three identities for $\cos 2\theta$.

40. PROOF Use the formula for $\sin(A + B)$ to derive the formula for $\sin 2\theta$, and use the formula for $\cos(A + B)$ to derive the formula for $\cos 2\theta$.

41. REASONING Derive the half-angle identities from the double-angle identities.

42. OPEN-ENDED Suppose a golfer consistently hits the ball so that it leaves the tee with an initial velocity of 115 feet per second and $d = \dfrac{2v^2 \sin \theta \cos \theta}{g}$. Explain why the maximum distance is attained when $\theta = 45°$.

Preparing for Assessment

43. Suppose θ is an angle in Quadrant IV of the coordinate plane such that $\sin \theta = -\frac{12}{13}$ and $\cos \theta = \frac{5}{13}$. What is $\sin 2\theta$? **MP 2, 6**

- A $-\frac{24}{26}$
- B $-\frac{120}{169}$
- C $-\frac{7}{13}$
- D $\frac{25}{169}$
- E $\frac{144}{169}$

44. An angle θ exists such that $270° < \theta < 360°$ and $\cos \theta = \frac{3}{5}$. What is the exact value of $\cos \frac{\theta}{2}$? **MP 2, 6**

- A $-\frac{2\sqrt{5}}{5}$
- B $-\frac{3}{10}$
- C $\frac{9}{25}$
- D $\frac{\sqrt{5}}{5}$
- E $\frac{6}{5}$

45. MULTI-STEP When the tree in the image casts a 12-foot shadow, the angle of elevation of the sun is θ. When the sun rises higher in the sky and the angle of elevation is 2θ, the length of shadow from the same tree is reduced to 4 feet. **MP 1, 2, 3, 6**

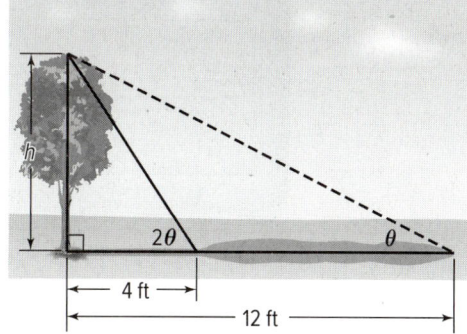

a. Which trigonometric identity can be used to solve this problem?

- A double-angle cosine identity
- B half-angle tangent identity
- C double-angle tangent identity
- D half-angle sine identity
- E half-angle cosine identity

b. Carey wants to calculate the height of the tree and performs the following algebra. Which lines contain errors? Choose all that apply.

- ☐ A $\tan 2\theta = \frac{h}{12}$
- ☐ B $\tan \theta = \frac{h}{4}$
- ☐ C $\tan 2\theta = \frac{2\tan \theta}{1 - \tan^2 \theta}$
- ☐ D $\frac{h}{4} = \frac{\frac{2h}{12}}{1 - \frac{h^2}{12^2}}$
- ☐ E $\frac{h}{4} = \frac{\frac{h}{6}}{\frac{144 - h^2}{144}}$
- ☐ F $\frac{h}{4}\left(\frac{144 - h^2}{144}\right) = \frac{h}{6}$
- ☐ G $144 - h^2 = 96$

c. To the nearest whole foot, what is the height h of the tree?

- A 5 ft
- B 7 ft
- C 8 ft
- D 10 ft
- E 16 ft

46. What is the exact value of $-\cos \frac{5\pi}{8}$? **MP 6**

47. For $\cos \theta = -\frac{4}{5}$, $90 < \theta < 180$, find the exact values. **MP 6**

a. $\sin 2\theta$
b. $\cos 2\theta$
c. $\tan 2\theta$
d. $\sin \frac{\theta}{2}$
e. $\cos \frac{\theta}{2}$
f. $\tan \frac{\theta}{2}$

EXPLORE 10-5

Graphing Technology Lab
Solving Trigonometric Equations

The graph of a trigonometric function is made up of points that represent all values that satisfy the function. To solve a trigonometric equation, you need to find all values of the variable that satisfy the equation. You can use a TI-83/84 Plus graphing calculator to solve trigonometric equations by graphing each side of the equation as a function and then locating the points of intersection.

Mathematical Practices
 5 Use appropriate tools strategically.

Activity 1 Real Solutions

Work cooperatively. Use a graphing calculator to solve $\sin x = 0.4$ if $0° \leq x < 360°$.

Step 1 Enter and graph related equations. Rewrite the equation as two equations, $Y1 = \sin x$ and $Y2 = 0.4$. Then graph the two equations. Because the interval is in degrees, set your calculator to degree mode.

KEYSTROKES: MODE ▼ ▼ ▶ ENTER
Y= SIN X,T,θ,n)
ENTER 0.4 ENTER GRAPH

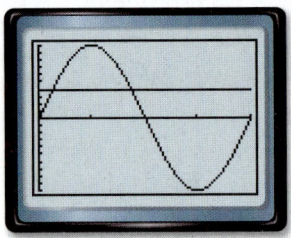

[0, 360] scl: 90 by [−1, 1] scl: 0.1

Step 2 Approximate the solutions. Based on the graph, you can see that there are two points of intersection in the interval $0° \leq x < 360°$. Use the **CALC** feature to determine the x-values at which the two graphs intersect.

The solutions are $x \approx 23.57°$ and $x \approx 156.4°$.

Activity 2 No Real Solutions

Work cooperatively. Use a graphing calculator to solve $\tan^2 x \cos x + 3 \cos x = 0$ if $0° \leq x < 360°$.

Step 1 Enter and graph related equations. The related equations to be graphed are $Y1 = \tan^2 x \cos x + 3 \cos x$ and $Y2 = 0$.

KEYSTROKES: Y= TAN X,T,θ,n) x^2 COS
X,T,θ,n) + 3 COS X,T,θ,n
) ENTER 0 ENTER

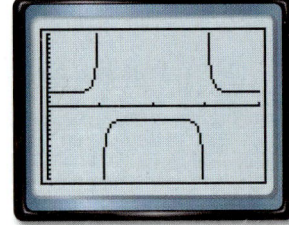

[0, 360] scl: 90 by [−15, 15] scl: 1

Step 2 These two functions do not intersect.

Therefore, the equation $\tan^2 x \cos x + 3 \cos x = 0$ has no real solutions.

Exercises

Work cooperatively. Use a graphing calculator to solve each equation for the values of x indicated.

1. $\sin x = 0.7$; $0° \leq x < 360°$
2. $\tan x = \cos x$; $0° \leq x < 360°$
3. $3 \cos x + 4 = 0.5$; $0° \leq x < 360°$
4. $0.25 \cos x = 3.4$; $-720° \leq x < 720°$
5. $\sin 2x = \sin x$; $0° \leq x < 360°$
6. $\sin 2x - 3 \sin x = 0$ if $-360° \leq x < 360°$

LESSON 5
Solving Trigonometric Equations

∴Then
- You verified trigonometric identities.

∴Now
1. Solve trigonometric equations.
2. Find extraneous solutions from trigonometric equations.

∴Why?
- When you ride a Ferris wheel that has a diameter of 40 meters and turns at a rate of 1.5 revolutions per minute, the height above the ground, in meters, of your seat after t minutes can be modeled by the equation

$$h = 21 - 20 \cos 3\pi t.$$

After the ride begins, how long is it before your seat is 31 meters above the ground for the first time?

New Vocabulary
trigonometric equations

Mathematical Practices
4 Model with mathematics.
6 Attend to precision.
7 Look for and make use of structure.

1 Solve Trigonometric Equations So far in this chapter, you have studied a special type of trigonometric equation called an identity. Trigonometric identities are equations that are true for all values of the variable for which both sides are defined. In this lesson, you will examine **trigonometric equations** that are true only for certain values of the variable. Solving these equations resembles solving algebraic equations.

Example 1   Solve Equations for a Given Interval

Solve $\sin \theta \cos \theta - \frac{1}{2} \cos \theta = 0$ if $0 \leq \theta \leq 180°$.

$\sin \theta \cos \theta - \frac{1}{2} \cos \theta = 0$ Original equation

$\cos \theta \left(\sin \theta - \frac{1}{2} \right) = 0$ Factor.

$\cos \theta = 0$ or $\sin \theta - \frac{1}{2} = 0$ Zero Product Property

$\theta = 90°$ $\sin \theta = \frac{1}{2}$

$\theta = 30°$ or $150°$

The solutions are 30°, 90°, and 150°.

CHECK You can check the answer by graphing $y = \sin \theta \cos \theta$ and $y = \frac{1}{2} \cos \theta$ in the same coordinate plane on a graphing calculator. Then find the points where the graphs intersect. You can see that there are infinitely many such points, but we are only interested in the points between 0° and 180°.

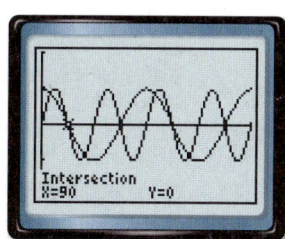

[0, 720] scl: 90 by [−1, 1] scl: 0.5

▸ Guided Practice

1. Find all solutions of $\sin 2\theta = \cos \theta$ if $0 \leq \theta \leq 2\pi$.

Trigonometric equations are usually solved for values of the variable between 0° and 360° or between 0 radians and 2π radians. There are solutions outside that interval. These other solutions differ by integral multiples of the period of the function.

connectED.mcgraw-hill.com   683

Example 2 Infinitely Many Solutions

Solve $\cos \theta + 1 = 0$ for all values of θ if θ is measured in radians.

$\cos \theta + 1 = 0$
$\cos \theta = -1$

Look at the graph of $y = \cos \theta$ to find solutions of $\cos \theta = -1$.

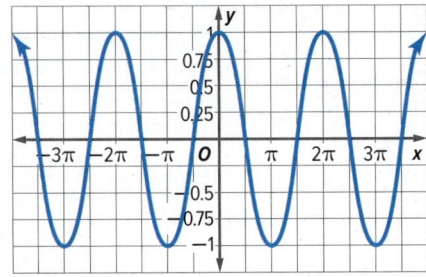

The solutions are π, 3π, 5π, and so on, and $-\pi$, -3π, -5π, and so on. The only solution in the interval 0 radians to 2π radians is π. The period of the cosine function is 2π radians. So the solutions can be written as $\pi + 2k\pi$, where k is any integer.

> **Study Tip**
>
> **Expressing Solutions as Multiples** The expression $\pi + 2k\pi$ includes all multiples of π in the solution set, so it is not necessary to list them separately.

Guided Practice

2A. Solve $\cos 2\theta + \cos \theta + 1 = 0$ for all values of θ if θ is measured in degrees.

2B. Solve $2 \sin \theta = -1$ for all values of θ if θ is measured in radians.

Trigonometric equations are often used to solve real-world problems.

Real-World Example 3 Solve Trigonometric Equations

AMUSEMENT PARKS Refer to the beginning of the lesson. How long after the Ferris wheel starts will your seat first be 31 meters above the ground?

$h = 21 - 20 \cos 3\pi t$	Original equation
$31 = 21 - 20 \cos 3\pi t$	Replace h with 31.
$10 = -20 \cos 3\pi t$	Subtract 21 from each side.
$-\dfrac{1}{2} = \cos 3\pi t$	Divide each side by -20.
$\cos^{-1}\left(-\dfrac{1}{2}\right) = 3\pi t$	Take the Arccosine.
$\dfrac{2\pi}{3} = 3\pi t$	The Arccosine of $-\dfrac{1}{2}$ is $\dfrac{2\pi}{3}$.
$\dfrac{2\pi}{3} + 2\pi k = 3\pi t$	k is any integer.
$\dfrac{2}{9} + \dfrac{2}{3}k = t$	Divide each term by 3π.

The least positive value for t is obtained by letting $k = 0$ in the first expression. Therefore, $t = \dfrac{2}{9}$ of a minute or about 13 seconds.

Guided Practice

3. How long after the Ferris wheel starts will your seat first be 41 meters above the ground?

2 Extraneous Solutions

Some trigonometric equations have no solution. For example, the equation $\cos \theta = 4$ has no solution because all values of $\cos \theta$ are between -1 and 1, inclusive. Thus, the solution set for $\cos \theta = 4$ is empty.

Example 4 Determine Whether a Solution Exists

Solve each equation.

a. $2 \sin^2 \theta - 3 \sin \theta - 2 = 0$ if $0 \leq \theta \leq 2\pi$

$2 \sin^2 \theta - 3 \sin \theta - 2 = 0$	Original equation
$(\sin \theta - 2)(2 \sin \theta + 1) = 0$	Factor.
$\sin \theta - 2 = 0$ or $2 \sin \theta + 1 = 0$	Zero Product Property
$\sin \theta = 2$ $2 \sin \theta = -1$	
This is not a solution because all values of $\sin \theta$ are between -1 and 1, inclusive. $\sin \theta = -\frac{1}{2}$	
$\theta = \frac{7\pi}{6}$ or $\frac{11\pi}{6}$	

The solutions are $\frac{7\pi}{6}$ or $\frac{11\pi}{6}$.

CHECK

$2 \sin \theta - 3 \sin \theta - 2 = 0$ $2 \sin^2 \theta - 3 \sin \theta - 2 = 0$

$2 \sin^2 \left(\frac{7\pi}{6}\right) - 3 \sin \left(\frac{7\pi}{6}\right) - 2 \stackrel{?}{=} 0$ $2 \sin^2 \left(\frac{11\pi}{6}\right) - 3 \sin \left(\frac{11\pi}{6}\right) - 2 \stackrel{?}{=} 0$

$2\left(\frac{1}{4}\right) - 3\left(-\frac{1}{2}\right) - 2 \stackrel{?}{=} 0$ $2\left(\frac{1}{4}\right) - 3\left(-\frac{1}{2}\right) - 2 \stackrel{?}{=} 0$

$\frac{1}{2} + \frac{3}{2} - 2 \stackrel{?}{=} 0$ $\frac{1}{2} + \frac{3}{2} - 2 \stackrel{?}{=} 0$

$0 = 0$ ✓ $0 = 0$ ✓

b. $\sin \theta = 1 + \cos \theta$ if $0° \leq \theta < 360°$

$\sin \theta = 1 + \cos \theta$	Original equation
$\sin^2 \theta = (1 + \cos \theta)^2$	Square each side.
$1 - \cos^2 \theta = 1 + 2 \cos \theta + \cos^2 \theta$	$\sin^2 \theta = 1 - \cos^2 \theta$
$0 = 2 \cos \theta + 2 \cos^2 \theta$	Set the left side equal to 0.
$0 = 2 \cos \theta (1 + \cos \theta)$	Factor.
$1 + \cos \theta = 0$ or $2 \cos \theta = 0$	Zero Product Property
$\cos \theta = -1$ $\cos \theta = 0$	
$\theta = 180$ $\theta = 90°$ or $270°$	

CHECK

$\sin \theta = 1 + \cos \theta$ $\sin \theta = 1 + \cos \theta$

$\sin 90° \stackrel{?}{=} 1 + \cos 90°$ $\sin 180° \stackrel{?}{=} 1 + \cos 180°$

$1 \stackrel{?}{=} 1 + 0$ $0 \stackrel{?}{=} 1 + (-1)$

$1 = 1$ ✓ $0 = 0$ ✓

$\sin \theta = 1 + \cos \theta$

$\sin 270° \stackrel{?}{=} 1 + \cos 270°$

$-1 \stackrel{?}{=} 1 + 0$

$-1 \neq 1$ ✗

The solutions are $90°$ and $180°$.

> **Problem-Solving Tip**
>
> **MP Regularity** Look for patterns in your solutions. Look for pairs of solutions that differ by exactly π or 2π and write your solutions with the simplest possible pattern.

Guided Practice

4A. $\sin^2 \theta + 2 \cos^2 \theta = 4$

4B. $\cos^2 \theta + 3 = 4 - \sin^2 \theta$

If an equation cannot be solved easily by factoring, try rewriting the expression using trigonometric identities. However, using identities and some algebraic operations, such as squaring, may result in extraneous solutions. So, it is necessary to check your solutions using the original equation.

> **Study Tip**
>
> **Solving Trigonometric Equations** Remember that *solving a trigonometric equation* means solving for all values of the variable.

Example 5 Solve Trigonometric Equations by Using Identities

Solve $2\sec^2\theta - \tan^4\theta = -1$ for all values of θ if θ is measured in degrees.

$2\sec^2\theta - \tan^4\theta = -1$ Original equation

$2(1 + \tan^2\theta) - \tan^4\theta = -1$ $\sec^2\theta = 1 + \tan^2\theta$

$2 + 2\tan^2\theta - \tan^4\theta = -1$ Distributive Property

$\tan^4\theta - 2\tan^2\theta - 3 = 0$ Set one side of the equation equal to 0.

$(\tan^2\theta - 3)(\tan^2\theta + 1) = 0$ Factor.

$\tan^2\theta - 3 = 0$ or $\tan^2\theta + 1 = 0$ Zero Product Property

$\tan^2\theta = 3$ $\tan^2\theta = -1$

$\tan\theta = \pm\sqrt{3}$ This part gives no solutions since $\tan^2\theta$ is never negative.

$\theta = 60° + 180°k$ and $\theta = -60° + 180°k$, where k is any integer. The solutions are $60° + 180°k$ and $-60° + 180°k$.

▶ **Guided Practice**

Solve each equation.

5A. $\sin\theta\cot\theta - \cos^2\theta = 0$ **5B.** $\dfrac{\cos\theta}{\cot\theta} + 2\sin^2\theta = 0$

Check Your Understanding

◯ = Step-by-Step Solutions begin on page R11.

Go Online! for a Self-Check Quiz

Example 1   **MP REGULARITY** Solve each equation if $0° \le \theta \le 360°$.

1. $2\sin\theta + 1 = 0$
2. $\cos^2\theta + 2\cos\theta + 1 = 0$
3. $\cos 2\theta + \cos\theta = 0$
4. $2\cos\theta = 1$
5. $\cos\theta = -\dfrac{\sqrt{3}}{2}$
6. $\sin 2\theta = -\dfrac{\sqrt{3}}{2}$
7. $\cos 2\theta = 8 - 15\sin\theta$
8. $\sin\theta + \cos\theta = 1$

Example 2   Solve each equation for all values of θ if θ is measured in radians.

9. $4\sin^2\theta - 1 = 0$
10. $2\cos^2\theta = 1$
11. $\cos 2\theta \sin\theta = 1$
12. $\sin\dfrac{\theta}{2} + \cos\dfrac{\theta}{2} = \sqrt{2}$
13. $\cos 2\theta + 4\cos\theta = -3$
14. $\sin\dfrac{\theta}{2} + \cos\theta = 1$

Solve each equation for all values of θ if θ is measured in degrees.

15. $\cos 2\theta - \sin^2\theta + 2 = 0$
16. $\sin^2\theta - \sin\theta = 0$
17. $2\sin^2\theta - 1 = 0$
18. $\cos\theta - 2\cos\theta\sin\theta = 0$
19. $\cos 2\theta \sin\theta = 1$
20. $\sin\theta\tan\theta - \tan\theta = 0$

Example 3   **21. LIGHT** The number of hours of daylight d in Hartford, Connecticut, may be approximated by the equation $d = 3\sin\dfrac{2\pi}{365}t + 12$, where t is the number of days after March 21.

     a. On what days will Hartford have exactly $10\dfrac{1}{2}$ hours of daylight?

     b. Using the results in part **a**, tell what days of the year have at least $10\dfrac{1}{2}$ hours of daylight. Explain how you know.

Examples 4–5 Solve each equation.

22. $\sin^2 2\theta + \cos^2 \theta = 0$
23. $\tan^2 \theta + 2 \tan \theta + 1 = 0$
24. $\cos^2 \theta + 3 \cos \theta = -2$
25. $\sin 2\theta - \cos \theta = 0$
26. $\tan \theta = 1$
27. $\cos 8\theta = 1$
28. $\sin \theta + 1 = \cos 2\theta$
29. $2 \cos^2 \theta = \cos \theta$

Practice and Problem Solving

Extra Practice is on page R10.

Example 1 Solve each equation for the given interval.

30. $\cos^2 \theta = \frac{1}{4};\ 0° \leq \theta \leq 360°$
31. $2 \sin^2 \theta = 1;\ 90° < \theta < 270°$
32. $\sin 2\theta - \cos \theta = 0;\ 0 \leq \theta \leq 2\pi$
33. $3 \sin^2 \theta = \cos^2 \theta;\ 0 \leq \theta \leq \frac{\pi}{2}$
34. $2 \sin \theta + \sqrt{3} = 0;\ 180° < \theta < 360°$
35. $4 \sin^2 \theta - 1 = 0;\ 180° < \theta < 360°$

Example 2 Solve each equation for all values of θ if θ is measured in radians.

36. $\cos 2\theta + 3 \cos \theta = 1$
37. $2 \sin^2 \theta = \cos \theta + 1$
38. $\cos^2 \theta - \frac{3}{2} = \frac{5}{2} \cos \theta$
39. $3 \cos \theta - \cos \theta = 2$

Solve each equation for all values of θ if θ is measured in degrees.

40. $\sin \theta - \cos \theta = 0$
41. $\tan \theta - \sin \theta = 0$
42. $\sin^2 \theta = 2 \sin \theta + 3$
43. $4 \sin^2 \theta = 4 \sin \theta - 1$

Example 3

44. **ELECTRONICS** The tallest structure in Tessa's hometown is a television transmitting tower with a height of 1626 feet. What is the measure of θ if the length of the shadow is 1 mile?

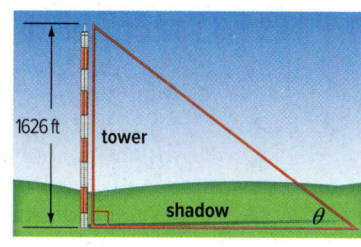

Examples 4–5 Solve each equation.

45. $2 \sin^2 \theta = 3 \sin \theta + 2$
46. $2 \cos^2 \theta + 3 \sin \theta = 3$
47. $\sin^2 \theta + \cos 2\theta = \cos \theta$
48. $2 \cos^2 \theta = -\cos \theta$

49. **MP SENSE-MAKING** Due to ocean tides, the depth y in meters of the River Thames in London varies as a sine function of x, the hour of the day. On a certain day that function was $y = 3 \sin\left[\frac{\pi}{6}(x - 4)\right] + 8$, where $x = 0, 1, 2, …, 24$ corresponds to 12:00 midnight, 1:00 A.M., 2:00 A.M., …, 12:00 midnight the next night.

  a. What is the maximum depth of the River Thames on that day?
  b. At what times does the maximum depth occur?

Solve each equation if θ is measured in radians.

50. $(\cos \theta)(\sin 2\theta) - 2 \sin \theta + 2 = 0$
51. $2 \sin^2 \theta + (\sqrt{2} - 1) \sin \theta = \frac{\sqrt{2}}{2}$

Solve each equation if θ is measured in degrees.

52. $\sin 2\theta + \frac{\sqrt{3}}{2} = \sqrt{3} \sin \theta + \cos \theta$
53. $1 - \sin^2 \theta - \cos \theta = \frac{3}{4}$

Solve each equation.

54. $2 \sin \theta = \sin 2\theta$

55. $\cos \theta \tan \theta - 2 \cos^2 \theta = -1$

56. DIAMONDS According to Snell's Law, $n_1 \sin i = n_2 \sin r$, where n_1 is the index of refraction of the medium the light is exiting, n_2 is the index of refraction of the medium the light is entering, i is the degree measure of the angle of incidence, and r is the degree measure of the angle of refraction.

  a. The index of refraction of a diamond is 2.42, and the index of refraction of air is 1.00. If a beam of light strikes a diamond at an angle of 35°, what is the angle of refraction?

  b. Explain how a gemologist might use Snell's Law to determine whether a diamond is genuine.

57. MP PERSEVERANCE A wave traveling in a guitar string can be modeled by the equation $D = 0.5 \sin (6.5x)° \sin (2500t)°$, where D is the displacement in millimeters at the position x millimeters from the left end of the string at time t seconds. Find the first positive time when the point 0.5 meter from the left end has a displacement of 0.01 millimeter.

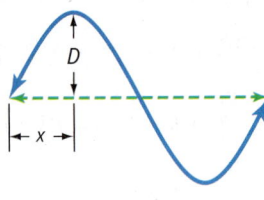

58. MULTIPLE REPRESENTATIONS Consider the trigonometric inequality $\sin \theta \geq \frac{1}{2}$.

  a. Tabular Construct a table of values for $0° \leq \theta \leq 360°$. For what values of θ is $\sin \theta \geq \frac{1}{2}$?

  b. Graphical Graph $y = \sin \theta$ and $y = \frac{1}{2}$ on the same graph for $0° \leq \theta \leq 360°$. For what values of θ is the graph of $y = \sin \theta$ above the graph of $y = \frac{1}{2}$?

  c. Analytic Based on your answers for parts **a** and **b**, solve $\sin \theta \geq \frac{1}{2}$ for all values of θ.

  d. Algebraic Solve each inequality if $0 \leq \theta \leq 360°$. Then solve each for all values of θ.

  i. $\cos \theta \geq \frac{\sqrt{2}}{2}$

  ii. $2 \sin \theta \leq \sqrt{3}$

  iii. $-\sin \theta \geq 0$

  iv. $\cos \theta - 1 < -\frac{1}{2}$

H.O.T. Problems Use Higher-Order Thinking Skills

59. CHALLENGE Solve $\sin 2x < \sin x$ for $0 \leq x \leq 2\pi$ without a calculator.

60. MP REASONING Compare and contrast solving trigonometric equations with solving linear and quadratic equations. What techniques are the same? What techniques are different? How many solutions do you expect?

61. WRITING IN MATH Why do trigonometric equations often have infinitely many solutions?

62. OPEN-ENDED Write an example of a trigonometric equation that has exactly two solutions if $0° \leq \theta \leq 360°$.

63. CHALLENGE How many solutions in the interval $0° \leq \theta \leq 360°$ should you expect for $a \sin (b\theta + c) = d$, if $a \neq 0$ and b is a positive integer?

Preparing for Assessment

64. The solution(s) for $\sec^2 \theta - 5 = -1$, in the interval $180° \leq \theta \leq 360°$, is given by which of the following? **MP** 2, 6

    I 120°
    II 210°
    III 240°
    IV 300°

- A I only
- B II only
- C I and II
- D III and IV
- E I, III, and IV

65. How many solutions does $1 + \sin \theta = \cos \theta$ have in the interval $0° \leq \theta < 360°$? **MP** 2, 6

- A 0
- B 1
- C 2
- D 4
- E infinitely many

66. Which of the following is a solution to $\tan \theta \sin \theta = \sin \theta$ in the interval $0° < \theta < 180°$? **MP** 2, 6

- A 30°
- B 45°
- C 60°
- D 90°
- E 135°

67. What is the solution to $4 \sin \theta - \sin \theta = 3$ on the interval $0 \leq \theta \leq 2\pi$? **MP** 2, 6

- A 0
- B $\frac{\pi}{2}$
- C π
- D $\frac{3\pi}{2}$
- E 2π

68. a. MULTI-STEP The graph of $y = \sin \theta$ is shown. Which of the following shows all of the solutions for $\sin \theta = -1$ in the interval $-\pi < \theta < \pi$? **MP** 2, 6

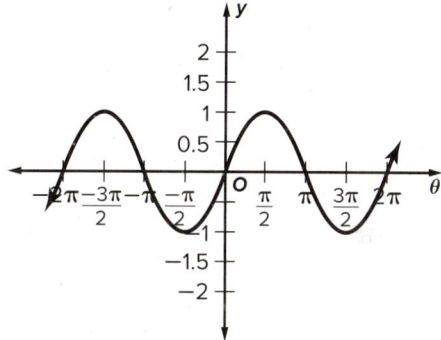

- A $-\pi$ and π
- B $-\frac{\pi}{2}$
- C 0
- D $-\frac{\pi}{2}$ and $\frac{3\pi}{2}$
- E $-\pi, -\frac{\pi}{2}, \frac{3\pi}{2}$ and π

b. How many solutions are there to $\sin \theta = 1$ for the domain $-2\pi < \theta < 2\pi$

- A 0
- B 1
- C 2
- D 4

CHAPTER 10
Study Guide and Review

Go Online! for Vocabulary Review Games and key vocabulary in 13 language

Study Guide

Key Concepts

Trigonometric Identities (Lessons 10-1, 10-2, and 10-5)
- Trigonometric identities describe the relationships between trigonometric functions.
- Trigonometric identities can be used to simplify, verify, and solve trigonometric equations and expressions.

Sum and Difference Identities (Lesson 10-3)
- For all values of A and B:
$$\cos(A \pm B) = \cos A \cos B \mp \sin A \sin B$$
$$\sin(A \pm B) = \sin A \cos B \pm \cos A \sin B$$
$$\tan(A \pm B) = \frac{\tan A \pm \tan B}{1 \mp \tan A \tan B}$$

Double-Angle and Half-Angle Identities (Lesson 10-4)
- Double-angle identities:
$$\sin 2\theta = 2 \sin \theta \cos \theta$$
$$\cos 2\theta = \cos^2 \theta - \sin^2 \theta$$
$$\cos 2\theta = 1 - 2 \sin^2 \theta$$
$$\cos 2\theta = 2 \cos^2 \theta - 1$$
$$\tan 2\theta = \frac{2 \tan \theta}{1 - \tan^2 \theta}$$
- Half-angle identities:
$$\sin \frac{\theta}{2} = \pm \sqrt{\frac{1 - \cos \theta}{2}}$$
$$\cos \frac{\theta}{2} = \pm \sqrt{\frac{1 + \cos \theta}{2}}$$
$$\tan \frac{\theta}{2} = \pm \sqrt{\frac{1 - \cos \theta}{1 + \cos \theta}}, \cos \theta \neq -1$$

FOLDABLES Study Organizer

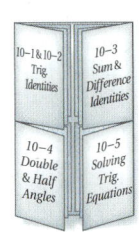

Use your Foldable to review the chapter. Working with a partner can be helpful. Ask for clarification of concepts as needed.

Key Vocabulary

cofunction identity (p. 655)

negative angle identity (p. 655)

Pythagorean identity (p. 655)

quotient identity (p. 655)

reciprocal identity (p. 655)

trigonometric equation (p. 683)

trigonometric identity (p. 655)

Vocabulary Check

Choose the correct term to complete each sentence.

1. The _____ can be used to find the sine or cosine of 75° if the sine and cosine of 90° and 15° are known.

2. The identities $\tan \theta = \frac{\sin \theta}{\cos \theta}$ and $\cot \theta = \frac{\cos \theta}{\sin \theta}$ are examples of _____.

3. The _____ can be used to find $\sin 60°$ using 30° as a reference.

4. The _____ identity can be used to find $\cos 22\frac{1}{2}°$.

5. The identities $\csc \theta = \frac{1}{\sin \theta}$ and $\sec \theta = \frac{1}{\cos \theta}$ are examples of _____.

6. The _____ can be used to find the sine or cosine of 120° if the sine and cosine of 90° and 30° are known.

7. $\cos^2 \theta + \sin^2 \theta = 1$ is an example of a(n) _____.

Concept Check

8. Explain how to verify a trigonometric identity.

9. Explain the difference between trigonometric identities and trigonometric equations.

Lesson-by-Lesson Review

10-1 Trigonometric Identities

Find the value of each expression.

10. $\sin \theta$, if $\cos \theta = \frac{\sqrt{2}}{2}$ and $270° < \theta < 360°$

11. $\sec \theta$, if $\cot \theta = \frac{\sqrt{2}}{2}$ and $90° < \theta < 180°$

12. $\tan \theta$, if $\cot \theta = 2$ and $0° < \theta < 90°$

13. $\cos \theta$, if $\sin \theta = -\frac{3}{5}$ and $180° < \theta < 270°$

14. $\csc \theta$, if $\cot \theta = -\frac{4}{5}$ and $270° < \theta < 360°$

15. **SOCCER** For international matches, the maximum dimensions of a soccer field are 110 meters by 75 meters. Find $\sin \theta$.

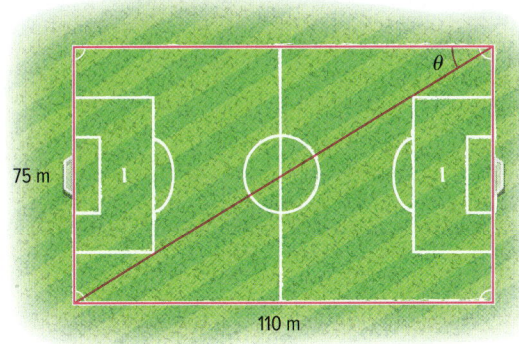

Simplify each expression.

16. $1 - \tan \theta \sin \theta \cos \theta$
17. $\tan \theta \csc \theta$
18. $\sin \theta + \cos \theta \cot \theta$
19. $\cos \theta (1 + \tan^2 \theta)$

Example 1
Find $\sin \theta$ if $\cos \theta = \frac{3}{4}$ and $0° < \theta < 90°$.

$\cos^2 \theta + \sin^2 \theta = 1$ — Trigonometric identity
$\sin^2 \theta = 1 - \cos^2 \theta$ — Subtract $\cos^2 \theta$ from each side.
$\sin^2 \theta = 1 - \left(\frac{3}{4}\right)^2$ — Substitute $\frac{3}{4}$ for $\cos \theta$.
$\sin^2 \theta = 1 - \frac{9}{16}$ — Square $\frac{3}{4}$.
$\sin^2 \theta = \frac{7}{16}$ — Subtract.
$\sin \theta = \pm \frac{\sqrt{7}}{4}$ — Take the square root of each side.

Because θ is in the first quadrant, $\sin \theta$ is positive.
Thus, $\sin \theta = \frac{\sqrt{7}}{4}$.

Example 2
Simplify $\cos \theta \sec \theta \cot \theta$.

$\cos \theta \sec \theta \cot \theta = \cos \theta \left(\frac{1}{\cos \theta}\right)\left(\frac{\cos \theta}{\sin \theta}\right)$
$= \cot \theta$

10-2 Verifying Trigonometric Identities

Verify that each of the following is an identity.

20. $\tan \theta \cos \theta + \cot \theta \sin \theta = \sin \theta + \cos \theta$

21. $\frac{\cos \theta}{\cot \theta} + \frac{\sin \theta}{\tan \theta} = \sin \theta + \cos \theta$

22. $\sec^2 \theta - 1 = \frac{\sin^2 \theta}{1 - \sin^2 \theta}$

23. **GEOMETRY** The right triangle shown at the right is used in a special quilt. Use the measures of the sides of the triangle to show that $\tan^2 \theta + 1 = \sec^2 \theta$.

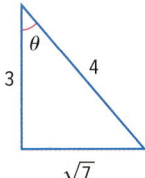

Example 3
Verify that $\frac{\cos \theta + 1}{\sin \theta} = \cot \theta + \csc \theta$ is an identity.

$\frac{\cos \theta + 1}{\sin \theta} \stackrel{?}{=} \cot \theta + \csc \theta$ — Original equation
$\frac{\cos \theta}{\sin \theta} + \frac{1}{\sin \theta} \stackrel{?}{=} \cot \theta + \csc \theta$ — Simplify.
$\cot \theta + \csc \theta = \cot \theta + \csc \theta$ ✓ — Simplify.

CHAPTER 10
Study Guide and Review Continued

10-3 Sum and Difference Identities

Find the exact value of each expression.

24. $\cos(-135°)$
25. $\cos 15°$
26. $\sin 210°$
27. $\sin 105°$
28. $\tan 75°$
29. $\cos 105°$

Verify that each of the following is an identity.

30. $\sin(\theta + 90) = \cos\theta$
31. $\sin\left(\dfrac{3\pi}{2} - \theta\right) = -\cos\theta$
32. $\tan(\theta - \pi) = \tan\theta$

Example 4

Find the exact value of $\sin 75°$.

Use $\sin(A + B) = \sin A \cos B + \cos A \sin B$.

$\sin 75° = \sin(30° + 45°)$
$= \sin 30° \cos 45° + \cos 30° \sin 45°$
$= \left(\dfrac{1}{2}\right)\left(\dfrac{\sqrt{2}}{2}\right) + \left(\dfrac{\sqrt{3}}{2}\right)\left(\dfrac{\sqrt{2}}{2}\right)$
$= \dfrac{\sqrt{2}}{4} + \dfrac{\sqrt{6}}{4}$ or $\dfrac{\sqrt{2} + \sqrt{6}}{4}$

10-4 Double-Angle and Half-Angle Identities

Find the exact values of $\sin 2\theta$, $\cos 2\theta$, $\sin\dfrac{\theta}{2}$, and $\cos\dfrac{\theta}{2}$ for each of the following.

33. $\cos\theta = \dfrac{4}{5}$; $0° < \theta < 90°$
34. $\sin\theta = -\dfrac{1}{4}$; $180° < \theta < 270°$
35. $\cos\theta = -\dfrac{2}{3}$; $\dfrac{\pi}{2} < \theta < \pi$

36. **BASEBALL** The infield of a baseball diamond is a square with side length 90 feet.
    a. Find the length of the diagonal.
    b. Write the ratio for $\sin 45°$ using the lengths of the baseball diamond.
    c. Use the formula $\sin\dfrac{\theta}{2} = \pm\sqrt{\dfrac{1 - \cos\theta}{2}}$ to verify the ratio you wrote in part **b**.

Example 5

Find the exact value of $\sin\dfrac{\theta}{2}$ if $\cos\theta = -\dfrac{3}{5}$ and θ is in the second quadrant.

$\sin\dfrac{\theta}{2} = \pm\sqrt{\dfrac{1 - \cos\theta}{2}}$ Half-angle identity

$= \pm\sqrt{\dfrac{1 - \left(-\dfrac{3}{5}\right)}{2}}$ $\cos\theta = -\dfrac{3}{5}$

$= \pm\sqrt{\dfrac{\dfrac{8}{5}}{2}}$ Subtract.

$= \pm\sqrt{\dfrac{4}{5}}$ Divide.

$= \pm\dfrac{2\sqrt{5}}{5}$ Simplify.

Because θ is in the second quadrant, $\sin\dfrac{\theta}{2} = \dfrac{2\sqrt{5}}{5}$.

10-5 Solving Trigonometric Equations

Find all solutions of each equation for the given interval.

37. $2\cos\theta - 1 = 0$; $0° \leq \theta < 360°$
38. $4\cos^2\theta - 1 = 0$; $0 \leq \theta < 2\pi$
39. $\sin 2\theta + \cos\theta = 0$; $0° \leq \theta < 360°$
40. $\sin^2\theta = 2\sin\theta + 3$; $0° \leq \theta < 360°$
41. $4\cos^2\theta - 4\cos\theta + 1 = 0$; $0 \leq \theta < 2\pi$

Example 6

Find all solutions of $\sin 2\theta - \cos\theta = 0$ if $0 \leq \theta < 2\pi$.

$\sin 2\theta - \cos\theta = 0$ Original equation
$2\sin\theta\cos\theta - \cos\theta = 0$ Double-angle identity
$\cos\theta(2\sin\theta - 1) = 0$ Factor.

$\cos\theta = 0$ or $2\sin\theta - 1 = 0$

$\theta = \dfrac{\pi}{2}, \dfrac{3\pi}{2}$ $\sin\theta = \dfrac{1}{2}$; $\theta = \dfrac{\pi}{6}, \dfrac{5\pi}{6}$

CHAPTER 10
Practice Test

1. **MULTIPLE CHOICE** Which expression is equivalent to $\sin \theta + \cos \theta \cot \theta$?
   A $\cot \theta$
   B $\tan \theta$
   C $\sec \theta$
   D $\csc \theta$

2. Verify that $\cos (30° - \theta) = \sin (60° + \theta)$ is an identity.

3. Verify that $\cos (\theta - \pi) = -\cos \theta$ is an identity.

4. **MULTIPLE CHOICE** What is the exact value of $\sin \theta$, if $\cos \theta = -\frac{3}{5}$ and $90° < \theta < 180°$?
   A $\frac{5}{3}$
   B $\frac{\sqrt{34}}{8}$
   C $-\frac{4}{5}$
   D $\frac{4}{5}$

Find the value of each expression.

5. $\cot \theta$, if $\sec \theta = \frac{4}{3}$; $270° < \theta < 360°$

6. $\tan \theta$, if $\cos \theta = -\frac{1}{2}$; $90° < \theta < 180°$

7. $\sec \theta$, if $\csc \theta = -2$; $180° < \theta < 270°$

8. $\cot \theta$, if $\csc \theta = -\frac{5}{3}$; $270° < \theta < 360°$

9. $\sec \theta$, if $\sin \theta = \frac{1}{2}$; $0° \le \theta < 90°$

Verify that each of the following is an identity.

10. $\sin \theta (\cot \theta + \tan \theta) = \sec \theta$

11. $\frac{\cos^2 \theta}{1 - \sin \theta} = \frac{\cos \theta}{\sec \theta - \tan \theta}$

12. $(\tan \theta + \cot \theta)^2 = \csc^2 \theta \sec^2 \theta$

13. $\frac{1 + \sec \theta}{\sec \theta} = \frac{\sin^2 \theta}{1 - \cos \theta}$

14. $\frac{\sin \theta}{1 - \cos \theta} = \csc \theta + \cot \theta$

15. **MULTIPLE CHOICE** What is the exact value of $\tan \frac{\pi}{8}$?
   A $\frac{\sqrt{2 - \sqrt{3}}}{2}$
   B $\sqrt{2} - 1$
   C $1 - \sqrt{2}$
   D $-\frac{\sqrt{2 - \sqrt{3}}}{2}$

16. **HISTORY** Some researchers believe that the builders of ancient pyramids, such as the Great Pyramid of Khufu, may have tried to build the faces as equilateral triangles. Later they had to change to other types of triangles. Suppose a pyramid is built such that a face is an equilateral triangle of side length 18 feet.

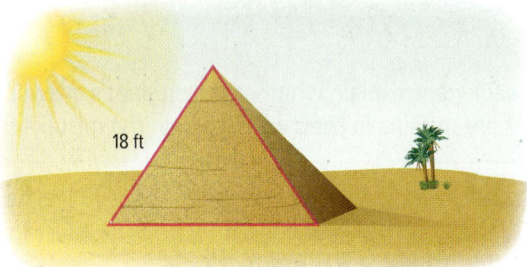

   a. Find the height of the equilateral triangle.

   b. Use the formula $\sin 2\theta = 2 \sin \theta \cos \theta$ and the measures of the equilateral triangle and its height to show that $\sin 2(30°) = \sin 60°$. Find the exact values.

Find the exact value of each expression.

17. $\cos (-225°)$
18. $\sin 480°$
19. $\cos 75°$
20. $\sin 165°$

21. **ROCKETS** A model rocket is launched with an initial velocity of 20 meters per second. The range of a projectile is given by the formula $R = \frac{v^2}{g} \sin 2\theta$, where R is the range, v is the initial velocity, g is acceleration due to gravity or 9.8 meters per second squared, and θ is the launch angle. What angle is needed in order for the rocket to reach a range of 25 meters?

Solve each equation for all values of θ if θ is measured in radians.

22. $2 \cos^2 \theta - 3 \cos \theta - 2 = 0$

23. $2 \sin 3\theta - 1 = 0$

Solve each equation for $0° \le \theta \le 360°$ if θ is measured in degrees.

24. $\cos 2\theta + \cos \theta = 2$

25. $\sin \theta \cos \theta - \frac{1}{2} \sin \theta = 0$

CHAPTER 10
Preparing for Assessment

Performance Task

Provide a clear solution to each part of the task. Be sure to show all of your work, include all relevant drawings, and justify your answers.

ERROR ANALYSIS A math teacher assigns her students a project that consists of grading and providing feedback on a hypothetical student's test.

For each item, explain whether the student's answer is correct. If the answer is incorrect, or if any steps are missing in their work, explain the mistake and provide a correction.

Part A
Find the exact value of $\cos \theta$ if $\sin = \frac{1}{2}$ and $90° < \theta < 180°$.

$$\cos^2 \theta + \sin^2 \theta = 1$$
$$\cos^2 \theta = 1 - \sin^2 \theta$$
$$\cos^2 \theta = 1 - \frac{1}{2}$$
$$\cos^2 \theta = \frac{1}{2}$$
$$\cos \theta = \frac{1}{\sqrt{2}}$$
$$\cos \theta = \frac{\sqrt{2}}{2}$$

Part B
Prove the identity $\cot(x) + \tan(x) = \sec(x)\csc(x)$.

$$\cot(x) + \tan(x) = \frac{\cos(x)}{\sin(x)} + \frac{\sin(x)}{\cos(x)}$$
$$= \frac{\cos^2(x)}{\sin(x)\cos(x)} + \frac{\sin^2(x)}{\sin(x)\cos(x)}$$
$$= \frac{\cos^2(x) + \sin^2(x)}{\sin(x)\cos(x)}$$
$$= \frac{1}{\sin(x)\cos(x)}$$
$$= \left(\frac{1}{\sin(x)}\right)\left(\frac{1}{\cos x}\right)$$
$$= \sec(x)\csc(x)$$

Part C
Find the exact value of $\cos(135°)$.

$$\cos(135°) = \cos(90°) + \cos(45°)$$
$$= \sin(90°)\cos(45°) + \cos(90°)\sin(45°)$$
$$= (1)\left(\frac{\sqrt{2}}{2}\right) + (0)\left(\frac{\sqrt{2}}{2}\right)$$
$$= \left(\frac{\sqrt{2}}{2}\right) + 0$$
$$= \frac{\sqrt{2}}{2}$$

Part D
Solve $\cos^2 \theta + \cos \theta = \sin^2 \theta$ for all values of θ if $0° \leq \theta < 360°$.

$$\cos^2 \theta + \cos \theta = \sin^2 \theta$$
$$\cos^2 \theta + \cos \theta = 1 - \cos^2 \theta$$
$$2\cos^2 \theta + \cos \theta - 1 = 0$$
$$(2\cos \theta - 1)(\cos \theta + 1) = 0$$
$$\cos \theta = \frac{1}{2} \text{ or } \cos \theta = -1$$
$$\theta = 60° \text{ and } 180°$$

Test-Taking Strategy

Example

Solve the problem below.

Simplify the trigonometric expression shown below by writing it in terms of $\sin \theta$.
$$\frac{\cos \theta}{\sec \theta + \tan \theta}$$

Step 1 Are there any mathematical operations you can apply? Are there are any laws or identities you can apply?
I can use trigonometric identities.

Step 2 How will you apply them?
I will rewrite $\sec \theta$ and $\tan \theta$ in terms of sine and cosine using the definitions of secant and tangent. Then I'll simplify the result and apply the Pythagorean identity to get cosine in terms of sine.

Step 3 What is the correct answer?
The answer is $1 - \sin \theta$.

> **Test-Taking Tip**
>
> **Strategies for Simplifying Expressions** Some problems, especially those involving trigonometric identities, require you to use the properties of algebra to simplify expressions. Follow the steps below to help prepare to solve these kinds of problems.

Apply the Strategy

Solve the problem.

Simplify $\dfrac{\sec \theta}{\cot \theta + \tan \theta}$ by writing it in terms of $\sin \theta$.

Answer the questions below.

a. Are there any mathematical operations you can apply? Are there are any laws or identities you can apply?

b. How will you apply them?

c. What is the correct answer?

CHAPTER 10
Preparing for Assessment
Cumulative Review

Read each question. Then fill in the correct answer on the answer document provided by your teacher or on a sheet of paper.

1. Suppose $180° < \theta < 270°$ and $\tan \theta = \frac{4}{3}$. What is $\cos \theta$?

   A $-\frac{3}{5}$

   B $-\frac{9}{25}$

   C $\frac{9}{25}$

   D $\frac{3}{4}$

   E $\frac{25}{9}$

2. When $P = 1$, which of the following trigonometric expressions, when defined, is considered an identity?

   $$\text{I} \quad -\sin^2 \theta - \cos^2 \theta = -P$$
   $$\text{II} \quad \tan^2 \theta - \sec^2 \theta = -P$$
   $$\text{III} \quad \frac{\sin \theta}{\sin (-\theta)} = P$$

   A I and II only

   B II only

   C III only

   D II and III only

   E I, II, and III

3. What is the remainder when $x^2 + 3x + 5$ is divided by $x + 1$?

   [   ]

Test-Taking Tip

Question 2 Because III is listed in three of the answer choices, first determine if III is an identity. If it is not, you can quickly eliminate choices C, D, and E.

4. A line segment extends from the origin of the coordinate plane $O(0, 0)$ to point $B(a, -b)$ in Quadrant IV, as shown in the figure below. If a, b, and c are positive real numbers, select all trigonometric ratios that represent the measure of the acute angle θ formed by $\overline{OB}$ and the x-axis.

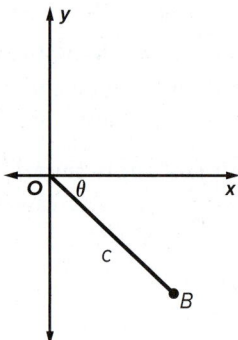

   A $\sin \theta = -\frac{b}{c}$

   B $\cos \theta = \frac{a}{c}$

   C $\sec \theta = -\frac{c}{a}$

   D $\tan \theta = \frac{a}{c}$

   E $\csc \theta = -\frac{c}{b}$

5. The time required to harvest the eggplant crop varies inversely with the size of the picking crew and directly with the number of rows planted. Last year a crew of 12 workers harvested 70 rows of eggplant in 2 days. This year the field was expanded to 100 rows. How many days will it take a crew of 8 workers to harvest this year's crop?

   A $\frac{14}{15}$

   B $1\frac{19}{21}$

   C $2\frac{1}{10}$

   D $4\frac{2}{7}$

6. What is the value of $\cos \theta$ when $\sin \theta = -\frac{\sqrt{3}}{2}$ and $180° < \theta < 270°$?

   [   ]

696 | Chapter 10 | Preparing for Assessment

7. Which shows a property of logarithms?

   A $\log b + \log c = \log(b+c)$

   B $\log_2 b = \dfrac{\log 2}{\log b}$

   C $\dfrac{\log a}{\log b} = \log a - \log b$

   D $\log b + \log c = \log(bc)$

   E $\log 2^b = b$

8. Which of the following trigonometric expressions has the same value as $\sin(270° + \theta)$?

   A $-\cos\theta$

   B $-\sin\theta$

   C $\cos\theta$

   D $-\cos\theta - \sin\theta$

   E $-\cos\theta + \sin\theta$

9. Which equation can be used to determine $\sin \angle DAF$, shown in the diagram below?

   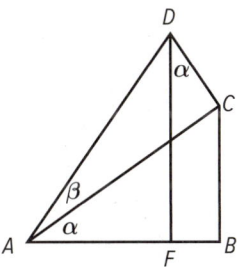

   A $\sin \angle DAF = \sin\alpha\cos\beta - \cos\alpha\sin\beta$

   B $\sin \angle DAF = \sin\alpha\sin\beta + \cos\alpha\cos\beta$

   C $\sin \angle DAF = \sin\alpha\cos\beta + \cos\alpha\sin\beta$

   D $\sin \angle DAF = \sin\alpha\sin\beta - \cos\alpha\cos\beta$

   E $\sin \angle DAF = \sin\alpha\sin\beta + \sin\alpha\sin\beta$

10. For a given angle θ such that $0° < \theta < 90°$, $\sin 2\theta = \dfrac{24}{25}$, and $\cos 2\theta = \dfrac{7}{25}$, the measure of $\tan 2\theta$ is defined by what numerical value?

    A $\dfrac{7}{24}$

    B $\dfrac{31}{25}$

    C $\dfrac{12}{7}$

    D $\dfrac{24}{7}$

    E $\dfrac{576}{49}$

11. When a stop sign casts an 8-foot shadow, the angle of elevation of the Sun is θ. When the angle of elevation is 2θ, the length of the shadow is 2 feet.

    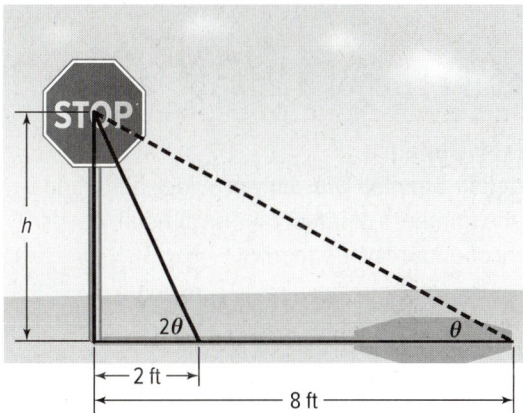

    a. What is the height h of the stop sign? Round to the nearest tenth of a foot.

    [____] ft

    b. **MP** What mathematical practice did you use to solve this problem?

12. For what value of θ, such that $0° < \theta < 360°$, is the equation $\sin^2\theta - \sin\theta = 2$ true?

    [____]°

Need Extra Help?

If you missed Question...	1	2	3	4	5	6	7	8	9	10	11	12
Go to Lesson...	10-1	10-2	4-3	10-1	7-5	10-1	6-7	10-3	10-3	10-4	10-4	10-5

Student Handbook

This **Student Handbook** can help you answer these questions.

What if I Need More Practice?

Extra Practice   R1
The **Extra Practice** section provides additional problems for each lesson so you have ample opportunity to practice new skills.

What if I Need to Check a Homework Answer?

Selected Answers and Solutions   R11
The answers to odd-numbered problems are included in **Selected Answers and Solutions**.

What if I Forget a Vocabulary Word?

Glossary/Glosario   R100
The **English-Spanish Glossary** provides definitions and page numbers of important or difficult words used throughout the textbook.

What if I Need to Find Something Quickly?

Index   R118
The **Index** alphabetically lists the subjects covered throughout the entire textbook and the pages on which each subject can be found.

What if I Forget a Formula?

Formulas, Symbols, and Parent Functions   Inside Back Cover
Inside the back cover of your math book is a list of **Formulas and Symbols** that are used in the book.

Extra Practice

CHAPTER 1 — Linear Equations

Solve each equation. Check your solution. (Lesson 1-1)

1. $a - 21 = 42$
2. $2x + 5 = -13$
3. $4(c - 5) - 2(3c + 5) = 6$

Solve each inequality. Then graph the solution set on a number line. (Lesson 1-2)

4. $d - 6 > -3$
5. $2c + 7 \geq 15$
6. $3x + 8 \leq 5$
7. $21 < -3(2y + 1)$

8. **TAXI** A taxi's rates are shown below. (Lesson 1-2)

Initial fare	$2.00
Mileage charge	$1.75

   a. Shannon wants to keep his fare under $35. How many miles can he travel in the taxi?
   b. Samantha has $21 for her taxi fare. If she plans on traveling 20 miles in the taxi, does she have enough money to cover the fare? Explain your answer.

Find the slope of the line that passes through each pair of points. (Lesson 1-3)

9. $(1, -7), (-3, 5)$
10. $(-1, 8), (3, -12)$
11. $(-2, -15), (4, 9)$

12. **ANIMALS** The table shows the weight in pounds of a Chihuahua for the first four weeks after its birth. (Lesson 1-3)

Time (wk)	Weight (lb)
0	2.0
1	2.5
2	3.25
3	4.7
4	5.6

    a. Find the average rate of change in the weight of the puppy between weeks 1 and 3.
    b. Find the average rate of change in the weight of the puppy between weeks 0 and 4.

Write an equation of the line passing through each pair of points. (Lesson 1-4)

13. $(-3, -14), (1, -2)$
14. $(-2, 11), (2, 3)$
15. $(-4, 2), (2, 5)$

Graph each inequality. (Lesson 1-5)

16. $y > -3$
17. $y < -2$
18. $2x - 5y \geq 10$

Solve each system of equations by using substitution or elimination. (Lesson 1-6)

19. $y = 3x + 4$
    $y = -2x - 6$
20. $y = -4x - 5$
    $y - 5x = -14$
21. $2a - 3b = -5$
    $2a + 3b = 13$
22. $-5a + 2b = 16$
    $4a - 5b = -6$

Solve each system of inequalities by graphing. (Lesson 1-7)

23. $y > 3x + 1$
    $y < -2x - 1$
24. $y \leq -x - 2$
    $y \leq 4x + 1$

25. **COLLEGE** The total score on a college entrance exam for the math and verbal sections is 2000. The college requires a math score of at least 875 and a verbal score of at least 800. (Lesson 1-7)
    a. Write and graph a system of inequalities to represent this situation.
    b. Give two examples of acceptable math and verbal scores for entrance to the college.

Graph each system of inequalities. Name the coordinates of the vertices of the feasible region. Find the maximum and minimum value of the given function for this region. (Lesson 1-8)

26. $x \geq -2$
    $y \leq 4$
    $y \geq x - 1$
    $f(x, y) = 2x - y$
27. $y \geq -3$
    $x \leq 5$
    $y \leq 2x + 4$
    $f(x, y) = -4x + 5y$

Solve each system of equations. (Lesson 1-9)

28. $2x - 3y + z = -4$
    $x + 2y - 3z = -9$
    $4x - y + 2z = -3$
29. $-3x + 4y - 2z = -14$
    $2x - 3y - 4z = 6$
    $-5x - 2y + 3z = 35$

CHAPTER 2 Relations and Functions

State the domain and range of each relation. Then determine whether each relation is a *function*. If it is a function, determine if it is *one-to-one*, *onto*, *both*, or *neither*. (Lesson 2-1)

1.

x	y
−2	4
−1	1
2	4
2	6

2.
0.5 → 2
1 → 3
2 → 5
2.5 → 6

State whether each function is a linear function. Write *yes* or *no*. Explain. (Lesson 2-2)

3. $f(x) = -2x + 1$
4. $3xy + 6 = 9$
5. $-\frac{7}{x} - 2 = y$
6. $4x - 5y = 18$

Describe the end behavior of each linear function graph. (Lesson 2-3)

7.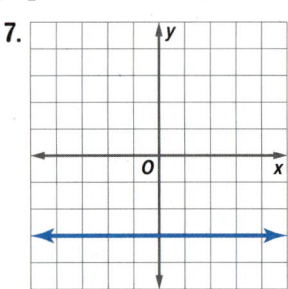
8.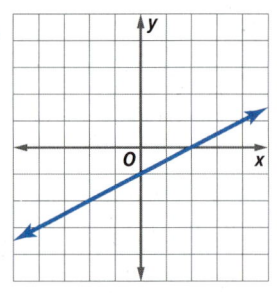

Describe the end behavior of each nonlinear function graph. (Lesson 2-3)

9.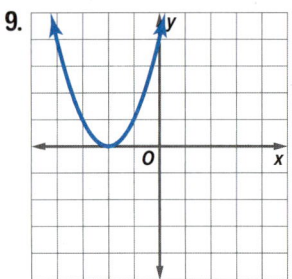
10.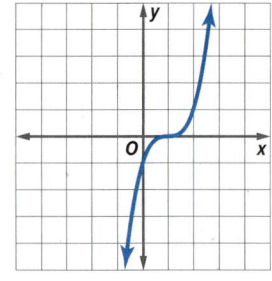

11. Use the given key features to sketch a nonlinear graph.

The function is continuous. The function is positive for −7 < x. The function has a relative maximum at (−3, 5) and a relative minimum at (5, 1). As $x \to \infty$, $f(x) \to \infty$ and as $x \to -\infty$, $f(x) \to -\infty$. (Lesson 2-4)

Graph each function. Identify the domain and range. (Lesson 2-5)

12. $g(x) = |-2x|$
13. $f(x) = [[x - 3]]$
14. $h(x) = 3|x| + 2$

Describe the translation in each function. Then graph the function. (Lesson 2-6)

15. $y = x^2 - 3$
16. $y = |x| + 5$
17. $y = (x + 2)^2$
18. $y = |x - 4|$

19. The graph of g(x) was obtained by translating the graph of f(x). Based on the graph, which equation can be used to describe g(x) in terms of f(x)? (Lesson 2-6)

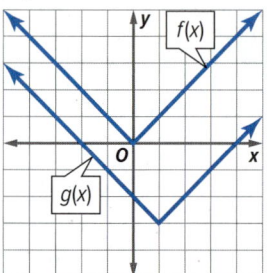

A $g(x) = f(x - 1) - 3$ C $g(x) = f(x - 3) - 1$
B $g(x) = f(x + 1) - 3$ D $g(x) = f(x - 3) + 1$

Find the *x*- and *y*-intercepts of the graph of each function. (Lesson 2-7)

20. $y = -2x + 2$
21. $y = -x^2 + 4x - 3$

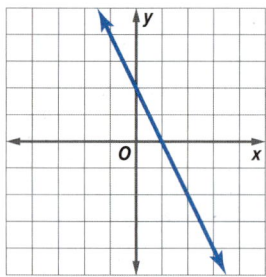

22. **ROCKET** A science teacher is conducting an experiment with a model rocket. The function $h = -16t^2 + 64t$ models the height of the rocket, where *h* is the height of the rocket in feet and *t* is the time in seconds after it is launched. Determine how many seconds the rocket is in the air. Solve by finding the zero of the related function. (Lesson 2-7)

CHAPTER 3 — Quadratic Functions

Complete parts a–c for each quadratic function.
a. Find the y-intercept, the equation of the axis of symmetry, and the x-coordinate of the vertex.
b. Make a table of values that includes the vertex.
c. Use this information to graph the function.
(Lesson 3-1)

1. $f(x) = x^2 - 3$
2. $f(x) = 2x^2 - 5x$

Use the related graph of each equation to determine its solutions. (Lesson 3-2)

3. $x^2 + x - 6 = 0$

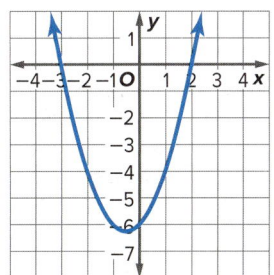

4. $x^2 - 2x - 3 = 0$

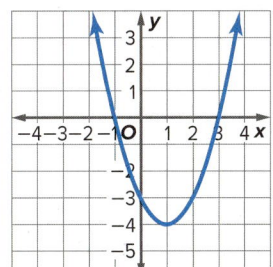

5. **ROCKET** A science teacher is conducting an experiment with a model rocket. The function $h = -16t^2 + 256t$ models the height of the rocket, where h is the height of the rocket in feet and t is the time in seconds after it is launched. Determine how many seconds the rocket is in the air. (Lesson 3-2)

Simplify. (Lesson 3-3)

6. $(10 + 3i) + (3 - 7i)$
7. $(2 + i)(2 - i)$
8. $\dfrac{5}{1 + 3i}$

Solve each equation by factoring. (Lesson 3-4)

9. $2x^2 + x - 10 = 0$
10. $2x^2 + x = 28$

11. **GAMES** Julio constructed a platform for a bean bag toss game. The plans for the original platform had dimensions of 3 feet by 5 feet. He made his platform larger by adding x feet to each side. The area of the new platform is 35 square feet. (Lesson 3-4)

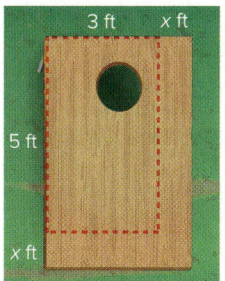

a. Write a quadratic equation that represents the area of his platform.
b. Find the dimensions of the platform Julio made.

Solve each equation by completing the square. (Lesson 3-5)

12. $x^2 - 2x - 24 = 0$
13. $x^2 - 2x = 35$
14. $5x^2 - 4x + 1 = 0$
15. $3x^2 + 6x + 10 = 0$

Solve each equation by using the Quadratic Formula. (Lesson 3-6)

16. $2x^2 - 3x - 9 = 0$
17. $4x^2 + 2x - 1 = 0$
18. $3x^2 + 4x - 2 = 0$
19. $3x^2 - 5x - 10 = 0$

Solve each inequality algebraically. (Lesson 3-7)

20. $x^2 + 3x < 18$
21. $x^2 - 3x \leq 28$
22. $2x^2 - 13x \geq -20$

CHAPTER 4 — Polynomials and Polynomial Functions

Simplify. (Lesson 4-1)

1. $(5x^2 + 3x - 7) - (3x^2 - x + 4)$
2. $(2y^2 - 4y - 5) + (3y^2 - 2y + 1)$
3. $2cd(3c - 4d) + 4d(c + 2d)$
4. $(r - s)(r + s)(3r - 2s)$

Find the indicated term of each expression. (Lesson 4-2)

5. third term of $(a + 2b)^6$
6. fourth term of $(y - 4x)^7$
7. **GEOMETRY** Suppose each of the dimensions of the cube shown below is increased by 0.25 centimeter. (Lesson 4-2)

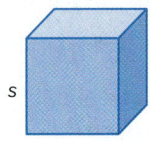

a. Write a binomial expression to represent the volume of the new cube.

b. Expand the binomial expression.

Simplify. (Lesson 4-3)

8. $\dfrac{4x^3y^4 - 10xy^5}{2xy}$
9. $(x^3 + 4x^2 - 4x - 7) \div (x + 1)$
10. **SHOES** The number of shoes sold by a store can be modeled by $-s^2 + 12s$, where s is the number of employees. Find the average number of shoes sold by each worker. (Lesson 4-3)

Find $p(-2)$ and $p(4)$ for each function. (Lesson 4-4)

11. $p(x) = -2x^2 + 5x - 3$
12. $p(x) = -3x^3 - x^2 + 2x - 8$
13. $p(x) = x^4 - 5x^3 + 4x + 6$

14. **CELL PHONES** The yearly sales of cell phones can be modeled by the function $s(t) = 0.45t^4 + 5.4t^3 - 15t^2 - 300t + 1000$, where $s(t)$ is the annual sales in millions of dollars and t is the number of years after 2005. (Lesson 4-5)

a. Graph the function for $0 \le t \le 10$.

b. Describe the turning points of the graph and its end behavior.

c. What trends in cell phone sales does this graph suggest?

d. Is it reasonable that the trend will continue indefinitely? Explain.

Factor completely. If the polynomial is not factorable, write *prime*. (Lesson 4-6)

15. $6x^4 - 5y^7$
16. $27a^3 - 125d^3$
17. $2ax^2 - 3bx^2 + cx^2 - 2ay^2 + 3by^2 - cy^2$

Transform each side of the equation to determine if it is an identity. (Lesson 4-7)

18. $(x^2 + 3)(x - 1)(x + 1) = (x^3 - x^2 + 3x - 3)(x + 1)$
19. $(4y + 5z)^2 = 16y^2 + 5z(4y + z)$

Given a polynomial and one of its factors, find the remaining factors of the polynomial. (Lesson 4-8)

20. $x^3 - 2x^2 - 5x + 6;\ x - 1$
21. $2x^3 - x^2 - 25x - 12;\ x + 3$
22. $6x^3 + 11x^2 + x - 4;\ 2x - 1$

Find all the zeros of each function. (Lesson 4-9)

23. $f(x) = x^3 - 4x^2 - 7x + 10$
24. $f(x) = x^4 - 8x^2 - 9$
25. $f(x) = x^4 + 3x^2 - 4$

26. **BUSINESS** The profit in hundreds of dollars for selling c calculators per day can be modeled by $R(c) = -0.005c^4 + 0.25c^3 + 0.01c^2 - 2.5c + 100$. (Lesson 4-9)

a. How many positive real zeros, negative real zeros, and imaginary zeros exist?

b. What is the meaning of the zeros in this situation?

CHAPTER 5 — Inverses and Radical Functions

Find $(f+g)(x)$, $(f-g)(x)$, $(f \cdot g)(x)$, and $\left(\dfrac{f}{g}\right)(x)$ for each $f(x)$ and $g(x)$. (Lesson 5-1)

1. $f(x) = x - 4$
   $g(x) = 2x + 3$

2. $f(x) = -x + 3$
   $g(x) = x^2 - 4$

Given $f(x) = x^2 + 1$ and $g(x) = x - 2$, find each value. (Lesson 5-2)

3. $[f \circ g](0)$
4. $[g \circ f](0)$
5. $[f \circ g](2)$
6. $[g \circ f](1)$
7. $[g \circ f](3)$
8. $[f \circ g](-1)$
9. $[g \circ f](-1)$
10. $[f \circ g](12)$
11. $[f \circ g](-5)$
12. $[f \circ g](-2)$
13. $[g \circ f](4)$
14. $[g \circ f](-9)$

Find $[f \circ g](x)$ and $[g \circ f](x)$, if they exist. (Lesson 5-2)

15. $f(x) = 3x + 2$
    $g(x) = x - 3$

16. $f(x) = 3x^2$
    $g(x) = x^2 + 5$

Determine whether each pair of functions are inverse functions. Write *yes* or *no*. (Lesson 5-3)

17. $f(x) = 2x - 8$
    $g(x) = \dfrac{1}{2}x + 4$

18. $f(x) = 3x - 4$
    $g(x) = 3x + 7$

19. **GEOMETRY** The formula for the volume of a sphere is $V = \dfrac{4}{3}\pi r^3$, where r is the radius of the sphere. (Lesson 5-3)
    a. Find the inverse of the function.
    b. Explain what purpose $V^{-1}(x)$ serves.

Identify the domain and range of each function. (Lesson 5-4)

20. $f(x) = \sqrt{2x} + 4$
21. $f(x) = \sqrt{x - 1} - 2$
22. $f(x) = -3\sqrt{x + 2}$

Graph each function. State the domain and range. (Lesson 5-4)

23. $f(x) = \sqrt{x} - 5$
24. $f(x) = 2\sqrt{x - 3}$
25. $f(x) = \dfrac{1}{2}\sqrt{x + 1} - 2$
26. $f(x) = \sqrt{2x - 5} + 3$

27. Write the square root function represented by the graph. (Lesson 5-4)

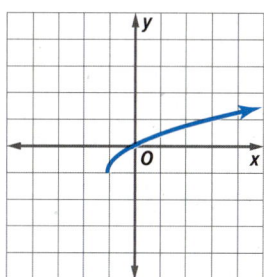

Graph each function. State key features of the graph. (Lesson 5-5)

28. $f(x) = \sqrt[3]{x} + 1$
29. $f(x) = 3\sqrt[3]{x - 2}$
30. $f(x) = \dfrac{1}{4}\sqrt[3]{x + 3} - 2$
31. $f(x) = -2\sqrt[3]{x - 1} + 3$

Solve each equation. (Lesson 5-6)

32. $\sqrt{5x + 14} - 2 = 6$
33. $\sqrt{4x + 15} = \sqrt{2x + 9}$
34. $\sqrt{x + 7} = \sqrt{x} - 1$

35. **WIND** The wind speed s in miles per hour can be modeled by the formula $s = 150\sqrt{\dfrac{p}{a}}$, where a is the area of a sail in square feet and p is the pressure of the wind on the sail in pounds per square feet. Determine the area of a sail if the wind speed is 75 mph and the pressure is 20 pounds per square feet. (Lesson 5-6)

CHAPTER 6 Exponential and Logarithmic Functions

Graph each function. State the domain and range. (Lesson 6-1)

1. $f(x) = 3(4)^x$
2. $f(x) = 2^{3x} - 3$

Solve each equation. (Lesson 6-2)

3. $2^{x-1} = 8^{x+3}$
4. $5^{2x+12} = 25^{10x-12}$

Write an equation for the *n*th term of each geometric sequence. (Lesson 6-3)

5. 0.75, 3, 12, …
6. 0.4, −2, 10, …

Evaluate each expression. (Lesson 6-4)

7. $\log_8 64$
8. $\log_7 1$
9. $\log_5 \frac{1}{25}$

10. **DOLPHINS** The number of dolphins living in an ocean region after t months can be approximated by $n(t) = 920 \log_{10}(t - 1)$. (Lesson 6-4)

    a. How many dolphins are in the ocean region after 4 months?
    b. How many dolphins are in the ocean region after 2 years?

11. **TAXI FARES** Diego collected data on the distance in miles and the fare in dollars for several taxi rides in his town. The ordered pairs are (2, 9.25), (5, 19.75), (8, 29.75), and (12, 44.50). (Lesson 6-5)

    a. What type of regression model works best with the data?
    b. If Diego rides a taxi for 10 miles, how much should he expect to pay?

Solve each equation. Check your solution. (Lesson 6-6)

12. $\log_4 3 + \log_4 x = \log_4 12$
13. $\log_7 6 - \log_7 2 = \log_7 x$
14. $4 \log_2 x = \log_2 81$
15. $\log_6 x + \log_6 (x + 5) = 2$

16. **POPULATION** The wolf population per year at a national park is listed in the table below. (Lesson 6-6)

Year	Population
2005	420
2006	399
2007	379
2008	360
2009	342
2020	?

a. Determine the annual percent of decrease of the population.
b. Write a logarithmic function for the time in years based upon population from 2005.
c. How many wolves will there be in the national park in 2020?

Solve each equation. Round to the nearest ten thousandth. (Lesson 6-7)

17. $5^x = 60$
18. $4^{x^2} = 21$
19. $3^{x-2} = 4^x$

Solve each equation. Round to the nearest ten thousandth. (Lesson 6-8)

20. $4e^x - 6 = 11$
21. $3e^{-x} + 4 = 17$
22. $2e^{3x} - 8 = 21$

Solve each equation. (Lesson 6-9)

23. $\log_{25} x = \frac{3}{2}$
24. $\log_9 81 = x$
25. $\log_6 (5x - 3) = \log_6 (x + 9)$
26. $\log_8 (x^2 + 6) = \log_8 (5x)$

27. **INTEREST** Stefanie deposited $2000 into a savings account paying 2.5% interest compounded continuously. (Lesson 6-10)

    a. Determine how long it will take Stefanie to have a balance of $2400.
    b. How long will it take Stefanie to double her investment?
    c. If Stefanie wanted to have $5000 after 10 years, how much should she have invested?

CHAPTER 7 Rational Functions

Simplify each expression. (Lesson 7-1)

1. $\dfrac{(x-2)(x+5)}{x^2+2x-8}$

2. $\dfrac{(x+3)(x-4)}{x^2+x-20}$

3. $\dfrac{x^2-16}{x^2-8x+12} \cdot \dfrac{x^2+3x-18}{x^2-7x+12}$

4. $\dfrac{x^2+10x+21}{x^2-5x+6} \div \dfrac{x^2+3x-28}{x^2-8x+15}$

5. **BUSINESS** The average hourly income I of a college student x years after graduation can be modeled by the function $I(x) = \dfrac{11x^2+33x+22}{x+2}$. Simplify the function. (Lesson 7-2)

Simplify each expression. (Lesson 7-2)

6. $\dfrac{2}{x^2-7x+12} + \dfrac{6}{x^2+x-20}$

7. $\dfrac{4}{x^2-11x+30} - \dfrac{3}{x^2-13x+40}$

Identify the asymptotes, domain, and range of each function. (Lesson 7-3)

8.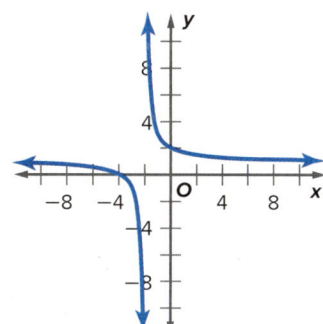

9.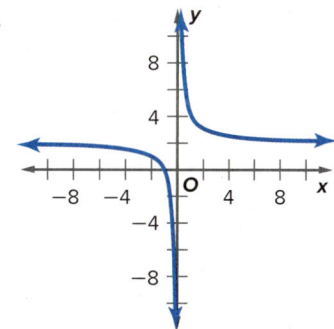

Identify the asymptotes, domain, and range of each function. (Lesson 7-3)

10.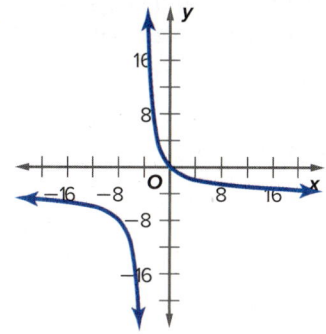

11. **FIELD TRIP** The American history class is planning a field trip to Washington, D.C. It costs $300 to rent the bus, and each ticket to the museum costs $12. (Lesson 7-3)

    a. If C represents the cost for each student and n represents the number of students, write an equation to represent the cost to each student as a function of how many students go on the field trip.

    b. Are there any limitations to the range or domain in this situation? Explain.

Graph each function. (Lesson 7-4)

12. $f(x) = \dfrac{2x}{x-3}$

13. $f(x) = \dfrac{x-4}{x+2}$

If a varies jointly as b and c, find a when $b = 2$ and $c = -3$, given each of the following. (Lesson 7-5)

14. $a = -12$ when $b = 4$ and $c = 6$

15. $a = -36$ when $b = 12$ and $c = -4$

Solve each equation. Check your solution. (Lesson 7-6)

16. $\dfrac{3}{x+4} + \dfrac{1}{x-1} = \dfrac{17}{24}$

17. $\dfrac{12}{x+4} - \dfrac{3}{x-3} = \dfrac{24}{x^2+x-12}$

18. $\dfrac{14}{x-5} - \dfrac{7}{x+5} = \dfrac{-7}{x^2-25}$

19. **EXERCISE** Natalie can bike 10 kilometers per hour faster than Aaron. By the time Natalie travels 60 kilometers, Aaron has gone 40 kilometers. (Lesson 7-6)

    a. Write an expression for Natalie's time.

    b. Write an expression for Aaron's time.

    c. Write and solve the rational equation to determine the speed of each biker.

CHAPTER 8 Statistics and Probability

Determine whether each situation describes a *survey*, an *experiment*, or an *observational study*. Then identify the sample, and suggest a population from which it may have been selected. (Lesson 8-1)

1. Every fifth person coming out of a football stadium is asked how often they go to a football game.

2. A company posts a controversial advertisement in their store and records the reactions of the customers.

3. A randomly selected group of people are chosen to test the effects of a new drug on energy levels.

4. A randomly selected group of people are asked their opinions on political topics.

Design and conduct a simulation using a probability model. Then report the results. (Lesson 8-2)

5. Natasha plays on the basketball team. This season she has made 80% of her free throws.

6. Pilar is playing a board game with eight different categories, each with the same number of questions that must be answered correctly in order to win.

For a sample survey of 120 soccer players, find the following population proportions. (Lesson 8-3)

7. 60 prefer weekend games.

8. 48 prefer to practice on Wednesdays.

9. 100 prefer to play outside.

10. 15 prefer to be goalie.

Compute the margin of error for each sample size. (Lesson 8-3)

11. 80

12. 250

Find the sample size for each margin of error. (Lesson 8-3)

13. ±1%

14. ±5%

15. **COLLEGE** The SAT scores of Mr. Williams's calculus class are given below. (Lesson 8-4)

SAT Scores				
2020	1250	1500	1700	1600
1830	1230	1330	1450	1250
1560	1450	1350	1430	1640
2210	1550	2200	2050	1750
2400	1900	2140	1280	1800

   a. Use a graphing calculator to create a histogram. Then describe the shape of the distribution.

   b. Describe the center and spread of the data using either the mean and standard deviation or the five-number summary. Justify your choice.

16. Ron wants to create a misleading histogram using the data in exercise **15**. Identify intervals along the horizontal axis that will make the data appear evenly distributed. (Lesson 8-5)

17. **ASSEMBLIES** The number of assemblies at West High School each year is normally distributed with a mean of 12.4 and a standard deviation of 1.6. (Lesson 8-6)

   a. What is the probability that there will be more than 10 assemblies in a given year?

   b. If the school has existed for 30 years, in how many of those years were there between 11 and 13 assemblies?

18. **STUDENT COUNCIL** The number of students that run for student council each year is normally distributed with a mean of 16.8 students and a standard deviation of 3.7. (Lesson 8-6)

   a. What is the probability that fewer than 10 students run in a given year?

   b. If the school has kept records for 20 years, in how many of those years were there between 15 and 20 students who ran for student council?

19. **FOOD PROCESSING** It is known that 0.1% of the cans produced at a food processing plant are an incorrect weight. An employee has a scale that is 98% accurate. The employee weighs some cans and discards any that the scale shows to be an incorrect weight. Do you think this is a good decision? Explain. (Lesson 8-7)

CHAPTER 9 Trigonometric Functions

Use a trigonometric function to find the value of x. Round to the nearest tenth. (Lesson 9-1)

1.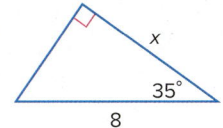

2.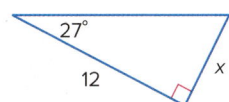

3. **FLAGPOLE** Tina is standing away from a flagpole. The angle of elevation from the ground to the top of the flagpole is 15°. The flagpole is 50 feet high. Determine how far Tina is standing from the flagpole. (Lesson 9-1)

4. **SAILS** One leg of a triangular sail of a sailboat is 25 feet long. The angle opposite this leg is 40°. What is the perimeter of the sail to the nearest tenth? (Lesson 9-1)

Rewrite each degree measure in radians and each radian measure in degrees. (Lesson 9-2)

5. 300°

6. −80°

7. $-\dfrac{2\pi}{3}$

8. $-\dfrac{\pi}{6}$

Find the exact value of each trigonometric function. (Lesson 9-3)

9. $\sin 120°$

10. $\csc \dfrac{3\pi}{4}$

The terminal side of angle θ in standard position intersects the unit circle at each point P. Find $\cos \theta$ and $\sin \theta$. (Lesson 9-4)

11. $P\left(\dfrac{35}{37}, \dfrac{12}{37}\right)$

12. $P\left(-\dfrac{\sqrt{3}}{2}, \dfrac{1}{2}\right)$

13. $P\left(\dfrac{\sqrt{2}}{2}, -\dfrac{\sqrt{2}}{2}\right)$

Determine the period of each function. (Lesson 9-4)

14.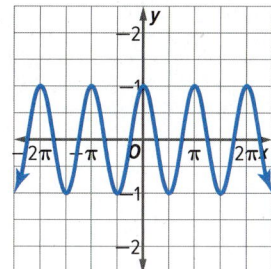

15.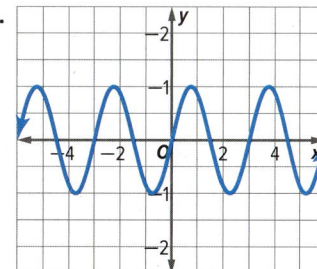

Find the exact value of each expression. (Lesson 9-4)

16. $\cos(-60)°$

17. $\sin\left(-\dfrac{3\pi}{4}\right)$

18. $\cos\left(\dfrac{13\pi}{4}\right)$

19. $\sin(-990°)$

Find the amplitude and period of each function. Then graph the function. (Lesson 9-5)

20. $y = 3\cos\theta$

21. $y = \dfrac{1}{2}\sin 3\theta$

State the amplitude, period, phase shift, and vertical shift for each function. Then graph the function. (Lesson 9-6)

22. $y = 2\tan(\theta + 30°) + 3$

23. $y = 4\cos\left(\theta - \dfrac{\pi}{2}\right) - 2$

CHAPTER 10 Trigonometric Identities and Equations

Find the exact value of each expression if $0° < \theta < 90°$. (Lesson 10-1)

1. If $\cos \theta = \frac{3}{5}$, find $\sin \theta$.

2. If $\tan \theta = 2$, find $\cot \theta$.

3. If $\sin \theta = \frac{\sqrt{5}}{3}$, find $\cos \theta$.

4. If $\csc \theta = \frac{3\sqrt{5}}{5}$, find $\tan \theta$.

Verify that each equation is an identity. (Lesson 10-2)

5. $\frac{\sin^2 \theta}{\cos^2 \theta} \cdot \csc^2 \theta = 1 + \tan^2 \theta$

6. $\sec \theta \cot \theta = \csc \theta$

7. $\sin \theta \cot \theta = \cos \theta$

8. $\frac{\sec^2 \theta}{\csc^2 \theta} = \tan^2 \theta$

9. **CONSTRUCTION** A window has the dimensions shown below. Use the measures of the sides of the triangle to show that $\sin^2 \theta + \cos^2 \theta = 1$. (Lesson 10-2)

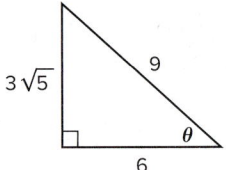

Find the exact value of each expression. (Lesson 10-3)

10. $\cos(-30°)$

11. $\sin(-120°)$

12. $\sin 135°$

13. $\cos 225°$

Find the exact value of each expression. (Lesson 10-4)

14. $\sin 15°$

15. $\cos \frac{\pi}{12}$

16. $\tan 67.5°$

17. $\sin 165°$

Solve each equation if $0° \leq \theta \leq 360°$. (Lesson 10-5)

18. $2 \sin \theta = 1$

19. $2 \cos \theta + 1 = 0$

20. $4 \cos^2 \theta - 1 = 0$

21. **TOYS** A toy's height h in inches can be modeled by the equation $h = 4 \sin \frac{11\pi}{12} t$, where t represents time in seconds. (Lesson 10-5)

    a. At what point will the toy be 4 inches above its resting position for the first time?

    b. At what point will the toy be 2 inches above its resting position for the first time?

22. **BUOYS** As waves move past a buoy, the height in feet of the buoy oscillates according to the equation $y = 6 \cos 30t$, where t is the time in seconds. If the buoy is at the top of the wave at 0 seconds, during what time intervals will the buoy be 3 feet below the midline? (Lesson 10-5)

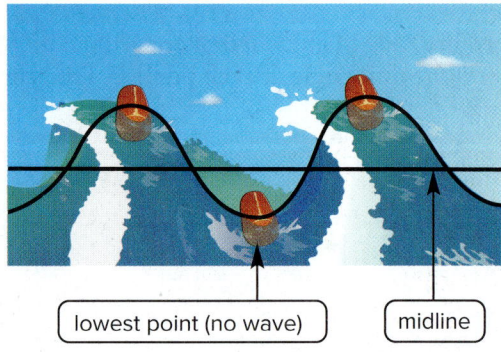

Selected Answers and Solutions

CHAPTER 0
Preparing for Advanced Algebra

Lesson 0-1

1. D = {1, 2, 3}, R = {6, 7, 10}; yes   **3.** D = {1, 2}, R = {5, 7, 9}; no   **5.** D = {−2, −1, 0, 3}, R = {−3, −2, 2}; yes   **7.** D = {−1, 0, 1, 2, 3}, R = {−3, −2, −1, 2, 3, 4}; no   **9.** I   **11.** none

Lesson 0-2

1. $a^2 + 6a + 8$ **3.** $h^2 - 16$ **5.** $b^2 + b - 12$ **7.** $r^2 - 5r - 24$ **9.** $p^2 + 16p + 64$ **11.** $2c^2 - 9c - 5$ **13.** $6m^2 - 7m - 20$ **15.** $2q^2 - 13q - 34$ **17a.** $n - 7$, $n + 2$ **17b.** $n^2 - 5n - 14$

Lesson 0-3

1. $4x(3x + 1)$ **3.** $4ab(2b - 3)$ **5.** $(y + 3)(y + 9)$ **7.** $(3y + 1)(y + 4)$ **9.** $(3x + 4)(x + 8)$ **11.** $(y - 4)(y - 1)$ **13.** $2(3a - b)(a - 8b)$ **15.** $(2x - 3y)(9x - 2y)$ **17.** $(3x - 4)^2$ **19.** $(x + 12)(x - 12)$ **21.** $(4y + 1)(4y - 1)$ **23.** $4(3y + 2)(3y - 2)$

Lesson 0-4

1. 60   **3.** 18   **5.** 120   **7.** 6   **9.** 6   **11.** permutations, 720   **13.** permutations, 5040   **15.** combinations, 15   **17.** permutations, 3024   **19a.** 2,176,782,336; 1,402,410,240   **19b.** 308,915,776; 712,882,560; The password with one digit is more secure, because the chance of someone guessing this password at random is $\frac{1}{712,882,560}$, which is less than the chance of someone guessing a 6-character password that contains only letters, $\frac{1}{308,915,776}$.

Lesson 0-5

1a.

Color	Frequency	Experimental Probability
red	6	0.12
blue	7	0.14
yellow	9	0.18
orange	12	0.24
purple	5	0.10
green	11	0.22

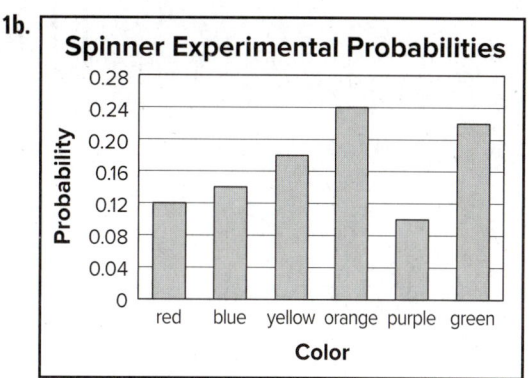

1b. Spinner Experimental Probabilities

1c.

Color	Frequency	Experimental Probability	Theoretical Probability
red	6	0.12	$0.1\overline{6}$
blue	7	0.14	$0.1\overline{6}$
yellow	9	0.18	$0.1\overline{6}$
orange	12	0.24	$0.1\overline{6}$
purple	5	0.10	$0.1\overline{6}$
green	11	0.22	$0.1\overline{6}$

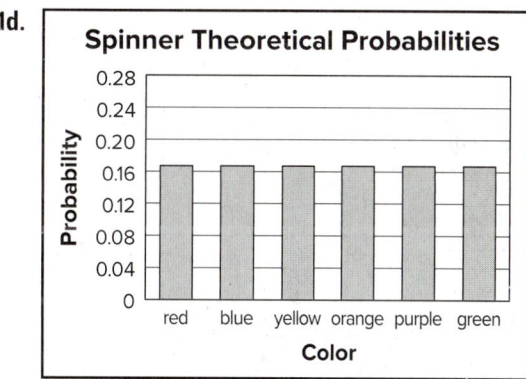

1d. Spinner Theoretical Probabilities

1e. Sample answer: Because all bars in the graph of the theoretical probabilities are the same height, the graph represents a uniform distribution. This means that in theory, the chance of landing on any one of the colors is equally likely. The graph of the experimental probabilities indicates that in practice, it is more likely that the spinner will land on orange or green than on any of the other colors, because the heights of those bars are taller than any others in the graph.

3a. mutually exclusive, $\frac{1}{2}$ **3b.** not mutually exclusive, $\frac{4}{13}$ **3c.** not mutually exclusive, $\frac{7}{13}$ **5a.** mutually exclusive, $\frac{11}{20}$ **5b.** not mutually exclusive, $\frac{29}{40}$ **5c.** not mutually exclusive, $\frac{2}{5}$

connectED.mcgraw-hill.com   **R11**

7. When events are mutually exclusive, $P(A \text{ and } B)$ will always equal 0, so the probability will simplify to $P(A \text{ or } B) = P(A) + P(B)$.   **9.** 1 to 3; 1 to 1

Lesson 0-6

1. independent; $\frac{1}{36}$ **3.** $\frac{1}{36}$ **5.** $\frac{25}{36}$ **7.** $\frac{5}{18}$ **9.** $\frac{5}{17}$ **11.** $\frac{1}{8}$
13. $\frac{1}{26}$ **15a.** $\frac{32}{41}$ **15b.** $\frac{2}{5}$ **15c.** $\frac{3}{7}$ **17a.** $\frac{91}{115}$ or about 79.1%   **17b.** $\frac{239}{308}$ or about 77.6%

Lesson 0-7

1. similar   **3.** neither   **5.** similar   **7.** 8; 21   **9.** 10.2; 13.6
11. 7.5 ft

Lesson 0-8

1. 39 ft   **3.** 8.3 cm   **5.** 5   **7.** 9.2   **9.** 8.5   **11.** yes   **13.** no
15. yes   **17.** about 2.66 m

Lesson 0-9

1. 451.8 pages, 399 pages, no mode   **3.** ≈111.3 text messages, 113 text messages, 125 text messages
5. Sample; Walk A: 47, ≈242.0, ≈15.6; Walk B: 92, ≈1115.4, ≈33.4; because the sample standard deviation of Walk B is greater than that of Walk A, there is more variability in the number of sponsors obtained by participants in Walk B than in Walk A.   **7.** 18, 23, 25, 27, 29; Sample answer: There are 18 students in the smallest math class at Central High and 29 students in the largest class. 25% of the classes have less than 23 students, 50% of the classes have less than 25 students, and 75% of the classes have less than 27 students.   **9.** 13; Sample answer: The interval beyond which any outliers would lie is $14.25 < x < 60.25$. Because $13 < 14.25$, it is an outlier.

Data Set	Mean	Median	Mode	Range	Standard Deviation
with outlier	≈35.8	36	36	38	≈9.3
without outlier	≈37.3	36	36	29	≈7.5

Removing the outlier did not affect the median or mode. However, the removal did affect the mean, standard deviation, and range. The mean increased, and the standard deviation and range decreased.

11a. Yes; sample answer: The interval beyond which any outliers would lie is $15.35 < x < 17.35$. Because $14.9 < 15.35$, it is an outlier.   **11b.** No; sample answer: The new interval beyond which any outliers would lie would be $15.4 < x < 17.4$. Because $17.4 > 17.35$, it would not be an outlier.   **11c.** Sample answers: data recording errors, manufacturing errors

CHAPTER 1
Linear Equations

Chapter 1 Get Ready

1. Multiply the numerators by the numerators and the denominators by the denominators.   **3.** Multiply (-1.5) by itself 3 times.   **5.** Find the square of 7 for the numerator and the square of 10 for the denominator.   **7.** yes   **9.** Insert 0 for x in the equation and solve for y.

Lesson 1-1

1. $12[x + (-3)]$ **3.** The sum of five times a number and seven equals 18.   **5.** The difference between five times a number and the cube of that number is 12.
7. Reflexive Property   **9.** 53   **11.** -8 **13.** -6 **15.** 3
17. 4   **19.** $q = \dfrac{8r - 3}{5}$ **21.** B   **23.** $8x^2$ **25.** $\dfrac{x}{4} + 5$
27. The quotient of the sum of three and a number and four is five.

(29) Let n = the number of home runs that Jacobs hit. Then $n + 6$ = the number of home runs that Cabrera hit.
$n + (n + 6) = 46$ Cabrera and Jacobs hit a combined total of 46 home runs.
$\quad 2n + 6 = 46$ Simplify.
$\quad\quad 2n = 40$ Subtract 6 from each side.
$\quad\quad\quad n = 20$ Divide each side by 2.
So, Jacobs hit 20 home runs and Cabrera hit $n + 6 = 20 + 6$ or 26 home runs.

31. substitution

33. multiplication $(=)$

35. 5   **37.** -3

(39) $5(-2x - 4) - 3(4x + 5) = 97$ Original equation
$-10x - 20 - 12x - 15 = 97$ Apply the Distributive Property.
$\quad\quad\quad -22x - 35 = 97$ Simplify the left side.
$\quad\quad\quad\quad\quad -22x = 132$ Add 35 to each side.
$\quad\quad\quad\quad\quad\quad x = -6$ Divide each side by -22.

41. -3 **43.** s = length of a side; $5s = 100$; 20 in.
45. $m = \dfrac{E}{c^2}$ **47.** $h = \dfrac{z}{\pi q^3}$ **49.** $a = \dfrac{y - bx - c}{x^2}$
51a. $V = \pi \cdot r \cdot r \cdot h$ **51b.** $h = \dfrac{V}{\pi r^2}$ **53.** -2 **55.** -4
57. $-\dfrac{117}{11}$ **59.** x = the cost of rent each month; $864 + 428 + 480 + 144 + 12x = 11{,}216$; $775 per month

(61) a. The integers from -5 to 5 are $-5, -4, -3, -2, -1, 0, 1, 2, 3, 4,$ and 5. Draw a number line and plot a point at each integer.

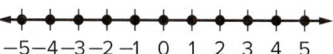

b. -5 and 5 are 5 units from zero, -4 and 4 are 4 units from zero, and so on.

Integer	Distance from Zero
-5	5
-4	4
-3	3
-2	2
-1	1
0	0
1	1
2	2
3	3
4	4
5	5

c. The points (x, y) = (integer, distance from zero) are $(-5, 5), (-4, 4), (-3, 3),$ $(-2, 2), (-1, 1), (0, 0), (1, 1),$ $(2, 2), (3, 3), (4, 4),$ and $(5, 5)$.

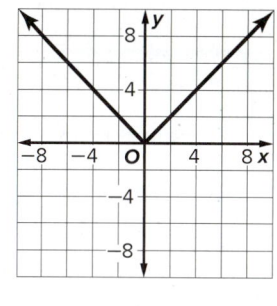

d. For positive integers, the distance from zero is the same as the integer. For negative integers, the distance is the integer with the opposite sign because distance is always positive.

63. $y_1 = y_2 - \sqrt{d^2 - (x_2 - x_1)^2}$ **65.** Sample answer: $3(x - 4) = 3x + 5$; $2(3x - 1) = 6x - 2$ **67.** B   **69.** B
71. C   **73a.** $2n + 2$ **73b.** $5n - 6$ **73c.** $n + 2n + 2 + 5n - 6 = 100$ **73d.** $n = 13$ **73e.** 13, 28, and 59   **73f.** $13 + 2(13) + 2 + 5(13) - 6 = 100$

Lesson 1-2

1. $b < 8$

3. $x \leq -6$

5. $w < 2$

(7) $s \geq \dfrac{s + 6}{5}$ Original inequality
$5s \geq s + 6$ Multiply each side by 5.
$4s \geq 6$ Subtract s from each side.
$s \geq 1.5$ Divide each side by 4.
The solution set is $\{s \mid s \geq 1.5\}$.

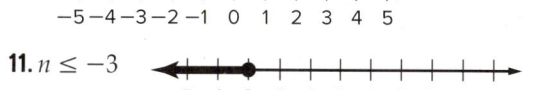

9. 40 bags

11. $n \leq -3$

13. $t \leq \dfrac{1}{2}$

15. $k < 27$

17. $z < 3$

connectED.mcgraw-hill.com   **R13**

19.
$12 < -4(3c - 6)$ Original inequality
$-3 > 3c - 6$ Divide each side by -4, reversing the inequality symbol.
$3 > 3c$ Add 6 to each side.
$1 > c$ Divide each side by 3.
The solution set is $\{c \mid c < 1\}$.

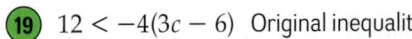

21. $z < 3$

23. $3x - 12 < 21; x < 11$ **25.** $5x - 6 > x; x > 1.5$
27. at least 8 hours
29. $x > -\frac{3}{4}$

31. $y > 18.75$

33. $v > -4.5$

35. $r > -\frac{3}{4}$

37a. $250 + 0.03(500a) \geq 700$ **37b.** $a \geq 30$; He must sell at least 30 advertisements. **39.** $\frac{x}{3} + 4 \leq 2x + 12; x \geq -4.8$

41. a. Let $d =$ the number of miles by which Jamie should increase her average daily run. Then $5 + d =$ her average daily distance after the increase.

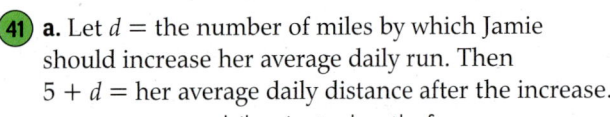

So, the inequality is $3(5 + d) \geq 26.2$.

b. $3(5 + d) \geq 26.2$ Original inequality
$5 + d \geq 8.73$ Divide each side by 3. Round to the nearest hundredth.
$d \geq 3.73$ Subtract 5 from each side.

In order to have enough endurance to run a marathon, Jamie should increase the distance of her average daily run by at least 3.73 miles.

43a. Sample answer:

Point	Resulting Statement	True or False
(0, 0)	$0 \geq 3$	False
(1, 1)	$1 \geq \frac{5}{2}$	False
(2, 2)	$2 \geq 2$	True
(3, 3)	$3 \geq \frac{3}{2}$	True
(4, 4)	$4 \geq 1$	True

43b. Sample answer:

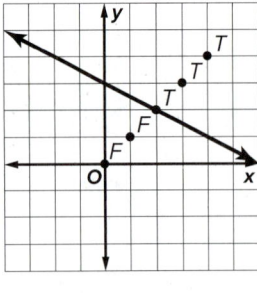

43c. Sample answer: The points on or above the line result in true statements, and the points below the line result in false statements. This is true for all points on the coordinate plane. **45.** No; Sample answer: Madlynn reversed the inequality sign when she added 1 to each side. Emilie did not reverse the inequality sign at all. **47.** Using the Triangle Inequality Theorem, we know that the sum of the lengths of any 2 sides of a triangle must be greater than the length of the remaining side. This generates 3 inequalities to examine.

$3x + 4 + 2x + 5 > 4x$ $3x + 4 + 4x > 2x + 5$
$x > -9$ $x > 0.2$

$2x + 5 + 4x > 3x + 4$
$x > -\frac{1}{3}$

In order for all 3 conditions to be true, x must be greater than 0.2. **49.** Sample answer: When one number is greater than another number, it is either more positive or less negative than that number. When these numbers are multiplied by a negative value, their roles are reversed. That is, the number that was more positive is now more negative than the other number. Thus, it is now *less than* that number and the inequality symbol needs to be reversed.

51. B **53a.** $\frac{5x}{-4} \geq 20$ **53b.** $x \leq -16$

53c.

53d. Substitute -15 and then -17 for x in $\frac{5x}{-4} \geq 20$. The inequality should be true for $x = -17$ and false for $x = -15$. You may choose other appropriate values to check the inequality. **55.** D **57.** $x \geq -3$

Lesson 1-3

1. 6 feet/min **3a.** about 6700 per year **3b.** about -5000 per year **3c.** The positive rate in part **a** represents an increase in the sales of digital cameras. The negative rate in part **b** represents a decrease in sales of film cameras. **5.** -3 **7.** $\frac{3}{5}$

9. Use the ordered pairs $(3, 20)$ and $(6, 40)$.

rate of change $= \dfrac{\text{change in } y}{\text{change in } x}$

$= \dfrac{\text{change in height}}{\text{change in time}}$ ← mm / ← days

$= \dfrac{40 - 20}{6 - 3}$

$= \dfrac{20}{3}$

The rate of change is $\dfrac{20}{3}$ mm/day.

11a. $0.15°/h$ **11b.** $-0.125°/h$; Yes; the number should be negative because her temperature is dropping. **11c.** Tuesday 8:00 A.M.–Tuesday 8:00 P.M. **13.** $\dfrac{14}{15}$
15. -2 **17.** $\dfrac{5}{3}$ **19.** 5

21. The line passes through $(0, 16)$ and $(10, 8)$.

$m = \dfrac{y_2 - y_1}{x_2 - x_1}$ Slope Formula

$= \dfrac{8 - 16}{10 - 0}$ $(x_1, y_1) = (0, 16), (x_2, y_2) = (10, 8)$

$= -\dfrac{8}{10}$ or -0.8 Simplify.

23. $\frac{4}{3}$ **25.** 3 **27.** $\frac{6}{5}$

29. slope = $\frac{\text{change in } y}{\text{change in } x}$

= $\frac{\text{change in vertical distance}}{\text{change in horizontal distance}}$

= $\frac{8.9}{2.8}$

≈ 3.2

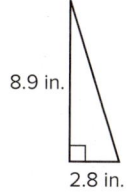

8.9 in.

2.8 in.

31. 9 **33.** 5 **35a.**

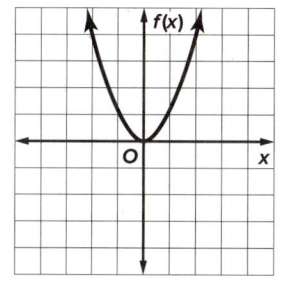

35b.

x	−4	−3	−2	−1	0	1	2	3	4
f(x)	16	9	4	1	0	1	4	9	16
slope		−7	−5	−3	−1	1	3	5	7

35c. Sample answer: The rate of change is not constant. The rate of change decreases as x approaches zero and then increases as x approaches infinity. **37.** Sample answer: Because the slope from (2, 3) to (5, 8) is the same as the slope from (5, 8) to (11, y), find the slope between each pair of points and set them equal to each other. Then solve for y.

$\frac{8 - 3}{5 - 2} = \frac{y - 8}{11 - 5}$

$\frac{5}{3} = \frac{y - 8}{6}$

$30 = 3(y - 8)$

$10 = y - 8$

$18 = y$

39. Sometimes; the slope of a vertical line is undefined.
41. E **43.** C **45.** A **47a.** 15 wins per month
47b. 14 wins per month **47c.** 27 wins per month
47d. The graph is steeper from July through September because the rate is higher from July through September.

Lesson 1-4

1. $y = 1.5x + 5$ **3.** $y = -2x + 11$ **5.** $y = -4x - 25$
7. $y = \frac{4}{3}x - \frac{2}{3}$

9. The slope of the given line is $\frac{7}{8}$. Lines that are parallel have the same slope, so the slope of the line parallel to the given line is $\frac{7}{8}$.

$y - y_1 = m(x - x_1)$ Point-slope form

$y - (-10) = \frac{7}{8}(x - 4)$ $(x_1, y_1) = (4, -10)$ and $m = \frac{7}{8}$

$y + 10 = \frac{7}{8}x - \frac{7}{2}$ Distributive Property

$y = \frac{7}{8}x - \frac{27}{2}$ Subtract 10 from each side.

11. $y = -\frac{1}{2}x + 5$ **13.** $y = 4.5x - 6.5$ **15.** $y = 4x - 15$
17. $y = -\frac{1}{4}x - 1$ **19.** $y = 2x - 2$

21. $y = -8x - 20$
23. $y = -0.5x + 3.35$ **25.** $y = \frac{1}{2}x$
27. $y = -\frac{1}{2}x + 6$
29. $y = 180x + 5900$
31. $y = -25x + 300$

33. First, find the slope. The line passes through (−6, 2) and (0, 6).

$m = \frac{y_2 - y_1}{x_2 - x_1}$ Slope Formula

$= \frac{6 - 2}{0 - (-6)}$ $(x_1, y_1) = (-6, 2), (x_2, y_2) = (0, 6)$

$= \frac{4}{6}$ or $\frac{2}{3}$ Simplify.

The graph intersects the y-axis at 6. So, $b = 6$. Substitute the values into the slope-intercept equation.

$y = mx + b$ Slope-intercept form

$y = \frac{2}{3}x + 6$ $m = \frac{2}{3}, b = 6$

35. 10 mi **37a.** $13.56 **37b.** She needs 142 ÷ 60 or 2.3 thread containers. Because she cannot buy a fractional container, round to the next highest integer, 3. Similarly, she needs 61 ÷ 20 or 4 baskets and 1 paint storage rack. The cost of her storage supplies is 3(14.99) + 4(15) + 49 or $153.97. This is how much money she needs to save after 2 months or about 9 weeks. The amount she needs to save per week is equal to the slope m between points (0, 32) and (9, 153.97) or $m = \frac{153.97 - 32}{9} = 13.56$.

39. Sample answer: Sometimes; while the two sets of parallel and perpendicular lines will always form a quadrilateral with four 90° angles, that figure will always be a rectangle, but not necessarily a square.

41. Sample answer: $y - 0 = a\left(x + \frac{b}{a}\right)$

43. Sample answer: $y - d = -\frac{d}{c}(x - 0)$
45. C **47.** C **49.** D **51a.** $y = 2$ **51b.** $x = 0$

Lesson 1-5

1. **3.**

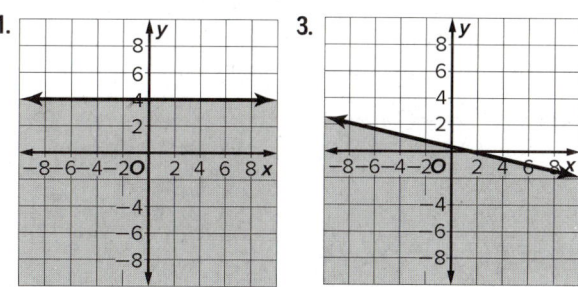

5a. $3.85g + 6.41q \leq 65$; g is the number of gallons of gas he buys. Gas is $3.85 a gallon; q is the number of quarts of oil he buys. Oil is $6.41 a quart. He only has $65 to spend so the inequality must be less than or equal to 65.

5b.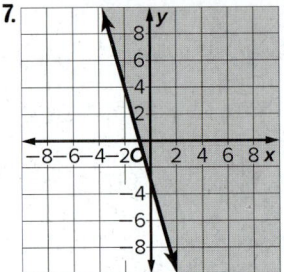

5c. No; (10, 8) is not in the shaded region.

7.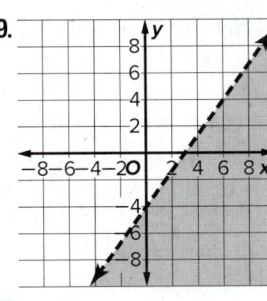

9.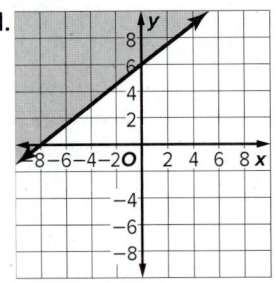

11. (graph shown)

13a. Let a be the number of hours Carlos works at the Main St. Deli. Then $8a$ is the total amount he earns from this job.

Let b be the number of hours Carlos works at babysitting. Then $6b$ is the total amount he earns from this job.

The total amount Carlos earns from both jobs is $8a + 6b$.

Because Carlos needs to earn at least \$700, the inequality is $8a + 6b \geq 700$.

13b. To graph the inequality, first graph the line $8a + 6b = 700$. Graph the line as a solid line because the boundary is included in the inequality. Because (0, 0) is not a solution of the inequality $8a + 6b \geq 700$, shade the region that does not include the origin.

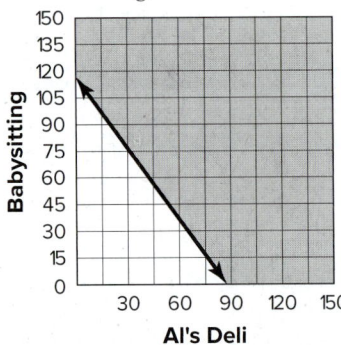

13c. If Carlos works 50 hours at each job, then $a = b = 50$; $8(50) + 6(50) = 400 + 300 = 700$, so Carlos will make enough money.

15. $y \geq -x + 2$ **17.** $y < -x$ **19.** $y \leq 4$

21. a. Sample answer: 8 ebooks and 10 movies; 15 e-books and 7 movies, 24 e-books and 3 movies
b. Sample answer: To begin, I wrote and graphed two inequalities, $7.99b + 17m \leq 250$ and $7.99b + 17m \geq 200$, to represent the situation. I selected points in the solution region to verify the total met the parameters of the problem.
c. I assumed that tax was already included in the cost of the e-books and movies. I assumed that there was no processing fee for her e-books.

23. (graph shown) **25.** Sample answer: $x > 3$

27. Paulo: $x - y \geq 2$ can be written as $y \leq x - 2$
29. The solution of the inequality $y < 2$ includes all points along the x-axis because the solution is the set of all points on the coordinate plane below the horizontal line $y = 2$. **31.** C, D **33.** C **35.** B
37a. Sample answer: $8n + 5e \geq 100$
37b. Sample answer: 10 necklaces and 8 earrings
37c. Sample answer: 11 necklaces and 4 earrings

Lesson 1-6

1. (3, 5) **3.** (3, −3) **5.** (6, 7)

7. Write each equation in slope-intercept form. Then graph.
$4x + 5y = -41 \rightarrow y = -\frac{4}{5}x - \frac{41}{5}$
$3y - 5x = 5 \rightarrow y = \frac{5}{3}x + \frac{5}{3}$

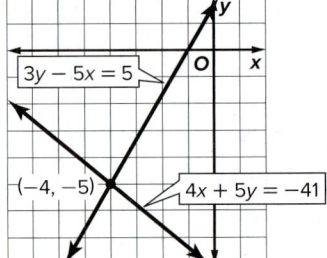

The graphs appear to intersect at (−4, −5). So, the solution is (−4, −5).

9a. $y = 0.15x + 2.70$, $y = 0.25x$ **9b.** \$6.75 for 27 photos
9c. You should use the online store if you are printing more than 27 digital photos, and the local store if you are printing fewer than 27 photos.

11.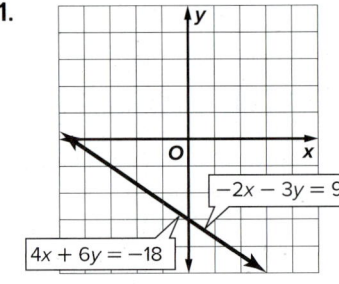

consistent and dependent

13. $(-2, 1)$

15. $2a + 8b = -8$ Multiply by 5. → $10a + 40b = -40$
$3a - 5b = 22$ Multiply by 8. → $24a - 40b = 176$

$10a + 40b = -40$ Equation 1 × 5
$\underline{(+) \; 24a - 40b = 176}$ Equation 2 × 8
$34a = 136$ Add the equations.
$ a = 4$ Divide each side by 34.

Substitute 4 for a into either original equation.
$2a + 8b = -8$ Equation 1
$2(4) + 8b = -8$ $a = 4$
$8 + 8b = -8$ Multiply.
$8b = -16$ Subtract 8 from each side.
$b = -2$ Divide each side by 8.

The solution is $(4, -2)$.

17. $(5, 1)$ **19.** $(-2, 7)$ **21.** $(-4, -3)$ **23.** no solution
25. $(2, -2)$ **27.** $(-2, 0)$ **29.** $(4, -1)$ **31.** 250 T-shirts **33.** $(-3, -4)$ **35.** infinite solutions
37. $(-1.5, -2)$

39. 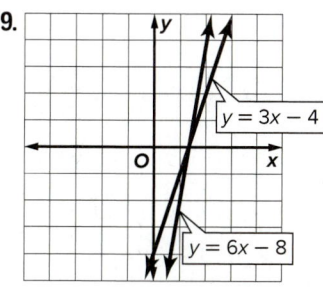 consistent and independent

41. 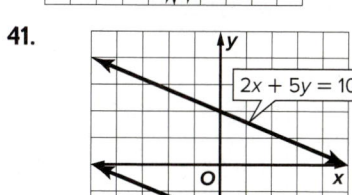 inconsistent

43. Write each equation in slope-intercept form. Then graph.
$-5x - 6y = 13$ → $y = -\frac{5}{6}x - \frac{13}{6}$
$12y + 10x = -26$ → $y = -\frac{5}{6}x - \frac{13}{6}$

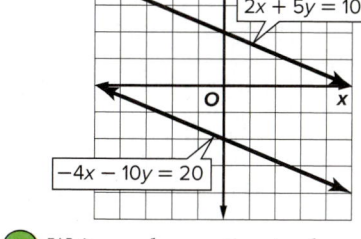 Because the equations are equivalent, their graphs are the same line. The system is consistent and dependent.

45. infinite solutions **47.** $(8, 4)$ **49.** $(-3, -1)$
51a. $x + y = 13$ and $4x + 2y = 38$
51b. 6 doubles matches and 7 singles matches
53. $(0, 4)$ **55.** no solution **57.** $(8, -6)$ **59.** $(5, 4)$
61. $(2.07, -0.39)$
63. $(15.03, 10.98)$ **65.** $(-5, 4)$ **67.** infinite solutions

69. $(16, -8)$ **71a.** Sample answer for men using $(0, 9.9)$ and $(44, 9.63)$: $y_m = -0.00613x + 9.9$; sample answer for women using $(0, 11.0)$ and $(44, 10.75)$: $y_w = -0.00568x + 11.0$.

71b.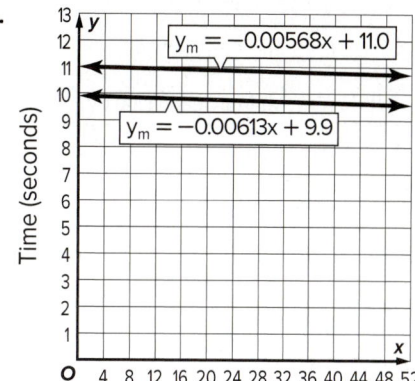

Based on these data, the women's performance will never catch up to the men's performance.

73. Let x represent the cost for an adult and y represent the cost for a student.

Van A: $2x + 5y = 77$
Van B: $2x + 7y = 95$

$2x + 5y = 77$ Multiply by -1. → $-2x - 5y = -77$

$-2x - 5y = -77$ Equation 1 × (−1)
$\underline{(+) \; 2x + 7y = 95}$ Equation 2
$2y = 18$ Add the equations.
$y = 9$ Divide each side by 2.

$2x + 7y = 95$ Equation 2
$2x + 7(9) = 95$ $y = 9$
$2x + 63 = 95$ Multiply.
$2x = 32$ Subtract 63 from each side.
$x = 16$ Divide each side by 2.

The solution is $(16, 9)$. So, the cost for an adult is $16 and the cost for a student is $9.

75. Find an equation for the diagonal that goes through $(6, 3)$ and $(2, 9)$.

Slope:
$m = \frac{y_2 - y_1}{x_2 - x_1}$
$= \frac{9 - 3}{2 - 6}$
$= \frac{6}{-4}$ or $-\frac{3}{2}$

Equation:
$y - y_1 = m(x - x_1)$
$y - 3 = -\frac{3}{2}(x - 6)$
$y - 3 = -\frac{3}{2}x + 9$
$y = -\frac{3}{2}x + 12$

Find an equation for the diagonal that goes through $(3, 4)$ and $(11, 18)$.

Slope:
$m = \frac{y_2 - y_1}{x_2 - x_1}$
$= \frac{18 - 4}{11 - 3}$
$= \frac{14}{8}$ or $\frac{7}{4}$

Equation:
$y - y_1 = m(x - x_1)$
$y - 4 = \frac{7}{4}(x - 3)$
$y - 4 = \frac{7}{4}x - \frac{21}{4}$
$y = \frac{7}{4}x - \frac{5}{4}$

Find the point of intersection of the diagonals.

$y = -\frac{3}{2}x + 12$ Equation 1

$\frac{7}{4}x - \frac{5}{4} = -\frac{3}{2}x + 12$ Substitute $\frac{7}{4}x - \frac{5}{4}$ for y.

$\frac{13}{4}x - \frac{5}{4} = 12$ Add $\frac{3}{2}x$ to each side.

$\frac{13}{4}x = \frac{53}{4}$ Add $\frac{5}{4}$ to each side.

$x = \frac{53}{13}$ Multiply each side by $\frac{4}{13}$.

$y = -\frac{3}{2}x + 12$ Equation 1

$y = -\frac{3}{2}\left(\frac{53}{13}\right) + 12$ $x = \frac{53}{13}$

$y = -\frac{159}{26} + 12$ Multiply.

$y = \frac{153}{26}$ Simplify.

The diagonals intersect at $\left(\frac{53}{13}, \frac{153}{26}\right)$.

77a.

Equation 1	
x	y
0	$\frac{16}{3}$
1	5
2	$\frac{14}{3}$
3	$\frac{13}{3}$
4	4

Equation 2	
x	y
0	−4
1	−2
2	0
3	2
4	4

Equation 3	
x	y
0	10
1	5
2	0
3	−5
4	−10

77b. Equations 1 and 2 intersect at (4, 4), equations 2 and 3 intersect at (2, 0), and equations 1 and 3 intersect at (1, 5); there is no solution that satisfies all three equations.

77c.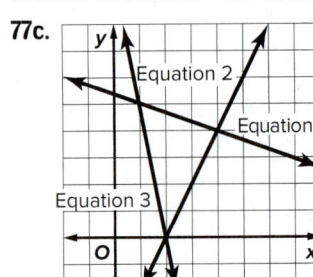

77d. If all three lines intersect at the same point, then the system has a solution. The system has no solution if the lines intersect at three different points, or if two or three lines are parallel.

79. $a \neq 0, b = 3$

81. Sample answer:
$4x + 5y = 21$ → $3(4x + 5y = 21)$
$3x - 2y = 10$ → $4(3x - 2y = 10)$

$12x + 15y = 63$ $4x + 5(1) = 21$
$(-)\ 12x - 8y = 40$ $4x + 5 = 21$
$23y = 23$ $4x = 16$
$y = 1$ $x = 4$

The solution is (4, 1).

83. D **85.** C **87.** 25 **89a.** $d + n = 27$
89b. $10d + 5n = 180$ **89c.** $n = 27 - d; d = 9$
89d. $9 + n = 27; n = 18$ **89e.** $5d + 5n = 135$
89f. $d = 9$ **89g.** $n = 18$
91. Still air: 525 mi/h and jetstream: 75 mi/h

Lesson 1-7

1.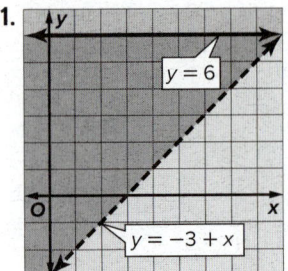

3.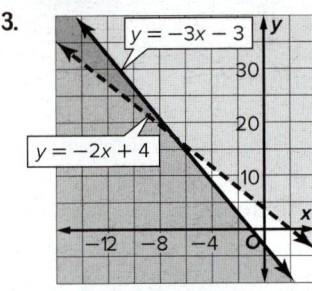

5. 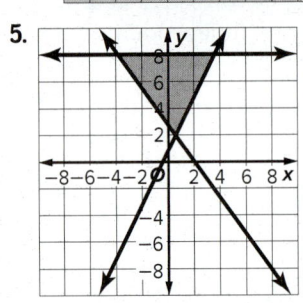 **7.**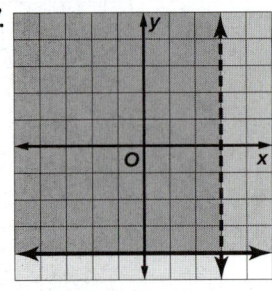

(3.5, 8), (−4, 8), (0.5, 2)

9 The solution of $y < -3x + 4$ is the region to the left of the boundary. The solution of $3y + x > -6$ is the region above the boundary. The intersection of the two regions is the solution of the system.

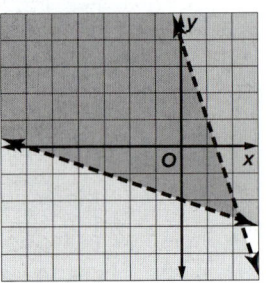

11. 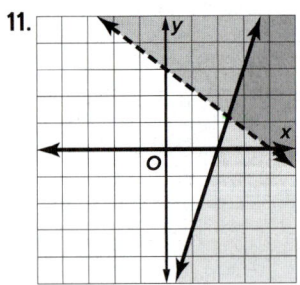 **13.**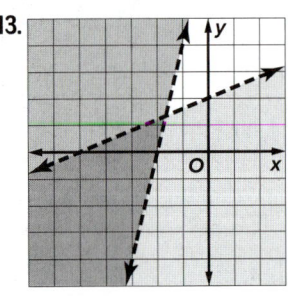

15.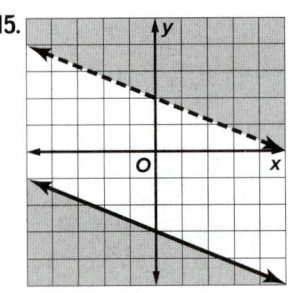

17.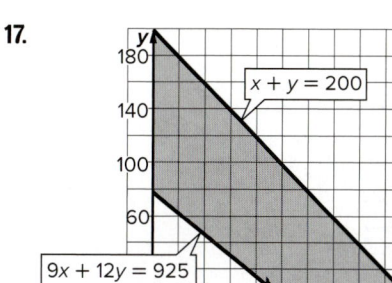

19. 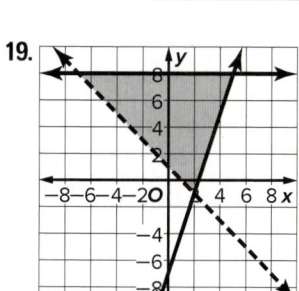 (2, −1), (5, 8), (−7, 8)

21. 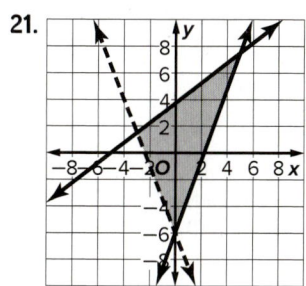 (−3, 1.5), (5, 7.5), (0, −6)

23. 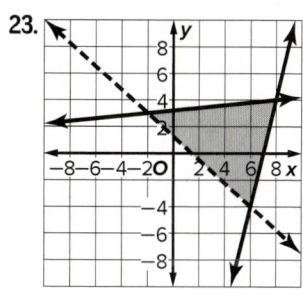 (8, 4), (6, −4), (−2, 3)

25. Let d represent the number of daytime minutes and n represent the number of nighttime minutes. Write a system of inequalities and then graph.

$d + n \leq 800$ Maximum number of minutes is 800.

$d \geq 2n$ At least twice as many daytime minutes as nighttime minutes

$n \geq 200$ At least 200 nighttime minutes

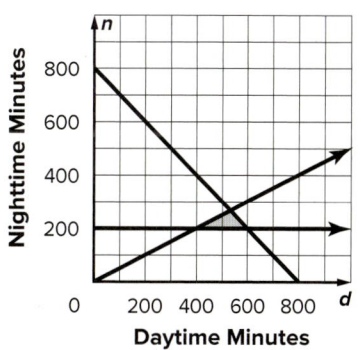

The intersection points are at (400, 200), (533.3, 266.7), and (600, 200).

$0.15d + 0.1n = 0.15(400) + 0.1(200)$ 400 daytime min, 200 nighttime min

$= 60 + 20$ or \$80   Simplify.

$0.15d + 0.1n = 0.15(533.3) + 0.1(266.7)$ 533.3 daytime min, 266.7 nighttime min

$\approx 80.0 + 26.7$ or \$106.70   Simplify.

$0.15d + 0.1n = 0.15(600) + 0.1(200)$ 600 daytime min, 200 nighttime min

$= 90 + 20$ or \$110   Simplify.

So, his maximum bill is \$110 and his minimum bill is \$80.

27a.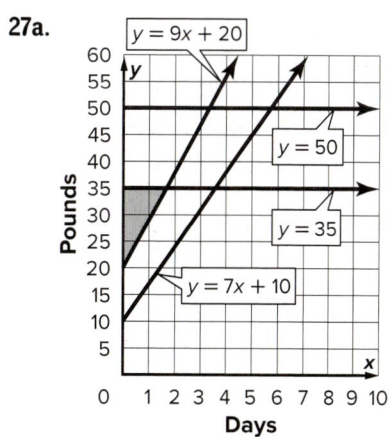

27b. $3\frac{1}{3}$ days   **27c.** Marc; Jessica could last about a quarter of a day longer than Marc.

29.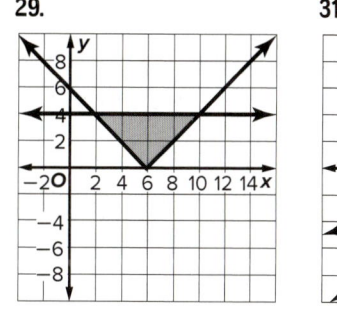

31.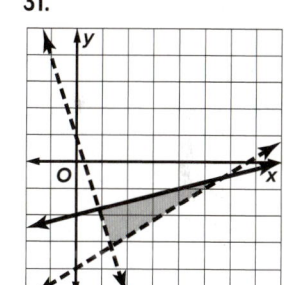

33.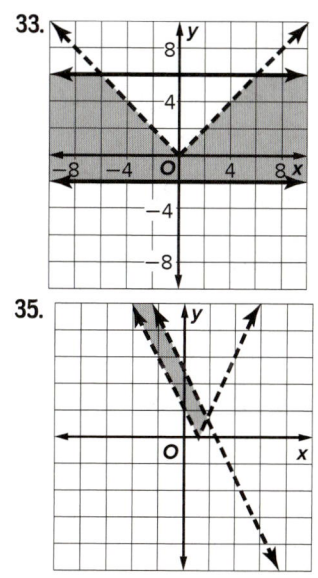

35.

37.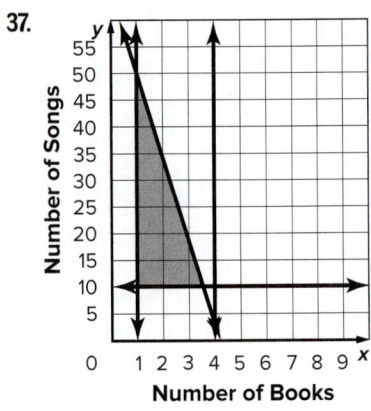

39. Let $w =$ the number of hours writing, and let $e =$ the number of hours exercising.
$w + e \leq 35$
$7 \leq e \leq 15$
$20 \leq w \leq 25$

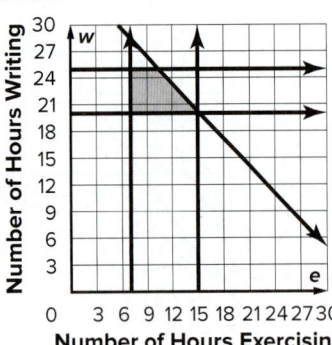

41. $(-6, -2)$, $\left(-3\frac{13}{17}, 6\frac{16}{17}\right)$, $\left(9\frac{1}{7}, 3\frac{5}{7}\right)$, $(0.8, -8.8)$

43. Let x represent the amount in the fund that pays 6% interest and y represent the amount in the fund that pays 10% interest. Write a system of inequalities and then graph.

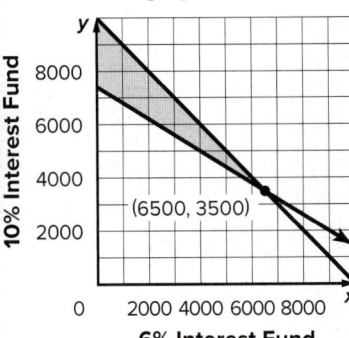

$x + y \leq 10{,}000$ Total amount invested is up to $10,000.
$0.06x + 0.10y \geq 740$ Total amount earned is at least $740.
The least amount Mr. Hoffman can invest in the risky fund, or the 10% interest fund, is $3500.

45. $(-3, -4)$, $(-6, 8)$, $(2, 6)$, $(5, 0)$; 75 units^2

47. Sample answer: $y \geq 2x - 6$, $y \leq -0.5x + 4$, $y \geq -3x - 6$; 47 **49.** Sample answer: Shade each inequality in their standard way, by shading above the line if $y >$ and shading below the line if $y <$ (or you can use test points). Once you determine where to shade for each inequality, the area where *every* inequality needs to be shaded is the actual solution. This is only the shaded area. **51.** B **53.** C

Lesson 1-8

1.

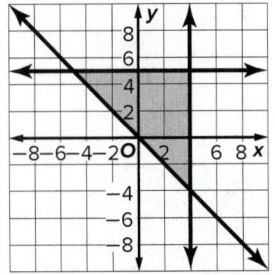

$(4, 5)$, $(4, -4)$, $(-5, 5)$; max $= 28$, min $= -35$

3.

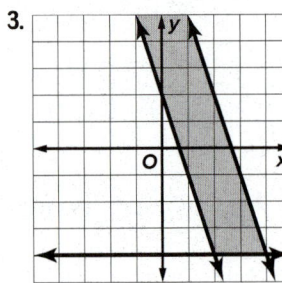

$(2, -4)$, $(4, -4)$; max does not exist, min $= -52$

5.

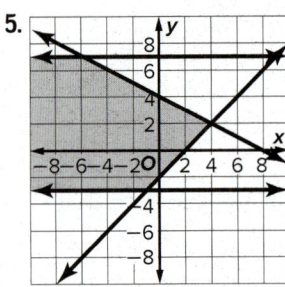

$(4, 2)$, $(-1, -3)$, $(-6, 7)$; max does not exist; min $= -30$

7a. $x \geq 10$, $y \geq 10$, $25x + 50y \leq 4200$, $3x + 5y \leq 480$

7b.

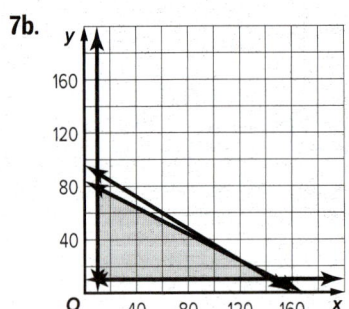

$(10, 10)$, $(10, 79)$, $\left(\frac{430}{3}, 10\right)$, $(120, 24)$

7c. $R(x, y) = 12.50x + 20y$ **7d.** 143 standard packages and 10 oversize packages; $1987.50; Sample answer: the vertex that maximizes the revenue function is $\left(\frac{430}{3}, 10\right)$. However, there cannot be a fractional package. Because there are weight and space constraints, $\frac{430}{3}$ or 143.3 needs to be rounded down to 143.

9. Graph the inequalities and locate the vertices.

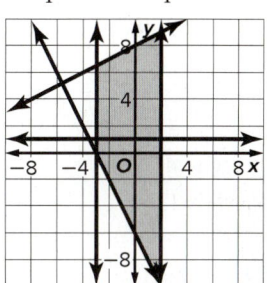

The vertices are at (2, −10), (−3, 0), (−3, 6.5), and (2, 9). Evaluate the function at each vertex.

(x, y)	−4x − 9y	f(x, y)
(2, −10)	−4(2) − 9(−10)	82
(−3, 0)	−4(−3) − 9(−0)	12
(−3, 6.5)	−4(−3) − 9(6.5)	−46.5
(2, 9)	−4(2) − 9(9)	−89

The maximum value is 82 at (2, −10). The minimum value is −89 at (2, 9).

11.

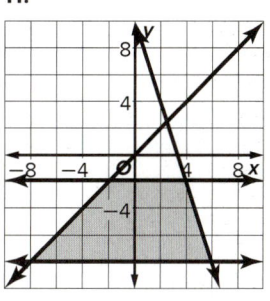

(6, −8), (4, −2), (−2, −2), (−8, −8); max = −8, min = −152

13.

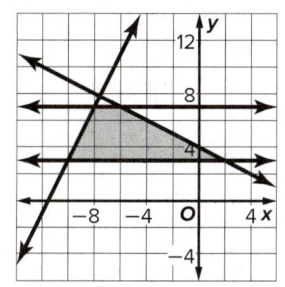

(−10, 3), (2, 3), (−6, 7), (−8, 7); max = 59, min = 9

15. Graph the inequalities and locate the vertices.

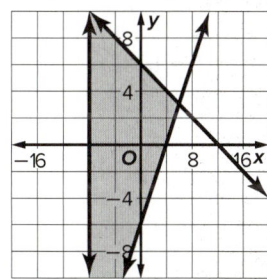

The vertices are at (6, 3), (−8, 10), and (−8, −18). Evaluate the function at each vertex.

(x, y)	10x − 6y	f(x, y)
(6, 3)	10(6) − 6(3)	42
(−8, 10)	10(−8) − 6(10)	−140
(−8, −18)	10(−8) − 6(−18)	28

The maximum value is 42 at (6, 3). The minimum value is −140 at (−8, 10).

17.

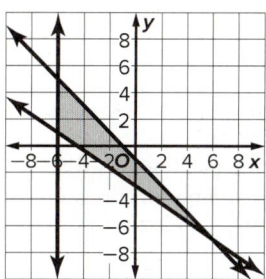

(−6, 1), (6, −7), (−6, 5); max = 48, min = 0

19.

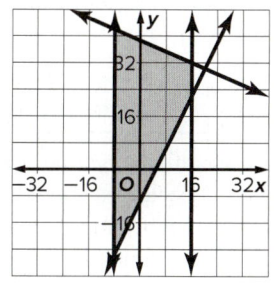

(−8, 44), (16, 32), (−8, −26), (16, 22); max = 672, min = −486

21.

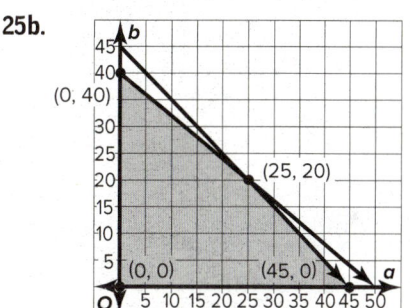

(5, −1), (1, 6), (−2, −8), (−4, −8), (−4, 6), max = 60, min = −112

23. 225 yellow cakes, 0 strawberry cakes

25a. $a \geq 0$, $b \geq 0$, $a + b \leq 45$, $\frac{2}{5}a + \frac{1}{2}b \leq 20$, $4a + 5b \leq 200$

25b.

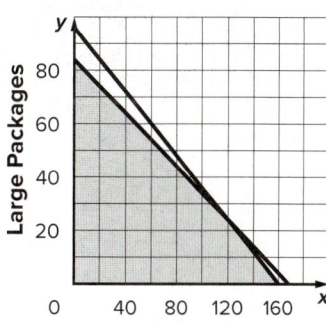

(0, 0), (0, 40), (25, 20), (45, 0)

25c. 25 sheds, 20 play houses
25d. $1250

27. a. Let x represent the number of small packages and y represent the number of large packages.

Write a system of inequalities. Then graph the inequalities and locate the vertices.

$x \geq 0$ number of small packages ≥ 0
$y \geq 0$ number of large packages ≥ 0
$25x + 50y \leq 4200$ weight of packages ≤ 4200 lb
$3x + 5y \leq 480$ capacity of packages ≤ 480 cu ft

The vertices are at (0, 84), (120, 24), and (160, 0). Evaluate the function $f(x, y) = 5x + 8y$ at each vertex.

(x, y)	$5x + 8y$	$f(x, y)$
(0, 84)	5(0) + 8(84)	672
(120, 124)	5(120) + 8(24)	792
(160, 0)	5(160) + 8(0)	800

To maximize revenue, 160 small packages and 0 large packages should be placed on a train car.
b. The maximum revenue per train car is $800.
c. No; if revenue is maximized, the company will not deliver any large packages, and customers with large packages to ship will probably choose another carrier.
29. Sample answer: $-2 \geq y \geq -6, 4 \leq x \leq 9$
31. b; The feasible region of Graph b is unbounded while the other three are bounded.
33. Sample answer: Even though the region is bounded, multiple maximums occur at A and B and all of the points on the boundary of the feasible region containing both A and B. This happened because that boundary of the region has the same slope as the function.
35. 110
37a.

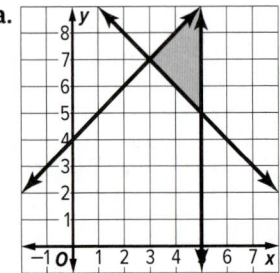

37b. (3, 7), (5, 5), (5, 9)
37c.

(x, y)	$5x - 2y$	$f(x, y)$
(3, 7)	5(3) − 2(7)	1
(5, 5)	5(5) − 2(5)	15
(5, 9)	5(5) − 2(9)	7

The maximum value is 15.
39. 6.4 million gallons of gasoline and 3.2 million gallons of fuel oil.

Lesson 1-9

1. A **3.** infinite solutions **5.** 2 sitcoms, 4 dramas, 1 talk show

7.
$-a + 4b + 2c = -13$ Multiply by 4. → $-4a + 16b + 8c = -52$
$-4a + 16b + 8c = -52$ Equation 3 (× 4)
$(+)\ 4a + 5b - 6c = 2$ Equation 1
$\overline{21b + 2c = -50}$
$-a + 4b + 2c = -13$ Multiply by −3. → $3a - 12b - 6c = 39$
$3a - 12b - 6c = 39$ Equation 3 × (−3)
$(+)\ -3a - 2b + 7c = -15$ Equation 2
$\overline{-14b + c = 24}$

The resulting system of two equations and two variables is shown below.
$21b + 2c = -50$ New equation 1

$-14b + c = 24$ New equation 2
$-14b + c = 24$ Multiply by −2. → $28b - 2c = -48$
$21b + 2c = -50$ New equation 1
$(+)\ 28b - 2c = -48$ New equation 2 × (−2)
$\overline{49b = -98}$ Add the equations.
$b = -2$ Divide each side by 49.
$-14b + c = 24$ New equation 2
$-14(-2) + c = 24$ Replace b with −2.
$28 + c = 24$ Multiply.
$c = -4$ Subtract 28 from each side.
$-a + 4b + 2c = -13$ Original equation 3
$-a + 4(-2) + 2(-4) = -13$ $b = -2$ and $c = -4$
$-a - 8 - 8 = -13$ Multiply.
$-a = 3$ Add 16 to each side.
$a = -3$ Multiply each side by −1.
The solution is $(-3, -2, -4)$.
9. $(-2, -1, 4)$ **11.** infinite solutions **13.** $(-4, -1, 6)$
15. no solution **17.** infinite solutions **19.** roller coasters: 5; bumper cars: 1; water slides: 4

21. $a =$ the amount invested in account A
$b =$ the amount invested in account B
$c =$ the amount invested in account C
$a + b + c = 100{,}000$ She invested a total of $100,000.
$a = c + 30{,}000$ She invested $30,000 more in account A than account C.
$0.04a + 0.08b + 0.1c = 6300$ The expected interest earned is $6300.

Substitute $a = c + 30{,}000$ in equations 1 and 3.
$a + b + c = 100{,}000$ Equation 1
$c + 30{,}000 + b + c = 100{,}000$ $a = c + 30{,}000$
$30{,}000 + b + 2c = 100{,}000$ Add.
$b + 2c = 70{,}000$ Simplify.
$0.04a + 0.08b + 0.1c = 6300$ Equation 3
$0.04(c + 30{,}000) + 0.08b + 0.1c = 6300$ $a = c + 30{,}000$
$0.04c + 1200 + 0.08b + 0.1c = 6300$ Distribute.
$1200 + 0.08b + 0.14c = 6300$ Add.
$0.08b + 0.14c = 5100$ Simplify.

Solve the system of two equations in two variables.
$b + 2c = 70{,}000$ Multiply by −0.08. → $0.08b + 0.14c = 5100$

$-0.08b - 0.16c = -5600$
$(+)\ 0.08b + 0.14c = 5100$
$\overline{-0.02c = -500}$
$c = 25{,}000$

Substitute to find b.
$b + 2c = 70{,}000$ Remaining equation in two variables
$b + 2(25{,}000) = 70{,}000$ $c = 25{,}000$
$b + 50{,}000 = 70{,}000$ Distribute.
$b = 20{,}000$ Simplify.

Substitute to find a.
$a + b + c = 100{,}000$ Equation 1
$a + 20{,}000 + 25{,}000 = 100{,}000$ $b = 20{,}000, c = 25{,}000$
$a + 45{,}000 = 100{,}000$ Add.
$a = 55{,}000$ Simplify.

The solution is (55,000, 20,000, 25,000). She invested $55,000 in account A, $20,000 in account B, and $25,000 in account C.

23. $y = -3x^2 + 4x - 6$; $a = -3$, $b = 4$, $c = -6$
25. Sample answer:
$3x + 4y + z = -17$
$2x - 5y - 3z = -18$;
$-x + 3y + 8z = 47$
$3x + 4y + z = -17$
$3(-5) + 4(-2) + 6 = -17$
$-15 + (-8) + 6 = -17$
$= -17$ ✓
$2x - 5y - 3z = -18$
$2(-5) - 5(-2) - 3(6) = -18$
$-10 + 10 - 18 = -18$
$-18 = -18$ ✓
$-x + 3y + 8z = 47$
$-(-5) + 3(-2) + 8(6) = 47$
$5 - 6 + 48 = 47$
$47 = 47$ ✓

27. Sample answer: First, combine two of the original equations using elimination to form a new equation with three variables. Next, combine a different pair of the original equations using elimination to eliminate the same variable and form a second equation with three variables. Do the same thing with a third pair of the original equations. You now have a system of three equations with three variables. Follow the same procedure you learned in this section. Once you find the three variables, you need to use them to find the eliminated variable. **29.** A **31a.** $3x + 3y = 6$
31b. Multiply the first equation by 2 to get $2x + 4y - 2z = -2$. Then subtract the third equation to get $3x + 3y - 2z = 2$. Adding these equations gives $-x + y = -4$.
31c. Multiply the equation from part **b** by 3 to get $-3x + 3y = -12$. Then add that to the equation in part **a** to get $y = -1$. Substitute into one of the original equations to solve for x: $x = 3$.
31d. $(3, -1, 2)$
33. $(1, 2, 1)$

Chapter 1 Study Guide and Review

1. unbounded   **3.** open sentence   **5.** solution   **7.** An independent system has exactly one solution, while a dependent system has an infinite number of solutions.
9. -7 **11.** $\frac{3}{2}$ **13.** $9.50   **15.** $m = \frac{r + 5}{pn}$ **17.** about 8 in.
19. $r > 55$

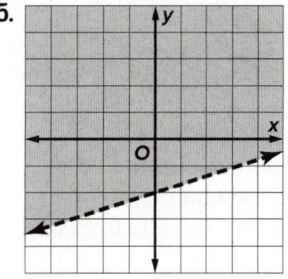

21. $p > -2$

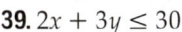

23. 18
25. -1
27. $y = \frac{2}{3}x - \frac{11}{3}$
29. $y = 5$
31. $y = \frac{3}{5}x + \frac{22}{5}$
33. $y = -\frac{3}{2}x + \frac{1}{2}$

35.

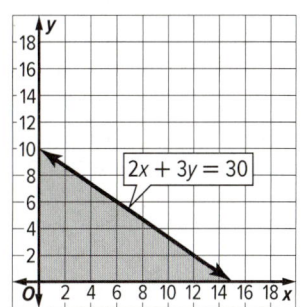

37.

(graph)

39. $2x + 3y \leq 30$

(graph)

41. infinitely many solutions
43. no solution
45. $(2, 4)$
47. $(5.25, -1.75)$
49. notebook: $2.50; pen $1.25
51.

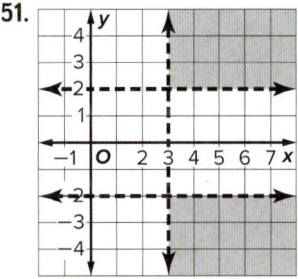

53.

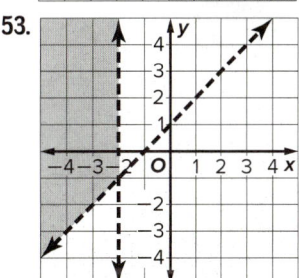

55. 126 simple and 63 grand
57. $(-23, -8, -6)$
59. hot dog: $3.25; popcorn $2.25; soda: $2.50

CHAPTER 2
Relations and Functions

Chapter 2 Get Ready

1. Follow a vertical line through the point to find the x-coordinate on the x-axis.   **3.** Quadrant III
5. not in quadrant; on y-axis   **7.** Subtract $3x$ from each side.   **9.** negative

Lesson 2-1

1. $D = \{5, 6, -2\}$, $R = \{3, -8, 1\}$; function; both

3. The domain is the set of x-values: $\{-2, 1, 4, 8\}$; the range is the set of y-values: $\{-4, -2, 6\}$. Because each element of the domain is paired with exactly one element of the range, the relation is a function. Because each element of the range corresponds to an element of the domain, the relation is an onto function.

5. 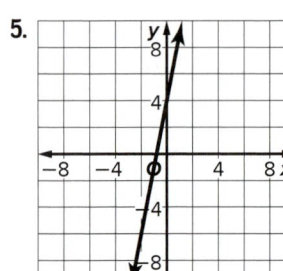 $D = \{\text{all real numbers}\}$, $R = \{\text{all real numbers}\}$; function; both; continuous

7. 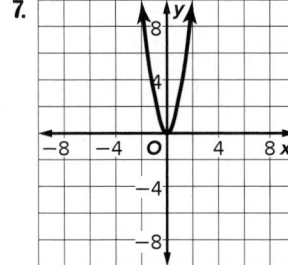 $D = \{\text{all real numbers}\}$, $R = \{y \mid y \geq 0\}$; function; onto; continuous

9a. $D = \{25, 26, 27, 28\}$; $R = \{22.9, 24.0, 23.8, 30.1\}$ **9b.** $\{(25, 22.9), (26, 24.0), (27, 23.8), (28, 30.1)\}$
9c. discrete
9d. yes

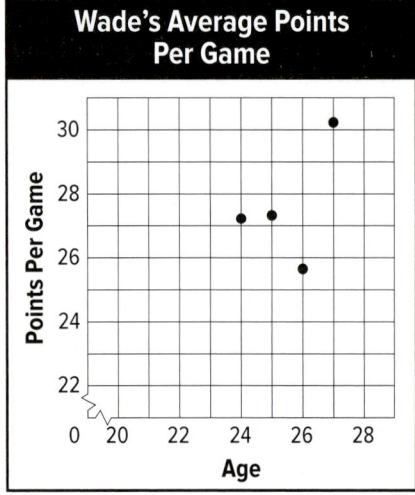

11. -69 **13.** Discrete; you cannot receive a fraction of a t-shirt.   **15.** $D = \{-8, 2, 4\}$, $R = \{-6, -4, 14\}$; not a function

17. 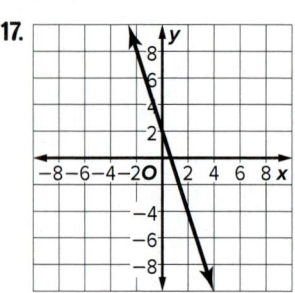 $D = \{\text{all real numbers}\}$, $R = \{\text{all real numbers}\}$; function; both; continuous

19. 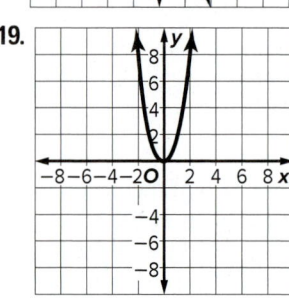 $D = \{\text{all real numbers}\}$, $R = \{y \mid y \geq 0\}$; function; onto; continuous

21. 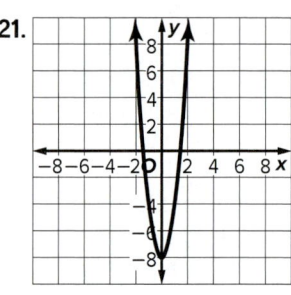 $D = \{\text{all real numbers}\}$, $R = \{y \mid y \geq -8\}$; function; onto; continuous

23.
$f(x) = 5x^3 + 1$ Original function
$f(-8) = 5(-8)^3 + 1$ Substitute -8 for each x.
$= 5(-512) + 1$ Evaluate $(-8)^3$.
$= -2560 + 1$ Multiply.
$= -2559$ Simplify.

25.a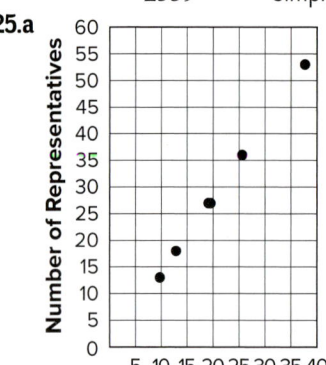

25b. $D = \{9.66, 12.87, 19.06, 19.47, 25.67, 37.69\}$, $R = \{13, 18, 27, 36, 53\}$ **25c.** discrete
25d. Yes; each domain value is paired with only one range value so the relation is a function.
27a. $\{(0, 1), (20, 1.6), (40, 2.2), (60, 2.8), (80, 3.4), (100, 4)\}$

27b.

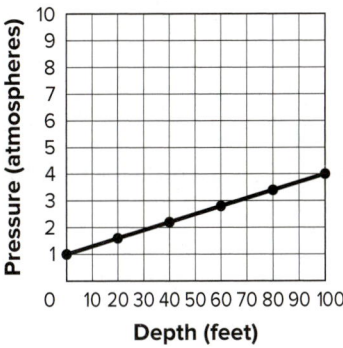

27c. D = $\{x \mid x \geq 0\}$, R = $\{y \mid y \geq 1\}$; continuous
27d. Yes; each domain value is paired with only one range value so the relation is a function.   **29.** 29
31. -72 **33.** -267 **35.** -4.5
37a. discrete; You cannot buy a fraction of a clothing item.
37b.

39. No; the domain of a discrete function can contain numbers other than integers as long as the domain is a set of individual values. For example, the domain could be a set of fractions or the set of multiples of five between 10 and 100.

41a.

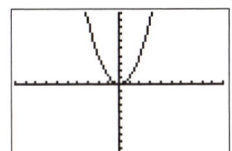

[−10, 10] scl: 1 by [−10, 10] scl: 1

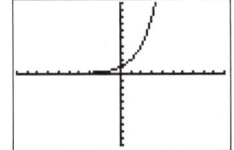

[−10, 10] scl: 1 by [−10, 10] scl: 1

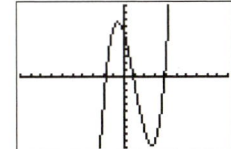

[−10, 10] scl: 1 by [−10, 10] scl: 1

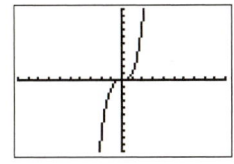

[−10, 10] scl: 1 by [−10, 10] scl: 1

41b.

Function	Possible Intersection Points
$f(x) = x^2$	0, 1, 2
$g(x) = 2^x$	0, 1
$h(x) = x^3 - 3x^2 - 5x + 6$	1, 2, 3
$j(x) = x^3$	1

41c. $g(x)$ and $j(x)$ are one-to-one, and $f(x)$ and $h(x)$ are not.

41d. $h(x)$ and $j(x)$ are onto, and $f(x)$ and $g(x)$ are not.

41e.

Function	One-to-one	Onto
$f(x) = x^2$	no	no
$g(x) = 2^x$	yes	no
$h(x) = x^3 - 3x^2 - 5x + 6$	no	yes
$j(x) = x^3$	yes	yes

43. Sample answer: $f(x) = 4x - 1$; $g(x) = 6x + 3$
45a. Sample answer:

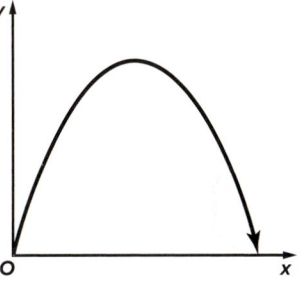

45b. Sample answer:

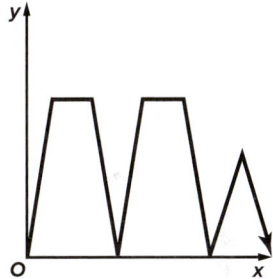

45c. Sample answer:

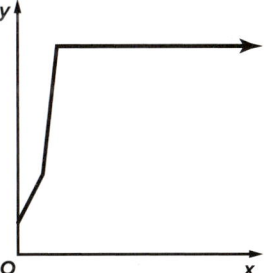

45d. Sample answer:

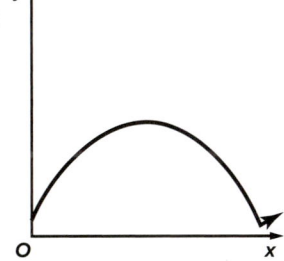

47a.

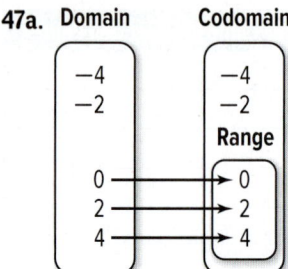

47b.

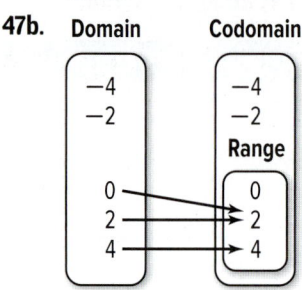

49. B **51.** A **53.** A

Lesson 2-2

1. Yes; it can be written as $f(x) = \frac{x}{5} + \frac{12}{5}$. **3.** No; x has an exponent that is not 1.

5. Selena: In the table representing Selena's savings, each increase of 1 week corresponds to an increase of $15 in her total savings. This means the rate of change is constant. This is a linear function. When the ordered pairs from the table are graphed, a single straight line can be drawn through all of the points.
Fiona: Place a straightedge against the points in the graph representing Fiona's savings. The points lie along the straightedge. This is a linear function. A single straight line can be drawn through all of the points on the graph.

7. line symmetry; $x = -2$ **9.** neither line nor point symmetry **11.** No; x has an exponent other than 1. **13.** No; x has an exponent other than 1. **15.** No; it cannot be written in $f(x) = mx + b$ form. **17.** No; it cannot be written in $y = mx + b$ form. **19.** Car A: Nonlinear function; the graph of the function is not a single straight line. Car B: Linear function; when the ordered pairs from the table are graphed, a single straight line can be drawn through all of the points.

21. There is no vertical line that divides the graph into two halves that are mirror images, so there is no line symmetry.
When the graph is rotated 180° about the point $(-1, -2)$, the rotation maps the original graph onto itself. The graph has point symmetry with $(-1, -2)$ as the point of symmetry.

23. point symmetry; all points on the line

25a. No; the equation of the function includes the variable x within absolute value symbols, so the equation cannot be written in the form $f(x) = mx + b$.

25b. $x = 12$; the line of symmetry passes through the peak of the roof and divides the roof into left and right halves that are mirror images of each other.

27. $y = -\frac{1}{6}x + \frac{7}{6}$; $m = -\frac{1}{6}$, $b = \frac{7}{6}$

29. $y = \frac{1}{18}x + \frac{31}{18}$; $m = \frac{1}{18}$, $b = \frac{31}{18}$

31. Rotate the given part of the graph 180° about the origin.

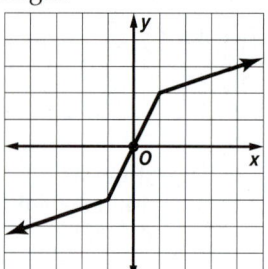

33. Sample answer: Never; the graph of $x = a$ is a vertical line. **35a.** Sample answer: When the earnings are determined by a constant hourly wage, the total earnings can be represented by $y = mx$ where m is the hourly wage. **35b.** Sample answer: the relationship between the cost and the number of gallons of gasoline purchased **37.** C **39.** C

Lesson 2-3

1. As $x \to +\infty$, $f(x) \to +\infty$ and as $x \to -\infty$ $f(x) \to -\infty$.
3. As $x \to +\infty$, $f(x) \to -\infty$ and as $x \to -\infty$, $f(x) \to +\infty$.
5. As $x \to +\infty$, $f(x) \to -\infty$ and as $x \to -\infty$, $f(x) \to -\infty$.
7. There is a zero near $x = -2$. There is a relative maximum near $x = -1$. There is a relative minimum near $x = 1$.
9. As $x \to +\infty$, $f(x) \to -\infty$ and as $x \to -\infty$ $f(x) \to +\infty$.
11. As $x \to +\infty$, $f(x) \to +\infty$ and as $x \to -\infty$ $f(x) \to -\infty$.
13. As $x \to +\infty$, $f(x) \to +\infty$ and as $x \to -\infty$ $f(x) \to +\infty$.

15. The values of $f(x)$ at $x = -1$ and $x = 0$ are less than at a point between them, indicating a relative maximum between -1 and 0. There must be a relative maximum near $x = -1$.
The values of $f(x)$ at $x = 0$ and $x = 1$ are greater than at a point between them, indicating a relative minimum between 0 and 1. There must be a relative minimum near $x = 1$.

17. Sample answer: Finn is correct. Even though the graph has a change in curvature at $(-2, 0)$, that point does not represent a relative minimum, relative maximum, or turning point because the function does not go from increasing to decreasing or from decreasing to increasing on either side of the point.

19. Sample answer: Graph B. As x approaches either positive or negative infinity, $f(x) = 2$.

21. Begin by choosing an end behavior that is possible for a linear function; for example: As $x \to +\infty$, $f(x) \to -\infty$ and as $x \to -\infty$, $f(x) \to +\infty$. Now sketch linear and nonlinear functions with this end behavior.

Sample answer:

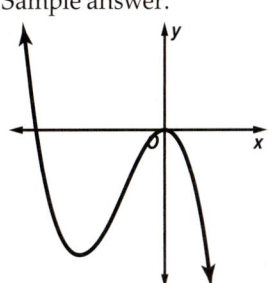 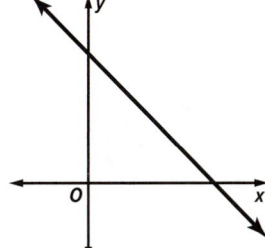

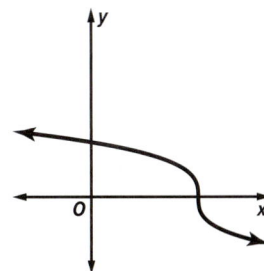 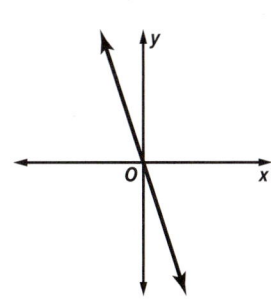

23. If a function has vertical symmetry, then the end behavior of f(x) as x → ∞ and as x → −∞ must be the same because both sides of the graph will approach the same value if the function is symmetric.

25. Sample answer: Point P is a relative maximum because if g is increasing as it approaches point P from the left, then P is higher than the points to its left. If g is decreasing as g moves away from P on the right, then P is higher than the points on its right. So, P must be a relative maximum. **27.** Sample answer: the end behavior stays the same for all even-powered functions, or the function goes to +∞ as x → +∞ and to +∞ as x → −∞. The end behavior stays the same for all odd-powered functions, or the function goes to +∞ as x → +∞ and to −∞ as x → −∞. **29a.** B, C, D **29b.** B, C **31a.** Relative minimum between x = 2 and x = 3. There are no nearby points that have a lesser value.
31b. As x → +∞, f(x) → +∞. As x → −∞, f(x) → +∞.
31c. Near x = 1 and x = 4

Lesson 2-4

1.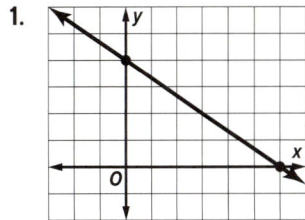

3. Sample answer: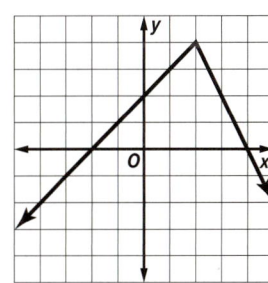

5.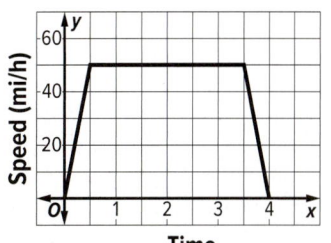

7. Because the domain of the function is −4 ≤ x ≤ 3, the graph will only appear on this interval. The range is −4 ≤ y ≤ 3. Because the graph is linear, this means the graph is a line segment that passes through the points (−4, 3) and (3, −4). Draw the segment between these points and check that the y-intercept is −1, as given.

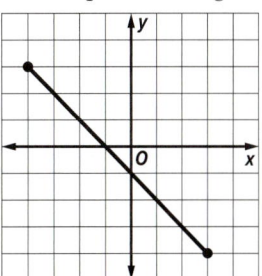

9. Sample answer: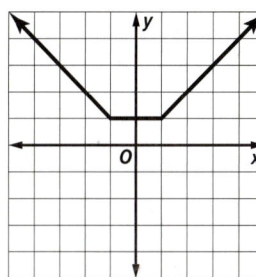

11. Sample answer: The y-intercept is 1; the function is increasing for all values of x; as x → ∞, f(x) → +∞ and as x → −∞, f(x) → −∞.

13. Write a function to represent the situation. Because they collect 10 cans every 20 minutes, they collect 30 cans per hour. After x hours, the total number of cans is y = 30x + 56. This is a linear function because it is of the form y = mx + b. Graph the function on the interval 0 ≤ x ≤ 4, because they collect cans for 4 hours.

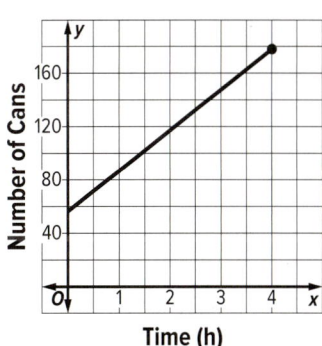

15. Sample answer:

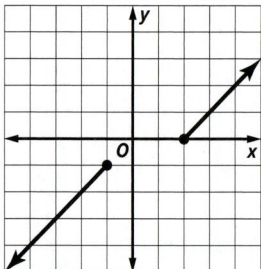

17. No; in order for the function to have the given end behavior, it must have negative values for some values of x and positive values for some values of x; because the function is continuous, the graph must cross the x-axis, which means the function has at least one x-intercept.

19. No; she did not draw a function with the correct end behavior.

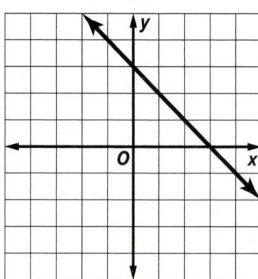

21.

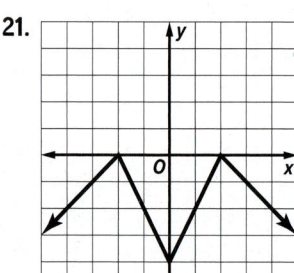

23. A **25.** B **27.** D

29a. Sample answer:

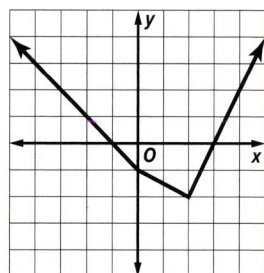

29b. Remain the same: continuous, nonlinear, end behavior; change: y-intercept, interval where negative, minimum **29c.** Sample answer based on sample answer for Part a:

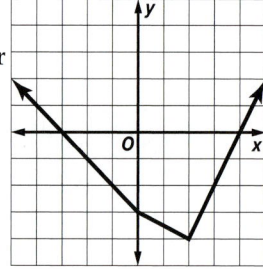

Lesson 2-5

1.

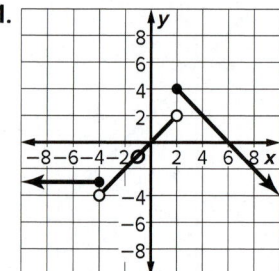

D = {all real numbers}; R = {y | y ≤ 4}

3. $g(x) = \begin{cases} x + 4 & \text{if } x < -2 \\ -3 & \text{if } -2 \leq x \leq 3 \\ -2x + 12 & \text{if } x > 3 \end{cases}$

5 If the number of tickets sold is greater than 0 but less than or equal to 250, then the drama club must do 1 performance. If the number of tickets sold is greater than 250 but less than or equal to 500, then the drama club must do 2 performances, and so on. You can use the pattern to make a table, where x is the number of tickets sold and P(x) is the number of performances. Then graph.

x	P(x)
0 < x ≤ 250	1
250 < x ≤ 500	2
500 < x ≤ 750	3
750 < x ≤ 1000	4
1000 < x ≤ 1250	5

7.

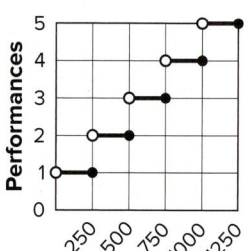

D = {all real numbers}; R = {all integers}

9.

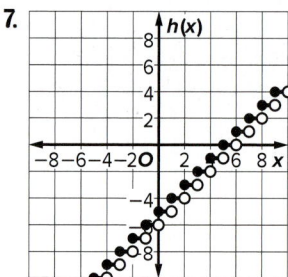

D = {all real numbers}; R = {f(x) | f(x) ≥ 0}

11.

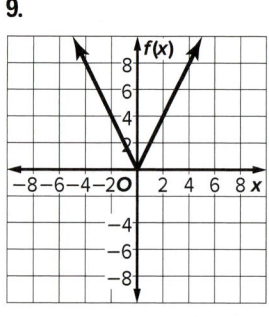

D = {all real numbers}; R = {s(x) | s(x) ≥ 6}

13.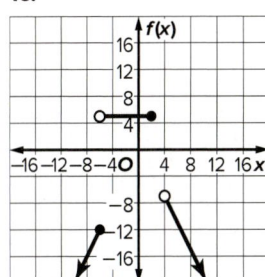
D = {x | x ≤ 2 or x > 4};
R = {f(x) | f(x) < −7, or f(x) = 5}

15.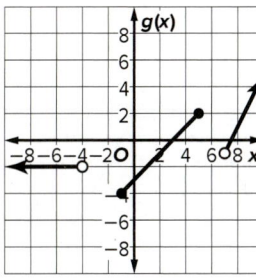
D = {x | x < −4, −1 ≤ x ≤ 5, or x > 7};
R = {g(x) | g(x) ≥ −4}

17. The left portion of the graph is the graph of g(x) = −x − 4. There is a circle at (−3, −1), so the linear function is defined for {x | x < −3}. The middle portion of the graph is the graph of g(x) = x + 1. There are dots at (−3, −2) and (1, 2), so the linear function is defined for {x | −3 ≤ x ≤ 1}. The right portion of the graph is the graph of g(x) = −6. There is a circle at (4, −6), so the linear function is defined for {x | x > 4}. Write the piecewise-defined function.

$$g(x) = \begin{cases} -x - 4 & \text{if } x < -3 \\ x + 1 & \text{if } -3 \leq x \leq 1 \\ -6 & \text{if } x > 4 \end{cases}$$

19. $g(x) = \begin{cases} 8 & \text{if } x \leq -1 \\ 2x & \text{if } 4 \leq x \leq 6 \\ 2x - 15 & \text{if } x > 7 \end{cases}$

21.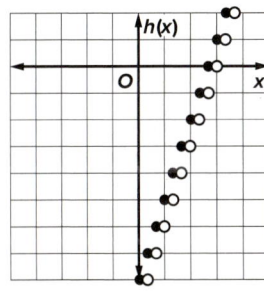
D = {all real numbers};
R = {all integers}

23.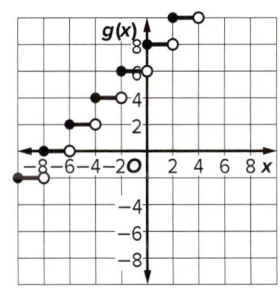
D = {all real numbers};
R = {all even integers}

25.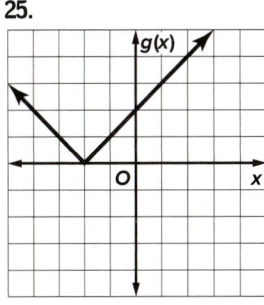
D = {all real numbers};
R = {g(x) | g(x) ≥ 0}

27.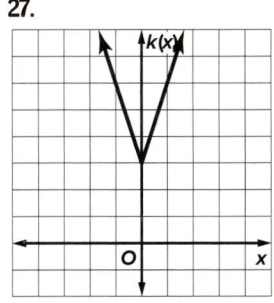
D = {all real numbers};
R = {k(x) | k(x) ≥ 3}

29.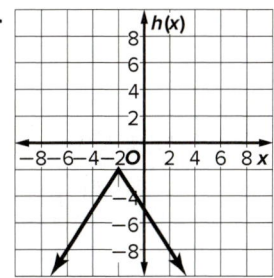
D = {all real numbers};
R = {h(x) | h(x) ≤ −2}

31a. f(a) = |a − 60|
31b. {a | a ≥ 0}
31c.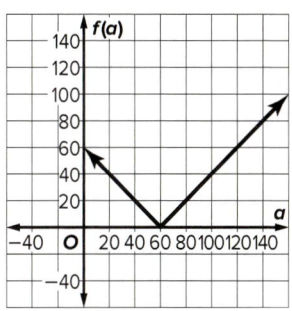

33. f(x) = 0.5x if x > 0, f(x) = 0 if x = 0, and f(x) = −0.5x if x < 0. So, according to the definition of absolute value, f(x) = |0.5x|.

35.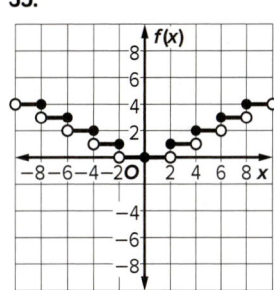
D = {all real numbers};
R = {all whole numbers}

37.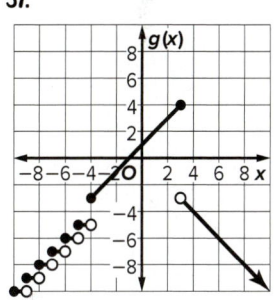
D = {all real numbers};
R = {g(x) | g(x) ≤ 4}

39a.
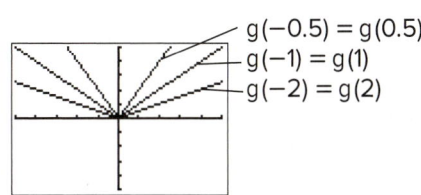
[−5, 5] scl: 1 by [−5, 5] scl: 1

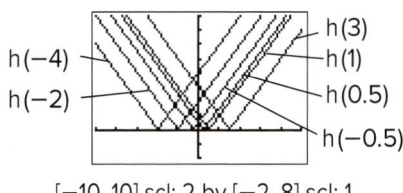
[−5, 5] scl: 1 by [−5, 5] scl: 1

[−10, 10] scl: 2 by [−2, 8] scl: 1

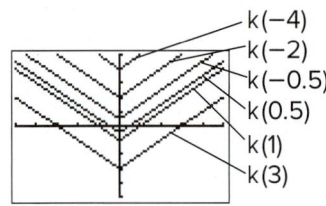

[−5, 5] scl: 1 by [−5, 5] scl: 1

39b. For $f(x)$, negative values of a cause the graph to be reflected in the x-axis. The greater the absolute value of a, the more the graph is vertically stretched. For $g(x)$, the greater the absolute value of b, the more the graph is horizontally stretched. For $h(x)$, as c increases, the graph shifts right. For $k(x)$, as d increases, the graph shifts up.

39c.

x	−2	−1	−0.5	0.5	2	4
$f(x) = -4|x|$	−8	−4	−2	−2	−8	−16
slope		4	4	0	−4	−4

x	−2	−1	−0.5	0.5	2	4
$f(x) = 4|x|$	8	4	2	2	−8	16
slope		−4	−4	0	−4	4

x	−2	−1	−0.5	0.5	2	4
$g(x) = |-4x|$	8	4	2	2	8	16
slope		−4	−4	0	4	4

x	−2	−1	−0.5	0.5	2	4
$g(x) = |4x|$	8	4	2	2	8	16
slope		−4	−4	0	4	4

x	−4	−2	−0.5	0.5	1	3
$h(x) = |x - (-4)|$	0	2	3.5	4.5	5	7
slope		1	1	1	1	1

x	−4	−2	−0.5	0.5	1	3
$h(x) = |x - 4|$	8	6	4.5	3.5	3	1
slope		−1	−1	−1	−1	−1

x	−4	−2	−0.5	0.5	1	3
$k(x) = |x| + (-4)$	0	−2	−3.5	−3.5	−3	−1
slope		−1	−1	0	1	1

x	−4	−2	−0.5	0.5	1	3
$k(x) = |x| + 4$	8	6	4.5	4.5	5	7
slope		−1	−1	0	1	1

39d. For $f(x)$, the slopes are opposite when the a-values are opposites. For $g(x)$, the slopes are the same when the b-values are opposites. For $h(x)$, the slopes are opposite when the c-values are opposites. For $k(x)$, the slopes are the same when the d-values are the opposites. The slope is constant for each section of the graph.

41.

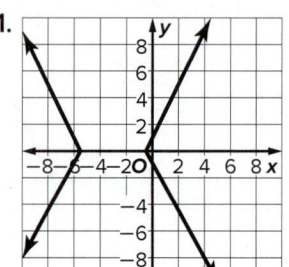

43. Sample answer: $f(x) = -|x - 2|$ **45.** B **47a.** C, D, F
47b. A **47c.** 10 **47d.** D

Lesson 2-6

1. translation of the graph of $y = x^2$ down 4 units

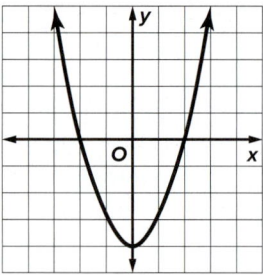

3. reflection of the graph of $y = |x|$ in the x-axis

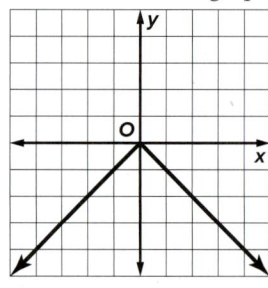

5. A transformation of the form $af(x)$, with $0 < a < 1$, compresses the graph vertically. So, $y = \frac{3}{5}x$ is a dilation of the graph of $y = x$ compressed vertically.

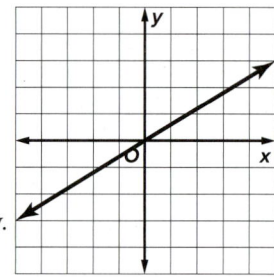

7. B

9. translation of the graph of $y = |x|$ down 3 units

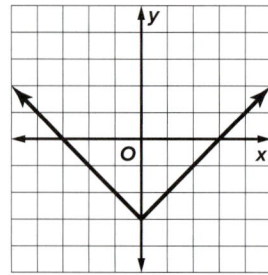

11. translation of the graph of $y = x$ up 2 units or left 2 units

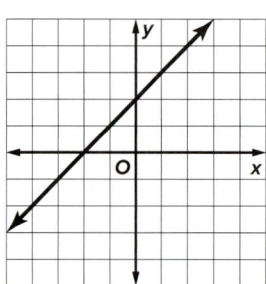

13. translation of the graph of $y = |x|$ left 6 units

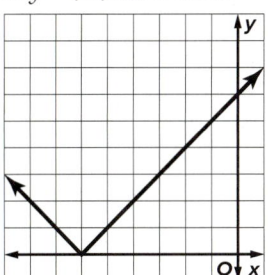

15. reflection of the graph of $y = x^2$ in the x-axis

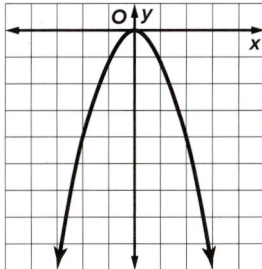

17. reflection of the graph of $y = |x|$ in the y-axis

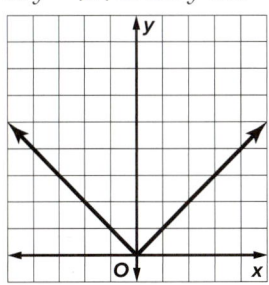

19. reflection of the graph of $y = x$ in the y-axis

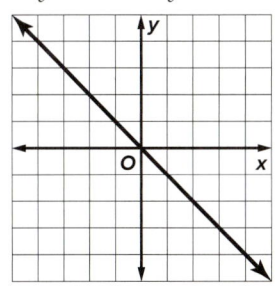

21. dilation of the graph of $y = x$ stretched vertically

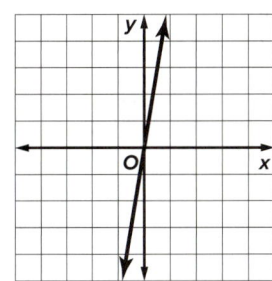

23. dilation of the graph of $y = |x|$ compressed horizontally

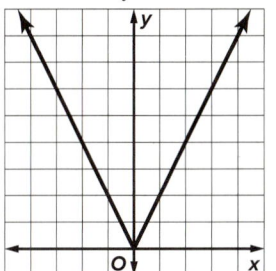

25. dilation of the graph of $y = x^2$ compressed vertically

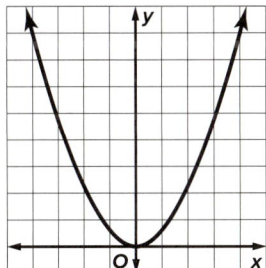

27. The parent graph has been translated up 7 units.

29. The parent graph has been translated right 5 units and reflected in the x-axis.

31. translation up 8 units or left 8 units

33. $y = x^2 + 1$ **35.** $y = x - 5$ **37.** $y = (x - 2)^2$

39 Since the lines have the same slope, the green line is a translation of the blue line down 2 units on the graph. Each unit on the graph on the vertical axis represents $1000, so this is a translation down $2000.

41 The graph is the graph of $y = x^2$ translated 4 units to the left and 6 units down, so the equation is $y = (x + 4)^2 - 6$ **43.** Sample answer: Because a vertical translation concerns only y-values and a horizontal translation concerns only x-values, order is irrelevant.

45. Sample graph:

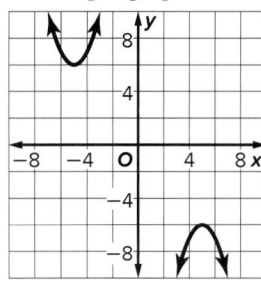

Sample answer: The figure in Quadrant II has been reflected and moved left 10 units.

47. Sample answer: It is not always true. When the axis of symmetry of the parabola is not along the y-axis, the graphs of the preimage and image will be different.
49. C   **51a.** A   **51b.** C   **51c.** B, E   **51d.** C

Lesson 2-7

1. x-intercept: $\frac{3}{2}$; y-intercept -3 **3.** x-intercept: 1; y-intercept: -3 **5.** x-intercept: $-\frac{8}{5}$; y-intercept: 8
7. x-intercept: 6; y-intercept: 4   **9.** $\frac{2}{3}$ **11.** -4
13 For a function in the form $f(x) = mx + b$, the value of b is the y-intercept. So the y-intercept of $f(x) = \frac{3}{4}x - 6$ is -6. To find the x-intercept, solve $0 = \frac{3}{4}x - 6$ for x. This shows that the x-intercept is 8.
15. x-intercept: -2; y-intercept: -8
17. x-intercept: 0.7; y-intercept: -2.8
19. x-intercept: 4.8; y-intercept: 4.8
21. x-intercept: $\frac{1}{4}$; y-intercept: 2   **23.** -3 **25.** 4
27. 16   **29.** Francesca must paint 21 faces to earn $80.
31. x-intercepts: -2, 1, 3; y-intercept: 6   **33.** $\frac{8}{3}$
35. -6 **37.** 3
39 a. Hernando has $40 saved, plus $30 from helping his father, for a total of $70.
39b. Hernando still needs $130 − $70 = $60.
39c. Let x be the number of weeks. Then $15x$ is the total amount Hernando earns from mowing the yard for x weeks. So, $f(x) = -60 + 15x$. To find the zero of the function, solve $0 = -60 + 15x$ to find that $x = 4$.
39d. Hernando must mow the neighbor's yard for 4 weeks.
41. The root and solution of $0 = 4x + 10$ is -2.5 because it is the value of x that makes the equation a true statement. The zero of $f(x) = 4x + 10$ is -2.5 because it is the value of x for which $f(x) = 0$.
43. No; the graph does not cross the x-axis.
45. C   **47.** D   **49.** B

Chapter 2 Study Guide and Review

1. one-to-one   **3.** identity   **5.** piecewise-defined
7. Polynomials with even degree have end behavior in the same direction, while polynomials with odd degree have end behavior in opposite directions.
9. D = {−4, −2, 1, 3}, R = {−4, 1, 3, 5}; not a function
11. D = {1, 2, 3, 4, 5}, R = {8.75, 14, 19.25, 24.50, 29.75}; a function; discrete

13. No; the variables have an exponent other than 1.
15. yes   **17.** No; x appears in a denominator.
19. line symmetry about the line $x = 1$; no point symmetry
21a. $f(x) \to -\infty$ as $x \to -\infty$, $f(x) \to +\infty$ as $x \to +\infty$
21b. Because the end behavior is in opposite directions, it is an odd-degree function.
21c. The graph intersects the x-axis at one point, so there is one real zero.
23. no rel max; rel. min: $x = -2.71$
25. Sample answer:

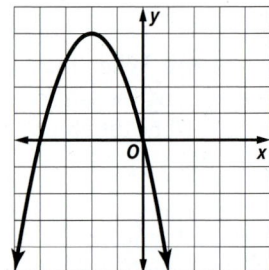

27.

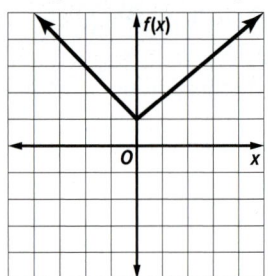

D = {all real numbers},
R = {$f(x) \mid f(x) \geq -7$}

29. D = {all real numbers},
R = {all integers}

31. D = {all real numbers}; R = {$f(x) \mid f(x) \geq 1$}

33. $y = x^2$ shifted down 3 units
35. $y = |x|$ reflected in the y-axis
37. $y = |x|$ expanded vertically
39. parabola   **41.** 1.29   **43.** 1, 5   **45.** -0.62, 1.62

R32

CHAPTER 3
Quadratic Functions

Chapter 3 Get Ready
1. substitute 2 for x in the equation
3. substitute 14 for x; $910
5. prime   7. $(x + 8)$ feet
9. Distributive Property

Lesson 3-1
1a. y-int $= 0$; axis of symmetry: $x = 0$; x-coordinate $= 0$

1b.
x	f(x)
−2	12
−1	3
0	0
1	3
2	12

1c.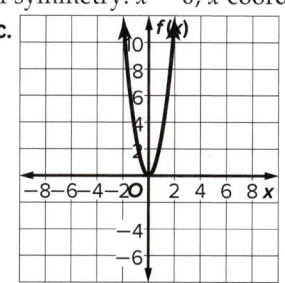

3a. y-int $= 0$; axis of symmetry: $x = 2$; x-coordinate $= 2$

3b.
x	f(x)
0	0
1	−3
2	−4
3	−3
4	0

3c.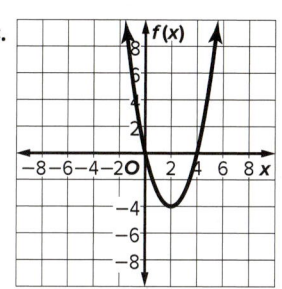

5a. y-int $= -3$; axis of symmetry: $x = 0.75$; x-coordinate $= 0.75$

5b.
x	f(x)
−1	7
0	−3
0.75	−5.25
1.5	−3
2.5	7

5c.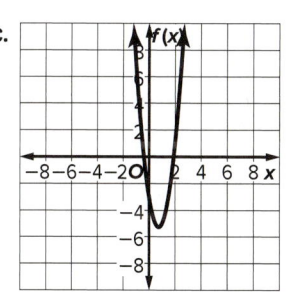

7. max $= 8$; D $=$ {all real numbers}, R $=$ {$f(x) \mid f(x) \leq 8$}

9. min $= -\frac{1}{3}$; D $=$ {all real numbers}, R $= \left\{ f(x) \mid f(x) \geq -\frac{1}{3} \right\}$ **11.** $28.75   **13a.** y-int $= 0$; axis of symmetry: $x = 0$; x-coordinate $= 0$

13b.
x	f(x)
−2	−8
−1	−2
0	0
1	−2
2	−8

13c.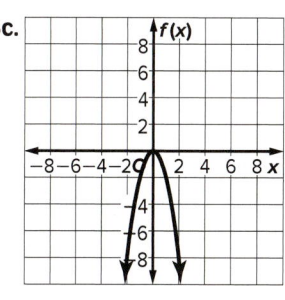

15a. y-int $= 3$; axis of symmetry: $x = 0$; x-coordinate $= 0$

15b.
x	f(x)
−2	7
−1	4
0	3
1	4
2	7

15c.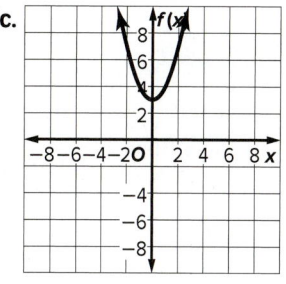

17a. y-int $= 5$; axis of symmetry: $x = 0$; x-coordinate $= 0$

17b.
x	f(x)
−2	−7
−1	2
0	5
1	2
2	−7

17c.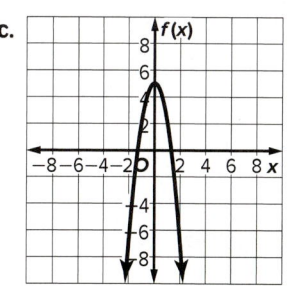

19 a. $f(x) = ax^2 + bx + c$
$\quad\quad\quad\;\downarrow\quad\;\downarrow\quad\;\downarrow$
$f(x) = 1x^2 - 3x - 10 \quad a = 1, b = -3, c = -10$

The y-intercept is $c = -10$.

$x = -\dfrac{b}{2a}$ Equation of the axis of symmetry

$\quad = -\dfrac{(-3)}{2(1)}$ $a = 1$ and $b = -3$

$\quad = \dfrac{3}{2}$ or 1.5   Simplify.

The equation of the axis of symmetry is $x = 1.5$. So, the x-coordinate of the vertex is 1.5.

b. Select five points, with the vertex in the middle and two points on either side of the vertex, including the y-intercept and its reflection.

x	f(x)	
0	−10	←y-intercept
1	−12	
1.5	−12.25	←vertex
2	−12	
3	−10	←reflection of y-intercept

c. Graph the five points from the table, connecting them with a smooth curve.

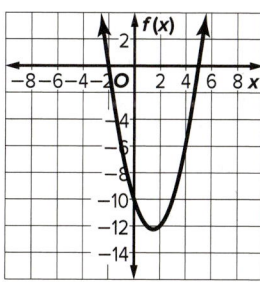

21a. y-int $= 9$; axis of symmetry: $x = 0.75$; x-coordinate $= 0.75$

21b.

x	f(x)
−1	4
0	9
0.75	10.125
1.5	9
2.5	4

21c.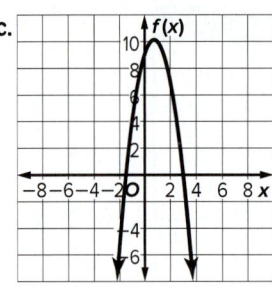

23. max $= -12$; D $=$ {all real numbers}, R $= \{f(x) \mid f(x) \leq -12\}$ **25.** max $= 13.25$; D $=$ {all real numbers}, R $= \{f(x) \mid f(x) \leq 13.25\}$ **27.** max $= 7$; D $=$ {all real numbers}, R $= \{f(x) \mid f(x) \leq 7\}$
29. min $= -9$; D $=$ {all real numbers}, R $= \{f(x) \mid f(x) \geq -9\}$ **31.** min $= -74$; D $=$ {all real numbers}, R $= \{f(x) \mid f(x) \geq -74\}$ **33a.** y-int $= -9$; axis of symmetry: $x = 1.5$; x-coordinate of vertex $= 1.5$

33b.

x	f(x)
0	−9
1	−13
1.5	−13.5
2	−13
3	−9

33c.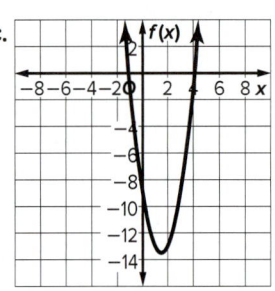

35a. y-int $= 0$; axis of symmetry: $x = \frac{5}{8}$; x-coordinate of vertex $= \frac{5}{8}$

35b.

x	f(x)
$-\frac{3}{4}$	−6
$\frac{1}{4}$	1
$\frac{5}{8}$	1.5625
1	1
2	−6

35c.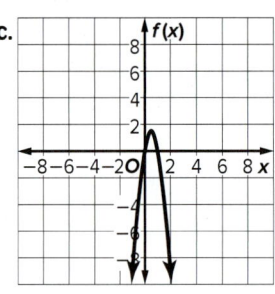

37a. y-int $= 4$; axis of symmetry: $x = -6$; x-coordinate of vertex $= -6$

37b.

x	f(x)
−10	−1
−8	−4
−6	−5
−4	−4
−2	−1

33c.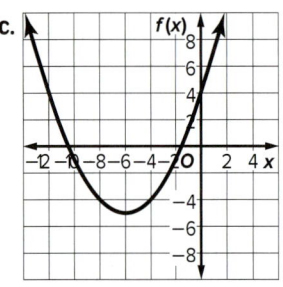

39a. y-int $= -2.5$; axis of symmetry: $x = -\frac{4}{3}$; x-coordinate of vertex $= -\frac{4}{3}$

39b.

x	f(x)
$-\frac{11}{3}$	3
$-\frac{8}{3}$	−2.5
$-\frac{4}{3}$	$-5\frac{1}{6}$
0	−2.5
1	3

39c.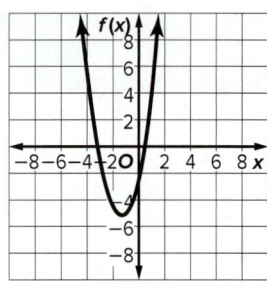

41a. $y = -x^2 + 6x + 475$ **41b.** D $= \{x \mid 0 \leq x \leq 25\}$, R $= \{y \mid 0 \leq y \leq 484\}$ **41c.** $11; Because the function has a maximum at $x = 3$, it is in the domain. Therefore, three $0.50 increases is reasonable. **41d.** $484
43. max $= 23$ **45.** max $= -0.10$ **47.** max $= -4.11$

49 $a = -5$, so the graph opens down and has a maximum value. The maximum value is the y-coordinate of the vertex.

$x = -\frac{b}{2a}$ Equation of the axis of symmetry

$= -\frac{4}{2(-5)}$ or 0.4 $a = -5$ and $b = 4$

The x-coordinate of the vertex is 0.4. Find the y-coordinate of the vertex by evaluating the function for $x = 0.4$.

$f(x) = -5x^2 + 4x - 8$ Original function
$= -5(0.4)^2 + 4(0.4) - 8$ $x = 0.4$
$= -7.2$ The maximum value of the function is -7.2.

The domain is all real numbers. The range is all real numbers less than or equal to the maximum value, or $\{f(x) \mid f(x) \leq -7.2\}$.

51. min $= -9.375$; D $=$ {all real numbers}, R $= \{f(x) \mid f(x) \geq -9.375\}$ **53.** min $= -23.5$; D $=$ {all real numbers}, R $= \{f(x) \mid f(x) \geq -23.5\}$
55. $f(x) = x^2 - 4x - 5$ **57.** $f(x) = x^2 - 6x + 8$

59 a. Sample answer: $795.63 **b.** Sample answer: He sells 600 at $1.50 per bottle, and 50 less for every price increase, so $(600 - 50x)(1.50 + 0.25x)$ represents his income as a function of x price increases. He incurs a fixed cost of $0.40 per bottle, so $(600 - 50x)(0.40)$ represents his fixed costs as a function of x price increases. He has to restock the machine once for every 50% of 400 or 200 bottles sold, so $10[(600 - 50x) \div 200]$ represents his variable costs as a function of x price increases. His profit p is equal to the income less the total costs. This can be represented by $p = (600 - 50x)(1.50 + 0.25x) - \{(600 - 50x)(0.40) + 10[(600 - 50x) \div 200]\}$.

$p = (600 - 50x)(1.50 + 0.25x) - \{(600 - 50x)(0.40) + 10[(600 - 50x) \div 200]\}$
$= 900 + 150x - 75x - 12.5x^2 - \{240 - 20x + 10[3 - 0.25x]\}$
$= 900 + 75x - 12.5x^2 - \{240 - 20x + 30 - 2.5x\}$
$= 900 + 75x - 12.5x^2 - \{270 - 22.5x\}$
$= 630 + 97.5x - 12.5x^2$

The maximum of the graph of this equation occurs at $(3.9, 820.15)$. His maximum profit is $820.15 per week at a price of 1.50 + 3.9(0.25) or about $2.48 per bottle. Because he cannot charge $2.48, round up to $2.50. At this rate, his profit is $795.63.   **61.** Madison; sample answer: $f(x)$ has a maximum of −2. $g(x)$ has a maximum of 1.   **63a.** $a = 22$; $b = 26$; $c = -6$; $d = 2$ **63b.** 0   **63c.** maximum   **65.** Sample answer: If the absolute value function has a minimum value, then it opens up. If it has a maximum value, then it opens down. The axis of symmetry passes through the maximum or minimum point. Thus, the absolute value function is reflected in the axis of symmetry.   **67.** 10.5   **69.** A   **71.** D

Lesson 3-2

1. no real solution

3. −4

5. Graph the related function $f(x) = x^2 - 3x - 18$. The equation of the axis of symmetry is $x = -\dfrac{(-3)}{2(1)}$ or 1.5. Make a table using x-values around 1.5. Then graph each point.

x	−3	0	1.5	3	6
f(x)	0	−18	−20.25	−18	0

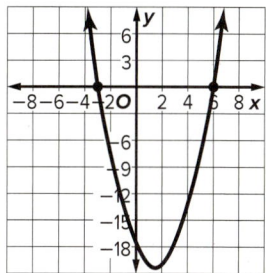

The zeros of the function are −3 and 6. So, the solutions of the equation are −3 and 6.

7.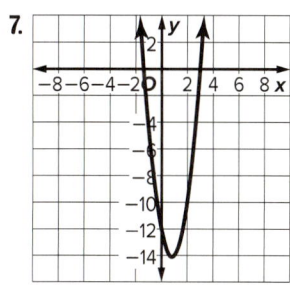
between −2 and −1, 3

9.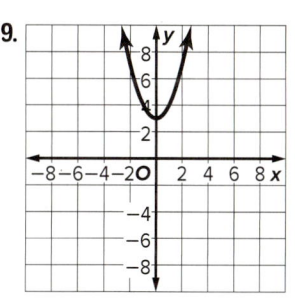
no real solution

11.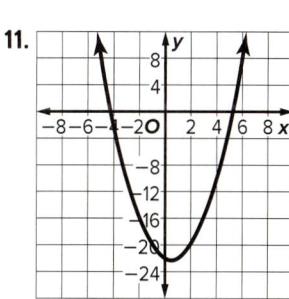
between −5 and −4, between 5 and 6

13. 5 seconds   **15.** no real solution   **17.** −2   **19.** −3, 4

21.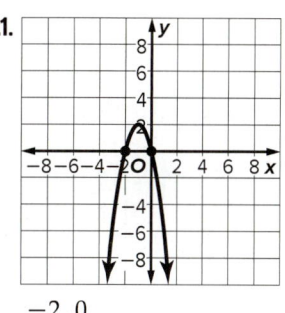
−2, 0

23.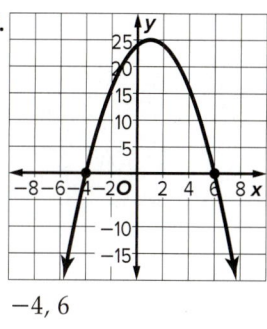
−4, 6

25.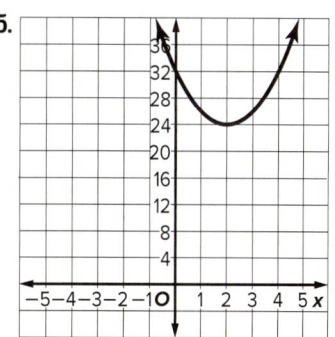
no real solution

27. between −1 and 0, between 1 and 2

29.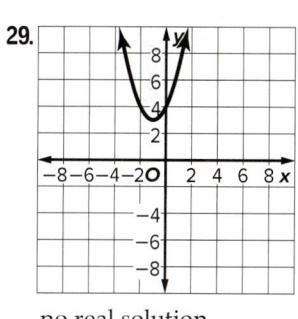
no real solution

31. between 0 and 1; between 2 and 3

33. Let x = one of the numbers. Then $-15 - x$ = the other number.

$x(-x - 15) = -54$ The product is −54.
$-x^2 - 15x = -54$ Distributive Property
$-x^2 - 15x + 54 = 0$ Add 54 to each side.

Graph the related function $f(x) = -x^2 - 15x + 54$. The equation of the axis of symmetry is $x = -\dfrac{(-15)}{2(-1)}$ or −7.5. Make a table using x-values around −7.5. Then graph each point.

x	−20	−15	−10	−7.5	−5	0	5
f(x)	−46	54	104	110.25	104	54	−46

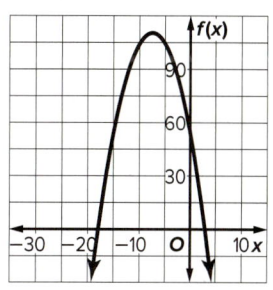

The zeros of the function are −18 and 3. So, the numbers are −18 and 3.

35. about −5.0 and 17.0 **37.** 11 and −19
39. about 3.4375 seconds

41.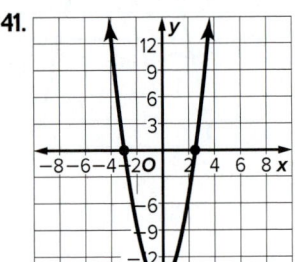
−3, between 2 and 3

43.
between −3 and −2, between 1 and 2

45.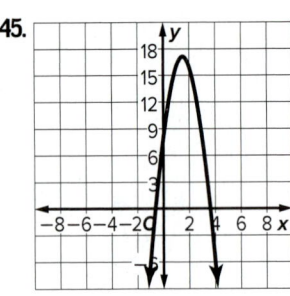
between −1 and 0, between 4 and 5

47.
between 3 and 4, between 8 and 9

49 Find t when $h_0 = 60$ and $h(t) = 0$.

$h(t) = -16t^2 + h_0$ Original equation
$0 = -16t^2 + 60$ $h(t) = 0$ and $h_0 = 60$

Graph the related function $f(t) = -16t^2 + 60$.

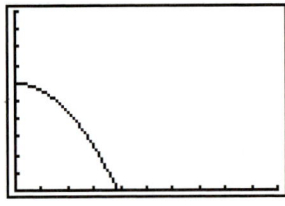

[0, 5] scl: 0.5 by [0, 100] scl: 10

Use the **Zero** feature in the **CALC** menu to find the positive zero of the function, because time cannot be negative. $x \approx 1.94$, so it would take the balloon about 1.94 seconds to hit the ground. Tony's brother should be 4.4 ft/s · 1.94 s or about 8.5 feet from the target when Tony lets go of the balloon.

51. 25 seconds   **53.** $k = 8$
55. $f(x) = -5x^2 + 30x + 80$
57. B   **59.** B   **61a.** C   **61b.** B, D, E, F, G
61c. −4 and 2   **61d.** −8
61e. See students' work.

Lesson 3-3

1. $9i$ **3.** 12   **5.** 1   **7.** $\pm 2i\sqrt{2}$ **9.** 3, −2   **11.** −3 + 2i
13. 70 − 60i **15.** $\frac{1}{2} - \frac{1}{2}i$ **17.** 12 + 6j amps   **19.** 13i
21. 9i **23.** −144i **25.** i **27.** −7   **29.** 9   **31.** 30 + 16i
33. 1 + i **35.** $\frac{1}{3} - \frac{5}{3}i$

37 $3x^2 + 48 = 0$ Original equation
   $3x^2 = -48$ Subtract 48 from each side.
   $x^2 = -16$ Divide each side by 3.
   $x = \pm\sqrt{-16}$ Square Root Property
   $x = \pm 4i$ $\sqrt{-16} = \sqrt{16} \cdot \sqrt{-1}$ or $4i$

39. $\pm i\sqrt{5}$ **41.** $\pm 4i$ **43.** 2, −3   **45.** $\frac{4}{3}$, 4   **47.** 25, −2
49. 4i **51.** 8   **53.** −21 + 15i **55.** $\frac{15}{13} + \frac{16}{13}i$
57. 11 + 23i **59.** $\frac{1}{7} - \frac{4\sqrt{3}}{7}i$

61 $V = C \cdot I$ Electricity formula
 $= (3 + 6j) \cdot (5 - j)$ $C = 3 + 6j$ and $I = 5 - j$
 $= 3(5) + 3(-j) + 6j(5) + 6j(-j)$ FOIL Method
 $= 15 - 3j + 30j - 6j^2$ Multiply.
 $= 15 - 3j + 30j - 6(-1)$ $j^2 = -1$
 $= 21 + 27j$ Simplify.

The voltage is 21 + 27j volts.

63. $(3 + i)x^2 + (-2 + i)x - 8i + 7$
65a. Sample answer: $x^2 + 9 = 0$
65b.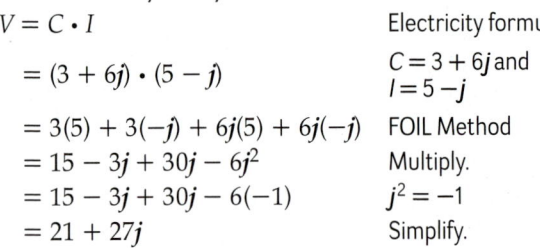

65c. Sample answer: $x^2 - 4x + 5 = 0$
65d.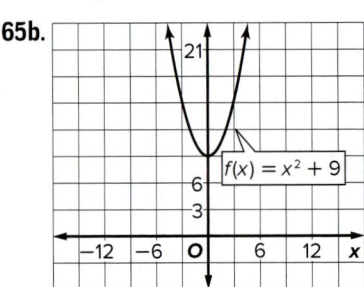

65e. Sample answer: A quadratic equation will have only complex solutions when the graph of the related function has no x-intercepts.   **67.** −11 − 2i
69. Sample answer: $(4 + 2i)(4 - 2i)$
71. C   **73.** C   **75.** 1   **77.** A, D, F

R36

Lesson 3-4

1. $x^2 + 3x - 40 = 0$ **3.** $6x^2 - 11x - 10 = 0$
5. $(6x - 1)(3x + 4)$ **7.** $(x - 7)(x + 3)$
9. $(2x - 11)(2x + 11)$ **11.** $0, \frac{3}{2}$ **13.** $0, 9$
15. 6 **17.** $x^2 - 14x + 49 = 0$
19. $5x^2 - 31x + 6 = 0$ **21.** $17c(3c^2 - 2)$
23. $3(x + 2)(x - 2)$

25. $48cg + 36cf - 4dg - 3df$ Original expression
$= (48cg + 36cf) + (-4dg - 3df)$ Group terms with common factors.
$= 12c(4g + 3f) + (-d)(4g + 3f)$ Factor the GCF from each group.
$= (12c - d)(4g + 3f)$ Distributive Property

27. $(x - 11)(x + 2)$ **29.** $(5x - 1)(3x + 2)$
31. $3(2x - 1)(3x + 4)$ **33.** $(3x + 5i)(3x - 5i)$
35. $-\frac{2}{5}, 6$ **37.** $0, 9$ **39.** $8, -3$ **41.** $2i, -2i$ **43.** $5, \frac{3}{4}$
45. 24 and 26 or -24 and -26 **47.** $x = 20$; 24 in. by 18 in. **49.** $-\frac{1}{2}, \frac{5}{6}$ **51.** $-\frac{3}{2}$ **53.** $6, -6$

55. To find the number of movie screens that produces a profit, first find the number of movie screens in which the profit is zero. Solve $-x^2 + 48x - 512 = 0$.
$ac = -1(-512)$ or 512
$m = 16$; $p = 32$ $mp = 512$ and $ac = 512$; $m + p = 48$ and $b = 48$
$-x^2 + 16x + 32x - 512 = 0$ Write the pattern.
$(-x^2 + 16x) + (32x - 512) = 0$ Group terms with common factors.
$-x(x - 16) + 32(x - 16) = 0$ Group terms with common factors.
$(-x + 32)(x - 16) = 0$ Distributive Property
$-x + 32 = 0$ or $x - 16 = 0$ Zero Product Property
$x = 32$ $x = 16$ Solve each equation.
The solutions are 16 and 32. When there are 16 or 32 movie screens, the profit is zero. Because a is negative, the graph of the function opens down and has a maximum value. So, $P(x)$ is nonnegative for $16 \leq x \leq 32$. When there are 16 to 32 movie screens, the company will not lose money.

57. $25x^2 - 100x + 51 = 0$ **59.** $-3, \frac{1}{2}$ **61.** $1, -\frac{5}{4}$
63. $-\frac{3}{2}, \frac{5}{6}$ **65.** $x^2 - 6^2$; $(x + 6)(x - 6)$ **67.** 20 in. by 15 in. **69.** 13 cm **71.** $2(3 - 4y)(3a + 8b)$

73. $6a^2b^2 - 12ab^2 - 18b^3$ Original expression
$= 6b^2(a^2 - 2a - 3b)$ Factor the GCF, $6b^2$.

75. $2(2x - 3y)(8a + 3b)$ **77.** $(x + y)(x - y)(5a + 2b)$
79. Sample answer: Morgan; Gwen did not have like terms in the parentheses in the third line.
81. $5x^2(2x - 3y)(4x^2 + 6xy + 9y^2)$
83. Sample answer:
$(x - p)(x - q) = 0$ Original equation
$x^2 - px - qx + pq = 0$ Multiply.
$x^2 - (p + q)x + pq = 0$ Simplify.
$x = -\frac{b}{2a}$ Formula for axis of symmetry
$x = -\frac{-(p+q)}{2(1)}$ $a = 1$ and $b = -(p + q)$
$x = \frac{p+q}{2}$ Simplify.
x is midway between p and q.   Definition of midpoint

85. Sample answer: Always; in order to factor using perfect square trinomials, the coefficient of the linear term, bx, must be a multiple of 2, or even. **87.** D
89. C **91.** $0, 4$ **93a.** $4(x + 4i)(x - 4i)$ **93b.** $i^2 = -1$
93c. see students' work

Lesson 3-5

1. $\{-8.45, -3.55\}$ **3.** $\{-12.87, -5.13\}$ **5.** 25 ft
7. $\{2 - i\sqrt{5}, 2 + i\sqrt{5}\}$ **9.** $\{-4.37, 1.37\}$ **11.** $\{-6.45, -1.55\}$
13. $y = -5(x - 2)^2 - 31$; $(2, -31)$ **15.** $\{-1.47, 7.47\}$
17. $\{-7.65, -2.35\}$ **19.** $\{-1, 3\}$ **21.** $\{4.67, 10.33\}$
23. $\{-0.95, 3.95\}$ **25.** $\{4, 5\}$ **27.** $\{-4.61, 2.61\}$
29. $\{1, 3\}$ **31.** $\left\{\frac{3 - i\sqrt{31}}{4}, \frac{3 + i\sqrt{31}}{4}\right\}$

33. $3x^2 - 6x - 9 = 0$ Original equation
$x^2 - 2x - 3 = 0$ Divide by the coefficient of the quadratic term, 3.
$x^2 - 2x = 3$ Add 3 to each side.
$x^2 - 2x + 1 = 3 + 1$ Because $\left(\frac{-2}{2}\right)^2 = 1$, add 1 to each side.
$(x - 1)^2 = 4$ Write the left side as a perfect square.
$x - 1 = \pm 2$ Square Root Property
$x = \pm 2 + 1$ Add 1 to each side.
$x = 2 + 1$ or $x = -2 + 1$ Write as two equations.
$= 3$ $= -1$ Simplify.
The solution set is $\{-1, 3\}$ or $\{x \mid x = -1, 3\}$.

35. $\{-2 - i\sqrt{7}, -2 + i\sqrt{7}\}$ **37.** $\{5 - 2i, 5 + 2i\}$
39. $\left\{\frac{7 - i\sqrt{47}}{4}, \frac{7 + i\sqrt{47}}{4}\right\}$ **41.** $\{2.65 - i\sqrt{1.5775}, 2.65 + i\sqrt{1.5775}\}$ **43.** $\{-0.89, 5.39\}$
45. $y = (x - 11)^2 + 4$; $(11, 4)$ **47.** $y = -8(x + 4)^2 - 18$; $(-8, -18)$ **49.** $\{2.38, 4.62\}$ **51.** $\{-1.26, 0.26\}$

53. a. The time in which the firework explodes is the t-coordinate of the vertex.
$x = -\frac{b}{2a}$ Equation of the axis of symmetry
$x = -\frac{(25)}{2(-1.5)}$ or $8\frac{1}{3}$ $a = -1.5$ and $b = 25$
So, the firework explodes after $8\frac{1}{3}$ seconds.
b. The time in which the firework explodes is the d-coordinate of the vertex.
$d = -1.5t^2 + 25t$ Original function
$d = -1.5\left(8\frac{1}{3}\right)^2 + 25\left(8\frac{1}{3}\right)$ $t = 8\frac{1}{3}$
$d \approx 104.2$ Simplify.
So, the firework explodes at a height of about 104.2 feet.

55. 2.56; $(x - 1.6)^2$

57a.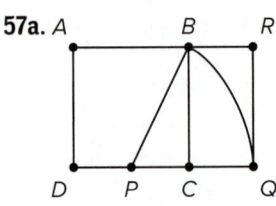

57b. $x = \dfrac{1 + \sqrt{5}}{2}$

57c.

CQ	x
2	$1 + \sqrt{5}$
3	$\dfrac{3 + 3\sqrt{5}}{2}$
4	$2 + 2\sqrt{5}$

57d. Sample answer: the x-values are multiples of $\dfrac{1 + \sqrt{5}}{2}$; $x = \dfrac{n(1 + \sqrt{5})}{2}$.   **59.** $x = \dfrac{-b}{2} \pm \sqrt{\dfrac{b^2}{4} - c}$

61. Sample answer: $x^2 - \dfrac{2}{3}x + \dfrac{1}{9} = \dfrac{1}{4}$; $\left\{\dfrac{5}{6}, -\dfrac{1}{6}\right\}$ **63.** B

65. $\dfrac{7}{6} + \dfrac{\sqrt{193}}{6}, \dfrac{7}{6} - \dfrac{\sqrt{193}}{6}$ **67.** $-3 \pm 2\sqrt{2}$

69a. 40   **69b.** B   **69c.** A, C   **71.** A, D, E

Lesson 3-6

1. $x = \dfrac{-b \pm \sqrt{b^2 - 4ac}}{2a}$ Quadratic Formula

$= \dfrac{-12 \pm \sqrt{12^2 - 4(1)(-9)}}{2(1)}$ $a = 1, b = 12,$ and $c = -9$

$= \dfrac{-12 \pm \sqrt{144 + 36}}{2}$ Multiply.

$= \dfrac{12 \pm \sqrt{180}}{2}$ Simplify.

$= \dfrac{-12 \pm 6\sqrt{5}}{2}$ $\sqrt{180} = 6\sqrt{5}$

$x = \dfrac{-12 + 6\sqrt{5}}{2}$ or $x = \dfrac{-12 - 6\sqrt{5}}{2}$ Write as two equations.

$= -6 + 3\sqrt{5}$ $= -6 - 3\sqrt{5}$ Simplify.

The solutions are $-6 + 3\sqrt{5}$ and $-6 - 3\sqrt{5}$.

3. $\left(\dfrac{5 + \sqrt{57}}{8}, \dfrac{5 - \sqrt{57}}{8}\right)$ **5.** $(1.5, -0.2)$

7. $\left(\dfrac{2 + 2\sqrt{7}}{3}, \dfrac{2 - 2\sqrt{7}}{3}\right)$ **9.** about 0.78 second

11a. -36 **11b.** 2 complex roots   **13a.** -76

13b. 2 complex roots   **15.** $\dfrac{-3 \pm \sqrt{15}}{2}$ **17.** $\dfrac{-7 \pm \sqrt{129}}{8}$

19. $\dfrac{-3 \pm i\sqrt{71}}{8}$

21. a. $b^2 - 4ac = 3^2 - 4(2)(-3)$ $a = 2, b = 3, c = -3$

$= 9 + 24$ or 33 Simplify.

b. The discriminate is positive and not a perfect square. So, there are 2 irrational roots.

c. $x = \dfrac{-b \pm \sqrt{b^2 - 4ac}}{2a}$ Quadratic Formula

$= \dfrac{-3 \pm \sqrt{3^2 - 4(2)(-3)}}{2(2)}$ $a = 2, b = 3,$ and $c = -3$

$= \dfrac{-3 \pm \sqrt{9 + 24}}{4}$ Multiply.

$= \dfrac{-3 \pm \sqrt{33}}{4}$ Simplify.

23a. 49   **23b.** 2 rational   **23c.** $\dfrac{1}{6}, -1$ **25a.** -87

25b. 2 complex   **25c.** $\dfrac{3 \pm i\sqrt{87}}{6}$ **27a.** 36

27b. 2 rational   **27c.** $1, -\dfrac{1}{5}$ **29a.** 1   **29b.** 2 rational

29c. $-1, -\dfrac{4}{3}$ **31a.** -16 **31b.** 2 complex   **31c.** $-1 \pm 2i$

33a. 0   **33b.** about 2.3 seconds   **35a.** 64

35b. 2 rational   **35c.** $0, -\dfrac{8}{5}$ **37a.** 160   **37b.** 2 irrational

37c. $\dfrac{-1 \pm \sqrt{10}}{6}$ **39a.** 13.48   **39b.** 2 irrational

39c. $\dfrac{-0.7 \pm \sqrt{3.37}}{0.6}$

41. a. $y = -0.26x^2 - 0.55x + 91.81$ Original equation

$= -0.26(17)^2 - 0.55(17) + 91.81$ Replace x with 17.

≈ 7.3 Simplify.

For 2017, the number of deaths per 100,000 is about 7.3. For 2015, the number is 25.06.

b. $y = -0.26x^2 - 0.55x + 91.81$ Original equation

$50 = -0.26x^2 - 0.55x + 91.81$ Replace y with 50.

$0 = -0.26x^2 - 0.55x + 41.81$ Subtract 50 from each side.

$x = \dfrac{-b \pm \sqrt{b^2 - 4ac}}{2a}$ Quadratic Formula

$= \dfrac{-(-0.55) \pm \sqrt{(-0.55)^2 - 4(-0.26)(41.81)}}{2(-0.26)}$ $a = -0.26, b = -0.55,$ and $c = 41.81$

$= \dfrac{0.55 \pm \sqrt{43.7894}}{-0.52}$ Simplify.

$x \approx -13.8$ or $x \approx 11.7$

Because the number of years after 2000 cannot be negative, the solution is 11.58. So, 11.7 years after 2000, or in 2011, the death rate will be 50 per 100,000.

c. $y = -0.26x^2 - 0.55x + 91.81$ Original equation

$0 = -0.26x^2 - 0.55x + 91.81$ Replace y with 0.

$x = \dfrac{-b \pm \sqrt{b^2 - 4ac}}{2a}$ Quadratic Formula

$= \dfrac{-(-0.55) \pm \sqrt{(-0.55)^2 - 4(-0.26)(91.81)}}{2(-0.26)}$ $a = -0.26, b = -0.55,$ and $c = 91.81$

$= \dfrac{0.55 \pm \sqrt{95.7849}}{-0.52}$ Simplify.

$x \approx -19.88$ or $x \approx 17.76$

Because the number of years after 2000 cannot be negative, the solution is 17.76. So, 17.76 years after 2000, or in 2017, the death rate will be 0 per 100,000. Sample answer: This prediction is not reasonable because the death rate from cancer will never be 0 unless a cure is found. If and when a cure will be found cannot be predicted.

43. Jonathan is correct; you must first write the equation in the form $ax^2 + bx + c = 0$ to determine the values of a, b, and c. Therefore, the value of c is -7, not 7.
45a. Sample answer: Always; when a and c are opposite signs, then ac will always be negative and $-4ac$ will always be positive. Because b^2 will also always be positive, then $b^2 - 4ac$ represents the addition of two positive values, which will never be negative. Hence, the discriminant can never be negative and the solutions can never be imaginary. **45b.** Sample answer: Sometimes; the roots will only be irrational if $b^2 - 4ac$ is not a perfect square.
47. -0.75 **49.** B **51.** $\left\{-\dfrac{3}{10}, 1\right\}$ **53.** $5 \pm \text{R38}$
55. two complex solutions

Lesson 3-7

1.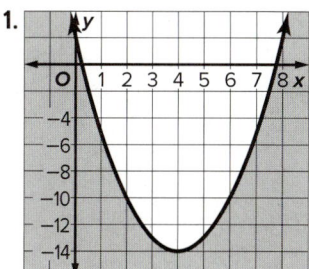

3.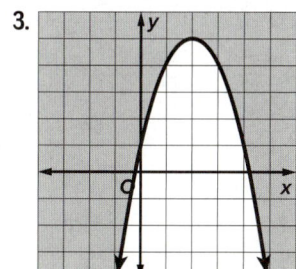

5. $\{x \mid -5 < x < -3\}$
7. $\{x \mid 0.29 \leq x \leq 1.71\}$
9. $\{x \mid -8 < x < 2\}$

11. $-x^2 + 12x = 28$ Related quadratic equation
$-x^2 + 12x - 28 = 0$ Subtract 28 from each side.
$x = \dfrac{-b \pm \sqrt{b^2 - 4ac}}{2a}$ Quadratic Formula
$= \dfrac{-12 \pm \sqrt{12^2 - 4(-1)(-28)}}{2(-1)}$ $a = -1$, $b = 12$, and $c = -28$
$x = \dfrac{-12 + \sqrt{32}}{-2}$ or $x = \dfrac{-12 - \sqrt{32}}{-2}$ Simplify and write as two equations.
≈ 3.17 ≈ 8.83 Simplify.

Plot 3.17 and 8.83 on a number line. Use dots because these values are solutions of the original inequality.

$x \leq 3.17$ | $3.17 \leq x \leq 8.83$ | $x \geq 8.83$

Test a value from each of the three intervals to see if it satisfies the original inequality.

$x \leq 3.17$ $\qquad$ $3.17 \leq x \leq 8.83$
Test $x = 0$. $\qquad$ Test $x = 5$.

$-x^2 + 12x \geq 28$ $\qquad$ $-x^2 + 12x \geq 28$
$-(0)^2 + 12(0) \geq 28$ $\qquad$ $-(5)^2 + 12(5) \geq 28$
$0 \ngeq 28$ $\qquad$ $35 \geq 28$

$x \geq 8.83$
Test $x = 10$.
$-x^2 + 12x \geq 28$
$-(10)^2 + 12(10) \geq 28$
$20 \ngeq 28$

The solution set is $\{x \mid 3.17 \leq x \leq 8.83\}$ or $[3.17, 8.83]$.

13.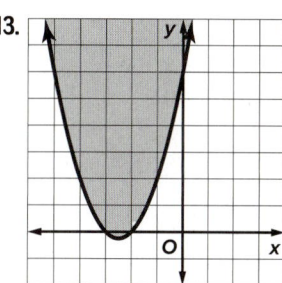

15.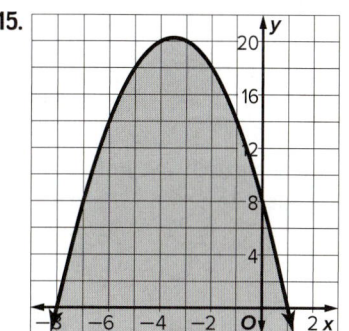

17.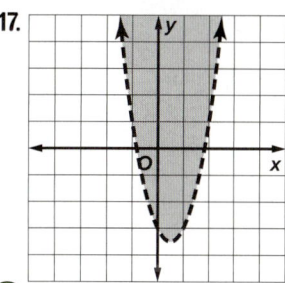

19. $\{x \mid 1.1 < x < 7.9\}$
21. {all real numbers}
23. $\{x \mid x < -1.42$ or $x > 8.42\}$ **25.** $\varnothing$
27. $\{x \mid x < -0.73$ or $x > 2.73\}$
29. $\{x \mid -0.5 \leq x \leq 2.5\}$

31. The function describes the height of the arch. You want to find the values of x for which $f(x) \geq 7$.
$f(x) \geq 7$ Original inequality
$-x^2 + 6x + 1 \geq 7$ $f(x) = -x^2 + 6x + 1$
$-x^2 + 6x - 6 \geq 0$ Subtract 7 from each side.
Graph the related function $y = -x^2 + 6x - 6$ using a graphing calculator.

At $x \approx 1.26$ and $x \approx 4.73$, $f(x) \geq 7$. So, at about 1.26 ft to 4.73 ft from the sides of the arch, the height is at least 7 ft.

[0, 5] scl: 0.5 by [0, 10] scl: 1

33. $\{x \mid 4 < x < 5\}$ **35.** $\{x \mid -1 < x < 2\}$ **37.** $\{x \mid x \leq -2.32$ or $x \geq 4.32\}$ **39.** $\{x \mid x \leq -1.58$ or $x \geq 1.58\}$
41. {all real numbers} **43.** $\{x \mid -2.84 < x < 0.84\}$

45a.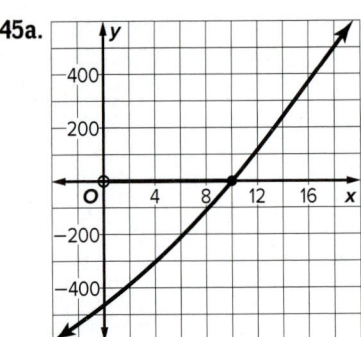

45b. greater than 0 ft but no more than 10.04 ft
47. $y \leq -x^2 + 2x + 6$
49. $\{x \mid x < -1.06 \text{ or } x > 7.06\}$

51 $11 = 4x^2 + 7x$ Related quadratic equation
$0 = 4x^2 + 7x - 11$ Subtract 11 from each side.
$x = \dfrac{-b \pm \sqrt{b^2 - 4ac}}{2a}$ Quadratic Formula
$x = \dfrac{-7 \pm \sqrt{7^2 - 4(4)(-11)}}{2(4)}$ $a = 4, b = 7,$ and $c = -11$
$x = \dfrac{-7 + \sqrt{255}}{8}$ or $x = \dfrac{-7 - \sqrt{255}}{8}$ Simplify and write as two equations.
$= 1$ $= -2.75$ Simplify.

Plot -2.75 and 1 on a number line. Use dots because these values are solutions of the original inequality.

Test a value from each of the three intervals to see if it satisfies the original inequality.

$x \leq -2.75$ $-2.75 \leq x \leq 1$
Test $x = -3.$ Test $x = 0.$
$11 \leq 4x^2 + 7x$ $11 \leq 4x^2 + 7x$
$11 \leq 4(-3)^2 + 7(-3)$ $11 \leq 4(0)^2 + 7(0)$
$11 \leq 15$ $11 \not\leq 0$
$x \geq 1$
Test $x = 2.$
$11 \leq 4x^2 + 7x$
$11 \leq 4(2)^2 + 7(2)$
$11 \leq 30$

The solution set is $\{x \mid x \leq -2.75 \text{ or } x \geq 1\}$.
53. $\{x \mid x < 0.61 \text{ or } x > 2.72\}$
55a. The graph of the profit function shifts down 25,000 units. The shift represents a $25,000 decrease in profit. **55b.** Sample answer: The manufacturer must sell more headsets to have profits of at least $100,000. However, the maximum number of headsets that can be sold decreases. The minimum number of headsets that must be sold increases by about 17,784, from about 29,549 to about 47,333. The maximum number of headsets that can be sold decreases by about 17,783, from 98,451 to 80,668. The range of the number of headsets that can be sold decreases from about 68,902 to about 33,335. **57a.** Sample answer: $x^2 + 2x + 1 \geq 0$

57b. Sample answer: $x^2 - 4x + 6 < 0$
59. No; the graphs of the inequalities intersect the x-axis at the same points.
61.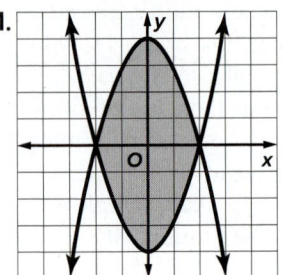

63. B   **65a.** outside   **65b.** C, E   **65c.** D   **67.** <

Chapter 3   Study Guide and Review

1. false, standard form   **3.** true   **5.** true   **7.** If $b^2 - 4ac < 0$, the equation has two complex roots. If $b^2 - 4ac = 0$, the equation has one real rational root. If $b^2 - 4ac > 0$, the equation has two real rational roots if $b^2 - 4ac$ is a perfect square and two real irrational roots if $b^2 - 4ac$ is not a perfect square.
9a. y-int: 12; $x = -\dfrac{5}{2}; -\dfrac{5}{2}$

9b.

x	y
-3	6
$-\dfrac{5}{2}$	$\dfrac{23}{4}$
-2	6
-1	8
0	12

9c.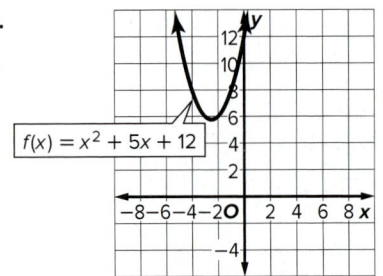

11a. y-int: -5; $x = \dfrac{9}{4}, \dfrac{9}{4}$

11b.

x	y
1	2
2	5
$\dfrac{9}{4}$	$\dfrac{41}{8}$
3	4
4	-1

11c.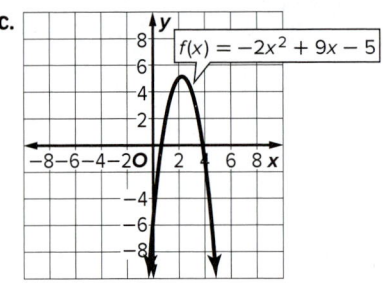

13. max; 1.25; D = {all real numbers}; R = $\{f(x) \mid f(x) \leq 1.25\}$ **15.** 75 T-shirts at $15 each
17. $\left\{-1, \dfrac{3}{2}\right\}$ **19.** 7.5 seconds   **21.** $15 + 3i$ **23.** $28 + 3i$
25. $x = \pm 5i$ **27.** $x = \pm i\sqrt{5}$ **29.** $x = \pm \dfrac{1}{2}i$
31. $x^2 + 10x + 21 = 0$ **33.** $3x^2 - x - 2 = 0$
35. $4x^2 + 5x + 1 = 0$ **37.** $\left\{-\dfrac{1}{2}, 3\right\}$ **39.** $\{-11i, 11i\}$
41. $x = 12$; 9 ft by 14 ft
43. 4; $(x - 2)^2$ **45.** 1.44; $(x + 1.2)^2$
47. $\dfrac{9}{25}$; $\left(x + \dfrac{3}{5}\right)^2$ **49.** $\{1 \pm i\sqrt{7}\}$ **51.** $\left\{1, -\dfrac{5}{2}\right\}$
53a. 0   **53b.** 1 real rational root   **53c.** {5}

R40

55a. 153 **55b.** 2 irrational real roots **55c.** $\left\{\dfrac{-3 \pm 3\sqrt{17}}{4}\right\}$

57a. -32 **57b.** 2 complex roots **57c.** $\{1 \pm 2i\sqrt{2}\}$

59a. -47 **59b.** 2 complex roots **59c.** $\left\{\dfrac{-5 \pm i\sqrt{47}}{4}\right\}$

61.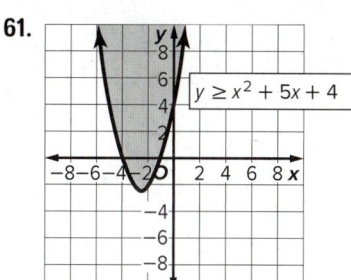

$y \geq x^2 + 5x + 4$

63.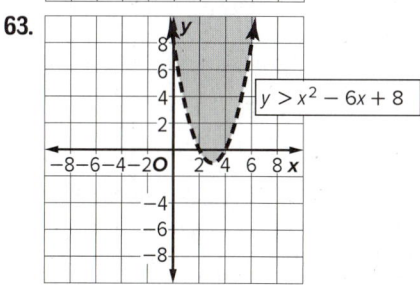

$y > x^2 - 6x + 8$

65. between 0 and 5 ft
67. {all real numbers}
69. $\{x \mid -1.69 < x < 0.44\}$

CHAPTER 4
Polynomials and Polynomial Functions

Chapter 4 Get Ready

1. $2xy + (-3) + (-z)$ 3. $-3a - 3b + 3c$
5. $a = 2, b = 8, c = 1$ 7. -0.13 and -3.87
9. the ordered pair for point E is two negative numbers

Lesson 4-1

1. $-8a^5b^2$ 3. $\dfrac{8a^6}{27b^3}$ 5. yes, 1 7. no 9. $-2x^2 - 6x + 3$
11. $8ab + 10a$ 13. $n^2 - 2n - 63$ 15. $750 - 2.5x$
17. $-8b^5c^3$ 19. $-yz^2$ 21. $\dfrac{a^2c^2}{2b^4}$ 23. z^{18} 25. yes; 3
27. no 29. $3b^2 + 6b - 5$ 31. $8x^3 + 4xy$

33. $(a + b)(a^3 - 3ab - b^2)$
$= a(a^3 - 3ab - b^2) + b(a^3 - 3ab - b^2)$ Distributive Property
$= a(a^3) - a(3ab) - a(b^2) + b(a^3) - b(3ab) - b(b^2)$
  Distributive Property
$= a^4 - 3a^2b - ab^2 + a^3b - 3ab^2 - b^3$ Multiply.
$= a^4 + a^3b - 3a^2b - 4ab^2 - b^3$ Simplify.

35. $10c^3 - c^2 + 4c$ 37. $12a^2b + 8a^2b^2 - 15ab^2 + 4b^2$
39. $4a^2x - 2a^2y + 10abx - 5aby + 6b^2x - 3b^2y$
41. $\dfrac{y^4}{81x^4}$ 43. $\dfrac{x^6}{16y^{14}}$ 45. b 47. $\dfrac{2}{5}cd^4$ 49. $\dfrac{1}{2}x^6y^3$

51. a. $d = rt$ distance = rate • time
$t = \dfrac{d}{r}$ Solve the formula for time.
$= \dfrac{2.367 \times 10^{21} \text{ m}}{3 \times 10^8 \text{ m/s}}$ ← Distance from Andromeda to Earth
  ← Speed of light
$= \dfrac{2.367}{3} \cdot \dfrac{10^{21}}{10^8} \cdot \dfrac{\text{m}}{\text{m/s}}$ Separate to get powers of the same base.
$\approx 0.789 \cdot 10^{21-8}$ s Subtract exponents.
$\approx 0.789 \times 10^{13}$ s Simplify.

It takes about 7.89×10^{12} seconds or about 250,190.26 years.

b. $t = \dfrac{d}{r}$ Write the formula.
$= \dfrac{2.28 \times 10^{11} \text{ m}}{3 \times 10^8 \text{ m/s}}$ ← Distance from the Sun to Mars
  ← Speed of light
$= \dfrac{2.28}{3} \cdot \dfrac{10^{11}}{10^8} \cdot \dfrac{\text{m}}{\text{m/s}}$ Separate to get powers of the same base.
$= 0.76 \times 10^{11-8}$ s Subtract exponents.
$= 0.76 \times 10^3$ s Simplify.
$= 760$ s Simplify.

It takes 760 seconds or about 12.67 minutes.

53. $2n^4 - 3n^3p + 6n^4p^4$ 55. $b^3 + \dfrac{b}{a} + \dfrac{1}{a^2}$
57. $2n^5 - 14n^3 + 4n^2 - 28$ 59. $64n^3 - 240n^2 + 300n - 125$
61a. $0.155x^2 + 8.818x + 835.8$
61b. $0.061x^2 - 10.57x + 112.4$ 63. 9 65. $\dfrac{1}{a^n} = \dfrac{a^0}{a^n} = a^{0-n} = a^{-n}$ 67. Sample answer: We would have a 0 in the denominator, which makes the expression undefined. 69. Sample answer: Astronomy deals with very large numbers that are sometimes difficult to work with because they contain so many digits. Properties of exponents make very large or very small numbers more manageable. As long as you know how far away a planet is from a light source, you can divide that distance by the speed of light to obtain how long it will take light to reach that planet. 71. D 73. A

Lesson 4-2

1. $c^5 + 5c^4d + 10c^3d^2 + 10c^2d^3 + 5cd^4 + d^5$
3. $x^6 - 24x^5 + 240x^4 - 1280x^3 + 3840x^2 - 6144x + 4096$
5. $x^5 + 15x^4 + 90x^3 + 270x^2 + 405x + 243$
7. $\dfrac{3}{32}$ or 0.09375

9. $(x + 3y)^8 = \sum_{k=0}^{8} \dfrac{8!}{k!(8-k)!} x^{8-k}(3y)^k$

$\dfrac{8!}{k!(8-k)!} x^{8-k}(3y)^k = \dfrac{8!}{4!(8-4)!} x^{8-4}(3y)^4$ For the fifth term, $k = 4$.
$= 70x^4(81y^4)$ $C(8, 4) = 70$, $(3y)^4 = 81y^4$
$= 5670x^4y^4$ Simplify.

11. $-108,864c^3d^5$ 13. $243a^5$ 15. $x^6 + 36x^5 + 540x^4 + 4320x^3 + 19,440x^2 + 46,656x + 46,656$ 17. $16a^4 + 128a^3b + 384a^2b^2 + 512ab^3 + 256b^4$ 19. $a^6 - 6a^5b + 15a^4b^2 - 20a^3b^3 + 15a^2b^4 - 6ab^5 + b^6$

21. Let w represent the number of women and m represent the number of men.
$(w + m)^{10} = \sum_{k=0}^{10} \dfrac{10!}{k!(10-k)!} w^{10-k}m^k$

To find the probability that 7 members are women, find the term in which the exponent of w is 7. Because $10 - 3 = 7$, find the term in which $k = 3$, the fourth term.

$\dfrac{10!}{k!(10-k)!} w^{10-k}m^k = \dfrac{10!}{3!(10-3)!} w^{10-3}m^3$ For the fourth term, $k = 3$.
$= 120w^7m^3$ $C(10, 3) = 120$

The probability of choosing a woman is $\dfrac{1}{2}$ and the probability of choosing a man is $\dfrac{1}{2}$.

$120w^7m^3 = 120\left(\dfrac{1}{2}\right)^7\left(\dfrac{1}{2}\right)^3$ $w = \dfrac{1}{2}$ and $m = \dfrac{1}{2}$
$= 120\left(\dfrac{1}{2}\right)^{10}$ Product of Powers Property
$= \dfrac{120}{1024}$ $\left(\dfrac{1}{2}\right)^{10} = \dfrac{1}{1024}$
$= \dfrac{15}{128}$ Simplify.

The probability that 7 of the members will be women is $\dfrac{15}{128}$, or about 0.117.

23. $84x^5z^2$ 25. $7168a^2b^6$ 27. $32,256x^5$ 29. $x^5 + \dfrac{5}{2}x^4 + \dfrac{5}{2}x^3 + \dfrac{5}{4}x^2 + \dfrac{5}{16}x + \dfrac{1}{32}$ 31. $32b^5 + 20b^4 + 5b^3 + \dfrac{5}{8}b^2 + \dfrac{5}{128}b + \dfrac{1}{1024}$ 33a. 0.121 33b. 0.121 33c. 0.309 35. Sample answer: While they have the same terms, the signs for $(x + y)^n$ will all be positive, while the signs for $(x - y)^n$ will alternate. 37. Sample answer: $\left(x + \dfrac{6}{5}y\right)^5$ 39. E 41. D 43. B, D

Lesson 4-3

1. $4y + 2x - 2$ **3.** $x - 8 - \dfrac{4}{x+2}$
5. $3z + 6 + \dfrac{6z - 15}{z^2 - 4z + 3}$ **7.** A **9.** $6a + 6 + \dfrac{21}{3a - 2}$
11. $3y + 5$ **13.** $x + 3y - 2$ **15.** $2a^2 + b - 3$
17. $3np - 6 + 7p$ **19.** $-w + 16 + \dfrac{1000}{w}$

21.
$$\begin{array}{r} b^2 - 5b + 6 \\ b+1\overline{)b^3 - 4b^2 + b - 2} \\ \underline{(-)\; b^3 + b^2} \\ -5b^2 + b \\ \underline{(-)\; -5b^2 - 5b} \\ 6b - 2 \\ \underline{(-)\; 6b + 6} \\ -8 \end{array}$$

The quotient is $b^2 - 5b + 6$, and the remainder is -8. So, the expression equals $b^2 - 5b + 6 - \dfrac{8}{b+1}$.

23. $x^4 + 4x^3 + 12x^2 + 52x + 208 + \dfrac{832}{x-4}$
25. $g^3 + 2g^2 + g + 2 - \dfrac{14}{g-2}$
27. $2x^4 + x^3 - x + \dfrac{2}{3} - \dfrac{2}{9x+3}$ **29.** $b - \dfrac{7}{2} + \dfrac{21b + 7}{4b^2 + 2b + 2}$
31. $5y^2 - 1 + \dfrac{y - 5}{5y^2 - y + 15}$ **33.** $V(t) = t^2 + 5t + 6$

35. a.
$$\begin{array}{r} 3500 \\ a^2 + 100\overline{)3500a^2} \\ \underline{(-)\; 3500a^2 + 350,000} \\ -350,000 \end{array}$$

The quotient is 3500, and the remainder is $-350,000$.
So, the expression equals $3500 - \dfrac{350,000}{a^2 + 100}$.

b. $n = 3500 - \dfrac{350,000}{a^2 + 100}$ Write the equation.

$= 3500 - \dfrac{350,000}{15^2 + 100}$ $a = 15$

$= 3500 - \dfrac{350,000}{325}$ Simplify.

≈ 2423 subscriptions

37. $\dfrac{4c^2d - 3d}{2}$ **39.** $n^2 - n - 1$
41. $3z^4 - z^3 + 2z^2 - 4z + 9 - \dfrac{13}{z+2}$ **43.** Sample answer: Sharon; Jamal actually divided by $x + 3$.
45. Sample answer: The degree of the quotient plus the degree of the divisor equals the degree of the dividend. **47.** $\dfrac{5}{x^2}$ does not belong with the other three. The other three expressions are polynomials. Because the denominator of $\dfrac{5}{x^2}$ contains a variable, it is not a polynomial. **49.** B **51.** C
53. Sample answer: It should just be -4 on the top, and it should be $-7x$, not $9x$, on the third line. The correct quotient is $8x - 4 + \dfrac{-7x + 6}{x^2 + 1}$.
55a. $2x - 1 + \dfrac{2x - 4}{2x^2 - 1}$
55b. $3x^2 - 5x + 6$
55c. $3x + 9 + \dfrac{25x - 7}{x^2 - 3x + 2}$
55d. $2x^2 - 2x + 6 - \dfrac{23}{x+3}$

Lesson 4-4

1. degree $= 6$, leading coefficient $= 11$ **3.** Not in one variable, because there are two variables, x and y.
5. $w(5) = -247$; $w(-4) = 104$ **7.** $4y^9 - 5y^6 + 2$
9. $1536a^3 - 426a^2 - 144a + 82$
11a. $f(x) \to -\infty$ as $x \to -\infty$. $f(x) \to +\infty$ as $x \to +\infty$.
11b. Because the end behavior is in opposite directions, it is an odd-degree function.
11c. The graph intersects the x-axis at three points, so there are three real zeros.

13.

x	y
0	2
-2	0
-1	9
-1.5	~17
0.5	0
1	0
0.75	-3
2	50

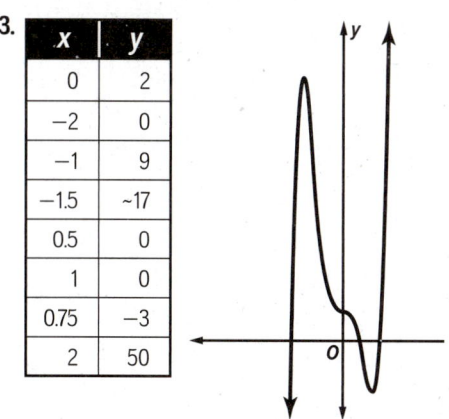

15. degree $= 6$, leading coefficient $= -12$
17. degree $= 4$, leading coefficient $= -5$
19. degree $= 2$, leading coefficient $= 3$
21. degree $= 9$, leading coefficient $= 2$
23. $p(-6) = 1227$; $p(3) = 66$
25. $p(-6) = -156$; $p(3) = 78$

27.
$p(x) = -x^3 + 3x^2 - 5$ Original function
$p(-6) = -(-6)^3 + 3(-6)^2 - 5$ Replace x with -6.
$= 216 + 108 - 5$ Simplify.
$= 319$ Simplify.

$p(x) = -x^3 + 3x^2 - 5$ Original function
$p(-6) = -(3)^3 + 3(3)^2 - 5$ Replace x with 3.
$= -27 + 27 - 5$ Simplify.
$= -5$ Simplify.

29. $18a^2 - 12a + 3$ **31.** $2b^4 - 4b^2 + 3$ **33.** $-64y^3 + 144y^2 - 104y + 25$ **35a.** $f(x) \to +\infty$ as $x \to -\infty$. $f(x) \to +\infty$ as $x \to +\infty$. **35b.** Because the end behavior is in the same direction, it is an even-degree function.
35c. The graph intersects the x-axis at four points, so there are four real zeros. **37a.** $f(x) \to -\infty$ as $x \to -\infty$. $f(x) \to +\infty$ as $x \to +\infty$. **37b.** Because the end behavior is in opposite directions, it is an odd-degree function.
37c. The graph intersects the x-axis at one point, so there is one real zero. **39a.** $f(x) \to -\infty$ as $x \to -\infty$. $f(x) \to -\infty$ as $x \to +\infty$. **39b.** Because the end behavior is in the same direction, it is an even-degree function.
39c. The graph intersects the x-axis at two points, so there are two real zeros.

41. Substitute for L and I in the formula and solve for W.
$W = 0.5LI^2$
$W = 0.5(24)(8^2)$
$W = 768$ joules
The amount of energy stored in an inductor with an inductance of 24 henries and a current of 8 amperes is 768 joules.

43. $p(-2) = -16$; $p(8) = 1024$ **45.** $p(-2) = -0.5$; $p(8) = 3112$ **47.** D **49.** A **51.** $3a^3 - 24a^2 + 240a + 66$ **53.** $5a^6 - 298a^2 + 1008a - 928$

55a.

x	p(x)
−7	−585
−6	0
−4	240
−3	135
−2	0
0	−144
1	−105
2	0
4	240
6	0
7	−585

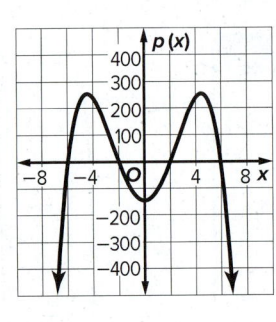

55b. −6, −2, 2, 6 **55c.** 2000 and 6000 items
55d. Sample answer: The negative values should not be considered because the company will not produce negative items.

57. The degree, 4, is even and the leading coefficient, −5, is negative. So, $f(x) \to -\infty$ as $x \to -\infty$ and $f(x) \to -\infty$ as $x \to +\infty$.
59. $h(x) \to +\infty$ as $x \to -\infty$; $h(x) \to -\infty$ as $x \to +\infty$
61. $g(x) \to -\infty$ as $x \to -\infty$; $g(x) \to +\infty$ as $x \to +\infty$
63. Sample answer: Shenequa is correct; the number of real zeros is equal to exactly the number of times the graph intersects the x-axis. **65.** Sample answer: $f(x) \to +\infty$ as $x \to -\infty$; $f(x) \to +\infty$ as $x \to +\infty$; $\frac{f(x)}{g(x)}$ will become a 2nd-degree function with a positive leading coefficient. **67.** Sometimes; a polynomial function with four real zeros may be a sixth-degree polynomial function with two imaginary zeros. A polynomial function that has four real zeros is at least a fourth-degree polynomial. **69.** C **71.** A **73a.** A, B, C, F **73b.** B **73c.** 2 **73d.** $2 + 3i$

Lesson 4-5

1. between −2 and −1

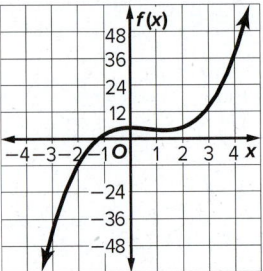

3. between 0 and 1 and between 2 and 3

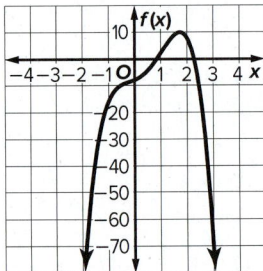

5. rel. max at $x \approx -1.8$; rel. min at $x \approx 1.1$; D = {all real numbers}, R = {all real numbers}

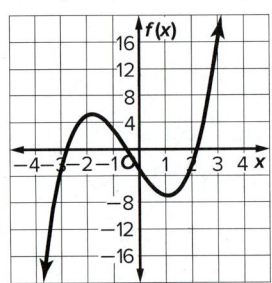

7. rel. max at $x \approx 2.4$; rel. min at $x \approx 0.3$; D = {all real numbers}, R = {all real numbers}

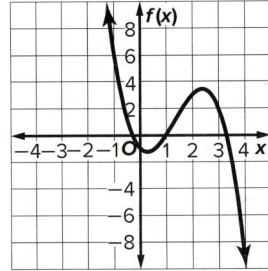

9a.

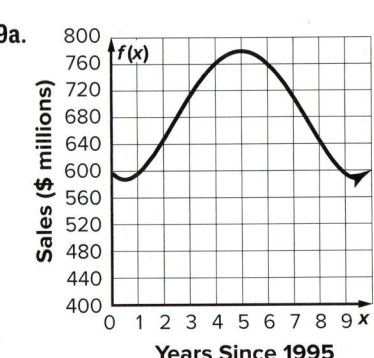

9b. Sample answer: Relative maximum at $x = 5$ and relative minimum at $x \approx 9.5$. $f(x) \to \infty$ as $x \to -\infty$ and $f(x) \to \infty$ as $x \to \infty$. The graph increases when $x < 5$ and $x > 9.5$ and decreases when $5 < x < 9.5$.

9c.
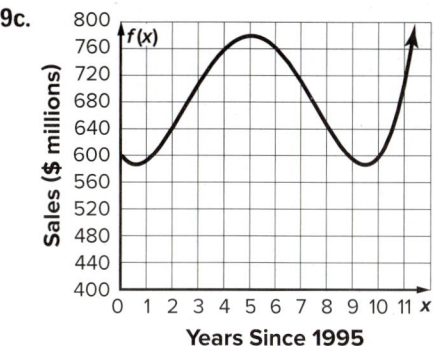

Sample answer: This suggests a dramatic increase in sales.

9d. Sample answer: No; with so many other forms of media on the market today, CD sales will not increase dramatically. In fact, the sales will probably decrease. The function appears to be accurate only until about 2005.

11 a. Because $f(x)$ is a third-degree polynomial function, it will have either 3 or 1 real zeros. Look at the values of $f(x)$ to locate the zeros. Then use the points to sketch the graph.

x	f(x)
−4	92
−3	41
−2	12
−1	−1
0	−4
1	−3
2	−4
3	−13
4	−36

← change in sign

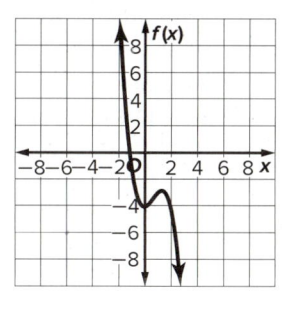

b. The value of $f(x)$ changes signs between $x = -2$ and $x = 1$. So, there is a zero between -2 and -1.

c. The value of $f(x)$ at $x = 0$ is less than the surrounding points, so there must be a relative minimum near $x = 0$. The value of $f(x)$ near $x = 1$ is greater than the surrounding points, so there must be a relative maximum near $x = 1$.

13a.

x	f(x)
−4	−155
−3	−80
−2	−33
−1	−8
0	1
1	0
2	−5
3	−8
4	−3
5	16

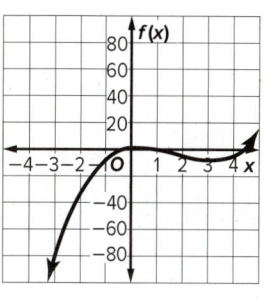

13b. at $x = 1$, between -1 and 0, and between $x = 4$ and $x = 5$ **13c.** rel. max: $x \approx \frac{1}{3}$, rel. min: $x = 3$

15a.

x	f(x)
−4	−176
−3	−77
−2	−22
−1	1
0	4
1	−1
2	−2
3	13
4	56

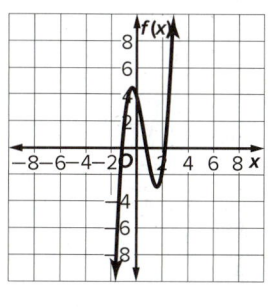

15b. between $x = -2$ and $x = -1$, between $x = 0$ and $x = 1$, and between $x = 2$ and $x = 3$

15c. rel. max: near $x = -0.3$; rel. min: near $x = 1.6$

17a.

x	f(x)
−4	372
−3	141
−2	36
−1	−3
0	−12
1	−3
2	36
3	141
4	372

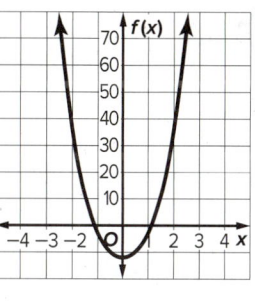

17b. between $x = -2$ and $x = -1$ and between $x = 1$ and $x = 2$ **17c.** min: near $x = 0$ **19.** rel. max: $x = -2.73$; rel. min: $x = 0.73$ **21.** rel. max: $x = 1.34$, no rel. min

23. Sample answer:

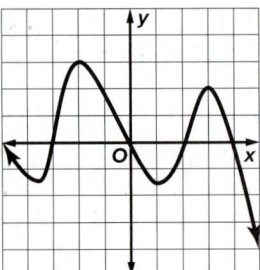

25. Sample answer:

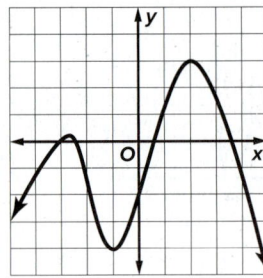

27. Sample answer:

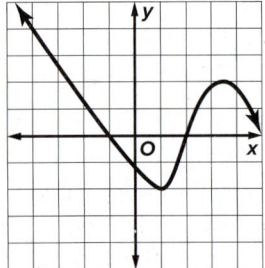

29. a.

x	d(x)
0	0
1	0.0145
2	0.056
3	0.1215
4	0.208
5	0.3125
6	0.432
7	0.5635
8	0.704
9	0.8505
10	1

b. Plot the points in the table and connect with a smooth curve.

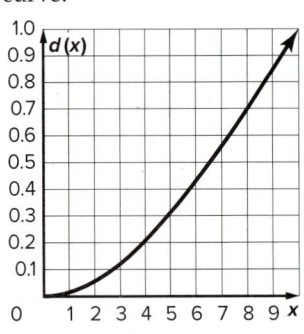

c. $d(x) \to +\infty$ as $x \to +\infty$; as x increases, $d(x)$ increases.

d. Sample answer: Because the diving board is only 10 feet long, x cannot be greater than 10. So, this trend cannot continue indefinitely.

31a. −2.5 (min), −0.5 (max), 1.5 (min) **31b.** −3.5, −1, 0, 3 **31c.** 4 **31d.** D = {all real numbers}; R = {$y \mid y \geq -3.1$} **33a.** −3.5 (min), −2.5 (max), −2 (min), −1 (max), 1 (min)
33b. −3.75, −3.25, −2, −1.75, −0.25, 2.9 **33c.** 6
33d. D = {all real numbers}; R = {$y \mid y \geq -5$}
35a. −2 (max), 1 (min) **35b.** −3, −0.5, 2 **35c.** 3
35d. D = {all real numbers}; R = {all real numbers}

37.

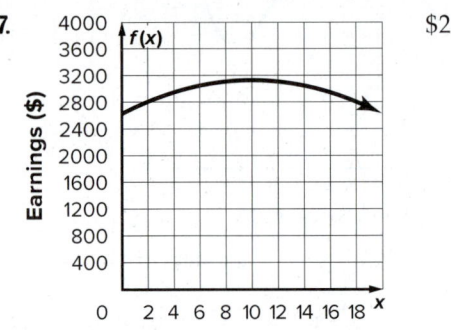

39a. zeros: $x \approx -1.73$ and 1.73; x-intercept: ≈ -1.73 and 1.73; y-intercept: 0; turning points: $x \approx -1.22$ and 1.22
39b. axis of symmetry: $x = 0$
39c. increasing: $x < -1.22$ and $0 < x < 1.22$; decreasing $-1.22 < x < 0$ and $x > 1.22$

41 a. Make a table of values and graph the function.

x	f(x)
−2	−16
−1.5	−4.5
−1	−4
−0.5	−4.6
0	−4
0.5	−3.3
1	−4
1.5	−3.5
2	8

← change in sign

The value of $f(x)$ changes signs between $x = 1.5$ and $x = 2$. So, there is a zero between these values, at approximately 1.75. The x-intercept ≈ 1.75. The y-intercept is at −4. The value of $f(x)$ at $x \approx -1.25$ and at $x \approx 0.5$ is greater than the surrounding points, so $x \approx -1.25$ and $x \approx 0.5$ are turning points. The value of $f(x)$ at $x \approx -0.5$ and at $x \approx 1.25$ is less than the surrounding points, so $x \approx -0.5$ and $x \approx 1.25$ are turning points.
b. The function does not have an axis of symmetry because the graph is not a parabola.
c. The function is increasing in the intervals $x \leq -1.25$, $-0.5 \leq x \leq 0.5$, and $x \geq 1.25$. The function is decreasing in the intervals $-1.25 \leq x \leq -0.5$ and $0.5 \leq x \leq 1.25$.

43a. no zeros, no x-intercepts, y-intercept: 5; no turning points **43b.** no axis of symmetry

43c. decreasing: $x \leq -4$; constant: $-4 < x \leq 0$; increasing: $x > 0$ **45.** As the x-values approach large positive or negative numbers, the term with the largest degree becomes more and more dominant in determining the value of $f(x)$.

47. Sample answer:

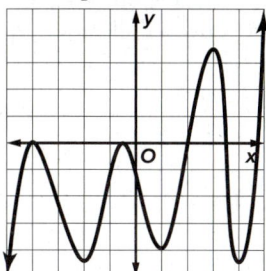

49. Sample answer: No; $f(x) = x^2 + x$ is an even degree, but $f(1) \neq f(-1)$.   **51.** Sample answer: From the degree, you can determine whether the graph is even or odd and the maximum number of zeros and turning points for the graph. You can create a table of values to help you find the approximate locations of turning points and zeros. The leading coefficient can be used to determine the end behavior of the graph, and, along with the degree, build the shape of the graph.
53. C   **55a.** C   **55b.** A, C, F   **55c.** D

Lesson 4-6

1. $(a + b)(3x + 2y - z)$ **3.** prime
5. $12q(w - q)(w^2 + qw + q^2)$
7. $x^2(a - b)(a^2 + ab + b^2)(a + b)(a^2 - ab + b^2)$
9. $(2c - 5d)(4c^2 + 10cd + 25d^2)$ **11.** $4, -4, \pm\sqrt{3}$
13. $-3, \dfrac{3 \pm 3i\sqrt{3}}{2}$ **15.** 2.5 ft   **17.** not possible
19. $\sqrt{6}, -\sqrt{6}, 2\sqrt{3}, -2\sqrt{3}$ **21.** $x(4x + y)(16x^2 - 4xy + y^2)$
23. $y^3(x^2 + y^2)(x^4 - x^2y^2 + y^4)$ **25.** prime
27. $(6x^2 - 5y^2)(2a - 3b + 4c)$
29. $8x^5 - 25y^3 + 80x^4 - x^2y^3 + 200x^3 - 10xy^3$ Original expression
   $= (8x^5 + 80x^4 + 200x^3) + (-25y^3 - x^2y^3 - 10xy^3)$
   Group to find a GCF.
   $= 8x^3(x^2 + 10x + 25) - y^3(25 + x^2 + 10x)$ Factor the GCF.
   $= 8x^3(x + 5)^2 - y^3(x + 5)^2$ Perfect squares
   $= (8x^3 - y^3)(x + 5)^2$ Distributive Property
   $= (2x - y)(4x^2 + 2xy + y^2)(x + 5)^2$ Difference of cubes
31. $6, -6, \pm 2i\sqrt{5}$ **33.** $\pm\sqrt{7}, \pm i\sqrt{13}$ **35.** $-\dfrac{1}{4}, \dfrac{1 \pm i\sqrt{3}}{8}$
37. $-15(x^2)^2 + 18(x^2) - 4$ **39.** not possible
41. $4(2x^5)^2 + 1(2x^5) + 6$ **43.** $\pm\sqrt{5}, \pm i\sqrt{2}$
45. $\pm\dfrac{2\sqrt{3}}{3}, \pm\dfrac{\sqrt{15}}{3}$ **47.** $\pm\dfrac{\sqrt{6}}{6}, \pm i\dfrac{\sqrt{3}}{2}$
49. $(x^2 + 25)(x + 5)(x - 5)$ **51.** $x(x + 2)(x - 2)(x^2 + 4)$
53. $(5x + 4y + 5z)(3a - 2b + c)$
55. $x(x + 3)(x - 3)(3x + 2)(2x - 5)$ **57.** $x = 8$; 8, 11, 5

59. $\pm\dfrac{2\sqrt{3}}{3}, \pm i\dfrac{\sqrt{2}}{2}$ **61.** $\pm\dfrac{1}{3}, \pm i\dfrac{\sqrt{10}}{2}$ **63.** $3, -3, \pm i\dfrac{\sqrt{15}}{3}$
65.
$x^6 - 26x^3 - 27 = 0$ Original equation
$(x^3)^2 - 26(x^3) - 27 = 0$ $(x^3)^2 = x^6$
$u^2 - 26u - 27 = 0$ Let $u = x^3$.
$(u - 27)(u + 1) = 0$ Factor.
$u - 27 = 0$ or $u + 1 = 0$ Zero Product Property
$x^3 - 27 = 0$ or $x^3 + 1 = 0$ Replace u with x^3.

$x^3 - 27 = 0$
$(x - 3)(x^2 + 3x + 9) = 0$ Difference of Two Cubes
$x - 3 = 0$ or $x^2 + 3x + 9 = 0$ Zero Product Property
$x = 3$ $x = \dfrac{-3 \pm \sqrt{3^2 - 4(1)(9)}}{2(1)}$
                $= \dfrac{-3 \pm 3i\sqrt{3}}{2}$

$x^3 + 1 = 0$
$(x + 1)(x^2 - x + 1) = 0$ Sum of Two Cubes
$x + 1 = 0$ or $x^2 - x + 1 = 0$ Zero Product Property
$x = -1$ $x = \dfrac{-(-1) \pm \sqrt{(-1)^2 - 4(1)(1)}}{2(1)}$
                $= \dfrac{1 \pm i\sqrt{3}}{2}$

The solutions are $-1, 3, \dfrac{-3 \pm 3i\sqrt{3}}{2}$, and $\dfrac{1 \pm i\sqrt{3}}{2}$.

67. $-1, 1, \pm\dfrac{1}{2}$ **69.** $\pm i\sqrt{5}, \pm i\sqrt{3}$ **71a.** Sample answer: First, determine how much Karen can invest each year. She allocates 7% of her income health, and retirement, so she has 93% of $50,000 or $46,500 before taxes. Her net income after taxes is $(1 - 0.28)(46,500)$ or $33,840 per year. Her annual expenses are $12(850 + 225 + 1400)$ or $29,700. Therefore, she can invest $33,840 - 29,700$ or $4140 each year.   **71b.** about 8%   **71c.** Each year, she also adds a new $4140 deposit. The table below shows the growth of each deposit.

Year	1st Deposit	2nd deposit	3rd Deposit	4th Deposit	5th Deposit
1	4140				
2	4140r	4140			
3	4140r^2	4140r	4140		
4	4140r^3	4140r^2	4140r	4140	
5	4140r^4	4140r^3	4140r^2	4140r	4140

Karen needs $(0.10)(200,000)$ or $20,000 at the end of year 5. Solve $4140r^4 + 4140r^3 + 4140r^2 + 4140r + 4140 = 20,000$ for the growth rate r. The growth rate is about 1.08, so the interest rate needs to be about 8% in order for Karen to have enough money at the end of Year 5.

71d. Karen does not get a pay increase or change jobs. Her rent or other expenses are constant. The interest rate for the CD remains constant.

73. a. $f(x) = (x + 6)[x + x + (x + 2) + (x + 2)] +$
$\qquad x[x + (x + 2) + (x + 2)] + x(x + 2)$
$\quad = (x + 6)(4x + 4) + x(3x + 4) + x(x + 2)$
$\quad = 4x^2 + 24x + 4x + 24 + 3x^2 + 4x + x^2 + 2x$
$\quad = 8x^2 + 34x + 24$

b. $f(x) = 8x^2 + 34x + 24$ Original function
$1366 = 8x^2 + 34x + 24$ Replace f(x) with 1366.
$0 = 8x^2 + 34x - 1342$ Subtract 1366 from each side.
$0 = 2(4x^2 + 17x - 671)$ Factor.
$0 = 2(4x + 61)(x - 11)$ Perfect squares
$4x + 61 = 0$ or $x - 11 = 0$ Zero Product Property
$x = -15.25$ $x = 11$ Simplify.
Because distance cannot be negative, $x = 11$ ft.

75. $(x + 2)^3(x - 2)^3$ **77.** $(x + y)^3(x - y)^3$
79. $(6x^n + 1)^2$ **81.** Sample answer: $a = 1, b = -1$
83. Sample answer: The factors can be determined by the x-intercepts of the graph. An x-intercept of 5 represents a factor of $(x - 5)$.   **85.** D   **87.** 18,600
89. Step 1: $4x, 4x$; Step 2: $-2x, 9x, 0$; Step 3: $-2x, 1, 2, 0$; Step 4: $-2x, 1, 2$; Step 5: $0, -\frac{1}{4}, -2$
91a. $\left\{-\frac{2}{3}, \frac{2}{3}, -\frac{2}{3}i, \frac{2}{3}i\right\}$
91b. $\{0, \sqrt{2}, -\sqrt{2}, \sqrt{3}, -\sqrt{3}\}$
91c. $\left\{0, 3, \frac{-3 - 3i\sqrt{3}}{2}, \frac{-3 + 3i\sqrt{3}}{2}\right\}$
93. $(x - 5)(x + 5)(x^2 + 2)$

Lesson 4-7

1. identity   **3.** identity   **5.** identity   **7.** not an identity
9. identity   **11.** identity   **13.** identity   **15.** identity

17. Multiply factors on the right side of the equation.
$(u + w)(u - w)(u^2 + uw + w^2)(u^2 - uw + w^2)$
$= (u^2 - w^2)(u^2 + uw + w^2)(u^2 - uw + w^2)$
$= (u^4 + u^3w + u^2w^2 - u^2w^2 - uw^3 - w^4)(u^2 - uw + w^2)$
$= (u^4 + u^3w - uw^3 - w^4)(u^2 - uw + w^2)$
$= u^6 - u^5w + u^4w^2 + u^5w - u^4w^2 + u^3w^3 - u^3w^3 +$
$\quad u^2w^4 - uw^5 - u^2w^4 + uw^5 - w^6$
$= u^6 - w^6$
This is equal to the expression on the left side of the equation, so the equation is an identity.

19.

	A	B	C
1	x	(x + 1)^2(x^3 − 6x^2 + 12x − 8)	(x^2 + 2x + 1)(x − 2)^3
2	0	-8	-8
3	1	-4	-4
4	2	0	0
5	3	16	16
6	4	200	200

21. Disagree; DeMarco chose values of x that are zeros of the expressions on either side of the equation, but the expressions are not equal for any other values of x.
23a. $x + y$ and $x - y$; $(x + y)(x - y)$
23b. The area of the red rectangle is equal to the area of the blue square minus the area of the small square in the lower right-hand corner; $x^2 - y^2$
23c. $x^2 - y^2 = (x + y)(x - y)$
25. Use the software to draw the graph of $y = (x + 1)^3$ and the graph of $y = x^3 + x^2 + x + 1$. The two graphs are different, which shows that the expressions on either side of the equation are not equal for all values of the variable.
27. $a = 1, b = 4, c = 6, d = 4, e = 1$
29. identity $(x + y + z)^2 = (x + y + z)(x + y + z)$
$= x^2 + xy + xz + yx + y^2 + yz + zx + zy + z^2$
$= x^2 + y^2 + z^2 + 2xy + 2xz + 2yz$
31. identity   **33a.** $x^4 + 2x^2y^2 + y^4$
33b. $x^4 + 2x^2y^2 + y^4$
33c. $x^2 - y^2$ and $2xy$ **35.** $8x$
37. This is an identity.   **39.** C   **41.** A, D

43a.

x	y	Left Side	Right Side
0	2	16	16
1	3	256	256
2	4	1296	1296
3	5	4096	4096
4	6	10,000	10,000

43b. It may be a polynomial identity.
45. B, C, D, F

Lesson 4-8

1. $58; -20$ **3.** 12,526   **5.** $x + 4, x - 4$ **7.** $x - 5, 2x - 1$
9. $71; -6$ **11.** $-435; -15$ **13.** $-4150; 85$ **15.** $647; -4$
17. $(x - 1)^2$ **19.** $x - 4, x + 1$ **21.** $x + 6, 2x + 7$
23. $x + 1, x^2 + 2x + 3$ **25.** $x - 4, 3x - 2$

27. a. Substitute the values for t into the function to determine the speed.
$f(1) = -0.001(1)^4 + 0.059(1)^3 - 1.51(1)^2 + 18.94(1) + 0.027$
$f(1) = -0.001 + 0.059 - 1.51 + 18.94 + 0.027$
$f(1) \approx 17.52$
$f(2) = -0.001(2)^4 + 0.059(2)^3 - 1.51(2)^2 + 18.94(2) + 0.027$
$f(2) = -0.016 + 0.472 - 6.04 + 37.88 + 0.027$
$f(2) \approx 32.32$
$f(3) = -0.001(3)^4 + 0.059(3)^3 - 1.51(3)^2 + 18.94(3) + 0.027$
$f(3) = -0.081 + 1.593 - 13.59 + 56.82 + 0.027$
$f(3) \approx 44.77$

b.
```
 8 | −0.001   0.059   −1.51    18.94    0.027
   |         −0.008   0.408   −8.816   80.992
   | −0.001   0.051  −1.102   10.124   81.019
```

The speed is about 81.02 ft/s. Assuming the first buoy is where the boat was at rest, the boat is traveling about 81 feet per second when it passes the second buoy. This speed converts to about 55 miles per hour. This type of acceleration may be possible for some watercraft.

29. $x + 2, x - 3, x^2 - x + 4$

31a. $g(x) = -9x^4 + 50x^3 + 51x^2 - 150x - 72$

31b.

x	g(x)
−5	−9922
−4	−4160
−3	−1242
−2	−112
−1	70
0	−72
1	−130
2	88
3	558
4	1040
5	1078
6	0

31c. There is a zero between $x = -2$ and $x = -1$ because $g(x)$ changes sign between the two values. There are also zeros between -1 and 0 and between $x = 1$ and $x = 2$ because $g(x)$ changes sign between the two values. There is also a zero at $x = 6$.

31d.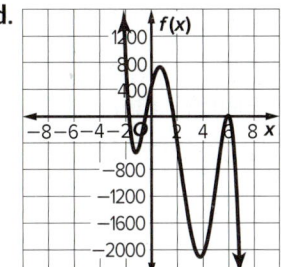

33.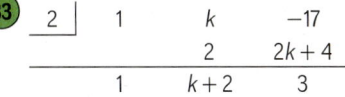

$-17 + 2k + 4 = 3$ Write an equation.
$-13 + 2k = 3$ Simplify.
$2k = 16$ Add 13 to each side.
$k = 8$ Divide each side by 2.

35. -3 **37.** $\pm\sqrt{6}, \pm\sqrt{3}$

39a. $x - c$ is a factor of $f(x)$.
39b. $x - c$ is not a factor of $f(x)$.   **39c.** $f(x) = x - c$
41. Sample answer: $f(x) = -x^3 + x^2 + x + 10$
43. Sample answer: A zero can be located using the Remainder Theorem and a table of values by determining when the output, or remainder, is equal to zero. For instance, if $f(6)$ leaves a remainder of 2 and $f(7)$ leaves a remainder of -1, then you know that there is a zero between $x = 6$ and $x = 7$.   **45.** B   **47.** C   **49.** A
51. 4   **53.** $c = -1, d = 3$

Lesson 4-9

1. $-2, 5$; 2 real   **3.** $-\frac{3}{2}, \frac{3}{2}, -\frac{3}{2}i, \frac{3}{2}i$; 2 real, 2 imaginary
5. 3 or 1; 0; 0 or 2   **7.** 1 or 3; 0 or 2; 0, 2, or 4
9. $-8, -2, 1$ **11.** $-4, 6, -4i, 4i$
13. $f(x) = x^3 - 9x^2 + 14x + 24$

15. $f(x) = x^4 - 3x^3 - x^2 - 27x - 90$
17. $-2, \frac{3}{2}$; 2 real   **19.** $-1, \frac{1 \pm i\sqrt{3}}{2}$; 1 real, 2 imaginary
21. $-\frac{8}{3}, 1$; 2 real   **23.** $-\frac{5}{2}, \frac{5}{2}, -\frac{5}{2}i, \frac{5}{2}i$; 2 real, 2 imaginary
25. $-2, -2, 0, 2, 2$; 5 real

27. Find the number of sign changes for $f(x)$ and $f(-x)$.
$f(x) = x^4 - 5x^3 + 2x^2 + 5x + 7$
    yes   yes   no   no
Because there are 2 sign changes, the function has 0 or 2 positive real zeros.
$f(-x) = x^4 + 5x^3 + 2x^2 - 5x + 7$
    no   no   yes   yes
Because there are 2 sign changes, the function has 0 or 2 negative real zeros.
Because $f(x)$ has degree 4, the function has 4 zeros. So, the function could have the following.
2 positive real zeros, 2 negative real zeros, 0 imaginary zeros
2 positive real zeros, 0 negative real zeros, 2 imaginary zeros
0 positive real zeros, 0 negative real zeros, 4 imaginary zeros

29. 0 or 2; 1, 2 or 4   **31.** 0 or 2; 0 or 2; 2, 4, or 6
33. $-6, -2, 1$
35. $-4, 7, -5i, 5i$ **37.** $4, 4, -2i, 2i$
39. $f(x) = x^3 - 2x^2 - 13x - 10$
41. $f(x) = x^4 + 2x^3 + 5x^2 + 8x + 4$
43. $f(x) = x^4 - x^3 - 20x^2 + 50x$

45a. Find the number of sign changes for $P(x)$ and $P(-x)$.
$P(x) = -0.006x^4 + 0.15x^3 - 0.05x^2 - 1.8x$
    yes   yes   no
Because there are 2 sign changes, the function has 0 or 2 positive real zeros.
$P(-x) = -0.006x^4 - 0.15x^3 - 0.05x^2 + 1.8x$
    no   no   yes
Because there is 1 sign change, the function has 1 negative real zero.
Because $f(x)$ has degree 4, the function has 4 zeros. So, the function could have the following.
2 positive real zeros, 1 negative real zero, 1 imaginary zero
0 positive real zeros, 1 negative real zero, 3 imaginary zeros

b. Nonnegative roots represent numbers of batches of frozen yogurt produced per day which lead to no profit for the manufacturer.

47. Because $f(x)$ is of degree 3, there are at most 3 zeros. For $f(x)$, there is one sign change, so there are 1 or 0 positive real zeros. For $f(-x)$, there are two sign changes, so there are 2 or 0 negative real zeros. You can find the zeros by factoring, dividing, or using synthetic division.

$f(x) = 4x^3 + 2x^2 - 4x - 2$
$= 2x^2(2x + 1) - 2(2x + 1)$
$= (2x^2 - 2)(2x + 1)$
$= 2(x^2 - 1)(2x + 1)$

The zeros are -1, $-\frac{1}{2}$, and 1. Plot all three zeros on a coordinate plane. You need to determine what happens to the graph between the zeros and at its extremes. Use a table to analyze values close to the zeros. Use these values to complete the graph.

x	−2	−$\frac{3}{4}$	0	3
y	−18	0.4375	−2	112

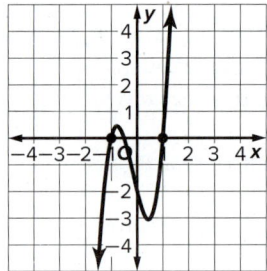

49.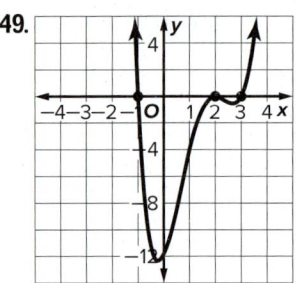

51. b **53a.** no more than 2.244%
53b. Sample answer: Interest is calculated before the payment is made, so the balance after the first month is $5000(1 + r) - 900$. Follow the pattern in the table to determine the balance after 6 months.

Month	Balance
1	$5000(1 + r) - 900$
2	$5000(1 + r)^2 - 900(1 + r) - 900$
3	$5000(1 + r)^3 - 900(1 + r)^2 - 900(1 + r) - 900$
4	$5000(1 + r)^4 - 900(1 + r)^3 - 900(1 + r)^2 - 900(1 + r) - 900$
5	$5000(1 + r)^5 - 900(1 + r)^4 - 900(1 + r)^3 - 900(1 + r)^2 - 900(1 + r) - 900$
6	$5000(1 + r)^6 - 900(1 + r)^5 - 900(1 + r)^4 - 900(1 + r)^3 - 900(1 + r)^2 - 900(1 + r) - 900$

Solve $5000(1 + r)^6 - 900(1 + r)^5 - 900(1 + r)^4 - 900(1 + r)^3 - 900(1 + r)^2 - 900(1 + r) - 900 = 0$. The monthly percentage rate r needs to be less than 2.244%. **55.** 1 positive, 2 negative, 2 imaginary; Sample answer: The graph crosses the positive x-axis once, and crosses the negative x-axis twice. Because the degree of the polynomial is 5, there are $5 - 3$ or 2 imaginary zeros.

57. Sample answer: $f(x) = (x + 2i)(x - 2i)(3x + 5)(x + \sqrt{5})(x - \sqrt{5})$; Use conjugates for the imaginary and irrational values.
59a. Sample answer: $f(x) = x^4 + 4x^2 + 4$
59b. Sample answer: $f(x) = x^3 + 6x^2 + 9x$ **61.** C **63.** C
65. $3i$ **67.** E **69a.** 1 **69b.** 4, 2, or 0 **69c.** $f(-2) = 0$
71a. 1 **71b.** 1 **71c.** 2 or 0

Chapter 4 Study Guide and Review

1. true **3.** false; depressed polynomial
5. true **7.** true
9. For a polynomial function $P(x)$, the number of positive real zeros is the same as the number of changes in sign of the coefficients of the terms of $P(x)$, or less than this by an even number, and the number of negative real zeros is the same as the number of changes in sign of the coefficients of the terms of $P(-x)$, or less than this by an even number.
11. $\frac{7x}{y^4}$
13. $r^2 + 8r - 5$
15. $m^3 - m^2p - mp^2 + p^3$
17. $a^3 + 3a^2b + 3ab^2 + b^3$
19. $-32z^5 + 240z^4 - 720z^3 + 1080z^2 - 810z + 243$
21. $x^5 - \frac{5}{4}x^4 + \frac{5}{8}x^3 - \frac{5}{32}x^2 + \frac{5}{256}x - \frac{1}{2024}$
23. $193{,}536x^2y^5$
25. $3x^3 + 2x^2y^2 - 4xy$
27. $a^3 + 3a^2 - 4a + 2$
29. $x^2 + 3x - 40$ units^2
31. This is not a polynomial in one variable. It has two variables, x and y.
33. $p(-2) = -3$; $p(x + h) = x^2 + 2xh + h^2 + 2x + 2h - 3$
35. $p(-2) = -25$; $p(x + h) = 3 - 5x^2 - 10xh - 5h^2 + x^3 + 3hx^2 + 3h^2x + h^3$
37a.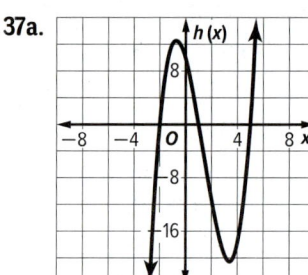

37b. The zeros are at -2, 1, and 5.
37c. rel. max: $x \approx -0.69$; rel. min: $x \approx 3.36$
39a.

39b. zeros at -2, 2, and 3

39c. rel. max: $x \approx -0.53$; rel. min: $x \approx 2.53$

41a.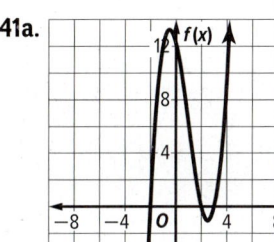

41b. between -1 and 0
41c. rel. max: $x \approx 0$; rel. min: $x \approx 0.80$
43. $(a - 2)(a + 2)(a^2 + 4)$
45. $2y(3x - 2y)(9x^2 + 6xy + 4y^2)$
47. $-7, 0, 5$
49. $x = 6$, length $= 9$ in., height $= 5$ in., width $= 7$ in.
51. $(a + b)^3 - (a^3 + b^3) = a^3 + 3a^2b + 3ab^2 + b^3 - a^3 - b^3$
$= 3a^2b + 3ab^2$
$= 3ab(a + b)$
53. $w(w + 2) + 3(w - 2) = w^2 + 2w + 3w - 6$
$= w^2 + 5w - 6$
$= (w + 6)(w - 1)$
55. $f(-2) = 18$; $f(4) = 0$
57. $f(-2) = 57$; $f(4) = 321$
59. $x^2 + 3x + 1$
61. positive real zeros: 3 or 1
negative real zeros: 0
imaginary zeros: 2 or 0
63. positive real zeros: 3 or 1
negative real zeros: 1
imaginary zeros: 4 or 2
65. positive real zeros: 2 or 0
negative real zeros: 2 or 0
imaginary zeros: 6, 4, or 2

CHAPTER 5
Inverses and Radical Functions

Chapter 5 Get Ready

1. Roots are the x-coordinates where the graph crosses the x-axis. **3.** 1 and between 0 and 1 **5.** Distributive Property **7.** Distributive Property **9.** $a = 4, b = 7, c = 3$

Lesson 5-1

1. $(f + g)(x) = 4x + 1$; $(f - g)(x) = -2x + 3$;
$(f \cdot g)(x) = 3x^2 + 5x - 2$; $\left(\dfrac{f}{g}\right)(x) = \dfrac{x + 2}{3x - 1}, x \neq \dfrac{1}{3}$

3a. $C(x) = 30.25x + 200$ **3b.** $P(x) = x + 3$
3c. $\left(\dfrac{C}{P}\right)(x) = \dfrac{39.25x + 200}{x + 3}$; this represents the dollar amount that each photographer will contribute.
3d. $46.73

5. $(f + g)(x) = 6x - 3$; $(f - g)(x) = -4x + 1$;
$(f \cdot g)(x) = 5x^2 - 7x + 2$; $\left(\dfrac{f}{g}\right)(x) = \dfrac{x - 1}{5x - 2}, x \neq \dfrac{2}{5}$

7.
$(f + g)(x) = f(x) + g(x)$ — Addition of functions
$= (3x) + (-2x + 6)$ — $f(x) = 3x$ and $g(x) = -2x + 6$
$= x + 6$ — Simplify.

$(f - g)(x) = f(x) - g(x)$ — Subtraction of functions
$= (3x) - (-2x + 6)$ — Substitution
$= 5x - 6$ — Simplify.

$(f \cdot g)(x) = f(x) \cdot g(x)$ — Multiplication of functions
$= (3x)(-2x + 6)$ — Substitution
$= -6x^2 + 18x$ — Simplify.

$\left(\dfrac{f}{g}\right)(x) = \dfrac{f(x)}{g(x)}$ — Division of functions
$= \dfrac{3x}{-2x + 6}, x \neq 3$ — Substitution

9. $(f + g)(x) = x^2 + x - 5$; $(f - g)(x) = x^2 - x + 5$;
$(f \cdot g)(x) = x^3 - 5x^2$; $\left(\dfrac{f}{g}\right)(x) = \dfrac{x^2}{x - 5}, x \neq 5$

11. $(f + g)(x) = 4x^2 - 8x$; $(f - g)(x) = 2x^2 + 8x - 8$;
$(f \cdot g)(x) = 3x^4 - 24x^3 + 8x^2 + 32x - 16$; $\left(\dfrac{f}{g}\right)(x) = \dfrac{3x^2 - 4}{x^2 - 8x + 4}, x \neq 4 \pm 2\sqrt{3}$

13a. $(f + g)(x) = 375x + 10$ **13b.** $(f - g)(x) = 25x + 40$
15. $2(g \cdot f) = 2x^3 - 4x^2 - 30x + 72$

17a.

x	$f(x) = x^2$	$g(x) = x$	$(f+g)(x) = x^2 + x$	$(f-g)(x) = x^2 - x$
-3	9	-3	6	12
-2	4	-2	2	6
-1	1	-1	0	2
0	0	0	0	0
1	1	1	2	0
2	4	2	6	2
3	9	3	12	6

17b.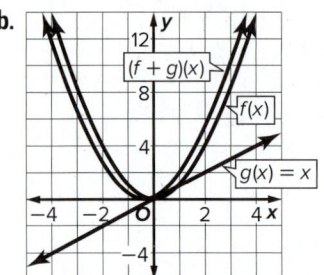

17c.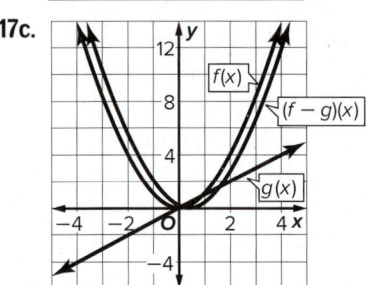

17d. Sample answer: For each value of x, the vertical distance between the graph of $g(x)$ and the x-axis is the same as the vertical distance between the graphs of $f(x)$ and $(f + g)(x)$ and between $f(x)$ and $(f - g)(x)$.

19. -2 **21.** -15 **23.** 1 **25.** 4
27. From the graph, $f(0) = 2$ and $g(0) = -1$.
$(f \cdot g)(0) = f(0) \cdot g(0) = 2 \cdot (-1) = -2$
29. -1 **31.** 1 **33.** 0 **35.** -1 **37.** 0

39a. Let $w(x)$ represent the function for women and $m(x)$ represent the function for men.
$(w + m)(x) = w(x) + m(x)$ — Addition of functions
$= (548.6x + 66{,}527)$
$\quad + (2090.7x + 62{,}243)$ — Substitution
$= 2639.3x + 128{,}770$ — Simplify.
The equation $y = 2639.3x + 128{,}770$ models the total number.
b. $(f - g)(x) = $ the number of men employed in the U.S. $-$ the number of women employed in the United States. So, the function represents the difference in the number of men and women employed in the United States.

41. 0 **43a.**

x	$(f \cdot g)(x)$	$\left(\dfrac{f}{g}\right)(x)$
-3	-60	$-\dfrac{5}{3}$
-2	-25	-1
-1	-8	$-\dfrac{1}{2}$
0	-3	$-\dfrac{1}{3}$
1	-4	-1
2	-5	-5
3	0	undef.

43b.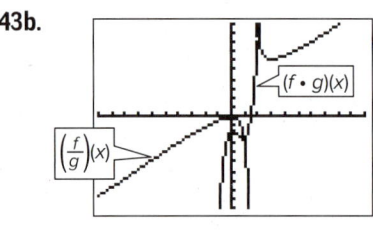

$[-20, 20]$ scl: 2 by $[-20, 20]$ scl: 2

43c. Sample answer: When x is 2 or 4, the functions are equal. **45.** Sample answer: $f(x) = 2x + 2$, $g(x) = x - 1$ **47a.** Never; because the functions are both linear functions, $(f + g)(x)$ is also a linear function and the domain is all real numbers. **47b.** Sometimes; if $f(x) = x$ and $g(x) = 5$, then $\left(\dfrac{f}{g}\right)(x) = \dfrac{x}{5}$ and the domain is all real numbers, but if $f(x) = 5$ and $g(x) = x$, then $\left(\dfrac{f}{g}\right)(x) = \dfrac{5}{x}$ and $x = 0$ is excluded from the domain. **49.** The domains of $(f + g)(x)$, $(f - g)(x)$, and $(f \cdot g)(x)$ are all real numbers because these are also polynomial functions. The domain of $\left(\dfrac{f}{g}\right)(x)$ is all real numbers except for any values of x for which $g(x) = 0$. **51.** B **53.** 76 **55.** A **57.** A

Lesson 5-2

1. 7 **3.** 103 **5.** 147 **7.** 3 **9.** 4
11. $f \circ g$ is undefined, D = ∅, R = ∅;
  $g \circ f = \{(0, 2)\}$, D = $\{0\}$, R = $\{2\}$.
13. $[f \circ g](x) = x^2 + 3x - 6$;
  $[g \circ f](x) = x^2 + 11x + 18$
15. $[f \circ g](10) = f(g(10)) = f(-2(10)^2) = f(-200) = -200 + 4 = -196$
17. 9 **19.** −50 **21.** −28 **23.** −128
25. $f \circ g = \{(6, 12)\}$, D = $\{6\}$, R = $\{12\}$; $g \circ f = \{(-7, 5), (4, 1), (-3, 8)\}$, D = $\{-7, -3, 4\}$, R = $\{1, 5, 8\}$
27. $f \circ g$ is undefined, D = ∅, R = ∅; $g \circ f = \{(-4, 9), (0, 1), (-6, 13), (2, -3)\}$, D = $\{-6, -4, 0, 2\}$, R = $\{-3, 1, 9, 13\}$.
29. $f \circ g = \{(-1, -2)\}$, D = $\{-1\}$, R = $\{-2\}$; $g \circ f$ is undefined, D = ∅, R = ∅.
31. $f \circ g$ is undefined, D = ∅, R = ∅; $g \circ f$ is undefined, D = ∅, R = ∅.
33. $f \circ g = \{(-2, 3), (-4, -1)\}$, D = $\{-4, -2\}$, R = $\{-1, 3\}$; $g \circ f = \{(12, -1), (9, 6), (8, 5)\}$, D = $\{8, 9, 12\}$, R = $\{-1, 5, 6\}$
35–41. D = {all real numbers}
35. $[f \circ g](x) = f(g(x)) = f(x + 5) = 2(x + 5) = 2x + 10$;
  $[g \circ f](x) = g(f(x)) = g(2x) = 2x + 5$
  Both compositions result in linear functions, so D = R = {all real numbers}.
37. $[f \circ g](x) = x^2 - 14$, R = $\{y \mid y \geq -14\}$;
  $[g \circ f](x) = x^2 - 8x + 6$, R = $\{y \mid y \geq -10\}$
39. $[f \circ g](x) = 32x^2 + 44x + 16$, R = $\{y \mid y \geq 0.875\}$;
  $[g \circ f](x) = 8x^2 - 4x + 7$, R = $\{y \mid y \geq 6.5\}$
41. $[f \circ g](x) = x^4 + 3x^2 + 1$, R = $\{y \mid y \geq 1\}$;
  $[g \circ f](x) = x^4 + 6x^3 + 11x^2 + 6x + 1$, R = $\{y \mid y \geq 0\}$
43a. $p(x) = 0.65x$; $t(x) = 1.0625x$
43b. Because $[p \circ t](x) = [t \circ p](x)$, either function represents the price.
43c. $1587.75 **45.** $5a^2 + 70a + 240$ **47.** 2 **49.** −4
51. −2 **53.** −2 **55.** −5 **57.** −3
59. $[f \circ (g \circ h)](2) = f(g \circ h)(2)) = f(g(h(2))) = f(g(2^2 - 2(2) + 1)) = f(g(1)) = f(-4(1) + 3) = f(-1) = -1 + 2 = 1$

61. 256
63.

x	f(x)	g(x)
1	0	1
2	2	4
3	4	7
4	6	10
5	9	13

65. $am^2x^2 + 2amxd + ad^2 + bmx + bd + c$
67. Each composition will have power pq. $(f \circ g)(x) = x^{qp} = x^{pq} (g \circ f)(x)$. **69.** Neither is correct; Denise multiplied the functions instead of finding the composition; Keiko did not apply the Distributive Property correctly; $[f \circ g](x) = 12x - 1$ **71a.** Always, because the range is dependent on the domain, the domain of $g[f(x)]$ is restricted by the domain of $f(x)$.
71b. Sometimes; when $f(x) = 4x$ and $g(x) = \sqrt{x}$, $g[f(x)] = \sqrt{4x}$, $x \geq 0$. The domain of $g(x)$ restricts the domain of $g[f(x)]$. When $f(x) = 4x^2$ and $g(x) = \sqrt{x}$, $g[f(x)] = \sqrt{4x^2}$. In this case, the domain of $g(x)$ does not restrict the domain of $g[f(x)]$.
73. C **75.** A, B, F **77.** 27 **79.** B
81a. $s(x) = x + 6.75$; $t(x) = 1.08x$
81b. $[s \circ t](x) = 1.08x + 6.75$; this represents the final cost of an order if sales tax is applied before the shipping fee.
81c. $[t \circ s](x) = 1.08x + 7.29$; this represents the final cost of an order if the shipping fee is applied before the sales tax.
81d. After; for any value of x, $1.08x + 6.75$ is less than $1.08x + 7.29$.

Lesson 5-3

1. $\{(10, -9), (-3, 1), (-5, 8)\}$
3. $f^{-1}(x) = -\dfrac{1}{3}x$

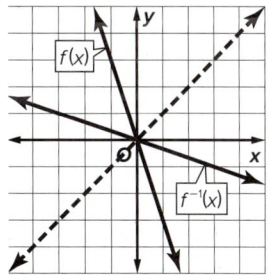

5. $h^{-1}(x) = \pm\sqrt{x + 3}$

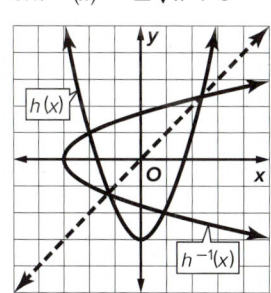

7. $(-\infty, 8]$,
$f^{-1}(x) = 8 + \sqrt{x + 1}$
and $[8, \infty)$,
$f^{-1}(x) = 8 - \sqrt{x + 1}$

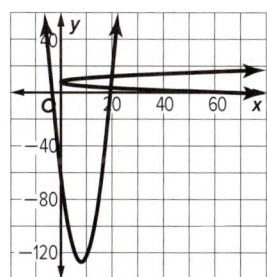

9. no **11.** $\{(6, -8), (-2, 6), (-3, 7)\}$
13. $\{(-1, 8), (-1, -8), (-8, -2), (8, 2)\}$

15. $f^{-1}(x) = x - 2$

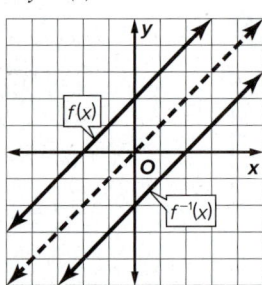

17. $f^{-1}(x) = -\frac{x}{2} + \frac{1}{2}$

19. $f^{-1}(x) = -\frac{3}{5}(x + 8)$

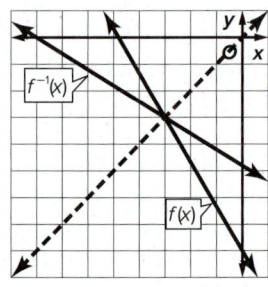

21. $h^{-1}(x) = \pm\sqrt{x - 4}$

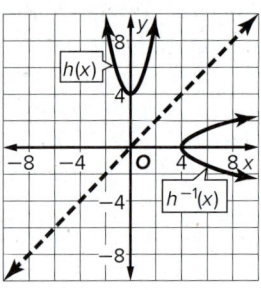

23. $f^{-1}(x) = \pm\sqrt{x - 3} - 1$

25. $(-\infty, +\infty)$, $f^{-1}(x) = -\frac{x}{8} + \frac{9}{8}$

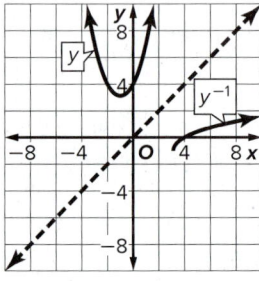
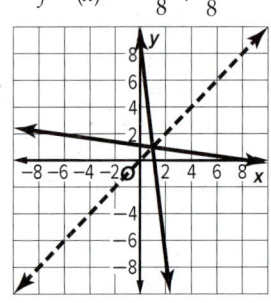

27. $(-\infty, -6]$, $f^{-1}(x) = -6 + \sqrt{x + 4}$ and $[-6, \infty)$, $f^{-1}(x) = -6 - \sqrt{x + 4}$

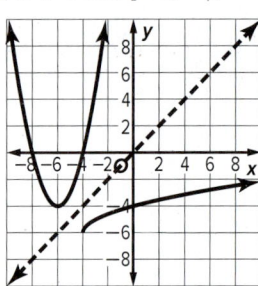

29. $(-\infty, 18]$, $f^{-1}(x) = 18 + \sqrt{x + 484}$ and $[18, \infty)$, $f^{-1}(x) = 18 - \sqrt{x + 484}$

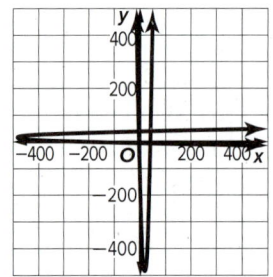

31. yes **33.** no **35.** yes

37. $[f \circ g](x) = f[g(x)]$ Composition of functions
$= f(\sqrt{x} - 6)$ Replace $g(x)$ with $\sqrt{x} - 6$.
$= [(\sqrt{x} - 6) + 6]^2$ Substitute $\sqrt{x} - 6$ for x in $f(x)$.
$= (\sqrt{x})^2$ Simplify.
$= x$ Evaluate.

$[g \circ f](x) = f[f(x)]$ Composition of functions
$= g[(x + 6)^2]$ Replace $f(x)$ with $(x + 6)^2$.
$= \sqrt{(x + 6)^2} - 6$ Substitute $(x + 6)^2$ for x in $g(x)$.
$= (x + 6) - 6$ Simplify the radical.
$= x$ Simplify.

The functions are inverses because $[f \circ g](x) = [g \circ f](x)$.

39a. $c(g) = 3.67g$ **39b.** $c(m) \approx 0.131m$

41. a. $A = \pi r^2$ Write the formula
$y = \pi x^2$ Write the formula using x and y.
$x = \pi y^2$ Exchange x and y in the equation.
$\frac{x}{\pi} = y^2$ Divide each side by π.
$\sqrt{\frac{x}{\pi}} = y$ Take the positive square root of each side.

So, replacing y with r and x with A, the inverse is $r = \sqrt{\frac{A}{\pi}}$.

b. $r = \sqrt{\frac{A}{\pi}}$ Write the inverse of the function.
$= \sqrt{\frac{36}{\pi}}$ $A = 36$
≈ 3.39 cm Simplify.

43. yes **45.** no **47.** no **49a.** $F^{-1}(x) = \frac{5}{9}(x - 32)$;

$F[F^{-1}(x)] = \frac{9}{5}\left[\frac{5}{9}(x - 32)\right] + 32 = x - 32 + 32 = x$;

$F^{-1}[F(x)] = \frac{5}{9}\left(\frac{9}{5}x + 32 - 32\right) = \frac{5}{9}\left(\frac{9}{5}x + 0\right) = x$.

49b. It can be used to convert Fahrenheit to Celsius.

51a.

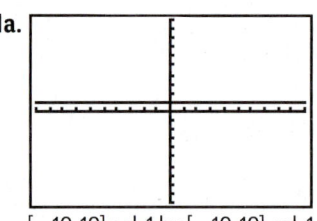

[−10, 10] scl: 1 by [−10, 10] scl: 1

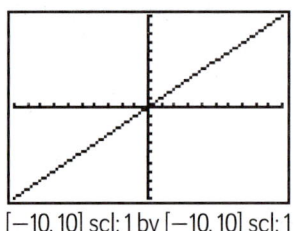
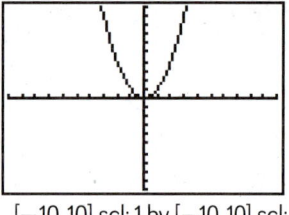

[−10, 10] scl: 1 by [−10, 10] scl: 1   [−10, 10] scl: 1 by [−10, 10] scl: 1

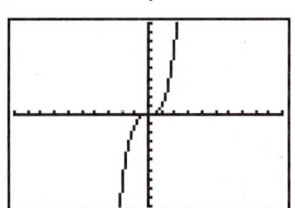
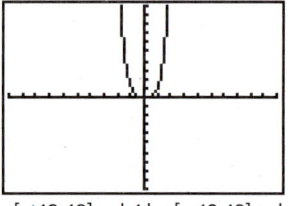

[−10, 10] scl: 1 by [−10, 10] scl: 1   [−10, 10] scl: 1 by [−10, 10] scl: 1

51b.

Function	Inverse a function?
$y = x^0$ or $y = 1$	no
$y = x^1$ or $y = x$	yes
$y = x^2$	no
$y = x^3$	yes
$y = x^4$	no

51c. n is odd.

53. Sample answer: $f(x) = 2x$, $f^{-1}(x) = 0.5x$; $f[f^{-1}(x)] = f^{-1}[f(x)] = x$

55. $y = \frac{1}{m}x - \frac{b}{m}$ **57.** A **59.** C **61.** B, D, E

63. $f^{-1}(x) = 4x^2$

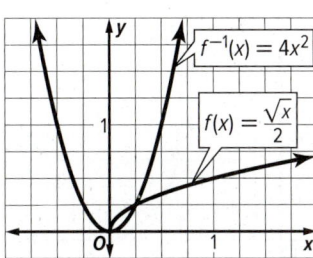

Lesson 5-4

1. D = $\{x \mid x \geq 0\}$; R = $\{f(x) \mid f(x) \geq 0\}$

3. D = $\{x \mid x \geq -8\}$; R = $\{f(x) \mid f(x) \geq -2\}$

5.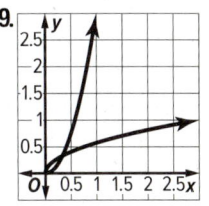
D = $\{x \mid x \geq 1\}$, $[1, \infty)$, or $\{x \geq 1\}$; R = $\{f(x) \mid f(x) \geq 0\}$, $[0, \infty)$, or $(f(x) \geq 0)$

7.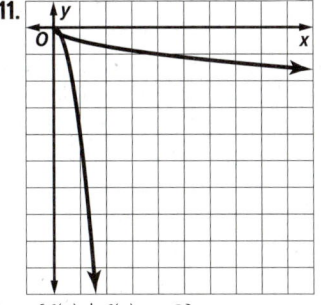
D = $\{x \mid x \geq \frac{5}{3}\}$, $[\frac{5}{3}, \infty)$, or $\{x \geq \frac{5}{3}\}$; R = $\{f(x) \mid f(x) \leq 5\}$, $(-\infty, 5]$, or $(f(x) \leq 5)$

9.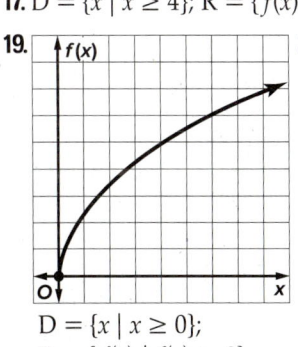

11.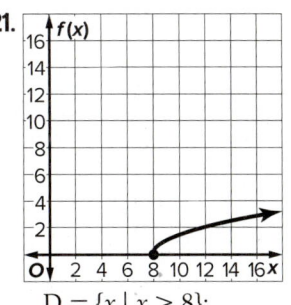

13. D = $\{x \mid x \geq 0\}$; R = $\{f(x) \mid f(x) \leq 2\}$

15. The domain only includes values for which the radicand is nonnegative.
$x - 2 \geq 0$ Write an inequality.
$x \geq 2$ Add 2 to each side.
The domain is $\{x \mid x \geq 2\}$. Find $f(2)$ to find the lower limit of the range.
$f(2) = 4\sqrt{2 - 2} - 8 = 0 - 8$ or -8
The range is $\{f(x) \mid f(x) \geq -8\}$.

17. D = $\{x \mid x \geq 4\}$; R = $\{f(x) \mid f(x) \geq -6\}$

19.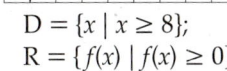
D = $\{x \mid x \geq 0\}$; R = $\{f(x) \mid f(x) \geq 0\}$

21. D = $\{x \mid x \geq 8\}$; R = $\{f(x) \mid f(x) \geq 0\}$

23.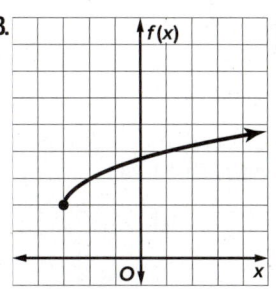
D = $\{x \mid x \geq -3\}$; R = $\{f(x) \mid f(x) \geq 2\}$

25.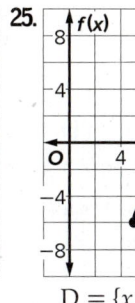
D = $\{x \mid x \geq 5\}$; R = $\{f(x) \mid f(x) \geq -6\}$

27.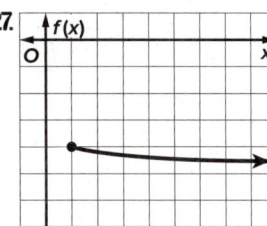
D = $\{x \mid x \geq 1\}$; R = $\{f(x) \mid f(x) \leq -4\}$

29. 1936 ft

31. $f^{-1}(x) = \sqrt{\frac{1}{2}x}, x \geq 0$
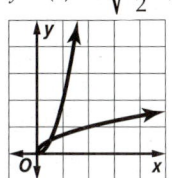

33. $f^{-1}(x) = \frac{1}{2}\sqrt{x}, x \geq 0$
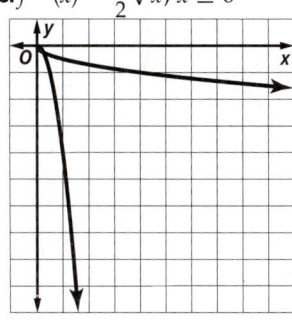

35. $f^{-1}(x) = -\sqrt{2x}, x \geq 0$
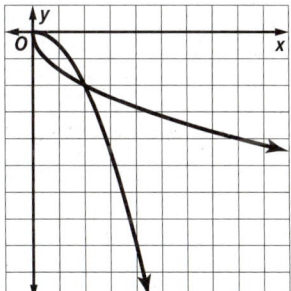

37. $f^{-1}(x) = \frac{1}{3}\sqrt{x + 4}, x \geq -4$
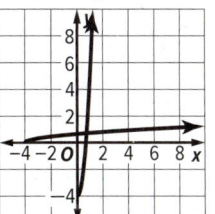

39a. $v = \sqrt{\frac{2E}{m}}$ **39b.** about 36.5 m/s **39c.** about 85,135 m/s **41.** $y = \sqrt{x - 4} - 6$; h is 4 and k is -6, use the equation $y = \sqrt{x - h} + k$ to get $y = \sqrt{x - 4} - 6$.

43. $y = -\sqrt{x + 6} - 6$

a. $T = 2\pi\sqrt{\dfrac{L}{g}}$ Original function

$T = 2\pi\sqrt{\dfrac{L}{32}}$ Replace g with 32.

Make a table of values for $0 \leq L \leq 10$. Graph the points and connect with a smooth curve.

L	T.
0	0
1	1.11
2	1.57
3	1.92
4	2.22
5	2.48
6	2.72
7	2.94
8	3.14
9	3.33
10	3.51

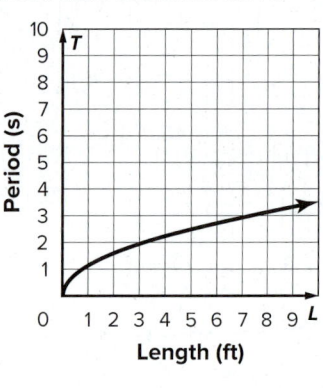

b. Use the table that you made in part **a**. (2, 1.57) means the period for a pendulum 2 feet long is about 1.57 seconds. (5, 2.48) means the period for a pendulum 5 feet long is about 2.48 seconds. (8, 3.14) means the period for a pendulum 8 feet long is about 3.14 seconds.

47. Sample answer: $y = -\sqrt{x+4} + 6$
49. Sample answer: $y = -\sqrt{x-8} + 14$
51. Molly; $y = \sqrt{5x+10}$ has an x-intercept of -2 and would be at the right of the given graph.
53a. Sample answer: The original is $y = x^2 + 2$ and inverse is $y = \pm\sqrt{x-2}$. **53b.** Sample answer: The original is $y = \pm\sqrt{x} + 4$ and inverse is $y = (x-4)^2$. **55a.** D **55b.** $\{f(x) \mid f(x) \geq -2\}$
57. B **59.** A, B

Lesson 5-5

1. $D = \{x \mid -\infty < x < +\infty\}$ or $(-\infty, +\infty)$; $R = \{f(x) \mid -\infty < f(x) < +\infty\}$ or $(-\infty, +\infty)$; End behavior: $f(x) \to +\infty$ as $x \to +\infty$ and $f(x) \to -\infty$ as $x \to -\infty$; Inflection point: (4, 0)
3. $D = \{x \mid -\infty < x < +\infty\}$ or $(-\infty, +\infty)$; $R = \{f(x) \mid -\infty < f(x) < +\infty\}$ or $(-\infty, +\infty)$; End behavior: $f(x) \to +\infty$ as $x \to +\infty$ and $f(x) \to -\infty$ as $x \to -\infty$; Inflection point: (0, −6)
5.

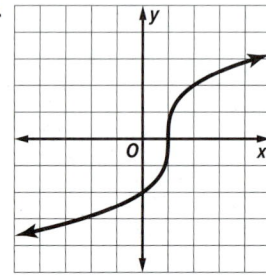

Graph of $f(x) = \sqrt[3]{x}$ stretched vertically by a factor of 2 and translated 1 unit right.
$D = \{x \mid -\infty < x < +\infty\}$ or $(-\infty, +\infty)$
$R = \{x \mid -\infty < f(x) < +\infty\}$ or $(-\infty, +\infty)$;
end behavior: $f(x) \to +\infty$ as $x \to +\infty$
$f(x) \to -\infty$ as $x \to -\infty$;
Inflection point: (1, 0)

7.

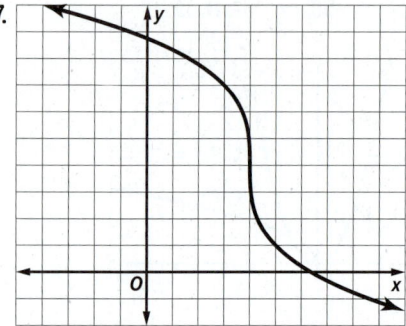

Graph of $f(x) = \sqrt[3]{x}$ stretched vertically by a factor of 3 and translated 4 units right and 4 units up; reflected in the line $x = 4$.
$D = \{x \mid -\infty < x < +\infty\}$ or $(-\infty, +\infty)$
$R = \{x \mid -\infty < f(x) < +\infty\}$ or $(-\infty, +\infty)$;
end behavior: $f(x) \to -\infty$ as $x \to +\infty$
$f(x) \to +\infty$ as $x \to -\infty$;
Inflection point: (4, 4)

9. 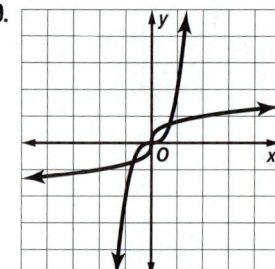 **11.**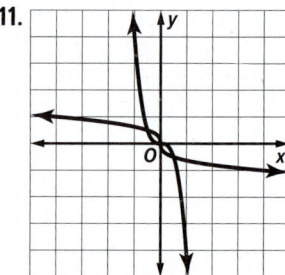

13. $D = \{x \mid -\infty < x < +\infty\}$ or $(-\infty, +\infty)$; $R = \{f(x) \mid -\infty < f(x) < +\infty\}$ or $(-\infty, +\infty)$; End behavior: $f(x) \to +\infty$ as $x \to +\infty$ and $f(x) \to -\infty$ as $x \to -\infty$; Inflection point: (−4, 0)
15. $D = \{x \mid -\infty < x < +\infty\}$ or $(-\infty, +\infty)$; $R = \{f(x) \mid -\infty < f(x) < +\infty\}$ or $(-\infty, +\infty)$; End behavior: $f(x) \to +\infty$ as $x \to +\infty$ and $f(x) \to -\infty$ as $x \to -\infty$; Inflection point: (2, 1)
17.

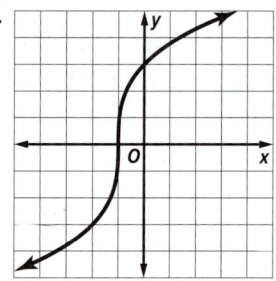

Graph of $f(x) = \sqrt[3]{x}$ stretched vertically by a factor of 3 and translated 1 unit left.
$D = \{x \mid -\infty < x < +\infty\}$ or $(-\infty, +\infty)$
$R = \{x \mid -\infty < f(x) < +\infty\}$ or $(-\infty, +\infty)$;
end behavior: $f(x) \to +\infty$ as $x \to +\infty$
$f(x) \to -\infty$ as $x \to -\infty$;
Inflection point: (−1, 0)

19.

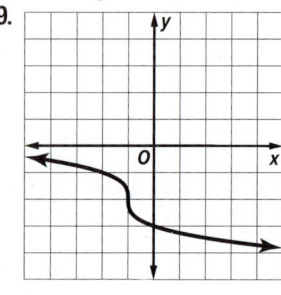

Graph of $f(x) = \sqrt[3]{x}$ and translated 1 unit left, 2 units down, and reflected in the line $x = -1$. $D = \{x \mid -\infty < x < +\infty\}$ or $(-\infty, +\infty)$; $R = \{f(x) \mid -\infty < f(x) < +\infty\}$ or $(-\infty, +\infty)$; End behavior: The values of y decrease as the values of x increase. Inflection point: (−1, −2)

21.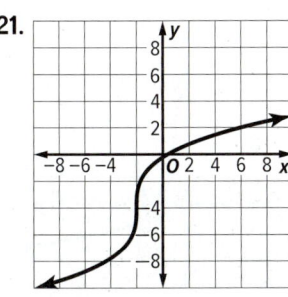

Graph of $f(x) = \sqrt[3]{x}$ stretched vertically by a factor of 3 and translated 2 units left and 4 units down. D = {x | $-\infty < x < +\infty$} or $(-\infty, +\infty)$; R = {$f(x)$ | $-\infty < f(x) < +\infty$} or $(-\infty, +\infty)$; End behavior: The values of y increase as the values of x increase. Inflection point: $(-2, -4)$

23. Make a table of values and graph the function.

x	y
−4	2
3	0
4	−2
5	−4
12	−6

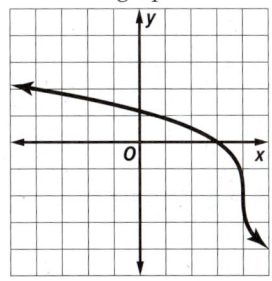

The graph is the same shape as $f(x) = \sqrt[3]{x}$, but stretched vertically by a factor of 2, translated 4 units to the right and 2 units down, and reflected in the line $x = 4$. D = {x | $-\infty < x < +\infty$} or $(-\infty, +\infty)$; R = {$f(x)$ | $-\infty < f(x) < +\infty$} or $(-\infty, +\infty)$; End behavior: The values of y decrease as the values of x increase. Inflection point: $(4, -2)$

25.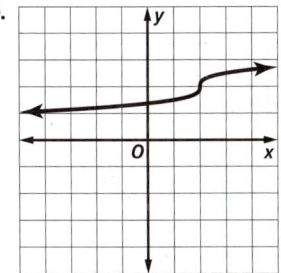

The graph is the same shape as $f(x) = \sqrt[3]{x}$, but compressed vertically by a factor of $\frac{1}{2}$, translated 2 units to the right and 2 units up. D = {x | $-\infty < x < +\infty$} or $(-\infty, +\infty)$; R = {$f(x)$ | $-\infty < f(x) < +\infty$} or $(-\infty, +\infty)$; End behavior: The values of y increase as the values of x increase. Inflection point: $(2, 2)$

27a.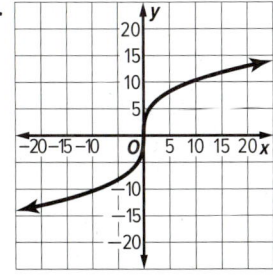

27b. 9.8 in.   **27c.** 4.6 oz

27d. Sample answer: The circumference of fruit depends on how the fruit grows, which in turn depends on many natural conditions that vary greatly, like the weather.
29a. Sample answer: The variable V is not really raised to the $\frac{1}{3}$ power to make the graph be of a standard cube root function. V is raised to the $\frac{2}{3}$ power, which gives a different graph entirely.   **29b.** about 165 in^2

31. The given graph is the graph of the parent function, $f(x) = \sqrt[3]{x}$, translated 1 unit to the left and stretched vertically be a factor of 2, so the function shown in the graph is $f(x) = 2\sqrt[3]{x+1}$.

33a.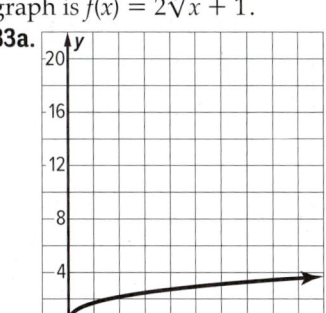

33b. D = {x | $0 \leq x < \infty$} or $[0, +\infty)$; R = {x | $0 \leq f(x) < \infty$} or $[0, +\infty)$; Because a cube cannot have a negative volume or negative side length, the domain and range include only nonnegative real numbers.
33c. 8 cm
35. Sample answer: The inverse function of the cube root of x is x cubed. and is an odd function. Therefore, the cube root of x is also an odd function.
37. Milo is correct. Sample answer: The graph of the function has an inflection point at (-2, 0), but $f(x) = \sqrt[3]{x}$, has an inflection point at the origin.
39a. C   **39b.** D = R = {$-\infty < f(x) < \infty$}
39c. A, B, C   **41.** 35.93   **43.** A

Lesson 5-6

1. 20   **3.** 13

5.
$$\sqrt[3]{x-2} = 3 \quad \text{Original equation}$$
$$(\sqrt[3]{x-2})^3 = 3^3 \quad \text{Raise each side to the third power.}$$
$$x - 2 = 27 \quad \text{Evaluate each side.}$$
$$x = 29 \quad \text{Add 2 to each side.}$$

7. 2   **9.** 49   **11.** $\frac{27}{2}$ **13a.** about 9.5 s   **13b.** about 324 feet   **15.** $-\frac{4}{3} \leq x \leq \frac{77}{3}$ **17.** $1 \leq y \leq 5$ **19.** $x > 1$
21. $x \leq -11$ **23.** 22   **25.** 3   **27.** no real solution   **29.** $\frac{1}{4}$
31. 9   **33.** $\frac{81}{16}$ **35.** 1 m   **37.** 3   **39.** 83   **41.** 61   **43.** 3
45. 18   **47.** 2   **49.** F   **51.** $x \geq 43$ **53.** no real solution

55.
$$\sqrt{d+3} + \sqrt{d+7} > 4 \quad \text{Original inequality}$$
$$\sqrt{d+3} > 4 - \sqrt{d+7} \quad \text{Subtract } \sqrt{d+7} \text{ from each side.}$$
$$d + 3 > 16 - 8\sqrt{d+7} + d + 7 \quad \text{Square each side.}$$
$$\frac{5}{2} < \sqrt{d+7} \quad \text{Simplify.}$$

$\frac{25}{4} < d + 7$ Square each side.

$-\frac{3}{4} < d$ Subtract 7 from each side.

$d > -\frac{3}{4}$ Rewrite inequality.

57. $-\frac{5}{2} \le y \le 2$ **59.** $a > 8$ **61.** $0 \le c < 3$

63. $W = \left(\frac{L}{0.55}\right)^3$

65. $A = \pi r^2$ Original formula
$250{,}000 = \pi r^2$ Replace A with 250,000.
$79{,}577.5 \approx r^2$ Divide each side by π.
$\sqrt{79{,}577.5} \approx r$ Take the square root of each side.
$282.1 \approx r$ Use a calculator.

The radius is about 282 ft.

67. $\sqrt{x+2} - 7 = -10$
69. never

$\frac{\sqrt{(x^2)^2}}{-x} = x$

$\frac{x^2}{-x} = x$

$x^2 = (x)(-x)$

$x^2 \ne -x^2$

71. They are the same number. **73.** 3
75. Sometimes; Sample answer: When the radicand is negative, then there will be extraneous roots.
77. 47.3 **79.** D **81.** $-1 < x \le 14$
83a. The expression under the root sign must be greater than zero. Then there is another part to solve the expression for x.
83b. $x \ge 11.2$, $x \le 12$ **83c.** yes

Chapter 5 Study Guide and Review

1. identity function **3.** radical equations
5. extraneous solution
7. Exchange x and y in the equation, then solve for y
9. $(f + g)(x) = x^2 + 4x + 10$
11. $(f \cdot g)(x) = 2x^3 + 13x^2 + 20x + 9$
13. $(f + g)(x) = x^3 + 10x - 8$
15. $(f \cdot g)(x) = 10x^4 - 80x$
17. $[f \circ g](x) = 8x - 9$
   $[g \circ f](x) = 8x - 1$
19. $[f \circ g](x) = 4x^2 - 4x + 5$
   $[g \circ f](x) = -2x^2 - 7$
21. $[f \circ g](x) = x^3 - 3x^2 + 3x - 1$
   $[g \circ f](x) = x^3 - 1$
23. $n = 36y$
25. $f^{-1}(x) = \frac{x+5}{-3}$

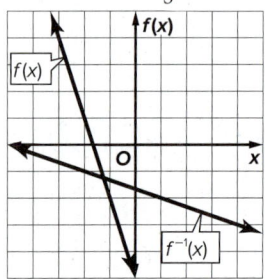

27. $f^{-1}(x) = \frac{5x-1}{4}$

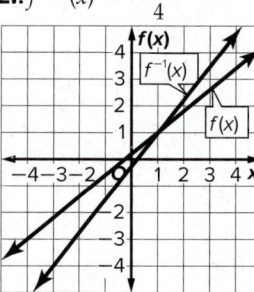

29. $f^{-1}(x) = \frac{-1 \pm \sqrt{x}}{2}$

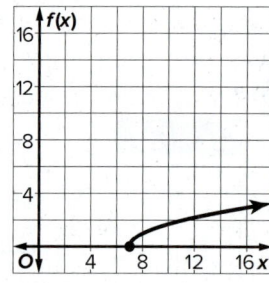

31. D = $\{x \mid x \ge 0\}$;
R = $\{f(x) \mid f(x) \ge 0\}$

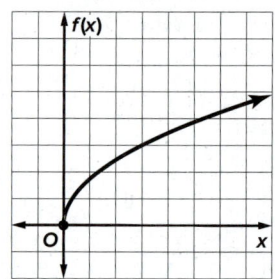

33. D = $\{x \mid x \ge 7\}$;
R = $\{f(x) \mid f(x) \ge 0\}$

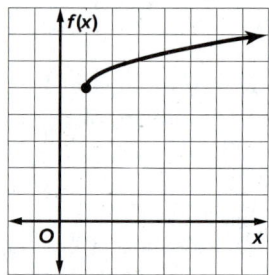

35. D = $\{x \mid x \ge 1\}$; R = $\{f(x) \mid f(x) \ge 5\}$

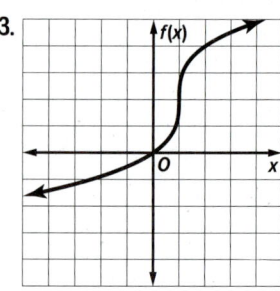

37. about 9.8 in.
39.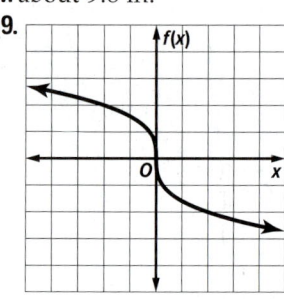
41.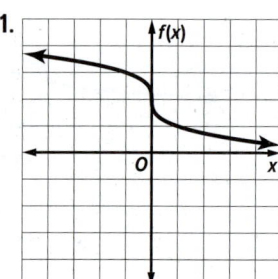

43.

45. 103 **47.** $\frac{17}{3}$ **49.** 3 **51.** 64
53. about 6.13 ft **55.** $x \ge 29$
57. $x \le -20$ **59.** $x \ge \frac{3}{2}$

R58

CHAPTER 6
Exponential and Logarithmic Functions

Chapter 6 Get Ready

1. add the exponent values together to simplify as a^{12}
3. Product of Powers Rule
5. no 7. line A is the function and line B is its inverse

Lesson 6-1

1. 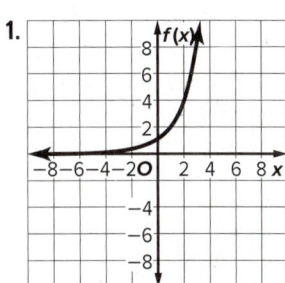 $D = (-\infty, \infty)$, {all real numbers}, or $\{-\infty < x < \infty\}$; $R = (0, \infty)$, $\{f(x) \mid f(x) > 0\}$, $R = (0, \infty)$, $\{f(x) \mid f(x) > 0\}$,

3. Make a table of values. Then plot the points, and sketch the graph.

x	$f(x) = 3^{x-2} + 4$
-2	$3^{-2-2} + 4 = 4\frac{1}{81}$
-1	$3^{-1-2} + 4 = 4\frac{1}{27}$
0	$3^{0-2} + 4 = 4\frac{1}{9}$
1	$3^{1-2} + 4 = 4\frac{1}{3}$
2	$3^{2-2} + 4 = 5$
3	$3^{3-2} + 4 = 7$
4	$3^{4-2} + 4 = 13$

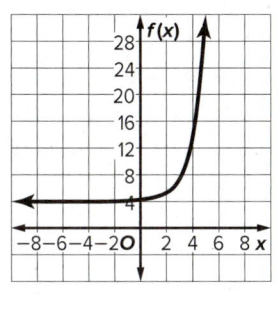

The domain is all real numbers and the range is all real numbers greater than 4.
$D = (-\infty, \infty)$, {all real numbers}, or $\{-\infty < x < \infty\}$; $R = (4, \infty)$, $\{f(x) \mid f(x) > 4\}$, or $\{4 < x < \infty\}$

5. 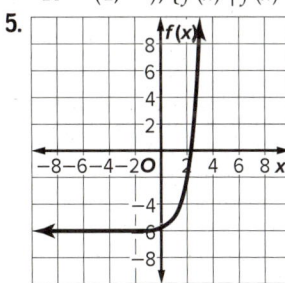 $D = (-\infty, \infty)$, {all real numbers}, or $\{-\infty < x < \infty\}$; $R = (-6, \infty)$, $\{f(x) \mid f(x) > -6\}$, or $\{-6 < x < \infty\}$

7.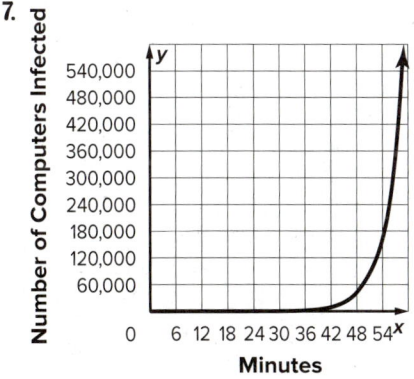

9.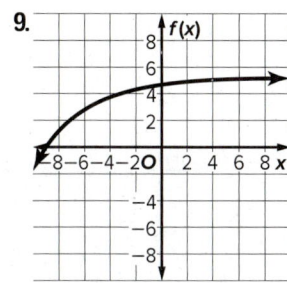
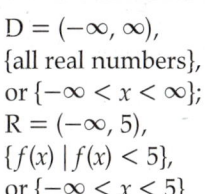
$D = (-\infty, \infty)$, {all real numbers}, or $\{-\infty < x < \infty\}$; $R = (-\infty, 5)$, $\{f(x) \mid f(x) < 5\}$, or $\{-\infty < x < 5\}$

11.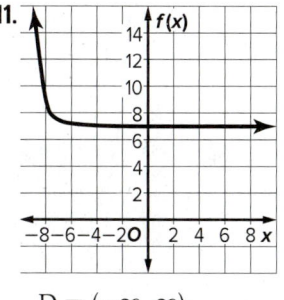
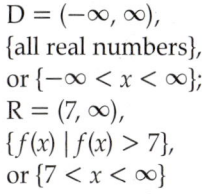
$D = (-\infty, \infty)$, {all real numbers}, or $\{-\infty < x < \infty\}$; $R = (7, \infty)$, $\{f(x) \mid f(x) > 7\}$, or $\{7 < x < \infty\}$

13. 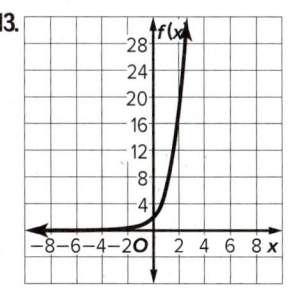 $D = (-\infty, \infty)$, {all real numbers}, or $\{-\infty < x < \infty\}$; $R = (0, \infty)$, $\{f(x) \mid f(x) > 0\}$, or $\{0 < x < \infty\}$

15. 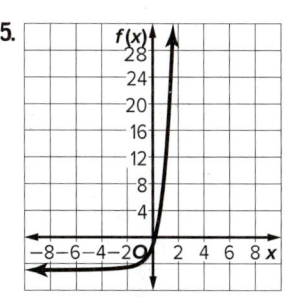 $D = (-\infty, \infty)$, {all real numbers}, or $\{-\infty < x < \infty\}$; $R = (-5, \infty)$, $\{f(x) \mid f(x) > -5\}$, or $\{-5 < x < \infty\}$

17. 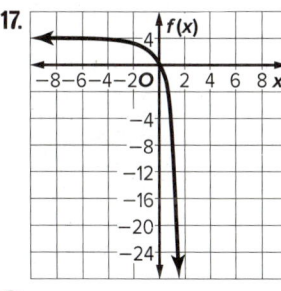 $D = (-\infty, \infty)$, {all real numbers}, or $\{-\infty < x < \infty\}$; $R = (-\infty, 4)$, $\{f(x) \mid f(x) < 4\}$, or $\{-\infty < x < 4\}$

19. $y = a(1 + r)^t$ Equation for exponential growth
$y = 65(1 + 0.3)^t$ $a = 65$ and $r = 0.3$
$y = 65(1.3)^t$ Simplify.

Make a table of values. Then plot the points, and sketch the graph.

t	$y = 65(0.3)^t$
0	$y = 65(1.3)^0 = 65$
2	$y = 65(1.3)^2 \approx 110$
4	$y = 65(1.3)^4 \approx 186$
6	$y = 65(1.3)^6 \approx 314$
8	$y = 65(1.3)^8 \approx 530$
10	$y = 65(1.3)^{10} \approx 896$

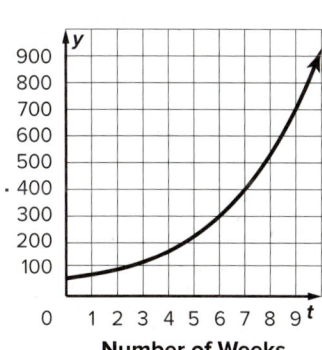

21.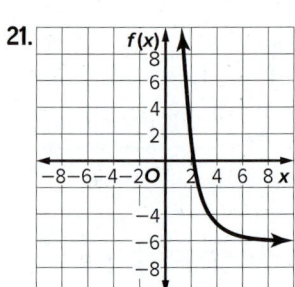
D = (−∞, ∞), {all real numbers}, or {−∞ < x < ∞}; R = (−6, ∞), {f(x) | f(x) > −6}, or {−6 < x < ∞}

23.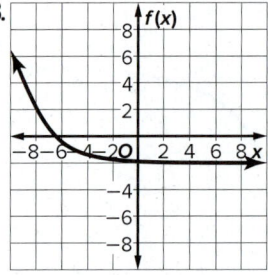
D = (−∞, ∞), {all real numbers}, or {−∞ < x < ∞}; R = (−2, ∞), {f(x) | f(x) > −2}, or {−2 < x < ∞}

25.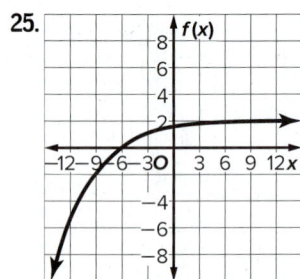
D = (−∞, ∞), {all real numbers}, or {−∞ < x < ∞}; R = (−∞, 2), {f(x) | f(x) < 2}, or {−∞ < x < 2}

27a. decay; 0.9

27b. The $P(x)$-intercept represents the number of inline skaters in 2004. The asymptote is the x-axis. The number of inline skaters can approach 0, but will never equal 0.

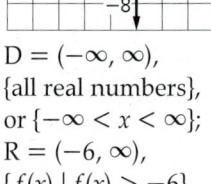

This makes sense as there will probably always be some who continue to skate. **29a.** $f(x) = 18(1.25)^{x-1}$

29b. growth; 1.25 **29c.** 134

31. $g(x) = 4(2)^{x-3}$ or $g(x) = \frac{1}{2}(2^x)$

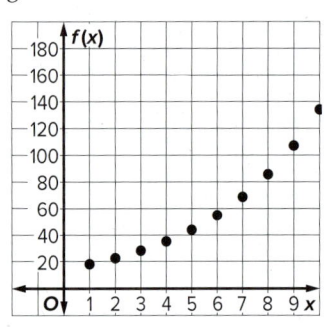

33. a.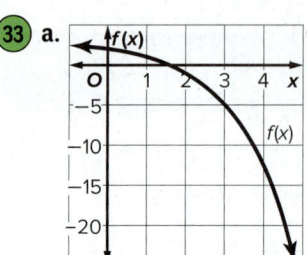

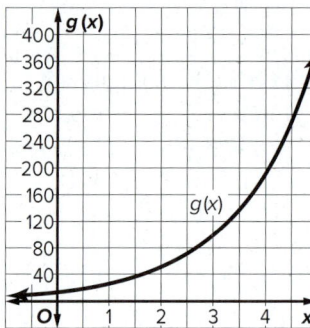 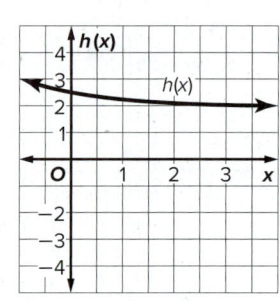

b. Sample answer: The graph of $f(x)$ appears to be the graph of $f(x) = b^x$ reflected across the x-axis. As the values of x increase, the output values decrease.

c. Sample answer: The graphs of $g(x)$ and $h(x)$ appear to be translated to the left.

d. Sample answer: $f(x)$ and $g(x)$ are growth and $h(x)$ is decay; the absolute value of the output is increasing for the growth functions and decreasing for the decay function.

35. Vince; the graphs of the function would be the same. **37.** Sample answer: 10 **39.** A **41.** B
43a. translated right 2 units, up 1 unit, and reflected in the x-axis. **43b.** increasing
43c. horizontal asymptote $y = 1$ **43d.** −8
43e.

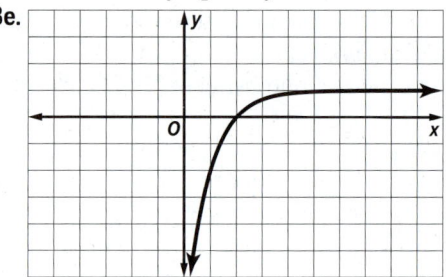

45. $1636.65

Lesson 6-2

1. 12 **3.** −10 **5a.** $c = 2^{\frac{t}{15}}$ **5b.** 16 cells **7.** $x \geq 4.5$ **9.** 0

11.
$81^{a+2} = 3^{3a+1}$	Original equation
$(3^4)^{a+2} = 3^{3a+1}$	Rewrite 81 as 3^4.
$3^{4(a+2)} = 3^{3a+1}$	Power of a Power
$3^{4a+8} = 3^{3a+1}$	Distributive Property
$4a + 8 = 3a + 1$	Property of Equality for Exponential Functions
$a + 8 = 1$	Subtract $3a$ from each side.
$a = -7$	Subtract 8 from each side.

13. $\frac{5}{3}$ **15a.** $y = 10{,}000(1.045)^x$ **15b.** about $26,336.52
17. $y = 256(0.75)^x$ **19.** $y = 144(3.5)^x$ **21.** $16,755.63

23. $55,085.44 **25.** $\{b \mid b > \frac{1}{5}\}$ **27.** $\{d \mid d \geq -1\}$
29. $\{w \mid w < \frac{2}{5}\}$
31. a. $\frac{a}{w^{1.31}} = \frac{170}{45^{1.31}}$ Write a proportion.
$a \cdot 45^{1.31} = w^{1.31} \cdot 170$ Cross Products Property
$a = \frac{w^{1.31} \cdot 170}{45^{1.31}}$ Divide each side by $45^{1.31}$.
$a = 1.16w^{1.31}$ Use a calculator.
b. $a = 1.16w^{1.31}$ Write the equation.
$= 1.16(430)^{1.31}$ $w = 430$
≈ 3268 yd^2 Use a calculator.
33. $\frac{1}{7}$ **35.** $-\frac{4}{13}$ **37.** 1 **39a.** $d = 1.30h^{\frac{3}{2}}$ **39b.** about 1001 cm **41a.** 2, 4, 8, 16

41b.

Cuts	Pieces
1	2
2	4
3	8
4	16

41c. $y = 2^x$ **41d.** $y = 0.003(2)^x$ **41e.** about 3,221,225.47 in.

43. Sample answer: Beth; Liz added the exponents instead of multiplying them when taking the power of a power. **45.** Reducing the term will be more beneficial. The multiplier is 1.3756 for the 4-year and 1.3828 for the 6.5%. **47.** Sample answer: $4^x \leq 4^2$
49. Sample answer: Divide the final amount by the initial amount. If n is the number of time intervals that pass, take the nth root of the answer. **51.** C
53. E **55a.** $-\frac{3}{7}$ **55b.** -1 **57a.** $x = 2$ **57b.** $x > 3$
57c. $x \geq 3$

Lesson 6-3

1. 1024 **3.** $a_n = 18 \cdot \left(\frac{1}{3}\right)^{n-1}$
5. $a_n = a_1 r^{n-1}$ nth term of a geometric sequence
$4 = a_1(3^{2-1})$ $a_n = 4, r = 3,$ and $n = 2$
$4 = a_1(3)$ Evaluate the power.
$\frac{4}{3} = a_1$ Divide each side by 3.
$a_n = a_1 r^{n-1}$ nth term of a geometric sequence
$a_n = \frac{4}{3}(3)^{n-1}$ $a_1 = \frac{4}{3}, r = 3$

7. $a_n = 12(-8)^{n-1}$ **9.** 1, 5, 25 or $-1, 5, -25$
11. 4095 **13.** $\frac{1}{16}$ **15.** 512 **17.** 93 in. **19.** 25
21. 512 **23.** $a_n = (-3)(-2)^{n-1}$ **25.** $a_n = (-1)(-1)^{n-1}$
27. $a_n = 8 \cdot \left(\frac{1}{4}\right)^{n-1}$ **29.** $a_n = 7(2)^{n-1}$
31. $a_n = \frac{1}{15,552}(6)^{n-1}$ **33.** $a_n = 648\left(\frac{1}{3}\right)^{n-1}$
35. 270, 90, 30 or $-270, 90, -30$
37. $\frac{7}{3}, \frac{14}{9}, \frac{28}{27}$ or $-\frac{7}{3}, \frac{14}{9}, -\frac{28}{27}$ **39.** 15 and 75
41. 99.19% **43.** 31.9375 **45.** 9707.82 **47.** 2188
49. $-87,381$
51. $S_n = \frac{a_1 - a_1 r^n}{1-r}$ Sum formula
$-2912 = \frac{a_1 - a_1(3^6)}{1-3}$ $S_n = -2912, r = 3,$ and $n = 6$
$-2912 = \frac{a_1(1-3^6)}{1-3}$ Distributive Property
$-2912 = \frac{-728a_1}{-2}$ Subtract.
$-2912 = 364a_1$ Simplify.
$-8 = a_1$ Divide each side by 364.

53. 64 **55.** 0.25
57. $S_n = \frac{a_1 - a_1 r^n}{1-r}$ Alternate sum formula
$= \frac{100 - 100(0.5)^5}{1-(0.5)}$ Substitution
$= 193.75$ ft Use a calculator.

59. 524,288 **61.** about 471 cm **63a.** $53.24, $64.42, $94.32 **63b.** $855.37 **63c.** Each payment made is rounded to the nearest cent, so the sum of the payments may be off by several cents.

65. $S_n = \frac{a_1 - a_n r}{1-r}$ Alternate sum formula
$a_n = a_1 \cdot r^{n-1}$ Formula for nth term
$\frac{a_n}{r^{n-1}} = a_1$ Divide both sides by r^{n-1}.
$S_n = \frac{\frac{a_n}{r^{n-1}} - a_n r}{1-r}$ Substitution.
$= \frac{\frac{a_n}{r^{n-1}} - \frac{a_n r \cdot r^{n-1}}{r^{n-1}}}{1-r}$ Multiply by $\frac{r^{n-1}}{1}$.
$= \frac{\frac{a_n(1-r^n)}{r^{n-1}}}{1-r}$ Simplify.
$= \frac{a_n(1-r^n)}{r^{n-1}(1-r)}$ Divide by $(1-r)$.
$= \frac{a_n(1-r^n)}{r^{n-1} - r^n}$ Simplify.

67. Sample answer: $n - 1$ needs to change to n, and the 10 needs to change to a 9. When this happens, the terms for both series will be identical (a_1 in the first series will equal a_0 in the second series, and so on), and the series will be equal to each other. **69.** 234
71. Sample answer: $4 + 8 + 16 + 32 + 64 + 128$
73. 78,732 **75.** 0.9375 **77.** C **79a.** 1000 **79b.** 16,000

Lesson 6-4

1. $8^3 = 512$ **3.** $\log_{11} 1331 = 3$ **5.** 2 **7.** 0
9. 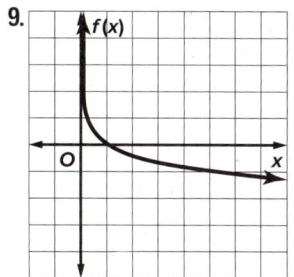 **11.**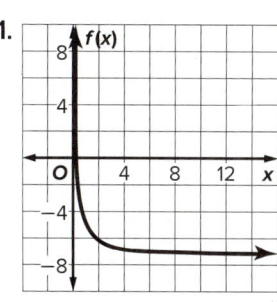

13. $2^4 = 16$ **15.** $9^{-2} = \frac{1}{81}$ **17.** $12^2 = 144$
19. $\log_9 \frac{1}{9} = -1$ **21.** $\log_2 256 = 8$ **23.** $\log_{27} 9 = \frac{2}{3}$
25. -2 **27.** 3 **29.** $\frac{1}{3}$ **31.** $\frac{1}{2}$
33. $\log_{\frac{1}{5}} 3125 = y$ Let the logarithm equal y.
$3125 = \frac{1}{5}^y$ Definition of logarithm

$5^5 = 5^{-1y}$ $\frac{1}{5} = 5^{-1}$

$5 = -1y$ Property of Equality for Exponential Functions

$-5 = y$ Divide each side by -1.

So, $\log_{\frac{1}{5}} 3125 = -5$.

35. 4

37.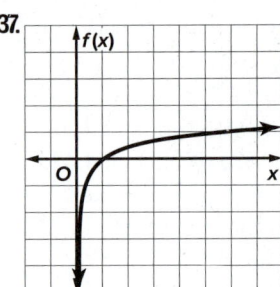

39.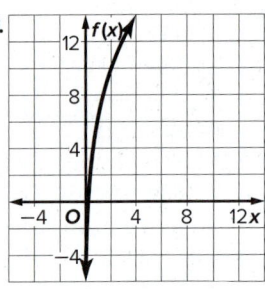

41.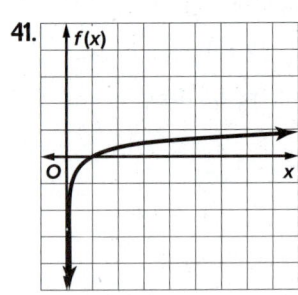

43.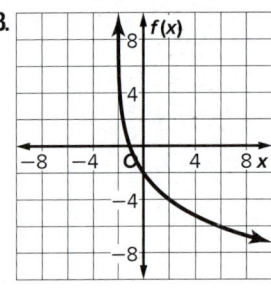

45.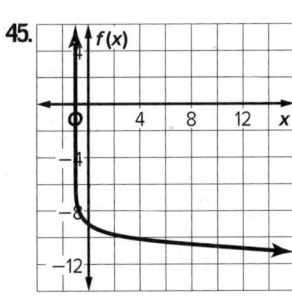

47.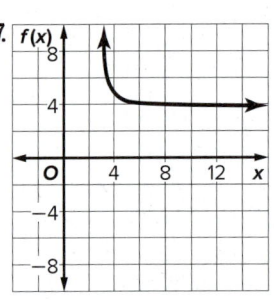

49a. 2   **49b.** 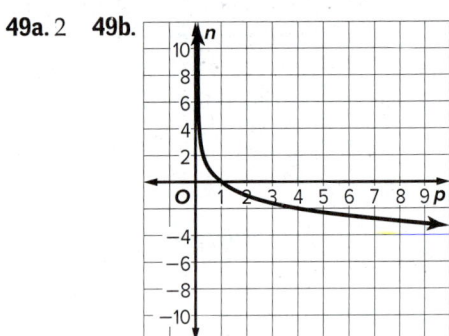   **49c.** $\frac{1}{8}$; less light

51. This represents a transformation of the graph of $f(x) = \log_2 x$.
$|a| = 4$: The graph expands vertically.
$h = 4$: The graph is translated 4 units to the right.
$k = 6$: The graph is translated 6 units up.

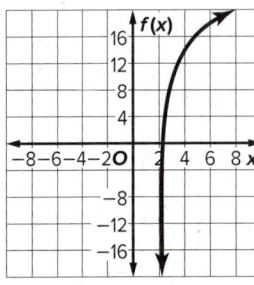

53.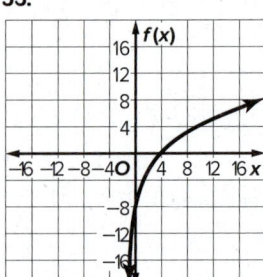

55.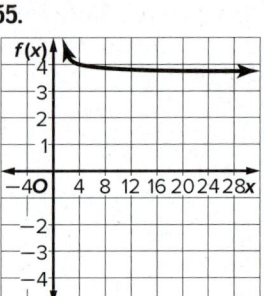

57a. $S(3) \approx 30$, $S(15) = 50$, $S(63) = 70$

57b. If $3000 is spent on advertising, $30,000 is returned in sales. If $15,000 is spent on advertising, $50,000 is returned in sales. If $63,000 is spent on advertising, $70,000 is returned in sales.

57c. Sales versus Money Spent on Advertising

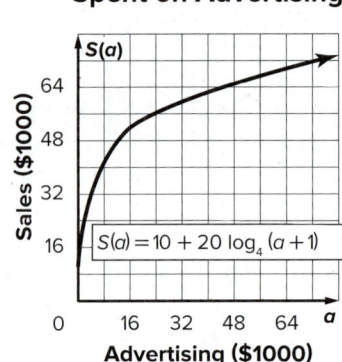

57d. Sample answer: Because eventually the graph plateaus, and no matter how much money you spend you are still returning about the same in sales.

59. a.
$\log_{\left(1 + \frac{0.24}{12}\right)} \frac{A}{2000} = 12t$ Original formula

$\log_{1.02} \frac{A}{2000} = 12t$ Simplify.

$\frac{A}{2000} = 1.02^{12t}$ Definition of logarithm

$A = 2000 \cdot 1.02^{12t}$ Multiply each side by 2000.

Make a table of values. Then plot the points, and sketch the graph.

t	$A = 2000 \cdot 1.02^{12t}$
0	$A = 2000 \cdot 1.02^{12(0)} = 2000$
2	$A = 2000 \cdot 1.02^{12(2)} \approx 3217$
4	$A = 2000 \cdot 1.02^{12(4)} \approx 5174$
6	$A = 2000 \cdot 1.02^{12(6)} \approx 8322$
8	$A = 2000 \cdot 1.02^{12(8)} \approx 13,386$
10	$A = 2000 \cdot 1.02^{12(10)} \approx 21,530$

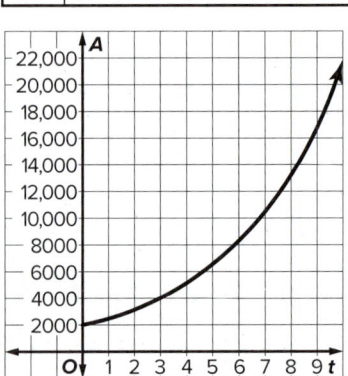

b. From the graph, $A = 4000$ at about $t = 3$. So, it will take approximately 3 years for the debt to double.

c. From the graph, $A = 6000$ at about $t = 4.5$. So, it will take approximately 4.5 years for the debt to triple.

61. Never; if zero were in the domain, the equation would be $y = \log_b 0$. Then $b^y = 0$. However, for any real number b, there is no real power that would let $b^y = 0$.

63. $\log_7 51$; Sample answer: $\log_7 51$ equals a little more than 2. $\log_8 61$ equals a little less than 2. $\log_9 71$ equals a little less than 2. Therefore, $\log_7 51$ is the greatest.

65. No; Elisa was closer. She should have $-y = 2$ or $y = -2$ instead of $y = 2$. Matthew used the definition of logarithms incorrectly. **67.** C **69.** B **71.** B, C, F
73. $a = 2$, $b = 3$

Lesson 6-5

1. nonlinear, positive, weak
3. linear, negative, strong
5. The x variable is sometimes referred to as the predictor variable because when you solve for x you are finding the solution to a prediction of unknown events.
7. linear; $y = \frac{1}{2}x + 3$
9. exponential
11. $x = 0$
13. $x = 2$
15. $x = 4$
17. 8 hours
19. nonlinear, no correlation
21. linear; $y = 1.5x + 4$
23. $x = 4$
25. $x = 2.68$
27. $x = 0$
29. power **31.** linear
33. linear, $y = 0.062756x + 0.32753$
35. weak **37.** perfect
39. Sample answer: If the right side of the graph is higher than the left side, the direction is positive. If the right side of the graph is lower than the left side, the direction is negative.

41. a. To find the values for the table, start with with $x = 0$ and a population of 502,000. Multiply the population by 1.032 (an increase of 3.2%) to find the population for each successive value of x.

x	Population
0	502,000
1	518,064
2	534,642
3	551,751
4	569,407

41b.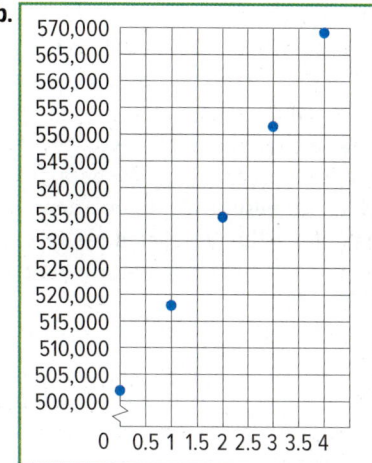

41c. Exponential growth is modeled by the equation $y = a(1 + r)^t$, where a is the initial amount and r is the growth rate. In this case, $a = 502,000$ and $r = 0.032$, so $y = 502,000(1 + 0.032)^t$. **41d.** exponential
41e. 2018 is 24 years after 1994, so let $t = 24$ in the equation $y = 502,000(1 + 0.032)^t$. $y \approx 1,069,095$
41f. Use technology to make a table of values and look for the first year in which the population is at least 1.5 million; find the corresponding value of x; 35 years.
43. Sample answer: It helps you eliminate possibilities and check your work. The linear regression model is very different then the quadratic regression model and by studying a scatterplot, you can sometimes see those differences.

45 a.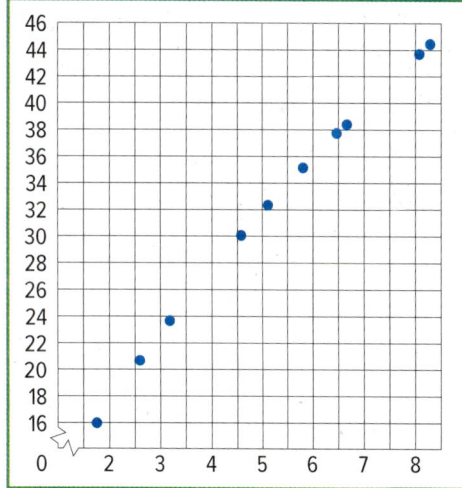

b. Sample answer: $0 \leq x \leq 10$ by 1 and $10 \leq y \leq 50$ by 5
c. Enter the data in a graphing calculator and try different regression models to find trhe one with the greatest value of r. A power regression is the best regression model.
d. Yes; the correlation coefficient is close to 1.
e. The power regression equation is $y = ax^b$, where $a \approx 11.0163$ and $b \approx 0.6595$. Solve $30.57 = 11.0163x^{0.6595}$. $x \approx 4.7$ cm.

47. Sample answer: We need to have the line of best fit to solve for x. If we choose the wrong model, the regression model will not have the highest correlation rate and the prediction will be innacurate.
49a. Sketching the graph could be more accurate if the data has a few outliers that are skewing the data.
49b-d. The sketched line of best fit would be linear, but with the graphing calculator the regression model would be a curve. So the graphing calculator would be more accurate in these cases.
51. B   **53.** A   **55a.** 0, 1, 2, 3, 4, 5, 6, 7, 8, 9
55b. 512, 256, 128, 64, 32, 16, 8, 4, 2, 1
55c. nonlinear, negative direction, strong correlation
55d. Exponential, because it has the strongest correlation.   **55e.** 16   **55f.** 2   **55g.** 9

Lesson 6-6

1. 2.085   **3.** 0.3685   **5.** Guadalupe Peak: 67,287.7 Pa; Emory Peak: 70,166.37 Pa; Anthony's Nose: 73,081.33 Pa; Panther Peak; 75,006.13 Pa; Buck Mountain: 76,662.09 Pa   **7.** 2.4182   **9.** 2   **11.** 13.4403   **13.** 2.1610

15. $\log_4 \frac{4}{3} = \log_4 4 - \log_4 3$ Quotient Property
$= 1 - \log_4 3$ Inverse Property of Exponents and Logarithms
$\approx 1 - 0.7925$ Replace $\log_4 3$ with 0.7925.
≈ 0.2075 Simplify.

17. 1.5   **19.** 2.1606   **21.** 3.4818   **23.** 8   **25.** 2

27. a. $P = \log_{10}\left(1 + \frac{1}{d}\right)$ Original equation
$10^P = 1 + \frac{1}{d}$ Definition of logarithm
$10^P - 1 = \frac{1}{d}$ Subtract 1 from each side.
$d(10^P - 1) = 1$ Multiply each side by d.
$d = \frac{1}{10^P - 1}$ Divide each side by $10^P - 1$.

b. $d = \frac{1}{10^P - 1}$ Write the formula.
$= \frac{1}{10^{0.097} - 1}$ $P = 0.097$
≈ 4 Use a calculator.

c. $P = \log_{10}\left(1 + \frac{1}{d}\right)$ Original equation
$= \log_{10}\left(1 + \frac{1}{1}\right)$ $d = 1$
$= \log_{10} 2$ Simplify.
≈ 0.30103 Replace $\log_{10} 2$ with 0.30103.

The probability is about 30.1%.

29. 2.1133   **31.** 0.1788   **33.** 1.7228   **35.** 2.0478   **37.** 3
39. 5   **41.** $85\frac{1}{3}$ **43.** $\left(\frac{x-2}{256}\right)^{\frac{1}{6}}$ **45.** $\sqrt{6}, -\sqrt{6}$ **47.** 5
49. 12   **51.** false   **53.** false   **55.** true   **57.** false
59a. Sample answer: about 6 yr   **59b.** Sample answer: She will earn 1050×12 or $12,600 from Social Security. She needs to earn $50,000 - $12,600 or $37,400 from interest. The principal in her CD account will need to be $37,400 \div 0.05$ or $748,000. Set up the exponential growth equation and solve for t.

$A = P(1+r)^t$ Exponential Growth
$748{,}000 = 320{,}000(1.15)^t$ $A = 748{,}000, P = 320{,}000,$ and $r = 0.15$
$2.3375 = 1.15^t$ Divide each side by 320,000.
$\log 2.3375 = \log 1.15^t$ Property of Equality of Logarithms
$\log 2.3375 = t \log 1.15$ Power Property of Logarithms
$\frac{\log 2.3375}{\log 1.15} = t$ Divide each side by log 1.15.
$6.07 \approx t$ Simplify.

59c. Sample answers: The annual increase in her account remains at 15%. She needs to pay taxes on Social Security. Each CD will earn 5% interest.
61a. Sample answer: $\log_b \frac{xz}{5} = \log_b x + \log_b z - \log_b 5$
61b. Sample answer: $\log_b m^4 p^6 = 4 \log_b m + 6 \log_b p$
61c. Sample answer: $\log_b \frac{j^8 k}{h^5} = 8 \log_b j + \log_b k - 5 \log_b h$
63a. $\log_b 1 = 0$, because $b^0 = 1$.   **63b.** $\log_b b = 1$, because $b^1 = b$.   **63c.** $\log_b b^x = x$, because $b^x = b^x$.
65. $\log_b 24 \neq \log_b 20 + \log_b 4$; all other choices are equal to $\log_b 24$.

67. $x^{3 \log_x 2 - \log_x 5} = x^{\log_x 2^3 - \log_x 5}$
$= x^{\log_x 8 - \log_x 5}$
$= x^{\log_x \frac{8}{5}}$
$= \frac{8}{5}$

69. C   **71.** C   **73.** 4.902   **75a.** all real numbers greater than or equal to 0   **75b.** all real numbers greater than or equal to 0   **75c.** 4.3   **75d.** 5.3   **75e.** 6.3   **75f.** Sample answer: The magnitude of each earthquake increases by 1 unit when the intensity increases by a factor of 10.

Lesson 6-7

1. 0.6990   **3.** -0.3979 **5.** 3.55×10^{24} ergs   **7.** 0.8442

9. $11^{b-3} = 5^b$ Original equation
$\log 11^{b-3} = \log 5^b$ Property of Equality for Logarithmic Functions
$(b-3) \log 11 = b \log 5$ Power Property of Logarithms
$b \log 11 - 3 \log 11 = b \log 5$ Distributive Property
$-3 \log 11 = b \log 5 - b \log 11$ Subtract $b \log 11$ from each side.
$-3 \log 11 = b(\log 5 - \log 11)$ Distributive Property
$\frac{-3 \log 11}{\log 5 - \log 11} = b$ Divide each side by log 5 − log 11.
$9.1237 \approx b$ Use a calculator.

11. $\{p \mid p \leq 4.4190\}$ **13.** $\frac{\log 23}{\log 4} \approx 2.2618$
15. $\frac{\log 5}{\log 2} \approx 2.3219$ **17.** 1.0414   **19.** 0.9138   **21.** -1.3979
23. 1.7740   **25.** 5.9647   **27.** ± 1.1691 **29.** $\{n \mid n > 0.6667\}$
31. $\{y \mid y \geq -3.8188\}$ **33.** $\frac{\log 18}{\log 7} \approx 1.4854$
35. $\frac{\log 16}{\log 2} = 4$ **37.** $\frac{\log 11}{\log 3} \approx 2.1827$

39. a.
$$n = 35[\log_4(t+2)] \quad \text{Original equation}$$
$$= 35[\log_4(10+2)] \quad t = 10$$
$$= 35[\log_4(12)] \quad \text{Simplify.}$$
$$= 35 \cdot \frac{\log_{10} 12}{\log_{10} 4} \quad \text{Change of Base Formula}$$
$$\approx 62.737 \quad \text{Use a calculator.}$$

In 2010, there are about 62.737 thousand, or 62,737 pet owners.

b.
$$n = 35[\log_4(t+2)] \quad \text{Original equation}$$
$$80 = 35[\log_4(t+2)] \quad n = 80$$
$$2.2857 \approx \log_4(t+2) \quad \text{Divide each side by 35.}$$
$$4^{2.2857} \approx t+2 \quad \text{Definition of logarithm}$$
$$22 \approx t$$

In 22 years after 2000, or in 2022, there will be 80,000 pet owners. An increase of about 20,000 pet owners since 2010 seems reasonable.

41. 3.3578 **43.** −0.0710 **45.** 4.7393 **47.** $\{x \mid x \geq 2.3223\}$
49. $\{x \mid x \leq 0.9732\}$ **51.** $\{p \mid p \leq 2.9437\}$
53. $\frac{\log 12}{\log 4} \approx 1.7925$ **55.** $\frac{\log 7.29}{\log 5} \approx 1.2343$
57. 113.03 cents
59. The graph of $g(x)$ is a dilation of the graph of $f(x)$, which is stretched vertically. This graph is also reflected across the x-axis.

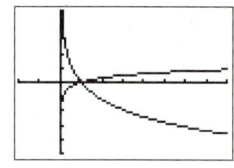

[−2, 8] scl: 1 by [−5, 5] scl: 1

61. The graph of $g(x)$ is a translation of the graph of $f(x)$, which is translated down 5 units.

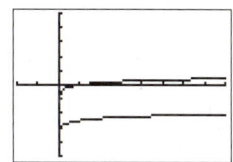

[−2, 8] scl: 1 by [−10, 10] scl: 2

63. The graph of $g(x)$ is a translation of the graph of $f(x)$, which is translated right 6 units.

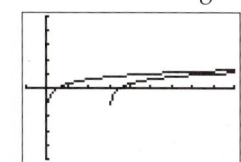

[−2, 18] scl: 2 by [−5, 5] scl: 1

65.
$$4^{x^2-3} = 16 \quad \text{Original equation}$$
$$4^{x^2-3} = 4^2 \quad \text{Rewrite 16 as } 4^2.$$
$$x^2 - 3 = 2 \quad \text{Property of Equality for Exponential Functions}$$
$$x^2 = 5 \quad \text{Add 3 to each side.}$$
$$x = \pm\sqrt{5} \quad \text{Take the square root of each side.}$$
$$\approx \pm 2.2361 \quad \text{Use a calculator.}$$

67. 3.5 **69.** −3.8188 **71a.** The solution is between 1.8 and 1.9. **71b.** (1.85, 13) **71c.** Yes; all methods produce the solution of 1.85. They all should produce the same result because you are starting with the same equation. If they do not, then an error was made.

73.
$$\log_{\sqrt{a}} 3 = \log_a x \quad \text{Original equation}$$
$$\frac{\log_a 3}{\log_a \sqrt{a}} = \log_a x \quad \text{Change of Base Formula}$$
$$\frac{\log_a 3}{\frac{1}{2}} = \log_a x \quad \sqrt{a} = a^{\frac{1}{2}}$$
$$2\log_a 3 = \log_a x \quad \text{Multiply numerator and denominator by 2.}$$
$$\log_a 3^2 = \log_a x \quad \text{Power Property of Logarithms}$$
$$3^2 = x \quad \text{Property of Equality for Logarithmic Functions}$$
$$9 = x$$

75. $\log_3 27 = 3$ and $\log_{27} 3 = \frac{1}{3}$; Conjecture: $\log_a b = \frac{1}{\log_b a}$

Proof:
$$\log_a b \stackrel{?}{=} \frac{1}{\log_b a} \quad \text{Original statement}$$
$$\frac{\log_b b}{\log_b a} \stackrel{?}{=} \frac{1}{\log_b a} \quad \text{Change of Base Formula}$$
$$\frac{1}{\log_b a} = \frac{1}{\log_b a} \quad \text{Inverse Property of Exponents and Logarithms}$$

77. B **79.** B **81.** B **83a.** 4491 **83b.** 8360
83c. 2079. Yes; Sample answer: The year seems reasonable based on the model of the equation.
83d. The domain is all real numbers greater than −5. These numbers allow for the value inside the parentheses to be positive.
83e. The range is all real numbers greater than or equal to 0. These numbers allow for a positive or 0 population.
83f. 1934 **85a.** $\log_2 30$ **85b.** $x < \log_2 30$
85c. $x > 2\log_4 7$ **85d.** $x \geq \log_2 10$ **85e.** $x \geq \log_{12} 0.05$
85f. $x \leq \log_{12} 0.05$ **85g.** $\log_{2.5}\left(\frac{4}{3}\right)$ **85h.** $x \geq \log_{2.5}\left(\frac{4}{3}\right)$

Lesson 6-8

1. $\ln 30 = x$ **3.** $\ln x = 3$ **5.** $7 \ln 2$ **7.** $\ln 17{,}496$
9. 0.1352 **11.** 993.6527 **13.** $\{x \mid -25.0855 < x < 15.0855, x \neq -5\}$ **15.** $\{x \mid x > 3.3673\}$ **17a.** about 243,092
17b. about 58 min **17c.** about 7 **19.** $\ln 0.1 = -5x$
21. $5.4 = e^x$ **23.** $e^{36} = x + 4$ **25.** $e^7 = e^x$ **27.** $7 \ln 10$
29.
$$7 \ln \tfrac{1}{2} + 5 \ln 2 = 7 \ln 2^{-1} + 5 \ln 2 \quad \text{Rewrite } \tfrac{1}{2} \text{ as } 2^{-1}.$$
$$= \ln 2^{-7} + \ln 2^5 \quad \text{Power Property of Logarithms}$$
$$= \ln (2^{-7})(2^5) \quad \text{Product Property of Logarithms}$$
$$= \ln 2^{-2} \quad \text{Simplify.}$$
$$= -2 \ln 2 \quad \text{Power Property of Logarithms}$$

31. $\ln 81x^6$ **33.** 3.7955 **35.** 0.6931 **37.** −0.5596
39. $\{x \mid x \leq 2.1633\}$ **41** $\{x \mid x > 8.0105\}$
43. $\{x \mid x < -239.8802 \text{ or } x > 239.8802\}$
45. The graph of $g(x)$ is a dilation of the graph of $f(x)$, which is compressed vertically by a factor of 0.5.

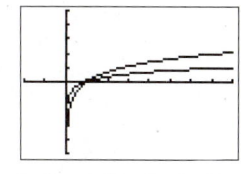

[−2, 8] scl: 1 by [−5, 5] scl: 1

47. The graph of $g(x)$ is a translation of the graph of $f(x)$, which is translated up 8 units.

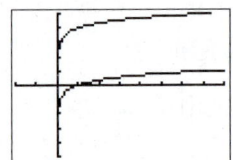

[−2, 8] scl: 1 by [−10, 10] scl: 2

49. The graph of $g(x)$ is a translation of the graph of $f(x)$, which is translated left 5 units.

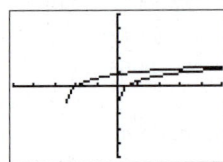

[−10, 10] scl: 2 by [−10, 10] scl: 2

51. **a.** $A = Pe^{rt}$ Exponential Formula
$= 800e^{(0.045)(5)}$ $P = 800, r = 0.045, t = 5$
$= 800e^{0.225}$ Simplify.
≈ 1001.86 Use a calculator.
About $1001.86 will be in the account.

b. $\ln \frac{A}{P} = rt$ Logarithmic Formula
$\ln 2 = 0.045t$ $A = 2P, r = 0.045$
$\frac{\ln 2}{0.045} = t$ Divide each side by 0.045.
$15.4 \approx t$ Use a calculator.

It would take about 15.4 years to double your money.

c. $\ln \frac{A}{P} = rt$ Logarithmic Formula
$\ln 2 = 9r$ $A = 2P, t = 9$
$\frac{\ln 2}{9} = r$ Divide each side by 9.
$0.077 \approx r$ Use a calculator.

You would need a rate of about 7.7%.

d. $A = Pe^{rt}$ Exponential Formula
$10{,}000 = Pe^{(0.0475)(12)}$ $A = 10{,}000, r = 0.0475, t = 12$
$10{,}000 = Pe^{0.57}$ Simplify.
$\frac{10{,}000}{e^{0.57}} = P$ Divide each side by $e^{0.57}$.
$5655.25 \approx P$ Use a calculator.

You would need to deposit about $5655.25.

53. $4 \ln 2 - 3 \ln 5$ **55.** $\ln x + 4 \ln y - 3 \ln z$

57. -0.8340 **59.** 1.1301

61a. 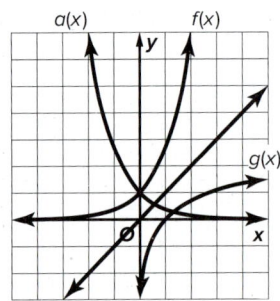   **61b.** y-axis; $a(x) = -e^x$

61c.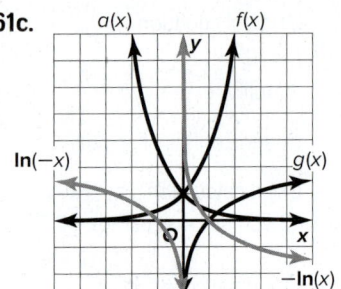

61d. Sample answer: No; these functions are reflections over $y = -x$, which indicates that they are not inverses.

63. Let $p = \ln a$ and $q = \ln b$. That means that $e^p = a$ and $e^q = b$.
$ab = e^p \times e^q$
$ab = e^{p+q}$
$\ln(ab) = (p + q)$
$\ln(ab) = \ln a + \ln b$

65. Sample answer: $e^{\ln 3}$ **67a.** $P(6) = 8$ billion
67b. $P(-16) = 6.01$ billion   **67c.** 53.319 years
67d. 1968   **69.** $x > e^{\frac{7}{9}}$ **71.** B   **73.** 0.403   **75.** B

Lesson 6-9

1. 16   **3.** C   **5.** $\left\{x \mid 0 < x \leq \frac{1}{64}\right\}$ **7.** $\left\{x \mid \frac{4}{3} < x < 2\right\}$

9. 3125   **11.** -2 **13.** 9

15. $\log_{12}(x^2 - 7) = \log_{12}(x + 5)$ Original equation
$x^2 - 7 = x + 5$ Property of Equality for Exponential Functions
$x^2 - x - 7 = 5$ Subtract x from each side.
$x^2 - x - 12 = 0$ Subtract 5 from each side.
$(x - 4)(x + 3) = 0$ Factor.
$x - 4 = 0$ or $x + 3 = 0$ Zero Product Property
$x = 4$ $x = -3$ Solve each equation.

17. 5   **19.** -3 **21.** 318 mph; The wind speed of 318 mph seems high because most tornadoes have wind speeds of about 110 mph.   **23.** $\{x \mid x \geq 256\}$

25. $\log_2 x \leq -2$ Original inequality
$0 < x \leq 2^{-2}$ Property of Inequality for Exponential Functions
$0 < x \leq \frac{1}{4}$ Simplify.

The solution is $\left\{x \mid 0 < x \leq \frac{1}{4}\right\}$ or $\left(0, \frac{1}{4}\right]$.

27. $\left\{x \mid 0 < x < \frac{1}{7}\right\}$ **29.** $\left\{x \mid \frac{1}{2} < x \leq 1\right\}$

31. $\left\{x \mid -\frac{5}{12} < x \leq 1\right\}$ **33.** $\{x \mid x \geq 8\}$

35a. 37   **35b.** 61

37. **a.** $\beta = 10 \log_{10}\left(\frac{I}{10^{-12}}\right)$ Original equation
$= 10 \log_{10}\left(\frac{1}{10^{-12}}\right)$ $I = 10$
$= 10 \log_{10} 10^{12}$ Write $\frac{1}{10^{-12}}$ as 10^{12}.
$= 10(12)$ or 120   Definition of logarithm

b. $\beta = 10 \log_{10}\left(\dfrac{I}{10^{-12}}\right)$ Original equation

$= 10 \log_{10}\left(\dfrac{10^{-2}}{10^{-12}}\right)$ $I = 10^{-2}$

$= 10 \log_{10} 10^{10}$ Quotient of Powers Property

$= 10(10)$ or 100 Definition of logarithm

c. Sample answer: The power of the logarithm only changes by 2. The power is the answer to the logarithm. That 2 is multiplied by the 10 before the logarithm. So we expect the decibels to change by 20.

39. $6\dfrac{17}{20}$ **41.** The logarithmic function of the form $y = \log_b x$ is the inverse of the exponential function of the form $y = b^x$. The domain of one of the two inverse functions is the range of the other. The range of one of the two inverse functions is the domain of the other.
43a. less than **43b.** less than **43c.** no **43d.** infinitely many **45a.** 19,952,623.2 **45b.** 665,087.4 **45c.** 5.8
47. D **49.** B **51.** C **53.** B

Lesson 6-10

1a. 5.545×10^{-10} **1b.** 1,578,843,530 yr
1c. about 30.48 mg **1d.** 3,750, 120,003 yr

3a.
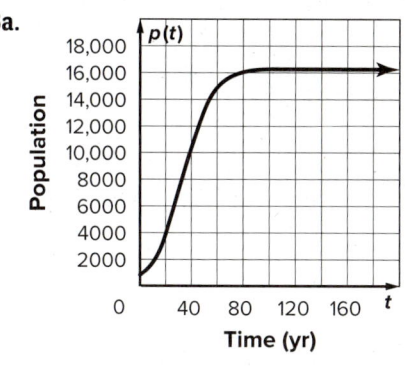

3b. $P(t) = 16{,}500$ **3c.** 16,500 **3d.** about 102 years

5 a.
$y = 80e^{kt}$ Original formula
$675 = 80e^{k(30)}$ $y = 675, t = 30$
$8.4375 = e^{30k}$ Divide each side by 80.
$\ln 8.4375 = \ln e^{30k}$ Property of Equality for Logarithmic Functions
$\ln 8.4375 = 30k$ $\ln e^x = x$
$\dfrac{\ln 8.4375}{30} = k$ Divide each side by 30.
$0.071 \approx k$ Use a calculator.

b.
$y = 80e^{kt}$ Original formula
$6000 = 80e^{(0.071)t}$ $y = 6000, k \approx 0.071$
$75 = e^{0.071t}$ Divide each side by 80.
$\ln 75 = \ln e^{0.071t}$ Property of Equality for Logarithmic Functions
$\ln 75 = 0.071t$ $\ln e^x = x$
$\dfrac{\ln 75}{0.071} = t$ Divide each side by 0.071.
$60.8 \approx t$ Use a calculator.

The bacteria will reach a population of 6000 cells in about 60.8 minutes.

c.
$35e^{0.0978t} > 80e^{0.071t}$ Formula for exponential growth
$\ln 35e^{0.0978t} > \ln 80e^{0.071t}$ Property of Inequality for Logarithms
$\ln 35 + \ln e^{0.0978t} > \ln 80 + \ln e^{0.071t}$ Product Property of Logarithms
$\ln 35 + 0.0978t > \ln 80 + 0.071t$ $\ln e^x = x$
$0.0268t > \ln 80 - \ln 35$ Subtract $(0.071t + \ln 35)$ from each side.
$t > \dfrac{\ln 80 - \ln 35}{0.0268}$ Divide each side by 0.0268.
$t > 30.85$ Use a calculator.

The number of cells of this bacteria exceed the number of cells in the other bacteria in about 30.85 minutes.

7.
$y = ae^{-0.00012t}$ Equation for the decay of Carbon-14
$0.85a = ae^{-0.00012t}$ $y = 0.85a$
$0.85 = e^{-0.00012t}$ Divide each side by a.
$\ln 0.85 = \ln e^{-0.00012t}$ Property of Equality for Logarithmic Functions
$\ln 0.85 = -0.00012t$ $\ln e^x = x$
$\dfrac{\ln 0.85}{-0.00012} = t$ Divide each side by -0.00012.
$1354 \approx t$ Use a calculator.

The bone is about 1354 years old.

9. about 14.85 billion yr **11.** about 20.1 yr

13a.
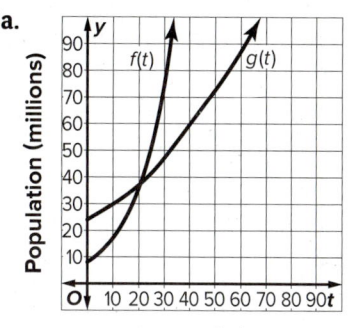

13b. The graphs intersect at $t = 20.79$. Sample answer: This intersection indicates the point at which both functions determine the same population at the same time.

13c. Sample answer: The logistic function $g(t)$ is a more accurate estimate of the country's population because $f(t)$ will continue to grow exponentially and $g(t)$ considers limitations on population growth such as food supply. **15.** $t \approx 113.45$ **17.** Sample answer: The spread of the flu throughout a small town. The growth of this is limited to the population of the town itself. **19.** 2.30 **21.** A **23.** C **25.** C **27a.** 100°C

27b. about -0.1099

27c. 24.6 min

Chapter 6   Study Guide and Review

1. logarithm   3. change of base formula
5. logarithmic function   7. Common logarithms have base 10, while natural logarithms have base e.

9.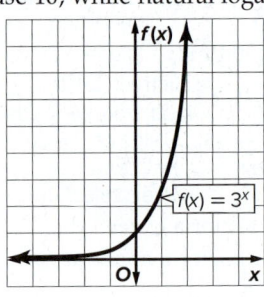
$D = (-\infty, \infty)$ or {all real numbers};
$R = (0, \infty)$ or $\{f(x) \mid f(x) > 0\}$

11.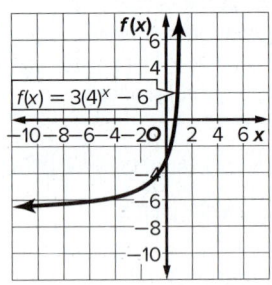
$D = (-\infty, \infty)$ or {all real numbers};
$R = (-6, \infty)$ or $\{f(x) \mid f(x) > -6\}$

13. 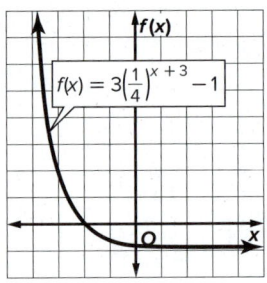 $D = (-\infty, \infty)$ or {all real numbers}; $R = (-1, \infty)$ or $\{f(x) \mid f(x) > -1\}$

15a. $f(x) = 120{,}000(0.97)^x$
15b. about 88,491
17. -7 19. $\frac{9}{41}$
21. -2 23. $x \leq -\frac{2}{5}$
25. $x \geq -1$ 27. 320   29. 8   31. 18, 54   33. 12, -36
35. 225   37. 215   39. 129   41. $2^{-4} = \frac{1}{16}$ 43. 4

45.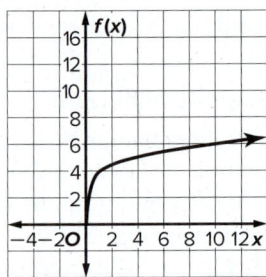

47. logarithmic; $y = -10.9503 + 26.2680\ln x$
49. exponential; $y = 0.8383(1.6997)x$
51. 2.5841   53. -1.2921 55. 30   57. $2\sqrt{3}$ 59. $\sqrt{\frac{5}{2}}$
61. 1.16 times   63. $x \approx \pm 1.3637$ 65. $r \approx 4.6102$
67. $\{x \mid x \leq -6.3013\}$ 69. $\{w \mid w \geq 3.3750\}$
71. 1.9769   73. 2.8424   75. 2.4935   77. 1.9459
79. 201.7144   81. $\{x \mid -3 < x < -0.2817\}$
83. $\{x \mid x < 1.9459\}$ 85. about 17.2 years   87. -6
89. $\left\{x \mid 0 < x \leq \frac{1}{125}\right\}$
91. -6 93. 1000   95. ≈ 5.8 days

CHAPTER 7
Rational Functions

Chapter 7 Get Ready
1. Sample answer: multiply each side by 8 **3.** The GCF of 72 and 77 is 1. **5.** $a = -4$, $b = 16$, $c = -1$ **7.** Associative Property **9.** Write the equation $5(11) = 8p$ and then solve for p.

Lesson 7-1
1. $\dfrac{x+3}{x+8}$; $x \neq -8$ **3.** D

5. $\dfrac{a^2x - b^2x}{by - ay} = \dfrac{x(a^2 - b^2)}{y(b-a)}$ Factor.

$= \dfrac{x(a-b)(a+b)}{y(b-a)}$ Factor.

$= \dfrac{-x(b-a)(a+b)}{y(b-a)}$ $a - b = -1(b-a)$

$= \dfrac{-x\cancel{(b-a)}(a+b)}{y\cancel{(b-a)}}$ Eliminate common factors.

$= \dfrac{-x(a+b)}{y}$ Simplify.

7. $\dfrac{2by}{3ax}$ **9.** $\dfrac{(a-b)(a+1)}{12(a-1)}$ **11.** 4 **13.** $\dfrac{x(x+6)}{x+4}$; $x \neq -4$ **15.** $\dfrac{(x+3)(x-z)}{4}$ **17.** $\dfrac{x(x+2)}{6(x+5)}$; $x \neq -5$ **19.** J **21.** $\dfrac{x^2}{x+6}$ **23.** $-\dfrac{c+4}{c+5}$ **25.** $\dfrac{c}{4ab^2f^2}$ **27.** $\dfrac{32b}{3ac^3f^2}$ **29.** $\dfrac{5a^4c}{3b}$

31. $\dfrac{y^2 + 8y + 15}{y-6} \cdot \dfrac{y^2 - 9y + 18}{y^2 - 9}$

$= \dfrac{(y+3)(y+5)}{y-6} \cdot \dfrac{(y-3)(y-6)}{(y-3)(y+3)}$ Factor.

$= \dfrac{\cancel{(y+3)}(y+5)}{\cancel{y-6}} \cdot \dfrac{\cancel{(y-3)}\cancel{(y-6)}}{\cancel{(y-3)}\cancel{(y+3)}}$

$= y + 5$ Simplify.

33. $\dfrac{(x+4)(x+2)}{2(x-5)}$ **35.** $\dfrac{(x-3)(x+1)}{6(x+7)}$ **37.** $\dfrac{-a^2(a+b)}{b^4}$ **39a.** $\dfrac{33}{121}$ **39b.** $\dfrac{33+m}{121+m}$ **41a.** $T(x) = \dfrac{0.4}{x+3}$

41b. about 3.9 mm thick **43.** $\dfrac{1}{4}$ **45.** $\dfrac{x(x+2)(x-1)}{(x+3)(x-7)}$

47. $\dfrac{20x^2y^6z^{-2}}{3a^3c^2} \cdot \left(\dfrac{16x^3y^3}{9acz}\right)^{-1}$

$= \dfrac{20x^2y^6z^{-2}}{3a^3c^2} \cdot \dfrac{9acz}{16x^3y^3}$ $\left(\dfrac{16x^3y^3}{9acz}\right)^{-1} = \dfrac{9acz}{16x^3y^3}$

$= \dfrac{20x^2y^6}{3a^3c^2z^2} \cdot \dfrac{9acz}{16x^3y^3}$ $z^{-2} = \dfrac{1}{z^2}$

$= \dfrac{2 \cdot 2 \cdot 5 \cdot x \cdot x \cdot y \cdot y \cdot y \cdot y \cdot y \cdot y \cdot 3 \cdot 3 \cdot a \cdot c \cdot z}{3 \cdot a \cdot a \cdot a \cdot c \cdot c \cdot z \cdot z \cdot 2 \cdot 2 \cdot 2 \cdot 2 \cdot x \cdot x \cdot x \cdot y \cdot y \cdot y}$ Factor.

$= \dfrac{\cancel{2} \cdot \cancel{2} \cdot 5 \cdot \cancel{x} \cdot \cancel{x} \cdot \cancel{y} \cdot \cancel{y} \cdot \cancel{y} \cdot y \cdot y \cdot y \cdot \cancel{3} \cdot 3 \cdot \cancel{a} \cdot \cancel{c} \cdot \cancel{z}}{\cancel{3} \cdot \cancel{a} \cdot a \cdot a \cdot \cancel{c} \cdot c \cdot \cancel{z} \cdot z \cdot \cancel{2} \cdot \cancel{2} \cdot 2 \cdot 2 \cdot \cancel{x} \cdot \cancel{x} \cdot x \cdot \cancel{y} \cdot \cancel{y} \cdot \cancel{y}}$ Eliminate common factors.

$= \dfrac{5 \cdot y \cdot y \cdot y \cdot 3}{a \cdot a \cdot c \cdot z \cdot 2 \cdot 2 \cdot x}$ Simplify.

$= \dfrac{15y^3}{4a^2cxz}$ Simplify.

49. $\dfrac{2(4x+1)(2x+1)}{5(2x-1)(x+2)}$ **51.** $\dfrac{2x+1}{-9x(x+2)}$ **53.** $\dfrac{x(x-2)(x+8)}{2(2x-1)(3x+1)}$ **55.** $\dfrac{-2(x-8)(x+4)(x-2)(x+1)}{(2x+1)^2(x^2+2x-6)}$

57a. 5 tracks $\cdot \dfrac{2 \text{ miles}}{1 \text{ track}} \cdot \dfrac{5280 \text{ feet}}{1 \text{ mile}} \cdot \dfrac{1 \text{ car}}{75 \text{ feet}}$

b. 5 tracks $\cdot \dfrac{2 \text{ miles}}{1 \text{ track}} \cdot \dfrac{5280 \text{ feet}}{1 \text{ mile}} \cdot \dfrac{1 \text{ car}}{75 \text{ feet}}$

$= \cancel{5} \text{ tracks} \cdot \dfrac{2 \text{ miles}}{1 \text{ track}} \cdot \dfrac{15 \cdot 352 \cancel{\text{ feet}}}{1 \cancel{\text{ mile}}} \cdot \dfrac{1 \text{ car}}{\cancel{5} \cdot 15 \cancel{\text{ feet}}}$

$= \dfrac{1 \cdot 2 \cdot 352 \cdot 1 \text{ car}}{1 \cdot 1 \cdot 1 \cdot 1}$

$= 704$ cars

c. 704 cars $\cdot \dfrac{8 \text{ attendants}}{1 \text{ car}} \cdot \dfrac{45 \text{ s}}{1 \text{ attendant}} \cdot \dfrac{1 \text{ min}}{60 \text{ s}} \cdot \dfrac{1 \text{ h}}{60 \text{ min}}$

$= 704$ cars $\cdot \dfrac{8 \text{ attendants}}{1 \cancel{\text{ car}}} \cdot \dfrac{45 \cancel{\text{ s}}}{1 \cancel{\text{ attendant}}} \cdot \dfrac{1 \text{ min}}{60 \cancel{\text{ s}}} \cdot \dfrac{1 \text{ h}}{60 \cancel{\text{ min}}}$

$= \dfrac{704 \cdot 8 \cdot 45 \cdot 1 \cdot 1 \text{ h}}{1 \cdot 1 \cdot 60 \cdot 60}$

$= 70.4$ hours

59. Sample answer: The two expressions are equivalent except that the rational expression is undefined at $x = 3$. **61.** $x^2 + x - 6$ **63.** Sample answer: Sometimes; with a denominator like $x^2 + 2$, in which the denominator cannot equal 0, the rational expression can be defined for all values of x. **65.** Sample answer: When the original expression was simplified, a factor of x was taken out of the denominator. If x were to equal 0, then this expression would be undefined. So, the simplified expression is also undefined for x. **67.** D **69.** D

Lesson 7-2
1. $80x^3y^3$ **3.** $3y(y-3)(y-5)$ **5.** $\dfrac{48y^4 + 25x^2}{20xy^3}$ **7.** $\dfrac{21b - 4}{36a}$ **9.** $\dfrac{9x + 15}{(x+3)(x+6)}$ **11.** $\dfrac{x - 11}{3(x+2)(x-2)}$ **13.** $\dfrac{14x - 10}{(x+1)(x-2)}$ **15.** $\dfrac{3y + 2}{y + 3}$ **17.** $\dfrac{2a + 5b}{3b - 8a}$ **19.** $180x^2y^4z^2$ **21.** $6(x+4)(2x-1)(2x+3)$ **23.** $\dfrac{28by^2z - 9bx}{105x^3y^4z}$ **25.** $\dfrac{20x^2y + 120y + 6x^2}{15x^3y}$ **27.** $\dfrac{15b^3 + 100ab^2 - 216a}{240ab^3}$

29. $\dfrac{6}{y^2 - 2y - 35} + \dfrac{4}{y^2 + 9y + 20}$

$= \dfrac{6}{(y-7)(y+5)} + \dfrac{4}{(y+4)(y+5)}$ Factor denominators.

$= \dfrac{6(y+4)}{(y-7)(y+5)(y+4)} + \dfrac{4(y-7)}{(y+4)(y+5)(y-7)}$

Multiply by missing factors.

$= \dfrac{6y + 24 + 4y - 28}{(y-7)(y+5)(y+4)}$ Add the numerators.

$= \dfrac{10y - 4}{(y-7)(y+5)(y+4)}$ Simplify.

31. $\dfrac{-10x - 10}{(2x-1)(x+6)(x-3)}$ **33.** $\dfrac{2x^2 + 32x}{3(x-2)(x+3)(2x+5)}$ **35.** $\dfrac{1000x + 800y}{x(x+2y)}$

37. $\dfrac{\dfrac{4}{x+5} + \dfrac{9}{x-6}}{\dfrac{5}{x-6} - \dfrac{8}{x+5}} = \dfrac{\dfrac{4(x-6)}{(x+5)(x-6)} + \dfrac{9(x+5)}{(x+5)(x-6)}}{\dfrac{5(x+5)}{(x+5)(x-6)} - \dfrac{8(x-6)}{(x+5)(x-6)}}$

$= \dfrac{\dfrac{4x - 24 + 9x + 45}{(x+5)(x-6)}}{\dfrac{5x + 25 - 8x + 48}{(x+5)(x-6)}}$ Simplify the numerator and denominator.

$= \dfrac{\dfrac{13x + 21}{(x+5)(x-6)}}{\dfrac{-3x + 73}{(x+5)(x-6)}}$ Combine like terms.

$= \dfrac{13x + 21}{(x+5)(x-6)} \div \dfrac{-3x + 73}{(x+5)(x-6)}$ Write as a division expression.

$= \dfrac{13x + 21}{(x+5)(x-6)} \cdot \dfrac{(x+5)(x-6)}{-3x + 73}$ Multiply by the reciprocal of the divisor.

$= \dfrac{13x + 21}{-3x + 73}$ Simplify.

39. $\dfrac{-x^2 + 33x + 16}{12x^2 + 11x - 27}$ **41.** $420x^5 y^4 z^3$

43. $(x+4)(x-4)(2x+1)(x-7)$ **45.** $\dfrac{360a^2 + 5a - 36}{60a^2}$

47. $\dfrac{42x + 41}{6(3x - 1)(x + 8)(2x + 3)}$ **49.** 0 **51.** $\dfrac{5a - 11}{6}$

53. $(x - 4)(x + 5)$ **55.** $-\dfrac{3}{2}$ **57.** -1

59a. $y = \dfrac{70x}{x - 70}$ **59b.** Sample answer: When the object is 70 mm away, y needs to be 0, which is impossible.

61. a. $P_0\left(\dfrac{s_0}{s_0 - x}\right) - P_0\left(\dfrac{s_0}{s_0 - y}\right) = \dfrac{P_0 s_0}{s_0 - x} - \dfrac{P_0 s_0}{s_0 - y}$

$= \dfrac{P_0 s_0 (s_0 - y)}{(s_0 - x)(s_0 - y)} - \dfrac{P_0 s_0 (s_0 - x)}{(s_0 - x)(s_0 - y)}$

$= \dfrac{P_0 s_0 (s_0 - y) - P_0 s_0 (s_0 - x)}{(s_0 - x)(s_0 - y)}$

$= \dfrac{P_0 s_0 s_0 - P_0 s_0 y - P_0 s_0 s_0 - P_0 s_0 x}{(s_0 - x)(s_0 - y)}$

$= \dfrac{P_0 s_0 x - P_0 s_0 y}{(s_0 - x)(s_0 - y)}$

b. $\dfrac{P_0 s_0 x - P_0 s_0 y}{(s_0 - x)(s_0 - y)} = \dfrac{(500)(332)(70) - (500)(332)(45)}{(332 - 70)(332 - 45)}$

$P_0 = 500, s_0 = 332, x = 70, y = 45$

$= \dfrac{4{,}150{,}000}{75{,}194}$ Simplify.

≈ 55.2 Hz Simplify.

63. $\dfrac{-3x^3 - 2x^2 + 16x - 5}{4x^3 + 18x^2 - 6x}$ **65.** Sample answer: $20a^4 b^2 c$, $15ab^6$, $9abc$ **67.** C **69.** B **71a.** C **71b.** C **71c.** B

Lesson 7-3

1. $x - 1 = 0$
$x = 1$
$f(x)$ is not defined when $x = 1$. So, there is a vertical asymptote at $x = 1$.
From $x = 1$, as x-values decrease, $f(x)$ values approach 0, and as x-values increase, $f(x)$ values approach 0. So there is a horizontal asymptote at $f(x) = 0$. The domain is all real numbers not equal to

1 or $D = \{x \mid x \neq 1\}$. The range is all real numbers not equal to 0 or $R = \{f(x) \mid f(x) \neq 0\}$.

3. 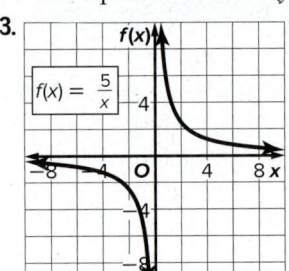 $D = \{x \mid x \neq 1\}$; $R = \{f(x) \mid f(x) \neq 0\}$

5. 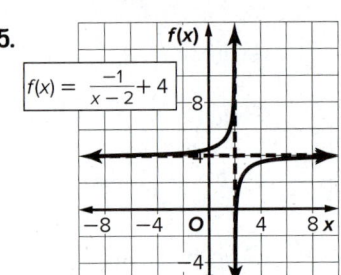 $D = \{x \mid x \neq 2\}$; $R = \{f(x) \mid f(x) \neq 4\}$

7. $x = -4, f(x) = 0; D = \{x \mid x \neq -4\}$; $R = \{f(x) \mid f(x) \neq 0\}$ **9.** $x = -6, f(x) = -2$; $D = \{x \mid x \neq -6\}; R = \{f(x) \mid f(x) \neq -2\}$

11. 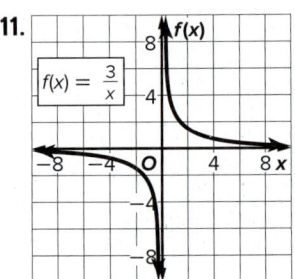 $D = \{x \mid x \neq 0\}$; $R = \{f(x) \mid f(x) \neq 0\}$

13. 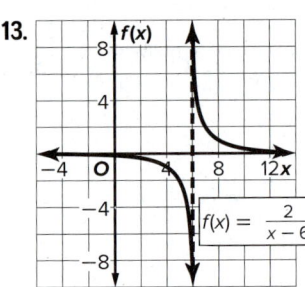 $D = \{x \mid x \neq 6\}$; $R = \{f(x) \mid f(x) \neq 0\}$

15. This represents a transformation of the graph of $f(x) = \dfrac{1}{x}$.
$a = 2$: The graph is stretched vertically.
$k = 3$: The graph is translated 3 units up. There is a horizontal asymptote at $f(x) = 3$. Domain: $D = \{x \mid x \neq 0\}$. Range: $R = \{f(x) \mid f(x) \neq 3\}$.

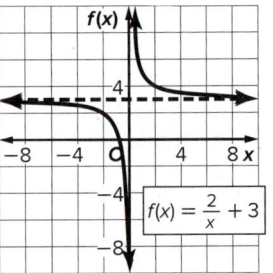

R70

17. 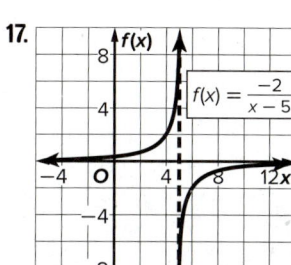 D = {x | x ≠ 5}; R = {f(x) | f(x) ≠ 0}

19. 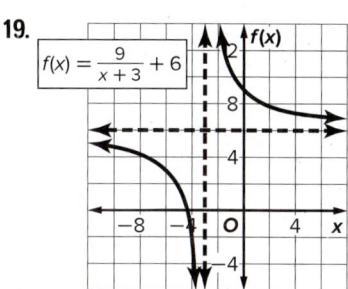 D = {x | x ≠ −3}; R = {f(x) | f(x) ≠ 6}

21. 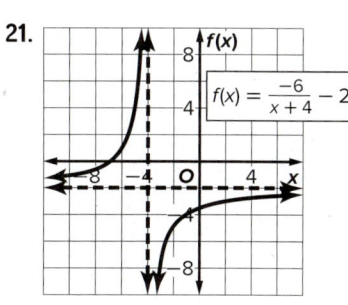 D = {x | x ≠ −4}; R = {f(x) | f(x) ≠ −2}

23a. $m = \dfrac{5000}{d}$

23b.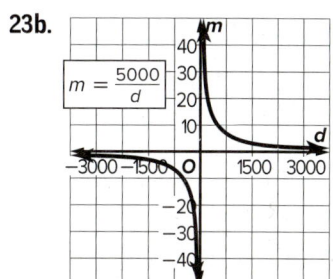

23c. 13.7 mi

25. 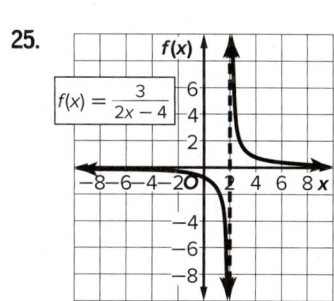 D = {x | x ≠ 2}; R = {f(x) | f(x) ≠ 0}

27. 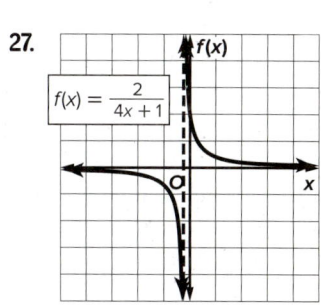 $D = \left\{x \,\middle|\, x \neq -\dfrac{1}{4}\right\}$; R = {f(x) | f(x) ≠ 0}

29. a.
$rt = d$ rate · time = distance
$rt = 60.5$ $d = 60.5$
$r = \dfrac{60.5}{t}$ Divide each side by t.

b. This represents a transformation of the graph of $f(x) = \dfrac{1}{x}$. There are asymptotes at $t = 0$ and $r = 0$. Because $a = 60.5$, the graph is stretched vertically.

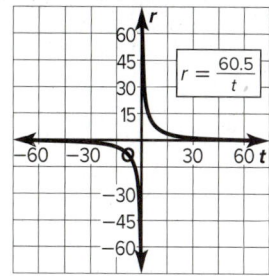

c. $r = \dfrac{60.5}{t}$ Write the equation.
$ = \dfrac{60.5}{0.48}$ $t = 0.48$
$ \approx 126$ ft/s Use a calculator.

31. D = {x | x ≠ −2}; R = {f(x) | f(x) ≠ −5}; x = −2, f(x) = −5

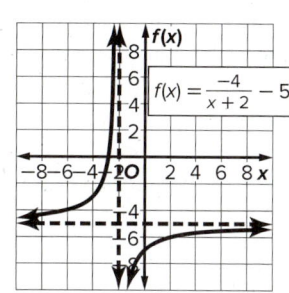

33. D = {x | x ≠ 4}; R = {f(x) | f(x) ≠ 3}; x = 4, f(x) = 3

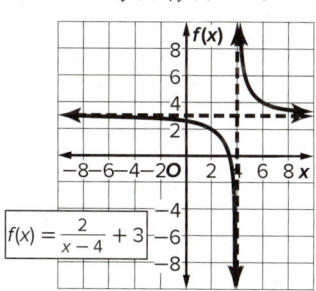

35. D = {x | x ≠ 7}; R = {f(x) | f(x) ≠ −8}; x = 7, f(x) = −8

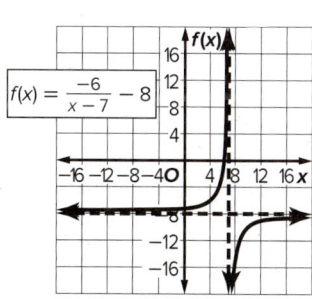

37a.

x	$a(x) = x^2$	$b(x) = x^{-2}$	$c(x) = x^3$	$d(x) = x^{-3}$
−4	16	$\frac{1}{16}$	−64	$-\frac{1}{64}$
−3	9	$\frac{1}{9}$	−27	$-\frac{1}{27}$
−2	4	$\frac{1}{4}$	−8	$-\frac{1}{8}$
−1	1	1	−1	−1
0	0	undefined	0	undefined
1	1	1	1	1
2	4	$\frac{1}{4}$	8	$\frac{1}{8}$
3	9	$\frac{1}{9}$	27	$\frac{1}{27}$
4	16	$\frac{1}{16}$	64	$\frac{1}{64}$

37b.
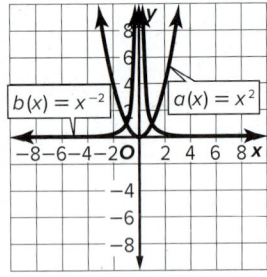

37c. $a(x)$: D = {all real numbers}, R = {$a(x) \mid a(x) \geq 0$}; as $x \to -\infty$, $a(x) \to \infty$, as $x \to \infty$, $a(x) \to \infty$; At $x = 0$, $a(x) = 0$, so there is a zero at $x = 0$. $b(x)$: D = {$x \mid x \neq 0$}, R {$b(x) \mid b(x) > 0$}; as $x \to -\infty$, $a(x) \to 0$, as $x \to \infty$, $a(x) \to 0$; At $x = 0$, $b(x)$ is undefined, so there is an asymptote at $x = 0$.

37d.
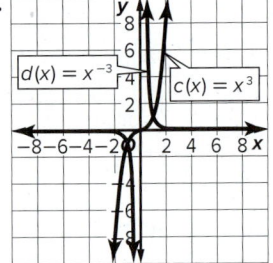

37e. $c(x)$: D = {all real numbers}, R = {all real numbers}; as $x \to -\infty$, $a(x) \to -\infty$, as $x \to \infty$, $a(x) \to \infty$; At $x = 0$, $a(x) = 0$, so there is a zero at $x = 0$. $d(x)$: D = {$x \mid x \neq 0$}, R {$b(x) \mid b(x) \neq 0$}; as $x \to -\infty$, $a(x) \to 0$, as $x \to \infty$, $a(x) \to 0$; At $x = 0$, $b(x)$ is undefined, so there is an asymptote at $x = 0$.

37f. For two power functions $f(x) = ax^n$ and $g(x) = ax^{-n}$, for every x, $f(x)$ and $g(x)$ are reciprocals. The domains are similar except that for $g(x)$, $x \neq 0$. Additionally, wherever $f(x)$ has a zero, $g(x)$ is undefined.

39a. The first graph has a vertical asymptote at $x = 0$ and a horizontal asymptote at $y = 0$. The second graph is translated 7 units up and has a vertical asymptote at $x = 0$ and a horizontal asymptote at $y = 7$. **39b.** Both graphs have a vertical asymptote at $x = 0$ and a horizontal asymptote at $y = 0$. The second graph is stretched by a factor of 4. **39c.** The first graph has a vertical asymptote at $x = 0$ and a horizontal asymptote at $y = 0$. The second graph is translated 5 units to the left and has a vertical asymptote at $x = -5$ and a horizontal asymptote at $y = 0$.

39d.
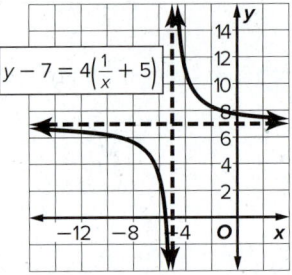

41. Sample answer: $f(x) = \frac{2}{x-3} + 4$ and $g(x) = \frac{5}{x-3} + 4$

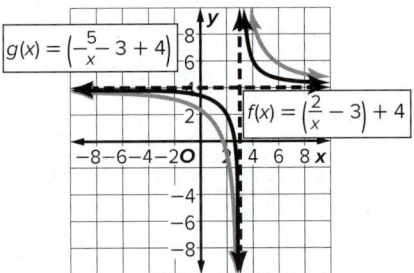

43. D **45a.** B, C, F **45b.** B

Lesson 7-4

1.
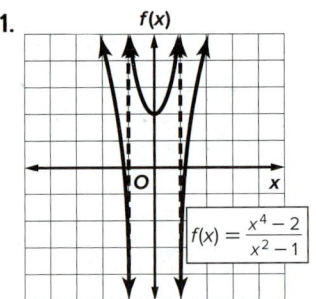

3a.
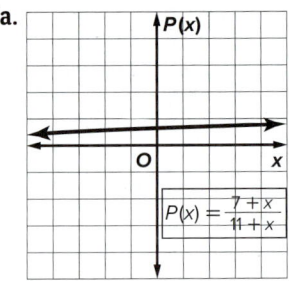

3b. The part in the first quadrant. **3c.** It represents his original field goal percentage of 63.6%. **3d.** $y = 1$; this represents 100% which he cannot achieve because he has already missed 4 field goals.

5 $x^2 + 8x + 20 = 0$ Set $a(x) = 0$.
Because $b^2 - 4ac = 8^2 - 4(1)(20)$ or -16, there are no real roots. So, there are no zeros.
$x + 2 = 0$ Set $b(x) = 0$.
$x = -2$ Subtract 2 from each side.
There is a vertical asymptote at $x = -2$. The degree of the numerator is greater than the degree of the denominator, so there is no horizontal asymptote. The difference between the degree of the numerator and the degree of the denominator is 1, so there is an oblique asymptote.

R72

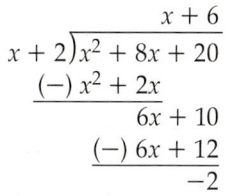

The oblique asymptote is $y = x + 6$.

7.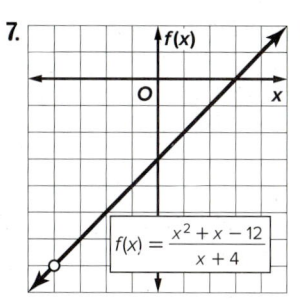

9.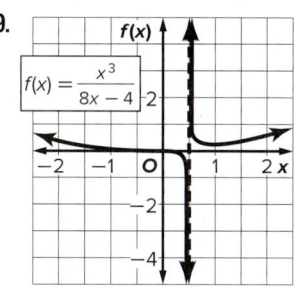

11.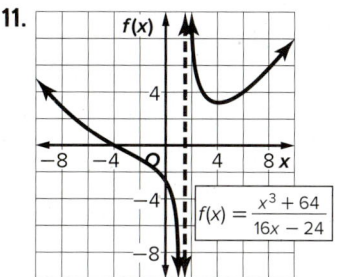

13.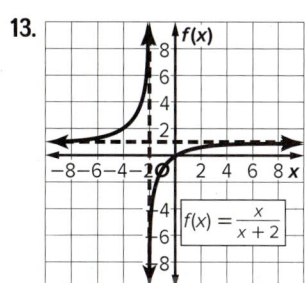

15. Because $a(x) = 4$, there are no zeros. The function is undefined for $x = 2$, so there is a vertical asymptote at $x = 2$. Because the degree of the numerator is less than the degree of the denominator, there is a horizontal asymptote at $f(x) = 0$. The difference between the degree of the numerator and the degree of the denominator is 2, so there is no oblique asymptote.

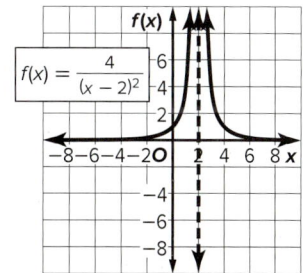

17.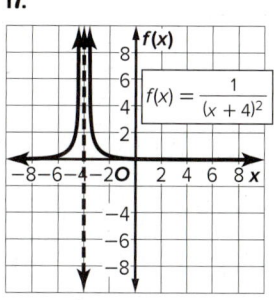

19.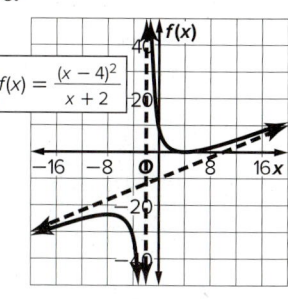

21.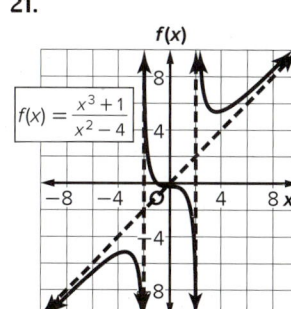

23.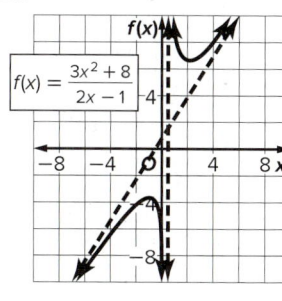

25.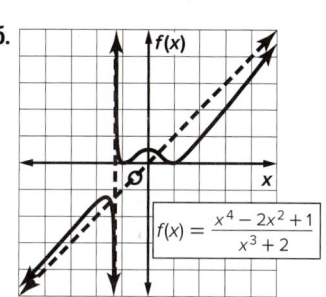

27a. $f(x) = \dfrac{x^2}{x+1}$ **27b.** Sample answer: I took the information I was given and made a visual. Then I set the denominator equal to 0 when x was -1, giving me $x + 1$. Because this is a rational function with no horizontal asymptotes I made the numerator x^2, meeting all the requirements I was given. **27c.** Sample answer: I made the assumption that there were no oblique asymptotes and the graph had no point discontinuity.

29.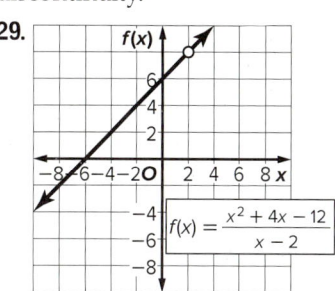

31.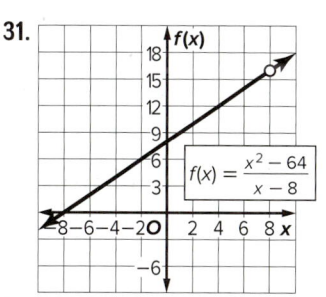

33.

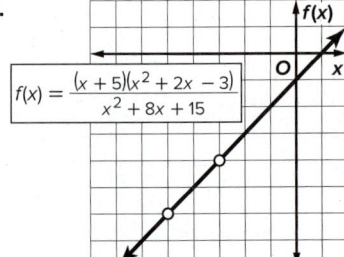

35.

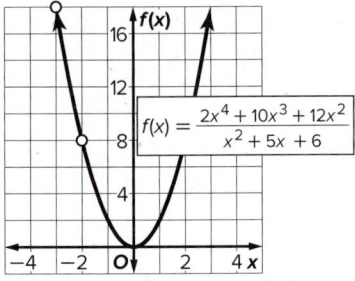

37. a. total cost = phone cost + monthly usage charge
$= 150 + 40x$

average monthly cost $= \dfrac{\text{total cost}}{\text{number of months}}$

$f(x) = \dfrac{150 + 40x}{x}$

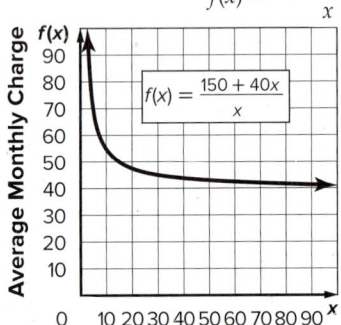

b. The vertical asymptote is $x = 0$. Because the degree of the numerator equals the degree of the denominator, the horizontal asymptote is at $f(x) = \dfrac{40}{1}$ or $f(x) = 40$.

c. Sample answer: The number of months and the average cost cannot have negative values.

d. $f(x) = \dfrac{150 + 40x}{x}$ Write the equation.

$45 = \dfrac{150 + 40x}{x}$ $f(x) = 45$

$45x = 150 + 40x$ Multiply each side by x.

$5x = 150$ Subtract $40x$ from each side.

$x = 30$ Divide each side by 5.

After 30 months, the average monthly charge will be $45.

39.

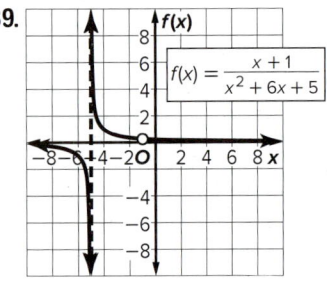

41.

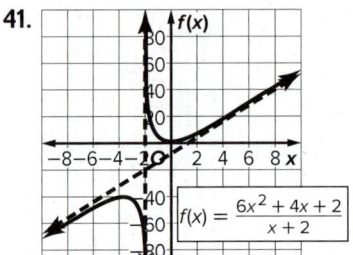

43. Similarities: Both have vertical asymptotes at $x = 0$. Both approach 0 as x approaches $-\infty$ and approach 0 as x approaches ∞. Differences: $f(x)$ has holes at $x = 1$ and $x = -1$, while $g(x)$ has vertical asymptotes at $x = \sqrt{2}$ and $x = -\sqrt{2}$. $f(x)$ has no zeros, but $g(x)$ has zeros at at $x = 1$ and $x = -1$.

45. $f(x) = \dfrac{x}{a-b} + \dfrac{c(a-b)}{a-b}$
$= \dfrac{x + ca - cb}{a-b}$

47. D   **49.** C   **51a.** C   **51b.** A   **51c.** B   **51d.** D

Lesson 7-5

1. 21   **3.** -32 **5.** -48 **7.** 1.5   **9.** -56 **11.** $m = \dfrac{1}{6}w$

13. $\dfrac{a_1}{b_1 c_1} = \dfrac{a_2}{b_2 c_2}$ Joint variation

$\dfrac{-60}{-5(4)} = \dfrac{a_2}{4(-3)}$ $a_1 = -60, b_1 = -5, c_1 = 4,$
$b_2 = 4, c_2 = -3$

$-60(4)(-3) = -5(4)(a_2)$ Cross multiply.

$720 = -20a_2$ Simplify.

$-36 = a_2$ Divide each side by -20.

15. -3 **17.** -22.5 **19.** 38   **21a.** $s = \dfrac{48}{t}$

21b. 3.2 hours   **23.** -10 **25.** direct   **27.** neither

29. $\dfrac{x_1}{y_2} = \dfrac{x_2}{y_1}$ Inverse variation

$\dfrac{16}{20} = \dfrac{x_2}{5}$ $x_1 = 16, y_1 = 5, y_2 = 20$

$16(5) = 20(x_2)$ Cross multiply.

$80 = 20x_2$ Simplify.

$4 = x_2$ Divide each side by 20.

31. -12 **33.** inverse; -2 **35.** combined; 10
37. direct; 4   **39.** direct; -2 **41.** inverse; 7   **43.** joint; 20

45a. $800 = rt$

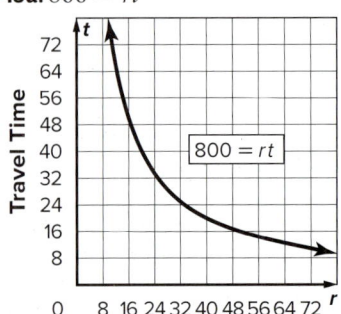

45b. $44.\overline{4}$ mph

47. a. $F = G\dfrac{m_1 m_2}{d^2}$ Law of Universal Gravitation
$= (6.67 \times 10^{-11})\dfrac{(7.36 \times 10^{22})(5.97 \times 10^{24})}{(3.84 \times 10^8)^2}$
$\approx 2 \times 10^{20}$ newtons

b. $F = G\dfrac{m_1 m_2}{d^2}$ Law of Universal Gravitation
$= (6.67 \times 10^{-11})\dfrac{(1.99 \times 10^{30})(5.97 \times 10^{24})}{(1.5 \times 10^{11})^2}$
$\approx 3.5 \times 10^{22}$ newtons

c. $F = G\dfrac{m_1 m_2}{d^2}$ Law of Universal Gravitation
$= (6.67 \times 10^{-11})\dfrac{(1000)(1000)}{(0.1)^2}$
$= 6.67 \times 10^{-3}$ newtons

49. a and g are directly related. **51.** Sample answer: The force of an object varies jointly as its mass and acceleration. **53a.** A **53b.** D

53c.

Number of Painters	Time to Paint the House (h)
10	3.75
15	2.5
20	1.86
25	1.5

53d.
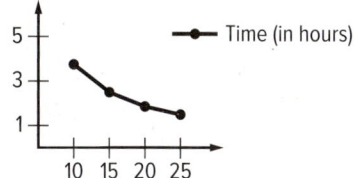

Time (in hours)

55. D **57.** C

Lesson 7-6

1. 11 **3.** 7

5. The LCD for the terms is $(x-5)(x-4)$.
$$\dfrac{8}{x-5} - \dfrac{9}{x-4} = \dfrac{5}{x^2 - 9x + 20}$$ Original equation

$\dfrac{(x-5)(x-4)(8)}{x-5} - \dfrac{(x-5)(x-4)(9)}{x-4} = \dfrac{(x-5)(x-4)(5)}{x^2-9x+20}$ Multiply by the LCD.

Divide common factors.

$(x-4)(8) - (x-5)(9) = 5$ Simplify.
$8x - 32 - 9x + 45 = 5$ Distribute.
$-x + 13 = 5$ Simplify.
$-x = -8$ Subtract 13 from each side.
$x = 8$ Divide each side by -1.

7. 14
9a. between 20 and 30 pounds

	Pounds	Price per Pound	Total Price
dried fruit	10	$6.25	6.25(10)
mixed nuts	m	$4.50	4.5m
trail mix	$10 + m$	$5.00	5(10 + m)

9b. $62.5 + 4.5m = 50 + 5m$ **9c.** 25; yes, the answer falls within the estimate **11a.** Kendal: $\dfrac{1}{60}$, Chandi: $\dfrac{1}{80}$
11b. Kendal: $\dfrac{x}{60}$, Chandi: $\dfrac{x}{80}$ **11c.** $\dfrac{x}{60} + \dfrac{x}{80} = 1$; between 30 and 40 mins **11d.** about 34.3 min; yes, the answer falls within the estimate **13.** $c < 0$, or $\dfrac{13}{18} < c$
15. $b < 0$, or $\dfrac{35}{12} < b$ **17.** 2 **19.** 1 **21.** ∅

23. cost of 3 pounds of bananas for \$0.90/pound = 0.9(3)
cost of x pounds of apples for \$1.25/pound = 1.25x
total weight = $3 + x$

$\dfrac{\text{total cost}}{\text{total weight}} = 1$ Write an equation.

$\dfrac{0.9(3) + 1.25x}{3 + x} = 1$ Substitute.

$\dfrac{2.7 + 1.25x}{3 + x} = 1$ Simplify the numerator.

$\dfrac{(3+x)(2.7 + 1.25x)}{3+x} = (3+x)(1)$ LCD is $(3+x)$. Multiply by the LCD.

$\dfrac{(3+x)(2.7+1.25x)}{3+x} = (3+x)(1)$ Divide out common factors.

$2.7 + 1.25x = 3 + x$ Simplify.
$2.7 + 0.25x = 3$ Subtract x from each side.
$0.25x = 0.3$ Subtract 2.7 from each side.
$x = 1.2$ Divide each side by 0.25.

She must purchase 1.2 pounds of apples.

25. $x < 0$ or $x > 1.75$ **27.** $x < -2$, or $2 < x < 14$
29. $x < -5$ or $4 < x < \dfrac{17}{3}$ **31.** 55.56 mph **33a.** 1; yes; 3

33b.
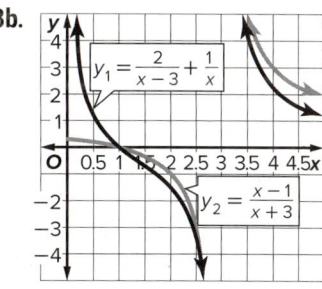

33c. 1; no

33d. Graph both sides of the equation. Where the graphs intersect, there is a solution. If they do not, then the possible solution is extraneous.

35. ∅ **37.** all real numbers except 5, −5, 0 **39.** Sample answer: Multiplying each sides of a rational inequality can produce extraneous solutions. Therefore, you should check all solutions to make sure that they satisfy the original equation or inequality. **41.** D **43.** D

45a.

	Distance (mi)	Avg Speed (mph)	Time (h)
With Current	4	$2+r$	$\frac{4}{2}+r$
Against Current	4	$2-r$	$\frac{4}{2}-r$

45b. $\frac{4}{2}+r+\frac{4}{2}-r=6$
45c. $(2+r)(2-r)$
45d. about 1.15 miles per hour

Chapter 7 Study Guide and Review

1. complex fraction **3.** rational equations
5. joint variation **7.** point discontinuity
9. If the degree of $a(x)$ is greater than the degree of $b(x)$, there is no horizontal asymptote. If the degree of $a(x)$ is less than the degree of $b(x)$, the horizontal asymptote is $y=0$. If the degree of $a(x)$ is equal to the degree of $b(x)$, the horizontal asymptote is $y=\dfrac{\text{leading coefficient of } a(x)}{\text{leading coefficient of } b(x)}$.

11. $\dfrac{10yz^2}{9x}$
13. $\dfrac{x-1}{x-2}$ **15.** $\dfrac{x-3}{x+1}$ **17.** $\dfrac{27b+10a^2}{12ab^2}$ **19.** $\dfrac{3xy^3+8y^3-5x}{6x^2y^2}$
21. $\dfrac{12x^2-10x+6}{2(x+2)(3x-4)(x+1)}$ **23.** $\dfrac{10x+20}{(x+6)(x+1)}$

25.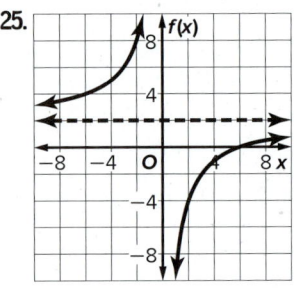
$D=\{x \mid x \neq 0\}$, $R=\{f(x) \mid f(x) \neq 2\}$

27.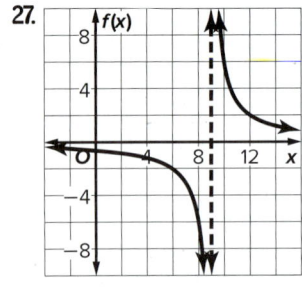
$D=\{x \mid x \neq 9\}$, $R=\{f(x) \mid f(x) \neq 0\}$

29.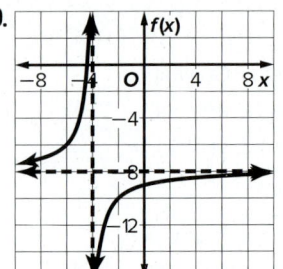
$D=\{x \mid x \neq -4\}$, $R=\{f(x) \mid f(x) \neq -8\}$

31. $x=-4, x=0$ **33.** $x=8$; hole: $x=-3$

35.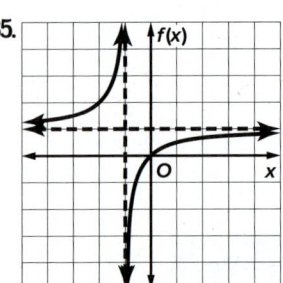

37.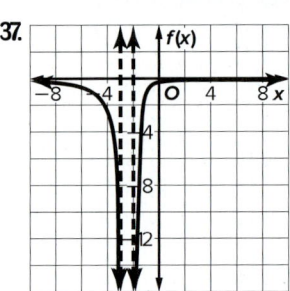

39. $a=15$ **41.** $y=-\dfrac{1}{3}$ **43.** $y=\dfrac{48}{5}$ **45.** $x=\dfrac{46}{17}$
47. $x=-7$ **49.** $x=8$ **51.** $-\dfrac{9}{10}<x<0$

CHAPTER 8
Statistics and Probability

Chapter 8 Get Ready

1. the average of the data; Add all the values together and divide by the total number of values.
3. the value that occurs most often in the set
5. Multiply the individual probabilities.
7. An outlier in a set of data will not affect the shape of the distribution plot.

Lesson 8-1

1. survey; sample: the students in the study; population: the student body
3. Observation study; sample answer: The scores of the participants are observed and compared without them being affected by the study. **5.** 909
7. The students' grades are being observed, so this is an observational study.
The sample consists of the physics students selected. The population is the set all college students that take a physics course.
9. Survey; sample answer: The data will be obtained from opinions given by members of the sample population. **11.** Observational study; sample answer: The eating habits of the participants will be observed and compared without them being affected by the study. **13.** 199 **15.** 448 **17.** 1674 **19.** 2721 **21.** There is no bias.
23. a. The sample group is the 8- to 18-year-olds who were actually surveyed. The population is represented by the sample, so the population is all 8- to 18-year-olds in the United States.
b. average time
c. Interpret the bar graph for each group. The red bar represents talking and the blue bar represents texting. Sample answer: The 8- to 10-year-old group talked for about 10 minutes a day and did not text at all. The 11- to 14-year-old group talked for about 30 minutes a day and texted for about 70 minutes a day. The 15- to 18-year-old group talked for about 40 minutes a day and texted for about 110 minutes a day.
d. Sample answer: A cell phone company might use a report like this to determine which age group to target in their ads.
25a. See students' work.
25b. Sample answer for Product A: ≈63.3%

Product A	
Number	Frequency
0–6	ʜʜ ʜʜ ʜʜ IIII
7–9	ʜʜ ʜʜ I

Sample answer for Product B: ≈76.7%

Product B	
Number	Frequency
0–7	ʜʜ ʜʜ ʜʜ ʜʜ III
8–9	ʜʜ II

25c. Sample answer: Yes; the probability that Product B is effective is 13.4% higher than that of Product A.
25d. Sample answer: It depends on what the product is and how it is being used. For example, if the product is a pencil sharpener, then the lower price may be more important than the effectiveness, and therefore, might not justify the price difference. However, if the product is a life-saving medicine, the effectiveness may be more important than the price, and therefore, might justify the price difference. **27.** true
29. An invalid sampling method and type of sample can produce bias. For example, if a sample is not random, the person conducting the study can influence the results by selecting a specific sample of people. Also, if an experiment is used when an observational study is the more logical type of study to be used, the study can be unreliable. For example, if someone wants to analyze the speeds of vehicles on a specific stretch of highway and decides to place an empty police car on the side of the road, the data will be affected by the police car. The results of this study will show lower speeds than are normally driven on the highway. Biased survey questions and incorrect procedures can affect the reliability of a study as well. A survey question that is poorly written may result in a response that does not accurately reflect the opinion of the participant.
31. D **33.** A, D **35.** D

Lesson 8-2

1. Sample answer: Use a spinner that is divided into two sectors, one containing 80% or 288° and the other containing 20% or 72°. Do 20 trials and record the results in a frequency table.

Outcome	Frequency
A	17
Below an A	3
Total	20

The probability of Clara getting an A on her next quiz is 0.85. The probability of earning any other grade is 1 − 0.85 or 0.15.

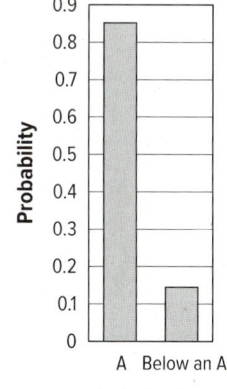

3. Sample answer: Use a random number generator to generate integers 1 through 20 where 1–9 represents tae kwon do, 10–15 represents yoga, 16–18 represents swimming, and 19–20 represents kickboxing. Do 20 trials and record the results in a frequency table.

Outcome	Frequency
tae kwon do	9
yoga	7
swimming	1
kickboxing	3
Total	20

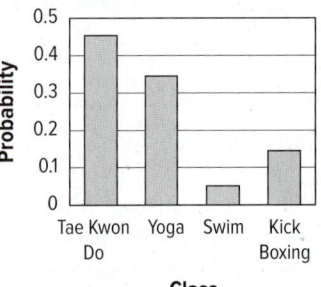

The probability of a customer taking the tae kwon do class is 0.45, taking yoga is 0.35, swimming is 0.05, and taking kickboxing is 0.15.

5. Sample answer: Use a spinner that is divided into two sectors. One sector should be 95% of the circle or $(0.95)360° = 342°$. The other should be 5% of the circle or $(0.05)360° = 18°$. Do 50 trials and record the results in a frequency table.

Outcome	Frequency
Sale	46
No Sale	4
Total	50

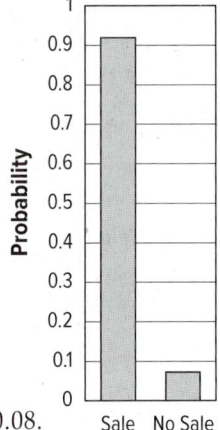

The probability of Ian selling a game is 0.92. The probability of not selling a game is $1 - 0.92$ or 0.08.

7. Sample answer: Use a random number generator to generate integers 1 through 20, where 1–8 represents drama, 9–14 represents mystery, 15–19 represents comedy, and 20 represents action. Do 20 trials and record the results in a frequency table.

Outcome	Frequency
drama	11
mystery	3
comedy	6
action	0
Total	20

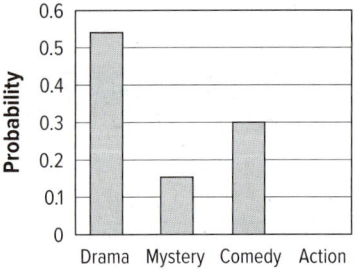

The probability of a customer choosing a drama is 0.55, choosing a mystery is 0.15, choosing a comedy is 0.3, and choosing an action film is 0. **9.** Sample answer: Use a random number generator to generate integers 1 through 20 where 1–9 represents Europe, 10–14 represents Asia, 15–17 represents South America, 18–19 represents Africa, and 20 represents Australia. Do 20 trials and record the results in a frequency table.

Outcome	Frequency
Europe	7
Asia	6
South America	5
Africa	2
Australia	0
Total	20

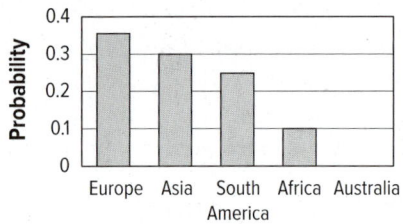

Travel Destinations

The probability of a customer traveling to Europe is 0.35, to Asia is 0.3, to South America is 0.25, to Africa is 0.1, and to Australia is 0.

11a. Sample answer: The theoretical probability that the player gets a hit is 30%, and the theoretical probability that he does not get a hit is 70%. Use a random number generator to generate integers 1 through 10. The integers 1–3 will represent a hit, and the integers 4–10 will represent the player not getting a hit. The simulation will consist of 50 trials.

11b. Sample answer: $P(\text{hit}) = 28\%$, $P(\text{not a hit}) = 72\%$

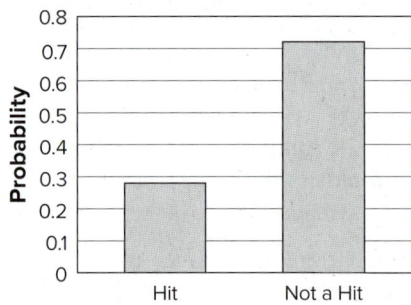

13a. Sample answer: The theoretical probability that a spin results in each prize is 20%. Use a random number generator to generate integers 1 through 5. The integer 1 will represent a hot pretzel, the integer 2 will represent a burger, the integer 3 will represent a large drink, the integer 4 will represent nachos, and the integer 5 will represent a small popcorn. The simulation will consist of 50 trials.

13b. Sample answer: $P(\text{hot pretzel}) = 20\%$, $P(\text{burger}) = 20\%$, $P(\text{large drink}) = 28\%$, $P(\text{nachos}) = 14\%$, $P(\text{small popcorn}) = 18\%$

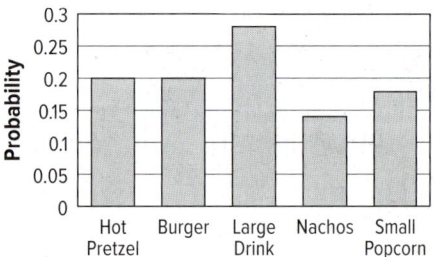

15a. Sample answer: 9, 10, 6, 6, 7, 9, 5, 9, 5, 7, 6, 5, 7, 3, 9, 7, 6, 7, 8, 7   **15b.** Sample answer: 4, 10, 5, 10, 6, 7, 12, 3, 7, 4, 7, 9, 3, 6, 4, 11, 5, 7, 5, 3

15c. Sample answer:

Trial	Sum of Roll	Sum of Output from Random Number Generator
1	9	4
2	10	10
3	6	5
4	6	10
5	7	6
6	9	7
7	5	12
8	9	3
9	5	7
10	7	4
11	6	7
12	5	9
13	7	3
14	3	6
15	9	4
16	7	11
17	6	5
18	7	7
19	8	5
20	7	3

15d. Sample answer:

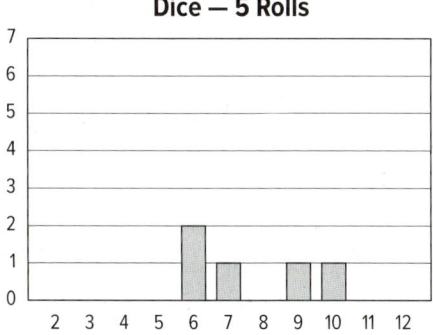

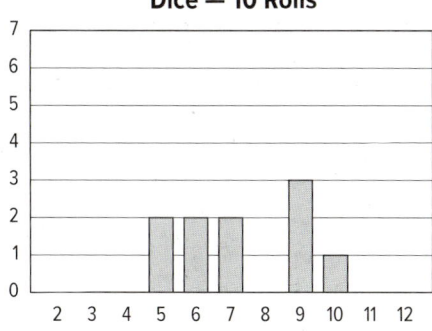

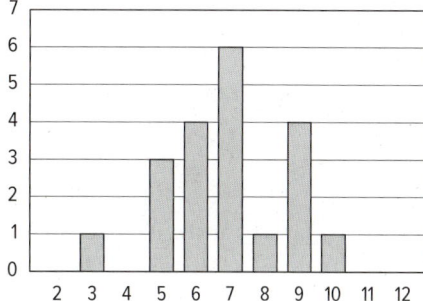

15e. Sample answer: The bar graph has more data points at the middle sums as more trials are added.

15f. Sample answer:

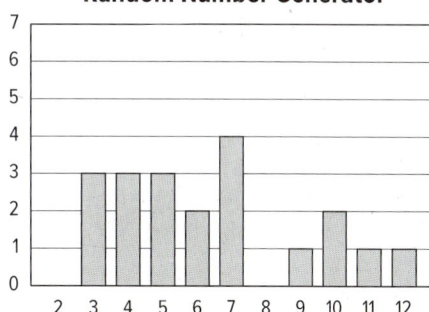

15g. Sample answer: They both have the most data points at the middle sums.   **17. Yes;** sample answer: If the spinner were going to be divided equally into three outcomes, each sector would measure 120. Because you only want to know the probability of outcome C, you can record spins that end in the red area as a success, or the occurrence of outcome C, and spins that end in the blue area as a failure, or an outcome of A or B.

19. Sample answer: You should consider the design of the simulation, how many trials were used, and whether the theoretical and experimental probabilities are reasonably close.   **21a.** Sample answer: Using a spinner with 8 equal regions   **21b.** Sample answer: Cat 1 – 0.14, Cat 2 – 0.15, Cat 3 – 0.12, Cat 4 – 0.10, Cat 5 – 0.13, Cat 6 – 0.16, Cat 7 – 0.09, Cat 8 – 0.11   **21c.** A   **23.** A, D, G

Lesson 8-3

1. 15   **3.** ±10%   **5.** $\mu = 52.5$ **7.** $p = 0.54$ **9.** $p = 0.60$
11. margin of error = ±5.77%
13. margin of error = ±3.64%
15. margin of error = ±2%
17. $\pm 3\% = \pm \frac{1}{\sqrt{n}}(100)$
$0.03\sqrt{n} = 1$
$\sqrt{n} = 33.333$
$n = 1111.11$
Round to the nearest individual; $n = 1111$.
19. $n = 204$ **21.** Fitness center A: 316.6, Fitness center B: 310.1. The claim is true.
23. Margin of error = $\pm \frac{1}{\sqrt{n}}(100)$ where n is the sample size, so $n = 250$.
margin of error = $\pm \frac{1}{\sqrt{250}}(100) \approx \pm 6.3 = \pm 6.3\%$
25. With a margin of error of ±0.7% the range is 85.3% to 86.7%.   **27.** Add the sum of all ratings and divide that by 100, the number of candy bars.

29. $p = \frac{250}{300} = 0.83$
31. $p = 0.75$ **33.** B **35.** A **37.** A, C
39a. ± 8.2% **39b.** $n = 473$ **39c.** For better results, Cassy's best option is to survey 473 homeowners with a margin of error of ± 4.6%.

Lesson 8-4

1 a. First, press STAT ENTER and enter each data value. Then, press 2nd [STAT PLOT] ENTER ENTER and choose ⬚. Finally, adjust the window to dimensions appropriate for the data.

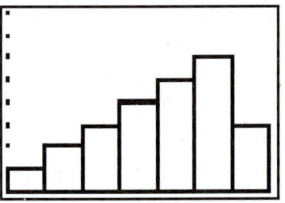
[4, 32] scl: 4 by [0, 8] scl: 1

Because the majority of the data is on the right and there is a tail on the left, the distribution is negatively skewed.

b. Sample answer: The distribution is skewed, so use the five-number summary. Press STAT ▶ ENTER ENTER and scroll down to display the statistics for the data set.

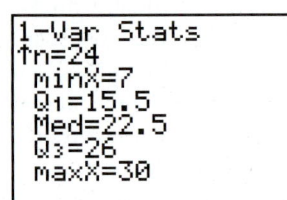

The range is 7 to 30 minutes. The median is 22.5 minutes, and half of the data are between 15.5 and 26 minutes.

3a. Mrs. Johnson's Class   Mr. Edmunds' Class

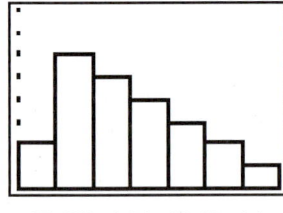 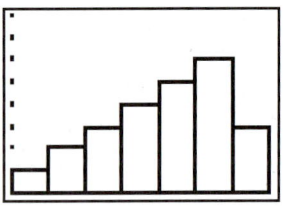
[5, 40] scl: 5 by [0, 8] scl: 1   [5, 40] scl: 5 by [0, 8] scl: 1

Mrs. Johnson's class, positively skewed; Mr. Edmunds' class, negatively skewed

3b. Sample answer: The distributions are skewed, so use the five-number summaries. The range for both classes is the same. However, the median for Mrs. Johnson's class is 17 and the median for Mr. Edmunds' class is 28. The lower quartile for Mr. Edmunds' class is 20. Because this is greater than the median for Mrs. Johnson's class, this means that 75% of the data from Mr. Edmunds' class is greater than 50% of the data from Mrs. Johnson's class. Therefore, we can conclude that the students in Mr. Edmunds' class had slightly higher sales overall than the students in Mrs. Johnson's class.

5a.

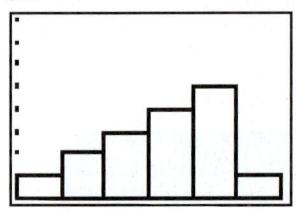 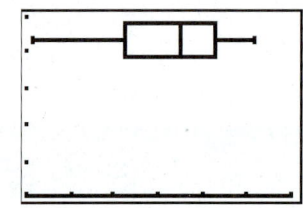
[50, 200] scl: 25 by [0, 8] scl: 1   [50, 200] scl: 25 by [0, 5] scl: 1

negatively skewed

5b. Sample answer: The distribution is skewed, so use the five-number summary. The range is 53 to 179 points. The median is 138.5 points, and half of the data are between 106.5 and 157 points.

7a. **Sophomore Year**

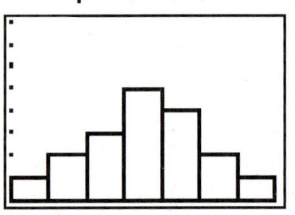
[1200, 1900] scl: 100 by [0, 8] scl: 1

Junior Year

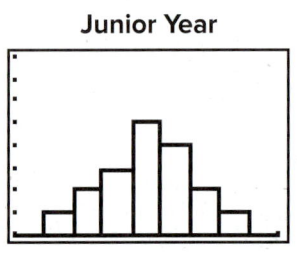
[1300, 2200] scl: 100 by [0, 8] scl: 1

both symmetric

7b. Sample answer: The distributions are symmetric, so use the means and standard deviations. The mean score for sophomore year is about 1552.9 with standard deviation of about 147.2. The mean score for junior year is about 1753.8 with standard deviation of about 159.1. We can conclude that the scores and the variation of the scores from the mean both increased from sophomore year to junior year.

9 a. Enter the tuitions for the public colleges as **L1**. Graph these data as **Plot1** by pressing 2nd [STAT PLOT] ENTER ENTER and choosing ⬚. Enter the tuitions for the private colleges as **L2**. Graph these data as **Plot2** by pressing 2nd [STAT PLOT] ▼ ENTER ENTER and choosing ⬚. For **Xlist**, enter L2. Adjust the window to dimensions appropriate for the data.

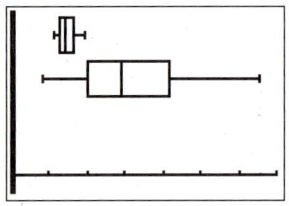
[0, 35,000] scl: 5000 by [−10, 100] scl: 1

For both sets of data, the whiskers are approximately equal, and the median is in the middle of the data. The distributions appear to be symmetric.
b. Sample answer: The distributions appear to be symmetric, so use the means and standard deviations. Press STAT ▶ ENTER ENTER to display the statistics for the public colleges.

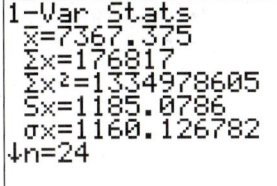

Press STAT ▶ ENTER 2nd [L1] ENTER to display the statistics for the private colleges.

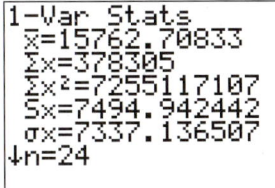

The mean for the public colleges is $7367.38 with standard deviation of about $1160.13. The mean for private colleges is about $15,762.71 with standard deviation of about $7337.14. We can conclude that not only is the average cost of private schools far greater than the average cost of public schools, but the variation of the costs from the mean is also much greater.

11a.

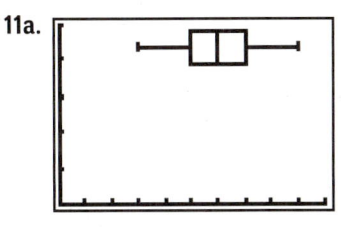

[0, 30] sc: 3 by [0, 5] scl: 1

Sample answer: The distribution is symmetric, so use the mean and standard deviation. The mean of the data is 18 with standard deviation of about 5.2 points.

11b.

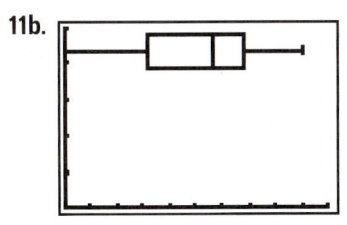

mean: 14.6; median: 17

[0, 30] sc: 3 by [0, 5] scl: 1

11c. Sample answer: Adding the scores from the first four games causes the shape of the distribution to go from being symmetric to being negatively skewed. Therefore, the center and spread should be described using the five-number summary.

13 a. Sample answer: Because the distribution is positively skewed, the median will be to the left of the mean closer to the majority of the data. An estimate for the median is 10. The mean will be more affected by the tail, and will be to the right of the majority of the data. An estimate for the mean is 14. **b.** Sample answer: Because the distribution is negatively skewed, the median will be to the right of the mean closer to the majority of the data. An estimate for the median is 24. The mean will be more affected by the tail, and will be to the left of the majority of the data. An estimate for the mean is 20. **c.** Sample answer: Because the distribution is symmetric, the mean and median will be approximately equal near the middle of the data. An estimate for the mean and median is 17.

15. Sample answer: The heights of the players on the Pittsburgh Steelers roster appear to represent a normal distribution.

Heights of the Players on the 2009 Pittsburgh Steelers Roster (inches)						
75	74	71	70	74	75	77	72
71	72	70	70	75	78	71	75
77	71	69	70	77	75	74	73
77	71	73	76	76	74	72	75
75	70	70	74	73	76	79	73
71	69	70	77	77	80	75	77
67	74	69	76	77	76		

The mean of the data is about 73.61 in. or 6 ft 1.61 in. The standard deviation is about 2.97 in.

The birth months of the players do not display central tendency.

[66, 82] scl: 2 by [0, 15] scl: 3

Birth Months of the Players on the 2009 Pittsburgh Steelers Roster						
1	12	10	3	11	1	10	5
4	8	9	11	1	1	11	5
8	6	11	4	3	4	8	5
3	7	2	1	11	4	3	2
1	1	6	1	6	8	11	9
3	3	1	6	9	1	9	9
6	5	10	11	11	12		

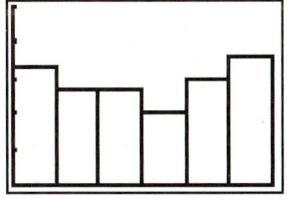

[0, 12] scl: 2 by [0, 15] scl: 3

17. C **19.** E **21a.** 9A: max 70, min 20; 9B: max 80, min 45
21b. 75% **21c.** 25% **21d.** 50%

Lesson 8-5

1. Sample answer: The conclusion may not be valid because the report is based on an observational study, so a cause-and-effect conclusion cannot be made. There may be other reasons why students who read a newspaper had higher test scores. For example, they may be more likely to read books or may spend more time studying.

3. Sample answer: The scale on the y-axis makes it appear as if the number of subscribers was almost constant, although the number of subscribers actually decreased by 25% from 2012 to 2016.

5. The conclusion is valid because the study was a controlled experiment with a large sample and individuals randomly assigned to a control group and a treatment group.

7. To analyze the graph, check the values on each axis. The least value on the y-axis is 30. This means the y-axis does not begin at 0.
The y-axis does not begin at a zero, which makes it appear as if sales revenue is increasing rapidly.

9. The conclusion is not valid because 6 out of 5235 people surveyed is approximately 0.1%; the population of the United States is about 320,000,000, and 0.1% of this population is about 320,000. The conclusion that millions of Americans know viral dance moves is not valid.

11. The formula for the margin of error is $\pm \frac{1}{\sqrt{n}}$ (100) where n is the sample size.
The sample size in this situation is 2050, because 2050 registered voters were surveyed.
So, the margin of error $= \pm \frac{1}{\sqrt{2050}}$ (100) ≈ 2.2.
The margin of error is about 2.2%, which means that it is unlikely that Rodriguez would receive less than 53% − 2.2% = 50.8% of the vote. Therefore, according to this survey, Rodrigues is likely to receive more than 50% of the vote, the election is not a toss-up, and Rodriguez is likely to win.

13. Disagree; it is possible to draw a cause-and-effect conclusion from a study if the study is a controlled experiment in which individuals are randomly assigned to a control group and to a treatment group.

15. Sample answer:

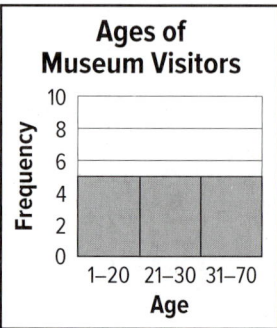

Lesson 8-6

17. B **19.** B, E **21.** D

1. $251 < X < 581$ **3a.** about 386 teens **3b.** 84%
5. 2.05; 2.05 standard deviations greater than the mean **7.** 3.08; 2.40 standard deviations less than the mean **9.** $X < 15.9$ or $X > 42.7$ **11a.** about 509
11b. 0.15% **13.** −1.33; 1.33 standard deviations less than the mean **15.** 177.7; 1.73 standard deviations greater than the mean

17. a. Find the z-values associated with 90,000 and 110,000. The mean μ is 113,627 and the standard deviation σ is 14,266.
$$z = \frac{X - \mu}{\sigma}$$
$$= \frac{90,000 - 113,627}{14,266}$$
$$\approx -1.656$$
$$z = \frac{X - \mu}{\sigma}$$
$$= \frac{110,000 - 113,627}{14,266}$$
$$\approx -0.254$$
Use a graphing calculator to find the area between the z-values.

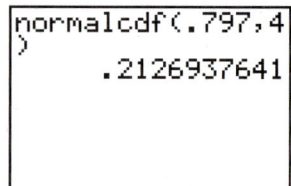

The value 0.35 is the percentage of batteries that will last between 90,000 and 110,000 miles. The total number of batteries in this group is 0.35 × 20,000 or 7000.

b. Find the z-value associated with 125,000.
$$z = \frac{X - \mu}{\sigma}$$
$$= \frac{125,000 - 113,627}{14,266}$$
$$\approx 0.797$$
We are looking for values greater than 125,000, so we can use a graphing calculator to find the area between $z = 0.797$ and $z = 4$.

```
normalcdf(.797,4
)
         .2126937641
```

The value 0.21 is the percentage of batteries that will last between more than 125,000 miles. The total number of batteries in this group is 0.21 × 20,000 or 4200.

c. Find the z-value associated with 100,000.

$$z = \frac{X - \mu}{\sigma}$$

$$= \frac{100{,}000 - 113{,}627}{14{,}266}$$

$$\approx -0.955$$

We are looking for values less than 100,000, so we can use a graphing calculator to find the area between $z = -4$ and $z = -0.955$.

```
normalcdf(-4,-.9
55)
         .1697571502
```

The probability that if you buy a car battery at random it will last less than 100,000 miles is about 17.0%.

19. a. The middle 68% represents all data values within one standard deviation of the mean. Add $\pm\$115$ to $829. The range of rates is $714 to $944.

b. Find the z-value associated with 1000.

$$z = \frac{X - \mu}{\sigma}$$

$$= \frac{1000 - 829}{115}$$

$$\approx 1.487$$

We are looking for values more than 1000, so we can use a graphing calculator to find the area between $z = 1.487$ and $z = 4$.

```
normalcdf(1.487,
4)
         .0684757505
```

The probability that a customer selected at random will pay more than $1000 is about 6.8%. Out of 900 people, about $0.068 \cdot 900$ or 62 people will pay more than $1000.

c. Sample answer: I would expect people with several traffic citations to lie to the far right of the distribution where insurance costs are highest, because I think insurance companies would charge them more.

d. Sample answer: I think auto insurance companies would charge younger people more than older people because they have not been driving as long. I think they would charge more for expensive cars and sports cars and less for cars that have good safety ratings. I think they would charge a person less if they have a good driving record and more if they have had tickets and accidents.

21. Sample answer: Hiroko; Monica's solution would work with a uniform distribution.   **23.** Sample answer:

True; according to the Empirical Rule, 99% of the data lie within 3 standard deviations of the mean. Therefore, only 1% will fall outside of three sigma. An infinitely small amount will fall outside of six-sigma.   **25.** Sample answer: The scores per team in each game of the first round of the 2010 NBA playoffs. The mean is 96.56 and the standard deviation is 11.06. The middle 68% of the distribution is $85.50 < X < 107.62$. The middle 95% is $74.44 < X < 118.68$. The middle 99.7% is $63.38 < X < 129.74$.   **27.** C   **29.** D   **31.** C   **33a.** 0.9332   **33b.** 0.7734   **33c.** 0.2956   **33d.** 0.9199

Lesson 8-7

1a. $P(\text{Alexa}) = \frac{5}{18}$, $P(\text{Gayle}) = \frac{4}{9}$, $P(\text{Mei}) = \frac{5}{18}$; because these probabilities are not equal, the method is not fair.   **1b.** Sample answer: If the sum is 2 through 5, Alexa is the Captain; if the sum is 6 or 8, Gayle is the captain; if the sum is 9 through 12, Mei is the captain; if the sum is 7, roll again.

3. The expected number of tokens from each roll is $20(0.02) + 10(0.1) + 5(0.18) + 2(0.3) + 1(0.4) = 3.3$. By playing 10 times, Latrell can expect to win $10(3.3) = 33$ tickets.
It takes 15 tickets to get a prize, so 33 tickets can be redeemed for 2 prizes, with 3 tickets leftover.
The value of the 2 prizes is $2(\$4.95) = \$9.90$, which is more than the cost of the 10 rolls, because $10(\$0.25) = \$2.50$.
The value of the prizes is worth more than the cost of playing the games, so this is a good decision.

5. Yes; $P(\text{Micah}) = P(\text{Carrie}) = \frac{1}{2}$

7. No; $P(\text{Micah}) = \frac{1}{3}$; $P(\text{Carrie}) = \frac{2}{3}$

9. No; the expected number on a box top is 2.16, so Ricardo can expect to have a sum of 17.28 with 8 boxes. This results in one prize worth $18.95, which is less than the cost of the 8 boxes ($29.20).

11. Sample answer: Latoya presents if the sum is 6 or less; David presents if the sum is 8 or more; roll again if the sum is 7.   **13.** Sample answer: Latoya presents if the number is 5 through 13; David presents if the number is 14 through 22; if the number is 23, choose a new number.

15. a. The probability that a pop is a berry pop is $\frac{1}{4}$ or 0.25.
The probability that a pop is not a berry pop is $1 - P(\text{berry}) = 1 - 0.25 = 0.75$.
The probability that none of the 6 pops is a berry pop is $(0.75)^6 \approx 17.8\%$.
This means that approximately 1 out of 5 boxes will not contain any berry pops, so it is not a very unlikely occurrence. For this reason, it makes sense to disagree with Brian's decision.

b. From part **a**, the probability that a box contains no berry pops is about 0.178.
A case contains 5 boxes. So, the probability of no berry pops in a case is $(0.178)^5 \approx 0.02\%$. Because this

probability is close to 0, this is very unlikely to happen by chance. For this reason, it makes sense to agree with Brian's decision in this case.
17. Disagree; Sample answer: Assign one option to the outcomes 1–3, assign the other option to the outcomes 4–6, and spin again if the outcome is 7.
19. $P(\text{red}) = 0.5$, but $P(\text{club}) = P(\text{spade}) = 0.25$, so the probabilities of the options are not all equal. Instead, he could assign the first option to the hearts. If the chosen card is a diamond, he should choose a new card until one of the options is chosen. **21.** B, C, E, F **23.** B **25.** It's a bad decision because the cost of cans is greater than the expected value of the prizes.

17. The conclusion may not be valid. The sample may not have been chosen using a simple random sample and may have been too small. This is an observational study, not an experiment. The cause-and-effect relationship cannot be determined.
19a. 68.5%
19b. about 2 games
21a. The president has double the chance of being chosen as everybody else. Everyone should have an equally likely chance of being chosen.
21b. The advisor should assign each member 1 number and if the other number is rolled, ignore it and roll again.

Chapter 8 Study Guide and Review

1. bias   **3.** Empirical Rule   **5.** A negatively skewed distribution has most of the data to the right of the mean and a positively skewed distribution has most of the data to the left of the mean.   **7.** survey; sample: every tenth shopper; population: all potential shoppers
9. survey; sample: every fifth person; population: student body
11a. Sample answer: The theoretical probability that the bus is late is 60%, and the theoretical probability that the bus is not late is 40%. Use a random number generator to generate integers 1 through 5. The integers 1–3 will represent the bus being late, and the integers 4–5 will represent the bus not being late. The simulation will consist of 50 trials.
11b. Sample answer: $P(\text{late}) = 58\%$

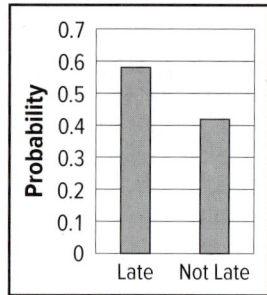

13. There is a population mean of 1.5 pets.
15a.

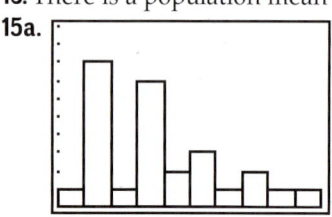

[8, 18] scl: 1 by [0, 10] scl: 1

positively skewed
15b. Sample answer: The distribution is skewed, so use the five-number summary. The values range from 8.7 to 18.1 days. The median is 11.55 days, and half of the data are between 9.55 and 13.3 days.

CHAPTER 9
Trigonometric Functions

Chapter 9 Get Ready
1. $a^2 + b^2 = c^2$ 3. Yes; they both measure 45°.
5. $x = 12$ 7. Each leg length will be $\frac{x\sqrt{2}}{2}$.

Lesson 9-1
1. $\sin \theta = \frac{4}{5}$; $\cos \theta = \frac{3}{5}$; $\tan \theta = \frac{4}{3}$; $\csc \theta = \frac{5}{4}$; $\sec \theta = \frac{5}{3}$; $\cot \theta = \frac{3}{4}$ 3. $\sin A = \frac{\sqrt{33}}{7}$, $\csc A = \frac{7\sqrt{33}}{33}$, $\cot A = \frac{4\sqrt{33}}{33}$, $\sec A = \frac{7}{4}$, $\tan A = \frac{\sqrt{33}}{4}$

5 The side opposite the 60° angle is given. The missing measure is the hypotenuse. Use the sine function to find x.

$\sin \theta = \frac{\text{opp}}{\text{hyp}}$ Sine function
$\sin 60° = \frac{22}{x}$ $\theta = 60°$ and opp = 22
$\frac{\sqrt{3}}{2} = \frac{22}{x}$ $\sin 60° = \frac{\sqrt{3}}{2}$
$\sqrt{3}x = 44$ Cross multiply.
$x = \frac{44}{\sqrt{3}}$ Divide each side by $\sqrt{3}$.
$x \approx 25.4$ Use a calculator.

7. 8.3   9. 25.4   11. about 274.7 ft   13. $\sin \theta = \frac{12}{13}$; $\cos \theta = \frac{5}{13}$; $\tan \theta = \frac{12}{5}$; $\csc \theta = \frac{13}{12}$; $\sec \theta = \frac{13}{5}$; $\cot \theta = \frac{5}{12}$ 15. $\sin \theta = \frac{\sqrt{51}}{10}$; $\cos \theta = \frac{7}{10}$; $\tan \theta = \frac{\sqrt{51}}{7}$; $\csc \theta = \frac{10\sqrt{51}}{51}$; $\sec \theta = \frac{10}{7}$; $\cot \theta = \frac{7\sqrt{51}}{51}$
17. $\sin A = \frac{8}{17}$, $\cos A = \frac{15}{17}$, $\csc A = \frac{17}{8}$, $\sec A = \frac{17}{15}$, $\cot A = \frac{15}{8}$ 19. $\sin B = \frac{3\sqrt{10}}{10}$, $\cos B = \frac{\sqrt{10}}{10}$, $\csc B = \frac{\sqrt{10}}{3}$, $\sec B = \sqrt{10}$, $\cot B = \frac{1}{3}$ 21. 12.7

23 $\tan \theta = \frac{\text{opp}}{\text{adj}}$ Tangent function
$\tan 30° = \frac{x}{18}$ Replace θ with 30°, opp with x, and adj with 18.
$\frac{\sqrt{3}}{3} = \frac{x}{18}$ $\tan 30° = \frac{\sqrt{3}}{3}$
$\frac{18\sqrt{3}}{3} = x$ Multiply each side by 18.
$10.4 \approx x$ Use a calculator.

25. 8.7   27. 132.5 ft   29. 30   31. 36.9   33. 32.5
35. 25.3 ft higher   37. $x = 21.9$, $y = 20.8$
39. $x = 19.3$, $y = 70.7$ 41. 54.9   43. 20.5   45. 11.5

47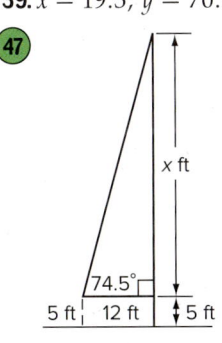

$\tan \theta = \frac{\text{opp}}{\text{adj}}$ Tangent function
$\tan 74.5° = \frac{x}{12}$ Replace θ with 74.5°, opp with x, and adj with 12.
$12 \cdot \tan 74.5° = x$ Multiply each side by 12.
$43 \approx x$ Use a calculator.

So, the height of the bird's nest is $43 + 5$ or 48 feet.

49a. about 647.2 ft   49b. about 239.4 ft
51. $m\angle A = 59°$, $a = 31.6$, $c = 36.9$ 53. $m\angle A = 38.7°$, $m\angle B = 51.3°$, $b = 7.5$, $c = 9.6$ 55. True; $\sin \theta = \frac{\text{opp}}{\text{hyp}}$ and the values of the opposite side and the hypotenuse of an acute triangle are positive, so the value of the sine function is positive.   57. Sample answer: The slope describes the ratio of the vertical rise to the horizontal run of the roof. The vertical rise is opposite the angle that the roof makes with the horizontal. The horizontal run is the adjacent side. So, the tangent of the angle of elevation equals the ratio of the rise to the run, or the slope of the roof; $\theta = 33.7°$.   59. C   61a. D   61b. D   61c. D   61d. C
61e. $28.3^2 \approx 15^2 + 24^2$

Lesson 9-2

1.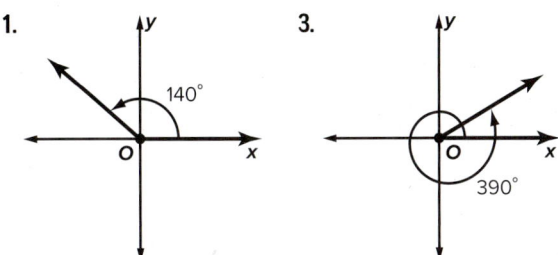

5 Sample answer:
positive angle: $175° + 360° = 535°$
negative angle: $175° - 360° = -185°$

7. 45°   9. $-\frac{2\pi}{9}$

11., 13.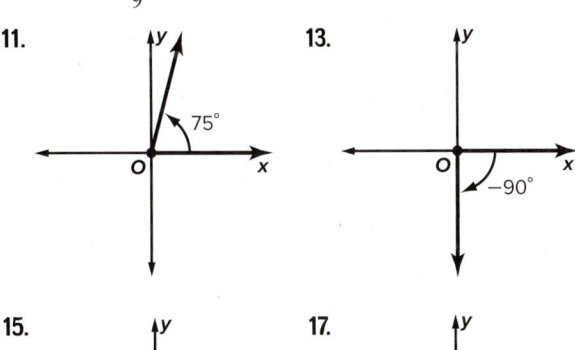

15., 17.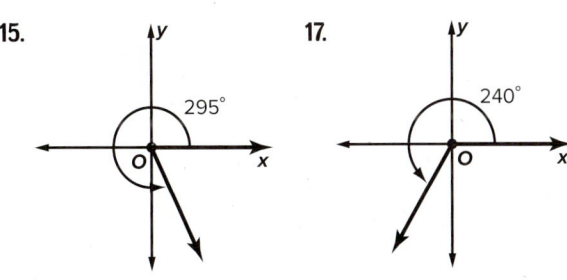

19. Sample answer: 410°, −310° **21.** Sample answer: 565°, −155° **23.** Sample answer: 280°, −440°

25. $330° = 330 \cdot \dfrac{\pi \text{ radians}}{180°}$
$= \dfrac{330\pi}{180}$ or $\dfrac{11\pi}{6}$ radians

27. −60° **29.** $\dfrac{19\pi}{18}$ **31.** about 12.6 ft **33.** 6.7 cm
35. 1 h 15 min **37.** Sample answer: 260°, −100°
39. Sample answer: $\dfrac{5\pi}{4}, -\dfrac{11\pi}{4}$

41. a.

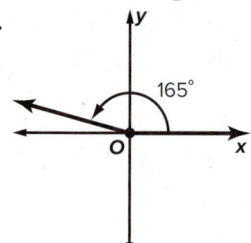

b. $165° = 165 \cdot \dfrac{\pi \text{ radians}}{180°}$
$= \dfrac{165\pi}{180}$ or $\dfrac{11\pi}{12}$ radians

c. $s = r\theta$ Formula for arc length
$= 6.5 \cdot \dfrac{11\pi}{12}$ $r = 6.5$ and $\theta = \dfrac{11\pi}{12}$
≈ 18.7 ft   Use a calculator.

d. The arc length would double. Because $s = r\theta$, if r is doubled and θ remains unchanged, then the value of s is also doubled.

43. 472.5° **45.** $-\dfrac{10\pi}{9}$ **47a.** 3 **47b.** 0.1 ft **49.** $x = 2$

51. Sample answer: 440° and −280°

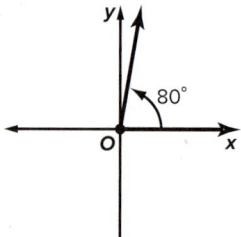

53. $\dfrac{\theta}{2\pi} = \dfrac{s}{2\pi r}$ Substitute.
$2\pi r\theta = 2\pi s$ Find the cross products.
$r\theta = s$ Divide each side by 2π.

One degree represents an angle measure that equals $\dfrac{1}{360}$ rotation around a circle. One radian represents the measure of an angle in standard position that intercepts an arc of length r. To change from degrees to radians, multiply the number of degrees by $\dfrac{\pi \text{ radians}}{180°}$. To change from radians to degrees, multiply the number of radians by $\dfrac{180°}{\pi \text{ radians}}$.

55. C **57a.** D **57b.** A **57c.** 142.5° **57d.** 9.9 ft
59a. 3 radians **59b.** $\dfrac{540}{\pi}$ degrees

Lesson 9-3

1. $\sin \theta = \dfrac{2\sqrt{5}}{5}, \cos \theta = \dfrac{\sqrt{5}}{5}, \tan \theta = 2, \csc \theta = \dfrac{\sqrt{5}}{2},$
$\sec \theta = \sqrt{5}, \cot \theta = \dfrac{1}{2}$ **3.** $\sin \theta = -1, \cos \theta = 0,$
$\tan \theta =$ undefined, $\csc \theta = -1, \sec \theta =$ undefined,
$\cot \theta = 0$

5. 65°

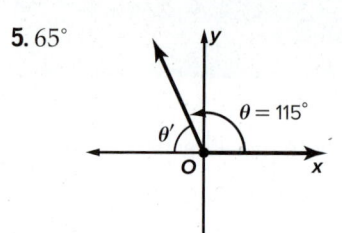

7. $\dfrac{\sqrt{2}}{2}$ **9.** −2

11a.

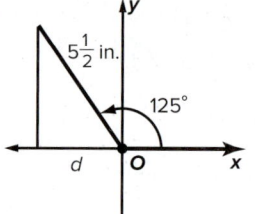

11b. 55°; $\cos 55° = \dfrac{d}{5\frac{1}{2}}$
11c. 3.2 in.

13.

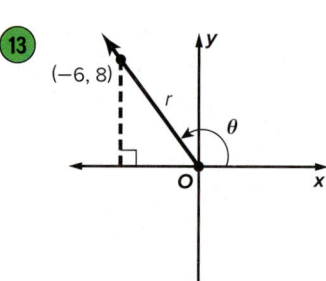

$r = \sqrt{x^2 + y^2}$
$= \sqrt{(-6)^2 + 8^2}$
$= \sqrt{100}$ or 10

Use $x = -6, y = 8,$ and $r = 10$.

$\sin \theta = \dfrac{y}{r}$ $\cos \theta = \dfrac{x}{r}$
$= \dfrac{8}{10}$ or $\dfrac{4}{5}$ $= \dfrac{-6}{10}$ or $-\dfrac{3}{5}$

$\tan \theta = \dfrac{y}{x}$ $\csc \theta = \dfrac{r}{y}$
$= \dfrac{8}{-6}$ or $-\dfrac{4}{3}$ $= \dfrac{10}{8}$ or $\dfrac{5}{4}$

$\sec \theta = \dfrac{r}{x}$ $\cot \theta = \dfrac{x}{y}$
$= \dfrac{10}{-6}$ or $-\dfrac{5}{3}$ $= \dfrac{-6}{8}$ or $-\dfrac{3}{4}$

15. $\sin \theta = -1, \cos \theta = 0, \tan \theta =$ undefined, $\csc \theta = -1, \sec \theta =$ undefined, $\cot \theta = 0$

17. $\sin \theta = -\dfrac{\sqrt{10}}{10}, \cos \theta = -\dfrac{3\sqrt{10}}{10}, \tan \theta = \dfrac{1}{3},$
$\csc \theta = -\sqrt{10}, \sec \theta = -\dfrac{\sqrt{10}}{3}, \cot \theta = 3$

19. 75° **21.** $\dfrac{\pi}{4}$

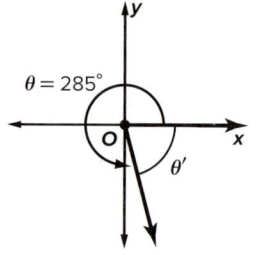

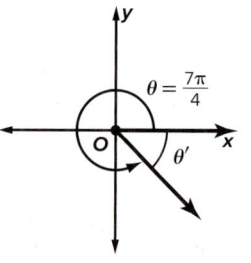

R86

23. 40° 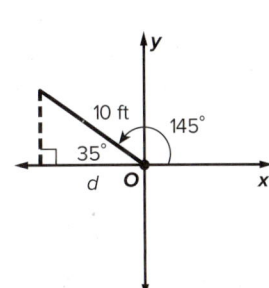 **25.** −1 **27.** −√2 **29.** $\frac{1}{2}$ **31.** $\frac{2\sqrt{3}}{3}$

33.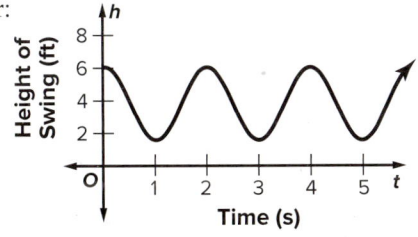
$\cos 35° = \frac{d}{10}$
$10 \cdot \cos 35° = d$
$8.2 \approx d$
The water reaches about 8.2 feet to the left of the sprinkler.

35. about 10.1 m **37.** $\cos \theta = -\frac{3}{5}$, $\tan \theta = -\frac{4}{3}$, $\csc \theta = \frac{5}{4}$, $\sec \theta = -\frac{5}{3}$, $\cot \theta = -\frac{3}{4}$ **39.** $\sin \theta = -\frac{15}{17}$, $\tan \theta = \frac{15}{8}$, $\csc \theta = -\frac{17}{15}$, $\sec \theta = -\frac{17}{8}$, $\cot \theta = \frac{8}{15}$
41. 0 **43.** $-\frac{1}{2}$ **45.** $\frac{\sqrt{3}}{2}$ **47.** No; for $\sin \theta = \frac{\sqrt{2}}{2}$ and $\tan \theta = -1$, the reference angle is 45°. However, for $\sin \theta$ to be positive and $\tan \theta$ to be negative, the reference angle must be in the second quadrant. So, the value of θ must be 135° or an angle coterminal with 135°. **49.** Sample answer: We know that $\cot \theta = \frac{x}{y}$, $\sin \theta = \frac{y}{r}$, and $\cos \theta = \frac{x}{r}$. Because $\sin 180 = 0$, it must be true that $y = 0$. Thus, $\cot \theta = \frac{x}{0}$, which is undefined. **51.** Sample answer: First, sketch the angle and determine in which quadrant it is located. Then use the appropriate rule for finding its reference angle θ'. A reference angle is the acute angle formed by the terminal side of θ and the x-axis. Next, find the value of the trigonometric function for θ'. Finally, use the quadrant location to determine the sign of the trigonometric function value of θ. **53.** E
55a. A, D, E, G **55b.** E **55c.** C, G

Lesson 9-4

1. $\cos \theta = \frac{15}{17}$, $\sin \theta = \frac{8}{17}$ **3.** 2 **5a.** 4 seconds
5b. Sample answer:

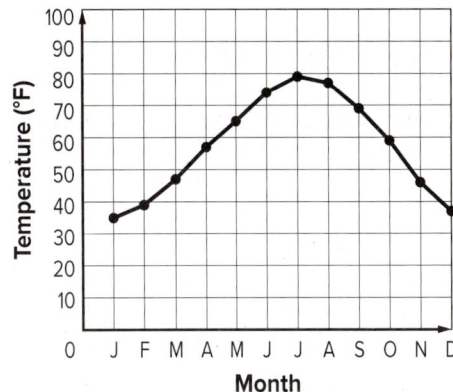

7. $-\frac{\sqrt{3}}{2}$ **9.** $\cos \theta = \frac{3}{5}$, $\sin \theta = -\frac{4}{5}$
11. $P\left(\frac{\sqrt{3}}{2}, \frac{1}{2}\right) = P(\cos \theta, \sin \theta)$

$\cos \theta = \frac{\sqrt{3}}{2}$ $\sin \theta = \frac{1}{2}$

13. 3 **15.** 12 **17.** 180°

19a.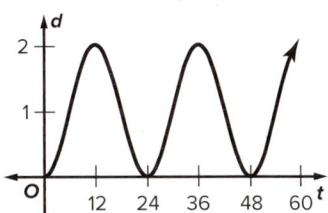

19b. 12 mo or 1 yr **21.** $\frac{1}{2}$ **23.** $\frac{\sqrt{2}}{2}$ **25.** $-\frac{\sqrt{3}}{2}$

27. a. The period is the time it takes to complete one rotation. So, the period is 60 seconds ÷ 2.5 or 24 seconds.
b. Because the siren is 1 mile from Ms. Miller's house and the beam of sound has a radius of 1 mile, then the minimum distance of the sound beam from the house is 0 miles and the maximum distance is 2 miles. Draw a sine curve with the pattern repeating every 24 seconds. Sample answer:

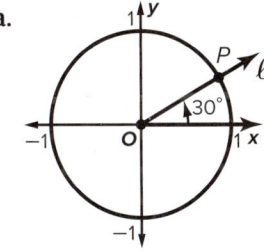

29a.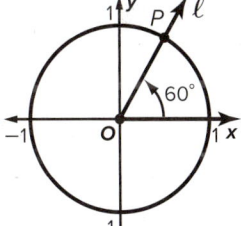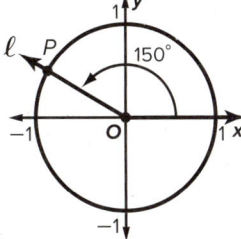

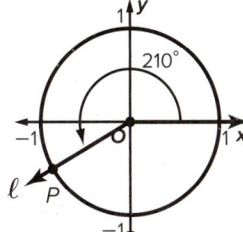 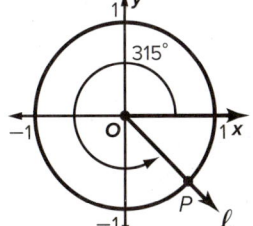

29b.

Angle	Slope
30	0.6
60	1.7
120	−1.7
150	−0.6
210	0.6
315	−1

29c. Sample answer: The slope corresponds to the tangent of the angle. For $\theta = 120°$, the x-coordinate of P is $-\frac{1}{2}$ and the y-coordinate is $\frac{\sqrt{3}}{2}$; slope = $\frac{\text{change in } y}{\text{change in } x}$. Because change in $x = -\frac{1}{2}$ and change in $y = \frac{\sqrt{3}}{2}$, slope = $\frac{\sqrt{3}}{2} \div \left(-\frac{1}{2}\right) = -\sqrt{3}$ or about -1.7.

31. $\cos 45° - \cos 30° = \frac{\sqrt{2}}{2} - \frac{\sqrt{3}}{2}$
$= \frac{\sqrt{2} - \sqrt{3}}{2}$

33. $-\frac{5\sqrt{3}}{2}$ **35.** 1 **37.** Benita; Francis incorrectly wrote $\cos \frac{-\pi}{3} = -\cos \frac{\pi}{3}$. **39.** Sometimes; the period of a sine curve could be $\frac{\pi}{2}$, which is not a multiple of π.
41. The period of a periodic function is the horizontal distance of the part of the graph that is nonrepeating. Each nonrepeating part of the graph is one cycle. **43.** C
45a. B, E, F **45b.** A **45c.** B, D

Lesson 9-5

1. amplitude: 4; period: 360°

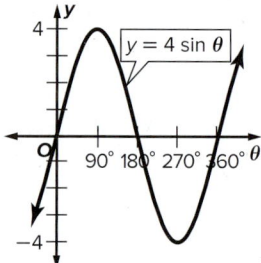

3. amplitude: 1; period: 180°

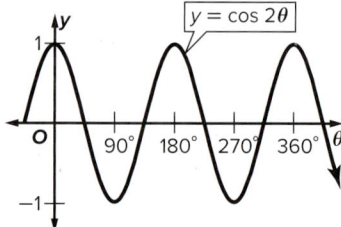

5a. $\frac{1}{14}$ or about 0.07 second

5b. $y = \sin 28\pi t$

7. period: 360°

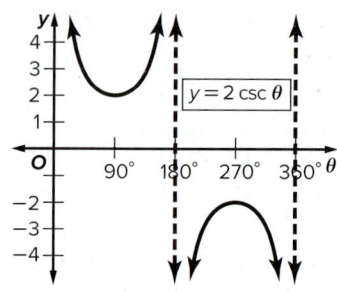

9. amplitude: 2; period: 360°

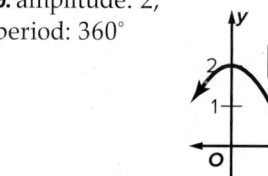

11. amplitude: 1; period: 180°

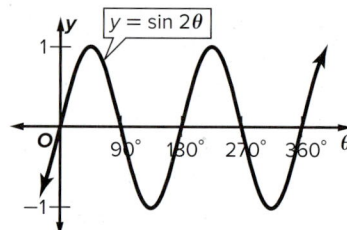

13. amplitude: 1; period: 720°

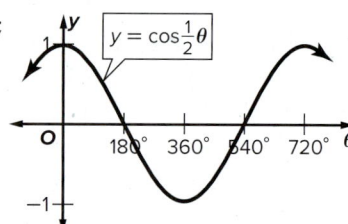

15. amplitude: $\frac{3}{4}$; period: 360°

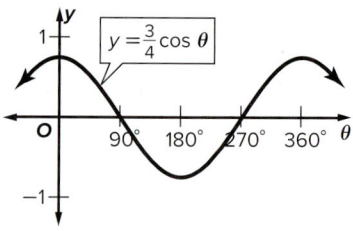

17. amplitude: $|a| = \left|\frac{1}{2}\right|$ or $\frac{1}{2}$
period: $\frac{360°}{|b|} = \frac{360°}{|2|}$
$= 180°$

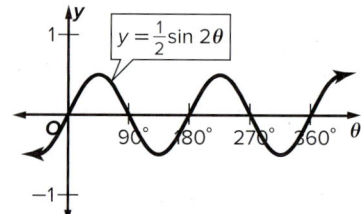

R88

19. amplitude: 3; period: 180°

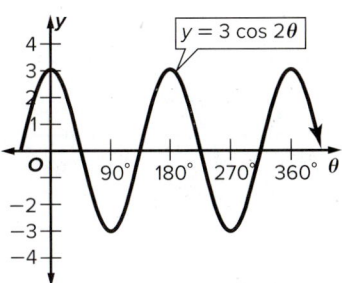

21a. $h = 4 \sin \frac{2}{3}\pi t$

21b.

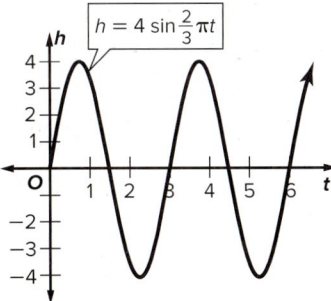

23. period: 360°

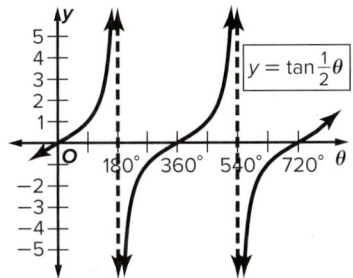

25. period: 180°

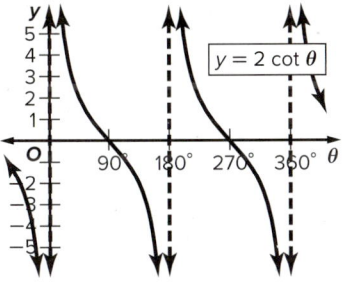

27. period: 180°

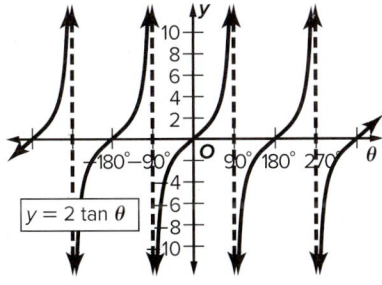

29. a. Because the frequency is 0.5, the period is $\frac{1}{0.5}$ or 2.

$$\text{period} = \frac{2\pi}{|b|} \quad \text{Write the relationship between the period and } b.$$
$$2 = \frac{2\pi}{|b|} \quad \text{Substitution}$$
$$b = \pi \quad \text{Solve for } b.$$

$y = a \sin b\theta$ General equation for the sine function
$h = 1 \sin \pi t$ Replace y with h, a with 1, b with π, and θ with t.
$h = \sin \pi t$ Simplify.

b.

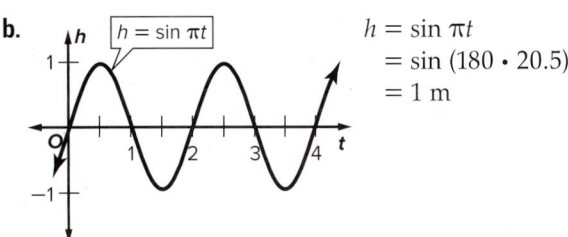

$h = \sin \pi t$
$= \sin (180 \cdot 20.5)$
$= 1$ m

31a. $y = \cos 260\pi t$
The amplitude remains the same. The period decreases because it is the reciprocal of the frequency.

31b. The amplitude remains the same. The period decreases because it is the reciprocal of the frequency.

33. amplitude: $\frac{1}{2}$; period: 480°

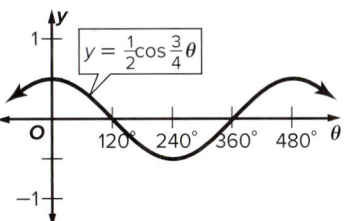

35. amplitude: does not exist; period: 450°

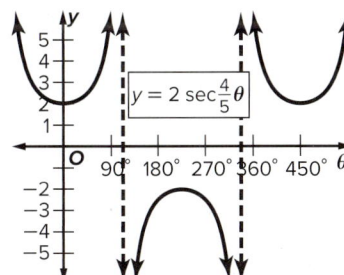

37. amplitude: does not exist; period: 30°

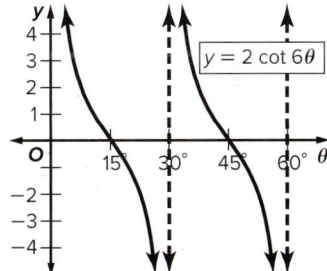

39. $180°$; $y = 5 \sin 2\theta$ **41.** The domain of $y = a \cos \theta$ is the set of all real numbers. The domain of $y = a \sec \theta$ is the set of all real numbers except the values for which $\cos \theta = 0$. The range of $y = a \cos \theta$ is $-a \leq y \leq a$. The range of $y = a \sec \theta$ is $y \leq -a$ and $y \geq a$.

43. Sample answer: $y = 3 \sin 2\theta$

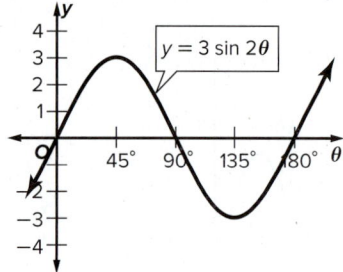

45. D   **47.** C

Lesson 9-6

1. 1; $360°$; $h = 180°$

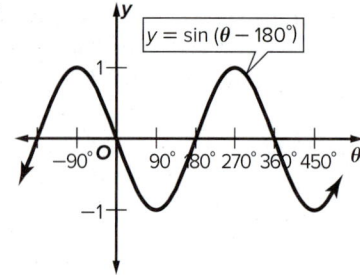

3. 1; 2π; $h = \dfrac{\pi}{2}$

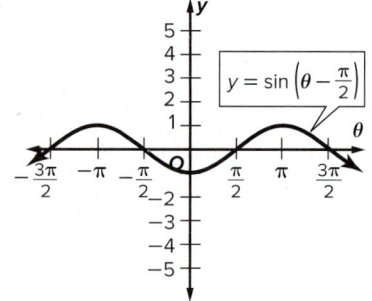

5. 1; $360°$; $k = 4$; $y = 4$

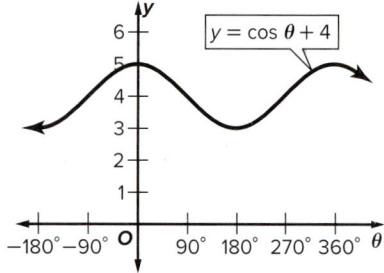

7. no amplitude; $180°$; $k = 1$; $y = 1$

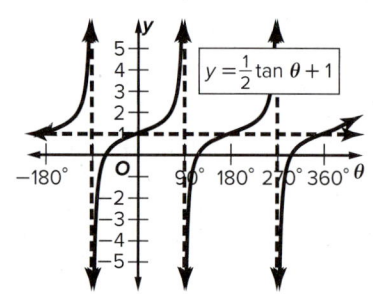

9. 2; $360°$; $h = -45°$; $k = 1$

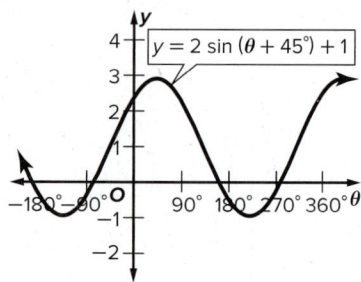

11. no amplitude; $90°$; $h = -30°$; $k = 3$

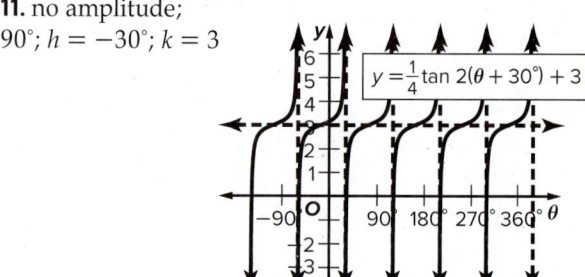

13. $P = 20 \sin 3\pi t + 110$

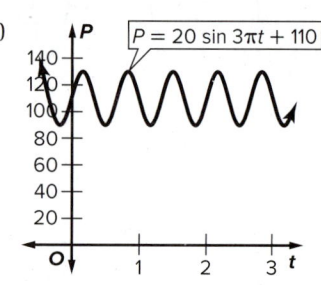

15. no amplitude; $180°$; $h = 90°$

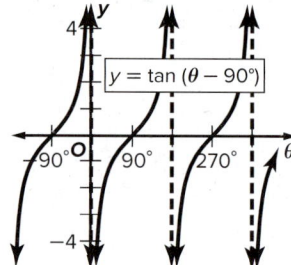

17. 2; 2π; $h = -\dfrac{\pi}{2}$

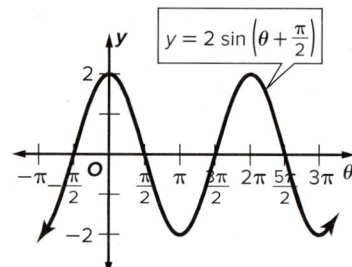

19. 3; 2π; $h = \dfrac{\pi}{3}$

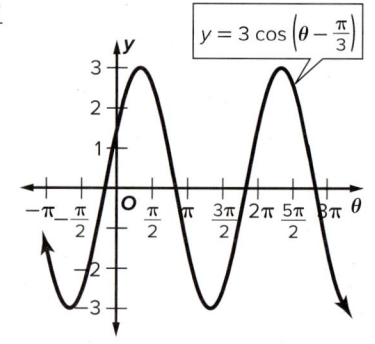

21. no amplitude; 180°; $k = -1$; $y = -1$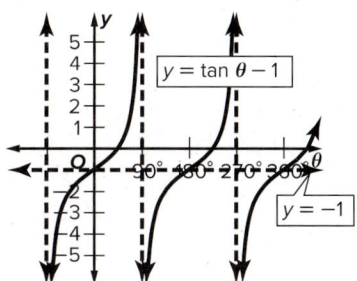

23. amplitude: $|a| = 2$
period: $\frac{360°}{|b|} = \frac{360°}{|1|}$ or 360°
vertical shift: $k = -5$
midline: $y = -5$
To graph $y = 2\cos\theta - 5$, first draw the midline. Then use it to graph $y = 2\cos\theta$ shifted 5 units down.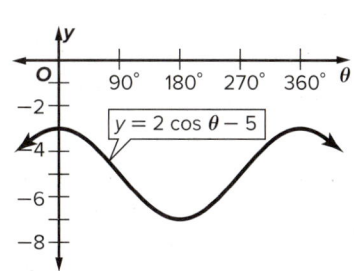

25. $\frac{1}{3}$; 360°; $k = 7$; $y = 7$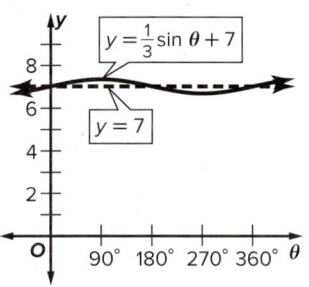

27. 1; 720°; $h = 90°$; $k = 2$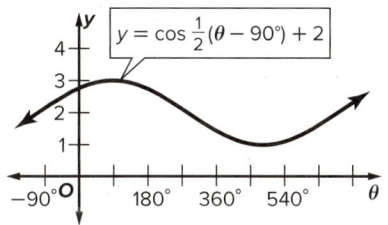

29. no amplitude; $\frac{\pi}{2}$; $h = -\frac{\pi}{4}$; $k = -5$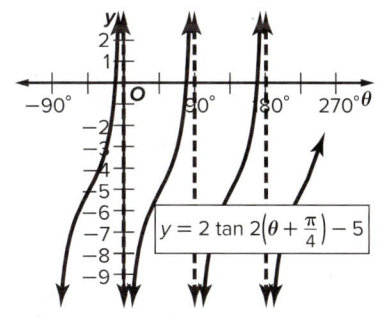

31. 1; 120°; $h = 45°$; $k = \frac{1}{2}$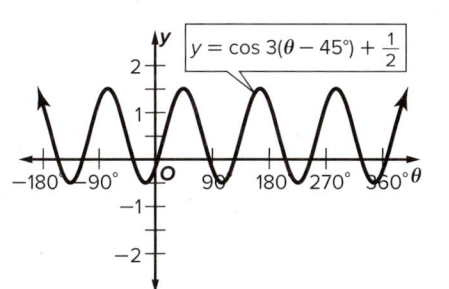

33. 3; 6π; $h = \frac{\pi}{2}$; $k = -2$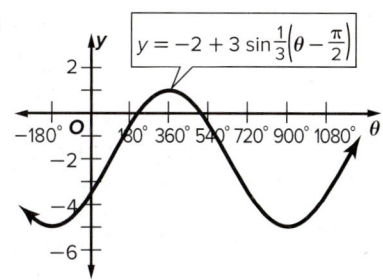

35. min: 10.2 ft; max: 13.8 ft
37. $y = \sin(x - 4) + 3$
39. $y = \tan(x - \pi) + 2.5$

41a. Sample answer: There are 8 groups of 4 horses, with each horse in a different location from other members of its group. Each group is represented by $f(t) = 2.5\sin\left(\frac{\pi}{10}t - \frac{g\pi}{4}\right) + 2.5$, where g represents a specific group, and $1 \leq g \leq 8$, $g \in W$.

41b. Sample answer: The value 2.5 represents the magnitude, where the bottom of the sine curve, at $f(x) = -2.5$, is when the horse is at its minimum height and the top of the curve, at $f(x) = 2.5$, is when the horse is at its maximum height. The 2.5 vertical shift was added at the end to set the minimum height at 0. There are 8 different maxima, so increments of $\frac{\pi}{4}$ to represent each group. Thus, each group is shifted from the others. The value g represents each of the 8 groups. The carousel does a complete revolution every 20 seconds, so the period is 20 and $b = \frac{2\pi}{20}$ or $\frac{\pi}{10}$.

41c. Sample answer: I assumed that the horses all went up and down at the same speed. I also assumed that there were 8 groups of 4 horses and each group was spaced out vertically by the same increment. This way the entire carousel can be modeled after one trigonometric function. I also assumed that the period of the horses coincided with the period of the carousel.

41d. Sample answer: In order to accomplish this, set the period of the horses so that it is not a multiple of the period of the carousel. Keep the function representing the horses the same, and set the period of the carousel to 23 seconds.

43. $\left(\frac{3\pi}{2}, 2\right)$ **45.** no maximum values **47.** The graphs are reflections of each other in the x-axis. **49.** The graphs are identical. **51.** 360°; Sample answer: $y = 2\cos(\theta + 90°)$

53. The midline lies halfway between the maximum and minimum values. So $y = \frac{4 + 2}{2}$ or 3. Because the midline is $y = 3$, the vertical shift is $k = 3$.

The amplitude is the difference between the midline value and the maximum value. So $|a| = |4 - 3|$ or 1.

Because the cycle repeats every 180°, the period is 180°.

$\text{period} = \frac{360°}{|b|}$ Write the relationship between the period and b.

$180° = \frac{360°}{|b|}$ Substitution

$b = 2$ Solve for b.

The graph is the sine curve shifted 45° to the right. So the phase shift is $h = 45°$.

$y = a \sin b(\theta - h) + k$ General equation for the sine function

$y = 1 \sin 2(\theta - 45°) + 3$ Replace a with 1, b with 2, h with 45°, and k with 3.

$y = \sin 2(\theta - 45°) + 3$ Simplify.

55. 180°; no phase shift; $k = 6$

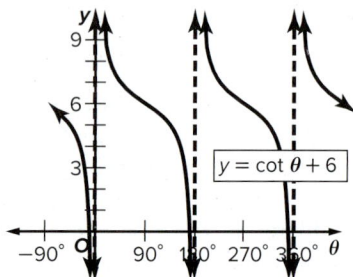

57. 120°; $h = 45°$; $k = 1$

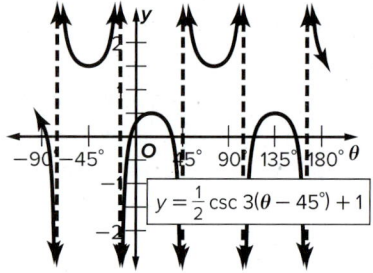

59. π; $h = -\frac{\pi}{2}$; $k = -3$

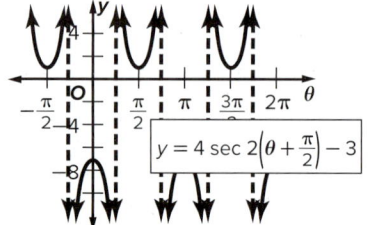

61. The graph of $y = 3 \sin 2\theta + 1$ has an amplitude of 3 rather than an amplitude of 1. It is shifted up 1 unit from the parent graph and is compressed so that it has a period of 180°.

63. Sample answer: $y = 2 \sin \theta - 3$

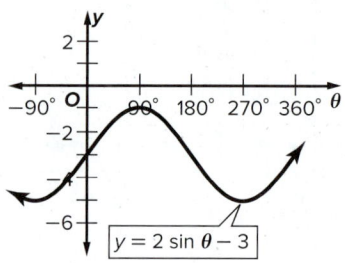

65. A **67.** C **69a.** B, D **69b.** B **69c.** C **69d.** A, C **69e.** E

Chapter 9 Study Guide and Review

1. true **3.** true **5.** Use the given trigonometric ratio to label two sides of a right triangle, and then use the Pythagorean Theorem to find the third side. Apply definitions to determine the other ratios. **7.** $a = 10.9$; $A = 65°$; $B = 25°$ **9.** $A = 15°$; $a = 4.0$; $c = 15.5$ **11.** $B = 55°$; $a = 12.6$; $b = 18.0$ **13.** about 8.8 feet **15.** 450° **17.** $-\frac{7\pi}{4}$ **19.** 295°, −425° **21.** $\frac{16}{\pi}$ **23.** $-\frac{\sqrt{3}}{3}$ **25.** 0 **27.** $\sin \theta = \frac{12}{13}$, $\cos \theta = \frac{5}{13}$, $\tan \theta = \frac{12}{5}$, $\csc \theta = \frac{13}{12}$, $\sec \theta = \frac{13}{5}$, $\cot \theta = \frac{5}{12}$ **29.** about 17.1 meters **31.** $\frac{-\sqrt{6}}{4}$ **33.** 0 **35.** 15 seconds **37.** amplitude: 1, period: 720°

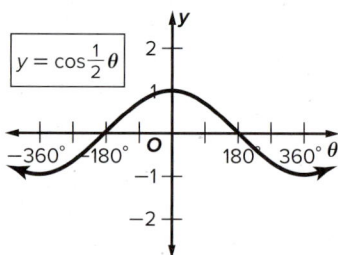

39. amplitude: not defined, period: 360°

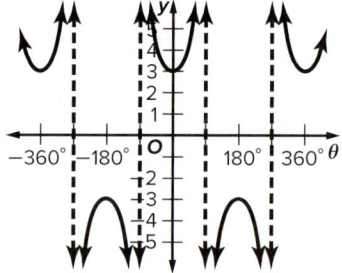

41. amplitude: not defined, period: 720°

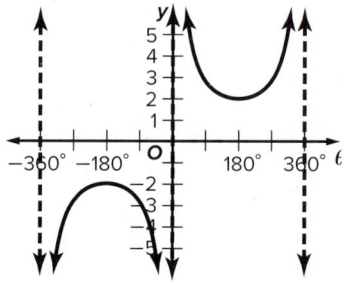

43. vertical shift: up 1; amplitude: 3; period 180°; phase shift: 90° right

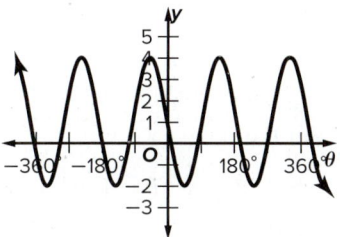

45. vertical shift: up 2
amplitude: not defined
period: $\dfrac{2\pi}{3}$
phase shift: right

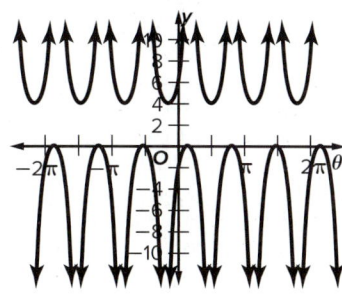

47. vertical shift: up 2
amplitude: $\dfrac{1}{3}$
period: 1080°
phase shift: 90° right

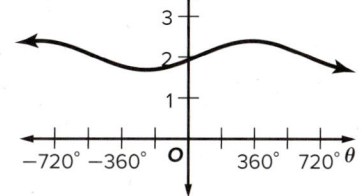

CHAPTER 10
Trigonometric Identities and Equations

Chapter 10 Get Ready
1. factorable; $-4a(4a - 1)$ 3. $x^2 + x - 42 = 0$
5. $180° - 135°$

Lesson 10-1
1. $\frac{1}{2}$ 3. $\frac{\sqrt{5}}{3}$ 5. $\sin\theta\cos\theta$ 7. $\cot^2\theta$ 9. $\frac{5}{4}$ 11. $\frac{4}{5}$
13. $-\frac{5}{4}$

15. $\cot^2\theta + 1 = \csc^2\theta$ Pythagorean Identity
$\left(\frac{1}{4}\right)^2 + 1 = \csc^2\theta$ Substitute $\frac{1}{4}$ for $\cot\theta$.
$\frac{1}{16} + 1 = \csc^2\theta$ Square $\frac{1}{4}$.
$\frac{17}{16} = \csc^2\theta$ Add.
$\pm\frac{\sqrt{17}}{4} = \csc\theta$ Take the square root of each side.
Because θ is in the third quadrant, $\csc\theta$ is negative.
So, $\csc\theta = -\frac{\sqrt{17}}{4}$.

17. $-\frac{12}{13}$ 19. $\frac{3}{5}$ 21. $\sec^3\theta$ 23. $\csc\theta$ 25. 1

27. $B = \frac{F\csc\theta}{I\ell}$ Original equation
$I\ell \cdot B = F\csc\theta$ Multiply each side by $I\ell$.
$I\ell B = F \cdot \frac{1}{\sin\theta}$ $\frac{1}{\sin\theta} = \csc\theta$
$I\ell B \sin\theta = F$ Multiply each side by $\sin\theta$.
The equation can be written as $F = I\ell B \sin\theta$.

29. $\sec\theta$ 31. 2   33. $2\cos^2\theta$ 35a. $\frac{\sqrt{65}}{9}$ 35b. $\frac{4\sqrt{65}}{65}$
35c. $\frac{4}{9}, -\frac{\sqrt{65}}{9}, -\frac{4\sqrt{65}}{65}$ 37. $\mu_k = \tan\theta$

39. $\frac{\cos\left(\frac{\pi}{2} - \theta\right) - 1}{1 + \sin(-\theta)} = \frac{\sin\theta - 1}{1 + \sin(-\theta)}$ $\cos\left(\frac{\pi}{2} - \theta\right) = \sin\theta$
$= \frac{\sin\theta - 1}{1 - \sin\theta}$ $\sin(-\theta) = -\sin\theta$
$= \frac{\sin\theta - 1}{-1(\sin\theta - 1)}$ $1 - \sin\theta = -1(\sin\theta - 1)$
$= \frac{1}{-1}$ or -1 Simplify.

41. $-\cot^2\theta$ 43. Sample answer: $x = 45°$ 45. The functions $\cos\theta$ and $\sin\theta$ can be thought of as the lengths of the legs of a right triangle, and the number 1 can be thought of as the measure of the corresponding hypotenuse.   47. Sample answer: $\frac{\sin\theta}{\cos\theta} \cdot \sin\theta$ and $\frac{\sin^2\theta}{\cos\theta}$ 49. $-\frac{4}{3}$ 51. E   53. D   55a. D

55b.
$\frac{\sin(90° - \theta)}{\cos(\theta - 90°)} = \frac{\sin(90° - \theta)}{\cos(90° - \theta)}$ Negative angle identity, $\cos(\theta - 90) = \cos(90 - \theta)$
$= \frac{\cos\theta}{\sin\theta}$ Substitute the cofunction identities.
$= \cot\theta$ Substitute the cotangent quotient identity.

57. $\sin^2\theta$ 59. C, D

Lesson 10-2
1. $\cot\theta + \tan\theta \stackrel{?}{=} \frac{\sec^2\theta}{\tan\theta}$
$\cot\theta + \tan\theta \stackrel{?}{=} \frac{\tan^2\theta + 1}{\tan\theta}$
$\cot\theta + \tan\theta \stackrel{?}{=} \frac{\tan^2\theta}{\tan\theta} + \frac{1}{\tan\theta}$
$\cot\theta + \tan\theta = \tan\theta + \cot\theta$ ✓

3. $\sin\theta \stackrel{?}{=} \frac{\sec\theta}{\tan\theta + \cot\theta}$
$\sin\theta \stackrel{?}{=} \frac{\frac{1}{\cos\theta}}{\frac{\sin\theta}{\cos\theta} + \frac{\cos\theta}{\sin\theta}}$
$\sin\theta \stackrel{?}{=} \frac{\frac{1}{\cos\theta}}{\frac{\sin^2\theta + \cos^2\theta}{\cos\theta\sin\theta}}$
$\sin\theta \stackrel{?}{=} \frac{\frac{1}{\cos\theta}}{\frac{1}{\cos\theta\sin\theta}}$
$\sin\theta \stackrel{?}{=} \frac{1}{\cos\theta} \cdot \frac{\cos\theta\sin\theta}{1}$
$\sin\theta = \sin\theta$ ✓

5. $\tan^2\theta\csc^2\theta \stackrel{?}{=} 1 + \tan^2\theta$
$\frac{\sin^2\theta}{\cos^2\theta} \cdot \frac{1}{\sin^2\theta} \stackrel{?}{=} \sec^2\theta$
$\frac{1}{\cos^2\theta} \stackrel{?}{=} \sec^2\theta$
$\sec^2\theta = \sec^2\theta$ ✓

7. $\frac{\tan^2\theta + 1}{\tan^2\theta} \stackrel{?}{=} \csc^2\theta$
$\frac{\sec^2\theta}{\tan^2\theta} \stackrel{?}{=} \csc^2\theta$
$\frac{\frac{1}{\cos^2\theta}}{\frac{\sin^2\theta}{\cos^2\theta}} \stackrel{?}{=} \csc^2\theta$
$\frac{1}{\cos^2\theta} \cdot \frac{\cos^2\theta}{\sin^2\theta} \stackrel{?}{=} \csc^2\theta$
$\frac{1}{\sin^2\theta} \stackrel{?}{=} \csc^2\theta$
$\csc^2\theta = \csc^2\theta$ ✓

9. $\sin^2\theta + \tan^2\theta \stackrel{?}{=} (1 - \cos^2\theta) + \frac{\sec^2\theta}{\csc^2\theta}$
$\sin^2\theta + \tan^2\theta \stackrel{?}{=} \sin^2\theta + \frac{\sec^2\theta}{\csc^2\theta}$
$\sin^2\theta + \tan^2\theta \stackrel{?}{=} \sin^2\theta + \frac{1}{\cos^2\theta} \div \frac{1}{\sin^2\theta}$
$\sin^2\theta + \tan^2\theta \stackrel{?}{=} \sin^2\theta + \frac{\sin^2\theta}{\cos^2\theta}$
$\sin^2\theta + \tan^2\theta \stackrel{?}{=} \sin^2\theta + \tan^2\theta$ ✓

11. $\cot\theta(\cot\theta + \tan\theta) \stackrel{?}{=} \csc^2\theta$
$\cot^2\theta + \cot\theta\tan\theta \stackrel{?}{=} \csc^2\theta$
$\cot^2\theta + \frac{\sin\theta}{\cos\theta} \cdot \frac{\cos\theta}{\sin\theta} \stackrel{?}{=} \csc^2\theta$
$\cot^2\theta + 1 \stackrel{?}{=} \csc^2\theta$
$\csc^2\theta = \csc^2\theta$ ✓

13. $\sin\theta\sec\theta\cot\theta \stackrel{?}{=} 1$
$\sin\theta \cdot \frac{1}{\cos\theta} \cdot \frac{\cos\theta}{\sin\theta} \stackrel{?}{=} 1$
$1 = 1$ ✓

15. $\frac{1 - 2\cos^2\theta}{\sin\theta\cos\theta} \stackrel{?}{=} \tan\theta - \cot\theta$
$\frac{(1 - \cos^2\theta) - \cos^2\theta}{\sin\theta\cos\theta} \stackrel{?}{=} \tan\theta - \cot\theta$

$\dfrac{\sin^2 \theta - \cos^2 \theta}{\sin \theta \cos \theta} \stackrel{?}{=} \tan \theta - \cot \theta$

$\dfrac{\sin^2 \theta}{\sin \theta \cos \theta} - \dfrac{\cos^2 \theta}{\sin \theta \cos \theta} \stackrel{?}{=} \tan \theta - \cot \theta$

$\dfrac{\sin \theta}{\cos \theta} - \dfrac{\cos \theta}{\sin \theta} \stackrel{?}{=} \tan \theta - \cot \theta$

$\tan \theta - \cot \theta = \tan \theta - \cot \theta$ ✓

17. $\cos \theta \stackrel{?}{=} \sin \theta \cot \theta$

$\cos \theta \stackrel{?}{=} \sin \theta \dfrac{\cos \theta}{\sin \theta}$

$\cos \theta = \cos \theta$ ✓

19. $\cos \theta \cos(-\theta) - \sin \theta \sin(-\theta) \stackrel{?}{=} 1$

$\cos \theta \cos \theta - \sin \theta (-\sin \theta) \stackrel{?}{=} 1$

$\cos^2 \theta + \sin^2 \theta \stackrel{?}{=} 1$

$1 = 1$ ✓

21. $\sec \theta - \tan \theta \stackrel{?}{=} \dfrac{1 - \sin \theta}{\cos \theta}$

$\dfrac{1}{\cos \theta} - \dfrac{\sin \theta}{\cos \theta} \stackrel{?}{=} \dfrac{1 - \sin \theta}{\cos \theta}$

$\dfrac{1 - \sin \theta}{\cos \theta} = \dfrac{1 - \sin \theta}{\cos \theta}$ ✓

23. $\sec \theta \csc \theta \stackrel{?}{=} \tan \theta + \cot \theta$

$\dfrac{1}{\cos \theta} \cdot \dfrac{1}{\sin \theta} \stackrel{?}{=} \dfrac{\sin \theta}{\cos \theta} + \dfrac{\cos \theta}{\sin \theta}$

$\dfrac{1}{\cos \theta \sin \theta} \stackrel{?}{=} \dfrac{\sin^2 \theta}{\sin \theta \cos \theta} + \dfrac{\cos^2 \theta}{\sin \theta \cos \theta}$

$\dfrac{1}{\cos \theta \sin \theta} \stackrel{?}{=} \dfrac{\sin^2 \theta + \cos^2 \theta}{\sin \theta \cos \theta}$

$\dfrac{1}{\cos \theta \sin \theta} = \dfrac{1}{\cos \theta \sin \theta}$ ✓

25. $(\sin \theta + \cos \theta)^2 \stackrel{?}{=} \dfrac{2 + \sec \theta \csc \theta}{\sec \theta \csc \theta}$

$(\sin \theta + \cos \theta)^2 \stackrel{?}{=} \dfrac{2 + \dfrac{1}{\cos \theta} \cdot \dfrac{1}{\sin \theta}}{\dfrac{1}{\cos \theta} \cdot \dfrac{1}{\sin \theta}}$

$(\sin \theta + \cos \theta)^2 \stackrel{?}{=} \left(2 + \dfrac{1}{\cos \theta \sin \theta}\right) \cdot \dfrac{\cos \theta \sin \theta}{1}$

$(\sin \theta + \cos \theta)^2 \stackrel{?}{=} 2 \cos \theta \sin \theta + 1$

$(\sin \theta + \cos \theta)^2 \stackrel{?}{=} 2 \cos \theta \sin \theta + \cos^2 \theta + \sin^2 \theta$

$(\sin \theta + \cos \theta)^2 = (\sin \theta + \cos \theta)^2$ ✓

27. $\csc \theta - 1 \stackrel{?}{=} \dfrac{\cot^2 \theta}{\csc \theta + 1}$

$\csc \theta - 1 \stackrel{?}{=} \dfrac{\csc^2 \theta - 1}{\csc \theta + 1}$

$\csc \theta - 1 \stackrel{?}{=} \dfrac{(\csc \theta - 1)(\csc \theta + 1)}{\csc \theta + 1}$

$\csc \theta - 1 = \csc \theta - 1$ ✓

29. $\sin \theta \cos \theta \tan \theta + \cos^2 \theta \stackrel{?}{=} 1$

$\sin \theta \cos \theta \cdot \dfrac{\sin \theta}{\cos \theta} + \cos^2 \theta \stackrel{?}{=} 1$

$\sin^2 \theta + \cos^2 \theta \stackrel{?}{=} 1$

$1 = 1$ ✓

31. $\csc^2 \theta \stackrel{?}{=} \cot^2 \theta + \sin \theta \csc \theta$

$\csc^2 \theta \stackrel{?}{=} \cot^2 \theta + \sin \theta \cdot \dfrac{1}{\sin \theta}$

$\csc^2 \theta \stackrel{?}{=} \cot^2 \theta + 1$

$\csc^2 \theta = \csc^2 \theta$ ✓

33. $\sin^2 \theta + \cos^2 \theta \stackrel{?}{=} \sec^2 \theta - \tan^2 \theta$

$1 \stackrel{?}{=} \tan^2 \theta + 1 - \tan^2 \theta$

$1 = 1$ ✓

35. yes

37. $\cot(-\theta) \tan(-\theta) = \dfrac{1}{\tan(-\theta)} \cdot \tan(-\theta)$ $\dfrac{1}{\tan(-\theta)} = \cot(-\theta)$

$= 1$ Simplify.

39. 1   **41.** 1   **43.** $\cos \theta$ **45.** 2   **47.** $\sin \theta$ **49.** 1

51. $y = -\dfrac{gx^2}{2v_0^2}(1 + \tan^2 \theta) + x \tan \theta$

53. a. $h = \dfrac{v_0^2 \sin^2 \theta}{2g} = \dfrac{47^2 \sin^2 \theta}{2(9.8)}$ Replace v_0 with 47 and g with 9.8.

$= \dfrac{2209 \sin^2 \theta}{19.6}$ Simplify.

$\dfrac{2209 \sin^2 30°}{19.6} \approx 28.2$ m $\theta = 30°$

$\dfrac{2209 \sin^2 45°}{19.6} \approx 56.4$ m $\theta = 45°$

$\dfrac{2209 \sin^2 60°}{19.6} \approx 84.5$ m $\theta = 60°$

$\dfrac{2209 \sin^2 90°}{19.6} \approx 112.7$ m $\theta = 90°$

b. Sample answer: Enter the equation $y = \dfrac{2209 (\sin \theta)^2}{19.6}$.

Use the window Xmin = 0, Xmax = 180, Xscl = 10, Ymin = 0, Ymax = 150, Yscl = 10, Xres = 1.

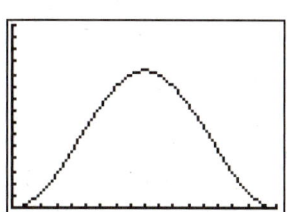

[0, 180] scl: 10 by [0, 150] scl: 10

c. $\dfrac{v_0^2 \tan^2 \theta}{2g \sec^2 \theta} \stackrel{?}{=} \dfrac{v_0^2 \sin^2 \theta}{2g}$

$\dfrac{v_0^2 \left(\dfrac{\sin^2 \theta}{\cos^2 \theta}\right)}{2g \left(\dfrac{1}{\cos^2 \theta}\right)} \stackrel{?}{=} \dfrac{v_0^2 \sin^2 \theta}{2g}$ $\tan^2 \theta = \dfrac{\sin^2 \theta}{\cos^2 \theta}$ and $\sec^2 \theta = \dfrac{1}{\cos^2 \theta}$

$\dfrac{v_0^2 \sin^2 \theta}{2g} = \dfrac{v_0^2 \sin^2 \theta}{2g}$ ✓ Simplify.

55. $\tan^2 \theta = \dfrac{\sin^2 \theta}{\cos^2 \theta} = \dfrac{1 - \cos^2 \theta}{\cos^2 \theta} = \dfrac{1}{\cos^2 \theta} - \dfrac{\cos^2 \theta}{\cos^2 \theta}$

$= \sec^2 \theta - 1$

57. Sample answer: counterexample 45°, 30°

59. Sample answer: Sine and cosine are the trigonometric functions with which most people are familiar, and all trigonometric expressions can be written in terms of sine and cosine. Also, by rewriting complex trigonometric expressions in terms of sine and cosine it may be easier to perform operations and to apply trigonometric properties.

61. Using the unit circle and the Pythagorean Theorem, we can justify $\cos^2 \theta + \sin^2 \theta = 1$.

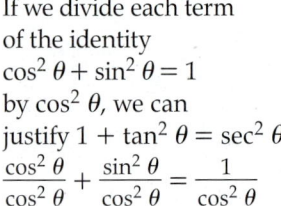

If we divide each term of the identity $\cos^2 \theta + \sin^2 \theta = 1$ by $\cos^2 \theta$, we can justify $1 + \tan^2 \theta = \sec^2 \theta$.

$$\frac{\cos^2 \theta}{\cos^2 \theta} + \frac{\sin^2 \theta}{\cos^2 \theta} = \frac{1}{\cos^2 \theta}$$

$$1 + \tan^2 \theta = \sec^2 \theta$$

If we divide each term of the identity $\cos^2 \theta + \sin^2 \theta = 1$ by $\sin^2 \theta$, we can justify $\cot^2 \theta + 1 = \csc^2 \theta$.

$$\frac{\cos^2 \theta}{\sin^2 \theta} + \frac{\sin^2 \theta}{\sin^2 \theta} = \frac{1}{\sin^2 \theta}$$

$$\cot^2 \theta + 1 = \csc^2 \theta$$

63. A **65a.** C

65b.
$$\frac{\sin^2 \theta + \cos^2 \theta}{\tan \theta \cos \theta} = \frac{1}{\tan \theta \cos \theta} \quad \text{Use the Pythagorean Identity.}$$

$$= \frac{1}{\left(\frac{\sin \theta}{\cos \theta}\right) \cos \theta} \quad \text{Substitute the quotient identity for } \tan \theta.$$

$$= \frac{1}{\sin \theta} \quad \text{Simplify.}$$

67. $\sin x \cos x \tan x = \sin x \cos x \left(\frac{\sin x}{\cos x}\right)$
$= \sin^2 x$
$= 1 - \cos^2 x$

69. D

Lesson 10-3

1. $\cos 165° = \cos (120° + 45°)$
$= \cos 120° \cos 45° - \sin 120° \sin 45°$
$= \left(-\frac{1}{2} \cdot \frac{\sqrt{2}}{2}\right) - \left(\frac{\sqrt{3}}{2} \cdot \frac{\sqrt{2}}{2}\right)$
$= -\frac{\sqrt{2}}{4} - \frac{\sqrt{6}}{4}$
$= -\frac{\sqrt{2} + \sqrt{6}}{4}$

3. $\frac{\sqrt{6} - \sqrt{2}}{4}$ **5.** $\frac{\sqrt{2}}{2}$ **7a.** 0 **7b.** The interference is destructive. The signals cancel each other completely.

9.
$$\cos\left(\frac{3\pi}{2} - \theta\right) \stackrel{?}{=} -\sin \theta$$
$$\cos\frac{3\pi}{2} \cos \theta + \sin\frac{3\pi}{2} \sin \theta \stackrel{?}{=} -\sin \theta$$
$$0 \cdot \cos \theta - 1 \cdot \sin \theta \stackrel{?}{=} -\sin \theta$$
$$-\sin \theta = -\sin \theta \checkmark$$

11.
$$\sin(\theta + \pi) \stackrel{?}{=} -\sin \theta$$
$$\sin \theta \cos \pi + \cos \theta \sin \pi \stackrel{?}{=} -\sin \theta$$
$$(\sin \theta)(-1) + (\cos \theta)(0) \stackrel{?}{=} -\sin \theta$$
$$-\sin \theta = -\sin \theta \checkmark$$

13. $-\frac{\sqrt{2}}{2}$ **15.** $\frac{\sqrt{6} - \sqrt{2}}{4}$ **17.** $\frac{\sqrt{2} + \sqrt{6}}{4}$

19.
$$\cos\left(\frac{\pi}{2} + \theta\right) \stackrel{?}{=} -\sin \theta$$
$$\cos\frac{\pi}{2} \cos \theta - \sin\frac{\pi}{2} \sin \theta \stackrel{?}{=} -\sin \theta$$
$$(0)(\cos \theta) - (1)(\sin \theta) \stackrel{?}{=} -\sin \theta$$
$$-\sin \theta = -\sin \theta \checkmark$$

21.
$$\cos (180° + \theta) \stackrel{?}{=} -\cos \theta$$
$$\cos 180° \cos \theta - \sin 180° \sin \theta \stackrel{?}{=} -\cos \theta$$
$$-1 \cdot \cos \theta - 0 \cdot \sin \theta \stackrel{?}{=} -\cos \theta$$
$$-\cos \theta = -\cos \theta \checkmark$$

23a. $y = 19.76 \sin\left(\frac{5\pi}{32}x - 1.79\right) + 64.27$ **23b.** The new function represents the middle average of the high and low temperatures for each month. **25.** $\sqrt{2} - \sqrt{6}$
27. $-2 + \sqrt{3}$ **29.** $2 - \sqrt{3}$

31. a. Let X be the endpoint of the segment that is 8 inches long.
$\sin (m\angle BAC)$
$= \sin (m\angle BAX + m\angle XAC)$
$= \sin (m\angle BAX) \cos (m\angle XAC) +$
$\quad \cos (m\angle BAX) \sin (m\angle XAC)$
$= \frac{8\sqrt{3}}{16} \cdot \frac{8}{10} + \frac{8}{16} \cdot \frac{6}{10} \quad \sin = \frac{\text{opp}}{\text{hyp}}$ and $\cos = \frac{\text{adj}}{\text{hyp}}$
$= \frac{4\sqrt{3}}{10} + \frac{3}{10} \quad$ Multiply.
$= \frac{3 + 4\sqrt{3}}{10} \quad$ Add.

b. Let X be the endpoint of the segment that is 8 inches long.
$\cos (m\angle BAC)$
$= \cos (m\angle BAX + m\angle XAC)$
$= \cos (m\angle BAX) \cos (m\angle XAC) -$
$\quad \sin (m\angle BAX) \sin (m\angle XAC)$
$= \frac{8}{16} \cdot \frac{8}{10} - \frac{8\sqrt{3}}{16} \cdot \frac{6}{10} \quad \sin = \frac{\text{opp}}{\text{hyp}}$ and $\cos = \frac{\text{adj}}{\text{hyp}}$
$= \frac{4}{10} - \frac{3\sqrt{3}}{10} \quad$ Multiply.
$\approx \frac{4 - 3\sqrt{3}}{10} \quad$ Add.

c. $\cos (m\angle BAC) = \frac{4 - 3\sqrt{3}}{10}$
$$m\angle BAC = \cos^{-1}\left(\frac{3 + 4\sqrt{3}}{10}\right)$$
$$\approx 96.9°$$

d. Because $m\angle BAC \neq 90$, the triangle formed is not a right triangle.

33a.

A	B	sin A	sin B	sin (A + B)	sin A + sin B
30°	90°	$\frac{1}{2}$	1	$\frac{\sqrt{3}}{2}$	$\frac{3}{2}$
45°	60°	$\frac{\sqrt{2}}{2}$	$\frac{\sqrt{3}}{2}$	$\frac{\sqrt{2} + \sqrt{6}}{4}$	$\frac{\sqrt{2} + \sqrt{3}}{2}$
60°	45°	$\frac{\sqrt{3}}{2}$	$\frac{\sqrt{2}}{2}$	$\frac{\sqrt{2} + \sqrt{6}}{4}$	$\frac{\sqrt{2} + \sqrt{3}}{2}$
90°	30°	1	$\frac{1}{2}$	$\frac{\sqrt{3}}{2}$	$\frac{3}{2}$

33b.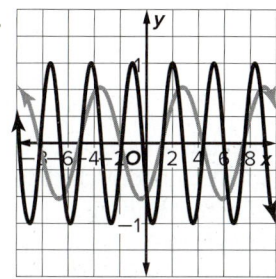

33c. No; a counterexample is: $\cos(30° + 45°) = \cos 30° + \cos 45°$, which equals $\frac{\sqrt{3}}{2} + \frac{\sqrt{2}}{2}$ or about 1.5731. Because a cosine value cannot be greater than 1, this statement must be false.

35.
$$\cos(A+B) \stackrel{?}{=} \frac{1 - \tan A \tan B}{\sec A \sec B}$$

$$\cos(A+B) \stackrel{?}{=} \frac{1 - \frac{\sin A}{\cos A} \cdot \frac{\sin B}{\cos B}}{\frac{1}{\cos A} \cdot \frac{1}{\cos B}}$$

$$\cos(A+B) \stackrel{?}{=} \frac{1 - \frac{\sin A \sin B}{\cos A \cos B}}{\frac{1}{\cos A \cos B}} \cdot \frac{\cos A \cos B}{\cos A \cos B} = 1$$

$$\cos(A+B) \stackrel{?}{=} \frac{\cos A \cos B - \sin A \sin B}{1} \quad \text{Simplify.}$$

$$\cos(A+B) = \cos(A+B) \checkmark \quad \text{Difference Identity}$$

37. $\sin(A+B)\sin(A-B) \stackrel{?}{=} \sin^2 A - \sin^2 B$
$(\sin A \cos B + \cos A \sin B)(\sin A \cos B - \cos A \sin B) \stackrel{?}{=} \sin^2 A - \sin^2 B$
$(\sin A \cos B)^2 - (\cos A \sin B)^2 \stackrel{?}{=} \sin^2 A - \sin^2 B$
$\sin^2 B \cos^2 B - \cos^2 A \sin^2 B \stackrel{?}{=} \sin^2 A - \sin^2 B$
$\sin^2 A \cos^2 B + \sin^2 A \sin^2 B - \sin^2 A \sin^2 B - \cos^2 A \sin^2 B \stackrel{?}{=} \sin^2 A - \sin^2 B$
$\sin^2 A (\cos^2 B + \sin^2 B) - \sin^2 B(\sin^2 A + \cos^2 A) \stackrel{?}{=} \sin^2 A - \sin^2 B$
$(\sin^2 A)(1) - (\sin^2 B)(1) \stackrel{?}{=} \sin^2 A - \sin^2 B$
$\sin^2 A - \sin^2 B = \sin^2 A - \sin^2 B \checkmark$

39. Sample answer: To determine wireless Internet interference, you need to determine the sine or cosine of the sum or difference of two angles. Interference occurs when waves pass through the same space at the same time. When the combined waves have a greater amplitude, constructive interference results. When the combined waves have a smaller amplitude, destructive interference results.

41. $d = \sqrt{(\cos A - \cos B)^2 + (\sin A - \sin B)^2}$
$d^2 = (\cos A - \cos B)^2 + (\sin A - \sin B)^2$
$d^2 = (\cos^2 A - 2\cos A \cos B + \cos^2 B) + (\sin^2 A - 2\sin A \sin B + \sin^2 B)$
$d^2 = \cos^2 A + \sin^2 A + \cos^2 B + \sin^2 B - 2\cos A \cos B - 2 \sin A \sin B$
$d^2 = 1 + 1 - 2\cos A \cos B - 2 \sin A \sin B \quad \begin{array}{l}\sin^2 A + \cos^2 A \\ = 1 \text{ and } \sin^2 B \\ + \cos^2 B = 1\end{array}$
$d^2 = 2 - 2\cos A \cos B - 2 \sin A \sin B$

Now find the value of d^2 when the angle having measure $A - B$ is in standard position on the unit circle, as shown in the figure below.

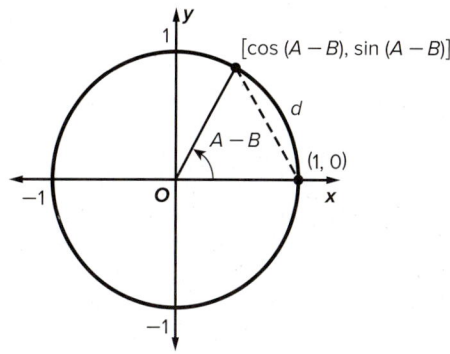

$d = \sqrt{[\cos(A-B) - 1]^2 + [\sin(A-B) - 0]^2}$
$d^2 = [\cos(A-B) - 1]^2 + [\sin(A-B) - 0]^2$
$= [\cos^2(A-B) - 2\cos(A-B) + 1] + \sin^2(A-B)$
$= \cos^2(A-B) + \sin^2(A-B) - 2\cos(A-B) + 1$
$= 1 - 2\cos(A-B) + 1$
$= 2 - 2\cos(A-B)$

43. D **45.** B **47.** D **49.** B, C, D

Lesson 10-4

1. $\frac{\sqrt{15}}{8}, \frac{7}{8}, \frac{\sqrt{8-2\sqrt{15}}}{4}, \frac{\sqrt{8+2\sqrt{15}}}{4}$ **3.** $-\frac{120}{169}, -\frac{119}{169}$, $\frac{3\sqrt{13}}{13}, \frac{2\sqrt{13}}{13}$ **5.** $-\frac{240}{289}, \frac{161}{289}, \frac{4\sqrt{17}}{17}, \frac{\sqrt{17}}{17}$ **7.** $\frac{\sqrt{2-\sqrt{2}}}{2}$

9a. $d = \frac{v^2 \sin 2\theta}{g}$ **9b.** ≈ 81 ft

11.
$(\sin\theta + \cos\theta)^2 \stackrel{?}{=} 1 + 2\sin\theta\cos\theta$
$(\sin\theta + \cos\theta)(\sin\theta + \cos\theta) \stackrel{?}{=} 1 + 2\sin\theta\cos\theta$
$\sin^2\theta + 2\sin\theta\cos\theta + \cos^2\theta \stackrel{?}{=} 1 + 2\sin\theta\cos\theta$
$1 + 2\sin\theta\cos\theta = 1 + 2\sin\theta\cos\theta \checkmark$

13. $\frac{240}{289}, -\frac{161}{289}, \frac{5\sqrt{34}}{34}, -\frac{3\sqrt{34}}{34}$

15. $\sin^2\theta = 1 - \cos^2\theta \quad \sin^2\theta + \cos^2\theta = 1$
$\sin^2\theta = 1 - \left(\frac{1}{5}\right)^2 \quad \cos\theta = \frac{1}{5}$
$\sin^2\theta = \frac{24}{25} \quad \text{Subtract.}$
$\sin\theta = \pm\frac{2\sqrt{6}}{5} \quad \text{Take the square root of each side.}$

Because θ is in the fourth quadrant, sine is negative. So,
$\sin\theta = -\frac{2\sqrt{6}}{5}$.

$\sin 2\theta = 2\sin\theta\cos\theta \quad \text{Double-angle identity}$
$= 2\left(-\frac{2\sqrt{6}}{5}\right)\left(\frac{1}{5}\right) \quad \sin\theta = -\frac{2\sqrt{6}}{5} \text{ and } \cos\theta = \frac{1}{5}$
$= -\frac{4\sqrt{6}}{25} \quad \text{Simplify.}$

$\cos 2\theta = 1 - 2\sin^2\theta \quad \text{Double-angle identity}$
$= 1 - 2\left(\frac{24}{25}\right) \quad \sin^2\theta = \frac{24}{25}$
$= -\frac{23}{25} \quad \text{Simplify.}$

$\sin\frac{\theta}{2} = \pm\sqrt{\frac{1-\cos\theta}{2}} \quad \text{Half-angle identity}$

$$= \pm\sqrt{\frac{1-\frac{1}{5}}{2}} \qquad \cos\theta = \frac{1}{5}$$

$$= \pm\sqrt{\frac{2}{5}} \qquad \text{Simplify.}$$

$$= \pm\frac{\sqrt{2}}{\sqrt{5}} \cdot \frac{\sqrt{5}}{\sqrt{5}} \text{ or } \pm\frac{\sqrt{10}}{5} \quad \text{Rationalize the denominator.}$$

If θ is between 270° and 360°, $\frac{\theta}{2}$ is between 135° and 180°. So, $\sin\frac{\theta}{2}$ is $\frac{\sqrt{10}}{5}$.

$$\cos\frac{\theta}{2} = \pm\sqrt{\frac{1+\cos\theta}{2}} \qquad \text{Half-angle identity}$$

$$= \pm\sqrt{\frac{1+\frac{1}{5}}{2}} \qquad \cos\theta = \frac{1}{5}$$

$$= \pm\sqrt{\frac{3}{5}} \qquad \text{Simplify.}$$

$$= \pm\sqrt{\frac{3}{5}} \cdot \frac{\sqrt{5}}{\sqrt{5}} \text{ or } \pm\frac{\sqrt{15}}{5} \quad \text{Rationalize the denominator.}$$

If θ is between 270° and 360°, $\frac{\theta}{2}$ is between 135° and 180°. So, $\cos\frac{\theta}{2}$ is $-\frac{\sqrt{15}}{5}$.

17. $-\frac{4}{5}, -\frac{3}{5}, \sqrt{\frac{\sqrt{5}+1}{2\sqrt{5}}}, \sqrt{\frac{\sqrt{5}-1}{2\sqrt{5}}}$ **19.** $\frac{\sqrt{2+\sqrt{2}}}{2}$

21. $-\sqrt{7-4\sqrt{3}}$ **23.** $\sqrt{3-2\sqrt{2}}$

25.
$$P = I_0^2 R \sin^2\theta t \qquad \text{Original equation}$$
$$P = I_0^2 R (\cos^2\theta t - \cos 2\theta t) \qquad \begin{array}{l}\sin^2\theta t = \cos^2 \\ \theta t - \cos 2\theta t\end{array}$$
$$P = I_0^2 R\left(\frac{1}{2}\cos 2\theta t + \frac{1}{2} - \cos 2\theta t\right) \qquad \begin{array}{l}\cos^2\theta t = \\ \frac{1}{2}\cos 2\theta t + \frac{1}{2}\end{array}$$
$$P = I_0^2 R\left(\frac{1}{2} - \frac{1}{2}\cos 2\theta t\right) \qquad \text{Simplify.}$$
$$P = \frac{1}{2}I_0^2 R - \frac{1}{2}I_0^2 R \cos 2\theta t \qquad \text{Distributive Property}$$

27.
$$1 + \frac{1}{2}\sin 2\theta \stackrel{?}{=} \frac{\sec\theta + \sin\theta}{\sec\theta}$$
$$\stackrel{?}{=} \frac{\frac{1}{\cos\theta} + \sin\theta}{\frac{1}{\cos\theta}}$$
$$\stackrel{?}{=} \frac{\frac{1}{\cos\theta} + \sin\theta}{\frac{1}{\cos\theta}} \cdot \frac{\cos\theta}{\cos\theta}$$
$$\stackrel{?}{=} 1 + \frac{1}{2} \cdot 2\sin\theta\cos\theta$$
$$\stackrel{?}{=} 1 + \frac{1}{2}\sin 2\theta \checkmark$$

29.
$$\tan\frac{\theta}{2} \stackrel{?}{=} \frac{\sin\theta}{1+\cos\theta}$$
$$\tan\frac{\theta}{2} \stackrel{?}{=} \frac{\sin 2\left(\frac{\theta}{2}\right)}{1+\cos 2\left(\frac{\theta}{2}\right)}$$
$$\tan\frac{\theta}{2} \stackrel{?}{=} \frac{2\sin\frac{\theta}{2}\cos\frac{\theta}{2}}{1 + 2\cos^2\frac{\theta}{2} - 1}$$
$$\tan\frac{\theta}{2} \stackrel{?}{=} \frac{2\sin\frac{\theta}{2}\cos\frac{\theta}{2}}{2\cos^2\frac{\theta}{2}}$$
$$\tan\frac{\theta}{2} \stackrel{?}{=} \frac{\sin\frac{\theta}{2}}{\cos\frac{\theta}{2}}$$

$$\tan\frac{\theta}{2} = \tan\frac{\theta}{2} \checkmark$$

31. $\frac{24}{25}, \frac{7}{25}, \frac{24}{7}$ **33.** $-\frac{3}{5}, -\frac{4}{5}, \frac{3}{4}$ **35.** $-\frac{4\sqrt{21}}{25}, \frac{17}{25}, -\frac{4\sqrt{21}}{17}$

37. No; Teresa incorrectly added the square roots, and Nathan used the half-angle identity incorrectly. He used sin 30° in the formula instead of first finding the cosine. **39.** If you are only given the value of cos θ, then cos $2\theta = 2\cos^2\theta - 1$ is the best identity to use. If you are only given the value of sin θ, then cos $2\theta = 1 - 2\sin^2\theta$ is the best identity to use. If you are given the values of both cos θ and sin θ, then cos $2\theta = \cos^2\theta - \sin^2\theta$ works just as well as the other two.

41. Find $\sin\frac{A}{2}$.

$$1 - 2\sin^2\theta = \cos 2\theta \qquad \text{Double-angle identity}$$
$$1 - 2\sin^2\frac{A}{2} = \cos A \qquad \begin{array}{l}\text{Substitute } \frac{A}{2} \text{ for } \theta \text{ and} \\ A \text{ for } 2\theta.\end{array}$$
$$\sin^2\frac{A}{2} = \frac{1-\cos A}{2} \qquad \text{Solve for } \sin^2\frac{A}{2}.$$
$$\sin\frac{A}{2} = \pm\sqrt{\frac{1-\cos A}{2}} \qquad \begin{array}{l}\text{Take the square root of} \\ \text{each side.}\end{array}$$

Find $\cos\frac{A}{2}$.

$$2\cos^2\theta - 1 = \cos 2\theta \qquad \text{Double-angle identity}$$
$$2\cos^2\frac{A}{2} - 1 = \cos A \qquad \text{Substitute } \frac{A}{2} \text{ for } \theta \text{ and } A \text{ for } 2\theta.$$
$$\cos^2\frac{A}{2} = \frac{1+\cos A}{2} \qquad \text{Solve for } \cos^2\frac{A}{2}.$$
$$\cos\frac{A}{2} = \pm\sqrt{\frac{1+\cos A}{2}} \qquad \text{Take the square root of each side.}$$

Find $\tan\frac{A}{2}$.

$$\tan\frac{A}{2} = \frac{\sin\frac{A}{2}}{\cos\frac{A}{2}} \qquad \text{Quotient Identity}$$

$$\tan\frac{A}{2} = \frac{\pm\sqrt{\frac{1-\cos A}{2}}}{\pm\sqrt{\frac{1+\cos A}{2}}} \qquad \text{Half-Angle Identities}$$

$$\tan\frac{A}{2} = \pm\sqrt{\frac{\frac{1-\cos A}{2}}{\frac{1+\cos A}{2}}} \qquad \text{Quotient Property of Radicals}$$

$$\tan\frac{A}{2} = \pm\sqrt{\frac{1-\cos A}{1+\cos A}} \qquad \text{Simplify.}$$

43. B **45a.** C **45b.** A, B **45c.** A **47a.** $-\frac{24}{25}$ **47b.** $\frac{7}{25}$ **47c.** $-\frac{24}{7}$ **47d.** $\frac{3\sqrt{10}}{10}$ **47e.** $\frac{\sqrt{10}}{10}$ **47f.** -3

Lesson 10-5

1. 210°, 330° **3.** 60°, 180°, or 300° **5.** 150°, 210°

7.
$$\cos 2\theta = 8 - 15\sin\theta \qquad \text{Original equation}$$
$$\cos 2\theta + 15\sin\theta - 8 = 0 \qquad \begin{array}{l}\text{Add } 15\sin\theta - 8 \\ \text{to each side.}\end{array}$$
$$1 - 2\sin^2\theta + 15\sin\theta - 8 = 0 \qquad \text{Double-angle identity}$$
$$-2\sin^2\theta + 15\sin\theta - 7 = 0 \qquad \text{Simplify.}$$
$$(-2\sin\theta + 1)(\sin\theta - 7) = 0 \qquad \text{Factor.}$$

R98

$-2\sin\theta + 1 = 0$ or $\sin\theta - 7 = 0$ Zero Product Property
$\sin\theta = \frac{1}{2}$ $\sin\theta = 7$
$\theta = 30°$ or $150°$ no solution because $0 \le \sin\theta \le 1$

9. $\pm\frac{\pi}{6} + 2k\pi$ or $\pm\frac{5\pi}{6} + 2k\pi$ **11.** $\frac{3\pi}{2} + 2k\pi$ **13.** $\pi + 2k\pi$
15. $90° + k \cdot 180°$ **17.** $45° + k \cdot 90°$ **19.** $270° + k \cdot 360°$
21a. There will be $11\frac{1}{2}$ hours of daylight 205 and 342 days after March 21; that is, on October 13 and February 26. **21b.** Every day from March 4 to October 13; sample explanation: Because the longest day of the year occurs around June 22, the days between February 26 and October 13 must increase in length until June 22 and then decrease in length until October 13.
23. $\frac{3\pi}{4} + \pi k$ **25.** $\frac{\pi}{2} + \pi k, \frac{\pi}{6} + 2k\pi, \frac{5\pi}{6} + 2k\pi$
27. $0° + k \cdot 45°$ or $0 + k \cdot \frac{\pi}{4}$
29. $\frac{\pi}{3} + 2k\pi, \frac{\pi}{2} + k\pi, \frac{5\pi}{3} + 2k\pi$
31. $135°, 225°$ **33.** $\frac{\pi}{6}$ **35.** $210°, 330°$

37.
$2\sin^2\theta = \cos\theta + 1$ Original equation
$2(1 - \cos^2\theta) = \cos\theta + 1$ $\sin^2\theta = 1 - \cos^2\theta$
$2 - 2\cos^2\theta = \cos\theta + 1$ Simplify.
$-2\cos^2\theta - \cos\theta + 1 = 0$ Subtract $\cos\theta + 1$ from each side and simplify.
$(-2\cos\theta + 1)(\cos\theta + 1) = 0$ Factor.
$-2\cos\theta + 1 = 0$ or $\cos\theta + 1 = 0$ Zero Product Property
$\cos\theta = \frac{1}{2}$ $\cos\theta = -1$

The solutions of $\cos\theta = \frac{1}{2}$ are $\frac{\pi}{3} + 2k\pi$ and $\frac{5\pi}{3} + 2k\pi$.
The solution of $\cos\theta = -1$ is $\pi + 2k\pi$.

39. $0 + 2k\pi$ **41.** $0° + k \cdot 180°$ **43.** $30° + k \cdot 360°, 150° + k \cdot 360°$ **45.** $\frac{7\pi}{6} + 2k\pi, \frac{11\pi}{6} + 2k\pi$ or $210° + k \cdot 360°, 330° + k \cdot 360°$ **47.** $0 + 2k\pi, \frac{\pi}{2} + k\pi$ or $0° + k \cdot 360°, 90° + k \cdot 180°$ **49a.** 11 m
49b. 7:00 A.M. and 7:00 P.M.
51. $\frac{\pi}{6} + 2\pi k, \frac{5\pi}{6} + 2\pi k, \frac{5\pi}{4} + 2\pi k, \frac{7\pi}{4} + 2\pi k$
53. $120° + 360°k, 240° + 360°k$
55. $\frac{\pi}{6} + 2\pi k, \frac{5\pi}{6} + 2\pi k$

57.
$D = 0.5 \sin(6.5x)\sin(2500t)$ Original equation
$0.01 = 0.5\sin(6.5 \cdot 500)\sin(2500t)$ $D = 0.01$ mm and $x = 0.5 \cdot 1000$ or 500 mm
$0.01 = 0.5\sin(3250)\sin(2500t)$ Simplify.
$0.1152 \approx \sin(2500t)$ Divide each side by $0.5\sin(3250)$.
$\sin^{-1}(0.1152) \approx 2500t$ Use the $\sin^{-1}$ function.
$6.6152 \approx 2500t$ Use a calculator.
$0.0026 \approx t$ Divide each side by 2500.
The time is about 0.0026 second.

59. $\frac{\pi}{3} < x < \pi$ or $\frac{5\pi}{3} < x < 2\pi$

61. Sample answer: All trigonometric functions are periodic. Therefore, once one or more solutions are found for a certain interval, there will be additional solutions that can be found by adding integral multiples of the period of the function to those solutions.

63. $0, b, 2b$ **65.** C **67.** B

Chapter 10   Study Guide and Review

1. difference of angles identity   **3.** double-angle identity   **5.** half-angle identity   **7.** Pythagorean identity   **9.** Trigonometric identities are equations that are true for all values of the variable for which both sides are defined. Trigonometric equations are true only for certain values of the variable.
11. $-\sqrt{3}$ **13.** $-\frac{4}{5}$ **15.** $\frac{15\sqrt{709}}{709}$
17. $\sec\theta$ **19.** $\sec\theta$

21.
$\dfrac{\cos\theta}{\cot\theta} + \dfrac{\sin\theta}{\tan\theta} \stackrel{?}{=} \sin\theta + \cos\theta$
$\cos\theta \div \dfrac{\cos\theta}{\sin\theta} + \sin\theta \div \dfrac{\sin\theta}{\cos\theta} \stackrel{?}{=} \sin\theta + \cos\theta$
$\cos\theta \cdot \dfrac{\sin\theta}{\cos\theta} + \sin\theta \cdot \dfrac{\cos\theta}{\sin\theta} \stackrel{?}{=} \sin\theta + \cos\theta$
$\sin\theta + \cos\theta = \sin\theta + \cos\theta$ ✓

23. $\tan^2\theta + 1 = \left(\dfrac{\sqrt{7}}{3}\right)^2 + 1 = \dfrac{7}{9} + 1 = \dfrac{7}{9} + \dfrac{9}{9} = \dfrac{16}{9}$;
$\sec^2\theta = \left(\dfrac{4}{3}\right)^2 = \dfrac{16}{9}$

25. $\dfrac{\sqrt{6} + \sqrt{2}}{4}$ **27.** $\dfrac{\sqrt{6} + \sqrt{2}}{4}$ **29.** $\dfrac{-\sqrt{6} + \sqrt{2}}{4}$

31.
$\sin\left(\dfrac{3\pi}{2} - \theta\right) \stackrel{?}{=} -\cos\theta$
$\sin\dfrac{3\pi}{2}\cos\theta - \cos\dfrac{3\pi}{2}\sin\theta \stackrel{?}{=} -\cos\theta$
$(-1)\cos\theta - (0)\sin\theta \stackrel{?}{=} -\cos\theta$
$-\cos\theta = -\cos\theta$ ✓

33. $\sin 2\theta = \dfrac{24}{25}, \cos 2\theta = \dfrac{7}{25}, \sin\dfrac{\theta}{2} = \dfrac{\sqrt{10}}{10}$, and $\cos\dfrac{\theta}{2} = \dfrac{3\sqrt{10}}{10}$

35. $\sin 2\theta = -\dfrac{4\sqrt{5}}{9}, \cos 2\theta = -\dfrac{1}{9}$, $\sin\dfrac{\theta}{2} = \dfrac{\sqrt{30}}{6}$, and $\cos\dfrac{\theta}{2} = \dfrac{\sqrt{6}}{6}$

37. $60°, 300°$

39. $90°, 210°, 270°, 330°$
41. $\dfrac{\pi}{3}, \dfrac{5\pi}{3}$

Glossary/Glosario

Multilingual eGlossary

Go to connectED.mcgraw-hill.com for a glossary of terms in these additional languages:

Arabic	Chinese	Hmong	Spanish	Vietnamese
Bengali	English	Korean	Tagalog	
Brazilian Portugese	Haitian Creole	Russian	Urdu	

English / Español

A

absolute value function (p. 120) A function written as $f(x) = |x|$, where $f(x) = \begin{cases} x \text{ if } x > 0 \\ 0 \text{ if } x = 0 \\ -x \text{ if } x < 0 \end{cases}$ values of x.

función del valor absoluto Una función que se escribe $f(x) = |x|$, donde $f(x) = \begin{cases} x \text{ si } x > 0 \\ 0 \text{ si } x = 0 \\ -x \text{ si } x < 0 \end{cases}$

amplitude (p. 627) For functions in the form $y = a \sin b\theta$ or $y = a \cos b\theta$, the amplitude is $|a|$.

amplitud Para funciones de la forma $y = a \text{ sen } b\theta$ o $y = a \cos b\theta$, la amplitud es $|a|$.

angle of depression (p. 598) The angle between a horizontal line and the line of sight from the observer to an object at a lower level.

ángulo de depresión Ángulo entre una recta horizontal y la línea visual de un observador a una figura en un nivel inferior.

angle of elevation (p. 598) The angle between a horizontal line and the line of sight from the observer to an object at a higher level.

ángulo de elevación Ángulo entre una recta horizontal y la línea visual de un observador a una figura en un nivel superior.

asymptote (p. 373) A line that a graph approaches.

asíntota Recta a la que se aproxima una gráfica.

axis of symmetry (p. 152) A line about which a figure is symmetric.

eje de simetría Recta respecto a la cual una figura es simétrica.

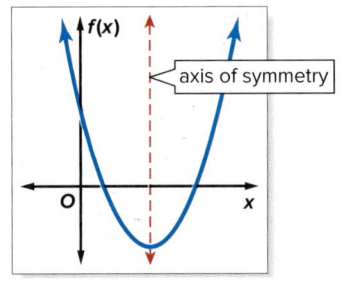

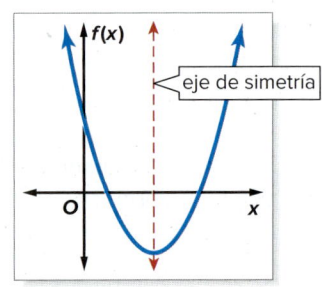

B

bias (p. 531) An error that results in a misrepresentation of members of a population.

Binomial Theorem (p. 237) If n is a nonnegative integer, then $(a + b)^n = 1a^n b^0 + \frac{n}{1} a^{n-1} b^1 + \frac{n(n+1)}{1 \cdot 2} a^{n-2} b^2 + \cdots + 1a^0 b^n$.

boundary (p. 35) A line or curve that separates the coordinate plane into two regions.

bounded (p. 60) A region is bounded when the graph of a system of constraints is a polygonal region.

sesgo Error que resulta en la representación errónea de los miembros de una población.

teorema del binomio Si n es un entero no negativo, entonces $(a + b)^n = 1a^n b^0 + \frac{n}{1} a^{n-1} b^1 + \frac{n(n+1)}{1 \cdot 2} a^{n-2} b^2 + \cdots + 1a^0 b^n$.

frontera Recta o curva que divide un plano de coordenadas en dos regiones.

acotada Una región está acotada cuando la gráfica de un sistema de restricciones es una región poligonal.

C

central angle (p. 607) An angle with a vertex at the center of the circle and sides that are radii.

Change of Base Formula (p. 425) For all positive numbers a, b, and n, where $a \neq 1$ and $b \neq 1$, $\log_a n = \frac{\log_b n}{\log_b a}$.

circular function (p. 620) A function defined using a unit circle.

combination (p. P11) An arrangement of objects in which order is not important.

combined variation (p. 503) When one quantity varies directly and/or inversely as two or more other quantities.

common logarithms (p. 423) Logarithms that use 10 as the base.

completing the square (p. 192) A process used to make a quadratic expression into a perfect square trinomial.

complex conjugates (p. 175) Two complex numbers of the form $a + bi$ and $a - bi$.

complex fraction (p. 470) A rational expression whose numerator and/or denominator contains a rational expression.

complex number (p. 173) Any number that can be written in the form $a + bi$, where a and b are real numbers and i is the imaginary unit.

composition of functions (p. 322) A function is performed, and then a second function is performed on the result of the first function. The composition of f and g is denoted by $f \circ g$, and $[f \circ g](x) = f[g(x)]$.

ángulo central Ángulo cuyo vértice es el centro del círculo y cuyos lados son radios.

fórmula del cambio de base Para todo número positivo a, b y n, donde $a \neq 1$ y $b \neq 1$, $\log_a n = \frac{\log_b n}{\log_b a}$.

funciones circulares Funciones definidas en un círculo unitario.

combinación Arreglo de elementos en que el orden no es importante.

variación combinada Cuando una cantidad varía directamente e inverso como dos o más otras cantidades.

logaritmos comunes El logaritmo de base 10.

completar el cuadrado Proceso mediante el cual una expresión cuadrática se transforma en un trinomio cuadrado perfecto.

conjugados complejos Dos números complejos de la forma $a + bi$ y $a - bi$.

fracción compleja Expresión racional cuyo numerador o denominador contiene una expresión racional.

número complejo Cualquier número que puede escribirse de la forma $a + bi$, donde a y b son números reales e i es la unidad imaginaria.

composición de funciones Se evalúa una función y luego se evalúa una segunda función en el resultado de la primera función. La composición de f y g se define con $f \circ g$, y $[f \circ g](x) = f[g(x)]$.

compound event (p. 14) Two or more simple events.

compound interest (p. 384) Interest paid on the principal of an investment and any previously earned interest.

conditional probability (p. P16) The probability of an event occurring given that another event has already occurred.

confidence level (p. 546) The probability that survey results from several samples of the same population will have a confidence interval that includes the actual value of the population parameter.

consistent (p. 43) A system of equations that has at least one ordered pair that satisfies both equations.

constant of variation (p. 500) The constant k used with direct or inverse variation.

constant term (p. 151) In $f(x) = ax^2 + bx + c$, c is the constant term.

constraints (p. 36) Conditions given to variables, often expressed as linear inequalities.

continuous relation (p. 88) A relation that can be graphed with a line or smooth curve.

correlation coefficient (p. 409) A measure that shows how well data are modeled by a linear equation.

cosecant (p. 594) For any angle, with measure α, a point $P(x, y)$ on its terminal side, $r = \sqrt{x^2 + y^2}$, $\csc \alpha = \frac{r}{y}$.

cosine (p. 594) For any angle, with measure α, a point $P(x, y)$ on its terminal side, $r = \sqrt{x^2 + y^2}$, $\cos \alpha = \frac{x}{r}$.

cotangent (p. 594) For any angle, with measure α, a point $P(x, y)$ on its terminal side, $r = \sqrt{x^2 + y^2}$, $\cot \alpha = \frac{x}{y}$.

coterminal angles (p. 605) Two angles in standard position that have the same terminal side.

cube root function (p. 345) A function described by the equation $f(x) = a\sqrt[3]{x - h} + k$.

cycle (p. 621) One complete pattern of a periodic function.

evento compuesto Dos o más eventos simples.

interés compuesto Interés obtenido tanto sobre la inversion inicial como sobre el interes conseguido.

probabilidad condicional Probabilidad de un evento dado que otro evento ya ha ocurrido.

nivel de confianza La probabilidad de que los resultados de un estudio basado en varias muestras de la misma población tengan un intervalo de confianza que incluye el valor real del parámetro de la población.

consistente Sistema de ecuaciones para el cual existe al menos un par ordenado que satusface ambas ecuaciones.

constante de variación La constante k que se usa en variación directa o inversa.

término constante En $f(x) = ax^2 + bx + c$, c es el término constante.

restricciones Condiciones a que están sujetas las variables, a menudo escritas como desigualdades lineales.

relación continua Relación cuya gráfica puede ser una recta o una curva suave.

coeficiente de correlación Una medida que demuestra cómo los datos bien son modelados por una ecuación linear.

cosecante Para cualquier ángulo de medida α, un punto $P(x, y)$ en su lado terminal, $r = \sqrt{x^2 + y^2}$, $\csc \alpha = \frac{r}{y}$.

coseno Para cualquier ángulo de medida α, un punto $P(x, y)$ en su lado terminal, $r = \sqrt{x^2 + y^2}$, $\cos \alpha = \frac{x}{r}$.

cotangente Para cualquier ángulo de medida α, un punto $P(x, y)$ en su lado terminal, $r = \sqrt{x^2 + y^2}$, $\cot \alpha = \frac{x}{y}$.

ángulos coterminales Dos ángulos en posición estándar que tienen el mismo lado terminal.

función de raíz cúbica Función que se describe con la ecuación $f(x) = a\sqrt[3]{x - h} + k$.

ciclo Un patrón completo de una función periódica.

D

decay factor (p. 376) In exponential decay, the base of the exponential expression, $1 - r$.

factor de decaimiento En decaimiento exponencial, la base de la expresión exponencial, $1 - r$.

degree of a polynomial (p. 231) The greatest degree of any term in the polynomial.

dependent (p. 43) A system of equations that has an infinite number of solutions.

dependent events (p. P16) Events in which the outcome of one event affects the outcome of another event.

dependent variable (p. 89) The other variable in a function, usually y, whose values depend on x.

depressed polynomial (p. 289) The quotient when a polynomial is divided by one of its binomial factors.

descriptive statistics (p. P24) The branch of statistics that focuses on collecting, summarizing, and displaying data.

dilation (p. 126) A transformation in which a geometric figure is enlarged or reduced.

dimensional analysis (p. 236) Performing operations with units.

direct variation (p. 500) y varies directly as x if there is some nonzero constant k such that $y = kx$. k is called the constant of variation.

discrete relation (p. 88) A relation in which the domain is a set of individual points.

discriminant (p. 202) In the Quadratic Formula, the expression $b^2 - 4ac$.

domain (p. P4) The set of all x-coordinates of the ordered pairs of a relation.

E

e (p. 430) The irrational number 2.71828…. e is the base of the natural logarithms.

elimination method (p. 45) Eliminate one of the variables in a system of equations by adding or subtracting the equations.

end behavior (pp. 103, 255) The behavior of the graph as x approaches positive infinity $(+\infty)$ or negative infinity $(-\infty)$.

equation (p. 5) A mathematical sentence stating that two mathematical expressions are equal.

experiment (p. 531) Something that is intentionally done to people, animals, or objects, and then the response is observed.

experimental probability (pp. P13, 538) What is estimated from observed simulations or experiments.

grado de un polinomio Grado máximo de cualquier término del polinomio.

dependiente Sistema de ecuaciones que posee un número infinito de soluciones.

evento dependiente Eventos en que el resultado de uno de los eventos afecta el resultado de otro de los eventos.

variable dependiente La otra variable de una función, por lo general y, cuyo valor depende de x.

polinomio reducido El cociente cuando se divide un polinomio entre uno de sus factores binomiales.

estadística descriptiva Rama de la estadística cuyo enfoque es la recopilación, resumen y demostración de los datos.

homotecia Transformación en que se amplía o se reduce un figura geométrica.

análisis dimensional Realizar operaciones con unidades.

variación directa y varía directamente con x si hay una constante no nula k tal que $y = kx$. k se llama la constante de variación.

relación discreta Relación en la cual el dominio es un conjunto de puntos individuales.

discriminante En la fórmula cuadrática, la expresión $b^2 - 4ac$.

dominio El conjunto de todas las coordenadas x de los pares ordenados de una relación.

e El número irracional 2.71828…. e es la base de los logaritmos naturales.

método de eliminación Eliminar una de las variables de un sistema de ecuaciones sumando o restando las ecuaciones.

comportamiento final El comportamiento de una gráfica a medida que x tiende a más infinito $(+\infty)$ o menos infinito $(-\infty)$.

ecuación Enunciado matemático que afirma la igualdad de dos expresiones matemáticas.

experimento Algo se hace intencionalmente poblar, los animales, o los objetos, y entonces la respuesta se observa.

probabilidad experimental Qué se estima de simulaciones o de experimentos observados.

exponential decay (p. 375) Exponential decay occurs when a quantity decreases exponentially over time.

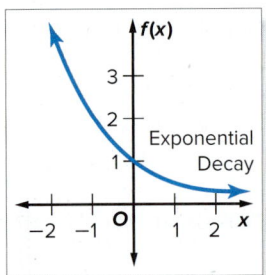

exponential equation (p. 383) An equation in which the variables occur as exponents.

exponential function (p. 373) A function of the form $y = ab^x$, where $a \neq 0$, $b > 0$, and $b \neq 1$.

exponential growth (p. 373) Exponential growth occurs when a quantity increases exponentially over time.

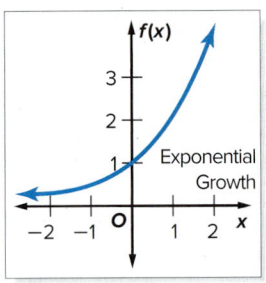

exponential inequality (p. 385) An inequality involving exponential functions.

extraneous solution (p. 352) A number that does not satisfy the original equation.

extrema (pp. 105, 263) The maximum and minimum values of a function.

desintegración exponencial Ocurre cuando una cantidad disminuye exponencialmente con el tiempo.

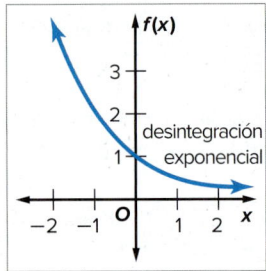

ecuación exponencial Ecuación en que las variables aparecen en los exponentes.

función exponencial Una función de la forma $y = ab^x$, donde $a \neq 0$, $b > 0$, y $b \neq 1$.

crecimiento exponencial El que ocurre cuando una cantidad aumenta exponencialmente con el tiempo.

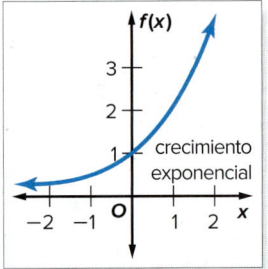

desigualdad exponencial Desigualdad que contiene funciones exponenciales.

solución extraña Número que no satisface la ecuación original.

extrema Son los valores máximos y mínimos de una función.

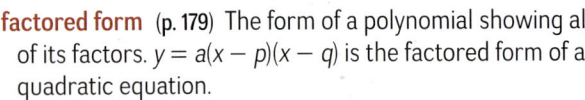

factored form (p. 179) The form of a polynomial showing all of its factors. $y = a(x - p)(x - q)$ is the factored form of a quadratic equation.

Factor Theorem (p. 289) The binomial $x - r$ is a factor of the polynomial $P(x)$ if and only if $P(r) = 0$.

fair (p. 573) An experiment or decision is fair if all possible outcomes or choices have the same probability of being chosen.

feasible region (p. 60) The intersection of the graphs in a system of constraints.

five-number summary (p. P26) The three quartiles and the maximum and minimum values in a set of data.

forma reducida La forma de un polinomio que demuestra todos sus factores. $y = a(x - p)(x - q)$ es la forma descompuesta en factores de una ecuación cuadrática.

teorema factor El binomio $x - r$ es un factor del polinomio $P(x)$ si $P(r) = 0$.

imparcial Un experimento, o una decisión, es imparcial si todos los resultados u opciones posibles tienen la misma probabilidad de ser elegidos.

región viable Intersección de las gráficas de un sistema de restricciones.

resumen de cinco números Los tres cuartiles y los valores máximo y mínimo de un conjunto de datos.

FOIL method (p. 179) The product of two binomials is the sum of the products of **F** the *first* terms, **O** the *outer* terms, **I** the *inner* terms, and **L** the *last* terms.

frequency (p. 628) The number of cycles in a given unit of time.

function (p. P4) A relation in which each element of the domain is paired with exactly one element in the range.

function notation (p. 89) An equation of y in terms of x can be rewritten so that $y = f(x)$. For example, $y = 2x + 1$ can be written as $f(x) = 2x + 1$.

método FOIL El producto de dos binomios es la suma de los productos de los primeros (**F**irst) términos, los términos exteriores (**O**uter), los términos interiores (**I**nner) y los últimos (**L**ast) términos.

frecuencia El número de ciclos en una unidad del tiempo dada.

función Relación en que a cada elemento del dominio le corresponde un solo elemento del rango.

notación funcional Una ecuación de y en términos de x puede escribirse en la forma $y = f(x)$. Por ejemplo, $y = 2x + 1$ puede escribirse como $f(x) = 2x + 1$.

G

geometric means (p. 391) The terms between any two nonsuccessive terms of a geometric sequence.

geometric series (p. 392) The sum of the terms of a geometric sequence.

greatest integer function (p. 119) A step function, written as $f(x) = [\![x]\!]$, where $f(x)$ is the greatest integer less than or equal to x.

growth factor (p. 375) In exponential growth, the base of the exponential expression, $1 + r$.

media geométrica Cualquier término entre dos términos no consecutivos de una sucesión geométrica.

serie geométrica La suma de los términos de una sucesión geométrica.

función del máximo entero Una función etapa que se escribe $f(x) = [\![x]\!]$, donde $f(x)$ es el meaximo entero que es menor que o igual a x.

factor del crecimiento En el crecimiento exponencial, la base de la expresión exponencial, $1 + r$.

H

horizontal asymptote (p. 491) A horizontal line which a graph approaches.

hyperbola (p. 483) The set of all points in the plane such that the absolute value of the difference of the distances from two given points in the plane, called foci, is constant.

asíntota horizontal Una linea horizontal a que un gráfico acerca.

hipérbola Conjunto de todos los puntos de un plano en los que el valor absoluto de la diferencia de sus distancias a dos puntos dados del plano, llamados focos, es constante.

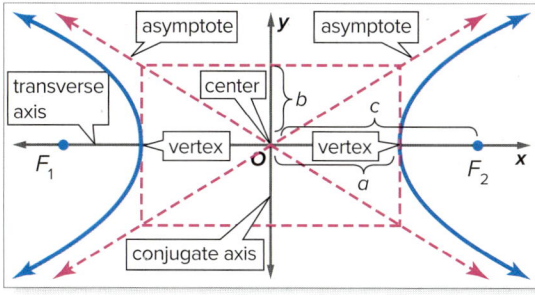
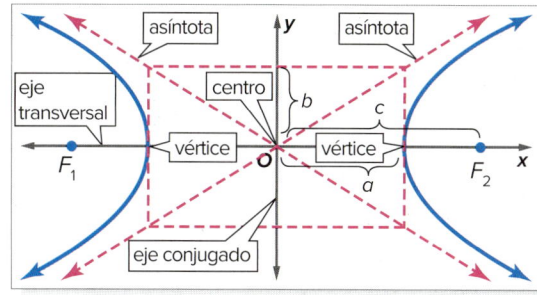

I

imaginary unit (p. 172) i, or the principal square root of -1.

inconsistent (p. 43) A system of equations with no ordered pair that satisfy both equations.

unidad imaginaria i, o la raíz cuadrada principal de -1.

inconsistente Un sistema de ecuaciones para el cual no existe par ordenado alguno que satisfaga ambas ecuaciones.

independent (p. 43) A system of equations with exactly one solution.

independent events (p. P16) Events in which the outcome of one event does not affect the outcome of another event.

independent variable (p. 89) In a function, the variable, usually x, whose values make up the domain.

inflection point (p. 345) A point of a curve at which a change in the direction of curvature happens.

infinity (p. 20) Without bound, or continues without end.

initial side (p. 604) The fixed ray of an angle.

independiente Un sistema de ecuaciones que posee una única solución.

eventos independientes Eventos en que el resultado de uno de los eventos no afecta el resultado de otro evento.

variable independiente En una función, la variable, por lo general x, cuyos valores forman el dominio.

punto de inflexión Punto de una curva en el que ocurre un cambio de dirección en su curvatura.

infinito Sin límite, o continúa sin extremo.

lado inicial de un ángulo El rayo fijo de un ángulo.

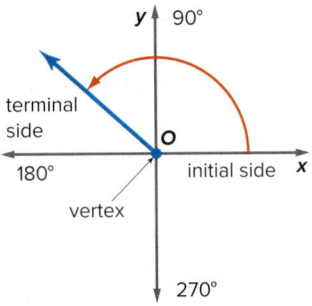

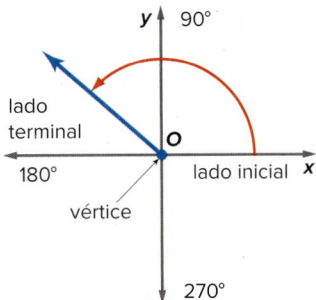

interquartile range (IQR) (p. P27) The range of the middle half of a set of data. It is the difference between the upper quartile and the lower quartile.

interval notation (p. 20) A way to describe the solution set of an inequality.

inverse function (p. 329) Two functions f and g are inverse functions if and only if both of their compositions are the identity function.

inverse relation (p. 329) Two relations are inverse relations if and only if whenever one relation contains the element (a, b) the other relation contains the element (b, a).

inverse variation (p. 502) y varies inversely as x if there is some nonzero constant k such that $xy = k$ or $y = \frac{k}{x}$, where $x \neq 0$ and $y \neq 0$.

amplitud intercuartílica Amplitud de la mitad central de un conjunto de datos. Es la diferencia entre el cuartil superior y el inferior.

notación del intervalo Una manera de describir el sistema de la solución de una desigualdad.

función inversa Dos funciones f y g son inversas mutuas si y sólo si las composiciones de ambas son la función identidad.

relaciones inversas Dos relaciones son relaciones inversas mutuas si y sólo si cada vez que una de las relaciones contiene el elemento (a, b), la otra contiene el elemento (b, a).

variación inversa y varía inversamente con x si hay una constante no nula k tal que $xy = k$ o $y = \frac{k}{x}$, donde $x \neq 0$ y $y \neq 0$.

J

joint variation (p. 501) y varies jointly as x and z if there is some nonzero constant k such that $y = kxz$.

variación conjunta y varía conjuntamente con x y z si hay una constante no nula k tal que $y = kxz$.

L

leading coefficient (p. 253) The coefficient of the term with the highest degree.

linear equation (p. 95) An equation that has no operations other than addition, subtraction, and multiplication of a variable by a constant.

linear function (p. 95) A function whose ordered pairs satisfy a linear equation.

linear inequality (p. 35) An inequality that describes a half-plane with a boundary that is a straight line.

linear programming (p. 60) The process of finding the maximum or minimum values of a function for a region defined by inequalities.

line of reflection (p. 126) The line over which a reflection flips a figure.

line of symmetry (p. 97) A line that divides a figure into two halves that are reflections of each other.

line symmetry (p. 97) Figures that match exactly when folded in half have line symmetry.

Location Principle (p. 262) Suppose $y = f(x)$ represents a polynomial function and a and b are two numbers such that $f(a) < 0$ and $f(b) > 0$. Then the function has at least one real zero between a and b.

linear term (p. 151) In the equation $f(x) = ax^2 + bx + c$, bx is the linear term.

logarithm (p. 397) In the function $x = b^y$, y is called the logarithm, base b, of x. Usually written as $y = \log_b x$ and is read "y equals log base b of x."

logarithmic equation (p. 437) An equation that contains one or more logarithms.

logarithmic function (p. 398) The function $y = \log_b x$, where $b > 0$ and $b \neq 1$, which is the inverse of the exponential function $y = b^x$.

logarithmic inequality (p. 438) An inequality that contains one or more logarithms.

logistic growth model (p. 448) A growth model that represents growth that has a limiting factor. Logistic models are the most accurate models for representing population growth.

lower quartile (p. 26) The median of the lower half of a set of data, indicated by LQ.

coeficiente líder Coeficiente del término de mayor grado.

ecuación lineal Ecuación sin otras operaciones que las de adición, sustracción y multiplicación de una variable por una constante.

función lineal Función cuyos pares ordenados satisfacen una ecuación lineal.

desigualdad lineal Desigualdad lineal es una desigualdad que describe un semiplano con un límite que es una línea recta.

programación lineal Proceso de hallar los valores máximo o mínimo de una función lineal en una región definida por las desigualdades.

línea de la reflexión La línea excedente que una reflexión mueve de un tirón una figura.

eje de simetría Línea que divide una figura en dos mitades que son reflejos mutuos.

simetría axial Las figuras que coinciden exactamente cuando se doblan por la mitad tienen simetría axial.

principio de ubicación Sea $y = f(x)$ una función polinómica con a y b dos números tales que $f(a) < 0$ y $f(b) > 0$. Entonces la función tiene por lo menos un resultado real entre a y b.

término lineal En la ecuación $f(x) = ax^2 + bx + c$, el término lineal es bx.

logaritmo En la función $x = b^y$, y es el logaritmo en base b, de x. Generalmente escrito como $y = \log_b x$ y se lee "y es igual al logaritmo en base b de x."

ecuación logarítmica Ecuación que contiene uno o más logaritmos.

función logarítmica La función $y = \log_b x$, donde $b > 0$ y $b \neq 1$, inversa de la función exponencial $y = b^x$.

desigualdad logarítmica Desigualdad que contiene uno o más logaritmos.

modelo logístico del crecimiento Un modelo del crecimiento que representa el crecimiento que tiene un factor limitador. Los modelos logísticos son los modelos más exactos para representar crecimiento de la población.

cuartil inferior Mediana de la mitad inferior de un conjunto de datos, se denota con CI.

M

mapping (p. P4) How each member of the domain is paired with each member of the range.

margin of error (p. 546) The limit on the difference between how a sample responds and how the total population would respond.

maximum value (p. 154) The y-coordinate of the vertex of the quadratic function $f(x) = ax^2 + bx + c$, where $a < 0$.

mean (p. P24) The sum of the values in a set of data divided by the total number of values in the set.

median (p. P24) The middle value or the mean of the middle values in a set of data when the data are arranged in numerical order.

midline (p. 636) A horizontal axis used as the reference line about which the graph of a periodic function oscillates.

minimum value (p. 154) The y-coordinate of the vertex of the quadratic function $f(x) = ax^2 + bx + c$, where $a > 0$.

mode (p. P24) The value or values that appear most often in a set of data.

mutually exclusive (p. P14) Two events that cannot occur at the same time.

transformaciones La correspondencia entre cada miembro del dominio con cada miembro del rango.

margen de error muestral Límite en la diferencia entre las respuestas obtenidas con una muestra y cómo pudiera responder la población entera.

valor máximo La coordenada y del vértice de la función cuadrática $f(x) = ax^2 + bx + c$, where $a < 0$.

media Suma de los valores en un conjunto de datos dividida entre el número total de valores en el conjunto.

mediana Valor del medio o la media de los valores del medio en un conjunto de datos, cuando los datos están ordenados numéricamente.

linea media Eje horizontal que se usa como recta de referencia alrededor de la cual oscila la gráfica de una función periódica.

valor mínimo La coordenada y del vértice de la función cuadrática $f(x) = ax^2 + bx + c$, donde $a > 0$.

moda Valor o valores que aparecen con más frecuencia en un conjunto de datos.

mutuamente exclusivos Dos eventos que no pueden ocurrir simultáneamente.

N

natural base, e (p. 430) An irrational number approximately equal to 2.71828... .

natural base exponential function (p. 430) An exponential function with base e, $y = e^x$.

natural logarithm (p. 430) Logarithms with base e, written ln x.

nonlinear function (p. 95) A function that is not modeled by a linear equation.

normal distribution (p. 566) A continuous, symmetric, bell-shaped distribution of a random variable.

base natural, e Número irracional aproximadamente igual a 2.71828... .

función exponencial natural La función exponencial de base e, $y = e^x$.

logaritmo natural Logaritmo de base e, el que se escribe ln x.

función no lineal Función que no se representa con una ecuación lineal.

distribución normal Distribución con forma de campana, simétrica y continua de una variable aleatoria.

Normal Distribution

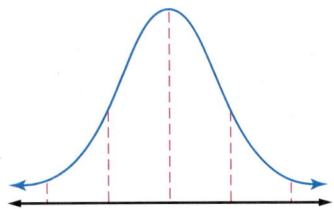

Distribución normal

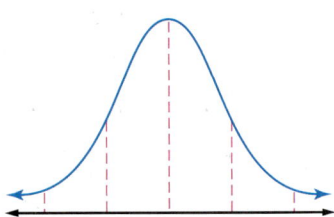

O

oblique asymptote (p. 493) An asymptote that is neither horizontal nor vertical and is sometimes called a *slant asymptote*.

observational study (p. 531) Individuals are observed and no attempt is made to influence the results.

odds (p. P15) A ratio that compares the number of ways an event can occur to the number of ways that it cannot occur.

one-to-one function (p. 87) **1.** A function where each element of the range is paired with exactly one element of the domain **2.** A function whose inverse is a function.

onto function (p. 87) Each element of the range corresponds to an element of the domain.

open sentence (p. 5) A mathematical sentence containing one or more variables.

optimize (p. 62) To seek the optimal price or amount that is desired to minimize costs or maximize profits.

ordered triple (p. 67) **1.** The coordinates of a point in space **2.** The solution of a system of equations in three variables x, y, and z.

outcome (p. P9) The results of a probability experiment or an event.

outlier (p. P27) A data point that does not appear to belong to the rest of the set.

asíntota oblicuo Una asíntota que es ni horizontal ni la vertical y a veces se llama una *asíntota inclinada*.

estudio de observación Observan a los individuos y no se hace ninguna tentativa de influenciar los resultados.

posibilidad Razón que compara el número de maneras en que puede ocurrir un evento al número de maneras en que no puede ocurrir dicho evento.

función biunívoca **1.** Función en la que a cada elemento del rango le corresponde sólo un elemento del dominio. **2.** Función cuya inversa es una función.

sobre la función Cada elemento de la gama corresponde a un elemento del dominio.

enunciado abierto Enunciado matemático que contiene una o más variables.

optimice Buscar el precio óptimo o ascender que se desea para reducir al mínimo costes o para maximizar de los beneficios.

triple ordenado **1.** Las coordenadas de un punto en el espacio **2.** Solución de un sistema de ecuaciones en tres variables x, y y z.

resultados Lo que produce un experimento o evento probabilístico.

valor atípico Dato que no parece pertenecer al resto el conjunto.

P

parabola (p. 151) The graph of a quadratic function. The set of all points in a plane that are the same distance from a given point, called the focus, and a given line, called the directrix.

parábola La gráfica de una función cuadrática. Conjunto de todos los puntos de un plano que están a la misma distancia de un punto dado, llamado foco, y de una recta dada, llamada directriz.

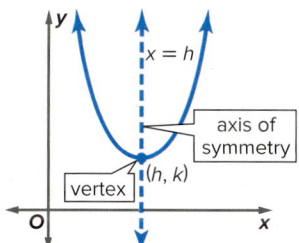

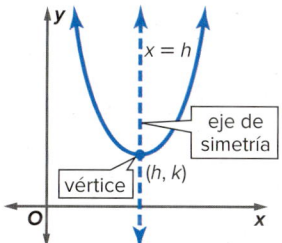

parallel lines (p. 30) Nonvertical coplanar lines with the same slope.

rectas paralelas Rectas coplanares no verticales con la misma pendiente.

parameter (p. 531) A measure that describes a characteristic of a population.

parámetro Una medida que describe una característica de una población.

Pascal's triangle (p. 237) A triangular array of numbers such that the $(n + 1)$th row is the coefficient of the terms of the expansion $(x + y)^n$ for $n = 0, 1, 2 \ldots$

period (p. 621) The least possible value of a for which $f(x) = f(x + a)$.

periodic function (p. 621) **1.** A function with y-values that repeat at regular intervals. **2.** A function is called periodic if there is a number a such that $f(x) = f(x + a)$ for all x in the domain of the function.

permutation (p. P9) An arrangement of objects in which order is important.

perpendicular lines (p. 30) In a plane, any two oblique lines, the product of whose slopes is -1.

phase shift (p. 635) A horizontal translation of a trigonometric function.

piecewise-defined function (p. 118) A function that is written using two or more expressions.

piecewise-linear function (p. 119) A function in which the equation for each interval is linear.

point discontinuity (p. 494) If the original function is undefined for $x = a$ but the related rational expression of the function in simplest form is defined for $x = a$, then there is a hole in the graph at $x = a$.

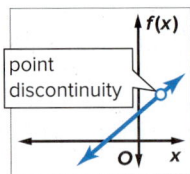

point of symmetry (p. 98) The common point of reflection for all points of a figure.

point-slope form (p. 29) An equation in the form $y - y_1 = m(x - x_1)$ where (x_1, y_1) are the coordinates of a point on the line and m is the slope of the line.

point symmetry (p. 98) Two distinct points P and P' are symmetric with respect to point M if and only if M is the midpoint of PP'. Point M is symmetric with respect to itself.

polynomial function (p. 254) A function that is represented by a polynomial equation.

polynomial identity (p. 282) A polynomial equation that is true for any values that are substituted for the variables.

polynomial in one variable (p. 253) $a_n x^n + a_{n-1} x^{n-1} + \cdots + a_2 x^2 + a_1 x + a_0$, where the coefficients $a_n, a_{n-1}, \ldots, a_0$ represent real numbers, and a_n is not zero and n is a nonnegative integer.

triángulo de Pascal Arreglo triangular de números tal que la fila de $(n + 1)$ está compuesta de proporciona los coeficientes de los términos de la expansión de $(x + y)^n$ para $n = 0, 1, 2 \ldots$

período El menor valor positivo posible para a, para el cual $f(x) = f(x + a)$.

función periódica **1.** Una función con y-valores aquella repetición con regularidad. **2.** Función para la cual hay un número a tal que $f(x) = f(x + a)$ para todo x en el dominio de la función.

permutación Arreglo de elementos en que el orden es importante.

rectas perpendiculares En un plano, dos rectas oblicuas cualesquiera cuyas pendientes tienen un producto igual a -1.

desplazamiento de fase Traslación horizontal de una función trigonométrica.

función por trozos-definida Una función se escribe que usando dos o más expresiones.

función lineal por partes Una función en que la ecuación para cada intervalo es lineal.

discontinuidad evitable Si la función original no está definida en $x = a$ pero la expresión racional reducida correspondiente de la función está definida en $x = a$, entonces la gráfica tiene una ruptura o corte en $x = a$.

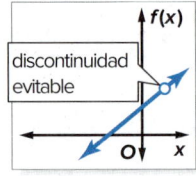

punto de simetría Punto común de reflexión para todos los puntos de una figura.

forma punto-pendiente Ecuación de la forma $y - y_1 = m(x - x_1)$ donde (x_1, y_1) es un punto en la recta y m es la pendiente de la recta.

simetría central Dos puntos distintos P y P' son simétricos respecto de un punto M, si y sólo si M es el punto medio de PP'. El punto M es simétrico respecto de sí mismo.

función polinomial Función representada por una ecuación polinomial.

identidad polinómica Ecuación polinómica que es verdadera para cualquier valor por el que se reemplazan las variables.

polinomio de una variable $a_n x^n + a_{n-1} x^{n-1} + \cdots + a_2 x^2 + a_1 x + a_0$, donde los coeficientes $a_n, a_{n-1}, \ldots, a_0$ son números reales, a_n no es nulo y n es un entero no negativo.

population (p. 24) An entire group of living things or objects.

population mean (p. 545) The mean, or average, of a characteristic for the entire population. The population mean can be represented by μ.

population proportion (p. 545) The ratio of the number of members in the population that have a particular characteristic to the number of members in the population.

positive correlation (p. 92) When the values in a scatter plot are closely linked in a positive manner.

power function (pp. 251, 254) An equation in the form $f(x) = ax^b$, where a and b are nonzero real numbers.

prime polynomial (p. 274) A polynomial that cannot be factored.

probability (p. P13) A measure of the chance that a given event will occur.

probability model (p. P13, 538) A mathematical model used to represent the outcomes of an experiment.

pure imaginary numbers (p. 172) The square roots of negative real numbers. For any positive real number b,
$$\sqrt{-b^2} = \sqrt{b^2} \cdot \sqrt{-1}, \text{ or } bi.$$

población Un grupo entero de cosas o de objetos vivos.

media de la población La media, o el promedio, de una característica de toda la población. La media de la población se puede representar con el símbolo μ.

proporción de la población La razón de la cantidad de miembros de la población que tienen una característica particular a la cantidad de miembros que hay en la población.

correlación positivo Cuando los valores en un diagrama de la dispersión se ligan de cerca de una manera positiva.

función potencia Ecuación de la forma $f(x) = ax^b$, donde a y b son números reales.

polinomio primero Un polinomio que no puede ser descompuesto en factores.

probabilidad Medida de la posibilidad de que ocurra un evento dado.

modelo de probabilidad Modelo matemático que se usa para representar los resultados de un experimento.

número imaginario puro Raíz cuadrada de un número real negativo. Para cualquier número real positivo b,
$$\sqrt{-b^2} = \sqrt{b^2} \cdot \sqrt{-1}, \text{ ó } bi.$$

Q

quadrantal angle (p. 613) An angle in standard position whose terminal side coincides with one of the axes.

quadrants (p. P4) The four areas of a Cartesian coordinate plane.

quadratic equation (p. 163) A quadratic function set equal to a value, in the form $ax^2 + bx + c = 0$, where $a \neq 0$.

quadratic form (p. 277) For any numbers a, b, and c, except for $a = 0$, an equation that can be written in the form $u^2 + u + c = 0$, where u is some expression in x.

Quadratic Formula (p. 199) The solutions of a quadratic equation of the form $ax^2 + bx + c = 0$, where $a \neq 0$, are given by the Quadratic Formula, which is
$$x = \frac{-b \pm \sqrt{b^2 - 4ac}}{2a}.$$

quadratic function (p. 151) A function described by the equation $f(x) = ax^2 + bx + c$, where $a \neq 0$.

quadratic inequality (p. 209) An inequality of the form $y > ax^2 + bx + c$, $y \geq ax^2 + bx + c$, $y < ax^2 + bx + c$, or $y \leq ax^2 + bx + c$.

quadratic term (p. 151) In the equation $f(x) = ax^2 + bx + c$, ax^2 is the quadratic term.

ángulo de cuadrante Ángulo en posición estándar cuyo lado terminal coincide con uno de los ejes.

cuadrantes Las cuatro regiones de un plano de coordenadas Cartesiano.

ecuación cuadrática Función cuadrática igual a un valor, de la forma $ax^2 + bx + c = 0$, donde $a \neq 0$.

forma de ecuación cuadrática Para cualquier número a, b, y c, excepto $a = 0$, una ecuación que puede escribirse de la forma $u^2 + u + c = 0$, donde u es una expresión en x.

fórmula cuadrática Las soluciones de una ecuación cuadrática de la forma $ax^2 + bx + c = 0$, donde $a \neq 0$, se dan por la fórmula cuadrática, que es
$$x = \frac{-b \pm \sqrt{b^2 - 4ac}}{2a}.$$

función cuadrática Función descrita por la ecuación $f(x) = ax^2 + bx + c$, donde $a \neq 0$.

desigualdad cuadrática Desigualdad cuadrática de la forma $y > ax^2 + bx + c$, $y \geq ax^2 + bx + c$, $y < ax^2 + bx + c$, o $y \leq ax^2 + bx + c$.

término cuadrático En la ecuación $f(x) = ax^2 + bx + c$, ax^2 el término cuadrático es ax^2.

connectED.mcgraw-hill.com

quartic function (p. 255) A fourth-degree function.

quartiles (p. P26) The values that divide a set of data into four equal parts.

quintic function (p. 255) A fifth-degree function.

R

radian (p. 606) The measure of an angle θ in standard position whose rays intercept an arc of length 1 unit on the unit circle.

radical equation (p. 352) An equation with radicals that have variables in the radicands.

radical function (p. 338) A function that contains the root of a variable.

radical inequality (p. 354) An inequality that has a variable in the radicand.

random sample (p. 531) A sample in which every member of the population has an equal chance of being selected.

range (pp. P4, P25) **1.** The set of all y-coordinates of a relation. **2.** The difference between the greatest and least values in a set of data.

rate of change (p. 21) How much a quantity changes on average, relative to the change in another quantity, over time.

rate of continuous decay (p. 445) The rate at which something decays continuously. Represented by a constant k in the exponential decay function $f(x) = ae^{-kt}$, where a is the initial value, and t is time in years.

rate of continuous growth (p. 445) The rate at which something grows continuously. The value of k in the exponential growth function, $f(x) = ae^{kt}$.

rational equation (p. 508) Any equation that contains one or more rational expressions.

rational expression (p. 467) A ratio of two polynomial expressions.

rational function (p. 491) An equation of the form $f(x) = \frac{p(x)}{q(x)}$, where $p(x)$ and $q(x)$ are polynomial functions, and $q(x) \neq 0$.

rational inequality (p. 513) Any inequality that contains one or more rational expressions.

reciprocal function (pp. 483, 595) **1.** A function of the form $f(x) = \frac{1}{a(x)}$, where $a(x)$ is a linear function and $a(x) \neq 0$. **2.** Trigonometric functions that are reciprocals of each other.

función quartic Una función del cuarto-grado.

cuartiles Valores que dividen un conjunto de datos en cuatro partes iguales.

función quintic Una función del quinto-grado.

radián Medida de un ángulo θ en posición normal cuyos rayos intersecan un arco de 1 unidad de longitud en el círculo unitario.

ecuación radical Ecuación con radicales que tienen variables en el radicando.

función radical Una función que contiene la raíz de una variable.

desigualdad radical Desigualdad que tiene una variable en el radicando.

muestra aleatoria Muestra en la cual cada miembro de la población tiene igual posibilidad de ser elegido.

rango 1. Conjunto de todas las coordenadas y de una relación. **2.** Diferencia entre el valor mayor y el menor en un conjunto de datos.

tasa de cambio Lo que cambia una cantidad en promedio, respecto al cambio en otra cantidad, por lo general el tiempo.

índice de desintegración continúa Ritmo al cual algo se desintegra continuamente. Representado por la constante k en la función de desintegración exponencial $f(x) = ae^{-kt}$, donde a es el valor inicial y t es el tiempo en años.

el índice del crecimiento continuo Es la tasa en la cual algo crece continuamente. El valor de k en la función exponencial del crecimiento, $f(x) = ae^{kt}$.

ecuación racional Cualquier ecuación que contiene una o más expresiones racionales.

expresión racional Razón de dos expresiones polinomiales.

función racional Ecuación de la forma $f(x) = \frac{p(x)}{q(x)}$, donde $p(x)$ y $q(x)$ son funciones polinomiales y $q(x) \neq 0$.

desigualdad racional Cualquier desigualdad que contiene una o más expresiones racionales.

funciones recíprocas 1. Una función de la forma $f(x) = \frac{1}{a(x)}$, donde $a(x)$ es una función linear y $a(x) \neq 0$. **2.** Funciones trigonométricas de la función que son reciprocals de uno a.

reference angle (p. 613) The acute angle formed by the terminal side of an angle in standard position and the x-axis.

reflection (p. 126) A transformation in which every point of a figure is mapped to a corresponding image across a line of symmetry.

regression curve (p. 408) A curve of best fit for a data set that is not modeled by a linear equation.

regression line (p. 408) A line of best fit.

relation (p. P4) A set of ordered pairs.

relative frequency (p. 538) The ratio of the number of observations in a category to the total number of observations.

relative maximum (pp. 105, 263) A point on the graph of a function where no other nearby points have a greater y-coordinate.

ángulo de referencia El ángulo agudo formado por el lado terminal de un ángulo en posición estándar y el eje x.

reflexión Transformación en que cada punto de una figura se aplica a través de una recta de simetría a su imagen correspondiente.

curva de regresión Curva de mejor ajuste para un conjunto de datos que no está representado por una ecuación lineal.

reca de regresión Una recta de óptimo ajuste.

relación Conjunto de pares ordenados.

frecuencia relativa Razón del número de observaciones en una categoría al número total de observaciones.

máximo relativo Punto en la gráfica de una función en donde ningún otro punto cercano tiene una coordenada y mayor.

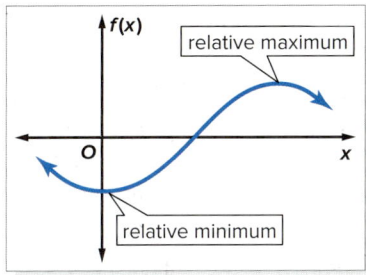

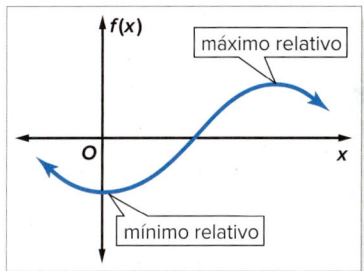

relative minimum (pp. 105, 263) A point on the graph of a function where no other nearby points have a lesser y-coordinate.

roots (p. 163) The solutions of a quadratic equation.

mínimo relativo Punto en la gráfica de una función en donde ningún otro punto cercano tiene una coordenada y menor.

raíz Las soluciones de una ecuación cuadrática.

S

sample (p. P24) A part of a population.

sample space (p. P9) The set of all possible outcomes of an experiment.

secant (p. 594) For any angle, with measure α, a point $P(x, y)$ on its terminal side, $r = \sqrt{x^2 + y^2}$, $\sec \alpha = \frac{r}{x}$.

set-builder notation (p. 15) The expression of the solution set of an inequality, for example $\{x \mid x > 9\}$.

simple event (p. P14) One event.

simplify (p. 229) To rewrite an expression without parentheses or negative exponents.

muestra Parte de una población.

espacio muestral Conjunto de todos los resultados posibles de un experimento probabilístico.

secante Para cualquier ángulo de medida α, un punto $P(x, y)$ en su lado terminal, $r = \sqrt{x^2 + y^2}$, $\sec \alpha = \frac{r}{x}$.

notación de construcción de conjuntos Escritura del conjunto solucion de una desigualdad, por ejemplo, $\{x \mid x > 9\}$.

evento simple Un solo evento.

reducir Escribir una expresión sin paréntesis o exponentes negativos.

simulation (p. 538) The use of a probability experiment to mimic a real-life situation.

sine (p. 594) For any angle, with measure α, a point $P(x, y)$ on its terminal side, $r = \sqrt{x^2 + y^2}$, $\sin \alpha = \frac{y}{r}$.

slope (p. 22) The ratio of the change in y-coordinates to the change in x-coordinates.

slope-intercept form (p. 28) The equation of a line in the form $y = mx + b$, where m is the slope and b is the y-intercept.

solution (p. 5) A replacement for the variable in an open sentence that results in a true sentence.

square root function (p. 338) A function that contains a square root of a variable.

Square Root Property (p. 173) For any real number n, if $x^2 = n$, then $x = \pm\sqrt{n}$.

standard deviation (p. P25) The square root of the variance.

standard form (p. 163) **1.** A linear equation written in the form $Ax + By = C$, where A, B, and C are integers whose greatest common factor is 1, $A \geq 0$, and A and B are not both zero. **2.** A quadratic equation written in the form $ax^2 + bx + c = 0$, where a, b, and c are integers, and $a \neq 0$.

standard normal distribution (p. 568) A normal distribution with a mean of 0 and a standard deviation of 1.

standard position (p. 604) An angle positioned so that its vertex is at the origin and its initial side is along the positive x-axis.

statistic (pp. P24, 531) A measure that describes a characteristic of a sample.

step function (p. 119) A function whose graph is a series of line segments.

substitution method (p. 44) A method of solving a system of equations in which one equation is solved for one variable in terms of the other.

survey (p. 531) Used to collect information about a population.

synthetic division (p. 244) A method used to divide a polynomial by a binomial.

synthetic substitution (p. 287) The use of synthetic division to evaluate a function.

simulación Uso de un experimento probabilístico para imitar una situación de la vida real.

seno Para cualquier ángulo de medida α, un punto $P(x, y)$ en su lado terminal, $r = \sqrt{x^2 + y^2}$, $\sin \alpha = \frac{y}{r}$.

pendiente La razón del cambio en coordenadas y al cambio en coordenadas x.

forma pendiente-intersección Ecuación de una recta de la forma $y = mx + b$, donde m es la pendiente y b la intersección.

solución Sustitución de la variable de un enunciado abierto que resulta en un enunciado verdadero.

función radical Función que contiene la raíz cuadrada de una variable.

Propiedad de la raíz cuadrada Para cualquier número real n, si $x^2 = n$, entonces $x = \pm\sqrt{n}$.

desviación estándar La raíz cuadrada de la varianza.

forma estándar **1.** Ecuación lineal escrita de la forma $Ax + By = C$, donde A, B, y C son enteros cuyo máximo común divisores 1, $A \geq 0$, y A y B no son cero simultáneamente. **2.** Una ecuación cuadrática escrita en la forma $ax^2 + bx + c = 0$, donde a, b, y c son números enteros, y $a \neq 0$.

distribución normal estándar Distribución normal con una media de 0 y una desviación estándar de 1.

posición estándar Ángulo en posición tal que su vértice está en el origen y su lado inicial está a lo largo del eje x positivo.

estadística Una medida que describe una característica de una muestra.

función etapa Función cuya gráfica es una serie de segmentos de recta.

método de sustitución Método para resolver un sistema de ecuaciones en que una de las ecuaciones se resuelve en una de las variables en términos de la otra.

encuesta Reunía información acerca de una población.

división sintética Método que se usa para dividir un polinomio entre un binomio.

sustitución sintética Uso de la división sintética para evaluar una función polinomial.

system of equations (p. 42) A set of equations with the same variables.

system of inequalities (p. 52) A set of inequalities with the same variables.

T
---

tangent (p. 594) **1.** A line that intersects a circle at exactly one point. **2.** For any angle, with measure α, a point $P(x, y)$ on its terminal side, $r = \sqrt{x^2 + y^2}$, $\tan \alpha = \frac{y}{x}$.

terminal side (p. 604) A ray of an angle that rotates about the center.

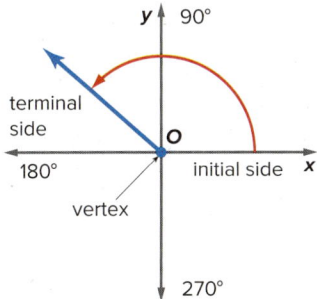

theoretical probability (pp. P13, 538) What should occur in a probability experiment.

translation (p. 125) A figure is moved from one location to another on the coordinate plane without changing its size, shape, or orientation.

tree diagram (p. P9) A diagram that shows all possible outcomes of an event.

trigonometric equations (p. 683) Equations containing at least one trigonometric function that is true for some but not all values of the variable.

trigonometric functions (p. 594) For any angle, with measure α, a point $P(x, y)$ on its terminal side, $r = \sqrt{x^2 + y^2}$, the trigonometric functions of α are as follows.

$\sin \alpha = \frac{y}{r}$ $\cos \alpha = \frac{x}{r}$ $\tan \alpha = \frac{y}{x}$
$\csc \alpha = \frac{r}{y}$ $\sec \alpha = \frac{r}{x}$ $\cot \alpha = \frac{x}{y}$

trigonometric identity (p. 655) An equation involving a trigonometric function that is true for all values of the variable for which the function is defined.

trigonometric ratio (p. 594) Compares the side lengths of a right triangle.

---

sistema de ecuaciones Conjunto de ecuaciones con las mismas variables.

sistema de desigualdades Conjunto de desigualdades con las mismas variables.

tangente **1.** Recta que interseca un círculo en un solo punto. **2.** Para cualquier ángulo, de medida α, un punto $P(x, y)$ en su lado terminal, $r = \sqrt{x^2 + y^2}$, $\tan \alpha = \frac{y}{x}$.

lado terminal Rayo de un ángulo que gira alrededor de un centro.

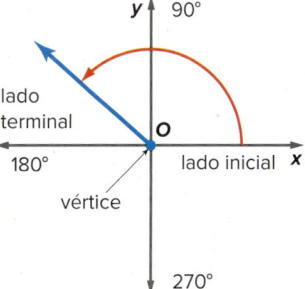

probabilidad teórica Lo que debería ocurrir en un experimento probabilístico.

traslación Se mueve una figura de un lugar a otro en un plano de coordenadas sin cambiar su tamaño, forma u orientación.

diagrama de árbol Diagrama que muestra todos los posibles resultados de un evento.

ecuación trigonométrica Ecuación que contiene por lo menos una función trigonométrica y que sólo se cumple para algunos valores de la variable.

funciones trigonométricas Para cualquier ángulo, de medida α, un punto $P(x, y)$ en su lado terminal, $r = \sqrt{x^2 + y^2}$, las funciones trigonométricas de α son las siguientes.

$\sin \alpha = \frac{y}{r}$ $\cos \alpha = \frac{x}{r}$ $\tan \alpha = \frac{y}{x}$
$\csc \alpha = \frac{r}{y}$ $\sec \alpha = \frac{r}{x}$ $\cot \alpha = \frac{x}{y}$

identidad trigonométrica Ecuación que involucra una o más funciones trigonométricas y que se cumple para todos los valores de la variable en que el función es definido.

razón trigonometric Compara las longitudes laterales de un triángulo derecho.

trigonometry (p. 594) The study of the relationships between the angles and sides of a right triangle.

turning point (pp. 105, 263) Point at which a graph turns. The location of relative maxima or minima.

trigonometría Estudio de las relaciones entre los lados y ángulos de un triángulo rectángulo.

momento crucial Un punto en el cual un gráfico da vuelta. La localización de máximos o de mínimos relativos.

U

unbounded (p. 60) A system of inequalities that forms a region that is open.

unit analysis (p. 236) The process of including unit measurement when computing.

unit circle (p. 620) A circle of radius 1 unit whose center is at the origin of a coordinate system.

no acotado Sistema de desigualdades que forma una región abierta.

análisis de la unidad Proceso de incluir unidades de medida al computar.

círculo unitario Círculo de radio 1 cuyo centro es el origen de un sistema de coordenadas.

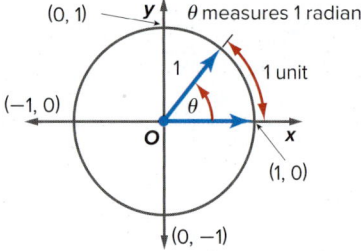

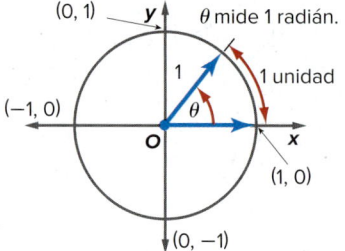

univariate data (p. 24) Data with one variable.

datos univariados Datos con una variable.

V

variable (p. P24) **1.** A characteristic of a population that can assume different values called *data*. **2.** A symbol, usually a letter, used to represent an unknown quantity.

variance (p. 25) The mean of the squares of the deviations from the arithmetic mean.

vertex (p. 152) **1.** Any of the points of intersection of the graphs of the constraints that determine a feasible region. **2.** The point at which the axis of symmetry intersects a parabola. **3.** The point on each branch nearest the center of a hyperbola.

vertical asymptote (p. 491) If the related rational expression of a function is written in simplest form and is undefined for $x = a$, then $x = a$ is a vertical asymptote.

vertical line test (p. 88) If no vertical line intersects a graph in more than one point, then the graph represents a function.

vertical shift (p. 636) When graphs of trigonometric functions are translated vertically.

variable **1.** Característica de una población que puede tomar diferentes valores llamados *datos*. **2.** Símbolo, generalmente una letra, que se usa para representar una cantidad desconocida.

varianza Media de los cuadrados de las desviaciones de la media aritmética.

vértice **1.** Cualqeiera de los puntos de intersección de las gráficas que los contienen y que determinan una región viable. **2.** Punto en el que el eje de simetría interseca una parábola. **3.** El punto en cada rama más cercano al centro de una hipérbola.

asíntota vertical Si la expresión racional que corresponde a una función racional se reduce y está no definida en $x = a$, entonces $x = a$ es una asíntota vertical.

prudba de la recta vertical Si ninguna recta vertical interseca una gráfica en más de un punto, entonces la gráfica representa una función.

cambio vertical Cuando los gráficos de funciones trigonométricas son translado verticales.

W

weighted average (p. 510) A method for finding the mean of a set of numbers in which some elements of the set carry more importance, or weight, than others.

promedio ponderado Un método para encontrar el medio de un sistema de los números en los cuales algunos elementos del sistema llevan más importancia, o peso, que otros.

X

x-intercept (p. 133) The x-coordinate of the point at which a graph crosses the x-axis.

intersección x La coordenada x del punto o puntos en que una gráfica interseca o cruza el eje x.

Y

y-intercept (p. 133) The y-coordinate of the point at which a graph crosses the y-axis.

intersección y La coordenada y del punto o puntos en que una gráfica interseca o cruza el eje y.

Z

zeros (pp. 134, 163) The x-intercepts of the graph of a function; the points for which $f(x) = 0$.

ceros Las intersecciones x de la gráfica de una función; los puntos x para los que $f(x) = 0$.

z-value (p. 568) The number of standard deviations that a given data value is from the mean.

valor Z Número de variaciones estándar que separa un valor dado de la media.

connectED.mcgraw-hill.com

Index

A

Absolute value, 188–189

Absolute value functions, 120–121

Activities. See Algebra Labs; Graphing Technology Labs; Spreadsheet Lab

Acute angles, 594–595

Addition
Associative Property of, 174
Commutative Property of, 174
of complex numbers, 174, 189
of cubes, 274–276
Distributive Property of, 174
of factors, P7–P8
of functions, 315–316
of polynomials, 231
of probabilities, P13–P15
of rational expressions, 477–478
in solving inequalities, 14
in solving systems of equations, 45, 46

Addition Property
of Equality, 6
of Inequality, 13

Algebraic expressions, 5, 120, 254

Algebra Labs
The Complex Plane, 188–189
Dimensional Analysis, 236
Interval and Set-Building Notation, 20
Quadratics and Rate of Change, 208

Amplitude, 627–630, 638

Angles
acute, 594–595
angle measures, 604–605, 606
central, 607
congruent, P20–P21
coterminal, 605
of depression, 598
double-angle identities, 675–676, 678
of elevation, 598
half-angle identities, 676–677, 678
of incline, 662
negative angle identities, 655
quadrantal, 613
radians and degrees, 606–607
reference, 613–615
of rotation, 605
in standard position, 604–605
trigonometric functions, 594–598, 612–613

Animations, 53, 120, 192, 263, 295, 471

Apothem, 603

Applications, P6, P9, P10, P11, P12, P15, P17, P19, P21, P24, P25, P26, P27, P28, P29, 9, 10, 16, 17, 21, 24, 25, 32, 33, 37, 40, 44, 48, 49, 55, 56, 62, 63, 64, 65, 71, 72, 76, 77, 78, 79, 90, 91, 92, 100, 101, 109, 114, 115, 123, 130, 135, 141, 156, 157, 161, 162, 205, 212, 213, 214, 217, 218, 219, 233, 240, 247, 248, 259, 265, 272, 278, 280, 307, 317, 318, 320, 324, 333, 334, 337, 340, 341, 357, 362, 378, 384, 393, 394, 395, 402, 407, 415, 419, 427, 434, 458, 474, 480, 481, 486, 487, 488, 490, 493, 496, 497, 504, 505, 506, 511, 512, 515, 521, 522, 523, 532, 534, 535, 541, 542, 543, 548, 549, 553, 556, 557, 558, 562, 563, 564, 570, 576, 577, 580, 581, 582, 584, 585, 598, 599, 600, 601, 608, 609, 616, 617, 623, 624, 625, 639, 640, 644, 645, 647, 650, 664, 665, 666, 671, 674, 679, 680, 686, 691, 692. See also Real-World Careers; Real-World Examples; Real-World Links

Archytas, 391

Arc length, 607

Area of parallelograms, 611

Argand plane, 188

Arithmetic means, 391

Aryabhatta, 657

Assessment. See also Guided Practice; Practice Test; Preparing for Assessment; Prerequisite Skills
ACT/SAT, 94, 124, 132, 159, 358, 489, 557
Multiple Choice, 9, 12, 19, 27, 34, 39, 40, 51, 58, 66, 73, 79, 80, 82–83, 94, 102, 110, 116, 124, 132, 137, 144–147, 159, 170, 178, 187, 190, 195, 207, 215, 221, 222–225, 235, 241, 247, 249, 261, 269, 272, 281, 286, 292, 300, 307, 308–311, 321, 328, 335, 344, 350, 355, 356, 358, 365, 366–369, 380, 389, 396, 404, 414, 415, 422, 429, 436, 442, 451, 459, 460–463, 471, 472, 475, 482, 489, 490, 498, 507, 516, 523, 524–527, 537, 544, 545, 550, 559, 565, 572, 578, 579, 585, 586–589, 610, 618, 619, 626, 633, 642, 647, 648–651, 661, 667, 673, 674, 681, 689, 693, 694–697

Associative Property
of Addition, 174
of Multiplication, 172, 174

Asymptotes, 373, 483–484, 491–495, 499, 629–630

Axis of symmetry, 152–153

B

Bari, Nina Karlovna, 46

Base
change of base formula, 425–426
logarithmic base, 398
natural base e, 430–431
natural base exponentials, 430–431

Bell curve, 566–569

Benford formula, 420

Bias, 531

Binomials
expansion of, 237–239
as factors of polynomials, 289
FOIL method, P6
multiplying, P6
Pascal's triangle and, 237

Binomial Theorem, 237–239

Boundary, 35

Bounded regions, 60–61

Box-and-whisker plots, 552–553, 555

Brahmagupta, 201

Break-even point, 42

C

Calculator. See Graphing calculators; Graphing Technology Labs

Carbon dating, 445–446

Careers. See Real-World Careers

Catenary curve, 430

Census, 375, 532

Centers
of circles, 607, 620

of distributions, 551–552, 566
of polygons, 603
statistical measures of, P24–P28

Central angle, 607

Central tendency. *See* Measures of center (central tendency)

Change of base formula, 425–426

Chapter 0
adding probabilities, P13–P15
congruent and similar figures, P20–P21
counting techniques, P9–P12
factoring polynomials, P7–P8
FOIL method, P6
measures of center, spread, and position, P24–P28
multiplying probabilities, P16–P19
posttest, P29
pretest, P3
Pythagorean Theorem, P22–P23
representing functions, P4–P5

Check for Reasonableness. *See* Study Tips

Circles
arc length, 607
central angle, 607
circumference, 606
degrees and radians, 606
unit, 620, 622

Circular functions, 620

Circumference of a circle, 606

Coefficients
correlation, 408–409
leading, 253
signs of, 294

Cofunction identities, 655

Combinations, P11–P12

Combined variation, 503

Common denominators, least (LCD), 476–478, 508–509

Common logarithms, 423–425

Common ratio, 391

Commutative Property
of Addition, 174
of Multiplication, 172, 174

Completing the square, 192–194, 198

Complex Conjugates Theorem, 297

Complex fractions, 470–471, 478

Complex numbers, 173–175
adding, 174, 189
conjugates, 175, 297
dividing, 175
graphing, 188–189
multiplying, 175
in quadratic equations, 202
subtracting, 174, 189

Complex plane, 188

Composition of functions, 322–324

Compound events, P14

Compound interest, 384–385, 433, 446
continuously compounded, 433, 446

Concept Summary
binomial expansion, 239
characteristics of linear systems, 44
degrees and radians, 606
designing, conducting, and reporting a simulation, 540
properties of exponents, 229
solving quadratic equations, 204
solving systems of equations, 46
transformations of functions, 128
zeros, factors, roots, and intercepts, 293

Conditional probability, P16–P17, P18

Confidence levels, 546–547

Congruence statements, P20

Congruent angles, P20–P21

Congruent figures, P20–P21

Congruent sides, P20

ConnectED. *See* Go Online!

Connections. *See* Applications; Interdisciplinary Connections

Consistent systems, 43–44

Constants
in translation, 125, 128
of variation, 500

Constant terms, 151

Constraints, 36, 62

Contingency tables, P18

Continuous functions, 112

Continuously compounded interest, 433, 446

Continuous probability distribution, 566

Continuous relations, 88–90

Contractions (dilations), 126–128

Converse of the Pythagorean Theorem, 23

Coordinate plane quadrants, P4

Corollary to Fundamental Theorem of Algebra, 294

Correlation coefficients, 408–409

Correlation of determination, 410

Correlations, 405–406

Cosecant, 594–595

Cosecant functions, 630

Cosine, 594–595, 655

Cosine functions, 622, 627–630

Cotangent, 594–595

Cotangent functions, 630

Coterminal angles, 605

Counting techniques, P9–P12

Cube root equations, 353–354

Cube root functions, 345–347, 351

Cubes, sum and difference of, 274–276

Cycles, 621–622

D

Data, P24. *See also* Graphs; Statistics; Tables
correlation of determination, 410
distributions of, 551–556
equations of best fit, 405–408
evaluating published reports, 561–562
identifying misleading uses of, 562
interquartile range, P27
mean, P24, 510, 545–546, 551, 554
median, P24, 553
mode, P24
modeling with polynomial functions, 270–271
modeling with quadratic functions, 160–162
normal distribution, 566–569
outliers, P27, 551, 555
range of, P4–P5, P25–P26, 87–89, 155, 329
regression lines and curves, 408–410
scatterplots of, 405–406
univariate, P24
variance of, P25–P26

Decay
  decay factor, 376
  exponential, 375–377, 407, 430, 445–446
Decisions
  analyzing, 574–575
  fair, 573–574
Degree of a polynomial, 231
Degrees, radians and, 606–607
Denominators, least common (LCD), 476–478, 508–509
Dependent events, P16–P17
Dependent systems, 43–44
Dependent variables, 89
Depressed polynomials, 289, 296
Depression, angles of, 598
Descartes, René, 294–295
Descartes' Rule of Signs, 295
Descriptive statistics, P24, 531
Diagnose readiness. See Prerequisite Skills
Diagrams
  tree, P9
  Venn, 174
Difference identities, 668–669, 670, 678
Difference of squares, P8, 180–181, 274–276
Dilations, 126–128
Dimensional analysis, 236
Diophantus of Alexandria, 6
Direct variation, 500–501
Discrete functions, 89–90
Discrete relations, 88–90
Discriminants, 202–204
Distance, 511, 514
Distributions of data, 551–556
  analyzing, 551–553
  comparing, 554–555
  normal, 566–569
Distributive Property, P7, 174, 179, 232
Division
  of complex numbers, 175
  of functions, 315–316
  of inequalities, 14
  of monomials, 229–230
  of polynomials, 242–246, 250, 287–288
  of rational expressions, 468–469

Division algorithm, 242
Division Property
  of Equality, 6
  of Inequality, 14
Domain
  of cube root functions, 345–346
  functions and, 87–89
  of inverses, 330–331
  of relations, P4–P5, 329
  of square root functions, 338
Doppler effect, 476, 481
Double-angle identities, 675–676, 678
Double zero, 256

E

e (natural base), 430–431
Elevation, angles of, 598
Elimination method, 45, 68
Empirical Rule, 566–568
End behavior, 103–104, 106–107, 255–256, 263–265
Enhanced Fujita scale, 437
Equality, properties of, 6–8
Equations. See also Quadratic equations
  absolute value, 120–121
  of best fit, 405–408
  cube root, 353–354
  definition, 5
  direct variation, 500–501
  equivalent, 43, 67, 430
  exponential, 383–385, 445–446
  exponential decay, 445, 446
  exponential growth, 445, 446
  hyperbolas, 483
  inverse functions, 329
  joint variation, 501
  linear, 5–11, 22–23, 28–31, 95–97, 134, 383
  logarithmic, 418, 430–431, 437–438, 443–444
  natural logarithm, 430–433
  parabolas, 72, 151–153, 160
  parallel lines, 31
  perpendicular lines, 31
  point-slope form, 29–30
  polynomial, 255–257, 263–265, 270–271, 287–288, 295–297, 301

  power, 251–252, 254, 340, 346–347, 408
  radical, 352–354, 359
  rational, 508–512, 517
  regression, 408–410
  roots of, 134
  slope-intercept form, 28–29
  solving
    with a calculator, 166
    by completing the square, 192–194, 204
    by elimination method, 45, 46
    by factoring, 180–183, 204
    by formula, 199–202
    by graphing, 42–44, 46, 133–134, 163–168, 171, 204, 273, 359–360, 381–382, 443–444, 517–518, 682
    with Square Root Property, 173, 191–192, 204
    systems of, 45, 46
    using least common multiple, 45–46
    using multiplication, 45–46
    using substitution, 45–46
    using tables, 45–46, 166
    square root, 354
    standard form of, 153, 163, 179, 383, 407
    systems of, 42–46, 45, 46
    trigonometric, 682, 683–686
Error, margin of, 546–547
Even-degree functions, 117, 140, 256–257
Events
  compound, P14
  dependent, P16–P17
  independent, P16
  mutually exclusive, P14
  simple, P14
Experimental probability, 13, 538, 540–541
Experiments, 531–532
Exponential decay, 375–377, 407, 430, 445–446
Exponential equations, 383–385, 445–446
Exponential functions. See also Functions
  converting to logarithmic functions, 407–408
  graphing, 373–377, 407
  natural base, 430–431
  regression, 408
  solving, 383–385
  standard form of, 407

Exponential growth, 373–375, 384, 407, 430, 445–447

Exponential inequalities, 385, 425

Exponents
 in polynomial division, 243–244
 properties of, 229–230

Expressions
 algebraic, 5, 120, 254
 equivalent, 243, 277, 310, 425, 430–431
 logarithmic, 397–398
 in quadratic form, 277
 radical, 352–354
 rational, 467–471, 477–478, 519
 simplifying, 431, 467–471
 verbal, 5

Extraneous solutions, 352–353, 685

Extrema of functions, 105–107, 112, 263–265

Factorial, P9

Factoring
 difference of cubes, 273–275
 general trinomials, 275
 greatest common factor, P7, 180, 275, 467
 grouping, 275–276
 least common denominator, 476–478, 508–509
 perfect squares, 180–181
 perfect square trinomials, P8, 180–182, 275
 polynomials, P7–P8, 274–276, 289, 293, 470, 476
 quadratic equations, 180–183, 204
 rational expressions, 470
 sum of cubes, 273–275

Factors, greatest common, P7, 180, 275, 467

Factor Theorem, 289

Fair, in decisions, 573–574

Families of graphs, 125–128, 138

Feasible regions, 60

Financial Literacy, 10, 32, 57, 72, 80, 157, 266, 290, 337, 375, 377, 387, 403, 421, 488, 497, 515, 571
 credit cards, 421
 income, 415, 557
 interest, 375, 384–385, 420, 433, 435, 446

 money, 17, 40, 79, 387
 savings, 99, 109, 455, 457

Five-number summary, P26

FOIL method, P6, 174, 179

Foldables™ Study Organizer, 74, 85, 138, 216, 302, 361, 453, 519, 579, 643, 690
 equations and inequalities, 3
 exponential and logarithmic functions, 371
 inverses and radical functions, 313
 polynomials and polynomial functions, 227
 quadratic functions, 149
 rational functions, 465
 relations and functions, 85
 statistics and probability, 529
 trigonometric functions, 591
 trigonometric identities and equations, 653

Formulas
 angle of incline, 662
 average speed, 493
 Benford, 420
 change of base, 425–426
 circumference of a circle, 606
 compound interest, 384
 continuously compounded interest, 433, 446
 distance, 511
 exponential function, 407
 exponential growth or decay, 407, 445, 446
 illuminance, 655
 loudness of sound, 424
 margin of error, 547
 natural exponential function, 407
 nth term of a geometric sequence, 390
 pitch, 338, 340
 population mean, 545
 population proportion, 545
 quadratic, 199–204
 simple interest, 375
 slope, 23–24
 standard form of an exponential function, 407
 temperature scale conversion, 334
 z-values, 568–569

Fractions, complex, 470–471, 478. See also Rational expressions

Frequency, 628–629

Frequency tables, P18, 575

Function notation, 61, 89

Functions. See also Transformations
 absolute value, 120–121
 arithmetic operations with, 315–316
 circular, 620
 combining, 317
 composition of, 322–324
 continuous, 89–90
 cosine, 622, 627–630
 cube root, 345–347, 351
 discrete, 89–90
 even- and odd-degree, 117, 140, 256–257
 exponential, 373–377, 383–385, 407–408, 430–431
 exponential decay, 375–377, 407, 430, 445–446
 exponential growth, 373–375, 384, 407, 430, 445–447
 extrema, 105–107, 112, 263–265
 finding zeros of, 134, 163, 293–297, 491
 greatest integer function, 119
 horizontal line test, 331
 identity, 332
 inverse, 329–332, 336, 340, 346–347
 linear, 95–97, 103–104, 111, 119–121, 408
 logarithmic, 397–398, 407–408
 logistic growth, 448, 450
 minimum and maximum values of, 60–61, 154, 263–265
 monomial, 251–252
 natural base exponential, 407, 430–431
 nonlinear, 95, 104–107, 112–113
 notation, 61, 89
 nth root, 351
 one-to-one, 87
 onto, 87
 periodic, 621–622, 627–630, 635–638
 piecewise-defined, 118–119
 power, 251–252, 254, 340, 346–347, 408
 quadratic, 127, 151–153, 160–166, 208, 408
 radical, 338–340, 345–347, 351
 rational, 491–495, 499
 reciprocal, 483–486, 502, 595, 630
 related, 134
 relations and, 87–90
 representing, P4–P5
 sine, 622, 627–630, 634–638
 square root, 338–340
 step, 119–120
 tangent, 629–630

translations, 125, 127, 128, 339, 346, 351, 374, 376, 399–400, 426, 485, 635–638
trigonometric, 594–598, 612–615, 622, 627–630, 635–638, 655–657, 682–686
turning points, 105–107, 263–265
variation, 500–503
vertical line test, 88–89, 330–332
zero, 106

Fundamental Counting Principle, P9–P12

Fundamental Theorem of Algebra, 293
Corollary to, 294

G

GCF (greatest common factor), P7, 180, 275, 467

Geometric means, 391

Geometric sequences, 390–391, 453

Geometric series, 392–393

Geometry
circles, 606–607, 620, 622
double-angle identities, 675–676, 678
hyperbolas, 483–486, 517
parabolas, 151–153
parallelograms, 611
polygons, 603
triangles, P22–P23, 237, 593–595, 643
trigonometric ratios, 594–598, 614, 620, 622

Geometry Labs
Areas of Parallelograms, 611
Regular Polygons, 603

Germain, Sophie, 275

Get Ready for the Chapter. See Prerequisite Skills

Get Ready for the Lesson. See Prerequisite Skills

Go Online!
ALEKS, 3, 45, 85, 149, 227, 313, 371, 529, 591, 653
Animations, 53, 120, 192, 263, 295, 471
Chapter Project, 3, 85, 149, 227, 313, 371, 465, 529, 591, 653
Coin Toss, 529
ConnectEd, 432, 438, 597
eStudent Edition, 230, 322, 418, 613, 656
eToolkit, 446, 484, 553, 628
Foldables™ Study Organizer, 3, 74, 85, 138, 149, 216, 227, 302, 313, 361, 371, 453, 465, 519, 529, 579, 591, 643, 653, 690

Geometer's Sketchpad, 3, 62, 85, 88, 149, 227, 313, 330, 371, 465, 494, 529, 568, 591, 653
Graphing Calculator, 276
Graphing Tools, 3, 23, 36, 44, 85, 152, 255, 371, 374, 399, 465, 591, 653
Interactive Student Guide, 3, 85, 149, 227, 313, 371, 465, 529, 591, 653
Learnsmart, 3, 85, 149, 227, 313, 371, 465, 529, 591, 653
Mapping Tool, P2
Personal Tutors, 8, 165, 212, 238, 339, 346, 385, 392, 510, 637, 664
Product Mat and Algebra Tiles, 149, 227, 313
Self-Check Quiz, 70, 183, 246, 355, 678
Spreadsheet Activity, 502
TI-Nspire Worksheet, 438
Tools, P2, 15
Vocabulary, P2, 30, 74, 86, 138, 150, 216, 302, 314, 361, 372, 453, 466, 519, 530, 579, 592, 643, 654, 690
Worksheets, 126

Graphing calculators, P26, 105, 155, 166, 175, 192, 264, 276, 403, 405, 409–410, 427, 434, 552, 569

Graphing Technology Labs
Analyzing Polynomial Functions, 301
Completing the Square, 198
Cooling, 452
Dividing Polynomials, 250
Graphing nth Root Functions, 351
Graphing Rational Functions, 499
Intersections of Graphs, 41
Inverse Functions and Relations, 336
Modeling Data with Polynomial Functions, 270–271
Modeling Data with Quadratic Functions, 160–162
Power Functions, 251–252
Solving Exponential Equations and Inequalities, 381–382
Solving Logarithmic Equations and Inequalities, 443–444
Solving Polynomial Equations by Graphing, 273
Solving Quadratic Equations by Graphing, 171
Solving Radical Equations, 359–360
Solving Rational Equations and Inequalities, 517–518
Solving Trigonometric Equations, 682
Systems of Linear Inequalities, 59
Trigonometric Graphs, 634

Graphs. See also Data; Graphing Technology Labs; Statistics
asymptotes, 373, 483–484, 491–495, 499
box-and-whisker plots, 552–553, 555
complex plane, 188–189
continuous, 90
cosecant, secant, and cotangent functions, 630
cube root functions, 345–347, 351
dilations, 126–128
discrete, 90
end behavior, 103–104, 256
exponential functions, 373–377, 381
exponential inequalities, 382
extrema, 105–107, 263–265
families of, 125–128, 138
feasible regions, 61–62
finding intercepts, 133
histograms, 552, 554
imaginary solutions, 194
intercepts, 28–29, 111–112, 133, 163, 293–294
intersections of, 41, 59
inverse functions, 330, 336, 397
inverse relations, 329, 336
inverse variation, 502
linear functions, 96–97, 101, 111
linear inequalities, 35, 52–54, 59
lines of best fit, 404–407
logarithmic functions, 398–400
misleading, 562
nonlinear functions, 112–113
normal distribution, 566
nth root functions, 351
on number lines, 14, 15, 355
periodic functions, 621–622, 635–638
phase shifts, 635
piecewise-defined functions, 116–117
point discontinuity, 494–495, 499
polynomial functions, 255–257, 273
quadrants, P4
quadratic functions, 127, 151–153
quadratic inequalities, 209
rational functions, 491–495, 499
reciprocal functions, 483–486, 502
reflections, 126–128
relations, P4
scatterplots, 405–406
sine and cosine functions, 622, 627–629, 634, 635–638
in solving equations, 133–134, 163–166
in solving quadratic equations, 163–166

R122

in solving radical equations, 359
in solving radical inequalities, 360
in solving systems of equations, 42–44
square root functions, 339–340
sums and differences of functions, 316
symmetry in, 96–98
tangent functions, 629–630
three-dimensional, 67
transformations, 125–129, 282–283, 339, 346, 351, 374–376, 399–400, 485–486, 635–638, 662–663
translations, 125, 128, 374, 485, 635–638
turning points, 263–265
verifying, 637
vertical shifts, 636–638

Greatest common factor (GCF), P7, 180, 275, 467

Greatest integer function, 119

Griddable, 81, 159, 358

Grouping, factoring by, 275–276

Growth
exponential, 373–375, 384, 407, 430, 445–447
factors, 374
logistic, 448, 450
population, 375, 447, 450, 454, 458, 459
rate of, 445

Guided Practice, 5, 6, 7, 8, 14, 15, 16, 21, 22, 23, 29, 30, 31, 35, 36, 42, 43, 45, 46, 52, 53, 54, 61, 62, 68, 70, 88, 89, 90, 96, 97, 98, 99, 104, 106, 107, 111, 112, 113, 118, 119, 120, 121, 125, 126, 127, 133, 134, 151, 153, 154, 155, 164, 165, 166, 172, 173, 174, 175, 179, 181, 182, 183, 191, 192, 193, 194, 200, 201, 203, 209, 210, 211, 212, 230, 231, 232, 237, 238, 239, 242, 243, 244, 245, 253, 254, 256, 257, 263, 264, 265, 274, 275, 276, 277, 282, 283, 288, 289, 294, 295, 296, 297, 315, 316, 322, 323, 324, 330, 332, 338, 339, 340, 345, 346, 347, 353, 355, 373, 374, 376, 377, 383, 384, 385, 390, 391, 392, 397, 398, 406, 408, 410, 416, 417, 418, 423, 424, 425, 426, 430, 431, 432, 433, 437, 438, 439, 445, 446, 447, 448, 467, 468, 469, 470, 471, 476, 477, 478, 483, 484, 485, 486, 493, 494, 495, 501, 502, 503, 508, 509, 510, 511, 512, 513, 532, 534, 539, 540, 541, 546, 547, 552, 553, 554, 555, 562, 567, 569, 574, 575, 594, 595, 596, 597, 598, 604, 605, 607, 612, 613, 614, 615, 620, 621, 622, 627, 628, 629, 630, 635, 636, 638, 656, 657, 662, 663, 664, 668, 669, 670, 675, 676, 677, 683, 684, 685

H

Half-angle identities, 676–677, 678

Half-lives, 445–446, 450

Hands-On. *See* Algebra Labs

Higher-Order Thinking Problems
Challenge, 11, 17, 38, 50, 55, 65, 71, 72, 93, 109, 115, 123, 131, 158, 169, 177, 186, 196, 206, 214, 234, 240, 248, 260, 280, 285, 291, 299, 320, 327, 334, 343, 349, 357, 379, 388, 395, 413, 421, 428, 435, 441, 450, 474, 481, 488, 497, 506, 515, 536, 558, 564, 571, 601, 609, 617, 625, 660, 666, 672, 688
Error Analysis, 11, 18, 26, 38, 50, 101, 106, 109, 115, 131, 136, 158, 177, 186, 196, 206, 214, 248, 260, 285, 327, 343, 349, 357, 379, 388, 403, 571, 577, 609, 660, 680
Make a Conjecture, 208, 336, 452, 603
Open-Ended, P23, 18, 33, 38, 57, 72, 93, 109, 115, 123, 136, 158, 169, 177, 186, 196, 206, 214, 234, 240, 248, 260, 268, 280, 285, 291, 299, 320, 327, 334, 343, 357, 379, 388, 395, 403, 413, 421, 435, 441, 450, 474, 481, 488, 497, 506, 515, 536, 543, 558, 564, 571, 577, 601, 609, 617, 625, 632, 641, 660, 672, 680, 688
Proof, 72, 234, 388, 395, 428, 435, 497, 660, 680
Reasoning, P15, 26, 109, 260, 342, 349, 641
Which One Doesn't Belong?, 65, 421, 474, 666
Writing in Math, 11, 18, 20, 26, 33, 38, 50, 57, 65, 72, 92, 101, 109, 115, 123, 131, 136, 158, 169, 177, 186, 196, 206, 214, 234, 240, 248, 260, 268, 280, 291, 299, 320, 327, 334, 343, 349, 357, 360, 379, 382, 388, 395, 403, 413, 421, 428, 435, 441, 450, 474, 481, 488, 497, 506, 515, 536, 543, 558, 564, 571, 577, 601, 609, 617, 625, 632, 641, 660,666, 672, 680, 688

Histograms, 552, 554

History Math Link. *See* Math History Link

Horizontal asymptotes, 483–484, 491–493

Horizontal line test, 331

H.O.T. Problems. *See* Higher-Order Thinking Problems

Hyperbolas, 483–486, 517

Hypotenuse, P22–P23, 594–595

I

Identities
cofunction, 655
difference, 668–669, 670, 678
double-angle, 675–676, 678
half-angle, 676–677, 678
identity function, 332
negative angle, 655
polynomial, 282–283
Pythagorean, 655
quotient, 655
reciprocal, 655
sum, 668, 670, 678
trigonometric, 655–656, 662–664, 670
verifying, 662–664, 670, 678

Identity functions, 332

Illuminance, 655

Imaginary axis, 188

Imaginary numbers
in complex numbers, 173–175, 194
graphing, 188–189, 194
pure, 172–173

Imaginary units, 172

Inclinometers, 596

Inconsistent systems, 43–44

Independent events, P16

Independent systems, 43–44

Independent variables, 89

Index of refraction, 688

Inequalities
addition of, 13–14
Addition Property of, 13
compound, 20
Division Property of, 14
with e and ln, 432–433
exponential, 385, 425
feasible regions, 60–61
graphing, 35, 52–54, 59, 74
interval notation, 20
linear, 13–18
logarithmic, 432–433, 438–439, 443–444
Multiplication Property of, 14
multistep, 15–16
negative numbers in, 425
one-step, 13
optimization, 62
Property of, for Exponential Functions, 385

Property of, for Logarithmic Functions, 438–439
quadratic, 209–212, 216
radical, 354–355, 360
rational, 513, 518–519
set-builder notation, 15, 20
solving, 52–54
Subtraction Property of, 13
systems of, 52–54, 74
vertices of an enclosed region, 54

Inferences, 532, 533–534

Inferential statistics, 531, 533–534

Infinity symbol, 20

Inflection point, 345–346

Infrasound, 629

Initial side of an angle, 604

Intercepts
finding, 133, 293–294
slope–intercept form, 28–29
x-intercepts, 133, 163
y-intercepts, 28–29, 111–112, 133

Interdisciplinary Connections. *See also* Applications
agriculture, 459
anthropology, 450
architecture, 100, 213
art, 535
astronomy, 234
biology, 26, 239, 279, 280, 290, 356, 394, 402, 447, 449, 450, 454, 458, 479, 481, 505, 539, 600, 601, 631
chemistry, 21, 317, 395, 510, 515
civil engineering, P21
computer science, 674
earth science, 400, 424, 632
engineering, P21, 535
environmental science, 428, 473
forestry, 449
geography, 679
health, 378
history, 674, 693
meteorology, 394
music, 427, 440, 503, 532, 632
paleontology, 449, 450
physical science, 190, 218, 219, 221, 334, 395, 504, 659, 666, 671, 680, 687
physics, 131, 162, 167, 175, 176, 177, 221, 247, 259, 342, 343, 349, 356, 364, 441, 450, 456, 506, 617, 629, 631, 638, 639, 665, 671, 672, 688, 693
science, 378, 384, 386, 395, 401, 415, 417, 419, 426, 440, 445, 449, 624

Interest, 375, 384–385, 420, 433, 435, 446

Interference, 668

Interquartile range (IQR), P27

Interval notation, 20

Inverse cosine, 597

Inverse functions
cube root, 346–347
equation for, 329
exponential, 400
graphs of, 330–332, 336
natural base and natural logarithm, 431
power, 340

Inverse Property, 330

Inverse relations, 329–332, 336

Inverse sine, 597

Inverse tangent, 597

Inverse trigonometric ratios, 597

Inverse variation, 502–503

IQR (interquartile range), P27

Irrational numbers, 192

Joint variation, 501

Key Concepts
absolute value of a complex number, 188
addition of rational expressions, 477
Addition Property of Inequality, 13
addition rules for probability, P14
angle measures, 604, 643
arc length, 607
Binomial Theorem, 238
box-and-whisker plots as distributions, 553
change of base formula, 425–426
circular and periodic functions, 643
combinations of n objects taken r at a time, P11
completing the square, 193
Complex Conjugates Theorem, 297
complex numbers, 173, 216
composition of functions, 322
compound interest, 384, 433, 446
conditional probability, P17
continuously compounded interest, 433, 446
convert between degrees and radians, 606
cosecant, secant, and cotangent functions, 630
Descartes' Rule of Signs, 295
direct variation, 500, 519
discriminant, 203
distributions of data, 551, 566, 568, 579
dividing rational expressions, 469
Division Property of Inequality, 14
double-angle identities, 675, 690
elimination method, 45
Empirical Rule, 566
end behavior of a polynomial function, 138, 255
evaluating published data, 579
evaluating trigonometric functions, 614
exponential functions, 453
exponential growth and decay, 445, 448
extrema and end behavior, 138
factorials, P10
Factor Theorem, 289, 302
feasible regions, 60
FOIL method for multiplying binomials, 179
functions, 87
functions and continuity, 138
functions on a unit circle, 620
Fundamental Counting Principle, P9
Fundamental Theorem of Algebra, 293
geometric sequences and series, 453
graphing linear functions, 111
graphing linear inequalities, 74
graphing nonlinear functions, 112
graphing quadratic functions, 152, 216
graphing quadratic functions – parabola, 152
graphing trigonometric functions, 637, 643
half-angle identities, 676, 690
horizontal line test, 331
important formulas, 407
inverse functions, 332, 361
inverse relations, 329–332
inverse trigonometric ratios, 597
inverse variation, 502, 519
joint variation, 501, 519
linear equations and inequalities, 74

linear equations and slope, 74
linear functions, 95, 111
linearity and symmetry, 138
linear programming, 74
Location Principle, 262–263
logarithms and logarithmic functions, 397, 453
logarithm with base b, 397
logistic growth function, 448
margin of error definition and formula, 547
maximum and minimum value, 154
measures of center, P24
measures of spread, P25
modeling data, 453
Multiplication Property of Inequality, 14
multiplying rational expressions, 469
natural base functions, 430
natural logarithms, 453
normal distributions, 566, 568, 579
nth term of a geometric sequence, 390
oblique asymptotes, 493
operations on functions, 315, 361
operations with polynomials, 302
optimization with linear programming, 62
parallel and perpendicular lines, 30
parent function of absolute value functions, 12
parent function of cube root functions, 345
parent function of exponential decay functions, 375
parent function of exponential growth functions, 373
parent function of logarithmic functions, 398–400
parent function of reciprocal functions, 483
parent function of square root functions, 338
partial sum of a geometric series, 392
permutations of n objects taken r at a time, P10
phase shifts, 635
point discontinuity, 495
point–slope form, 29–30
polynomial functions and graphs, 302
population mean and population proportion, 545
population parameters, 579
Power Property of Logarithms, 418

powers of binomials, 302
probability of dependent events, P16
probability of independent events, P16
Product Property of Logarithms, 416
Properties of Equality, 6, 437
Property of Equality for Logarithmic Functions, 437
Property of Inequality for Exponential Functions, 385
Property of Inequality for Logarithmic Functions, 438–439
property of inverses, 330
quadrantal angles, 613
Quadratic Formula, 200
quadratic functions, 152, 216
quadratic inequalities, 216
Quotient Property of Logarithms, 417
random sampling, 579
rational equations and inequalities, 519
rational expressions, 469, 477, 519
reciprocal functions, 519
reference angles, 613
Remainder Theorem, 287, 302
right triangle trigonometry, 594, 627, 643
roots and zeros, 302
simplifying monomials, 230
sine and cosine functions, 627
sketching graphs of functions, 138
slope–intercept form, 28
slope of a line, 23
solutions of a quadratic equation, 164
solving equations by graphing, 138
solving polynomial equations, 302
solving quadratic equations, 216
solving quadratic equations by graphing, 216
solving radical equations, 352, 361
solving radical inequalities, 354
solving rational inequalities, 513
Solving Systems of Inequalities, 52
special functions and parent functions, 138
square root and cube root functions, 361
standard form of a linear equation, 383
statistical study types, 531
subtraction of rational expressions, 477
Subtraction Property of Inequality, 13
sum and difference identities, 668, 690

sum and difference of cubes, 274–276
symmetric and skewed distributions, 551
synthetic division, 245
systems of equations and inequalities, 74
systems of equations in three variables, 74
tangent functions, 629
theoretical and experimental probability, P13
transformations of cube root functions, 346
transformations of exponential functions, 374
transformations of logarithmic functions, 399
transformations of reciprocal functions, 485
transformations of square root functions, 339
translations of trigonometric graphs, 643
trignometric values for special angles, 595
trigonometric functions in right triangles, 594, 627, 643
trigonometric functions of general angles, 612–613, 643
trigonometric identities, 655, 690
types of regression equations and models, 408
using exponential and logarithmic functions, 453
using probability to make decisions, 579
using statistical experiments, 579
verifying identities, 662, 663
vertical and horizontal asymptotes, 491
vertical line test, 88
vertical shifts, 636
Zero Product Property, 180
zeros of even- and odd-degree functions, 256
z-value formula, 568

Kilometers, 5

L

Labs. *See* Algebra Labs; Geometry Labs; Graphing Technology Labs; Spreadsheet Lab

Law of Cooling, 435
Law of Diminishing Returns, 169
Law of Large Numbers, 541
Leading coefficient, 253
Least common denominators (LCD), 476–478, 508–509
Least common multiple (LCM), 476
Linear equations, 95. *See also* Equations
  point-slope form, 29–30
  roots of, 134
  slope, 22–23
  slope-intercept form, 28–29
  solving, 5–11
  standard form of, 383
  writing, 28–31
Linear functions. *See also* Functions
  end behavior, 103–104
  identifying, 95–97
  piecewise-linear functions, 119–121
  regression, 408
  sketching graphs of, 111
Linear inequalities, 13–18
  applying, 36
  graphing, 14, 15, 35, 52–54, 59
  multistep, 15–16
  one-step, 13–15
Linear programming, 60–65
Linear regression lines and curves, 408–410
Linear terms, 151
Line of reflection, 126
Line of symmetry, 97–98
Lines
  of best fit, 404–409
  horizontal line test, 331
  parallel, 30–31
  perpendicular, 30–31
  of reflection, 126
  regression, 408–410
  slope, 22–23
  of symmetry, 97–98
  vertical, 29
  vertical line test, 88–89, 330–332
Ln (natural logarithm), 430–431. *See also* Logarithms
The Location Principle, 262–263
Logarithmic equations, 418, 430–431, 437–438, 443–444

Logarithmic expressions, 397–398
Logarithmic functions, 398–400, 407–408
Logarithmic inequalities, 438–439, 443–444
Logarithms
  change of base formula, 425–426
  common, 423–425
  functions, 397–400
  natural, 430–431
  natural base e, 430–431, 432
  properties of, 416–418
  solving equations, 418, 424–425
  solving exponential inequalities, 425
Logistic growth model, 448, 450
Lower quartile, 26

M

Mapping, P4
Marginal cost, 169
Margin of error, 546–547
Mathematical Practice Standards
  Construct Arguments, 18, 65, 186, 196, 268, 299, 334, 388, 450, 481, 488, 534, 536, 548, 558, 601, 617, 641, 680
  Critique Arguments, 57, 93, 123, 284, 403, 413, 428, 441, 474, 506, 548–549, 564, 577, 625, 660
  Justify Arguments, 543
  Modeling, 16, 18, 26, 31, 33, 36, 38, 47, 48, 49, 50, 56, 57, 65, 72, 100, 136, 155, 156, 168, 195, 204, 234, 259, 325, 340, 341, 347, 348, 349, 386, 402, 408, 435, 440, 480, 487, 504, 536, 548, 557, 600, 614, 659, 670
  Perseverance, 11, 61, 185, 213, 222, 258, 291, 348, 394, 419, 449, 479, 496, 608, 632, 658, 688
  Precision, 7, 9, 10, 29, 32, 64, 68, 134, 167, 222, 231, 257, 275, 289, 374, 401, 408, 413, 427, 477, 505, 546, 564, 664, 679
  Reasoning, 7, 24, 26, 29, 33, 38, 46, 50, 90, 93, 101, 115, 121, 131, 136, 158, 169, 177, 206, 240, 247, 248, 267, 268, 280, 283, 290, 291, 298, 299, 320, 327, 334, 340, 342, 346, 355, 357, 378, 379, 395, 403, 412, 420, 428, 435, 441, 472, 474, 488, 495, 497, 502, 506, 571, 607, 609, 616, 617, 625, 629, 632, 640, 660, 666, 671, 672, 680, 688
  Regularity, 128, 233, 393, 639, 685, 686

Sense-Making, 104, 112, 119, 122, 128, 130, 131, 184, 205, 212, 240, 266, 279, 280, 326, 330, 333, 356, 377, 426, 434, 471, 486, 497, 546, 564, 577, 599, 624, 665, 670, 687
  Structure, 25, 33, 88, 89, 91, 92, 97, 98, 115, 181, 320, 439, 484, 514, 623
  Tools, 45, 105, 157, 284, 418, 448, 515, 554, 556, 570, 574, 577, 595, 636
Math History Link
  Archytas, 391
  Aryabhatta, 657
  Bari, Nina Karlovna, 46
  Brahmagupta, 201
  Diophantus of Alexandria, 6
  Germain, Sophie, 275
  Napier, John, 425
  Taylor, Brook, 509
  Weierstrass, Karl, 120
  Zhang Heng, 438
Maximum values
  of polynomial functions, 261–262
  of quadratic functions, 154
  relative, 105–107, 263–265
Means
  arithmetic, 391
  of data, P24, 510, 545–546, 551, 554
  geometric, 391
Measures, angle, 604, 643
Measures of center (central tendency), P24–P28
  means, P24, 391, 510, 545–546, 551, 554
  median, P24, 553
  mode, P24
Measures of spread (variation), P25–P27
  range, P4–P5, P25–P26, 87–89, 155, 329, 338, 345–346
  standard deviations, P25–P26, 551, 553, 554, 568
  variance, P25–P26
Median, P24, 553
Mid-Chapter Quiz, 40, 117, 190, 272, 337, 415, 490, 560, 619, 674
Midlines, 636–638
Miles, 5
Minimum values, 60–61
  of polynomial functions, 261–264
  of quadratic functions, 154
  relative, 105–107
Mode, P24

Modeling
  equations of best fit, 405–409
  logistic growth model, 448, 450
  probability, P13
  simulations, 538–541
  visualizing problems by, 669

Monomial functions, 251–252

Monomials
  division of, 229–230
  least common denominator, 476
  multiplication of, 229–230
  simplifying, 230

Multiple Choice, 9, 12, 19, 27, 34, 39, 40, 51, 58, 66, 73, 79, 80, 82–83, 94, 102, 110, 116, 124, 132, 137, 144–147, 159, 170, 178, 187, 190, 195, 207, 215, 221, 222–225, 235, 241, 247, 249, 261, 269, 272, 281, 286, 292, 300, 307, 308–311, 321, 328, 335, 344, 350, 355, 356, 358, 365, 366–369, 380, 389, 396, 404, 414, 415, 422, 429, 436, 442, 451, 459, 460–463, 471, 472, 475, 482, 489, 490, 498, 507, 516, 523, 524–527, 537, 544, 545, 550, 559, 565, 572, 578, 579, 585, 586–589, 610, 618, 619, 626, 633, 642, 647, 648–651, 661, 667, 673, 674, 681, 689, 693, 694–697

Multiple representations, 11, 18, 26, 50, 93, 123, 158, 177, 185, 196, 234, 248, 259, 268, 291, 319, 320, 327, 334, 379, 388, 428, 435, 440, 450, 474, 481, 488, 515, 536, 543, 609, 625, 659, 672, 688

Multiples, least common (LCM), 476

Multiplication
  Associative Property of, 172, 174
  of binomials, P6
  Commutative Property of, 172, 174
  of complex numbers, 175
  Distributive Property of, 172, 174
  FOIL method, P6, 174, 179
  of functions, 315–316
  of imaginary numbers, 172–173
  of inequalities, 14
  of monomials, 229–230
  of polynomials, 232
  of probabilities, P16–P19
  of rational expressions, 468–469
  in solving systems of equations, 45–46

Multiplication Property
  of Equality, 6
  of Inequality, 14

Multiplicative inverse, 229

Multistep inequalities, 15–16

Multi-Step problems, 12, 19, 27, 33, 34, 38, 39, 51, 58, 64, 66, 73, 102, 116, 124, 132, 136, 137, 158, 159, 169, 170, 178, 187, 235, 240, 241, 249, 261, 269, 279, 286, 299, 320, 321, 328, 334, 335, 358, 380, 387, 389, 396, 404, 413, 414, 420, 422, 429, 436, 442, 473, 475, 482, 489, 507, 516, 526, 537, 544, 550, 559, 560, 565, 572, 578, 602, 610, 617, 618, 619, 633, 640, 642, 661, 667, 673, 681, 689

Mutually exclusive events, P14

Napier, John, 425

Natural base e, 430–431

Natural base exponential function, 430

Natural exponential function, 407

Natural logarithm, 430–431

Negative angle identities, 655

Negative correlations, 405–406

Negative Exponent Property, 229

Negatively skewed distributions, 551

Newton's Law of Cooling, 435

Nonlinear functions, 95
  end behavior, 103–104, 106–107, 255–256, 263–265
  extrema, 105–107, 112, 263–265
  sketching graphs of, 112–113

Normal distribution, 566–569

Notation
  composition of functions, 323
  function, 61, 89
  in Fundamental Counting Principle, P9
  imaginary units, 174
  inequality, 14
  infinity, 20
  interval, 20
  phase and vertical shifts, 636
  set-builder, 15, 20, 164
  sigma, 392
  triangles, 595
  trigonometric functions, 594

Nth roots, 351

Nth terms, 390

Number lines, 14, 15, 355

Numbers
  complex, 173–175, 188–189, 194, 202
  complex conjugates, 175, 297
  imaginary, 172–175, 194
  irrational, 192
  random, 539, 540

Number Theory, 165, 167, 168, 184, 206

Oblique asymptotes, 493–494

Observational studies, 531–532

Odd-degree functions, 117, 140, 256–257

Odds, P15

One-to-one function, 87

Onto functions, 87

Open Ended. See Higher-Order Thinking Problems

Open sentence, 5

Optimization, 62

Ordered pairs, P4, 329–332

Ordered triples, 67

Outcomes, P9, P13. See also Probability

Outliers, P27, 551, 555

Palermo scale, 397

Parabolas
  axis of symmetry, 152–153
  equation for, 72, 151–153, 160
  graphs of, 127, 151–153, 160–161
  maximum values, 154
  minimum values, 154
  in quadratic inequalities, 209–211
  transformations of, 127
  vertex, 152–153

Parallelograms, 611

Parameters, 531

Parent functions
  of absolute value functions, 12
  of cube root functions, 345
  of exponential decay functions, 375
  of exponential growth functions, 373
  of logarithmic functions, 398–400
  of reciprocal functions, 483
  of square root functions, 338
  transformations of, 125–128, 138

Pascal's triangle, 237

Perfect squares, 180–181

Perfect square trinomials, P8, 180–182, 275

Periodic functions, 621–622, 635–638
  cosecant, 630
  cosine, 622, 627–630
  cotangent, 630
  phase shifts, 635
  secant, 630
  sine, 622, 627–630
  tangent, 629–630
  translations of, 635–638
  vertical shifts of, 636–638

Periods, 621, 627, 638

Permutations, P9–P11

Personal Tutors. *See* Go Online!

Phase shifts, 635–638

Piecewise-defined functions, 118–119

Piecewise-linear functions, 119–121

Planes, complex, 188

Point discontinuity, 494–495, 499

Points
  break-even, 42
  center, 551–552, 566, 603, 607
  inflection, 345–346
  relative maximum, 105–107, 263–265
  relative minimum, 105–107, 263–265
  symmetry, 98–99
  turning, 105–107, 263–265

Point-slope form, 29–30

Point symmetry, 98–99

Polygons
  apothems, 603
  regular, 603

Polynomial functions
  end behavior, 255–256
  even- and odd-degree, 117, 140, 256–257
  finding zeros of, 295–297
  graphing, 255–257, 301
  maximum and minimum points, 263–265
  modeling data with, 270–271
  in one variable, 253
  synthetic substitution, 287–288, 296
  turning points, 105–107, 263–265
  writing, from zeros, 297

Polynomial identities, 282–283

Polynomials
  adding, 231

  binomials, P6, 237–239, 289
  degrees of, 231
  depressed, 289, 296
  difference of cubes, 274–276
  dividing, 242–246, 250, 287–288
  factoring, P7–P8, 274–276, 289, 293, 470, 476
  least common denominator of, 476
  Location Principle, 262–263
  long division, 242–244
  monomials, 229–230, 476
  multiplying, 232
  in one variable, 253
  perfect square trinomials, P8
  prime, 274
  quadratic form of, 277
  simplifying, 231–232
  solving by graphing, 273
  subtracting, 231
  sum of cubes, 274–276
  synthetic division, 244–246, 287
  synthetic substitution, 287–288, 296

Population, P24
  growth, 375, 447, 450, 454, 458, 459
  means, 545–546
  proportions, 545–546

Positive correlations, 405–406

Positively skewed distributions, 551

Power functions, 251–252, 254, 340, 346–347, 408

Power of a Power Property, 229

Power of a Product Property, 229

Power of a Quotient Property, 229

Power Property of Logarithms, 418

Practice Test, 79, 143, 221, 307, 365, 459, 523, 585, 647, 693

Predictions, 410

Preparing for Advanced Algebra. *See* Chapter 0

Preparing for Assessment, 12, 19, 27, 34, 39, 51, 58, 66, 73, 80, 82–83, 94, 102, 110, 116, 124, 132, 137, 144–147, 159, 170, 178, 187, 195, 207, 215, 222–225, 235, 241, 249, 261, 269, 281, 286, 292, 300, 308–311, 321, 328, 335, 344, 350, 358, 366–369, 380, 389, 396, 404, 414, 422, 429, 436, 442, 451, 460–463, 475, 482, 489, 498, 507, 516, 524–527, 537, 544, 545, 550, 559, 565, 572, 578, 579, 586–589, 610, 618, 626, 633, 642, 648–651, 661, 667, 673, 681, 689, 694–697.

Prerequisite Skills. *See also* Chapter 0
  Get Ready for the Chapter, 4, 45, 46, 86, 150, 314, 372, 530, 592, 654

Prime polynomials, 274

Probability
  addition of, P13–P15
  analyzing decisions, 574–575
  Benford formula, 420
  combinations, P11, P11–P12
  compound events, P14
  conditional, P16–P17, P18
  definition, P13
  dependent events, P16–P17
  experimental, P9, P13, 13, 538
  fair decisions, 573–574
  Fundamental Counting Principle, P9–P12
  independent events, P16
  Law of Large Numbers, 541
  models, P13, 538
  multiplication of, P16–P19
  mutually exclusive events, P14
  odds, P15
  outcomes, P9, P13
  permutations, P9–P11
  sample spaces, P9
  simple events, P14
  simple probability, P13
  theoretical, 13, 538
  tree diagrams, P9
  uniform model, P13

Problem-solving. *See also* Problem-Solving Tips
  dimensional analysis, 236
  four-step plan, 69

Problem-Solving Tips, 340, 347, 447, 539, 669, 685

Product of Powers Property, 229

Product Property of Logarithms, 416

Products. *See* Multiplication

Properties
  Addition Property of Equality, 6
  Addition Property of Inequality, 13
  Associative Property of Addition, 174
  Associative Property of Multiplication, 172, 174
  Commutative Property of Addition, 174
  Commutative Property of Multiplication, 172, 174
  Distributive, P7
  Distributive Property of Addition, 174

Distributive Property of Multiplication, 172, 174
Division Property of Equality, 6
Division Property of Inequality, 14
of Equality, 6
of Equality for Logarithmic Functions, 437
of Exponents, 229
of Inequality for Exponential Functions, 385
of Inequality for Logarithmic Functions, 438–439
of Inverses, 330
Multiplication Property of Equality, 6
Multiplication Property of Inequality, 14
Negative Exponent, 229
Power of a Power, 229
Power of a Product, 229
Power of a Quotient, 229
Power Property of Logarithms, 418
Product of Powers, 229
Product Property of Logarithms, 416
Quotient of Powers, 229
Quotient Property of Logarithms, 417
Reflexive Property of Equality, 6
Square Root, 173, 191–192
Substitution Property of Equality, 6
Subtraction Property of Equality, 6
Subtraction Property of Inequality, 13
Symmetric Property of Equality, 6
Transitive Property of Equality, 6
Zero Power Property, 229
Zero Product, 180

Proportional sides, P20

Proportions
with inverse variations, 502
population, 545–546

Pure imaginary numbers, 172–173

Pythagorean identities, 655

Pythagorean Theorem, P22–P23

Pythagorean triples, 285

Q

Quadrantal angles, 613

Quadrants, P4, 656

Quadratic equations
estimating solutions, 165–166
factored form, 179
regression, 408
roots of, 163–166, 200–204
solving
  with a calculator, 166
  by completing the square, 192–194, 204
  by factoring, 180–183, 204
  by formula, 199–202
  by graphing, 163–166, 171, 204
  with Square Root Property, 173, 191–192, 204
standard form, 153, 163, 179
zeros of, 163

Quadratic form, of polynomials, 277

Quadratic Formula, 199–204

Quadratic functions
graphs of, 127, 151–153
modeling data with, 160–162
rate of change and, 208
regression, 408
solving by graphing, 163–166, 171, 204
standard quadratic form, 153, 163, 179
zeros of, 163

Quadratic inequalities, 209–212

Quadratic terms, 151

Quartiles, P26

Quotient identities, 655

Quotient of Powers Property, 229

Quotient Property of Logarithms, 417

Quotients. See Division

R

Radians, 606–607

Radical equations, 352–354, 359

Radical expressions, 352–354

Radical functions, 338–340, 345–347, 351

Radical inequalities, 354–355, 360

Radicands, 200, 202

Random number generators, 539, 540

Random samples, 531

Range, P4–P5
of cube root functions, 345–346
of a data set, P25–P26
of functions, 87–89, 154
of inverse functions, 329–330, 332
of quadratic functions, 154–155
of square root functions, 338

Rates
of change, 21–23
of continuous decay, 445
of continuous growth, 445
decay factor, 376
exponential, 375–377, 407, 430, 445–446
logistic, 448, 450

Rational equations, 508–512, 517

Rational expressions, 467
addition of, 477–478
complex fractions, 470–471, 478
simplifying, 467–471, 656–657
subtraction of, 477–478

Rational functions, 491–495, 499

Rational inequalities, 513, 518

Ratios
rates of change, 21–23
rational expressions, 467
slope, 22–23
trigonometric, 594–595

Reading Math
angle of rotation, 605
conditional probability, P16
extraneous, 509
factorials, P9
feasible regions, 61
function notation, 61, 89
geometric means, 391
half-angle identities, 677
imaginary units, 174
intersections, 316
inverse functions, 330
inverse trigonometric ratios, 597
labeling triangles, 595
logarithms, 397
permutations, P10
radian measures, 606
repeated roots, 294
set-builder notation, 15
translation, 126
zero of the function, 134

Real axis, 188

Real-World Careers
acoustical engineer, 424
assemblers, 540
electrical engineers, 175

electrician, 676
music management, 70
operations manager, 62
pilot, 486
recreation workers, 36
truck driver, 232
U.S. Coast Guard Boatswain's Mate, 493

Real-World Examples, P18, 16, 21, 22, 36, 44, 53, 70, 96, 106, 113, 120, 134, 155, 166, 175, 183, 211, 232, 237, 254, 288, 317, 324, 340, 347, 375, 377, 384, 390, 392, 400, 417, 424, 445, 446, 447, 448, 486, 493, 503, 511, 512, 533–534, 538, 547, 552, 561, 567, 569, 573, 574–575, 596, 615, 621, 629, 638, 657, 669, 678, 684

Real-World Links, 16, 53, 90, 96, 155, 166, 183, 211, 254, 265, 288, 317, 324, 340, 347, 375, 377, 384, 400, 417, 424, 433, 446, 448, 503, 552, 567, 569, 596, 598, 605, 615, 621, 629, 638, 678

Real-World Problems, 69

Reasonableness, Check for. *See* Study Tips

Reciprocal functions
graphing, 483–486, 502
transformations of, 485–486
of trigonometric functions, 595, 630

Reciprocal identities, 655

Reference angles, 613–615

Reflection, 126–128
graphing, 152
line of reflection, 126, 152
line of symmetry, 97–98

Reflectional symmetry, 112

Reflexive Property of Equality, 6

Regression lines and curves, 408–410

Relations, P4
continuous, 88–89
discrete, 88–89
domain of, P4–P5
functions and, 87–90
horizontal line test, 331
inverse, 329–332, 336
mapping, P4
range of, P4–P5
vertical line test, 88–89, 330–332

Relative frequency, 538

Relative maximum, 105–106, 263–265

Relative minimum, 105–106, 263–265

Remainder Theorem, 287–288

Review. *See* Guided Practice; Preparing for Assessment; Prerequisite Skills; Review Vocabulary; Study Guide and Review

Review Vocabulary
absolute value, 314
combination, 530
complex conjugates, 297
domain, 150, 372
equation, 86
evaluate, 4
formula, 654
function, 86, 150, 372, 466, 592
identity, 654
inequality, 4
inequality symbols, 14
infinity, 104, 255
inverse function, 592
least common multiple, 45, 466
perfect square, 180
permutation, 530
power, 4
radicand, 200
random, 530
range, 150, 372
rational number, 314, 466
relation, 86, 314
relative frequency, P18
symmetry, 152
trigonometric functions, 654
zero, 256

Richter scale, 400, 424

Right triangles
45°–45°–90°, 593
identifying, P23
Pythagorean Theorem, P22–P23
trigonometric functions for acute angles, 594–595

Roots, 134
complex numbers, 194, 202
discriminant and, 202–204
extraneous, 509
imaginary, 293
irrational, 192
number of, 293–295
of quadratic equations, 163–166, 200–204
rational, 191
types of, 293–295

S

Sample, of a population, P24

Sample size, 547

Sample spaces, P9

Scatterplots, 405–406

Secant, 594–595

Secant functions, 630

Sense-Making. *See* Mathematical Practice Standards

Sequences, geometric, 390–391, 453

Set-builder notation, 15, 20, 164

Sides
congruent, P20
proportional, P20
of a right triangle, P22–23
terminal, 604–605

Sigma notation, 392

Similar figures, P20–P21

Simple events, P14

Simple probability model, P13

Simplifying
with e and ln, 431
expressions, 467–471, 695
logarithmic expressions, 431
monomials, 229–230
polynomials, 231–232
rational expressions, 467–471
with trigonometric functions, 656–657

Simulations, 538–541

Sine, 594–595

Sine functions, 622, 627–630, 634–638

Skewness, 551–555

Slant asymptotes, 493–494

Slope, 22–23
in direct variation, 500
of parallel lines, 30–31
of perpendicular lines, 30–31
of vertical lines, 29

Slope–intercept form, 28–29

Snell's Law, 688

SOH-CAH-TOA, 595

Solutions
extraneous, 352–353, 685
imaginary, 173, 194
infinitely many, 67, 69, 684
irrational, 201

no, 69
no real, 165, 682
one rational, 200–201
one real, 164, 682
of the open sentence, 5
two rational, 200
two real, 164

Spreadsheet Lab
Investigating Special Right Triangles, 593

Squared differences, P25

Square Root Property, 173, 191–192

Square roots
equations, 353–354
functions, 338–340
inequalities, 354–355
of negative numbers, 172
perfect squares, 180
positive, 23
principle, 354

Standard deviations, P25–P26, 551, 553, 554, 568

Standard form
of equations, 153, 163, 179, 383, 407
of exponential functions, 407
of linear equations, 383
of quadratic equations, 153, 163, 179

Standard normal distribution, 568–569

Standard position, angles in, 604–605

Statistic, 531

Statistics, P24–P28, 531. *See also* Data; Graphs; Measures of center
descriptive, P24, 531
five-number summary, P26
inferential, 531, 533–534
interquartile range, P27
margin of error, 546–547
mean, P24, 510, 545–546, 551, 554
measures of center, P24–P28
measures of spread, P25–P27
median, P24, 553
mode, P24
outliers, P27, 551, 555
parameters, 531
population means, 545–546
population proportions, 545–546
quartiles, P26
range, P4–P5, P25–P26
simulations, 538–541

standard deviations, P25–P26, 551, 553, 568
types of studies, 531–532, 531–534
variance, P25–26
weighted average, 510

Step functions, 119–120

Study Guide and Review, 74–78, 138–142, 216–220, 302–306, 361–364, 453–458, 519–522, 579–584, 643–646, 690–692

Study Organizer. *See* Foldables™ Study Organizer

Study Tips
adding and subtracting equations, 45, 46
adding and subtracting rational expressions, 477
alternative methods, 231
amplitude, 628, 629
angles of elevation and depression, 598
asymptotes, 485
boundaries, 54
carbon dating, 446
census, 532
checking solutions, P8, 275, 353, 418
choosing the sign, 677
combinations, 238
comparing data sets, 554, 562
complex numbers, 174, 175, 202
composition of functions, 323
confirming answers, 68
cycles, 621
decimals and percentages, 546
degree of the function, 263
deriving formulas, 676
discrete and continuous relations, 88
distance problems, 511
domain and range, 154, 155, 339
double zero, 256
eliminating choices, 468
eliminating common factors, 469
end behavior, 374, 400
equations of best fit, 406
evaluating permutations and combinations, P11
exponential formulas, 407, 408
exponential function formula, 407
expressing solutions as multiples, 684
factoring polynomials, 289

factoring trinomials, 181, 182
fair decisions, 574
finding intersects, 448
formula for simple interest, 375
fractions, 153
graphing calculators, 238
graphing equations, 29
graphing inequalities, 14
graphing polynomial functions, 257
graphing vertical shifts, 636
greatest integer function, 119
grouping 6 or more terms, 276
imaginary solutions, 194
imaginary units, 174
independent events, P17
independent quantities, 22
infinite and no solutions, 69
interquartile range, P27
inverse functions, 330
inverse variation, 502
joint and marginal frequencies, P18
joint variation, 501
leading coefficient and degree, 255
line of best fit, 406
line symmetry, 97
margin of error, 546
maximum and minimum, 263
modeling data sets, 408
mutually and not mutually exclusive events, P14
natural base and log expressions, 432
negative numbers in inequalities, 425
normal distributions, 567, 568
oblique asymptotes, 494
odd functions, 263
optional graphs, 164
order of operations, 283
outliers, 555
periods, 628
phase and vertical shifts, 636
piecewise-defined functions, 119
plotting reflections, 152
point symmetry, 98
power of 1, 231
precision, 7
quadrantal angles, 613
quadrants, 656

quadratic form, 153, 277
radians, 606
radical inequalities, 354–355
random number generators, 539
range of z-values, 569
rational inequalities, 513
rational values, 106, 264
read the question, 8
reasonableness, P21
reasoning, 7
reciprocal functions, 630
related linear functions, 134
with repetition, P10
roots, 203
rounding, 433
set-builder notation, 164
simplifying, 431, 470, 657
simplifying logarithmic expressions, 431
simplifying rational expressions, 470
sine and cosine functions, 22
sketching angles, 614
sketching graphs, 112
slope and classifying systems, 43
slope is constant, 23
SOH-CAH-TOA, 595
solving trigonometric equations, 686
squared differences, P25
Square Root Property, 192
square roots, 181
standard normal distribution, 568
study notebook, 204
substitute values, 354
sum and difference identities, 670
symmetry, 568
systems of equations, 43
tangent function, 630
testing for zeros, 296
transformations, 128
trigonometric functions, 596, 686
unbiased estimator, P25
undefined terms, 478
unit circles, 622
using tables, 510
using your text, 6
verifying a graph, 637
vertical asymptotes, 484
why subtract?, P14
zero at the origin, 295
zero exponent, 399
zero on calculators, 296

z-values, 569
Substitution method, 44–45, 68
Substitution Property of Equality, 6
Subtraction
of complex numbers, 174, 189
of cubes, 274–276
of functions, 315–316
of polynomials, 231
of rational expressions, 477–478
solving inequalities, 14
in solving systems of equations, 45, 46
Subtraction Property
of Equality, 6
of Inequality, 13
Sum. See Addition
Sum identities, 668, 670, 678
Sum of squares, 180, 274–276
Surveys, 531–532
Symbols. See also Notation
function, 61, 89
in Fundamental Counting Principle, P9
imaginary units, 174
inequality, 14
infinity, 20
interval, 20
sigma, 392
triangles, 595
Symmetric Property of Equality, 6
Symmetry
axis of, 152–153
in distributions, 551–553
in graphs, 96–98
line, 97–98, 111–112
point, 98–99
reflectional, 112
Synthetic division, 244–246, 287
Synthetic substitution, 287–288, 296
Systems of equations, 42–46
classifying, 43–44
quadratic functions, 160–161
solve
using elimination method, 45, 46
using graphs, 42–44, 46
using multiplication and least common multiple, 45–46
using substitution, 44–45, 46
using tables, 42, 46
in three variables, 67–70
Systems of inequalities, 52–54, 59

T

Tables
in graphing, 151, 153, 257
in solving quadratic equations, 166
two-way frequency, P18, 575
Tangent, 594–595, 655
Tangent functions, 629–630
Taylor, Brook, 509
Taylor's Theorem, 509
Teaching Tips
continuous graphs, 90
end behavior, 104
graphing calculators, 105
strategies for working backward, 367
transformations of cube root functions, 346
Technology. See Applications; Graphing Technology Labs; Spreadsheet Lab
Technology Tips, 540
Terminal side of an angle, 604–605
Terms
constant, 151
grouping, 276
linear, 151
nth, 390
quadratic, 151
undefined, 478
Test Practice, 79, 143, 307, 365, 459, 523, 585, 647, 693
Test-Taking Strategies, 81, 145, 223, 309, 367, 461, 525, 587, 649, 695
Test-Taking Tips, 83, 147, 310, 368, 526, 588, 650, 696
for drawing a picture, 309
for guessing and checking, 525
for reading math problems, 145
for simplifying expressions, 695
for solving griddable problems, 81
for solving multi-step problems, 587
for solving using a graph, 223
for using a scientific calculator, 649
for using technology, 461
Theorems
Binomial, 237–239
Complex Conjugates, 297
converse of the Pythagorean, 23
Corollary to Fundamental Theorem of Algebra, 294
Factor, 289

Fundamental Theorem of Algebra, 293
Pythagorean, P22–P23
Remainder, 287–288
Taylor's, 509

Theoretical probability, 13, 538

Transformations, 125–129
of cube root functions, 346, 351
dilations, 126–128
of each side of an equation, 663–664
of exponential decay, 375–376
of exponential growth, 374
of logarithmic functions, 399–400
polynomial identities, 282–283
of reciprocal functions, 485–486
reflections, 97–98, 126–128, 152
of square root functions, 339
translations, 125, 127, 128, 339–400, 346, 351, 374, 376, 426, 485, 635–638
of trigonometric graphs, 635–638
of trigonometric identities, 662–663

Transitive Property of Equality, 6

Translations, 128
of cube root functions, 346
of exponential functions, 374, 376
of logarithmic functions, 399–400, 426
of n*th* root functions, 351
of quadratic functions, 127
of reciprocal functions, 485
of square root functions, 339
of trigonometric graphs, 635–638

Tree diagrams, P9

Triangles
45°–45°–90°, 593
labeling, 595
Pascal's, 237
Pythagorean Theorem, P22–P23
right, P22–P23, 593–595, 643

Trigonometric functions, 594–598
for acute angles, 594–595
extraneous solutions, 685
of general angles, 612–613
graphs of, 622, 627–630, 635–638
reciprocal, 630
with reference angles, 613–615
simplifying expressions with, 656–657
solving, 682, 683–686
translations of, 635–638
trigonometric identities and, 655–656

Trigonometric identities
identifying, 655–656
in solving equations, 685–686

verifying, 662–664

Trigonometric ratios, 594–598, 614, 620, 622

Trigonometry, 594–598
angle measures, 604–605, 606
circular functions, 620
cosecant, 594–595, 630
cosine, 594–595, 622, 627–630, 655
cotangent, 594–595, 630
equations, 682, 683–686
graphs, 622, 627–630, 634, 635–638
half-angle identities, 676–677, 678
identities, 655–656, 662–664, 670
periodic functions, 621–622, 635–638
quadrantal angles, 613
reference angles, 613–615
regular polygons, 603
right triangles, P22–P23, 593–595, 643
secants, 594–595, 630
tangent, 594–595, 629–630, 655
unit circles, 620, 622

Trinomials
factoring, 181–182
general, 275
perfect square, P8, 180–182, 275

Turning points, 105–107, 263–265

Two-way frequency tables, P18, 575

Unbiased estimator, P25
Unbounded regions, 60–61
Uniform probability model, P13
Unit analysis, 236
Unit circles, 620, 622
Univariate data, P24
Upper quartile, 26

Variables, P24
dependent, 89
independent, 89
polynomials in one, 253
systems of equations in three, 67–70

Variance, P25–P26. *See also* Measures of spread

Variation
constants of, 500
direct, 500–501
functions, 500–503
inverse, 502–503
joint, 501

Verbal expressions, 5
Vertex, 54, 152–153
Vertical asymptotes, 483–484, 491–493
Vertical lines, 97, 111–112
Vertical line test, 88–89, 330–332
Vertical shifts, 636–638
Vocabulary Check, 74, 138, 216, 302, 361, 453, 519, 579, 690
Volume of a rectangular prism, 471

W

Wakeboarding, 605
Watch Out!
arc length, 29, 607
completing the square, 193
equation of a vertical line, 29
evaluating the function, 317
functions, P5, 317, 638
graph of speed versus time, 113
inverse functions, 332
logarithmic base, 398
maxima and minima, 154
parent functions, 638
percents, 385
point discontinuities, 495
positive square roots, P23
sigma notation, 392
standard deviations, 553
synthetic division, 245, 246
synthetic substitution, 289
two-way tables, 575
verify identities, 663
zeros, 165
zeros *vs.* vertical asymptotes, 492

Weierstrass, Karl, 120
Weighted average, 510
Which One Doesn't Belong?. *See* Higher-Order Thinking Problems
Whisker. *See* Box-and-whisker plots
Work problems, 512, 514
Writing in Math, 117. *See also* Higher-Order Thinking Problems

X

x-coordinates, P4, 22–23
x-intercepts, 133, 163

Y

y-coordinates, P4, 22–23
y-intercepts, 28–29, 111–112, 133

Z

Zero functions, 106
Zero Power Property, 229
Zero Product Property, 180
Zeros
  double, 256
  of even- and odd-degree functions, 256–257
  of functions, 134, 163, 293, 294–297, 491
  locating, 263–265
  location principle, 262–263
  positive and negative, 295–296
  signs of coefficients and, 294
Zhang Heng, 438
z-values, 568–569

Symbols

$f(x) = \{$	piecewise-defined function	Σ	sigma, summation
$f(x) = |x|$	absolute value function	$\bar{x}$	mean of a sample
$f(x) = [\![x]\!]$	function of greatest integer not greater than x	μ	mean of a population
$f(x, y)$	f of x and y, a function with two variables, x and y	s	standard deviation of a sample
$\overrightarrow{AB}$	vector AB	σ	standard deviation of a population
i	the imaginary unit	$P(B|A)$	the probability of B given that A has already occurred
$[f \circ g](x)$	f of g of x, the composition of functions f and g	$_nP_r$	permutation of n objects taken r at a time
$f^{-1}(x)$	inverse of $f(x)$	$_nC_r$	combination of n objects taken r at a time
$b^{\frac{1}{n}} = \sqrt[n]{b}$	nth root of b	$\text{Sin}^{-1} x$	Arcsin x
$\log_b x$	logarithm base b of x	$\text{Cos}^{-1} x$	Arccos x
$\log x$	common logarithm of x	$\text{Tan}^{-1} x$	Arctan x
$\ln x$	natural logarithm of x		

Parent Functions

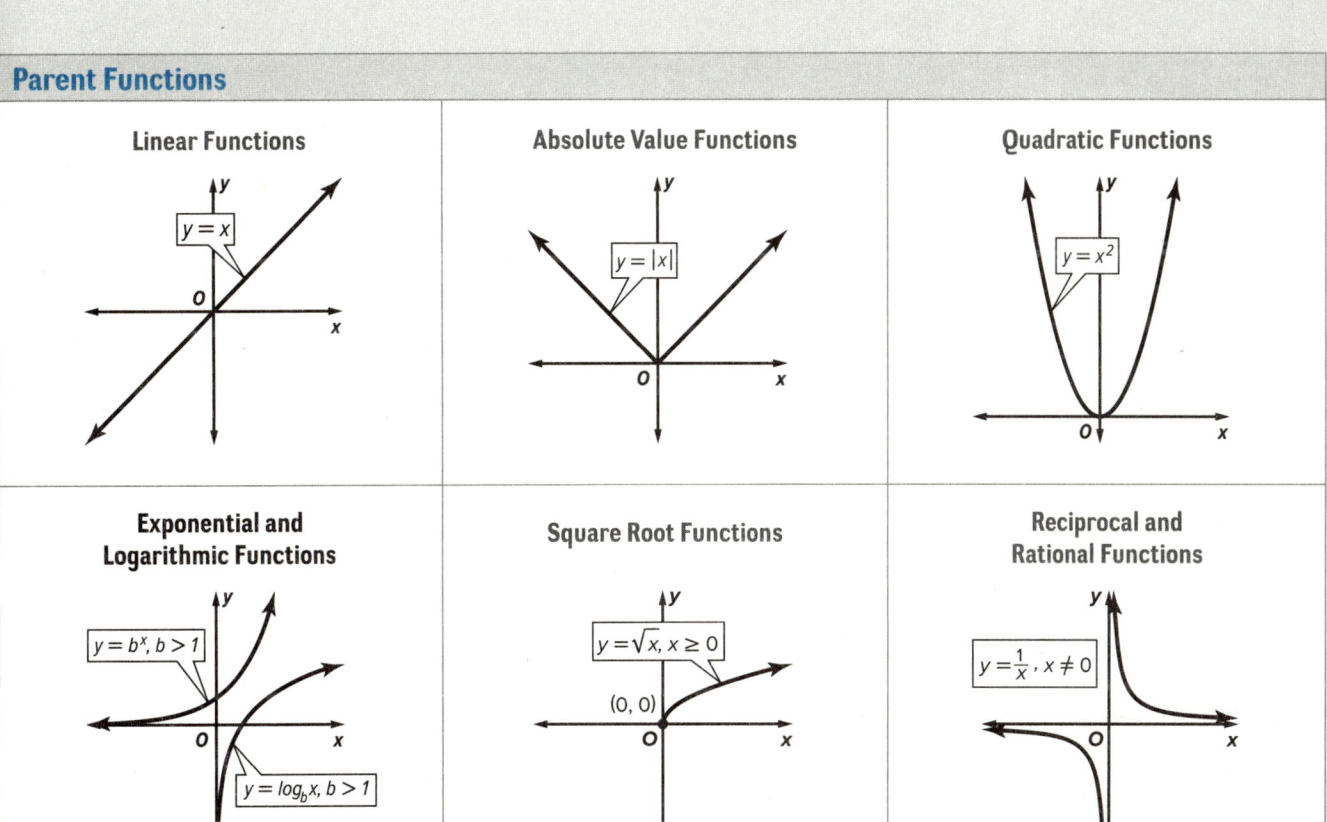

Formulas

Coordinate Geometry

Midpoint	$M = \left(\dfrac{x_1 + x_2}{2}, \dfrac{y_1 + y_2}{2}\right)$	**Distance**	$d = \sqrt{(x_2 - x_1)^2 + (y_2 - y_1)^2}$
		Slope	$m = \dfrac{y_2 - y_1}{x_2 - x_1}, x_2 \neq x_1$

Matrices

Adding	$\begin{bmatrix} a & b \\ c & d \end{bmatrix} + \begin{bmatrix} e & f \\ g & h \end{bmatrix} = \begin{bmatrix} a+e & b+f \\ c+g & d+h \end{bmatrix}$	**Multiplying by a Scalar**	$k\begin{bmatrix} a & b \\ c & d \end{bmatrix} = \begin{bmatrix} ka & kb \\ kc & kd \end{bmatrix}$
Subtracting	$\begin{bmatrix} a & b \\ c & d \end{bmatrix} - \begin{bmatrix} e & f \\ g & h \end{bmatrix} = \begin{bmatrix} a-e & b-f \\ c-g & d-h \end{bmatrix}$	**Multiplying**	$\begin{bmatrix} a & b \\ c & d \end{bmatrix} \cdot \begin{bmatrix} e & f \\ g & h \end{bmatrix} = \begin{bmatrix} ae+bg & af+bh \\ ce+dg & cf+dh \end{bmatrix}$

Polynomials

Quadratic Formula	$x = \dfrac{-b \pm \sqrt{b^2 - 4ac}}{2a}, a \neq 0$	**Square of a Difference**	$(a - b)^2 = (a - b)(a - b) = a^2 - 2ab + b^2$
Square of a Sum	$(a + b)^2 = (a + b)(a + b) = a^2 + 2ab + b^2$	**Product of Sum and Difference**	$(a + b)(a - b) = (a - b)(a + b) = a^2 - b^2$

Logarithms

Product Property	$\log_x ab = \log_x a + \log_x b$	**Power Property**	$\log_b m^p = p \log_b m$
Quotient Property	$\log_x \dfrac{a}{b} = \log_x a - \log_x b, b \neq 0$	**Change of Base**	$\log_a n = \dfrac{\log_b n}{\log_b a}$

Conic Sections

Parabola	$y = a(x - h)^2 + k$ or $x = a(y - k)^2 + h$	**Ellipse**	$\dfrac{x^2}{a^2} + \dfrac{y^2}{b^2} = 1$ or $\dfrac{y^2}{a^2} + \dfrac{x^2}{b^2} = 1, a, b \neq 0$
Circle	$x^2 + y^2 = r^2$ or $(x - h)^2 + (y - k)^2 = r^2$	**Hyperbola**	$\dfrac{x^2}{a^2} - \dfrac{y^2}{b^2} = 1$ or $\dfrac{y^2}{a^2} - \dfrac{x^2}{b^2} = 1, a, b \neq 0$

Sequences and Series

nth term, Arithmetic	$a_n = a_1 + (n - 1)d$	**nth term, Geometric**	$a_n = a_1 r^{n-1}$
Sum of Arithmetic Series	$S_n = n\left(\dfrac{a_1 + a_n}{2}\right)$ or $S_n = \dfrac{n}{2}[2a_1 + (n - 1)d]$	**Sum of Geometric Series**	$S_n = \dfrac{a_1 - a_1 r^n}{1 - r}$ or $S_n = \dfrac{a_1 - a_n r}{1 - r}, r \neq 1$

Trigonometry

Law of Sines	$\dfrac{\sin A}{a} = \dfrac{\sin B}{b} = \dfrac{\sin C}{c}, a, b, c \neq 0$		
Law of Cosines	$a^2 = b^2 + c^2 - 2bc \cos A$	$b^2 = a^2 + c^2 - 2ac \cos B$	$c^2 = a^2 + b^2 - 2ab \cos C$
Trigonometric Functions	$\sin \theta = \dfrac{\text{opp}}{\text{hyp}}$ $\csc \theta = \dfrac{\text{hyp}}{\text{opp}} = \dfrac{1}{\sin \theta}$	$\cos \theta = \dfrac{\text{adj}}{\text{hyp}}$ $\sec \theta = \dfrac{\text{hyp}}{\text{adj}} = \dfrac{1}{\cos \theta}$	$\tan \theta = \dfrac{\text{opp}}{\text{adj}} = \dfrac{\sin \theta}{\cos \theta}$ $\cot \theta = \dfrac{\text{adj}}{\text{opp}} = \dfrac{\cos \theta}{\sin \theta}$
Pythagorean Identities	$\cos^2 \theta + \sin^2 \theta = 1$	$\tan^2 \theta + 1 = \sec^2 \theta$	$\cot^2 \theta + 1 = \csc^2 \theta$